SYSTEMS ANALYSIS AND DESIGN

FIFTH EDITION

KENNETH E. KENDALL
Rutgers University
School of Business–Camden
Camden, New Jersey

JULIE E. KENDALL
Rutgers University
School of Business–Camden
Camden, New Jersey

Prentice-Hall International, Inc.

Acquisitions Editor: Robert Horan
Vice President and Publisher: Natalie Anderson
Associate Editor: Lori Cerreto
Editorial Assistant: Erika Rusnak
Marketing Manager: Sharon Turkovich
Marketing Assistant: Jason Smith
Managing Editor (Production): John Roberts
Production Assistant: Dianne Falcone
Permissions Coordinator: Suzanne Grappi
Associate Director, Manufacturing: Vincent Scelta
Production Manager: Arnold Vila
Manufacturing Buyer: Diane Peirano
Design Manager: Patricia Smythe
Senior Designer: Kevin Kall
Interior Design: Dorothy Bungert
Cover Design: Karen Quigley, QT Designs
Cover Art: Bette Ridgeway, *Voyage of the Spirit*, 1999.
Composition and Full-Service Project Management: UG / GGS Information Services, Inc.
Printer/Binder: R. R. Donnelley–Willard

Credits and acknowledgments borrowed from other sources and reproduced, with permission, in this textbook appear on appropriate page within text.

10 9 8 7 6 5 4 3 2 1
ISBN 0-13-042365-3

To Julia A. Kendall and the memory of Edward J. Kendall,
whose lifelong example of working together will inspire us forever

COMPANY TRADEMARKS

Apple and Macintosh are registered trademarks of Apple Computer. Dragon Naturally Speaking is a registered trademark of Lernout and Hauspie. FormFlow99 is a registered trademark of JetForm Corporation. Dreamweaver, Macromedia Flash, and Likeminds are trademarks of Macromedia. HyperCase is a registered trademark of Raymond J. Barnes, Richard L. Baskerville, Julie E. Kendall, and Kenneth E. Kendall. Lotus 1–2-3, Freelance Graphics, Organizer, and Smartmasters are registered trademarks of Lotus Corporation. Micrografx Designer, Flowcharter, WebCharter, and Graphics Suite are registered trademarks of Micrografx Corporation. Microsoft Windows, Microsoft Access, Microsoft Word, Microsoft FrontPage, Microsoft PowerPoint, Microsoft Project, Microsoft Excel, Microsoft Visio Professional are registered trademarks of Microsoft Corporation. Netscape Communicator and Netscape Navigator are registered trademarks of Netscape Communications Corp. OmniPage is a trademark of Scansoft. ProModel, ProcessModel, and Service Model are registered trademarks of PROMODEL Corporation. Visible Analyst is a registered trademark of Visible Systems Corporation. Web Strategy Pro and Business Plan Pro 4.0 are trademarks of Palo Alto Software. WinFax Pro, and Norton Internet Security are registered trademarks of Symantec. Workflow BPR is a trademark of Holosofx, Inc. Other product and company names mentioned herein may be the trademarks of their respective owners. Companies, names, and/or data used in screens and sample output are fictitious unless otherwise noted.

BRIEF CONTENTS

CONTENTS

PART I SYSTEMS ANALYSIS FUNDAMENTALS

PREFACE

The fifth edition of *Systems Analysis and Design* includes many new and updated features. In particular:

- Full color with **meaningful color coding** of DFDs, E-Rs, structure charts, and UML diagrams
- More than **70 Consulting Opportunities** including many new mini-cases focused on designing for ecommerce, performing workflow analysis, and modeling with UML
- Expanded **Web-based design**
- New approaches for designing **ecommerce** Web sites
- Expanded coverage of **graphical user interface** (GUI) design
- New ideas on ecommerce project management
- New design approaches for **wireless technologies, ERP**, and **Web-based systems**
- Expanded design for **intranets and extranets**, including easy onscreen navigation approaches
- New **UML** section of the object-oriented chapter with new consulting opportunities, diagrams, and problems
- New coverage on implementing **Web site security and privacy** safeguards including firewalls, corporate privacy policies, PKI, SSL, SET, VPN, URL filters, and email filtering
- Expanded coverage of software to monitor Web traffic, perform audience profiling, and promote corporate Web sites to ensure the **effectiveness of new ecommerce systems**
- Updated, ongoing **CPU case** for use with **Visible Analyst** and **Microsoft Access**
- Updated **HyperCase 2.5**, a graphical organizational simulation for the Web that allows students to apply their skills

DESIGN FEATURES

Figures take on a stylized look in order to help students more easily grasp the subject matter.

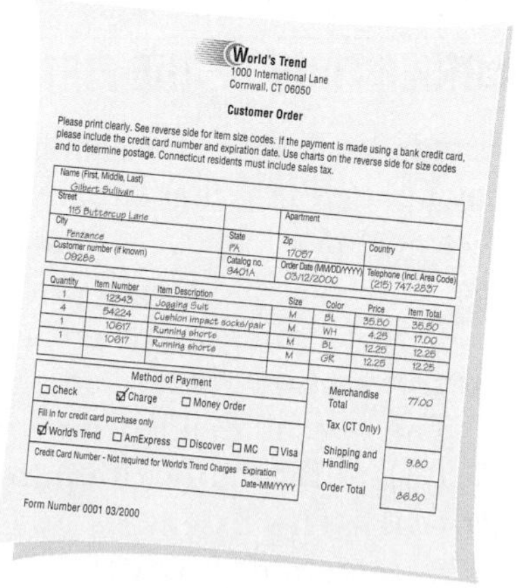

Paper forms are used throughout to show input and output design as well as the design of questionnaires. Blue ink is always used to show writing or data input, thereby making it easier to identify what was filled in by users. Although most organizations have computerization of manual processes as their eventual goal, much data capture is still done using paper forms. Improved form design enables analysts to ensure accurate and complete input and output. Better forms can also help streamline new internal workflows that result from newly automated business-to-consumer (B2C) applications for ecommerce on the Web.

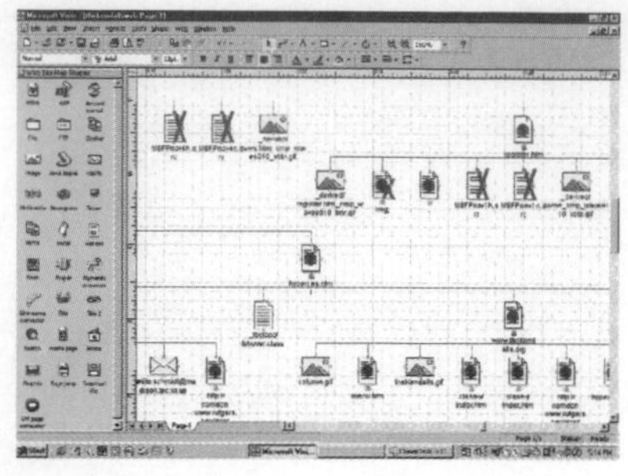

Computer display screens demonstrate important features of software that are useful to the analyst. This example shows how a Web site can be evaluated for broken links by using a package such as Microsoft Visio. Actual screen shots show important aspects of design. Analysts are continuously seeking to improve the appearance of the screens and Web pages they design. Colorful examples help to illustrate why some screen designs are particularly effective.

Conceptual diagrams are used to introduce the many tools that systems analysts have at their disposal. This example shows the differences between logical data flow diagrams and physical data flow diagrams. Conceptual diagrams are color-coded so that students can distinguish easily among them, and their functions are clearly indicated. Many other important tools are illustrated, including entity-relationship diagrams, structure charts, and structured English.

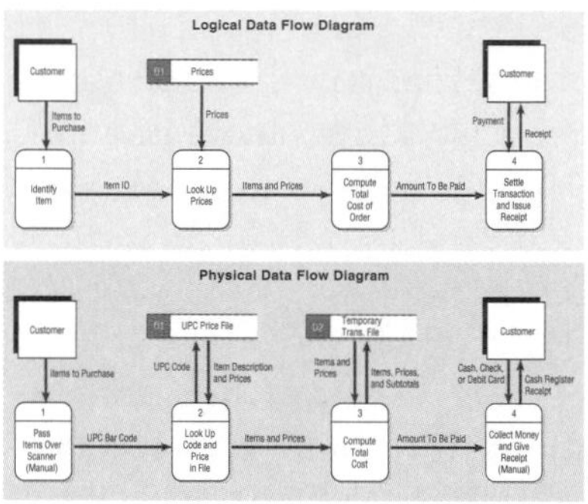

Tables are used when an important list needs special attention, or when information needs to be organized or classified. In addition, tables are used to supplement the understanding of the reader in a way that is different from the way material is organized in the narrative portion of the text. Most analysts find tables a useful way to organize numbers and text into a meaningful "snapshot."

This example of a table from Chapter 3 shows how analysts can refine their activity plans for analysis by breaking them down into smaller tasks and then estimating how much time it will take to complete them. The underlying philosophy of our book is that systems analysis and design is a process that integrates the use of many tools with the unique talents of the systems analyst to systematically improve business through the implementation or modification of computerized information systems. Systems analysts can grow in their work by taking on new IT challenges and keeping current in their profession through the application of new techniques and tools.

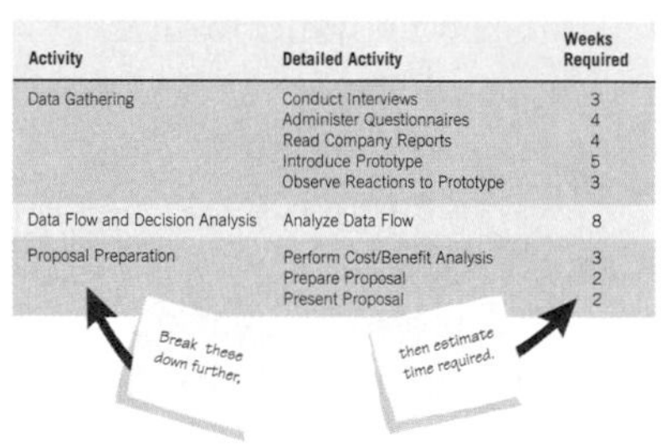

Activity	Detailed Activity	Weeks Required
Data Gathering	Conduct Interviews	3
	Administer Questionnaires	4
	Read Company Reports	4
	Introduce Prototype	5
	Observe Reactions to Prototype	3
Data Flow and Decision Analysis	Analyze Data Flow	8
Proposal Preparation	Perform Cost/Benefit Analysis	3
	Prepare Proposal	2
	Present Proposal	2

Break these down further.

then estimate time required.

A BRIEF TOUR OF THE FIFTH EDITION

Systems analysis and design is typically taught in one or two semesters. Our book may be used in either situation. The text is appropriate for undergraduate (junior or senior) curricula at a four-year university, graduate school, or community college. The level and length of the course can be varied and supplemented by using real-world projects, HyperCase, or other materials available on the instructor's resource section of our Companion Web site.

The text is divided into five major parts: Systems Analysis Fundamentals (Part I), Information Requirements Analysis (Part II), The Analysis Process (Part III), The Essentials of Design (Part IV), and Software Engineering and Implementation (Part V).

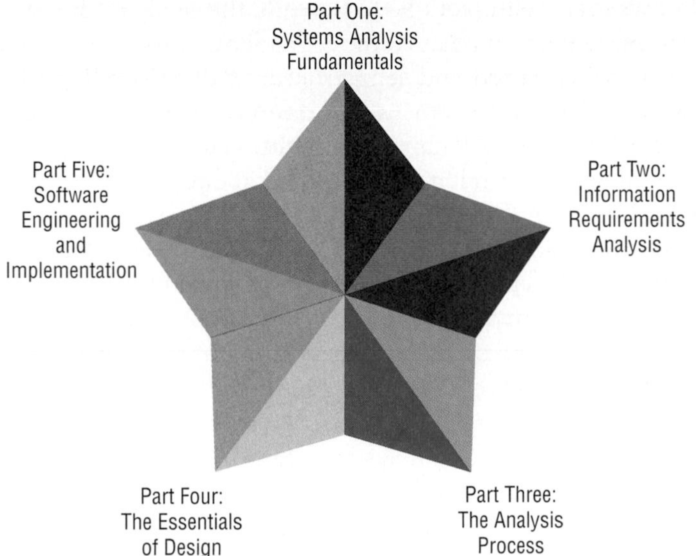

Part I (Chapters 1–3) stresses the basics that students need to know about what an analyst does; how a variety of information systems, including handheld, wireless technologies and ERP systems, fit into organizations; how to determine whether a systems project is worthy of commitment; new coverage of ecommerce project management; and how to manage a systems project using special software tools. There is new material on virtual teams and virtual organizations. Techniques for drawing entity-relationship diagrams and context-level data flow diagrams when first entering the organization are introduced. Chapter 3 introduces a new tool, the Feasibility Impact Grid, to assess impacts of the development of new systems at both strategic and operational levels. Alternative systems analysis and design methods such as ETHICS are introduced. The three roles of the systems analyst as consultant, supporting expert, and agent of change are also introduced, and there are updated ideas on ethical issues and professional guidelines for serving as a systems consultant.

Part II (Chapters 4–8) emphasizes the use of systematic and structured methodologies. Attention to analysis helps analysts ensure that they are addressing the correct problem before designing the system. The presentation of each methodology (sampling, investigating hard data, interviewing, questionnaires, and observation) moves students closer to understanding what information users need and how those needs may best be ascertained. Chapter 4 introduces a new software tool for doing workflow analysis that helps in the integration of ecommerce into the traditional business processes. Chapter 5 includes material on joint application design (JAD) for ascertaining information requirements in concert with users. Chapter 7 is especially innovative and goes well beyond the typical text in showing how to accomplish systematic observation of decision-makers. Chapter 8 is unique in its treatment of prototyping as another data gathering technique that enables the analyst to solve the right problem by getting users involved from the start. This chapter also includes new material on rapid application development (RAD), which is conceptually close to prototyping. RAD provides an accelerated approach to the SDLC that is particularly suitable for designing ecommerce applications.

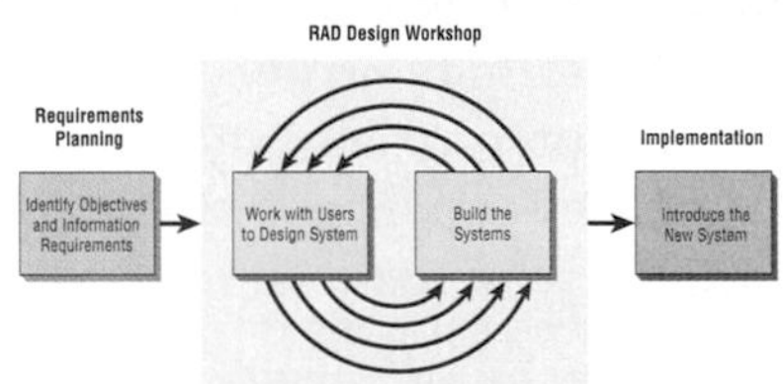

Part III (Chapters 9–14) details the analysis process. It builds on the previous two parts to move students into analysis of data flows as well as structured and semistructured decisions. It provides step-by-step details on how to use structured techniques to draw data flow diagrams (DFDs). Chapter 9 provides coverage of how to create child diagrams; how to develop both logical and physical data flow diagrams; and how to partition data flow diagrams. A new section discussing the object-oriented approach of use cases and data flow diagrams is included. The object-oriented approach in Chapter 10 features material on the data repository and vertical balancing of data flow diagrams. Chapter 11 includes material on developing process specifications. A discussion of both logical and physical process specifications shows how to use process specifications for horizontal balancing.

Part III also covers how to diagram structured decisions with the use of structured English, decision tables, and decision trees. Students then progress to a consideration of semistructured decisions that are featured in decision support systems. New material in Chapter 12 gives practical guidelines to analysts about choosing decision support system methods and software. New approaches to supporting decision making include expert systems, neural nets, using the analytic hierarchy process (AHP), the use of recommendation systems, Web-based systems, and the use of simulations such as Promodel and ServiceModel to aid in decision-making. In addition, push technologies are introduced.

New material on evaluating vendor support when choosing hardware and software for a new system is included in Chapter 13. In addition, students are taught several methods for forecasting costs and benefits, which are necessary to the discussion of acquiring software and hardware. Chapter 14 stresses the importance of a professionally prepared written and oral presentation of the systems proposal.

Part IV (Chapters 15–19) covers the essentials of design. It begins with designing output, since many practitioners believe systems to be output driven. The design of Web-based forms is covered in detail. Particular attention is paid to relating output method to content, the effect of output on users, and designing good forms and screens. Chapter 15 compares advantages and disadvantages of output, including Web screens, audio, CD-ROM, DVD, and electronic output such as e-mail, faxes, and bulletin boards. Chapter 16 includes innovative material on designing Web-based input forms as well as other electronic form design. Also included is computer-assisted form design.

Chapter 16 also includes expanded coverage of Web site design, including guidelines on when designers should add video, audio, and animation to Web site designs. New material introduces the uses of Web push and pull technologies for output design. Also new is expanded consideration of how to create effective graphics for corporate Web sites and designing effective onscreen navigation for Web site users.

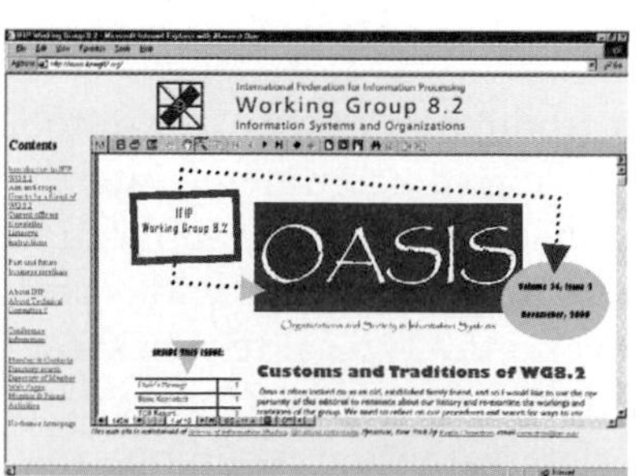

New material includes expanded coverage of intranet and extranet page design. Consideration of database integrity constraints has been included, and how the user interacts with the computer and how to design an appropriate interface are also covered. The importance of user feedback and correct ergonomic design of computer workstations are also found in Part IV. How to design accurate data-entry procedures that take full advantage of computer and human capabilities to assure entry of quality data is emphasized here.

Chapter 17 demonstrates how to use the entity-relationship diagram to determine record keys, as well as providing guidelines for file/database relation design. Students are shown the relevance of database design for the overall usefulness of the system, and how users actually use databases. Chapter 18 features new material on designing easy onscreen navigation for Web site visitors. It also features updated material on important aspects of data mining and data warehousing. Innovative approaches to searching on the Web are also presented. Material on GUI design is also highlighted and innovative approaches to designing dialogs are provided. Chapter 19 includes new material on managing the supply chain through the effective design of business-to-business (B2B) ecommerce systems.

Part V (Chapters 20–22) introduces students to structured software engineering and documentation techniques as ways to implement a quality system. Chapter 20 includes a section on the important concepts of code generation and design re-engineering. We also cover developments in structured techniques while also teaching students which techniques are appropriate for particular situations.

The material on structure charts includes details on how to use data flow diagrams to draw structure charts. In addition, material on system security and firewalls is included. Testing, auditing, and maintenance of systems are discussed in the context of total quality management. Chapter 21 presents innovative tools for modeling networks, which can be done with popular tools such as Microsoft Visio. A discussion of groupware is also included. Part V also introduces the student to designing client/server systems and designing distributed systems.

New material on security and privacy in relation to designing ecommerce applications is included. Expanded coverage on security, specifically firewalls, gateways, public key infrastructure (PKI), secure electronic translation (SET), secure socket layering (SSL), virus protection software, URL filtering products, email filtering products, and virtual private networks (VPN) has been added. Additionally, new topics of interest to designers of ecommerce applications including the development of audience profiling, and the development and posting of corporate privacy policies are covered.

New coverage of how the analyst can promote and then monitor a corporate Web site is included in this section, which features Web activity monitoring, Web site promotion, Web traffic analysis, and audience profiling to ensure the effectiveness of new ecommerce systems. Techniques for evaluating the completed information systems project are covered systematically as well.

Part V concludes with Chapter 22 on object-oriented systems analysis and design, which includes a new in-depth section on using the unified modeling language (UML). Through several examples and Consulting Opportunities, this chapter demonstrates how to use an object-oriented approach. UML is introduced as an agreed-upon standard notation for object-oriented analysis and design. New consulting opportunities, diagrams, and problems enable students to learn and use UML to model systems from an object-oriented perspective.

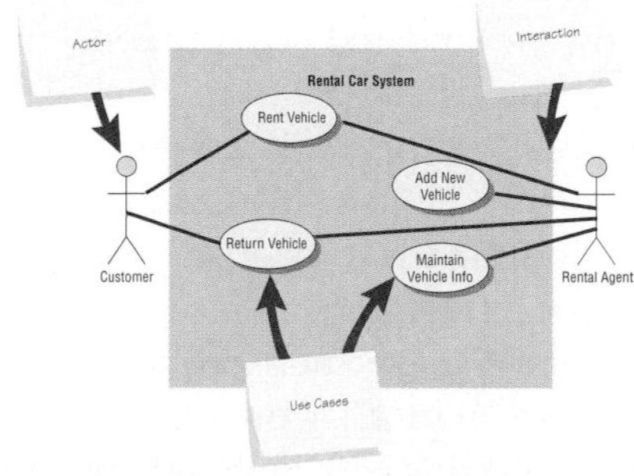

The fifth edition contains a **Glossary** of terms and a separate list of **Acronyms** used in the book and the systems analysis and design field.

PEDAGOGICAL FEATURES

Chapters in the Fifth Edition contain:

- **End-of-chapter Summaries** that tie together the salient points of each chapter while providing an excellent review source for exams
- **Keywords and Phrases**
- **Review Questions**
- **Problems**
- **Group Projects** that help students work together in a systems team to solve important problems that are best solved through group interaction
- **Consulting Opportunities**—now more than 70 minicases throughout the book
- **HyperCase Experiences**
- **CPU Episodes**—parts of an ongoing case threaded throughout the book

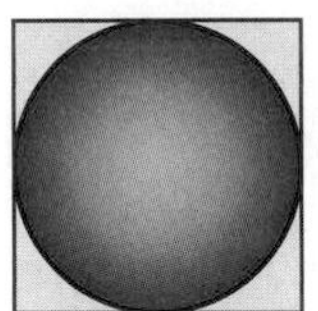

CONSULTING OPPORTUNITIES

The fifth edition presents more than 70 Consulting Opportunities, and many of them address new topics that have arisen in the field, including designing ecommerce applications for the Web, workflow analysis, and using UML to model information systems from an object-oriented perspective. Consulting Opportunities can be used for stimulating in-class discussions or assigned as homework or take-home exam questions. Since not all systems are extended two- or three-year projects, our book contains many Consulting Opportunities that can be solved quickly in 20 to 30 minutes of group discussion or individual writing. These mini-cases, written in a humorous manner to enliven the material, require students to synthesize what they have learned up to that point in the course, ask students to mature in their professional and ethical judgment, and expect students to articulate the reasoning that led to their systems decisions.

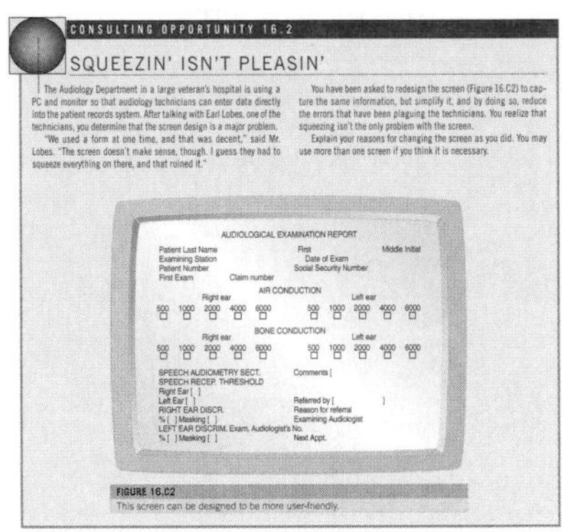

CONSULTING OPPORTUNITY 16.2

SQUEEZIN' ISN'T PLEASIN'

The Audiology Department in a large veteran's hospital is using a PC and monitor so that audiology technicians can enter data directly into the patient records system. After talking with Earl Lobes, one of the technicians, you determine that the screen design is a major problem.

"We used a form at one time, and that was decent," said Mr. Lobes. "The screen doesn't make sense, though. I guess they had to squeeze everything on there, and that ruined it."

You have been asked to redesign the screen (Figure 16.C2) to capture the same information, but simplify it, and by doing so, reduce the errors that have been plaguing the technicians. You realize that squeezing isn't the only problem with the screen.

Explain your reasons for changing the screen as you did. You may use more than one screen if you think it is necessary.

FIGURE 16.C2
This screen can be designed to be more user-friendly.

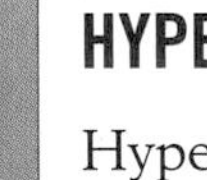

HYPERCASE EXPERIENCES

HyperCase Experiences that pose challenging student exercises are present in each chapter. HyperCase version 2.5 is now available on the Web. HyperCase now has updated organizational problems featuring state-of-the-art technological systems as well. HyperCase is an original virtual organization that allows students who access it to become immediately immersed in organizational life. Students will interview people, observe office environments, analyze their prototypes, and review the documentation of their existing systems. HyperCase version 2.5 is Web-based interactive software that presents an organization called Maple Ridge Engineering in a colorful, three-dimensional graphics environment. HyperCase permits

HYPERCASE® EXPERIENCE 3

"I hope everyone you've encountered at MRE has treated you well. Here's a short review of some of the ways you can access our organization through HyperCase. The reception area at MRE contains the key links to the rest of our organization. Perhaps you've already discovered these on your own, but I wanted to remind you of them now, because I don't want to get so engrossed in the rest of our organizational problems that I forget to mention them.

"The telephone on the receptionist's desk has instructions about how to answer the phone in the rest of the organization. You have my permission to pick up the phone if it is ringing and no one else answers it.

"The empty doorway you see is a link to the next room, which we call the East Atrium. You have probably noticed that all open doorways are links to adjacent rooms. Notice the building map displayed in the reception area. You are free to go to public areas such as the canteen, but as you know, you must have an employee escort you into a private office. You cannot go there on your own.

"By now you have probably noticed the two documents and the computer on the small table in the reception area. The little one is the MRE internal phone directory. Just click on an employee name, and if that person is in, he or she will grant you an interview and a tour of the office. I leave you to your own devices in figuring out what the other document is.

"The computer on the table is on and displays the World Wide Web home page for MRE. You should take a look at the corporate page and visit all the links. It tells the story of our company and the people who work here. We're quite proud of it and have gotten positive feedback about it from visitors.

"If you have had a chance to interview a few people and see how our company works, I'm sure you are becoming aware of some of the politics involved. We are also worried, though, about more technical issues, such as what constitutes feasibility for a training project and what does not."

FIGURE 3.HC1
The reception room resembles a typical corporation. While you are in this HyperCase screen, find the directory if you want to visit someone.

professors to begin approaching the systems analysis and design class with exciting multimedia material. Carefully watching their use of time and managing multiple methods, students use the hypertext characteristics of HyperCase on the Web to create their own individual paths through the organization.

Maple Ridge Engineering (MRE) is drawn from the actual consulting experiences of the authors of the original version (Raymond Barnes, Richard Baskerville, Julie E. Kendall, and Kenneth E. Kendall). Allen Schmidt joined the project for Version 2.0. Peter Schmidt was the HTML programmer and Jason Reed created the images for the Web version.

In each chapter, there are special HyperCase Experiences that include assignments (and even some clues) to help students solve the difficult organizational problems they encounter at MRE. HyperCase has been fully classroom-tested, and was an award winner in the Decision Sciences Institute Innovative Instruction competition.

CPU CASE EPISODES

In keeping with our belief that a variety of approaches is important, we have once again integrated the Central Pacific University (CPU) Case into every chapter of the fifth edition. The CPU case makes use of the popular CASE tool Visible Analyst by Visible Systems, Inc., as well as Microsoft Access for the example screen shots and the student exercises.

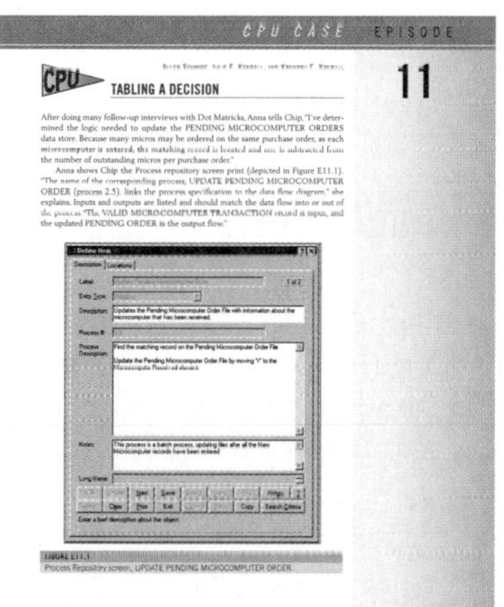

CPU CASE EPISODE 11

TABLING A DECISION

After doing many follow-up interviews with Dot Matricks, Anna tells Chip, "I've determined the logic needed to update the PENDING MICROCOMPUTER ORDERS data store. Because many micros may be ordered on the same purchase order, as each microcomputer is entered, the matching record is located and one is subtracted from the number of outstanding micros per purchase order."

Anna shows Chip the Process repository screen print (depicted in Figure E11.1). "The name of the corresponding process, UPDATE PENDING MICROCOMPUTER ORDER (process 2.5), links the process specification to the data flow diagram," she explains. Inputs and outputs are listed and should match the data flow into or out of the process. "The VALID MICROCOMPUTER TRANSACTION record is input, and the updated PENDING ORDER is the output flow."

The CPU Case takes students through all phases of the systems development life cycle, demonstrating the capabilities of Visible Analyst, a student edition of which can be bundled with our book. This CASE tool gives students an opportunity to solve problems on their own, using Visible Analyst and data that users of the book can download from the Web containing Visible Analyst exercises specifically keyed to each chapter of the book. Additionally, partially completed exercises in Microsoft Access files are also available for student use on the Web. The CPU Case has been fully classroom-tested with a variety of students over numerous terms. The case is detailed, rigorous, and rich enough to stand alone as a systems analysis and design project spanning one or two terms. Alternatively, the CPU Case can be used as a way to teach the use of CASE tools in conjunction with the assignment of a one- or two-term real-world project outside the classroom.

EXPANDED WEB SUPPORT

Kendall & Kendall's *Systems Analysis and Design*, Fifth Edition, adds expanded Web-based support to solid yet lively pedagogical techniques in the information systems field.

The Web site, located at **www.prenhall.com/kendall/**, contains a wealth of critical learning and support tools, which keep class discussions exciting.

★ **HyperCase version 2.5**, an award-winning, virtually interactive organization game. Students are encouraged to interview people in the organization, analyze problems, modify data flow diagrams and data dictionaries, react to prototypes, and design new input and output. HyperCase now has a distinctive 3-D look.

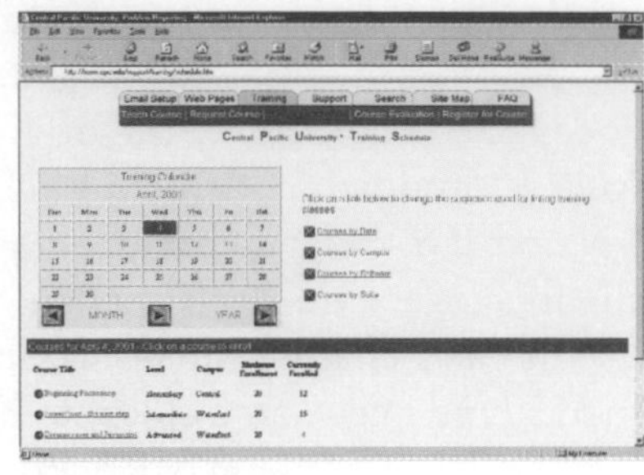

★ **Student Exercises based on the ongoing CPU Case**, with partially solved problems and examples stored in Visible Analyst files and Microsoft Access files, so students can develop a Web-based computer management system.
★ **Interactive Study Guide,** featuring true/false and multiple-choice questions for each chapter. Students receive automatic grading and feedback upon completing each quiz.
★ **Instructor's Manual** (within a secure faculty section) with answers to problems, solutions to cases, and suggestions for approaching the subject matter.
  - A complete set of PowerPoint presentation slides for use in lectures, which includes *all* of the technical figures from the fifth edition
  - Sample Course Outlines for one- or two-semester or quarter courses
  - Solutions to Student Exercises based on the ongoing CPU Case, with solutions and examples stored in Visible Analyst files and Microsoft Access files
  - The *HyperCase Users' Guide to the Corporation*, an instructor's guide to interpreting HyperCase and suggested approaches for classroom use.

EXPANDED CD-ROM SUPPLEMENTAL SUPPORT

New to the fifth edition package is an **Instructor's Resource CD-ROM**, featuring all of the instructor supplements in one convenient place. Resources include:

★ A complete set of **PowerPoint presentation slides** for use in lectures
★ **Image Library**, a collection of all text art organized by chapter
★ **Instructor's Manual** in Microsoft Word
★ **Test Item File** in Microsoft Word
★ **Windows PH Test Manager**, a comprehensive suite of tools for testing and assessment that allows instructors to easily create and distribute tests
★ **Solutions to Student Exercises** based on the ongoing CPU Case, with solutions and examples stored in Visible Analyst files and Microsoft Access files

ACKNOWLEDGMENTS

When we began writing the fifth edition of *Systems Analysis and Design*, we were heartened and motivated by the rapid and widespread adoption of the Web for IT and other social and personal uses. What this means, in part, is that the demand for systems analysts is expected to grow for the near future. More people than ever before are expected to be systems designers, and even more people are supposed to be sophisticated consumers of Web-based systems and information. Users interact, critique, and respond to what they see and hear. Good systems analysts and designers use both art and science to reply to the feedback they receive in order to develop systems that are in tune with their users, their environments, and even society.

Our cover artist, Bette Ridgeway, tells us that her painting, *Voyage of the Spirit* (which we first viewed on her Web site), was a work created through love, inspiration, and hard work. It is a product of controlled chaos, restrained and disciplined spontaneity. We think you will see that its creation is similar to what happens in the creation of new information systems. You learn and apply numerous structured techniques, methods, tools, and approaches. But when the time comes to interpret what is happening in the organization and to develop meaningful information systems from the application of rules to your analysis, your training combines with creativity to produce a system that is both in keeping with the character of the organization and uniquely reflective of you as a systems analyst.

As with any new edition, our students deserve applause for helping us to improve the book. We appreciate all of their comments and insights. We want to thank our co-author, Allen Schmidt, for all of the ability, interest, and cheerfulness he brings to our collaborations. He is a unique person. Our enthusiastic thanks also go to Peter Schmidt and Jason Reed for their dedication to the HyperCase project. Their potential seems to know no bounds. We also want to thank the other two original authors of HyperCase, Richard Baskerville and Raymond Barnes, who contributed so much.

We would also like to thank Mike Anderson, Wayne Huang, John Janney, and Ping Zhang and for their insightful contributions to the fifth edition. We would like to thank our editor, Robert Horan, who knows that maintaining a vision while working hard usually results in rich rewards. Lori Cerreto was also a joy to work with. Kudos to Kevin Kall, who inspired us by pulling together the stunning design of this text, and Sharon Turkovich, who kept us going with the marketing campaign. Their knowledge, insight, and enthusiasm have kept this project moving toward our shared goals.

Sandra Gormley, our production editor at UG / GGS Information Services, Inc., also deserves an enormous amount of praise for maintaining a calm presence and helping us with the difficult task of setting and staying with priorities. Because of her, this edition was an absolute joy to work on. Finally, there were many people we did not meet in person at Prentice Hall, UG / GGS Information Services, Inc., and elsewhere who helped us by designing the book, drawing the art, composing the pages, and securing permissions. We thank them all.

Many of our reviewers, colleagues, and friends have encouraged us through the process of writing this book. We thank them for their comments on our work. They include: Ayman Abu Hamdieh; Nora Braun; Catherine Brooker; Jim and Jan Buffington; Jim, Lorraine, and Rob Ciupek; Charles J. Coleman; Melissa Covelli; Gordon Davis; Dorothy Dologite; Phillip Ein-Dor; Bruce Fanning; Alma Gallagher; Paul Gray; Maurice Green; Varun Grover; Nancy V. Gulick; Andy and Pam Hamingson; Chung Kwong Han; Mike Jones; Rajiv Kishore; Sarah Klammer; Ken and Nancy Kopecky; Art and Joan Kraft; Lee and Judi Krajewski; Deborah LaBelle; Ken and Jane Laudon; Richard Madigan; Emily McMullen; Marlyn Milne; Robert Mockler; Brian Modena; Nancy Omaha Boy; Ciaran O'Mearain; M. Owen Lee; JoAnn Peach; Joel and Bobbie Porter; Suzy Ramirez; Jane Van Saun; Caryn Schmidt; Marc and Jill Schniederjans; Scott Dixon Smith; Mikaela Sullivan; Eric and Tisha Stahl; Cheryl Vargo; Malcolm Warner; Mun Yi; Shaker and Patricia Zahra; Joachim Zoundi; and all of our friends and colleagues in the Association for Information Systems, the Decision Sciences Institute, IFIP Working Group 8.2, and all those involved in the KPMG Ph.D Project.

Our heartfelt thanks go to Julia A. Kendall and to the memory of Edward J. Kendall. Their belief that love, goals, and hard work are an unbeatable combination continues to infuse our every endeavor.

1

ASSUMING THE ROLE OF THE SYSTEMS ANALYST

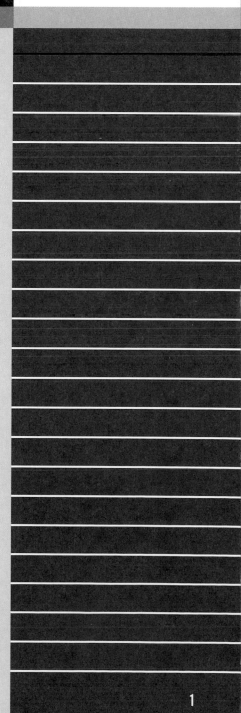

INFORMATION AS AN ORGANIZATIONAL RESOURCE

Organizations have long recognized the importance of managing key resources such as labor and raw materials. Information has now moved to its rightful place as a key resource. Decision makers now understand that information is not just a byproduct of conducting business; rather, it fuels business and can be the critical factor in determining the success or failure of a business.

MANAGING INFORMATION AS A RESOURCE

To maximize the usefulness of information, a business must manage it correctly, just as it manages other resources. Managers need to understand that costs are associated with the production, distribution, security, storage, and retrieval of all information. Although information is all around us, it is not free, and its strategic use for positioning a business competitively should not be taken for granted.

MANAGING COMPUTER-GENERATED INFORMATION

The ready availability of networked computers, along with access to the Internet and the World Wide Web, has created an information explosion throughout society in general and business in particular. Managing computer-generated information differs in significant ways from handling manually produced data. Usually there is a greater quantity of computer information to administer. Costs of organizing and maintaining it can increase at alarming rates, and users often treat it less skeptically than information obtained in different ways. This chapter examines the fundamentals of different kinds of information systems, the varied roles of systems analysts, and the phases in the systems development life cycle; it also introduces Computer-Aided Software Engineering (CASE) tools.

TYPES OF SYSTEMS

Information systems are developed for different purposes, depending on the needs of the business. Transaction processing systems (TPS) function at the operational level of the organization; office automation systems (OAS) and knowledge work systems (KWS) support work at the knowledge level. Higher-

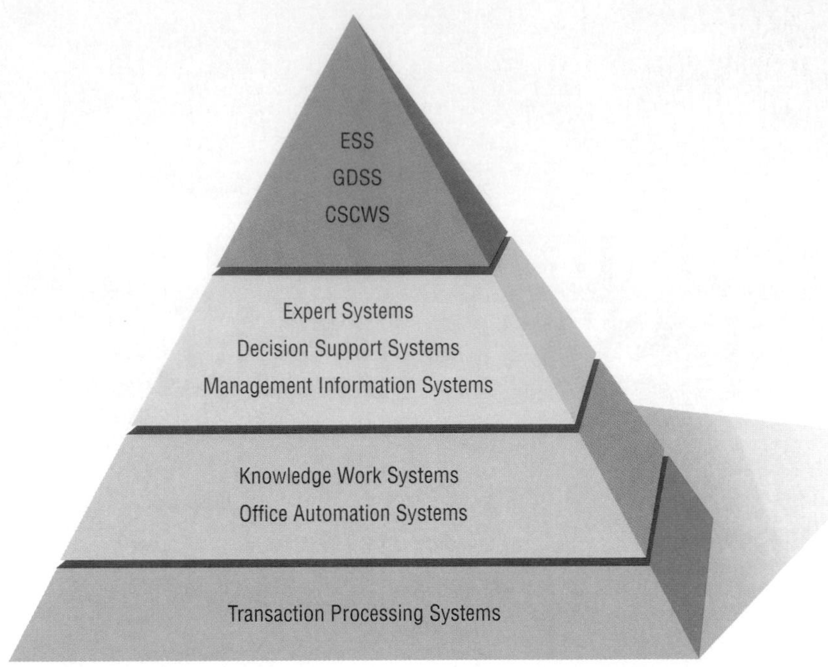

FIGURE 1.1

A systems analyst may be involved with any of or all these systems.

level systems include management information systems (MIS) and decision support systems (DSS). Expert systems apply the expertise of decision makers to solve specific, structured problems. On the strategic level of management we find executive support systems (ESS). Group decision support systems (GDSS) and the more generally described computer supported collaborative work systems (CSCWS) aid group-level decision making of a semistructured or unstructured variety.

The variety of information systems that analysts may develop are shown in Figure 1.1. Notice that the figure presents these systems from the bottom up, indicating that the operational, or lowest level of the organization, is supported by TPS and the highest, or strategic level of semistructured and unstructured decisions, is supported by ESS, GDSS, and CSCWS at the top. This text uses the terms *management information systems, information systems* (IS), *computerized information systems*, and *computerized business information systems* interchangeably to denote computerized information systems that support the broadest range of business activities through the information they produce.

TRANSACTION PROCESSING SYSTEMS

Transaction processing systems (TPS) are computerized information systems that were developed to process large amounts of data for routine business transactions such as payroll and inventory. A TPS eliminates the tedium of necessary operational transactions and reduces the time once required to perform them manually, although people must still input data to computerized systems.

Transaction processing systems are boundary-spanning systems that permit the organization to interact with external environments. Because managers look to the data generated by the TPS for up-to-the-minute information about what is happening in their companies, it is essential to the day-to-day operations of business that these systems function smoothly and without interruption.

OFFICE AUTOMATION SYSTEMS AND KNOWLEDGE WORK SYSTEMS

At the knowledge level of the organization are two classes of systems. Office automation systems (OAS) support data workers, who do not usually create new knowledge but rather analyze information so as to transform data or manipulate it in some way before sharing it with, or formally disseminating it throughout, the organization and, sometimes, beyond. Familiar aspects of OAS include word processing, spreadsheets, desktop publishing, electronic scheduling, and communication through voice mail, email (electronic mail), and video conferencing.

Knowledge work systems (KWS) support professional workers such as scientists, engineers, and doctors by aiding them in their efforts to create new knowledge and by allowing them to contribute it to their organization or to society at large.

MANAGEMENT INFORMATION SYSTEMS

Management information systems (MIS) do not replace transaction processing systems; rather, all MIS include transaction processing. MIS are computerized information systems that work because of the purposeful interaction between people and computers. By requiring people, software (computer programs), and hardware (computers, printers, etc.) to function in concert, management information systems support a broader spectrum of organizational tasks than transaction processing systems, including decision analysis and decision making.

To access information, users of the management information system share a common database. The database stores both data and models that help the user interpret and apply that data. Management information systems output information that is used in decision making. A management information system can also help unite some of the computerized information functions of a business, although it does not exist as a singular structure anywhere in the business.

DECISION SUPPORT SYSTEMS

A higher-level class of computerized information systems is decision support systems (DSS). The DSS is similar to the traditional management information system because they both depend on a database as a source of data. A decision support system departs from the traditional management information system because it emphasizes the support of decision making in all its phases, although the actual decision is still the exclusive province of the decision maker. Decision support systems are more closely tailored to the person or group using them than is a traditional management information system.

EXPERT SYSTEMS AND ARTIFICIAL INTELLIGENCE

Artificial intelligence (AI) can be considered the overarching field for expert systems. The general thrust of AI has been to develop machines that behave intelligently. Two avenues of research of AI are understanding natural language and analyzing the ability to reason through a problem to its logical conclusion. Expert systems use the approaches of AI reasoning to solve the problems put to them by business (and other) users.

Expert systems are a very special class of information system that have been made practicable for use by business as a result of widespread availability of hardware and software such as personal computers (PCs) and expert system shells. An expert system (also called a knowledge-based system) effectively captures and uses the knowledge of an expert for solving a particular problem experienced in an organization. Notice that unlike DSS, which leave the ultimate judgment to the decision maker, an expert system selects the best solution to a problem or a specific class of problems.

The basic components of an expert system are the knowledge base, an inference engine connecting the user with the system by processing queries via languages such as SQL (structured query language), and the user interface. People called knowledge engineers capture the expertise of experts, build a computer system that includes this expert knowledge, and then implement it. It is entirely possible that building and implementing expert systems will be the future work of many systems analysts.

GROUP DECISION SUPPORT SYSTEMS AND COMPUTER-SUPPORTED COLLABORATIVE WORK SYSTEMS

When groups need to work together to make semistructured or unstructured decisions, a group decision support system may afford a solution. Group decision support systems (GDSS), which are used in special rooms equipped in a number of different configurations, permit group members to interact with electronic support—often in the form of specialized software—and a special group facilitator. Group decision support systems are intended to bring a group together to solve a problem with the help of various supports such as polling, questionnaires, brainstorming, and scenario creation. GDSS software can be designed to minimize typical negative group behaviors such as lack of participation due to fear of reprisal for expressing an unpopular or contested viewpoint, domination by vocal group members, and "group think" decision making. Sometimes GDSS are discussed under the more general term *computer supported collaborative work* (CSCW), which might include software support called "groupware" for team collaboration via networked computers.

EXECUTIVE SUPPORT SYSTEMS

When executives turn to the computer, they are often looking for ways to help them make decisions on the strategic level. Executive support systems (ESS) help executives organize their interactions with the external environment by providing graphics and communications support in accessible places such as boardrooms or personal corporate offices. Although ESS rely on the information generated by TPS and MIS, executive support systems help their users address unstructured decision problems, which are not application-specific, by creating an environment that is conducive to thinking about strategic problems in an informed way. ESS extend and support the capabilities of executives, permitting them to make sense of their environments.

INTEGRATING TECHNOLOGIES FOR SYSTEMS

As new technologies are adopted and diffused, some of the systems analyst's work will be devoted to integrating traditional systems with new ones as shown in Figure 1.2. This section describes some of the new information technologies systems analysts will be using as businesses work to integrate their ecommerce applications into their traditional businesses or as they begin entirely new ebusinesses.

ECOMMERCE APPLICATIONS AND WEB SYSTEMS

Many of the systems discussed here can be imbued with greater functionality if they are migrated to the World Wide Web or if they are originally conceived and implemented as Web-based technologies. A recent survey found that in the com-

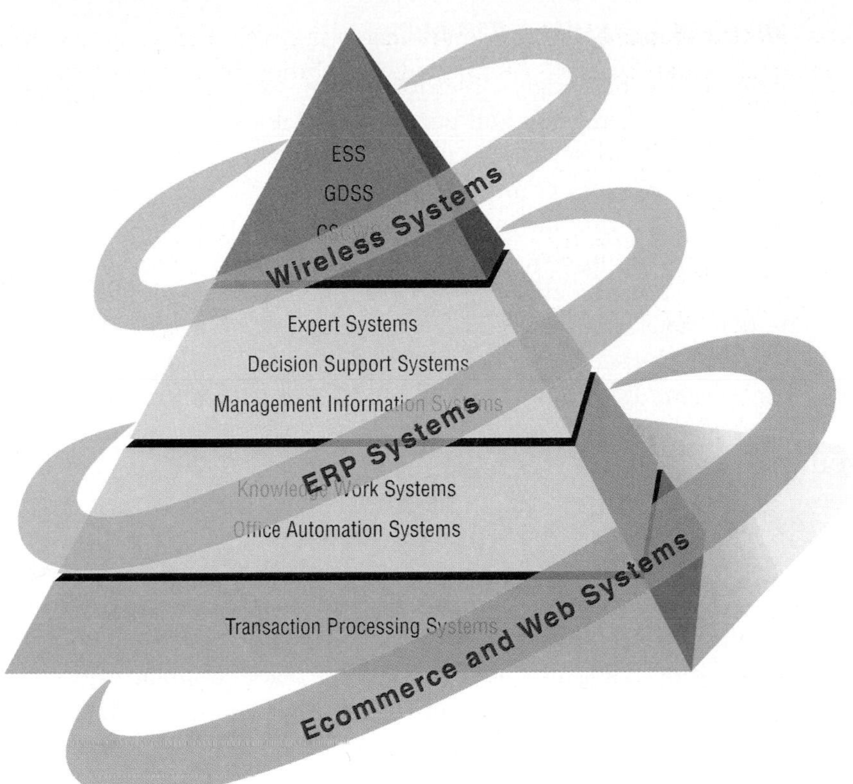

FIGURE 1.2
Systems analysts need to be aware that integrating technologies affect all types of systems.

ing year, half of all small- to mid-sized businesses responded that the Internet was their most favored strategy to pursue business growth. This response was more than twice the number of those who said they would be pursuing a strategic alliance as a way to grow. There are many benefits to mounting an application on the Web:

1. Increasing awareness of the availability of a service, product, industry, person, or group.
2. The possibility of 24-hour access for users.
3. Standardizing the design of the interface.
4. Creating a system that can extend globally rather than remain local, thus reaching people in remote locations without worry of the time zone in which they are located.

ENTERPRISE RESOURCE PLANNING SYSTEMS

Many organizations envision potential benefits from the integration of many information systems existing on different management levels and within different functions. Enterprise resource planning (ERP) systems are designed to perform this integration. Instituting ERP requires enormous commitment and organizational change. Often systems analysts serve as consultants to ERP endeavors that use proprietary software. Popular ERP software includes SAP and mysap as well as PeopleSoft, and packages from Oracle, Bään, and JD Edwards. Some of these packages are targeted toward moving enterprises onto the Web. Typically, analysts as well as some users require vendor training, support, and maintenance to be able to properly design, install, maintain, update, and use a particular ERP package.

SYSTEMS FOR WIRELESS AND HANDHELD DEVICES

Analysts are being called to design a plethora of new systems and applications, including many for wireless devices and personal digital assistants (PDAs) such as the popular Palm handheld series. In addition, analysts may find themselves designing standard or wireless communications networks that integrate voice, video, and email into organizational intranets or industry extranets. Wireless ecommerce is referred to as mcommerce (for mobile commerce).

In more advanced settings, analysts may be called upon to design intelligent agents, software that can assist users in tasks where the software learns preferences of the users over time and then acts on those preferences. For example, in the use of pull technology an intelligent agent would search the Web for stories of interest to the user having observed the users' behavior patterns with information over time and would conduct searches on the Web without continual prompting from the user.

One example of this type of software is that being developed by Microsoft based on Bayesian statistics (using statistics to infer probabilities) and decision-making theory in combination with monitoring a user's behavior concerning the handling of incoming information (such as a message from home, a phone call from a client, a call on the cell phone, or updated analysis of one's stock portfolio). The result is notification manager software that also places a dollar value to each piece of incoming information from a variety of sources and how it should best be displayed. For instance, based on decision theory, probability statistics, and the user's own previous behavior, a phone call from home could be valued at a $1.00 and could pop up on their computer screen, whereas a cold sales call could be valued at 20 cents (i.e., lower value) and could appear as a note on a pager.

OPEN SOURCE SOFTWARE

An alternative to traditional software development where proprietary code is hidden from the users is called open source software. It stands for a development model and philosophy of distributing software free and publishing its source code. In this way the code, or computer instructions, can be studied, shared, and modified by many users and programmers. Rules of this community include the idea that any program modifications must be shared with all the people on the project. Examples include the Linux operating system and Apache software used for servers that host Web sites.

If software is distributed free, how will software companies make money? They will provide a service, customizing programs for users and then following up with continued support. In an open source software world, systems development would continue its evolution into a service industry. It would move away from a manufacturing model where products are shrink-wrapped and packaged in eye-catching boxes and shipped out the door, just as any manufactured product might be.

Open source development is useful for handheld devices and communication equipment. Its use may encourage progress in creating standards for devices to communicate more easily. It is not certain if or when open source software would be widely accepted on high-end computers. It is conceivable that if it were, some of the severe shortages of programmers would be assuaged, and some large problems could be solved through extensive collaboration.

NEED FOR SYSTEMS ANALYSIS AND DESIGN

Systems analysis and design, as performed by systems analysts, seeks to analyze data input or data flow systematically, processing or transforming data, data storage, and information output within the context of a particular business.

"And that's why we need a computer."

Furthermore, systems analysis and design is used to analyze, design, and implement improvements in the functioning of businesses that can be accomplished through the use of computerized information systems.

Installing a system without proper planning leads to great dissatisfaction and frequently causes the system to fall into disuse. Systems analysis and design lends structure to the analysis and design of information systems, a costly endeavor that might otherwise have been done in a haphazard way. It can be thought of as a series of processes systematically undertaken to improve a business through the use of computerized information systems. A large part of systems analysis and design involves working with current and eventual users of information systems.

Anyone who interacts with an information system in the context of his or her work in the organization can be called an end user. Over the years the distinctions among users have become blurred. Further, any categories of users employed should not be thought of as exclusive.

However end users are classified, one fact about them remains pertinent to the systems analyst: Some kind of user involvement throughout the systems project is critical to the successful development of computerized information systems. Systems analysts, whose roles in the organization are discussed next, are the other essential component in developing useful information systems.

ROLES OF THE SYSTEMS ANALYST

The systems analyst systematically assesses how businesses function by examining the inputting and processing of data and the outputting of information with the intent of improving organizational processes. Many improvements involve better

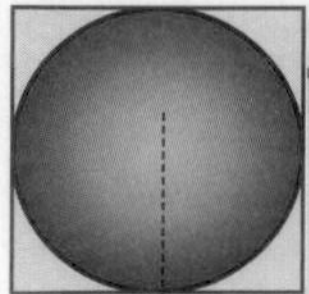

CONSULTING OPPORTUNITY 1.1

HEALTHY HIRING: ECOMMERCE HELP WANTED

"You'll be happy to know that we made a strong case to management that we should hire a new systems analyst to specialize in ecommerce development," says Al Falfa, a systems analyst for the multi-outlet international chain of Marathon Vitamin Shops. He is meeting with his large team of systems analysts to decide on the qualifications that their new team member should possess. Al continues, saying, "In fact, they were so excited by the possibility of our team helping to move Marathon into an ecommerce strategy that they've said we should start our search now and not wait until the fall.

Ginger Rute, another analyst, agrees, saying, "As long as the economy is healthy, the demand for Web site developers is far outstripping the supply. We should move quickly. I think our new person should be knowledgeable in CASE tools, Visual Basic, and Java Script, just to name a few."

Al looks surprised at Ginger's long list of languages, but then replies, "Well, that's certainly one way we could go. But I would also like to see a person with some business savvy. Most of the people coming out of school will have solid programming skills, but they should know about accounting, inventory, and distribution of goods and services, too."

The newest member of the system analysis group, Vita Minn, finally breaks into the discussion. She says, "One of the reasons I chose to come to work with all you was that I thought we all got along quite well together. Because I had some other opportunities, I looked very carefully at what the atmosphere was here. From what I've seen, we're a friendly group. Let's be sure to hire someone who has a good personality and who fits in well with us."

Al concurs, continuing, "Vita's right. The new person should be able to communicate well with us, and with business clients too. We are always communicating in some way, through formal presentations, drawing diagrams, or interviewing users. If they understand decision making, it will make their job easier, too. Also, Marathon is interested in integrating ecommerce into the entire business. We need someone who at least grasps the strategic importance of the Web. Page design is such a small part of it."

Ginger injects again with a healthy dose of practicality, saying, "Leave that to management. I still say the new person should be a good programmer." Then she ponders aloud, "I wonder how important UML will be?"

After listening patiently to everyone's wish list, one of the senior analysts, Cal Siem, speaks up, jokingly saying, "We'd better see if Superman is available!"

As the group shares a laugh, Al sees an opportunity to try for some consensus, saying, "We've had a chance to hear a number of different qualifications. Let's each take a moment and make a list of the qualifications we personally think are essential for the new ecommerce development person to possess. We'll share them and continue discussing until we can describe the person in enough detail to turn a description over to the human resources group for processing."

What qualifications should the systems analysis team be looking for when hiring their new ecommerce development team member? Is it more important to know specific languages or to have an aptitude for picking up languages and software packages quickly? How important is it that the person being hired have some basic business understanding? Should all team members possess identical competencies and skills? What personality or character traits are desirable in a systems analyst who will be working in ecommerce development?

support of business functions through the use of computerized information systems. This definition emphasizes a systematic, methodical approach to analyzing—and potentially improving—what is occurring in the specific context created by a business.

Our definition of a systems analyst is necessarily broad. The analyst must be able to work with people of all descriptions and be experienced in working with computers. The analyst plays many roles, sometimes balancing several at the same time. The three primary roles of the systems analyst are consultant, supporting expert, and agent of change.

SYSTEMS ANALYST AS A CONSULTANT

The systems analyst frequently acts as a systems consultant to a business and thus may be hired specifically to address information systems issues within a business. Such hiring can be an advantage because outside consultants can bring with them a fresh perspective that other members of an organization do not possess. It also means that outside analysts are at a disadvantage because the true organizational culture can never be known to an outsider. As an outside consultant, you will rely heavily on the systematic methods discussed throughout this text to analyze and design appropriate information systems for a particular business. In addition, you will rely on information system users to help you understand the organizational culture from others' viewpoints.

SYSTEMS ANALYST AS SUPPORTING EXPERT

Another role that you may be required to play is that of supporting expert within a business where you are regularly employed in some systems capacity. In this role the analyst draws on professional expertise concerning computer hardware and software and their uses in the business. This work is often not a full-blown systems project, but rather it entails a small modification or decision affecting a single department.

As the support expert, you are not managing the project; you are merely serving as a resource for those who are. If you are a systems analyst employed by a manufacturing or service organization, many of your daily activities may be encompassed by this role.

SYSTEMS ANALYST AS AGENT OF CHANGE

The most comprehensive and responsible role that the systems analyst takes on is that of agent of change, whether internal or external to the business. As an analyst, you are an agent of change whenever you perform any of the activities in the systems development life cycle (discussed in the next section) and are present in the business for an extended period (from two weeks all the way to more than a year). An agent of change can be defined as a person who serves as a catalyst for change, develops a plan for change, and works with others in facilitating that change.

Your presence in the business changes it. As a systems analyst, you must recognize this fact and use it as a starting point for your analysis. Hence, you must interact with users and management (if they are not one and the same) from the very beginning of your project. Without their help you cannot understand what is happening in an organization, and real change cannot take place.

If change (that is, improvements to the business that can be realized through information systems) seems warranted after analysis, the next step is to develop a plan for change along with the people who must enact the changes. Once a consensus is reached on the change that is to be made, you must constantly interact with those who are changing. You facilitate change by using your expertise with humans as well as with computers to bring about their integration in a human–machine information system.

As a systems analyst acting as an agent of change, you advocate a particular avenue of change involving the use of information systems. In addition, you teach users the process of change, because you are aware that changes in the information system do not occur independently but cause changes in the rest of the organization as well.

QUALITIES OF THE SYSTEMS ANALYST

From the foregoing descriptions of the roles the systems analyst plays, it is easy to see that the successful systems analyst must possess a wide range of qualities. Many different kinds of people are systems analysts, so any description is destined to fall short in some way. There are some qualities, however, that most systems analysts seem to display.

Above all, the analyst is a problem solver. He or she is a person who views the analysis of problems as a challenge and who enjoys devising workable solutions. When necessary, the analyst must be able to systematically tackle the situation at hand through skillful application of tools, techniques, and experience. The analyst must also be a communicator capable of relating meaningfully to other people over extended periods of time. Systems analysts need enough computer experience to program, understand the capabilities of computers, glean information requirements from users, and communicate what is needed to programmers.

It is appropriate at this time for you to reflect on the personal and professional ethics you bring to a consulting relationship. Most probably your values have been shaped by your parents, your education, and perhaps formal spiritual training. When you begin interacting with and advising others, you must be aware of the ethical framework that will shape your responses. Work to clarify the values that you embrace as you build relationships (and create actual and psychological contracts) with users and team members. Self-knowledge can help you to become a better analyst, and the codes of conduct of professional groups such as the Association for Computing Machinery (ACM) provide a reasonable context for examining your ethical beliefs. You can access the ACM Code of Conduct at http://www.acm.org/constitution/code.html.

The systems analyst must be a self-disciplined, self-motivated individual who is able to manage and coordinate innumerable project resources, including other people. Systems analysis is a demanding career but, in compensation, an ever-changing and always challenging one.

THE SYSTEMS DEVELOPMENT LIFE CYCLE

Throughout this chapter we have referred to the systematic approach analysts take to the analysis and design of information systems. Much of this is embodied in what is called the systems development life cycle (SDLC). The SDLC is a phased approach to analysis and design that holds that systems are best developed through the use of a specific cycle of analyst and user activities.

Analysts disagree on exactly how many phases there are in the systems development life cycle, but they generally laud its organized approach. Here we have divided the cycle into seven phases, as shown in Figure 1.3. Although each phase is presented discretely, it is never accomplished as a separate step. Instead, several activities can occur simultaneously, and activities may be repeated. It is more useful to think of the SDLC as accomplished in phases (with activities in full swing overlapping with others and then tapering off) and not in separate steps.

FIGURE 1.3
The seven phases of the systems development life cycle.

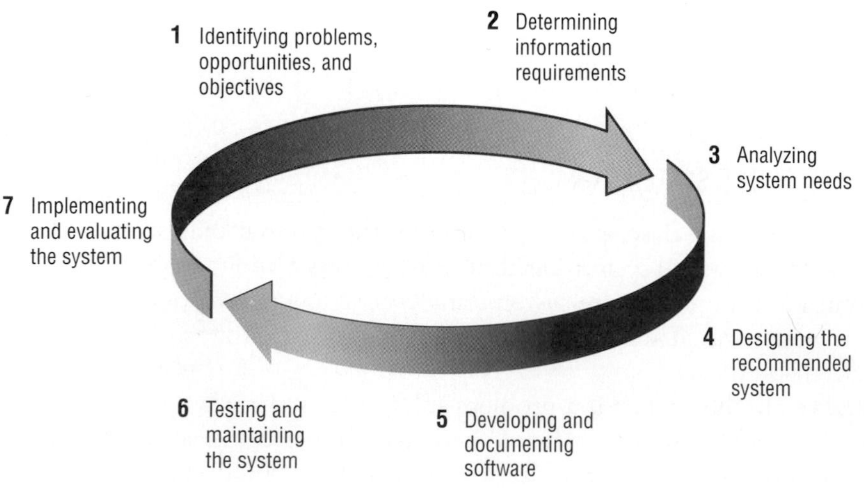

IDENTIFYING PROBLEMS, OPPORTUNITIES, AND OBJECTIVES

In this first phase of the systems development life cycle, the analyst is concerned with identifying problems, opportunities, and objectives. This stage is critical to the success of the rest of the project, because no one wants to waste subsequent time addressing the wrong problem.

The first phase requires that the analyst look honestly at what is occurring in a business. Then, together with other organizational members, the analyst pinpoints problems. Often, these problems will be brought up by others, and they are the reason the analyst was initially called in. Opportunities are situations that the analyst believes can be improved upon through the use of computerized information systems. Seizing opportunities may allow the business to gain a competitive edge or set an industry standard.

Identifying objectives is also an important component of the first phase. First, the analyst must discover what the business is trying to do. Then the analyst will be able to see if some aspect of information systems applications can help the business reach its objectives by addressing specific problems or opportunities.

The people involved in the first phase are the users, analysts, and systems managers coordinating the project. Activities in this phase consist of interviewing user management, summarizing the knowledge obtained, estimating the scope of the project, and documenting the results. The output of this phase is a feasibility report containing a problem definition and summarizing the objectives. Management must then make a decision on whether to proceed with the proposed project. If the user group does not have sufficient funds in its budget or wishes to tackle unrelated problems, or if the problems do not require a computer system, a manual solution may be recommended, and the systems project does not proceed any further.

DETERMINING INFORMATION REQUIREMENTS

The next phase that the analyst enters is that of determining information requirements for the particular users involved. Among the tools used to define information requirements in the business are sampling and investigating hard data, interviewing, questionnaires, observing decision makers' behavior and office environments, and even prototyping.

Rapid application development (RAD) is an object-oriented approach to systems development that includes a method of development (including the generating of information requirements) as well as software tools. In this text it is paired with prototyping in Chapter 8, because the philosophical approach it takes is similar to prototyping, even though the method for creating a design quickly and getting rapid feedback from users differs somewhat. (There is more about object-oriented approaches in Chapter 22).

In the information requirements phase of the SDLC, the analyst is striving to understand what information users need to perform their jobs. You can see that several of the methods for determining information requirements involve interacting directly with users. This phase serves to fill in the picture that the analyst has of the organization and its objectives. Sometimes only the first two phases of the systems development life cycle are completed. This kind of study may have a different purpose and is typically carried out by a specialist called an information analyst (IA).

The people involved in this phase are the analysts and users, typically operations managers and operations workers. The systems analyst needs to know the details of current system functions: the who (the people who are involved), what (the business activity), where (the environment in which the work takes place), when (the timing), and how (how the current procedures are performed) of the business under study. The analyst must then ask why the business uses the current system. There may be good reasons for doing business using the current methods, and these should be considered when designing any new system.

If the reason for current operations is that "it's always been done that way," however, the analyst may wish to improve on the procedures. Business process reengineering may be of help in framing an approach for rethinking the business in a creative way. At the completion of this phase, the analyst should understand how the business functions and have complete information on the people, goals, data, and procedures involved.

ANALYZING SYSTEM NEEDS

The next phase that the systems analyst undertakes involves analyzing system needs. Again, special tools and techniques help the analyst make requirement determinations. One such tool is the use of data flow diagrams to chart the input, processes, and output of the business's functions in a structured graphical form. From the data-flow diagrams, a data dictionary is developed that lists all the data items used in the system, as well as their specifications: whether they are alphanumeric or text and how much space they take up when printed.

During this phase the systems analyst also analyzes the structured decisions made. Structured decisions are those for which the conditions, condition alternatives, actions, and action rules can be determined. There are three major methods for analysis of structured decisions: structured English, decision tables, and decision trees.

Not all decisions in organizations are structured, but it is still important for the systems analyst to understand them. Semistructured decisions (decisions made under risk) are often supported by decision support systems. When analyzing semistructured decisions, the analyst examines the decisions based on the degree of decision-making skill required, the degree of problem complexity, and the number of criteria considered when the decision is made.

Analysis of multiple-criteria decisions (decisions where many factors must be balanced) is also part of this phase. Many techniques are available for analyzing multiple-criteria decisions, including the tradeoff process and the use of weighting methods.

At this point in the systems development life cycle, the systems analyst prepares a systems proposal that summarizes what has been found, provides cost/benefit analyses of alternatives, and makes recommendations on what (if anything) should be done. If one of the recommendations is acceptable to management, the analyst proceeds along that course. Each systems problem is unique, and there is never just one correct solution. The manner in which a recommendation or solution is formulated depends on the individual qualities and professional training of each analyst.

DESIGNING THE RECOMMENDED SYSTEM

In the design phase of the systems development life cycle, the systems analyst uses the information collected earlier to accomplish the logical design of the information system. The analyst designs accurate data-entry procedures so that data going into the information system are correct. In addition, the analyst provides for effective input to the information system by using techniques of good form and screen design.

Part of the logical design of the information system is devising the user interface. The interface connects the user with the system and is thus extremely impor-

"You can't just punch in 'let there be light' without writing the code underlying the user interface functions."

tant. Examples of user interfaces include a keyboard (to type in questions and answers), onscreen menus (to elicit user commands), and a variety of Graphical User Interfaces (GUIs) that use a mouse or touch screen.

The design phase also includes designing files or databases that will store much of the data needed by decision makers in the organization. A well-organized database is the basis for all information systems. In this phase the analyst also works with users to design output (either onscreen or printed) that meets their information needs.

Finally, the analyst must design controls and backup procedures to protect the system and the data and to produce program specification packets for programmers. Each packet should contain input and output layouts, file specifications, and processing details; it may also include decision trees or tables, dataflow diagrams, a system flowchart, and the names and functions of any prewritten code routines.

DEVELOPING AND DOCUMENTING SOFTWARE

In the fifth phase of the systems development life cycle, the analyst works with programmers to develop any original software that is needed. Some of the structured techniques for designing and documenting software include structure charts, Nassi-Shneiderman charts, and pseudocode. The systems analyst uses one or more of these devices to communicate to the programmer what needs to be programmed.

During this phase the analyst also works with users to develop effective documentation for software, including procedure manuals, online help, and Web sites featuring Frequently Asked Questions (FAQ), on "Read Me" files shipped with new software. Documentation tells users how to use software and what to do if software problems occur.

Programmers have a key role in this phase because they design, code, and remove syntactical errors from computer programs. If the program is to run in a mainframe environment, job control language (JCL) must be created. To ensure quality, a programmer may conduct either a design or a code walkthrough, explaining complex portions of the program to a team of other programmers.

TESTING AND MAINTAINING THE SYSTEM

Before the information system can be used, it must be tested. It is much less costly to catch problems before the system is signed over to users. Some of the testing is completed by programmers alone, some of it by systems analysts in conjunction with programmers. A series of tests to pinpoint problems is run first with sample data and eventually with actual data from the current system.

Maintenance of the system and its documentation begins in this phase and is carried out routinely throughout the life of the information system. Much of the programmer's routine work consists of maintenance, and businesses spend a great deal of money on maintenance. Some maintenance, such as program updates, can be done automatically via a vendor site on the World Wide Web. Many of the systematic procedures the analyst employs throughout the systems development life cycle can help ensure that maintenance is kept to a minimum.

IMPLEMENTING AND EVALUATING THE SYSTEM

In this last phase of system development, the analyst helps implement the information system. This phase involves training users to handle the system. Some training is done by vendors, but oversight of training is the responsibility of the systems analyst. In addition, the analyst needs to plan for a smooth conversion from the old system to the new one. This process includes converting files from old formats to new ones or building a database, installing equipment, and bringing the new system into production.

Evaluation is shown as part of this final phase of the systems development life cycle mostly for the sake of discussion. Actually, evaluation takes place during every phase. A key criterion that must be satisfied is whether the intended users are indeed using the system.

It should be noted that systems work is often cyclical. When an analyst finishes one phase of system development and proceeds to the next, the discovery of a problem may force the analyst to return to the previous phase and modify the work done there. For example, during the testing phase, the programmer may discover that the program does not work correctly, either because code was not written to support certain portions of the system design or the design was incomplete. In either event the programs must be modified, and the analyst may have to change some of the system design materials. In turn, it may be necessary for the analyst to meet with the user and reinvestigate how a specific business activity functions.

THE IMPACT OF MAINTENANCE

After the system is installed, it must be maintained, meaning that the computer programs must be modified and kept up to date. Figure 1.4 illustrates the average amount of time spent on maintenance at a typical MIS installation. Estimates of the time spent by departments on maintenance have ranged from 48 to 60 percent of the total time spent developing systems. Very little time remains for new system

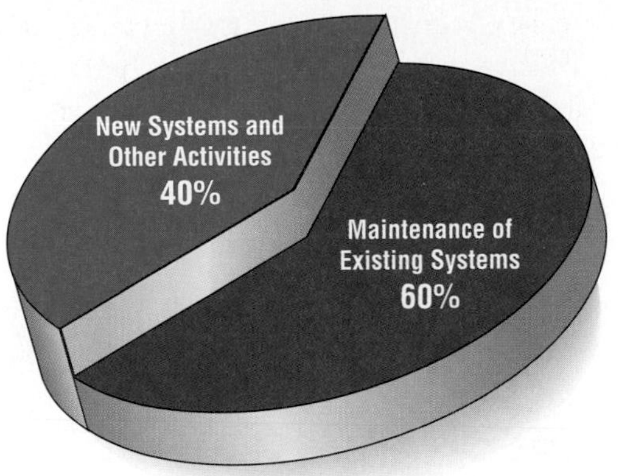

FIGURE 1.4
Some researchers estimate that the amount of time spent on system maintenance may be as much as 60 percent of the total time spent on systems projects.

development. As the number of programs written increases, so does the amount of maintenance they require.

Maintenance is performed for two reasons. The first of these is to correct software errors. No matter how thoroughly the system is tested, bugs or errors creep into computer programs. Bugs in commercial PC software are often documented as "known anomalies" and are corrected when new versions of the software are released or in an interim release. In customized software, bugs must be corrected as they are detected.

The other reason for performing system maintenance is to enhance the software's capabilities in response to changing organizational needs, generally involving one of the following three situations:

1. *Users often request additional features after they become familiar with the computer system and its capabilities.* These requested features may be as simple as displaying additional totals on a report or as complicated as developing new software.
2. *The business changes over time.* Software must be modified to encompass changes as new government or corporate reporting requirements arise, as new client information needs to be produced, and so on.
3. *Hardware and software are changing at an accelerated pace.* A system that uses older technology may be modified to use the capabilities of newer technology. An example of such a change is replacing a mainframe with a client/server system in which client PCs house some data and processing and servers provide data and applications by splitting up tasks efficiently via a local area network (LAN).

Figure 1.5 illustrates the amount of resources—usually time and money—spent on system development and maintenance. The area under the curve represents the total dollar amount spent. You can see that over time the total cost of maintenance is likely to exceed that of system development. At a certain point it becomes more feasible to perform a new systems study, because the cost of continued maintenance is clearly greater than that of creating an entirely new information system.

In summary, maintenance is an ongoing process over the life cycle of an information system. After the information system is installed, maintenance usually takes the form of correcting previously undetected program errors. Once these are corrected, the system approaches a steady state, providing dependable service to its users. Maintenance during this period may consist of removing a few previously

FIGURE 1.5

Resource consumption over the system life.

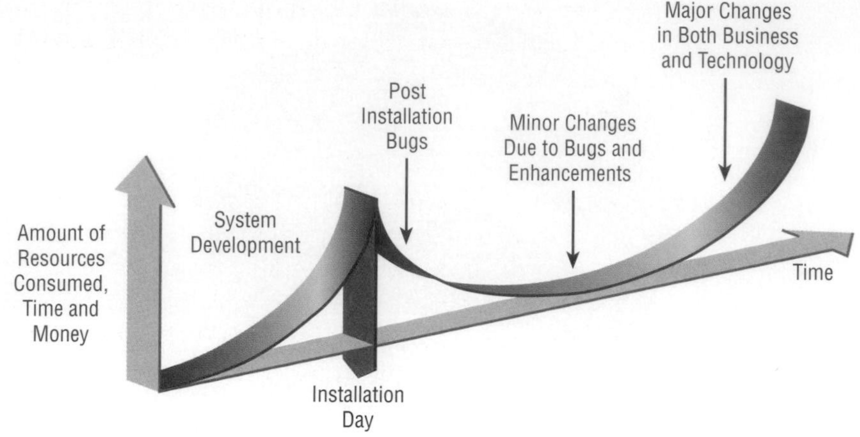

undetected bugs and updating the system with a few minor enhancements. As time goes on and the business and technology change, however, the maintenance effort increases dramatically.

USING CASE TOOLS

Throughout this book we emphasize the need for a systematic, thorough approach to the analysis, design, and implementation of information systems. We recognize that to be productive, systems analysts must be organized, accurate, and complete in what they set out to do. Since the early 1990s, analysts have begun to benefit from productivity tools, called Computer-Aided Software Engineering (CASE) tools, that have been created explicitly to improve their routine work through the use of automated support. In a recent study it was found that larger IS departments, those with more than 10 employees, were most likely than smaller IS departments to adopt CASE tools. Organizational systems, procedures, and management practices may constrain the spread of CASE tools. Therefore, systems analysts who want to foster the use of CASE tools should work closely with the organization to ensure that it is supportive of CASE tool adoption and usage in all these aspects. Analysts rely on CASE tools to increase productivity, communicate more effectively with users, and integrate the work that they do on the system from the beginning to the end of the life cycle.

INCREASING ANALYST PRODUCTIVITY

Visible Analyst allows its users to draw and modify diagrams easily. By our definition the analyst can thus become more productive simply by reducing the considerable time typically spent in manually drawing and redrawing data-flow diagrams until they are acceptable.

A tools package such as Visible Analyst also enhances group productivity by allowing analysts to share work easily with other team members, who can simply access the file on their PCs and review or modify what has been done. Such sharing reduces the time necessary to reproduce and distribute data-flow diagrams among team members. Thus, rather than mandating a strict distribution and feedback response schedule, such a tools package further allows members of the systems analysis team to work with the diagrams whenever they have the time.

CASE tools also facilitate interaction among team members by making diagramming a dynamic, iterative process rather than one in which changes are cumbersome and therefore tend to become a drain on productivity. In this instance the CASE tool for drawing and recording data-flow diagrams affords a record of the team's changing thinking regarding data flows.

IMPROVING ANALYST–USER COMMUNICATION

For the proposed system to come into being and actually be used, excellent communication among analysts and users throughout the systems development life cycle is essential. The success of the eventual system implementation rests on the capability of analysts and users to communicate in a meaningful way. So far, it has been the experience of analysts currently using CASE tools that their use fosters greater, more meaningful communication among users and analysts.

Providing a Means of Communication. Analysts and users alike report that CASE tools afford them a means of communication about the system during its conceptualization. Through the use of automated support featuring onscreen output, clients can readily see how data flows and other system concepts are depicted, and they can then request corrections or changes that would have taken too much time with a manual system.

Whether a particular diagram will be adjudged useful by users or analysts at the end of the project is questionable. What is important is that such automated support for many life cycle design activities serves as a means to an end by acting as a catalyst for analyst–user interaction. The same arguments used to support CASE tools' role in increasing productivity are equally valid in this arena; that is, the manual tasks of drawing, reproducing, and distributing take much less time, so work in progress can be shared more easily with users.

INTEGRATING LIFE CYCLE ACTIVITIES

The third reason for using CASE tools is to integrate activities and provide continuity from one phase to the next throughout the systems development life cycle.

CASE tools are especially useful when a particular phase of the life cycle requires several iterations of feedback and modification. Recall that user involvement can be important in each of the phases. Integration of activities through the underlying use of technologies makes it easier for users to understand how all the life cycle phases are interrelated and interdependent.

ACCURATELY ASSESSING MAINTENANCE CHANGES

The fourth and possibly one of the most important reasons for using CASE tools is that they enable users to analyze and assess the impact of maintenance changes. For example, the size of an element such as a customer number may need to be made larger. The CASE tool will cross-reference every screen, report, and file within which the element is used, leading to a comprehensive maintenance plan.

UPPER AND LOWER CASE

CASE tools are classified as lower CASE, upper CASE, and integrated CASE, which combines both upper and lower CASE in one toolset. Although experts disagree about what precisely constitutes an upper CASE tool versus a lower CASE tool, it might be helpful to conceptualize upper CASE tools on the basis of whom they support. Upper CASE tools primarily help analysts and designers. Lower CASE tools are used more often by programmers and workers who must implement the systems designed via upper CASE tools.

UPPER CASE TOOLS

An upper CASE tool allows the analyst to create and modify the system design. All the information about the project is stored in an encyclopedia called the CASE repository, a large collection of records, elements, diagrams, screens, reports, and other information. (See Figure 1.6.) Analysis reports may be produced using the repository information to show where the design is incomplete or contains errors.

Upper CASE tools can also help support the modeling of an organization's functional requirements, assist analysts and users in drawing the boundaries for a given project, and help them visualize how the project meshes with other parts of the organization. In addition, some upper CASE tools can support prototyping of screen and report designs.

FIGURE 1.6
The repository concept.

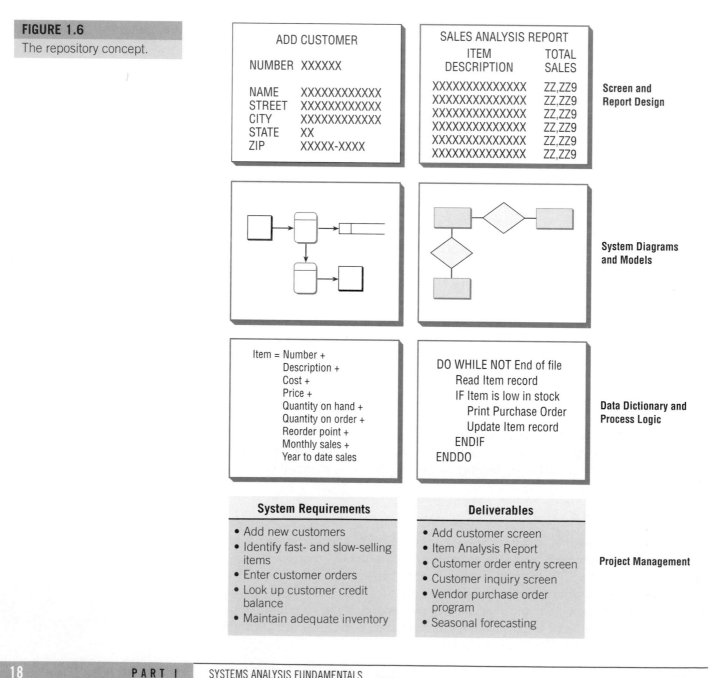

Screen and Report Design

```
ADD CUSTOMER

NUMBER  XXXXXX

NAME    XXXXXXXXXXXX
STREET  XXXXXXXXXXXX
CITY    XXXXXXXXXXXX
STATE   XX
ZIP     XXXXX-XXXX
```

```
SALES ANALYSIS REPORT
     ITEM             TOTAL
  DESCRIPTION         SALES

XXXXXXXXXXXXXX        ZZ,ZZ9
XXXXXXXXXXXXXX        ZZ,ZZ9
XXXXXXXXXXXXXX        ZZ,ZZ9
XXXXXXXXXXXXXX        ZZ,ZZ9
XXXXXXXXXXXXXX        ZZ,ZZ9
XXXXXXXXXXXXXX        ZZ,ZZ9
```

System Diagrams and Models

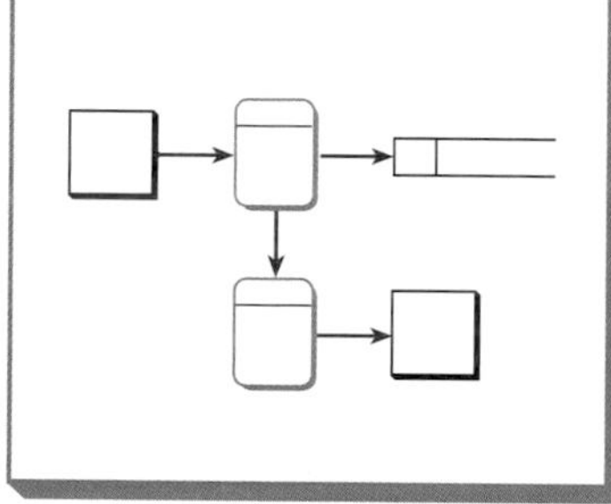

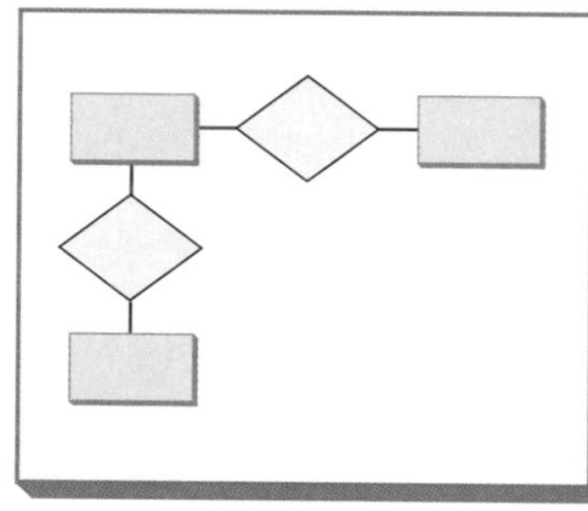

Data Dictionary and Process Logic

```
Item = Number +
       Description +
       Cost +
       Price +
       Quantity on hand +
       Quantity on order +
       Reorder point +
       Monthly sales +
       Year to date sales
```

```
DO WHILE NOT End of file
   Read Item record
   IF Item is low in stock
      Print Purchase Order
      Update Item record
   ENDIF
ENDDO
```

Project Management

System Requirements
- Add new customers
- Identify fast- and slow-selling items
- Enter customer orders
- Look up customer credit balance
- Maintain adequate inventory

Deliverables
- Add customer screen
- Item Analysis Report
- Customer order entry screen
- Customer inquiry screen
- Vendor purchase order program
- Seasonal forecasting

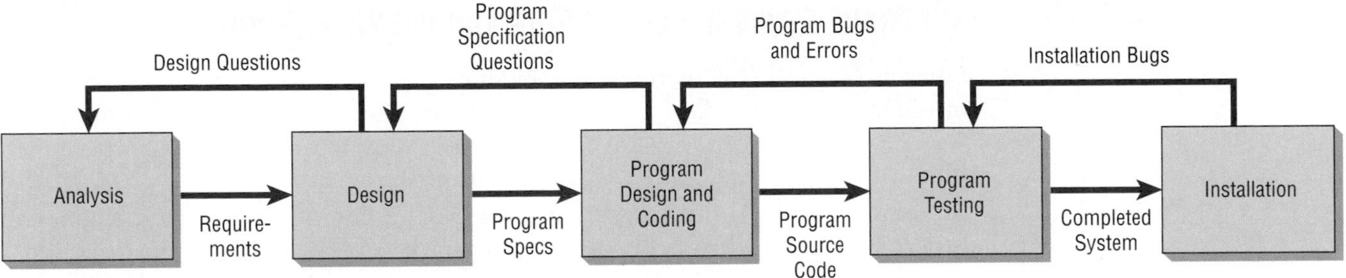

Traditional Systems Development Life Cycle

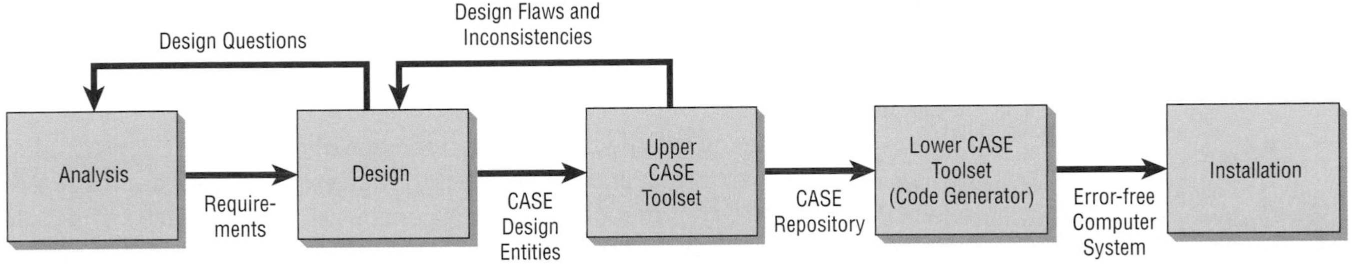

CASE Systems Development Life Cycle

FIGURE 1.7
Traditional versus CASE systems development life cycles.

LOWER CASE TOOLS

Lower CASE tools are used to generate computer source code, eliminating the need for programming the system. Code generation has several advantages:

1. The system may be produced more quickly than by writing computer programs. Becoming familiar with the methodology used by the code generator, however, often takes a great deal of time, so program generation may initially be slower. In addition, the design must be thoroughly entered into the toolset, which may require a lengthy period of time.
2. The amount of time spent on maintenance decreases with code generation. There is no need to modify, test, and debug computer programs. Instead, the CASE design is modified, and the code is regenerated. Decreased time spent on maintenance results in more time to develop new systems and helps to relieve a backlog of projects under consideration for development.
3. Code may be generated in more than one computer language, so it is easier to migrate systems from one platform, such as a mainframe, to another, perhaps a PC. For example, VA Corporate Edition, Version 7.5, can generate full ANSI, COBOL, or C-language 3 GL source code.
4. Code generation provides a cost-effective way of tailoring systems purchased from third-party vendors to the needs of the organization. Often, modifying purchased software requires such great effort that the cost of doing so exceeds that of the software. With code generation software, purchasing CASE design and repository for the application enables the analyst to modify the design and generate the revised computer system.
5. Generated code is free of computer program errors. The only potential errors are design errors, which may be minimized by running CASE analysis reports to ensure that the system design is complete and correct.

Figure 1.7 illustrates the traditional systems development life cycle and the CASE life cycle. Notice that the program coding, testing, and debugging portions of the cycle are eliminated from the CASE life cycle.

SOFTWARE REVERSE ENGINEERING AND REENGINEERING

Software reverse engineering and reengineering are methods for extending the life of older programs, called legacy software. Both approaches use computer-assisted reengineering CARE software to analyze and restructure existing computer code. Several reverse engineering toolsets are available.

Note that the term *reengineering* is used in a number of different engineering, programming, and business contexts. Often, it is used to mean "business process reengineering," which is a way of reorienting an organization around key processes. Systems analysts can play an important role in business process reengineering, because many of the necessary changes are possible only through the availability of innovative information technology. One company developing software to aid the analyst in business process reengineering is Holosofx. Through the use of these tools and others, many analysts are now being educated in ways to facilitate organizational change.

Reverse engineering is the opposite of code generation. The computer source code is examined, analyzed, and converted into repository entities, as illustrated in Figure 1.8. The first step in software reverse engineering is to load existing computer programs' code (as written in COBOL, C, or another high-level language) into the toolset. Depending on the reverse engineering toolset used, the code is analyzed and the toolset produces some or all the following:

1. Data structures and elements, describing the files and records stored by the system
2. Screen designs, if the program is on line
3. Report layouts for batch programs
4. A structure chart showing the hierarchy of the modules in the program
5. Database design and relationships

The design stored in the repository may be modified or incorporated into other CASE project information. When all the modifications are complete, the new system code may be regenerated. Reengineering refers to the complete process of converting program code to the CASE design, modifying the design, and regenerating the new program code.

The advantages of using a reverse engineering toolset are numerous:

1. The time required for system maintenance is reduced, freeing up time for new development.
2. Documentation, which may have been nonexistent or minimal for older programs, is produced.

FIGURE 1.8
Reverse engineering concepts.

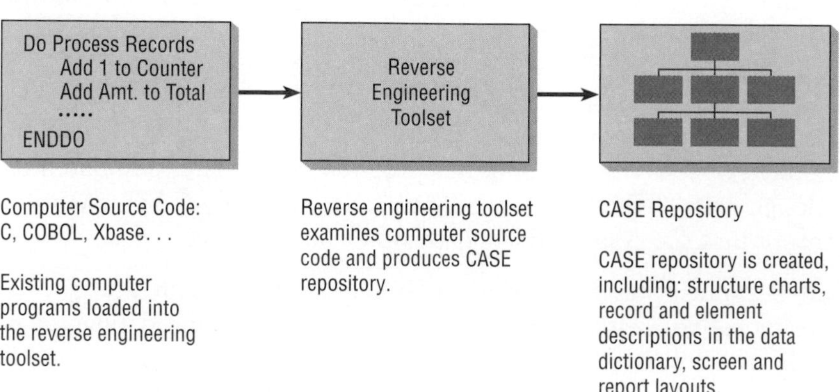

3. Structured programs are created from unstructured or loosely structured computer code.
4. Future maintenance changes are easier to make, because changes may be made at the design level rather than at the code level.
5. Analysis may be performed on the system to eliminate unused portions of computer code, code that may still exist in older programs even though it was made obsolete by revisions of the program throughout the years.

OBJECT-ORIENTED SYSTEMS ANALYSIS AND DESIGN

A very different approach to systems development is object-oriented (O-O) systems analysis and design. Object-oriented techniques, which are based on object-oriented programming concepts, can help analysts respond to organizational demands for new systems that must undergo continuous maintenance, adaptation, and redesign. In object-oriented programming, objects are created that include not only code about data, but also instructions about the operations to be performed on it.

Operational prototypes (discussed in Chapter 8) are frequently used during the design phase, often done through rapid application development (RAD). Chapter 22 provides a practical explanation of object-oriented analysis (OOA) and design (OOD) using the standardized, graphical notation unified modeling language (UML), a standardized language in which objects that are created include not only code about data but also instructions about the operations to be performed on it. That chapter takes the structured analysis and design presented in the preceding 21 chapters as its point of departure. UML is used to describe, envision, and then document an object-oriented system that is being developed.

NEED FOR STRUCTURED ANALYSIS AND DESIGN

Structured analysis and design provides a systematic approach to designing and building quality computer systems. Throughout the phases of analysis and design, the analyst should proceed step by step, obtaining feedback from users and analyzing the design for omissions and errors. Moving too quickly to the next phase may require the analyst to rework portions of the design that were produced earlier.

Figure 1.9 illustrates the cost of correcting an error detected in each of the phases. Notice that considerably more effort is required to correct an error in each

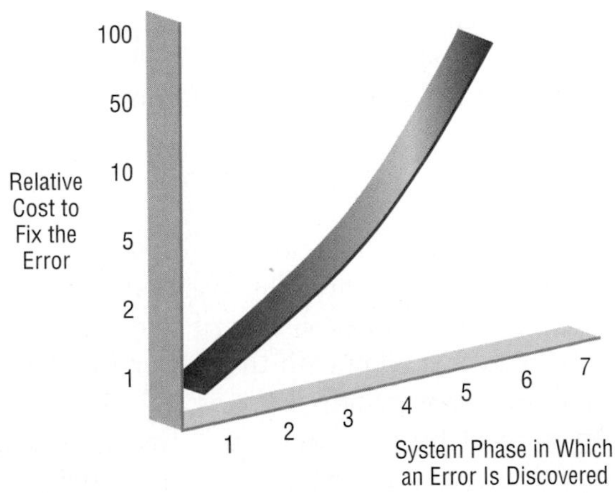

FIGURE 1.9
Cost to fix an error and the phase where the error is detected.

succeeding phase. For example, suppose several elements were omitted when the analyst was examining the details of the data used within the system. If the analyst learned that these elements were missing after programs were written, the file, report, and screen layouts would have to be modified, as would the test data files, programs, and documentation. These corrections might take 100 hours, whereas if the elements had been part of the original design, it might have taken only 4 hours at the outset to add them to the design materials and programs. Look for suggestions throughout the following chapters that indicate where analysis should take place and how to determine whether the systems design is accurate and complete.

Analysts have at their disposal a number of PC software tools that may be used to assist them in the development of systems. In addition to CASE and reverse engineering tools, the following may be used at various parts of the analysis and design life cycle: management software to optimize the allocation of people and project resources, prototyping software to rapidly create screens and reports that users can review and modify, form design tools to help design forms or source documents, and graphics and presentation software to assist in creating illustrations and producing a professional presentation to users.

ALTERNATIVE METHODOLOGIES

Although this text focuses on the most widely used approach in practice, at times the analyst will recognize that the organization could benefit from an alternative approach. Perhaps a systems project using a structured approach has recently failed, or perhaps the organizational subcultures, composed of several different user groups, seem more in step with an alternative method. We cannot do justice to these methods in a small space; each deserves and has inspired its own books and research. By mentioning these approaches here, however, we hope to help you become aware that under certain circumstances, your organization may want to consider an alternative or supplement to structured analysis and design and to the systems development life cycle.

Among the most popular alternatives are prototyping (distinct from the prototyping we discuss in Chapter 8), ETHICS, the project champion approach, Soft Systems Methodology, and Multiview. Prototyping, established in other disciplines and applied to IS (information systems), was offered as a response to the long development times associated with the systems development life cycle approach and to the uncertainty often surrounding user requirements. ETHICS was introduced as a sociotechnical methodology combining social and technical solutions. The project champion approach, a concept borrowed from marketing, adopts the strategy of involving one key person from each area affected by the system to ensure the system's success. Soft Systems Methodology was envisioned as a way to model a world that is often chaotic by using "rich pictures," ideographs that capture characteristic organizational narratives. Multiview was proposed as a way to organize and use elements of several competing methodologies.

SUMMARY

Information can be viewed as an organizational resource. As such, it must be managed carefully, just as other resources are. The availability of affordable computer power to organizations has meant an explosion of information, and consequently, more attention must be paid to coping with the information generated.

All computerized information systems have as their basis a database that stores the data necessary to support business functions. Transaction processing systems (TPS) support large-volume, routine business transactions such as payroll and

HYPERCASE® EXPERIENCE 1

"Welcome to Maple Ridge Engineering, what we call MRE. We hope you'll enjoy serving as a systems consultant for us. Although I've worked here five years in different capacities, I've just been reassigned to serve as an administrative aide to Snowden Evans, the head of the new Training and Management Systems Department. We're certainly a diverse group. As you make your way through the company, be sure to use all your skills, both technical and people-oriented, to understand who we are and to identify the problems and conflicts that you think should be solved regarding our information systems.

"To bring you up to date, let me say that Maple Ridge Engineering is a medium-sized medical engineering company. Last year, our revenues exceeded $287 million. We employ about 335 people. There are about 150 administrative employees as well as management and clerical staff like myself; approximately 75 professional employees, including engineers, physicians, and systems analysts; and about 110 trade employees, such as drafters and technicians.

"There are four offices. You will visit us through HyperCase in our home office in Maple Ridge, Tennessee. We have three other branches in the southern United States as well: Atlanta, Georgia; Charlotte, North Carolina; and New Orleans, Louisiana. We'd love to have you visit when you're in the area.

"For now, you should explore HyperCase using either Netscape Navigator or Microsoft Internet Explorer.

"To learn more about Maple Ridge Engineering as a company or to find out how to interview our employees, who will use the systems you design, and how to observe their offices in our company, you may want to start by going to the World Wide Web Internet site found at the location www.prenhall.com/kendall.

"This Web page contains useful information about the project as well as files that may be downloaded to your computer. One file is a set of Visible Analyst data files that match HyperCase. They contain a partially constructed series of data-flow diagrams, entity-relationship diagrams, and repository information. The HyperCase Web site also contains additional exercises that may be assigned. HyperCase is designed to be explored, and you should not overlook any object or clue on a Web page. To start consulting immediately, click the mouse on the MRE Reception area link and read more about HyperCase in Chapter 2."

inventory. Office automation systems (OAS) support data workers who use word processing, spreadsheets, and so on to analyze, transform, or manipulate data. Knowledge work systems (KWS) support professionals such as scientists and engineers who create new knowledge. Management information systems (MIS) are computerized information systems that support a broader range of business functions than do transaction processing systems. Most often, MIS output reports to decision makers. Decision support systems (DSS) are information systems whose output is tailored to their users' needs and that help support decision makers in making semistructured decisions. Expert systems capture the expertise of decision makers for use in solving a problem or a class of problems. Analysts may be called upon to design a variety of new systems, including recommendation systems, that combine intelligent agents, expert systems, and other Web-based technologies that permit interactivity with sophisticated filtering and polling. Group decision support systems (GDSS) and computer supported collaborative work systems (CSCWS) bring together group members in special electronic ways to help groups solve semistructured or unstructured problems. Executive support systems (ESS) help executives organize their interactions with the external environment by providing graphics and communications support in accessible locations.

Many applications are either originating on, or moving to, the Web to support ecommerce. The advantages of establishing a Web presence include increasing the awareness of a service or product, industry, person, or group; permitting round-the-clock access; designing and employing a standard user interface; and reaching remote people via global rather than local systems. Analysts are also designing networks and applications for wireless systems. When ecommerce moves to a wireless implementation it is called mcommerce (for mobile commerce).

When organizations want to integrate their many diverse functions and levels of functionality and management they may ask the analyst to assess the potential for developing an enterprise requirements planning (ERP) system. ERP requires substantial organizational change and commitment and is often implemented by specially trained consultants versed in one of several popular proprietary packages such as SAP, PeopleSoft, or JD Edwards.

A departure from traditional software development and distribution is called open source software, a development model and philosophy of distributing software free and publishing its source code (computer instructions) so that it can be studied, shared, and modified by many users and programmers.

Systems analysis and design is a systematic approach to identifying problems, opportunities, and objectives; to analyzing the information flows in organizations; and to designing computerized information systems to solve a problem. As information proliferates, a systematic, planned approach to the introduction, modification, and maintenance of information systems is essential. Systems analysis and design provide such an approach.

Systems analysts are required to take on many roles in the course of their work. Some of these roles are (1) an outside consultant to business, (2) supporting expert within a business, and (3) agent of change in both internal and external situations.

Analysts possess a wide range of skills. First and foremost, the analyst is a problem solver, someone who enjoys the challenge of analyzing a problem and devising a workable solution. Systems analysts require communication skills that allow them to relate meaningfully to many different kinds of people on a daily basis, as well as computer skills. End-user involvement is critical to their success.

Analysts proceed systematically. The framework for their systematic approach is provided in what is called the systems development life cycle (SDLC). This life cycle can be divided into seven sequential phases, although in reality the phases are interrelated and are often accomplished simultaneously. The seven phases are identifying problems, opportunities, and objectives; determining information requirements; analyzing system needs; designing the recommended system; developing and documenting software; testing and maintaining the system; and implementing and evaluating the system.

Automated, PC-based software packages for systems analysis and design are called CASE (Computer-Aided Software Engineering) tools. The four reasons for adopting CASE tools are increasing analyst productivity, improving communication among analysts and users, integrating life cycle activities, and analyzing and assessing the impact of maintenance changes.

Analysts also use CARE (computer-assisted reengineering) approaches to do software reverse engineering and reengineering to extend the life of legacy software.

A different approach to systems development is object-oriented analysis (OOA) and object-oriented design (OOD). These techniques are based on object-oriented programming concepts that have become codified in UML, a standardized modeling language in which objects that are created include not only code about data but also instructions about the operations to be performed on it.

When the organizational situation demands it, the analyst may depart from the SDLC to try an alternative methodology such as prototyping, ETHICS, the project champion approach, Soft Systems Methodology, or Multiview.

KEYWORDS AND PHRASES

- agent of change
- analyst's ethical framework
- artificial intelligence (AI)
- CARE (computer-assisted reengineering)
- CASE repository
- CASE tools
- code generation maintenance
- computer-generated information
- computer-supported collaborative work systems (CSCWS)
- decision support systems (DSS)
- ecommerce applications
- enterprise resource planning (ERP) systems
- ETHICS
- executive support systems (ESS)
- expert systems
- groupware
- group decision support systems (GDSS)
- knowledge work systems (KWS)
- legacy software
- management information systems (MIS)
- mcommerce (mobile commerce)
- migrate systems
- Multiview
- object-oriented analysis (OOA)
- object-oriented design (OOD)
- object-oriented (O-O) systems analysis and design
- office automation systems (OAS)
- open source software
- personal digital assistant (PDA)
- project champion approach
- prototyping
- rapid application development (RAD)
- reengineering
- Soft Systems Methodology
- software reverse engineering
- supporting expert
- systems analysis and design
- systems analyst
- systems consultant
- systems development life cycle (SDLC)
- transaction processing systems (TPS)
- unified modeling language (UML)
- Web systems

REVIEW QUESTIONS

1. Describe why information is most usefully thought of as an organizational resource rather than as an organizational byproduct.
2. Define what is meant by a transaction processing system.
3. Explain the difference between office automation systems (OAS) and knowledge work systems (KWS).
4. Compare the definition of management information systems (MIS) with the definition of decision support systems (DSS).
5. Define the term *expert systems*. How do expert systems differ from decision support systems?
6. List the problems of group interaction that group decision support systems (GDSS) and computer supported collaborative work systems (CSCWS) were designed to address.
7. Which is the more general term, CSCWS or GDSS? Explain.
8. Define the term *mcommerce*.
9. List the advantages of mounting applications on the World Wide Web.
10. What is the overarching reason for designing ERP systems?

11. Define what is meant by open source software.
12. List the advantages of using systems analysis and design techniques in approaching computerized information systems for business.
13. List three roles that the systems analyst is called upon to play. Provide a definition for each one.
14. What personal qualities are helpful to the systems analyst? List them.
15. List and briefly define the seven phases of the systems development life cycle (SDLC).
16. What is rapid application development (RAD)?
17. Define software reverse engineering and reengineering as they apply to CARE (computer-assisted reengineering).
18. List the four reasons for adopting CASE tools.
19. Define the terms *object-oriented analysis* and *object-oriented design*.
20. What is UML?

SELECTED BIBLIOGRAPHY

Alavi, M. "An Assessment of the Prototyping Approach to Information Systems." *Communications of the ACM*, Vol. 26, No. 6, June 1984, pp. 556–63.

Avison, D. E., and A. T. Wood-Harper. *Multiview: An Exploration in Information Systems Development*. Oxford: Blackwell Scientific Publications, 1990.

Beath, C. M. "Supporting the Information Technology Champion." *MIS Quarterly*, Vol. 15, No. 3, September 1991, pp. 355–72.

Business2.0. "Challenges to Small Business on the Net," June 26, 2000.

Checkland, P. B. *Systems Thinking, Systems Practice*. Chichester, U.K.: Wiley, 1981.

———. "Soft Systems Methodology." *Human Systems Management*, Vol. 8, No. 4, 1989, pp. 271–89.

Coad, P., and E. Yourdon. *Object-Oriented Analysis*, 2d ed. Englewood Cliffs, NJ: Prentice Hall, 1991.

Davis, G. B., and M. H. Olson. *Management Information Systems: Conceptual Foundation, Structure, and Development*, 2d ed. New York: McGraw-Hill, 1985.

Holsapple, C. W., and A. B. Whinston. *Business Expert Systems*. Homewood, IL: Irwin, 1987.

http://www.telelogic.com/solution/lanuage/uml.asp

Jackson, M. A. *Systems Development*. Englewood Cliffs, NJ: Prentice Hall, 1983.

Kendall, K. E. "Behavioral Implications for Systems Analysis and Design: Prospects for the Nineties." *Journal of Management Systems*, Vol. 3, No. 1, 1991, pp. 1–4.

Kendall, J. E., and K. E. Kendall. "Systems Analysis and Design." In *International Encyclopedia of Business and Management*, Vol. 5, edited by M. Warner, pp. 4749–59. London: Routledge, 1996.

———. "Information Delivery Systems: An Exploration of Web Push and Pull Technologies." *Communications of AIS*, Article 14, April 23, 1999. Vol. 1.

Kendall, J. E., K. E. Kendall, R. Baskerville, and R. Barnes. "An Empirical Comparison of a Hypertext-based Systems Analysis Case with Conventional Cases and Role Playing." *The DATA BASE for Advances in Information Systems*, Vol. 27, No. 1, Winter 1996, pp. 58–77.

Laudon, K. C., and J. P. Laudon. *Management Information Systems*, 4th ed. Upper Saddle River, NJ: Prentice Hall, 1996.

Lohr, S. "Code Name: Mainstream." *New York Times*, August 28, 2000, pp. C1, C7.

Markoff, J. "Microsoft Sees Software 'Agent' as Way to Avoid Distractions." *New York Times*, July 17, 2000, pp. C1, C4.

Mumford, E., and M. Weir. *Computer Systems in Work Design—the ETHICS Method*. London: Associated Business Press, 1979.

Naumann, J. D., and A. M. Jenkins. "Prototyping The New Paradigm for Systems Development." *MIS Quarterly*, Vol. 6, No. 3, September 1982, pp. 29–44.
Sahami, M. S. Dumais, D. Heckerman, and E. Horvitz. "A Bayesian Approach to Filtering Junk E-Mail." AAAI Technical Report WS-98–05. AAAI Workshop on Learning for Text Categorization, July 1998, Madison, WI.
Schmidt, A. *Working with Visible Analyst for Windows*. Upper Saddle River, NJ: Prentice Hall, 1996.
Sharma, S. and A. Rai. "CASE Deployment in IS Organizations." *Communications of the ACM*, Vol. 43, No. 1, January 2000, pp. 80–88.
Stohr, E. A., and S. Viswanathan. "Recommendation Systems." In *Emerging Information Technologies: Improving Decisions, Cooperation, and Infrastructure*, edited by K. E. Kendall, pp. 21–44. Thousand Oaks, CA: Sage Publications, 1999.
Visible Analyst (VA), Corporate Edition, Version 7.5, Product and Pricing Information. Waltham, MA: Visible Systems Corporation, 2000.
Whitten, J. L., L. D. Bentley, and V. M. Barlow. *Systems Analysis and Design Methods*, 3d ed. Homewood, IL: Irwin, 1994.
Wood-Harper, A. T., S. Corder, J. R. G. Wood, and H. Watson. "How We Profess: The Ethical Systems Analyst." *Communications of the ACM*, Vol. 39, No. 3, March 1996, pp. 69–77.
Yourdon, E. *Modern Structured Analysis*. Englewood Cliffs, NJ: Prentice Hall, 1989.

1

Allen Schmidt, Julie E. Kendall, and Kenneth E. Kendall

THE CASE OPENS

On a warm, sunny day in late October, Chip Puller parks his car and walks into his office at Central Pacific University. It felt good to be starting as a systems analyst, and he was looking forward to meeting the other staff.

In the office, Anna Liszt introduces herself. "We've been assigned to work as a team on a new project. Why don't I fill you in with the details, and then we can take a tour of the facilities."

"That sounds good to me," Chip replies. "How long have you been working here?"

"About five years," answers Anna. "I started as a programmer analyst, but the last few years have been dedicated to analysis and design. I'm hoping we'll find some ways to increase our productivity," Anna continues.

"Tell me about the new project," Chip says.

"Well," Anna replies, "like so many organizations, we have a large number of microcomputers with different software packages installed on them. In the 1980s there were few microcomputers and a scattered collection of software, but there has been a rapid increase in recent years. The current system used to maintain software and hardware has been overwhelmed."

"What about the users? Who should I know? Who do you think will be important in helping us with the new system?" Chip asks.

"You'll meet everyone, but there are key people I've recently met, and I'll tell you what I've learned so you'll remember them when you meet them.

"Dot Matricks is manager of all microcomputer systems at Central Pacific. We seem to be able to work together well. She's very competent. She'd really like to be able to improve communication among users and analysts."

"It will be a pleasure to meet her," Chip speculates.

"Then there's Mike Crowe, the micromaintenance expert. He really seems to be the nicest guy, but too busy. We need to help lighten his load. The software counterpart to Mike is Cher Ware. She's a free spirit, but don't get me wrong, she knows her job," Anna says.

"She could be fun to work with," Chip says.

"Could be," Anna agrees. "You'll meet the financial analyst, Paige Prynter, too. I haven't figured her out yet."

"Maybe I can help," Chip says.

"Last, you should—I mean, you will—meet Hy Perteks, who does a great job running the Information Center. He'd like to see us be able to integrate our life cycle activities."

"It sounds promising," Chip says. "I think I'm going to like it here."

EXERCISE

E-1. From the introductory conversation Chip and Anna shared, which elements mentioned might suggest the use of CASE tools?

UNDERSTANDING ORGANIZATIONAL STYLE AND ITS IMPACT ON INFORMATION SYSTEMS

2

ORGANIZATIONAL FUNDAMENTALS

To analyze and design appropriate information systems, systems analysts need to comprehend the organizations they work in as systems shaped through the interactions of three main forces: the levels of management, design of organizations, and organizational cultures.

Organizations are large systems composed of interrelated subsystems. The subsystems are influenced by three broad levels of management decision makers (operations, middle management, and strategic management) that cut horizontally across the organizational system. Organizational cultures and subcultures all influence the way people in subsystems interrelate. These topics and their implications for information systems development are considered in this chapter.

ORGANIZATIONS AS SYSTEMS

Organizations are usefully conceptualized as systems designed to accomplish predetermined goals and objectives through people and other resources that they employ. Organizations are composed of smaller, interrelated systems (departments, units, divisions, etc.) serving specialized functions. Typical functions include accounting, marketing, production, data processing, and management. Specialized functions (smaller systems) are eventually reintegrated through various mechanisms to form an effective organizational whole.

The significance of conceptualizing organizations as complex systems is that systems principles allow insight into how organizations work. To ascertain information requirements properly and to design appropriate information systems, it is of primary importance to understand the organization as a whole. All systems are composed of subsystems (which include information systems); therefore, when studying an organization, we also examine how smaller systems are involved and how they function.

INTERRELATEDNESS AND INTERDEPENDENCE OF SYSTEMS

All systems and subsystems are interrelated and interdependent. This fact has important implications both for organizations and for those systems analysts who seek to help them better achieve their goals. When any element of a system is changed or eliminated, the rest of the system's elements and subsystems are also impacted.

"The E-Mail isn't functioning - pass it on."

For example, suppose that the administrators of an organization decide not to hire personal secretaries any longer and to replace their functions with networked PCs. This decision has the potential to impact not only the secretaries and the administrators but also all the organizational members who built up communications networks with the now-departed secretaries.

SYSTEMS PROCESSES

All systems process inputs from their environments. By definition, processes change or transform inputs into outputs. Whenever you examine a system, check to see what is being changed or processed. If nothing is changed, you may not be identifying a process. Typical processes in systems include verifying, updating, and printing.

ORGANIZATIONAL BOUNDARIES

Another aspect of organizations as systems is that all systems are contained by boundaries separating them from their environments. Organizational boundaries exist on a continuum ranging from extremely permeable to almost impermeable. To continue to adapt and survive, organizations must be able first to import people, raw materials, and information through their boundaries (inputs) and then to exchange their finished products, services, or information with the outside world (outputs).

SYSTEM FEEDBACK FOR PLANNING AND CONTROL

Feedback is one form of system control. As systems, all organizations use planning and control to manage their resources effectively. Figure 2.1 shows how system outputs are used as feedback that compares performance with goals. This comparison in turn helps administrators formulate more specific goals as inputs. An example is a manufacturing company that produces red-white-and-blue weight-training sets as well as gun-metal colored sets. The company finds that one year after the Olympics, very few red-white-and-blue sets are pur-

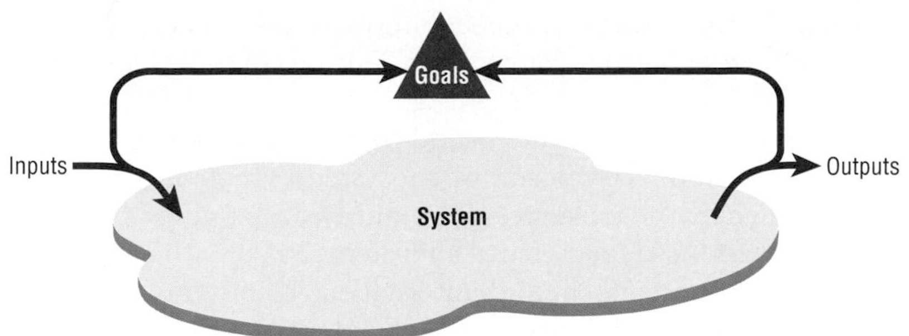

FIGURE 2.1
System outputs serve as feedback that compares performance to goals.

chased. Production managers use that information as feedback to make decisions about what quantities of each color to produce. Feedback in this instance is useful for planning and control.

The ideal system, however, is one that self-corrects or self-regulates in such a way that decisions on typical occurrences are not required. An example is a computerized information system for production planning that takes into account current and projected demand and formulates a proposed solution as output. An Italian knitwear manufacturer that markets its clothing in the United States has just such a system. This company produces most of its sweaters in white, uses its computerized inventory information system to find out what colors are selling best, and then dyes sweaters in hot-selling colors immediately before shipping them. With a self-regulating information system, a manager can review the proposed production figures and make interventions only when exceptional factors, not accounted for in the computer's software formulation of the problem, are present.

ENVIRONMENTS FOR ORGANIZATIONAL SYSTEMS

Feedback is received from within the organization and from the outside environments around it. Anything external to an organization's boundaries is considered to be an environment. Numerous environments, with varying degrees of stability, constitute the milieu in which organizations exist.

Among these environments are (1) the environment of the community where the organization is physically located, which is shaped by the size of its population and its demographic profile, including factors such as education and average income; (2) the economic environment, influenced by market factors, including competition; and (3) the political environment, controlled through state and local governments. Although changes in environmental status can be planned for, they often cannot be directly controlled by the organization.

OPENNESS AND CLOSEDNESS IN ORGANIZATIONS

Related and similar to the concept of external boundary permeability is the concept of internal openness or closedness of organizations. Openness and closedness also exist on a continuum, because there is no such thing as an absolutely open or completely closed organization.

Openness refers to the free flow of information within the organization. Subsystems such as creative or art departments often are characterized as open, with a free flow of ideas among participants and very few restrictions on who gets what information at what time when a creative project is in its infancy.

At the opposite end of the continuum might be a defense department unit assigned to work on top-secret defense planning affecting national security. Each person needs to receive clearance, timely information is a necessity, and access to information is only on a "need to know" basis. This sort of unit is limited by numerous rules.

Using a systems overlay to understand organizations allows us to acknowledge the idea of systems composed of subsystems; their interrelatedness and their interdependence; the existence of boundaries that allow or prevent interaction between various departments and elements of other subsystems and environments; and the existence of internal environments characterized by degrees of openness and closedness, which might differ across departments, units, or even projects.

VIRTUAL ORGANIZATIONS AND VIRTUAL TEAMS

Not all organizations or parts of organizations are visible in a physical location. Entire organizations or units of organizations can now possess virtual components that permit them to change configurations to adapt to changing project or marketplace demands. Virtual enterprises use networks of computers and communications technology to bring people with specific skills together electronically to work on projects that are not physically located in the same place. Information technology enables coordination of these remote team members. Often virtual teams spring up in already-established organizations; in some instances, however, organizations of remote workers have been able to succeed without the traditional investment in infrastructure.

There are several potential benefits to virtual organizations, such as the possibility of reducing costs of physical facilities, more rapid response to customer needs, and helping virtual employees to fulfill their familial obligations to children or aging parents. Just how important it will be to meet the social needs of virtual workers is still open to research and debate. One example of a need for tangible identification with a culture arose when students who were enrolled in an online virtual university, with no physical campus, kept requesting items such as sweatshirts, coffee mugs, and pennants with the virtual university's logo imprinted on them. These items are meaningful cultural artifacts that traditional brick and mortar schools have long provided.

Many systems analysis and design teams are now able to work virtually, and in fact, many of them marked the path for other types of employees to follow in

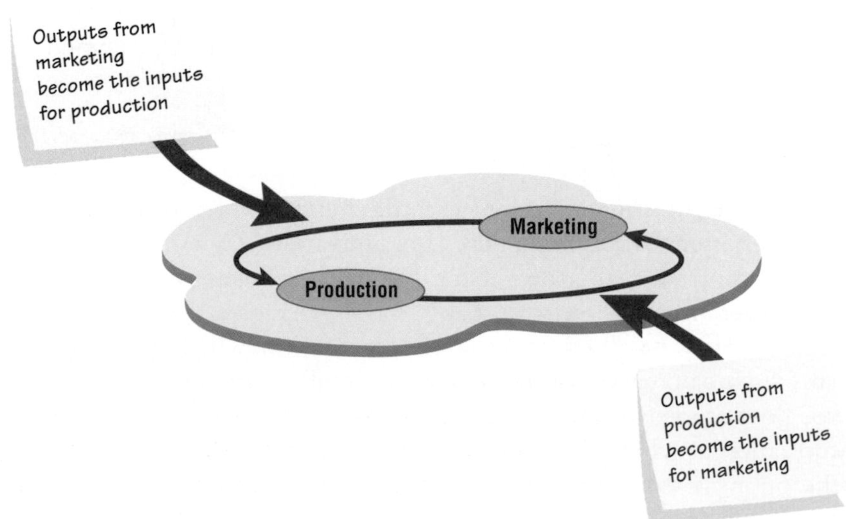

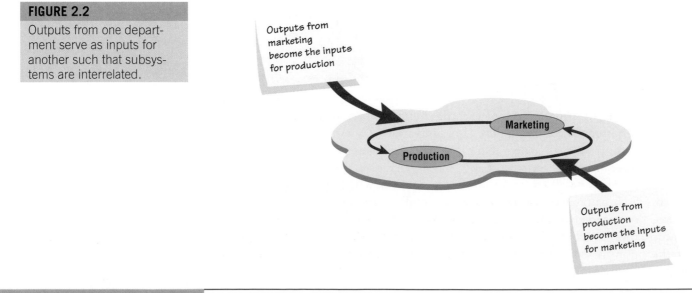

FIGURE 2.2
Outputs from one department serve as inputs for another such that subsystems are interrelated.

How a marketing manager may view the organization

How a production manager may see the organization

FIGURE 2.3

A depiction of the personal perspective of functional managers shows that they feature their own functional area as central to the organization.

accomplishing work virtually. Some applications permit analysts who are providing technical assistance over the Web to "see" the software and hardware configuration of the user requesting help, in this way creating an ad hoc virtual team composed of the analyst and user.

TAKING A SYSTEMS PERSPECTIVE

Taking a systems perspective allows systems analysts to start broadly clarifying and understanding the various businesses with which they will come into contact. It is important that members of subsystems realize that their work is interrelated. Notice in Figure 2.2 that the outputs from the production subsystems serve as inputs for marketing and that the outputs of marketing serve as new inputs for production. Neither subsystem can properly accomplish its goals without the other.

Problems occur when each manager possesses a different picture of the importance of his or her own functional subsystem. In Figure 2.3 you can see that the marketing manager's personal perspective shows the business as driven by marketing, with all other functional areas interrelated but not of central importance. By the same token the perspective of a production manager positions production at the center of the business, with all other functional areas driven by it.

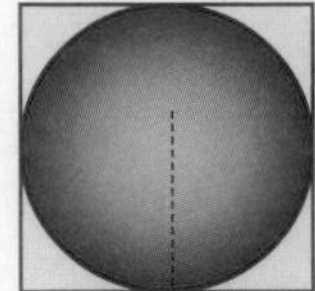

CONSULTING OPPORTUNITY 2.1

THE E IN VITAMIN E STANDS FOR ECOMMERCE

"Our retail shops and mail-order division are quite healthy," says Bill Berry, one of the owners of Marathon Vitamin Shops, "but to be competitive, we must establish an ecommerce Web site." His father, and coowner, exclaims, "I agree, but where do we start?" The elder Berry knew, of course, that it wasn't a case of setting up a Web page and asking customers to email their orders to the retail store. He identified eight different parts to ecommerce and realized that they were all part of a larger system. In other words, all the parts had to work together to create a strong package. His list of elements essential to ecommerce included the following:

1. Attracting customers to an ecommerce Web site
2. Informing customers about products and services offered
3. Allowing customers to customize products online
4. Completing transactions with the customers
5. Accepting payment from customers in a variety of forms
6. Supporting customers after the sale via the Web site
7. Arranging for the delivery of goods and services
8. Personalizing the look and feel of the Web site for different customers

Bill Berry read the list and contemplated it for a while. "It is obvious that ecommerce is more complex than I thought," he says. You can help the owners of Marathon Vitamin Shops in the following ways:

1. Make a list of the elements that are interrelated or interdependent. Then write a paragraph stating why it is critical to monitor these elements closely.
2. Decide on the boundaries of the system. That is, write a paragraph expressing an opinion on which elements are critical for Marathon Vitamin Shops and which elements can be explored at a later date.
3. Suggest which elements should be handled in-house and which should be outsourced to another company that may be better able to handle the job. Justify your suggestions in two paragraphs, one for the in-house jobs and one for the outsourced tasks.

The relative importance of functional areas as revealed in the personal perspectives of managers takes on added significance when managers rise to the top through the ranks, becoming strategic managers. They can create problems if they overemphasize their prior functional information requirements in relation to the broader needs of the strategic manager.

For example, if a production manager continues to stress production scheduling and performance of line workers, the broader aspects of forecasting and policy making may suffer. This tendency is a danger in all sorts of businesses: where engineers work their way up to become administrators of aerospace firms, college professors move from their departments to become deans, or programmers advance to become executives of software firms. Their tunnel vision often creates problems for the systems analyst trying to separate actual information requirements from desires for a particular kind of information.

ENTERPRISE RESOURCE PLANNING: VIEWING THE ORGANIZATION AS A SYSTEM

An enterprise resource planning system, or ERP, is a term used to describe an integrated organizational (enterprise) information system. ERP is software that helps the flow of information between the functional areas in the organization. It is a customized system that, rather than being developed in-house, is usually purchased from one of the software development companies well known for its ERP packages, such as SAP, Bään, Oracle, PeopleSoft, or J.D. Edwards. The product is then customized to fit the requirements of a particular company. Typically, the vendor requires an organizational commitment in terms of specialized user or analyst training. Many ERP packages are designed to run on the Web.

ERP evolved from materials requirements planning (MRP), the information systems designed to improve manufacturing in general and assembly in particular. ERP systems now include manufacturing components and thus help with capacity planning, material production scheduling, and forecasting. Beyond manufacturing (and its service counterpart), ERP includes sales and operations planning, distribution, and managing the supply chain. It therefore impacts all the areas within the organization including accounting, finance, management, marketing, and information systems. An organization can take advantage of ERP systems that use electronic data interchange (EDI) to communicate with its business partners.

Implementing an ERP solution may be frustrating because it is difficult to analyze a system currently in use and then fit the ERP model to that system. Furthermore, companies tend to design their business processes before ERP is implemented. This redesign is called business process reengineering (BPR). Unfortunately, this process is often rushed and the proposed business model does not always match the ERP functionality. The result is further customizations, extended implementation time frames, higher costs, and often the loss of user confidence. Analysts need to be aware of the magnitude of the problem they are tackling when trying to implement ERP packages.

DEPICTING SYSTEMS GRAPHICALLY

A system or subsystem as it exists within the corporate organization may be graphically depicted in several ways. The various graphical models show the boundaries of the system and the information used within the system.

SYSTEMS AND THE CONTEXT-LEVEL DATA FLOW DIAGRAM

The first model is the context-level data flow diagram (also called an environmental model). Data flow diagrams focus on the data flowing into and out of the system and the processing of the data. These basic components of every computer program may be described in detail and used to analyze the system for accuracy and completeness.

As shown in Figure 2.4, the context-level diagram employs only three symbols: (1) a rectangle with rounded corners, (2) a square with two shaded edges,

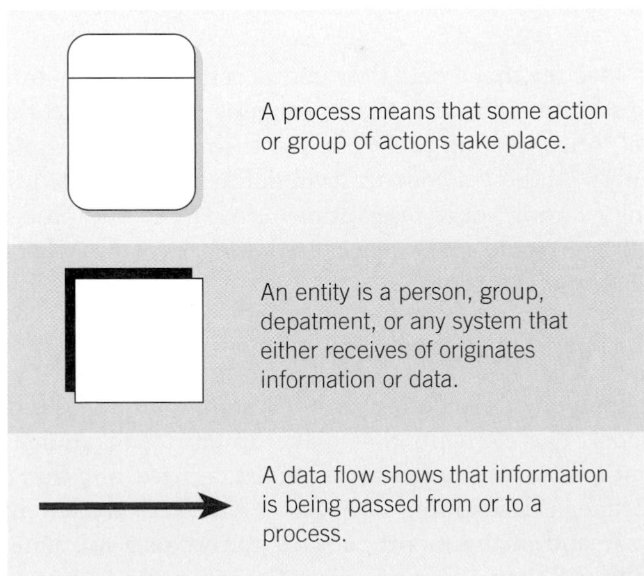

A process means that some action or group of actions take place.
An entity is a person, group, depatment, or any system that either receives of originates information or data.
A data flow shows that information is being passed from or to a process.

FIGURE 2.4
The basic symbols of a data flow diagram.

FIGURE 2.5

A context-level flow diagram for an airline reservation system.

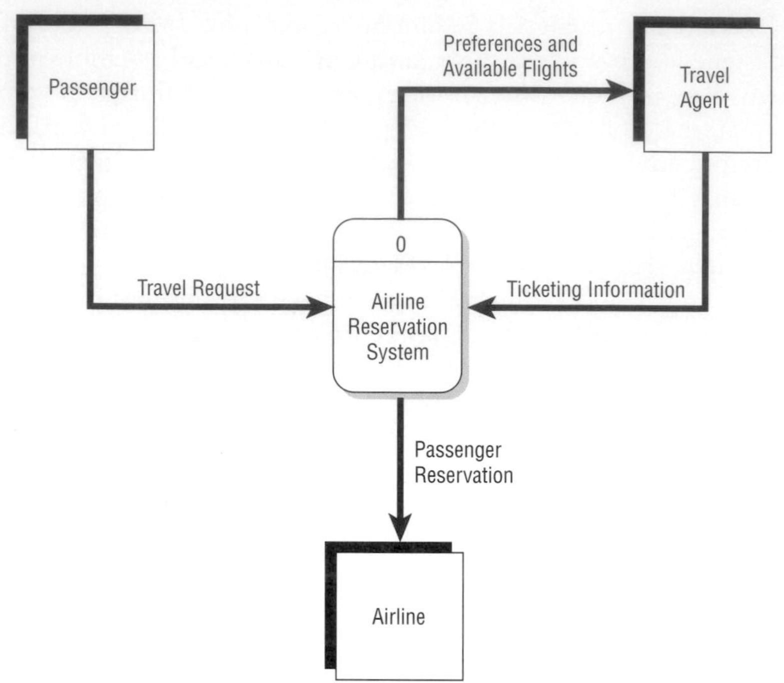

and (3) an arrow. Processes transform incoming data into outgoing information, and the content level has only one process, representing the entire system. The external entity represents any entity that supplies or receives information from the system but is not a part of the system. This entity may be a person, a group of people, a corporate position or department, or other systems. The lines that connect the external entities to the process are called data flows, and they represent data.

An example of a context-level data flow diagram is found in Figure 2.5. In this example, the most basic elements of an airline reservation system are represented.

The passenger (an entity) initiates a travel request (data flow). The context-level diagram doesn't show enough detail to indicate exactly what happens (it isn't supposed to), but we can see that the passenger's preferences and the available flights are sent to the travel agent, who sends ticketing information back to the process. We can also see that the passenger reservation is sent to the airline.

In Chapter 9 we see that a data flow contains much information. For example, the passenger reservation contains the passenger's name, airline, flight number(s), date(s) of travel, price, seating preference, and so on. For now, however, we are concerned mainly with how a context level defines the boundaries of the system. In the preceding example, only reservations are part of the process. Other decisions that the airline would make (e.g., purchasing airplanes, changing schedules, pricing) are not part of this system.

SYSTEMS AND THE ENTITY-RELATIONSHIP MODEL

One way a systems analyst can define proper system boundaries is to use an entity-relationship model. The elements that make up an organizational system can be referred to as entities. An entity may be a person, a place, or a thing, such as a passenger on an airline, a destination, or a plane. Alternatively, an entity may be an event, such as the end of the month, a sales period, or a machine breakdown. A relationship is the association that describes the interaction among the entities.

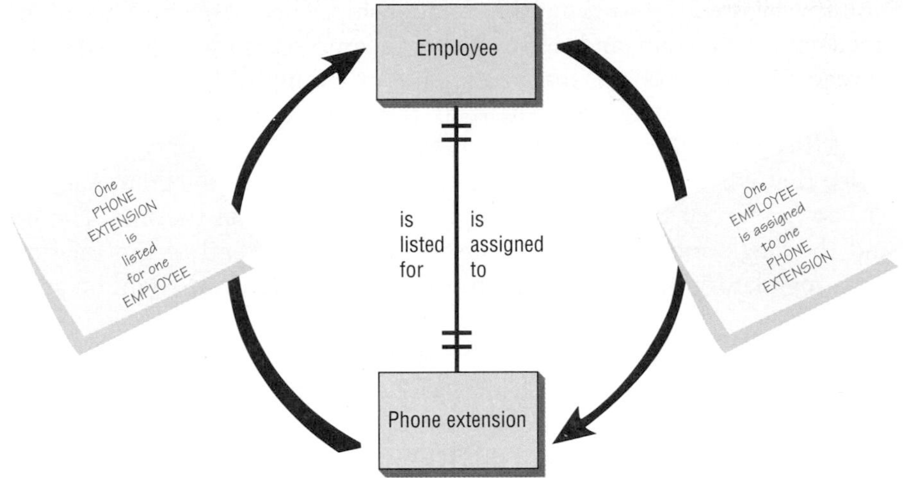

FIGURE 2.6
An entity relationship diagram showing a one-to-one relationship.

There are many different conventions for drawing entity-relationship, or E-R, diagrams (with names like crow's foot, Arrow, or Bachman notation). In this book, we use crow's foot notation. For now, we assume that an entity is a plain rectangular box.

Figure 2.6 shows a simple entity-relationship diagram. Two entities are linked together by a line. In this example, the end of the line is marked with two short parallel marks (I I), signifying that this relationship is one-to-one. Thus, exactly one employee is assigned to one phone extension. No one shares the phone extension in this office.

The colored arrows are not part of the entity-relationship diagram. They are present to demonstrate how to read the entity-relationship diagram. The phrase on the right side of the line is read from top to bottom as follows: "One EMPLOYEE is assigned to one PHONE EXTENSION." On the left side, as you read from bottom to top, the arrow says, "One PHONE EXTENSION is listed for one EMPLOYEE."

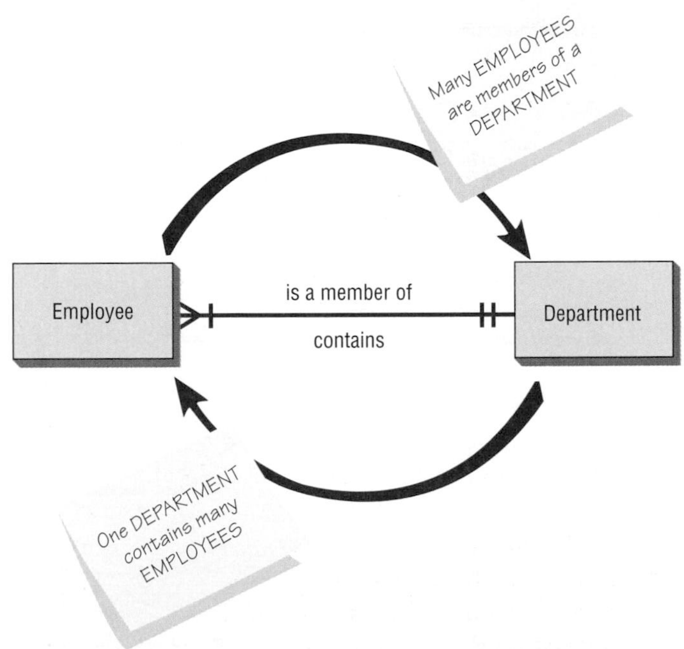

FIGURE 2.7
An entity relationship diagram showing a many-to-one relationship.

Similarly, Figure 2.7 shows another relationship. The crow's foot notation (>+) is obvious on this diagram, and this particular example is a many-to-one example. As you read from left to right, the arrow signifies, "Many EMPLOYEES are members of a DEPARTMENT." As you read from right to left, it implies, "One DEPARTMENT contains many EMPLOYEES."

Notice that when a many-to-one relationship is present, the grammar changes from "is" to "are" even though the singular "is" is written on the line. The crow's foot and the single mark do not literally mean that this end of the relationship must be a mandatory "many." Instead, they imply that this end could be anything from one to many.

Figure 2.8 elaborates on this scheme. Here we have listed a number of typical entity relationships. The first, "An EMPLOYEE is assigned to an OFFICE," is a one-to-one relationship. The second one is a one-to-many relationship: "One CARGO AIRCRAFT will serve one or more DISTRIBUTION CENTERs." The third one is slightly different because it has a circle at one end. It can be read as "A SYSTEMS ANALYST may be assigned to MANY PROJECTS," meaning that the analyst can be assigned to no projects [that is what the circle (**O**), for zero, is for], one, or many projects. Likewise, the circle (**O**) indicates that none is possible in the next relationship. Recall that the short mark means one. Therefore, we can read it as follows: "A MACHINE may or may not be undergoing SCHEDULED MAINTE-

FIGURE 2.8

Examples of different types of relationships in E-R diagrams.

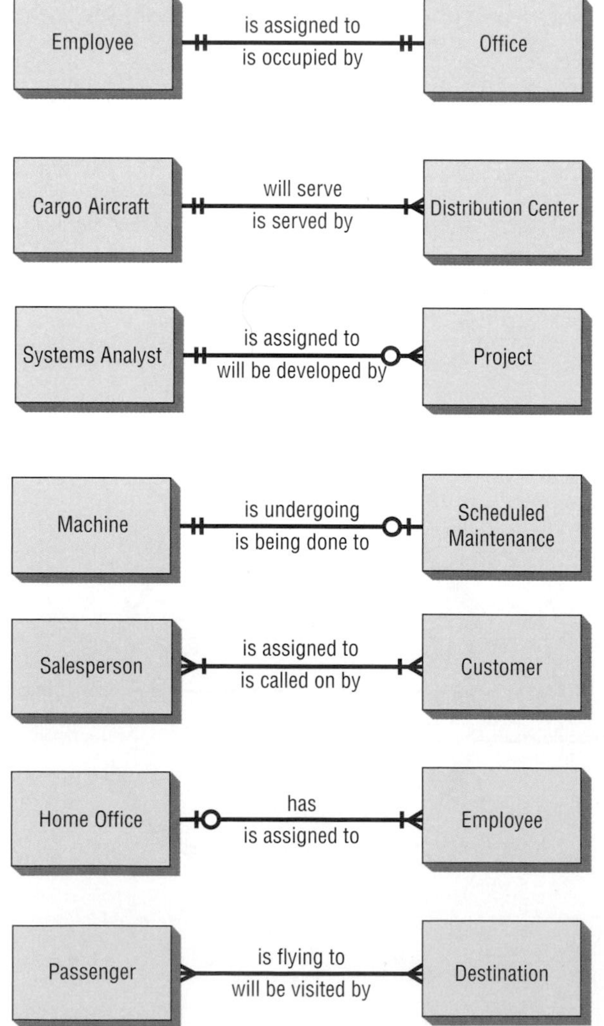

NANCE." Notice that the line is written as "is undergoing," but the end marks on the line indicate that either no maintenance (**O**) or maintenance (**|**) is actually going on.

The next relationship states, "One or many SALESPEOPLE (plural of SALESPERSON) are assigned to one or more CUSTOMERs." It is the classic many-to-many relationship. The next relationship can be read as follows: "The HOME OFFICE can have one or many EMPLOYEEs," or "One or more EMPLOYEEs may or may not be assigned to the HOME OFFICE." Once again, the **|** and **O** together imply a Boolean situation, in other words, zero or one.

The final relationship shown here can be read as, "Many PASSENGERs are flying to many DESTINATIONs." This symbol ⤙ is preferred by some to indicate a mandatory "many" condition. (Would it ever be possible to have only one passenger or only one destination?) Even so, some CASE tools such as Visible Analyst do not offer this possibility, because the optional one-or-more condition as shown in the SALESPERSON-CUSTOMER relationship above will do.

Up to now we have modeled all our relationships using just one simple rectangle and a line. This method works well when we are examining the relationships of real things like real people, places, and things. Sometimes, though, we create new items in the process of developing an information system. Some examples are invoices, receipts, files, and databases. When we want to describe how a person relates to a receipt, for example, it becomes convenient to indicate the receipt in a different way, as shown in Figure 2.9 as an associative entity.

An associative entity can only exist if it is connected to at least two other entities. For that reason, some call it a gerund, a junction, an intersection, or a concatenated entity. This wording makes sense because a receipt wouldn't be necessary unless there were a customer and a salesperson making the transaction.

Another type of entity is the attributive. When an analyst wants to show data that are completely dependent on the existence of a fundamental entity, an attributive entity should be used. For example, if a video store had multiple copies of the same video title, an attributive entity could be used to designate which copy of the tape is being checked out. The attributive entity is useful for showing repeating groups of data. For example, suppose we are going to model the relationships that exist when a patron gets tickets to a concert or show. The entities seem obvious at first: "a PATRON and a CONCERT/SHOW," as shown in Figure 2.10. What sort of relationship exists? At first glance the PATRON gets a reserva-

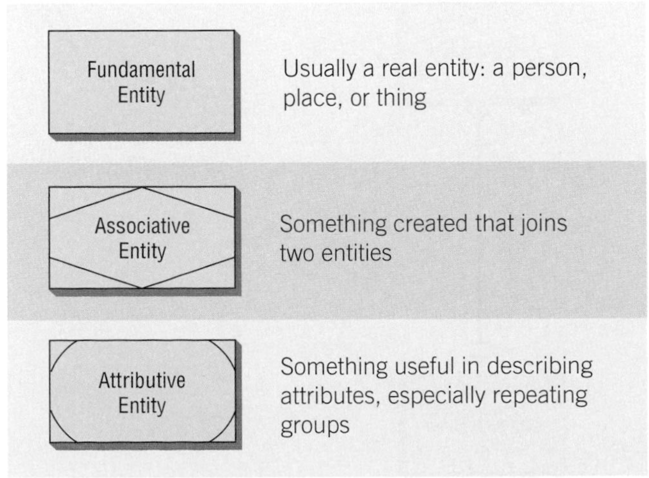

FIGURE 2.9
Three different types of entities used in E-R diagrams.

FIGURE 2.10

Starting to draw an E-R diagram; the first attempt.

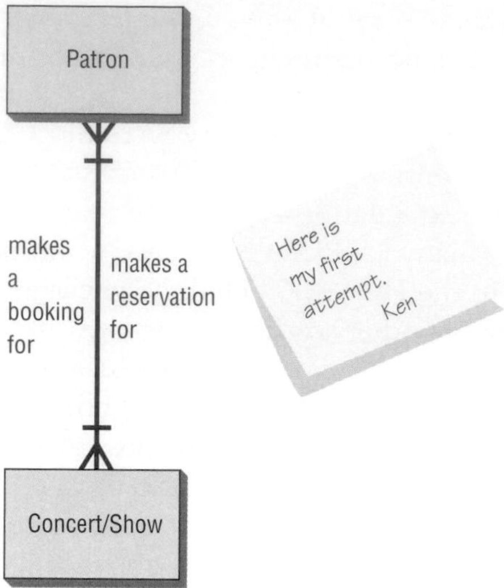

tion for a CONCERT/SHOW, and the CONCERT/SHOW can be said to have made a booking for a PATRON.

The process isn't that simple of course, and the E-R diagram need not be that simple either. The PATRON actually makes a RESERVATION, as shown in Figure 2.11. The RESERVATION is for a CONCERT/SHOW. The CONCERT/SHOW holds the RESERVATION, and the RESERVATION is in the name of the PATRON. We added an associative entity here because a RESERVATION was

FIGURE 2.11

Improving the E-R diagram by adding an associative entity called RESERVATION.

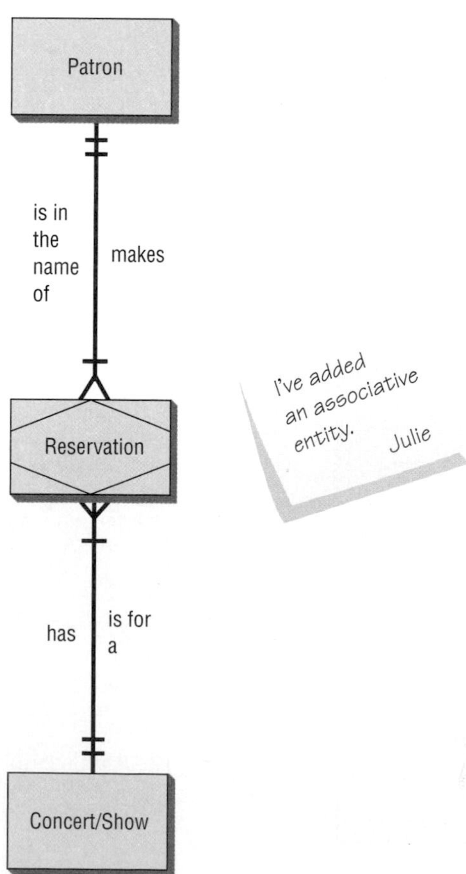

created due to the information system required to relate the PATRON and the CONCERT/SHOW.

Again this process is quite simple, but because concerts and shows have many performances, the entity relationship diagram is drawn once more in Figure 2.12. Here we add an attributive entity to handle the many performances of the CONCERT/SHOW. In this case the RESERVATION is made for a particular PERFORMANCE, and the PERFORMANCE is one of many that belong to a specific CONCERT/SHOW. In turn the CONCERT/SHOW has many performances, and one PERFORMANCE has a RESERVATION that is in the name of a particular PATRON.

To the right of this E-R diagram is a set of data attributes that make up each of the entities in this E-R diagram. Some entities may have attributes in common. The attributes that are underlined can be searched for. The attributes are referred to as keys and are discussed in Chapter 17.

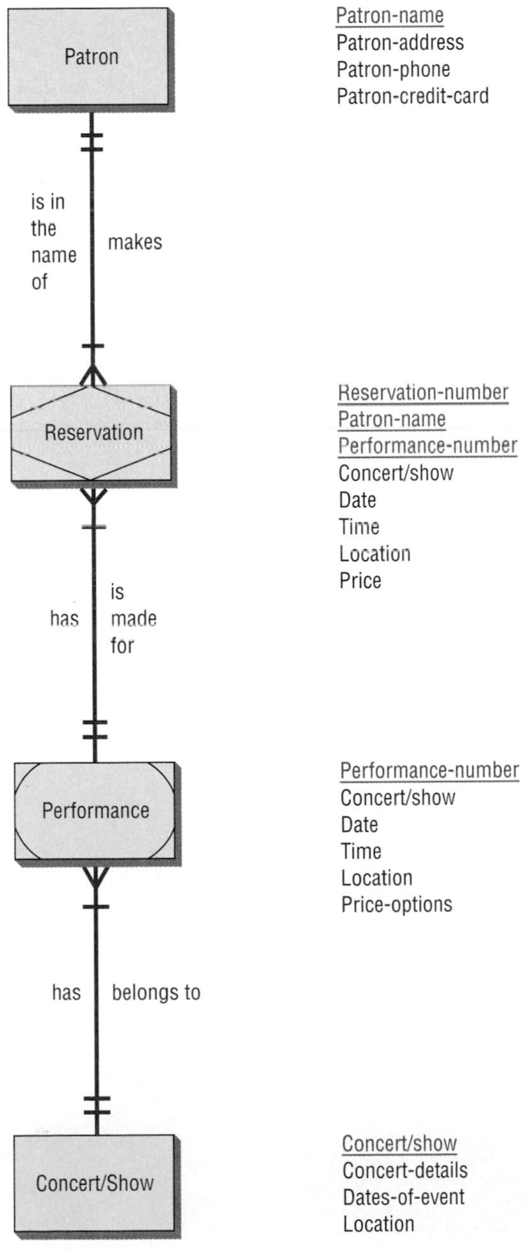

FIGURE 2.12
A more complete E-R diagram showing data attributes of the entities.

Entity-relationship diagrams are often used by systems designers to help model the file or database. It is even more important, however, that the systems analyst understand early both the entities and relationships in the organizational system. In sketching out some basic E-R diagrams, the analyst needs to:

1. List the entities in the organization to gain a better understanding of the organization.
2. Choose key entities to narrow the scope of the problem to a manageable and meaningful dimension.
3. Identify what the primary entity should be.
4. Confirm the results of steps 1 through 3 through other data-gathering methods (investigation, interviewing, administering questionnaires, observation, and prototyping), as discussed in Chapters 4 through 8.

It is critical that the systems analyst begin to draw E-R diagrams upon entering the organization rather than waiting until the database needs to be designed, because E-R diagrams help the analyst understand what business the organization is actually in, determine the size of the problem, and discern whether the right problem is being addressed. The E-R diagrams need to be confirmed or revised as the data-gathering process takes place.

Organizational factors that influence the analysis and design of information systems include levels of management and organizational culture. We discuss each of these factors and their implications for the analysis and design of information systems in the remaining sections of this chapter.

LEVELS OF MANAGEMENT

Management in organizations exists on three broad, horizontal levels: operational control, managerial planning and control (middle management), and strategic management, as shown in Figure 2.13. Each level carries its own responsibilities, and all work toward achieving organizational goals and objectives in their own ways.

OPERATIONS MANAGEMENT

Operational control forms the bottom tier of three-tiered management. Operations managers make decisions using predetermined rules that have predictable outcomes when implemented correctly.

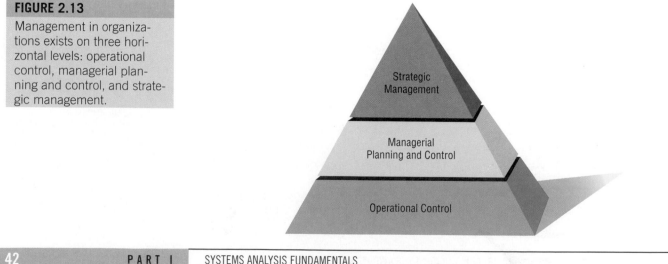

FIGURE 2.13
Management in organizations exists on three horizontal levels: operational control, managerial planning and control, and strategic management.

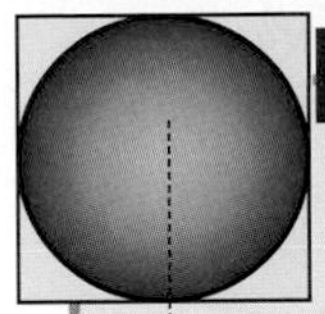

CONSULTING OPPORTUNITY 2.2

WHERE THERE'S CARBON, THERE'S A COPY

"I don't know what we do with the pink ones yet," Richard Russell admitted. "They're part of a quadruplicate form that rips apart. All I know is that we keep them for the filing clerk, and he files them when he has time."

Richard is a newly hired junior account executive for Carbon, Carbon & Rippy, a brokerage house. You are walking through the steps he takes in making a stock purchase "official" because his boss has asked you to streamline the process whereby stock purchase information is stored in the computer and retrieved.

After you leave, Richard continues thinking about the pink forms. He tells his clerk, Harry Schultz, "In my two months here, I haven't seen anyone use those. They take up my time and yours, not to mention all the filing space. Let's pitch them."

Richard and Harry proceed to open all the old files kept by Richard's predecessor and throw out the filed pink forms, along with those accumulated but not yet filed. It takes hours, but they make a lot of room. "Definitely worth the time," Richard reassures Harry.

Three weeks later, an assistant to Richard's boss, Carol Vaness, appears. Richard is happy to see a familiar face, greeting her with, "Hi, Carol. What's new?"

"Same old thing," Carol sighs. "Well, I guess it isn't old to you, because you're the newcomer. But I need all those pesky pink forms."

Almost in shock, Richard exchanges looks with Harry, then mumbles, "You're kidding, of course."

Carol looks more serious than Richard ever thought possible, replying, "No joke. I summarize all the pink forms from all the brokers, and then my totals are compared with computerized stock purchase information. It's part of our routine, three-month audit for transaction accuracy. My work depends on yours. Didn't Ms. McCue explain that to you when you started?"

What systems concept did Richard and Harry ignore when tossing out the pink forms? What are the possible ramifications for systems analysts if general systems concepts are ignored?

Operations managers are the decision makers whose work is the most clear-cut because of the high degree of certainty in their decision-making environment. They make decisions that affect implementation in work scheduling, inventory control, shipping, receiving, and control of processes such as production. Operations managers oversee the operating details of the organization, ensuring that the basic tasks of the organization are accomplished on time and in accordance with organizational constraints.

MIDDLE MANAGEMENT

Middle management forms the second, or intermediate, tier of the three-tiered management system. Middle managers make short-term planning and control decisions about how resources may best be allocated to meet organizational objectives.

Middle managers experience very little certainty in their decision-making environment. Their decisions range all the way from forecasting future resource requirements to solving employee problems that threaten productivity. Few of the decisions that middle managers make are as structured as those made by operations managers. The decision-making domain of middle managers can usefully be characterized as partly operational and partly strategic, with constant fluctuations.

STRATEGIC MANAGEMENT

Strategic management is the third level of three-tiered management control. Strategic managers look outward from the organization to the future, making decisions that will guide middle and operations managers in the months and years ahead.

Strategic managers work in a highly uncertain decision-making environment. Through statements of goals and the determination of strategies and policies to achieve them, strategic managers actually define the organization as a whole.

Theirs is the broad picture, wherein the company decides to develop new product lines, divest itself of unprofitable ventures, acquire other compatible companies, or even allow itself to be sold.

There are sharp contrasts among the decision makers on many dimensions. For instance, strategic planners have multiple decision objectives, whereas operations managers have single ones. It is often difficult for high-level managers to identify problems, but it is easy for operations managers to do so. Strategic planners are faced with semistructured problems, whereas lower-level managers deal mostly with structured problems. The alternative solutions to a problem facing the strategic managers are often difficult to articulate, but the alternatives that operations managers work with are usually easy to enumerate. The nature of the decisions made by the various levels of management is entirely different. Strategic managers most often make one-time decisions, whereas the decisions made by operations managers tend to be repetitive. Finally, the decision style of strategic managers tends toward the heuristic, whereas operations managers tend to be mainly analytic in their style.

IMPLICATIONS FOR INFORMATION SYSTEMS DEVELOPMENT

Each of the three management levels holds differing implications for developing management information systems. Some of the information requirements for managers are clear-cut, whereas others are fuzzy and overlap.

Operations managers need internal information that is of a repetitive, low-level nature. They are highly dependent on information that captures current performance, and they are large users of online, real-time information resources. The need of operations managers for past performance information and periodic information is only moderate. They have little use for external information that allows future projections or creation of what-if scenarios.

Information systems designed for operations managers are valuable if they can provide information to help in controlling operations in a timely manner. Much of the information needed for operations is already being generated or can be easily pinpointed, but it would be more useful if it were accessible from an online system.

On the next management level, middle managers, who both plan and control, are in need of both short- and longer-term information. Due to the troubleshooting nature of their jobs, middle managers experience extremely high needs for information in real time. To control properly, they also need current information on performance as measured against set standards.

Middle managers are highly dependent on internal information. In contrast to operations managers, they have a high need for historical information, along with information that allows prediction of future events and simulation of numerous possible scenarios.

Strategic managers differ somewhat from both middle and operations managers in their information requirements. They are highly dependent on information from external sources that supply news of market trends and the strategies of competing corporations.

Because the task of strategic managing demands projections into the uncertain future, strategic managers have a high need for information of a predictive nature and information that allows creation of many different what-if scenarios. Strategic managers also exhibit strong needs for periodically reported information as they seek to adapt to fast-moving changes.

Strategic planners need general, summarized information rather than the highly detailed, raw data required by low-level managers. Information for strategic

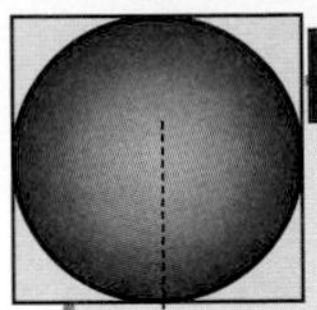

CONSULTING OPPORTUNITY 2.3

PYRAMID POWER

"We really look up to you," says Paul LeGon. As a systems analyst, you have been invited to help Pyramid, Inc., a small, independent book publishing firm that specializes in paperback books outside of the publishing mainstream.

Paul continues, "We deal with what some folks think are fringe topics. You know, pyramid power, end-of-the-world prophecies, and healthier living by thinking of the color pink. Sometimes when people see our books, they just shake their heads and say, 'Tut—uncommon topic.' But we're not slaves to any particular philosophy, and we've been very successful. So much so that because I'm 24, people call me the 'boy king.'" Paul pauses to decipher your reaction.

Paul continues, "I'm at the top as president, and functional areas such as editorial, accounting, production, and marketing are under me."

Paul's assistant, Ceil Toom, who has been listening quietly up to now, barges in with her comments: "The last systems experts that did a project for us recommended the creation of liaison committees of employees between accounting, production, and marketing so that we could share newly computerized inventory and sales figures across the organization. They claimed that committees like that would cut down on needless duplication of output, and each functional area would be better integrated with all the rest."

Paul picks up the story, saying, "It was fair—oh, for a while—and the employees shared information, but the reason you're here is that the employees said they didn't have time for committee meetings and were uncomfortable sharing information with people from other departments who were farther up the ladder than they were here at Pyramid."

According to Paul and Ceil, what were the effects of installing a management information system at Pyramid, Inc., that required people to share information in ways that were not consistent with their structure? Propose some general ways to resolve this problem so that Pyramid employees can still get the sales and inventory figures they need.

planners may be older and estimated, whereas operational managers need current, accurate information. Finally, the strategic planner needs qualitative information, mainly from external sources, rather than the quantitative information from internal sources required by the operations manager.

ORGANIZATIONAL CULTURE

Organizational culture is a burgeoning area of research that has grown remarkably since the mid-1980s. Just as it is appropriate to think of organizations as including many technologies, it is similarly appropriate to see them as hosts to multiple, often competing subcultures.

Because the area is relatively new, there is little agreement on what precisely constitutes an organizational subculture. It is agreed, however, that competing subcultures may be in conflict, attempting to gain adherents to their vision of what the organization should be. Research is in progress to determine the effects of virtual organizations and virtual teams on the creation of subcultures when members do not share a physical workspace.

Rather than thinking about culture as a whole, it is more useful to think about the researchable determinants of subcultures, such as shared verbal and nonverbal symbolism. Verbal symbolism includes shared language used to construct, convey, and preserve subcultural myths, metaphors, visions, and humor. Nonverbal symbolism includes shared artifacts, rites, and ceremonies such as clothing; the use, placement, and decoration of offices; and rituals for celebrating members' birthdays, promotions, and retirements.

Subcultures coexist within "official" organizational cultures. The officially sanctioned culture may prescribe a dress code, suitable ways to address superiors and coworkers, and proper ways to deal with the outside public. Subcultures may be powerful determinants of information requirements, availability, and use.

Organizational members may belong to one or more subcultures within the organization. Subcultures may exert a powerful influence on member behavior, including sanctions for or against the use of information systems.

Understanding and recognizing predominant organizational subcultures may help the systems analyst overcome the resistance to change that arises when a new information system is installed. For example, the analyst might devise user training to address specific concerns of organizational subcultures. Identifying subcultures may also help in the design of decision support systems that are tailored for interaction with specific user groups.

SUMMARY

There are three broad organizational fundamentals to consider when analyzing and designing information systems: the concept of organizations as systems, the various levels of management, and the overall organizational culture.

Organizations are complex systems composed of interrelated and interdependent subsystems. In addition, systems and subsystems are characterized by their internal environments on a continuum from open to closed. An open system allows free passage of resources (people, information, materials) through its boundaries; closed systems do not permit free flow of input or output. Organizations and teams can also be organized virtually with remote members connected electronically who are not in the same physical workspace. Enterprise resource planning (ERP) systems are integrated organizational (enterprise) information systems developed with customized, proprietary software that helps the flow of information between the functional areas in the organization. They support a systems view of the organization.

Entity relationship diagrams help the systems analyst understand the entities and relationships that comprise the organizational system. The four different kinds of E-R diagrams are a one-to-one relationship, a one-to-many relationship, a many-to-one relationship, and a many-to-many relationship.

The three levels of managerial control are operational, middle management, and strategic. The time horizon of decision making is different for each level.

Organizational cultures and subcultures are important determinants of how people use information and information systems. By grounding information systems in the context of the organization as a larger system, it is possible to realize that numerous factors are important and should be taken into account when ascertaining information requirements and designing and implementing information systems.

KEYWORDS AND PHRASES

- associative entity
- attributive entity
- closedness
- crow's foot notation
- enterprise resource planning (ERP)
- entity (fundamental entity)
- entity-relationship (ER) diagrams
- environment
- feedback
- interdependent
- interrelatedness
- organizational boundaries
- organizational culture
- openness
- operations management
- middle management
- strategic management
- systems
- virtual enterprise
- virtual organization
- virtual team

"You seem to have already made a good start at MRE. Even though I can tell you a lot about the company, remember that there are a number of ways to orient yourself within it. You will want to interview users, observe their decision-making settings, and look at archival reports, charts, and diagrams. To do so, you can click on the telephone directory to get an appointment with an interviewee, click on the building map to go to a particular location, or click on organizational charts that show you the functional areas and formal hierarchical relationships at MRE.

"Many of the rules of corporate life apply within the MRE HyperCase. For instance, there are many public areas in which you are free to walk. If you want to tour a private corporate office, however, you must first book an appointment with one of our employees. Some secure areas are strictly off limits to you, just because you are an outsider and could pose a security risk.

"I don't think you'll find us excessively secretive, however, because you may assume that any employee who grants you an interview will also grant you access to the archival material in his or her files as well as to current work. You'll be able to go about your consulting freely in most cases. If you get too curious or invade our privacy in some way, we'll let you know. We're not afraid to tell you what the limits are.

"Unfortunately, some people in the company never seem to make themselves available to consultants. If you need to know more about these hard-to-get interviewees, I suggest you be persistent. There are lots of ways to find out about the people and the systems of MRE, but much of the time creativity is what pays off. You will notice that the systems consultants who follow their hunches, sharpen their technical skills, and never stop thinking about piecing together the puzzles here at MRE are the ones who are the best.

"Remember to use multiple methods—interviewing, observation, and investigation—to understand what we at MRE are trying to tell you. Sometimes actions, documents, and offices actually speak louder than words!"

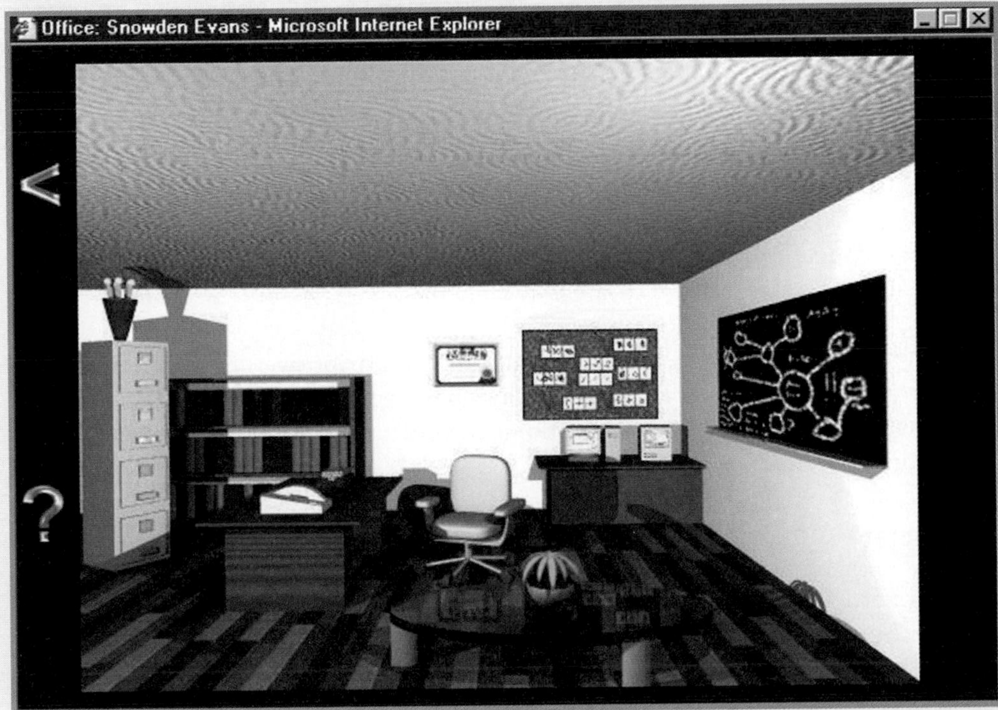

FIGURE 2.HC1

Click on key words in HyperCase and find out more detail.

(Continued)

HYPERCASE QUESTIONS

1. What major organizational change recently took place at MRE? What department(s) were involved, and why was the change made?
2. What does the Management Systems Unit at MRE do? Who are its clients?
3. What are the goals and strategies of the Engineering and Systems division at MRE? What are the goals of the Training and Management Systems Department?
4. Would you categorize MRE as a service industry, a manufacturer, or both? What kind of "products" does MRE "produce" (i.e., does it offer material goods, services, or both)? Suggest how the type of industry MRE is in affects the information systems it uses.
5. What type of organizational structure does MRE have? What are the implications of this structure for MIS?
6. Describe in a paragraph the "politics" of the Training and Management Systems Department at MRE. Who is involved, and what are some of the main issues?

REVIEW QUESTIONS

1. What are the three groups of organizational fundamentals that carry implications for the development of information systems?
2. What is meant by saying that organizational subsystems are interrelated and interdependent?
3. Define the term *organizational boundary*.
4. What are the two main purposes for feedback in organizations?
5. Define openness in an organizational environment.
6. Define closedness in an organizational environment.
7. What is the difference between a traditional organization and a virtual one?
8. What are the potential benefits and a drawback of a virtual organization?
9. Give an example of how systems analysts could work with users as a virtual team.
10. What is ERP, and what is its purpose?
11. What problems do analysts often encounter when they try to implement an ERP package?
12. What is meant by the term *entity-relationship diagram*?
13. What symbols are used to draw E-R diagrams?
14. List the types of E-R diagrams.
15. How do an entity, an associative entity, and an attributive entity differ?
16. List the three broad, horizontal levels of management in organizations.
17. How can understanding organizational subcultures help in the design of information systems?

PROBLEMS

1. "It's hard to focus on what we want to achieve. I look at what our real competitors, the convenience stores, are doing and think we should copy that. Then a hundred customers come in, and I listen to each of them, and they say we should keep our little store the same, with friendly clerks and old-fashioned cash registers. Then, when I pick up a copy of SuperMarket News, they say

that the wave of the future is super grocery stores, with no individual prices marked and UPC scanners replacing clerks. I'm pulled in so many directions I can't really settle on a strategy for our grocery store," admits Geoff Walsham, owner and manager of Jiffy Geoff's Grocery Store.

In a paragraph, apply the concept of permeable organizational boundaries to analyze Geoff's problem in focusing on organizational objectives.

2. Write seven sentences explaining the right-to-left relationships in Figure 2.8.
3. Draw an entity-relationship diagram of a patient–doctor relationship.
   a. Which of the types of E-R diagrams is it?
   b. In a sentence or two, explain why the patient–doctor relationship is diagrammed in this way.
4. You began drawing E-R diagrams soon after your entry into the health maintenance organization for which you're designing a system. Your team member is skeptical about using E-R diagrams before design of the database is begun. In a paragraph, persuade your team member that early use of E-R diagrams is worthwhile.
5. Sandy works as a manager for Arf-Arf Dog Food Company. Because there are several different suppliers of ingredients and their prices fluctuate, she has come up with several different formulations for the same dog food product, depending on the availability of particular ingredients from particular suppliers. She then orders ingredients accordingly. Even though she cannot predict when ingredients will become available at a particular price, her ordering of supplies can be considered routine.
   a. On what level of management is Sandy working? Explain in a paragraph.
   b. What attributes of her job would have to change before you would categorize her as working on a different level of management? List them.
6. Many of the people who work at Arf-Arf Dog Food Company laugh about Arf-Arf and about how silly it is that making dog food is such an important enterprise in the United States. Other groups in the company, however, are extremely proud of Arf-Arf's products and reputation for producing quality dog chow, and they happily display industry awards they have received. In a paragraph, identify and give descriptive names to organizational subcultures that seem apparent at Arf Arf.

GROUP PROJECTS

1. Break up into groups of five. Assign one person to be the Web site designer, one to write copy for a company's product, one to keep track of customer payments, one to worry about distribution, and one to satisfy customers that have questions about using the product. Then select a simple product (one that does not have too many versions of the product). Good examples are a camera, a VCR, a box of candy, and a specialty travel hat. Now spend 20 minutes trying to explain to the Web site designer what to put on the Web site. Describe in about three paragraphs what experience your group had in coordination. Elaborate of the interrelatedness of subsystems within the organization (your group).
2. With your group, draw a context-level diagram of your school's or university's registration system. Label each entity and process. Discuss why there appear to be different ways to draw the diagram. Reach consensus as a group about the best way to draw the diagram and defend your choice in a paragraph. Now, working with your group's members, follow the appropriate steps for developing an E-R diagram and create one for your school or university registration system. Make sure your group indicates whether the relationship you depict is one-to-one, one-to-many, many-to-one, or many-to-many.

SELECTED BIBLIOGRAPHY

Bleeker, S. E. "The Virtual Organization." *Futurist*, Vol. 28, No. 2, 1994, pp. 9–14.

Burch, J. G., F. R. Strater, and G. Grudnitski. *Information Systems: Theory and Practice*, 3d ed. New York: Wiley, 1983.

Chen, P. "The Entity-Relationship Model—Towards a Unified View of Data." *ACM Transactions on Database Systems*, Vol. 1, March 1976, pp. 9–36.

Ching, C., C. W. Holsapple, and A. B. Whinston. "Toward IT Support for Coordination in Network Organizations." *Information Management*, Vol. 30, No. 4, 1996, pp. 179–99.

Davis, G. B., and M. H. Olson. *Management Information Systems, Conceptual Foundations, Structure, and Development*, 2d ed. New York: McGraw-Hill, 1985.

Galbraith, J. R. *Organizational Design*. Reading, MA: Addison-Wesley, 1977.

Kendall, K. E., J. R. Buffington, and J. E. Kendall. "The Relationship of Organizational Subcultures to DSS User Satisfaction." *Human Systems Management*, March 1987, pp. 31–39.

PeopleSoft. Available: ⟨http://peoplesoft.com⟩. Last updated 1999.

Warkentin, M., L. Sayeed, and R. Hightower. "Virtual Teams versus Face-to-Face Teams; An Exploratory Study of a Web-Based Conference System." In *Emerging Information Technologies: Improving Decisions, Cooperation, and Infrastructure*, edited by K. E. Kendall, pp. 241–62. Thousand Oaks, CA: Sage Publications, 1999.

Yager, S. E. "Everything's Coming Up Virtual." Available: ⟨http://info.acm.org/crossroads/xrds4-1/organi.html⟩. Last updated March 6, 2000.

Allen Schmidt, Julie E. Kendall, and Kenneth E. Kendall

PICTURING THE RELATIONSHIPS

"So the project involves more than simply performing maintenance work on the current programs," Chip says. "Are we using a formal methodology for analyzing and designing the new system?"

"Yes," replies Anna. "We are also using a CASE tool, Visible Analyst, to analyze and design the system.[1] We've recently installed the product on our microcomputer in the office." Anna motions to a large computer work area containing a microcomputer with attached laser printer and mainframe terminal.

"That's great!" exclaims Chip. "Have you used Visible Analyst on other projects?"

"No," Anna says. "This is the pilot project involving Visible Analyst. I've attended training sessions, and I'm somewhat familiar with its operation. Are you familiar with Visible Analyst?"

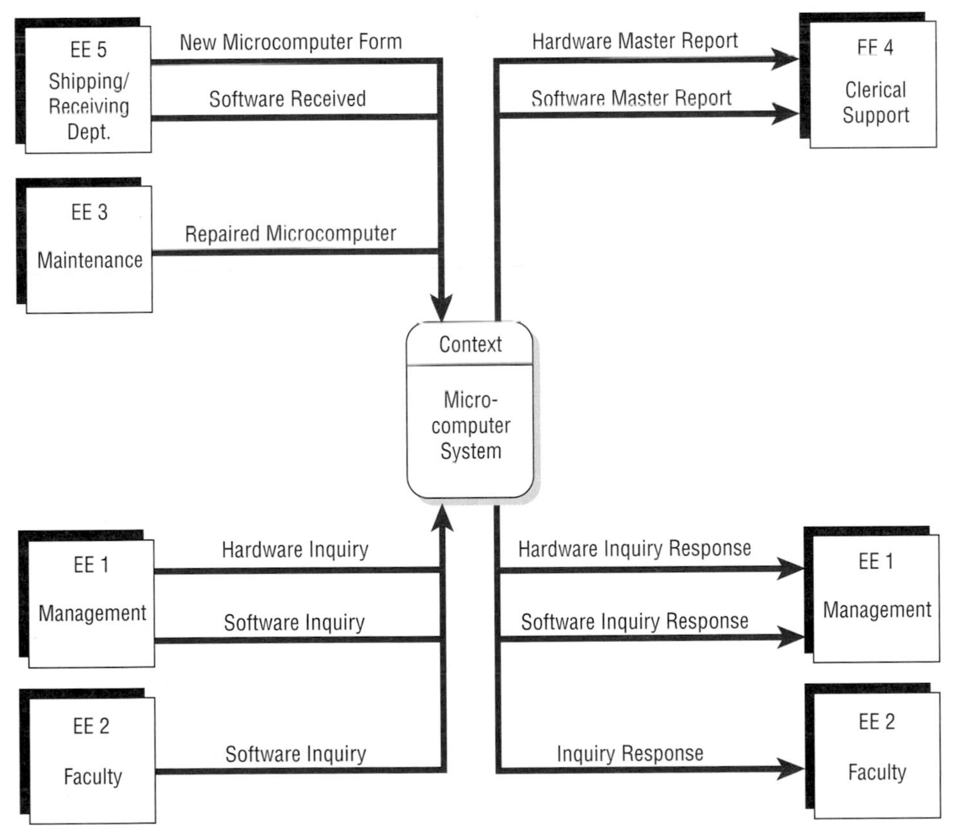

FIGURE E2.1
Context-level data flow diagram, Current System.

[1]For more details on how to begin using Visible Analyst, see Allen Schmidt, *Working with Visible Analyst Workbench for Windows* (Upper Saddle River, NJ: Prentice Hall, 1996).

The Central Pacific University case can be adapted to other CASE tools. Other tools include Microsoft Visio. Alternatively, many of the exercises can be accomplished manually if CASE tools are unavailable.

2

"I've had some experience with it at school. I understand that's one of the reasons that I was hired for this position," answers Chip. "Let me see what you've done so far."

They both move to the microcomputer area, where Anna loads Visible Analyst. With a few easy mouse clicks she comes to a context-level data flow diagram (See Figure E2.1). "It's very useful to begin thinking of the system this way," Anna says as they look at the diagram on the screen.

Chip agrees, saying, "I can very easily see what you think is happening with the system. For instance, I see that the external entity Management supplies hardware and software inquiries and receives the corresponding response in return. It shows the system within the larger organization."

"I've also drawn an E-R diagram of the system so I could start understanding the relationships between the data structures," Anna says as she brings up the entity-relationship diagram on the screen. (See Figure E2.2.)

"Yes, the many-to-many and one-to-many relationships are very clear when you look at this," Chip says, viewing the screen. "You've got a good start here," Chip continues. "Let's get to work and see what needs to be done next."

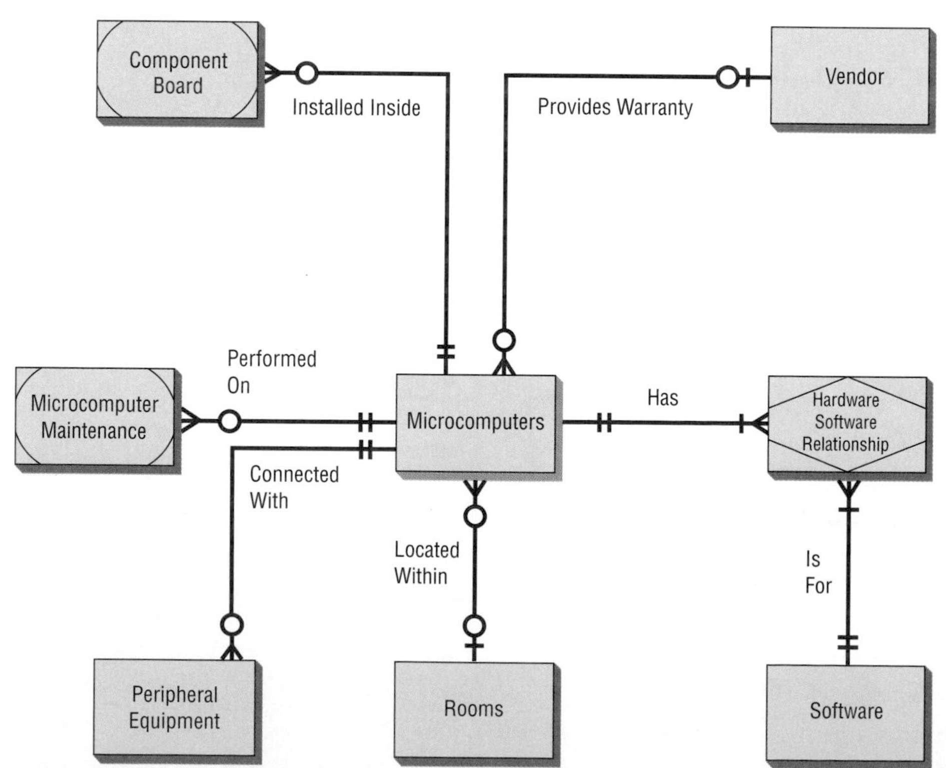

FIGURE E2.2
Entity-relationship diagram, Microcomputer System.

EXERCISES*

E-1. Use Visible Analyst to view and print the context-level data flow diagram for the microcomputer systems, as Chip and Anna did.

E-2. Use the Repository feature to view the entry for the central process.

E-3. Use Visible Analyst to view and print the entity-relationship diagram for the microcomputer system.

E-4. Explain why the external entities on the context-level diagram are not found on the entity-relationship diagram.

E-5. Explain why the entities MANAGEMENT and FACULTY are found on both sides of the process on the context-level diagram.

*Exercises preceded by a CD-ROM icon require the program Visible Analyst or another CASE tool. A CD-ROM containing Visible Analyst examples is provided free of charge to any professor adopting this book. The examples on the disk may be imported into Visible Analyst and then used by students.

DETERMINING FEASIBILITY AND MANAGING ANALYSIS AND DESIGN ACTIVITIES

3

PROJECT FUNDAMENTALS

Initiating projects, determining project feasibility, scheduling projects, and planning and then managing activities and team members for productivity are all important capabilities for the systems analyst to master. As such, they are considered project fundamentals.

A systems project begins with problems or with opportunities for improvement within a business that often come up as the organization adapts to change. The increasing popularity of ecommerce means that some fundamental changes are occurring as businesses either originate their enterprises on, or move their internal operations as well as external relationships to, the Internet. Changes that require a systems solution occur in the legal environment as well as in the industry's environment. Once a project is suggested, the systems analyst works quickly with decision makers to determine whether it is feasible. If a project is approved for a full systems study, the project activities are scheduled through the use of tools such as Gantt charts and Program Evaluation and Review Techniques (PERT) diagrams so that the project can be completed on time. Part of assuring the productivity of systems analysis team members is effectively managing their scheduled activities. This chapter is devoted to a discussion of these project fundamentals.

PROJECT INITIATION

Systems projects are initiated by many different sources for many reasons. Some of the projects suggested will survive various stages of evaluation to be worked on by you (or you and your team); others will not and should not get that far. Businesspeople suggest systems projects for two broad reasons: (1) to experience problems that lend themselves to systems solutions and (2) to recognize opportunities for improvement through upgrading, altering, or installing new systems when they occur. Both situations can arise as the organization adapts to and copes with natural, evolutionary change.

PROBLEMS WITHIN THE ORGANIZATION

Managers do not like to conceive of their organization as having problems, let alone talk about them or share them with someone from outside. Good managers, however, realize that recognizing symptoms of problems or, at a later stage, diag-

FIGURE 3.1

Checking output, observing employee behavior, and listening to feedback are all ways to help the analyst pinpoint systems problems and opportunities.

To Identify Problems	Look for These Specific Signs:
Check output against performance criteria	• Too many errors • Work completed slowly • Work done incorrectly • Work done incompletely • Work not done at all
Observe behavior of employees	• High absenteeism • High job dissatisfaction • High job turnover
Listen to external feedback from: Vendors Customers Suppliers	• Complaints • Suggestions for improvement • Loss of sales • Lower sales

nosing the problems themselves and then confronting them are imperative if the business is to keep functioning at its highest potential.

Problems surface in many different ways. One way of conceptualizing what problems are and how they arise is to think of them as situations where goals have never been met or are no longer being met. Useful feedback gives information about the gap between actual and intended performance. In this way feedback spotlights problems.

In some instances problems that require the services of systems analysts are uncovered because performance measures are not being met. Problems (or symptoms of problems) with processes that are visible in output and that could require the help of a systems analyst include excessive errors and work performed too slowly, incompletely, incorrectly, or not at all. Other symptoms of problems become evident when people do not meet baseline performance goals. Changes in employee behavior such as unusually high absenteeism, high job dissatisfaction, or high worker turnover should alert managers to potential problems. Any of these changes, alone or in combination, might be sufficient reason to request the help of a systems analyst.

Although difficulties such as those just described occur within the organization, feedback on how well the organization is meeting intended goals may come from outside, in the form of complaints or suggestions from customers, vendors, or suppliers and lost or unexpectedly lower sales. This feedback from the external environment is extremely important and should not be ignored.

A summary of symptoms of problems and approaches useful in problem detection is provided in Figure 3.1. Notice that checking output, observing or researching employee behavior, and listening to feedback from external sources are all valuable in problem finding. When reacting to accounts of problems within the organization, the systems analyst plays the roles of consultant, supporting expert, and as agent of change, as discussed in Chapter 1. As you might expect, roles for the systems analyst shift subtly when projects are initiated because the focus is on opportunities for improvement rather than on the need to solve problems.

SELECTION OF PROJECTS

Projects come from many different sources and for many reasons. Not all should be selected for further study. You must be clear in your own mind about the reasons for recommending a systems study on a project that seems to address a problem or

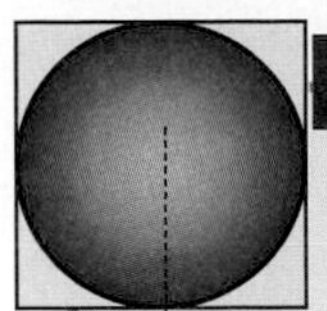

CONSULTING OPPORTUNITY 3.1

THE SWEETEST SOUND I'VE EVER SIPPED

Felix Straw, who represents one of the many U.S. distributors of the European soft drink Sipps, gazes unhappily at a newspaper weather map, which is saturated with dark red, indicating that most of the United States is experiencing an early spring heat wave with no signs of a letup. Pointing to the paper as he speaks, he tells your systems group, "It's the best thing that could happen to us, or at least it should be. But when we had to place our orders three months ago, we had no idea that this spring monster heat wave was going to devour the country this way!" Nodding his head toward a picture of their European plant on the wall, he continues. "We need to be able to tell them when things are hot over here so we can get enough product. Otherwise, we'll miss out every time. This happened two years ago and it just about killed us.

"Each of us distributors meets with our district managers to do three-month planning. When we agree, we fax our orders into European headquarters. They make their own adjustments, bottle the drinks, and then we get our modified orders about 9 to 15 weeks later. But we need ways to tell them what's going on now. Why, we even have some new superstores that are opening up here. They should know we have extra-high demand."

Corky, his assistant agrees, saying, "Yeah, they should at least look at our past sales around this time of year. Some springs are hot, others are just average."

Straw concurs, saying, "It would be music to my ears, it would be really sweet, if they would work with us to spot trends and changes—and then respond quickly."

Stern's, based in Blackpool, England, is a European beverage maker and the developer and producer of Sipps. Sipps is a sweet, fruit-flavored, nonalcoholic, noncarbonated drink, which is served chilled or with ice, and it is particularly popular when the weather is hot. Selling briskly in Europe and growing in popularity in the United States since its introduction five years ago, Sipps has had a difficult time adequately managing inventory and keeping up with U.S. customer demand, which is affected by seasonal temperature fluctuations. Places with year-round temperature climates and lots of tourists (such as Florida and California) have large standing orders, but other areas of the country could benefit from a less cumbersome, more responsive order-placing process. Sipps is distributed by a network of local distributors located throughout the United States and Canada.

As one of the systems analysts assigned to work with the U.S. distributors of Sipps, begin your analysis by listing some of the key symptoms and problems you have identified after studying the information flows, ordering process, and inventory management and after interviewing Mr. Straw and his assistant. In a paragraph describe which problems might indicate the need for a systems solution.

Note: This consulting opportunity is loosely based on J. C. Perez, "Heineken's HOPS Software Keeps A-Head on Inventory," *PC Week*, Vol. 14, No. 2, January 13, 1997, pp. 31 and 34.

could bring about improvement. Consider the motivation that prompts a proposal on the project. You need to be sure that the project under consideration is not being proposed simply to enhance your own political reputation or power or that of the person or group proposing it, because there is a high probability that such a project will be ill-conceived and eventually ill-accepted.

As outlined in Chapter 2, prospective projects need to be examined from a systems perspective in such a way that you are considering the impact of the proposed change on the entire organization. Recall that the various subsystems of the organization are interrelated and interdependent, so a change to one subsystem might affect all the others. Even though the decision makers directly involved ultimately set the boundaries for the systems project, a systems project cannot be contemplated or selected in isolation from the rest of the organization.

Beyond these general considerations are five specific criteria for project selection:

1. Backing from management
2. Appropriate timing of project commitment
3. Possibility of improving attainment of organizational goals
4. Practical in terms of resources for the systems analyst and organization
5. Worthwhile project compared with other ways the organization could invest resources

First and foremost is backing from management. Absolutely nothing can be accomplished without the endorsement of the people who eventually will foot the bill. This statement does not mean that you lack influence in directing the project or that people other than management can't be included, but management backing is essential.

Another important criterion for project selection includes timing for you and the organization. Ask yourself and the others who are involved if the business is presently capable of making a time commitment for installation of new systems or improvement to existing ones. You must also be able to commit all or a portion of your time for the duration.

A third criterion is the possibility of improving attainment of organizational goals. The project should put the organization on target, not deter it from its ultimate goals.

A fourth criterion is selecting a project that is practicable in terms of your resources and capabilities as well as those of the business. Some projects will not fall within your realm of expertise, and you must be able to recognize them.

Finally, you need to come to a basic agreement with the organization about the worthiness of the systems project relative to any other possible project being considered. Remember that when a business commits to one project, it is committing resources that thereby become unavailable for other projects. It is useful to view all possible projects as competing for the business resources of time, money, and people.

DETERMINING FEASIBILITY

Once the number of projects has been narrowed according to the criteria discussed previously, it is still necessary to determine if the selected projects are feasible. Our definition of feasibility goes much deeper than common usage of the term, because systems projects feasibility is assessed in three principal ways: operationally, technically, and economically. The feasibility study is not a full-blown systems study. Rather, the feasibility study is used to gather broad data for the members of management that in turn enables them to make a decision on whether to proceed with a systems study.

Data for the feasibility study can be gathered through interviews, which are covered in detail in Chapter 5. The kind of interview required is directly related to the problem or opportunity being suggested. The systems analyst typically interviews those requesting help and those directly concerned with the decision-making process, typically management. Although it is important to address the correct problem, the systems analyst should not spend too much time doing feasibility studies, because many projects will be requested and only a few can or should be executed. The feasibility study must be highly time-compressed, encompassing several activities in a short span of time.

DEFINING OBJECTIVES

The systems analyst serves as catalyst and supporting expert primarily by being able to see where processes can be improved. Optimistically, opportunities can be conceived of as the obverse of problems; yet, in some cultures, crisis also means opportunity. What looms as a disturbing problem for a manager might be turned into an opportunity for improvement by an alert systems analyst.

Improvements to systems can be defined as changes that will result in incremental yet worthwhile benefits. There are many possibilities for improvements, including:

1. Speeding up a process
2. Streamlining a process through the elimination of unnecessary or duplicated steps
3. Combining processes
4. Reducing errors in input through changes of forms and display screens
5. Reducing redundant storage
6. Reducing redundant output
7. Improving integration of systems and subsystems

It is well within the systems analyst's capabilities to notice opportunities for improvements. People who come into daily contact with the system, however, may be even better sources of information about improvements that should be made. If improvements have already been suggested, your expertise is needed to help determine whether the improvement is worthwhile and how it can be implemented.

It is worthwhile for an analyst to create a feasibility impact grid (FIG) for understanding and assessing what impacts (if any) improvements to existing systems can make. Figure 3.2 maps out such a grid. The labels on the far left-hand side describe a variety of systems that exist currently or are proposed. They are categorized into three system types: ecommerce systems, management information systems (MIS), and transaction processing systems (TPS). Listed at the top are the seven process objectives. Red arrows in the grid show that a positive impact can be made when a system improvement is made. Green arrows indicate that the system has been implemented and that the improvement positively impacted the process objective.

Notice that the transaction processing systems show a positive effect on the process objectives in almost every case. Traditional management information systems may help make better decisions, but sometimes they do not help the efficient collecting, storing, or retrieving of data. Thus, there are fewer arrows in that part of the grid. When entering the world of ecommerce, the analyst needs to be aware of how each system improvement may effect process objectives. Notice that the analyst who completed this grid recognized that although some process objectives were effected, others were not.

Of equal importance is how corporate objectives are affected by improvements to information systems. These corporate objectives include:

1. Improving corporate profits
2. Supporting the competitive strategy of the organization
3. Improving cooperation with vendors and partners
4. Improving internal operations support so that goods and services are produced efficiently and effectively
5. Improving internal decision support so that decisions are more effective
6. Improving customer service
7. Increasing employee morale

Once again a feasibility impact grid (FIG) can be drawn to increase awareness of the impacts made on the achievement of corporate objectives. The grid shown in Figure 3.3 is similar to the process objectives grid described earlier, but it emphasizes the point that improvements to the management information system greatly affect corporate objectives. You may recall that traditional MIS did not affect many process objectives, but on the other hand, they do affect most of the corporate objectives.

FIGURE 3.2

An analyst can use a feasibility impact grid to show how each system component affects process objectives.

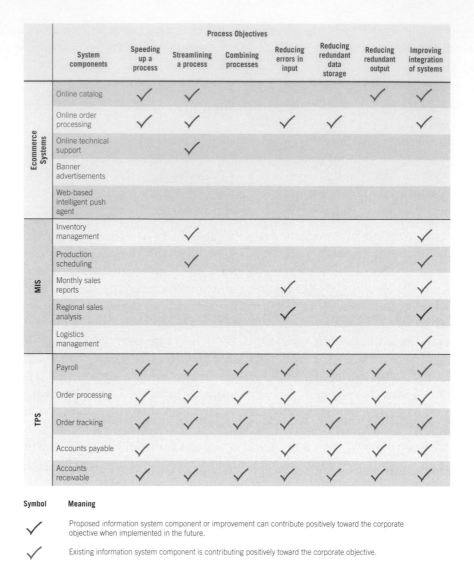

	System components	Process Objectives: Speeding up a process	Streamlining a process	Combining processes	Reducing errors in input	Reducing redundant data storage	Reducing redundant output	Improving integration of systems
Ecommerce Systems	Online catalog	✓	✓				✓	✓
	Online order processing	✓	✓		✓	✓		✓
	Online technical support		✓					
	Banner advertisements							
	Web-based intelligent push agent							
MIS	Inventory management		✓					✓
	Production scheduling		✓					✓
	Monthly sales reports				✓			✓
	Regional sales analysis				✓			✓
	Logistics management					✓		✓
TPS	Payroll	✓	✓	✓	✓	✓	✓	✓
	Order processing	✓	✓	✓	✓	✓	✓	✓
	Order tracking	✓	✓	✓	✓	✓	✓	✓
	Accounts payable	✓			✓	✓	✓	✓
	Accounts receivable	✓	✓	✓	✓	✓	✓	✓

Symbol	Meaning
✓	Proposed information system component or improvement can contribute positively toward the corporate objective when implemented in the future.
✓	Existing information system component is contributing positively toward the corporate objective.

It is essential that analysts systematically go through the steps in developing feasibility impact grids. By understanding process and corporate objectives, an analyst realizes why he or she is building systems and comprehends what the importance of designing efficient and effective systems might be. Analysts can communicate those impacts to the decision makers evaluating (and paying for) the project.

An analyst should be aware that there also are some unacceptable objectives for systems projects. As mentioned before, they include undertaking a project solely to prove the prowess of the systems analysis team or purely to assert the superiority of one department over another in terms of its power to command internal resources. Without a consideration of its true contribution to the achievement of the organization's goals, it is also unacceptable to automate manual procedures for the sake of automation alone or to invest in new technology because of infatuation with the "bells and whistles" it provides over and above what the present system offers.

The objectives of the project need to be cleared formally on paper as well as informally by talking to people in the business. Find out what problem they believe the systems project would solve or what situation it would improve and what their expectations are for the proposed system.

	System components	Speeding up a process	Streamlining a process	Combining processes	Reducing errors in input	Reducing redundant data storage	Reducing redundant output	Improving integration of systems
		Corporate Objectives						
Ecommerce Systems	Online catalog		✔				✔	✔
Ecommerce Systems	Online order processing		✔	✔			✔	✔
Ecommerce Systems	Online technical support		✔				✔	✔
Ecommerce Systems	Banner advertisements	✔		✔			✔	
Ecommerce Systems	Web-based intelligent push agent						✔	
MIS	Inventory management	✓	✔		✓	✓	✓	
MIS	Production scheduling	✓	✔		✓	✓	✓	✓
MIS	Monthly sales reports	✓	✔		✓	✓		✓
MIS	Regional sales analysis	✔			✔	✔		✔
MIS	Logistics management	✓	✔		✓	✓		
TPS	Payroll				✓			✓
TPS	Order processing	✓			✓		✓	
TPS	Order tracking	✓			✓		✓	
TPS	Accounts payable			✓	✓		✓	
TPS	Accounts receivable			✓	✓		✓	

Symbol	Meaning
✔	Proposed information system component or improvement can contribute positively toward the corporate objective when implemented in the future.
✓	Existing information system component is contributing positively toward the corporate objective.

FIGURE 3.3
An analyst can use a feasibility impact grid to show how each system component affects corporate objectives.

DETERMINING RESOURCES

Resource determination for the feasibility study follows the same broad pattern discussed previously and will be revised and reevaluated if and when a formal systems study is commissioned. A project must be feasible in all three ways to merit further development, as shown in Figure 3.4. Resources are discussed in relationship to three areas of feasibility: technical, economic, and operational.

Technical Feasibility. A large part of determining resources has to do with assessing technical feasibility. The analyst must find out whether current technical resources can be upgraded or added to in a manner that fulfills the request under consideration. Sometimes, however, "add-ons" to existing systems are costly and not worthwhile, simply because they meet needs inefficiently. If existing systems cannot be added onto, the next question becomes whether there is technology in existence that meets the specifications.

At this point the expertise of systems analysts is beneficial, because by using their own experience and their contact with vendors, systems analysts will be able

FIGURE 3.4

The three key elements of feasibility include technical, economic, and operational feasibility.

The Three Key Elements of Feasibility
Technical Feasibility Add on to present system Technology available to meet users' needs
Economic Feasibility Systems analysts' time Cost of systems study Cost of employees' time for study Estimated cost of hardware Cost of packaged software/software development
Operational Feasibility Whether the system will operate when installed Whether the system will be used

to answer the question of technical feasibility. Usually the response to whether a particular technology is available and capable of meeting the users' requests is "yes," and then the question becomes an economic one.

Economic Feasibility. Economic feasibility is the second part of resource determination. The basic resources to consider are your time and that of the systems analysis team, the cost of doing a full systems study (including time of employees you will be working with), the cost of the business employee time, the estimated cost of hardware, and the estimated cost of software or software development.

The concerned business must be able to see the value of the investment it is pondering before committing to an entire systems study. If short-term costs are not overshadowed by long-term gains or produce no immediate reduction in operating costs, the system is not economically feasible and the project should not proceed any further.

Operational Feasibility. Suppose for a moment that technical and economic resources are both judged adequate. The systems analyst must still consider the operational feasibility of the requested project. Operational feasibility is dependent on the human resources available for the project and involves projecting whether the system will operate and be used once it is installed.

If users are virtually wed to the present system, see no problems with it, and generally are not involved in requesting a new system, resistance to implementing the new system will be strong. Chances for it ever becoming operational are low.

Alternatively, if users themselves have expressed a need for a system that is operational more of the time, in a more efficient and accessible manner, chances are better that the requested system will eventually be used. Much of the art of determining operational feasibility rests with the user interfaces that are chosen, as we see in Chapter 18.

At this point, determining operational feasibility requires creative imagination on the part of the systems analyst as well as the powers of persuasion to let users know which interfaces are possible and which will satisfy their needs. The systems analyst must also listen carefully to what users really want and what it seems they will use. Ultimately, however, assessing operational feasibility largely involves educated guesswork.

JUDGING FEASIBILITY

From the foregoing discussion, it is evident that judging the feasibility of systems projects is never a clear-cut or easy task. Furthermore, project feasibility is a decision to be made not by the systems analyst but instead by management. Decisions are based on feasibility data expertly and professionally gathered and presented by the analyst.

The systems analyst needs to be sure that all three areas of technical, economic, and operational feasibility are addressed in the preliminary study. The study of a requested systems project must be accomplished quickly so that the resources devoted to it are minimal, the information output from the study is solid, and any existing interest in the project remains high. Remember that this is a preliminary study, which precedes the system study, and it must be executed rapidly and competently.

Projects that meet the criteria discussed in the project selection subsection earlier in the chapter, as well as the three criteria of technical, economic, and operational feasibility, should be chosen for a detailed systems study. At this point the systems analyst must act as a supporting expert, advising management that the requested systems project meets all the selection criteria and has thus qualified as an excellent candidate for further study. Remember that a commitment from management now means only that a systems study may proceed, not that a proposed system is accepted. Generally, the process of feasibility assessment is effective in screening out projects that are inconsistent with the business's objectives, technically impossible, or economically unprofitable. Although it is painstaking, studying feasibility is worthwhile and saves businesses and systems analysts a good deal of time and money in the end.

ACTIVITY PLANNING AND CONTROL

Systems analysis and design involves many different types of activities that together make up a project. The systems analyst must manage the project carefully if the project is to be successful. Project management involves the general tasks of planning and control.

Planning includes all the activities required to select a systems analysis team, assign members of the team to appropriate projects, estimate the time required to complete each task, and schedule the project so that tasks are completed in a timely fashion. Control means using feedback to monitor the project, including comparing the plan for the project with its actual evolution. In addition, control means taking appropriate action to expedite or reschedule activities to finish on time while motivating team members to complete the job properly.

ESTIMATING TIME REQUIRED

The systems analyst's first decision is to determine the amount of detail that goes into defining activities. The lowest level of detail is the systems development life cycle itself, whereas the highest extreme is to include every detailed step. The optimal answer to planning and scheduling lies somewhere in between.

A structured approach is useful here. In Figure 3.5 the systems analyst beginning a project has broken the process into three major phases: analysis, design, and implementation. Then the analysis phase is further broken down into data gathering, data flow and decision analysis, and proposal preparation. Design is broken down into data-entry design, input and output design, and data organization. The implementation phase is divided into implementation and evaluation.

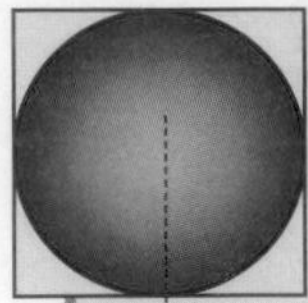

CONSULTING OPPORTUNITY 3.2

FOOD FOR THOUGHT

We could really make some changes. Shake up some people. Let them know we're with it. Technologically, I mean," said Malcolm Warner, vice president for AllFine Foods, a wholesale dairy products distributor. "That old system should be overhauled. I think we should just tell the staff that it's time to change."

"Yes, but what would we actually be improving?" Kim Han, assistant to the vice president, asks. "I mean, there aren't any substantial problems with the system input or output that I can see."

Malcolm snaps, "Kim, you're purposely not seeing my point. People out there see us as a stodgy firm. A new computer system could help change that. Change the look of our invoices. Send jazzier reports to the food store owners. Get some people excited about us as leaders in wholesale food distributing and computers."

"Well, from what I've seen over the years," Kim replies evenly, "a new system is very disruptive, even when the business really needs it. People dislike change, and if the system is performing the way it should, maybe there are other things we could do to update our image that wouldn't drive everyone nuts in the process. Besides, you're talking big bucks for a new gimmick."

Malcolm says, "I don't think just tossing it around here between the two of us is going to solve anything. Check on it and get back to me. Wouldn't it be wonderful?"

A week later Kim enters Malcolm's office with several pages of interview notes in hand. "I've talked with most of the people who have extensive contact with the system. They're happy, Malcolm. And they're not just talking through their hats. They know what they're doing."

"I'm sure the managers would like to have a newer system than the guys at Quality Foods," Malcolm replies. "Did you talk to them?"

Kim says, "Yes. They're satisfied."

"And how about the people in systems? Did they say the technology to update our system is out there?" Malcolm inquires insistently.

"Yes. It can be done. That doesn't mean it should be," Kim says firmly.

As the systems analyst for AllFine Foods, how would you assess the feasibility of the systems project Malcolm is proposing? Based on what Kim has said about the managers, users, and systems people, what seems to be the operational feasibility of the proposed project? What about the economic feasibility? What about the technological feasibility? Based on what Kim and Malcolm have discussed, would you recommend that a full-blown systems study be done?

In subsequent steps the systems analyst needs to consider each of these tasks and break them down further so that planning and scheduling can take place. Figure 3.6 shows how the analysis phase is described in more detail. For example, data gathering is broken down into five activities, from conducting interviews to observing reactions to the prototype. This particular project requires data flow analysis but not decision analysis, so the systems analyst has penciled in "analyze data flow" as the single step in the middle phase. Finally, proposal preparation is broken down into three steps: perform cost/benefit analysis; prepare proposal, and present proposal.

FIGURE 3.5
Beginning to plan a project by breaking it into three major activities.

Phase	Activity
Analysis	Data Gathering Data Flow and Decision Analysis Proposal Preparation
Design	Data-Entry Design Input Design Output Design Data Organization
Implementation	Implementation Evaluation

Break apart the major activities into smaller ones.

Activity	Detailed Activity	Weeks Required
Data Gathering	Conduct Interviews	3
	Administer Questionnaires	4
	Read Company Reports	4
	Introduce Prototype	5
	Observe Reactions to Prototype	3
Data Flow and Decision Analysis	Analyze Data Flow	8
Proposal Preparation	Perform Cost/Benefit Analysis	3
	Prepare Proposal	2
	Present Proposal	2

FIGURE 3.6

Refining the planning and scheduling of analysis activities by adding detailed tasks and establishing the time required to complete these tasks.

The systems analyst, of course, has the option to break down steps further. For instance, the analyst could specify each of the persons to be interviewed. The amount of detail necessary depends on the project, but all critical steps need to appear in the plans.

Sometimes the most difficult part of project planning is the crucial step of estimating the time it takes to complete each task or activity. When quizzed about reasons for lateness on a particular project, project team members cited poor scheduling estimates that hampered the success of projects from the outset. There is no substitute for experience in estimating time requirements, and systems analysts who have had the opportunity of an apprenticeship are fortunate in this regard.

Planners have attempted to reduce the inherent uncertainty in determining time estimates by projecting most likely, pessimistic, and optimistic estimates and then using a weighted average formula to determine the expected time an activity will take. This approach offers little more in the way of confidence, however. Perhaps the best strategy for the systems analyst is to adhere to a structured approach in identifying activities and describing these activities in sufficient detail. In this manner, the systems analyst will at least be able to limit unpleasant surprises.

USING GANTT CHARTS FOR PROJECT SCHEDULING

A Gantt chart is an easy way to schedule tasks. It is a chart on which bars represent each task or activity. The length of each bar represents the relative length of the task.

Figure 3.7 is an example of a two-dimensional Gantt chart where time is indicated on the horizontal dimension and a description of activities makes up the vertical dimension. In this example the Gantt chart shows the analysis or information-gathering phase of the project. Notice on the Gantt chart that conducting interviews will take three weeks, administering the questionnaire will take four weeks, and so on. These activities overlap part of the time. In the chart the special symbol ▲ signifies that it is week 9. The bars with color shading represent projects or parts of projects that have been completed, telling us that the systems analyst is

FIGURE 3.7

Using a two-dimensional Gantt chart for planning activities that can be accomplished in parallel.

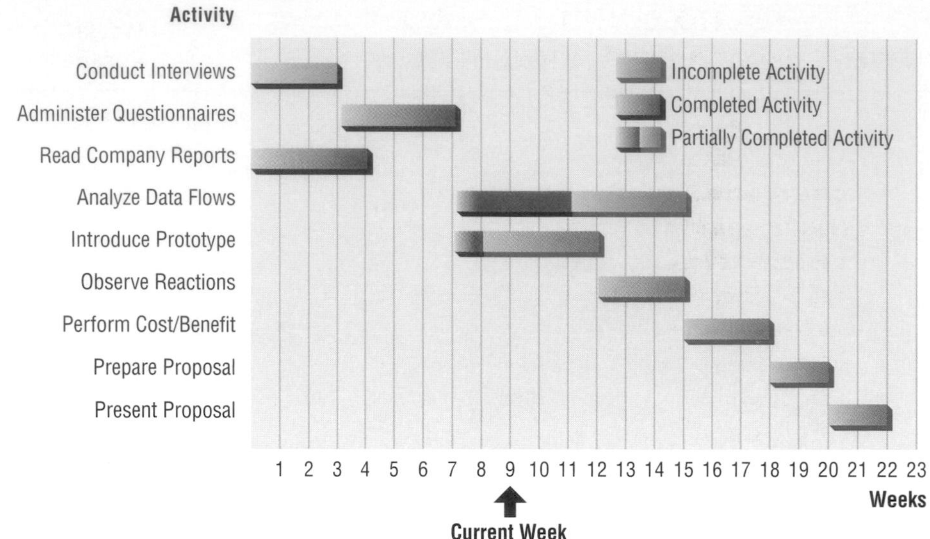

behind in introducing prototypes but ahead in analyzing data flows. Action must be taken on introducing prototypes soon so that other activities or even the project itself will not be delayed as a result.

The main advantage of the Gantt chart is its simplicity. The systems analyst will find not only that this technique is easy to use but also that it lends itself to worthwhile communication with end users. Another advantage of using a Gantt chart is that the bars representing activities or tasks are drawn to scale; that is, the size of the bar indicates the relative length of time it will take to complete each task.

USING PERT DIAGRAMS

PERT is an acronym for Program Evaluation and Review Techniques. A program (a synonym for a project) is represented by a network of nodes and arrows that are then evaluated to determine the critical activities, improve the schedule if necessary, and review progress once the project is undertaken. PERT was developed in the late 1950s for use in the U.S. Navy's Polaris nuclear submarine project. It reportedly saved the U.S. Navy two years' development time.

PERT is useful when activities can be done in parallel rather than in sequence. The systems analyst can benefit from PERT by applying it to systems projects on a smaller scale, especially when some team members can be working on certain activities at the same time that fellow members are working on other tasks.

Figure 3.8 compares a simple Gantt chart with a PERT diagram. The activities expressed as bars in the Gantt chart are represented by arrows in the PERT diagram. The length of the arrows has no direct relationship with the activity durations. Circles on the PERT diagram are called events and can be identified by numbers, letters, or any other arbitrary form of designation. The circular nodes are present to (1) recognize that an activity is completed and (2) indicate which activities need to be completed before a new activity may be undertaken (precedence).

In reality activity C may not be started until activity A is completed. Precedence is not indicated at all in the Gantt chart, so it is not possible to tell whether activity C is scheduled to start on day 4 on purpose or by coincidence.

A project has a beginning, a middle, and an end; the beginning is event 10 and the end is event 50. To find the length of the project, each path from beginning to

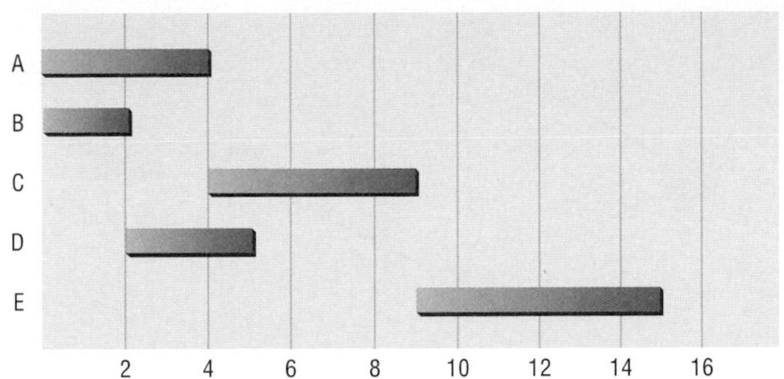

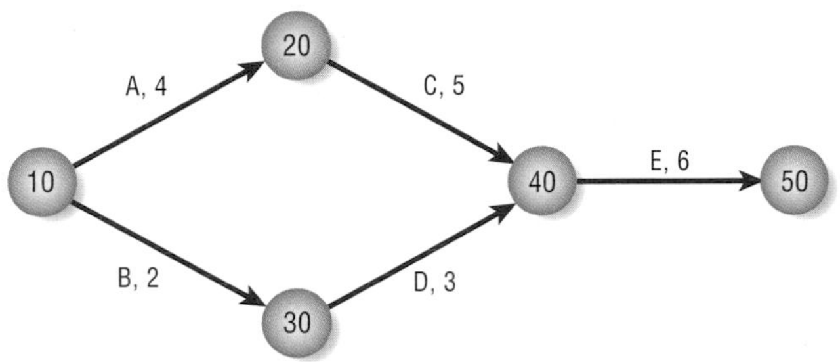

FIGURE 3.8

Gantt charts compared with PERT diagrams for scheduling activities.

end is identified, and the length of each path is calculated. In this example path 10–20–40–50 has a length of 15 days, whereas path 10–30–40–50 has a length of 11 days. Even though one person may be working on path 10–20–40–50 and another on path 10–30–40–50, the project is not a race. The project requires that both sets of activities (or paths) be completed; consequently, the project takes 15 days to complete.

The longest path is referred to as the critical path. Although the critical path is determined by calculating the longest path, it is defined as the path that will cause the whole project to fall behind if even one day's delay is encountered on it. Note that if you are delayed one day on path 10–20–40–50, the entire project will take longer, but if you are delayed one day on path 10–30–40–50, the entire project will not suffer. The leeway to fall behind somewhat on noncritical paths is called slack time.

Occasionally, PERT diagrams need pseudo-activities, referred to as dummy activities, to preserve the logic of or clarify the diagram. Figure 3.9 shows two PERT diagrams with dummies. Project 1 and project 2 are quite different, and the way the dummy is drawn makes the difference clear. In project 1 activity C can only be started if both A and B are finished, because all arrows coming into a node must be completed before leaving the node. In project 2, however, activity C requires only activity B's completion and can therefore be under way while activity A is still taking place.

Project 1 takes 14 days to complete, whereas project 2 takes only 9 days. The dummy in project 1 is necessary, of course, because it indicates a crucial precedence relationship. The dummy in project 2, on the other hand, is not required, and activity A could have been drawn from 10 to 40 and event 20 may be eliminated completely.

Therefore, there are many reasons for using a PERT diagram over a Gantt chart. The PERT diagram allows

1. Easy identification of the order of precedence
2. Easy identification of the critical path and thus critical activities
3. Easy determination of slack time

FIGURE 3.9
Precedence of activities is important in determining the length of the project when using a PERT diagram.

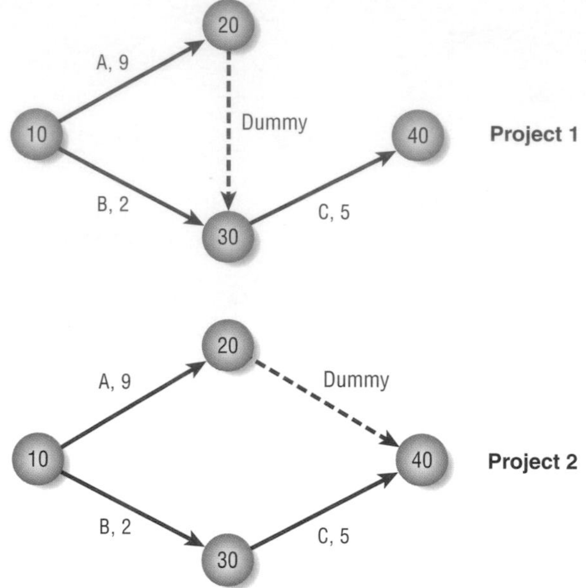

A PERT Example. Suppose a systems analyst is trying to set up a realistic schedule for the data-gathering and proposal phases of the systems analysis and design life cycle. The systems analyst looks over the situation and lists activities that need to be accomplished along the way. This list, which appears in Figure 3.10, also shows that some activities must precede other activities. The time estimates were determined as discussed in an earlier section of this chapter.

Drawing the PERT Diagram. In constructing the PERT diagram, the analyst looks first at those activities requiring no predecessor activities, in this case A (conduct interviews) and C (read company reports). In the example in Figure 3.11, the analyst chose to number the nodes 10, 20, 30, and so on, and he or she drew two arrows out of the beginning node 10. These arrows represent activities A and C and are labeled as such. Nodes numbered 20 and 30 are drawn at the end of these respective arrows. The next step is to look for any activity requiring only A as a predecessor; task B (administer questionnaires) is the only one, so it can be represented by an arrow drawn from node 20 to node 30.

Because activities D (analyze data flow) and E (introduce prototype) require both activities B and C to be finished before they are started, arrows labeled D and

FIGURE 3.10
Listing activities for use in drawing a PERT diagram.

Activity		Predecessor	Duration
A	Conduct Interviews	None	3
B	Administer Questionnaires	A	4
C	Read Company Reports	None	4
D	Analyze Data Flow	B, C	8
E	Introduce Prototype	B, C	5
F	Observe Reactions to Prototype	E	3
G	Perform Cost/Benefit Analysis	D	3
H	Prepare Proposal	G	2
I	Present Proposal	H	2

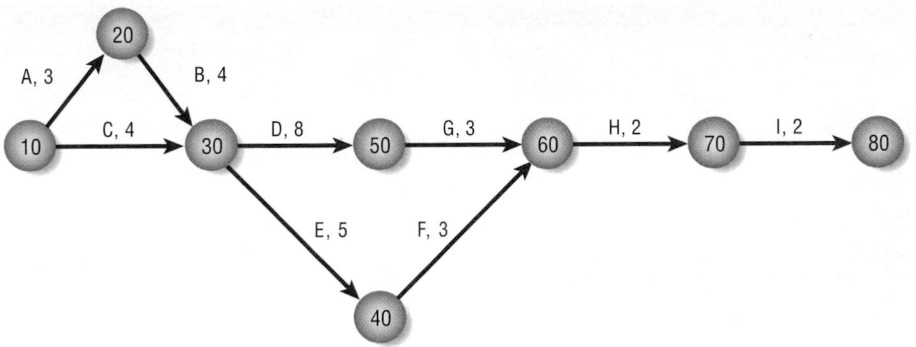

FIGURE 3.11
A completed PERT diagram for the analysis phase of a systems project.

E are drawn from node 30, the event that recognizes the completion of both B and C. This process is continued until the entire PERT diagram is completed. Notice that the entire project ends at an event called "node 80."

Identifying the Critical Path. Once the PERT diagram is drawn, it is possible to identify the critical path by calculating the sum of the activity times on each path and choosing the longest path. In this example, there are four paths: 10–20–30–50–60–70–80, 10–20–30–40–60–70–80, 10 30 50 60 70 80, and 10–30–40–60–70–80. The longest path is 10 20 30 50 60–70–80, which takes 22 days. It is essential that the systems analyst carefully monitor the activities on the critical path so as to keep the entire project on time or even shorten the project length if warranted.

COMPUTER-BASED PROJECT SCHEDULING

Using PCs for project scheduling has now become practical and straightforward. Microsoft Project, Symantec's Timeline, and Computer Associates' CA-Super Project are three good examples of powerful programs.

An example of project management from Microsoft Project can be found in Figure 3.12. Here you can see that the expediting problem completed earlier in this chapter is entered into Microsoft Project. New tasks can be entered into either the top or the bottom part of the screen, whichever is easier for the user. Let's assume we want to enter the task "Analyze data flow" on the bottom half of the screen. First, we enter the name of the activity, then its duration, 8d (including a qualifier: d for day, w for week, etc.), and the ID for any predecessors (in this case there are two). The ID, or identifier, is simply the number of the task. We don't have to enter a start date if we want the computer program to schedule it for us (as soon as possible, given the predecessors). The upper part of the table lists the activities in the order in which we entered them. Following is a Gantt chart. The lighter bars appear in orange on the screen and indicate the critical path. The darker bars are shaded blue and are noncritical.

Figure 3.13 is another screen from Microsoft Project. The bottom half of the screen is essentially the same, but the top half now shows a PERT diagram. Computer programs take the liberty of representing the tasks or activities with rectangles rather than with arrows. Although this goes against the traditional conventions used in PERT diagrams, the software authors feel that it is easier to read tasks in rectangular boxes than to read them on arrows. A darker line, shown in red on the actual screen, indicates the critical path. Once activities are drawn on the screen, they can be repositioned using a mouse to enhance readability and

FIGURE 3.12

This screen from Microsoft Project illustrates a Gantt chart.

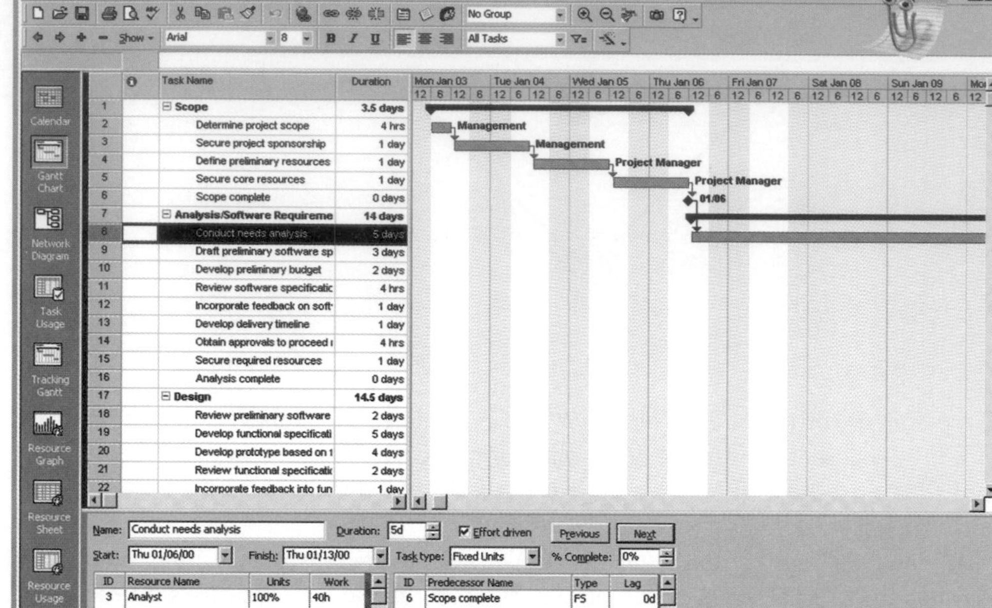

communication with others. The dark box indicates that we are looking at that activity now. The dotted vertical line on the right side of the screen shows the user where the page break will occur. The icons at the top of the page are familiar to anyone who uses Word for Windows, Excel, or PowerPoint, three other Microsoft products.

FIGURE 3.13

Computer-based project management programs often show activities as rectangles, not arrows.

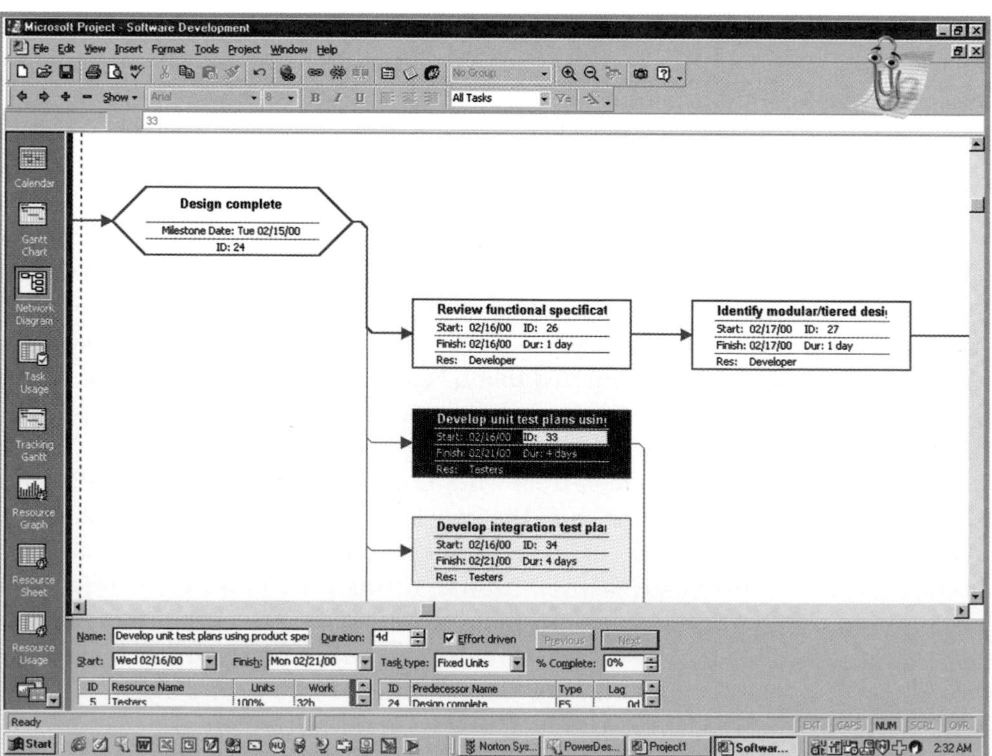

TIMEBOXING

A recent development in project management is the concept of timeboxing. Traditionally, a project is broken down into phases, milestones, and tasks, but the timeboxing approach uses an absolute due date for the project and whatever has been accomplished by that due date is implemented. It is important to set a reasonable due date given the project size and goals. It is also important to prioritize the project goals so that the most important ones are delivered to the users by the due date. Lesser goals may be implemented later in the project. An example of timeboxing is creating a Web site that contains the most important features with some of the minor pages containing an "Under Construction" image.

Other approaches to scheduling include integrated personal information managers, or PIMs. Some examples of a PIM include Microsoft Outlook and Organizer by Lotus Corporation. These PIMs are useful because they are a repository for phone and fax numbers of business associates; for daily, weekly, or monthly planners; and for to-do lists. Some PIMs are designed to be shells that enable you to launch other programs and even allow you to store similar data from word-processing and spreadsheet programs in "folders" organized around a particular topic. Some are good at sharing data with other programs, whereas others include Gantt charts to aid in project management. Most PIMs can be synchronized with PIMs in Palm computers and other handheld devices, cell phones, and watches, allowing for excellent, wireless portability.

MANAGING ANALYSIS AND DESIGN ACTIVITIES

Along with managing time and resources, systems analysts must also manage people. Management is accomplished primarily by communicating accurately to team members who have been selected for their competency and compatibility. Goals for project productivity must be set, and members of systems analysis teams must be motivated to achieve them.

COMMUNICATION STRATEGIES FOR MANAGING TEAMS

Teams have their own personalities, a result of combining each individual team member with every other in a way that creates a totally new network of interactions. A way to organize your thinking about teams is to visualize them as always seeking a balance between accomplishing the work at hand and maintaining the relationships among team members.

In fact, teams will often have two leaders, not just one. Usually one person will emerge who leads members to accomplish tasks, and another person will emerge who is concerned with the social relationships among group members. Both are necessary for the team. These individuals have been labeled by other researchers as, respectively, task leader and socioemotional leader. Every team is subject to tensions that are an outgrowth of seeking a balance between accomplishing tasks and maintaining relationships among team members.

For the team to continue its effectiveness, tensions must be continually resolved. Minimizing or ignoring tensions will lead to ineffectiveness and eventual disintegration of the team. Much of the tension release necessary can be gained through skillful use of feedback by all team members. All members, however, need to agree that the way they interact (i.e., process) is important enough to merit some time. Productivity goals for processes are discussed in a later section.

Securing agreement on appropriate member interaction involves creating explicit and implicit team norms (collective expectations, values, and ways of behaving) that guide members in their relationships. A team's norms belong to it

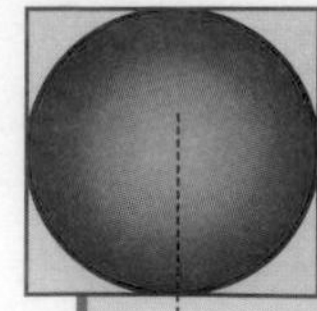

CONSULTING OPPORTUNITY 3.3

GOAL TENDING

"Here's what I think we can accomplish in the next five weeks," says Hy, the leader of your systems analysis team as he confidently pulls out a schedule listing each team member's name alongside a list of short-term goals. Just a week ago your systems analysis team went through an intense meeting on expediting their project schedule for the Kitchener, Ontario, Redwings, a hockey organization whose management is pressuring you to produce a prototype.

The three other members of the team look at the chart in surprise. Finally, one of the members, Rip, speaks: "I'm in shock. We each have so much to do as it is, and now this."

Hy replies defensively, "We've got to aim high, Rip. They're in the off-season. It's the only time to get them. If we set our goals too low, we won't finish the systems prototype, let alone the system itself, before another hockey season passes. The idea is to give the Kitchener Redwings the fighting edge through the use of their new system."

Fiona, another team member, enters the discussion, saying, "Goodness knows their players can't give them that!" She pauses for the customary groan from the assembled group, then continues. "But seriously, these goals are killers. You could have at least asked us what we thought, Hy. We may even know better than you what's possible."

"This is a pressing problem, not a tea party, Fiona," Hy replies. "Polite polling of team members was out of the question. Something had to be done quickly. So I went ahead with these. I say we submit our schedule to management based on this. We can push back deadlines later if we have to. But this way they'll know we're committed to accomplishing a lot during the off-season."

As a fourth team member listening to the foregoing exchange, formulate three suggestions that would help Hy improve his approach to goal formation and presentation. How well-motivated do you think the team will be if they share Fiona's view of Hy's goals? What are the possible ramifications of supplying management with overly optimistic goals? Write one paragraph devoted to short-term effects and another one discussing the long-term effects of setting unrealistically high goals.

and will not necessarily transfer from one team to another. These norms change over time and are better thought of as a team process of interaction rather than a product.

Norms can be functional or dysfunctional. Just because a particular behavior is a norm for a team does not mean it is helping the team to achieve its goals. For example, an expectation that junior team members should do all project scheduling may be a team norm. By adhering to this norm, the team is putting extreme pressure on new members and not taking full advantage of the experience of the team. It is a norm that, if continued, could make team members waste precious resources.

Team members need to make norms explicit and periodically assess whether norms are functional or dysfunctional in helping the team achieve its goals. The overriding expectation for your team must be that change is the norm. Ask yourself whether team norms are helping or hindering the team's progress.

SETTING PROJECT PRODUCTIVITY GOALS

When you have worked with your team members on various kinds of projects, you or your team leader will acquire acumen for projecting what the team can achieve within a specific amount of time. Using the hints discussed in the earlier section in this chapter on methods for estimating time required and coupling them with experience will enable the team to set worthwhile productivity goals.

Systems analysts are accustomed to thinking about productivity goals for employees who show tangible outputs such as the number of blue jeans sewn per hour, the number of entries keyed in per minute, or the number of items scanned per second. As manufacturing productivity rises, however, it is becoming clear that managerial productivity must keep pace. It is with this aim in mind that productivity goals for the systems analysis team are set.

Goals need to be formulated and agreed to by the team, and they should be based on team members' expertise, former performance, and the nature of the specific project. Goals will vary somewhat for each project undertaken, because sometimes an entire system will be installed, whereas other projects might involve limited modifications to a portion of an existing system.

MOTIVATING PROJECT TEAM MEMBERS

Although motivation is an extremely complex topic, it is a good one to consider, even if briefly, at this point. To oversimplify, recall that people join organizations to provide for some of their basic needs such as food, clothing, and shelter. All humans, however, also have higher-level needs that include affiliation, control, independence, and creativity. People are motivated to fulfill unmet needs on several levels.

Team members can be motivated, at least partially, through participation in goal setting, as described in the previous section. The very act of setting a challenging but achievable goal and then periodically measuring performance against the goal seems to work in motivating people. Goals act almost as magnets in attracting people to achievement.

Part of the reason goal setting motivates people is that team members know prior to any performance review exactly what is expected of them. The success of goal setting for motivating can also be ascribed to it affording each team member some autonomy in achieving the goals. Although a goal is predetermined, the means to achieve it may not be. In this instance team members are free to use their own expertise and experience to meet their goals.

Setting goals can also motivate team members by clarifying for them and others what must be done to get results. Team members are also motivated by goals because goals define the level of achievement that is expected of them. This use of goals simplifies the working atmosphere, but it also electrifies it with the possibility that what is expected can indeed be done.

MANAGING ECOMMERCE PROJECTS

Many of the approaches and techniques discussed earlier are transferable to ecommerce project management. You should be cautioned, however, that although there are many similarities, there are also many differences. One difference is that the data used by the ecommerce systems are scattered all over the organization. Therefore, you are not just managing data within a self-contained department or even one solitary unit. Hence, many organizational politics can come into play, because often units feel protective of the data they generate and do not understand the need to share them across the organization.

Another stark difference is that ecommerce project teams typically need more staff with a variety of skills, including developers, consultants, and system integrators, from across the organization. Neatly defined, stable project groups that exist within a cohesive IS group or systems development team will be the exception rather than the rule. In addition, because so much help may be required initially, ecommerce project managers need to build partnerships externally and internally well ahead of the implementation, perhaps sharing talent across projects to defray costs of ecommerce implementations and to muster the required numbers of people with the necessary expertise. The potential for organizational politics to drive a wedge between team members is very real.

One way to prevent politics from sabotaging a project is for the ecommerce project manager to emphasize the integration of the ecommerce with the organization's internal systems and in so doing emphasize the organizational aspect embedded in the ecommerce project. As one ecommerce project manager told us, "Designing the front end [what the consumer sees] is the easy part of all this. The real challenge comes from integrating ecommerce strategically into all the organization's systems."

A fourth difference between traditional project management and ecommerce project management is that because the system will be linking with the outside world via the Internet, security is of the utmost importance. Developing and implementing a security plan before the new system is in place is a project in and of itself and must be managed as such.

AVOIDING PROJECT FAILURES

The early discussions you have with management and others requesting a project, along with the feasibility studies you do, are usually the best defenses possible against taking on projects that have a high probability of failure. Your training and experience will improve your ability to judge the worthiness of projects and the motivations that prompt others to request projects. If you are part of an in-house systems analysis team, you must keep current with the political climate of the organization as well as with financial and competitive situations.

You can also learn from the wisdom gained by people involved in earlier project failures. When asked to reflect on why projects had failed, professional programmers cited the setting of impossible or unrealistic dates for completion by management, belief in the myth that simply adding more people to a project would expedite it (even though the original target date on the project was unrealistic), and management behaving unreasonably by forbidding the team to seek professional expertise from outside of the group to help solve specific problems.

Remember that you are not alone in the decision to begin a project. Although apprised of your team's recommendations, management will have the final say about whether a proposed project is worthy of further study (that is, further investment of resources). The decision process of your team must be open and stand up to scrutiny from those outside of it. The team members should consider that their reputation and standing in the organization are inseparable from the projects they accept.

SUMMARY

The five major project fundamentals that the systems analyst must handle are (1) project initiation, (2) determining project feasibility, (3) activity planning and control, (4) project scheduling, and (5) managing systems analysis team members. Projects may be requested by many different people within the business or by systems analysts themselves.

Selecting a project is a difficult decision, because more projects will be requested than can actually be done. Five important criteria for project selection are (1) that the requested project be backed by management, (2) that it be timed appropriately for a commitment of resources, (3) that it move the business toward attainment of its goals, (4) that it be practical, and (5) that it be important enough to be considered over other possible projects.

If a requested project meets these criteria, a feasibility study of its operational, technical, and economic merits can be done. Through the feasibility study, systems analysts gather data that enable management to decide whether to proceed with a full systems study. Project planning includes the estimation of time required for each of the analyst's activities, scheduling them, and expediting them if necessary to ensure that a project is completed on time. One technique available to the systems analyst for scheduling tasks is the Gantt chart, which displays activities as bars on a graph.

3

"I hope everyone you've encountered at MRE has treated you well. Here's a short review of some of the ways you can access our organization through HyperCase. The reception area at MRE contains the key links to the rest of our organization. Perhaps you've already discovered these on your own, but I wanted to remind you of them now, because I don't want to get so engrossed in the rest of our organizational problems that I forget to mention them.

"The telephone on the receptionist's desk has instructions about how to answer the phone in the rest of the organization. You have my permission to pick up the phone if it is ringing and no one else answers it.

"The empty doorway you see is a link to the next room, which we call the East Atrium. You have probably noticed that all open doorways are links to adjacent rooms. Notice the building map displayed in the reception area. You are free to go to public areas such as the canteen, but as you know, you must have an employee escort you into a private office. You cannot go there on your own.

"By now you have probably noticed the two documents and the computer on the small table in the reception area. The little one is the MRE internal phone directory. Just click on an employee name, and if that person is in, he or she will grant you an interview and a tour of the office. I leave you to your own devices in figuring out what the other document is.

"The computer on the table is on and displays the World Wide Web home page for MRE. You should take a look at the corporate page and visit all the links. It tells the story of our company and the people who work here. We're quite proud of it and have gotten positive feedback about it from visitors.

"If you have had a chance to interview a few people and see how our company works, I'm sure you are becoming aware of some of the politics involved. We are also worried, though, about more technical issues, such as what constitutes feasibility for a training project and what does not."

FIGURE 3.HC1
The reception room resembles a typical corporation. While you are in this HyperCase screen, find the directory if you want to visit someone.

(Continued)

HYPERCASE QUESTIONS

1. What criteria does the Training Unit use to judge the feasibility of a new project? List them.
2. List any changes or modifications to these criteria that you would recommend.
3. Snowden Evans has asked you to help prepare a proposal for a new project tracking system for the Training Unit. Briefly discuss the technical, economic, and operational feasibility of each alternative for a proposed project tracking system for the Training Unit.
4. Which option would you recommend? Use evidence from HyperCase to support your decision.

Another technique, called PERT (for Program Evaluation and Review Techniques), displays activities as arrows on a network. PERT helps the analyst determine the critical path and slack time, which is the information required for effective project control. The timeboxing approach uses an absolute due date for the project, and whatever has been accomplished by that due date is implemented.

Computer-based project scheduling using PCs is now practical. In addition, personal information managers (PIMs) can be used by analysts to do planning, create repositories for phone and fax numbers, or even launch other programs. Most PIMs can be synchronized with PIMs in Palm computers and other handheld devices allowing for excellent portability.

Once a project has been judged feasible, the systems analyst must manage the team members and their activities, time, and resources. Such management is accomplished by communicating with team members. Teams are constantly seeking a balance between working on tasks and maintaining relationships within the team. Tensions arising from attempting to achieve this balance must be addressed. Often two leaders of a team will emerge, a task leader and a socioemotional leader. Members must periodically assess team norms to ensure that the norms are functional rather than dysfunctional for the attainment of team goals.

Managing ecommerce projects is similar to managing traditional IS projects in a number of ways, but there are four ways in which it departs significantly from these practices. The first is that the data you will be coordinating are scattered all over the organization (which has political ramifications), another is that specialized team members are drawn from across the organization (so organizational politics may also loom), a third is that the ecommerce project manager should be emphasizing strategic integration of ecommerce into all the organization's systems; and the fourth is that security concerns must be managed first when establishing an ecommerce site.

It is important that the systems analysis team set reasonable productivity goals for tangible outputs and process activities. Unrealistic management deadlines, adding unneeded personnel to a project that is trying to meet an unrealistic deadline, and not permitting developer teams to seek expert help outside their immediate group were cited by programmers as reasons projects had failed. Project failures can usually be avoided by examining the motivations for requested projects as well as your team's motives for recommending or avoiding a particular project.

KEYWORDS AND PHRASES

- computer-based project scheduling
- critical path
- economic feasibility
- ecommerce project management
- feasibility impact grid (FIG)
- Gantt chart
- operational feasibility
- PERT diagrams
- personal information managers (PIMs)
- productivity goals
- socioemotional leader
- task leader
- team motivation
- team norms
- team process
- technical feasibility
- timeboxing

REVIEW QUESTIONS

1. What are the five major project fundamentals?
2. List three ways to find out about problems or opportunities that might call for a systems solution.
3. List the five criteria for systems project selection.
4. Examine the feasibility impact grid shown in Figure 3.2. List the corporate objectives that seem to be affected positively by ecommerce systems.
5. Define technical feasibility.
6. Define economic feasibility.
7. Define operational feasibility.
8. When is a two-dimensional Gantt chart more appropriate than a one-dimensional Gantt chart?
9. When is a PERT diagram useful for systems projects?
10. List three advantages of a PERT diagram over a Gantt chart for scheduling systems projects.
11. Define the term *critical path*.
12. Define the technique of timeboxing.
13. List the functions of PC-based project scheduling that are available in common software packages.
14. List the functions of some commonly used personal information manager (PIM) software.
15. List the two types of team leaders.
16. What is meant by dysfunctional team norm?
17. What is meant by team process?
18. What are three reasons that goal setting seems to motivate systems analysis team members?
19. What are four ways in which ecommerce project management differs from traditional project management?
20. What are three reasons programmers cited for project failure?

PROBLEMS

1. Dressman's Chocolates of St. Louis makes an assortment of chocolate candy and candy novelties. The company has six in-city stores, five stores in major metropolitan airports, and a small mail-order branch. Dressman's has a small, computerized information system that tracks inventory in its plant, helps schedule production, and so on, but this system is not tied directly into any of its retail outlets. The mail-order system is handled manually. Recently, several Dressman's stores experienced a rash of complaints from mail-order customers that the candy was spoiled upon arrival, that it did not

come when promised, or that it never arrived; the company also received several letters complaining that candy in various airports tasted stale. Finally, a few sales clerks in company stores reported being asked whether the firm would be willing to market a new, dietetic form of chocolate made with sugar-free, artificial sweetener.

Description	*Task*	*Must Follow*	*Expected Time (Days)*
Draw data flow	P	None	9
Draw decision tree	Q	P	12
Revise tree	R	Q	3
Write up project	S	R, Z	7
Organize data dictionary	T	P	11
Do output prototype	X	None	8
Revise output design	Y	X	14
Design database	Z	T, Y	5

You had been working for two weeks with Dressman's on some minor modifications for its inventory information system when you overheard two managers discussing these occurrences. List the possible opportunities or problems among them that might lend themselves to systems projects.

2. Where is most of the feedback on problems with Dressman's products coming from in problem 1? How reliable are the sources? Explain in a paragraph.
3. After getting to know them better, you have approached Dressman's management people with some of your ideas on possible systems improvements that could address some of the problems or opportunities given in problem 1.
   a. In two paragraphs, provide your suggestions for systems projects. Make any realistic assumptions necessary.
   b. Are there any problems or opportunities discussed in problem 1 that are not suitable? Explain your response.
4. The systems analysis consulting firm of Flow Associates is working on a systems design project for Wind and Waves Waterbeds, Inc.
   a. Using the data from the table in problem 1, draw a Gantt chart to help Flow Associates organize its design project.
   b. When is it appropriate to use a Gantt chart? What are the disadvantages? Explain in a paragraph.
5. Figure 3.EX1 is a PERT diagram based on the data from problem 4. List all paths and calculate and identify the critical path.
6. Recently, two analysts just out of college have joined your systems analyst group at the newly formed company, Mega Phone. When talking to you about the group, they mention that some things strike them as odd. One is

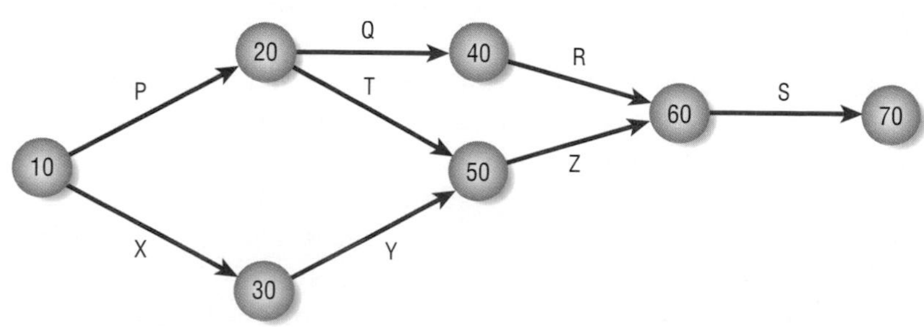

FIGURE 3.EX1
The PERT diagram from Flow Associates.

that group members seem to look up to two group leaders, Bill and Penny, not just one.

Their observation is that Bill seems pretty relaxed, whereas Penny is always planning and scheduling activities. They have also observed that everyone "just seems to know what to do" when they get into a meeting, even though no instructions are given. Finally, they have remarked on the openness of the group in addressing problems as they arise, instead of letting things get out of hand.

a. By way of explanation to the new team members, label the types of leaders Bill and Penny appear to be, respectively.
b. Explain the statement that "everyone just seems to know what to do." What is guiding their behavior?
c. What concept best describes the openness of the group that the new team members commented on?

GROUP PROJECT

1. With your group members, explore project manager software such as Microsoft Project. What features are available? Work with your group to list them. Have your group evaluate the usefulness of the software for managing a systems analysis and design team project. In a paragraph, state whether the software you are evaluating facilitates team member communication and management of team activities, time, and resources. State which particular features support these aspects of any project. Note whether the software falls short of these criteria in any regard.

SELECTED BIBLIOGRAPHY

Adam, E. E., Jr., and R. J. Ebert. *Production and Operations Management*, 3d ed. Englewood Cliffs, NJ: Prentice Hall, 1986.

Bales, R. F. *Personality and Interpersonal Behavior*. New York: Holt, Rinehart and Winston, 1970.

Gildersleeve, T. R. *Successful Data Processing for Systems Analysis*. Englewood Cliffs, NJ: Prentice Hall, 1977.

Glass, R. "Evolving a New Theory of Project Success." *Communications of the ACM*, Vol. 42, No. 11, 1999, pp. 17–19.

Linberg, K. R. "Software Perceptions about Software Project Failure: A Case Study." *Journal of Systems and Software*, Vol. 49, Nos. 2 and 3, 1999, pp. 177–92.

Merry, U., and M. E. Allerhand. *Developing Teams and Organizations*. Reading, MA: Addison-Wesley, 1977.

Schein, E. H. *Process Consultation: Its Role in Organization Development*. Reading, MA: Addison-Wesley, 1969.

Walsh, B. "Your Network's Not Ready for E-Commerce." *Network Computing*. Available: (http://www.networkcomputing.com/922/922colwalsh.html). Last updated October 19, 1999.

Weinberg, G. M. *Rethinking Systems Analysis and Design*. Boston: Little, Brown, 1982.

3

ALLEN SCHMIDT, JULIE E. KENDALL, AND KENNETH E. KENDALL

GETTING TO KNOW U

Chip enters Anna's office one day saying, "I think the project will be a good one, even though it's taking some long hours to get started."

Anna looks up from her screen and smiles. "I like what you've done in getting us organized," she says. "I hadn't realized Visible Analyst could help us this much with project management. I've decided to do a PERT diagram for the data-gathering portion of the project. It should help us plan our time and work as a team on parallel activities."

"Can I take a look at the PERT diagram?" asks Chip.

Anna shows him a screen with a PERT diagram on it (see Figure E3.1) and remarks, "This will help immensely. It is much easier than planning haphazardly."

"I notice that you have Gather Reports, Gather Records and Data Capture Forms, and Gather Qualitative Documents as parallel tasks," notes Chip, gazing at the screen.

"Yes," replies Anna. "I thought that we would split up the time that it takes to gather the information. We can also divide up the task of analyzing what we have learned."

"I notice that you have a rather large number of days allocated for interviewing the users," notes Chip.

"Yes," replies Anna. "This activity also includes creating questions, sequencing them, and other tasks, such as taking notes of the office environment and analyzing them. I've also assumed a standard of six productive hours per day."

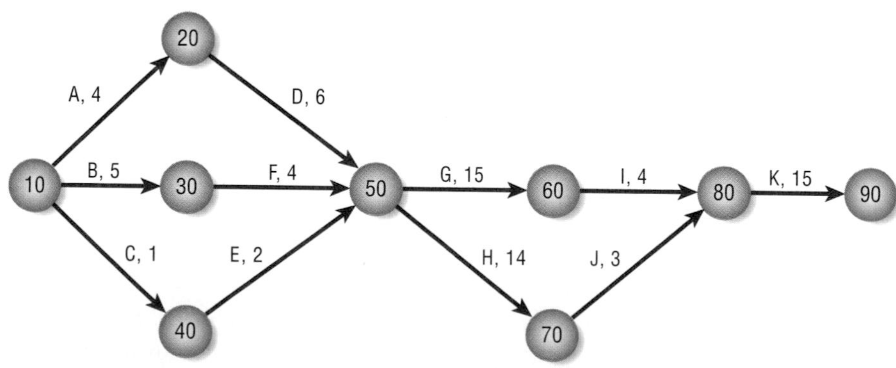

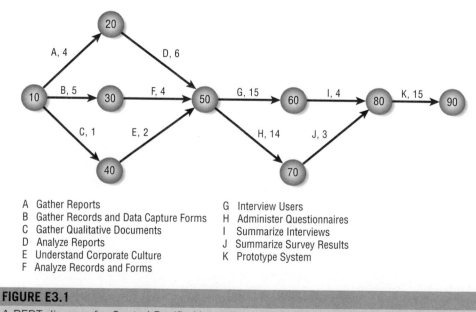

A Gather Reports
B Gather Records and Data Capture Forms
C Gather Qualitative Documents
D Analyze Reports
E Understand Corporate Culture
F Analyze Records and Forms
G Interview Users
H Administer Questionnaires
I Summarize Interviews
J Summarize Survey Results
K Prototype System

FIGURE E3.1
A PERT diagram for Central Pacific University that is used for the gathering information phase.

Anna glances at her watch. "But now it's getting late. I think we've made a lot of progress in setting up our project. Let's call it a day, or should I say evening? Remember, I got us tickets for the football game."

Chip replies, "I haven't forgotten. Let me get my coat, and we'll walk over to the stadium together."

Walking across campus later, Chip says, "I'm excited. It's my first game here at CPU. What's the team mascot, anyway?"

"Chipmunks, of course," says Anna.

"And the team colors?" Chip asks, as they enter the stadium.

"Blue and white," Anna replies.

"Oh, that's why everyone's yelling, 'Go Big Blue!'" Chip says, listening to the roar of the crowd.

"Precisely," says Anna.

EXERCISES*

Use Visible Analyst to define the following User entities and User Requirements.

E-1. Use Visible Analyst to view the Gathering Information PERT diagram.

E-2. List all paths and calculate and determine the critical path for the Gathering Information PERT diagram.

E-3. Use Visible Analyst to create the PERT diagram shown in Figure E3.2. It represents the activities involved in interviewing the users and observing their offices.

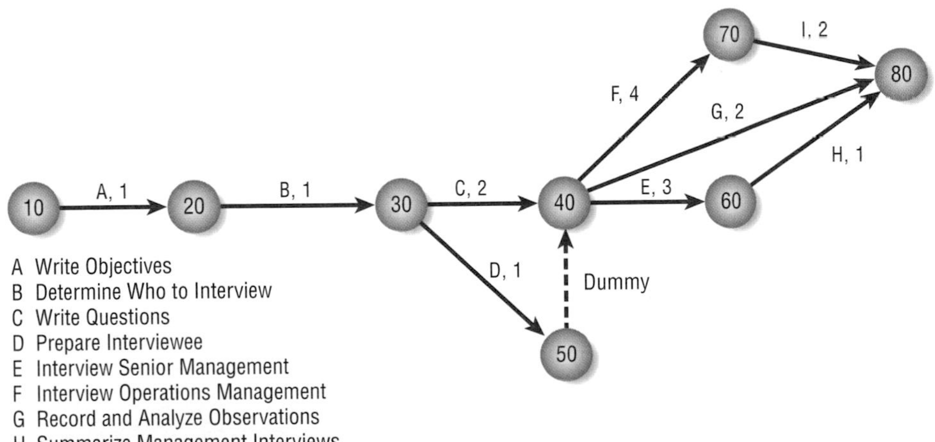

FIGURE E3.2
A PERT diagram for Central Pacific University that is used for interviewing users.

*Exercises preceded by a CD-ROM icon require the program Visible Analyst or another CASE tool. A CD-ROM containing Visible Analyst examples is provided free of charge to any professor adopting this book. The examples on the disk may be imported into Visible Analyst and then used by students.

Activity	Predecessor	Duration
A Determine Overall Prototype Screens and Reports	None	2
B Determine Report and Screen Contents	A	4
C Create Report Prototypes	B	3
D Create Screen Prototypes	B	4
E Obtain Report Prototype Feedback	C	1
F Obtain Screen Prototype Feedback	D	2
G Modify Report Prototypes	E	2
H Modify Screen Prototypes	F	4
I Obtain Final Approval	G, H	2

FIGURE E3.3

A list of activities and estimated duration times for the CPU project.

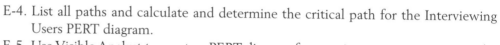

E-4. List all paths and calculate and determine the critical path for the Interviewing Users PERT diagram.

E-5. Use Visible Analyst to create a PERT diagram for creating system prototypes. The activity information is shown in Figure E3.3.

SAMPLING AND INVESTIGATING HARD DATA

4

Underlying all the data-gathering methods of investigation, interviewing, and observing are the crucial decisions regarding what to examine and whom to question or observe. The systems analyst can make these decisions based on a structured approach called sampling.

SAMPLING

Sampling is the process of systematically selecting representative elements of a population. When these selected elements are examined closely, it is assumed that the analysis will reveal useful information about the population as a whole.

The systems analyst has to make a decision on two key issues. First, there are many reports, forms, output documents, memos, and Web sites that have been generated by members of the organization. Which of these should the systems analyst pay attention to, and which should the systems analyst ignore?

Second, a great many employees can be affected by the proposed information system. Which people should the systems analyst interview, seek information from via questionnaires, or observe in the process of carrying out their decision-making roles?

THE NEED FOR SAMPLING

There are many reasons a systems analyst would want to select either a representative sample of data to examine or representative people to interview, question, or observe. They include

1. Containing costs
2. Speeding up the data gathering
3. Improving effectiveness
4. Reducing bias

Examining every scrap of paper, talking with everyone, and reading every Web page from the organization would be far too costly for the systems analyst. Copying reports, asking employees for valuable time, and duplicating unnecessary surveys would result in much needless expense.

Sampling helps accelerate the process by gathering selected data rather than all data for the entire population. In addition, the systems analyst is spared the burden of analyzing data from the entire population.

Effectiveness in data gathering is an important consideration as well. Sampling can help improve effectiveness if information that is more accurate can be obtained. Such sampling is accomplished, for example, by talking to fewer employees but asking them questions that are more detailed. In addition, if fewer people are interviewed, the systems analyst can afford the time to follow up on missing or incomplete data, thus improving the effectiveness of data gathering.

Finally, data-gathering bias can be reduced by sampling. When the systems analyst interviews an executive of the corporation, for example, the executive is involved with the project, because this person has already given a certain amount of time to the project and would like it to succeed. When the systems analyst asks an opinion about a permanent feature of the installed information system, the executive interviewed may provide a biased evaluation, because there is little possibility of changing it.

SAMPLING DESIGN

A systems analyst must follow four steps to design a good sample:

1. Determine the data to be collected or described.
2. Determine the population to be sampled.
3. Choose the type of sample.
4. Decide on the sample size.

These steps are described in detail in the following subsections.

Determining the Data to Be Collected or Described. The systems analyst needs a realistic plan about what will be done with the data once they are collected. If irrelevant data are gathered, then time and money are wasted in the collection, storage, and analysis of useless data.

The duties and responsibilities of the systems analyst at this point are to identify the variables, attributes, and associated data items that need to be gathered in the sample. The objectives of the study must be considered as well as the type of data-gathering method (investigation, interviews, questionnaires, observation) to be used. The kinds of information sought when using each of these methods are discussed in more detail in this and subsequent chapters.

Determining the Population to Be Sampled. Next, the systems analyst must determine what the population is. In the case of hard data, the systems analyst needs to decide, for example, if the last two months are sufficient, or if an entire year's worth of reports are needed for analysis.

Similarly, when deciding whom to interview, the systems analyst has to determine whether the population should include only one level in the organization or all the levels, or maybe the analyst should even go outside of the system to include the reactions of customers, vendors, suppliers, or competitors. These decisions are explored further in the chapters on interviewing, questionnaires, and observation.

Choosing the Type of Sample. The systems analyst can use one of four main types of samples, as pictured in Figure 4.1. They are convenience, purposive, simple, and complex. Convenience samples are unrestricted, nonprobability samples. A sample could be called a convenience sample if, for example, the systems analyst posts a notice on the company's intranet asking for everyone interested in the new sales performance reports to come to a meeting at 1 P.M. on Tuesday the 12th. Obviously, this sample is the easiest to arrange, but it is also the most unreliable. A purposive

	Not Based on Probability	Based on Probability
Sample elements are selected directly without restrictions	Convenience	Simple random
Sample elements are selected according to specific criteria	Purposive	Complex random (systematic, stratified, and cluster)

The systems analyst should use a complex random sample if possible.

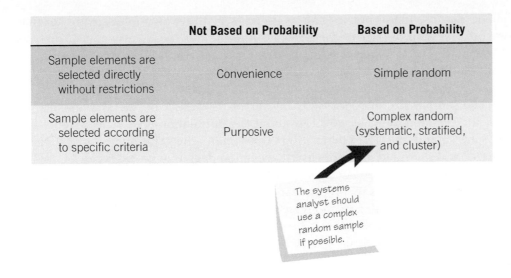

FIGURE 4.1
Four main types of samples the analyst has available.

sample is based on judgment. A systems analyst can choose a group of individuals who appear knowledgeable and who are interested in the new information system. Here the systems analyst bases the sample on criteria (knowledge about and interest in the new system), but it is still a nonprobability sample. Thus, purposive sampling is only moderately reliable. If you choose to perform a simple random sample, you need to obtain a numbered list of the population to ensure that each document or person in the population has an equal chance of being selected. This step often is not practical, especially when sampling involves documents and reports. The complex random samples that are most appropriate for the systems analyst are (1) systematic sampling, (2) stratified sampling, and (3) cluster sampling.

In the simplest method of probability sampling, systematic sampling, the systems analyst would, for example, choose to interview every *k*th person on a list of company employees. This method has certain disadvantages, however. You would not want to use it to select every *k*th day for a sample because of the potential periodicity problem. Furthermore, a systems analyst would not use this approach if the list were ordered (for example, a list of banks from the smallest to the largest), because bias would be introduced.

Stratified samples are perhaps the most important to the systems analyst. Stratification is the process of identifying subpopulations, or strata, and then selecting objects or people for sampling within these subpopulations. Stratification is often essential if the systems analyst is to gather data efficiently. For example, if you want to seek opinions from a wide range of employees on different levels of the organization, systematic sampling would select a disproportionate number of employees from the operational control level. A stratified sample would compensate for this. Stratification is also called for when the systems analyst wants to use different methods to collect data from different subgroups. For example, you may want to use a survey to gather data from middle managers, but you might prefer to use personal interviews to gather similar data from executives.

Sometimes the systems analyst must select a group of documents or people to study. This process is referred to as cluster sampling. Suppose an organization had 20 helpdesks scattered across the country. You may want to select one or two of these helpdesks under the assumption that they are typical of the remaining ones.

Deciding on the Sample Size. Obviously, if everyone in the population viewed the world the same way or if each of the documents in a population contained exactly the same information as every other document, a sample size of one would be sufficient. Because that is not the case, it is necessary to set a sample size greater than one but less than the size of the population itself.

It is important to remember that the absolute number is more important in sampling than the percentage of the population. We can obtain satisfactory results sampling 20 people in 200 or 20 people in 2,000,000.

The sample size depends on many factors, some set by the systems analyst, some determined by what we know about the population itself. The systems analyst can choose the acceptable interval estimate (i.e., the degree of precision desired) and the standard error (by choosing the degree of confidence).

Furthermore, the characteristics of the population may change the sample size. If the sales figures in a report range from $10,000 to $15,000, a small sample size would be sufficient to give you an accurate estimate of average sales. If the sales ranged from $1,000 to $100,000, however, a much larger sample size would be required. Sample size is discussed in greater detail in the following section.

THE SAMPLE SIZE DECISION

The sample size often depends on the cost involved or the time required by the systems analyst, or even the time available by people in the organization. This subsection gives the systems analyst some guidelines for determining the required sample size under ideal conditions.

Determining Sample Size When Sampling Data on Attributes. Sometimes the systems analyst might want to find out what proportion of people in an organization think a certain way or have certain characteristics. Other times the analyst may need to know what percentage of input forms contain errors. This type of data can be referred to as attribute data.

The systems analyst needs to follow seven steps, some of which are subjective judgments, to determine the required sample size:

1. Determine the attribute you will be sampling.
2. Locate the database or reports where the attribute can be found.
3. Examine the attribute. Estimate p, the proportion of the population having the attribute.
4. Make the subjective decision regarding the acceptable interval estimate, i.
5. Choose the confidence level and look up the confidence coefficient (z value) in a table.
6. Calculate σ_p, the standard error of the proportion as follows:

$$\sigma_p = \frac{i}{z}$$

7. Determine the necessary sample size, n, using the following formula:

$$n = \frac{p(1-p)}{\sigma_p^2} + 1$$

The first step, of course, is to determine which attribute you will be sampling. Once this is done, you can find out where this data is stored, perhaps in a database, on a form, or in a report.

It is important to estimate p, the proportion of the population having the attribute so that you set the appropriate sample size. Many textbooks on systems

First decide on the confidence level,

Confidence Level	Confidence Coefficient (z value)
99%	2.58
98	2.33
97	2.17
96	2.05
95	1.96
90	1.65
80	1.28
50	0.67

then look up the z value.

FIGURE 4.2

A table of area under a normal curve can be used to look up a value once the systems analyst decides on the confidence level.

analysis suggest using a heuristic of 0.25 for $p(1 - p)$. This value almost always results in a sample size larger than necessary because 0.25 is the maximum value of $p(1 - p)$, which occurs only when $p = 0.50$. When $p = 0.10$, as is more often the case, $p(1 - p)$ becomes 0.09, resulting in a much smaller sample size.

Steps 4 and 5 are subjective decisions. The acceptable interval estimate of ±0.10 means that you are willing to accept an error of no more than 0.10 in either direction from the actual proportion, p. The confidence level is the desired degree of certainty, say, for example, 95 percent. Once the confidence level is chosen, the confidence coefficient (also called a z value) can be looked up in a table like the one found in Figure 4.2.

Steps 6 and 7 complete the process by taking the parameters found or set in steps 3 through 5 and entering them into two equations to eventually solve for the required sample size.

Example

The foregoing steps can best be illustrated by an example. Suppose the A. Sembly Company, a large manufacturer of shelving products, asks you to determine what percentage of orders contain mistakes. You agree to do this job and perform the following steps. You

1. Determine that you will be looking for orders that contain mistakes in names, addresses, quantities, or model numbers.
2. Locate copies of order forms from the past six months.
3. Examine some of the order forms and conclude that only about 5 percent (0.05) contain errors.
4. Make a subjective decision that the acceptable interval estimate will be ±0.02.
5. Choose a confidence level of 95 percent. Look up the confidence coefficient (z value) in Figure 4.2. The z value equals 1.96.
6. Calculate σ_p as follows:

$$\sigma_p = \frac{i}{z}$$

$$= \frac{0.02}{1.96} = 0.0102$$

7. Determine the necessary sample size, n, as follows:

$$n = \frac{p(1-p)}{\sigma_p^2} + 1$$

$$= \frac{0.05(0.95)}{(0.0102)(0.0102)} + 1 = 458$$

The conclusion, then, is to set the sample size at 458. Obviously, a greater confidence level or a smaller acceptable interval estimate would require a larger sample size, as shown below. If we keep the acceptable interval estimate the same but increase the confidence level to 99 percent (with a z value of 2.58), the standard error of the proportion is

$$\sigma_p = \frac{i}{z}$$

$$= \frac{0.02}{2.58} = 0.0078$$

and the necessary sample size is

$$n = \frac{p(1-p)}{\sigma_p^2} + 1$$

$$= \frac{(0.05)(0.95)}{(0.0078)(0.0078)} + 1 = 782$$

If the confidence level stays at 95 percent (z value = 1.96) but the acceptable interval estimate is set to 0.01, the standard error of the proportion is

$$\sigma_p = \frac{i}{z}$$

$$= \frac{0.01}{1.96} = 0.0051$$

and the necessary sample size is

$$n = \frac{p(1-p)}{\sigma_p^2} + 1$$

$$= \frac{(0.05)(0.95)}{(0.0051)(0.0051)} + 1 = 1827$$

Determining Sample Size When Sampling Data on Variables. A systems analyst may sometimes need to gather information on actual numbers such as gross sales, amount of items returned, or number of mistakes keyed in. Data of this type are referred to as variables.

The steps in determining the necessary sample size for variables is similar to the steps for attribute data. The steps are as follows:

1. Determine the variable you will be sampling.
2. Locate the database or reports where the variable can be found.
3. Examine the variable to gain some idea about its magnitude and dispersion. Ideally, it would be useful to know the mean to determine a more appropriate

acceptable interval estimate and the standard deviation, *s*, to determine sample size (in step 7).
4. Make a subjective decision regarding the acceptable interval estimate, *i*.
5. Choose a confidence level and look up the confidence coefficient (*z* value) in a table.
6. Calculate $\sigma_{\bar{x}}$ the standard error of the mean as follows:

$$\sigma_{\bar{x}} = \frac{i}{z}$$

7. Determine the necessary sample size, *n*, using the following formula:

$$n = \left(\frac{s}{\sigma_{\bar{x}}}\right)^2 + 1$$

Step 3 is difficult to do precisely. To estimate the mean and standard deviation accurately, you would actually have to sample the population. This dilemma is, of course, a chicken-or-egg problem because the systems analyst does not know which comes first. Consequently, it is not necessary to know precisely the mean and standard deviation. A reasonable estimate will do.

The importance of estimating the mean and standard deviation becomes obvious when you attempt to set the acceptable interval estimate. Suppose you were asked to determine the average gross sales for a small greeting card store ($3,500 per week) and a large grocery store ($70,000 per week). A desired interval estimate of $350 for the small greeting card store would be inappropriate for the large grocery store.

Furthermore, the estimated dispersion is important because the smaller the dispersion (as measured by the standard deviation), the smaller the necessary sample size. For example, suppose you were trying to determine the average age of purchasers of music on CDs (compact discs). If you make the assumption that anyone from 10 to 100 years of age purchases CDs, you would need a very large sample size. If, however, you estimate that generally the ages of CD purchasers are between 13 and 23, you can get by with a much smaller sample size.

The formulas for determining the necessary sample size for variables are different from the formulas for attribute data. Because the systems analyst must take more care in estimating the magnitude and dispersion of a variable, determining the sample size for a variable is slightly more difficult than determining the sample size for attribute data.

Example

Another example from the A. Semble Company can be used to illustrate how to find the necessary sample size for variables. Now you are asked to determine the average dollar amount for an order. You agree to do this task and perform the following steps:

1. Determine that you will be looking for the mean dollar amount for orders on shelving.
2. Locate copies of order forms from the past six months.
3. Examine some of the order forms and conclude that orders average about $1,500, with a standard deviation, *s*, of about $100.
4. Make a subjective decision that the acceptable interval estimate will be $5.00.
5. Chose a confidence level of 96 percent. Look up the confidence coefficient (*z* value) in Figure 4.2. The *z* value equals 2.05.

6. Calculate $\sigma_{\bar{x}}$ as follows:

$$\sigma_{\bar{x}} = \frac{i}{z}$$

$$= \frac{5.00}{2.05} = 2.44$$

7. Determine the necessary sample size, n, as follows:

$$n = \left(\frac{s}{\sigma_{\bar{x}}}\right)^2 + 1$$

$$= \left(\frac{100}{2.44}\right)^2 + 1$$

$$= 1{,}680 + 1$$

$$= 1{,}681$$

You therefore would want to sample 1,681 orders to determine the mean dollar amount of orders.

As with sampling attribute data, a greater confidence level or a smaller acceptable interval estimate would require a larger sample size. If we keep the acceptable interval estimate the same but increase the confidence level to 99 percent (with a z value of 2.58), the standard error of the mean is

$$\sigma_{\bar{x}} = \frac{i}{z}$$

$$= \frac{5.00}{2.58} = 1.94$$

and the necessary sample size is

$$n = \left(\frac{s}{\sigma_{\bar{x}}}\right)^2 + 1$$

$$= \left(\frac{100}{1.94}\right)^2 + 1$$

$$= 2{,}658$$

If the confidence level stays at 96 percent (z value = 2.05) but the acceptable interval estimate is set to \$1.00, the standard error of the mean is

$$\sigma_{\bar{x}} = \frac{i}{z}$$

$$= \frac{1.00}{2.05} = 0.488$$

and the necessary sample size is

$$n = \left(\frac{s}{\sigma_{\bar{x}}}\right)^2 + 1$$

$$= \left(\frac{100}{0.488}\right)^2 + 1$$

$$= 41{,}992$$

This result implies that accuracy to the nearest dollar would be far too costly in sampling, because sampling over 41,000 of anything is excessive.

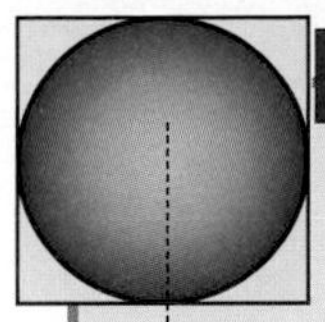

CONSULTING OPPORTUNITY 4.1

TRAPPING A SAMPLE

"Real or fake? Fake or real? Who would have thought it, even five years ago?" howls Sam Pelt, a furrier who owns stores in New York, Washington, D.C., Beverly Hills, and Copenhagen. Sylva Foxx, a systems analyst with her own consulting firm, is talking with Sam for the first time. Currently, P & P, Ltd. (which stands for Pelt and Pelt's son) is using a PC that supports package software for a select customer mailing list, accounts payable and accounts receivable, and payroll.

Sam is interested in making some strategic decisions that will ultimately affect the purchasing of goods for his four fur stores. He feels that although the computer might help, other approaches should also be considered.

Sam continues, "I think we should talk to all of the customers when they come in the door. Get their opinions. You know, some of them are getting very upset about wearing fur from endangered species. They're very environmentally minded. They prefer fake to real, if they can save a baby animal. Some even like faking better, calling them 'fun furs.' And I can charge almost the same for a good look-alike.

It's a very fuzzy proposition, though. If I get too far away from my suppliers of pelts, I may not get what I want when I need it. They see the fake fur people as worms, worse then moths! If I deal with them, the real fur men might not talk to me. They can be animals. On the other hand, I feel strange showing fakes in my stores. All these years, we've prided ourselves on having only the genuine article."

Sam continues, in a nearly seamless monologue, "I want to talk to each and every employee, too."

Sylva glances at him furtively and begins to interrupt. "But that will take months, and purchasing may come apart at the seams unless they know soon what—"

Pelt interrupts, "I don't care how long it takes, if we get the right answers. But they have to be right. Not knowing how to solve this dilemma about fake furs is making me feel like a leopard without its spots."

Sylva talks to Sam Pelt a bit longer and then ends the interview by saying, "I'll talk it all over with the other analysts at the office and let you know what we come up with. I think we can outfox the other furriers if we use the computer to help us sample opinions, rather than trapping unsuspecting customers into giving an opinion. But I'll let you know what they say. This much is for sure: If we can sample and not talk to everybody before making a decision, every coat you sell will have a silver lining."

As one of the systems analysts who is part of Sylva Foxx's firm, suggest some ways that Sam Pelt can use the PC he has to sample adequately the opinions of his customers, store managers, buyers, and any others you feel will be instrumental in making the strategic decision regarding the stocking of fake furs in what has always been a real fur store. Suggest a type of sample for each group and justify it. The constraints you are subject to include the need to act quickly so as to remain competitive, the need to retain a low profile so that competing furriers are unaware of your fact-gathering, and the need to keep costs of data gathering to a reasonable level.

Determining Sample Size When Sampling Qualitative Data. A good deal of information cannot be obtained by searching through files. That information can best be obtained by interviewing people in the organization.

There are no magic formulas to help the systems analyst set the sample size for interviewing. The overriding variable that determines how many people the systems analyst should interview in depth is the time an interview takes. A true in-depth interview and follow-up interview is very time-consuming for both the interviewer and the participant.

A good rule of thumb is to interview at least three people on every level of the organization and at least one from each of the organization's functional areas (as described in Chapter 2) who will be directly involved with a new or updated system. Remember also that one does not have to interview more people just because it is a larger organization. If the stratified sample is done properly, a small number of people will adequately represent the entire organization.

KINDS OF INFORMATION SOUGHT IN INVESTIGATION

The systems analyst seeks facts and figures, financial information, organizational contexts, and document types and problems through the sampling and investigation of hard data, as shown in Figure 4.3. Hard data accumulated in records supply information that cannot be obtained through other methods such as interviewing or observation.

FIGURE 4.3
Kinds of information sought in investigation.

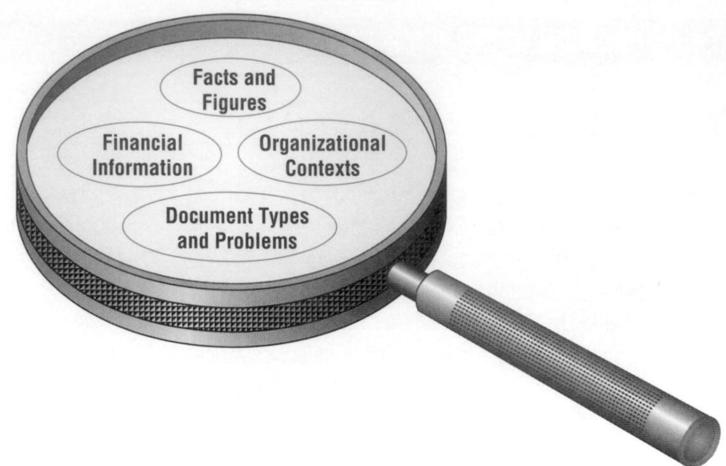

Although hard data are produced by the organization as a generic product (much like "to whom it may concern" material), the analyst needs to remember that meanings taken from hard data are constructed personally by organizational members. So it becomes important to ask: Who were the documents produced for originally, and why they have been kept? In other words, the role of the document in the organization needs to be understood. Generally, well-documented organizations may be more rigid than businesses with less documentation, because documentation may function to create one-way centralized control.

The analyst also needs to be aware that documents in an organization can serve as persuasive messages, because they offer information about how people are expected to behave. It also follows that organizational change is facilitated through the changing of documents.

TYPES OF HARD DATA

As the systems analyst works to understand the organization and its information requirements, it will become important to examine different types of hard data that offer information unavailable through any other method of data gathering. Hard data reveal where the organization has been and where its members believe it is going. To piece together an accurate picture, the analyst needs to examine both quantitative and qualitative hard data.

Analyzing Quantitative Documents. Many quantitative documents are available for interpretation in any business, and they include reports used for decision making, performance reports, records, and a variety of forms. All these documents have a specific purpose and audience for which they are targeted.

Reports used for decision making A systems analyst needs to obtain some of the documents that are used in running the business. These documents are often paper reports regarding the status of inventory, sales, or production. Many of these reports are not complex, but they serve mainly as feedback for quick action. For example, a sales report may summarize the amount sold and the type of sales. In addition, sales reports might include graphical output comparing revenue and income over a set number of periods. Such reports enable the decision maker to spot trends easily.

Production reports include recent costs, current inventory, recent labor, and plant information. Beyond these key reports, many summary reports are used by decision makers to provide background information, spot exceptions to normal occurrences, and afford strategic overviews of organizational plans.

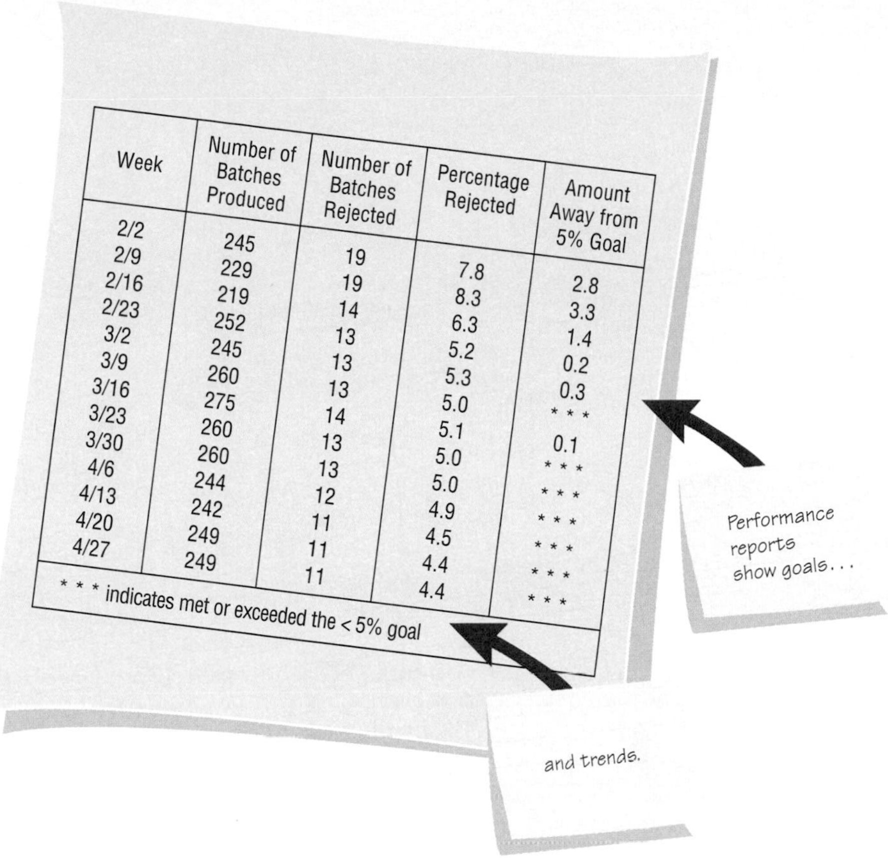

Week	Number of Batches Produced	Number of Batches Rejected	Percentage Rejected	Amount Away from 5% Goal
2/2	245	19	7.8	2.8
2/9	229	19	8.3	3.3
2/16	219	14	6.3	1.4
2/23	252	13	5.2	0.2
3/2	245	13	5.3	0.3
3/9	260	13	5.0	* * *
3/16	275	14	5.1	0.1
3/23	260	13	5.0	* * *
3/30	260	13	5.0	* * *
4/6	244	12	4.9	* * *
4/13	242	11	4.5	* * *
4/20	249	11	4.4	* * *
4/27	249	11	4.4	* * *
* * * indicates met or exceeded the < 5% goal				

Performance reports show goals. . .
and trends.

FIGURE 4.4
A performance report showing improvement.

Performance reports Most performance reports take on the general form of actual versus intended performance. One important function of performance reports is to assess the size of the gap between actual and intended performance. It is also important to be able to determine if that gap is widening or narrowing as an overall trend in whatever performance is being measured. Figure 4.4 shows a clear improvement in sales performance over two to three months. The analyst will want to note if performance measurement is available and adequate for key organizational areas.

Records Records provide periodic updates of what is occurring in the business. If the record is updated in a timely fashion by a careful recorder, it can provide much useful information to the analyst. Figure 4.5 is a manually completed payment record for apartment rental. There are several ways that the analyst can inspect a record:

1. Checking for errors in amounts and totals
2. Looking for opportunities for improving the recording form design
3. Observing the number and type of transactions
4. Watching for instances where the computer can simplify the work (i.e., calculations and other data manipulation)

Data capture forms Before you set out to change the information flows in the organization, you need to be able to understand the system that is currently in place. You or one of your team members may want to collect and catalog a blank copy of each form (official or unofficial) that is in use. (Sometimes businesses have a person already charged with forms management, who would be your first source for forms in use.)

FIGURE 4.5

A manually completed payment record.

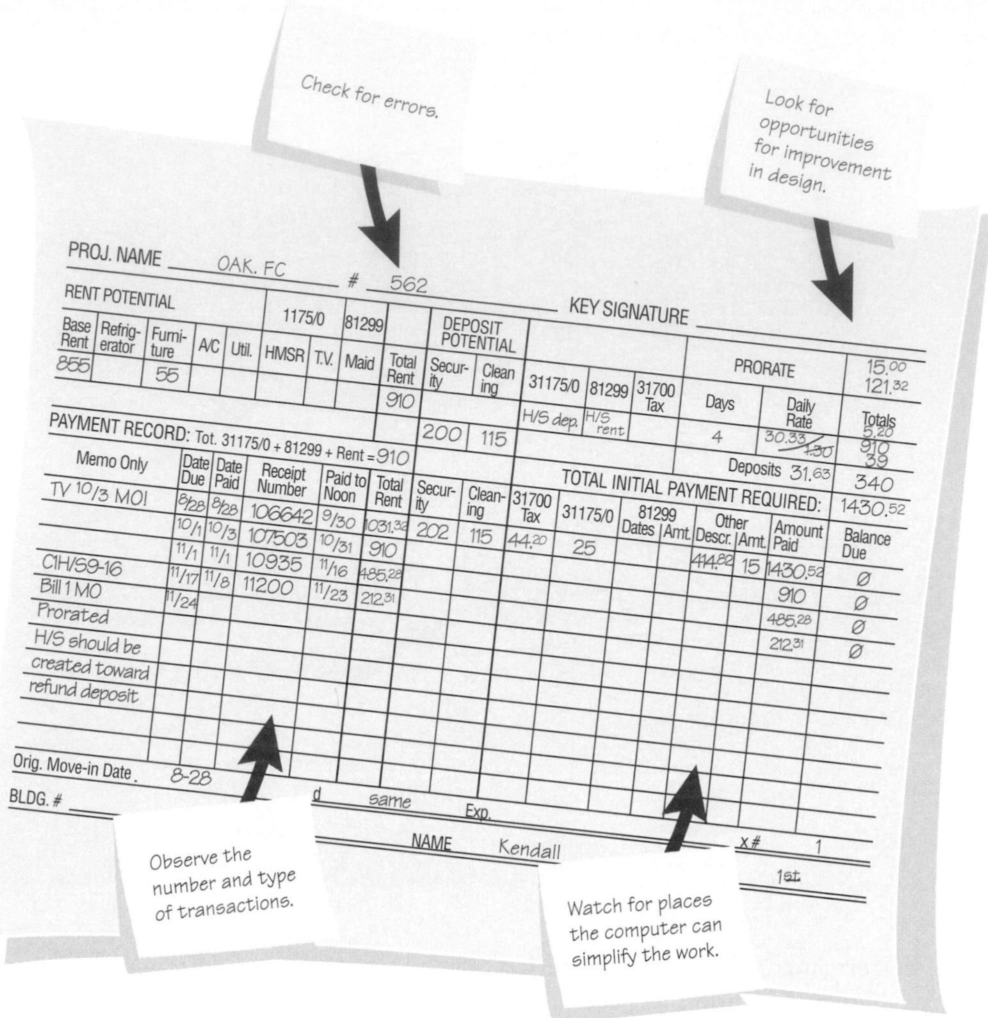

PROJ. NAME OAK. FC # 562 KEY SIGNATURE

RENT POTENTIAL 1175/0 81299 DEPOSIT POTENTIAL

Base Rent 855 | Refrig-erator | Furni-ture 55 | A/C | Util. | HMSR | T.V. | Maid | Total Rent 910 | Secur-ity 200 | Clean ing 115

31175/0 H/S dep. | 81299 H/S rent | 31700 Tax | PRORATE | Days 4 | Daily Rate 30.33 | Totals 15.00 121.32 5.20 910 39

PAYMENT RECORD: Tot. 31175/0 + 81299 + Rent = 910

Deposits 31.63 340

TOTAL INITIAL PAYMENT REQUIRED: 1430.52

Memo Only	Date Due	Date Paid	Receipt Number	Paid to Noon	Total Rent	Secur-ity	Clean-ing	31700 Tax	31175/0	81299 Dates	Amt.	Other Descr.	Amt.	Amount Paid	Balance Due
TV 10/3 MO!	8/28	8/28	106642	9/30	1031.32	202	115	44.20	25			414.82	15	1430.52	Ø
	10/1	10/3	107503	10/31	910									910	Ø
C1H/S9-16	11/1	11/1	10935	11/16	485.28									485.28	Ø
Bill 1 MO	11/17	11/8	11200	11/23	212.31									212.31	Ø
Prorated	11/24														
H/S should be created toward refund deposit															

Orig. Move-in Date 8-28

BLDG. # d same Exp.

NAME Kendall X# 1

1st

Blank forms, along with their instructions for completion and distribution, can be compared with filled-in forms to see if any data items are consistently left blank on the forms; whether the people who are supposed to receive the forms actually do get them; and if they follow standard procedures for using, storing, and discarding them. Remember to print out any Web-based forms that require users to print them. Alternatively, electronic versions that can be submitted via the Web or email can be identified and stored in a database for later inspection.

To proceed when creating a catalog of forms to help you understand the information flow currently in use in the business:

1. Collect examples of all the forms in use, whether officially sanctioned by the business or not (official versus bootleg forms).
2. Note the type of form (whether printed in-house, handwritten, computer-generated in-house, online forms, Web fill-in forms, printed externally and purchased, etc.).
3. Document the intended distribution pattern.
4. Compare the intended distribution pattern with who actually receives the form.

Although this procedure is time-consuming, it is useful. Another approach is to sample data capture forms that have already been completed. Remember to check databases that store consumer data when sampling input from ecommerce transactions. The analyst must keep in mind many particular questions, as illustrated in Figure 4.6. They include the following:

1. Is the form filled out in its entirety? If not, what items have been omitted, and are they consistently omitted? Why?
2. Are there forms that are never used? Why? (Check the design and appropriateness of each form for its purported function.)
3. Are all copies of forms circulated to the proper people or filed appropriately? If not, why not? Can people who must access online forms do so?
4. If there is a paper form that is offered as an alternative to a Web-based form, compare the completion rates for both.
5. Are "unofficial" forms being used on a regular basis? (Their use might indicate a problem in standard procedures or may indicate political battles within the organization.)

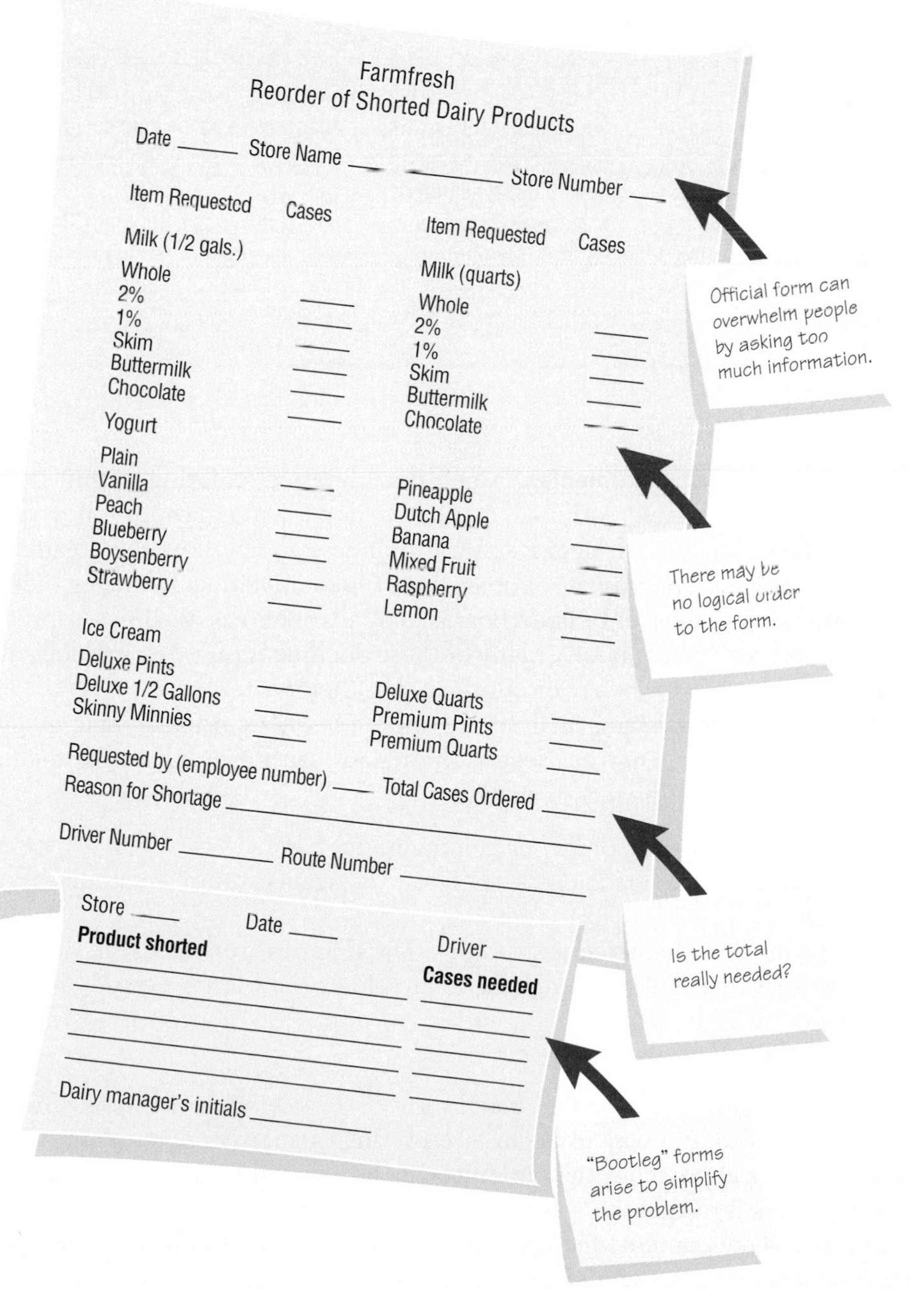
Farmfresh
Reorder of Shorted Dairy Products

Date ____ Store Name ____ Store Number ____

Item Requested	Cases	Item Requested	Cases
Milk (1/2 gals.)		Milk (quarts)	
Whole	____	Whole	____
2%	____	2%	____
1%	____	1%	____
Skim	____	Skim	____
Buttermilk	____	Buttermilk	____
Chocolate	____	Chocolate	____
Yogurt			
Plain	____	Pineapple	____
Vanilla	____	Dutch Apple	____
Peach	____	Banana	____
Blueberry	____	Mixed Fruit	____
Boysenberry	____	Raspberry	____
Strawberry	____	Lemon	____
Ice Cream			
Deluxe Pints	____	Deluxe Quarts	____
Deluxe 1/2 Gallons	____	Premium Pints	____
Skinny Minnies	____	Premium Quarts	____

Requested by (employee number) ____ Total Cases Ordered ____
Reason for Shortage ____
Driver Number ____ Route Number ____

Store ____ Date ____ Driver ____

Product shorted	Cases needed
____	____
____	____
____	____
____	____
____	____

Dairy manager's initials ____

FIGURE 4.6
Questions to ask about official and bootleg forms that are already filled out.

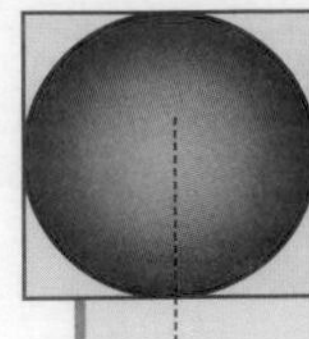

CONSULTING OPPORTUNITY 4.2

A ROSE BY ANY OTHER NAME . . . OR QUALITY, NOT QUANTITIES

"I think we have everything we need. I've sampled financial statements, sales figures for each branch, wastage for each shop—we have it all. With all these numbers, we should be able to figure out how to keep Fields in the green, or at least at the forefront of the flower business. We can even show Seymour Fields himself how his new computer system can make it all happen," says Rod Golden, a junior systems analyst working for a medium-sized consulting group.

The firm, under the supervision of its head systems analyst, Clay Potts, has been working on a systems project for the entire chain of 15 successful florist shops and indoor floral markets called Fields. Each of three Midwestern cities has five Fields outlets.

"Although it's just a budding enterprise now, eventually we want to grow with offshoots to half a dozen states," says Seymour Fields, the owner. "I want to reap the benefits of all the happiness we've sown so far. I think we can do it by playing my hunches about what is the best time to purchase flowers at each European market we buy from, and then we should prune back our purchases.

"Over the past three years, I've written lots of memos to our managers about this plan. They've written some good ones back, too. I think we're ready to stake out some territory on this soon," continues Seymour, painting a rosy picture of Fields's future.

"I agree," says Rod. "When I come back from my analysis of these figures," he says, indicating a large stack of material he has unearthed from Fields field offices, "we'll be able to deliver."

Three weeks later, Rod returns to Clay with wilting confidence. "I don't know what to make of all this. I can't seem to get at what's causing the company's growth, or how it's managed. They've been expanding, but I've been through all the figures, and nothing really seems to make sense yet."

Clay listens empathetically, then says, "You've given me a germ of an idea. What we need is some cross-pollination, a breath of fresh air. We need to dig a little deeper. Did you examine anything but their bottom line?"

Rod looks startled and replies, "No, I—uh—what do you mean?"

How can Clay Potts tactfully explain to Rod Golden that examination of qualitative as well as quantitative documents could be important to delivering an accurate assessment of Fields's potential to be a more fruitful enterprise? In a paragraph, recommend some specific documents that should be read. List the specific steps Rod should follow in evaluating qualitative documents obtained from Fields. Write a paragraph to explain how qualitative documents help in presenting an overall account of Fields's success.

Analyzing Qualitative Documents. Many documents circulating within organizations are not quantitative. Although qualitative documents may not follow a predetermined form, analysis of them is critical to understanding how the organizational members engage in the process of organizing. Qualitative documents include email messages, memos, signs on bulletin boards and in work areas, Web pages, procedure manuals, and policy handbooks. Many of these documents are rich in details revealing the expectations for behavior of others that their writers hold.

Although many systems analysts are apprehensive about analyzing qualitative documents, they need not be. Several guidelines can help analysts to take a systematic approach to this sort of analysis:

1. Examine documents for key or guiding metaphors.
2. Look for insiders versus outsiders or an "us against them" mentality in documents.
3. List terms that characterize good or evil and appear repeatedly in documents.
4. Look for the use of meaningful graphics, logos, and icons posted on common areas or on Web pages.
5. Recognize a sense of humor, if present.

Each guideline is explained briefly in the following paragraphs. Guidelines should be viewed as a way to begin interpreting qualitative documents. Certainly there are other approaches that go much further in analysis than necessary for the systems analyst's purposes.

Examining documents for key or guiding metaphors is done because language shapes behavior; thus, the metaphors we employ are critical. For example, an organization that discusses employees as "part of a great machine" or "cogs in a wheel" might be taking a mechanistic view of the organization. Notice that the guiding metaphor in the memo in Figure 4.7 is, "We're one big happy family." The analyst can use this information to predict the kinds of metaphors that will be persuasive in the organization as well as to understand how a new system might be characterized metaphorically.

When the analyst finds language that pits one group or department against another or sets the business as a whole against competitors, it is possible to understand more clearly the "us against them" politics that exist. Obviously, if one department is battling another, it may be impossible to gain any cooperation on a systems project until the politics are resolved in a satisfactory manner. Similarly, if the business is fiercely competitive in its remarks about other businesses in the industry, part of the thrust for the systems project may come from wanting to hold the competitive edge.

Listing the terms found in qualitative documents that characterize actions, groups, or events as good or evil allows the analyst to see what is considered good in the business and what is deemed bad or wrong. This knowledge allows insights into what group values are being espoused. For instance, if accumulating information without apparent reason is described in negative terms (as depicted in Figure 4.8), the analyst will want to ensure that system output is actually used and is not being accumulated "just in case."

The fourth guideline reminds the analyst to look for extra information that can be gained from noticing the use of graphics, logos, and icons posted on common areas shared by employees in the company or displayed on Web pages. With the explosion in the use of color monitors and printing, the ease of use of graphics and drawing packages, and the widespread access to the Internet, the use of graphics,

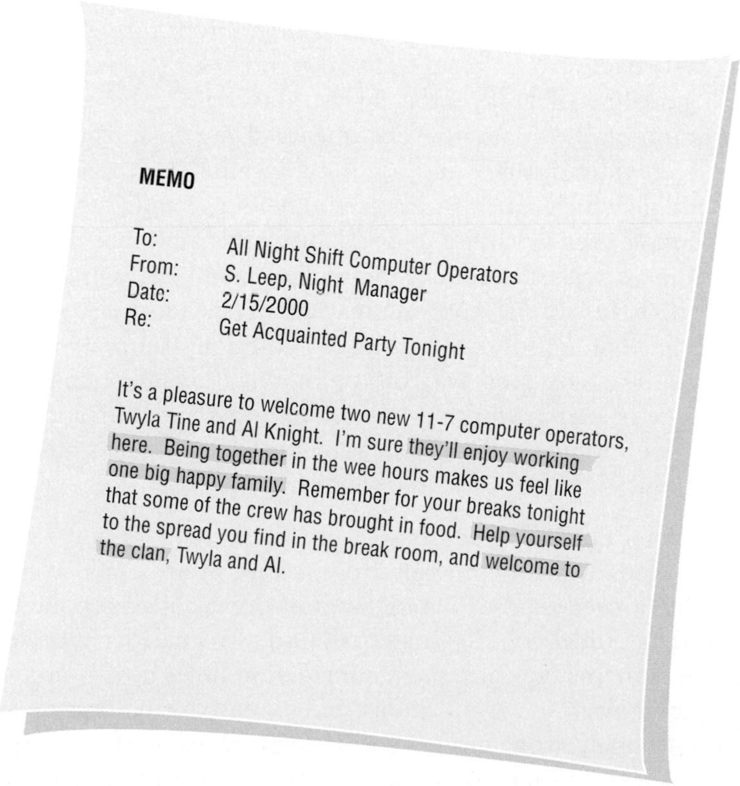

MEMO

To: All Night Shift Computer Operators
From: S. Leep, Night Manager
Date: 2/15/2000
Re: Get Acquainted Party Tonight

It's a pleasure to welcome two new 11-7 computer operators, Twyla Tine and Al Knight. I'm sure they'll enjoy working here. Being together in the wee hours makes us feel like one big happy family. Remember for your breaks tonight that some of the crew has brought in food. Help yourself to the spread you find in the break room, and welcome to the clan, Twyla and Al.

FIGURE 4.7
Analysis of memos provides insights into the metaphors that guide the organization's thinking.

FIGURE 4.8

Analysis of memos reveals values, attitudes, and beliefs of organizational members regarding the use of information.

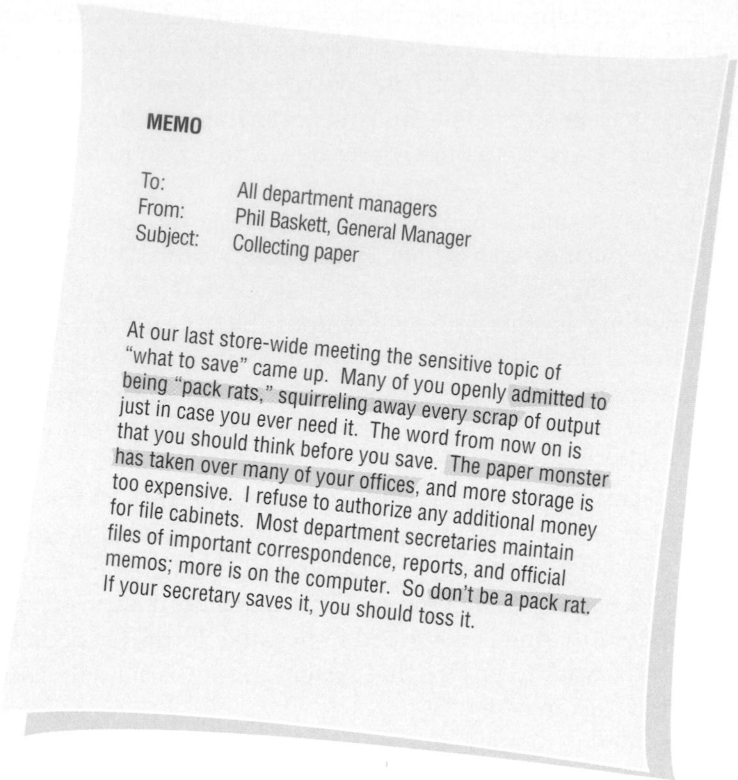

MEMO

To: All department managers
From: Phil Baskett, General Manager
Subject: Collecting paper

At our last store-wide meeting the sensitive topic of "what to save" came up. Many of you openly admitted to being "pack rats," squirreling away every scrap of output just in case you ever need it. The word from now on is that you should think before you save. The paper monster has taken over many of your offices, and more storage is too expensive. I refuse to authorize any additional money for file cabinets. Most department secretaries maintain files of important correspondence, reports, and official memos; more is on the computer. So don't be a pack rat. If your secretary saves it, you should toss it.

logos, and icons to convey a message has also increased. They are yet another form of information and expression that the analyst can use to understand the culture, values, and preferences of organizational members. In addition, when analyzed and understood they may permit the analyst to incorporate some of the design sensibilities, metaphors, and themes of the organization into future ecommerce sites or information systems design.

The fifth guideline encourages the analyst to recognize the existence of a sense or senses of humor in the qualitative documents of the organization. Assessing the use of humor provides a quick and accurate barometer of many organizational variables, including which subculture a person belongs to and what kind of morale exists. For example, conservative, traditionalist organizational members tend to relate formula jokes, with a beginning, a middle, and an ending punch line. People who are sensitive to others' communication needs tend to relate anecdotal, humorous stories that are self-deprecating. In addition, if morale is low, "gallows" humor tends to be a common way of coping with organizational uncertainties. Cartoons that are posted in workspaces, coffee rooms, and corridors are helpful in revealing a sense of humor and some areas of frustration or tension among organizational members.

Memos When possible, analysts should sample memos sent throughout the business. Sometimes, however, memos are not kept or are made available only to those who have "a need to know" as defined in organizational policy. Along with the five preceding guidelines, the analyst should also consider who sends memos and who receives them. Typically, most information flows downward and horizontally rather than upward in organizations, and extensive email systems mean messages are sent to many work groups and individuals, too. Memos in print or on email reveal a lively, continuing dialogue in the organization. Analysis of memo

Source: © 2000 The New Yorker Collection from cartoonbank.com. All Rights Reserved.

content will provide you with a clear idea of the values, attitudes, and beliefs of organizational members.

Signs or posters on bulletin boards or in work areas Although signs may seem incidental to what is happening in the organization, they serve as subtle reinforcers of values to those who read them, as depicted in Figure 4.9. Signs such as "Quality Is Forever" or "Safety First" give the analyst a feel for the official organizational culture. It is also instructive to note for whom signs are intended and to find out through interviews whether organizational members are held accountable for acting on the information posted.

Corporate Web Sites Web sites used for business-to-consumer (B2C) ecommerce as well as those used for business-to-business (B2B) transactions should also be viewed by the analyst. Examine the contents for metaphors, humor, use of design features (such as color, graphics, animation, and hyperlinks), and the meaning and clarity of any messages provided. Think about the Web site from three dimensions: technical, aesthetic, and managerial. Are there discrepancies between the stated goals of the organization and what is presented to the intended viewer? How much customization of the Web site is available for each user? How much personalization of the Web site is possible? If you are not designing ecommerce sites for the organization, how does what you see on their Web sites affect the systems you are investigating? Remember to note the level of interactivity of the Web site or sites, the accessibility of the messages, and the apparent security.

Manuals Other qualitative documents the analyst should examine are organizational manuals, including manuals for computer operating procedures and online manuals. Manuals should be analyzed following the five guidelines spelled out previously. Writers of manuals are allowed more elaboration in making a point than is typ-

FIGURE 4.9

Posted signs reveal the official organizational culture.

ically accorded to those writing memos or posting signs. Remember that manuals present the "ideal", the way machines and people are expected to behave. Examining manuals, both printed and online, systematically will give you a picture of the way things ought to happen. It is important to recall that printed manuals are rarely kept current and are sometimes relegated to a shelf, unused. Check to see if the manuals you are looking at are current and whether they are followed or forgotten.

Policy handbooks The last type of qualitative document we consider is the policy handbook. Although these documents typically cover broad areas of employee and corporate behavior, you can be primarily concerned with those that address policies about computer services, use, access, and charges. Some computer policies may appear on screen or on a Web site whenever a particular program is used.

Policies are larger guidelines that spell out the organization's ideal of how members should conduct themselves to achieve strategic goals, as depicted in Figure 4.10. Once again, members may not even be aware of particular policies, or policies may be purposely sidestepped in the name of efficiency or simplicity. Examining policies, however, allows the systems analyst to gain an awareness of the values, attitudes, and beliefs guiding the corporation.

WORKFLOW ANALYSIS

You can learn many things about an organization's information flow by examining the forms it uses. Some signs to watch for that may be symptomatic of larger problems include the following:

1. Data or information doesn't flow as intended (too many people, too few people, or the wrong people receive it).
2. Bottlenecks in the processing of forms result in slowed or stopped work.
3. Access to online forms is cumbersome. Web forms must be printed out and faxed or sent rather than submitted electronically.
4. Unnecessary duplication of work occurs because employees are unaware that information is already in existence on another form that they do not receive.
5. Employees lack understanding about the interrelatedness of information flow (i.e., they do not know that their output work serves as input for another person).

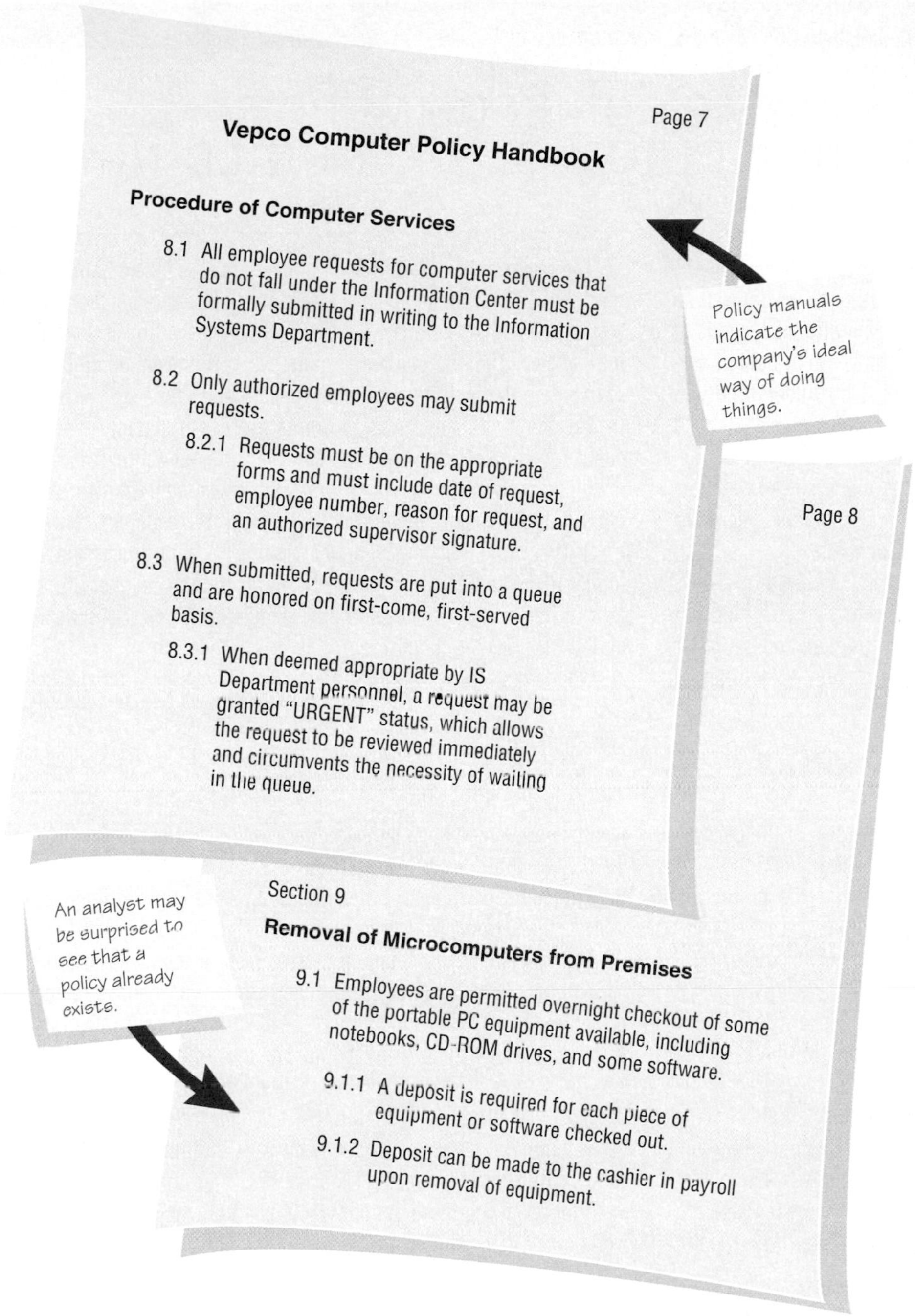

Page 7

Vepco Computer Policy Handbook

Procedure of Computer Services

8.1 All employee requests for computer services that do not fall under the Information Center must be formally submitted in writing to the Information Systems Department.

8.2 Only authorized employees may submit requests.

8.2.1 Requests must be on the appropriate forms and must include date of request, employee number, reason for request, and an authorized supervisor signature.

8.3 When submitted, requests are put into a queue and are honored on first-come, first-served basis.

8.3.1 When deemed appropriate by IS Department personnel, a request may be granted "URGENT" status, which allows the request to be reviewed immediately and circumvents the necessity of waiting in the queue.

Policy manuals indicate the company's ideal way of doing things.

Page 8

Section 9

Removal of Microcomputers from Premises

9.1 Employees are permitted overnight checkout of some of the portable PC equipment available, including notebooks, CD-ROM drives, and some software.

9.1.1 A deposit is required for each piece of equipment or software checked out.

9.1.2 Deposit can be made to the cashier in payroll upon removal of equipment.

An analyst may be surprised to see that a policy already exists.

FIGURE 4.10
Policy manuals provide insights into the prescribed way of behaving in the organization.

One approach to analyzing a system is to diagram the flow of information, paperwork, and decisions. Unlike playscript analysis (introduced in Chapter 7) that focuses on people as the important element in the system, diagrams of workflow tend to treat the work itself as being the important element. A business process reengineering (BPR) tool called Workflow BPR™ from Holosofx, Inc. <www.holosofx.com> is actually a set of tools that include modeling tools; analysis measures; simulation for analyzing bottlenecks; and integration with CASE tools, financial systems, or human resource management systems. Workflow BPR is unique because it allows analysts to take the process models entered into Workflow BPR and export them to other software packages.

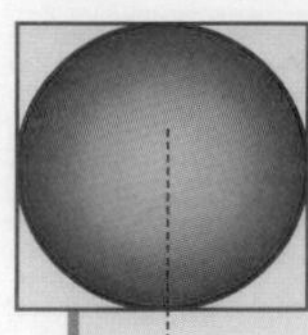

CONSULTING OPPORTUNITY 4.3

WORKFLOW ANALYSIS FOR ECOMMERCE INTEGRATION: BUILDING A SITE OR BUILDING A VISION?

"You'd think this should be easier. We have a killer ecommerce site and our customers love it. Unfortunately, although we can sell an insurance policy over the Web to a customer in a matter of seconds, it takes us 30 days to actually issue the darn thing. We're being hammered in our marketing research. Customers don't understand why they are waiting so long to get their policy," says Ryan Taylor, senior director of marketing ecommerce for a large insurance company on the East Coast. The ecommerce site he is describing is used by insurance agents, business owners, individuals, and company sales representatives to quote and request new insurance.

Ryan continues, "We've tried to integrate this before but it always gets bogged down in politics. There are currently three groups at odds with each other here: marketing, underwriting, and operations." Each of these departments has established a number of manual processes and audits that are part of issuing new insurance policies received from the Web site.

Ryan continues, explaining to Chandler, the recently hired director of the operations department, "Marketing reviews each new policy after the initial sale, to make sure that the submission of the policy is complete. The underwriting department reviews the request for completeness and accuracy, in effect reauditing what the marketing department looked at. Your department, operations, reaudits what underwriting does and then keys the information into the mainframe."

Chandler cuts in, "I realize I have not been here very long, but it seems to me the problem lies in that no one owns the entire process. Each department functions as a stand-alone silo. Why don't we just get everyone together and agree on a stream-lined process?"

Ryan explains, "Twice we've tried to bring these three factions together, and twice we've failed. In fact, the last time, marketing and operations got in a shouting match. The meeting ended when one of your managers threw up their hands and stormed out of the meeting saying, 'This is a waste of time. It didn't work before, and it won't work this time'".

"It's got to be different this time," Ryan exclaims. He is charged with the challenging task of bringing together the warring factions to create a way to process policy issuances electronically within five days.

Chandler looks puzzled. "You already have a functioning ecommerce site," he says. "What are the departments' concerns?"

Ryan replies, "Marketing is worried about how the customer will be treated by other departments. They are also concerned about job security and the elimination of jobs if the system becomes fully automated. Underwriting is worried about loss of control and wonders if an automated system can be as effective as human underwriter in evaluating the quality of each insurance policy. They are worried that they could be taken out of the process completely. Operations people are also concerned that their jobs are at stake if an automated system is put in place."

The release of the initial ecommerce site was in the hands of the marketing department. From all accounts of people on the team, it was successful. Said one, "It was relatively easy, and it was rather inexpensive. The real challenge came when the ecommerce site had to be integrated with the entire organization."

The main challenge was to create a cross-functional team that could reach a consensus on the creation of a fully integrated policy issuance ecommerce site. The first step Ryan went through was to identify key decision makers in each of the three departments and bring them together into a project team. He also brought in eight highly respected yet disinterested organizational members to serve as a buffer as the project team navigated stormy political seas associated with moving to an integrated, electronic system.

When the project team first met, Ryan, who is from the marketing department, turned to them and said, "I have permission from our chief marketing executive to eliminate all the jobs in marketing that are involved with issuing policies if that's what the solution calls for. As one of the three major stakeholders in this process, we abide by whatever is agreed upon by the project team." In this way he diffused suspicions that he would be partial to a solution that saved marketing jobs to the detriment of the other groups or to the detriment of the improved electronic processing in general.

The first step the ecommerce project team took was to examine the vendors, technologies, and consultants who could provide an integrated suite of tools to support the electronic processing of insurance policies. Software that could provide an end-to-end solution was considered. The team also considered whether to have the entire processing system built for them. They took the opportunity to see other companies that were successful in building sites, and they were excited with the possibilities.

After reviewing these options, their eventual decision was to hire a workflow design company that could perform business process reengineering. The consulting company gathered all the data and documented the current process from start to finish. The insurance company provided project management and business experts. After analyzing the current workflow, the consultants jointly developed an automated process that allowed the computer to perform most of the audits. In addition, the new workflow made use of imaging and optical character recognition to cut down on the need for data-entry personnel.

Because the process is now automated, people who submit incomplete applications are not permitted to move off the application screen. As the application moves through the company, the customer is notified via email about the progress of the request. Because 85 percent of submissions are clean because of up-front electronic audits, only 15 percent need to be reviewed by underwriters. Clean submissions are sent directly to the mainframe, and policies are issued in three to five days, rather than 25 to 30 days as before.

(Continued)

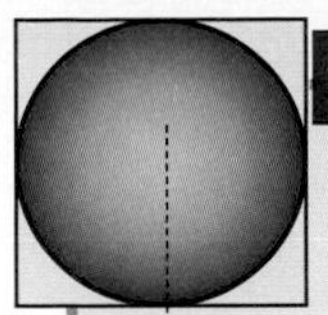

CONSULTING OPPORTUNITY 4.3 (CONTINUED)

When an analyst performs workflow analysis what sort of data are collected? Try sketching (by hand) the workflow depicted for this insurance company's policy issuance process, from the initial price inquiry on the Web to issuing a new policy. (You can use symbols as shown in the BPR Workflow section of this chapter.) Marketing released the ecommerce site two and a half years before this project team was formed. How do consumers view a business that establishes a Web presence but is unable to live up to expectations of rapid delivery influenced by their other Web experiences? Write a paragraph describing the organizational barriers to the integration of electronic policy processing that the project team faced. Compare and contrast managing the front-end development of an ecommerce site development and managing strategic integration of ecommerce into an organization's systems.

MODELING

In Workflow BPR the modeling feature can be used to model the existing system or to model a new system. When used to analyze the existing system, the analyst needs to observe what happens in the organization on a typical day.

Process modeling in Workflow BPR is built around objects. These models show how objects are transferred and where they are going. The process must unfold chronologically, and depending on what particular conditions or decisions exist, the object may take different paths.

In the medical insurance application example shown in Figure 4.11, a small bitmap graphic image of a form on a clipboard is used to show the application form as it appears in the system. A rounded rectangle is used to show a process, in this case a process to analyze risk factors. When the results of the analysis are obtained, different events can occur given different outcomes. The branching that is based on different outcomes is illustrated by the dark blue diamond in the workflow diagram. Diamonds have traditionally been used to show a decision that can be made by a human or that can be automated.

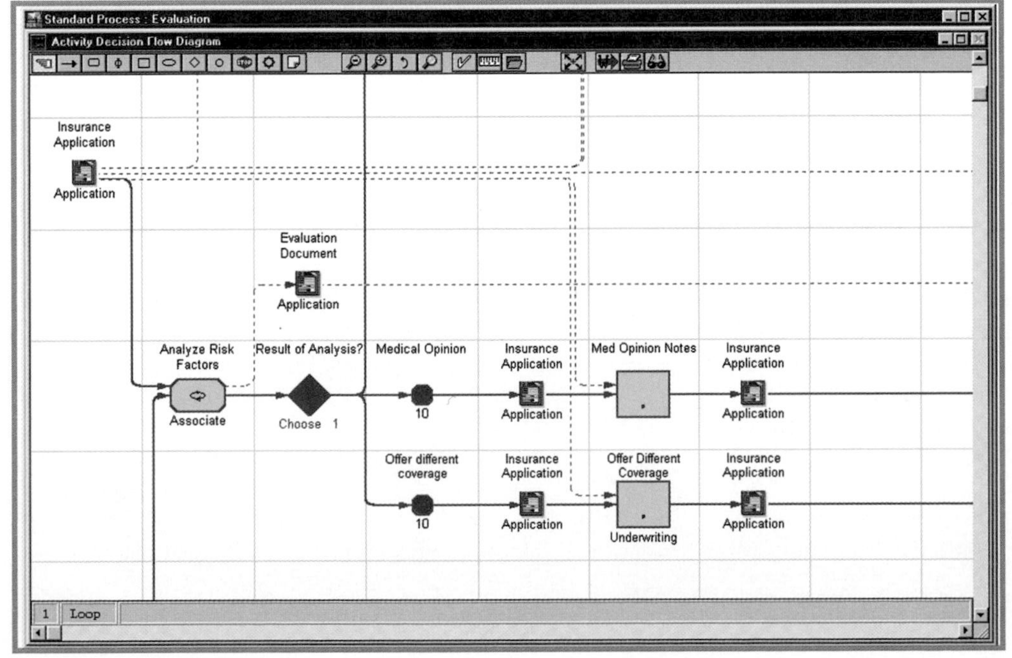

FIGURE 4.11
An analyst can study the workflow of an organization by diagramming the processes, as shown in this activity decision flow diagram of a medical insurance application process. The software used is Workflow BPR by Holosofx, Inc.

If additional analysis is desired, it is necessary to determine the probabilities of each branch extending from the diamond. These probabilities are shown in red. Shown is only a portion of the workflow diagram, but the entire diagram will reveal that the probabilities add up to 100 percent.

ANALYSIS

Any process in a manufacturing or service organization can have multiple outcomes as decisions are made (e.g., a form is routed to different departments depending on the action required). In Workflow BPR these outcomes are called process cases. Workflow BPR measures each of these possible outcomes based on the probabilities that the outcome will occur and then summarizes the results. Figure 4.12 shows the detailed results from some of the cases generated in the medical insurance application model. The analyst can use this method to double check the accuracy and completeness of this model.

SIMULATION

Another useful feature of Workflow BPR is the simulation tool. By varying rates of input, activities can be simulated so that short-term performance issues such as bottlenecks in a process can be identified and addressed. By analyzing the simulation results, a systems analyst can develop procedures to successfully plan for and manage uncontrollable variations of the inputs.

Figure 4.13 shows the results of a simulation for the medical insurance application model. Ten simulations were run, and averages were then calculated. The analyst can use such accurate information to determine paperwork processing time and average cost of processing applications. Recently, analysts have been using BPR Workflow Analysis to integrate front-end ecommerce Web sites with a variety of internal business processes. The real power of ecommerce becomes apparent when internal business processes are linked to electronic processes. In doing so, long-neglected processes are often streamlined, automated, and speeded up to save time, money, and employees. In addition, the focus of ecommerce for the company becomes strategic rather than transaction-oriented.

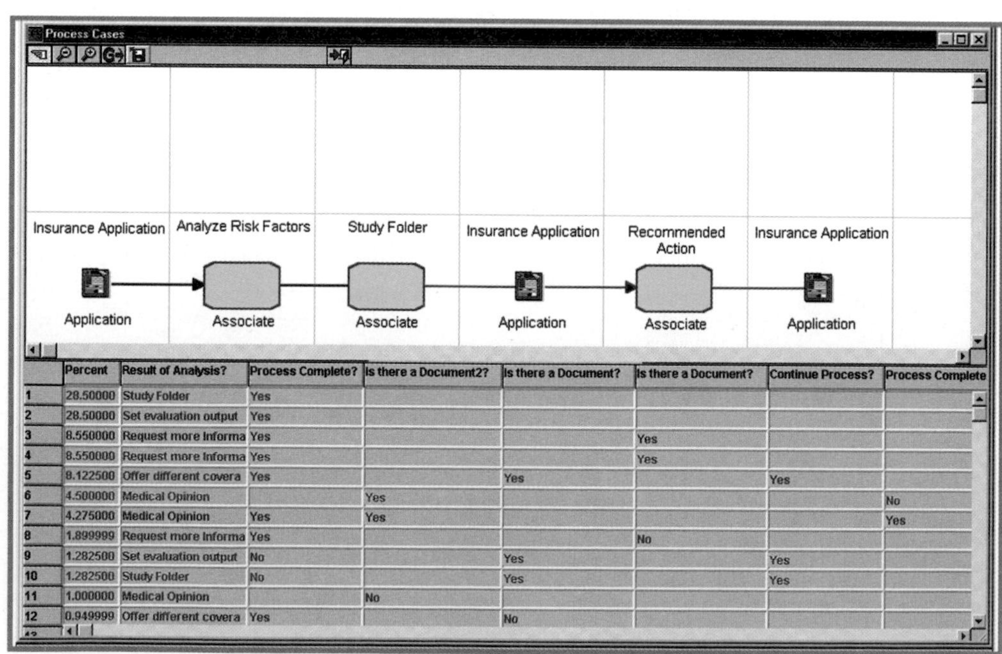

	Percent	Result of Analysis?	Process Complete?	Is there a Document2?	Is there a Document?	Is there a Document?	Continue Process?	Process Complete
1	28.50000	Study Folder	Yes					
2	28.50000	Set evaluation output	Yes					
3	8.550000	Request more Informa	Yes			Yes		
4	8.550000	Request more Informa	Yes			Yes		
5	8.122500	Offer different covera	Yes		Yes		Yes	
6	4.500000	Medical Opinion		Yes				No
7	4.275000	Medical Opinion	Yes	Yes				Yes
8	1.899999	Request more Informa	Yes			No		
9	1.282500	Set evaluation output	No		Yes		Yes	
10	1.282500	Study Folder	No		Yes		Yes	
11	1.000000	Medical Opinion		No				
12	0.949999	Offer different covera	Yes		No			

FIGURE 4.12

Once the process is diagrammed, it can be checked for accuracy by enumerating all the cases. The software used is Workflow BPR by Holosofx, Inc.

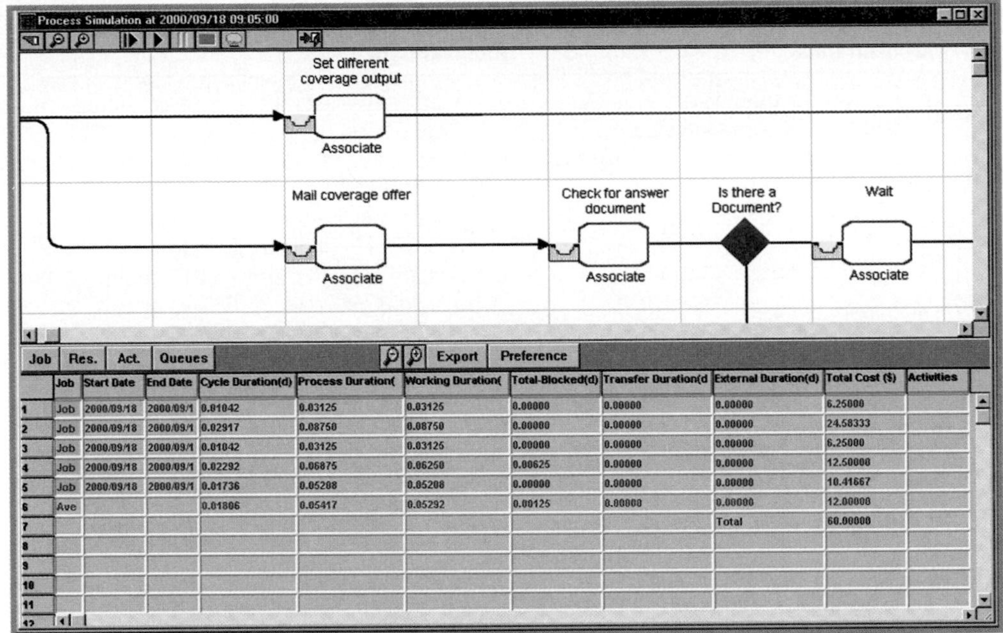

FIGURE 4.13

Simulation can be used to analyze cycle times and costs. This example of a medical insurance application process was modeled using Workflow BPR by Holosofx, Inc.

ABSTRACTING DATA FROM ARCHIVAL DOCUMENTS

Much of the data, both quantitative and qualitative, that you will need will not be in current use. Rather, they will be stored in archives of some sort. Some material is kept because of government regulations, some because of accounting practices, and some because the business may have formerly been called upon, in lawsuits or for other purposes, to produce historical information.

Examples of archival data that may be of interest to the systems analyst include actuarial records, budgets, and sales reports. One common characteristic that all archival data share is that their collection was paid for by someone other than the systems analyst. This fact may or may not influence the content, but as a result, archival data are relatively low-cost informational sources. Another advantage of archival data is that there is no overt changing of data, because the producer is unaware that he or she is being studied.

One disadvantage of using archival data is that the analyst may be uncertain about their meaning if only a limited subset of the data originally created exists. Another disadvantage of using archival data is that the records that survive (or that someone decided to keep) may not be the most important or meaningful ones. A third disadvantage of using archival data is that there is a large degree of built-in bias because someone has decided, for whatever reasons, to file the original data as well as to preserve some data and not others. In addition, new data are difficult to obtain from the same or equivalent samples. Finally, some archival data are inaccessible because they have not been converted into a format or medium that is not currently in use, and to do so would be an expensive undertaking. Advantages and disadvantages of using archival data are summarized in Figure 4.14.

Even with all the preceding cautions, the following guidelines can make abstracting data from archives and archival databases worthwhile:

1. Fragment the data into subclasses and make cross-checks to reduce errors.
2. Compare reports on the same phenomenon by different analysts (or maybe even someone within the organization).

FIGURE 4.14
Advantages and disadvantages of the use of archival data by the systems analyst.

Advantages of Using Archival Data	Disadvantages of Using Archival Data
• Archival documents are already paid for—by someone else. • Data are not changed because the producer is unaware of being studied.	• Uncertainty exists if the data are only a subset of original data. • The records that survive may not be the most important or meaningful. • The data may be biased because someone originally decided what to file. • New data are difficult to obtain from equivalent samples.

3. Realize the inherent bias associated with original decisions to file, keep, or destroy reports.
4. Use other methods such as interviewing and observation to fill out your organizational picture and supply cross-checks.

SUMMARY

The process of systematically selecting representative elements of a population is called sampling. The purpose of sampling is to select and study documents such as invoices, sales reports, and memos or perhaps to select and interview, give questionnaires to, or observe members of the organization. Sampling can reduce cost, speed data gathering, potentially make the study more effective, and possibly reduce the bias in the study.

A systems analyst must follow four steps in designing a good sample. First, there is a need for determining the population itself. Second, the type of sample must be decided. Third, the sample size is calculated. Finally, the data that need to be collected or described must be planned.

The types of samples useful to a systems analyst are convenience samples, purposive samples, simple random samples, and complex random samples. The last type includes the subcategories of systematic sampling and stratified sampling. There are several guidelines to follow when determining sample size. The systems analyst can make a subjective decision regarding acceptable interval estimates, then a confidence level is chosen and the necessary sample size can be calculated.

Workflow analysis enables the analyst to diagram the flow of information, paperwork, and decisions that comprise the business processes in an organization. With that data the analyst can model what exists or what the new system may look like as they build models around objects. In addition, workflow can be simulated so that bottlenecks in processes can be identified and cleared. Workflow BPR has recently been used to integrate ecommerce business processes with internal business processes in a strategic manner.

Systems analysts need to investigate current and archival data and forms, including reports, documents, financial statements, content of corporate Web sites, Web forms meant to be printed out and those that are electronically submitted, procedure manuals, and email content and memos. Current and archival data and forms reveal where the organization has been and where its members believe it is going. Both quantitative and qualitative documents need to be analyzed. Because documents are persuasive messages, it must be recognized that changing them might well change the organization.

There are many ways to analyze both quantitative and qualitative documents, but it is important to remember that the investigation of archival data has drawbacks as well as advantages. Because many of these drawbacks can be overcome, archival investigation is worthwhile.

HYPERCASE® EXPERIENCE

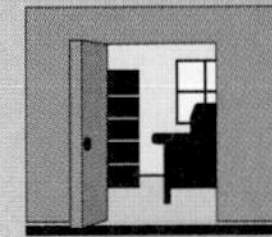

4

"We're glad you find MRE an interesting place to consult. According to the grapevine, you've been busy exploring the home office. I know, there's so much going on. We find it hard to keep track of everything ourselves. One thing we've made sure of over the years is that we try to use the methods that we believe in. Have you seen any of our reports? How about the data that were collected on one of Snowden's questionnaires? He seems to favor questionnaires over any other method. Some people resent them, but I think you can learn a lot from the results. Some people have been good about cooperating on these projects. Have you met Kathy Blandford yet?"

HYPERCASE QUESTIONS

1. Use clues from the case to evaluate the Training Unit's computer experience and its staff's feeling about a computerized project tracking system. What do you think the consensus is in the Training Unit toward a computerized tracking system?
2. What reports and statements are generated by the Training Unit during project development? List each with a brief description.
3. According to the interview results, what are the problems with the present project tracking system in the Training Unit?
4. Describe the "project management conflict" at MRE. Who is involved? Why is there a conflict?
5. How does the Management Systems Unit keep track of project progress? Briefly describe the method or system.

KEYWORDS AND PHRASES

acceptable interval estimate
archival data
attribute data
business-to-business (B2B) ecommerce
business-to-consumer (B2C) ecommerce
business process reengineering (BPR)
cluster sampling
complex random sample
confidence level
convenience sample
corporate Web sites
dispersion
hard data
purposive sample
qualitative data
quantitative data
sample population
sampling
simple random sample
stratified sampling
systematic sampling
variables
workflow analysis

REVIEW QUESTIONS

1. Define what is meant by sampling.
2. List four reasons why the systems analyst would want to sample data or select representative people to interview.
3. What are the four steps to follow to design a good sample?
4. Give an example of a convenience sample.
5. Define what is meant by taking a purposive sample.
6. Why is it often impractical to use a simple random sample for sampling documents and reports?

7. List the three approaches to complex random sampling.
8. Define what is meant by stratification of samples.
9. Give an example of attribute data.
10. What two changes can cause the analyst to use a larger sample size when sampling for attributes?
11. In what way does determining the sample size for variables differ from determining the sample size for attribute data?
12. What effect on sample size does using a greater confidence level have when sampling attribute data?
13. What is the overriding variable that determines how many people the systems analyst should interview in depth?
14. List three of the quantitative documents that the analyst should analyze when attempting to understand the organization.
15. List six qualitative sources that the analyst should analyze when attempting to understand the organization.
16. What are four guidelines for the systematic analysis of qualitative documents?
17. Define what is meant by workflow analysis using a business process reengineering tool.
18. List two uses of workflow business process reengineering software.
19. What are two advantages of using archival data as a source for understanding the organization?
20. What are five disadvantages of using archival data as a source for understanding the organization?

PROBLEMS

1. Dee Fektiv is concerned that too many forms are being filled out incorrectly. She feels that about 10 percent of all the forms have an error.
   a. How large a sample size should Dee use to be 99 percent certain she will be within 0.02?
   b. Suppose Dee will accept a confidence level of 95 percent that she will be within 0.02. What will the sample size of forms be now?
2. Rhea Fund has asked you to determine the average number of rebates mailed on a daily basis. You examine some records and believe the average to be about 200, with a standard deviation of 20. You want to be more certain, so you decide to sample.
   a. How large should the sample size be if you want to be 99 percent certain that the number will be within 5 of the mean?
   b. How large should the sample size be if you change the confidence limit to 95 percent?
3. Phil Ittup, a member of your systems analysis group, has been assigned the task of interviewing organizational members for your systems study. The business, Fall Back Industries, has five layers of management. Also, production, accounting, marketing, systems, logistics, and top management are the functional areas that will be affected by the proposed system. Each level has about 40 people. Production has 80 people total, accounting has 35, marketing has 42, systems has 10, and logistics has 28. There are five people in top management.
   a. Draw up an interviewing sample for Phil to follow in choosing who to interview. Justify why you have recommended that he interview the number of people you have chosen. What factors influenced your decision?
   b. While waiting for you to make up your sample, Phil decided that, just to be on the safe side, he would skip sampling and interview each person. In

a paragraph, explain to him why it is unwise for your systems analysis team not to follow a sampling strategy.

c. Phil has finally asked you for a hard-and-fast rule for how many people to interview in the organization. If you know of one, state it. If there isn't one, explain why one does not exist.

4. "I see that you have quite a few papers there. What all do you have in there?" asks Betty Kant, head of the MIS task force that is the liaison group between your systems group and Sawder's Furniture Company. You are shuffling a large bundle of papers as you prepare to leave the building.

"Well, I've got some financial statements, production reports from the last six months, and some performance reports that Sharon got me that cover goals and work performance over the last six months," you reply as some of the papers fall to the floor. "Why do you ask?"

Betty takes the papers from you and puts them on the nearest desk. She answers, "Because you don't need all this junk. You're here to do one thing, and that's talk to us, the users. Bet you can't read one thing in there that'll make a difference."

a. The only way to convince Betty of the importance of each document is to tell her what you are looking for in each one. Use a paragraph to explain what each kind of document contributes to the systems analyst's understanding of the business.

b. While you are speaking with Betty, you realize you actually need other quantitative documents as well. List any you are missing.

5. Figures 4.EX1 and 4.EX2 are two filled-in pages that comprise a larger computer usage log that the administrator of the information center has been using.

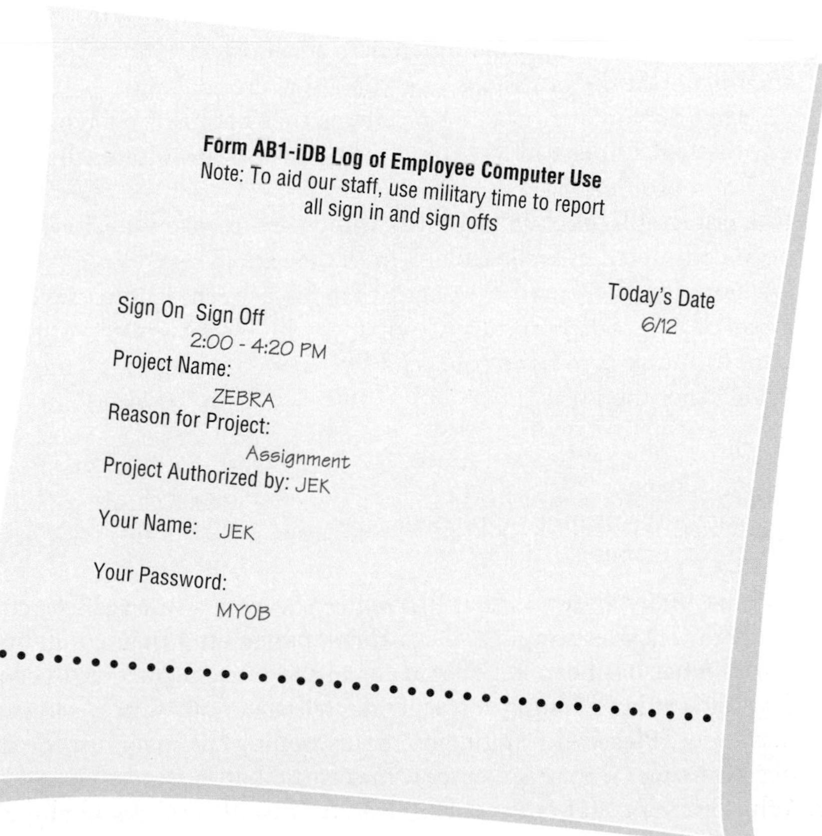

Form AB1-iDB Log of Employee Computer Use
Note: To aid our staff, use military time to report all sign in and sign offs

Today's Date
6/12

Sign On Sign Off
2:00 - 4:20 PM

Project Name:
ZEBRA

Reason for Project:
Assignment

Project Authorized by: JEK

Your Name: JEK

Your Password:
MYOB

FIGURE 4EX.1
One user's idea of how to complete the computer usage log.

FIGURE 4EX.2

A second user's idea of how to complete the computer usage log.

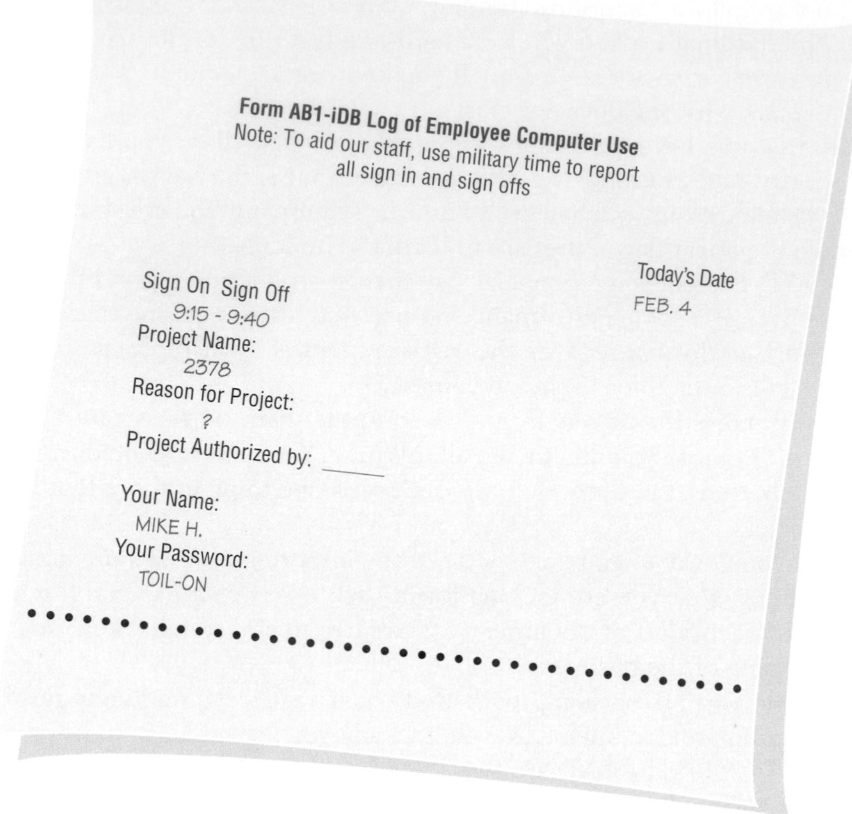

Form AB1-iDB Log of Employee Computer Use
Note: To aid our staff, use military time to report all sign in and sign offs

Today's Date
FEB. 4

Sign On Sign Off
9:15 - 9:40

Project Name:
2378

Reason for Project:
?

Project Authorized by: _____

Your Name:
MIKE H.

Your Password:
TOIL-ON

a. Analyze the way the log has been completed. Compare the data collected with the data the information center administrator intended to get from the form. List the omissions or errors users are making.

b. To what do you attribute the problems that people are having with filling out the log? Explain in a sentence or two the reason for each problem you found in problem 5a.

c. In a paragraph, explain why it is important to examine forms that have been completed as well as blank ones that are in use.

6. You've sampled the email messages that have been sent to several middle managers of Sawder's Furniture Company, which ships build-your-own particleboard furniture across the country. Here is one that repeats a message found in several other memos.

   To: Sid, Ernie, Carl
   From: Imogene
   Re: computer/printer supplies
   Date: November 10, 2000

   It has come to my attention that I have been waging a war against requests for computer and printer supplies (disks, toner, paper, etc.) that are all out of proportion to what has been negotiated for in the current budget. Because we're all good soldiers here, I hope you will take whatever our supply sergeant says is standard issue. Please, no "midnight requisitioning" to make up for shortages. Thanks for being GI in this regard; it makes the battle easier for us all.

   a. What metaphor(s) is (are) being used? List the predominant metaphor and other phrases that play on that theme.

b. If you found repeated evidence of this idea in other email messages, what interpretation would you have? Use a paragraph to explain.
c. In a paragraph, describe how the people in your systems analysis group can use the information from the email messages to shape their systems project for Sawder's.
d. In interviews with Sid, Ernie, and Carl, there has been no mention of problems with obtaining enough computer and printer supplies. In a paragraph, discuss why such problems may not come up in interviews and discuss the value of examining email messages and other memos in addition to interviewing.

7. "Here's the main policy manual we've put together over the years for system users," says Al Bookbinder, as he blows the dust off the manual and hands it to you. Al is a document keeper for the systems department of Prechter and Gumbel, a large manufacturer of health and beauty aids. "Everything any user of any part of the system needs to know is in what I call the Blue Book. I mean it's chock a block with policies. It's so big, I'm the only one with a complete copy. It costs too much to reproduce it." You thank Al and take the manual with you. When you read through it, you are astonished at what it contains. Most pages begin with a message such as: "This page supersedes page 23.1 in manual Vol. II. Discard previous inserts; do not use."
   a. List your observations about the frequency of use of the Blue Book.
   b. How user friendly are the updates in the manual? Write a sentence explaining your answer.
   c. Write a paragraph commenting on the wisdom of having all-important policies for all systems users in one book.
   d. Suggest a solution that incorporates the use of online policy manuals for some users.
8. Arch Ives, a newly hired and rather shy systems analyst with your systems group, has come to you with an idea that he assures you will save the team time, and hence the business you're consulting with will save money. Arch proposes that rather than doing time-consuming interviews or collecting current quantitative and qualitative documents, the team should rely primarily on what it discovers in an archival database. "They've kept everything. They even have their old reports from earlier systems projects. We don't have to talk with anyone; we can access the database all day, without seeing a soul."
   a. List the merits of what Arch is proposing.
   b. List the problems with what Arch is proposing.
   c. Can business process reengineering using workflow analysis be performed on archival data? Answer yes or no, and defend your answer in a paragraph.
   d. In a paragraph, explain tactfully to Arch why the systems team uses many different approaches to gathering data.

GROUP PROJECT

1. Assume your group will serve as a systems analysis and design team for a project designed to computerize or enhance the computerization of all business aspects of a 15-year-old, national U.S. trucking firm called Maverick Transport. Maverick is a less-than-a-truckload (LTL) carrier. The people in management work from the philosophy of just in time (JIT), in which they have created a partnership that includes the shipper, the receiver, and the carrier (Maverick Transport) for the purpose of transporting and delivering the materials required just in time for their use on the production line. Maverick maintains

626 tractors for hauling freight and has 45,000 square feet of warehouse space and 21,000 square feet of office space.

a. Along with your group members, develop a list of sources of archival data that should be checked when analyzing the information requirements of Maverick.
b. When this list is complete, devise a sampling scheme that would permit your group to get a clear picture of the company without having to read each document generated in its 15-year history.

SELECTED BIBLIOGRAPHY

Babbie, R. R. *Survey Research Methods.* Belmont, CA: Wadsworth, 1973.

Bormann, E. G. *Discussion and Group Methods: Theory and Practice,* 2d ed. New York: Harper and Row, 1975.

Emory, C. W. *Business Research Methods,* 3d ed. Homewood, IL: Irwin, 1985.

Holosofx, Inc. on the Web, <www.holosofx.com>, El Segundo, CA: 1997–2000.

Johnson, B. M. *Communication—The Process of Organizing.* Boston: Allyn and Bacon, 1977.

Kendall, J. E., and K. E. Kendall. "Metaphors and Methodologies: Living Beyond the Systems Machine." *MIS Quarterly,* Vol. 17, No. 2, June 1993, pp. 149–71.

———. "Metaphors and Their Meaning for Information Systems Development." *European Journal of Information Systems,* 1994, pp. 37–47.

Markus, M. L., and Lee, A. S. "Special Issue on Intensive Research in Information Systems: Using Qualitative, Interpretive, and Case Methods to Study Information Technology—Second Installment." *MIS Quarterly,* Vol. 24, No. 1, March 2000, p. 1.

Sano, D. *Designing Large-Scale Web Sites: A Visual Methodology.* New York: Wiley Computer Publishing, 1996.

Schultze, U. "A Confessional Account of an Ethnography about Knowledge Work." *MIS Quarterly,* Vol. 24, No. 1, March 2000, pp. 3–41.

Webb, E. J., D. T. Campbell, R. D. Schwartz, and L. Sechrest. *Unobtrusive Measures: Nonreactive Research in the Social Sciences.* Chicago: Rand McNally College Publishing, 1966.

Allen Schmidt, Julie E. Kendall, and Kenneth E. Kendall

MINDING THE MEMOS

"Anna, I've been busy gathering some memos to help us understand what's behind some of the changes people have requested to the system," says Chip as he hands over a sheaf of papers to Anna.

"Good idea," she replies, "but do you have any way to make them meaningful to us?"

Chip carefully rereads the memos shown in Figures E4.1 through E4.4.

"They tell an interesting story," Chip replies. "I can't wait for you to take a look."

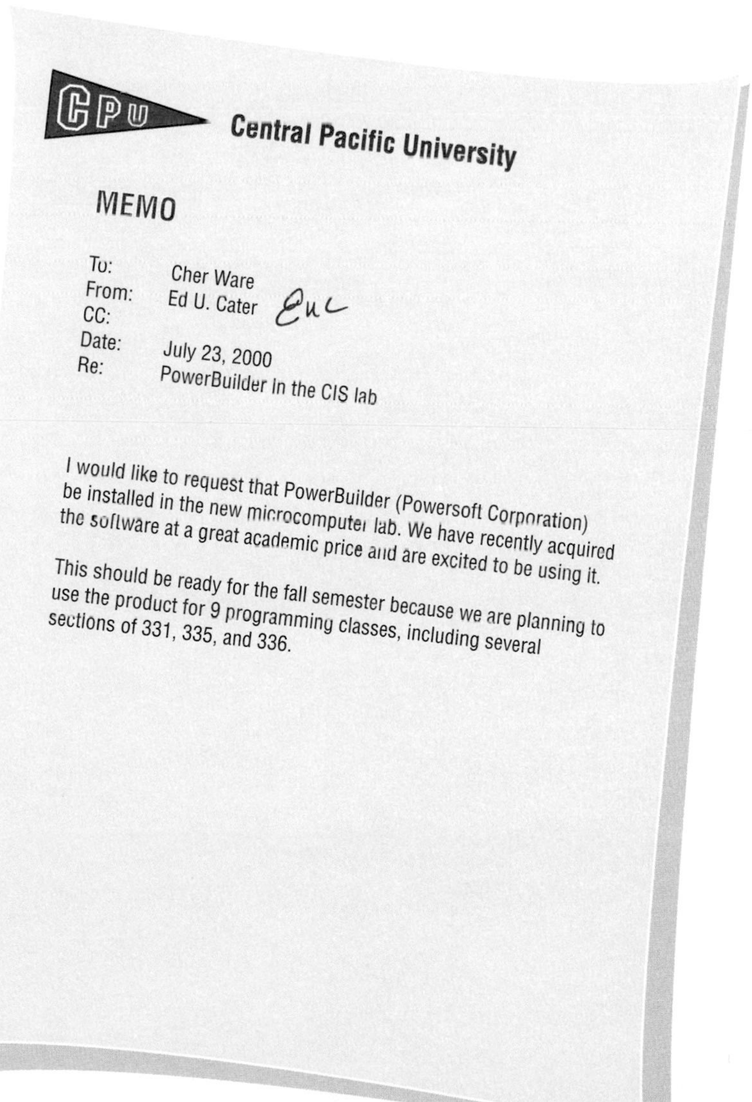

Central Pacific University

MEMO

To: Cher Ware
From: Ed U. Cater
CC:
Date: July 23, 2000
Re: PowerBuilder in the CIS lab

I would like to request that PowerBuilder (Powersoft Corporation) be installed in the new microcomputer lab. We have recently acquired the software at a great academic price and are excited to be using it.

This should be ready for the fall semester because we are planning to use the product for 9 programming classes, including several sections of 331, 335, and 336.

FIGURE E4.1
July 23d memo from Ed U. Cater.

4

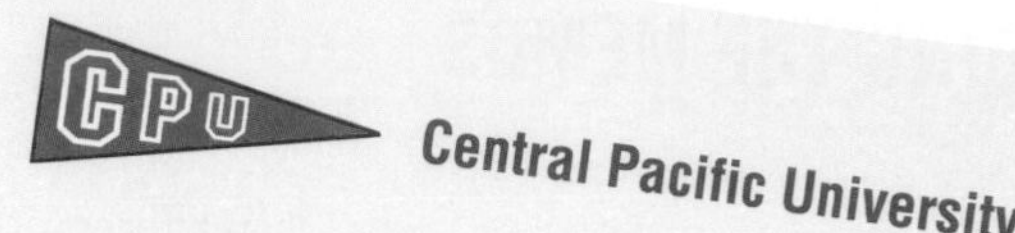

MEMO

To: Cher Ware
From: Ed U. Cater
CC: Dot Matricks
Date: July 30, 2000
Re: PowerBuilder install problems

Thanks for your recent memo. I went into the new lab to check out PowerBuilder and found out that it does not run correctly and that many of the features are not installed.

I tried to reinstall PowerBuilder on one of the machines and apparently there is not enough hard drive space to include the full product. In addition, the amount of main memory is not enough to work smoothly with the product. Is there a means of adding more memory and increasing the amount of hard drive space? Time is running short for us to set up projects and make student handouts.

Extreme urgency is needed in fixing the problem!!

FIGURE E4.2
July 30th memo from Ed U. Cater.

Central Pacific University

MEMO

To: Ed U. Cater
From: Cher Ware CW
CC: Dot Matricks, Mike Crowe
Date: August 1, 2000
Re: PowerBuilder install problems

I have checked with Mike on the problem with main memory and additional hard drive space. He has informed me that there are many other requests for hardware installation prior to the start of the semestor and doubts that the machines will be upgraded until the middle of October.

You may still use PowerBuilder with the minimal configuration, but the help system and other features will not be available.

We are evaluating budgetary concerns and looking for options to obtain the funds to upgrade the machines.

Another potential solution is to run some of the software from the installation CD. Let us know what you decide.

FIGURE E4.3
August 1st memo from Cher Ware.

EPISODE 4 CPU CASE

MEMO

To: Cher Ware
From: Ed U. Cater
CC: Dot Matricks, Mike Crowe
Date: August 19, 2000
Re: PowerBuilder install problems

Thanks for your recent memo on the PowerBuilder installation problems. I have thought over your suggestions and considered using CD-ROM drives; these are slow, however, being 20 speed. Further, we feel that the CD-ROMs may be mishandled, causing read problems, or may become lost.

The help system is essential for students who will be writing programs, and therefore it must be installed. Is there any other computer lab that has the available hard drive space and main memory? These labs should be on the main campus—we hope in the Computer Sciences building or one nearby.

Another solution is to delete all unnecessary software from the machines that are in the CIS lab. I realize that this will severely affect other instructors and courses, but perhaps their software is also installed in other labs.

I don't mean to sound panicked, but time is running short. The students need this to complete the course. Please respond quickly.

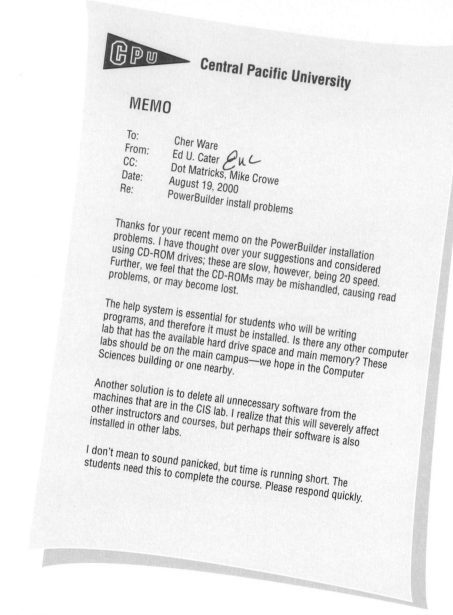

FIGURE E4.4
August 19th memo from Ed U. Cater.

EXERCISES

E-1. Analyze the four memos depicted in Figures E4.1 through E4.4. Discuss how you analyzed them and, in a paragraph, describe what you have found.

E-2. In a paragraph, answer the following questions: Is any information missing from the memos? What is it? Where would you get it?

E-3. In a paragraph, discuss who you would contact to better understand the memos. In addition, discuss how you would bring up the content of the memos so that they can be incorporated into the requirements determination phase.

INTERVIEWING

5

Before you interview someone else, you must in effect interview yourself. You need to know your biases and how they will affect your perceptions. Your education, intellect, upbringing, emotions, and ethical framework all serve as powerful filters for what you will be hearing in your interviews.

You need to think through the interview thoroughly before you go. Visualize why you are going, what you will ask, and what will make it a successful interview in your eyes. The other half is the individual you will interview. You must anticipate how to make the interview fulfilling for him or her as well.

KINDS OF INFORMATION SOUGHT

An information-gathering interview is a directed conversation with a specific purpose that uses a question-and-answer format. In the interview you want to get the opinions of the interviewee and his or her feelings about the current state of the system, organizational and personal goals, and informal procedures, as shown in Figure 5.1.

Above all, seek the opinions of the person you are interviewing. Opinions may be more important and more revealing than facts. For example, imagine asking the owner of a traditional store who has recently added an online store how many customer refunds she typically gives for Web transactions each week. She replies, "About 20 to 25 a week." When you monitor the transactions and discover that the average is only 10.5 per week, you might conclude that the owner is overstating the facts and the problem.

Imagine instead that you ask the owner what her major concerns are and that she replies, "In my opinion, customer returns of goods purchased over the Web are way too high. We must strive to get it right the first time." By seeking opinions rather than facts, you discover a key problem that the owner wants addressed.

In addition to opinions, you should try to capture the feelings of the interviewee. Remember that the interviewee knows the organization better than you do. You can understand the organization's culture more fully by listening to the feelings of the respondent. You can also determine the existing degree of optimism.

Expressed feelings help capture emotion and attitudes. If the owner of the store tells you, "I feel encouraged that you are working on this project," you can take it as a positive sign that the project will go well. This information is available only through asking about feelings.

FIGURE 5.1
Kinds of information sought in interviewing.

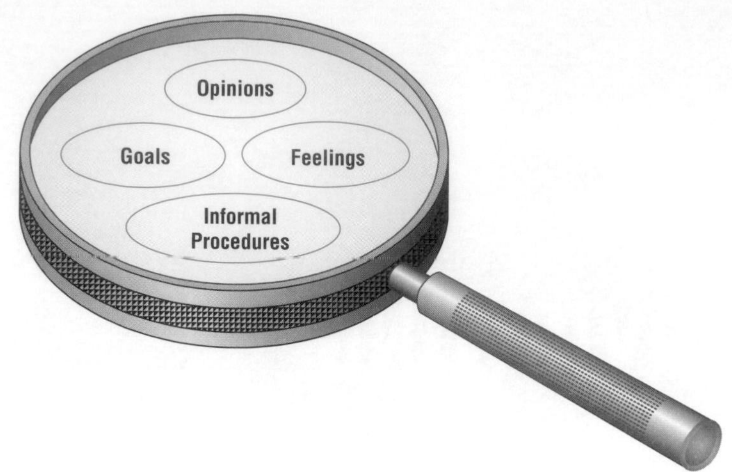

Goals are important information that can be gleaned from interviewing. Facts that you obtain from hard data may explain past performance, but goals project the organization's future. Try to find out as many of the organization's goals as possible from interviewing. You may not be able to determine goals through any other data-gathering methods.

In the interview you are setting up a relationship with someone who is probably a stranger to you. You need to build trust and understanding quickly, but at the same time you must maintain control of the interview. You also need to sell the system by providing needed information to your interviewee. Do so by planning for the interview before you go so that conducting it is second nature to you. Fortunately, effective interviewing can be learned. As you practice, you will see yourself improving. Later in the chapter we discuss Joint Application Design (JAD) (pronounced as one word jăd, rhymes with add), which can serve as an alternative to one-on-one interviewing in certain situations.

PLANNING THE INTERVIEW

FIVE STEPS IN INTERVIEW PREPARATION

The five major steps in interview preparation are shown in Figure 5.2. These steps include a range of activities from gathering basic background material to deciding who to interview.

Read Background Material. Read and understand as much background information about the interviewees and their organization as possible. This material can often be obtained by a quick call to your contact person to ask for a corporate Web site, a current annual report, a corporate newsletter, or any publications sent out to explain the organization to the public. Check the Internet for any corporate information such as that in Standard and Poor's.

As you read through this material, be particularly sensitive to the language the organizational members use in describing themselves and their organization. What you are trying to do is build up a common vocabulary that will eventually enable you to phrase interview questions in a way that is understandable to your interviewee. Another benefit of researching your organization is to maximize the time you spend in interviews; without such preparation you may waste time asking general background questions.

Steps in Planning the Interview	
1	Read Background Material
2	Establish Interviewing Objectives
3	Decide Whom to Interview
4	Prepare the Interviewee
5	Decide on Question Types and Structure

FIGURE 5.2
Steps the systems analyst follows in planning the interview.

Establish Interviewing Objectives. Use the background information you gathered as well as your own experience to establish interview objectives. There should be four to six key areas concerning information-processing and decision-making behavior about which you will want to ask questions. These areas include information sources, information formats, decision-making frequency, qualities of information, and decision-making style.

Decide Whom to Interview. When deciding whom to interview, include key people at all levels who will be affected by the system in some manner. As discussed in Chapter 4, it is important to sample organizational members. Strive for balance so that as many user's needs are addressed as possible. Your organizational contact will also have some ideas about whom should be interviewed.

Prepare the Interviewee. Prepare the person to be interviewed by calling ahead or sending an email message and allowing the interviewee time to think about the interview. If you are doing an in-depth interview, it is permissible to email your questions ahead of time to allow your interviewee time to think over their responses. Because there are many objectives to fulfill in the interview (including building trust and observing the workplace), however, interviews should typically be conducted in person and not via email. Arrange time for phone calls and meetings. Interviews should be kept to 45 minutes, an hour at the most. No matter how much your interviewees seem to want to extend the interview beyond this limit, remember that when they spend time with you, they are not doing their work. If interviews go over an hour, it is likely that the interviewees will resent the intrusion, whether or not they articulate their resentment.

Decide on Question Types and Structure. Write questions to cover the key areas of decision making that you discovered when you ascertained interview objectives. Proper questioning techniques are the heart of interviewing. Questions have some basic forms you need to know. The two basic question types are open-ended and closed. Each question type can accomplish something a little different from the other, and each has benefits and drawbacks. You need to think about the effect each question type will have.

It is possible to structure your interview in three different patterns: a pyramid structure, a funnel structure, or a diamond structure. Each is appropriate under different conditions and serves a different function, and each one is discussed later in this chapter.

The following discussion describes in detail some of the important decisions the interviewer must make, such as which questions to ask and how, whether to structure the interview, and how to document the interview.

QUESTION TYPES

Open-Ended Questions. Open-ended questions include those such as, "What do you think about putting all the managers on an intranet?" and "Please explain how you make a scheduling decision." Consider the term *open-ended*. "Open" actually describes the interviewee's options for responding. They are open. The response can be two words or two paragraphs. Some examples of open-ended questions are found in Figure 5.3.

The benefits of using open-ended questions are numerous and include the following:

1. Putting the interviewee at ease
2. Allowing the interviewer to pick up on the interviewee's vocabulary, which reflects his or her education, values, attitudes, and beliefs
3. Providing richness of detail
4. Revealing avenues of further questioning that may have gone untapped
5. Making it more interesting for the interviewee
6. Allowing more spontaneity
7. Making phrasing easier for the interviewer
8. Using them in a pinch if the interviewer is caught unprepared

As you can see, there are several advantages to using open-ended questions. There are, however, also many drawbacks:

1. Asking questions that may result in too much irrelevant detail
2. Possibly losing control of the interview
3. Allowing responses that may take too much time for the amount of useful information gained
4. Potentially seeming that the interviewer is unprepared
5. Possibly giving the impression that the interviewer is on a "fishing expedition" with no real objective for the interview

You must carefully consider the implications of using open-ended questions for interviewing.

Closed Questions. The alternative to open-ended questions is found in the other basic question type: closed questions. Such questions are of the basic form, "How many subordinates do you have?" The possible responses are closed to the interviewees, because they can only reply with a finite number such as "None," "One," or "Fifteen." Some examples of closed questions can be found in Figure 5.4.

A closed question limits the response available to the interviewee. You may be familiar with closed questions through multiple-choice exams in college. You are given a question and five responses, but you are not allowed to

FIGURE 5.3
Open-ended interview questions allow the respondent open options for responding. The examples were selected from different interviews and are not shown in any particular order.

Open-Ended Interview Questions

- What's your opinion of the current state of business-to-business ecommerce in your firm?
- What are the critical objectives of your department?
- Once the data are submitted via the Web site, how are they processed?
- Describe the monitoring process that is available online.
- What are some of the common data-entry errors made in this department?
- What are the biggest frustrations you've experienced during the transition to ecommerce?

Closed Interview Questions

- How many times a week is the project repository updated?
- On average, how many calls does the call center receive monthly?
- Which of the following sources of information is most valuable to you:
    - Completed customer complaint forms
    - Email complaints from consumers who visit the Web site
    - Face-to-face interaction with customers
    - Returned merchandise
- List your top two priorities for improving the technology infrastructure.
- Who receives this input?

FIGURE 5.4
Closed interview questions limit the options the respondent has for responding. The examples were selected from different interviews and are not shown in any particular order.

write down your own response and still be counted as having correctly answered the question.

A special kind of closed question is the bipolar question. This type of question limits the interviewee even further by only allowing a choice on either pole, such as yes or no, true or false, agree or disagree. Examples of bipolar questions can be found in Figure 5.5.

The benefits of using closed questions of either type include the following:

1. Saving time
2. Easily comparing interviews
3. Getting to the point
4. Keeping control over the interview
5. Covering lots of ground quickly
6. Getting to relevant data

The drawbacks of using closed questions are substantial, however. They include the following:

1. Being boring for the interviewee
2. Failing to obtain rich detail (because the interviewer supplies the frame of reference for the interviewee)
3. Missing main ideas for the preceding reason
4. Failing to build rapport between interviewer and interviewee

Thus, as the interviewer, you must think carefully about the question types you will use.

Both open-ended and closed questions have advantages and drawbacks, as shown in Figure 5.6. Notice that choosing one question type over the other actually involves a trade-off; although an open-ended question affords breadth and depth of reply, responses to open-ended questions are difficult to analyze.

Bipolar Interview Questions

- Do you use the Web to provide information to vendors?
- Do you agree or disagree that ecommerce on the Web lacks security?
- Do you want to receive a printout of your account status every month?
- Does your Web site maintain a FAQ page for employees with payroll questions?
- Is this form complete?

FIGURE 5.5
Bipolar interview questions are a special kind of closed question. The examples were selected from different interviews and are not shown in any particular order.

FIGURE 5.6

Attributes of open-ended and closed questions.

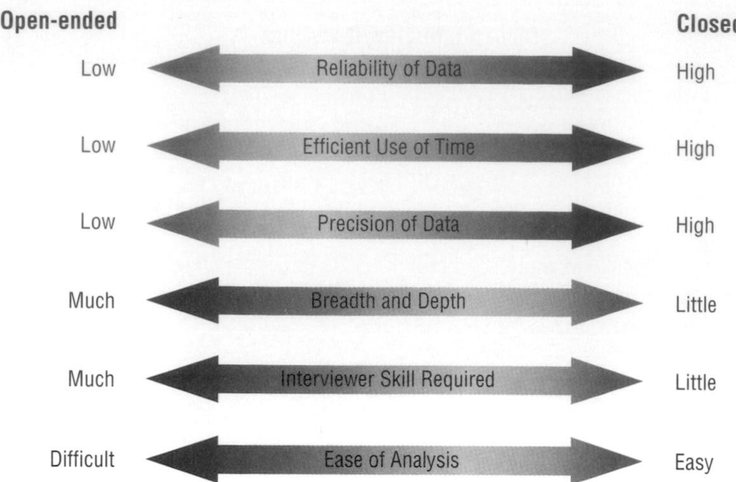

Probes. A third type of question is the probe or follow-up. The strongest probe is the simplest: the question, "Why?" Other probes are "Can you give me an example?" and "Will you elaborate on that for me?" Some examples of probing questions can be found in Figure 5.7. The purpose of the probe is to go beyond the initial answer to get more meaning, to clarify, and to draw out and expand on the interviewee's point. Probes may be either open-ended or closed questions.

It is essential to probe. Most beginning interviewers are reticent about probing and consequently accept superficial answers. They are usually grateful that employees have granted interviews and feel somewhat obligated to accept unqualified statements politely.

If done in a systematic and determined manner, your probing will be acknowledged as a sign that you are listening to what's being said, thinking it through, and responding appropriately. This response can only help the situation. Rather than using a tough "investigative-reporter" approach, you should probe in a way that exhibits your thoroughness and your desire to comprehend the interviewee's responses. Formulating a sound probe indicates that you are listening.

QUESTION PITFALLS

By wording your questions beforehand, you are able to correct any poor questions that you have written. Watch for troublesome question types that can ruin your data. Such questions are called "leading questions" and "double-barreled questions."

FIGURE 5.7

Probes allow the systems analyst to follow up questions to get more detailed responses. The examples were selected from different interviews and are not shown in any particular order.

Probes
• Why? • Give an example of how ecommerce has been integrated into your business processes. • Please give an illustration of the security problems you are experiencing with your online bill payment system. • You mentioned both an intranet and an extranet solution. Please give an example of how you think each differs. • What makes you feel that way? • Tell me step by step what happens after a customer hits the "Submit" button on the Web registration form.

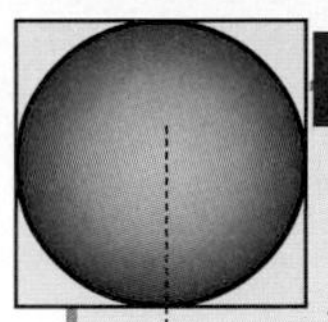

CONSULTING OPPORTUNITY 5.1

STRENGTHENING YOUR QUESTION TYPES

Strongbodies, a large, local chain of sports clubs, has experienced phenomenal growth in the past five years. Management would like to refine its decision-making process for purchasing new body-building equipment. Currently, managers listen to customers, attend trade shows, look at advertisements, and put in requests for new equipment purchases based on their subjective perceptions. These are then approved or denied by Harry Mussels.

Harry is the first person you will interview. He is a 37-year-old division manager who runs five area clubs. He travels all over the city to their widespread locations. He keeps an office at the East location, although he is there less than a quarter of the time.

In addition, when Harry is present at a club, he is busy answering business-related phone calls, solving on-the-spot problems presented by managers, and interacting with club members. His time is short, and to compensate for that he has become an extremely well-organized, efficient divisional manager. He cannot grant you a lot of interview time. Yet his input is important, and he feels he would be the main beneficiary of the proposed system.

What types of interview questions might be most suitable for your interview with Harry? Why are they most appropriate? How will your choice of question type affect the amount of time you spend in preparation for interviewing Harry? Write 5 to 10 questions of this type. What other techniques might you use to supplement information unavailable through that type of question? Write a paragraph to explain.

Avoiding Leading Questions. Leading questions tend to lead the interviewee into a response that you seem to want. The response is then biased because you are setting up a kind of trap. An example is, "You agree with other managers that inventory control should be computerized, don't you?" You have made it very uncomfortable to disagree. An alternative, preferred phrasing is, "What do you think of computerizing inventory control?" Your data will be more reliable and more valid, and hence easier to understand and more useful, with such preferred wording.

Avoiding Double-Barreled Questions. Double-barreled questions are those that use only one question mark for what are actually two separate questions. A question such as, "What decisions are made during a typical day and how do you make them?" is an example of a double-barreled question. If your interviewee responds to this type of question, your data may suffer.

A double-barreled question is a poor choice because interviewees may answer only one question (purposely or not), or you may mistake which question they are answering and draw the wrong conclusion. If you are lucky enough to discover your error, it still means orally retracing steps and straightening out the misunderstanding, a process that takes extra time. Problems can be avoided by phrasing questions carefully beforehand.

ARRANGING QUESTIONS IN A LOGICAL SEQUENCE

Just as there are two generally recognized ways of reasoning—inductive and deductive—there are two similar ways of organizing your interviews. A third way combines both inductive and deductive patterns.

Using a Pyramid Structure. Inductive organization of interview questions can be visualized as having a pyramid shape. Using this form, the interviewer begins with very detailed, often closed, questions. The interviewer then expands the topics by allowing open-ended questions and more generalized responses, as shown in Figure 5.8.

A pyramid structure should be used if you believe your interviewee needs to warm up to the topic. It is also useful if the interviewee seems reluctant to address

FIGURE 5.8

The pyramid structure for interviewing goes from specific to general questions.

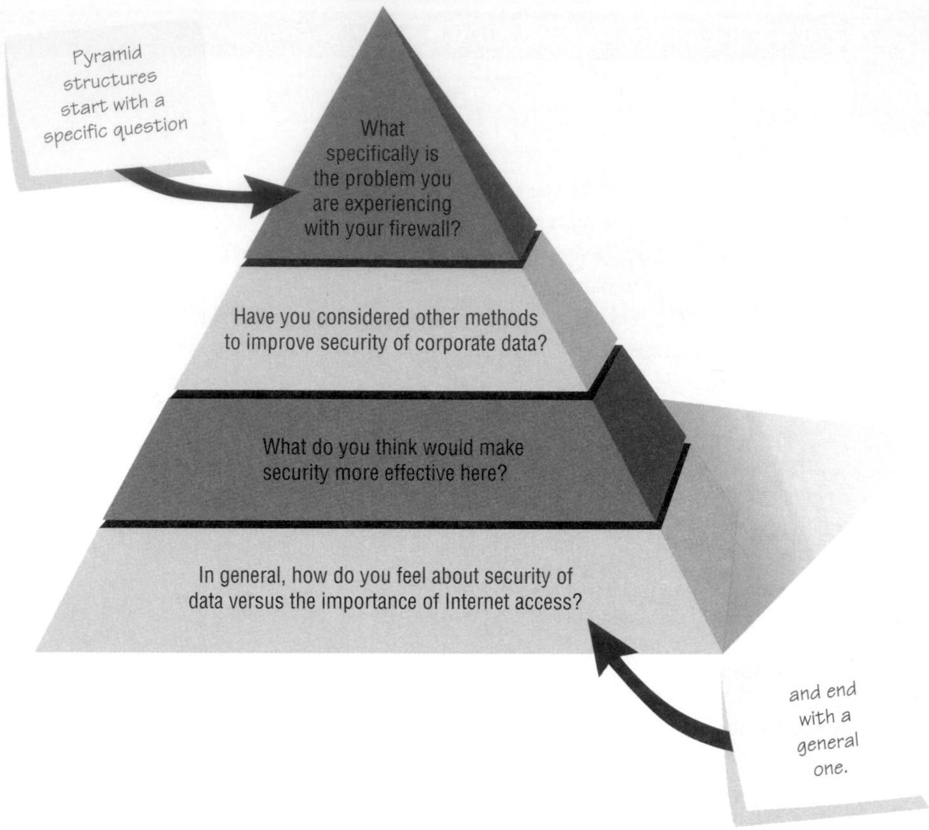

the topic. For example, if you are interviewing someone who has told you over the phone that he or she does not need to talk with you because that person already knows what is wrong with the forecasting model, you should probably structure the interview as a pyramid.

Using a pyramid structure for question sequencing is also useful when you want an ending determination about the topic. Such is the case in the final question, "In general, how do you feel about forecasting?"

Using a Funnel Structure. In the second kind of structure, the interviewer takes a deductive approach by beginning with generalized, open-ended questions and then narrowing the possible responses by using closed questions. This interview structure can be thought of as funnel-shaped, like that depicted in Figure 5.9.

Using the funnel structure method provides an easy, nonthreatening way to begin an interview. Respondents will not feel pressured that they are giving a "wrong" response to an open-ended question. A funnel-shaped question sequence is also useful when the interviewee feels emotional about the topic and needs freedom to express those emotions. A benefit of using a funnel structure is that organizing the interview in such a manner may elicit so much detailed information that long sequences of closed questions and probes are unnecessary.

Using a Diamond-Shaped Structure. Often a combination of the two above structures, one resulting in a diamond-shaped interview structure, is best. This structure entails beginning in a very specific way, then examining general issues, and finally coming to a very specific conclusion, as shown in Figure 5.10.

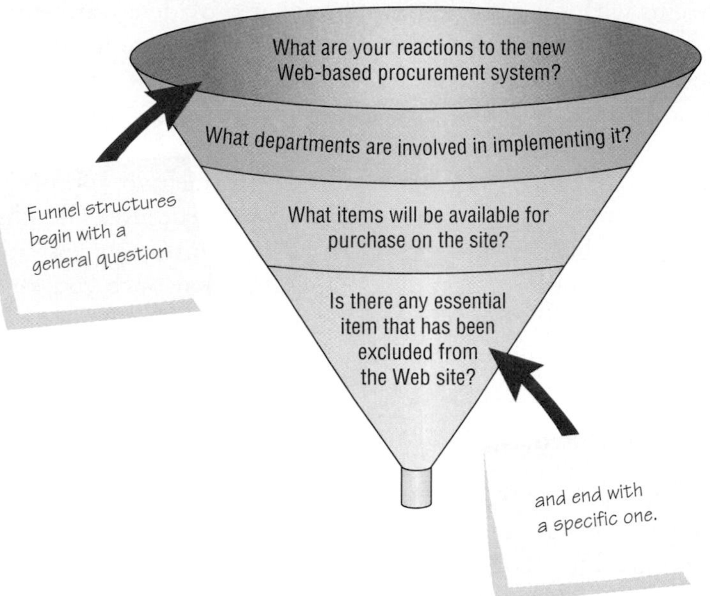
What are your reactions to the new
Web-based procurement system?
What departments are involved in implementing it?
What items will be available for
purchase on the site?
Is there any essential
item that has been
excluded from
the Web site?
Funnel structures
begin with a
general question
and end with
a specific one.

FIGURE 5.9

The funnel structure for interviewing begins with broad questions and then funnels to specific questions.

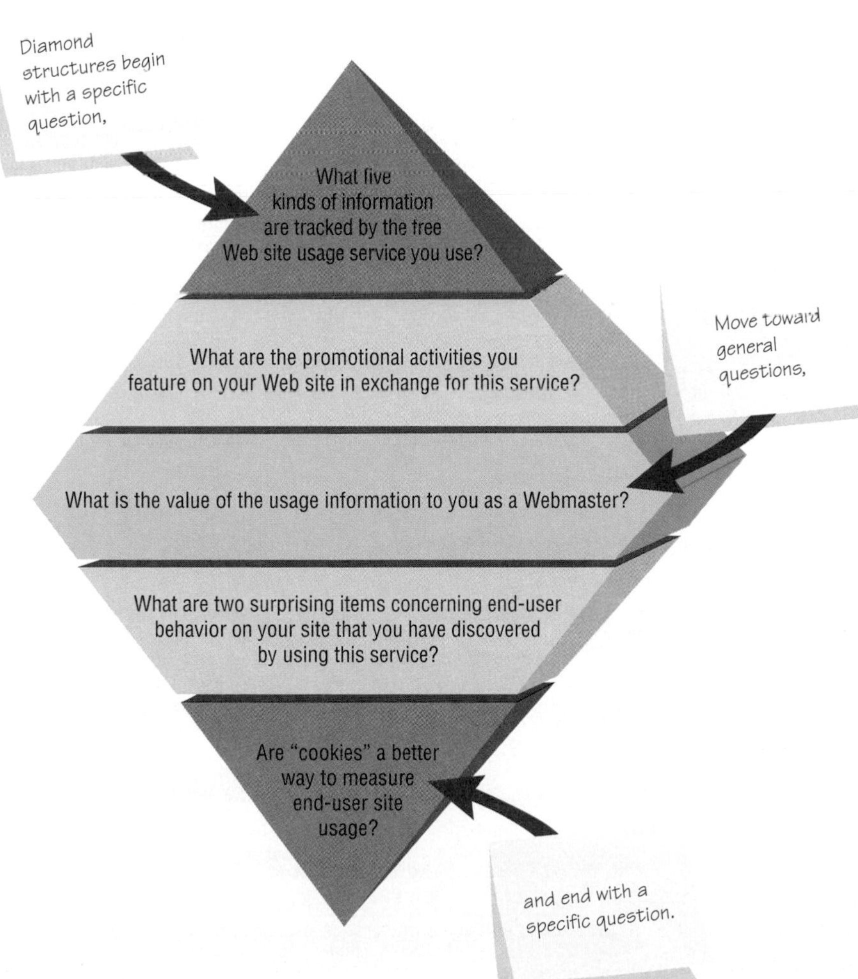
Diamond
structures begin
with a specific
question,
What five
kinds of information
are tracked by the free
Web site usage service you use?
What are the promotional activities you
feature on your Web site in exchange for this service?
Move toward
general
questions,
What is the value of the usage information to you as a Webmaster?
What are two surprising items concerning end-user
behavior on your site that you have discovered
by using this service?
Are "cookies" a better
way to measure
end-user site
usage?
and end with a
specific question.

FIGURE 5.10

The diamond-shaped structure for interviewing combines the pyramid and funnel structures.

The interviewer begins with easy, closed questions that provide a warm-up to the interview process. In the middle of the interview, the interviewee is asked for opinions on broad topics that obviously have no "right" answer. The interviewer then narrows the questions again to get specific questions answered, thus providing closure for both the interviewee and the interviewer.

The diamond structure combines the strengths of the other two approaches but has the disadvantage of taking longer than either other structure. The chief advantage of using a diamond-shaped structure is keeping your interviewee's interest and attention through a variety of questions. Remember that once you know how to ask the right questions at the right time, you have many options for sequencing them.

STRUCTURED VERSUS UNSTRUCTURED INTERVIEWS

Many beginning interviewers believe that because interviews are similar to conversations, they would be better off not structuring the questions or question sequences in their interviews. For a completely structured interview, everything is planned out and the plan is strictly adhered to. Closed questions are at the core of a completely structured interview.

There are explicit trade-offs involved for each of 10 variables, as shown in Figure 5.11. Notice that although it is difficult to evaluate an unstructured interview, it is easier to evaluate one that is structured; greater contact time is required to conduct an unstructured interview than a structured one; much training is needed to conduct a successful unstructured interview; limited training is required to conduct a completely structured interview; and so on.

FIGURE 5.11
Attributes of unstructured and structured interviews to consider when deciding on an interview format.

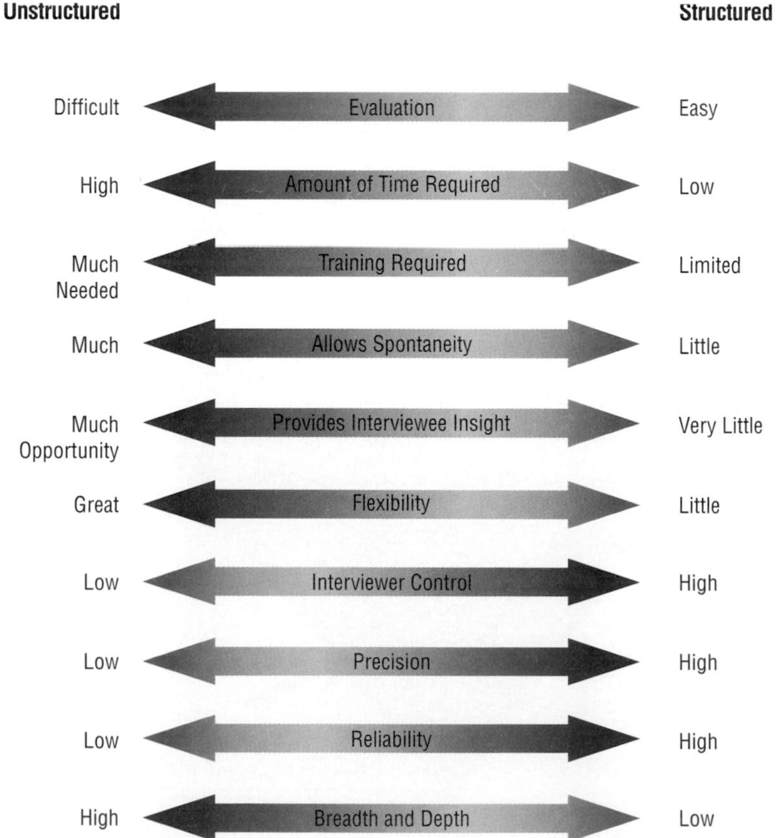

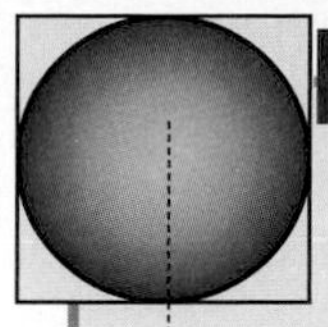

CONSULTING OPPORTUNITY 5.2

SKIMMING THE SURFACE

You are about to leave SureCheck Dairy after a preliminary tour when another member of your systems analysis team calls you at the dairy to say he cannot make his interview appointment with the plant manager because of illness. The plant manager is extremely busy, and you want to keep his enthusiasm for the project going by doing things as scheduled. You also realize that without the initial interview data, the rest of your data gathering will be slowed. Although you have no interview questions prepared, you make the decision to go ahead and interview the plant manager on the spot.

You have learned that SureCheck is interested in processing its own data on quantities and kinds of dairy products sold so that its people can use that information to better control production of the company's large product line (it includes whole, skim, 2 percent, and 1 percent milk, half-and-half, cottage cheese, yogurt, and frozen novelties). Sales managers are currently sending their sales figures to corporate headquarters, 600 miles away, and processing turnaround seems slow. You will base your ad-libbed questions on what you have just found out on the tour.

In the few minutes before your interview begins, decide on a structure for it: funnel, pyramid, or diamond. In a paragraph, justify why you would proceed with the interview structure you have chosen based on the unusual context of this interview. Write a series of questions and organize them in the structure you have chosen.

Awareness of the trade-offs between structured and unstructured interviews will enable you to make a better decision about which kind of interview is more appropriate for a particular situation. If you decide to follow the unstructured route, you should still prepare for the interview as discussed in the previous steps. For the unstructured approach we advocate, at minimum, a brief outline (including many questions worded as precisely as they will be asked). Remember that unstructured here refers only to the order in which the questions are asked and does not imply a lack of other preparation.

The only way to tell if your question is appropriate is to word it precisely beforehand and anticipate possible responses and how you will follow up on them. It is necessary for you to project what the other person might say. This step takes considerable time and thought and is in itself an excellent argument for preparing several potential questions that will allow you to take various avenues during the course of the interview. This approach is actually a branching program: If the answer is "yes," you proceed one way; if "no," you take the other path of questioning.

MAKING A RECORD OF THE INTERVIEW

Record the most important aspects of your interview. Either you can use a tape recorder or you can rely on taking notes with pen and paper, but it is important to make a permanent record during the actual interview.

Whether you take notes or make an audio recording depends in part on whom you are interviewing and what you will do with the information once the interview is over. In addition, there are advantages and disadvantages inherent in each recording methods.

Making an Audio Recording. Consider your interviewee when deciding how to record your interview. When you make an appointment, tell the interviewee that you would like to make an audio recording of the interview. Mention what you will do with the tape: Either it will be listened to by you and by the other team members and then destroyed, or it will be transcribed and used as information for system development. Be truthful about your intentions and reassuring about the confidentiality of any of the interviewee's remarks. If your interviewee refuses to allow you permission to make an audio recording, graciously accept that constraint.

Audio recording has advantages and disadvantages. The advantages are that such recording accomplishes the following:

1. It provides a completely accurate record of what each person said.
2. It frees the interviewer to listen to what is said and to respond more rapidly.
3. It allows better eye contact and hence better development of rapport between the interviewer and the interviewee.
4. It allows a replay of the interview for other team members.

The disadvantages of audio recording are also numerous. They include the following:

1. The interviewee may possibly be nervous and less apt to respond freely.
2. The interviewer may possibly be less apt to listen, because it's all being recorded.
3. It is difficult to locate important passages on a long tape.
4. The costs of data gathering are increased because of the need to transcribe tapes.

The decision to make an audio recording of interviews is a professional one that you will have to make based on what you know about interviewing, the interviewee's outlook on taping, and the particular project. Evaluate audio recording each time it is chosen, just as you would evaluate any other procedure.

Notetaking. Notetaking may be your only way to record the interview if your interviewee refuses your request to make an audio tape. It is important that you somehow record the interview as it takes place. Notetaking includes the following advantages:

1. It keeps the interviewer alert.
2. It aids recall of important questions.
3. It helps recall of important interview trends.
4. It shows interviewer interest in the interview.
5. It demonstrates the interviewer's preparedness.

There are some very good reasons for notetaking, but it is not without drawbacks. The disadvantages include the following:

1. Losing vital eye contact (and therefore rapport) between the interviewer and interviewee.
2. Losing the train of the conversation.
3. Making the interviewee hesitant to speak when notes are being made.
4. Causing excessive attention to facts and too little attention to feelings and opinions.

Even though the technology (and temptation) exists to take notes on your laptop or notebook computer during the interview, we do not recommend it. Our experience shows that although your accuracy of reporting may increase, your rapport with interviewees and your general concentration decrease. Focus on listening to the interviewee.

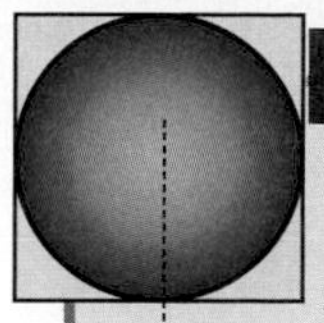

CONSULTING OPPORTUNITY 5.3

HITTING THE HIGH NOTE

Late in the afternoon, you see Gabriel Garcia, manager of Chumco's infant toy line, turning a corner in the corridor to return to his office. Gabriel, who is walking rather slowly, looks baffled. Your team member, Arthur Brown, just completed an interview with him about the possibility of having your systems analysis team design a decision support system to support Gabriel's process of deciding which new toys to produce. As project leader, you are curious about uncharacteristic behavior and want to learn his impression of the interview.

You easily catch up with Gabriel and ask if everything went all right in the interview. He replies, "Yes, it was okay, I guess. We covered a lot of territory. It was a little strange, though. I'm not even sure I'll recognize the guy who interviewed me if I see him again. All I saw was the top of his head, his pad, and his pencil. He wrote down every word I said! I felt like I was in court. What are you going to do with all that information? Publish quotations? I hope it doesn't get back to the wrong people."

What important notetaking guidelines did Arthur apparently forget to follow? Make a list. As project team leader, what assurances can you give Gabriel now? Respond in a paragraph.

To prepare for your interview properly, you must understand yourself, your own biases, and your values. Then research your interviewee and his or her organization, sketch out key areas for questioning, contact interviewees, write interview questions, and formulate an interview plan.

BEFORE THE INTERVIEW

The day preceding your interview, contact your interviewees via phone or email to confirm times and places of interviews. Coordinate appointments with any other team members and gather necessary materials.

When you conduct interviews, dress appropriately, perhaps wearing what you would wear to interview for a job in the same organization. Because you will be controlling the interview, you must dress in a credible manner. Failing to dress appropriately could result in poor data gathering. The interviewee's response to you are geared to their initial perceptions.

Arrive a little early for your interview. You can use the extra time to review your notes or start making observations about the organization. (See Chapter 7 for observation techniques.) Affirm with the interviewee that you are present and ready to begin the interview.

CONDUCTING THE ACTUAL INTERVIEW

BEGINNING THE INTERVIEW

As you enter, firmly shake your interviewee's hand, whether you are male or female. As in any other business situation, a handshake helps establish your credibility and trustworthiness. Remind the interviewee of your name and briefly outline once more why you are there and why you chose to interview him or her.

As you sit down, immediately take out your cassette recorder or note pad. Remind your interviewee that you will record important points. Tell the interviewee what you will do with the data you collect and be reassuring about its confidentiality.

Now is the time to check whether your audio recorder and microphone are working properly. Some beginning interviewers are too embarrassed to take the time to check equipment, but the result of assuming that you're taping when you're not is disastrous. Your interviewee will respect your professionalism in guaranteeing that everything is in working order before you begin.

Depending on the structure you are following in your interview, you may begin with some very general, nonthreatening, open-ended questions. Opening an interview this way helps to relax you and your interviewee. It also provides a frame of reference for you that is helpful in tailoring your later questions. By listening closely to early responses, you can pick up on vocabulary and jargon (maybe the organization doesn't have "departments," for example, but instead calls them "units") and the relevant metaphors such as, "The EDP department is a zoo," "Our scheduling system works like a well-oiled machine," or "We're one big happy family here." Recall from Chapters 2 and 4 that metaphors help reveal membership in organizational subcultures.

Early open-ended responses can also reveal the attitudes, values, and beliefs of the interviewee that will help you understand how he or she uses information and how he or she feels toward others in the organization. You must listen and respond appropriately to what your interviewee is saying.

As you proceed through your interview schedule, mention to your respondent what sort of detail you would like to receive in answers. For instance, if you feel you need depth on a question, encourage your interviewee to provide an example. If you have only passing interest in a topic, tell your respondent a "yes" or "no" is sufficient. You are in control of time usage in this kind of interview, and providing guidelines for length of response is helpful in maintaining interview balance.

All interview material should be covered in 45 minutes to an hour, and by now you are well aware of the planning and managing necessary to accomplish this goal. Closing the interview appropriately is just as important as opening it well.

During the interview, reflect back to some of your interviewee's responses through paraphrasing or summarizing to make sure you understand his or her meaning. If at any time you are unsure, you must ask for definitions or other clarification. Feigning knowledge only works against your ultimate objectives. The close of the interview is not the place to bring up all these concerns. Rather, they are more naturally handled as they arise.

The end of the interview is, however, a natural place to ask one key question: "Is there anything we haven't touched on that you feel is important for me to know?" Considered a formula question by the interviewee most of the time, the response will often be "No." You are interested in the other times, when this question opens the proverbial floodgates and much new (and often surprising) data are presented, though.

As you conclude the interview, there are other procedures to follow. Summarize and provide feedback on your overall impressions. Inform the interviewee about the subsequent steps to take and what you and other team members will do next. You may wish to ask the interviewee with whom you should talk next. Set up future appointment times for follow-up interviews, thank the interviewee for his or her time, and shake hands.

WRITING THE INTERVIEW REPORT

Although the interview itself is complete, your work on the interview data is just beginning. You need to capture the essence of the interview through a written report. It is imperative that you write the interview report as soon as possible after the interview. This step is another way you can ensure quality of interview data.

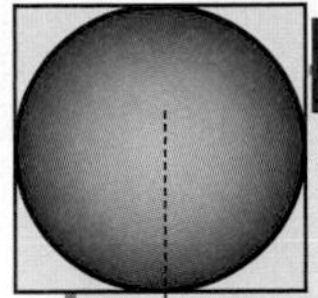

CONSULTING OPPORTUNITY 5.4

A SYSTEMS ANALYST, I PRESUME?

"Know what I think of the work the last systems analyst team did? The printouts created are a jungle. To figure out the cost of raw materials to us, I have to cut my way through the overgrowth of data, hacking my path with a pen. I cross out everything that's irrelevant. Sometimes I physically rip out the excess vegetation until I reach the numbers I need," says Henry Stanley, accounting supervisor for Zenith Glass Company. As you interview him, he points unhappily to an untidy stack of mutilated printouts sprouting beside his desk.

Identify the overriding metaphor Henry is using to describe the printouts he is receiving and the accessibility of information in them. In a paragraph, describe how this step helps you understand Henry's attitude toward any work proposed by your systems analysis team. In a paragraph adopt Henry's metaphor and extend it in a more positive sense during your interview with him.

The longer you wait to write up your interview, the more suspect the quality of your data becomes.

After this initial summary, go into more detail, noting main points of the interview and your own opinions. Figure 5.12 shows a sample interview report form that will aid you in capturing the interview essentials. It will help you in planning for your next interview as well.

Review the interview report with the respondent at a follow-up meeting. This step helps clarify the meaning the interviewee had in mind and lets the interviewee know that you are interested enough to take the time to understand his or her point of view and perceptions.

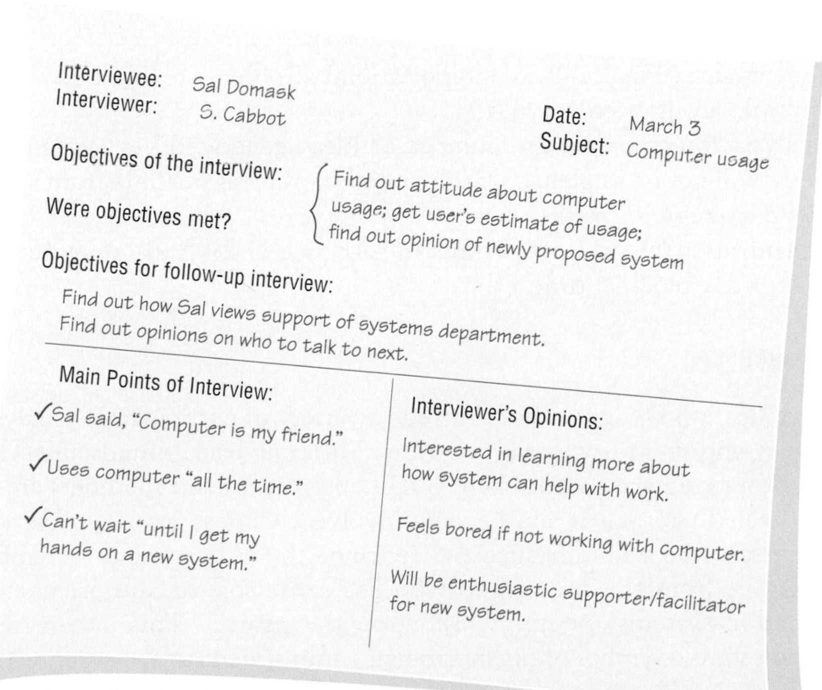

Interviewee: Sal Domask
Interviewer: S. Cabbot
Date: March 3
Subject: Computer usage

Objectives of the interview:
Were objectives met?
Find out attitude about computer usage; get user's estimate of usage; find out opinion of newly proposed system

Objectives for follow-up interview:
Find out how Sal views support of systems department.
Find out opinions on who to talk to next.

Main Points of Interview:	Interviewer's Opinions:
✓Sal said, "Computer is my friend."	Interested in learning more about how system can help with work.
✓Uses computer "all the time."	Feels bored if not working with computer.
✓Can't wait "until I get my hands on a new system."	Will be enthusiastic supporter/facilitator for new system.

FIGURE 5.12
An interviewer's follow-up report documenting data gathered as well as the interviewer's reaction to them.

JOINT APPLICATION DESIGN

No matter how adept you become as an interviewer, you will inevitably experience situations where one-on-one interviews do not seem to be as useful as you would like. Personal interviews are time-consuming and subject to error, and their data are prone to misinterpretation. An alternative approach to interviewing users one by one, called Joint Application Design (JAD), was developed by IBM. The motivation for using JAD is to cut the time (and hence the cost) required by personal interviews, to improve the quality of the results of information requirements assessment, and to create more user identification with new information systems as a result of the participative processes.

Although JAD can be substituted for personal interviews at any appropriate juncture during the systems development life cycle, it has usually been employed as a technique that allows you, as a systems analyst, to accomplish requirements analysis and to design the user interface jointly with users in a group setting. Sometimes group decision support systems (GDSS) can be used, but it is not mandatory. The many intricacies of this approach can only be learned in a paid seminar demonstrating proprietary methods. We can, however, convey enough information about JAD here to make you aware of some of its benefits and drawbacks in comparison with one-on-one interviews.

Joint Application Design requires some specialized skills on the part of the analyst and many skills and a firm commitment on the part of the organization and users who undertake to use this approach. In certain situations, however, JAD can be very effective and should be considered as an alternative to more traditional methods of systems analysis.

CONDITIONS THAT SUPPORT THE USE OF JAD

The following list of conditions will help you decide when the use of JAD may be fruitful. Consider using Joint Application Design when:

1. User groups are restless and want something new, not a standard solution to a typical problem.
2. The organizational culture supports joint problem-solving behaviors among multiple levels of employees.
3. Analysts forecast that the number of ideas generated via one-on-one interviews will not be as plentiful as the number of ideas possible from an extended group exercise.
4. Organizational workflow permits the absence of key personnel during a two-to-four-day block of time.

WHO IS INVOLVED

Joint Application Design sessions include a variety of participants—analysts, users, executives, and so on—who will contribute differing backgrounds and skills to the sessions. Your primary concern here is that all project team members are committed to the JAD approach and become involved. Choose an executive sponsor, a senior person who will introduce and conclude the JAD session. Preferably, select an executive from the user group who has some sort of authority over the IS (information systems) people working on the project. This person will be an important, visible symbol of organizational commitment to the systems project.

At least one IS analyst should be present, but the analyst usually takes a passive role, unlike traditional interviewing where the analyst controls the interaction. As the project analyst, you should be present during JAD to listen to what users say and what they require. In addition, you will want to give an expert opinion about any disproportionate costs of solutions proposed during the JAD session itself. Without this kind of immediate feedback, unrealistic solutions with excessive costs may creep into the proposal and prove costly to discourage later on.

From eight to a dozen users can be chosen from any rank to participate in JAD sessions. Try to select users above the clerical level who can articulate what information they need to perform their jobs as well as what they desire in a new or improved computer system. Some of the ideas on sampling from Chapter 4 can be put to good use here, because your goal is to get a representative sample of users without forming such a large group that it becomes unwieldy during group interactions.

The session leader should not be an expert in systems analysis and design but rather someone who has excellent communication skills to facilitate appropriate interactions. Consider having a full-time training department member serve as session leader. Note that you do not want to use a session leader who reports to another person in the group. To avoid this possibility, an organization may want to retain an outside management consultant to serve as session leader. The point is to get a person who can bring the group's attention to bear on important systems issues, satisfactorily negotiate and resolve conflicts, and help group members reach consensus rather than rely on simple majority rule to make decisions.

Your JAD session should also include one or two observers who are analysts or technical experts from other functional areas to offer technical explanations and advice to the group during the sessions. In addition, one scribe from the IS department should attend the JAD sessions to formally write down everything that is done. Ensure that the scribe publishes the record of JAD results rapidly once the group has met. Slow publication of results risks losing the time savings and momentum that are the prime motivations for using JAD in the first place. Consider selecting a second scribe from a user department as well. The responsibility for recording systems issues can then be left to the IS scribe, and the scribe from the user group can note business-related content.

PLANNING FOR THE JAD SESSION

One of the keys to a successful JAD workshop is laying the groundwork through advance study and planning. The session leader chosen will work with an executive sponsor to determine the scope of the project that JAD must cover. Sometimes the project requires more than one JAD workshop, but some of the benefits of JAD erode if too many independent sessions are planned. Following the definition of the project scope, participants will be selected, and the session leader will learn the application by doing some interviews with key users. The purpose of the interviews is to gather information to facilitate understanding of what is happening in the business.

WHERE TO HOLD JAD MEETINGS

If at all possible, we recommend holding the two-to-four-day sessions offsite, away from the organization, in comfortable surroundings. Some groups use executive centers or even group decision support facilities that are available at major universities. The idea is to minimize the daily distractions and responsibilities of the participants' regular work. The room itself should comfortably hold the 20 or so people invited. Minimal presentation support equipment includes two overhead projectors, a whiteboard, a flip chart, and easy access to a copier. Group decision support rooms will also provide networked PCs, a projection system, and software written to facilitate group interaction while minimizing unproductive group behaviors.

Give adequate thought to the creature comforts of participants as well; a JAD session will be an intense experience, quite different in nature from a typical day's work for most people. Plan for sufficient food as well as ample refreshments during scheduled breaks before lunch and again in late afternoon.

Schedule your JAD session when all participants can commit to attending. Do not hold the sessions unless everyone who has been invited can actually attend. This rule is critical to the success of the sessions. Ensure that all participants receive an agenda before the meeting, and consider holding an orientation meeting for a half-day one week or so before the workshop so that those involved know what is expected of them. Such a premeeting allows you to move rapidly and act confidently once the actual meeting is convened.

ACCOMPLISHING A STRUCTURED ANALYSIS OF PROJECT ACTIVITIES

IBM recommends that the JAD sessions examine these points in the proposed systems project: planning, receiving, receipt processing/tracking, monitoring and assigning, processing, recording, sending, and evaluating. For each topic, the questions who, what, how, where, and why should also be asked and answered. Clearly, ad hoc interactive systems such as decision support systems and other types of systems dependent on decision-maker style (including prototype systems) are not as easily analyzed with the structured approach of JAD.

As the analyst involved with the JAD sessions, you should receive the notes of the scribes and prepare a specifications document based on what happened at the meeting. Systematically present the management objectives as well as the scope and boundaries of the project. Specifics of the system, including details on screen and report layouts, should also be included. For a guide for what to include, see Chapter 13, which details the composition of a systems proposal. You can also look to the client organization for other guidelines to be followed in preparing this document.

POTENTIAL BENEFITS OF USING JAD IN PLACE OF TRADITIONAL INTERVIEWING

There are four major potential benefits that you, the users, and your systems analysis team should consider when you weigh the possibilities of using Joint Application Design. The first potential benefit is time savings over traditional one-on-one interviews. Some organizations have estimated that JAD sessions have provided a 15 percent time savings over the traditional approach.

Hand-in-hand with time savings is the rapid development possible via JAD. Because user interviews are not accomplished serially over a period of weeks or months, the development can proceed much more quickly.

A third benefit to weigh is the possibility of improved ownership of the information system. As analysts, we are always striving to involve users in meaningful ways and to encourage users to take early ownership of the systems we are designing. Due to its interactive nature and high visibility, JAD helps users become involved early in systems projects and treats their feedback seriously. Working through a JAD session eventually helps reflect user ideas in the final design.

A final benefit of participating in JAD sessions is the creative development of designs. The interactive character of JAD has a great deal in common with brainstorming techniques that generate new ideas and new combinations of ideas because of the dynamic and stimulating environment. Designs can evolve through facilitated interactions, rather than in relative isolation.

POTENTIAL DRAWBACKS OF USING JAD

There are three drawbacks or pitfalls that you should also weigh when making a decision on whether to do traditional one-on-one interviews or to use Joint Application Design. The first drawback is that JAD requires the commitment of a large block of time from all 18 to 20 participants. Because JAD requires a two-to-four-day commitment, it is not possible to do any other activities concurrently or to time-shift any activities, as is typically done in one-on-one interviewing.

A second pitfall occurs if preparation for the JAD sessions is inadequate in any regard or if the follow-up report and documentation of specifications is incomplete. In these instances resulting designs could be less than satisfactory. Many variables need to come together correctly for JAD to be successful. Conversely, many things can go wrong. The success of designs resulting from JAD sessions is less predictable than that achieved through standard interviews.

Finally, the necessary organizational skills and organizational culture may not be sufficiently developed to enable the concerted effort required to be productive in a JAD setting. In the end you will have to judge whether the organization is truly committed to, and prepared for, this approach.

SUMMARY

This chapter has covered the process of interviewing, which is one method systems analysts use for collecting data on information requirements. Systems analysts listen for goals, feelings, opinions, and informal procedures in interviews with organizational decision makers. They also sell the system during interviews. Interviews are preplanned question-and-answer dialogues between two people. The analyst uses the interview to develop their relationship with a client, to observe the workplace, and to collect data regarding information requirements. Although email can be used to prepare the interviewee by posing questions prior to meeting, interviews should typically be conducted in person and not electronically.

There are five steps to be taken in preplanning the interview:

1. Read background material.
2. Establish interviewing objectives.
3. Decide whom to interview.
4. Prepare the interviewee.
5. Decide on question types and structure.

Questions are of two basic types: open-ended or closed. Open-ended questions leave open all response options for the interviewee. Closed questions limit the possible options for response. Probes or follow-up questions can be either open-ended or closed, but they ask the respondent for a more detailed reply.

Interviews can be structured in three basic structures: pyramid, funnel, or diamond. Pyramid structures begin with detailed, closed questions and broaden to more generalized questions. Funnel structures begin with open-ended, general questions and then funnel down to more specific, closed questions. Diamond-shaped structures combine the strengths of the other two structures, but they take longer to conduct. There are trade-offs involved when deciding how structured to make interview questions and question sequences.

Interviews should be documented with audio recordings or notes. After the interview the interviewer should write a report listing the main points provided as well as opinions about what was said. It is extremely important to document the interview soon after it has taken place.

"Well, I did warn you that things weren't always smooth here at MRE. By now you've met many of our key employees and are starting to understand the 'lay of the land.' Who would have thought that some innocent decisions about hardware, like whether to buy a COMTEX or Shiroma, would cause such hostility? Well, live and learn, I always say. At least now you'll know what you're up against when you have to start recommending hardware!

"It's funny that not all questions are created equal. I myself favor asking open-ended questions, but when I have to answer them, it is not always easy. Have you been taking the opportunity to view people's offices when you've been in there to do your interviews? You can learn a lot more by using a structured observation method such as STROBE."

HYPERCASE QUESTIONS

1. Using the interview questions posed in HyperCase, give five examples of open-ended questions and five examples of closed questions. Explain why your examples are correctly classified as either open-ended or closed question types.
2. List three probing questions that are part of the HyperCase interviews. In particular, what did you learn by following up on the questions you asked Snowden Evans?

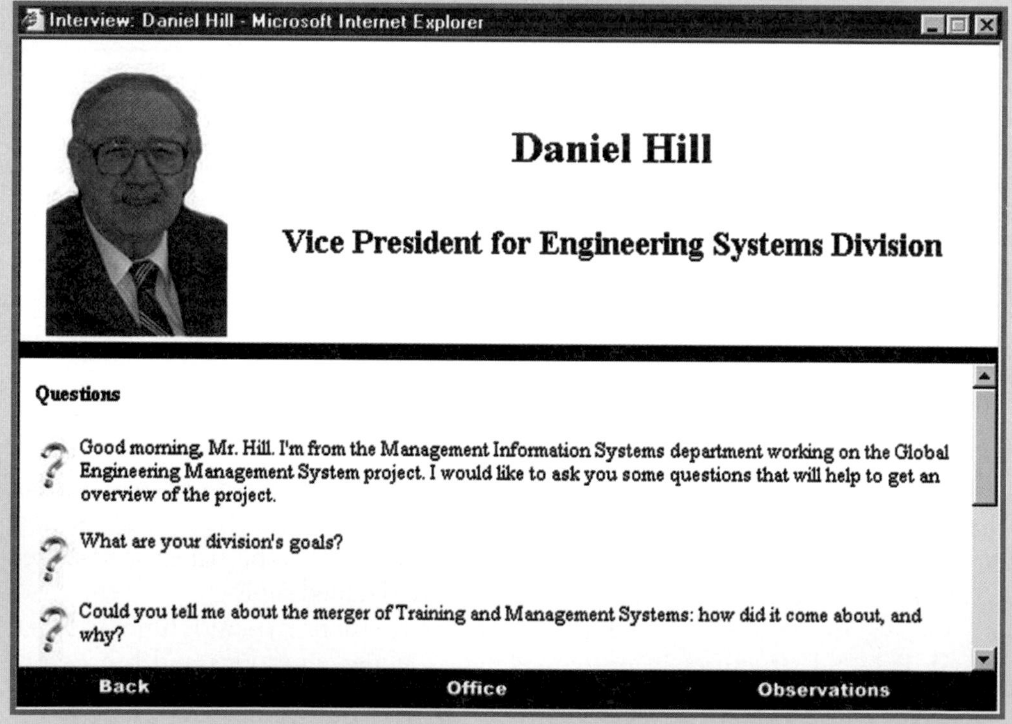

FIGURE 5.HC1

Pointing to a question in HyperCase will reveal an answer.

To cut both the time and cost of personal interviews, analysts may want to consider Joint Application Design (JAD) as an alternative. Using JAD, analysts can both analyze requirements and design a user interface with users in a group setting. Careful assessment of the particular organizational setting will help the analyst judge whether JAD is a suitable alternative.

KEYWORDS AND PHRASES

- bipolar closed questions
- closed questions
- diamond-shaped structure
- double-barreled questions
- funnel structure
- informal procedures
- interviewee feelings
- interviewee goals
- interviewee opinions
- Joint Application Design (JAD)
- leading questions
- open-ended questions
- probes
- pyramid structure
- structured interviews
- unstructured interviews

REVIEW QUESTIONS

1. What kinds of information should be sought in interviews?
2. List the five steps in interview preparation.
3. Define what is meant by open-ended interview questions. Give eight benefits and five drawbacks of using them.
4. When are open-ended questions appropriate for use in interviewing?
5. Define what is meant by closed interview questions. Give six benefits and four drawbacks of using them.
6. When are closed questions appropriate for use in interviewing?
7. What is a probing question? What is the purpose of using a probing question in interviews?
8. What are leading questions? Why should they be avoided in interviews?
9. What are double-barreled questions? Why should they be avoided in interviews?
10. Define what is meant by pyramid structure. When is it useful to employ it in interviews?
11. Define what is meant by funnel structure. When is it useful to employ it in interviews?
12. Define what is meant by diamond-shaped structure. When is it useful to employ it in interviews?
13. What are the 10 variables that become trade-offs between structured and unstructured interviews?
14. What are four advantages and four disadvantages of making audio recordings of interviews?
15. What are five advantages and four disadvantages of notetaking during interviews?
16. Why are you discouraged from using a laptop computer to record an interview which you are conducting?
17. Define Joint Application Design.
18. List the situations that warrant use of JAD in place of personal organizational interviews.
19. List the potential benefits of using Joint Application Design.
20. List the three potential drawbacks of using JAD as an alternative to personal interviews.

PROBLEMS

1. While going over your interview schedule, you notice several questions that seem inadequate. Here are the original questions for the sales manager of Sampson Paper Products, whose company has expressed a desire to put some of its sales information on the Web and to allow managers to comment on it interactively to refine their sales projections. Rewrite the questions in a more appropriate manner.
   1. Your subordinates told me that you have a high level of computer anxiety. Is that true?
   2. I'm new to this. What did I leave out?
   3. What are your most used sources of information on sales figures, and how frequently do you use them?
   4. Do you agree with other sales managers that putting some monthly sales on the Web and then doing trend analysis would be a major improvement?
   5. Isn't there a better way to project sales than the antiquated one you're using now?
2. As part of your systems analysis project to update the automated accounting functions for Chronos Corporation, a maker of digital watches, you will be interviewing Harry Straiter, the chief accountant. Write four to six interview objectives covering his use of information sources, information formats, decision-making frequency, desired qualities of information, and decision-making style.
   a. In a paragraph, write down how you will approach Harry to set up an interview.
   b. State which structure you will choose for this interview. Why?
   c. Harry has three subordinates who also use the system. Would you interview them also? Why or why not?
   d. Write three open-ended questions that you will email to Harry prior to your interview. Write a sentence explaining why it is preferred to conduct an interview in person rather than via email.
3. Here are five questions written by one of your systems analysis team members. Her interviewee is the local manager of LOWCO, an outlet of a national discount chain, who has asked you to work on a management information system to provide inventory information. Review these questions for your team member.
   1. When was the last time you thought seriously about your decision-making process?
   2. Who are the trouble makers in your store, I mean the ones who will show the most resistance to changes in the system that I have proposed?
   3. Are there any decisions you need more information about to make them?
   4. You don't have any major problems with the current inventory control system, do you?
   5. Tell me a little about the output you'd like to see.
      a. Rewrite each question to be more effective in eliciting information.
      b. Order your questions in either a pyramid, funnel, or diamond-shaped structure and label the structure used.
      c. What guidelines can you give your team member for improving her interviewing questions for the future? Make a list of them.
4. Ever since you entered the door, your interviewee, Max Hugo, has been shuffling papers, looking at his watch, and lighting and snuffing out cigarettes. Based on what you know about interviews, you guess that Max is nervous because of the other work he needs to do. In a paragraph, describe how you

would deal with this situation so that the interview can be accomplished with Max's full attention. (Max cannot reschedule the interview for a different day.)

5. Write a series of six closed questions that cover the subject of decision-making style for the accountant described in problem 2.
6. Write a series of open-ended questions that cover the subject of decision-making style for the accountant described in problem 2.
7. Examine the interview structure presented in the sequencing of the following questions:
   1. How long have you been in this position?
   2. What are your key responsibilities?
   3. What reports do you receive?
   4. How do you view the goals of your department?
   5. How would you describe your decision-making process?
   6. How can that process best be supported?
   7. How frequently do you make those decisions?
   8. Who is consulted when you make a decision?
   9. What is the one decision you make that is essential to departmental functioning?
      a. What structure is being used? How can you tell?
      b. Restructure the interview by changing the sequence of the questions (you may omit some if necessary). Label the structure you have used.
8. The following is the first interview report filed by one of your systems analysis team members: "In my opinion, the interview went very well. The subject allowed me to talk with him for an hour and a half. He told me the whole history of the business, which was very interesting. The subject also mentioned that things have not changed all that much since he has been with the firm, which is about 16 years. We are meeting again soon to finish the interview, because we did not have time to go into the questions I prepared."
   a. In two paragraphs, critique the interview report. Assume you asked the team member to use the report form provided in Figure 5.12. What critical information is missing?
   b. What information is extraneous to the interview report?
   c. If what is reported actually occurred, what three suggestions do you have to help your teammate conduct a better interview next time?

GROUP PROJECTS

1. With your group members, role play a series of interviews with various end users at Maverick Transport (first introduced in the Chapter 4 Group Project). Each member of your group should choose one of the following roles: company president, information technology director, dispatcher, customer service agent, or truck driver. Those group members playing roles of Maverick Transport employees should attempt to briefly describe their job responsibilities, goals, and informational needs.

   Remaining group members should play the roles of systems analysts and devise interview questions for each employee. If there are enough people in your group, each analyst may be assigned to interview a different employee. Those playing the roles of systems analysts should work together to develop common questions that they will ask as well as questions tailored to each individual employee. Be sure to include open-ended, closed, and probing questions in your interviews.

Maverick Transport is attempting to change from outdated and unreliable technology to more state-of-the-art, dependable technology. The company is seeking to move from dumb terminals attached to a mainframe because it wants to use PCs in some way and is also interested in investigating a satellite system for tracking freight and drivers. In addition, the company is interested in pursuing ways to cut down on the immense storage requirements and difficult access of the troublesome handwritten, multipart forms that accompany each shipment.

2. Conduct all five interviews in a role-playing exercise. If there are more than 10 people in your group, permit two or more analysts to ask questions.
3. Debrief from the interviews, using the form provided in this chapter.
4. Formulate follow-up questions for second interviews based on what you found out from the debriefing reports. Your group should produce a written list of follow-up questions for each employee interviewed.
5. With your group write a plan for a JAD session that takes the place of personal interviews. Include relevant participants, suggested setting, and so on.

SELECTED BIBLIOGRAPHY

Ackroyd, S., and J. A. Hughes. *Data Collecting in Context*, 2d ed. New York: Longman, 1992.

Cash, C. J., and W. B. Stewart Jr. *Interviewing Principles and Practices*, 4th ed. Dubuque, IA: Wm. C. Brown, 1986.

Cooper, D. R., and P. S. Schindler. *Business Research Methods*, 6th ed. New York: Irwin/McGraw Hill, 1998.

Deetz, S. *Transforming Communication, Transforming Business: Building Responsive and Responsible Workplaces.* Cresskill, NJ: Hampton Press, 1995.

Di Salvo, V. *Business and Professional Communication.* Columbus, OH: Merrill, 1977.

Emerick, D. , K. Round, and S. Joyce. *Exploring Web Marketing and Project Management*, Upper Saddle River, NJ: Prentice Hall PTR, 2000.

Gane, C. *Rapid Systems Development.* New York: Rapid Systems Development, 1987.

Joint Application Design. *GUIDE Publication GPP-147.* Chicago: GUIDE International, 1986.

Strauss, J., and R. Frost. *E-Marketing*, 2d ed. Upper Saddle River, NJ: Prentice Hall, 2001.

ALLEN SCHMIDT, JULIE E. KENDALL, AND KENNETH E. KENDALL

TELL ME MORE, I'LL LISTEN

"I've scheduled preliminary interviews with five key people. Because you've been so busy with Visible Analyst, I decided to do the first round of interviews myself," Anna tells Chip as they begin their morning meeting.

"That's fine with me," Chip says. "Just let me know when I can fill in. Who will you be talking to first? Dot?"

"No secret there, I guess," replies Anna. "She's critical to the success of the system. Her word is it when it comes to whether a project will fly or not."

"Who else?" asks Chip.

"I'll see who Dot refers me to, but I set up appointments with Mike Crowe, the maintenance expert; Cher Ware, the software specialist; and Paige Prynter, CPU's financial analyst."

"Don't forget Hy Perteks," says Chip.

"Right. The Information Center will be important to our project," says Anna. "Let me call and see when he's available."

After a brief phone conversation with Hy, Anna turns once again to Chip.

"He'll meet with me later today," Anna confirms.

"I can't wait to hear what they're thinking about the new system," says Chip. "Good luck."

INTERVIEWS WITH CPU STAFF MEMBERS

Interview One
Respondent: Dorothy (Dot) Matricks, manager, Personal Computers
Interviewer: Anna Liszt
Location: Dorothy's office

Anna: (Extending her hand as she enters Dorothy's office) Hello, Dorothy. It's good to see you once again. I think we last saw each other when they had the reception for the new president.

Dot: (Rising from her desk, shakes hands with Anna) Please, call me Dot. And I remember that reception, too. It was fun. Please, have a seat (she indicates a chair beside her desk) while I call Pat to put a hold on my phone calls. I didn't know at the time we'd be working together. But (she continues with a laugh), it seems sooner or later computer people find each other. I've heard through the grapevine that your group is contemplating helping us out of our quagmire here.

Anna: I'm not sure it's a "quagmire," but the administration has requested that the systems and programming group help you to manage your PCs with a system of your own.

Dot: (Sits back in her chair with a chuckle) I couldn't be more delighted. That means that my efforts to get some help—or should I say my unabashed pleading-has not fallen on deaf ears. Tell me more.

Anna: I thought I'd keep this first interview short, about a half an hour to forty-five minutes (glancing at her watch). My overall objective is to find out about usage of PCs on campus currently, from your perspective. Later we can get into the system you use to manage the PCs and its strengths and weaknesses.

Dot: It's easy enough to give an overview, because it's something I often communicate to people. Let me begin with a little history so you can understand where I'm coming from. We started getting involved with PCs in the early eighties. We thought we were very high-tech to be buying them as fast as they were produced.

Anna: Yes. Not many schools or even businesses had a plan for implementing personal computer systems.

Dot: Don't be misled by the haze of history. We purchased some PCs early, mostly for the accounting area. But we didn't have a plan, except that we reacted to demand. The beginning was slow, but once hardware and software became available, we grew explosively. Once a marketing professor saw what accounting was doing, he or she would say, "What's new for me? We don't want to deal with the mainframe unless we have to." And so we grew and grew. We now have about 1,620 PCs with three or four pieces of equipment attached to each one. Do you believe there are more than 6,000 tagged items in our inventory? By the end of next fall, 200 more PCs will be added. You have probably noticed that we have a mix of brands. IBM is used primarily in the business courses, and Macs are used in the art department. Science areas like to use both IBMs and Macs, it seems. Many of our computers are IBM compatibles. By the way, we sometimes use the term *microcomputer* or *personal computer* to represent any desktop or notebook computer. Both are kept in inventory. We use the term *hardware* to include not only computers but scanners, digital cameras, printers, and other digital equipment.

Anna: (Nodding as she absorbs Dot's response) That's a lot for anyone to manage. But you seem to be up-to-date on what you have in inventory. What system are you using now to keep track of it all?

Dot: I've been here since the beginning, and I honestly believe we've done better than most, but still the database system we use is inadequate. Again, my analysis is that just like we outgrew old machines, we have outgrown our management system. But we are reluctant to fiddle with it too much.

Anna: Why?

Dot: Well, I guess it's because something is better than nothing, even if something is antiquated. We try to be resourceful. It all started in the late eighties. I, and the few other people who were assigned to the PCs (on a part-time basis, back then), began to realize that something had to be done to prevent chaos. People were asking us for things, and we couldn't lay our hands on them. At the same time, computers were breaking down, and we did not know who had the service contract on them, if they had ever been serviced, and so on. We sensed disaster in the making. From our old mainframe experience, we started trying to organize what we had in a logical way. We all agreed to get a system running with a little off-the-shelf database package (which was not user-friendly, may I be the first to point out). We had very much of a family atmosphere at the time.

Anna: What functions did the package perform?

Dot: It was very basic information, which is what we desperately needed at the time. It sounds simple now, but we took some time and we were thrilled to have inventory information, including the type of equipment, the manufacturer, initial cost, room number where equipment was located, serial number, and equipment purchase date. Even that took extensive updating.

Anna: Is it the same system you run today?

Dot: Yes and no. Same software, but we've been through three updates.

Anna: What other improvements were made?

Dot: After using the system for a couple of years, we knew what we wanted. We added fields to capture the memory size on each PC, and graphics boards were installed in the machines.

Anna: And is this the same system you are using now?

Dot: Yes. Yes it is. We can print a number of reports and some summary information. But don't get the wrong impression. The system as it stands is clearly inadequate. You won't get much argument on that point from anyone you interview. But we did what we could with what we had at the time. In fact, there are still a few members of the original group here. I will admit that it's been interesting to watch the PCs grow. It makes me feel like I had a hand in helping develop a very important area.

Anna: What do you see as the strengths of the current system?

Dot: I was in it from the start, so I find it fairly easy to use. And it's flexible enough to produce a variety of reports. It's brought us quite far and it does provide the elementary information required to manage the PCs.

Anna: Earlier you alluded to some limitations of the system. What are the specific weaknesses you were speaking about?

Dot: (Reflecting before she speaks) In a way they are not weaknesses of the system per se, but they are changes in the types of PCs we are seeing that the system was not far-sighted enough to accommodate. For instance, the number of internal boards and disk drives has increased markedly. A few machines have modems, some have different graphics boards, such as simple SVGA graphics in many desktops, and others contain a graphics accelerator board for fast image manipulation. The laptops have their own EGA graphics boards, but don't ask me anything about the technical differences! Many have single hard drives, but some have two hard drives. Most of the machines have $3\frac{1}{2}$ inch diskette drives, but some have the new super drives. We have no information on file regarding the components, and we are asked many questions each week, such as, "Where can I find a machine with a graphics accelerator card and a CD-ROM burner?" Another problem is that we do not have a concise report of which peripherals are connected to the machines, so we do not have a handle on what type of keyboard, printer, or external CD burner each unit uses. Memory is also a mixed bag. Some machines have 4–64 megabytes, some have 8–128 meg, and some have 256 megabytes of memory installed. You can imagine what it is like to try and find the right memory to run a particular software package. Sometimes we run the software and run into problems.

Anna: How do you track what software is installed on which machine?

Dot: Well, unfortunately, we don't have a good grasp of that. We started to track it, but as I mentioned, the whole PC area exploded campuswide. We spend so much time putting out brush fires that we're losing the battle to keep track of the information we have.

Anna: Are there similar problems with maintaining the equipment?

Dot: You're getting the idea, now. We fix whatever we can as soon as possible. Some of the machines are shipped out on warranty. We don't even dream of preventive maintenance, even though we are agreed that it is important.

Anna: Really, I am interested in what you are dreaming about for the new system. What would you like it to accomplish?

Dot: That's easy for me to summarize. All the weaknesses I've out-lined should be addressed. I'd like it to have a dossier about each machine, its internal components, the peripherals attached to it. I would also like to have good cost and repair information maintained. As changes occur, we need to keep the files up to date.

Anna: Anything specifically related to software?

Dot: Software cross-referencing is a must.

Anna: One of the items I picked up on during our conversation today has been your dissatisfaction with the capability of the old system to keep pace with the PCs' growth. How would any future plans affect the system we develop?

Dot: Certainly we will be adding significant numbers of machines every year. The requests for machines far exceed the budget for several years. We expect that new technology will be adding new components, such as optical disks, that must be added to the system. Also, the use of notebooks that actually leave the premises with users will probably grow. Already they are checked out by individuals for teaching remote classes, research, curriculum development, and the like.

Anna: (Glancing at her notes) We've covered quite a bit, I think I am beginning to understand what the old system does and doesn't do, what you'd like to see in the new system, and what you project for growth in the coming years. Is there anything else that you think is important for me to know that I haven't asked?

Dot: It's an oversight of mine, but I should mention that we also have PCs on our four satellite campuses in outlying areas. Those machines and all they entail need to be included in our system plans.

Anna: I know there are several people who can help with this project. Is there anyone in particular you would recommend that I talk to?

Dot: There are several people that you will want to seek out. They will be very useful to us. Mike Crowe is our maintenance expert. He's been here almost as long as I have. You will enjoy him very much. His counterpart in software is Cher Ware. She's easy to talk with. Don't forget to touch base with Paige Prynter. She's in charge of financial information about the PCs. She'll have what you need in that area. And Hy Perteks runs the Information Center for us. Certainly you will want to see him before you're done.

Anna: Yes. In fact, they are already on my schedule. We must be thinking alike. As I summarize our interview for the systems team, I may have some follow-up questions for you.

Dot: I'm delighted to be a part of this. Call me anytime.

Anna: (Standing up and extending her hand to Dot) Thanks very much for your time. The information you provided gave me a solid start. I will be back in touch.

Dot: (Shaking hands and standing) Let me know how I can help. My door is always open.

Interview Two
Respondent: Mike Crowe, maintenance expert, PCs
Interviewer: Anna Liszt
Location: Mike's workroom, at a workbench

Anna: (Entering his workroom and extending her hand) Hello, Mike. I'm Anna from systems and programming. The administration has asked my team to develop a system to keep track of maintenance costs, preventive maintenance, and other information about the PCs. Dot said you'd be a good person to talk to.

Mike: (Shakes her hands heartily. Clears a spot for Anna to sit beside the workbench) Hello Anna. What do you need to know?

Anna: I'd like to ask you some questions regarding the maintenance of the PCs.

Mike: (Looks around a room strewn with open computers, cables, parts that defy description, tools, and general clutter) As you can see, we're constantly working on the machines that have problems. Some of these are breakdowns, but a lot of the work is upgrading machines to include new capabilities. These things on the workbench are getting memory expansion; those over there are having graphics accelerator boards installed. Really sharp images, with high-speed 24-bit resolution. The computers stacked in boxes behind me are to be installed in Room 472. They'll be linked using Ethernet software and a Maxus XZ server with 256 meg of RAM. A T1 line is connected to the backbone, and soon all the computer lab rooms will have really fast SLIP server access and Internet speed.

Anna: Tell me about your preventive maintenance program.

Mike: (Laughs) When we have time. We would like to blow the dust off every machine, keyboard, and printer, and vacuum the CPUs periodically. The disk drives should also be cleaned once in a while. Often we simply don't have the time to accomplish this work.

Anna: How many people are working on maintenance?

Mike: Me, my assistant, and quite a few students who work part-time for me.

Anna: How do you know when to perform the periodic maintenance?

Mike: Well, we don't have any exact way to do that. Normally, we go from room to room, as we have time. When a room is completed, we write it on a list. Let me show you the clipboard we use. We just keep it hanging on the wall. Because the students do a lot of the preventive maintenance, I'm not directly involved in each room. I spot-check their work. You know how kids are, though. Sometimes they forget to write which rooms they've completed, and I have to get after them. But I rely on them. They're good, for the most part.

Anna: What aspects of maintenance would you like the new system to help you with?

Mike: I've seen some systems at other places, and they can get pretty fancy. I don't think I need anything that complicated. I would like to know which machines are still under warranty. That's a big one. Right now, if a machine breaks down I have to look through stacks of information to find out the warranty period and when we bought the thing.

Anna: (Nodding as she makes a note) What else would be useful?

Mike: I'd like to know which machines are lemons, which ones are constantly breaking down, I mean. I'd take those right out of the high-usage areas. Knowing how often we had to repair the machine would be useful. It would be great to have a list of machines showing which ones need preventive maintenance the most. That would probably cut down on the number sent in for repairs.

Anna: Do all the machines have the same preventive maintenance interval?

Mike: No. (Mike's portable pager sounds. In response, he goes to the phone and has a short conversation about a computer problem.) Now, where were we? Oh, yes. The interval between maintenance. There is different timing for each machine. It would be good to keep that information on file somewhere. It also depends on whether it's a desktop or notebook. The notebooks need less in the line of periodic maintenance.

Anna: Is there anything else you would like to add that I haven't covered?

Mike: Let me say it again, loud and clear. We need warranty information. Also, it would be good to get a report saying which machine needs preventive maintenance and when. Putting it in order of room numbers where the machines are located would make it easy to find the things. We would like to have a Web site problem reporting that would enable folks to report problems when they occur, assuming they can use a different machine. The Web site would have a form to complete that would capture the problem information and send it to us as an email.

Anna: (Standing and extending her hand to shake Mike's) Thanks for your time, Mike. May I get back to you with any further questions and also have you review my interview summary?

Mike: (Standing and shaking hands) Sure. Just have the office page me and leave your number. If you build me a system like the one I just described, I'll put it to good use.

Interview Three

Respondent: Cher Ware, software specialist, personal computer systems
Interviewer: Anna Liszt
Location: Cher's office

Anna: (Entering the open office and extending her hand to the woman perched on an old sofa to one side of the office) How do you do? I'm Anna from systems and programming. Dot mentioned that you would be an important person to talk to about building a new system for managing the PCs. I'd like to ask you some questions about the systems you have for managing software.

Cher: (Gesturing to a spot on the sofa next to her and shaking hands with Anna) Sure. I was expecting you. Dot told me all about you. She keeps us going. She is order itself. I'm happy to talk to you because I know that we need a system for managing our software. It's not like I haven't been trying to keep a grasp on all this, but we've had a fantastic explosion of software. It's growing like *The Thing That Ate Sacramento.* I feel like we're living in a science fiction flick half the time. The software is clearly trying to gobble up our database capacity. Just to track it takes a lot.

Anna: (Laughing) Well, what are some of the basics about *The Thing,* the software that's in use here at CPU?

5

Cher: Well, it started around the early eighties. Or was it 1985? I lose track. The seventies were the best for me personally, but the whole millennium was good too. What was your question?

Anna: How many software packages did you have in the eighties through the nineties?

Cher: Early on, there were only a few. Some simple packages. A word processor, a database, a spreadsheet. Boy, when you think of how things have changed!

Anna: Can you contrast that to the number of packages that are in use now?

Cher: A humongous variety now. Lots and lots of versions of each of them, too. You don't see plain vanilla packages anymore, either. There are several word processors, several databases, spreadsheet, and graphics packages. Then we have software for the science and math programs and Mac packages for the art department. And they're pretty jazzy, too. Not to mention the grammar checkers for the English profs. And everybody wants Internet software, not only access to the Web, like browsers and email, but the kind you use to create Web pages and that cool animation stuff. You know, I don't actually know how many packages there are.

Anna: How does the current system work for managing software?

Cher: It's a simple database system that was developed years ago. We never expected to see the growth that we have. Almost since the beginning, our system hasn't been able to maintain all the information we need. Information is also missing. Let me say it better. Not all the software packages have been captured by our system. Many times a professor gets software for class or research and forgets to tell us about it. I wish that I wasn't the last person on earth to know about nifty updates and new packages. If there was just one process for registering software with us that everyone had to follow, life would be rosier.

Anna: What process is followed when your office receives a new piece of software?

Cher: We inventory it, inform the professor that it has arrived, and key the information into the database. Then it is delivered to the lab or to the prof who requested it.

Anna: What happens to the older version of the software, if there is one?

Cher: Chaos city. I mean we are talking nightmare time. Golden oldies should be deleted from the hard disks and scratched, but that's not the case. Often we have several versions of the same software in several different labs and campuses, even though we try not to let it happen. Microsoft Word® is a good example. We have Word 95, Word 97, and Word 2000. The same is true for our versions of PowerPoint® and many other packages. Really, though, sometimes there's a good reason for having multiple versions, because not all the equipment in all the labs is upgraded to run new software. Also, some software is used by more than one course and the students should not have to switch versions from one course to another.

Anna: Do site licenses add further complications?

Cher: You guessed it. Recently, we've been getting site licenses for some of the most commonly used software. Some of it is used on a LAN where there are many workstations and only one copy of the software. If there is no site license, we need to know how many copies of a particular package we have and on which machines they're located.

Anna: How do you determine which machines or labs will have a new package installed?

Cher: We like to think we have that under control. The normal situation is to use the labs that are designated for that application. For example, Photoshop® is installed in Room 320, the art department's lab. It is an awesome package, by the way. Visible Analyst is installed in the information sciences lab. Some packages, like Word, are installed in several labs. There are some exceptions to this, though. For instance, some of the scientific software now requires a graphics accelerator board and 8–128 MB of RAM. Typically, scientific packages would be installed in the science- and math-complex labs, but only the machines in the information sciences lab have enough memory in them. So that's where the new scientific packages wind up.

Anna: Describe the problems you encounter when locating a machine for installation of new software.

Cher: Sometimes we have requirements for graphics and printers and don't really know which machines have the specific configuration needed. That kind of information is not maintained by either our system or the hardware system. Sometimes the machines don't have enough speed, especially the older ones. Sometimes the hard disks are simply full. We usually investigate these cases, though. Lots of times students put their own games and stuff on machines. We take them off when it's noticed.

Anna: Explain what happens when you receive a request for the location of a particular software package.

Cher: We have a sorted listing of software by its name, which also contains the room number. We can't trust it completely, though, because it's often outdated and incomplete. Not all software is registered with our area, as I mentioned earlier. For example, last week a professor asked me where he could use the language "C++." We informed him of the labs where it was supposed to be located, and later he called to tell us that he found it close to his office, on machines that weren't supposed to have it.

Anna: Do you currently keep financial information pertaining to software?

Cher: Not on the same database. I know this is critical information that should be maintained, though. It would be extremely useful to know the total costs of each software package and category, such as word processors. It would also be great to have the total cost available for an upgrade. The upgrades seem to happen so often that we can hardly install all the packages and provide training before a new release is announced.

Anna: From what you've said today, I can see that you have an incredibly complex operation. You've been very helpful explaining how the software is managed and giving me ideas about what you would like to see the new system do. Is there anything we have not covered that you would like to mention?

Cher: Well, talking to you reminded me of a lot of things I hadn't thought about for a while. I hope the new system can help out, especially in getting all software registered with our office. We'd also like to be able to tie into the hardware system to determine which machines will actually run the software we have.

Anna: (Rises and extends her hand to shake Cher's) I will be back in touch with you as the project continues. We should be able to help out. I'll ask you to review an interview summary in a few days. Thanks very much for your time.

Cher: (Shaking hands and rising as Anna leaves) My pleasure. We have a lot of fun here, even though it's crazy. I'm happy to help any way I can.

Interview Four
Respondent: Paige Prynter, financial analyst
Interviewer: Anna Liszt
Location: Paige's office

Anna: (Knocking on Paige's door, and as door opens, extends her hand to shake Paige's) Hello Ms. Prynter. I'm Anna Liszt from systems and programming.

Paige: (Shaking hands, then showing Anna to a chair opposite her desk) I've been expecting you. Please have a seat. Dot told me you would be contacting me.

Anna: The administration has asked my group to help build a system to manage the PCs, and I am doing a series of interviews with key people who will use the information provided by the system.

Paige: The system is sorely needed. What do you need to know from me?

Anna: I'd like to ask you some questions about the financial needs regarding the PCs used at CPU. More specifically, what types of reports are you currently receiving?

Paige: We get a report listing the cost of all the PCs and a total. That is about the extent of our financial information right now.

Anna: Would it be useful to have subtotals added to the reports?

Paige: Yes. That would be extremely useful for costs on each type of machine.

Anna: Do you receive financial information on software?

Paige: You've touched on an issue that is controversial these days. We receive absolutely no computerized information on software. And, of course, the software has just snowballed. We have no idea of the total amount invested. We scrape by with outdated requisitions for software. We desperately need more information about software purchases to formulate better controls and put together reasonable budgets. We need subtotals by product and by category of software, such as word processing.

Anna: How do site licenses fit into the picture?

Paige: We would like to have the figure for the site license as a total and then not have to calculate the amount for each copy.

Anna: Your current needs are clearly pressing. But is there anything you would add to the system for the future?

Paige: Yes. We would like to input the cost of an upgrade to a particular software package and have the computer tell us how much it would cost for all the currently installed software. We also need subtotals by product for both hardware and software, along with totals. It would also be useful to have totals for each of the satellite campuses.

Anna: You've answered all the questions I have right now. Is there anything you would like to add?

Paige: Yes. We need to take an accurate inventory of the hardware periodically. Machines have an annoying habit of moving from one room to another over the course of a semester, but we need to know what we have and precisely where it is. You can imagine how time-consuming the inventory process is. Automating it would be highly desirable.

Anna: May I get back to you with a summary of our interview and also any further questions I might have?

Paige: Certainly. Just set up an appointment ahead of time again, and I will be glad to speak with you.

Anna: (Rising and extending her hand to shake Paige's) Thank you very much for your time. Your input will be valuable in helping to put together the new system.

Paige: (Shaking Anna's hand and rising from her desk) I hope the new system will provide us with the vital information we need. I don't like to complain, but it's been a long time in coming. Please close the door behind you.

Interview Five

Respondent: Hy Perteks, director, Information Center
Interviewer: Anna Liszt
Location: Hy's office in the Information Center

Anna: (Walking past several students and faculty working on PCs to get to Hy's open door; pausing at the entrance) Hello, Hy. I'm Anna.

Hy: (Getting up from a crowded desk to welcome her and extending his hand to shake Anna's) I remember you. We were at that conference together about two years ago. Please come in and pull up a chair. I got your message, and Dot said you'd be by, too.

Anna: (Sitting down in a chair by the side of Hy's desk) She probably told you that the administration has asked my group to help with designing a system to manage the PCs. I've been doing a series of interviews with key people, and it's time for me to find out about the needs you are experiencing here in the Information Center.

Hy: What we need so badly that I can taste it is a centralized bank of information on the PCs and the software we have.

Anna: Who is served by the Information Center?

Hy: Our clients come from every level of the university. We serve administrators, both on the managerial and support staff levels, and we also serve the needs of faculty for teaching and research. Our clients are pretty evenly split between those groups, with maybe a little lean toward the faculty.

Anna: What services do you provide them?

Hy: We do a lot of training. A lot. Every so often we offer classes in popular software packages. You've probably gotten some of our flyers announcing those classes. We answer, or try to answer, tons of technical questions as well. They usually are specific as to how to perform advanced tasks with the software. A third category of service we perform is helping users to adapt software to their particular application. We also help in figuring out which software would be most effective in solving their problems.

Anna: It seems that the staff here has to be fairly well-versed in many different areas. Do you have an expert in each software package?

Hy: No, although I personally am familiar with the basics of all the packages. I receive training in the operation of the software and work on small projects just to familiarize myself with new programs. When the nitty-gritty detailed technical questions arise, I dive into the manuals and other reference materials we have here in the IC library. But I'm not alone. I often call on specialists who are on the faculty to help with particularly thorny problems. At times I even call software vendors for help.

Anna: How often do you add new software packages?

Hy: More often than you would believe. We do upgrades, as well as altogether new software. For example, we recently updated a math word processor and are scheduling classes for training. We are being absolutely deluged with requests from the math and science faculty. They are so tickled that the package is here and running. There's never a dull moment. That's why I love the job.

Anna: What are some of the problems you are experiencing that might be dealt with more effectively via an improved system?

Hy: I know there's a lot that can be done that we haven't, just because no one has taken the time. For one, we need to know what version of software a person is using. You wouldn't believe how critical that is in determining the solution to a technical question, such as how to create a macro or transfer files from one package to another. Lots of times the users just aren't sure of what version they're using. We'd also like to know the phone numbers of software providers so we can quickly phone for assistance. Sometimes it's difficult to find an expert around the university for a particular program. I meet a lot of interesting people searching for help, though.

Anna: Other people have mentioned the need to know which software is located on which machine in which room. Is this important to you?

Hy: You bet. We get requests for training, or someone wants to use software that isn't in the Information Center. If it's an oddball package, we don't know where it's located without lots of calling around.

Anna: You've told me a good deal in a short time. Is there anything else that you'd like to add that we haven't covered?

Hy: I know I talk a lot, but I mean well. Really, I think we've covered all the bases. Let me just add that I would like to have a sense of how many people would be interested in training on software that we have now. I'd also like to know what software people want us to purchase. We can't get everything, but if they don't ask, I'll never know. That's a tall order, though. I'm not sure how to get that without surveying the entire university.

Anna: (Rising and extending her hand to shake Hy's) We'll try to include as much of what you've said as we can. Thanks for your time. May I get back to you with follow-up questions if they are needed?

Hy: (Shaking her hand and rising to accompany her to the door) No problem. Service is what we're here for. If you end up using a package for the PC as part of the solution, I can get involved even more.

EXERCISES

E-1. Assume Chip conducted all the initial interviews rather than Anna. In small groups, role play the part of Chip and the interviewees. Assign students to analyze and comment on the differences in the results.

E-2. Analyze the five interviews. In a paragraph, discuss what type of structure each interview had.

E-3. List each interview, 1 through 5, and then write a paragraph for each, discussing ways that Anna might improve on her interviews for next time.

E-4. Analyze the questions used in the five interviews. In a paragraph, discuss the question types used and whether they were appropriate for getting needed information.

USING QUESTIONNAIRES

6

KINDS OF INFORMATION SOUGHT

The use of questionnaires is an information-gathering technique that allows systems analysts to study attitudes, beliefs, behavior, and characteristics of several key people in the organization who may be affected by the current and proposed systems, as shown in Figure 6.1. Attitudes are what people in the organization say they want (in a new system, for instance), beliefs are what people think is actually true, behavior is what organizational members do, and characteristics are properties of people or things.

Responses gained through questionnaires (also called surveys) using closed questions can be quantified. If you are surveying people via email or the Web, you can use software to turn electronic responses directly into data tables for use in a spreadsheet or other statistical software for analysis. Responses to questionnaires using open-ended questions are analyzed and interpreted in other ways. Answers to questions on attitudes and beliefs are notably sensitive to the wording chosen by the systems analyst.

Through the use of questionnaires, the analyst may be seeking to quantify what was found in interviews. In addition, questionnaires may be used to determine how widespread or limited a sentiment expressed in an interview really is. Conversely, questionnaires can be used to survey a large sample of system users to sense problems or raise important issues before interviews are scheduled.

Throughout this chapter, we compare and contrast questionnaires with interviews, which were covered in Chapter 5. There are many similarities between the two techniques, and perhaps the ideal would be to use them in conjunction with each other, either following up unclear questionnaire responses with an interview or designing the questionnaire based on what is discovered in the interview. Each technique, however, has its own specific functions, and it is not always necessary or desirable to use both.

PLANNING FOR THE USE OF QUESTIONNAIRES

At first glance questionnaires may seem to be a quick way to gather massive amounts of data about how users assess the current system, about what problems they are experiencing with their work, and about what people expect from a new or modified system. Although it is true that you can gather a lot of information through questionnaires without spending time in face-to-face interviews, develop-

FIGURE 6.1
Kinds of information sought when using questionnaires.

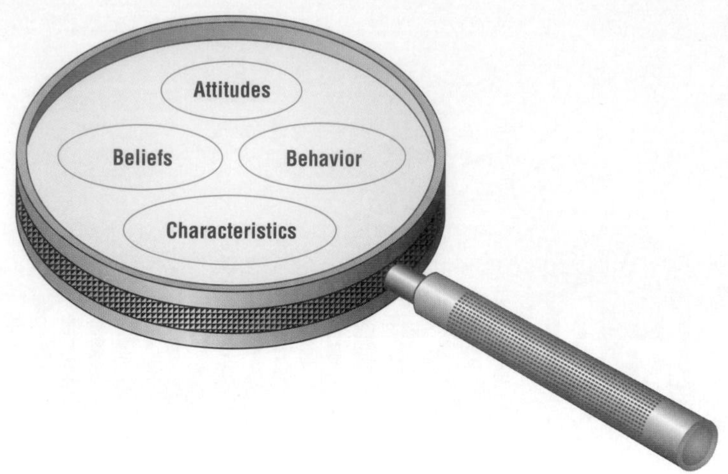

ing a useful questionnaire takes extensive planning time in its own right. When you decide to survey users via email or the Web, you face additional planning considerations concerning confidentiality, authentication of identity, and problems of multiple responses.

You must first decide what you are attempting to gain through using a survey. For instance, if you want to know what percentage of users prefers a FAQ page as a means of learning about new software packages, a questionnaire might be the right technique. If you want an in-depth analysis of a manager's decision-making process, an interview is a better choice.

Here are some guidelines to help you decide whether use of questionnaires is appropriate. Consider using questionnaires if:

1. The people you need to question are widely dispersed (different branches of the same corporation).
2. A large number of people are involved in the systems project, and it is meaningful to know what proportion of a given group (for example, management) approves or disapproves of a particular feature of the proposed system.
3. You are doing an exploratory study and want to gauge overall opinion before the systems project is given any specific direction.
4. You wish to be certain that any problems with the current system are identified and addressed in follow-up interviews.

Once you have determined that you have good cause to use a questionnaire and have pinpointed the objectives to be fulfilled through its use, you can begin formulating questions.

WRITING QUESTIONS

The biggest difference between the questions used for most interviews and those used on questionnaires is that interviewing permits interaction between the questions and their meanings. In an interview the analyst has an opportunity to refine a question, define a muddy term, change the course of questioning, respond to a puzzled look, and generally control the context.

Few of these opportunities are possible on a questionnaire. Thus, for the analyst, questions must be transparently clear, the flow of the questionnaire cogent, the respondent's questions anticipated, and the administration of the questionnaire planned in detail. (A respondent is the person who responds to or answers the questionnaire.)

The basic question types used on the questionnaire are open-ended and closed, as discussed for interviewing. Due to the constraints placed on questionnaires, some additional discussion of question types is warranted.

Open-Ended Questions. Recall that open-ended questions (or statements) are those that leave all possible response options open to the respondent. For example, open-ended questions on a questionnaire might read, "Describe any problems you are currently experiencing with output reports" or "In your opinion, how helpful are the user manuals for the current system's accounting package?"

When you write open-ended questions for a questionnaire, anticipate what kind of response you will get. It is important that responses you receive are capable of correct interpretation. Otherwise, many resources will have been wasted in the development, administration, and interpretation of a useless questionnaire.

For instance, if you ask a question such as, "How do you feel about the system?" the responses are apt to be too broad for accurate interpretation or comparison. Therefore, even when you write an open-ended question, it must be narrow enough to guide respondents to answer in a specific way. (Examples of open-ended questions can be found in Figure 6.2.)

For instance, if you really want to gather feelings toward the current system, you might couch your questions in the context of satisfaction versus dissatisfaction with the system. Furthermore, you might suggest some system features to prompt respondents in recalling which features are of interest.

Open-ended questions are particularly well suited to situations in which you want to get at organizational members' opinions about some aspect of the system, whether product or process. In such cases you will want to use open-ended questions when it is impossible to list effectively all the possible responses to the question.

In addition, open-ended questions are useful in exploratory situations. These situations occur when the systems analyst is not able (because of diversity of opinion or far-flung employees) to determine precisely what problems plague the current system. Responses to the open-ended questions will then be used to focus on cited problems more narrowly, via interviews with a handful of key decision makers.

Closed Questions. Recall that closed questions (or statements) are those that limit or close the response options available to the respondent. For example, in Figure 6.3 the statement in question 23 ("Below are the six software packages currently available in the Information Center. Please check the package you personally use most frequently") is closed. Notice that respondents are not asked why the package is preferred, nor are they asked to select more than one, even if that is a more representative response.

Closed questions should be used when the systems analyst is able to list effectively all the possible responses to the question and when all the listed responses are mutually exclusive, so that choosing one precludes choosing any of the others.

Use closed questions when you want to survey a large sample of people. The reason becomes obvious when you start imagining how the data you are collecting will look. If you use only open-ended questions for hundreds of people, correct analysis and interpretation of their responses becomes impossible without the aid of a computerized content analysis program.

There are trade-offs involved in choosing either open-ended or closed questions for use on questionnaires. Figure 6.4 summarizes these trade-offs. Notice that responses to open-ended questions can help analysts gain rich, exploratory insights

FIGURE 6.2
Open-ended questions used for questionnaires.

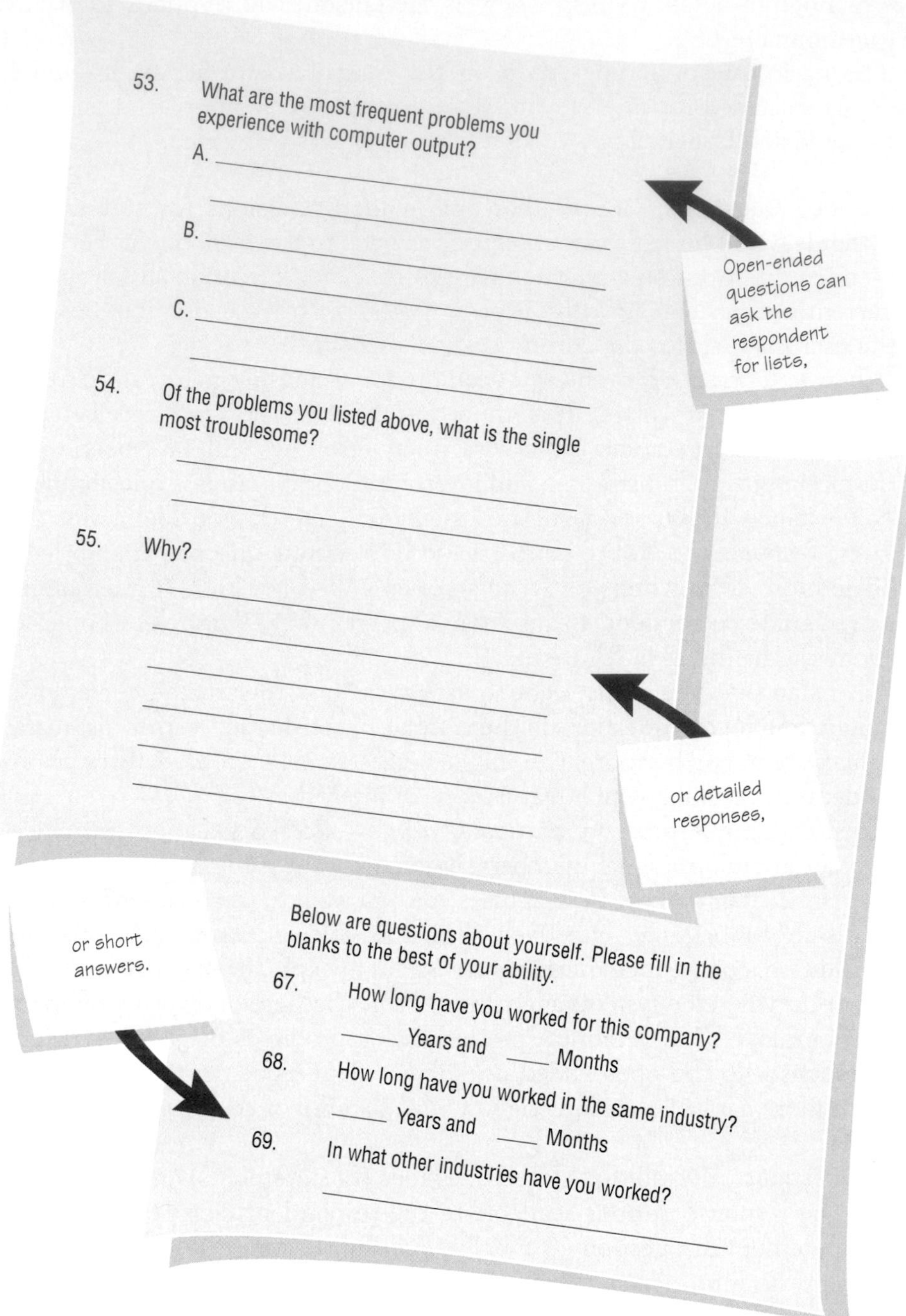

as well as breadth and depth on a topic. Although open-ended questions can be written easily, responses to them are difficult and time-consuming to analyze.

53. What are the most frequent problems you experience with computer output?
A.
B.
C.
54. Of the problems you listed above, what is the single most troublesome?
55. Why?
Open-ended questions can ask the respondent for lists,
or detailed responses,
or short answers.
Below are questions about yourself. Please fill in the blanks to the best of your ability.
67. How long have you worked for this company?
Years and Months
68. How long have you worked in the same industry?
Years and Months
69. In what other industries have you worked?

When we refer to the writing of closed questions with either ordered or unordered answers, we often refer to the process as scaling. The use of scales in surveys is discussed in detail in a later section.

Choice of Words. Just as with interviews, the language of questionnaires is an extremely important aspect of their effectiveness. Even if the systems analyst has a standard set of questions concerning system development, it is wise to write them to reflect the business's own terminology.

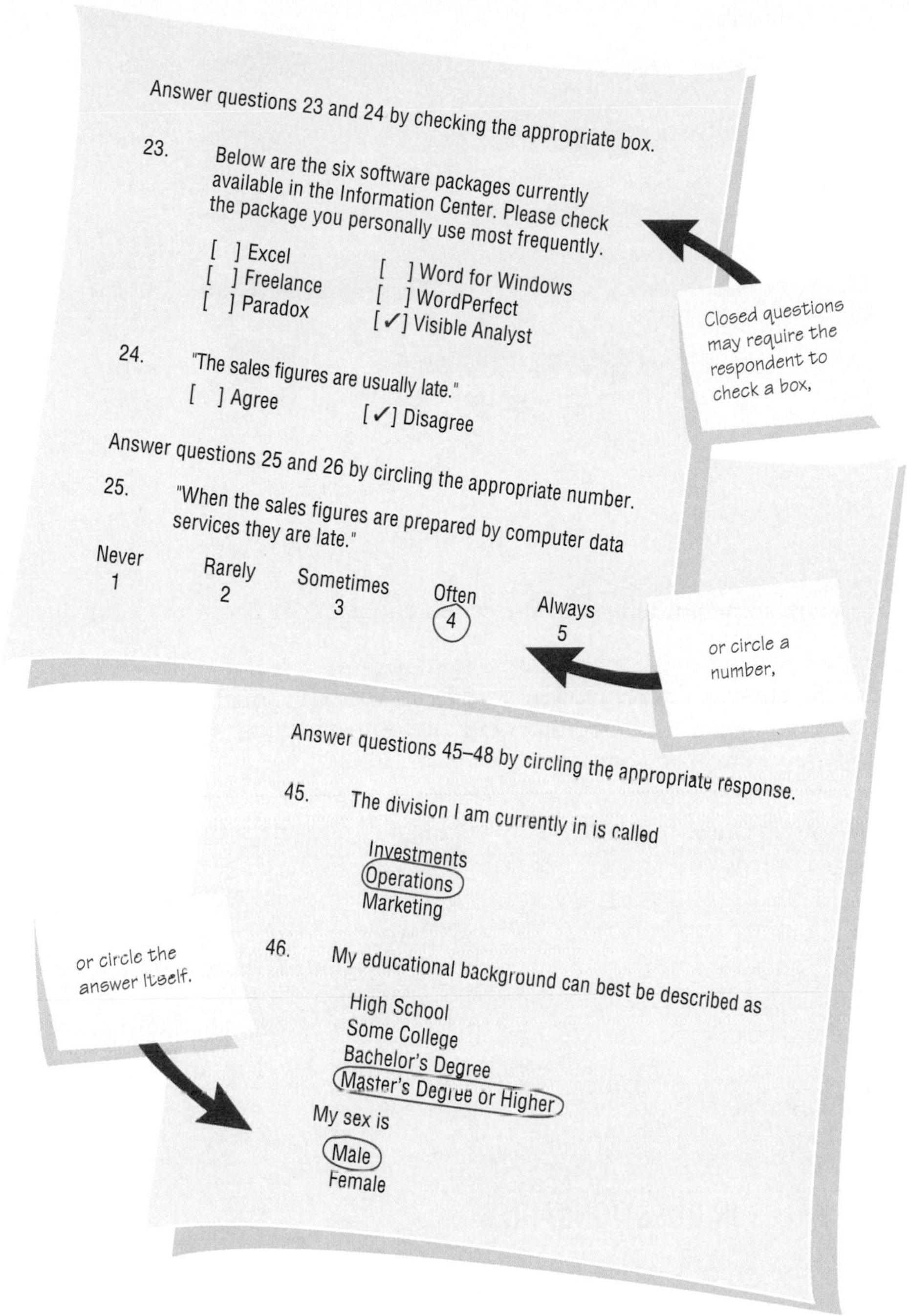

Answer questions 23 and 24 by checking the appropriate box.

23. Below are the six software packages currently available in the Information Center. Please check the package you personally use most frequently.

[ ] Excel
[ ] Freelance
[ ] Paradox
[ ] Word for Windows
[ ] WordPerfect
[✓] Visible Analyst

24. "The sales figures are usually late."

[ ] Agree
[✓] Disagree

Answer questions 25 and 26 by circling the appropriate number.

25. "When the sales figures are prepared by computer data services they are late."

Never	Rarely	Sometimes	Often	Always
1	2	3	(4)	5

Answer questions 45–48 by circling the appropriate response.

45. The division I am currently in is called

Investments
(Operations)
Marketing

46. My educational background can best be described as

High School
Some College
Bachelor's Degree
(Master's Degree or Higher)

My sex is

(Male)
Female

Closed questions may require the respondent to check a box,

or circle a number,

or circle the answer itself.

FIGURE 6.3
Closed questions on questionnaires can help ensure responses.

Respondents appreciate the efforts of someone who bothers to write a questionnaire reflecting their own language usage. For instance, if the business uses the term *supervisors* instead of *managers,* or *units* rather than *departments,* incorporating the preferred terms into the questionnaire helps respondents relate to the meaning of the questions. Responses will be easier to interpret accurately, and respondents will be more enthusiastic overall.

To check whether language used on the questionnaire is that of the respondents, try some sample questions on a pilot (test) group. Ask them to pay particular attention to the appropriateness of the wording and to change any words that do not ring true.

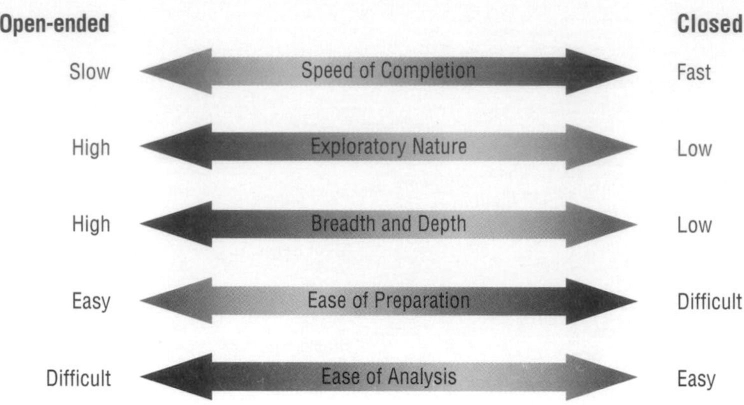

FIGURE 6.4
Trade-offs between the use of open-ended and closed questions on questionnaires.

Here are some guidelines to use when choosing language for your questionnaire:

1. Use the language of respondents whenever possible. Keep wording simple.
2. Work at being specific rather than vague in wording. Avoid overly specific questions as well.
3. Keep questions short.
4. Do not patronize respondents by talking down to them through low-level language choices.
5. Avoid bias in wording. Avoiding bias also means avoiding objectionable questions.
6. Target questions to the correct respondents (that is, those who are capable of responding). Don't assume too much knowledge.
7. Ensure that questions are technically accurate before including them.
8. Use software to check whether the reading level is appropriate for the respondents.

USING SCALES IN QUESTIONNAIRES

Scaling is the process of assigning numbers or other symbols to an attribute or characteristic for the purpose of measuring that attribute or characteristic. Scales are often arbitrary and may not be unique. For example, temperature is measured in a number of ways; the two most common are the Fahrenheit scale (where water freezes at 32 degrees and boils at 212 degrees) and the Celsius scale (where freezing occurs at 0 degrees and boiling at 100 degrees).

SCALING FUNDAMENTALS

Reasons for Scaling. The systems analyst may want to design scales either (1) to measure the attitudes or characteristics of the people answering the questionnaire or (2) to have the respondents judge the subjects of the questionnaire. Let's look at how each of these types of scales can be applied. For our purposes, each will use the same set of questions about a number of sample monthly statement printouts for the Never Fail Bank of America.

If the analyst wants to measure attitudes or characteristics of the respondents, the responses can be combined or grouped to reflect that information. A number of people might fall into a group that doesn't want to change the monthly printouts at any cost, another group may want cleaner output, and a third group might want to add features such as sorting checks by number and category. Here we are trying to measure the differences among respondents, and it doesn't matter how each one of the sample printouts was rated. Other examples of measuring characteristics or attitudes are discussed later in this chapter.

If the systems analyst were interested in how each of the sample monthly statements fared, the respondents would serve as judges. In this case, it wouldn't matter how much respondents differ in their attitudes.

Measurement. There are four different forms of measurement scales, each form offering different degrees of accuracy. The form of measurement also dictates how to analyze the data collected. The four forms of measurement are

1. Nominal
2. Ordinal
3. Interval
4. Ratio

Nominal scales are used to classify things. A question such as:

What type of software do you use the most?

1 = A Word Processor
2 = A Spreadsheet
3 = A Database
4 = An Email Program

uses a nominal scale. Obviously, nominal scales are the weakest of the forms of measurement. Generally, all the analyst can do with them is obtain totals for each classification.

Ordinal scales, like nominal scales, allow classification. The difference, however, is that the ordinal scale also implies rank ordering. In this example a systems analyst asks an end user to circle one of the numbers:

The support staff of the Technical Support Group is:

1. Extremely Helpful
2. Very Helpful
3. Moderately Helpful
4. Not Very Helpful
5. Not Helpful At All

Ordinal scales are useful because one class is greater or less than another class. On the other hand, no assumption can be made that the difference between choices 1 and 2 is the same as the difference between choices 3 and 4.

Interval scales possess the characteristic that the intervals between each of the numbers are equal. Due to this characteristic, mathematical operations can be performed on the questionnaire data, resulting in a more complete analysis.

Examples of interval scales are the Fahrenheit and Celsius scales, which measure temperature.

The foregoing example of the Information Center is definitely not that of an interval scale, but by anchoring the scale on either end, the analyst may want to assume the respondent perceives the intervals to be equal:

How useful is the support given by the Technical Support Group?

Not Useful At All				Extremely Useful
1	2	3	4	5

If the systems analyst makes this assumption, more quantitative analysis is possible.

Ratio scales are similar to interval scales insofar as the intervals between numbers are assumed to be equal. Ratio scales, however, have an absolute zero. An example of a ratio scale is length as measured by a ruler. Another example is the following:

Approximately how many hours do you spend on the Internet daily?

0 2 4 6 8

Ratio scales will be used less often by the systems analyst than other scales.

As a guideline, a systems analyst should use:

1. A ratio scale when the intervals are equal and there is an absolute zero
2. An interval scale when it can be assumed that the intervals are equal but that there is no absolute zero
3. An ordinal scale when it is impossible to assume the intervals are equal but when the classes can be ranked
4. A nominal scale if the systems analyst wants to classify things but if they cannot be ranked

Validity and Reliability. There are two measures of performance in constructing scales: validity and reliability. The systems analyst should be aware of these concerns.

Validity is the degree to which the question measures what the analyst intends to measure. For example, if the purpose of the questionnaire is to determine whether the organization is ready for a major change in computer operations, do the questions measure that?

Reliability measures consistency. If the questionnaire was administered once and then again under the same conditions and if the same results were obtained both times, the instrument is said to have external consistency. If the questionnaire contains subparts and these parts have equivalent results, the instrument is said to have internal consistency. Both external and internal consistency is important.

CONSTRUCTING SCALES

The actual construction of scales is a serious task. Careless construction of scales can result in one of the following problems:

1. Leniency
2. Central tendency
3. Halo effect

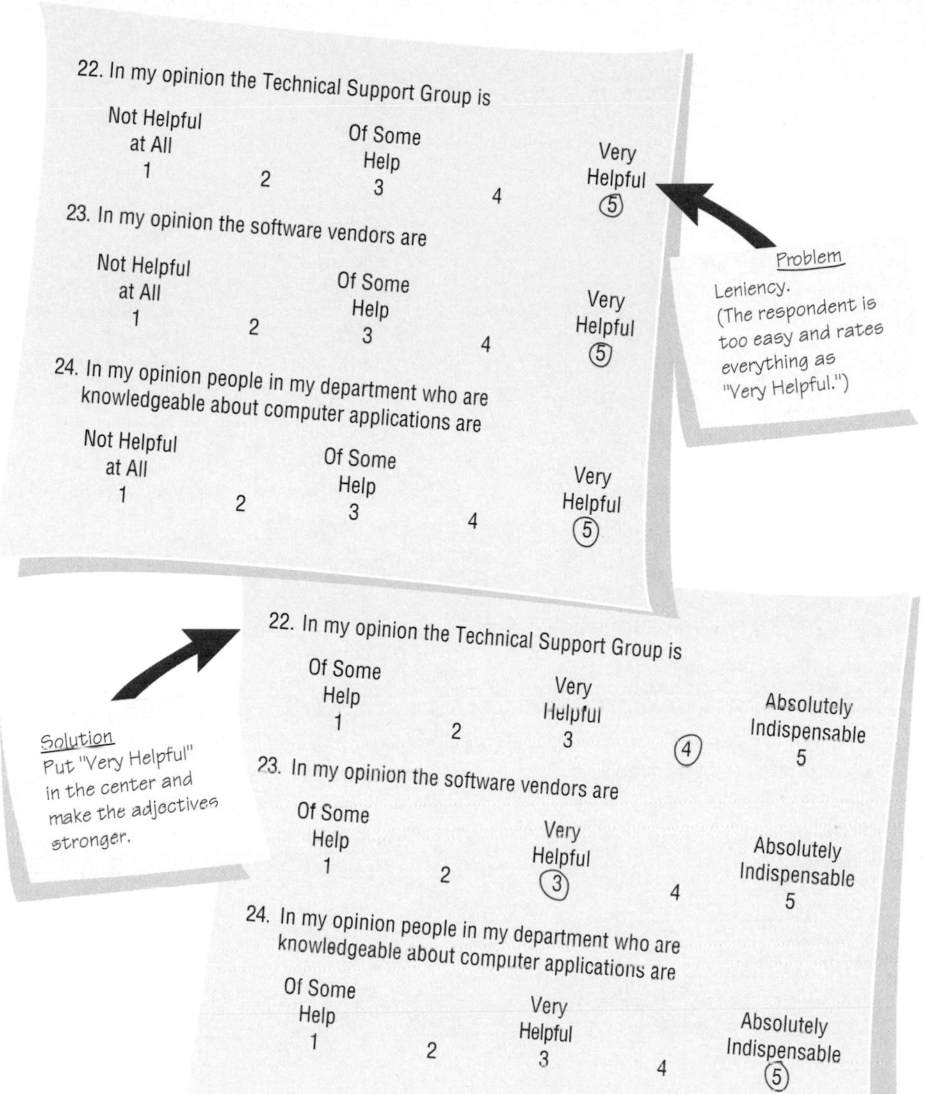

FIGURE 6.5

Correcting the problem of respondent's leniency.

Leniency is a problem caused by respondents who are easy raters. A systems analyst can avoid the problem of leniency by moving the "average" category to the left (or right) of center, as shown in Figure 6.5.

Central tendency is a problem that occurs when respondents rate everything as average. The analyst can improve the scale (1) by making the differences smaller at the two ends, (2) by adjusting the strength of the descriptors, or (3) by creating a scale with more points. An example of correcting for central tendency can be found in Figure 6.6.

The halo effect is a problem that arises when the impression formed in one question carries into the next question. For example, if you are rating an employee about whom you have a very favorable impression, you may give a high rating in every category or trait, regardless of whether or not it is a strong point of the

FIGURE 6.6
Correcting the problem of respondent's central tendency.

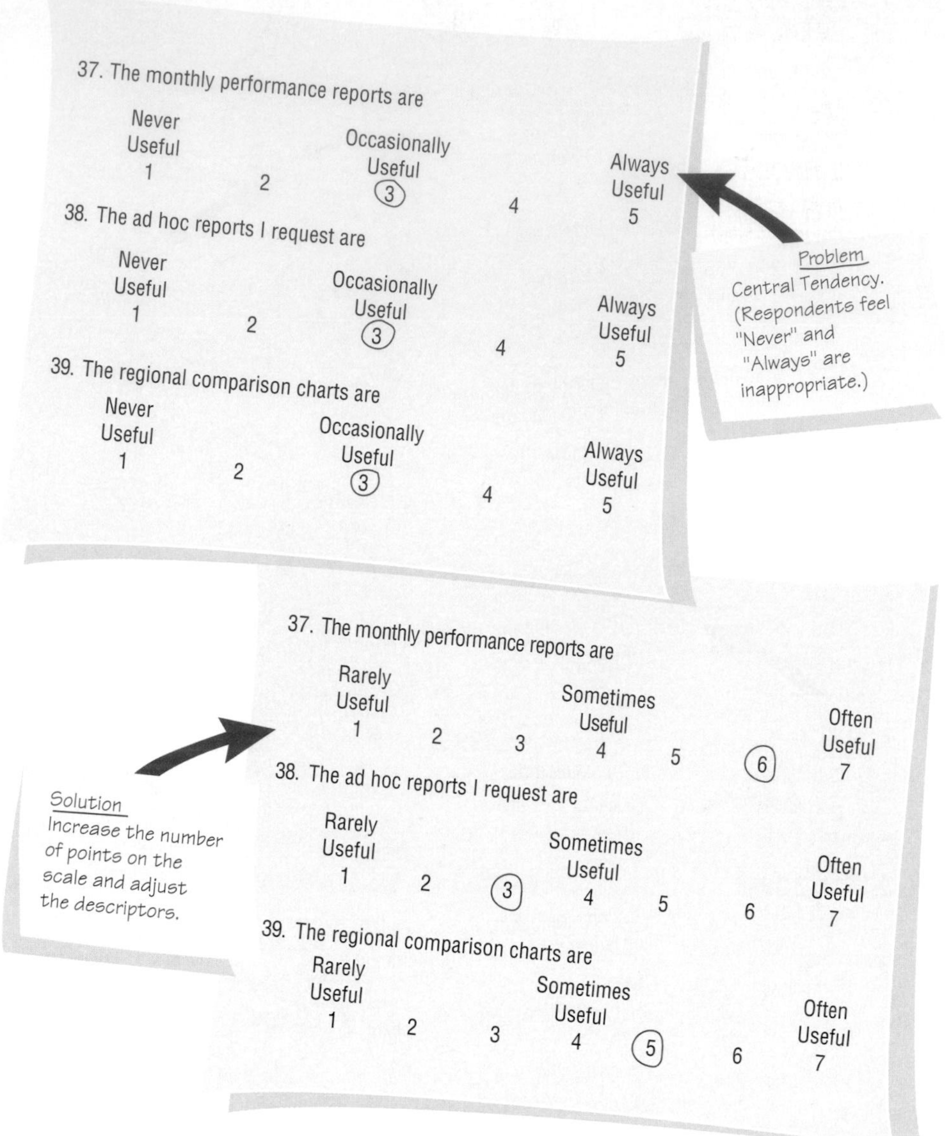

employee's. The solution is to place one trait and several employees on each page, rather than one employee and several traits on a page. An example of correcting for the halo effect can be found in Figure 6.7.

DESIGNING THE QUESTIONNAIRES

Many of the same principles that are relevant to the design of forms for data input (as covered in Chapter 16) are important here as well. Although the intent of the questionnaire is to gather information on attitudes, beliefs, behavior, and characteristics whose impact may substantially alter users' work, respondents are not always motivated to respond. Remember that organizational members as a whole tend to receive too many surveys, many of which are often ill conceived and trivial.

A well-designed, relevant questionnaire can help overcome some of this resistance to respond. This section discusses the stylistic considerations that can help improve the response rate to questionnaires. It also presents the guidelines for ordering the content for the best results.

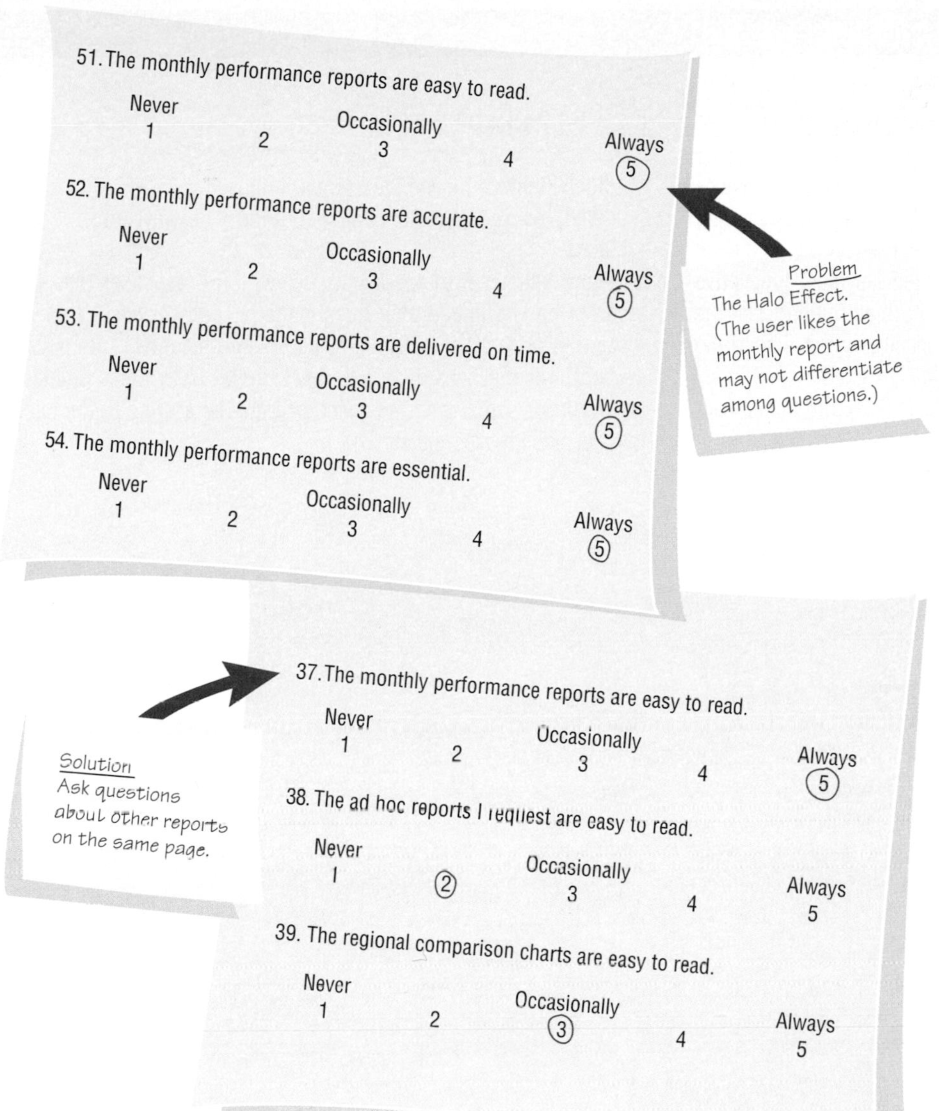

FIGURE 6.7

Correcting the problem of respondent's answering everything favorably due to a halo effect.

QUESTIONNAIRE FORMAT

Allow Ample White Space. The most important consideration in designing the questionnaire format is to allow enough white space so that the respondent is drawn into the form. White space refers to blank space surrounding text on a page or screen. A questionnaire that is squeezed together without benefit of adequate white space is not as likely to be completed, even though it will take less paper to print. To further increase the response rate, use only white or off-white paper for printed questionnaires. When designing Web surveys, make sure that the display is easy to follow, and if the form continues onto several screens, make it easy to scroll to other sections of the form.

Allow Adequate Space for Responses. Beyond allowing white space to set off the printed material, allow adequate space for responses as well. If you expect respondents to write a paragraph in response to an open-ended question, you must leave three to five blank lines for them to do so. Your Web form should permit a generous amount of text to be typed into the response space for open-ended questions.

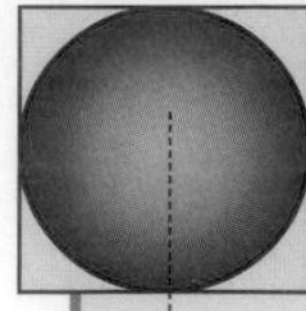

CONSULTING OPPORTUNITY 6.1

THE UNBEARABLE QUESTIONNAIRE

"I'm going to go into a depression or at least a slump if someone doesn't figure this out soon," say Penny Stox, office manager for Carbon, Carbon, & Rippy, a large brokerage firm. Penny is sitting across a conference table from you and two of her most productive account executives, By Lowe and Sal Hy. You are all mulling over the responses to a questionnaire that has been distributed among the firm's account executives, which is shown in Figure 6.C1.

"We need a crystal ball to understand these," By and Sal call out together.

"Maybe it reflects some sort of optimistic cycle, or something," Penny says as she reads more of the responses. "Who designed this gem, anyway?"

"Rich Kleintz," By and Sal call out in unison.

"Well, as you can see, it's not telling us anything," Penny exclaims.

Penny and her staff are dissatisfied with the responses they have received on the unbearable questionnaire, and they feel that the responses are unrealistic reflections of the amount of information account executives want. In a paragraph state why these problems are occurring. On a separate sheet, change the scaling of the questions to avoid these problems.

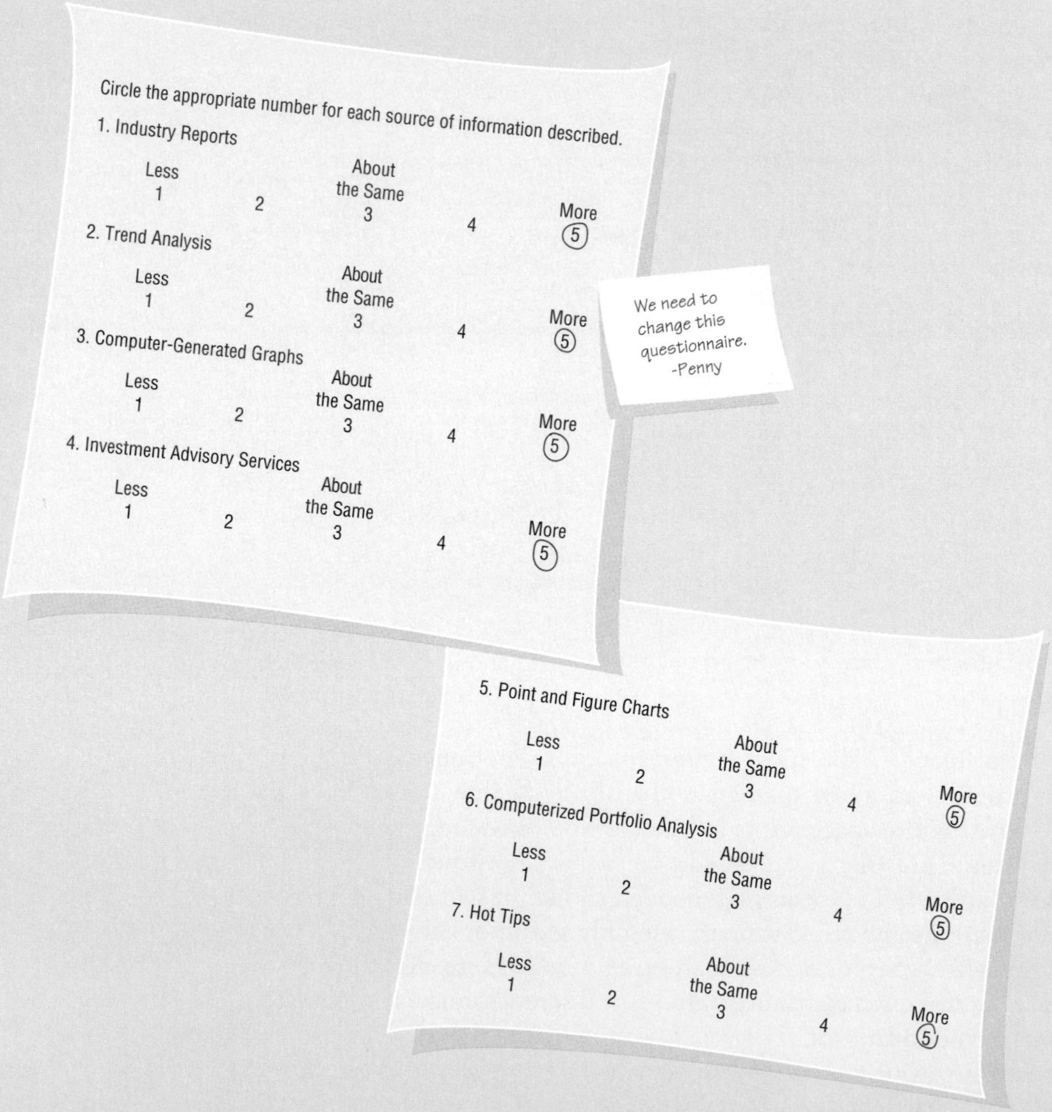

Circle the appropriate number for each source of information described.

1. Industry Reports

Less		About the Same		More
1	2	3	4	(5)

2. Trend Analysis

Less		About the Same		More
1	2	3	4	(5)

3. Computer-Generated Graphs

Less		About the Same		More
1	2	3	4	(5)

4. Investment Advisory Services

Less		About the Same		More
1	2	3	4	(5)

We need to change this questionnaire.
-Penny

5. Point and Figure Charts

Less		About the Same		More
1	2	3	4	(5)

6. Computerized Portfolio Analysis

Less		About the Same		More
1	2	3	4	(5)

7. Hot Tips

Less		About the Same		More
1	2	3	4	(5)

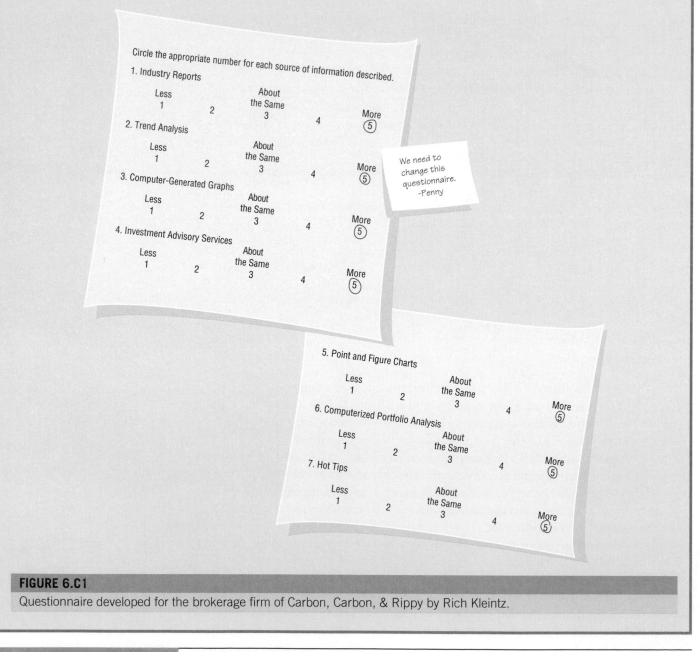

FIGURE 6.C1
Questionnaire developed for the brokerage firm of Carbon, Carbon, & Rippy by Rich Kleintz.

Name	Appearance	Purpose
One-line text box	[ ]	Used to obtain a small amount of text and limit the answer to a few words.
Scrolling text box	[ ]	Used to obtain one or more paragraphs of text.
Check box	[ ]	Used to obtain a Yes-No answer (e.g., Do you wish to be included on the mailing list?).
Radio button	( )	Used to obtain a yes-no or true-false answer.
Drop-down menu	[▾]	Used to obtain more consistent results. Respondent is able to choose the appropriate answer from a predetermined list (e.g., a list of state abbreviations).
Push button	Button	Most often used for an action (e.g., a respondent pushes a button marked "Submit" or "Clear").

FIGURE 6.8
When designing a Web survey, keep in mind that there are different ways to capture responses.

Even if a blank response space looks small, respondents should be able to scroll down within the box to enter a longer answer. To see what a scrolling text box looks like, examine Figure 6.8. It shows the six basic types of response entry that you can incorporate on a Web survey form.

Ask Respondents to Clearly Mark Their Answers. Another good practice for capturing a response correctly is to request that respondents circle their answers (or numbers if you are employing a scale). Figure 6.9 shows what can happen if respondents aren't encouraged to circle numbers. In this example, the analyst would have a difficult time trying to determine if the respondent meant to choose a 3 or 4.

On a printed survey, sometimes it is permissible to have respondents check a box [ ] created by a pair of brackets or check a space ( ) created with a pair of parentheses. Care must be taken to allow enough space for people who make large check marks, however. Data tabulation and analysis become difficult if respondents inadvertently check through several boxes. The researchers at the Georgia Tech Graphic, Visualization, and Usability Center <www.cc.gatech.edu/gvu/> have found that setting off a line with an asterisk on either end helps mark the beginning and end of a space intended for a typed response on an email survey.

On an email form, you can ask respondents to replace a number on a scale with an X. On a Web survey, you can ask respondents bipolar questions by having them click on a check box or radio button to indicate a "Yes" or "No" preference or a "True" or "False" preference.

For posing closed questions that are not bipolar, you can use a drop-down box to list alternatives. In this way you can ensure more consistent results, because the user chooses an appropriate answer from a predetermined list.

Use Objectives to Help Determine Format. Before you design the questionnaire, you need to articulate your objectives. For instance, if your objective is to poll as many organizational members as possible concerning an identified list of problems

FIGURE 6.9

Improving the accuracy of responses by asking a respondent to circle a number.

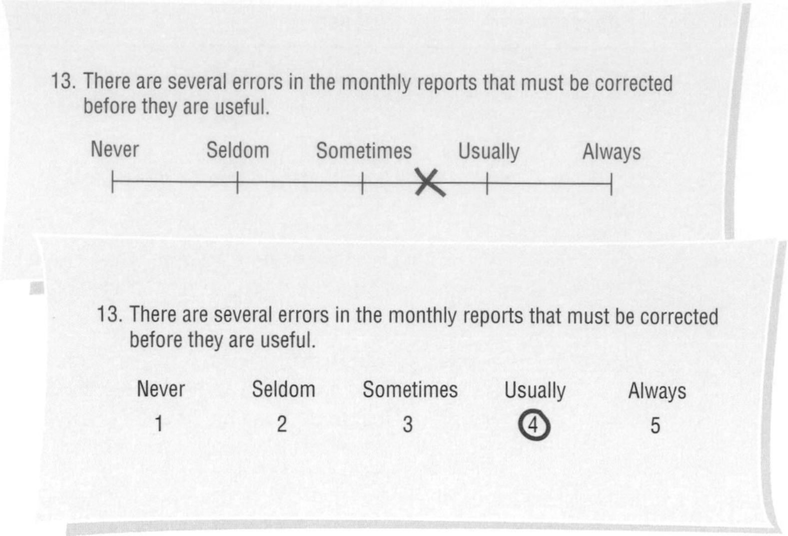

in the current system, it is probably best to use a response form that is machine-readable. If so, how you design the questionnaire and what kinds of instructions you will include are affected.

Alternatively, if you desire written responses, you need to calculate the amount of space that will be needed for the length of response you want, and then you will have to be sure to include that space on the form or on a separate answer sheet. You might need to plan for both numerical and written responses.

You may also wish to assign someone other than the respondents to enter the responses on the questionnaire. Although doing so presents a greater possibility of errors in interpretation, it does prevent the kind of mechanical data-entry errors that inexperienced respondents might make. In addition, keep in mind that questionnaire sheets on which respondents can write directly are often easier for them to complete correctly than machine-readable response forms.

Be Consistent in Style. Organize the questionnaire consistently throughout. Put instructions in the same place in relation to question subsections so that respondents always know where to find instructions.

Be consistent, as shown in Figure 6.10. If you use shaded boxes, use them in the same way from question to question. Following this format consistently allows respondents to get through the questionnaire quickly and reduces the chance of error.

Another major part of designing the questionnaire is deciding the order in which questions should appear. You will often need the input of a pilot group to help decide the most appropriate question order.

ORDER OF QUESTIONS

There is no best way to order questions on the questionnaire. Once again, as you order questions, you must think about your objectives in using the questionnaire and then determine the function of each question in helping you to achieve your objectives. It is also important to see the questionnaire through the respondent's eyes. If you are without the help of a pilot group, always ask yourself how respondents will feel about the order and placement of particular questions as well as whether that is indeed the reaction you want.

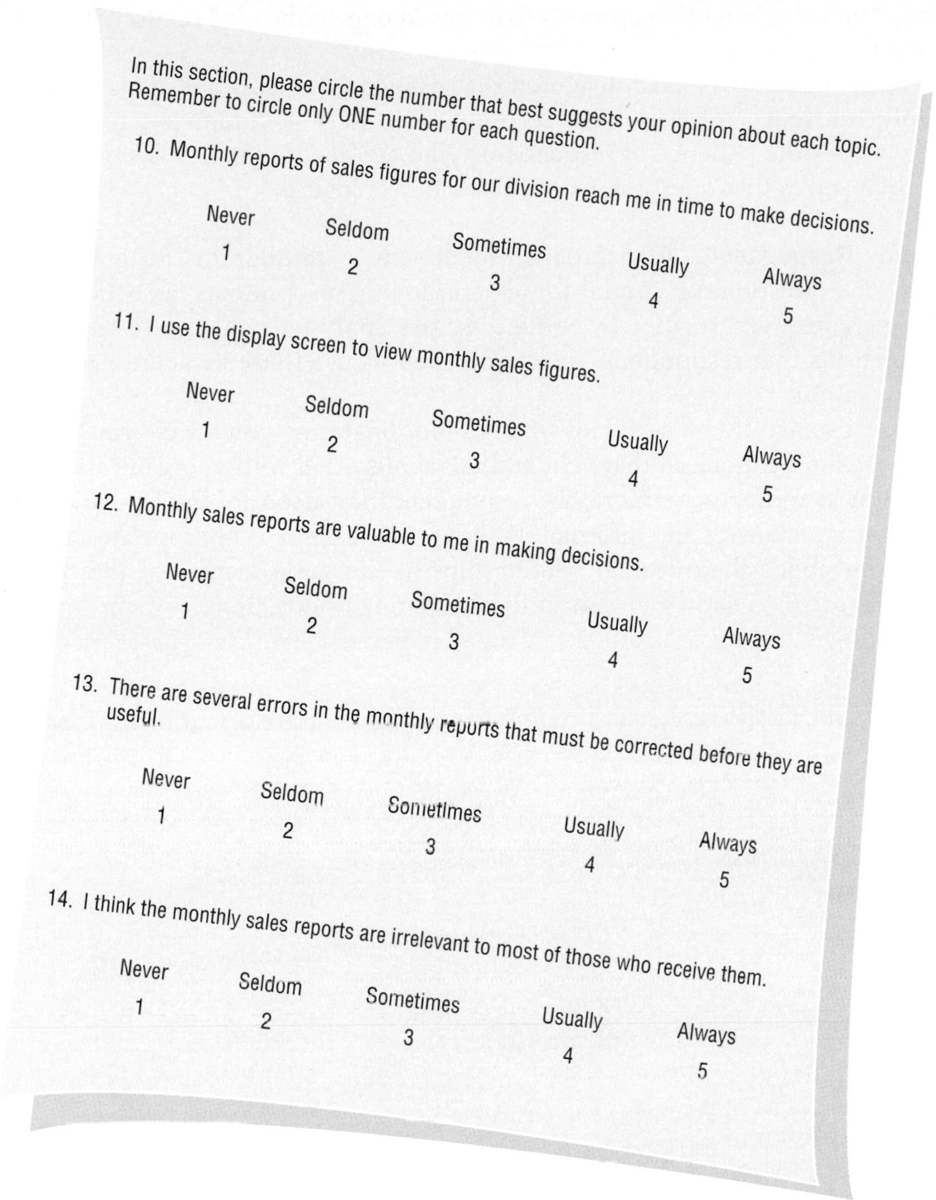

In this section, please circle the number that best suggests your opinion about each topic. Remember to circle only ONE number for each question.

10. Monthly reports of sales figures for our division reach me in time to make decisions.

Never	Seldom	Sometimes	Usually	Always
1	2	3	4	5

11. I use the display screen to view monthly sales figures.

Never	Seldom	Sometimes	Usually	Always
1	2	3	4	5

12. Monthly sales reports are valuable to me in making decisions.

Never	Seldom	Sometimes	Usually	Always
1	2	3	4	5

13. There are several errors in the monthly reports that must be corrected before they are useful.

Never	Seldom	Sometimes	Usually	Always
1	2	3	4	5

14. I think the monthly sales reports are irrelevant to most of those who receive them.

Never	Seldom	Sometimes	Usually	Always
1	2	3	4	5

FIGURE 6.10

Using a consistent format in presenting questions and responses.

Questions of Importance to Respondents Go First. The first questions should deal with subjects that respondents view as being important. This approach makes for an intriguing start to the questionnaire, and it is a technique to get people involved quickly. Respondents should feel that by answering each question and completing the form, they can cause a change or have some impact. For example, if organizational members are keen on rating potential software packages for the proposed system, begin with questions on this subject.

This technique is quite different from the typical novice's approach, which is to begin by asking for demographic information such as job title, years with the company, number of subordinates, gender, and years in school. Some people will find this style boringly similar to other forms they routinely fill out, whereas others may actually find it threatening, especially if you stressed confidentiality earlier but are now requesting data that will clearly allow identification of the respondent.

Cluster Items of Similar Content Together. When you build a frame of reference for respondents, take advantage of it by placing questions related to each other close together on the questionnaire. For example, all questions dealing with end-user computing might be grouped together in one subheaded section of the questionnaire.

Some researchers have suggested that data are cleaner if questions appear randomly, but that approach is not recommended here. Randomization in this case merely tries the patience of respondents who are likely to prefer being able to see for themselves that questionnaire construction is logical.

Employ Respondents' Associational Tendencies. Similar to the guideline for clustering questions on similar topics, employing respondents' associational tendencies goes even further by reminding the analyst to anticipate the kinds of associations that respondents make and then to use these associations in ordering questions.

For instance, if you ask, "How many subordinates do you have?" you will probably want to continue in that vein and ask about other formal organizational relationships as well. Furthermore, the respondent may associate the formal organizational structure with the informal. If that is the case, it is appropriate to include questions about the informal relationships in the same section of the questionnaire. Figure 6.11 shows an associational ordering of questions.

FIGURE 6.11
Using an associational ordering of questions to enable respondents to think about unrelated subjects.

Below are questions about data-entry procedures. Please answer the questions in the space provided.

21. List the methods you feel are appropriate for correcting erroneous entries.

22. List the methods that currently are in use to prevent data-entry errors.

23. Which of the methods you listed are effective? Why?

24. Which of the methods you listed are ineffective? Why?

25. What controls would you like to see added to improve the quality of data entry?

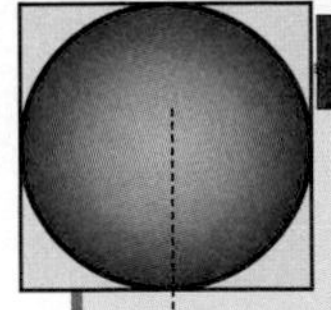

CONSULTING OPPORTUNITY 6.2

ORDER IN THE COURTS

"I love my work," Tennys says, beginning the interview with a volley. "It's a lot like a game. I keep my eye on the ball and never look back," he continues. Tennyson "Tennys" Courts is a manager for Global Health Spas, Inc., which has popular health and recreation spas worldwide.

"Now that I've finished my M.B.A., I feel like I'm on top of the world with Global," Tennys says. "I think I can really help this outfit shape up with its computers and health spas."

Tennys is attempting to help your systems group, which is developing a system to be used by all 80 outlets (where currently each group handles its paperwork in its own way). "Can I bounce this off you?" he asks Terri Towell, a member of your team of systems analysts. "It's a questionnaire I designed for distribution to all spa managers."

Ever the good sport, Terri tells Tennys that she'd love to take a look at the form. But back in the office, Terri puts the ball in your court. Systematically critique Tennys' technique as depicted in Figure 6.C2, and explain to him point by point what it needs to be a matchless questionnaire with a winning form. Building on your critique, tell Tennys what he should do to rewrite the form as an email survey instead.

QUESTIONNAIRE FOR ALL MANAGERS OF HEALTH SPAS

URGENTFILL OUT IMMEDIATELY AND RETURN PERSONALLY TO YOUR DIVISION MANAGER. YOUR NEXT PAYCHECK WILL BE WITHHELD UNTIL IT IS CONFIRMED THAT YOU HAVE TURNED THIS IN.

In ten words or fewer, what complaints have you lodged about the current computer system in the last six months to a year?

Are there others who feel the same way in your outlet as you do? Who? List their names and positions.

1. 2.

3. 4.

5. 7.

What is the biggest problem you have when communicating your information requirements to headquarters? Describe it briefly.

How much computer downtime did you experience last year?

1 - 2 - 3 - 4 - 5 - 6 - 7 - 8 - 9 - 10 -

Is there any computer equipment you never use?

Description Serial Number

Do you want it removed? Agree Neutral Disagree

In your opinion, what's next as far as computers and Global Spas are concerned?

Thanks for filling this out. • • • • • • • • • • • • • • • •

Terri
Please help me improve this form.
Tennys

FIGURE 6.C2
Questionnaire developed for managers of Global Health Spas by Tennys Court.

Bring up Less Controversial Items First. In your preliminary assessment of what is happening in the business, you will have run into some issues that are, for one reason or another, divisive to particular groups. If you believe that those issues still must be examined, try to put less controversial items before divisive or inflammatory items in a questionnaire.

For example, suppose you realize that moving key tasks to the Web has long been a sore point with some employees, but you want to find out how widespread this sentiment is. In this case the questions should still be asked, but they should follow less upsetting ones.

Because your overall objective is to gather data on attitudes, beliefs, behavior, and characteristics, you expect and seek some diversity among respondents; otherwise, a questionnaire would be extraneous. You want respondents to feel as unthreatened by and interested in the questions being asked as possible, however, without getting overwrought about a particular issue.

ADMINISTERING QUESTIONNAIRES

RESPONDENTS

Deciding who will receive the questionnaire is handled in conjunction with the task of setting up objectives for its results. Sampling, which was covered in Chapter 4, helps the systems analyst to determine what sort of representation is necessary and hence what kind of respondents should receive the questionnaire.

Recipients are often chosen as representative because of their rank, length of service with the company, job duties, or special interest in the current or proposed system. Be sure to include enough respondents to allow for a reasonable sample in the event that some questionnaires are not returned or some response sheets are incorrectly completed and thus must be discarded.

METHODS OF ADMINISTERING THE QUESTIONNAIRE

The systems analyst has several options for administering the questionnaire, and the choice of administration method is often determined by the existing business situation. Options for administering the questionnaire include the following:

1. Convening all concerned respondents together at one time
2. Personally handing out blank questionnaires and taking back completed ones
3. Allowing respondents to self-administer the questionnaire at work and drop it in a centrally located box
4. Mailing questionnaires to employees in branch or satellite sites and supplying a deadline, instructions, and return postage
5. Administering the questionnaire electronically either via email or on the Web

Each of these five methods has advantages and disadvantages. Collecting questionnaire data from a group gathered in one place at one time is helpful because there is no wait time (except whatever time it takes to complete the form) involved before getting back the data. In addition, the analyst is better able to control the data collection situation by guaranteeing that everyone receives the same instructions and that 100 percent of the forms will be returned.

A disadvantage of group data collection is that not all employees in the sample will be free at the scheduled time. In addition, there may be some resentment at being asked to focus on the task of filling out the questionnaire when other work seems more pressing. Peer pressure in this context can work for or against the completion of the questionnaire. If key respondents seem favorable, the majority will catch on and react in the same way, and vice versa.

The systems analyst can also guarantee a good response rate by personally handing out and collecting questionnaires, but analyst time becomes a problem when a large or widely dispersed group is sampled. In addition, respondents may be skeptical that even though the questionnaire stressed confidentiality, the analyst is all too aware of who is turning in which form.

Frequently, respondents are allowed to self-administer the questionnaire. Response rates with this method are a little lower than with the other methods, because people may forget about the form, lose it, or purposely ignore it. Self-administration, however, allows people to feel that their anonymity is ensured and may result in less guarded answers from some respondents. Both email and Web surveys fall into the category of self-administered questionnaires.

One way to increase the response rate on self-administered forms is to set up a central drop box at an employee's desk and ask him or her to cross off the names of respondents who return a form. That way, a particular person is not associated with a particular response form, but there is still subtle pressure to return the form.

The response rate for mailing, the fourth method of questionnaire administration, is notably poorer than other methods. Mailing a questionnaire does not involve the respondent in a personal way with the survey, but it is often important to include remote organizational members just because they are not as involved with life at headquarters. In addition, they will likely have a different perspective on current and prospective computer systems that should be taken into account.

Administering the questionnaire electronically, either via email or posted on the Web, is one way to quickly reach current system users. Costs of duplication are minimized. In addition, responses can be made at the convenience of the respondent and then can be automatically collected and stored electronically. Reminders to respondents can be easily and inexpensively sent via email, as can notifications to the analyst about when the respondent has opened the email. Some software now turns email data into data tables for use in spreadsheet or statistical software analysis. Research has not definitively determined response rates for electronic surveys, so some caution is warranted. In addition, respondents may question the confidentiality of responses that are given via email.

New research, however, shows that respondents are willing to respond to questions about highly sensitive matters via the Internet. Thus, questions that may be difficult to pose in person regarding systems problems may be acceptable to ask on a Web survey. The analyst can provide a password system for the Web survey. Such a system will improve confidentiality as well as respondents' perception of confidentiality. If done correctly, it can also help authenticate the respondent's identity, verifying that only the respondent surveyed actually answered the survey and that person has responded only once to the questionnaire. Remember to include both a "Submit" button and a "Clear" button at the end of the Web survey form. One additional concern with electronic surveys is the possibility that the analyst could miss potential users who do not currently use email or who do not have access to the Web. If this is an issue, traditional paper-and-pencil methods should be used.

SUMMARY

By using questionnaires, systems analysts can gather data on attitudes, beliefs, behavior, and characteristics from key people in the organization. Questionnaires are useful if people in the organization are widely dispersed, many people are involved with the systems project, exploratory work is necessary before recommending alternatives, or there is a need for problem sensing before interviews are conducted.

Once objectives for the questionnaire are articulated, the analyst can begin writing either open-ended or closed questions. Choice of wording is extremely important and should reflect the language of the organizational members. Ideally, the questions should be simple, specific, short, free of bias, not patronizing, technically accurate, addressed to those who are knowledgeable, and at an appropriate reading level.

Scaling is the process of assigning numbers or other symbols to an attribute or characteristic. The systems analyst may want to use scales either to measure the attitudes or characteristics of respondents or to have respondents act as judges for the subject of the questionnaire.

The four forms of measurement are nominal, ordinal, interval, and ratio scales. The form of measurement is often dictated by the data, and the analysis of data is in turn dictated to some degree by the form of measurement.

HYPERCASE® EXPERIENCE 6

"You've probably noticed by now that not everyone enjoys filling out questionnaires at MRE. We seem to get more questionnaires than most organizations. I think it's because many of the employees, especially those from the old Training Unit, value the contributions of questionnaire data in our work with clients. When you examine the questionnaire that Snowden distributed, you'll probably want not only to look at the results, but also to critique it from a methods standpoint. I always feel strongly that we can improve our internal performance so that eventually we can better serve our clients. The next time we construct a questionnaire, we want to be able to improve three things: the reliability of the data, the validity of the data, and the response rate we get."

HYPERCASE QUESTIONS

1. What evidence of questionnaires have you found at MRE? Be specific about what you have found and where.
2. Critique the questionnaire that Snowden circulated. What can be done to it to improve its reliability, validity, and response rate? Provide three practical suggestions.
3. Write a short questionnaire to follow up on some aspects of the merger between Management Systems and the Training Unit at MRE that are still puzzling you. Be sure to observe all the guidelines for good questionnaire design.
4. Redesign the questionnaire you wrote in problem 3 so that it can be used as a Web survey.

Systems analysts need to be concerned with validity and reliability. Validity means that the questionnaire measures what the systems analyst intended to measure. Reliability means that the results are consistent.

Analysts should be careful to avoid problems such as leniency, central tendency, and the halo effect when constructing scales.

Consistent control of the questionnaire format and style can result in a better response rate. Web surveys can be designed to encourage consistent responses by including radio buttons and drop-down lists and scrolling text boxes to pose both open-ended and closed questions. In addition, the meaningful ordering and clustering of questions is important for helping respondents understand the questionnaire. Surveys can be administered in a variety of ways, including (but not limited to) electronically via email or the Web or with the analyst present in a group of users.

KEYWORDS AND PHRASES

- central tendency
- check box
- closed questions
- drop-down box
- halo effect
- interval scale
- leniency
- nominal scale
- open-ended questions
- ordinal scale
- questionnaire
- radio button
- ratio scale
- reliability
- scaling
- scrolling text box
- survey respondents
- validity

REVIEW QUESTIONS

1. What kinds of information is the systems analyst seeking through the use of questionnaires or surveys?
2. List four situations that make the use of questionnaires appropriate.
3. What are the two basic question types used on questionnaires?
4. List two reasons why a systems analyst would use a closed question on a questionnaire.
5. List two reasons why a systems analyst would use an open-ended question on a questionnaire.
6. What are the seven guidelines for choosing language for the questionnaire?
7. Define what is meant by scaling.
8. What are two kinds of information that can be gained by the use of scales on questionnaires?
9. What are nominal scales used for?
10. What is the diffenence between nominal and ordinal scales?
11. Give and example of an interval scale.
12. When should the analyst use interval scales?
13. Describe the difference between interval and ratio scales.
14. Define reliability as it refers to the construction of scales.
15. Define validity as it refers to the construction of scales.
16. List three problems that can occur because of careless construction of scales.
17. What are four actions that can be taken to ensure that the questionnaire format is conducive to a good response rate?
18. Which questions should be placed first on the questionnaire?
19. Why should questions on similar topics be clustered together?
20. What is an appropriate placement of controversial questions?
21. List five methods for administering the questionnaire.
22. What considerations are necessary when questionnaires are Web-based?

PROBLEMS

1. Cab Wheeler is a newly hired systems analyst with your group. Cab has always felt that questionnaires are a waste. Now that you will be doing a systems project for MegaTrucks, Inc., a national trucking firm with branches and employees in 130 cities, you want to use a questionnaire to elicit some opinions about the current and proposed systems.
   a. Based on what you know about Cab and MegaTrucks, give three persuasive reasons why he should use a survey for this study.
   b. Given your careful arguments, Cab has agreed to use a questionnaire but strongly urges that all questions be open ended so as not to constrain the respondents. In a paragraph, persuade Cab that closed questions are useful as well. Be sure to point out trade-offs involved with each question type.
2. "Everytime we get consultants in here, they pass out some goofy questionnaire that has no meaning to us at all. Why don't they bother to personalize it, at least a little?" asks Ray Dient, head of emergency systems. You are discussing the possibility of beginning a systems project with Pohattan Power Company (PPC) of Far Meltway, N.J.
   a. What steps will you follow to personalize a standardized questionnaire?
   b. What are the advantages of adapting a questionnaire to a particular organization? What are the disadvantages?

3. A sample question from the draft of the Pohattan Power Company questionnaire reads:
   I have been with the company:

   20–upwards years
   10–15 years upwards
   5–10 years upwards
   less than a year
   Check one that most applies.
   a. What kind of a scale is the question's author using?
   b. What errors have been made in the construction of the question, and what might be the possible responses?
   c. Rewrite the question to achieve clearer results.
   d. Where should the question you've written appear on the questionnaire?
4. Also included on the PPC questionnaire is this question: When residential customers call, I always direct them to our Web site to get an answer.

Sometimes	*Never*	*Always*	*Usually*
1	2	3	4

   a. What type of scale is this one intended to be?
   b. Rewrite the question and possible responses to achieve better results.
5. Another question used on the PPC draft questionnaire reads: My satisfaction with access time to the customer database is:

Displeased				*Satisfaction*
0%	5%	25%	50%	100%

   a. What kind of scale does the question's author intend to use?
   b. Is the intended scale appropriate for the question? Why or why not?
   c. Rewrite the question to achieve clearer results.
6. Figure 6.EX1 is a questionnaire designed by an employee of Green Toe Textiles, which specializes in manufacturing men's socks. Di Wooly wrote the

FIGURE 6.EX1
Questionnaire developed by Di Wooly.

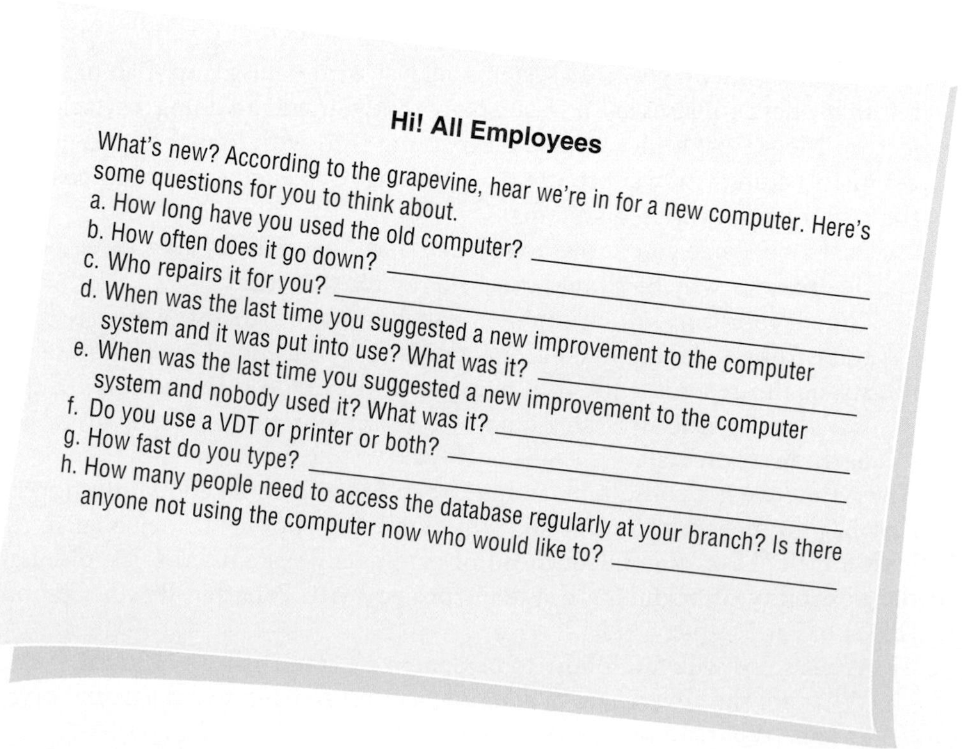

Hi! All Employees

What's new? According to the grapevine, hear we're in for a new computer. Here's some questions for you to think about.

a. How long have you used the old computer? __________
b. How often does it go down? __________
c. Who repairs it for you? __________
d. When was the last time you suggested a new improvement to the computer system and it was put into use? What was it? __________
e. When was the last time you suggested a new improvement to the computer system and nobody used it? What was it? __________
f. Do you use a VDT or printer or both? __________
g. How fast do you type? __________
h. How many people need to access the database regularly at your branch? Is there anyone not using the computer now who would like to? __________

questionnaire because, as the office manager at headquarters in Juniper, Tennessee, she is concerned with the proposed purchase and implementation of a new computer system.

   a. Provide a one-sentence critique for each question given.
   b. In a paragraph, critique the layout and style in terms of white space used, room for responses, ease of responding, and so on.

7. Based on what you surmise Ms. Wooly is trying to get through the questionnaire, rewrite and reorder the questions (use both open-ended and closed questions) so that they follow good practice and result in useful information for the systems analysts. Indicate next to each question that you write whether it is open-ended or closed, and write a sentence indicating why you have written the question this way.
8. Redesign the questionnaire you created for problem 7 for use on email. Write a paragraph saying what changes were necessary to accommodate users on email.
9. Redesign the questionnaire you created for problem 7 as a Web survey. Write a paragraph saying what changes were necessary to accommodate users on the Web.
10. Frieda Forall, head of nursing, is insistent that employees at her for-profit hospital should complete questionnaires about the proposed computer system at home, on their own time, not the hospital's.
    a. Write a paragraph to convince Frieda that administering the questionnaire in a way other than self-administration off the premises would garner better results and be a worthwhile use of employees' time.
    b. Nurse Forall is wavering in her position on self-administration but as yet remains unconvinced. List some incentives that can motivate managers to administer a questionnaire properly the first time.

GROUP PROJECTS

1. Using the interview data you gained from the group exercise on Maverick Transport in Chapter 5, meet with your group to brainstorm the design of a questionnaire for the hundreds of truck drivers that Maverick Transport employs. Recall that Maverick is interested in implementing a satellite system for tracking freight and drivers. There are other systems that may affect the drivers as well. As your group constructs the questionnaire, consider the drivers' likely level of education and any time constraints the drivers are under for completing such a form.
2. Using the interview data you gained from the group exercise on Maverick Transport in Chapter 5, your group should meet to design an email or Web questionnaire for surveying the company's 20 programmers (15 of whom have been hired in the past year) about their skills, ideas for new or enhanced systems, and so on. As your group constructs the programmer survey, consider what you have learned about users in the other interviews as well as what vision the director of information technology holds for the company.

SELECTED BIBLIOGRAPHY

Babbie, E. R. *Survey Research Methods.* Belmont, CA: Wadsworth, 1973.

Cooper, D. R., and P. S. Schindler. *Business Research Methods*, 6th ed. New York: Irwin/McGraw Hill, 1998.

Dillman, D. A. *Mail and Telephone Surveys*. New York: Wiley, 1978.

Emory, C. W. *Business Research Methods*, 3d ed. Homewood, IL: Irwin, 1985.

Georgia Tech's Graphic, Visualization, and Usability Center. "WWW User Surveysat." Available: <www.cc.gatech.edu/gvu/> 1998.

Hessler, R. M. *Social Research Methods*. New York: West, 1992.

Peterson, R. A. *Constructing Effective Questionnaires*. Thousand Oaks, CA: Sage Publications, 1999.

Strauss, J., and R. Frost. *E-Marketing*, 2d ed. Upper Saddle River, NJ: Prentice Hall, 2001.

Sudman, S., and N. M. Bradburn. *Asking Questions: A Practical Guide to Questionnaire Design*. San Francisco: Jossey-Bass, 1988.

Allen Schmidt, Julie E. Kendall, and Kenneth E. Kendall

THE QUEST CONTINUES

Anna sits at her desk, reviewing the interview summaries and the memos that were gathered during the summer. Several stacks of papers are neatly filed in expansion folders.

"We have so much information," she remarks to Chip, "yet I sense that we are only working with the tip of the iceberg. I don't have a solid feeling for the difficulties of faculty members and research staff. Are they experiencing some of the problems that came out in the memos and interviews? Are there additional problems we haven't heard about?"

Chip looks up from his work of trying to extract key points for defining the problems. "I wonder if we should do more interviews, or perhaps gather more documents," he says.

"That would be a good idea, but how many interviews should we conduct and who should we interview?" Anna replies. "Suppose we interview several staff members and base the new system on the results. We could interview the wrong people and design a system to satisfy only their needs, missing key problems that the majority of faculty and staff need to have solved."

"I see what you mean," Chip answers. "Perhaps we should design a questionnaire and survey the faculty and research staff. Most of them would return the survey if it was easy to complete, especially people with major concerns."

"Great idea!" Anna says. "How should we decide which questions to include on the survey?"

"Let's speak with some key people and base the survey on the results. A good starting point would be Hy Perteks, because he is always talking with the faculty and staff. I'll give him a call and arrange a meeting," Chip says.

Chip arranged the meeting for the following morning. It would be held in a conference room adjacent to the Information Center.

"Thanks for meeting with us on such short notice," Chip opens. "We're thinking about surveying the faculty and research staff to obtain additional information that will help us define the system concerns."

"I think it's a tremendous idea," Hy replies. "I would also like some information that would give me a clue as to what type of software should be available in the Information Center and the type of training we should provide."

"What type of software information do you think we should obtain?" asks Anna.

"Certainly the major package types used," Hy answers. "Word-processing software is extremely popular. We should find out which package each user likes and, equally important, which version of the package. I know that many are using Microsoft Word and others are using WordPerfect. I have a few folks using some minor word-processing software they are comfortable with. Database software also varies. Many are using Access, and others are involved with Paradox, File Maker Pro, or FoxPro. Same for spreadsheets, with Lotus, Excel, and Quattro Pro being the most popular.

"Another consideration would be what type of specialized software is being used by groups of faculty members," muses Hy. "Many of the people in the math department are using Exp, a math word processor. Others are using various software packages for a number of courses. For instance, the information science people are using Visible Analyst, but a few are using Visio. I've also heard that we're getting some biology and astronomy software. And the art department uses Macs almost exclusively for full-color production. Many of the faculty are getting heavily into software for constructing Web sites, such as Dreamweaver and Front Page."

"Other than software packages and versions, what types of information should we capture?" asks Chip.

"I would like to know what level of expertise each person has," responds Hy. "No doubt, some are beginners, whereas others have a good knowledge but have not mastered all the features of a particular package. Some are experts, without question. They know the software inside and out. I'm interested in the beginners and intermediate users, because we should be providing different training for them. And I would really like to know who the experts are. Then I could contact them and ask if they would be willing to conduct a training session or serve as a resource if someone has a problem with an advanced feature of the software."

"Is there anything else you feel we should find out about in the survey?" asks Chip.

"The only other thing that I worry about are problems that result in a faculty or staff member not using the software," Hy replies.

"What do you mean?" asks Chip.

"Well, suppose a person has the software but it is installed incorrectly or displays a message, such as 'Not enough memory to run,' or 'This wizard isn't installed,'" replies Hy. "I've had some inquiries about this matter recently. One person said that they couldn't use Access except for simple tasks because he always got a message saying that the wizards weren't installed. It turned out that the system was not configured correctly to run over a network. It was a simple matter to fix the problem, but it had been going on for a long time! Other such problems must exist, and I'd like to know what they are. That would make our whole staff more productive and comfortable with using PCs."

"Do you know of a representative faculty or staff member whom we should interview?" Anna inquires.

"Yes, there's a faculty member in math, Rhoda Booke, who has consistently shown interest in hardware and software issues. I've helped her a number of times, and she's always friendly and grateful."

"Thanks once again for all your help," says Chip. "We'll get back to you later with the results of the survey."

Anna arranges a meeting with Rhoda and explains the nature of the project and why she was selected as a faculty representative. The meeting was held in a small conference room in the math department.

"Thanks for meeting with us," says Anna after introductions had been made. "We'd like to have the faculty perspective on problems encountered with PCs and the associated software. Our goal is to provide the faculty with the best possible resources with the least number of problems."

"I'm really glad to be a part of the project," exclaims Rhoda. "I've been using software for about five years, and what a learning experience it has been! Thank goodness that Hy is available as a resource person. I've taken hours of his time, and it's been well worth the effort. I feel much more productive, and the students are using software that helps them grasp the material more thoroughly than simply doing math exercises and reading the text."

"That's good news. But are there some difficulties that you've been experiencing?" asks Chip.

"Well, becoming familiar with the software is a major hurdle. I spent a good portion of last summer, when I wasn't working on my book, learning how to use some of the classroom software for both algebra and calculus. The stuff's great, but I got stuck several times and had to call for help. It's necessary to understand the software to prepare lesson plans and explain to the students how to use it."

"How about problems with installing the software or hardware?" Anna asks.

"Oh, yes!" exclaims Rhoda. "I tried to install the software, and it went smoothly until the part where the screen asked questions about a number of import file formats for graphics, such as PSD and PNG. I didn't even know what those letters meant," laughs Rhoda.

"Then there were setup problems," Rhoda continues. "I needed to figure out what to install on the network and what to include on the local hard drive. What a learning experience that was. Some of the computers in the student lab gave us 'Not enough memory' error messages, and we learned that they had been installed with minimum memory. I've heard that the physics faculty had the same problem."

"Are there any other concerns you have or feel that we should include on our survey to the faculty and research staff?" Chip asks.

"It would be useful to know who is using the same software in different departments and what software is supplied by which vendor. Perhaps if we have many packages from one vendor, we could get a larger discount for software. The department software budget is already overwhelmed with requests," Rhoda says.

"Thanks for all your help," Anna says. "If you think of any additional questions we should include on the survey, please do not hesitate to call us."

Back in their office, the analysts start compiling a list of the issues to be contained on the survey.

"We certainly need to ask about the software in use and about training needs," remarks Anna. "We should also address the problems that are occurring."

"Agreed," replies Chip. "I feel that we should include questions on software packages, vendors, versions, level of expertise, and training concerns. What I'm not so sure about is how to obtain information on problems the faculty and staff are encountering. How should we approach these issues?"

"Well," replies Anna, "we should focus on matters with which they are familiar. We might ask questions about the type of problems that are occurring, but certainly not technical ones that they wouldn't find to be of interest. For example, we should not ask, 'How much main memory is available on your machine?' because they may not know or care. They probably would not be familiar with the Windows files or network access rights for the software either. And the survey should not ask any questions that we could easily look up answers to, such as 'Who is the vendor for the software?'"

"I see," replies Chip. "Let's divide the questions into categories. Some would be closed questions and some would be open ended. Then there's the matter of which structure to use.

EXERCISES

E-1. Based on the dialogue among Chip, Anna, Hy, and Rhoda, make a detailed list of concerns raised about hardware and software for faculty computing.

E-2. From the list of concerns, select the issues that would best be phrased as closed questions.
E-3. From the list of concerns, select the issues that would best be phrased as open-ended questions.
E-4. On the basis of problems 2 and 3, design a questionnaire to be sent to the faculty and research staff.
E-5. Pilot your questionnaire by having other students in class fill it out. On the basis of their feedback and your capability to analyze the data you receive, revise your questionnaire.

OBSERVING DECISION-MAKER BEHAVIOR AND THE OFFICE ENVIRONMENT

7

KINDS OF INFORMATION SOUGHT

Observing the decision maker and the decision maker's physical environment are important information-gathering techniques for the systems analyst. Through observing activities of decision makers, the analyst seeks to gain insight about what is actually done, not just what is documented or explained. In addition, through observation of the decision maker, the analyst attempts to see firsthand the relationships that exist between decision makers and other organizational members.

By observing the office environment, the systems analyst seeks the symbolic meaning of the work context for decision makers. The analyst examines physical elements of the decision maker's workspace for their influence on decision-making behavior. Furthermore, through observation of the physical elements over which the decision maker has control (clothing, desk position, and so on), the analyst works to understand what messages the decision maker is sending. Finally, through observation, the analyst works to comprehend the influence of the decision maker on others in the organization. All these kinds of information are summarized in Figure 7.1.

OBSERVING A DECISION MAKER'S BEHAVIOR

Systems analysts use observation for many reasons. One reason is to gain information about decision makers and their environments that is unavailable through any other method. Observing also helps to confirm or negate and reverse what has been found through interviewing, questionnaires, and other methods.

Observation must be structured and systematic if the findings are to be interpretable. Thus, it is of utmost importance that the systems analyst know what is being observed. Great care and thought must go into what and who will be observed as well as when, where, why, and how. It is not enough simply to be aware of the need for observation.

Many observational schemes are available, each with its own objectives. Analysts are encouraged to draw from research as well as from their own experience to devise observational schemes that are workable.

OBSERVING A TYPICAL MANAGER'S DECISION-MAKING ACTIVITIES

Managers' workdays have been described as a series of interruptions punctuated by short bursts of work. In other words, pinning down what a manager "does" is a slippery proposition even under the best of circumstances. For the systems analyst

FIGURE 7.1
Kinds of information sought when observing decision-maker behavior and the office environment.

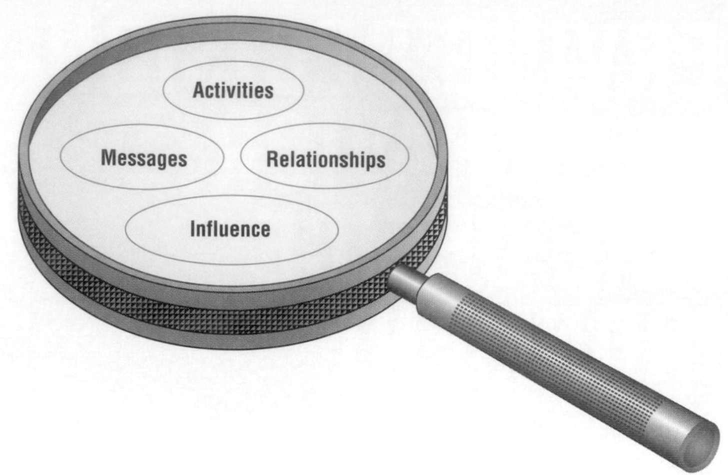

to grasp adequately how managers characterize their work, interviews and questionnaires are used, as discussed in Chapters 5 and 6. Observation, however, allows the analyst to see firsthand how managers gather, process, share, and use information to get work done.

The following steps aid in observing a manager's typical decision-making activities:

1. Decide what is to be observed (activities).
2. Decide at what level of concreteness activities are to be observed (that is, will the analyst observe that "The manager freely shared information with subordinates" or make a much more concrete observation such as, "Manager sends a copy of the same memo to three subordinates"?). Determining the level of concreteness of observation will also dictate the amount of inference in each observation and subsequently the amount of interpretation needed once observations are made.
3. Create categories that adequately capture key activities.
4. Prepare appropriate scales, checklists, or other materials for observation.
5. Decide when to observe.

Deciding when to observe is covered in the next subsection.

TIME AND EVENT SAMPLING

Each approach to when to observe has its own advantages and trade-offs. Time sampling allows the analyst to set up specific intervals at which to observe managers' activities. For example, time sampling might specify observing a decision maker during five randomly chosen 10-minute intervals throughout seven 8-hour days. The advantages of time sampling include cutting down on the bias that might otherwise enter into observations made "just anytime." Time sampling also allows for a representative view of activities that occur fairly frequently.

The drawbacks of time sampling include gathering observational data in a piecemeal fashion that may not allow sufficient time for an event such as a decision to unfold in its entirety. A second problem with time sampling when gathering observational data is that rare or infrequent but important events (for example, a strategic decision on a five-year investment in a new management information system) may not be represented in the time that is sampled. Nonetheless, the decision is important and will have an impact.

	Time Sampling	Event Sampling
Advantages	• Cuts down on bias with randomization of observations • Allows a representative view of frequent activities	• Allows observation of behavior as it unfolds • Allows observation of an event designated as important
Disadvantages	• Gathers data in a fragmented fashion that doesn't allow time for a decision to unfold • Misses infrequent but important decisions	• Takes a great deal of analyst's time • Misses a representative sample of frequent decisions

FIGURE 7.2
Advantages and disadvantages of time versus event sampling.

Event sampling addresses both these concerns by purposefully sampling entire events such as "a board meeting" or "a user training session" rather than sampling time periods randomly. Event sampling provides for observation of an integral behavior in its natural context. A drawback of event sampling is that it may not be possible to achieve a representative sample of frequent occurrences.

In light of the pros and cons of both approaches, analysts who are interested in observation are encouraged to combine time and event sampling when deciding on the what, when, why, and how of decision-maker activities. A comparison and contrast of time versus event sampling is given in Figure 7.2.

As discussed in Chapter 2, decision making occurs on the operational, managerial, and strategic levels of the organization. The preceding discussion assumed that decision makers at all levels of the organization will interact with the information system and hence should be observed.

OBSERVING A DECISION MAKER'S BODY LANGUAGE

The systems analyst subconsciously observes body language during the interviews and other interactions. This discussion is intended to bring that awareness to the conscious level, where it can be recognized and used by the analyst. Understanding body language enables the analyst to understand better the information requirements of the decision maker by adding dimension to what is being said. Nevertheless, although it is important to observe decision makers' body language, precise interpretation of it, movement for movement, is immensely difficult and also varies across cultures.

Adjective Pairs and Categories. Adjective pairs have become a popular way to record behavior. An example of decision-making behavior described in adjective pairs is decisive/indecisive, confident/not confident, assertive/unassertive, and so on, as depicted in Figure 7.3.

Category systems for recording decision-making behavior were discussed briefly before. The analyst determines activity categories before observations are undertaken. An example of a concrete category is "Accesses database personally." Examples of a category system that asks for more analyst inference in recording observation are "Uses internal sources of data" or "Shows initiative in accessing

FIGURE 7.3

A sample adjective-pairs sheet for observing a decision maker.

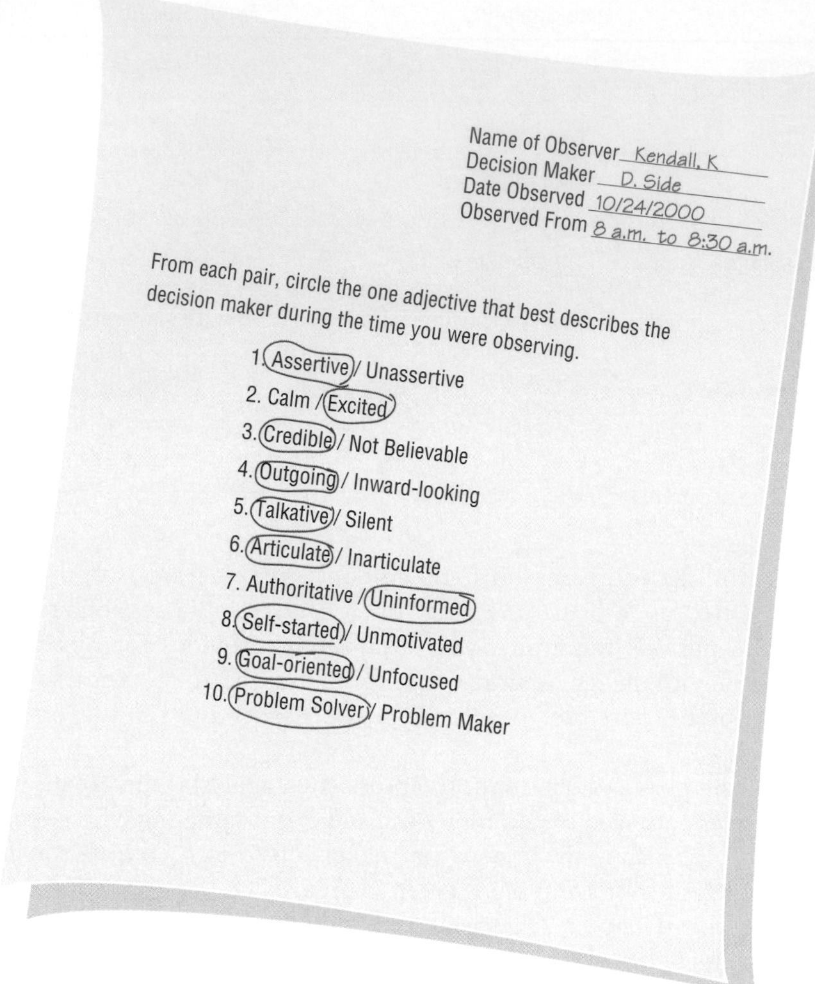
Name of Observer Kendall, K
Decision Maker D. Side
Date Observed 10/24/2000
Observed From 8 a.m. to 8:30 a.m.

From each pair, circle the one adjective that best describes the decision maker during the time you were observing.

1. Assertive / Unassertive
2. Calm / Excited
3. Credible / Not Believable
4. Outgoing / Inward-looking
5. Talkative / Silent
6. Articulate / Inarticulate
7. Authoritative / Uninformed
8. Self-started / Unmotivated
9. Goal-oriented / Unfocused
10. Problem Solver / Problem Maker

data." Forms listing categories are then copied in sufficient number and taken along to be completed as the analyst observes. A sample form illustrating the category system is shown in Figure 7.4.

The Analyst's Playscript. Systems analysts can also use a technique called playscripting to record observed behavior. With this technique the "actor" is the decision maker who is observed "acting" or making decisions. In setting up a playscript, the actor is listed in the left-hand column and all his or her actions are listed in the right-hand column, as shown in Figure 7.5. All activities are recorded with action verbs so that a decision maker would be described as "talking," "sampling," "corresponding," and "deciding."

Playscript is an organized and systematic approach that demands the analyst be able to understand and articulate the action taken by each observed decision maker. This approach eventually assists the systems analyst in determining what information is required for major or frequent decisions made by the observed people. For instance, from the quality assurance manager example in the playscript, it becomes clear that even though this decision maker is on the middle management level, he or she still requires a fair amount of external information to perform the required activities of this specific job.

Name of Observer Kim McCabe
Decision Maker A.K. Stratton
Date Observed 6/10/2000
Observied From 8:30 to 12:00
(Start time to end time)

Each time the decision maker is observed engaging anew in the behavior listed, put 1 mark in the box beside the appropriate category. Mark only when you actually observe the decision maker in action. Fill out the first column ONLY when you are in the presence of the decision maker you are observing.

Behavior	Number of Times Behavior Occurs	Total	Percentage of Total
Instructs Subordinates	///	3	5
Instructs Peers	//	2	4
Instructs Superiors	/	1	2
Questions Subordinates	//	2	4
Questions Peers	///	3	5
Questions Superiors	/	1	2
Reprimands Subordinates	—	0	0
Reprimands Peers	—	0	0
Reprimands Superiors	—	0	0
Opens Mail	𝍸	5	9
Answers Phone	𝍸 //	7	13
Makes Phone Call	𝍸 𝍸	10	18
Reads External Information	𝍸 𝍸 ///	13	23
Reads Internal Information	///	3	5
Processes Own Information	//	2	4
Asks Others to Process Information	////	4	7
		56	≈ 100%

FIGURE 7.4
A sample observation form employing categories of decision-maker behavior.

OBSERVING THE PHYSICAL ENVIRONMENT

Observing the activities of decision makers is just one way to assess their information requirements. Observing the physical environment in which decision makers work also reveals much about their information requirements. Most often, such observing means systematically examining the offices of decision makers, because offices constitute their primary workplace. Decision makers influence and are in turn influenced by their physical environments.

STRUCTURED OBSERVATION OF THE ENVIRONMENT (STROBE)

Film critics sometimes use a structured form of criticism called mise-en-scène analysis to systematically assess what is in a single shot of the film. They look at editing, camera angle, set decor, and the actors and their costumes to find out how

FIGURE 7.5

A sample page from the analyst's playscript describing decision making.

Playscript Analysis Company: Solid Steel Shelving Analyst: L. Bracket Scenario: Quality Assurance Date: 9/3/2000

Decision Maker (Actor)	Information-Related Activity (Script)
Quality Assurance Manager	Asks shop floor supervisor for the day's production report
Shop Floor Supervisor	Prints out daily computerized production report
	Discusses recurring problems in production runs with quality assurance (QA) manager
Quality Assurance Manager	Reads production report
	Compares current report with other reports from the same week
	Inputs data from daily production run into QA model on micro
	Observes onscreen results of QA model
	Calls steel suppliers to discuss deviations from quality standards
Shop Floor Supervisor	Attends meeting on new quality specifications with quality assurance manager and vice president of production
Quality Assurance Manager	Drafts letter to inform suppliers on new quality specifications agreed upon in meeting
	Sends draft to vice president via email
Vice President of Production	Reads drafted letter
	Returns corrections and comments via email
Quality Assurance Manager	Reads corrected letter on email
	Rewrites letter to reflect changes

they are shaping the meaning of the film as intended by the director. Sometimes the film's mise-en-scène will contradict what is said in the dialogue. For information requirements analysis, the systems analyst can take on a role similar to that of the film critic. It often is possible to observe the particulars of the surroundings that will confirm or negate the organizational narrative (or dialogue) that is found through interviews or questionnaires.

The method for *STR*uctured *OB*servation of the *E*nvironment is referred to as STROBE. It is systematic because (1) it provides a standard methodology and standard classification for the analysis of those organizational elements that influence decision making, (2) it allows other systems analysts to apply the same analytic framework to the same organization, and (3) it limits analysis to the organization as it exists during the current stage in its life cycle.

A correspondence exists between the elements of analysis for film criticism and those used in the STROBE assessment of a decision maker's information requirements. This correspondence makes it possible for the systems analyst to remember easily the analogy, as pictured in Figure 7.6. A more detailed discussion of each of these elements follows. There are seven concrete elements that are easily observable by the systems analyst. These elements can reveal much about the way a decision maker gathers, processes, stores, and shares information as well as about the decision maker's credibility in the workplace. The seven observable elements are described in the following paragraphs.

Office Location. One of the first elements a systems analyst should observe is the location of a particular decision maker's office with respect to other offices. Accessible offices tend to increase interaction frequency and informal messages, whereas inaccessible offices tend to decrease the interaction frequency and increase task-oriented messages. Offices distributed along the perimeter of the building usually result in a report or memo being held up in one of the offices, whereas office clusters encourage information sharing. It is also likely that the people whose offices are separated from others may tend to view the organization differently and so drift farther apart from other organization members in their objectives.

Placement of the Decision Maker's Desk. Placement of a desk in the office can provide clues to the exercise of power by the decision maker. Executives who enclose a visitor in a tight space with the visitor's back to the wall while allowing themselves a lot of room put themselves into the strongest possible power position. An executive who positions his or her desk facing the wall with a chair at the side for a visitor is probably encouraging participation and equal exchanges. The systems analyst should notice the arrangement of the office furniture and in particular the placement of the desk.

Filmic Elements	Organizational Elements
Set location	Office location
People positioned within a frame	Decision maker's placement in an office (i.e., desk placement)
Stationary objects	File cabinets, bookshelves, and equipment for storing information
Props (movable objects)	Calculators, PCs, and other items used for processing information
External objects (brought in from other scenes)	Trade journals, newspapers, and items used for external information
Lighting and color	Office lighting and color
Costumes	Clothing worn by decision makers

FIGURE 7.6
The analogy between filmic elements for observation and elements of STROBE.

FIGURE 7.7

Observe a decision maker's office for clues concerning his or her personal storage, processing, and sharing of information.

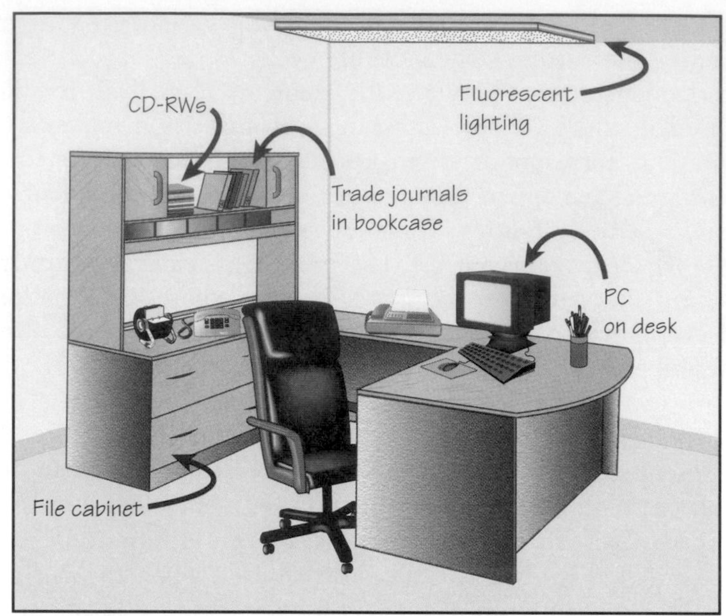

Stationary Office Equipment. File cabinets, bookshelves, and other large equipment for storing items are all included in the category of stationary office equipment. If there is no such equipment, it is likely the decision maker stores very few items of information personally. If there is an abundance of such equipment, it is presumed the decision maker stores and values much information.

Props. The term *props* (an abbreviation of the stage/film term *properties*) refers to all the small equipment used to process information, including calculators, PCs, pens, pencils, and rulers. The presence of calculators and PCs suggests that a decision maker who possesses such equipment is more likely to use it personally than one who must leave the room to use it.

Trade Journals and Newspapers. A systems analyst needs to know what type of information is used by the decision maker. Observation of the type of publications stored in the office can reveal whether the decision maker is looking for external information (found in trade journals, newspaper clippings about other companies in the industry, and so on) or relies more on internal information (company reports, intraoffice correspondence, policy handbooks).

Office Lighting and Color. Lighting and color play an important role in how a decision maker gathers information. An office lighted with warm, incandescent lighting indicates a tendency toward more personal communication. An executive in a warmly lit office will gather more information informally, whereas another organizational member working in a brightly lit, brightly colored office may gather information through more formal memos and official reports. Figure 7.7 shows a decision maker's office equipment, props, trade journals, office lighting, and color.

Clothing Worn by Decision Makers. Much has been written about the clothing worn by executives and others in authority. The systems analyst can gain an understanding of the credibility exhibited by managers in the organization by observing the clothing they wear on the job. The formal three-piece suit for a man or the skirted suit for a woman represents the maximum authority, according to some

Characteristics of Decision Makers	Corresponding Elements in the Physical Environment
Gathers information informally	Warm, incandescent lighting and colors
Seeks extraorganizational information	Trade journals present in office
Processes data personally	PCs, calculators present in office
Stores information personally	Equipment/files present in office
Exercises power in decision making	Desk placed for power
Exhibits credibility in decision making	Wears authoritative clothing
Shares information with others	Office easily accessible

FIGURE 7.8
A summary of decision-maker characteristics that correspond to observable elements in the physical environment.

researchers who have studied perceptions of executive appearance. Casual dressing by leaders tends to open the door for more participative decision making, but such attire often results in some loss of credibility in the organization if the predominant culture values traditional, conservative clothing.

Through the use of STROBE, the systems analyst can gain a better understanding of how managers gather, process, store, and use information. A summary of the characteristics exhibited by decision makers and the corresponding observable elements is shown in Figure 7.8. The following subsection examines options available to the systems analyst for recording and documenting observations.

APPLYING STROBE

Analysts may choose among many application strategies when using the STROBE approach. These strategies vary from very structured (such as taking photographs for later analysis) to unstructured. Four strategies are described in the following paragraphs.

Analysis of Photographs. Photographing the environments of decision makers and then analyzing the photographs for elements of STROBE is most closely allied with the original use of mise-en-scène for film criticism. Interestingly, this application has parallels in much earlier management work; at the turn of the century, Frank Gilbreth used film in his famous time-motion studies, analyzing frame-by-frame what motions were necessary to complete a task.

Photographic applications of STROBE have some distinct advantages. One is that a document is made that can be referred to repeatedly, which can be extremely helpful when organizational visits must be limited due to time, distance, or expense. A second advantage is that the photographer can focus specifically on pertinent elements of STROBE and thereby exclude extraneous elements. In addition, using photography for STROBE allows a side-by-side comparison of organizations, because the limitations of time and space are overcome by photography. A fourth advantage is that a photograph can supply detail that is easily overlooked during personal contact, when the systems analyst is not only observing but also conducting an interview or investigating hard data. A fifth advantage is that a photograph can be scanned into a Web document for use by all team members if the organization so agrees to this use.

There are also drawbacks to using photography for implementing STROBE. First and foremost may be the problem of deciding what to photograph. In contrast to the human eye, photographs are very limited as to what they can aim at and "take in." The second drawback is that photography, although it may prove

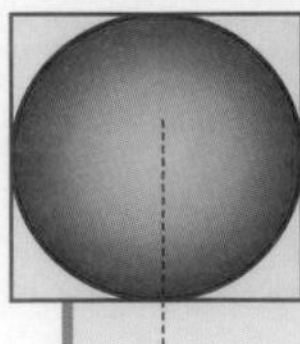

CONSULTING OPPORTUNITY 7.1

DON'T BANK ON THEIR SELF-IMAGE OR NOT EVERYTHING IS REFLECTED IN A MIRROR

"I don't want any power here," demurs Dr. Drew Charles, medical director of the regional blood center where your systems group has just begun a project. "I'm up to my neck in work just keeping the regional physicians informed so they follow good bloodbanking practices," he says, as he shields his eyes from the bright sunlight streaming into his office. He clicks off the monitor connected to his PC and turns his attention to you and the interview.

Dr. Charles is dressed in a conservative, dark wool suit and is wearing a red-striped silk necktie. He continues, "In fact, I don't make decisions. I'm here purely in a positive support role." He pulls out the organizational chart shown here to illustrate his point. "It is as clear as a fracture. The administrator is the expert on all administrative matters. I am the medical consultant only."

Dr. Charles's office is stacked high not only with medical journals such as *Transfusion* but also with *BYTE* magazine and *Business Week*. Each is opened to a different page, as if the doctor were in the process of devouring each new morsel of information. The overflow journals, however, are not stored meticulously on metal bookshelves as expected. In sharp contrast to the gleaming, new equipment you saw being used in the donor rooms, the journals are piled a foot high on an old blood-donating bed that has been long retired from its intended use.

Next, you decide to interview the chief administrator, Craig Bunker, to whom Dr. Charles has alluded. Fifteen minutes after the scheduled start of your appointment, Bunker's secretary, Dawn Upshaw, finally allows you to enter his office. Bunker, who has just finished a phone call, is dressed in a light-blue sport coat, checkered slacks, light-blue shirt, and a necktie. "How are you doing? I've just been checking around to see how everything's perking along," Bunker says by way of introduction. He is outgoing and very friendly.

As you glance around the room, you notice that there are no filing cabinets, nor is there a PC such as the one Dr. Charles was using. There are lots of photos of Craig Bunker's family, but the only item resembling a book or magazine is the center's newsletter, *Bloodline*. As the interview begins in earnest, Bunker cheerfully launches into stories about the Pennsylvania Blood Center, where he held the position of assistant administrator six years ago.

Finally, you descend the stairs to the damp basement level of the Heath Lambert Mansion. The bloodmobiles have just returned, and processed blood has been shipped to area hospitals. You decide to talk with Sang Kim, a bloodmobile driver; Jenny McLaughlin, the distribution manager; and Roberta Martin, a lab technician who works the night shift.

Roberta begins, "I don't know what we'd do without the doctor." In the same vein, though, Sang feeds the conversation by remarking, "Yeah, he helped us by thinking up a better driving schedule last week."

Jenny adds, "Dr. Charles is invaluable in setting the inventory levels for each hospital, and if it wasn't for him, we wouldn't have word processing yet, let alone our new computer."

As one of the systems analysis team members assigned to the blood center project, develop an anecdotal checklist using STROBE to help you systematically interpret the observations you made about the offices of Dr. Charles and Craig Bunker. Consider any disparities between a decision maker's clothing, what a decision maker states, and what is said by others; between office location and what is stated; and between office equipment and policies stated. In addition, in a paragraph, suggest possible follow-up interviews and observations to help settle any unresolved questions.

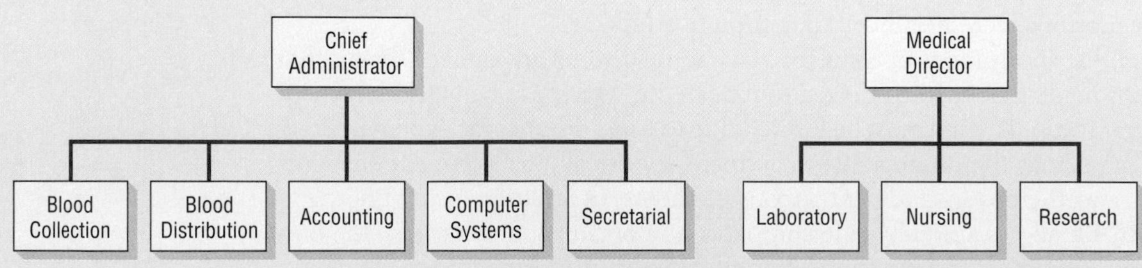

FIGURE 7.C1

Organization chart of the regional blood center.

Source: © 2000 The New Yorker Collection from the cartoonbank.com. All Rights Reserved.

unobtrusive in the long term, is initially obtrusive. The systems analyst will face problems of decision makers who wind up posing as well as intentionally or unintentionally changing their environments in an attempt to make them somehow more acceptable to the analyst.

Checklist/Likert Scale Approach. A second application of STROBE is a less-structured technique than photography: a checklist/Likert scale approach. Researchers developed five-point Likert-type scales relating to seven decision-maker characteristics that were observable through physical elements in decision makers' organizational environments, as shown in Figure 7.9.

In the original study using this scale to assess 16 blood administrators and medical directors from the United States and Canada, researchers found convergent and discriminant validity of the information gained through the STROBE scales and of the information gained through interviewing and behavioral scales. The same Likert-type scales are recommended to systems analysts, who can use them as an application of STROBE in conjunction with more traditional methods.

Anecdotal List (With Symbols). A third and even less-structured way to implement STROBE is through the use of an anecdotal checklist with meaningful shorthand symbols. This approach to STROBE was useful in ascertaining the information requirements for four key decision makers in a Midwestern blood center.

As can be seen in Figure 7.10, five shorthand symbols were used by the systems analysts to evaluate how observation of the elements of STROBE compared with the organizational narrative generated through interviews. The five symbols are as follows:

1. A checkmark means the narrative is confirmed.
2. An "X" means the narrative is reversed.
3. An oval or eye-shaped symbol serves as a cue for the systems analyst to look further.
4. A square means observation of the elements of STROBE modifies the narrative.
5. A circle means the narrative is supplemented by what is observed.

FIGURE 7.9

Likert-type scales for use in observing the physical environment of decision makers with STROBE.

STROBE Scales for Observing the Physical Environment

1. Office lighting, walls, paintings, and graphics are warm-toned, creating an informal arena for information exchange.

Fluorescent lights, cool-colored walls, no decorations				*Incandescent lights, warm-colored wall, warm graphics*
1	2	3	4	5

2. Office contains various forms of information brought from outside the organization, including trade journals, association newsletters, and business newspapers.

No outside sources of information				*Four or more journals or newspapers*
1	2	3	4	5

3. Aids for processing of information are present in the office and are easily accessible.

No calculators or PCs visible				*Calculator or PCs accessible without leaving a chair*
1	2	3	4	5

4. Office houses many pieces of equipment used for storing information.

No storage cabinets in office				*Four or more file cabinets or shelves*
1	2	3	4	5

5. Desk is placed to maximize territory for administrator and limit visitor space.

Desk placed against wall				*Desk used as barrier with little space for visitor*
1	2	3	4	5

6. Wears authoritative business suits rather than casual or sporty clothing.

Wears casual or sporty clothing				*Wears conservative business suits*
1	2	3	4	5

7. Administrator's office is easily accessible.

Office located on separate floor from subordinates				*Office within 50 feet of subordinates*
1	2	3	4	5

When STROBE is implemented in this manner, the first step is to determine key organizational themes growing out of interviews. Next, the elements of STROBE are systematically observed, and then a matrix is constructed that lists major ideas from the organizational narrative about information gathering, processing, storing, and sharing on one axis and elements of STROBE on the other. When narrative and observations are compared, one of the five appropriate symbols is then used to characterize the relationship between the narrative and the relevant element observed. The analyst thus creates a table that first documents and then aids in the analysis of observations.

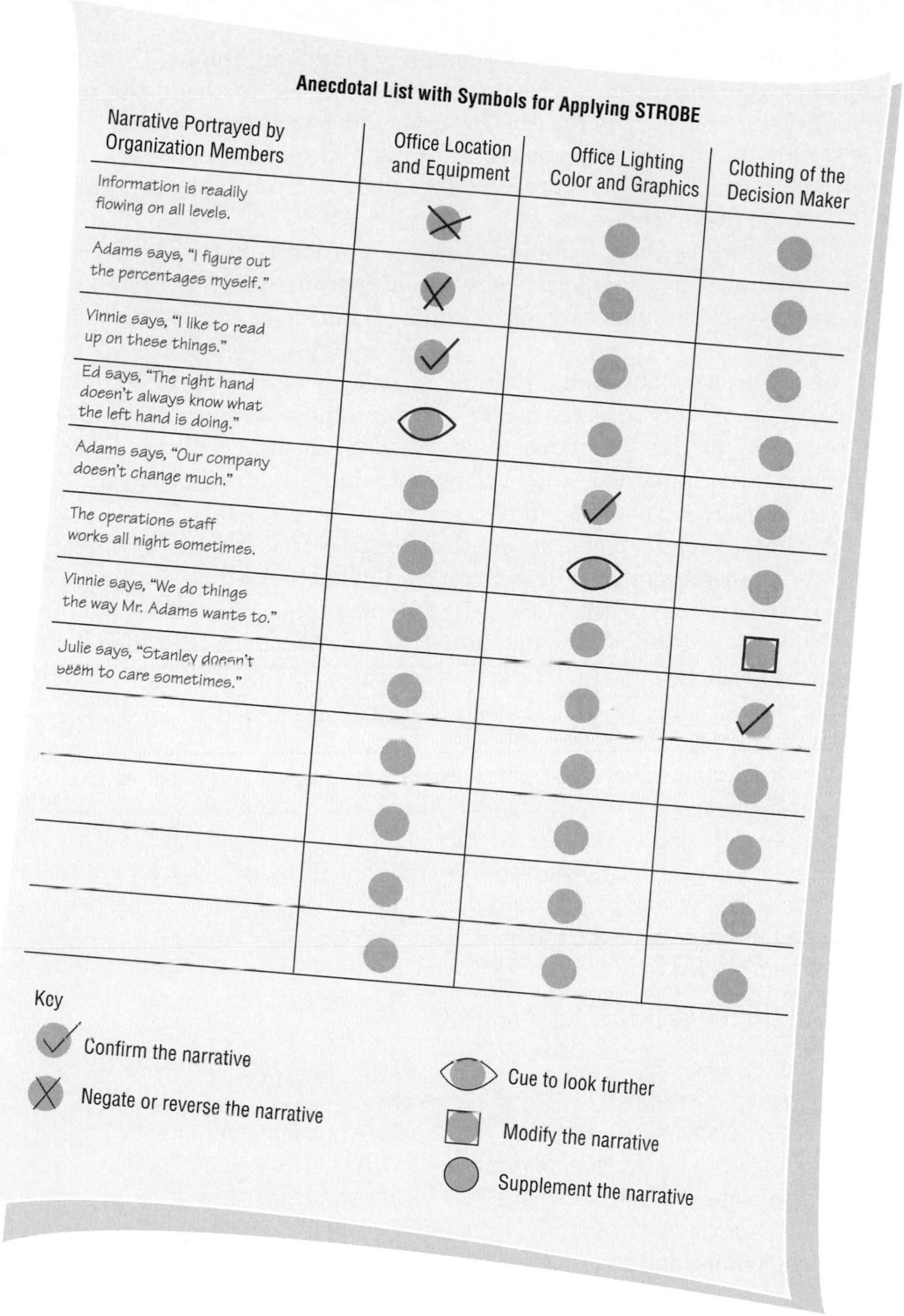

Anecdotal List with Symbols for Applying STROBE

Narrative Portrayed by Organization Members	Office Location and Equipment	Office Lighting Color and Graphics	Clothing of the Decision Maker
Information is readily flowing on all levels.	Negate	Supplement	Supplement
Adams says, "I figure out the percentages myself."	Negate	Supplement	Supplement
Vinnie says, "I like to read up on these things."	Confirm	Supplement	Supplement
Ed says, "The right hand doesn't always know what the left hand is doing."	Cue	Supplement	Supplement
Adams says, "Our company doesn't change much."	Supplement	Confirm	Supplement
The operations staff works all night sometimes.	Supplement	Cue	Supplement
Vinnie says, "We do things the way Mr. Adams wants to."	Supplement	Supplement	Modify
Julie says, "Stanley doesn't seem to care sometimes."	Supplement	Supplement	Confirm
	Supplement	Supplement	Supplement
	Supplement	Supplement	Supplement
	Supplement	Supplement	Supplement
	Supplement	Supplement	Supplement

Key

- Confirm the narrative
- Negate or reverse the narrative
- Cue to look further
- Modify the narrative
- Supplement the narrative

FIGURE 7.10

An anecdotal list with symbols for use in applying STROBE.

Observation/narrative Comparison. The fourth way to implement STROBE is also the least structured method. Although filmgoers rarely attend a film with a mise-en-scène checklist in hand, few of its elements fail to make at least a subconscious impact on them. As long as the systems analyst is aware of the elements of mise-en-scène and as long as they are consciously observed, valuable insights can be gained, even without the aid of a checklist. Examining the organization from a heightened awareness of the elements of STROBE creates a base for making structured observations. These observations can be used later in assessing information requirements.

SUMMARY

Analysts use observation as an information-gathering technique. Through observation they gain insight into what is actually done, see firsthand the relationships among decision makers in the organization, understand the influence of the physical setting on the decision maker, interpret the messages sent by the decision maker through clothing and office arrangement, and comprehend the influence of the decision maker on others.

Using time or event sampling, the analyst observes typical decision-maker activities and body language. There are numerous systems for recording such observations, including category systems, checklists, scales, field notes, and playscripts.

In addition to observing a decision maker's behavior, the systems analyst should observe the decision maker's surroundings. A method for *STR*uctured *OB*servation of the *E*nvironment is called STROBE. A systems analyst uses STROBE in the same way that a film critic uses a method called mise-en-scène analysis to analyze a shot in a film.

Several concrete elements in the decision maker's environment can be observed and interpreted. These elements include (1) office location, (2) placement of the decision maker's desk, (3) stationary office equipment, (4) props such as calculators and PCs, (5) trade journals and newspapers, (6) office lighting and color, and (7) clothing worn by the decision maker. STROBE can be used to gain a better understanding of how decision makers actually gather, process, store, and share information.

There are a number of alternatives for applying STROBE in an organization. These include analysis of photographs that can be mounted on a team Web site, using a checklist based on Likert scales, adopting an anecdotal list with symbols, and simply writing up an observation/narrative comparison. Each method has certain advantages as well as drawbacks that the analyst needs to weigh when choosing one alternative over the other.

KEYWORDS AND PHRASES

adjective pairs
analyst's playscript
category systems
clothing worn by decision makers
decision maker's body language
event sampling
office lighting and color
office location
placement of decision maker's desk
props
stationary office equipment
STROBE
systematic observation
time sampling
trade journals and newspapers

REVIEW QUESTIONS

1. List three reasons why observation is useful to the systems analyst in the organization.
2. Why is it important that observation of decision makers be structured and systematic?
3. List five steps to help the analyst observe the decision maker's typical activities.
4. What are the advantages of using time sampling of observations?
5. What are the disadvantages of using event sampling of observations?
6. Compare the use of adjective pairs versus the analyst's playscript.

"We're proud of our building here in Tennessee. In fact, we used the architectural firm of I. M. Paid to carry the same theme, blending into the local landscape while still reaching out to our clients, throughout all the branches. We get lots of people coming through just to admire the building once they catch on to where it is exactly. In fact, by Tennessee standards we get so many sightseers that it might as well be the pyramids! Well, you can see for yourself as you go through. The East Atrium is my favorite place: plenty of light, lots of louvered blinds to filter it. Yet it has always fascinated me that the building and its furnishings might tell a story quite different from the one its occupants tell.

"Sometimes employees complain that the offices all look the same. The public rooms are spectacular, though. Even the lunchroom is inviting. Most people can't say that about their cafeterias at work. You'll notice that we all personalize our offices, anyway. So even if the offices were of the "cookie cutter" kind, their occupants' personalities seem to take over as soon as they have been here a while. What have you seen? Was there anything that surprised you so far?"

HYPERCASE QUESTIONS

1. Use STROBE to compare and contrast Snowden Evans's and Ketcham's offices. What sort of conclusion about each person's use of information technology can you draw from your observations? How compatible do Evans and Ketcham seem in terms of the systems they use? What other clues to their storage, use, and sharing of information can you discover based on your observations of their offices?

FIGURE 7.HC1
There are hidden clues in HyperCase. Use STROBE.

(Continued)

2. Carefully examine Kathy Blandford's office. Use STROBE to confirm, reverse, or negate what you have learned during your interview with her. List anything you found out about Ms. Blandford from observing her office that you did not know from the interview.
3. Carefully examine the contents of the MRE reception area using STROBE. What inferences can you make about the organization? List them. What interview questions would you like to ask, based on your observations of the reception area? Make a list of people you would like to interview and the questions you would ask each of them.
4. Describe in a paragraph the process you would go through in applying STROBE to observing an MRE office setting. List all elements in the MRE offices that seem important to understanding the users' decision-making behavior.

7. What three attributes make STROBE a systematic approach to observing the decision maker's physical environment?
8. List the seven concrete elements of the decision maker's physical environment that can be observed by the systems analyst using STROBE.
9. What are four different application strategies for using STROBE?

PROBLEMS

1. "I think I'll be able to remember most everything he does," says Ceci Awll. Ceci is about to interview Biff Welldon, vice president of strategic planning of OK Corral, a steak restaurant chain with 130 stores. "I mean, I've got a good memory. I think it's much more important to listen to what he says than to observe what he does anyway." As one of your systems analysis team members, Ceci has been talking with you about the desirability of writing down her observations of Biff's office and activities during the interview.
   a. In a paragraph, persuade Ceci that listening is not enough in interviews and that observing and recording those observations are also important.
   b. Ceci seems to have accepted your idea that observation is important but still doesn't know what to observe. Make a list of items and behaviors to observe, and in a sentence beside each behavior, indicate what information Ceci should hope to gain through observation of it.
   c. Ceci is uncomfortable writing down observations during her interviews. In a paragraph, suggest two methods for recording observations that do not require that they be used as the observation occurs. Now recommend one of the methods, and in a sentence or two, justify why you feel it would be a good method for Ceci in particular to try.
2. "We're a progressive company, always looking to be ahead of the power curve. We'll give anything a whirl if it'll put us ahead of the competition, and that includes every one of us," says I. B. Daring, an executive with Michigan Manufacturing (2M). You are interviewing him as a preliminary step in a systems project, one in which his subordinates have expressed interest. As you listen to I. B., you look around his office to see that most of the information he has stored on shelves can be classified as internal procedures manuals. In addition, you notice a PC on a back table of I. B.'s office. The monitor's screen is

covered with dust, and the manuals stacked beside the PC are still encased in their original shrink-wrap. Even though you know that 2M uses an intranet, no cables are visible going to or from I. B.'s PC. You look up behind I. B.'s massive mahogany desk to see on the wall five framed oil portraits of 2M's founders, all clustered around a gold plaque bearing the corporate slogan, which states, "Make sure you're right, then go ahead."

a. What is the organizational narrative or storyline as portrayed by I. B. Daring? Rephrase it in your own words.
b. List the elements of STROBE that you have observed during your interview with I. B.
c. Next to each element of STROBE that you have observed, write a sentence on how you would interpret it.
d. Construct a matrix with the organizational storyline down the left-hand side of the page and the elements of STROBE across the top. Using the symbols from the "anecdotal list" application of STROBE, indicate the relationship between the organizational storyline as portrayed by I. B. and each element you have observed (that is, indicate whether each element of STROBE confirms, reverses, causes you to look further, modifies, or supplements the narrative).
e. Based on your observations of STROBE and your interview, state in a paragraph what problems you are able to anticipate in getting a new system approved by I. B. and others. In a sentence or two, discuss how your diagnosis might have been different if you had only talked to I. B. over the phone or had read his written comments on a systems proposal.

GROUP PROJECTS

1. Arrange to visit a local organization that is expanding or otherwise enhancing its information systems. To allow your group to practice the various observation methods described in this chapter, assign one of the following techniques to each group member: time sampling, event sampling, observing decision makers' body language, developing the analyst's playscript, and using STROBE. Many of these strategies can be employed during one-on-one interviews, whereas some require formal organizational meetings. Try to accomplish several objectives during your visit to the organization by scheduling it at an appropriate time, one that permits all team members to try their assigned method of observation. Using multiple methods such as interviewing and observation (often simultaneously) is the only cost-effective way to get a true, timely picture of the organization's information requirements.
2. The members of your group should meet and discuss their findings. Were there any surprises? Did the information garnered through observation confirm, reverse, or negate what was learned in interviews? Were any of the findings from the observational methods in direct conflict with each other? Work with your group to develop a list of ways to address any puzzling information (for example, by doing follow-up interviews).

SELECTED BIBLIOGRAPHY

Edwards, A., and R. Talbot. *The Hard-Pressed Researcher*. New York: Longman, 1994.

Kendall, K. E., and J. E. Kendall. "Observing Organizational Environments: A Systematic Approach for Information Analysts." *MIS Quarterly*, Vol. 5, No. 1, 1981, pp. 43–55.

———. "STROBE: A Structured Approach to the Observation of the Decision-Making Environment." *Information and Management*, Vol. 7, No. 1, 1984, pp. 1–11.

———. "Structured Observation of the Decision-Making Environment: A Validity and Reliability Assessment." *Decision Sciences*, Vol. 15, No. 1, 1984, pp. 107–18.

Runkel, P. J., and J. E. McGrath. *Research on Human Behavior: A Systematic Guide to Method*. New York: Holt, Rinehart and Winston, 1972.

Shultis, R. L. " 'Playscript'-A New Tool Accountants Need." *NAA Bulletin*, Vol. 45, No. 12, August 1964, pp. 3–10.

Weick, K. E. "Systematic Observational Methods." In *The Handbook of Social Psychology*, 2d ed., Vol II, edited by G. Lindzey and E. Aronson, Reading, MA: Addison-Wesley, 1968.

Allen Schmidt, Julie E. Kendall, and Kenneth E. Kendall

SEEING IS BELIEVING

"Chip, I know the interviews took a long time, but they were worth it," Anna says defensively as Chip enters her office with a worried look on his face.

"I'm sure of that," Chip says. "You really made a good impression on them. People have stopped me in the hall and said they're glad we're working on the new system. I'm not worried about the interviews themselves. But I was concerned that we didn't have time to discuss observations before you did them."

"Rest assured, I was all eyes," Anna laughs. "I used a technique called STROBE, or Structured observation of the environment, to see our decision maker's habitats systematically. You'll be interested in these notes I wrote up for each person I interviewed," says Anna, as she hands Chip her written, organized observations from each interview.

OBSERVATIONS OF DECISION MAKERS' OFFICES

DECISION MAKER	DOT MATRICKS
Office location	An enclosed office in the Administrative Data Processing area. The door is usually open. Large windows with a beautiful view are located opposite the door.
Placement of desk	Center of the room with a chair across the desk from Dot and another chair at the side of the desk.
Stationary equipment	Two large bookcases containing a variety of books. One is ceiling height and another is desk height.
Decorations	On the lower bookcase are pictures of Dot's children. There are several pictures on the walls. One is a farm scene of a horse and buggy and two riders trotting up a dirt road.
Props	A computer workstation, turned on with a network sign-on screen displayed. A stack of reports is on the left side. Several pens and a printer calculator are above the reports. A notebook computer is in the shorter bookcase.
Trade journals and newspapers	Several copies of *Computerworld* and the *Journal of Management Systems* are on top of the bookcase. The latest *Computerworld* issue is on the desk.
Office lighting/color	Brightly lit, warm tan walls with a brown accent stripe.
Clothing	A dress, authoritatively jacketed with a navy blazer.

DECISION MAKER	MIKE CROWE
Office location	Workroom near the mainframe and computer network complex. A desk is in a cubicle partitioned at one end of the room.

Placement of desk	Against the wall. A chair is at the side of the desk and is piled high with technical manuals.
Stationary equipment	A file cabinet and a half-size bookcase. A long, low workbench covered with computers and parts.
Decorations	Posters of a magnified microchip showing the circuitry. Several posters of railroad trains.
Props	The bookcase is stacked high with papers, magazines, manuals, software packages, and diskettes. A Pentium 860 is on the desk. Displayed on it are several open windows overlapping each other.
Trade journals and newspapers	There are numerous catalogs for parts and a stack of *PC Magazines* are in evidence.
Office lighting/color	The office is well lit with large overhead fluorescent lights as well as desk and work-area lighting.
Clothing	Dark slacks, a lightly striped shirt, and a dark tie accenting the shirt stripes.

DECISION MAKER	CHER WARE
Office location	Within Administrative Data Processing. Cubicle, near the center of the PC area.
Placement of desk	Facing a wall of the cubicle with a chair behind the desk and at the side of the desk. An old sofa is against another wall.
Stationary equipment	A bookcase and a file cabinet. The bookcase contains a variety of books referencing PC software and hardware.
Decorations	Posters of mountain scenery, tranquil lakes, a forest, and one that says "Flower Power."
Props	The desk is rather cluttered with paper, pencils, coffee cups, and the like. A personal computer is on the desk, with a Web page displaying a link to download a software update.
Trade journals and newspapers	Several internal reports. The latest copy of *PC World* magazine is stacked next to two other PC magazines. Each is folded back, exposing a product review page.
Office lighting/color	The cubicle has burgundy walls and is warmly lit with ceiling and desk lighting.
Clothing	A flowered pastel blouse and a denim skirt.

DECISION MAKER	PAIGE PRYNTER
Office location	Administration building, an enclosed office. Located near other decision makers. The office door is normally closed. Thin vertical windows are part of the wall to the hallway.

Placement of desk	Close to the door. Expansive space behind the desk. Chair of visitor is against the wall and directly across the desk from Paige's chair.
Stationary equipment	A file cabinet is in the corner. A bookcase contains bound sets of books neatly organized.
Decorations	A framed print of an English landscape is on the wall.
Props	A personal computer sits on the desk. It is turned off. A gold pen set is the only other object on the desk.
Trade journals and newspapers	A copy of the *Wall Street Journal* and some educational journals are visible on a small bookcase. No journals or reports are on the desk.
Office lighting/color	Lighting is fluorescent. The colors of the office are grey and mauve.
Clothing	Skirted suit with white blouse.

DECISION MAKER	**HY PERTEKS**
Office location	In the Information Center. A faculty and staff resource room with PC and several mainframe workstations. The office is partitioned off at one end of the room.
Placement of desk	The desk has its side against the partition wall with a chair to the side of it. The space behind and in front of the desk are fairly balanced.
Stationary equipment	Two desk-height bookcases and several four-drawer file cabinets.
Decorations	The walls are covered with numerous posters and with artwork that his children have created. The posters are of astronomical objects, distant countries, and interesting surrealistic-looking computer artwork, some depicting fractals.
Props	The desk has a Pentium 700 displaying a Web site. There are pens and some Post-it notes on the desk. Several of CD holders with a variety of CDs in them are on a shelf. There is a large box of read/write CDs on another shelf.
Trade journals and newspapers	Several issues of *PC Magazine* and a Macintosh journal are on the bookcase. The latest issue of *Training* is on the desk.
Office lighting/color	The office is well lit with desk and overhead fluorescent lamps. The color is a warm off-white.
Clothing	A sports jacket with a pale yellow shirt and dark tie. Dress slacks that coordinate with his jacket.

7

EXERCISES

E-1. Based on Anna's written observation of Dot's office and clothing, use STROBE to analyze Dot as a decision maker. In two paragraphs, compare and contrast what you learned in Dot's interview (Chapter 5) and what you learned via STROBE.

E-2. After examining Anna's written observations about Mike Crowe's office, use STROBE to analyze Mike as a decision maker. What differences (if any) did you see between Mike in his interview (Chapter 5) and Mike in Anna's observations? Use two paragraphs to answer.

E-3. Use STROBE to analyze Anna's written observations about Cher Ware and Paige Prynter. Use two paragraphs to compare and contrast the decision-making style of each person as it is revealed by their offices and clothing.

E-4. Use STROBE to analyze Anna's written observations about Hy Perteks. Now compare your analysis with Hy's interview in Chapter 5. Use two paragraphs to discuss whether STROBE confirms, negates, reverses, or serves as a cue to look further in Hy's narrative. (Include any further questions you would ask Hy to clarify your interpretation.)

PROTOTYPING AND RAPID APPLICATION DEVELOPMENT

8

KINDS OF INFORMATION SOUGHT

Prototyping of information systems is a worthwhile technique for quickly gathering specific information about users' information requirements. As becomes evident in the second section of this chapter, there are four basic approaches to prototyping. Generally speaking, effective prototyping should come early in the systems development life cycle, during the requirements determination phase. Prototyping, however, is a complex technique that requires knowledge of the entire systems development life cycle before it is successfully accomplished.

Prototyping is included at this point in the text to underscore its importance as an information-gathering technique. When using prototyping in this way, the systems analyst is seeking initial reactions from users and management to the prototype, user suggestions about changing or cleaning up the prototyped system, possible innovations for it, and revision plans detailing which parts of the system need to be done first or which branches of an organization to prototype next. Figure 8.1 shows the four kinds of information that analysts seek during prototyping.

One special instance of prototyping that uses an object-oriented approach is called rapid application development, or RAD.

INITIAL USER REACTIONS

As the systems analyst presenting a prototype of the information system, you are keenly interested in the reactions of users and management to the prototype. You want to know in detail how they react to working with the prototype and how good the fit is between their needs and the prototyped features of the system. Reactions are gathered through observation, interviews, and feedback sheets (possibly questionnaires) designed to elicit each person's opinion about the prototype as he or she interacts with it. Through such user reactions, the analyst discovers many perspectives on the prototype, including whether users seem happy with it and whether there will be difficulty in selling or implementing the system.

USER SUGGESTIONS

The analyst is also interested in user and management suggestions about refining or changing the prototype that has been presented. Suggestions are garnered from those experiencing the prototype as they work with it for a specified period. The

FIGURE 8.1

Kinds of information sought when prototyping.

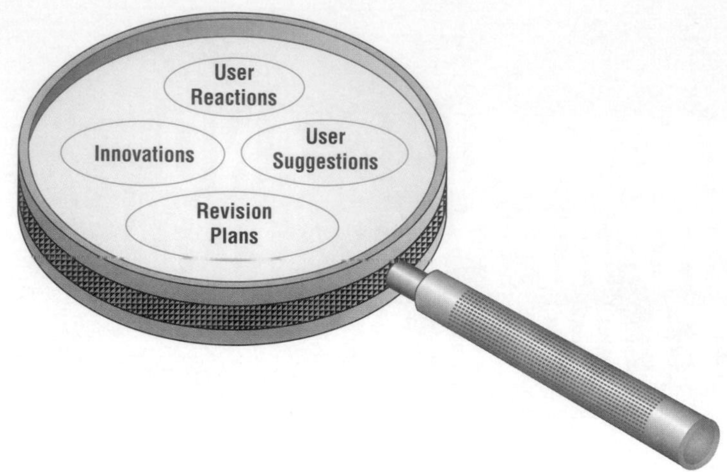

time that users spend with the prototype is usually dependent on their dedication to and interest in the systems project.

Suggestions are the product of users' interaction with the prototype as well as their reflection on that interaction. The suggestions obtained from users should point the analyst toward ways of refining, changing, or "cleaning up" the prototype so that it better suits users' needs.

INNOVATIONS

Innovations for the prototype (which, if successful, will be part of the finished system) are part of the information sought by the systems analysis team. Innovations are new system capabilities that have not been thought of prior to the time when users began to interact with the prototype. These innovations go beyond the current prototyped features by adding something new and innovative.

REVISION PLANS

Prototypes preview the future system. Revision plans help identify priorities for what should be prototyped next. In situations where many branches of an organization are involved, revision plans help to determine which branches to prototype next.

Information gathered in the prototyping phase allows the analyst to set priorities and redirect plans inexpensively, with a minimum of disruption. Because of this feature, prototyping and planning go hand in hand.

APPROACHES TO PROTOTYPING

KINDS OF PROTOTYPES

The word *prototype* is used in many different ways. Rather than attempting to synthesize all these uses into one definition or trying to mandate one correct approach to the somewhat controversial topic of prototyping, we illustrate how each of several conceptions of prototyping may be usefully applied in a particular situation, as shown in Figure 8.2.

Patched-Up Prototype. The first kind of prototyping has to do with constructing a system that works but is patched up or patched together. In engineering this approach is referred to as breadboarding: creating a patched-together, working model of an (otherwise microscopic) integrated circuit.

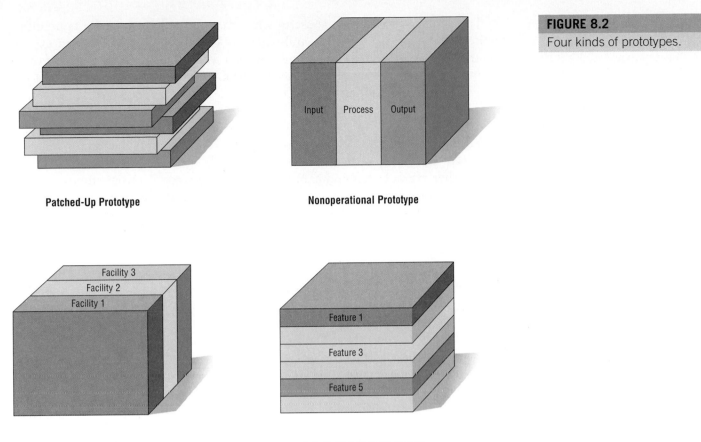

FIGURE 8.2
Four kinds of prototypes.

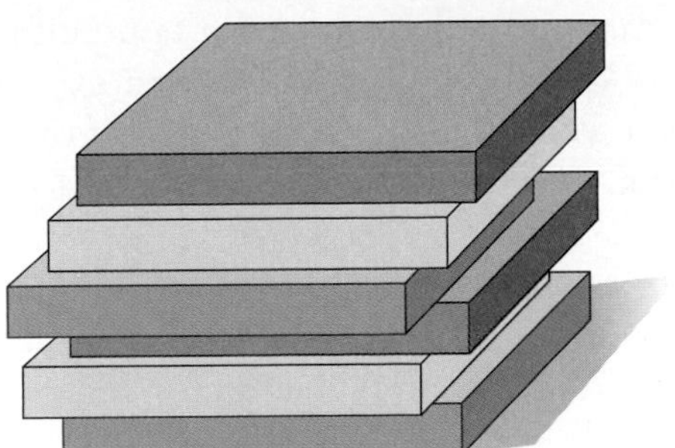

Patched-Up Prototype

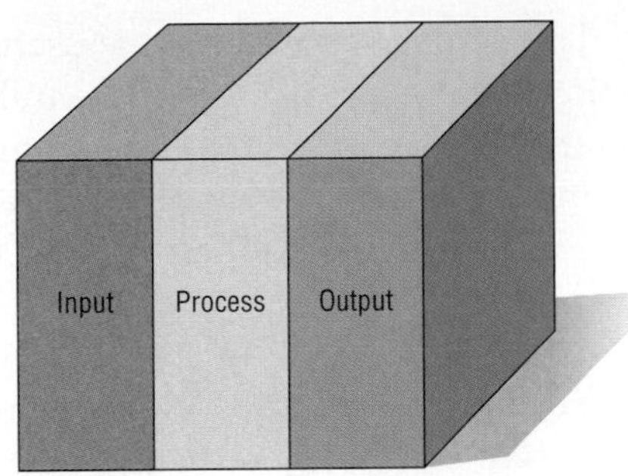

Nonoperational Prototype

Facility 3
Facility 2
Facility 1

First-of-a-Series Prototype

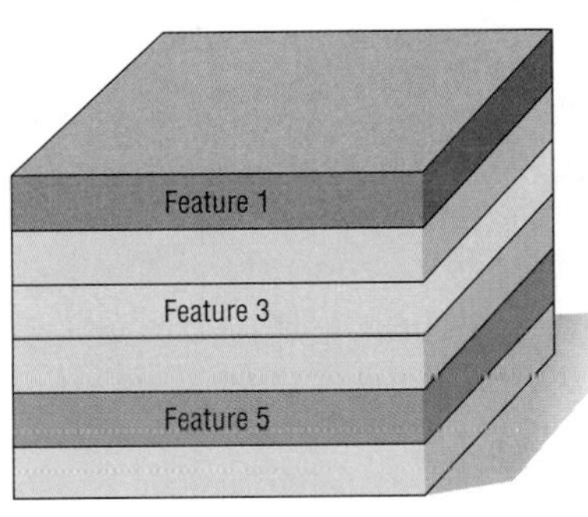

Selected Features Prototype

An example in information systems is a working model that has all the necessary features but is inefficient. In this instance of prototyping, users can interact with the system, getting accustomed to the interface and types of output available. The retrieval and storage of information may be inefficient, however, because programs were written rapidly with the objective of being workable rather than efficient.

Another example of a patched-up prototype is an information system that has all the proposed features but is really a basic model that will eventually be enhanced.

Nonoperational Prototype. The second conception of a prototype is that of a nonworking scale model that is set up to test certain aspects of the design. An example of this approach is a full-scale model of an automobile that is used in wind tunnel tests. The size and shape of the auto are precise, but the car is not operational. In this case only features of the automobile essential to wind tunnel testing are included.

A nonworking scale model of an information system might be produced when the coding required by the applications is too extensive to prototype but when a useful idea of the system can be gained through the prototyping of the input and output only. In this instance processing, because of undue cost and time, would not be prototyped. Some decisions on the utility of the system, however, could still be made based on prototyped input and output.

First-of-a-Series Prototype. A third conception of prototyping involves creating a first full-scale model of a system, often called a pilot. An example is prototyping the first airplane of a series. The prototype is completely operational and is a realization of what the designer hopes will be a series of airplanes with identical features.

This type of prototyping is useful when many installations of the same information system are planned. The full-scale working model allows users to experience realistic interaction with the new system, yet it minimizes the cost of overcoming any problems that it presents. Creation of a working model is one of the types of prototyping done with RAD.

For example, when a retail grocery chain intends to use EDI (electronic data interchange) to check in suppliers' shipments in a number of outlets, a full-scale model might be installed in one store so as to work through any problems before the system is implemented in all the others.

Another example is found in banking installations for electronic funds transfer. A full-scale prototype is installed in one or two locations first, and if successful, duplicates are installed at all locations based on customer usage patterns and other key factors.

Selected Features Prototype. A fourth conception of prototyping concerns building an operational model that includes some of, but not all, the features that the final system will have. An analogy would be a new retail shopping mall that opens before the construction of all shops is complete.

In a newly opened retail mall, essential functions such as being able to purchase some goods, eating in a fast-food restaurant, and parking nearby are possible, although not all space is occupied and not all goods that will ultimately be for sale are available when the complex first opens. Nonetheless, from initial contact with the retail complex, it is possible to gain a good understanding of what future visits will be like.

When prototyping information systems in this way, some, but not all, essential features are included. For example, a system menu may appear on screen that lists six features: add a record, update a record, delete a record, search a record for a key word, list a record, or scan a record. In the prototyped system, however, only three of the six may be available for use, so that the user may add a record (feature 1), delete a record (feature 3), and list a record (feature 5).

When this kind of prototyping is done, the system is accomplished in modules so that if the features that are prototyped are evaluated as successful, they can be incorporated into the larger, final system without undertaking immense work in interfacing. Prototypes done in this manner are part of the actual system. They are *not* just a mock-up as in the first definition of prototyping considered previously. A selected features prototype is an additional type of prototype built when using RAD.

PROTOTYPING AS AN ALTERNATIVE TO THE SYSTEMS DEVELOPMENT LIFE CYCLE

Some analysts argue that prototyping should be considered as an alternative to the systems development life cycle (SDLC). Recall that the SDLC, introduced in Chapter 1, is a logical, systematic approach to follow in the development of information systems.

Complaints about going through the SDLC center around two interrelated main concerns. The first concern is the extended time required to go through the development life cycle. As the investment of analyst time increases, the cost of the delivered system rises proportionately.

The second concern about using the SDLC is that user requirements change over time. During the long interval between the time that user requirements are analyzed and the time that the finished system is delivered, user requirements are evolving. Thus, because of the extended development cycle, the resulting system may be criticized for inadequately addressing current user information requirements.

It is apparent that the concerns are interrelated, because they both pivot on the time required to complete the SDLC and the problem of falling out of touch with user requirements during subsequent development phases. If a system is developed in isolation from users (after initial requirements analysis is completed), it will not meet their expectations.

A corollary of the problem of keeping up with user information requirements is the suggestion that users cannot really know what they do or do not want until they see something tangible. In addition, in the traditional SDLC, it often is too late to change an unwanted system once it is delivered.

To overcome these problems, some analysts propose that prototyping be used as an alternative to the systems development life cycle. When prototyping is used in this way, the analyst effectively shortens the time between ascertainment of information requirements and delivery of a workable system. In addition, using prototyping instead of the traditional SDLC might overcome some of the problems of accurately identifying user information requirements.

With a prototype, users can actually see what is possible and how their requirements translate into hardware and software. Any of the four kinds of prototyping discussed earlier might be used.

Drawbacks to supplanting the SDLC with prototyping include prematurely shaping a system before the problem or opportunity being addressed is thoroughly understood. Also, using prototyping as an alternative may result in producing a system that is accepted by specific groups of users but that is inadequate for overall system needs.

The approach we advocate here is to use prototyping as a part of the traditional SDLC. In this view prototyping is considered as an additional, specialized method for ascertaining users' information requirements.

DEVELOPING A PROTOTYPE

In this section guidelines for developing a prototype are advanced. The term *prototyping* is taken in the sense of the last definition that was discussed, that is, a selected-features prototype that will include some but not all features, one that, if successful, will eventually be part of the larger, final system that is delivered.

FIGURE 8.3
Certain factors determine whether a system is more or less suitable for prototyping.

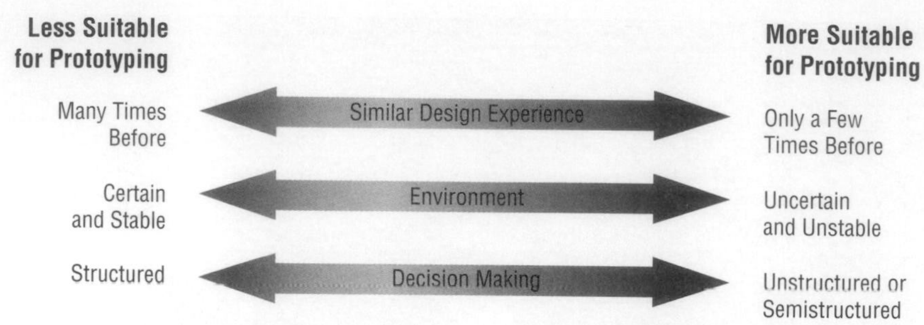

When deciding whether to include prototyping as a part of the SDLC, the systems analyst needs to consider what kind of problem is being solved and in what way the system presents the solution. Different types of systems and their suitability for prototyping are depicted in Figure 8.3. A straightforward payroll or inventory system that solves a highly structured problem in a traditional manner is not a good candidate for prototyping, because the outcome of the system as a solution is well-known and predictable.

Rather, consider the novelty and complexity of the problem and its solution. A novel and complex system that addresses unstructured or semistructured problems in a nontraditional way is a perfect candidate for prototyping. Decision support systems, which are the subject of Chapter 12, are personalized information systems that support users in semistructured decision making. As such, DSSs are well-suited to prototyping.

The systems analyst must also evaluate the environmental context for the system when deciding whether to prototype. If the system will exist in an environment that is stable for long periods, prototyping may be unnecessary. If the environment for the system changes rapidly, however, prototyping should be seriously considered. By their nature prototypes are evolutionary and can absorb many revisions.

The prototype system is actually an operational portion of the eventual system that you will build. It is not a complete system, because you will strive to build it quickly; only some essential functions will be included in the model. It is important, however, to envision and then build the prototype as part of the actual system with which the user will interact. It must incorporate enough representative functions to allow users to understand that they are interacting with a real system.

Prototyping is a superb way to elicit feedback about the proposed system and about how readily it is fulfilling the information needs of its users, as depicted in Figure 8.4. The first step of prototyping is to estimate the costs involved in building a module of the system. If costs of programmers' and analysts' time as well as equipment costs are within the budget, building of the prototype can proceed. Prototyping is an excellent way to facilitate the integration of the information system into the larger system of the organization.

GUIDELINES FOR DEVELOPING A PROTOTYPE

Once the decision to prototype has been made, four main guidelines must be observed when integrating prototyping into the requirements determination phase of the SDLC:

1. Work in manageable modules.
2. Build the prototype rapidly.
3. Modify the prototype in successive iterations.
4. Stress the user interface.

As you can see, the guidelines suggest ways of proceeding with the prototype that are necessarily interrelated. Each guideline is explained in the following subsections.

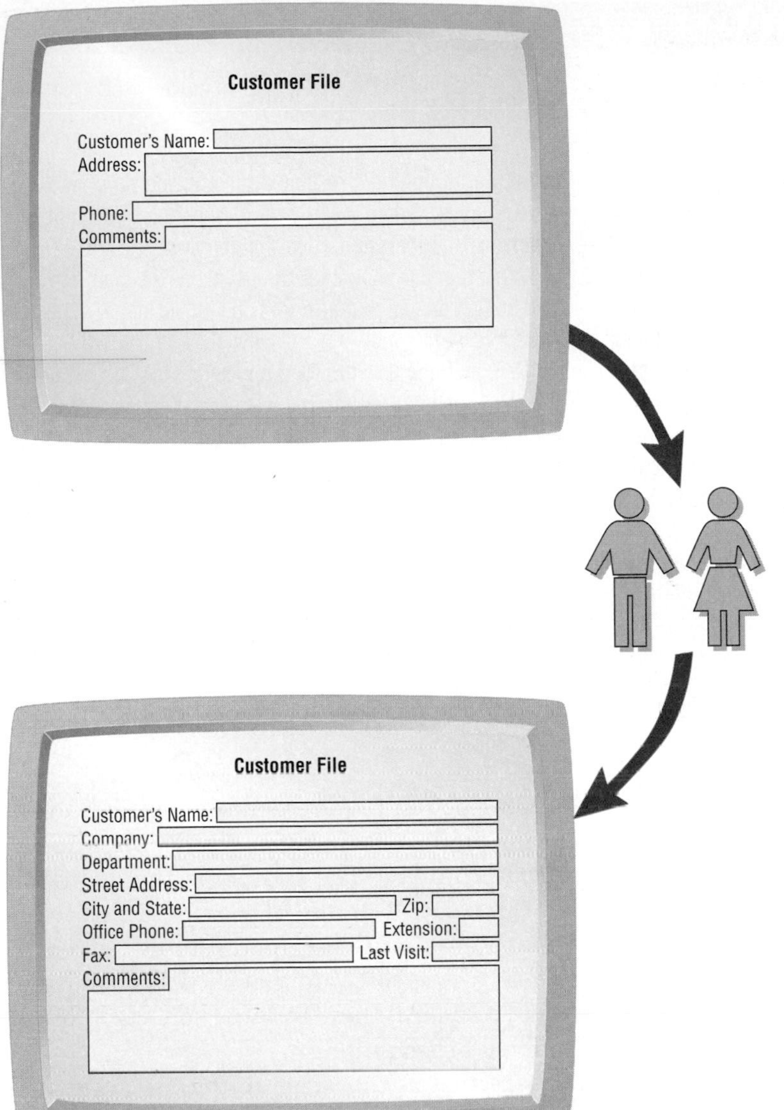

FIGURE 8.4
Eliciting user feedback results in improved screens that better address user requirements.

Working in Manageable Modules. When prototyping some of the features of a system into a workable model, it is imperative that the analyst work in manageable modules. One distinct advantage of prototyping is that it is not necessary or desirable to build an entire working system for prototype purposes.

A manageable module is one that allows users to interact with its key features yet can be built separately from other system modules. Module features that are deemed less important are purposely left out of the initial prototype.

Building the Prototype Rapidly. Speed is essential to the successful prototyping of an information system. Recall that one complaint voiced against following the traditional SDLC is that the interval between requirements determination and delivery of a complete system is far too long to address evolving user needs effectively.

Analysts can use prototyping to shorten this gap by using traditional information-gathering techniques to pinpoint salient information requirements, and then they can quickly make decisions that bring forth a working model. In effect the user sees and uses the system very early in the SDLC instead of waiting for a finished system to gain hands-on experience.

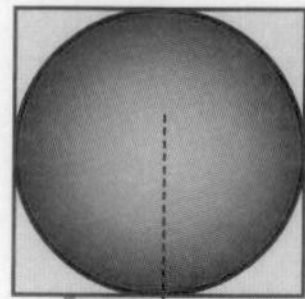

CONSULTING OPPORTUNITY 8.1

IS PROTOTYPING KING?

"As you know, we're an enthusiastic group. We're not a dynasty yet, but we're working on it," Paul Le Gon tells you. Paul (introduced in Consulting Opportunity 2.2), at 24 years of age, is the "boy king" of Pyramid, Inc., a small but successful, independent book publishing firm that specializes in paperback books outside of the publishing mainstream. As a systems analyst, you have been hired by Pyramid, Inc., to help develop a computerized warehouse inventory and distribution information system.

"We're hiring lots of workers," Paul continues, as if to convince you of the vastness of Pyramid's undertaking. "And we feel Pyramid is positioned perfectly as far as our markets in the north, south, east, and west are concerned.

"My assistant, Ceil Toom, and I have been slaving away, thinking about the new system. And we've concluded that what we really need is a prototype. As a matter of fact, we've tunneled through a lot of material. Our fascination with the whole idea has really pyramided."

As you formulate a response to Paul, you think back over the few weeks you've worked with Pyramid, Inc. You think that the business problems that its information system must resolve are very straight forward. You also know that the people in the company are on a limited budget and cannot afford to spend like kings. Actually, the entire project is quite small.

Ceil, building on what Paul has said, tells you, "We don't mean to be too wrapped up with it, but we feel prototyping represents the new world. And that's where we all want to be. We know we need a prototype. Have we convinced you?"

Based on Paul and Ceil's enthusiasm for prototyping and what you know about Pyramid's needs, would you support construction of a prototype? Why or why not? Formulate your decision and response in a letter to Paul Le Gon and Ceil Toom. Present a justification for your decision based on overall criteria that should be met to justify prototyping.

After a brief analysis of information requirements using traditional methods, such as interviewing, observing, and researching archival data, the analyst constructs working models for the prototype. The prototype should take less than a week to put together; two or three days is preferable and possible. Remember that to build a prototype this quickly, you must use special tools, such as an existing database management system, as well as software that allows generalized input and output, interactive systems, and so on. All these tools permit speed of construction that is impossible with traditional programming.

It is important to emphasize that at this stage in the life cycle, the analyst is still gathering information about what users need and want from the information system. The prototype becomes a valuable extension of traditional requirements determination. The analyst assesses user feedback about the prototype to get a better picture of overall information needs.

Putting together an operational prototype both rapidly and early in the SDLC allows the analyst to gain valuable insight into how the remainder of the project should go. By showing users very early in the process how parts of the system actually perform, rapid prototyping guards against overcommitting resources to a project that may eventually become unworkable. Later, when RAD is discussed you again see the importance of rapid systems building.

Modifying the Prototype. A third guideline for developing the prototype is that its construction must support modifications. Making the prototype modifiable means creating it in modules that are not highly interdependent. If this guideline is observed, less resistance is encountered when modifications in the prototype are necessary.

The prototype is generally modified several times, going through several iterations. Changes in the prototype should move the system closer to what users say is important. Each modification necessitates another evaluation by users.

As with the initial development, modifications must be accomplished swiftly, usually in a day or two, to keep the momentum of the project going. The exact

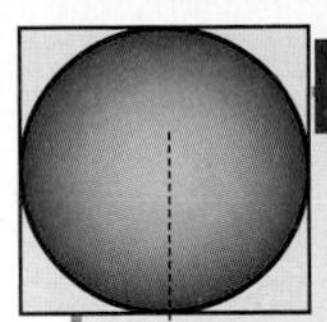

CONSULTING OPPORTUNITY 8.2

CLEARING THE WAY FOR CUSTOMER LINKS

World's Trend (see Chapter 9 for a detailed corporate description) is building a Web site on which to sell clearance merchandise usually sold through the Web and through its catalog operation. As a newly hired Web consultant, Lincoln Cerf finds himself in a very cold, wintry city, fighting his way through several inches of snow to meet with one of the system team members, Mary Maye, at World's Trend headquarters.

Mary welcomes Lincoln, saying "At least the weather doesn't seem to effect our Web sales! They're brisk no matter what." Lincoln groans appreciatively at her weak attempt at humor, smiles, and says, "I gathered from your email last week that you are trying to determine the type of information that needs to be displayed on our clearance Web site." Mary replies, "Yes, I'm trying to get it organized in the best possible way. Our customers are all so busy. I know photos of all our merchandise can take a long time to appear on the page if a customer is accessing the Web via a slower modem from home."

Mary continues by saying, "Linc, I'm not even that concerned about how to design our clearance site at this time. I am worried, though, about how much information we need to include on a page. For example, when items are on clearance, not all colors and sizes are available. Which do you think is better, to include some basic information and let the customer click a button to ask for more information or to be as complete as possible on one page? If I use the linking method, then I could fit more items on the screen . . . but it might be too orderly. Customers like the look and feel of a sale where merchandise is kind of jumbled together."

Linc continues her line of thought, saying, "Yeah, I wonder how customers want the information organized? Have you actually watched them use the Web? I mean, do they look for shoes when they buy a suit? If so, should shoes appear on the suit page or be linked in some way?"

Mary comments, "Those are my questions, too. Then I wonder if we should just try this approach for men's clothes first, before we implement it for women's clothing. What if men's and women's approaches to shopping on the Web are different?"

As a third member of the World's Trend Web site development group, respond in a brief written report to Lincoln and Mary about whether you should use a prototype to elicit recommendations from potential customers about the proposed Web site. What type of prototype is appropriate? Consider each form of prototype and explain why each type would apply (or would not apply) to this problem. Devote a paragraph to each explanation.

timing of modifications, however, depends on how dedicated users are to interacting with modified prototypes. Systems analysts must encourage users to do their share by evaluating changes rapidly.

The prototype is not a finished system. Entering the prototyping phase with the idea that the prototype will require modification is a helpful attitude that demonstrates to users how necessary their feedback is if the system is to improve.

Stressing the User Interface. The user's interface with the prototype (and eventually the system) is very important. Because what you are really trying to achieve with the prototype is to get users to further articulate their information requirements, they must be able to interact easily with the system's prototype. For many users the interface *is* the system. It should not be a stumbling block.

For example, at this stage the goal of the analyst is to design an interface that both allows the user to interact with the system with a minimum of training and allows a maximum of user control over represented functions. Although many aspects of the system will remain undeveloped in the prototype, the user interface must be well developed enough to enable users to pick up the system quickly and not be put off. Online, interactive systems using GUI interfaces are ideally suited to prototypes. Chapter 18 describes in detail the considerations that are important in designing the user interface.

Many of the intricacies of interfaces must be streamlined or ignored altogether in the prototyping phase. If prototype interfaces are not what users need or want or if systems analysts find that the interfaces do not adequately allow system access, however, they too are candidates for modification.

DISADVANTAGES OF PROTOTYPING

As with any information-gathering technique, there are several disadvantages to prototyping. The first is that it can be quite difficult to manage prototyping as a project within the larger systems effort. The second disadvantage is that users and analysts may adopt a prototype as a completed system when it is in fact inadequate and was never intended to serve as a finished system.

The analyst needs to weigh these disadvantages against the known advantages when deciding whether to prototype, when to prototype, and how much of the system to prototype.

Managing the Project. All the systems analysts' management skills that you learned in Chapter 3 come into play again as your systems analysis team constructs and modifies a prototype. All the possible problems that project management is subject to are relevant here.

Although several iterations of the prototype may be necessary, extending the prototype indefinitely also creates problems. It is important that the systems analysis team devise and then carry out a plan regarding how feedback on the prototype will be collected, analyzed, and interpreted. Set up specific time periods during which you and management decision makers will use feedback to evaluate how well the prototype is performing. Even though the prototype is prized for its evolutionary nature, the analyst cannot permit prototyping to overtake other phases in the SDLC.

Elicit feedback from users periodically, not just once, and ask them if previous suggestions for improvements or changes have been acted upon satisfactorily. Feedback is directed to the members of the systems analysis team for their reaction and possible modification of the prototype to better fit user needs. Recall that modifications to the prototype should be managed on a tight schedule of only a day or two each throughout the successive iterations.

Adopting an Incomplete System as Complete. A second major disadvantage of prototyping is that if a system is needed badly and welcomed readily, the prototype may be accepted in its unfinished state and pressed into service without the necessary refinements. Although superficially this method may seem to be an appealing way to shortcut the development effort, it works to the business' and team's disadvantage.

Users will develop interaction patterns with the prototype system that are not compatible with what will actually occur with the complete system. In addition, a prototype will not perform all necessary functions. Eventually, when users discover the deficiencies, user backlash may develop if the prototype has been mistakenly adopted and integrated into the business as if it were a complete system.

ADVANTAGES OF PROTOTYPING

Prototyping is not necessary or appropriate in every systems project, as we have seen. The advantages, however, should also be given consideration when deciding whether to prototype. The three major advantages of prototyping are the potential for changing the system early in its development, the opportunity to stop development on a system that is not working, and the possibility of developing a system that more closely addresses users' needs and expectations. All three advantages are interrelated.

Changing the System Early in Its Development. Successful prototyping depends on early and frequent user feedback, which can be used to help modify the system and make it more responsive to actual needs. As with any systems effort, early changes are less expensive than changes made late in the project's development.

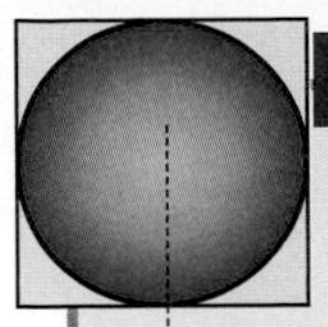

CONSULTING OPPORTUNITY 8.3

TO HATCH A FISH

"Just be a little patient. I think we need to add a few more features before we turn it over to them. Otherwise, this whole prototype will sink, not swim," says Sam Monroe, a member of your systems analysis team. All four members of the team are sitting together in a hurriedly called meeting, and they are discussing the prototype that they are developing for an information system to help managers monitor and control water temperature, number of fish released, and other factors at a large, commercial fish hatchery.

"They've got plenty to do already. Why, the system began with four features and we're already up to nine. I feel like we're swimming upstream on this one. They don't need all that. They don't even want it," argues Belle Uga, a second member of the systems analysis team. "I don't mean to carp, but just give them the basics. We've got enough to tackle as it is."

"I think Monroe is more on target," volunteers Wally Ide, a third member of the team, baiting Belle a little. "We have to show them our very best, even if it means being a few weeks later in hatching our prototype than we promised."

"Okay," Belle says warily, "but I want the two of you to tell the managers at the hatchery why we aren't delivering the prototype. I don't want to. And I'm not sure they'll let you off the hook that easily."

Monroe replies, "Well, I guess we could, but we probably shouldn't make a big deal out of being later than we wanted. I don't want to rock the boat."

Wally chimes in, "Yeah. Why point out our mistakes to everyone? Besides, when they see the prototype, they'll forget any complaints they had. They'll love it."

Belle finds a memo in her notebook from their last meeting with the hatchery managers and reads it aloud. "Agenda for meeting of September 22. 'Prototyping—the importance of rapid development, putting together the user analyst team, getting quick feedback for modification . . .'" Belle's voice trails off, omitting the last few agenda items. In the wake of her comments, Monroe and Ide look unhappily at each other.

Monroe speaks first. "I guess we did make a try to get everyone primed for receiving a prototype quickly, and to be involved from day one." Noting your silence up until now, Monroe continues, "But still waters run deep. What do you think we should do next?" he asks you.

As the fourth member of the systems analysis team, what actions do you think should be taken? In a one- or two-paragraph email message to your teammates, answer the following questions: Should more features be added to the hatchery system prototype before giving it to the hatchery managers to experiment with? How important is the rapid development of the prototype? What are the trade-offs involved in adding more features to the prototype versus getting a more basic prototype to the client when it was promised? Complete your message with a recommendation.

Because the prototype can be changed many times and because flexibility and adaptation are at the heart of prototyping, the feedback that calls for a change in the system is often the action taken. Feedback will help tell you if changes are warranted in the input, process, or output areas or if all three need adjustment.

When changing a prototype, analysts do not need to worry about wasting many labor-hours of their efforts and those of programmers who have developed a full-blown system only to find that it needs modifications. Although the prototype represents an investment of time and money, it is always considerably less expensive than a completed system. Concomitantly, system problems and oversights are much easier to trace and detect in a prototype with limited features and limited interfaces than they are in a complex system.

Scrapping Undesirable Systems. A second advantage of using prototyping as an information-gathering technique is the possibility of scrapping a system that is just not what users and analysts had hoped it would be. Once again, the issue of time and money arises. A prototype represents much less of an investment than a completely developed system.

Permanently removing the prototype system from use is done when it becomes apparent that the system is not useful and does not fulfill the information requirements (and other objectives) that have been set. Although scrapping the prototype is a difficult decision to make, it is infinitely better than putting increasing sums of time and money into a project that is plainly unworkable.

Designing a System for Users' Needs and Expectations. A third advantage of prototyping is that the system being developed should be a better fit with users' needs and expectations. Many studies of failed information systems indict the long interval between requirements determination and the presentation of the finished system; these systems failed precisely because it is common for systems analysts to develop systems while sequestered away from users during this critical period.

It is a better practice to interact with users throughout the SDLC. If your team makes a commitment to ongoing user involvement in all phases of the project, the prototype can be used as an interactive tool that shapes the final system to accurately reflect users' requirements.

Users who take early ownership of the information system work to ensure its success. One way to foster early user support is to involve users actively in prototyping.

If your evaluation of the prototype indicates that the system is functioning well and within the guidelines that have been set, the decision should be to keep the prototype going and continue expanding it to include other functions as planned. Then, it is considered an operational prototype. The decision is made to keep the prototype functioning if the prototype is within the budget set for programmers' and analysts' time, if users find the system worthwhile, and if it is meeting the information requirements and objectives that have been set. A list comparing disadvantages and advantages of prototyping is given in Figure 8.5.

USERS' ROLE IN PROTOTYPING

The users' role in prototyping can be summed up in two words: honest involvement. Without user involvement there is little reason to prototype. The precise behaviors necessary for interacting with a prototype can vary, but it is clear that the user is pivotal to the prototyping process. Realizing the importance of the user to the success of the process, the members of the systems analysis team must encourage and welcome input and guard against their own natural resistance to changing the prototype.

INTERACTION WITH THE PROTOTYPE

There are three main ways a user can be of help in prototyping:

1. Experimenting with the prototype
2. Giving open reactions to the prototype
3. Suggesting additions to or deletions from the prototype

All the above stem from the users' initial and successive interactions with the prototype.

Experimenting with the Prototype. Users should be free to experiment with the prototype. In contrast to a mere list of systems features, the prototype allows users the reality of hands-on interaction. Mounting a prototype on an inter-

FIGURE 8.5
Disadvantages and advantages of prototyping.

Disadvantages to Prototyping	Advantages to Prototyping
• Difficult to manage prototyping as a project within a larger systems effort • Users and analysts may adopt a prototype as a completed system when it is inadequate	• Potential exists for changing the system early in its development • Opportunity exists to stop development on a system that is not working • May address user needs and expectations more closely

active Web site is one way to facilitate this interaction. Limited functionality along with the capability to send comments to the systems team can be included.

Users need to be encouraged to experiment with the prototype. The final system will be delivered with documentation stating how the system is to be used, and this in effect constrains experimentation. In the prototyping stage, however, the user is free from all but minimal instruction on how to use the system. When that is the case, experimentation becomes necessary to make the prototype work. Experimentation is an important part of RAD.

Analysts need to be present at least part of the time when experimentation is occurring. They can then observe users' interactions with the system, and they are bound to see interactions they never planned. A form for observing user experimentation with the prototype is shown in Figure 8.6. Some of the variables you should observe include user reactions to the prototype, user suggestions for changing or expanding the prototype, user innovations for using the system in completely new ways, and any revision plans for the prototype that aid in setting priorities. When revising the prototype, analysts should circulate their recorded observations among team members so that everyone is fully informed.

Giving Open Reactions to the Prototype. Another aspect of the users' role in prototyping requires that they give open reactions to the prototype. Unfortunately, these reactions are not something that occur on demand. Rather, making users secure enough to give an open reaction is part of the relationship between analysts and users that your team works to build.

Prototype Evaluation Form

Observer Name		Date	
System or Project Name		Company or Location	
Program Name or Number		Version	

	User 1	User 2	User 3	User 4
User Name				
Period Observed				
User Reactions				
User Suggestions				
Innovations				
Revision Plans				

FIGURE 8.6
An important step in prototyping is to properly record user reactions, user suggestions, innovations, and revision plans.

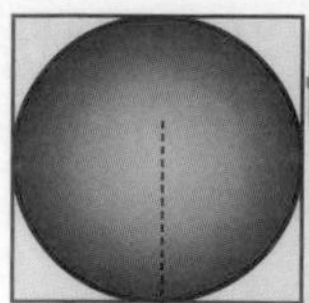

CONSULTING OPPORTUNITY 8.4

THIS PROTOTYPE IS ALL WET

"It can be changed. It's not a finished product, remember," affirms Sandy Beach, a systems analyst for RainFall, a large manufacturer of fiberglass bathtub and shower enclosures for bathrooms. Beach is anxiously reassuring Will Lather, a production scheduler for RainFall, who is poring over the first hard-copy output produced for him by the prototype of the new information system.

"Well, it's okay," Lather says quietly. "I wouldn't want to bother you with anything. Let's see, . . . yes, *here* they are," he says as he finally locates the monthly report summarizing raw materials purchased, raw materials used, and raw materials in inventory.

Lather continues paging through the unwieldy computer printout. "This will be fine." Pausing at a report, he remarks, "I'll just have Miss Fawcett copy this part over for the people in Accounting." Turning a few more pages, he says, "And the guy in Quality Assurance should really see this column of figures, although the rest of it isn't of much interest to him. I'll circle it and make a copy of it for him. Maybe I should phone part of this into the warehouse, too."

As Sandy prepares to leave, Lather bundles up the pages of the reports, commenting, "The new system will be a big help. I'll make sure everybody knows about it. Anything will be better than the 'old monster' anyway. I'm glad we've got something new."

Sandy leaves Will Lather's office feeling a little lost at sea. Thinking it over, he starts wondering why Accounting, QA, and the warehouse aren't getting what Will thinks they should. Beach phones a few people, and he confirms that what Lather has told him is true. They need the reports and they're not getting them.

Later in the week Sandy approaches Lather about rerouting the output as well as changing some of the features of the system. These modifications would allow Lather to get onscreen answers regarding what-if scenarios about changes in the prices suppliers are charging; changes in the quality rating of the raw materials available from suppliers (or both) as well as allow him to see what would happen if a shipment was late.

Lather is visibly upset with Sandy's suggestions for altering the prototype and its output. "Oh, don't do it on my account. It's okay really. I don't mind taking the responsibility for routing information to people. I'm always showering them with stuff anyway. Really, this is working pretty well. I would hate to have you take it away from us at this point. Let's just leave it in place."

Sandy is pleased that Lather seems so satisfied with the prototyped output, but he is concerned about Lather's unwillingness to change the prototype, because he has been encouraging users to think of it as an evolving product, not a finished one.

Write a brief report to Sandy listing changes to the prototype prompted by Will's reactions. In a paragraph, discuss ways that Sandy can calm Lather's fears about having the prototype "taken away." Discuss in a paragraph some actions that can be taken *before* a prototype is tried out to prepare users for its evolutionary nature.

In addition, if users feel wary about commenting on or criticizing what may be a pet project of organizational superiors or peers, it is unlikely that open reactions to the prototype will be forthcoming. Providing a private (relatively unsupervised) period for users to interact with and respond to the prototype is one way to insulate them from unwanted organizational influences. An exclusive Web site set up for users and analysts can also help in this regard.

Suggesting Changes to the Prototype. A third aspect of the users' role in prototyping is their willingness to suggest additions to or deletions from the features being tried. The analysts' role is to elicit such suggestions by assuring users that the feedback they provide is taken seriously, by observing users as they interact with the system, and by conducting short, specific interviews with users concerning their experiences with the prototype.

Although users will be asked to articulate suggestions and innovations for the prototype, in the end it is the analyst's responsibility to weigh this feedback and translate it into workable changes where necessary. Users need to be encouraged to brainstorm about possibilities and to be reminded that their input during the prototyping phase helps determine whether to save, scrap, or modify the system. In other words, users should never be resigned to accepting in the prototype stage something less than what they want. Systems analysts must remember to stress to users and management alike that prototyping is the most appropriate time for system changes.

To facilitate the prototyping process, the analyst must clearly communicate the purposes of prototyping to users, along with the idea that prototyping is valuable only when users are meaningfully involved.

RAPID APPLICATION DEVELOPMENT

Rapid application development (RAD) is an object-oriented approach to systems development that includes a method of development as well as software tools. It makes sense to discuss RAD and prototyping in the same chapter, because they are conceptually very close. Both have as their goal the shortening of time typically needed in a traditional SDLC between the design and implementation of the information system. Ultimately, both RAD and prototyping are trying to meet rapidly changing business requirements more closely. Once you have learned the concepts of prototyping, it is much easier to grasp the essentials of RAD, which can be thought of as a specific implementation of prototyping.

Some developers are looking at RAD as a helpful approach in new ecommerce, Web-based environments where so-called first-mover status of a business might be important. In other words, to deliver an application to the Web before their competitors, businesses may want their development team to experiment with RAD.

PHASES OF RAD

There are three broad phases to RAD that engage both users and analysts in assessment, design, and implementation. Figure 8.7 depicts these three phases. Notice that RAD involves users in each part of the development effort, with intense participation in the business part of the design.

Requirements Planning Phase. In the requirements planning phase, users and analysts meet to identify objectives of the application or system and to identify information requirements arising from those objectives. This phase requires intense involvement from both groups; it is not just signing off on a proposal or document. In addition, it may involve users from different levels of the organization (as covered in Chapter 2). In the requirements planning phase, when information requirements are still being addressed, you may be working with the CIO (if it is a large organization) as well as with strategic planners, especially if you are working with an ecommerce application that is meant to further the strategic aims of the organization. The orientation in this phase is

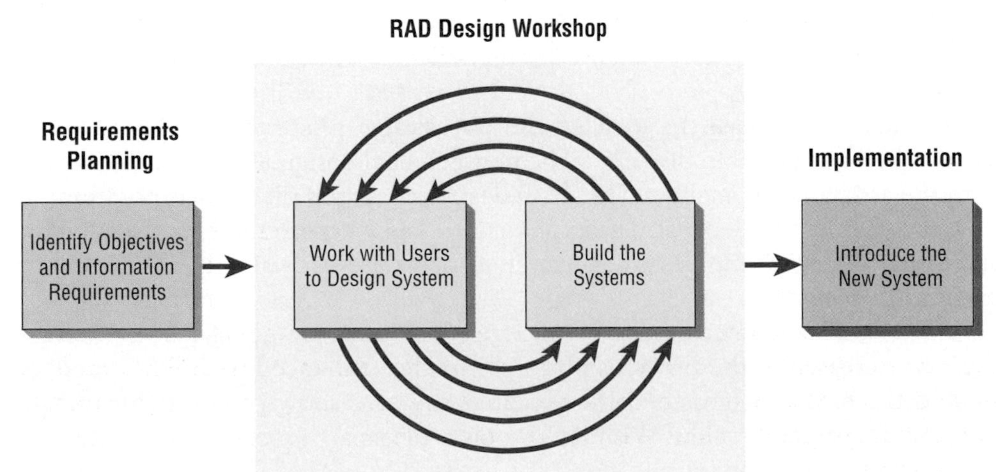

FIGURE 8.7
The RAD Design Workshop is the heart of the interactive development process.

toward solving business problems. Although information technology and systems may even drive some of the solutions proposed, the focus will always remain on reaching business goals.

RAD Design Workshop. The RAD design workshop phase is a design and refine phase that can best be characterized as a workshop. When you imagine a workshop, you know that participation is intense, not passive, and that it is typically hands on. Usually participants are seated at round tables or in a U-shaped configuration of chairs with attached desks where each person can see the other and where there is space to work on a notebook computer. If you are fortunate enough to have a group decision support systems (GDSS) room (as covered in Chapter 12) available at the company or through a local university, use it to conduct at least part of your RAD design workshop.

In some ways a GDSS decision room is an ideal setting, because analysts and users can agree on designs and then some of the systems analysts and programmers can be working to build and show visual representations of the designs and workflow to users. RAD design workshops can take place over a series of days, but extended blocks of time (approximately three day-long workshops, depending on the size of the system) are useful.

During the RAD design workshop, users respond to actual working prototypes and analysts refine designed modules (using some of the software tools mentioned later) based on user responses. The workshop format is very exciting and stimulating, and if you have experienced users and analysts, there is no question that this creative endeavor can propel development forward at an accelerated rate.

Implementation Phase. In figure shown earlier, you can see that analysts are working with users intensely during the workshop to design the business or nontechnical aspects of the system. As soon as these aspects are agreed upon and the systems are built and refined, the new systems or part of systems are tested and then introduced to the organization. Because RAD can be used to create new ecommerce applications for which there is no old system, there is often no need for (and no real way to do) running the old and new systems in parallel before implementation.

By this time, the RAD design workshop will have generated excitement, user ownership, and acceptance of the new application. Typically, change brought about in this manner is far less wrenching than when a system is delivered with little or no user participation.

Martin's Pioneering Approaches to RAD. In Figure 8.8 you can see our conceptualization of James Martin's original RAD phases. In the first phase Martin discusses requirements planning. Here high-level users decide on what functions the application should feature.

In the second phase, here called the user design phase, Martin characterizes users as being engaged in discussing the nontechnical design aspects of the system, with the assistance of analysts. The RAD design workshop phase incorporates both the user phase and the construction phase into one, because the highly interactive and visual nature of the design and refine process are occurring in an interactive, participative way.

In the construction phase many different activities are going on. Any designs that were created in the previous phase are further enhanced with RAD tools. As soon as the new functions become available, they are shown to users for interaction, comments, and review. With RAD tools analysts are able to make continuous changes in the design of applications.

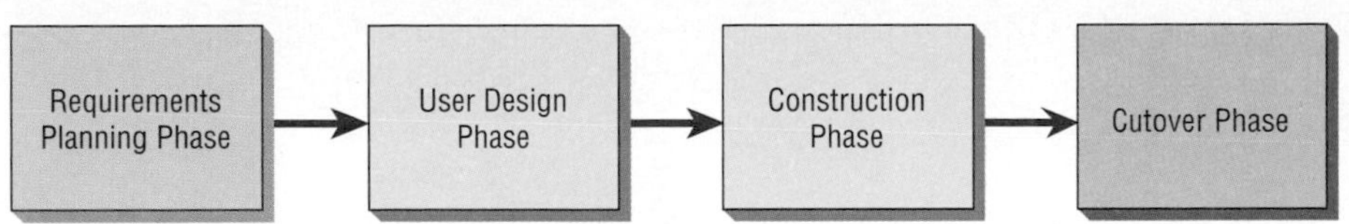

FIGURE 8.8
Martin's phases of RAD.

In Martin's fourth and final phase, the cutover phase, the newly developed application will replace the old one. While it is being run in parallel with the old application, the new one is tested, users are trained, and organizational procedures are changed before the cutover occurs.

Software Tools for RAD. As you would expect, RAD software tools are often newer, often object-oriented tools. They include such familiar programs as Microsoft Access, Microsoft Visual Basic 6, Symantec's Visual Café, Visual C++, and even RAD with Java using JBuilder 3.

One way the tools differ from one another is in their capabilities to support client/server applications (for example, MS Access does not, Visual Basic does) as well as their ease of use and the amount of programming skill that is required. Most RAD applications have stayed on the small, PC side, although their true power may be for client/server applications that need to run across multiple platforms.

The tools for developing Java applications with a RAD approach have come into play more slowly. The result is that organizations (and their IT departments) must decide what trade-offs are involved with the decision to wait until RAD tools are more user friendly. If companies decide to use RAD tools that are complicated and difficult to learn, they run the risk that inexperienced programmers may take longer than expected to learn and use the tools. In addition, any gains anticipated from RAD may be diminished.

Although there are almost as many different phases of RAD identified as there are analysts, the four phases proposed by Martin—requirements planning, the user design phase, the construction phase, and the cutover phase—are useful. Let's examine each in a little more detail, comparing and contrasting them to the features of classic prototyping and the traditional SDLC.

COMPARING RAD TO THE SDLC

In Figure 8.9 you can compare the phases of the SDLC with those detailed for RAD at the beginning of the section. Notice that the ultimate purpose of RAD is to shorten the SDLC and in this way respond more rapidly to dynamic information requirements of organizations. The SDLC takes a more methodical, systematic approach that ensures completeness and accuracy and has as its intention the creation of systems that are well integrated into standard business procedures and culture.

The RAD design workshop phase is a departure from the standard SDLC design phases, because RAD software tools are used to generate screens and to exhibit the overall flow of the running of the application. Thus, when users approve this design, they are signing off on a visual model representation, not just a conceptual design represented on paper, as is traditionally the case.

The implementation phase of RAD is in many ways less stressful than others, because the users have helped to design the business aspects of the system and are well aware of what changes will take place. There are few surprises, and the change is something that is welcomed. Often when using the SDLC, there is a lengthy time during development and design when analysts are separated from users. During this period, requirements can change and users can be caught off guard if the final product is different than anticipated over many months.

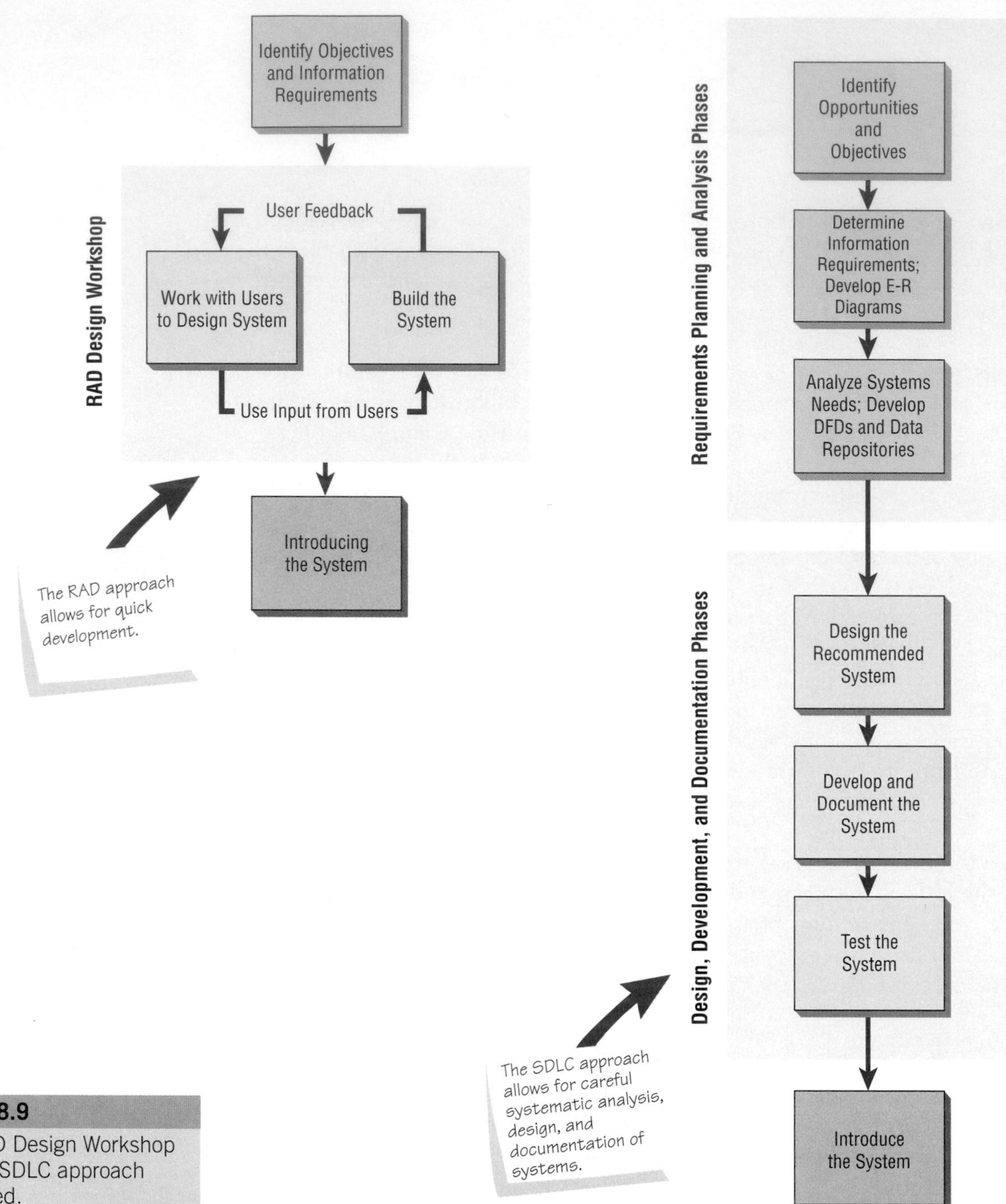

FIGURE 8.9

The RAD Design Workshop and the SDLC approach compared.

When to Use RAD. As an analyst you want to learn as many approaches and tools as possible to facilitate getting your work done in the most appropriate way. Certain applications and system work will call forth certain methodologies. Consider using RAD when:

1. Your team includes programmers and analysts who are experienced with it, and,
2. There are pressing business reasons for speeding up the portion of an application development, or
3. When you are working with a novel ecommerce application and your development team believes that the business can sufficiently benefit over their

competitors from being an innovator if this application is the first or among the first to appear on the Web, or

4. When users are sophisticated, and highly engaged with the organizational goals of the company.

How to Use RAD within the SDLC. We recommend that when the situation calls for it, you should consider using RAD. Rather than considering it a distinct or complete methodology that stands on its own, however, we believe that it is most powerful when used in conjunction with the SDLC, much the same way you would incorporate prototyping into the SDLC as discussed earlier. RAD can be used as an incisive and targeted tool to update, improve, or innovate selected portions of a systems effort. If you use it in combination with the SDLC, you are taking advantage of two very powerful approaches.

Disadvantages of RAD. The difficulties with RAD, like other types of prototyping, arise because systems analysts try to hurry the project too much. Suppose two carpenters are hired to build two storage sheds for two neighbors. The first carpenter follows the SDLC philosophy, whereas the second follows the RAD philosophy.

The first carpenter is systematic, inventorying every tool, lawn mower, and piece of patio furniture to determine the correct size for the shed, designing a blueprint of the shed, and writing specifications for every piece of lumber and hardware. The carpenter builds the shed with little waste and has precise documentation about how the shed was built if anyone wants to build another just like it, repair it, or paint it using the same color.

The second carpenter jumps right into the project by estimating the size of the shed, getting a truckload of lumber and hardware, building a frame and discussing it with the owner of the property as modifications are made if certain materials are not available, and making a return trip to return the lumber not used. The shed gets built faster, but if a blueprint is not drawn, the documentation never exists.

Other disadvantages include some of the trade-offs that have now become familiar between time and attention to detail. If an application is delivered quickly using RAD but does not address the pressing business problems it was developed to address, RAD is not an improvement over slower, more systematic, methods.

Another drawback is the potentially steep learning curve for programmers inexperienced with RAD tools. Hoped-for gains in delivering an application may not materialize if programmers find it difficult to learn and apply new tools. This disadvantage is not exclusive to RAD tools of course, but is a more general concern whenever programmers and analysts are faced with acquiring new skills at the same time they are doing development work.

SUMMARY

Prototyping is an information-gathering technique useful for supplementing the traditional systems development life cycle. When systems analysts use prototyping, they are seeking user reactions, suggestions, innovations, and revision plans to make improvements to the prototype and thereby modify system plans with a minimum of expense and disruption. Systems that support semistructured decision making (as decision support systems do) are prime candidates for prototyping.

The term *prototyping* carries several different meanings, four of which are commonly used. The first definition of prototyping is that of constructing a patched-up prototype. A second definition of prototyping is a nonoperational prototype that is used to test certain features of the design. A third conception of prototyping is creating the first-of-a-series prototype that is fully operational. This kind of prototype is useful when many installations of the same information system (under similar conditions) are planned. The fourth kind of prototyping is a selected features prototype that has some of, but not all, the essential system features. It uses self-contained modules as building blocks so that if prototyped features are successful, they can be kept and incorporated into the larger, finished system.

The four major guidelines for developing a prototype are to (1) work in manageable modules, (2) build the prototype rapidly, (3) modify the prototype, and (4) stress the user interface.

One disadvantage of prototypes is that managing the prototyping process is difficult because of the rapidity of the process and its many iterations. A second disadvantage is that an incomplete prototype may be pressed into service as if it were a complete system.

Although prototyping is not always necessary or desirable, it should be noted that there are three main, interrelated advantages to using it: (1) the potential for changing the system early in its development, (2) the opportunity to stop development on a system that is not working, and (3) the possibility of developing a system that more closely addresses users' needs and expectations.

Users have a distinct role to play in the prototyping process. Their main concern must be to interact with the prototype through experimentation. Systems analysts must work systematically to elicit and evaluate users reactions to the prototype. Sometimes reactions can be sought by mounting the prototype on an interactive Web site that permits users to see limited functionality and to enter comments for the systems team. Analysts then work to incorporate worthwhile user suggestions and innovations into subsequent modifications.

One particular use of prototyping is rapid application development, or RAD. It is an object-oriented approach with three phases: requirements planning, RAD design workshop, and implementation.

KEYWORDS AND PHRASES

building the prototype rapidly
first-of-a-series prototype
implementation
modifying the prototype
nonoperational prototype
patched-up prototype
prototype
RAD Design Workshop
rapid application development (RAD)
requirements planning phase
selected features prototype
stressing the user interface
user involvement with prototyping
working in manageable modules

REVIEW QUESTIONS

1. What four kinds of information is the analyst seeking through prototyping?
2. What is meant by the term *patched-up prototype*?
3. Define a prototype that is a nonworking scale model.
4. Give an example of a prototype that is a first full-scale model.
5. Define what is meant by a prototype that is a model with some, but not all, essential features.
6. List the advantages and disadvantages of using prototyping to *replace* the traditional systems development life cycle.
7. Describe how prototyping can be used to augment the traditional systems development life cycle.
8. What are the criteria for deciding whether a system should be prototyped?
9. List four guidelines the analyst should observe in developing a prototype.
10. What are the two main problems identified with prototyping?

"Thank goodness it's the time of year when everything is new. I love spring; it's the most exhilarating time here at MRE. The trees are so green, with leaves in so many different shades. So many new projects to do, too; so many new clients to meet. It's really exciting. It reminds me of prototyping. Or what I know about prototyping, anyway. It's something new and fresh, a quick way to find out what's happening.

"In fact, I believe that we have a few prototypes already going on here. The best thing about them is that they can change. I don't know anyone who's really been satisfied with a first pass at a prototype. But it is fun to be involved with something that is happening fast, and something that will change."

HYPERCASE QUESTIONS

1. Locate the prototype currently proposed for use in one of MRE's departments. Suggest a few modifications that would make this prototype even more responsive to the unit's needs.
2. Using a word processor, construct a nonoperational prototype for a Training Unit Project Reporting System. If you have a hypertext program available, attempt to create partial functionality by making the menus functional. *Hint*: See sample screens in Chapters 15 and 16 to help you in your design.

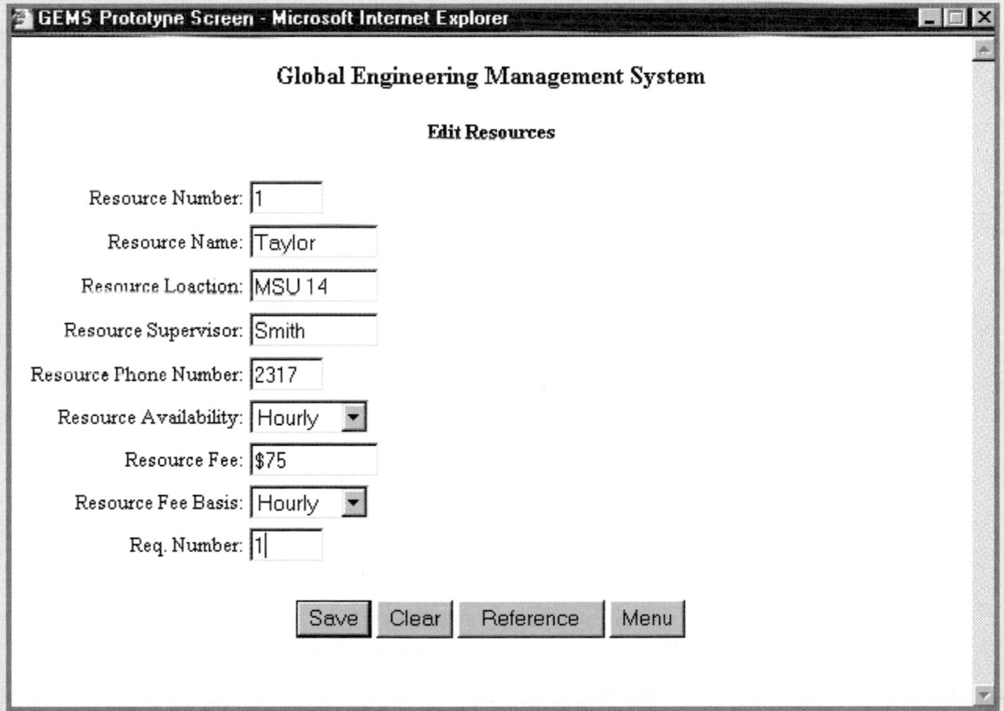

FIGURE 8.HC1
One of the many prototype screens found in HyperCase.

11. List the three main advantages in using prototyping.
12. How can a prototype mounted on an interactive Web site facilitate the prototyping process? Answer in a paragraph.
13. What are three ways that a user can be of help in the prototyping process?
14. Define what is meant by RAD.
15. What are the three phases of RAD according to your text?

PROBLEMS

1. As part of a larger systems project, Clone Bank of Clone, Colorado, wants your help in setting up a new monthly reporting form for its checking and savings account customers. The president and vice presidents are very attuned to what customers in the community are saying. They think that their customers want a checking account summary that looks like the one offered by the other three banks in town. They are unwilling, however, to commit to that form without a formal summary of customer feedback that supports their decision. Feedback will not be used to change the prototype form in any way. They want you to send a prototype of one form to one group and to send the old form to another group.
   a. In a paragraph discuss why it probably is not worthwhile to prototype the new form under these circumstances.
   b. In a second paragraph discuss a situation under which it would be advisable to prototype a new form.
2. C. N. Itall has been a systems analyst for Tun-L-Vision Corporation for many years. When you came on board as part of the systems analysis team and suggested prototyping as part of the SDLC for a current project, C. N. said, "Sure, but you can't pay any attention to what users say. They have no idea what they want. I'll prototype, but I'm not 'observing' any users."
   a. As tactfully as possible, so as not to upset C. N. Itall, make a list of the reasons that support the importance of observing user reactions, suggestions, and innovations in the prototyping process.
   b. In a paragraph describe what might happen if part of a system is prototyped and no user feedback about it is incorporated into the successive system.
3. "Every time I think I've captured user information requirements, they've already changed. It's like trying to hit a moving target. Half the time, I don't think they even know what they want themselves," exclaims Flo Chart, a systems analyst for 2 Good 2 Be True, a company that surveys product use for the marketing divisions of several manufacturing companies.
   a. In a paragraph, explain to Flo Chart how prototyping can help her to better define users' information requirements.
   b. In a paragraph, comment on Flo's observation: "Half the time, I don't think they even know what they want themselves." Be sure to explain how prototyping can actually help users better understand and articulate their own information requirements.
   c. Suggest how an interactive Web site featuring a prototype might address Flo's concerns about capturing user information requirements. Use a paragraph.
4. Harold, a district manager for the multioutlet chain of Sprocket's Gifts, thinks that building a prototype can mean only one thing: a nonworking scale model. He also believes that this way is too cumbersome to prototype information systems and thus is reluctant to do so.
   a. Briefly (in two or three paragraphs) compare and contrast the other three kinds of prototyping that are possible so that Harold has an understanding of what prototyping can mean.

b. Harold has an option of implementing one system, trying it, and then having it installed in five other Sprocket locations if it is successful. Name a type of prototyping that would fit well with this approach, and in a paragraph defend your choice.

5. "I've got the idea of the century!" proclaims Bea Kwicke, a new systems analyst with your systems group. "Let's skip all this SDLC garbage and just prototype everything. Our projects will go a lot more quickly, we'll save time and money, and all the users will feel as if we're paying attention to them instead of going away for months on end and not talking to them."
   a. List the reasons you (as a member of the same team as Bea) would give her to dissuade her from trying to scrap the SDLC and prototype every project.
   b. Bea is pretty disappointed with what you have said. To encourage her, use a paragraph to explain the situations you think would lend themselves to prototyping.
6. The following remark was overheard at a meeting between managers and a systems analysis team at the Fence-Me-In fencing company: "You told us the prototype would be finished three weeks ago. We're still waiting for it!"
   a. In a paragraph, comment on the importance of rapid delivery of a portion of a prototyped information system.
   b. List three elements of the prototyping process that must be controlled to ensure prompt delivery of the prototype.
   c. What are some elements of the prototyping process that are difficult to manage? List them.
7. Nordic Designs, a chain of stores specializing in contemporary furniture from Scandinavia, has been circulating a corporate newsletter that brags about the prototype of its shipping information system. The newsletter story proclaims, "Our shipping information system prototype was put into service as soon as it was delivered. With absolutely no changes necessary, managers say it's the perfect solution to tracking furniture shipments. Watch for the prototype in your store soon."
   a. How has the writer of the story apparently misunderstood the concept of prototyping? Explain in a paragraph.
   b. List the problems faced by designers of prototypes if users expect that "absolutely no changes are necessary."

GROUP PROJECTS

1. Divide your group into two smaller subgroups. Have group 1 follow the processes specified in this chapter for creating prototypes. Using a CASE tool or a word processor, group 1 should devise two nonworking prototype screens using the information collected in the interviews with Maverick Transport employees accomplished in the group exercise in Chapter 5. Make any assumptions necessary to create two screens for truck dispatchers. Group 2 (playing the roles of dispatchers) should react to the prototype screens and provide feedback about desired additions and deletions.
2. The members of group 1 should revise the prototype screens based on the user comments they received. Those in group 2 should respond with comments about how well their initial concerns were addressed with the refined prototypes.
3. As a united group, write a paragraph discussing your experiences with prototyping for ascertaining information requirements.

SELECTED BIBLIOGRAPHY

Alavi, M. "An Assessment of the Prototyping Approach to Information Systems Development." *Communications of the ACM,* Vol. 27, No. 6, June 1984, pp. 556–63.

Avison, D., and D. N. Wilson. "Controls for Effective Prototyping." *Journal of Management Systems,* Vol. 3, No. 1, 1991.

Billings, C., M. Billings, and J. Tower. *Rapid Application Development with Oracle Designer/2000*. Reading, MA: Addison-Wesley, 1996.

Davis, G. B., and M. H. Olson. *Management Information Systems: Conceptual Foundations, Structure, and Development,* 2d ed. New York: McGraw-Hill, 1985.

Dearnley, P., and P. Mayhew. "In Favour of System Prototypes and Their Integration into the Systems Development Cycle." *Computer Journal,* Vol. 26, February 1983, pp. 36–42.

Ghione, J. "A Web Developer's Guide to Rapid Application Development Tools and Techniques." *Netscape World,* June 1997.

Gremillion, L. L., and P. Pyburn. "Breaking the Systems Development Bottleneck." *Harvard Business Review,* March–April 1983, pp. 130–37.

Harrison, T. S. "Techniques and Issues in Rapid Prototyping." *Journal of Systems Management*. Vol. 36, No. 6, June 1985, pp. 8–13.

Liang, D. *Rapid Java Application Development Using JBuilder 3*. Upper Saddle River, NJ: Prentice Hall, 2000.

McMahon, D. *Rapid Application Development with Visual Basic 6 (Enterprise Computing)*. New York: McGraw-Hill Professional Publishing, 1999.

——— *Rapid Application Development with Visual* C++. New York: McGraw-Hill Professional Publishing, 1999.

Naumann, J. D., and A. M. Jenkins. "Prototyping: The New Paradigm for Systems Development." *MIS Quarterly*, September 1982, pp. 29–44.

Sano, D. *Designing Large-Scale Web Sites*. New York: John Wiley, 1996.

ALLEN SCHMIDT, JULIE E. KENDALL, AND KENNETH E. KENDALL

REACTION TIME

"We need to get a feel for some of the output needed by the users," Anna comments. "It will help to firm up some of our ideas on the information they require."

"Agreed," replies Chip. "It will also help us determine the necessary input. From that we can design corresponding data-entry screens. Let's create prototype reports and screens and get some user feedback. Why don't we use Microsoft Access to quickly create screens and reports? I'm quite familiar with the software."

Anna starts by developing the PREVENTIVE MAINTENANCE REPORT prototype. Based on interview results, she sets to work creating the report she feels Mike Crowe will need.

"This report should be used to predict when machines should have preventive maintenance," Anna thinks. "It seems to me that Mike would need to know *which* machine needs work performed as well as *when* the work should be scheduled. Now let's see, what information would identify the machine clearly? The inventory number, brand name, and model would identify the machine. I imagine the room and campus should be included to quickly locate the machine. A calculated maintenance date would tell Mike when the work should be completed. What sequence should the report be in? Probably the most useful would be by location."

The PREVENTIVE MAINTENANCE REPORT prototype showing the completed report is shown in Figure E8.1. Notice that Xxxxxxx's and generic dates are used to indicate where data should be printed. Realistic Campus and Room Locations as well as Inventory Numbers are included. They are necessary for Access to accomplish group printing.

The report prototype is soon finished. After printing the final copy, Anna takes the report to both Mike Crowe and Dot Matricks. Mike Crowe is enthusiastic about the project and wants to know when the report will be in production. Dot is similarly impressed.

Several changes come up. Mike wants an area to write in the Completion Date of the preventive maintenance so the report can be used to reenter the dates into the computer. Dot wants the report number assigned by data control to appear at the top of the form for reference purposes. She also suggests that the report title be changed to WEEKLY PREVENTIVE MAINTENANCE REPORT. The next steps are to modify the prototype report to reflect the recommended changes and then have both Mike and Dot review the result.

The report is easily modified and printed. Dot is pleased with the final result. "This is really a fine method for designing the system," she comments. "It's so nice to feel that we are a part of the development process and that our opinions count. I'm starting to feel quite confident that the final system will be just what we've always wanted."

Mike has similar praise, observing, "This will make our work so much smoother. It eliminates the guesswork about which machines need to be maintained. And sequencing them by room is a fine idea. We won't have to spend so much time returning to rooms to work on machines."

Chip makes a note about each of these modifications on the Prototype Evaluation Form shown in Figure E8.2. This form gets Chip organized and documents the prototyping process.

Preventive Maintenance Report

Week Of 1/11/2001

1/11/01

Page 1 of 1

Campus Location	Room Location	Inventory Number	Brand Name	Model	Last Preventive Maintenance Date	Done
Central Administration	11111	84004782	Xxxxxxxxxxxxxxxx	Xxxxxxxxxxxxxxxxxxxx	11/4/00	___
Central Administration	11111	90875039	Xxxxxxxxxxxxxx	Xxxxxxxxxxxxxxxx	10/24/00	___
Central Administration	11111	93955411	Xxxxxxxxxxxxxxxx	Xxxxxxxxxxxxxxxxxxxx	11/4/00	___
Central Administration	11111	99381373	Xxxxxxxxxxxxxxxx	Xxxxxxxxxxxxxxxxxxxx	10/24/00	___
Central Administration	22222	10220129	Xxxxxxxxxxxxxx	Xxxxxxxxxxxxxxxxxx	10/24/00	___
Central Administration	99999	22838234	Xxxxxxxxxxxxxx	Xxxxxxxxxxxxxxxx	10/24/00	___
Central Administration	99999	24720952	Xxxxxxxxxxxxxx	Xxxxxxxxxxxxxxxxxx	10/24/00	___
Central Administration	99999	33453403	Xxxxxxxxxxxxxxxx	Xxxxxxxxxxxxxxxx	11/4/00	___
Central Administration	99999	34044449	Xxxxxxxxxxxx	Xxxxxxxxxxxxxxxxxx	11/4/00	___
Central Administration	99999	40030303	Xxxxxxxxxxxxxxxx	Xxxxxxxxxxxxxxxxxxxx	11/4/00	___
Central Administration	99999	47403948	Xxxxxxxxxxxxxxxx	Xxxxxxxxxxxxxxxx	10/24/00	___
Central Administration	99999	56620548	Xxxxxxxxxxxxxxxx	Xxxxxxxxxxxxxxxx	11/4/00	___
Central Computer Science	22222	34589349	Xxxxxxxxxxxx	Xxxxxxxxxxxxxxxxxx	10/24/00	___
Central Computer Science	22222	38376910	Xxxxxxxxxxxx	Xxxxxxxxxxxxxxxxxx	10/24/00	___
Central Computer Science	22222	94842282	Xxxxxxxxxxxxxxxx	Xxxxxxxxxxxxxxxxxxxx	10/24/00	___
Central Computer Science	99999	339393	Xxxxxxxxxxxxxx	Xxxxxxxxxxxxxxxx	11/4/00	___
Central Zoology	22222	11398423	Xxxxxxxxxxxxxxxx	Xxxxxxxxxxxxxxxxxxxx	10/24/00	___
Central Zoology	22222	28387465	Xxxxxxxxxxxxxxxx	Xxxxxxxxxxxxxxxxxxxx	11/4/00	___
Central Zoology	99999	70722533	Xxxxxxxxxxxxxx	Xxxxxxxxxxxxxxxx	10/24/00	___
Central Zoology	99999	99481102	Xxxxxxxxxxxxxxxx	Xxxxxxxxxxxxxxxxxxxx	10/24/00	___

FIGURE E8.1

Prototype for PREVENTIVE MAINTENANCE REPORT. This report needs to be revised.

Chip and Anna next turned their attention to creating screen prototypes. "Because I like the hardware aspect of the system, why don't I start working on the ADD NEW MICROCOMPUTER screen design?" asks Chip.

"Sounds good to me," Anna replies. "I'll focus on the software aspects."

Chip analyzes the results of detailed interviews with Dot and Mike. He compiles a list of elements that each user would need when adding a computer. Other elements, such as location and maintenance information, would update the MICROCOMPUTER MASTER later, after the machine was installed.

The ADD NEW MICROCOMPUTER prototype screen created with the Access form feature is shown in Figure E8.3. Placed on the top of the screen are the current date and time as well as a centered screen title. Field captions are placed on the screen, with the left characters aligned. Check boxes are included for several fields, as well as pull-down lists for the type of Monitor, Printer, and Internet connections. A small

Prototype Evaluation Form

Observer Name	Chip Puller		Date 1/06/2000
System or Project Name Microcomputer System		Company or Location Central Pacific University	
Program Name or Number	Prev. Maint.	Version	1

	User 1	User 2	User 3	User 4
User Name	Mike C.	Dot M.		
Period Observed	1/06/2000 A.M.	1/06/2000 A.M.		
User Reactions	Generally favorable, got excited about project	Excellent!		
User Suggestions	Add the date when maintenance was performed.	Place a form number on top for reference. Place word WEEKLY in title.		
Innovations				
Revision Plans	Modify on 1/08/2000. Review with Dot and Mike.			

FIGURE E8.2
Prototype evaluation form.

Board Code table is included to add several internal boards for one microcomputer. An Add Record and Print button are included.

"Having the database tables defined sure helps to make quick prototypes," Chip comments. "It didn't take very long to complete the screen. Would you like to watch me test the prototype?"

"Sure," replies Anna. "This is my favorite part of prototyping."

Chip executes the screen design as Anna, Mike, and Dot watch. The pull-down list boxes and check boxes make it easy to enter accurate data.

"I really like this," Dot says. "May I try adding some data?"

"Be my guest," replies Chip. "Try to add both invalid and valid data. And notice the help messages that appear at the bottom of the screen to indicate what should be entered."

Dot is plainly enjoying herself as she enters data and tests the screen. Mike also spends time testing the screen. Both users state that they have a good understanding of the system and how it will operate when complete. Enthusiasm for the project is taking on a life of its own.

Anna returns to her desk and creates the ADD SOFTWARE RECORD screen design.

When Anna completes the screen design, she asks Cher to test the prototype. Cher keys information in, checks the pull-down list values, and views help messages.

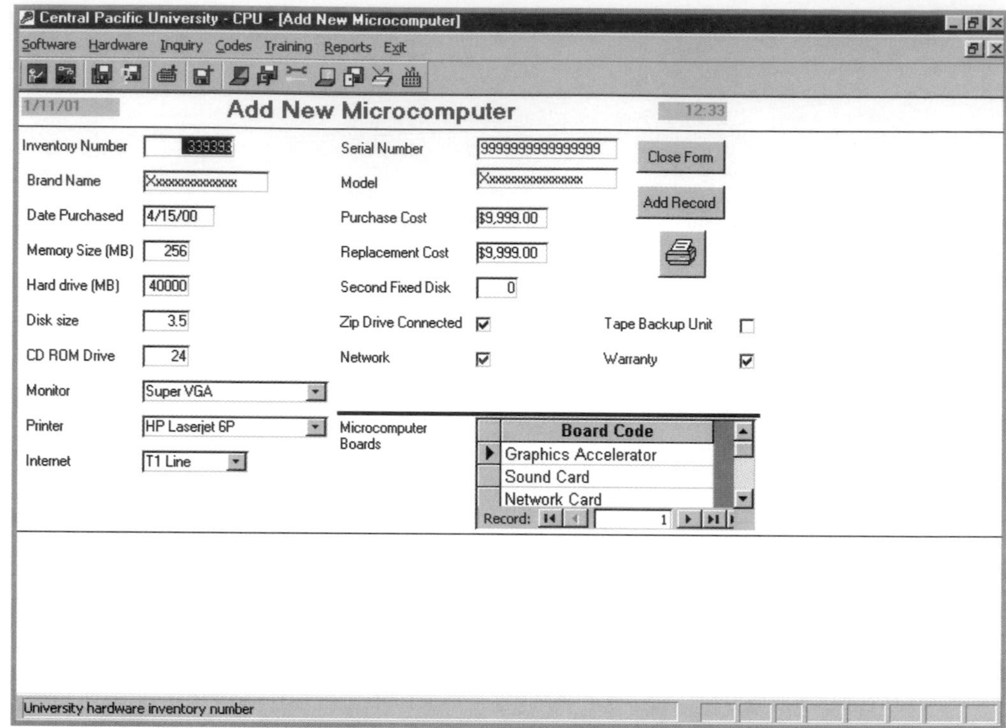

FIGURE E8.3

Prototype for the ADD NEW MICROCOMPUTER screen. Microsoft Access was used as the prototyping tool. Improvements can be made at this stage.

"I really like the design of this screen and how it looks," remarks Cher. "It lacks some of the fields that would normally be included when a software package is entered, though, like the computer brand and model that the software runs on, the memory required, monitor, and the printer or plotter required. I would also like buttons to save the record and exit the screen."

"Those are all doable. I'll make the changes and get back to you," replies Anna, making some notes to herself.

A short time later, Cher again tests the ADD SOFTWARE RECORD screen. It includes all the features that she requires. The completed screen design is shown in Figure E8.4. Notice that there is a line separating the software information from the hardware entries.

"Chip, I was speaking with Dot and she mentioned that there has been funding for putting some of the information on the Web, as part of a unified Web site for technology support at CPU," comments Anna, looking up from her computer. "I have been busy creating a prototype for the Web page menus and the first screen, one to report technology problems. Because solving problems is Mike's area, I have invited him and Dot to review the prototype. Care to join the session?"

"Sure," replies Chip. "I am interested in working on the design of some of the Web pages."

A short time later Mike, Dot and Chip are gathered around Anna as she demonstrates the Web page, illustrated in Figure E8.5.

"I really like the menu style," comments Dot. "The main feature tabs on the top are easy to use, and I like the way they change color when one is clicked."

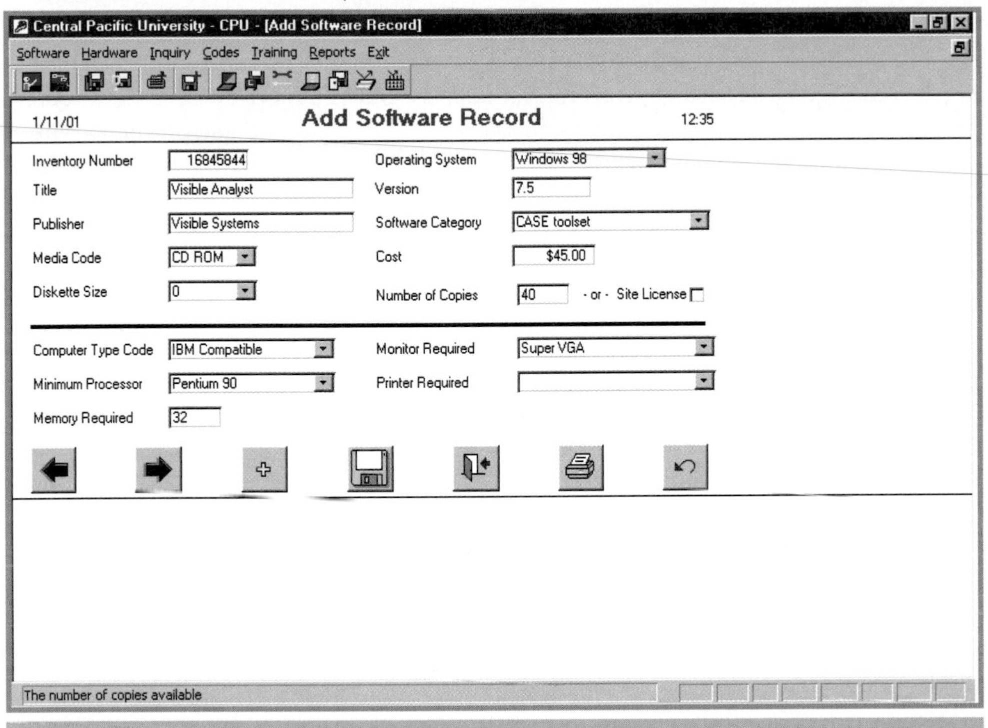

FIGURE E8.4

Prototype for the ADD SOFTWARE RECORD screen. Microsoft Access was used as the prototyping tool. This screen can be improved.

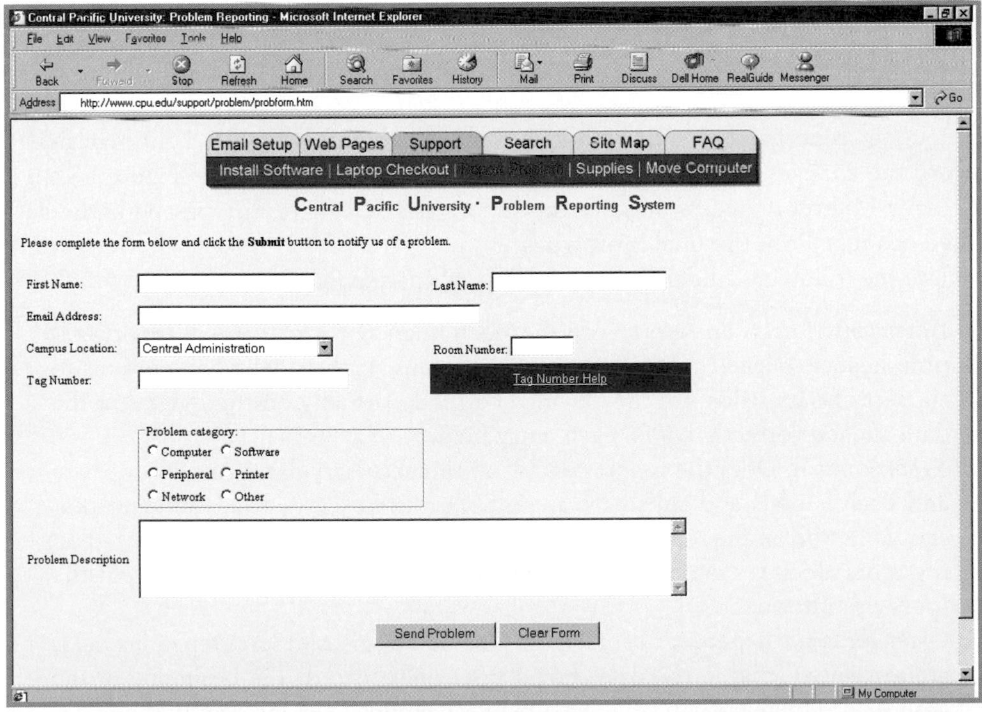

FIGURE E8.5

Prototype for the PROBLEM REPORTING SYSTEM Web page. This Web page needs some improvement.

8

"Yes, and having submenus underneath the main one for the features of each tab makes it easy to find what you are looking for," adds Mike. "I do have some suggestions for the Web page for reporting problems, though. It would be more useful if the Problem Category selection area were moved to the top of the page. Each problem type is assigned to a different technician, one who is more or less an expert in that area. We need an additional check box to identify if it is Macintosh- or IBM-compatible equipment or software we are working with. The Tag Number help is a great idea. Many people do not realize that each piece of equipment has a small metal identifying tag on it with a unique inventory number. Hmmm. . . . That large blue area seems to stand out too much. After all, it is just help. I think that it would be better to replace it with a small graphic image."

"I think that these changes will be easy to do," remarks Anna.

"Great," replies Mike. "It would also be useful to include the tech support hotline phone number on the Web page. If it's a real emergency, it might speed up our resolution to the problem. We should add an entry field for their phone number as well. Of course, we could always look it up, but the person reporting the problem may be in a computer lab or another location away from their office."

"Good idea!" exclaims Dot. "This is going to be extremely helpful to the faculty and staff. I think that we should prototype all the Web pages for the site. I realize that Web pages are supposed to change from time to time, but let's get these as good as possible from the start!"

Anna glances at Chip and grins. "I guess you'll be working on Web page design sooner than you think!"

Anna and Chip continued to work on prototypes by designing, obtaining user feedback, and modifying the design to accommodate user changes. Now that the work is complete, they have a solid sense of the requirements of the system.

EXERCISES*

Critique the report and screen prototypes for the problems below (E-1 through E-10). Record the changes on a copy of the Prototype Evaluation Form (see Figure E8.2). If you have Microsoft Access, modify the report and screen prototypes with the suggested changes. Print the final prototypes.

Use the following guidelines to help in your analysis:

1. *Alignment of fields on reports.* Are the fields aligned correctly? Are report column headers aligned correctly over the columns? If the report has captions to the left of data fields, are they aligned correctly (usually on the left)? Are the data aligned correctly *within* each entry field?
2. *Report content.* Does the report contain all the necessary data? Are appropriate and useful totals and subtotals present? Are there extra totals or data that should not be on the report? Are codes or the meaning of the codes printed on the report (codes should be avoided because they may not clearly present the user with information)?
3. *Check the visual appearance of the report.* Does it look pleasing? Are repeating fields group printed (that is, the data should print only once, at the beginning of the group)? Are there enough blank lines between groups to easily identify them?
4. *Screen data and caption alignment.* Are the captions correctly aligned on the screens? Are the data fields correctly aligned? Are the data *within* a field correctly aligned?

5. *Screen visual appearance.* Does the screen have a pleasing appearance? Is there enough vertical spacing between fields? Is there enough horizontal spacing between columns? Are the fields logically grouped together? Are features, such as buttons and check boxes, grouped together?
6. *Does the screen contain all the necessary functional elements?* Look for missing buttons that would help the user work smoothly with the screen; also look for missing data, extra unnecessary data, or fields that should be replaced with a check box or drop-down list.

E-1. The HARDWARE INVENTORY LISTING is shown in Figure E8.6. It shows all personal computers, sorted by campus and room.

01/11/2001

Hardware Inventory Listing

Page 1 of 1

Campus	Room Location	Inventor Number	Brand Name	Model	Present
Central Administration	11111	84004782	Xxxxxxxxxx	Xxxxxxxxxxxxxxxxx	
Central Administration	11111	90875039	Xxxxxxxxxx	Xxxxxxxxxxxxxxxx	
Central Administration	11111	93955411	Xxxxxxxxxx	Xxxxxxxxxxxxxxxx	
Central Administration	11111	99381373	Xxxxxxxxxx	Xxxxxxxxxxxxxxxx	
		Total number of machines in room is		4	
Central Administration	22222	10220129	Xxxxxxxxxx	Xxxxxxxxxxxxxxxx	
		Total number of machines in room is		1	
Central Administration	99999	22838234	Xxxxxxxxxx	Xxxxxxxxxxxxxxxx	
Central Administration	99999	24720952	Xxxxxxxxxx	Xxxxxxxxxxxxxxxx	
Central Administration	99999	33453403	Xxxxxxxxxx	Xxxxxxxxxxxxxxxx	
Central Administration	99999	34044449	Xxxxxxxxxx	Xxxxxxxxxxxxxxxx	
Central Administration	99999	40030303	Xxxxxxxxxx	Xxxxxxxxxxxxxxxx	
Central Administration	99999	47403948	Xxxxxxxxxx	Xxxxxxxxxxxxxxxx	
Central Administration	99999	56620548	Xxxxxxxxxx	Xxxxxxxxxxxxxxxx	
		Total number of machines in room is		7	
		Total number of machines at campus is		12	
Central Computer Science	22222	34589349	Xxxxxxxxxx	Xxxxxxxxxxxxxxxx	
Central Computer Science	22222	38376910	Xxxxxxxxxx	Xxxxxxxxxxxxxxxx	
Central Computer Science	22222	94842282	Xxxxxxxxxx	Xxxxxxxxxxxxxxxx	
		Total number of machines in room is		3	
Central Computer Science	99999	339393	Xxxxxxxxxx	Xxxxxxxxxxxxxxxx	
		Total number of machines in room is		1	
		Total number of machines at campus is		4	
Central Zoology	22222	11398423	Xxxxxxxxxx	Xxxxxxxxxxxxxxxx	
Central Zoology	22222	28387465	Xxxxxxxxxx	Xxxxxxxxxxxxxxxx	
		Total number of machines in room is		2	
Central Zoology	99999	70722533	Xxxxxxxxxx	Xxxxxxxxxxxxxxxx	
Central Zoology	99999	99481102	Xxxxxxxxxx	Xxxxxxxxxxxxxxxx	
		Total number of machines in room is		2	
		Total number of machines at campus is		4	
		Total number of machines is		24	

FIGURE E8.6

Prototype for the HARDWARE INVENTORY LISTING. Many improvements can be made to this report.

E-2. The SOFTWARE INVESTMENT REPORT is illustrated in Figure E8.7. It is used to calculate the total amount invested in software.

E-3. One page of the INSTALLED MICROCOMPUTER REPORT is shown in Figure E8.8. It shows the information for a single machine. (The full report would be many pages long.)

1/11/01

Software Investment Report

Page 1 of 2

Title	Version	Operating System	Site License	Number of Copies	Unit Cost	Software Cost
Acrobat	4	Windows 98	No	12	$100.00	$1,200.00
After Effects	4.1	Windows 98	No	20	$259.00	$5,180.00
AutoCAD	2000	Windows 98	No	40	$375.00	$15,000.00
Corel Draw!	9	Windows 98	No	20	$150.00	$3,000.00
Dreamweaver	3	Windows 98	No	40	$198.00	$7,920.00
Eudora Pro	4	Windows 98	No	60	$45.00	$2,700.00
Fireworks	3	Windows 98	No	20	$99.00	$1,980.00
Front Page	2000	Windows 98	No	40	$35.00	$1,400.00
In Design	1.5	Windows 98	No	20	$359.00	$7,180.00
Internet Explorer	5.5	Windows 98	Yes		$999.99	$999.99
Netscape Navigator	7.5	Windows 95	Yes		$999.99	$999.99
Office	2000	Windows 98	No	400	$35.00	$14,000.00
Pagemaker	6.5	Windows 98	No	20	$279.00	$5,580.00
Photo Delux	3	Windows 98	No	60	$25.00	$1,500.00
Photoshop	5.5	Windows 98	No	20	$289.00	$5,780.00
Quark Express	4	Windows 98	No	12	$340.00	$4,080.00
Visible Analyst	7.1	Windows 95	No	40	$45.00	$1,800.00
Visible Analyst	7.5	Windows 98	No	40	$45.00	$1,800.00
Visio	4.0	Windows 98	No	40	$120.00	$4,800.00
Visual Basic	6.0	Windows 98	No	40	$35.00	$1,400.00
Windows Draw	6	Windows 98	No	30	$39.99	$1,199.70
Word Perfect	6	Unix	No	20	$35.00	$700.00
WordPerfect Suite	9	Windows 98	No	120	$35.00	$4,200.00
					Total Software Cost	$97,199.67

FIGURE E8.7

Prototype for the SOFTWARE INVESTMENT REPORT. A number of changes can be made to improve this report.

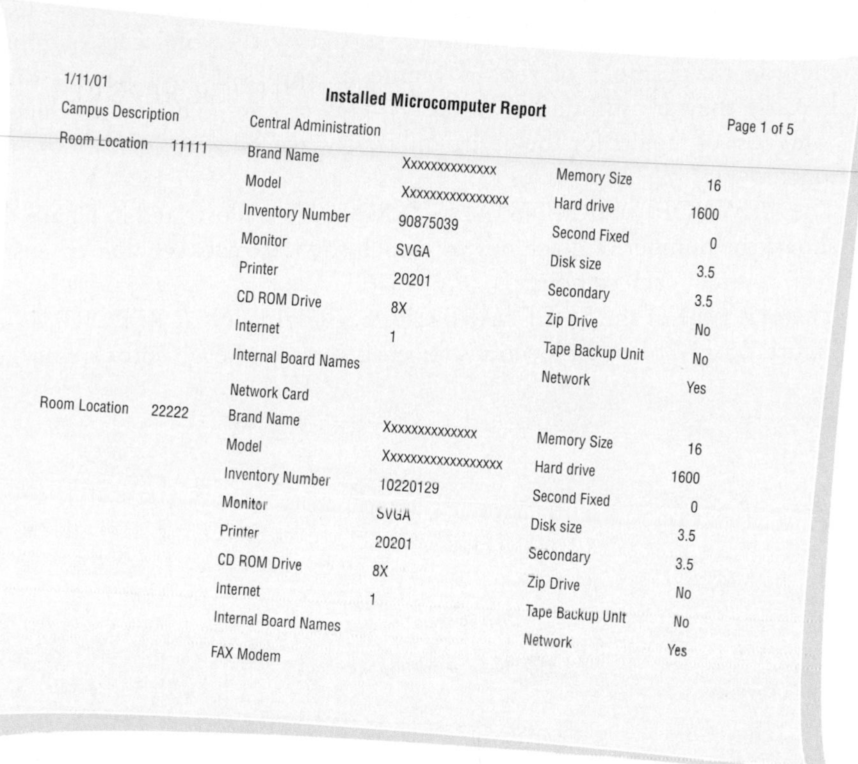

1/11/01

Installed Microcomputer Report

Campus Description Central Administration

Page 1 of 5

Room Location 11111

Brand Name	Xxxxxxxxxxxxxx	Memory Size	16
Model	Xxxxxxxxxxxxxxxx	Hard drive	1600
Inventory Number	90875039	Second Fixed	0
Monitor	SVGA	Disk size	3.5
Printer	20201	Secondary	3.5
CD ROM Drive	8X	Zip Drive	No
Internet	1	Tape Backup Unit	No
Internal Board Names		Network	Yes
Network Card			

Room Location 22222

Brand Name	Xxxxxxxxxxxxxx	Memory Size	16
Model	Xxxxxxxxxxxxxxxxxx	Hard drive	1600
Inventory Number	10220129	Second Fixed	0
Monitor	SVGA	Disk size	3.5
Printer	20201	Secondary	3.5
CD ROM Drive	8X	Zip Drive	No
Internet	1	Tape Backup Unit	No
Internal Board Names		Network	Yes
FAX Modem			

FIGURE E8.8

Prototype for the INSTALLED MICROCOMPUTER REPORT. Many improvements can be made to this report.

01/11/2001

Microcomputer Problem Report

Page 1 of 1

Inventory Number	Brand Name	Model	Cost of Repairs	Warranty
40030303	Xxxxxxxxxxxx	Xxxxxxxxxxxxx	$955.25	Yes
34589349	Xxxxxxxxxxxx	Xxxxxxxxxxxxx	$720.00	Yes
56620548	Xxxxxxxxxxxx	Xxxxxxxxxxxxx	$487.22	Yes
84004782	Xxxxxxxxxxxx	Xxxxxxxxxxxxx	$376.90	Yes
24720952	Xxxxxxxxxxxx	Xxxxxxxxxxxxx	$290.00	Yes
93955411	Xxxxxxxxxxxx	Xxxxxxxxxxxxx	$242.55	Yes
33453403	Xxxxxxxxxxxx	Xxxxxxxxxxxxx	$155.50	Yes
99381373	Xxxxxxxxxxxx	Xxxxxxxxxxxxx	$40.00	Yes
10220129	Xxxxxxxxxxxx	Xxxxxxxxxxxxx	$20.00	Yes

FIGURE E8.9

Prototype for the MICROCOMPUTER PROBLEM REPORT. Users might ask for changes to this report based on this prototype.

E-4. Figure E8.9 shows the prototype for the MICROCOMPUTER PROBLEM REPORT. It should list all machines sorted by the total cost of repairs and include the number of repairs (some machines do not have a high cost, because they are still under warranty). This prototype is used to calculate the total cost of repairs for the entire university, as well as to identify the problem machines.

E-5. The NEW SOFTWARE INSTALLED REPORT is illustrated in Figure E8.10. It shows the number of machines with each software package that are installed in each room of each campus.

E-6. The first page of the SOFTWARE CROSS-REFERENCE REPORT is shown in Figure E8.11. It lists all locations for each version of each software package.

1/11/01

New Software Installed Report

Page 1 of 4

Software Category Title	CASE Version	Operating System	Publisher	Brand / Model	Campus Room	Number of Copies
Visible Analyst	6.2	9	Visible Systems	Xxxxxxxxxxxxxxxx	1001	1
Visible Analyst	7.1	9	Visible Systems	Xxxxxxxxxxxxxxxxxxxx Xxxxxxxxxxxx	11111 1007	1
Visible Analyst	7.1	9	Visible Systems	Xxxxxxxxxxxxxxxxxx Xxxxxxxxxxxxxx	22222 1001	1
Visible Analyst	7.1	9	Visible Systems	Xxxxxxxxxxxxxxxxx Xxxxxxxxxxxxxxxx	99999 1001	1
Visible Analyst	7.1	9	Visible Systems	Xxxxxxxxxxxxxxxx Xxxxxxxxxxxxxxxx	99999 1018	1
Visible Analyst	7.5	8	Visible Systems	Xxxxxxxxxxxxxxxxxxxx Xxxxxxxxxxxxxxxx	22222 1018	1
Visio	4.0	8	Microsoft	Xxxxxxxxxxxxxxxxxxxx Xxxxxxxxxxxxxxxx Xxxxxxxxxxxxxxxxxxxx	22222 1018 22222	1

Software Category Title	DPUB Version	Operating System	Publisher	Brand / Model	Campus Room	Number of Copies
In Design	1.5	8	Adobe	Xxxxxxxxxxxx	1001	1
In Design	1.5	8	Adobe	Xxxxxxxxxxxxxxxxx Xxxxxxxxxxxxxxxx	99999 1018	1
Pagemaker	6.5	8	Adobe	Xxxxxxxxxxxxxxxxxxxx Xxxxxxxxxxxxxx	22222 1001	1
Pagemaker	6.5	8	Adobe	Xxxxxxxxxxxxxxxxx Xxxxxxxxxxxxxxxx Xxxxxxxxxxxxxxxxxxxx	99999 1018 22222	2

FIGURE E8.10

Prototype for the NEW SOFTWARE INSTALLED REPORT. Modifications can improve this.

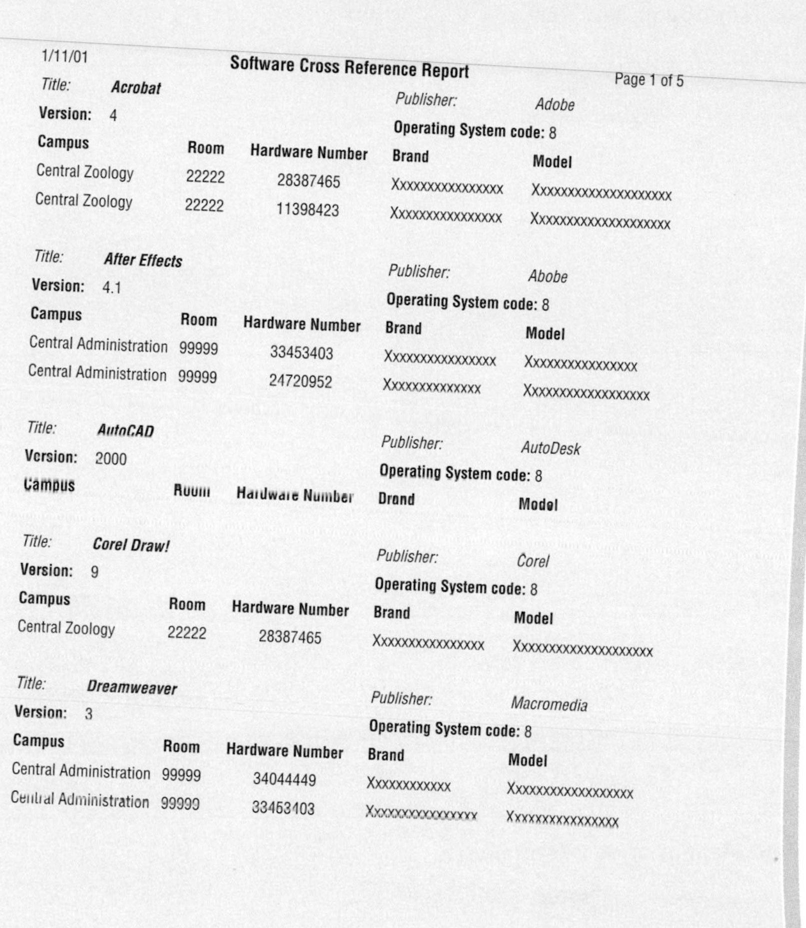

1/11/01 — Software Cross Reference Report — Page 1 of 5

Title: Acrobat — Publisher: Adobe
Version: 4 — Operating System code: 8

Campus	Room	Hardware Number	Brand	Model
Central Zoology	22222	28387465	Xxxxxxxxxxxxxxxx	Xxxxxxxxxxxxxxxxxxxx
Central Zoology	22222	11398423	Xxxxxxxxxxxxxxxx	Xxxxxxxxxxxxxxxxxxxx

Title: After Effects — Publisher: Abobe
Version: 4.1 — Operating System code: 8

Campus	Room	Hardware Number	Brand	Model
Central Administration	99999	33453403	Xxxxxxxxxxxxxxxx	Xxxxxxxxxxxxxxxx
Central Administration	99999	24720952	Xxxxxxxxxxxxxx	Xxxxxxxxxxxxxxxxxx

Title: AutoCAD — Publisher: AutoDesk
Version: 2000 — Operating System code: 8

Campus	Room	Hardware Number	Brand	Model

Title: Corel Draw! — Publisher: Corel
Version: 9 — Operating System code: 8

Campus	Room	Hardware Number	Brand	Model
Central Zoology	22222	28387465	Xxxxxxxxxxxxxxxx	Xxxxxxxxxxxxxxxxxxxx

Title: Dreamweaver — Publisher: Macromedia
Version: 3 — Operating System code: 8

Campus	Room	Hardware Number	Brand	Model
Central Administration	99999	34044449	Xxxxxxxxxxxx	Xxxxxxxxxxxxxxxxxx
Central Administration	99999	33453403	Xxxxxxxxxxxxxxxx	Xxxxxxxxxxxxxxxx

FIGURE E8.11
Prototype for the SOFTWARE CROSS REFERENCE REPORT. Many improvements can be made to this report.

E-7. Figure E8.12 illustrates the DELETE MICROCOMPUTER RECORD screen. The entry area is the Hardware Inventory Number field. The other fields are for display only, to identify the machine. The users would like the ability to print each record before they delete it. They also want to scroll to the next and previous records. *Hint*: Examine the fields shown in the HARDWARE INVENTORY LISTING report.

E-8. Mike Crowe needs a screen to enable him to change maintenance information about personal computers. Sometimes these are routine changes, such as the LAST PREVENTIVE MAINTENANCE DATE or the NUMBER OF REPAIRS, but other changes may occur only sporadically, such as the expiration of a warranty. The HARDWARE INVENTORY NUMBER is entered, and the matching

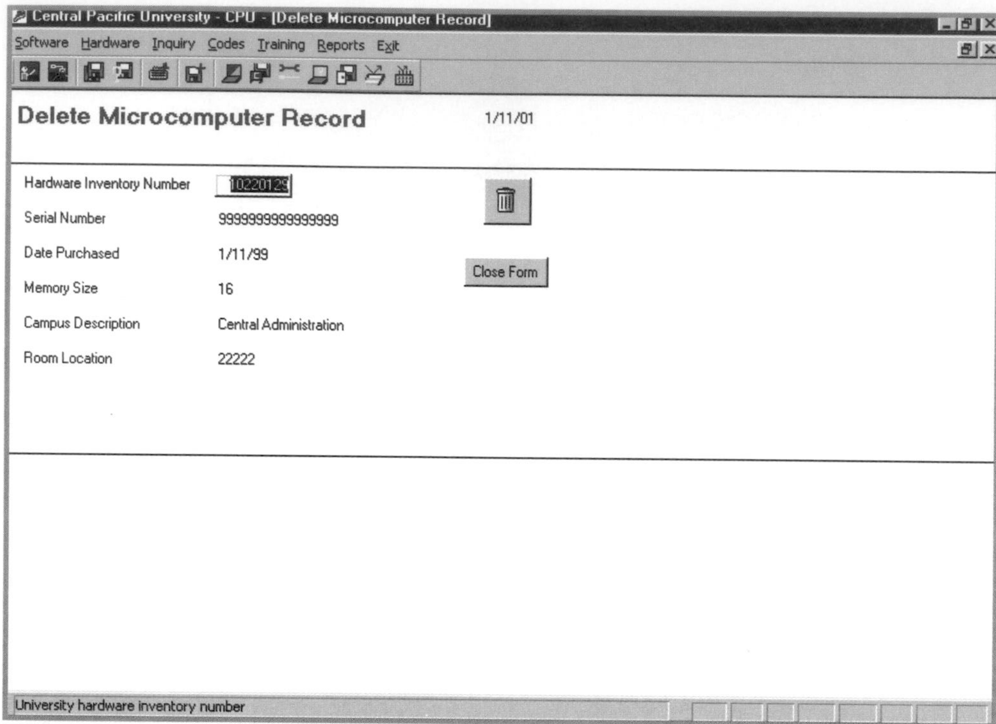

FIGURE E8.12

Prototype for the DELETE MICROCOMPUTER RECORD SCREEN. Changes can improve it.

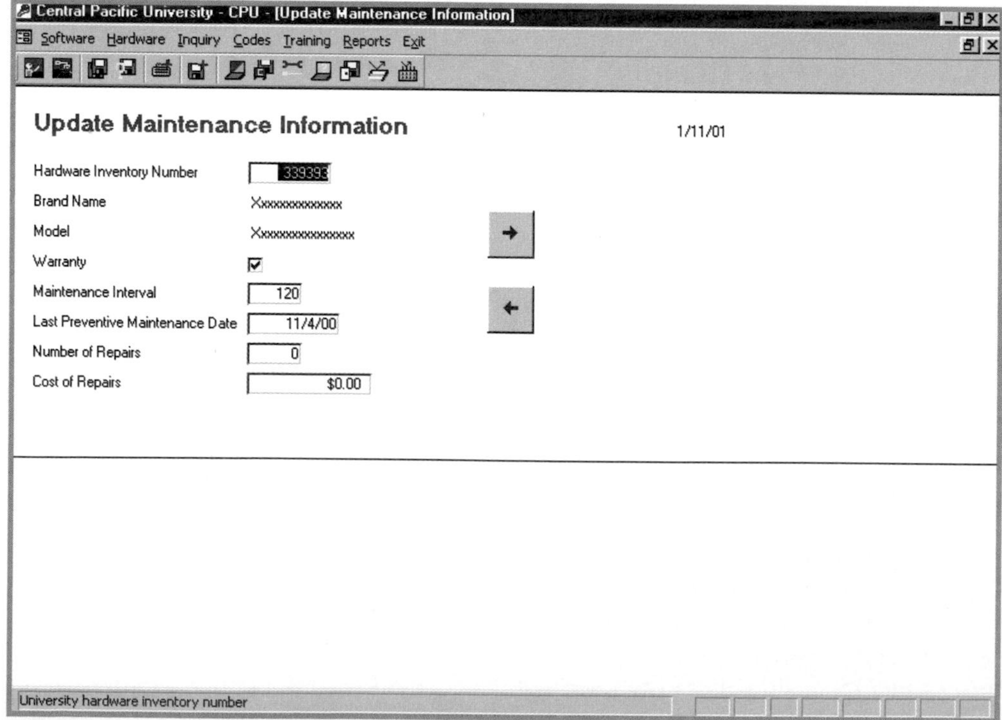

FIGURE E8.13

Prototype for the UPDATE MAINTENANCE INFORMATION SCREEN. Modifications can make the screen better.

MICROCOMPUTER RECORD is found. The BRAND and MODEL are displayed for feedback. The operator may then change the WARRANTY, MAINTENANCE INTERVAL, NUMBER OF REPAIRS, LAST PREVENTIVE MAINTENANCE DATE, and TOTAL COST OF REPAIRS fields. Mike would like to print the screen information, as well as undo any changes, easily. The screen is illustrated in Figure E8.13.

E-9. The SOFTWARE LOCATION INQUIRY screen is shown in Figure E8.14. The TITLE, VERSION NUMBER, and OPERATING SYSTEM are entered. The output portion of the screen should show the CAMPUS LOCATION, ROOM LOCATION, HARDWARE INVENTORY NUMBER, BRAND NAME, and MODEL. Buttons allow the user to move to the next record, the previous record, and to close and exit the screen.

E-10. Figure E8.15 illustrates the HARDWARE CHARACTERISTIC INQUIRY screen design. The operator enters a BRAND NAME and CD-ROM DRIVE type. The MONITOR and PRINTER CODE fields have drop-down lists to select the appropriate codes. The display portion of the inquiry screen consists of CAMPUS, ROOM, and INVENTORY NUMBER.

*Exercises preceded by a CD ROM icon require the program Visible Analyst or another CASE tool. A CD-ROM containing Visible Analyst examples is provided free of charge to any professor adopting this book. The examples on the disk may be imported into Visible Analyst and then used by students.

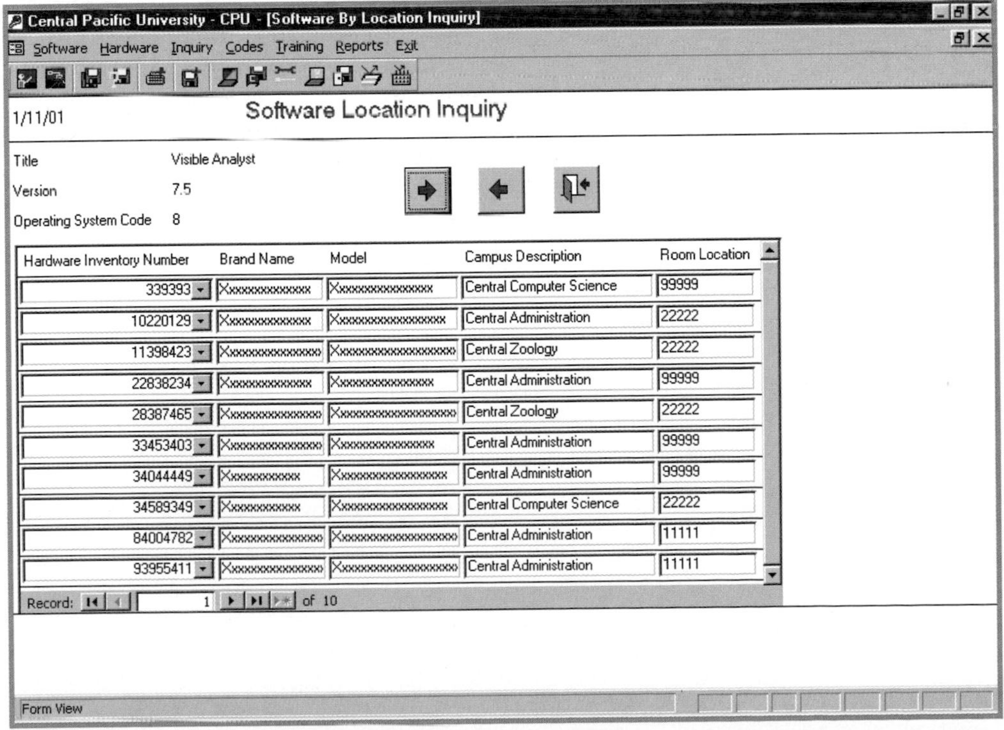

FIGURE E8.14
Prototype for the SOFTWARE LOCATION INQUIRY SCREEN. Many improvements can be made.

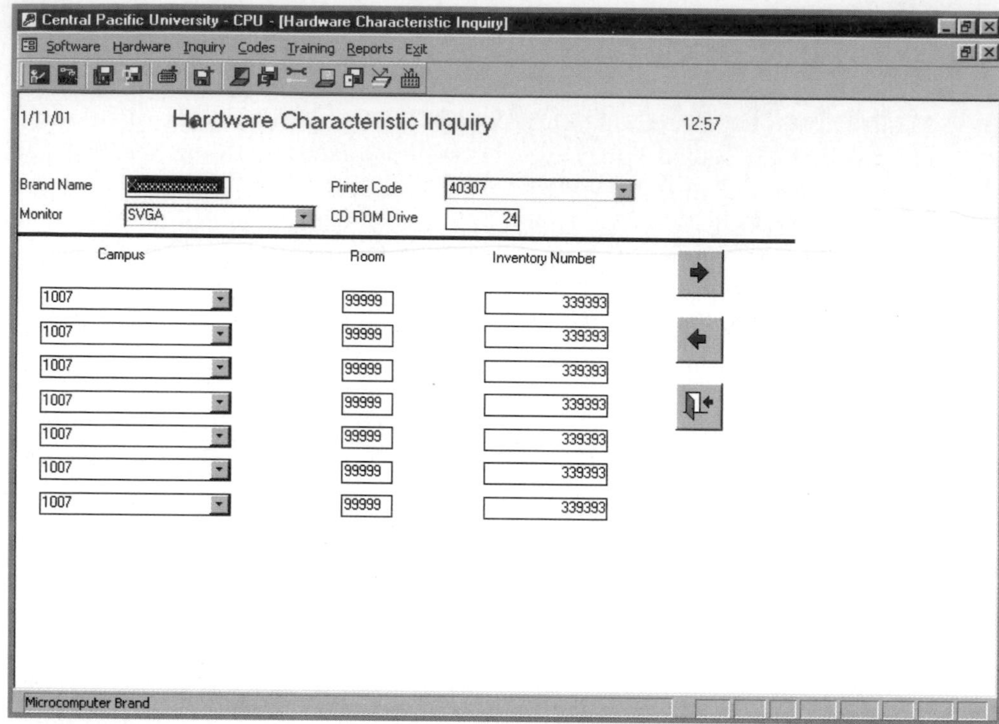

FIGURE E8.15

Prototype for the HARDWARE CHARACTERISTIC SCREEN. Modifications will improve the screen.

USING DATA FLOW DIAGRAMS

9

The systems analyst needs to make use of the conceptual freedom afforded by data flow diagrams, which graphically characterize data processes and flows in a business system. In their original state data flow diagrams depict the broadest possible overview of system inputs, processes, and outputs, which correspond to those of the general systems model discussed in Chapter 2. A series of layered data flow diagrams may also be used to represent and analyze detailed procedures within the larger system.

THE DATA FLOW APPROACH TO REQUIREMENTS DETERMINATION

When systems analysts attempt to understand the information requirements of users, they must be able to conceptualize how data move through the organization, the processes or transformation that the data undergo, and what the outputs are. Although interviews and the investigation of hard data provide a verbal narrative of the system, a visual depiction can crystallize this information in a useful way.

Through a structured analysis technique called data flow diagrams (DFDs), the systems analyst can put together a graphical representation of data processes throughout the organization. The data flow approach emphasizes the logic underlying the system. By using combinations of only four symbols, the systems analyst can create a pictorial depiction of processes that will eventually provide solid system documentation.

ADVANTAGES OF THE DATA FLOW APPROACH

The data flow approach has four chief advantages over narrative explanations of the way data move through the system.

1. Freedom from committing to the technical implementation of the system too early.
2. Further understanding of the interrelatedness of systems and subsystems.
3. Communicating current system knowledge to users through data flow diagrams.
4. Analysis of a proposed system to determine if the necessary data and processes have been defined.

Perhaps the biggest advantage lies in the conceptual freedom found in the use of the four symbols (covered in the upcoming subsection on DFD conventions). (You will recognize three of the symbols from Chapter 2.) None of the symbols specifies the physical aspects of implementation. For instance, although an analyst will signify that data are stored at a particular point, the data flow approach does not dictate specifying the medium for storage. Thus, systems analysts can conceptualize necessary data flows and avoid committing too quickly to their technical realization.

The data flow approach has the additional advantage of serving as a useful exercise for systems analysts, enabling them to better understand the interrelatedness of the system and its subsystems. Recall that in Chapter 2 we stressed the importance of being able to differentiate the system from its environment by locating its boundaries. It requires discipline and true understanding to conceptualize the system in a broad overview and then explode it into its functional subsystems.

A third advantage of the data flow approach is that it can be used as a tool to interact with users. An interesting use of DFDs is to show them to users as incomplete representations of the analyst's understanding of the system. Users can then be asked to comment on the accuracy of the analyst's conceptualization, and the analyst can incorporate changes that more accurately reflect the system from the users' perspectives.

Although many texts tout the ease of communicating to users through data flow diagrams, this benefit does not occur automatically. If you want to use DFDs for interaction, you must assume responsibility for educating users about their purposes. Necessary background must be provided to users before data flow diagrams will be meaningful rather than confusing.

The last advantage of using data flow diagrams is that they allow analysts to describe each component used in the diagram. Analysis can then be performed to ensure that all necessary output may be obtained from the input data and that processing logic is reflected in the diagram. Detecting and correcting errors and design flaws of this nature in the earlier stages of the systems development life cycle is far less costly than in the later phases of programming, testing, and implementation.

Recall that in Chapter 4 we examined workflow analysis that visually depicts work in the organization, in manual or automated form. In addition, in workflow analysis, work must unfold chronologically. By contrast, DFDs emphasize the processing of data or the transforming of data as they move through a variety of processes. In logical DFDs, there is no distinction between manual or automated processes. Neither are the processes graphically depicted in chronological order. Rather, processes are eventually grouped together when further analysis dictates that it makes sense to do so. Manual processes are put together, and automated processes can also be paired with each other.

CONVENTIONS USED IN DATA FLOW DIAGRAMS

Four basic symbols are used to chart data movement on data flow diagrams: a double square, an arrow, a rectangle with rounded corners, and an open-ended rectangle (closed on the left side and open-ended on the right), as shown in Figure 9.1. An entire system and numerous subsystems can be depicted graphically with these four symbols in combination.

The double square is used to depict an external entity (another department, a business, a person, or a machine) that can send data to or receive data from the system. The external entity, or just entity, is also called a source or destination of data, and it is considered to be external to the system being described. Each entity is labeled with an appropriate name. Although it interacts with the system, it is considered as outside the boundaries of the system. Entities should be named with a noun. The same entity may be used more than once on a given data flow diagram to avoid crossing data flow lines.

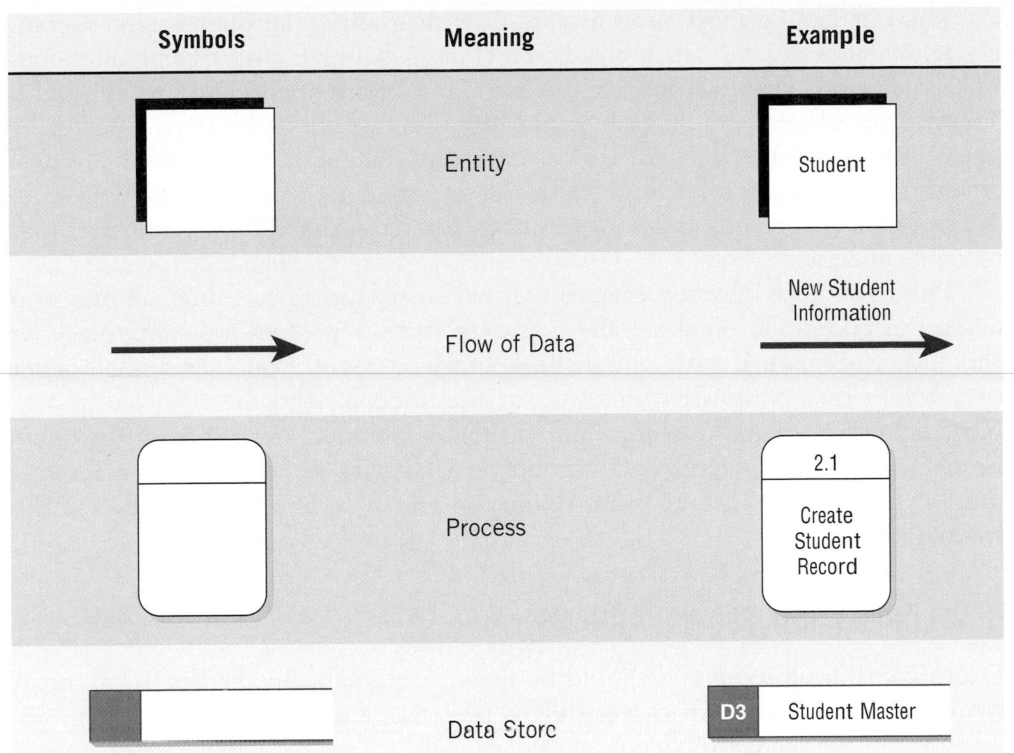

Symbols
Meaning
Example
Entity
Student
Flow of Data
New Student Information
Process
2.1
Create Student Record
Data Store
D3
Student Master

FIGURE 9.1

The four basic symbols used in data flow diagrams, their meanings, and examples.

The arrow shows movement of data from one point to another, with the head of the arrow pointing toward the data's destination. Data flows occurring simultaneously can be depicted doing just that through the use of parallel arrows. Because an arrow represents data about a person, place, or thing, it too should be described with a noun.

A rectangle with rounded corners is used to show the occurrence of a transforming process. Processes always denote a change in or transformation of data; hence, the data flow leaving a process is *always* labeled differently from the one entering it. Processes represent work being performed within the system and should be named using one of the following formats. A clear name makes it easier to understand what the process is accomplishing.

1. Assign the name of the whole system when naming a high-level process. An example is INVENTORY CONTROL SYSTEM.
2. To name a major subsystem, use a name such as INVENTORY REPORTING SUBSYSTEM or INTERNET CUSTOMER FULFILLMENT SYSTEM.
3. Use a verb-adjective-noun format for detailed processes. The verb describes the type of activity, such as COMPUTE, VERIFY, PREPARE, PRINT, or ADD. The noun indicates what the major outcome of the process is, such as REPORT or RECORD. The adjective illustrates which specific output, such as BACK-ORDERED or INVENTORY, is produced. Examples of complete process names are COMPUTE SALES TAX, VERIFY CUSTOMER ACCOUNT STATUS, PREPARE SHIPPING INVOICE, PRINT BACK-ORDERED REPORT, SEND CUSTOMER EMAIL CONFIRMATION, VERIFY CREDIT CARD BALANCE, and ADD INVENTORY RECORD.

A process must also be given a unique identifying number indicating its level within the diagram. This organization is discussed later in this chapter. Several data flows may go into and out of each process. Examine processes with a single flow in and out for missing data flows.

The last basic symbol used in data flow diagrams is an open-ended rectangle, which represents a data store. The rectangle is drawn with two parallel lines that are closed by a short line on the left side and are open-ended on the right. These symbols are drawn only wide enough to allow identifying lettering between the parallel lines. In logical data flow diagrams, the type of physical storage (for example, tape, diskette) is not specified. At this point the data store symbol is simply showing a depository for data that allows addition and retrieval of data.

The data store may represent a manual store, such as a filing cabinet, or a computerized file or database. Because data stores represent a person, place, or thing, they are named with a noun. Temporary data stores, such as scratch paper or a temporary computer file, are not included on the data flow diagram. Neither are any blank forms or blank diskettes included, even though they may be necessary for a business activity. Give each data store a unique reference number, such as D1, D2, D3, and so on, to identify its level as described in the following section.

DEVELOPING DATA FLOW DIAGRAMS

Data flow diagrams can and should be drawn systematically. Figure 9.2 summarizes the steps involved in successfully completing data flow diagrams. First, the systems analyst needs to conceptualize data flows from a top-down perspective.

FIGURE 9.2
Steps in developing data flow diagrams.

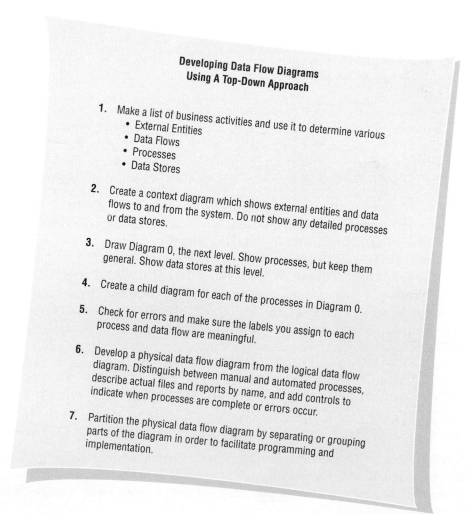

Developing Data Flow Diagrams Using A Top-Down Approach

1. Make a list of business activities and use it to determine various
   - External Entities
   - Data Flows
   - Processes
   - Data Stores
2. Create a context diagram which shows external entities and data flows to and from the system. Do not show any detailed processes or data stores.
3. Draw Diagram 0, the next level. Show processes, but keep them general. Show data stores at this level.
4. Create a child diagram for each of the processes in Diagram 0.
5. Check for errors and make sure the labels you assign to each process and data flow are meaningful.
6. Develop a physical data flow diagram from the logical data flow diagram. Distinguish between manual and automated processes, describe actual files and reports by name, and add controls to indicate when processes are complete or errors occur.
7. Partition the physical data flow diagram by separating or grouping parts of the diagram in order to facilitate programming and implementation.

To begin a data flow diagram, collapse the organization's system narrative into a list with the four categories of external entity, data flow, process, and data store. This list in turn helps determine the boundaries of the system you will be describing. Once a basic list of data elements has been compiled, begin drawing a context diagram.

CREATING THE CONTEXT DIAGRAM

With a top-down approach to diagramming data movement, the diagrams move from general to specific. Although the first diagram helps the systems analyst grasp basic data movement, its general nature limits its usefulness. The initial context diagram should be an overview, one including basic inputs, the general system, and outputs. This diagram will be the most general one, really a bird's-eye view of data movement in the system and the broadest possible conceptualization of the system.

The context diagram is the highest level in a data flow diagram and contains only one process, representing the entire system. The process is given the number zero. All external entities are shown on the context diagram as well as major data flow to and from them. The diagram does not contain any data stores and is fairly simple to create, once the external entities and the data flow to and from them are known to analysts from interviews with users and as a result of document analysis.

DRAWING DIAGRAM 0 (THE NEXT LEVEL)

More detail than the context diagram permits is achievable by "exploding the diagrams." Inputs and outputs specified in the first diagram remain constant in all subsequent diagrams. The rest of the original diagram, however, is exploded into close-ups involving three to nine processes and showing data stores and new lower-level data flows. The effect is that of taking a magnifying glass to view the original data flow diagram. Each exploded diagram should use only a single sheet of paper. By exploding DFDs into subprocesses, the systems analyst begins to fill in the details about data movement. The handling of exceptions is ignored for the first two or three levels of data flow diagramming.

Diagram 0 is the explosion of the context diagram and may include up to nine processes. Including more processes at this level will result in a cluttered diagram that is difficult to understand. Each process is numbered with an integer, generally starting from the upper left-hand corner of the diagram and working toward the lower right-hand corner. The major data stores of the system (representing master files) and all external entities are included on Diagram 0. Figure 9.3 schematically illustrates both the context diagram and Diagram 0.

Because a data flow diagram is two-dimensional (rather than linear), you may start at any point and work forward or backward through the diagram. If you are unsure of what you would include at any point, take a different external entity, process, or data store, and then start drawing the flow from it. You may:

1. Start with the data flow from an entity on the input side. Ask questions such as: "What happens to the data entering the system?" "Is it stored?" "Is it input for several processes?"
2. Work backwards from an output data flow. Examine the output fields on a document or screen. (This approach is easier if prototypes have been created.) For each field on the output, ask: "Where does it come from?" or "Is it calculated or stored on a file?" For example, when the output is a PAYCHECK, the EMPLOYEE NAME and ADDRESS would be located on an EMPLOYEE file, the HOURS WORKED would be on a TIME RECORD, and the GROSS PAY and DEDUCTIONS would be calculated. Each file and record would be connected to the process that produces the paycheck.

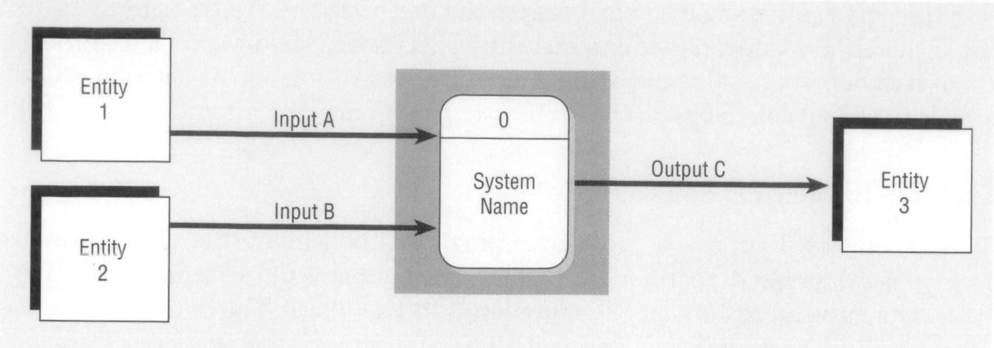

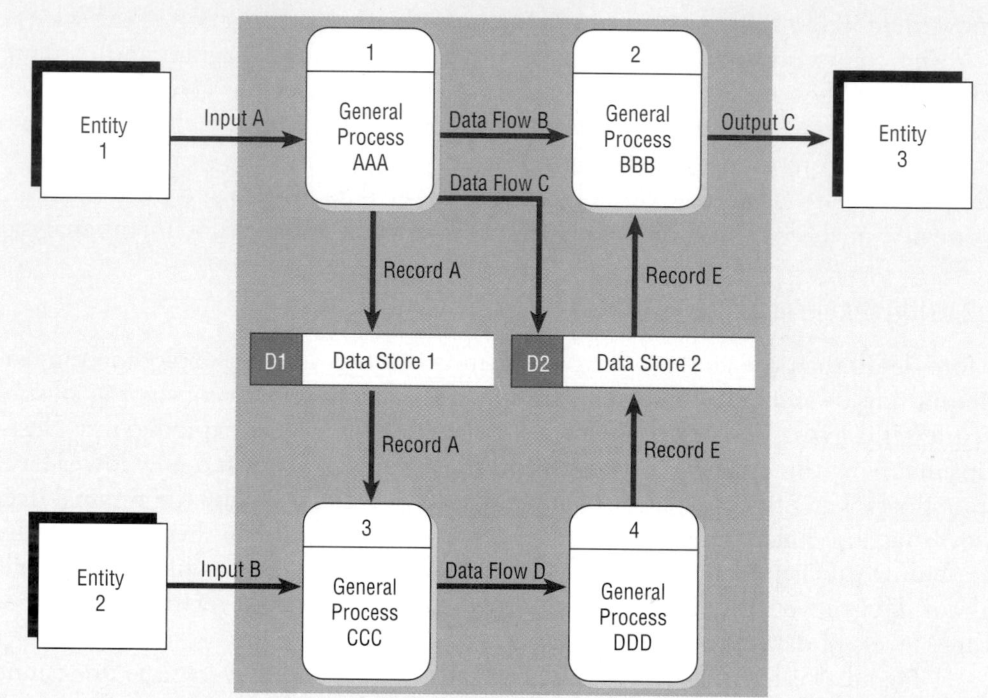

FIGURE 9.3

Context diagrams (above) can be "exploded" into Diagram 0 (below). Note the greater detail in Diagram 0.

3. Examine the data flow to or from a data store. Ask: "What processes put data into the store?" or "What processes use the data?" Note that a data store used in the system you are working on may be produced by a different system. Thus, from your vantage point, there may not be any data flow into the data store.
4. Analyze a well-defined process. Look at what input data the process needs and what output it produces. Then connect the input and output to the appropriate data stores and entities.
5. Take note of any fuzzy areas where you are unsure of what should be included or of what input or output is required. Awareness of problem areas will help you formulate a list of questions for follow-up interviews with key users.

CREATING CHILD DIAGRAMS (MORE DETAILED LEVELS)

Each process on Diagram 0 may in turn be exploded to create a more detailed child diagram. The process on Diagram 0 that is exploded is called the *parent process*, and the diagram that results is called the *child diagram*. The primary rule for creating child diagrams, vertical balancing, dictates that a child diagram cannot

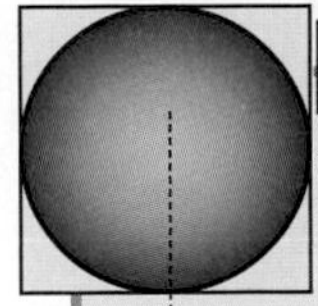

CONSULTING OPPORTUNITY 9.1

GO WITH THE FLOW

"Let's see. We've got a clerk adding up the day's receipts from the cash register tape by calculator. After she adds them initially, she separates them into separate departments, including juvenile, maternity, and infants. Then she gives her departmental subtotals and total on scratch paper to me," says Luis Asperilla.

Pamela Coburn, a systems analyst who is working with a group of 26 franchise clothing stores called Bonton's, is talking to the South Street store's manager, Luis, and she is trying to understand the data flows within the store. Luis continues the narrative. "Then I recheck the day's receipts, looking for any discrepancies. Next, I enter the day's breakdown of the day's receipts, their departments, and the total day's receipts into the ledger, and I fill out the deposit for the bank. All daily receipt information is stored in one place, in the ledger in my office."

Pamela asks, "Do you keep a copy anywhere?" Luis pauses, then replies, "Well, there is a weekly report that summarizes all the weekly information for the head of franchising in New York. The people there enter it into their computers and we get sent a printout at the end of the month. So if I wait five weeks, I do in effect get a copy back, except that I keep each printout and reconcile it against my own monthly summary that I do by hand. You'd be surprised at how often there is a mistake in what is sent back. Then I write a letter and try to get it corrected so my six-month inventories come out right. I keep copies of all correspondence to the people in New York in a file drawer. I'm always writing to them on something they've screwed up. And I need a copy to prove I sent in a correction."

Luis continues, "Computers in New York seem worth it, I suppose, but I think they introduce an awful lot of errors if you don't use common sense when you enter the numbers in. But the ledger book does get heavy to pull down from the shelf by the end of the year."

"I keep a lot of what happens in the store in my head, too," Luis adds thoughtfully. "It's so hard to write everything down, like which customers are allowed layaway privileges and information like that, because we get so busy. I keep a few notes in my desk. I think you'll find that I'm really organized compared with the other managers in town."

Draw a data flow diagram of Luis's description of the store's data flows. What are some of the specific physical barriers to implementation that Pamela can overcome by representing the store's data flow in a data flow diagram? Describe them in a paragraph.

produce output or receive input that the parent process does not also produce or receive. All data flow into or out of the parent process must be shown flowing into or out of the child diagram.

The child diagram is given the same number as its parent process in Diagram 0. For example, process 3 would explode to Diagram 3. The processes on the child diagram are numbered using the parent process number, a decimal point, and a unique number for each child process. On Diagram 3, the processes would be numbered 3.1, 3.2, 3.3, and so on. This convention allows the analyst to trace a series of processes through many levels of explosion. If Diagram 0 depicts processes 1, 2, and 3, the child diagrams 1, 2, and 3 are all on the same level.

Entities are usually not shown on the child diagrams below Diagram 0. Data flow that matches the parent flow is called an *interface data flow* and is shown as an arrow from or into a blank area of the child diagram. If the parent process has data flow connecting to a data store, the child diagram may include the data store as well. In addition, this lower-level diagram may contain data stores not shown on the parent process. For example, a file containing a table of information, such as a tax table, or a file linking two processes on the child diagram may be included. Minor data flow, such as an error line, may be included on a child diagram but not on the parent.

Processes may or may not be exploded, depending on their level of complexity. When a process is not exploded, it is said to be functionally primitive and is called a *primitive process*. Logic is written to describe these processes and is discussed in detail in Chapter 11. Figure 9.4 illustrates detailed levels within a child data flow diagram.

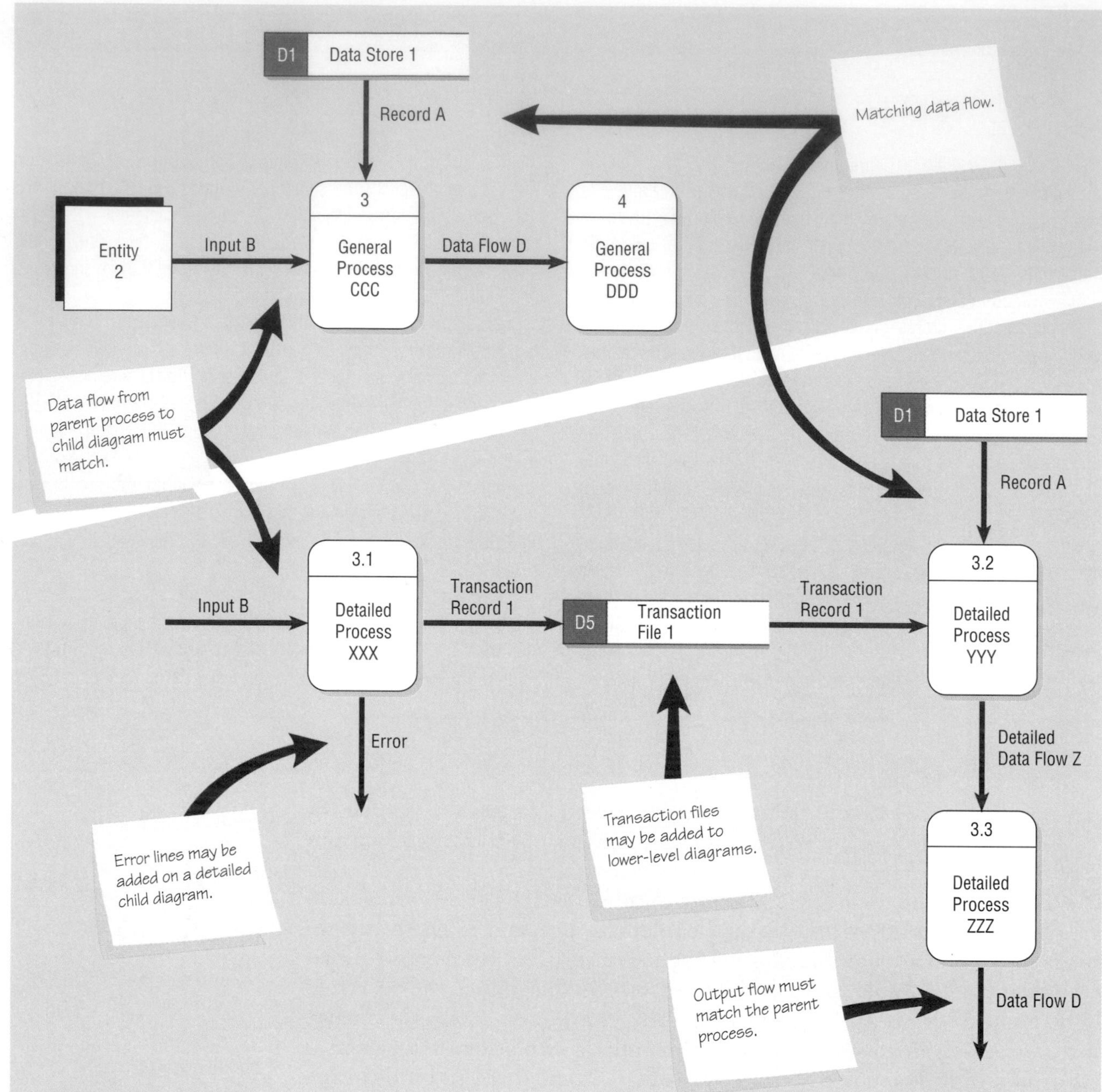

FIGURE 9.4
Differences between the parent diagram (above) and the child diagram (below).

CHECKING THE DIAGRAMS FOR ERRORS

Numerous errors may occur when drawing data flow diagrams. Some of the more common mistakes are shown in Figure 9.5.

It is useful to see how mistakes can come about in a data flow diagram. Figure 9.6 is an example of a data flow diagram that, if implemented, would produce an employee paycheck with many flaws. Several common errors made when drawing data flow diagrams are as follows.

1. Forgetting to include a data flow or pointing an arrowhead in the wrong direction. An example is a drawn process showing all its data flow as input or as output. Each process transforms data and must receive input and pro-

Data flows should not split into two or more different data flows.

All data flows must EITHER originate or terminate at a process.

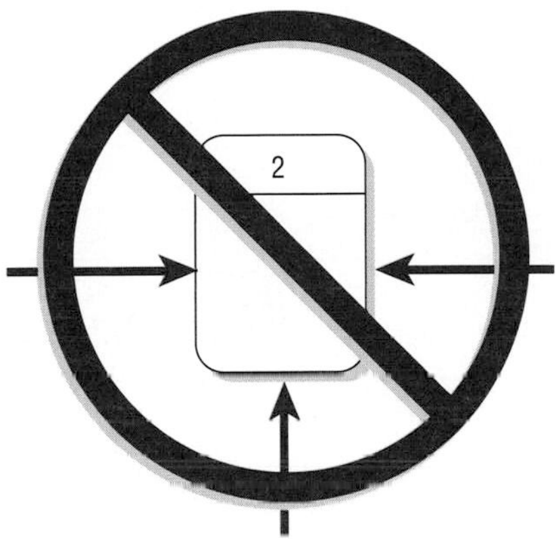

Processes need to have at least one input data flow and one output data flow.

FIGURE 9.5
Don't break the rules when drawing data flow diagrams.

duce output. This type of error usually occurs when the analyst has forgotten to include a data flow or has placed an arrowhead pointing in the wrong direction. Process 1 in Figure 9.6 contains only input because the GROSS PAY arrow is pointing in the wrong direction. This error also affects process 2, CALCULATE WITHHOLDING AMOUNT, which is in addition missing a data flow representing input for the withholding rates and the number of dependents.

2. Connecting data stores and external entities directly to each other. Data stores and entities may not be connected to each other; data stores and external entities must connect only with a process. A file does not interface with another file without the help of a program or a person moving the data, so EMPLOYEE MASTER in Figure 9.6 cannot directly produce the CHECK RECONCILIATION file. External entities do not directly work with files. For example, you would not want a customer rummaging around in the customer master file. Thus, in Figure 9.6 the EMPLOYEE does not create the EMPLOYEE TIME FILE. Two external entities directly connected indicate that they wish to communicate with each other. This connection is not included on the data flow diagram unless the system is facilitating the communication. Producing a report is an instance of this sort of communication. A process must still be interposed between the entities to produce the report, however.
3. Incorrectly labeling processes or data flow. Inspect the data flow diagram to ensure that each object or data flow is properly labeled. A process should indicate the system name or use the verb-adjective-noun format. Each data flow should be described with a noun.

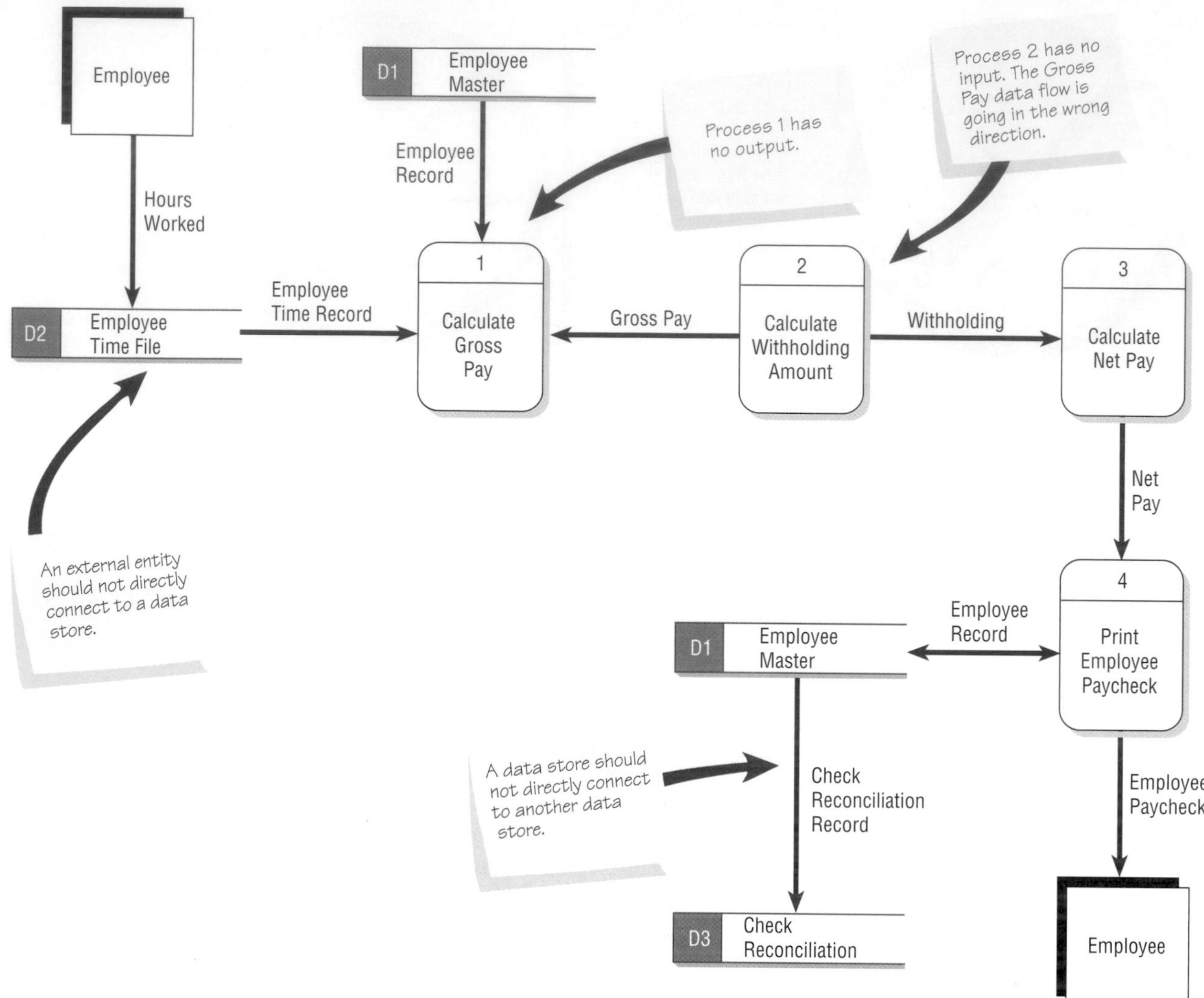

FIGURE 9.6
Typical errors that can occur in a data flow diagram (payroll example).

4. Including more than nine processes on a data flow diagram. Having too many processes creates a cluttered diagram that is confusing to read and hinders rather than enhances communication. If more than nine processes are involved in a system, group some of the processes that work together into a subsystem and place them in a child diagram.
5. Omitting data flow. Examine your diagram for linear flow, that is, data flow where each process has only one input and one output. Except in the case of very detailed child data flow diagrams, linear data flow is somewhat rare. Its presence usually indicates that the diagram has missing data flow. For instance, in Figure 9.6 the process CALCULATE WITHHOLDING AMOUNT needs the number of dependents that an employee has and the WITHHOLDING RATES as input. In addition, NET PAY cannot be calculated solely from the WITHHOLDING, and the EMPLOYEE PAYCHECK cannot be created from the NET PAY alone; it also needs to include an EMPLOYEE NAME and the current and year-to-date payroll and WITHHOLDING AMOUNT figures.
6. Creating unbalanced decomposition in child diagrams. Each child diagram should have the same input and output data flow as the parent process. An exception to this rule is minor output, such as error lines, which are included only on the child diagram. The data flow diagram in Figure 9.7 is correctly drawn. Note that although the data flow is not linear, you can clearly follow a path directly from the source entity to the destination entity.

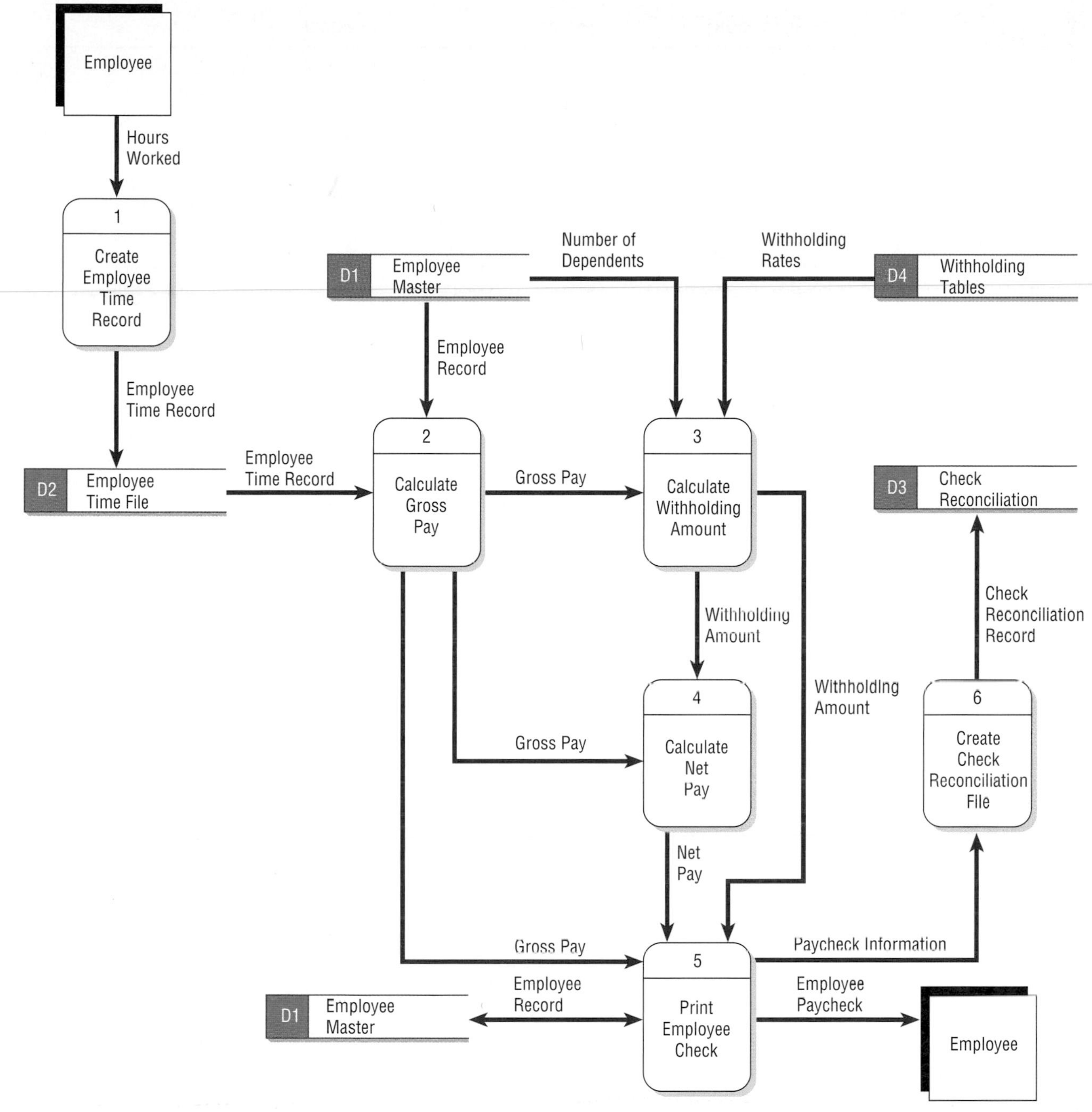

FIGURE 9.7
The correct data flow diagram for the payroll example.

LOGICAL AND PHYSICAL DATA FLOW DIAGRAMS

Data flow diagrams are categorized as either logical or physical. A logical data flow diagram focuses on the business and how the business operates. It is not concerned with how the system will be constructed. Instead, it describes the business events that take place and the data required and produced by each event. Conversely, a physical data flow diagram shows how the system will be implemented, including the hardware, software, files, and people involved in the system. The chart shown in Figure 9.8 contrasts the features of logical and physical models. Notice that the logical model reflects the business, whereas the physical model depicts the system.

Ideally, systems are developed by analyzing the current system (the current logical DFD) and then adding features that the new system should include (the

FIGURE 9.8
Features common of logical and physical data flow diagrams.

Design Feature	Logical	Physical
What the model depicts	How the business operates	How the system will be implemented (or how the current system operates)
What the processes represent	Business activities	Programs, program modules, and manual procedures
What the data stores represent	Collections of data regardless of how the data are stored	Physical files and databases, manual files
Type of data stores	Show data stores representing permanent data collections	Master files, transition files. Any processes that operate at two different times must be connected by a data store.
System controls	Show business controls	Show controls for validating input data, for obtaining a record (record found status), for ensuring successful completion of a process, and for system security (example: journal records)

proposed logical DFD). Finally, the best methods for implementing the new system should be developed (the physical DFD).

Developing a logical data flow diagram for the current system affords a clear understanding of how the current system operates and thus a good starting point for developing the logical model of the current system. This time-consuming step is often omitted so as to go straight to the proposed logical DFD. An example of one type of logical model is the navigation charts created for Web sites when using Microsoft's Frontpage.

One argument in favor of taking the time to construct the logical data flow diagram of the current system is that it can be used to create the logical data flow diagram of the new system. Processes that will be unnecessary in the new system may be dropped, and new features, activities, output, input, and stored data may

FIGURE 9.9
The progression of models from logical to physical.

Derive the logical data flow diagram for the current system by examining the physical data flow diagram and isolating unique business activities.

Create the logical data flow diagram for the new system by adding the input, output, and processes required in the new system to the logical data flow diagram for the current system.

Derive the physical data flow diagram by examining processes on the new logical diagram. Determine where the user interfaces should exist, the nature of the processes, and necessary data stores.

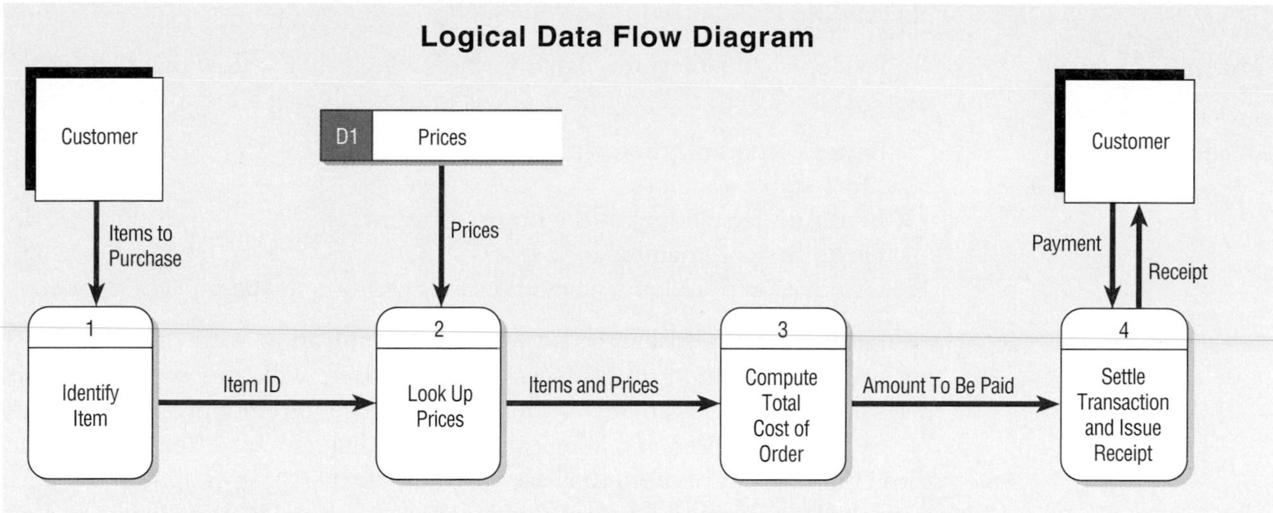

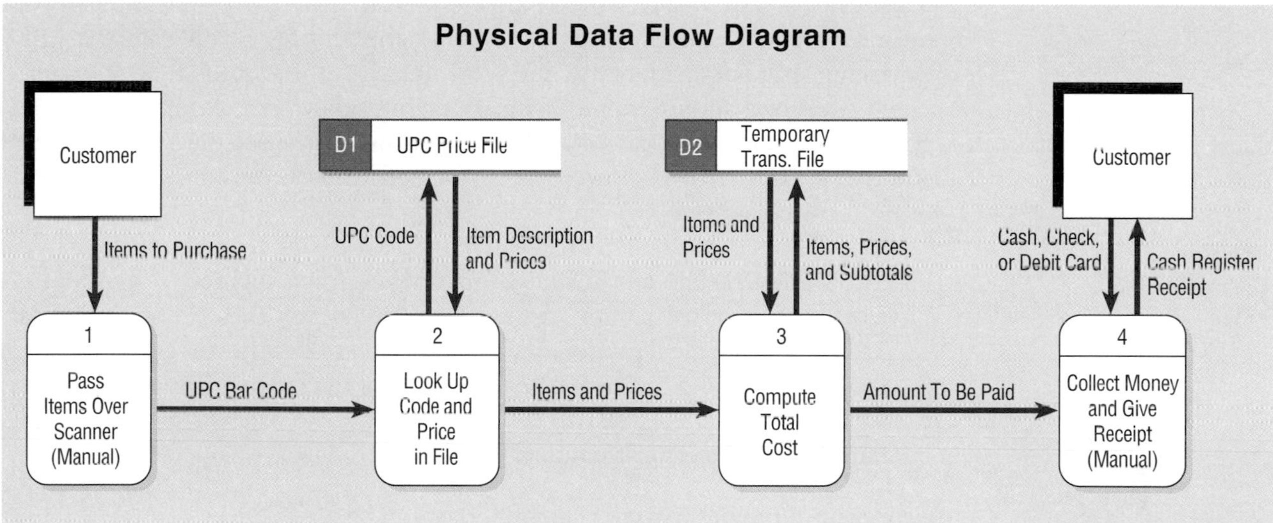

FIGURE 9.10

The physical data flow diagram (below) shows certain details not found on the logical data flow diagram (above).

be added. This approach provides a means of ensuring that the essential features of the old system are retained in the new system. In addition, using the logical model for the current system as a basis for the proposed system provides for a gradual transition to the design of the new system. After the logical model for the new system has been developed, it may be used to create a physical data flow diagram for the new system. The progression of these models is illustrated in Figure 9.9.

Figure 9.10 shows a logical data flow diagram and a physical data flow diagram for a grocery store cashier. The CUSTOMER brings the ITEMS to the register; PRICES for all ITEMS are LOOKED UP and then totaled; next, PAYMENT is given to the cashier; finally, the CUSTOMER is given a RECEIPT. The logical data flow diagram illustrates the processes involved without going into detail about the physical implementation of activities. The physical data flow diagram shows that a bar code—the universal product code (UPC) BAR CODE found on most grocery store items—is used. In addition, the physical data flow diagram mentions manual processes such as scanning, explains that a temporary file is used to keep a subtotal of items, and indicates that the PAYMENT could be made by CASH, CHECK, or DEBIT CARD. Finally, it refers to the receipt by its name, CASH REGISTER RECEIPT.

DEVELOPING LOGICAL DATA FLOW DIAGRAMS

To develop such a diagram, first construct a logical data flow diagram for the current system. There are a number of advantages to using a logical model, including:

1. Better communication with users.
2. More stable systems.
3. Better understanding of the business by analysts.
4. Flexibility and maintenance.
5. Elimination of redundancies and easier creation of the physical model.

A logical model is easier to use when communicating with users of the system because it is centered on business activities. Users will thus be familiar with the essential activities and many of the information requirements of each activity.

Systems formed using a logical data flow diagram are often more stable than those that are not because they are based on business events and not on a particular technology or method of implementation. Logical data flow diagrams represent features of a system that would exist no matter what the physical means of doing business are. For example, activities such as applying for a video store membership card, checking out a videotape, and returning the tape would all occur whether the video store had an automated, manual, or hybrid system. A logical data flow diagram has a business emphasis and helps the analyst understand the business being studied, grasp why procedures are performed, and determine the expected result of performing a task.

The new system will be more flexible and easier to maintain if its design is based on a logical model. Business functions are not subject to frequent change. Physical aspects of the system change more frequently than do business functions.

Examining a logical model may help you to create a better system by eliminating redundancies and inefficient methods that exist in the current system. In addition, the logical model is easy to create and simpler to use because it does not often contain data stores other than master files or a database.

DEVELOPING PHYSICAL DATA FLOW DIAGRAMS

When the logical model of the new system is complete, it may be used to create a physical data flow diagram for the new system. The physical data flow diagram shows how the system will be constructed. Just as logical data flow diagrams have certain advantages, physical data flow diagrams have others, including:

1. Clarifying which processes are manual and which are automated.
2. Describing processes in more detail than do logical DFDs.
3. Sequencing processes that have to be done in a particular order.
4. Identifying temporary data stores.
5. Specifying actual names of files and printouts.
6. Adding controls to ensure the processes are done properly.

Figure 9.11 lists the contents of physical data flow diagrams. Notice that the list includes manual processes such as opening mail orders, creating a batch of forms for keying, and visually inspecting a form. Also included are processes for adding, deleting, changing, and updating records. Each master file should link to one corresponding process for each of these tasks. Physical data flow diagrams are often more complex than logical data flow diagrams simply because of the many data stores present in a system. The acronym CRUD is often used for Create, Read, Update, and Delete, the activities that must be present in a system for each master file. A CRUD matrix is a tool to represent where each of these processes occurs

Contents of Physical Data Flow Diagrams
• Manual processes • Processes for adding, deleting, changing, and updating records • Data entry and verifying processes • Validation processes for ensuring accurate data input • Sequencing processes to rearrange the order of records • Processes to produce every unique system output • Intermediate data stores • Actual file names used to store data • Controls to signify completion of tasks or error conditions

FIGURE 9.11
Physical data flow diagrams contain many items not found in logical data flow diagrams.

within a system. Figure 9.12 is a CRUD matrix for an Internet storefront. Notice that some of the processes include more than one activity. Data-entry processes such as keying (either batch or online) and verifying are also part of physical data flow diagrams.

Because much of the work performed in a system involves validation, such processes for ensuring accurate input should be included. It is estimated that 50 to 90 percent of program code is related to validation. Sequencing processes such as sorting and merging may also be included. Processes to produce every unique system output should be added, because each report or screen should be produced by a separate process.

Physical data flow diagrams also have intermediate data stores, often, a transaction file or a temporary database table. Transaction or master files or a database is required to link any two processes that operate at different times. For example, a program may be used to process customer orders on a minute-by-minute basis. The information that this program generates may then have to be stored in a monthly file for sending customer bills and in an annual file for producing an annual sales summary report. Also part of physical data flow diagrams are physical

Activity	Customer	Item	Order	Order Detail
Customer Logon	R			
Item Inquiry		R		
Item Selection		R	C	C
Order Checkout	U	U	U	R
Add Account	C			
Add Item		C		
Close Customer Account	D			
Remove Obsolete Item		D		
Change Customer Demographics	RU			
Change Customer Order	RU	RU	RU	CRUD
Order Inquiry	R	R	R	R

FIGURE 9.12
A CRUD matrix for an Internet storefront. This tool can be used to represent where each of four processes (Create, Read, Update and Delete) occurs within a system.

data stores. The data stores are designated by the actual names of the files or database (for example, CUSTOMER MASTER FILE rather than the label CUSTOMERS used on a logical data flow diagram), and they may be further described by including the data set name, number of records, and other key attributes.

Controls are also included in physical data flow diagrams. Among them are editing input data, "record found" status when accessing a file or database, password verification when accessing a secure Web site, security and backup controls such as a journal record, and batch update controls to ensure that files produced by one process are correctly transmitted to the next process. In the physical data flow diagram, make distinctions between which processes are manual and which are automated.

Manual processes should be documented with written procedures that instruct employees on how to accomplish the tasks at hand. Automated procedures require computer programs, either written in-house or purchased from a vendor. Processes that are automated should be described in the diagram either as online or batch.

Batch means that a group of records are processed automatically. Batch and online programs are used on multiple platforms, such as a mainframe or midrange computer, on the Internet, in a client/server system, or in a distributed system. A distributed system uses a local area network or a midrange computer at a regional office, perhaps a store or manufacturing plant. Data are periodically transmitted to and from a centralized computer called the host computer. Some systems use a combination of both batch and online systems for a task. Figure 9.13 illustrates the use of batch and online programs for different platforms.

FIGURE 9.13

Batch and on-line programs used for different platforms.

Type of Platform	Batch	Online Entry, Batch Update	Online
Mainframe/ midrange	Billing, bulk data entry, report production, file archival.	Keying data that are stored on transaction files for overnight update.	Entering data on a screen with immediate update of database tables.
Distributed	Transmitting data from centralized computer to local computer.	Entering data at local site online and transmitting them to the host computer. Updating the host computer databases.	Online queries and data entry at local site.
Personal computer	Batch update of databases with data received in delimited text format. Report production.	Entering sales on a portable computer by sales representatives. Consolidation of data and updating server database with daily orders.	Queries and data entry performed on a personal computer.
Internet	Sending periodic email newsletter to members of a group. Purchasing email addresses for marketing. Analyzing sales figures.	Filling out and submitting a Web form that is sent as email to the business site. Batch update of databases at the central site.	A Web form is completed and submitted. Database tables are immediately updated. Sending a confirmation email after an order is received.

Timing information may also be included. For example, an edit program must be run before an update program. Updates must be performed before producing a summary report, or an order must be entered on a Web site before the amount charged to a credit card may be verified with the financial institution. Note that because of such considerations, a physical data flow diagram may appear more linear than a logical model.

Intermediate data stores often consist of transaction files used to store data between processes. Because most processes that require access to a given set of data are unlikely to execute at the same instant in time, transaction files must hold the data from one process to the next. An easily understood example of this concept is found in the everyday experience of grocery shopping. The activities are:

1. Selecting items from shelves.
2. Checking out and paying the bill.
3. Transporting the groceries home.
4. Preparing a meal.
5. Eating the meal.

Each of these five activities would be represented by a separate process on a physical data flow diagram, and each one occurs at a different time. For example, you would not typically transport the groceries home and eat them at the same time. Therefore, a "transaction data store" is required to link each task. When you are selecting items, the transaction data store is the shopping cart. After the next process (checking out), the cart is unnecessary. The transaction data store linking checking out and transporting the groceries home is the shopping bag (cheaper than letting you take the cart home!). Bags are an inefficient way of storing the groceries once they are home, so cupboards and a refrigerator are used as a transaction data store between the activity of transporting the goods home and preparing the meal. Finally, a plate, bowl, and cup constitute the link between preparing and eating the meal.

Create the physical data flow diagram for a system by analyzing its output and input. When creating a physical data flow diagram, input data flow from an external entity is sometimes called a trigger because it starts the activities of a process, and output data flow to an external entity is sometimes called a response because it is sent as the result of some activity. Determine which data fields or elements need to be keyed. These fields are called *base elements* and must be stored in a file. Elements that are not keyed but are rather the result of a calculation or logical operation are called *derived elements*. When examining the output, determine whether the information must immediately be displayed or made available to numerous users. The processes that produce such output are usually online. Processes that involve a high volume of transactions, such as billing or check processing, bulk email transmission, or a large number of records that need to be summarized, are usually batch processes, meaning that the documents are keyed as a group, edited as a group, or printed as a group. Printed reports are usually produced by batch processes, and screens tend to be online processes.

Analyze the output data flows and ask the question, "Is the information output coming from base elements on input flows or from calculations?" Often, the answer is easier after you have an understanding of the project data dictionary, which is discussed in Chapter 10. Regardless, create a process for each distinct output. If the stored information necessary for the report or screen is located in several files, show each file as an input data flow. If the output data needs to appear in

Event	Source	Trigger	Activity	Response	Destination
Customer logon.	Customer	Customer number and password	Find customer record and verify password. Send Welcome Web page.	Welcome Web page.	Customer
Customer browses items at Web storefront.	Customer	Item information	Find item price and quantity available. Send Item Response Web page.	Item Response Web page	Customer
Customer places item into shopping basket at Web storefront.	Customer	Item purchase (item number and quantity)	Store data on Order Detail Record. Calculate shipping cost using shipping tables. Update customer total. Update item quantity on hand.	Items Purchased Web page	Customer
Customer Checks Out.	Customer	Clicks Check Out button on Web page	Display Customer Order Web page.	Verification Web page	
Obtain Customer Payment.	Customer	Credit card information	Verify credit card amount with credit card company. Send.	Credit card data Customer feedback	Credit card company Customer
Send Customer email.		Temporal, hourly	Send customer an email confirming shipment.		Customer

FIGURE 9.14

An event response table for an Internet storefront.

a specific sequence, check to see if the files need to be sorted or indexed to match the sequence. Sorting is usually included on a lower-level child diagram as a separate process. Also analyze the input. In a lower-level diagram, include processes for keying, input record validation, and verification. Finally, be sure to add processes for updating master files with input data.

Sometimes it is not clear how many processes to place in one diagram and when to create a child diagram. One suggestion is to examine each process and count the number of data flows entering and leaving it. If the total is greater than four, the process is a good candidate for a child diagram. Physical data flow diagrams are illustrated in the example later in this chapter.

EVENT MODELING AND DATA FLOW DIAGRAMS

Another approach to creating physical data flow diagrams is to create a simple data flow diagram fragment for each unique system event. An event is an input to the system and happens at a specific time and place. Events cause the system to do something and act as a trigger to the system. Triggers start activities and processes, which in turn use data or produce output. An example of an event is a customer reserving a flight on the Web. As each Web form is submitted, processes are activated, such as validating and storing the data and formatting and displaying the next Web page. Events fall into two categories: external events and temporal events. External events are those that occur outside the system and come from an external entity. Temporal events occur at certain fixed times. For example, a temporal event might be the weekly emailing of a newsletter with product and marketing information or sending out billing statements each month. Temporal events tend to be batch processes.

Events are usually summarized in an event table. An example of an event table for an Internet storefront business is illustrated in Figure 9.14. Notice that, with the exception of the event column, the columns contain components found on data flow diagrams, such as the external entities that are represented by source and destination. A data flow diagram fragment is represented by each row in the table. Each fragment is a single process on a data flow diagram along with the data flow, data stores, and external entities that interact with the process. All the fragments are combined to form Diagram 0. If there are more than seven fragments, some are grouped into subsystems. The trigger and response columns become the input and output data flows, and the activity becomes the process. The analyst must determine the data stores required for the process by examining the input and output data flow. Figure 9.15 illustrates a portion of the data flow diagram for the first three rows of the event table shown in Figure 9.14.

The advantage of building data flow diagrams based on events is that the users are familiar with the events that take place within their business area and know how the events drive other activities.

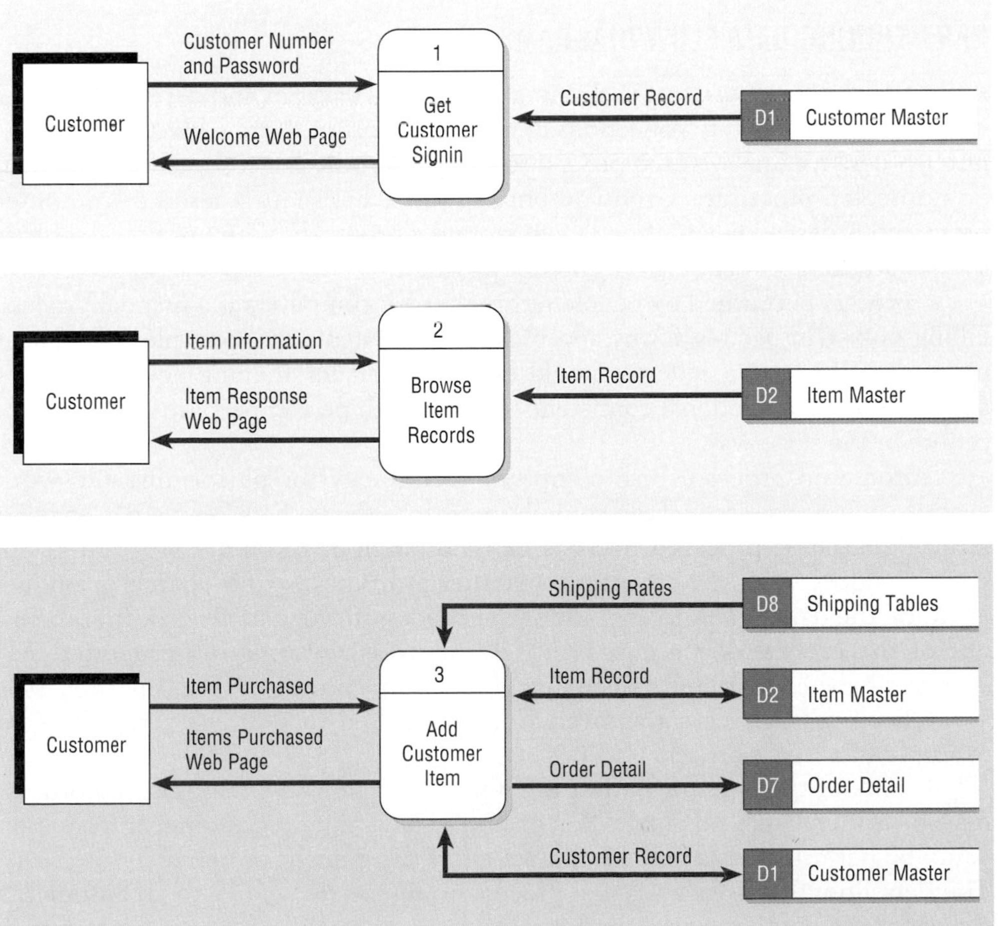

FIGURE 9.15
Data flow diagrams for the first three rows of the Internet storefront event response table.

USE CASES AND DATA FLOW DIAGRAMS

1. Another approach to creating data flow diagrams adapted from Unified Modeling Language (UML) (explained in Chapter 22) is to develop use cases. A use case summarizes an event and has a similar format to process specifications (described in Chapter 11). Each use case defines one activity and its trigger, input, and output. Figure 9.16 illustrates a use case for Process 3, Add Customer Item (shown previously in Figure 9.15). Notice that the activities performed determine the data stores required. This approach allows the analyst to work with users to understand the nature of the processes and activities and then create a single data flow diagram fragment. When creating use cases, first make an initial attempt to define the use cases without going into detail. This step provides an overview of the system and leads to the creation of Diagram 0. Decide what the names should be and provide a brief description of the activity. List the activities, inputs, and outputs for each one.
2. Document the steps used in each use case. These should be in the form of business rules that list or explain the activities completed for each use case. If at all possible, list them in the sequence that they would normally be executed.
3. Determine the data used by each step. This step is easier if a data dictionary has been completed or if prototypes have been approved. It may involve meeting with the users for a follow-up interview or analyzing system documents, such as forms, screens, or reports.
4. Critique the use cases. Have the users review and suggest modifications of the use cases. It is important that the use cases are written clearly.

PARTITIONING DATA FLOW DIAGRAMS

Partitioning is the process of examining a data flow diagram and determining how it should be divided into collections of manual procedures and collections of computer programs. Analyze each process to determine whether it should be a manual or automated procedure. Group automated procedures into a series of computer programs. A dashed line is often drawn around a process or group of processes that should be placed into a single computer program.

A process performed by people rather than by computers is a manual process. Filling out or inspecting forms and picking order items are examples of manual processes. Written procedures should be developed for training new employees and developing operational consistency so that each person performs a given procedure in the same way.

Automated processes use computer technology for performing the work even if such processes include some human activity, such as keying or verifying input data. These processes become either a batch or an online program as the system is developed. To determine whether a process is to be batch or online, examine the data flow into and out of the process. If the data flow both into and out of the process is composed entirely of stored information generated and accessed by the computer, thus requiring no human intervention, the process is a batch process. Using a transaction or master file to produce a report is an example of a batch process.

If some of the input or output is entered or examined by people, the process may be either batch or online. For example, keying new customer information could be performed as a batch by a data-entry department or online by the users. The data flow that links a manual process or an external entity to an automated process represents a person–computer interaction that requires a user interface, a means for an individual to work with the information technology. This interface is

Use case name: Add Customer Item			
Description: Adds an item for a customer Internet order.			Process ID: 3
Trigger: Customer places an order item in the shopping basket.			
Trigger type: External ■ Temporal □			
Input Name	**Source**	**Output Name**	**Destination**
Item Purchased (Item Number and Quantity)	Customer	Items Purchased Confirmation Web Page	Customer

Steps Performed	Information for Steps
1. Find Item Record using the Item Number. If the item is not found, place a message on the Items Purchased Web page.	Item Number, Item Record
2. Store item data on Order Detail Record.	Order Detail Record
3. Use the Customer Number to find the Customer Record.	Customer Number, Customer Record
4. Calculate Shipping Cost using shipping tables. Using the Item Weight from the Item Record and the Zip Code from the Customer Record, look up the Shipping Cost in the Shipping Tables.	Zip Code, Item Weight, Shipping Table
5. Modify the Customer Total using the Quantity Purchased and the Item Price. Add the Shipping Cost. Update the Customer Record.	Item Record, Quantity Purchased, Shipping Cost, Customer Record
6. Modify the item Quantity On Hand and update the Item Record.	Quantity Ordered, Item Record

FIGURE 9.16
A Use Case form for the Internet storefront describes the Add Customer Item activity and its triggers, input, and output.

typically online, and it may consist of a screen, a report, or a handheld optical scanner like those commonly used in retail stores.

Batch processes are usually used when programs process a high volume of data. For example, a large amount of data may need to be entered when processing a mailbag containing customer orders in a mail-order company. Batch processes are also used when a large amount of data must be read and summarized—or otherwise processed—to produce output. An example is reading an entire file of a bank's customers to determine overdrawn accounts.

Another consideration the analyst must address in the physical data flow diagram is whether several batch processes should be combined into one computer program or job stream. A *job stream* is several programs written separately but running back to back. Online programs are usually reserved for low-volume transactions or inquiries or cases in which an employee is working directly with a customer, such as a telephone inquiry about the current status of a bank account.

To develop a collection or group of computer programs and manual procedures, you should examine each process and ask questions about the nature of the work being done. An element of experimentation or play may enter the design process at this time, as you think about each process in the data flow diagram and see if it could be both batch and online. Reflect on which option would be better for the user community. Describe the processes as manual, batch, or online.

There are six reasons for partitioning data flow diagrams:

1. *Different user groups.* Are the processes performed by several different user groups, often at different physical locations within the company? If so, they should be partitioned into different computer programs. An example is the need to process customer returns and customer payments in a department store. Both processes involve obtaining financial information that is used to adjust customer accounts (subtracting from the amount the customer owes), but they are performed by different user groups at different locations. The counter handling items returned by customers is usually located at a desk near the store entrance. The payment counter is located somewhere in the interior of the store (for security) and is staffed by security personnel. Each group needs a different screen for recording the particulars of the transaction, either a credit screen or a payment screen.
2. *Timing.* Examine the timing of the processes. If two processes execute at different times, they cannot be grouped into one program. Timing issues may also involve how much data is presented at one time on a Web page. If an ecommerce site has rather lengthy Web pages for ordering data or making an airline reservation, the Web pages may be partitioned into separate programs that format and present the data.
3. *Similar tasks.* If two processes perform similar tasks and both are batch processes, they may be grouped into one computer program. For example, in a monthly run to adjust customer balances, both return item credits and customer payments are subtracted from the customer balance due. These two adjustment processes may readily be combined into one program.
4. *Efficiency.* Several batch processes may be combined into one program for efficient processing. For example, if a series of reports needs to use the same large input files, producing them from the same batch program may save considerable computer run time.
5. *Consistency of data.* Processes may be combined into one program for consistency of data. For example, an accounts receivable report needs to be printed periodically to show the amount due from each customer. The same figures must also be included on bills sent to the customers. If these two distinct outputs were produced in separate computer runs and if the customer master file was updated between the runs, the bills would have different data. The result is an unreliable, inconsistent system.
6. *Security.* Processes may be partitioned into different programs for security reasons. An example is a system with a process for adding a new customer and a process for changing customer financial information. Each process should have

a separate program because, whereas only one person (or several in a large organization) should have password access to the program that changes customer financial data, many persons may be needed and authorized to add new customer records. Another use of partitioning is to show security in an Internet system. A dashed line may be placed around Web pages that are on a secure server to separate them from those Web pages on a server that is not secured. A Web page that is used for obtaining the customer user's identification and password is usually partitioned from order entry or other business pages.

A DATA FLOW DIAGRAM EXAMPLE

The corporation in our example is FilmMagic, a video rental chain founded by three people with expertise in the video rental business. The plan is to have a series of stores scattered strategically around a metropolitan area. The company has also adopted a unique policy of giving free rentals and videos to its high-volume customers in an attempt to gain a large market share. According to one of the company's owners, "If the airlines can have frequent flyer programs, our video stores can have a recurrent rental program." Consequently, a monthly customer bonus program will be part of the system.

CREATING THE CONTEXT DIAGRAM

A summary of the business activities obtained from interviews with the owners of FilmMagic is illustrated in Figure 9.17. An alternative approach would be to expand the list to a table or a set of use cases. The context-level data flow diagram, representing an overview of the entire system, appears in Figure 9.18. Because the system must keep track of the number of videos a customer has rented, the external CUSTOMER has the most data flow to and from it. Note that the context diagram is relatively simple.

DRAWING DIAGRAM 0

Diagram 0, shown in Figure 9.19, depicts the major activities for the FilmMagic video rental system. Note that there is one process for each major activity. Each process is analyzed to determine the data required and the output produced. Process 1, RENT VIDEO ITEMS, summarizes the main function of the system and is thus a complex process. Notice the many input and output data flows.

To draw the data flow diagram correctly, questions must be asked such as, "What information is needed to rent a video?" A VIDEO RENTAL ITEM (which may be either a videocassette or a video game), a PAYMENT, and a CUSTOMER ID (a rental card) are required from the CUSTOMER. The VIDEO RENTAL ITEM is used to find matching information about the video, such as the price and description. The process creates a CASH TRANSACTION, which will eventually produce information about the total cash received. The CUSTOMER RECORD is obtained and updated with the total amount of the rental. A double-headed arrow indicates that the CUSTOMER RECORD is obtained from and replaced in the same file location. The RENTAL RECEIPT and video are given to the CUSTOMER. RENTAL INFORMATION, such as the date and the item rented, is produced for later use to generate MANAGEMENT REPORTS.

The other processes are simpler, with less input and output. Process 3, CHECK IN CUSTOMER VIDEO RETURN, updates the CUSTOMER data store indicating that items are no longer checked out. New customers must be added to the CUSTOMER data store before a video may be checked out. Process 5, ADD NEW CUSTOMER, takes NEW CUSTOMER INFORMATION and issues the

FIGURE 9.17
Start with a list of business activities, which will help you identify processes, external entities, and data flows.

Summary of Business Activities
Customer Rental System

1. Customers apply for a video rental card. They fill out a form and provide a means of verifying their identity. They are issued a video rental card.
2. Customers rent videos by giving the clerk their video rental card and the video cassettes or video games. The clerk totals the amount of the rental, which is received from the customers. Customers are given a receipt with the due date on it. A record is created for each item rented.
3. Customers return video cassettes or video games. If the video is returned late, a note and the amount of the late fee is made on the record.
4. If a customer has a late fee, he or she is required to pay the amount the next time an item is rented.
5. The company has several special policies designed to provide a competitive edge in the video rental market. Once a month, the customer rental records are reviewed for customers who have rented more than the bonus level, currently set at $50. Bonus customers are sent a letter thanking them for their business as well as issuing them several free rental coupons (depending on the amount of rental for the month).
6. Once a year, the customer records are examined for customers who have rented more than a yearly bonus level (currently set at $250). A letter, free rental coupons, and a certificate for a free video (if a customer has rented over two times the bonus level) are sent to the customer.

FIGURE 9.18
Context level diagram for the FilmMagic video rental stores.

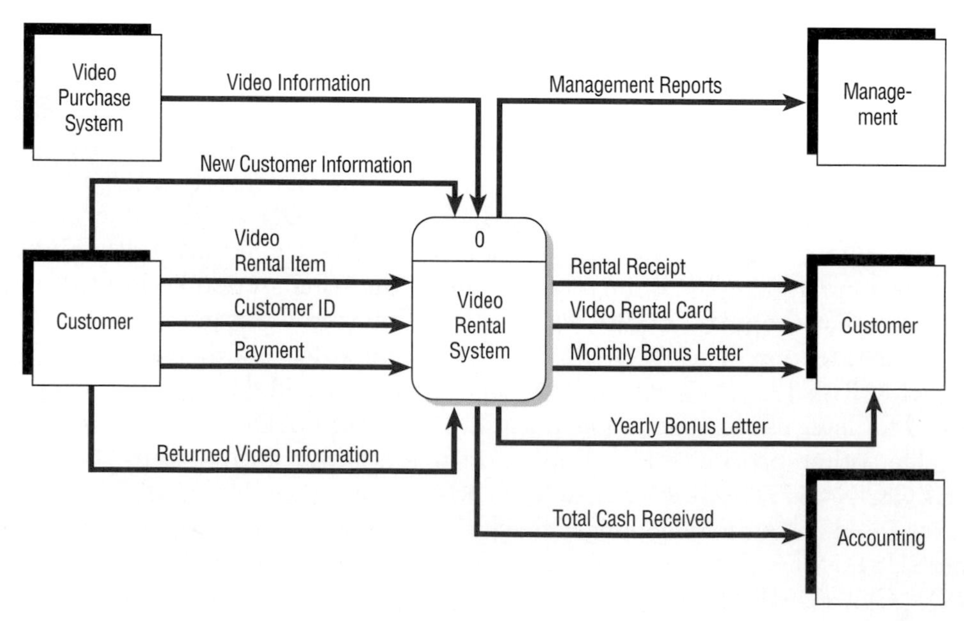

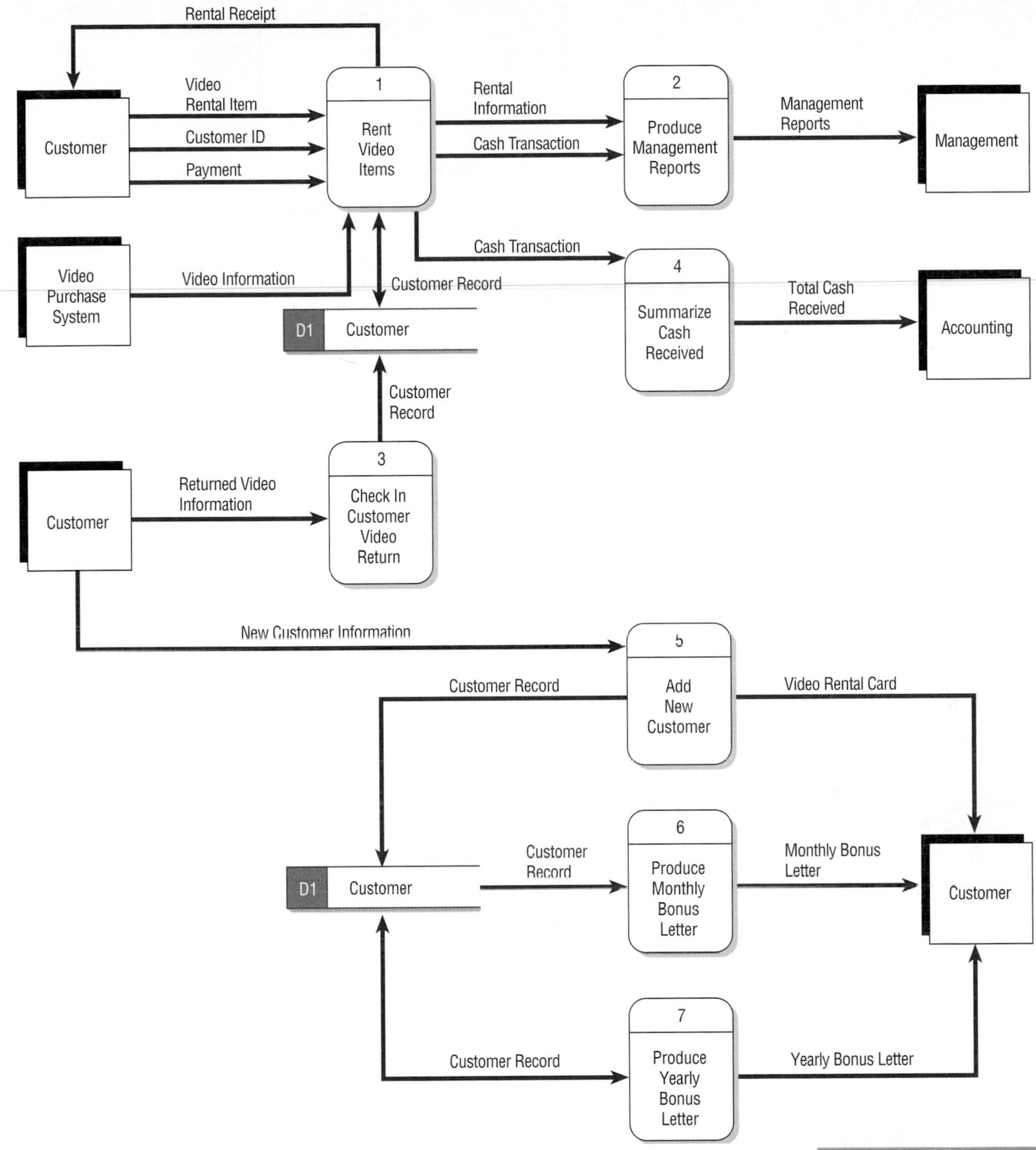

FIGURE 9.19

Diagram 0 for the FilmMagic video rental system shows seven major processes. This logical DFD tells us what the system does, what is stored, who or what provides the inputs, and who receives the output.

customer a VIDEO RENTAL CARD. The card must be presented each time a customer wishes to check out a video.

Processes 2 and 4 produce useful information for managing the business and making decisions, such as when to lower the price of videos that are in demand and when to advertise to draw more customers, thereby increasing cash flow. Processes 6 and 7 use CUSTOMER data store information to PRODUCE MONTHLY and YEARLY CUSTOMER BONUS LETTERS. Notice that the names of the data flows going out of the processes are different, indicating that something has transformed input data to produce output. All processes start with a verb such as RENT, PRODUCE, CHECK IN, SUMMARIZE, or ADD.

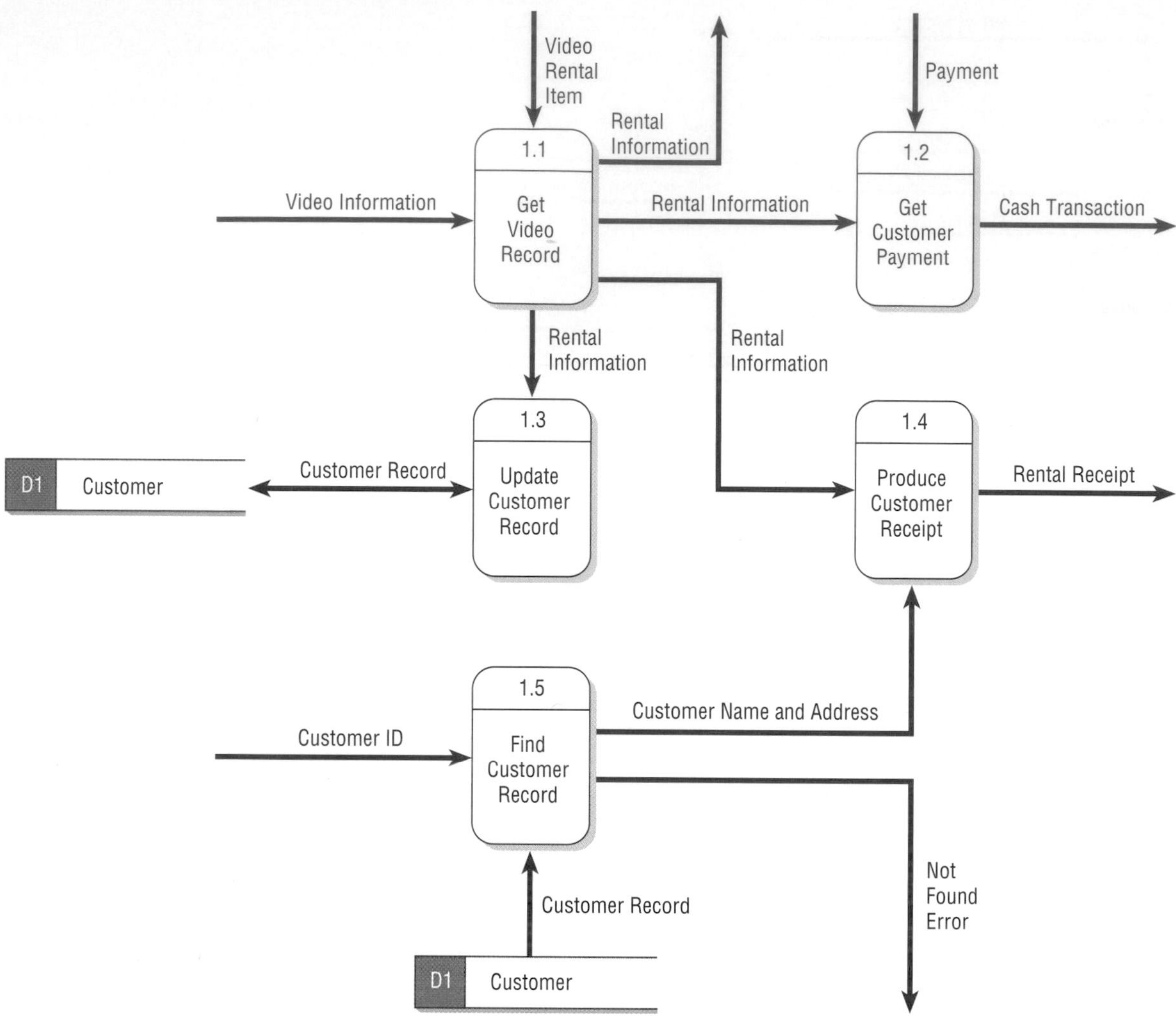

FIGURE 9.20

The child diagram for process 1 shows more detail than does Diagram 0. Process 1.1 is GET VIDEO RECORD. The logical DFD tells us what is accomplished but not how to do it.

CREATING A CHILD DIAGRAM

Figure 9.20 is the child diagram of process 1, RENT VIDEO ITEMS, in the FilmMagic example. The input data flow VIDEO INFORMATION is connected only to the process GET VIDEO RECORD. The source of this input is a blank area in the drawing. This incomplete interface flow matches the flow into process 1 on Diagram 0. The same is true for VIDEO RENTAL, PAYMENT, and CUSTOMER ID.

The CUSTOMER RECORD is also an interface data flow, but it is connected on Diagram 1 to the CUSTOMER data store because data stores in the parent diagram may also be included on the child diagram. The output data flows CASH TRANSACTION and RENTAL RECEIPT are interface flows that match the parent process output. The flow NOT FOUND ERROR is not depicted in the parent process because an error line is considered a minor output.

Child diagram processes are more detailed, illustrating the logic required to produce the output. The process GET VIDEO RECORD uses VIDEO RENTAL, which indicates what video the customer wishes to rent, to find the matching VIDEO INFORMATION (title, price, and so on). Process 1.5, FIND CUSTOMER RECORD, uses the CUSTOMER ID on the video rental card to locate the CUSTOMER record. The CUSTOMER NAME AND ADDRESS are printed on the RENTAL RECEIPT printed from process 1.4.

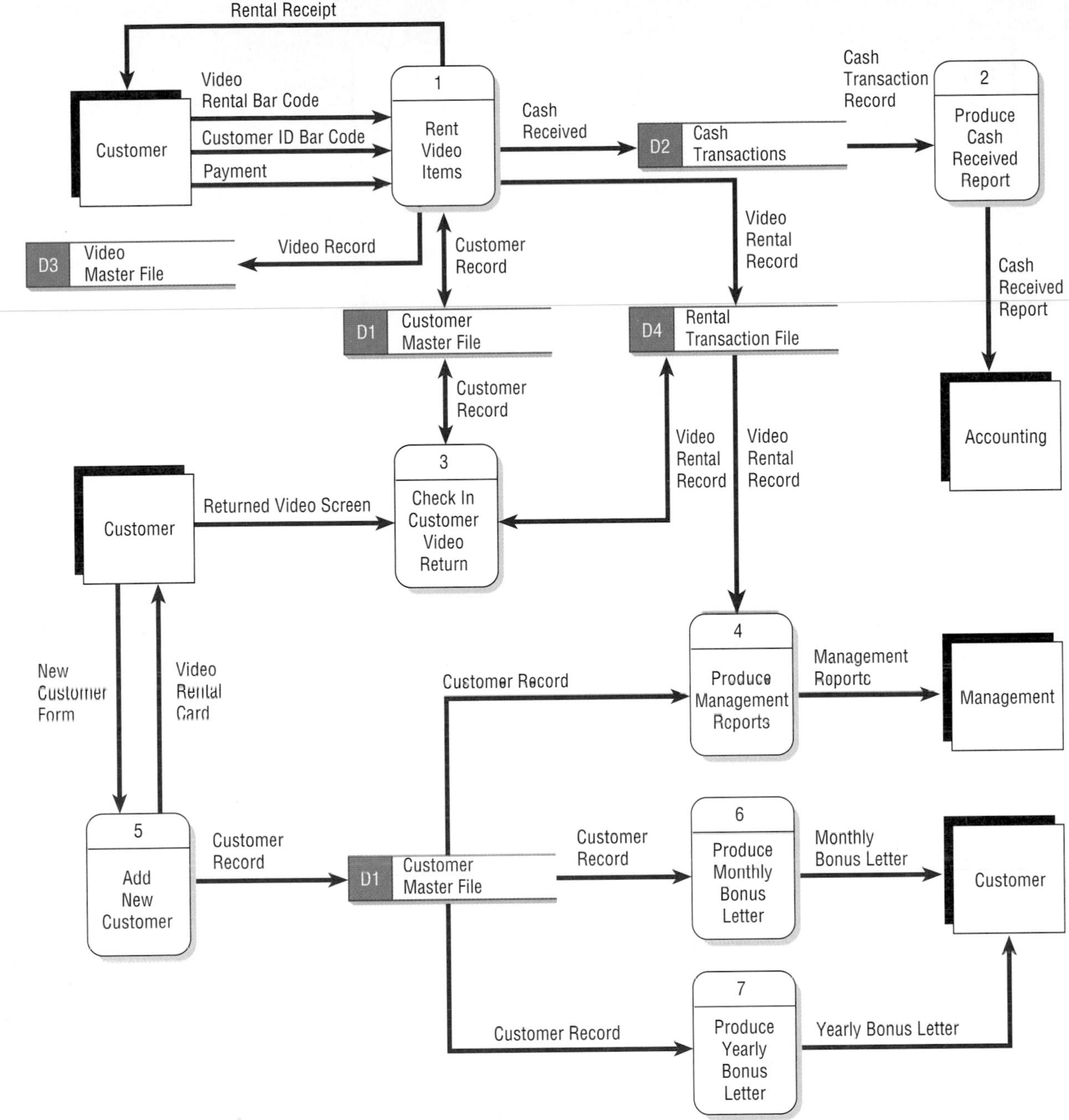

FIGURE 9.21
This physical data flow diagram corresponds to logical Diagram 0. Note the subtle differences. CUSTOMER ID is now CUSTOMER ID BAR CODE, emphasizing physical implementation.

CREATING A PHYSICAL DATA FLOW DIAGRAM

Figure 9.21 is the physical data flow diagram corresponding to the FilmMagic logical Diagram 0. Data flow names have been changed to reflect implementation. The customer now supplies a VIDEO RENTAL BAR CODE and a CUSTOMER ID BAR CODE to process 1, RENT VIDEO ITEMS. The entity VIDEO PURCHASE SYSTEM has been replaced with a VIDEO MASTER FILE because files are used to communicate between systems. There are now two transaction files. The RENTAL TRANSACTION FILE is used to store information from the time the videos are rented until they are returned. The CASH TRANSACTIONS file is necessary because videos are rented throughout the day while the CASH RECEIVED REPORT is produced once a week. Data are entered using the RETURNED VIDEO SCREEN (and any late charges, are calculated in process 3, CHECK IN CUSTOMER VIDEO RETURN). New customers fill out the NEW

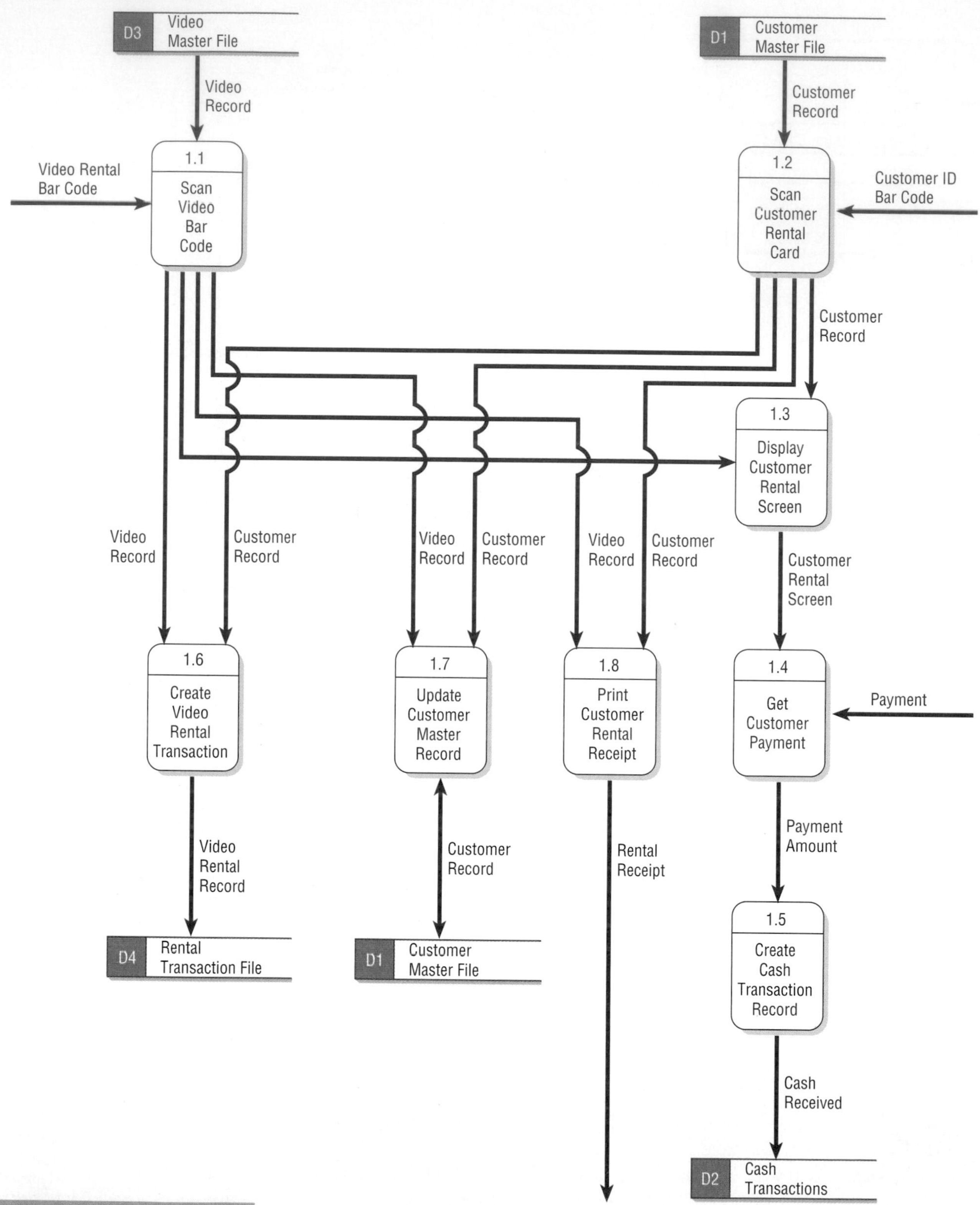

FIGURE 9.22

This physical child data flow diagram shows details about real-world implementation. The logical diagram process 1.1 was GET VIDEO RECORD, but the physical diagram process 1.1 tells us how (SCAN VIDEO BAR CODE).

CUSTOMER FORM, whereas on the logical data flow diagram this step is simply called NEW CUSTOMER INFORMATION.

An example of a physical child data flow diagram is Diagram 1 of the FilmMagic example, which is illustrated in Figure 9.22. Notice that there are processes for scanning bar codes, displaying screens, locating records, and creating and updating files. The sequence of activities is important here, because the emphasis is on how the system will work and in what order events happen.

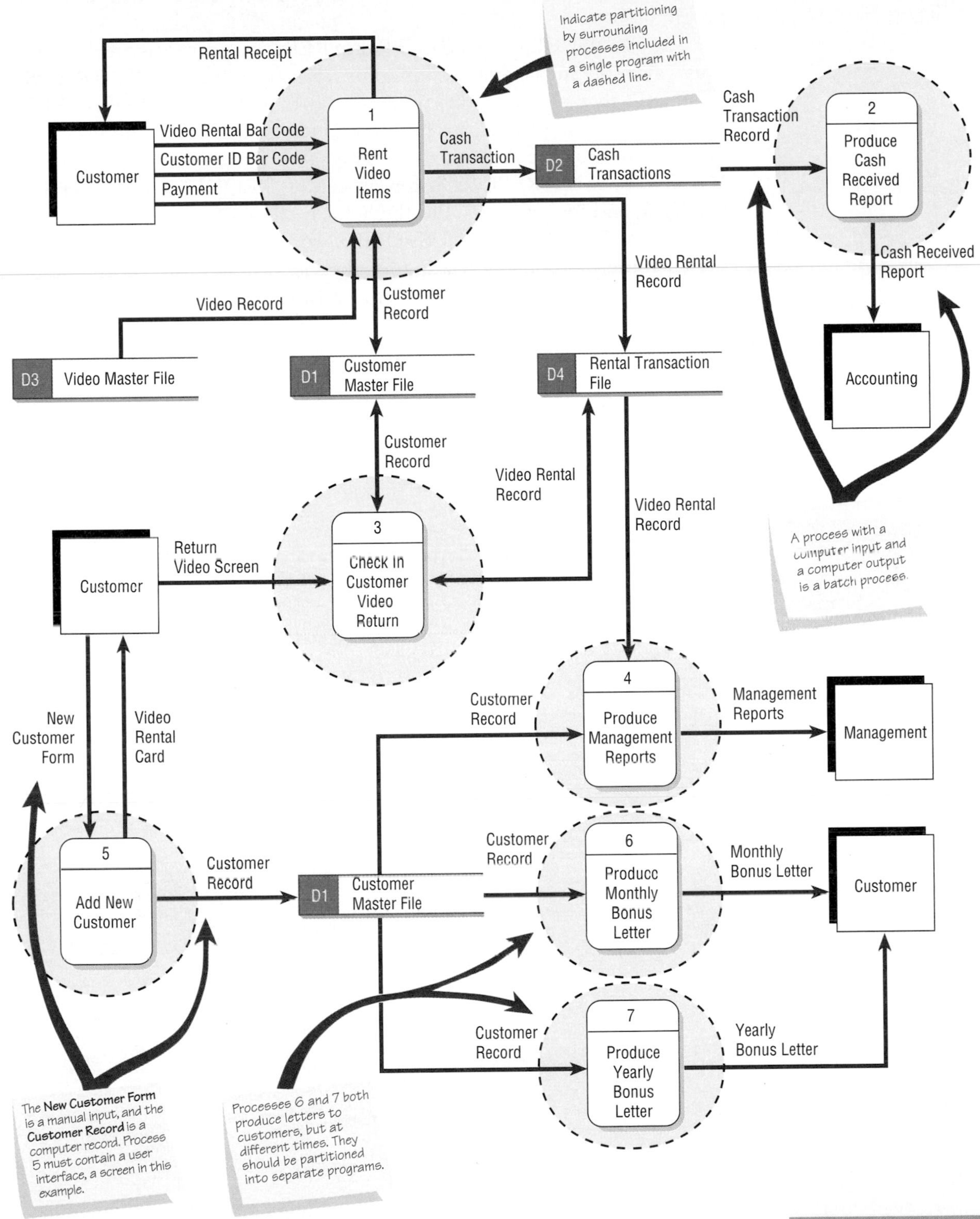

FIGURE 9.23
Partitioning the FilmMagic physical data flow diagram. Partitioning takes the physical DFD and makes programming and implementation manageable.

PARTITIONING THE DATA FLOW DIAGRAM

Figure 9.23 illustrates partitioning for the FilmMagic physical data flow diagram. Notice the use of dotted lines to indicate which processes should be in separate programs. The process RENT VIDEO ITEMS operates on a minute-by-minute basis. The process CHECK IN CUSTOMER VIDEO RETURN also operates on a minute-by-minute basis. Returns, however, are handled at a time later than the rental process, and both procedures should thus be in separate programs.

The PRODUCE CASH RECEIVED REPORT process is weekly and therefore must also be in a separate program. Because the CASH TRANSACTION RECORD that goes into this process and the CASH RECEIVED REPORT that comes out of the process are both computer information, the process should be implemented as a batch program. The same is true for process 4, PRODUCE MANAGEMENT REPORTS; for process 6, PRODUCE MONTHLY BONUS LETTER; and for process 7, PRODUCE YEARLY BONUS LETTER.

Process 5, ADD NEW CUSTOMER, could be either batch or online. Because the customer is probably waiting for the video rental card on the other side of a counter, an online process would provide the best customer service.

A SECOND DATA FLOW DIAGRAM EXAMPLE

Often, a person's first exposure to data flow diagrams seems confusing because there are so many new concepts and definitions. This next example is intended to illustrate the development of a data flow diagram by selectively looking at each of

FIGURE 9.24

A summary of business activities for World's Trend Catalog Division.

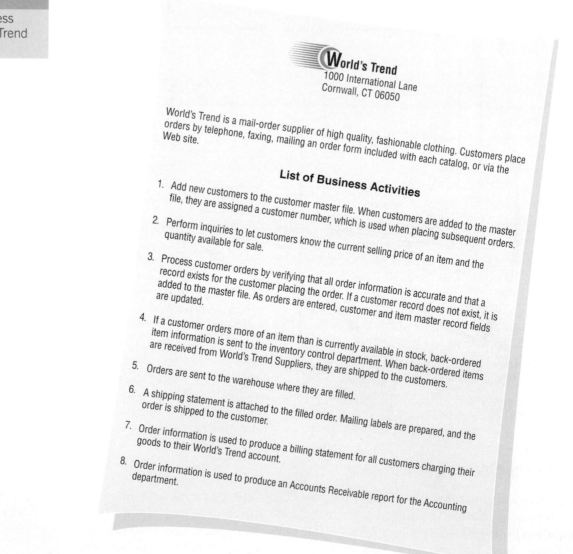

1000 International Lane
Cornwall, CT 06050

World's Trend is a mail-order supplier of high quality, fashionable clothing. Customers place orders by telephone, faxing, mailing an order form included with each catalog, or via the Web site.

List of Business Activities

1. Add new customers to the customer master file. When customers are added to the master file, they are assigned a customer number, which is used when placing subsequent orders.
2. Perform inquiries to let customers know the current selling price of an item and the quantity available for sale.
3. Process customer orders by verifying that all order information is accurate and that a record exists for the customer placing the order. If a customer record does not exist, it is added to the master file. As orders are entered, customer and item master record fields are updated.
4. If a customer orders more of an item than is currently available in stock, back-ordered item information is sent to the inventory control department. When back-ordered items are received from World's Trend Suppliers, they are shipped to the customers.
5. Orders are sent to the warehouse where they are filled.
6. A shipping statement is attached to the filled order. Mailing labels are prepared, and the order is shipped to the customer.
7. Order information is used to produce a billing statement for all customers charging their goods to their World's Trend account.
8. Order information is used to produce an Accounts Receivable report for the Accounting department.

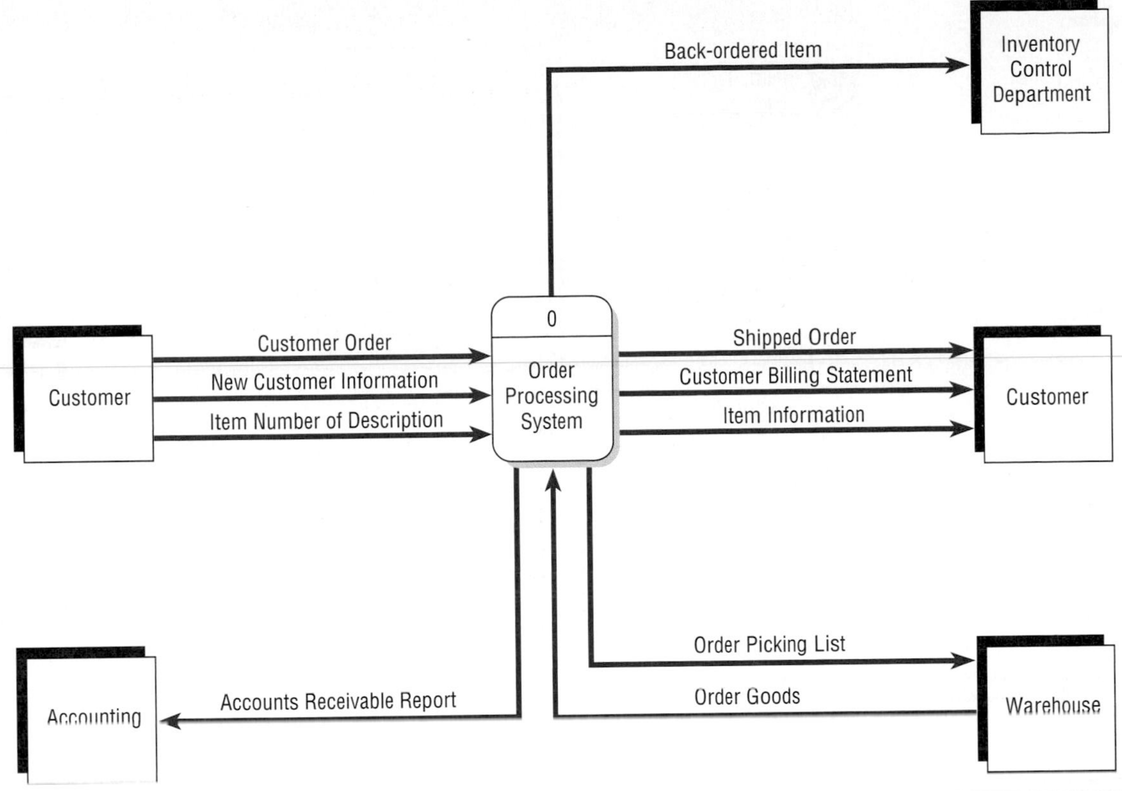

FIGURE 9.25
A context-level data flow diagram for the order processing system at World's Trend.

the components explored earlier in this chapter. The example, called "World's Trend Catalog Division," will also be used to illustrate concepts covered in Chapters 10 and 11.

A list of business activities for World's Trend can be found in Figure 9.24. You could develop this list using information obtained through interviews, investigation, and observation. The list can be used to identify external entities such as CUSTOMER, ACCOUNTING, and WAREHOUSE as well as data flows such as ACCOUNTS RECEIVABLE REPORT and CUSTOMER BILLING STATEMENT. Later (when developing level 0 and child diagrams), the list can be used to define processes, data flows, and data stores.

Once this list of activities is developed, create a context diagram as shown in Figure 9.25. This diagram shows the ORDER PROCESSING SYSTEM in the middle (no processes are described in detail in the context level diagram) and five external entities (the two entities called CUSTOMER are really one and the same). The data flows that come from and go to the external entities are shown as well (for example, CUSTOMER ORDER and ORDER PICKING LIST).

Next, go back to the activity list and make a new list of as many processes and data stores as you can find. You can add more later, but start making the list now. If you think you have enough information, draw a level 0 diagram such as the one found in Figure 9.26. Call this Diagram 0 and keep the processes general so as not to overcomplicate the diagram. Later, you can add detail. When you are finished drawing the seven processes, draw data flows between them and to the external entities (the same external entities shown in the context diagram). If you think there needs to be a data store such as ITEM MASTER or CUSTOMER MASTER, draw those in and connect them to processes using data flows. Now take the time to number the processes and data stores. Pay particular attention to making the labels meaningful. Check for errors and correct them before moving on.

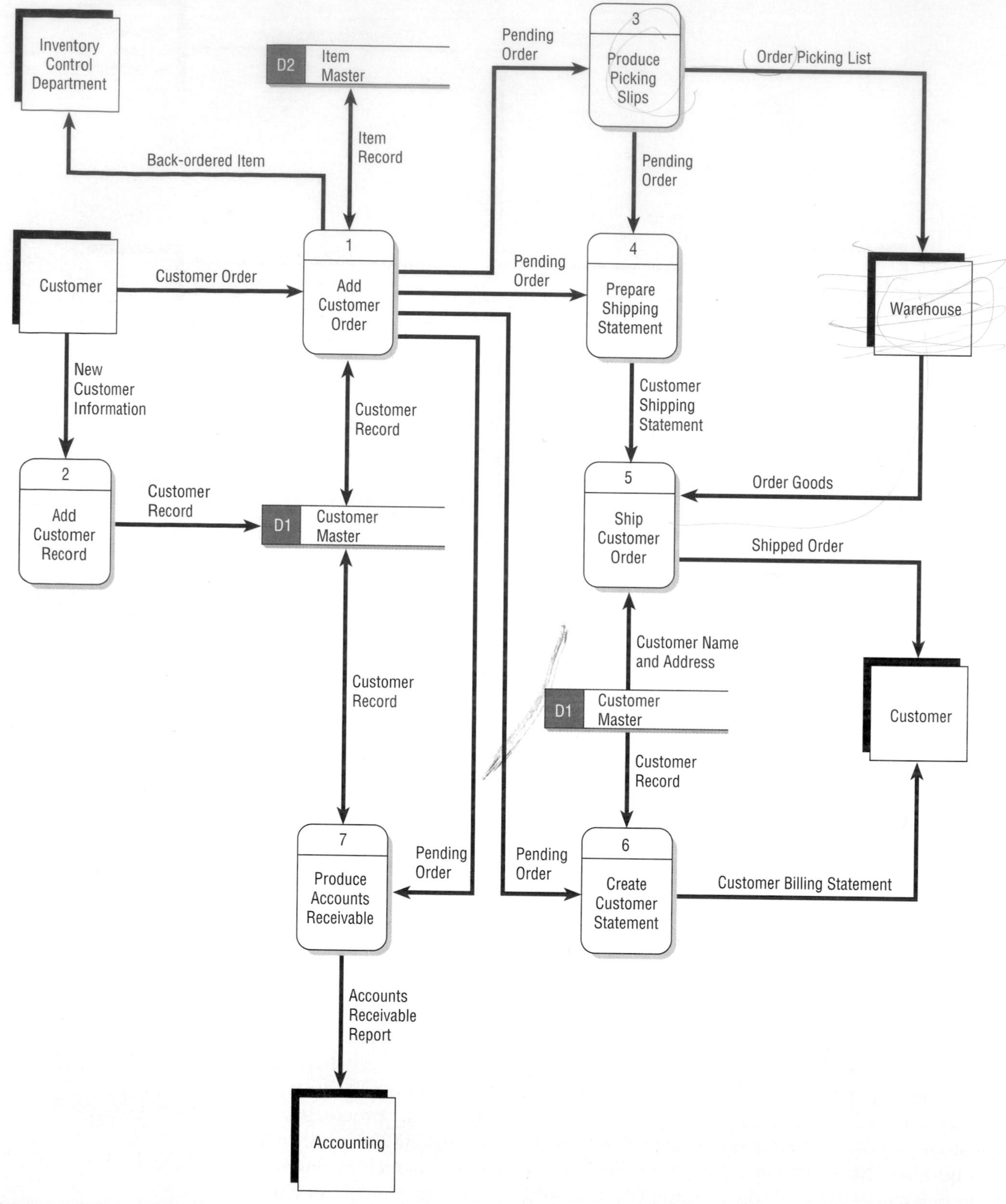

FIGURE 9.26
Diagram 0, of the order processing system for World's Trend Catalog Division.

At this point try to draw a child diagram (sometimes also called a level 1 diagram) like the one in Figure 9.27. Number your child diagrams Diagram 1, Diagram 2, and so on, in accordance with the number you assigned to each process in the level 0 diagram. When you draw Diagram 1, make a list of subprocesses first. A process such as ADD CUSTOMER ORDER can have subprocesses (in this case, there are seven). Connect these subprocesses to one another and also to data stores

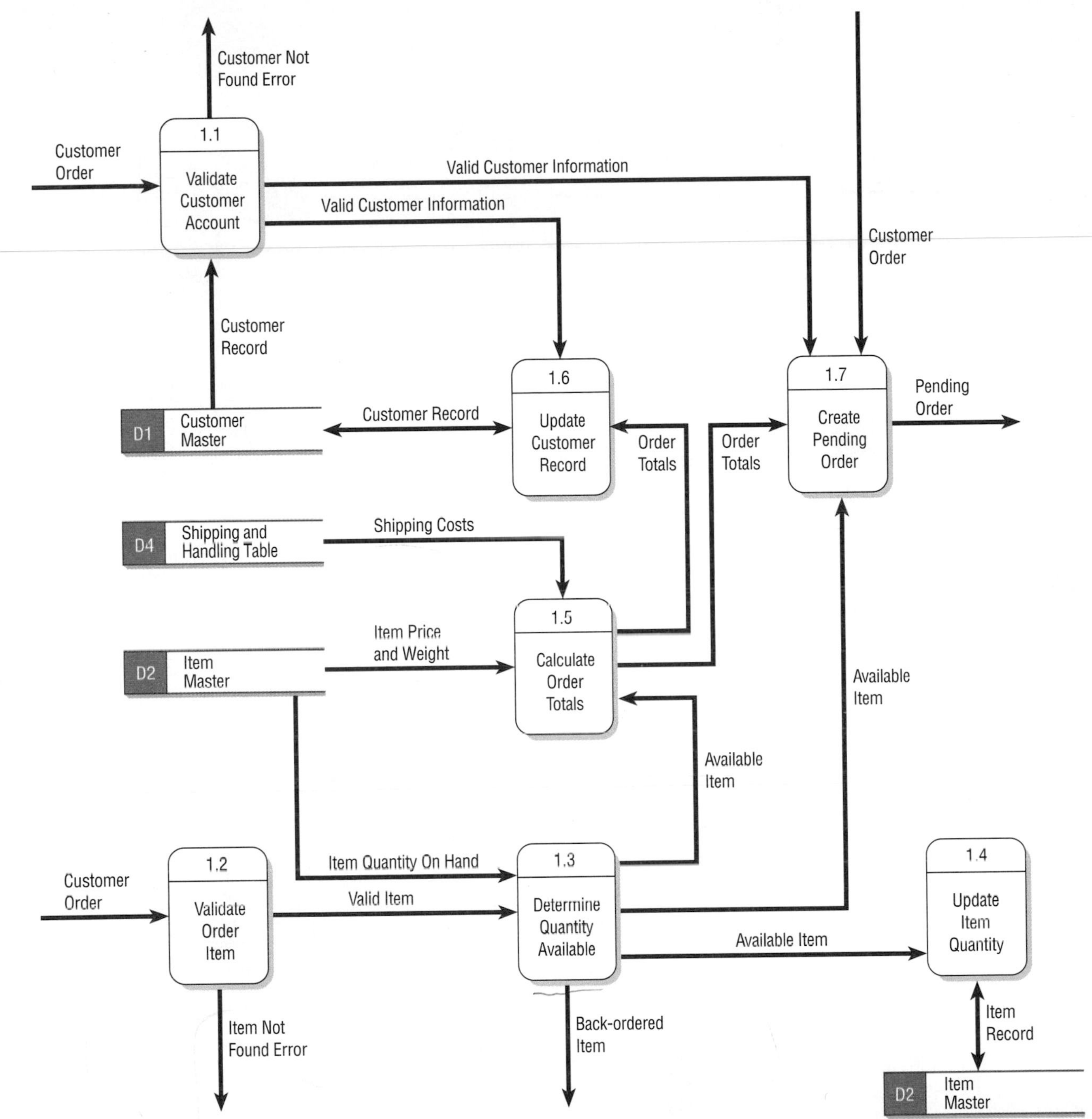

FIGURE 9.27
Diagram 1, of the order processing system for World's Trend Catalog Division.

when appropriate. Subprocesses do not have to be connected to external entities, because we can always refer to the parent (or level 0) data flow diagram to identify these entities. Label the subprocesses 1.1, 1.2, 1.3, and so on. Take the time to check for errors and make sure the labels make sense.

If you want to go beyond the logical model and draw a physical model as well, look at Figure 9.28, which is an example of a physical data flow child diagram of process 3, PRODUCE PICKING SLIPS. When you label a physical model, take care to describe the process in great detail. For example, subprocess 3.3 in a logical model could simply be SORT ORDER ITEM, but in the physical model, a better label is SORT ORDER ITEM BY LOCATION WITHIN

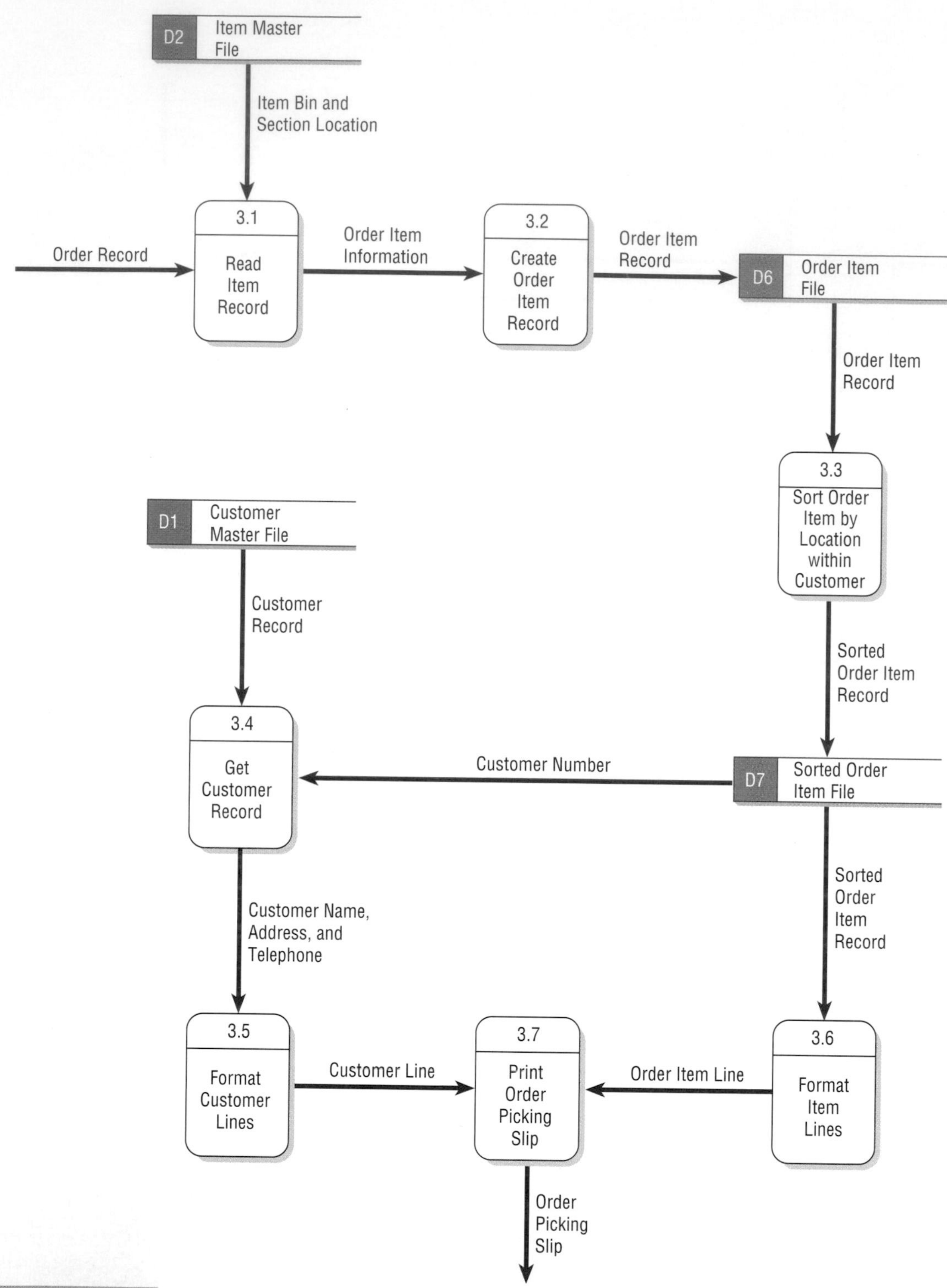

FIGURE 9.28

A physical data flow child diagram for World's Trend Catalog Division.

CUSTOMER. When you write a label for a data store, refer to the actual file or database, such as CUSTOMER MASTER FILE or SORTED ORDER ITEM FILE. When you describe data flows, describe the actual form, report, or screen. For example, when you print a slip for order picking, call the data flow ORDER PICKING SLIP.

Finally, take the physical data flow diagram and suggest partitioning through combining or separating the processes. As stated earlier, there are many reasons for partitioning: identifying distinct processes for different user groups, separating

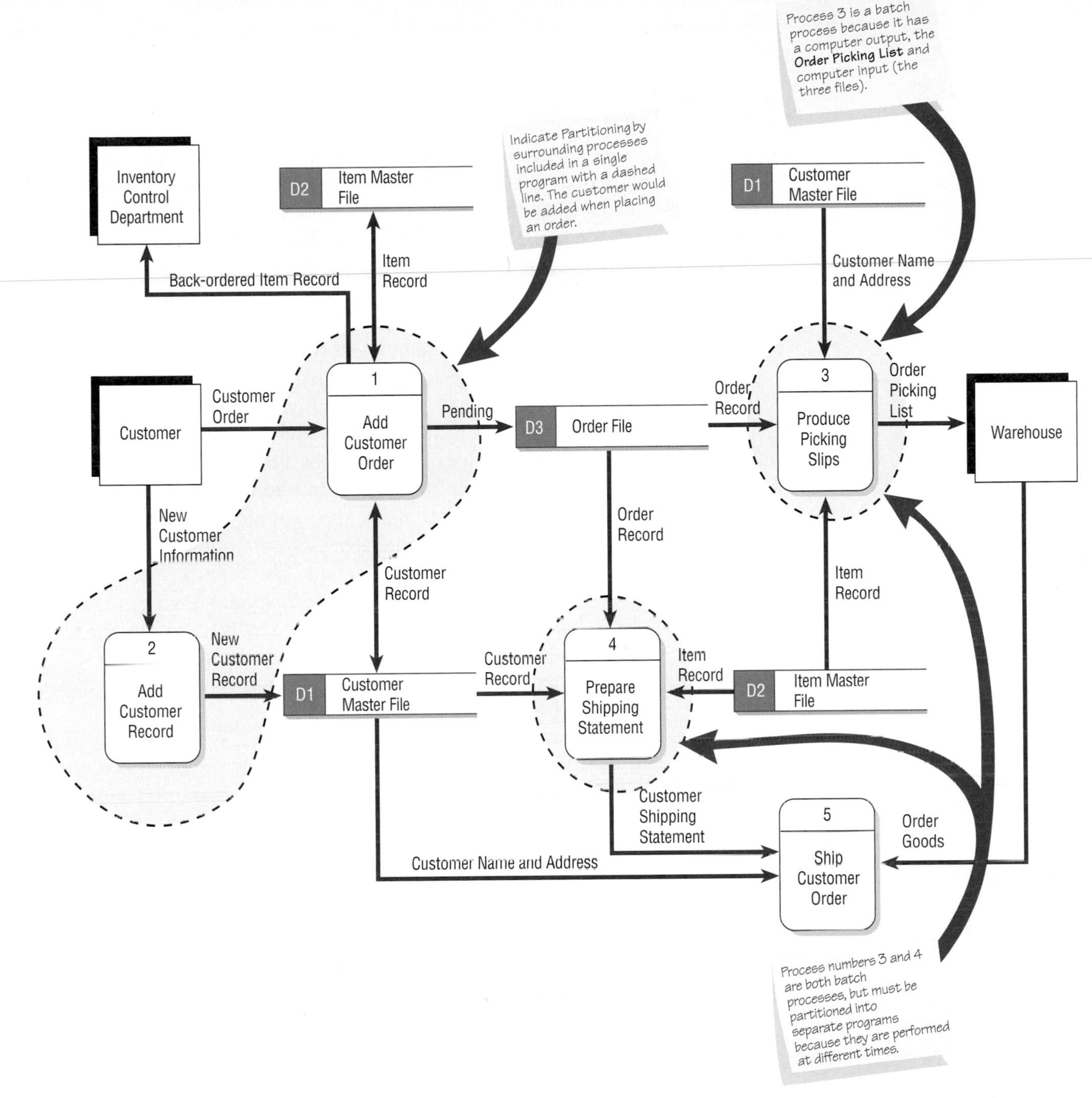

FIGURE 9.29

Partitioning the data flow diagram (Showing part of Diagram 0).

processes that need to be performed at different times, grouping similar tasks, grouping processes for efficiency, combining processes for consistency, or separating them for security. Figure 9.29 shows that partitioning is useful in the case of World's Trend Catalog Division. You would first group processes 1 and 2 because it would make sense to add new customers at the same time their first order was placed. You would then put processes 3 and 4 in two separate partitions. Although both are batch processes, they must be done at different times from each other and thus cannot be grouped into a single program.

The process of developing a data flow diagram is now completed from the top down, first drawing a companion physical data flow diagram to accompany the logical data flow diagram, then partitioning the data flow diagram by grouping or separating the processes. The World's Trend example is used again in Chapters 10 and 11.

USING DATA FLOW DIAGRAMS

Data flow diagrams are useful throughout the analysis and design process. Use original, unexploded data flow diagrams early when ascertaining information requirements. At this stage they can help provide an overview of data movement through the system, lending a visual perspective unavailable in narrative data.

Tradeoffs are involved in deciding how far the data streams should be exploded. Time may be wasted and understandability sacrificed if data flow diagrams are overly complex. On the other hand, if the data flow diagrams are underexploded, errors of omission could eventually occur that might affect the system that is being developed.

If data flow diagrams are used as a tool to solicit more specific information requirements from users, they should not be highly exploded or finalized in any medium before users have had a chance to walk through them with the systems analyst. Changes need to be incorporated after getting users' input. Overly exploded diagrams may not be helpful to users. In addition, if data flow diagrams are too complex before presentation to users, the systems analyst is more likely to defend the representation than to welcome any user corrections.

After exploding the original data flow diagrams, use them as a tool for further interaction with users. At this stage the data flow diagram shows your own conceptualization of the business data streams. Educate key users about the conventions used in data flow diagrams, and then go through the successive levels with them. Ask for changes that they may suggest to clarify processes or to make the diagrams more accurate in some way.

After user modifications are added, users and the systems analysis team approve the data flow diagrams as accurate reflections of the organization's data flows. The data flow diagrams can then be finalized and drawn using a CASE tool, as shown in Figure 9.30.

FIGURE 9.30
A Visible Analyst screen showing a data flow diagram.

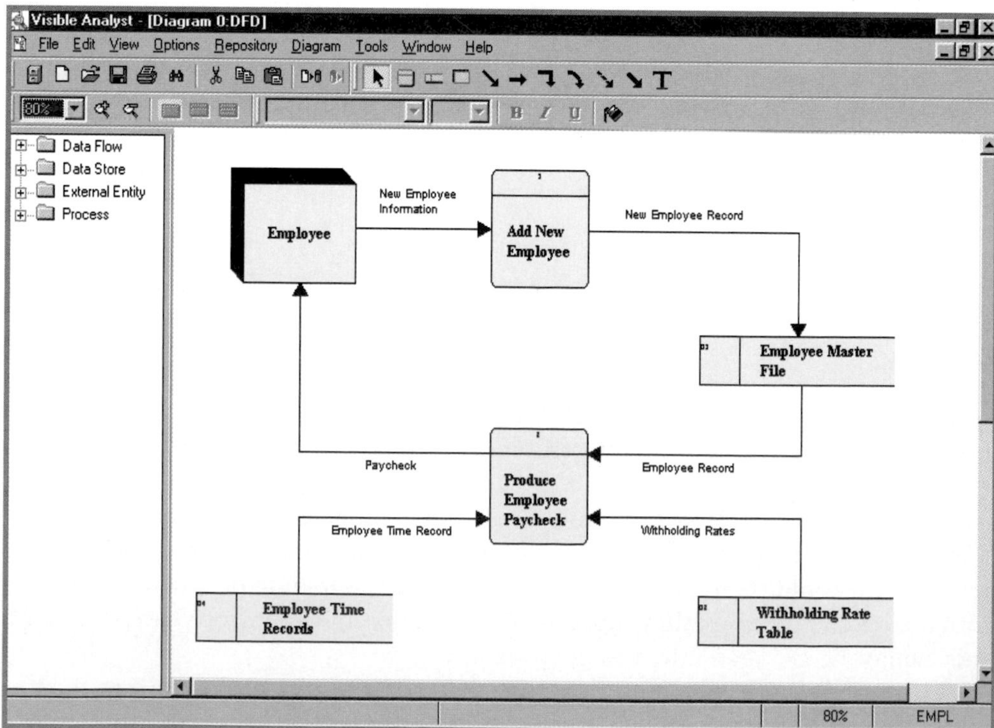

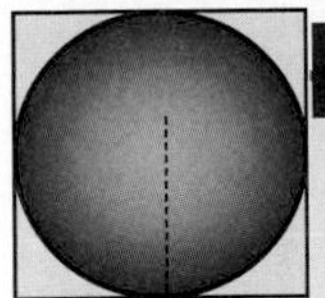

CONSULTING OPPORTUNITY 9.2

THERE'S NO BUSINESS LIKE FLOW BUSINESS

The phone at Merman's rings, and Annie Oaklea, head of costume inventory, picks it up and answers a query by saying, "Let me take a look at my inventory cards. Sorry, it looks as if there are only two male bear suits in inventory, with extra growly expressions at that. We've had a great run on bear. When do you need them? Perhaps one will be returned. No, can't do it, sorry. Would you like these two sent, regardless? The name of your establishment? Manhattan Theatre Company? London branch? Right. Delightful company! I see by our account card that you've rented from us before. And how long will you be needing the costumes?"

Figure 9.C1 is a data flow diagram that sets the stage for processing of costume rentals from Merman's. It shows rentals like the one Annie is doing for Manhattan Theatre Company.

After conversing for another few moments about shop policy on alterations, Annie concludes her conversation by saying, "You are very lucky to get the bears on such short notice. I've got another company reserving them for the first week in July. I'll put you down for the bear suits, and they'll be taken to you directly by our courier. As always, prompt return will save enormous trouble for us all."

(Continued)

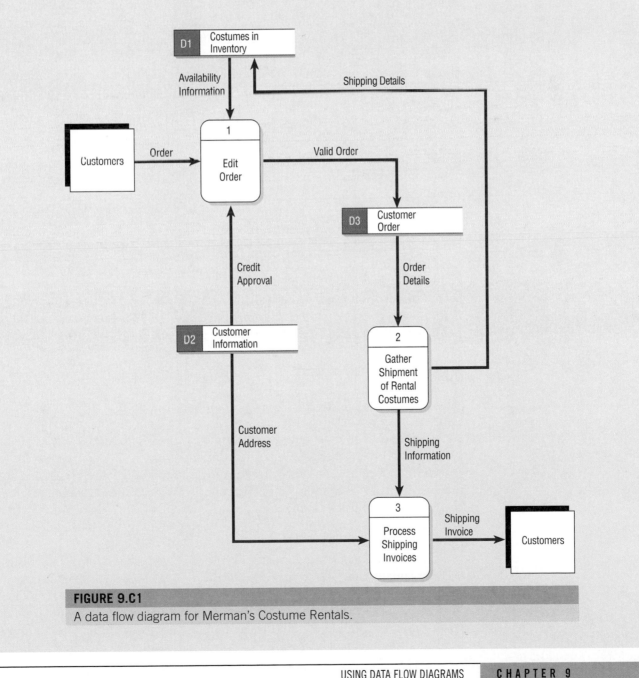

FIGURE 9.C1
A data flow diagram for Merman's Costume Rentals.

Merman's costume rental enterprise is located in London's world-famous West End theatre district. When a theatre or television production company lacks the resources (either time or expertise) to construct a costume in its own shop, the cry goes up to "Ring up Merman's!" and it proceeds to rent what it needs with a minimum of fuss.

The shop (more aptly visualized as a warehouse) goes on for three floors full of costume racks, holding thousands of costumes hung together by historical period, then grouped as to whether they are for men or women, and then by costume size.[1] Most theatre companies are able to locate precisely what they need through Annie's capable assistance.

Now tailor-make the *rental return* portion of the data flow diagram given earlier. Remember that timely returns are critical for keeping the spotlight on costumes rented from Merman's.

[1]Western Costume Company in Hollywood, California, is said to have more than 1 million costumes worth about $40 million.

Once their content is clarified, the data flow diagrams need to be redrawn and relabeled in a meaningful way. The systems analyst might be quite competent at sketching through the logic of the data stream for data flow diagrams, but to make the diagrams truly communicative, meaningful labels for all data components are also required. Labels should not be generic, because then they do not tell enough about the situation at hand. All general systems models bear the configuration of input, process, and output, so labels for a data flow diagram need to be more specific than that.

Consider effective naming as a top priority so that someone unfamiliar with the system will be able to pick up a data flow diagram and, with a little training, understand what it depicts. Make labels as specific yet concise as possible. Try to avoid using the same term to mean two different things. Conversely, consolidate terms wherever possible, using only a few terms for the data item. Part of the reason data flow diagrams are effective is that they are consistent from page to page (recall the requirement that inputs and outputs remain constant between diagrams). The same sort of consistency should be evident in labeling.

Finally, remember that data flow diagrams are used to document the system. Assume that data flow diagrams will be around longer than the people who drew them, which is, of course, always true if an external consultant is drawing them. Data flow diagrams can be used for documenting high or low levels of analysis and helping to substantiate the logic underlying the data flows of the organizations.

SUMMARY

To better understand the logical movement of data throughout a business, the systems analyst draws data flow diagrams (DFDs). Data flow diagrams are structured analysis and design tools that allow the analyst to comprehend the system and subsystems visually as a set of interrelated data flows.

Graphical representations of data movement storage and transformation are drawn with the use of four symbols: a rounded rectangle to depict data processing or transformations, a double square to show an outside data entity (source or receiver of data), an arrow to depict data flow, and an open-ended rectangle to show a data store.

The systems analyst extracts data processes, sources, stores, and flows from early organizational narrative and uses a top-down approach to first draw a context diagram of the system within the larger picture. Then a level 0 logical data flow diagram is drawn. Processes are shown and data stores are added. Next, the analyst creates a child diagram for each of the processes in Diagram 0. Inputs and outputs remain constant, but the data stores and sources change. Exploding the original data flow diagram allows the systems analyst to focus on ever more detailed depictions of data movement within the system. The analyst then develops a physical data flow diagram from the logical data flow diagram, partitioning it to facilitate programming. Each process is analyzed to determine whether it should be a manual or automated procedure. Automated procedures are subsequently grouped into a series of computer programs, designated as either batch or online. Alternative methods of creating data flow diagrams include using event response tables and use cases. Six considerations for partitioning data flow diagrams include whether processes are performed by different user groups, processes execute at the same times, processes perform similar tasks, batch processes can be combined for efficient processing, processes may be combined into one program for consistency of data, or processes may be partitioned into different programs for security reasons.

The advantages of data flow diagrams include the simplicity of notation; using them to gain clearer information from users, allowing the systems analyst to conceptualize necessary data flows without being tied to a particular physical implementation, allowing analysts to better conceptualize the interrelatedness of the system and its subsystems, and analyzing a proposed system to determine if the necessary data and processes have been defined.

KEYWORDS AND PHRASES

- base elements
- batch processes
- child diagram
- context diagram
- data flow diagram fragment
- data flow diagrams
- data-oriented systems
- data store
- derived elements
- event modeling
- event response tables
- events triggers
- exploding
- external entity (source or destination)
- external events
- functionally primitive
- interface data flow
- level 0 diagram
- logical model
- online processes
- parent process
- partitioning
- physical data stores
- physical model
- primitive process
- temporal events
- top-down approach
- transaction data store
- transforming process
- UML (unified modeling language)
- use case
- vertical balancing

REVIEW QUESTIONS

1. What is one of the main methods available for the analyst to use when analyzing data-oriented systems?
2. What are the four advantages of using a data flow approach over narrative explanations of data movement?
3. What are the four data items that can be symbolized on a data flow diagram?
4. What is a context diagram? Contrast it to a level 0 DFD.
5. Define the top-down approach as it relates to drawing data flow diagrams.
6. Describe what "exploding" data flow diagrams means.
7. What are the trade-offs involved in deciding how far data streams should be exploded?
8. Why is labeling data flow diagrams so important? What can effective labels on data flow diagrams accomplish for those unfamiliar with the system?

HYPERCASE® EXPERIENCE

"You take a very interesting approach to the problems we have here at MRE. I've seen you sketching diagrams of our operation almost since the day you walked in the door. I'm actually getting used to seeing you doodling away now. What did you call those? Oh, yes. Context diagrams. And flow charts? Oh, no. Data flow diagrams. That's it, isn't it?"

HYPERCASE QUESTIONS

1. Find the data flow diagrams already drawn in MRE. Make a list of those you found and add a column to show where in the organization you found them.
2. Draw a context diagram modeling the Training Unit Project Development process, one that is based on case interviews with relevant Training Unit staff. Then draw a level 0 diagram detailing the process.

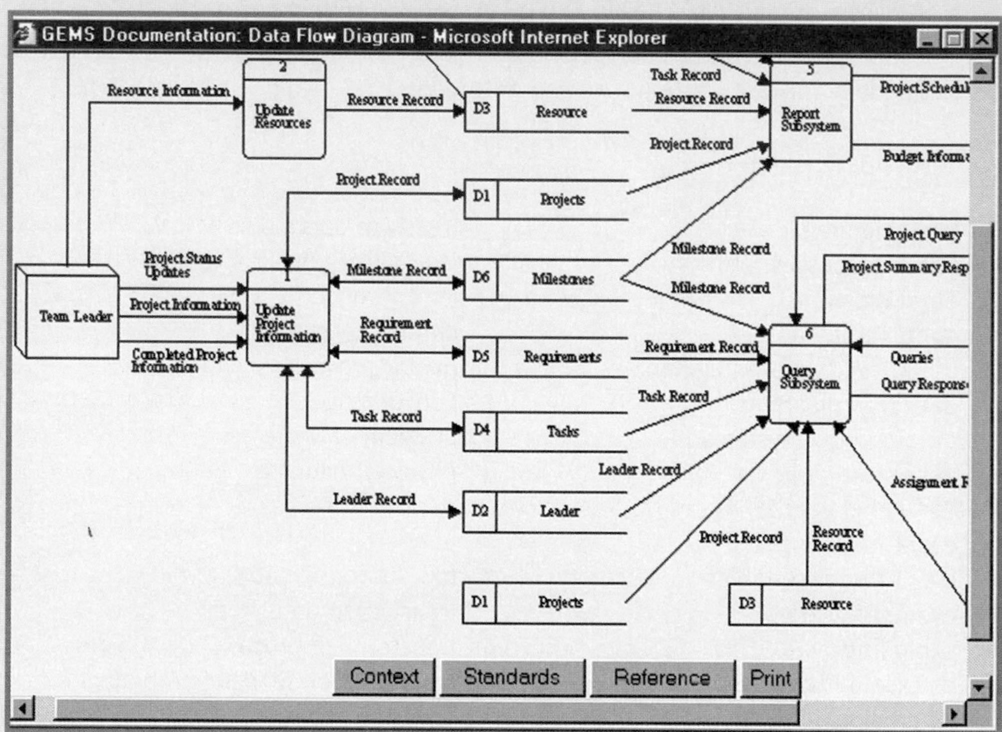

FIGURE 9.HC1
In HyperCase you can click on elements in a data flow diagram.

9. What is the difference between a logically and physically oriented data flow diagram?
10. List three reasons for creating a logically oriented data flow diagram.
11. List five characteristics found on a physical data flow diagram that are not on a logical data flow diagram.
12. When are transaction files required in the system design?
13. How can an event table be used to create a data flow diagram?
14. List the major sections of a use case.
15. How can a use case be used to create a data flow diagram?
16. What is partitioning, and how is it used?
17. How can an analyst determine when a user interface is required?
18. List three ways of determining partitioning on a data flow diagram.
19. List three ways to use completed data flow diagrams.

PROBLEMS

1. Pamela Coburn, a systems analyst, has worked for some time with Luis Asperilla, the manager of the South Street Bonton's clothing store, observing him in action during the day and talking with him whenever business slows. Pamela is feeling fairly sure of the store's processes now and wants to capture what she's learned on paper. (Consulting Opportunity 9.1 provides the background Pamela has gained.)
   a. Draw a context diagram for Bonton's.
   b. Draw a level 0 data flow diagram of data movement at Bonton's on South Street, as Pamela might.
   c. Explode one of the processes from your first-level diagram into more detail, adding data stores and data flows. Make reasonable assumptions about the operation of a retail clothing store, if necessary, to finish the diagrams.
   d. In a paragraph write a description of the process you exploded in problem 1c. What assumptions, if any, did you have to make to draw the second-level diagram?
2. In two paragraphs defend the statement, "An advantage of logical data flow diagrams is that they free the systems analyst from premature commitment to technical implementation of the system." Use an example to support what you write.
3. Up to this point you seem to have had excellent rapport with Kathy Kline, one of the managers who will use the system you are proposing. When you showed her the data flow diagrams you drew, however, she did not understand them.
   a. In a paragraph write down in general terms how to explain to a user what a data flow diagram is. Be sure to include a list of symbols and what they mean.
   b. It takes some effort to educate users about data flow diagrams. Is it worthwhile to share them with users? Why or why not? Defend your response in a paragraph.
4. One common experience that students in every college and university share is enrolling in a college course.
   a. Draw a first-level data flow diagram of data movement for enrollment in a college course. Use a single sheet and label each data item clearly.
   b. Explode one of the processes in your original data flow diagram into subprocesses, adding data flows and data stores.
   c. List the parts of the enrollment process that are "hidden" to the outside observer and about which you have had to make assumptions to complete a second-level diagram.
5. Figure 9.EX1 is a level 1 data flow diagram of data movement in a Niagara Falls tour agency called Marilyn's Tours. Read it over, checking for any inaccuracies.
   a. List and number the errors that you have found in the diagram.
   b. Redraw and label the data flow diagram of Marilyn's so that it is correct. Be sure that your new diagram employs symbols properly so as to cut down on repetitions and duplications where possible.
6. Perfect Pizza wants to install a system to record orders for pizza and chicken wings. When regular customers call Perfect Pizza on the phone, they are asked their phone number. When the number is typed into a computer, the name, address, and last order date is automatically brought up on the screen. Once the order is taken, the total, including tax and delivery, is calculated. Then the

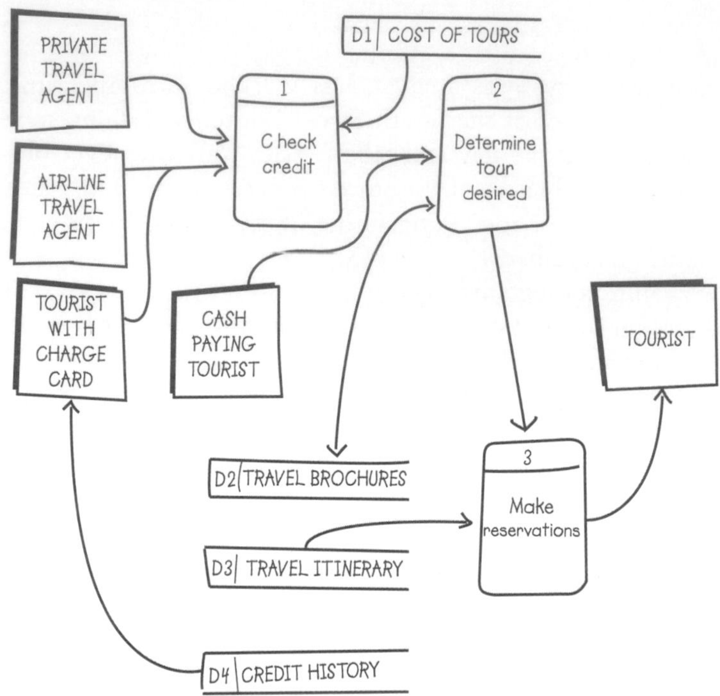

FIGURE 9.EX1
A hand-sketched data flow diagram for Marilyn's Tours.

order is given to the cook. A receipt is printed. Occasionally, special offers (coupons) are printed so the customer can get a discount. Drivers who make deliveries give customers a copy of the receipt and a coupon (if any). Weekly totals are kept for comparison with last year's performance. Write a summary of business activities for taking an order at Perfect Pizza.

7. Draw a context diagram for Perfect Pizza (problem 6).
8. Explode the context-level diagram in problem 7 showing all the major processes. Call this Diagram 0. It should be a logical data flow diagram.
9. Draw a logical child diagram for Diagram 0 in problem 8 for the process that adds a new customer if he or she is not currently in the database (has never ordered from Perfect Pizza before).
10. Draw a physical data flow diagram for problem 8.
11. Draw a physical data flow diagram for problem 9.
12. Partition the physical data flow diagram in problem 8, grouping and separating processes as you deem appropriate. Explain why you partitioned the data flow diagram in this manner. (Remember that you do not have to partition the entire diagram, only the parts that make sense to partition.)
13. a. Draw a logical child diagram for process 6 in Figure 9.26.
    b. Draw a physical child diagram for process 6 in Figure 9.26.
14. Draw a physical data flow diagram for process 1.1 in Figure 9.27.
15. Create a context diagram for a real estate agent trying to create a system that matches up buyers with potential houses.
16. Draw a logical data flow diagram showing general processes for problem 15. Call it Diagram 0.
17. Create a context diagram for billing in a dental office. External entities include the patients and insurance companies.
18. Draw a logical data flow diagram showing general processes for problem 17. Call it Diagram 0.

19. Create a use case for the list of six activities for the FilmMagic Video Rental System. Refer to Diagram 0 for the Video Rental System.
20. Create an event table for the six activities of the FilmMagic Video Rental System.
21. Create an event table for the activities listed for World's Trend Order Processing System.
22. Create a use case for the list of seven processes for the World's Trend Order Processing System.
23. Create a CRUD matrix for FilmMagic.
24. Create a CRUD matrix for the files of World's Trend.
25. Create a data flow diagram for the following situation:

Technical Temporaries

Technical Temporaries is a company that specializes in placing employees in businesses for short periods of time. The company specializes in "temporaries" who have a high degree of proficiency in working with PC software, such as word processing and spreadsheets, as well as other technical areas. Employees must pass proficiency tests for areas in which they wish to be certified. The system described below is responsible for matching employees with short-term openings that are available.

List of Business Activities

a. Businesses telephone the company to request temporaries to fill specific positions. The requests are used to create a Temporary Employment Request record. If the business requesting the temporary employee is not on the Employer Master file, a record is created for it.
b. Employees are selected to fill the temporary positions based on employee qualifications and availability. The Temporary Employee Master and the Temporary Employment Request files are used to list all qualified candidates.
c. Contracts are sent to the selected temporaries. Information is printed from the Employee Master, Employer Master, and Temporary Employment Request files.
d. Returned contracts are used to update the Employee Master file. The Temporary Employment Request file is updated with scheduling and personnel information.
e. Monthly schedules are printed for each employee. They contain information from the Employee Master, the Employer Master, and the Temporary Employment Request files, and they are sequenced by employment date for each employee.
f. Notification is sent to the business requesting the temporary employees confirming the date and qualifications of the workers as well as their names.

26. Create a physical child data flow diagram for the following situation: The Fastbase Corporation develops PC database software products that are sold for both the domestic and international market. Customers receiving the products are sent a set of additional fonts if they return a warranty registration card that is included with the software and documentation. The diagram represents a batch process and is the child of process 5, ADD CUSTOMER REGISTRATION. The following tasks are included:
    a. Inspect the Warranty Registration Card received from the customer to ensure that the information is complete and accurate. Incomplete cards are placed in a reject box.

b. Data entry operators key the Warranty Registration Card, thus creating a Warranty Registration File.
c. A different data entry operator verifies the keyed data by reentering the Warranty Registration Card information. The data entry terminal compares the data previously keyed with the entry made by the second operator. Discrepancies are displayed.
d. The Warranty Registration File is input to a batch edit program. Each record is checked for accuracy. Errors are printed on a Warranty Validation Report, and valid records are placed on a Valid Warranty Registration File.
e. The Valid Warranty Registration File is used as input, along with the Customer Master File, into the Customer Warranty Update Program. Records are added or updated, depending on whether the customer already exists on the Customer Master File.
f. The Valid Warranty Registration File is used to print a series of mailing labels for sending the font software to the customer.

27. Use the principles of partitioning to determine which of the processes in problem 20 should be included in separate programs.

28. Create a physical child data flow diagram for the following situation: The local Personal Computer Users Group holds meetings once a month with informative speakers, door prizes, and sessions for special interest groups. A laptop computer is taken to the meetings, and it is used to add the names of new members to the group. The diagram represents an online process and is the child of process 1, ADD NEW MEMBERS. The following tasks are included:
    a. Key the new member information.
    b. Validate the information. Errors are displayed on the screen.
    c. When all the information is valid, a confirmation screen is displayed. The operator visually confirms that the data are correct and either accepts the transaction or cancels it.
    d. Accepted transactions add new members to the Membership Master file, which is stored on the laptop hard drive.
    e. Accepted transactions are written to a Membership Journal file, which is stored on a diskette.

GROUP PROJECTS

1. Meet with your group to develop a context diagram for Maverick Transport (first introduced in Chapter 5). Use any data you have subsequently generated with your group about Maverick Transport. (*Hint*: Concentrate on one of the company's functional areas rather than try to model the entire organization.)
2. Using the context diagram developed in problem 1, develop with your group a level 0 logical data flow diagram for Maverick Transport. Make any assumptions necessary to draw it. List them.
3. With your group, choose one key process and explode it into a logical child diagram. Make any assumptions necessary to draw it. List follow-up questions and suggest other methods to get more information about processes that are still unclear to you.
4. Use the work your group has done to date to create a physical data flow diagram of a portion of the new system you are proposing for Maverick Transport.

SELECTED BIBLIOGRAPHY

Colter, M. "A Comparative Examination of Systems Analysis Techniques." *MIS Quarterly*, Vol. 8, No. 1, June 1984, pp. 51–66.

Davis, G. B., and M. H. Olson. *Management Information Systems, Conceptual Foundations, Structure, and Development*, 2d ed. New York: McGraw-Hill, 1985.

Gane, C., and T. Sarson. *Structured Systems Analysis and Design Tools and Techniques*. Englewood Cliffs, NJ: Prentice Hall, 1979.

Gore, M., and J. Stubbe. *Elements of Systems Analysis*, 3d ed. Dubuque, IA: William C. Brown, 1983.

Leeson, M. *Systems Analysis and Design*. Chicago: Science Research Associates, 1985.

Lucas, H. *Information Systems Concepts for Management*, 3d ed. New York: McGraw-Hill, 1986.

Martin, J. *Strategic Data-Planning Methodologies*. Englewood Cliffs, NJ: Prentice Hall, 1982.

McFadden, F. R., and J. A. Hoffer. *Data Base Management*. Menlo Park, CA: Benjamin/Cummings, 1985.

Senn, J. A. *Analysis and Design of Information Systems*. New York: McGraw-Hill, 1984.

Sprague, R. H., and E. D. Carlson. *Building Effective Decision Support Systems*. Englewood Cliffs, NJ: Prentice Hall, 1982.

EPISODE CPU CASE

9

ALLEN SCHMIDT, JULIE E. KENDALL, AND KENNETH E. KENDALL

JUST FLOWING ALONG

After the results of interviews, questionnaires, and prototyping are gathered and analyzed, Anna and Chip move to the next step, modeling the system. Their strategy is to create a layered set of data flow diagrams and then describe the components.

Modeling starts with analyzing the context diagram of the current microcomputer system. This diagram is simple to create and is the foundation for successive levels because it describes the external entities and the major data flow.

"Shall we create a physical data flow diagram of the current system?" asks Chip.

Anna replies, "No, it's fairly simple to understand, and we wouldn't gain any significant new knowledge of how the system operates. Let's start by creating a logical model of the current system."

The logical data flow diagrams are completed within a few days. Anna and Chip hold an afternoon meeting to review the diagrams and give each other feedback. "These look good," remarks Chip. "We can clearly see the business events that comprise the current system."

Anna replies, "Yes, let's take the current logical data flow diagrams and add all the requirements and desired features of the new system. We can also eliminate any of the unnecessary features that wouldn't be implemented in the new system."

Anna takes the context-level diagram (shown in Chapter 2) and adds many of the reports, inquiries, and other information included in the new system. The finished context-level diagram is shown in Figure E9.1. Notice there the many new data flows. The maintenance department will receive reports that currently are not available. One report, for example, helps to automate the installation of new microcomputers, the INSTALLATION LISTING, and another report intended for management shows which software is located on which machines, the SOFTWARE CROSS-REFERENCE REPORT.

Chip reviews the finished diagram, commenting, "This is more art than science. It looks like all the requirements of the new system are included. But it is far more complex than I originally thought it would be."

Anna replies, "Let's expand this to Diagram 0 for the new system. This will be a logical data flow diagram because we want to focus on the business needs. Perhaps it would be best if we work in a team for this diagram."

After working for several hours that afternoon and a good portion of the next morning, they complete the diagram. It is reviewed and modified with some minor changes. The finished Diagram 0 is shown in Figures E9.2 and E9.3. Because it is a logical diagram, it shows no keying or validation operations, nor does it show any temporary data stores or transaction files. Timing is not a consideration (an example is the ADD NEW MICROCOMPUTER process, where it appears that orders are updated and reports simultaneously produced).

"This finally looks right," muses Chip. "All the major processes, data flow, and data stores are accounted for. And the overall diagram doesn't look too complicated."

"Putting all of the inquiries into one subsystem and all the reports into another helped. Remember how complex the original diagram was?" asks Anna.

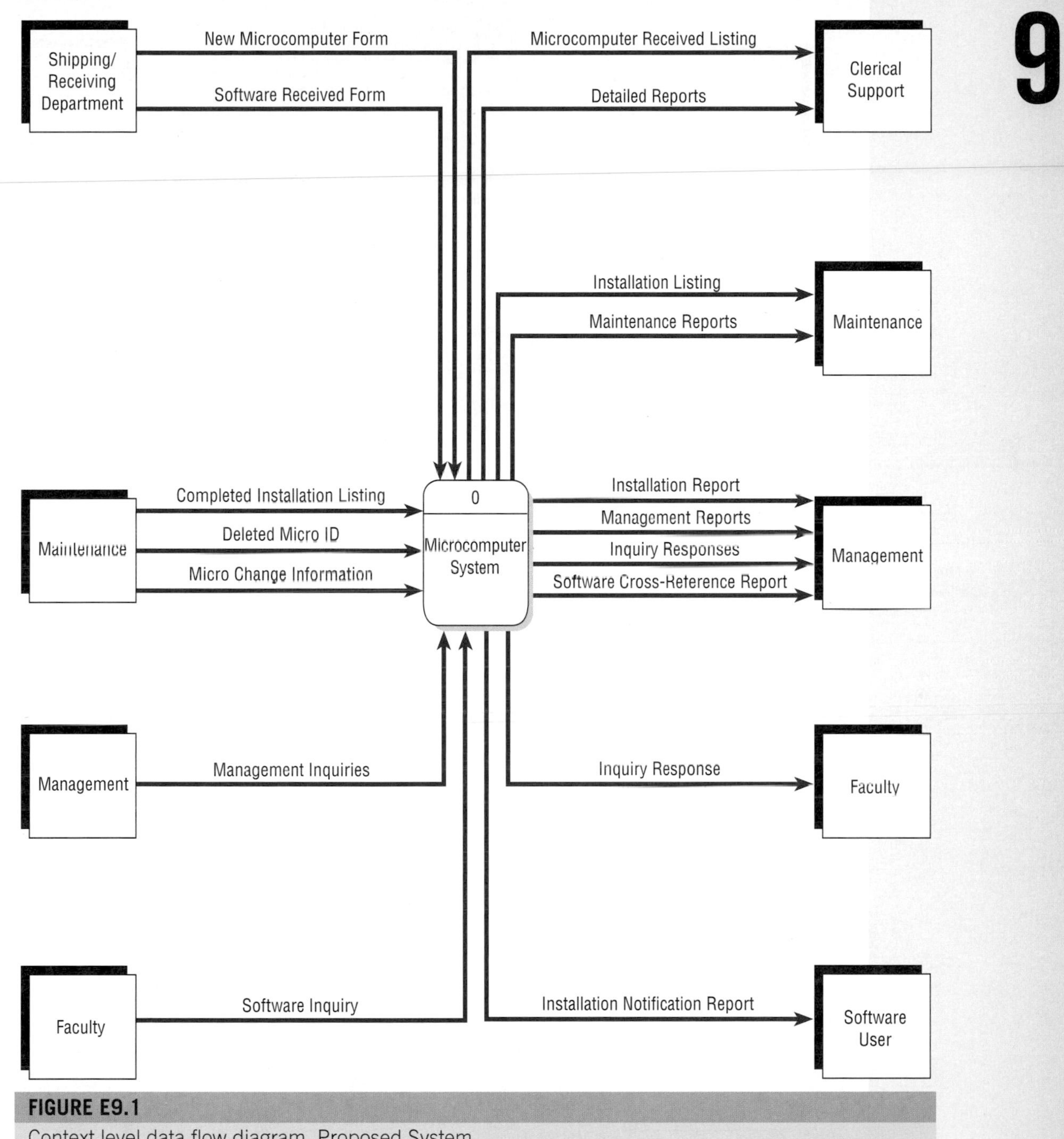

FIGURE E9.1
Context level data flow diagram, Proposed System.

"I sure do," Chip replies. "I started to think we were tackling too much at once with this system. At least it's more manageable now. Because this is finished, what's the next step?"

"We need to decide how to implement the data flow diagram into a series of steps, which are shown on the physical data flow diagram," Anna says. "This logical data flow diagram shows the business tasks, or *what* should be accomplished. Now we need to

9

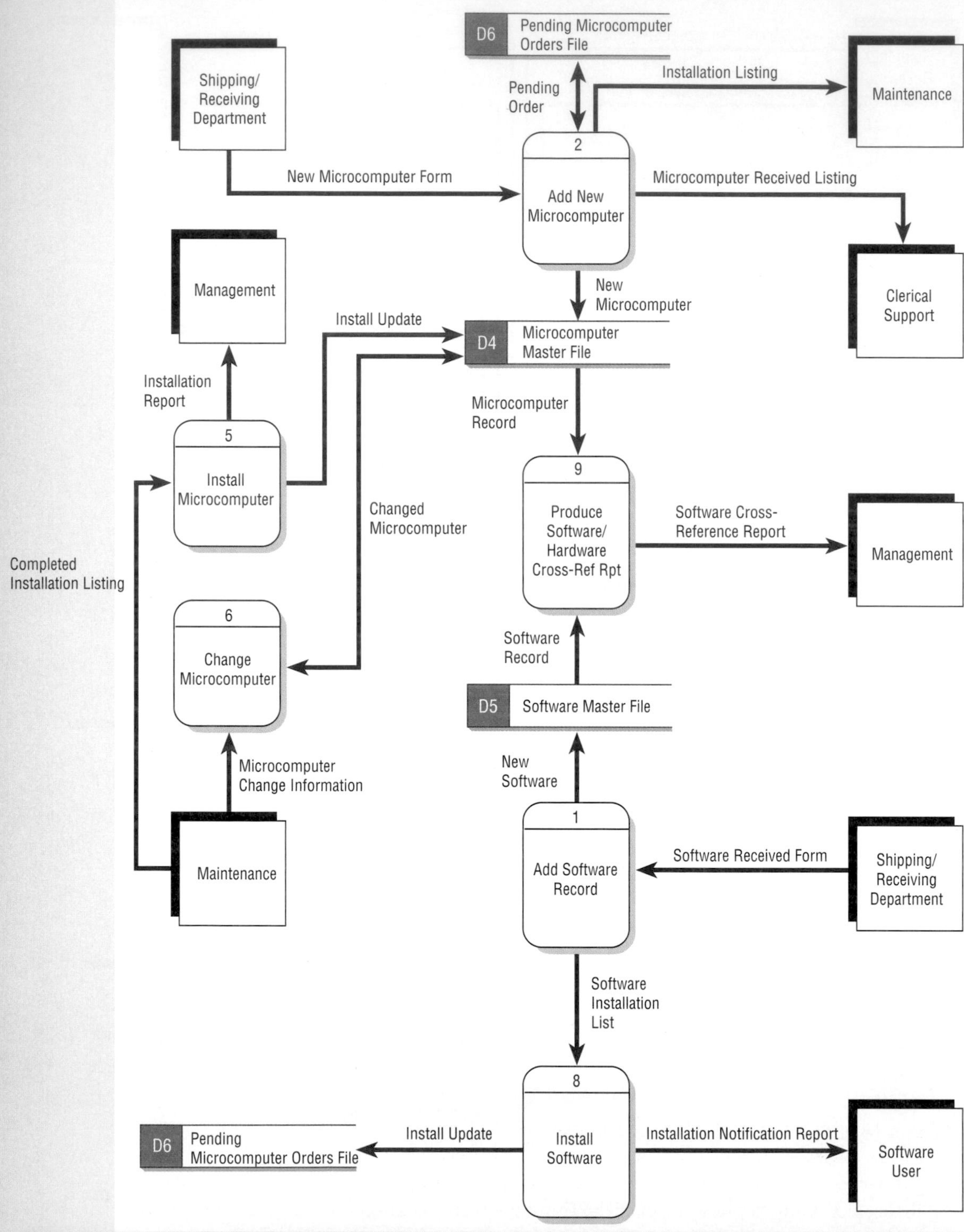

FIGURE E9.2
Diagram 0: Proposed Microcomputer System (part 1).

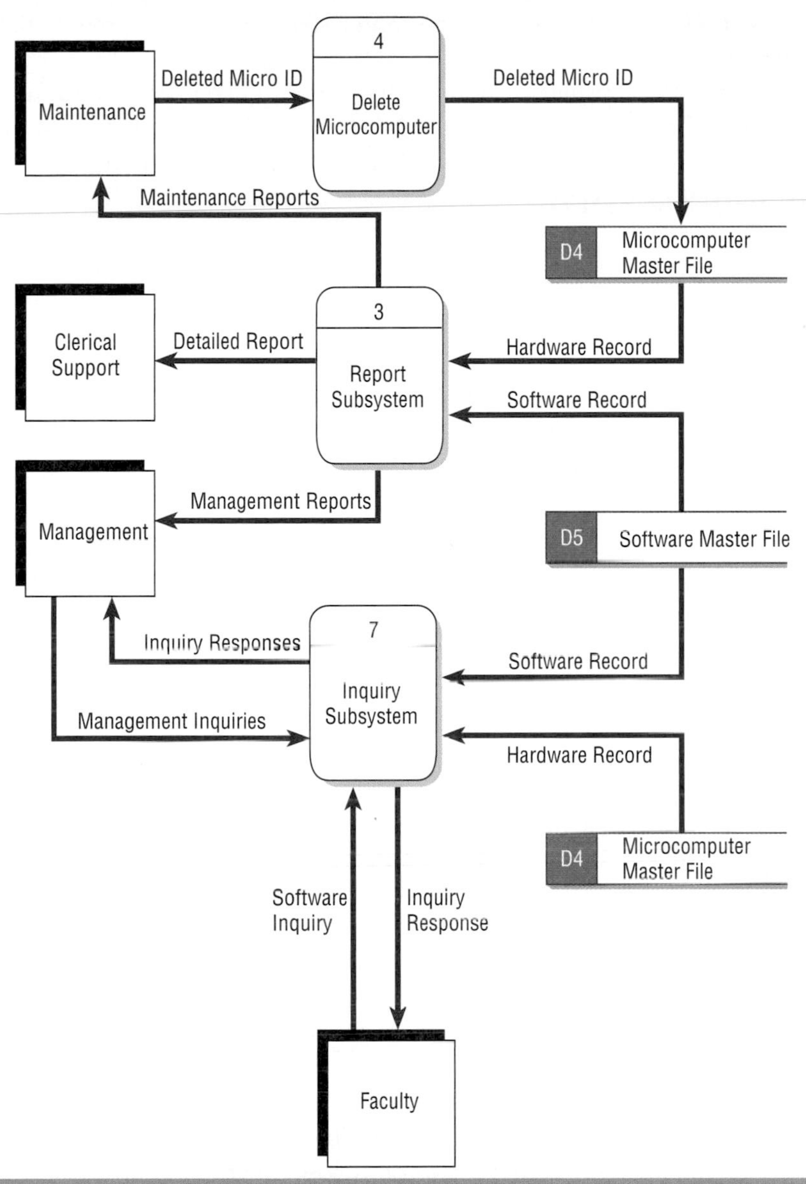

FIGURE E9.3
Diagram 0: Proposed Microcomputer System (part 2).

show *how* the system will work. Keying, validation, information on whether programs are online or batch, and transaction files need to be added."

Chip and Anna divide up the work by major tasks to be accomplished. Chip starts working on the ADD MICROCOMPUTER process.

When Chip draws the diagrams, he sees that he is drawing a level 0 diagram and then exploding it into many level 1 diagrams. Just as a parent may have many children, there may be many level 1 diagrams for a specific level 0 diagram. For this reason some analysts refer to them as parent and child diagrams.

Chip and Anna decide to abbreviate level 0 diagram as Diagram 0. The details are shown on a child diagram, Diagram 2. The external entities do not appear on the diagram, because they are only shown on the context diagram and Diagram 0, the context diagram explosion.

Figure E9.4 is the finished version of Diagram 2, a batch process for adding new microcomputers. The NEW MICROCOMPUTER FORM is an input interface flow that matches the parent diagram. The MICROCOMPUTER RECEIVED LISTING and INSTALLATION LISTING are output interface flows. Notice that several transaction files are needed to hold data between each process that runs at a different time. Errors are also shown, as a minor interface flow, which need not be present on the parent diagram. Keying, validation, and sorting processes are included, because they are necessary for implementation of the design.

Chip has some difficulty creating the diagram. The starting point is the input flow, NEW MICROCOMPUTER FORM. This flow has to be keyed, and because it is a batch process, the forms should be rekeyed by a separate operator to catch any keying errors. The results are stored on a transaction file.

Chip is unsure what activities will take place next so he decides to work backward from the MICROCOMPUTER MASTER FILE data store. The records must be added to the master file. Thus, an update program is necessary. "Would the input to the ADD MICROCOMPUTER RECORD process be the keyed transactions?" Chip wonders to himself. "No. The data must be edited to ensure validity. The edit program should check all records for syntax errors and confirm that a record for the same microcomputer does not already exist on the master file."

"Should the edit and update activities be accomplished in the same process?" Chip wonders. After he gives it some thought, the answer becomes clear. "We need to have a file of the new transactions for printing the INSTALLATION LISTING and the MICROCOMPUTER RECEIVED LISTING. It makes sense to separate the programs and create a VALID MICROCOMPUTER TRANSACTION file out of the edit program."

Chip then decides to work backward from the INSTALLATION LISTING interface flow. "It needs to be printed, but what is the sequence of the prototype report? Ah, here it is!" he exclaims softly. Because the INSTALLATION LISTING needs to be in sequence by the manufacturer and model, it needs to be sorted. A sort process was added with the MICROCOMPUTER TRANSACTION RECORD as input. This same sequence was needed for the MICROCOMPUTER RECEIVED LISTING. "All done," thinks Chip. "One more review and . . . whoops, forgot about updating the PENDING MICROCOMPUTER ORDERS FILE data store." After one last change, the diagram was completed.

Anna reviews the diagram for omissions and errors. "I can see that you've put a great deal of thought into this," she exclaims. "It's really well designed. I've been working on Diagram 1, an explosion of process 1, ADD SOFTWARE RECORD. Perhaps you would like to review the finished result."

"Sure," replies Chip. "I'll check it for omissions and errors."

Diagram 1 is shown in Figure E9.5. Because this diagram is an online process, there are no key and verify operations. Instead, NEW SOFTWARE INFORMATION is keyed and edited by the same program. Errors are reported on the screen and corrected by the operator. After all errors have been corrected, the operator has a chance to sight-verify the data. If correct, the operator presses a key to accept the data; otherwise the transaction may be canceled or corrected.

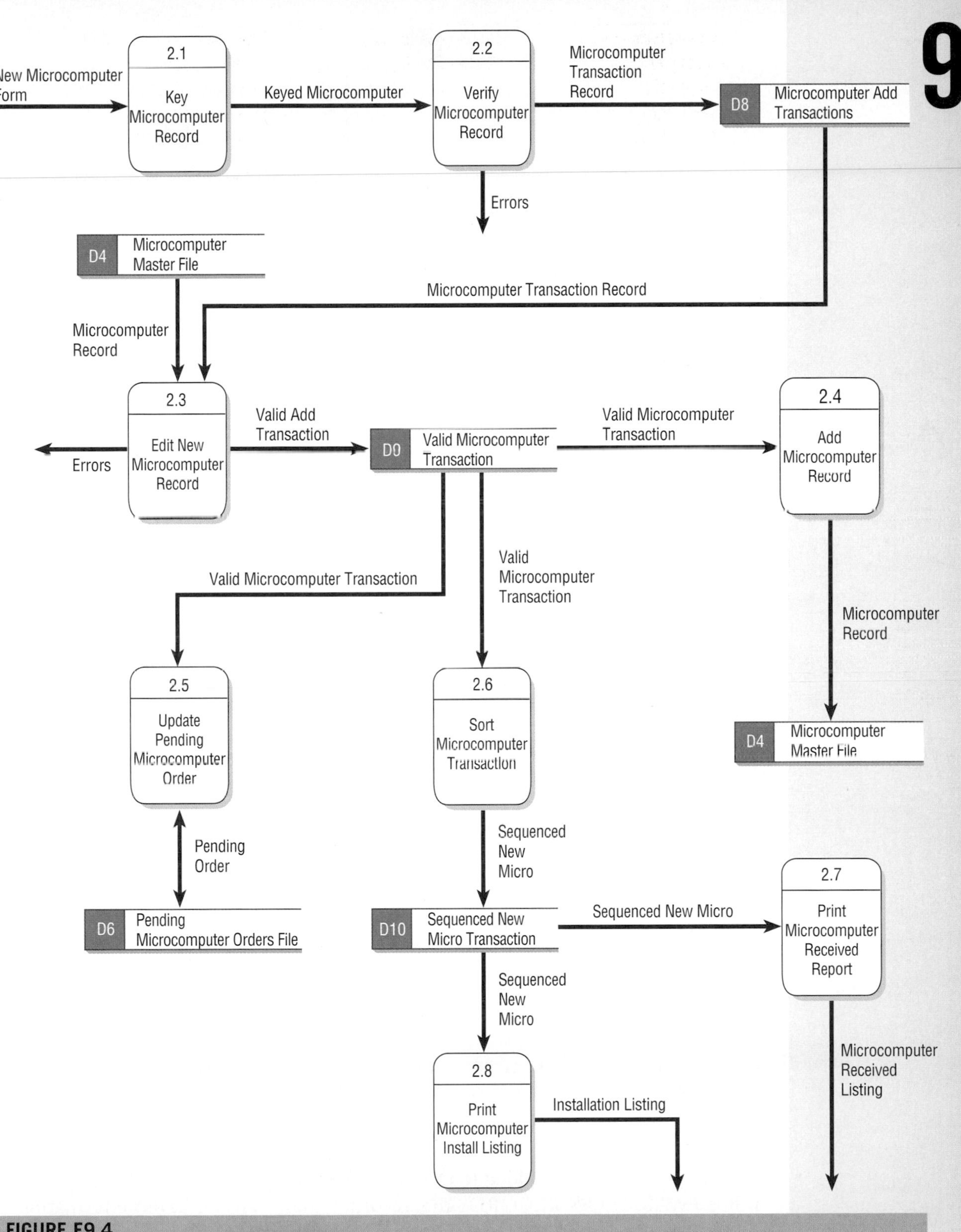

FIGURE E9.4

Diagram 2: Proposed Microcomputer System.

FIGURE E9.5

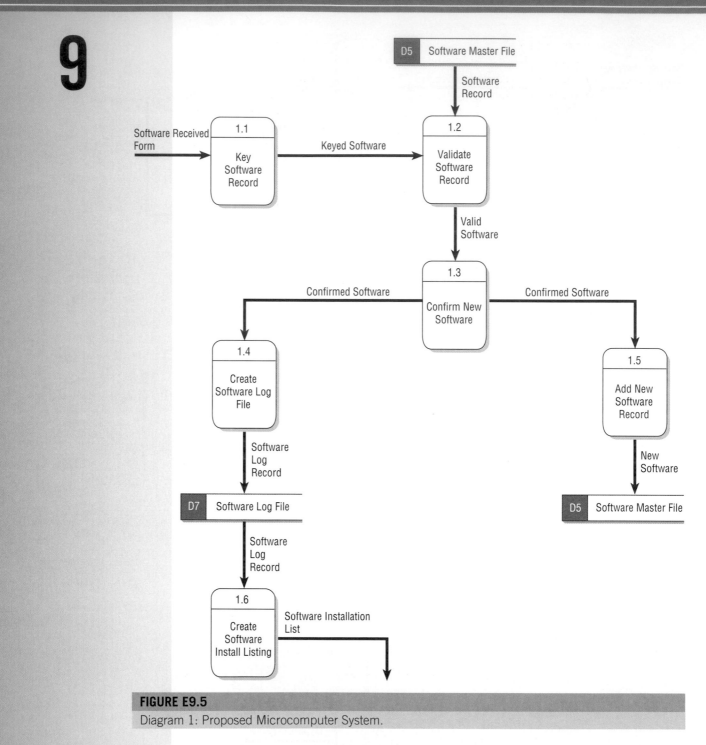

Diagram 1: Proposed Microcomputer System.

Confirmed data are added to the SOFTWARE MASTER file and used to create a SOFTWARE LOG RECORD. This record contains all the keyed information as well as the date, time, and user ID of the person entering the transaction. In the ADD SOFTWARE diagram, this record is used to create the SOFTWARE INSTALLATION LIST as well as to provide a backup of all new transactions and an audit trail of entries.

Because Chip and Anna are using Visible Analyst to create the data flow diagrams, all the components of the diagram may be described in the Visible Analyst repository. Chip starts by working on the ADD MICROCOMPUTER data flow diagram.

The description for process 2.5, UPDATE PENDING MICROCOMPUTER ORDER, is shown in Figure E9.6. The **Label** area contains the text that appears on the diagram. The Process Description entry is one of the most important entry areas. In the example shown Chip outlines the detailed logic for updating the PENDING MICROCOMPUTER ORDER file.

A second description screen, which is shown in Figure E9.7, shows the diagram that the process is located on.

Similarly, Chip describes process 1 from the parent data flow diagram. This process is printed as a Web page using the Report Query feature of Visible Analyst, and it is illustrated in Figure E9.8. Notice that when a repository entity is previewed using a Web browser, the information from several screens is included on one Web page and the analyst may bookmark or print this page. Large previewed reports may be viewed by scrolling up and down the Web page.

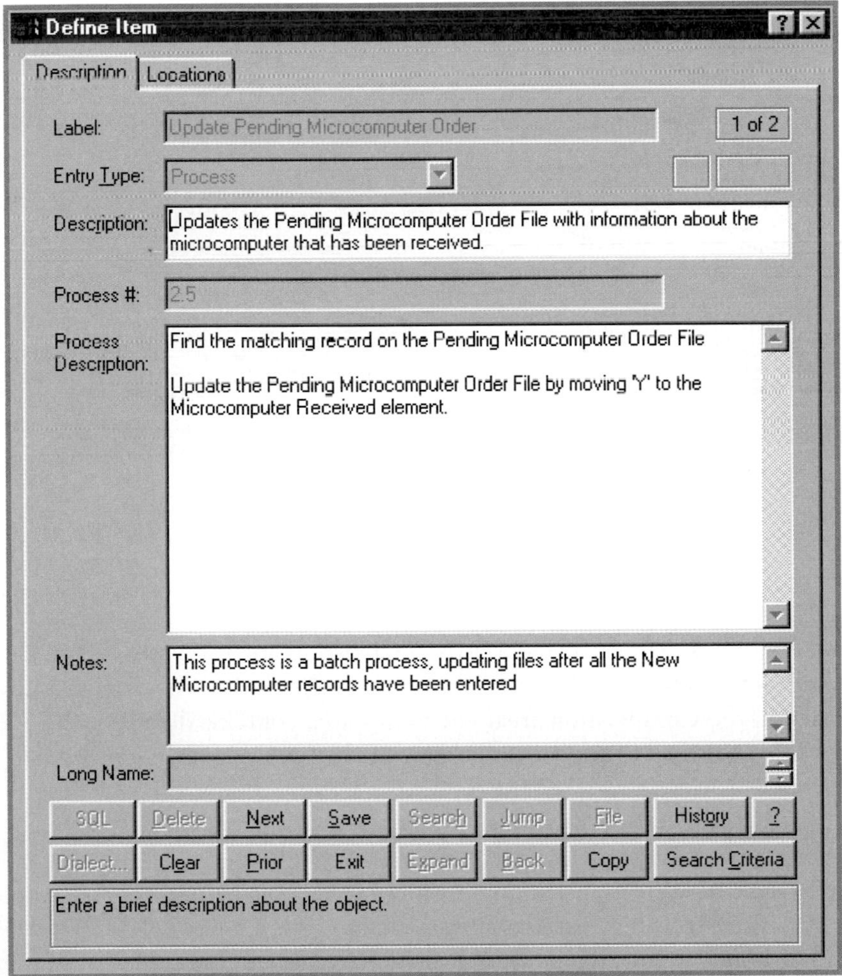

FIGURE E9.6
Process description screen, UPDATE PENDING MICROCOMPUTER ORDER.

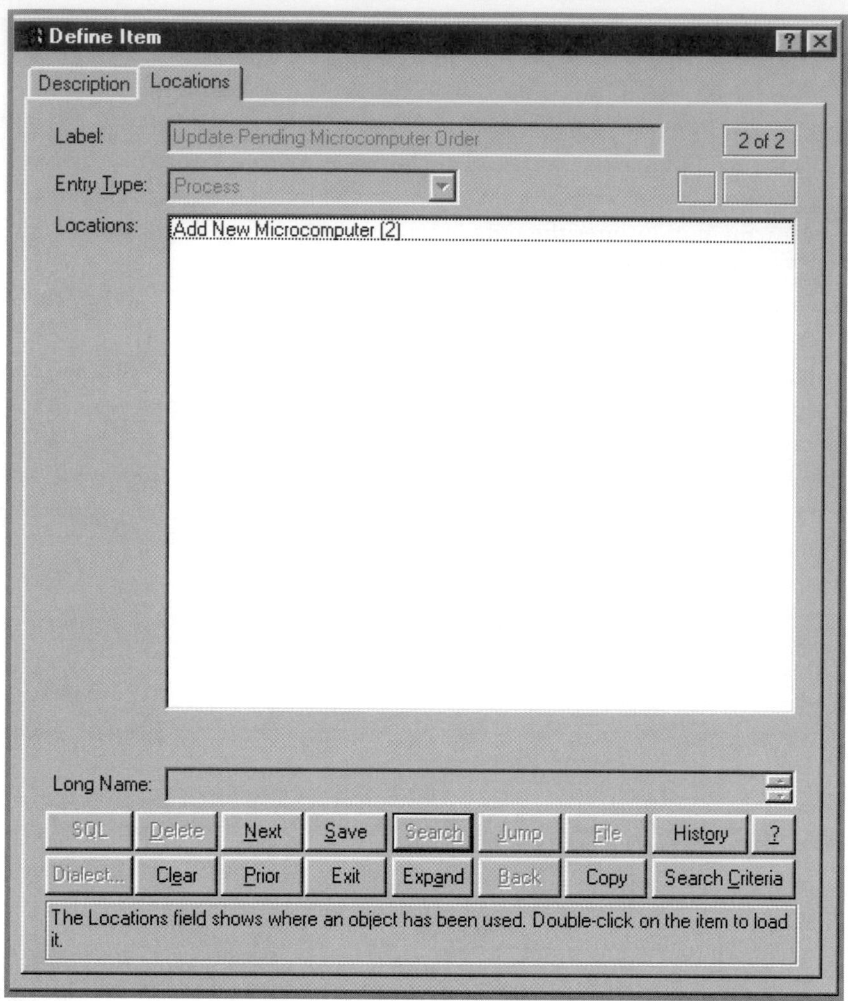

FIGURE E9.7

Process description screen, UPDATE PENDING MICROCOMPUTER ORDER, second screen.

Anna describes the SOFTWARE MASTER data store, shown in Figure E9.9. This data store has the structure of the record stored in the SOFTWARE RECORD, which is indicated in the **Composition** area. The **Notes** area contains details of the index elements or key fields and the approximate size of the file in records.

The NEW MICROCOMPUTER FORM data flow designed by Chip is shown in Figure E9.10. The **Alias** area contains the name of the computer screen that will be used to enter the data from the form. This data flow has the NEW MICROCOMPUTER FORM RECORD in its **Composition** field. This record contains the details, such as structures and elements, that are the details of the NEW MICROCOMPUTER FORM. The **Notes** area contains some of the details of how the form is represented on a screen. A second screen, illustrated in Figure E9.11, indicates the diagrams that contain the data flow.

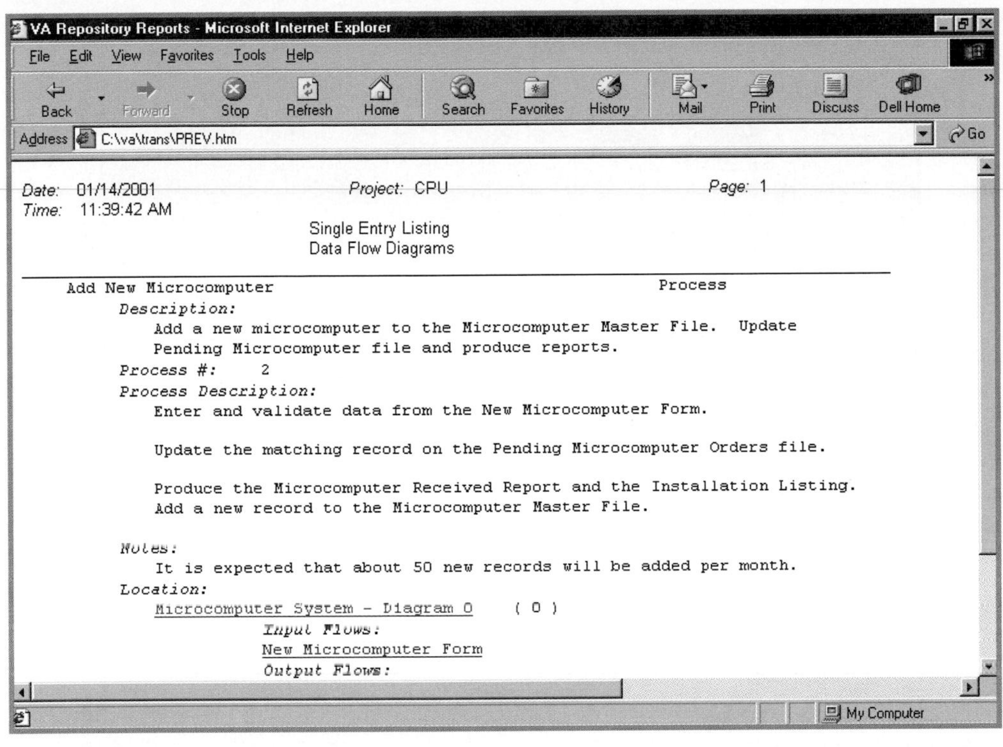

FIGURE E9.8
Web-based output for process 2.

It takes time to enter descriptions for all the objects, but once the entries are complete, Visible Analyst will provide analysis of the design. The Analyze feature for data flow diagrams provides several important features for validating the data flow diagram, the explosion diagrams, and the connections.

When a specified data flow diagram is analyzed, the resultant report may reveal that any of the following data flow diagram syntax errors exist in that DFD:

1. The data flow diagram must have at least one process and must not have any freestanding objects or objects connected to themselves.
2. A process must receive at least one data flow and create at least one data flow. Processes with all input or all output should not occur.
3. A data store should be connected to at least one process.
4. External entities should not be connected to each other. Although they communicate independently, that communication is not part of the system being designed.

Visible Analyst does not show the following errors or check the standards set by Chip and Anna for the project:

1. Data flow names into and out of a process should change (with exceptions).
2. Linear flow (several processes with only one input and output) is rarely found. Except in very low level processes, it is a warning sign that some of the processes may be missing input or output flow.

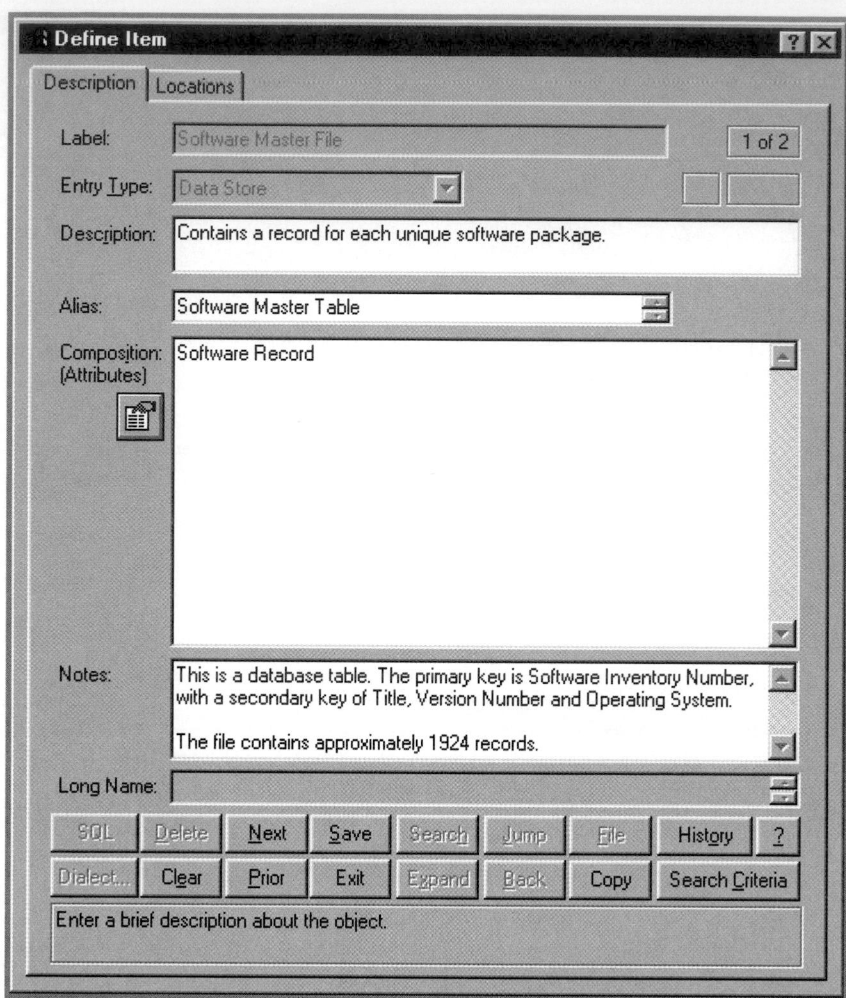

FIGURE E9.9
Data store description screen, SOFTWARE MASTER.

3. External entities should not be connected directly to data stores. For example, you would not let an employee rummage through the Employee master file!
4. Process names should contain a verb describing the work being performed (with exceptions, such as INQUIRY SUBSYSTEM). Data flow names should be nouns.

Chip and Anna both use Visible Analyst to verify that the data flow diagram syntax is correct. The analysis report is shown in Figure E9.12. Notice that a descriptive error message is given for each error, with the diagram object that relates to the problem indicated with quotation marks. This error report was generated because of the syntactical errors in the data flow diagram shown in Figure E9.13.

Visible Analyst will also check that the levels balance among data flow diagram processes and the child diagrams. Inputs and outputs that do not match are shown.

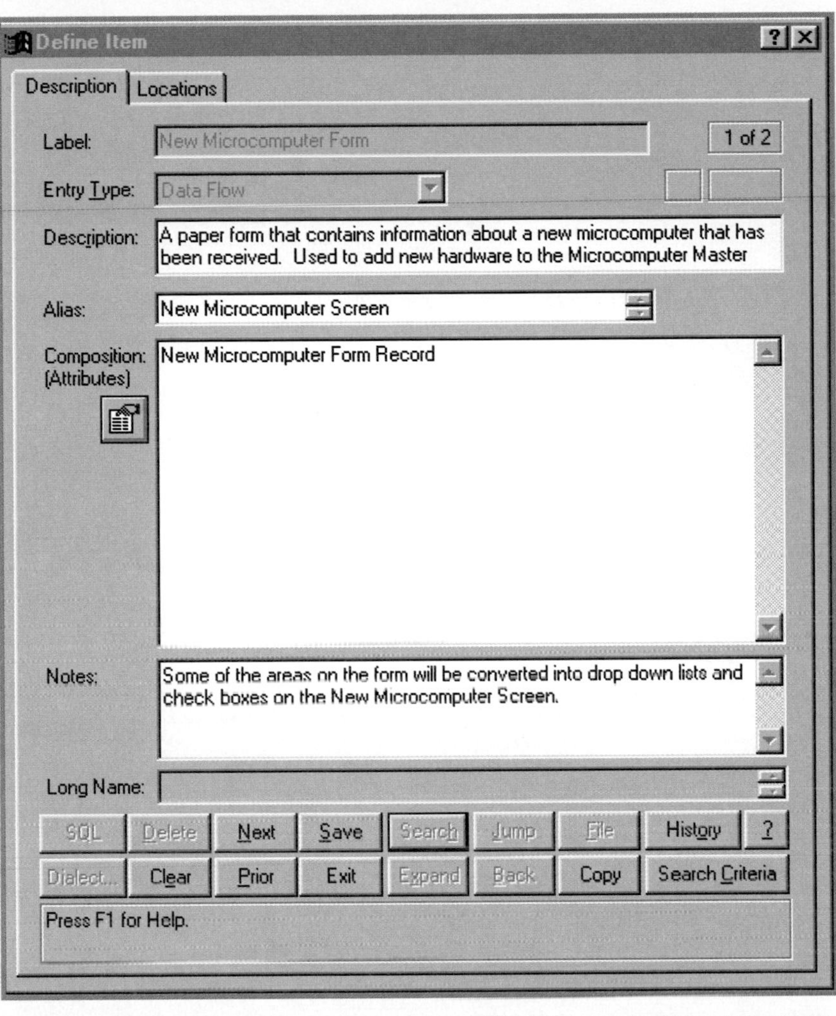

FIGURE E9.10
Data flow description screen, NEW MICROCOMPUTER form.

EXERCISES*

E-1. Use Visible Analyst to view the context diagram for the proposed microcomputer system. Experiment with the **Zoom** controls on the lower toolbar to change from a global to a detailed view of the diagram. Double click on the central process to examine the Repository entry for it. Click Exit to return to the diagram. Right click on the central process to display the object menu for the central process. Use the Explode option to display Diagram 0, representing the details of the central process. Maximize the window and double click on some of the data stores and data flows to examine their repository entries. Press Exit to return to the diagram. Zoom to 100 percent and scroll around the screen to view different regions of the diagram; then print the diagram using a landscape orientation. Click FILE, NEST, and PARENT to return to the context-level diagram. Maximize the window.

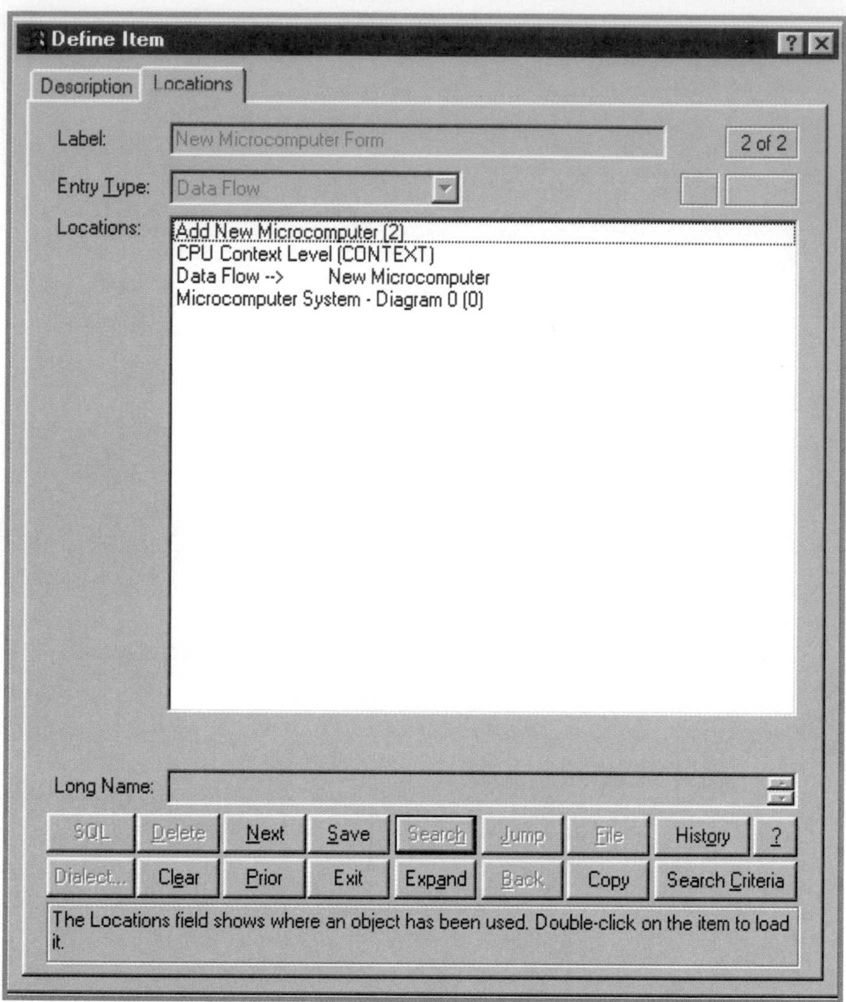

FIGURE E9.11
Second data flow description screen, NEW MICROCOMPUTER form.

E-2. Modify Diagram 0 of the proposed microcomputer system. Add process 10, UPDATE SOFTWARE RECORD. You will have to move the Management external entity lower in the diagram; place it to the left of process 7, Inquiry Subsystem. Create a Repository entry for the process and then press **Exit** to return to the diagram. Print the diagram using a landscape orientation.

Input:	1.	Software Change Data, from Clerical Support
	2.	Software Delete ID, from Management
Output:	1.	Software Record, an update from the Software Master data store

E-3. Explode to Diagram 10, UPDATE SOFTWARE RECORD. Maximize the window and create the diagram illustrated in Figure E9.14. Connect to the SOFTWARE MASTER FILE using a double-headed arrow. (*Hint*: Right click on the data flow, select Change Item, then select Change Type, and Terminator Type, Double Filled.) Print the final diagram.

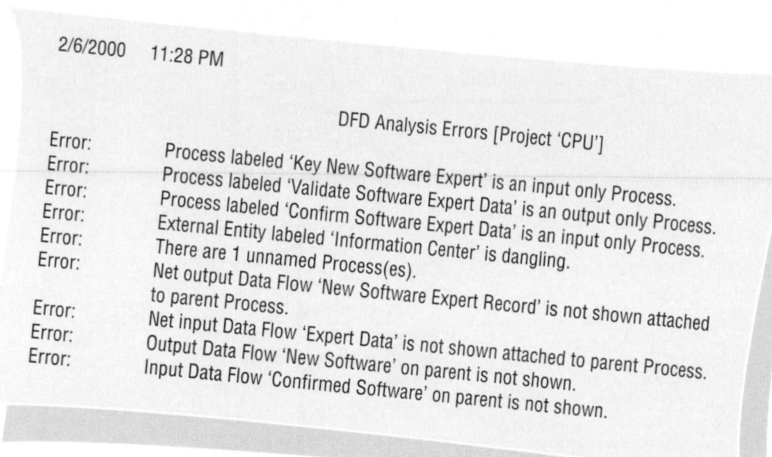
2/6/2000 11:28 PM

DFD Analysis Errors [Project 'CPU']

Error: Process labeled 'Key New Software Expert' is an input only Process.
Error: Process labeled 'Validate Software Expert Data' is an output only Process.
Error: Process labeled 'Confirm Software Expert Data' is an input only Process.
Error: External Entity labeled 'Information Center' is dangling.
Error: There are 1 unnamed Process(es).
Error: Net output Data Flow 'New Software Expert Record' is not shown attached to parent Process.
Error: Net input Data Flow 'Expert Data' is not shown attached to parent Process.
Error: Output Data Flow 'New Software' on parent is not shown.
Error: Input Data Flow 'Confirmed Software' on parent is not shown.

FIGURE E9.12
Data flow diagram error report.

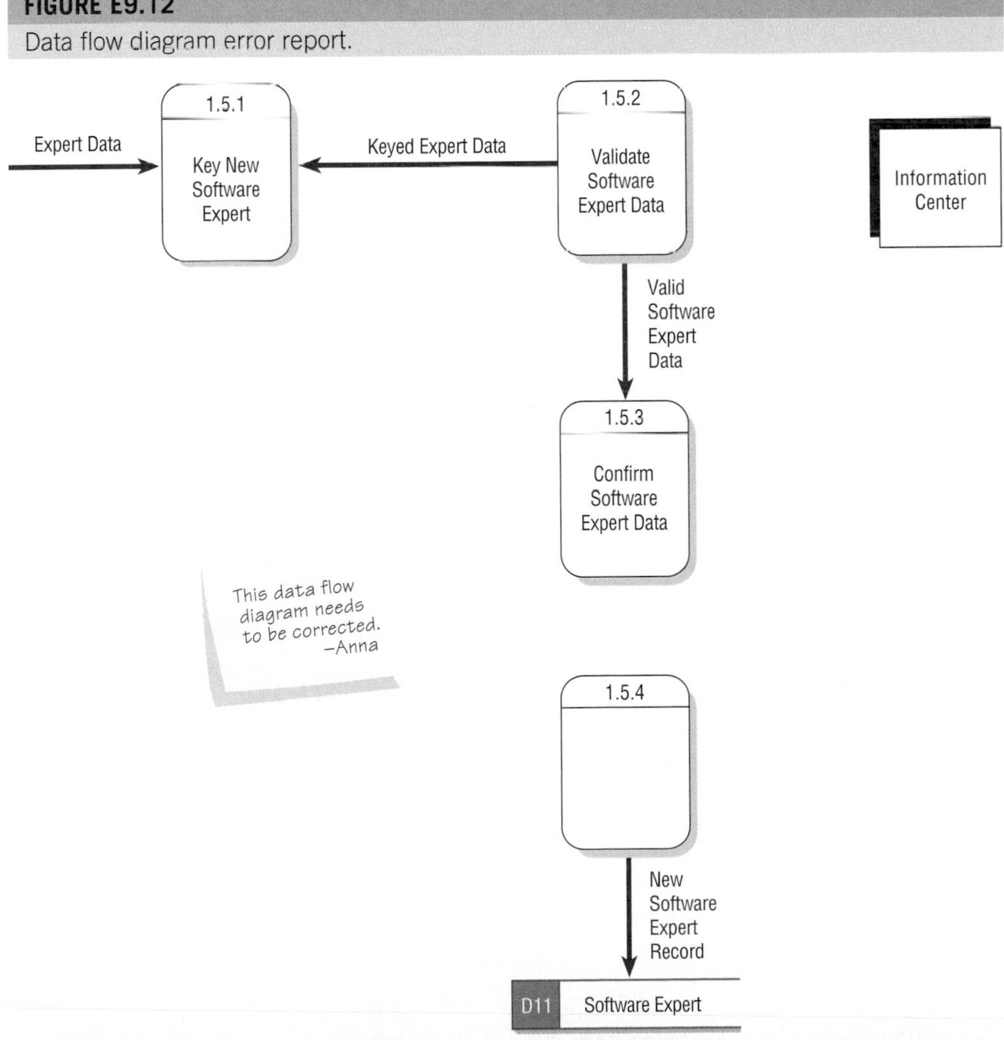

FIGURE E9.13
Data flow diagram with errors.

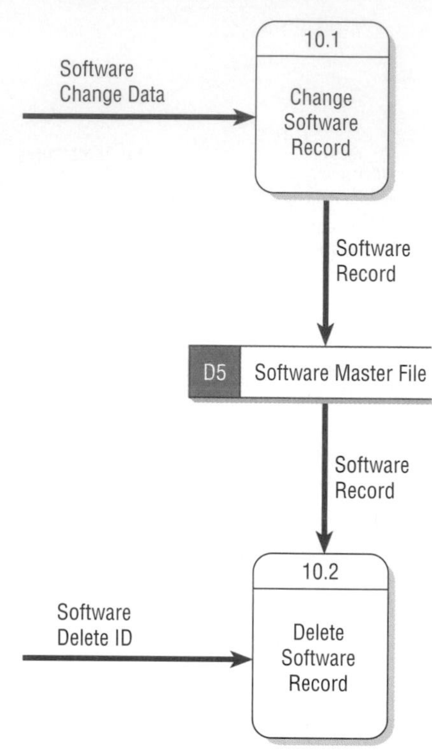

FIGURE E9.14
Data flow diagram, UPDATE SOFTWARE RECORD.

E-4. Modify Diagram 8, INSTALL SOFTWARE. Add the following processes, describing each in the Repository. Zoom to 100 percent and scroll around the screen, checking your diagram for a professional appearance. Print the final result.

Process:		8.2 Install Microcomputer Software
Description:		Manual process, place software on machine
Input:	1.	Microcomputer Location, from process 8.1
	2.	Software Title and Version, from process 8.1
Output:	1.	Installed software form
Process:		8.3 Create Installed Software Transaction
Description:		Batch data-entry process for creating installed software transactions, including validation
Input:	1.	Installed software form
Output:	1.	Installed Software transaction, to Installed Software data store
Process:		8.4 Update Software Master
Description:		Random update of the SOFTWARE MASTER FILE data store with update information

Input:	1.	Installed Software Transaction
Output:	1.	Software Master, update
Process:		8.5 Produce Installation Notification
Description:		Produce an installation notification informing users onto which machines the software has been installed
Input:	1.	Installed Software Transaction
	2.	Software Master, from the SOFTWARE MASTER FILE data store
	3.	Hardware Master, from the MICROCOMPUTER MASTER FILE data store
Output:	1.	Installation Notification listing, an interface flow

E-5. Modify Diagram 6, CHANGE MICROCOMPUTER RECORD, which is shown in Figure E9.15. This is an interactive, online program to change microcomputer

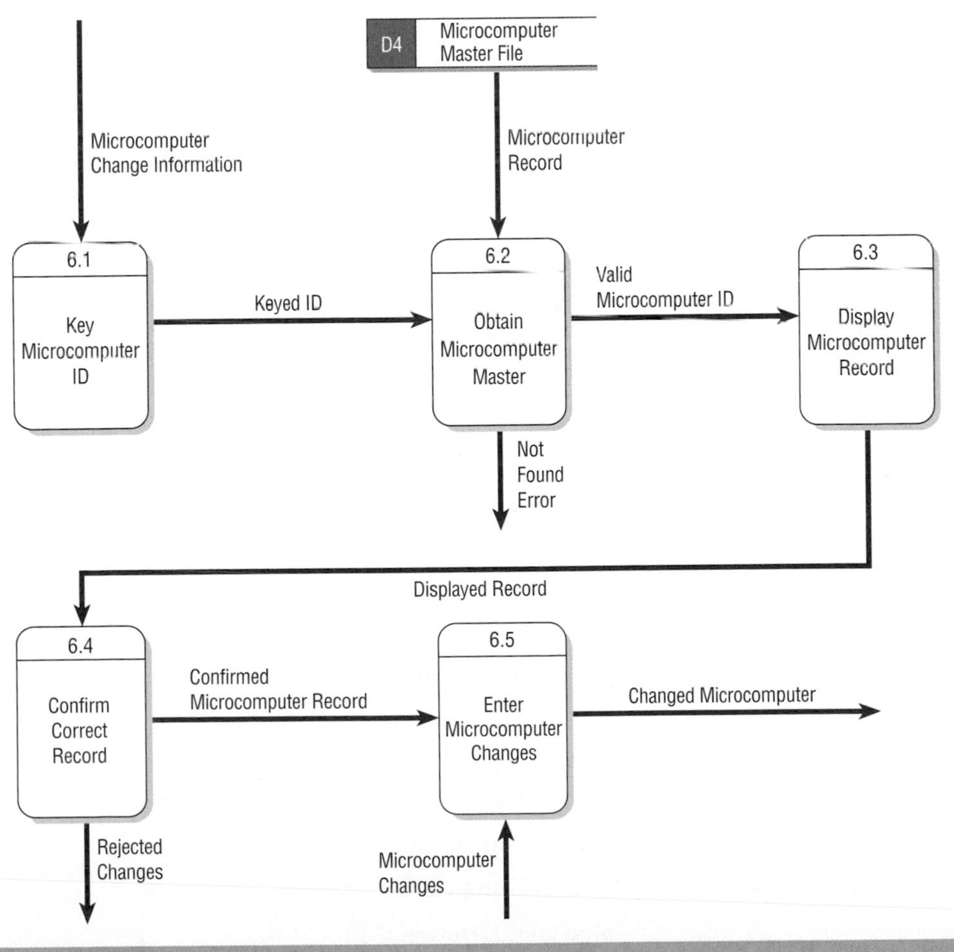

FIGURE E9.15
Data flow diagram, CHANGE MICROCOMPUTER RECORD.

information. Add the following three processes. Create Repository entries for each of the processes as well as the data flow. When completed, zoom to 100 percent and change any data flow arrows that are not straight and move data flow labels for a professional-looking graph. Print the diagram using landscape orientation.

a. Process 6.6, VALIDATE CHANGES. This process edits each change field for validity. The input is the KEYED CHANGES. The output fields are CHANGE ERRORS (interface flow) and VALID CHANGES (to process 6.7).
b. Process 6.7, CONFIRM CHANGES. This process is a visual confirmation of the changes. The operator has a chance to reject the changes or accept them. Input is the VALID CHANGES. The output fields are REJECTED CHANGES (interface flow) and CONFIRMED CHANGES (to process 6.8).
c. Process 6.8, REWRITE MICROCOMPUTER MASTER. This process is a rewrite of the Microcomputer Master record with the changes on the record. Input is the CONFIRMED CHANGES. Output flow is the MICROCOMPUTER MASTER record, to the MICROCOMPUTER MASTER FILE data store.

E-6. Create the explosion data flow diagram for process 4, DELETE MICROCOMPUTER. The following table summarizes input, process, and output. Describe each process and data flow in the Repository. When completed, zoom to 100 percent, move any data flow lines that are not aligned correctly, move the data flow labels for a professional-looking graph, and print the diagram.

Process:	4.1	Key Delete ID
Description:		The microcomputer ID is keyed interactively
Input:	1.	Deleted Micro ID
Output:	1.	Keyed Delete
Process:	4.2	Obtain Microcomputer Record
Description:		Microcomputer Master record is read to ensure that it exists
Input:	1.	Keyed Delete (interface)
	2.	Microcomputer record, from the Microcomputer Master data store
Output:	1.	Not Found Error (interface)
	2.	Valid Microcomputer record
Process:	4.3	Confirm Microcomputer Deletion
Description:		The microcomputer information is displayed on the screen for operator confirmation or rejection
Input:	1.	Valid Microcomputer record
Output:	1.	Rejected Deletion (interface)
	2.	Confirmed Deletion
Process:	4.4	Delete Microcomputer Record

Description: The microcomputer record is *logically* (not physically) deleted from the Microcomputer master file by rewriting the record with an I for inactive in the Record Code field

Input: 1. Confirmed Deletion

Output: 2. Deleted Microcomputer, a double-headed arrow to the Microcomputer Master data store

E-7. Run the data flow diagram analysis feature (select Diagram Analyze and select Current Diagram). Print the report for each of the data flow diagrams described in previous problems. Examine the diagrams and note the problems detected.

E-8. Run the syntax checking report (select Repository and Syntax Check) to produce the syntax check report for the diagrams. Examine and interpret the information provided.

E-9. Create an Undescribed Repository Entities report for the diagrams produced in previous problems. Select Repository and Reports and make the selections illustrated in Figure E9.16. Print the report and make a note of the corrections that need to be made for the design material to be complete.

E-10. Create a matrix showing all the data stores for all diagrams. Select Repository and Reports and make the selections illustrated in Figure E9.17. Preview the report using your Internet browser and examine the output. Write a short paragraph stating what the report is illustrating and how it would be useful to the analyst.

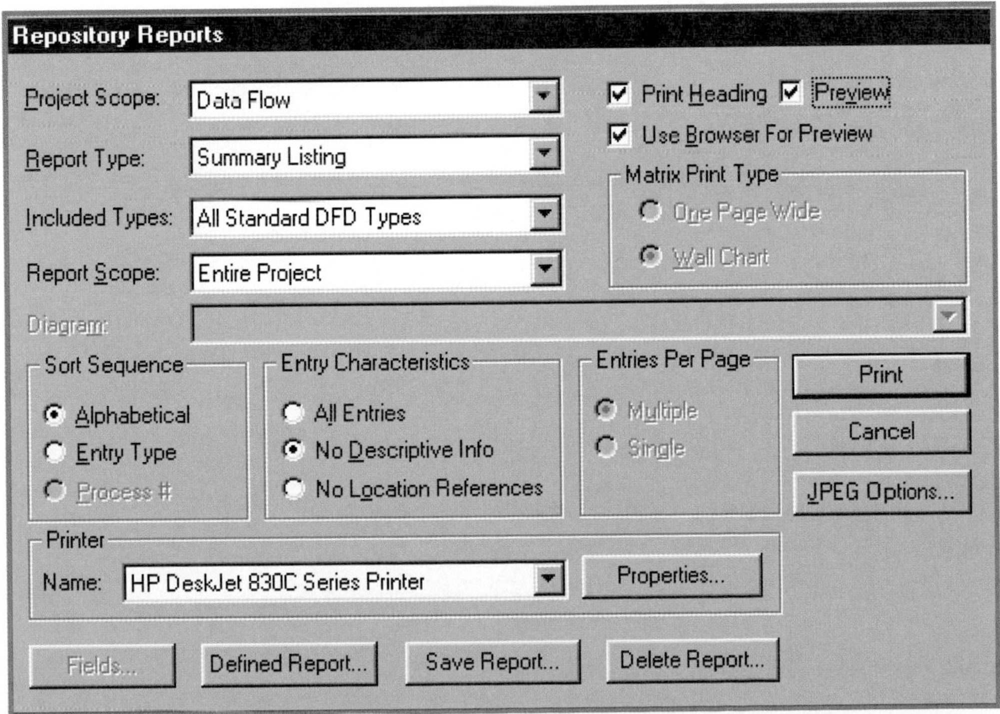

FIGURE E9.16

Repository report inquiry screen used to create a summary listing report.

9

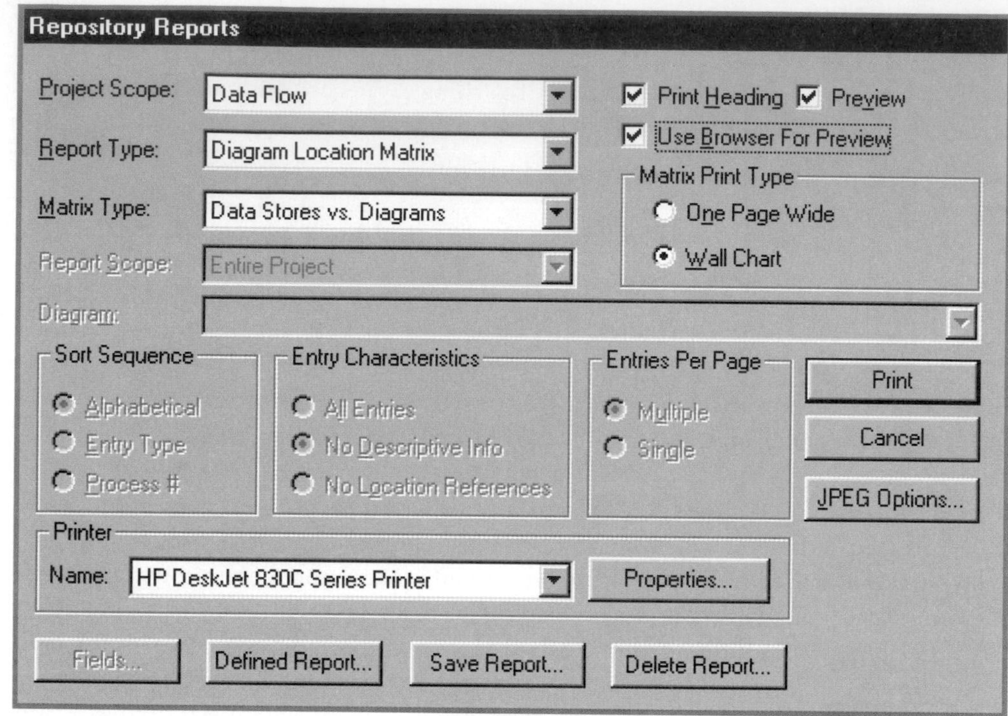

FIGURE E9.17

Repository report inquiry screen used to create a diagram location matrix report.

*Exercises preceded by a CD-ROM icon require the program Visible Analyst or another CASE tool. A CD-ROM containing Visible Analyst examples is provided free of charge to any professor adopting this book. The examples on the disk may be imported into Visible Analyst and then used by students.

ANALYZING SYSTEMS USING DATA DICTIONARIES

10

After successive levels of data flow diagrams are complete, systems analysts use them to help catalog the data processes, flows, stores, structures, and elements in a data dictionary. Of particular importance are the names used to characterize data items. When given an opportunity to name components of data oriented systems, the systems analyst needs to work at making the name meaningful yet exclusive of other existing data component names. This chapter covers the data dictionary, which is another method to aid in the analysis of data-oriented systems.

THE DATA DICTIONARY

The data dictionary is a specialized application of the kinds of dictionaries used as references in everyday life. The data dictionary is a reference work of data about data (that is, *metadata*), one that is compiled by systems analysts to guide them through analysis and design. As a document, the data dictionary collects and coordinates specific data terms, and it confirms what each term means to different people in the organization. The data flow diagrams covered in Chapter 9 are an excellent starting point for collecting data dictionary entries.

Systems analysts must be aware of and catalog different terms that refer to the same data item. This awareness helps them avoid duplication of effort, allows better communication between organizational departments sharing a database, and makes maintenance more straightforward. The data dictionary can also serve as a consistent standard for data elements.

Automated data dictionaries (also part of the CASE tools mentioned earlier) are valuable for their capacity to cross-reference data items, thereby allowing necessary program changes to all programs sharing a common element. This feature supplants changing programs on a haphazard basis, or it prevents waiting until the program won't run because a change has not been implemented across all programs sharing the updated item. Clearly, automated data dictionaries become important for large systems that produce several thousand data elements requiring cataloging and cross-referencing.

NEED FOR UNDERSTANDING THE DATA DICTIONARY

Many database management systems now come equipped with an automated data dictionary. These dictionaries can be either elaborate or simple. Some computerized data dictionaries automatically catalog data items when programming is done;

others simply provide a template to prompt the person filling in the dictionary to do so in a uniform manner for every entry.

Despite the existence of automated data dictionaries, understanding what data compose a data dictionary, the conventions used in data dictionaries, and how a data dictionary is developed are issues that remain pertinent for the systems analyst during the systems effort. Understanding the process of compiling a data dictionary can aid the systems analyst in conceptualizing the system and how it works. The upcoming sections allow the systems analyst to see the rationale behind what exists in automated as well as manual data dictionaries.

In addition to providing documentation and eliminating redundancy, the data dictionary may be used to:

1. Validate the data flow diagram for completeness and accuracy.
2. Provide a starting point for developing screens and reports.
3. Determine the contents of data stored in files.
4. Develop the logic for data flow diagram processes.

THE DATA REPOSITORY

Although the data dictionary contains information about data and procedures, a larger collection of project information is called a repository. The repository concept is one of the many impacts of CASE tools and may contain the following:

1. Information about the data maintained by the system, including data flow, data stores, record structures, and elements.
2. Procedural logic.
3. Screen and report design.
4. Data relationships, such as how one data structure is linked to another.
5. Project requirements and final system deliverables.
6. Project management information, such as delivery schedules, achievements, issues that need resolving, and project users.

The data dictionary is created by examining and describing the contents of the data flow, data stores, and processes, as illustrated by Figure 10.1. Each data store and data flow should be defined and then expanded to include the details of the elements it contains. The logic of each process should be described using the data

FIGURE 10.1
How data dictionaries relate to data flow diagrams.

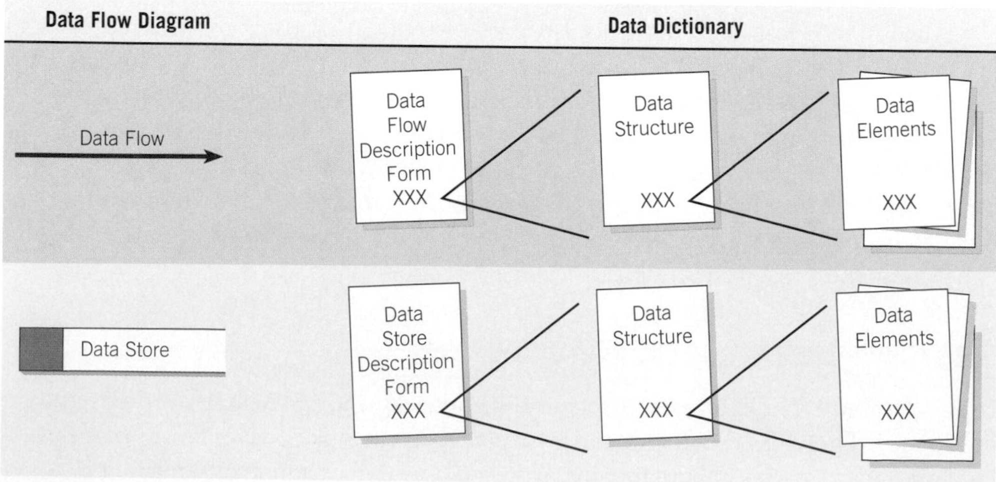

World's Trend
1000 International Lane
Cornwall, CT 06050

Customer Order

Please print clearly. See reverse side for item size codes. If the payment is made using a bank credit card, please include the credit card number and expiration date. Use charts on the reverse side for size codes and to determine postage. Connecticut residents must include sales tax.

Name (First, Middle, Last): Gilbert Sullivan
Street: 115 Buttercup Lane
Apartment:
City: Penzance
State: PA
Zip: 17057
Country:
Customer number (if known): 09288
Catalog no.: 9401A
Order Date (MM/DD/YYYY): 03/12/2000
Telephone (Incl. Area Code): (215) 747-2837

Quantity	Item Number	Item Description	Size	Color	Price	Item Total
1	12343	Jogging Suit	M	BL	35.50	35.50
4	54224	Cushion impact socks/pair	M	WH	4.25	17.00
1	10617	Running shorts	M	BL	12.25	12.25
1	10617	Running shorts	M	GR	12.25	12.25

Method of Payment
☐ Check ☑ Charge ☐ Money Order

Fill in for credit card purchase only
☑ World's Trend ☐ AmExpress ☐ Discover ☐ MC ☐ Visa

Credit Card Number - Not required for World's Trend Charges   Expiration Date-MM/YYYY

Merchandise Total	77.00
Tax (CT Only)	
Shipping and Handling	9.80
Order Total	86.80

Form Number 0001 03/2000

FIGURE 10.2
An order form from World's Trend Catalog Division.

flowing into or out of the process. Omissions and other design errors should be noted and resolved.

The four data dictionary categories—data flows, data structures, data elements, and data stores—should be developed to promote understanding of the data of the system. Procedural logic is presented in Chapter 11.

To illustrate how data dictionary entries are created, we use an example for World's Trend Catalog Division. This company sells clothing and other items by mail order (or faxing the mail-order form), using a toll-free phone order system, and via the Internet using customized Web forms. Regardless of the origin of the order, the underlying data captured by the system are the same for all three methods.

The World's Trend order form shown in Figure 10.2 gives some clues about what to enter into a data dictionary. First, you need to capture and store the name, address, and telephone number of the person placing the order. Then you need to address the details of the order: the item description, size, color, price, quantity, and so on. The customer's method of payment must also be determined. Once you have done this, these data may be stored for future use. This example is used throughout this chapter to illustrate each part of the data dictionary.

FIGURE 10.3

An example of a data flow description from World's Trend Catalog Division.

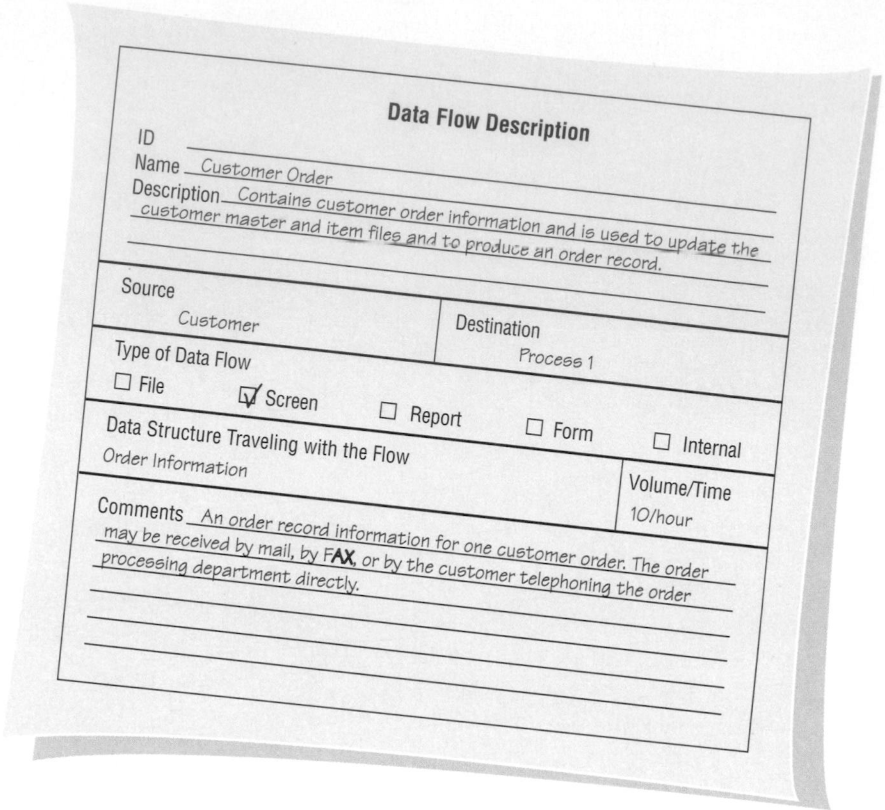

Data Flow Description

ID

Name Customer Order

Description Contains customer order information and is used to update the customer master and item files and to produce an order record.

Source: Customer

Destination: Process 1

Type of Data Flow

☐ File ☑ Screen ☐ Report ☐ Form ☐ Internal

Data Structure Traveling with the Flow: Order Information

Volume/Time: 10/hour

Comments An order record information for one customer order. The order may be received by mail, by FAX, or by the customer telephoning the order processing department directly.

DEFINING THE DATA FLOW

Data flow is usually the first component to be defined. System inputs and outputs are determined from interviewing, observing users, and analyzing documents and other existing systems. The information captured for each data flow may be summarized using a form containing the following information:

1. ID, an optional identification number. Sometimes the ID is coded using a scheme to identify the system and the application within the system.
2. A unique descriptive name for this data flow. This name is the text that should appear on the diagram and be referenced in all descriptions using the data flow.
3. A general description of the data flow.
4. The source of the data flow. The source could be an external entity, a process, or a data flow coming from a data store.
5. The destination of the data flow (same items listed under the source).
6. An indication of whether the data flow is a record entering or leaving a file or a record containing a report, form, or screen. If the data flow contains data that are used between processes, it is designated as *internal*.
7. The name of the data structure describing the elements found in this data flow. For a simple data flow, it could be one or several elements.
8. The volume per unit of time. The data could be records per day or any other unit of time.
9. An area for further comments and notations about the data flow.

Once again we can use our World's Trend Catalog Division example from Chapter 9 to illustrate a completed form. Figure 10.3 is an example of the data flow description representing the screen used to add a new CUSTOMER ORDER and to update the customer and item files. Notice that the external entity CUSTOMER is the input and that PROCESS 1 is the destination, providing linkage

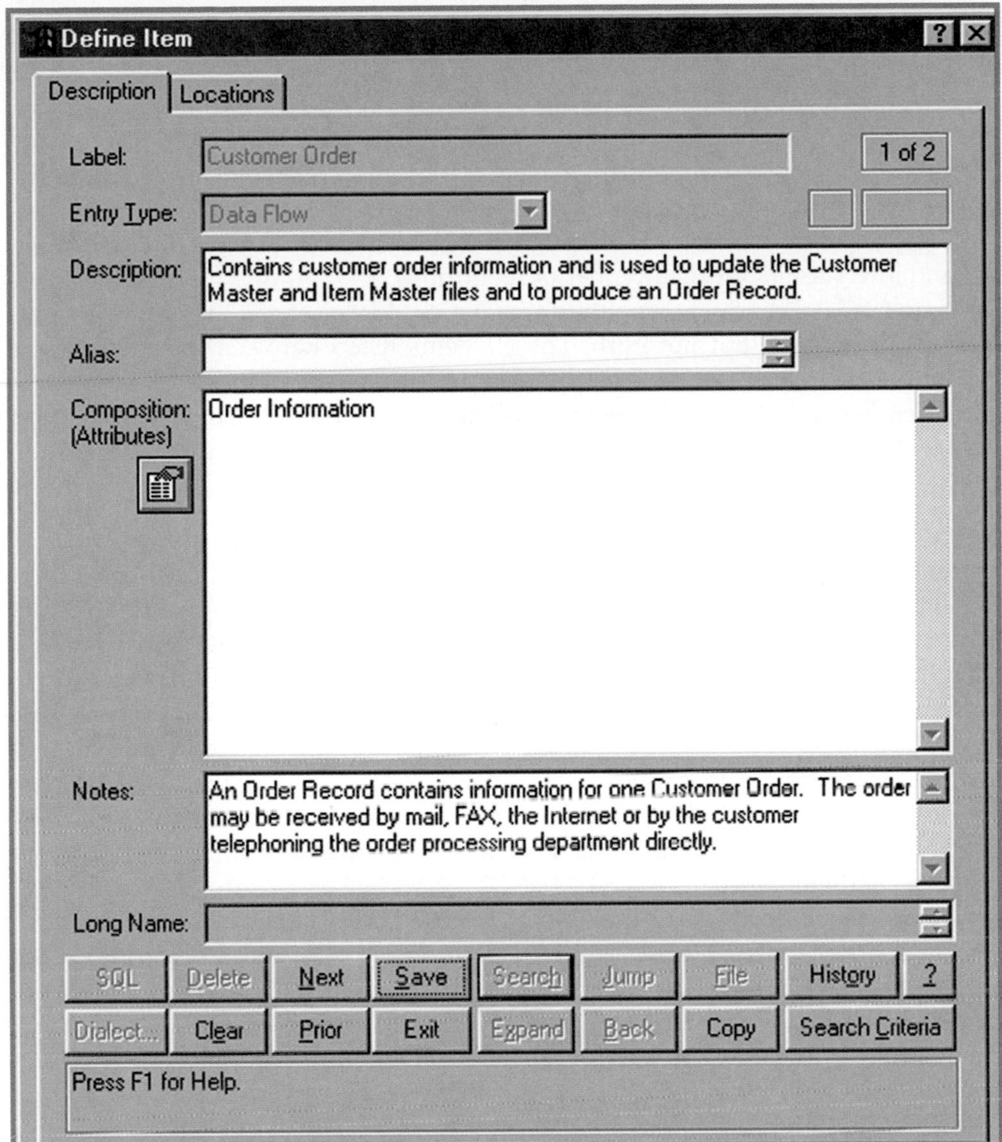

FIGURE 10.4
A Visible Analyst screen showing a data flow description.

back to the data flow diagram. The checked box for "Screen" indicates that the flow represents an input screen. It could be any screen, such as a mainframe, Graphical User Interface (GUI), or Web page. The detailed description of the data flow could appear on this form, or it could be represented as a data structure.

Data flow for all input and output should be described first, because it usually represents the human interface, followed by the intermediate data flow and the data flow to and from data stores. The detail of each data flow is described using elements, sometimes called fields, or a data structure, a group of elements.

A simple data flow may be described using a single element, such as a customer number used by an inquiry program to find the matching customer record. An example of an electronic form is shown in Figure 10.4. Visible Analyst was used to create the form.

DESCRIBING DATA STRUCTURES

Data structures are usually described using algebraic notation. This method allows the analyst to produce a view of the elements that make up the data structure along with information about those elements. For instance, the analyst will denote whether there are many of the same element within the data structure (a repeating group) or whether two elements may exist mutually exclusive of each other. The algebraic notation uses the following symbols:

1. An equal sign (=) means "is composed of."
2. A plus sign (+) means "and."
3. Braces { } indicate repetitive elements, also called repeating groups or tables. There may be one repeating element or several within the group. The repeating group may have conditions, such as a fixed number of repetitions or upper and lower limits for the number of repetitions.
4. Brackets [ ] represent an either/or situation. Either one element may be present or another, but not both. The elements listed between the brackets are mutually exclusive.
5. Parentheses ( ) represent an optional element. Optional elements may be left blank on entry screens and may contain spaces or zeros for numeric fields on file structures.

Figure 10.5 is an example of the data structure for adding a customer order at World's Trend Catalog Division. Each NEW CUSTOMER SCREEN consists of the entries found on the right side of the equal signs. Some of the entries are elements, but others, such as CUSTOMER NAME, ADDRESS, and TELEPHONE, are groups of elements or structural records. For example, CUSTOMER NAME is made up of FIRST NAME, MIDDLE INITIAL, and LAST NAME. Each structural record must be further defined until the entire set is broken down into its component elements. Notice that following the definition for the customer order screen are definitions for each structural record. Even a field as simple as the TELEPHONE NUMBER is defined as a structure so that the area code may be processed individually.

Structural records and elements that are used within many different systems are given a nonsystem specific name, such as street, city, and zip, that does not reflect the functional area within which they are used. This method allows the analyst to define these records once and use them in many different applications. For example, a city may be a customer city, supplier city, or employee city. Notice the use of parentheses to indicate that (MIDDLE INITIAL), (APARTMENT), and (ZIP EXPANSION) are optional ORDER information (but not more than one). Indicate the OR condition by enclosing the options in square brackets and separating them with the symbol ¦.

LOGICAL AND PHYSICAL DATA STRUCTURES

When data structures are first defined, only the data elements that the user would see, such as a name, address, and balance due, are included. This stage is the logical design, showing what data the business needs for its day-to-day operations. Using the logical design as a basis, the analyst then designs the physical data structures, which include additional elements necessary for implementing the system. Examples of physical design elements are the following:

1. Key fields used to locate records in a file. An example is an item number, which is not required for a business to function but is necessary for identifying and locating computer records.
2. Codes to identify the status of master records, such as whether an employee is active (currently employed) or inactive. Such codes can be maintained on files that produce tax information.
3. Transaction codes are used to identify types of records when a file contains different record types. An example is a credit file containing records for returned items as well as records of payments.

Customer Order = Customer Number +
Customer Name +
Address +
Telephone +
Catalog Number +
Order Date +
{Available Order Items} +
Merchandise Total +
(Tax) +
Shipping and Handling +
Order Total +
Method of Payment +
(Credit Card Type) +
(Credit Card Number) +
(Expiration Date)

Customer Name = First Name +
(Middle Initial) +
Last Name

Address = Street +
(Apartment) +
City +
State +
Zip +
(Zip Expansion) +
(Country)

Telephone = Area Code +
Local Number

Available Order Items = Quantity Ordered +
Item Number +
Item Description +
Size +
Color +
Price +
Item Total

Method of Payment = [Check ¦ Charge ¦ Money Order]

Credit Card Type = [World's Trend ¦ American Express ¦ MasterCard ¦ Visa]

FIGURE 10.5
Data structure example for adding a customer order at World's Trend Catalog Division.

4. Repeating group entries containing a count of how many items are in the group.
5. Limits on the number of items in a repeated group.
6. A password used by a customer accessing a secure Web site.

Figure 10.6 is an example of the data structure for a Customer Billing Statement, one showing that the Order Line is both a repeating item and a structural record. The Order Line limits are from 1 to 5, indicating that the customer may order from one to five items on this screen. Additional items would appear on subsequent orders.

The repeating group notation may have several other formats. If the group repeats a fixed number of times, that number is placed next to the opening brace, as in 12 {Monthly Sales}, where there are always 12 months in the year. If no num-

FIGURE 10.6
Physical elements added to a data structure.

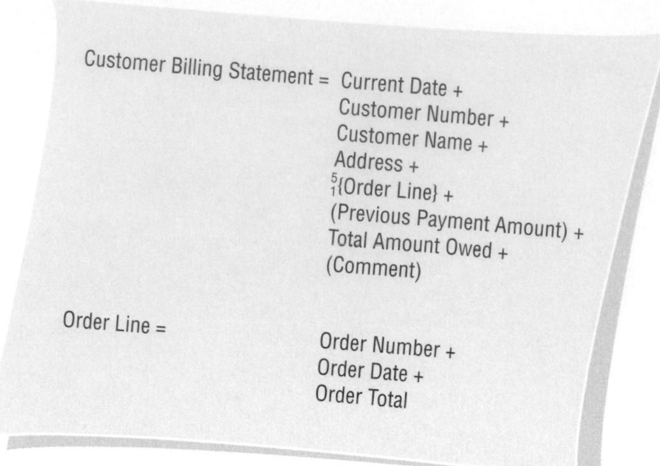

ber is indicated, the group repeats indefinitely. An example is a file containing an indefinite number of records, such as Customer Master File = {Customer Records}.

Customer Billing Statement = Current Date +
Customer Number +
Customer Name +
Address +
$^{5}_{1}${Order Line} +
(Previous Payment Amount) +
Total Amount Owed +
(Comment)
Order Line =
Order Number +
Order Date +
Order Total

The number of entries in repeating groups may also depend on a condition, such as an entry on the Customer Master Record for each item ordered. This condition could be stored in the data dictionary as {Items Purchased} 5, where 5 is the number of items.

DATA ELEMENTS

Each data element should be defined once in the data dictionary and may also be entered previously on an element description form, such as the one illustrated in Figure 10.7. Characteristics commonly included on the element description form are the following:

1. Element ID. This optional entry allows the analyst to build automated data dictionary entries.
2. The name of the element. The name should be descriptive, unique, and based on what the element is commonly called in most programs or by the major user of the element.
3. Aliases, which are synonyms or other names for the element. Aliases are names used by different users within different systems. For example, a CUSTOMER NUMBER may also be called a RECEIVABLE ACCOUNT NUMBER or a CLIENT NUMBER.
4. A short description of the element.
5. Whether the element is base or derived. A base element is one that is initially keyed into the system, such as a customer name, address, or city. Base elements must be stored on files. Derived elements are created by processes as the result of calculations or logic. An example is the total amount that a customer owes or an employee's gross pay. Analysis of base and derived elements differ, and this difference provides a means of determining areas of the system that may need further work.
6. The length of an element. This value should be the *stored* length of the item. The on-screen and printed lengths of the item may differ from this value, but the programs responsible for displaying the item on the screen or printing it in a report will insert any additional formatting characters required. An important consideration is how long to make an element. Some elements have standard lengths. In the United States, for example, lengths for state name

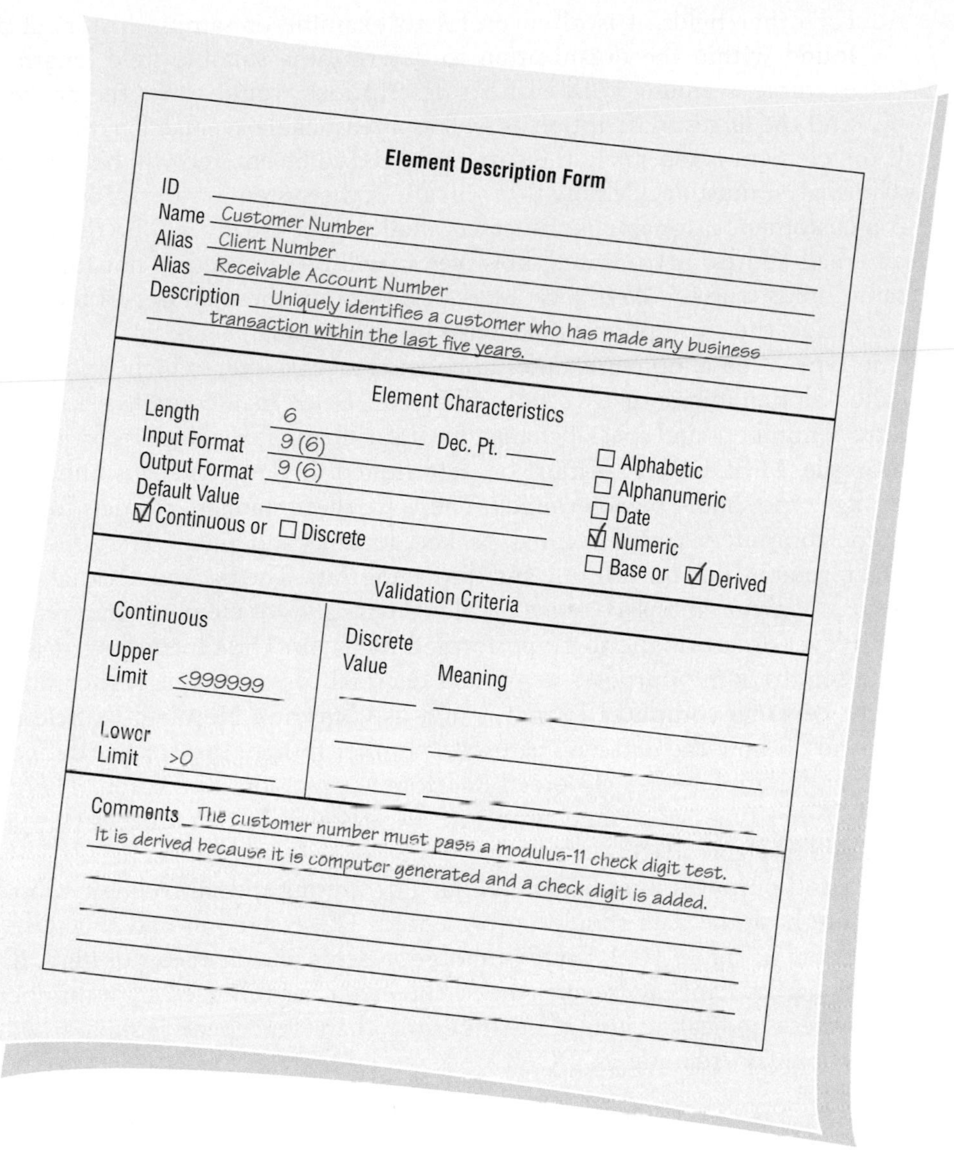

Element Description Form

ID
Name Customer Number
Alias Client Number
Alias Receivable Account Number
Description Uniquely identifies a customer who has made any business transaction within the last five years.

Element Characteristics

Length 6
Input Format 9 (6) Dec. Pt.
Output Format 9 (6)
Default Value
☑ Continuous or ☐ Discrete

☐ Alphabetic
☐ Alphanumeric
☐ Date
☑ Numeric
☐ Base or ☑ Derived

Validation Criteria

Continuous
Upper Limit <999999
Lower Limit >0

Discrete
Value Meaning

Comments The customer number must pass a modulus-11 check digit test. It is derived because it is computer generated and a check digit is added.

FIGURE 10.7
An element description form example from World's Trend Catalog Division.

abbreviations, zip codes, and telephone numbers are all standard. For other elements, the lengths may vary, and the analyst and user community must jointly decide the final length based on the following considerations:

a. Numeric amount lengths should be determined by figuring the largest number the amount will probably contain and then allowing reasonable room for expansion. Lengths designated for totals should be large enough to accommodate the sum of the numbers accumulated in them.
b. Name and address fields may be given lengths based on the following table. For example, a last name field of 11 characters will accommodate 98 percent of the last names in the United States.

Field	*Length*	*Percent of Data That Will Fit (U.S.)*
Last Name	11	98
First Name	18	95
Company Name	20	95
Street	18	90
City	17	99

c. For other fields, it is often useful to examine or sample historical data found within the organization to determine a suitable field length. For example, scanning a list of item descriptions would allow the analyst to find the largest description as well as a reasonable average length.

If the element is too small, the data that need to be entered will be truncated. The analyst must decide how that will affect the system outputs. For example, if a customer's last name is truncated, mail would usually still be delivered; if an email address is truncated, however, it will be returned as not found. The same holds true for Web page addresses. Because these addresses tend to be very large, the corresponding elements must have a large size.

7. The type of data: numeric, date, alphabetic, or character, which is sometimes called alphanumeric or text data. Character fields may contain a mixture of letters, numbers, and special characters. If the element is a date, its format—for example, MMDDYYYY—must be determined. If the element is numeric, its storage type should be determined. There are three standard formats for mainframe computers: zoned decimal, packed decimal, and binary. The zoned decimal format is used for printing and displaying data. The packed decimal format is commonly used to save space on file layouts and for elements that require a high level of arithmetic to be performed on them. The binary format is suitable for the same purposes as packed decimal format but is less commonly used. Personal computer formats, such as Currency, Number, or Scientific, depend on how the data will be used. Number formats are further defined as integer, long integer, single precision, double precision, and so on. There are many other types of formats used with PC systems. Several of these are summarized in Figure 10.8.
8. Input and output formats should be included, using special coding symbols to indicate how the data should be presented. These symbols and their use are illustrated in Figure 10.9. Each symbol represents one character or digit. If the same character repeats several times, the character followed by a number in parentheses indicating how many times the character repeats is substituted for the group. For example, XXXXXXXX would represent as X(8).

FIGURE 10.8
Some examples of data formats used in PC systems.

Data Type	Meaning
Bit	A value of 1 or 0, a true/false value
Char, varchar, text	Any alphanumeric character
Datetime, smalldatetime	Alphanumeric data, several formats
Decimal, numeric	Numeric data that are accurate to the least significant digit; can contain a whole and decimal portion
Float, real	Floating-point values that contain an approximate decimal value
Int, smallint, tinyint	Only integer (whole digit) data
Currency, money, smallmoney	Monetary numbers accurate to four decimal places
Binary, varbinary, image	Binary strings (sound, pictures, video)
Cursor, timestamp, uniqueidentifier	A value that is always unique within a database
Autonumber	A number that is always incremented by one when a record is added to a database table

Formatting Character	Meaning
X	May enter or display / print any character
9	Enter or display only numbers
Z	Display leading zeros as spaces
,	Insert commas into a numeric display
.	Insert a period into a numeric display
/	Insert slashes into a numeric display
-	Insert a hyphen into a numeric display
V	Indicate a decimal position (when the decimal point is not included)

FIGURE 10.9
Format character codes.

9. Validation criteria for ensuring that accurate data are captured by the system. Elements are either discrete, meaning they have certain fixed values, or continuous, with a smooth range of values. Here are common editing criteria:
   a. A range of values is suitable for elements that contain continuous data. For example, in the United States a student grade point average may be from 0.00 through 4.00. If there is only an upper or lower bound to the data, a limit is used instead of a range.
   b. A list of values is indicated if the data are discrete. Examples are codes representing the colors of items for sale in World's Trends' catalog.
   c. A table of codes is suitable if the list of values is extensive (for example, state abbreviations, telephone country codes, or U.S. telephone area codes).
   d. For key or index elements, a check digit is often included.
10. Any default value the element may have. The default value is displayed on entry screens and is used to reduce the amount of keying that the operator may have to do. Usually, several fields within each system have default values. When using GUI lists or drop-down lists, the default value is the one currently selected and highlighted. When using radio buttons, the option for the default value is selected and when using check boxes, the default value (either "yes" or "no") determines whether or not the check box will have an initial check in it.
11. An additional comment or remarks area. This might be used to indicate the format of the date, special validation that is required, the check digit method used (explained in Chapter 19), and so on.

An example of a Visible Analyst data element description form can be found in Figure 10.10. As shown on the form, the CUSTOMER NUMBER may be called CLIENT NUMBER elsewhere in the system (perhaps old code written with this alias needs to be updated). The form is also useful because we can tell from it that the element is a numeric variable with a length of six characters. This variable can be as large as 999999 but cannot be less than zero.

Another kind of data element is an alphabetic element. In the case shown in Figure 10.11, the element is a discrete variable assigned certain codes. At World's Trend Catalog Division, codes are used to describe colors: BL for blue, WH for white, and GR for green. When this element is implemented, a table will be needed for users to look up the meanings of these codes. (Coding is discussed further in Chapter 19.)

DATA STORES

All base elements must be stored within the system. Derived elements, such as the employee year-to-date gross pay, may also be stored in the system. Data stores are created for each different data entity being stored. That is, when data flow base

FIGURE 10.10

Visible Analyst screens showing an element description. Two pages are required to define an element.

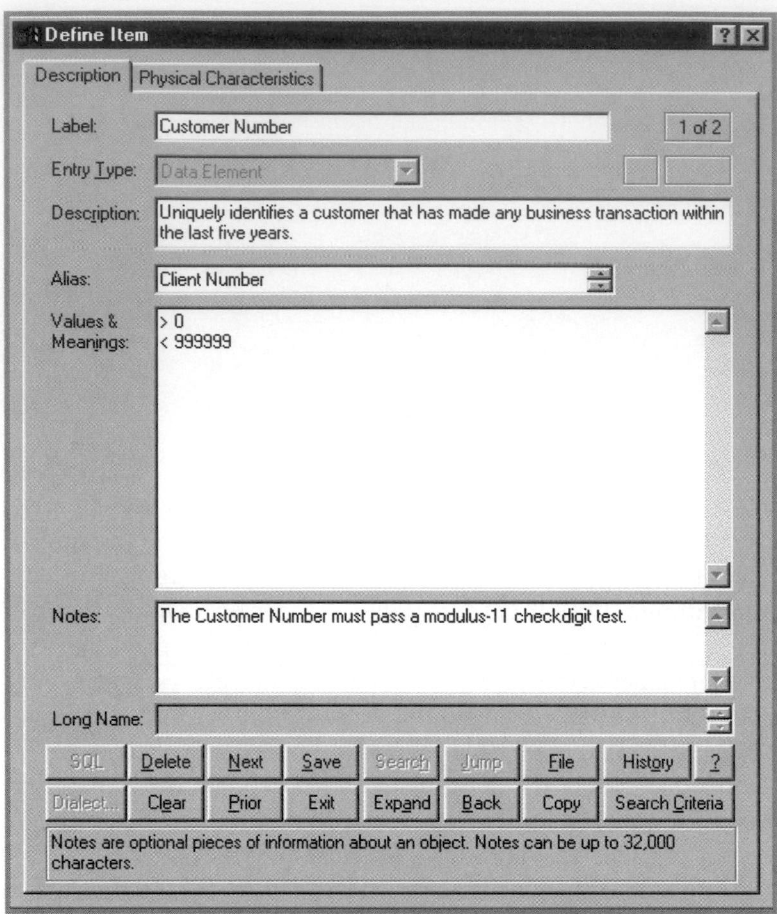

Define Item
Description
Physical Characteristics
Label:
Customer Number
1 of 2
Entry Type:
Data Element
Description:
Uniquely identifies a customer that has made any business transaction within the last five years.
Alias:
Client Number
Values & Meanings:
> 0
< 999999
Notes:
The Customer Number must pass a modulus-11 checkdigit test.
Long Name:
SQL
Delete
Next
Save
Search
Jump
File
History
?
Dialect...
Clear
Prior
Exit
Expand
Back
Copy
Search Criteria
Notes are optional pieces of information about an object. Notes can be up to 32,000 characters.

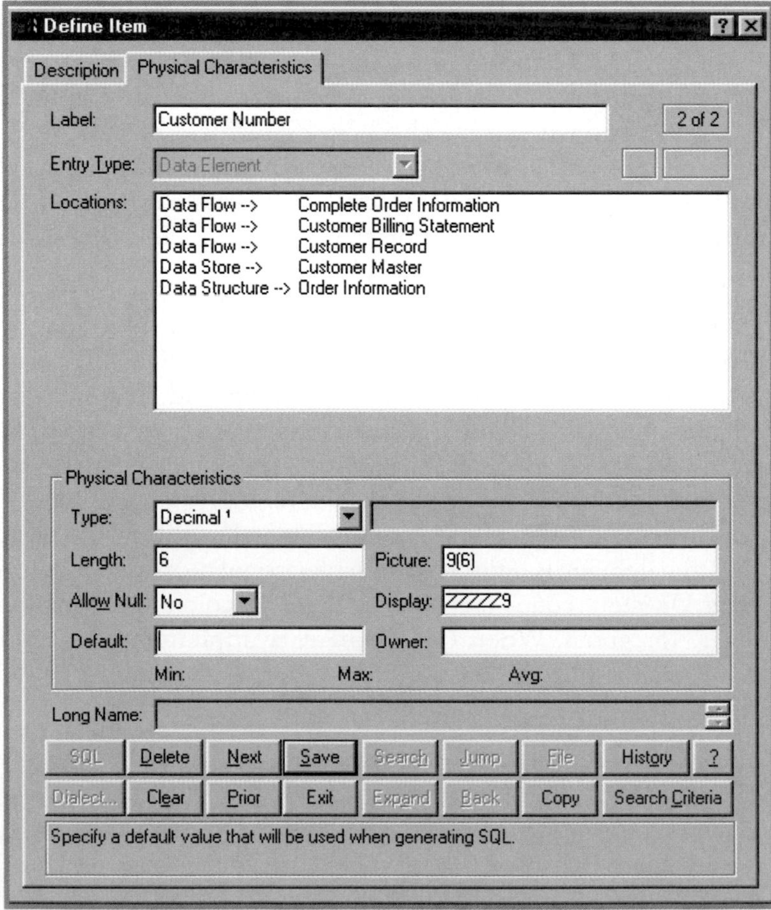

Define Item
Description
Physical Characteristics
Label:
Customer Number
2 of 2
Entry Type:
Data Element
Locations:
Data Flow --> Complete Order Information
Data Flow --> Customer Billing Statement
Data Flow --> Customer Record
Data Store --> Customer Master
Data Structure --> Order Information
Physical Characteristics
Type:
Decimal
Length:
6
Picture:
9(6)
Allow Null:
No
Display:
ZZZZZ9
Default:
Owner:
Min:
Max:
Avg:
Long Name:
SQL
Delete
Next
Save
Search
Jump
File
History
?
Dialect...
Clear
Prior
Exit
Expand
Back
Copy
Search Criteria
Specify a default value that will be used when generating SQL.

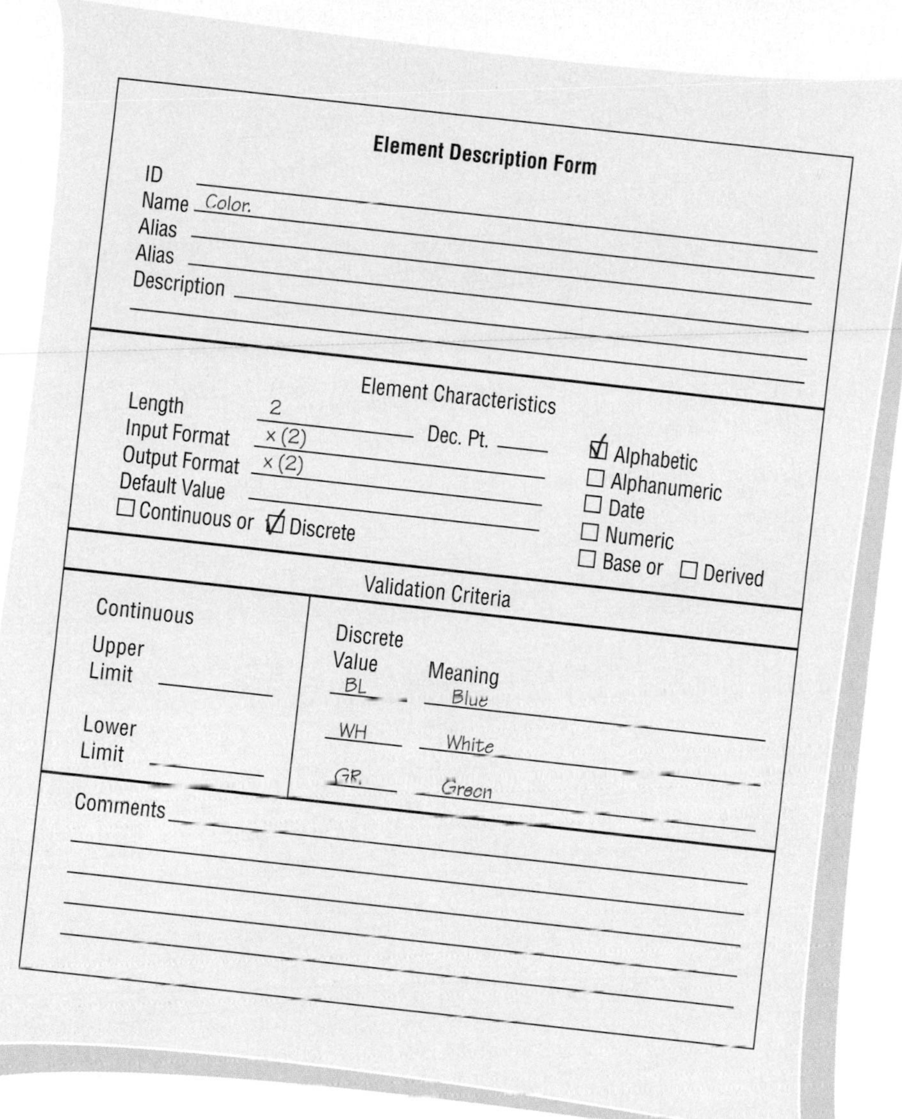
Element Description Form

ID
Name Color.
Alias
Alias
Description

Element Characteristics

Length 2
Input Format x (2) Dec. Pt.
Output Format x (2)
Default Value
☐ Continuous or ☑ Discrete

☑ Alphabetic
☐ Alphanumeric
☐ Date
☐ Numeric
☐ Base or ☐ Derived

Validation Criteria

Continuous
Upper Limit
Lower Limit

Discrete Value	Meaning
BL	Blue
WH	White
GR.	Green

Comments

FIGURE 10.11

An example of an alphabetic element description form from World's Trend Catalog Division.

elements are grouped together to form a structural record, a data store is created for each unique structural record.

Because a given data flow may only show part of the collective data that a structural record contains, you may have to examine many different data flow structures to arrive at a complete data store description. For example, when adding a customer, you may initially include only information known when the record is first created. Running balances, transaction dates, and other information added to the CUSTOMER data store only after business has progressed would be on different data flows.

Figure 10.12 is a typical form used to describe a data store. The information included on the form is as follows:

1. The Data Store ID. The ID is often a mandatory entry to prevent the analyst from storing redundant information. An example would be D1 for the CUSTOMER MASTER FILE.
2. The Data Store Name, which is descriptive and unique.
3. An Alias for the file, such as CLIENT MASTER FILE for the CUSTOMER MASTER FILE.

FIGURE 10.12

A data store form example for World's Trend Catalog Division.

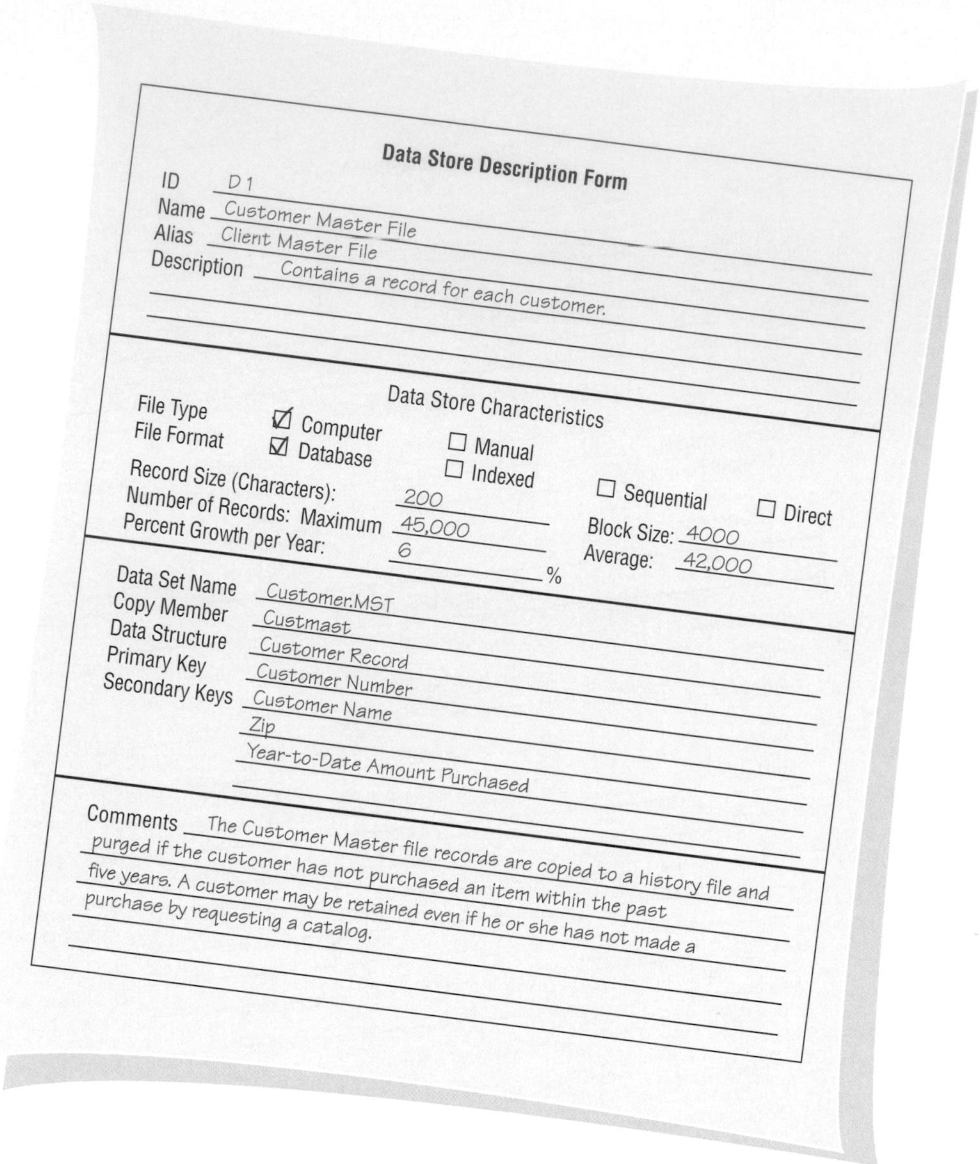

Data Store Description Form

ID D1
Name Customer Master File
Alias Client Master File
Description Contains a record for each customer.

Data Store Characteristics

File Type ☑ Computer ☐ Manual
File Format ☑ Database ☐ Indexed ☐ Sequential ☐ Direct
Record Size (Characters): 200 Block Size: 4000
Number of Records: Maximum 45,000 Average: 42,000
Percent Growth per Year: 6 %

Data Set Name Customer.MST
Copy Member Custmast
Data Structure Customer Record
Primary Key Customer Number
Secondary Keys Customer Name
Zip
Year-to-Date Amount Purchased

Comments The Customer Master file records are copied to a history file and purged if the customer has not purchased an item within the past five years. A customer may be retained even if he or she has not made a purchase by requesting a catalog.

4. A short description of the data store.
5. The file type, either manual or computerized.
6. If the file is computerized, the file format designates whether the file is a database file called a table or if it has the format of a traditional flat file. (File formats are detailed in Chapter 17.)
7. The maximum and average number of records on the file as well as the growth per year. This information helps the analyst to predict the amount of disk space required for the application and is necessary for hardware acquisition planning.
8. The data set name specifies the file name, if known. In the initial design stages, this item may be left blank. An electronic form produced using Visible Analyst is shown in Figure 10.13. This example shows that the CUSTOMER MASTER file is stored on a computer in the form of a database with a maximum number of 45,000 records. (Records and the keys used to sort the database are explained in Chapter 17.)
9. The data structure should use a name found in the data dictionary, providing a link to the elements for this data store. Alternatively, the data elements could be described on the data store description form or on the CASE tool screen for the data store. Primary and secondary keys must be elements (or a combina-

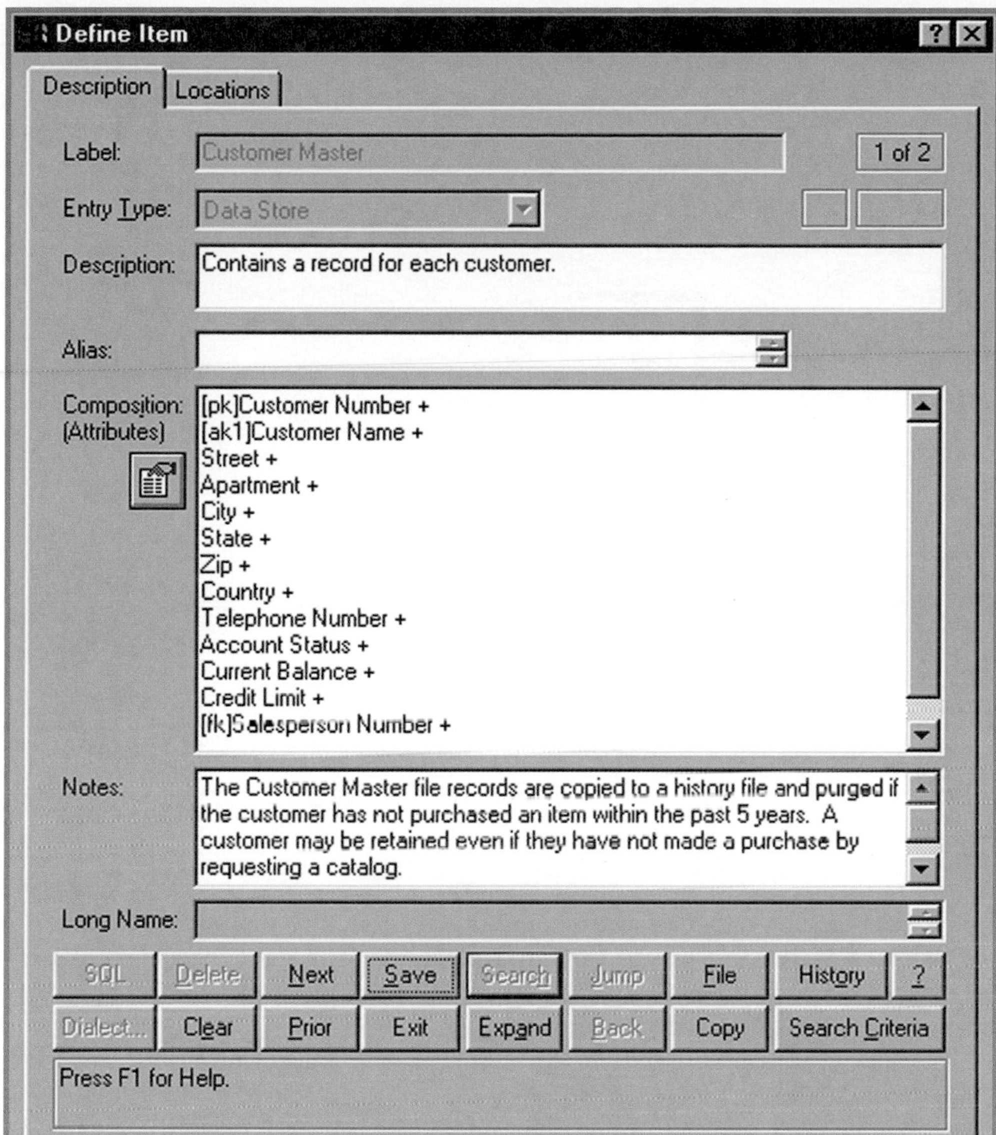

FIGURE 10.13
Visible Analyst screen showing a data store description.

tion of elements) found within the data structure. In the example the CUSTOMER NUMBER is the primary key and should be unique. The CUSTOMER NAME, ZIP, and YEAR-TO-DATE AMOUNT PURCHASED are secondary keys used to control record sequencing on reports and to locate records directly. (Keys are discussed in Chapter 17.) Comments are used for information that does not fit into any of the above categories. They may include update or backup timing, security, or other considerations.

CREATING THE DATA DICTIONARY

Data dictionary entries may be created after the data flow diagram has been completed, or they may be constructed as the data flow diagram is being developed. The use of algebraic notation and structural records allows the analyst to develop the data dictionary and the data flow diagrams using a top-down approach. For instance, the analyst may create a data flow Diagram 0 after the first few interviews and, at the same time, make the preliminary data dictionary entries. Typically, these entries consist of the data flow names found on the data flow diagram and their corresponding data structures. After several additional interviews

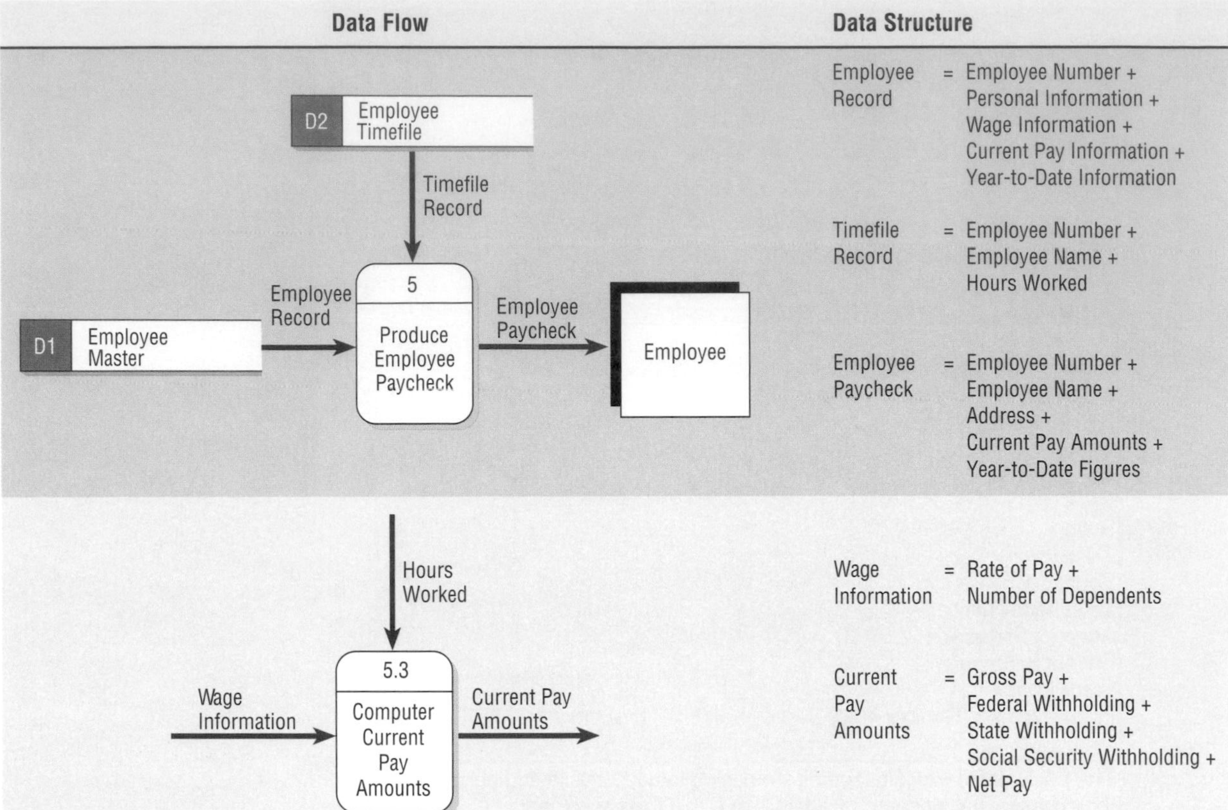

FIGURE 10.14
Two data flow diagrams and corresponding data dictionary entries for producing an employee paycheck.

have been conducted to learn the details of the system, the analyst will expand the data flow diagram and create the child diagrams. The data dictionary is then modified to include the new structural records and elements gleaned from further interviews, observation, and document analysis.

Each level of a data flow diagram should use data appropriate for the level. Diagram 0 should include only forms, screens, reports, and records. As child diagrams are created, the data flow into and out of the processes becomes more and more detailed, including structural records and elements. Thus, each data flow diagram has data appropriate for the level of detail it is depicting. The data dictionary is needed because you would not want to show records and screens on a detailed child data flow diagram nor scores of elements on a high-level data flow diagram.

Figure 10.14 illustrates a portion of two data flow diagram levels and corresponding data dictionary entries for producing an employee paycheck. Process 5, found on Diagram 0, is an overview of the production of an EMPLOYEE PAYCHECK. The corresponding data dictionary entry for EMPLOYEE RECORD shows the EMPLOYEE NUMBER and four structural records, the view of the data obtained early in the analysis. Similarly, TIMEFILE RECORD and the EMPLOYEE PAYCHECK are also defined as series of structures.

If prototypes or sample documents are available, the complete data dictionary and set of data flow diagrams may be developed. If you have not previously obtained complete information, use the preliminary documentation to formulate questions such as "How is the payroll produced?" and "What fields are found on the paycheck?" for a second series of interviews with systems users. After each interview, the emerging system is documented by creating child diagrams and completing the corresponding data structures (process 5.3, for example). This procedure is repeated until all data structures are completely defined.

As mentioned in Chapter 9, data flow diagrams must be balanced vertically between the parent process and the child diagram, yet each diagram should use meaningful names for its level of data flow. Balancing data flow diagram levels is accomplished by using the data dictionary structures. Names do not have to match between the parent process and a corresponding child diagram. What is important is that the data flow names on the child data flow diagram are contained as elements or structural records within the data flow on the parent process. Returning to the example, WAGE INFORMATION (input into process 5.3, COMPUTE CURRENT PAY AMOUNTS) is a structural record contained within the EMPLOYEE RECORD (input to process 5). Similarly, GROSS PAY (output from process 5.3.4 a lower level process not shown in the figure) is contained within the structural record CURRENT PAY AMOUNTS (output from the parent process 5.3, COMPUTE CURRENT PAY AMOUNTS).

ANALYZING INPUT AND OUTPUT

An important step in creating the data dictionary is to identify and categorize system input and output data flow. Input and output analysis forms, such as the one example shown in Figure 10.15, may be used to organize the information obtained

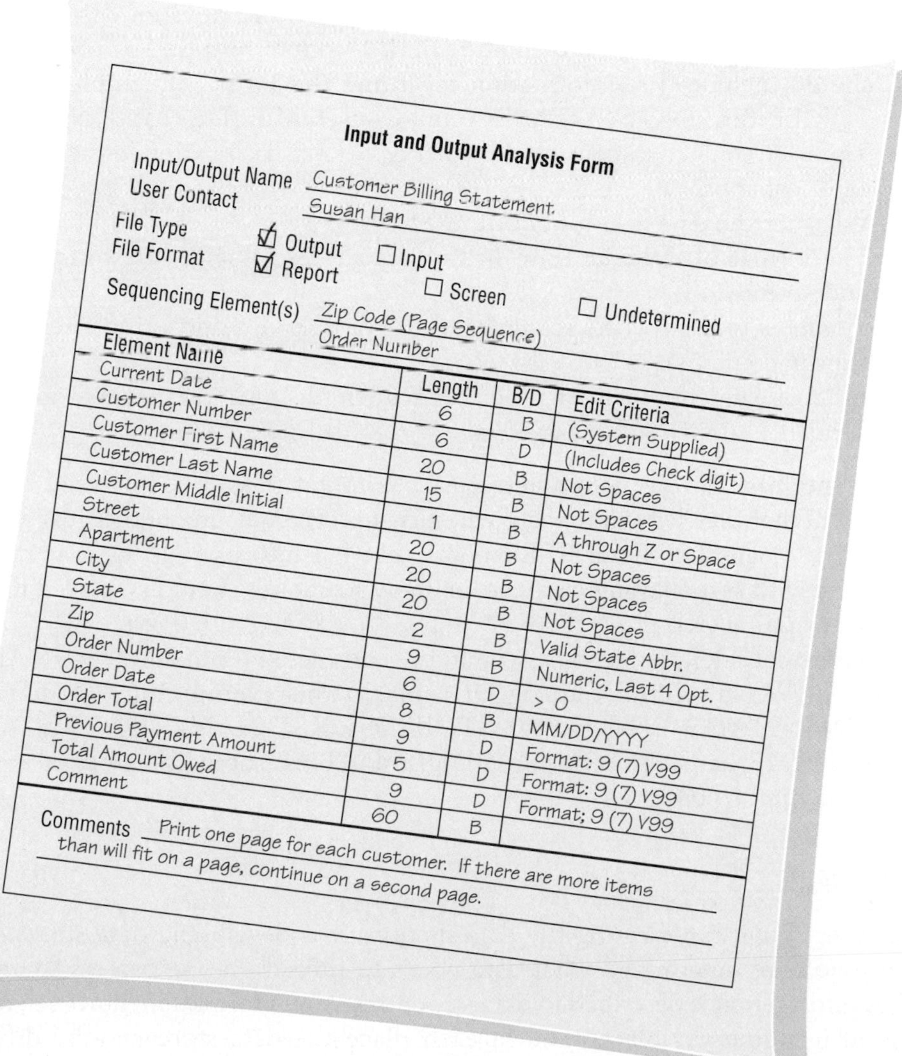

Input and Output Analysis Form

Input/Output Name: *Customer Billing Statement*
User Contact: *Susan Han*
File Type: ☑ Output ☐ Input
File Format: ☑ Report ☐ Screen ☐ Undetermined
Sequencing Element(s): *Zip Code (Page Sequence)*, *Order Number*

Element Name	Length	B/D	Edit Criteria
Current Date	6	B	(System Supplied)
Customer Number	6	D	(Includes Check digit)
Customer First Name	20	B	Not Spaces
Customer Last Name	15	B	Not Spaces
Customer Middle Initial	1	B	A through Z or Space
Street	20	B	Not Spaces
Apartment	20	B	Not Spaces
City	20	B	Not Spaces
State	2	B	Valid State Abbr.
Zip	9	B	Numeric, Last 4 Opt.
Order Number	6	D	> 0
Order Date	8	B	MM/DD/YYYY
Order Total	9	D	Format: 9 (7) V99
Previous Payment Amount	5	D	Format: 9 (7) V99
Total Amount Owed	9	D	Format: 9 (7) V99
Comment	60	B	

Comments: *Print one page for each customer. If there are more items than will fit on a page, continue on a second page.*

FIGURE 10.15
An input/output analysis form example for World's Trend Catalog Division.

CONSULTING OPPORTUNITY 10.1

WANT TO MAKE IT BIG IN THE THEATRE? IMPROVE YOUR DICTION(ARY)!

As you enter the door of Merman's, Annie Oaklea greets you warmly, saying, "I'm delighted with the work you have done on the data flow diagrams. I would like you to keep playing the role of systems analyst for Merman's and see if you can eventually get a new information system for our costume inventory sewn up. Unfortunately, some of the terms you're using don't come off very well in the language of Shakespeare. Bit of a translation problem, I suspect."

Clinging to Annie's initial praise, you are undaunted by her exit line. You determine that a data dictionary based on the rental and return data flow diagrams would make a big hit.

Begin by writing entries for a manual system in as much detail as possible. Prepare two data process entries, two data flow entries, two data store entries, one data structure entry, and four data element entries using the formats in this chapter. Portraying interrelated data items with preciseness will result in rave reviews. (Refer to Consulting Opportunity 9.2.)

from interviews and document analysis. Notice that this form contains the following commonly included fields:

1. A descriptive name for the input or output. If the data flow is on a logical diagram, the name should identify what the data are (for example, CUSTOMER INFORMATION). If the analyst is working on the physical design or if the user has explicitly stated the nature of the input or output, however, the name should include that information regarding the format. Examples are CUSTOMER BILLING STATEMENT and CUSTOMER DETAILS INQUIRY.
2. The user contact responsible for further details clarification, design feedback, and final approval.
3. Whether the data is input or output.
4. The format of the data flow. In the logical design stage, the format may be undetermined.
5. Elements indicating the sequence of the data on a report or on a screen (perhaps in columns).
6. A list of elements, including their names, lengths, and whether they are base or derived, and their editing criteria.

Once the form has been completed, each element should be analyzed to determine whether the element repeats, whether it is optional, or whether it is mutually exclusive of another element. Elements that fall into a group or that regularly combine with several other elements in many structures should be placed together into a structural record.

These considerations can be seen in the completed Input and Output Analysis Form for World's Trend Catalog Division. In this example of a CUSTOMER BILLING STATEMENT, the CUSTOMER FIRST NAME, the CUSTOMER LAST NAME, and CUSTOMER MIDDLE INITIAL should be grouped together in a structural record.

DEVELOPING DATA STORES

Another activity in creating the data dictionary is developing data stores. Up to now, we have determined what data needs to flow from one process to another. This information is described in data structures. The information, however, may be stored in numerous places, and in each place the data store may be different. Whereas data flows represent data in motion, data stores represent data at rest.

For example, when an order arrives at World's Trend, it contains information of a temporary nature, that is, the information needed to fill that particular order. Meanwhile, some of the information on the order form might be stored permanently. Examples of the latter include information about customers (so catalogs can be sent to them) and information about items (because these items will appear on many other customer's orders). Figure 10.16 shows that from a series of customer orders, information can be captured and stored into two data stores called CUSTOMER MASTER FILE and the ITEM MASTER FILE. An example of these data stores can be found in Figure 10.17.

Data stores contain information of a permanent or semipermanent nature. An ITEM NUMBER, DESCRIPTION, and ITEM COST are examples of information that is relatively permanent. So is the TAX RATE. When the ITEM COST is multiplied by the TAX RATE, however, the TAX CHARGED is calculated (or derived). Derived values do not have to be stored in a data store. When developing data stores, it is acceptable to start with some information and then add more information to the data store when you analyze more data flows and realize more information needs to be added.

When data stores are created for only one report or screen, we refer to them as "user views," because they represent the way that the user wants to see the information. A systems designer has to determine whether to set up individual files representing user views or set up a database representing many user views instead. The trade-offs associated with these two approaches are discussed in Chapter 17.

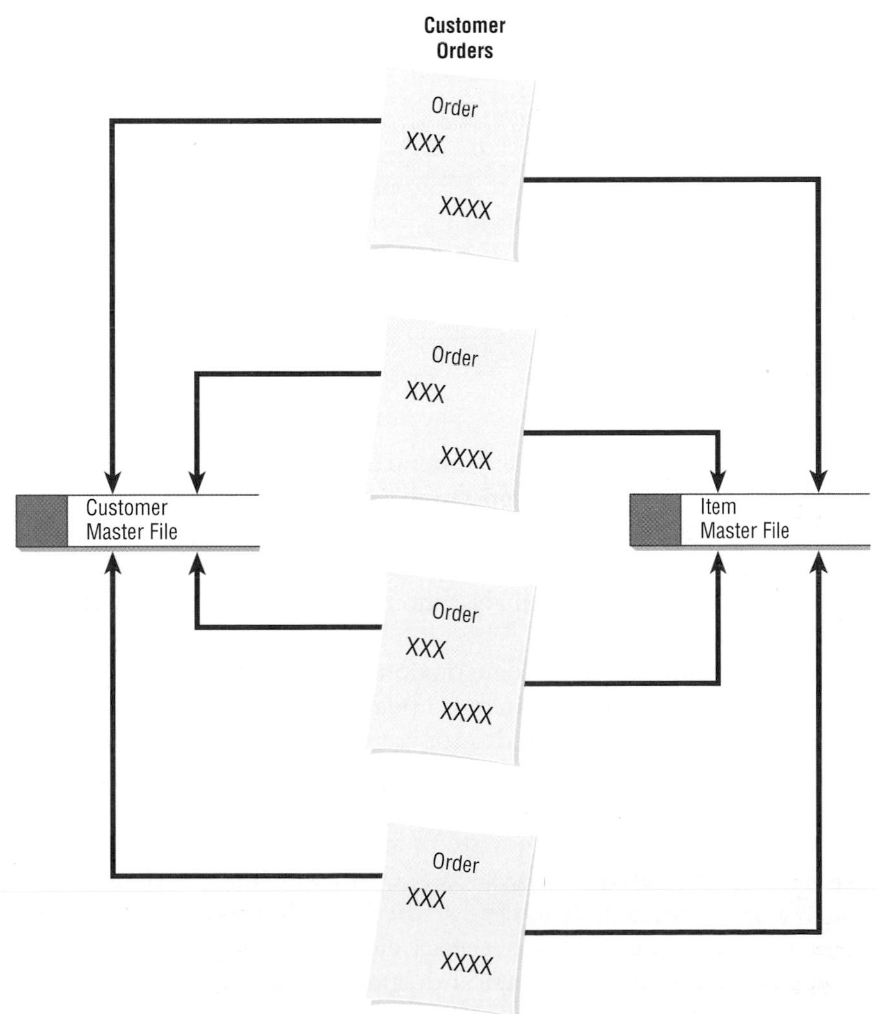

FIGURE 10.16
Information that comes from customer orders may find its way into different data stores.

FIGURE 10.17
Data stores derived from a pending order at World's Trend Catalog Division.

```
Customer Master =        Customer Number +
                         Customer Name +
                         Address +
                         Telephone +
                         Corporate Credit Card Number +
                         Expiration Date
Item Master =            Item Number +
                         Price +
                         Quantity on Hand
Order Record =           Customer Number +
                         Catalog Number +
                         Order Date +
                         {Available Order Items} +
                         Merchandise Total +
                         (Tax) +
                         Shipping and Handling +
                         Order Total +
                         Method of Payment +
                         (Credit Card Type) +
                         (Credit Card Number) +
                         (Expiration Date)
Available Order Items =  Item Number +
                         Quantity Ordered +
                         Quantity Shipped +
                         Current Price
Method of Payment =      [Check ¦ Charge ¦ Money Order]
Credit Card Type =       [World's Trend ¦ American Express ¦ MasterCard ¦ Visa]
```

USING THE DATA DICTIONARY

The ideal data dictionary is automated, interactive, online, and evolutionary. As the systems analyst learns about the organization's systems, data items are added to the data dictionary. On the other hand, the data dictionary is not an end in itself and must never become so. To avoid becoming sidetracked with the building of a complete data dictionary, the systems analyst should view it as an activity that parallels systems analysis and design.

To have maximum power, the data dictionary should be tied into a number of systems programs so that when an item is updated or deleted from the data dictionary, it is automatically updated or deleted from the database. The data dictionary becomes simply a historical curiosity if it is not kept current. Automated data dictionaries allow dramatic improvements in the upkeep of documentation. In doing so they also change the work of the systems analyst.

The data dictionary may also be used to create screens, reports, and forms. For example, examine the data structure for the World's Trend ORDER PICKING SLIP in Figure 10.18. Because the necessary elements and their lengths have been defined, the process of creating physical documents consists of arranging the elements in a pleasing and functional way using design guidelines and common sense.

FIGURE 10.18

Data structure for an order picking slip at World's Trend Catalog Division.

```
Order Picking Slip =      Order Number +
                          Order Date +
                          Customer Number +
                          Customer Name +
                          Customer Address +
                          Customer Telephone +
                          {Order Item Selection} +
                          Number of Items

Order Item Selection =    Item Number +
                          Item Description +
                          Size Description +
                          Color Description +
                          Warehouse Section +
                          Shelf Number +
                          Quantity Ordered +
                          Quantity Picked

Customer Name =           First Name +
                          (Middle Initial) +
                          Last Name

Address =                 Street +
                          (Apartment) +
                          City +
                          State +
                          Zip +
                          (Zip Expansion) +
                          (Country)

Telephone =               Area Code +
                          Local Number
```

Repeating groups become columns and structural records are grouped together on the screen, report, or form. The report layout for the World's Trends ORDER PICKING SLIP is shown in Figure 10.19. Notice that FIRST NAME and LAST NAME are grouped together in NAME and that QUANTITY (PICKED and ORDERED), SECTION, SHELF NUMBER, ITEM NUMBER, ITEM DESCRIPTION, SIZE, and COLOR form a series of columns, because they are the repeating elements.

The data structure and elements for a data store are commonly used to generate corresponding computer source language code, which is then incorporated into computer programs. The data dictionary may be used in conjunction with a data flow diagram to analyze the system design, detecting flaws and areas that need clarification. Some considerations are:

1. All base elements on an output data flow must be present on an input data flow to the process producing the output. Base elements are keyed and should never be created by a process.
2. A derived element must be created by a process and should be output from at least one process into which it is not input.

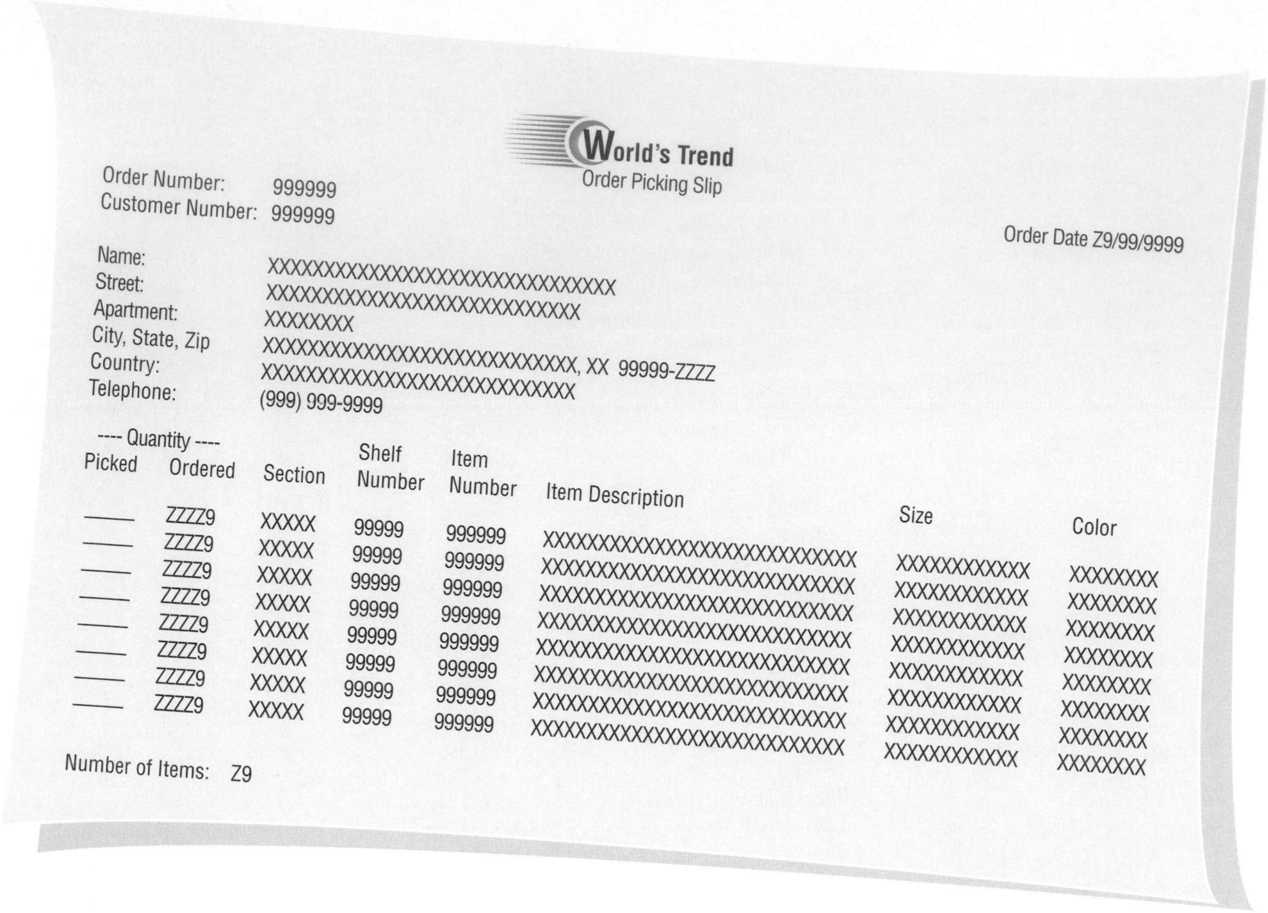

World's Trend
Order Picking Slip

Order Number: 999999
Customer Number: 999999

Order Date Z9/99/9999

Name: XXXXXXXXXXXXXXXXXXXXXXXXXXXXXX
Street: XXXXXXXXXXXXXXXXXXXXXXXXXXX
Apartment: XXXXXXXX
City, State, Zip XXXXXXXXXXXXXXXXXXXXXXXXXXXX, XX 99999-ZZZZ
Country: XXXXXXXXXXXXXXXXXXXXXXXXXXX
Telephone: (999) 999-9999

---- Quantity ----							
Picked	Ordered	Section	Shelf Number	Item Number	Item Description	Size	Color
____	ZZZZ9	XXXXX	99999	999999	XXXXXXXXXXXXXXXXXXXXXXXXXXX	XXXXXXXXXXXX	XXXXXXXX
____	ZZZZ9	XXXXX	99999	999999	XXXXXXXXXXXXXXXXXXXXXXXXXXX	XXXXXXXXXXXX	XXXXXXXX
____	ZZZZ9	XXXXX	99999	999999	XXXXXXXXXXXXXXXXXXXXXXXXXXX	XXXXXXXXXXXX	XXXXXXXX
____	ZZZZ9	XXXXX	99999	999999	XXXXXXXXXXXXXXXXXXXXXXXXXXX	XXXXXXXXXXXX	XXXXXXXX
____	ZZZZ9	XXXXX	99999	999999	XXXXXXXXXXXXXXXXXXXXXXXXXXX	XXXXXXXXXXXX	XXXXXXXX
____	ZZZZ9	XXXXX	99999	999999	XXXXXXXXXXXXXXXXXXXXXXXXXXX	XXXXXXXXXXXX	XXXXXXXX
____	ZZZZ9	XXXXX	99999	999999	XXXXXXXXXXXXXXXXXXXXXXXXXXX	XXXXXXXXXXXX	XXXXXXXX
____	ZZZZ9	XXXXX	99999	999999	XXXXXXXXXXXXXXXXXXXXXXXXXXX	XXXXXXXXXXXX	XXXXXXXX

Number of Items: Z9

FIGURE 10.19
Order picking slip created from the data dictionary.

3. The elements that are present on a data flow coming into or going out of a data store must be contained within the data store.

Even though the trend is toward online automated data dictionaries, it is important to appreciate the importance of compiling even a rudimentary data dictionary that is common to the organization. If begun early, a data dictionary can save many hours of time in the analysis and design phases. The data dictionary is the one common source in the organization for answering questions and settling disputes about any aspect of data definition. A current data dictionary can serve as an excellent reference for maintenance efforts on unfamiliar systems. Automated data dictionaries can serve as references for both people and programs.

SUMMARY

Using a top-down approach, the systems analyst uses data flow diagrams to begin compiling a data dictionary, which is a reference work containing data about data, or *metadata*, on all data processes, stores, flows, structures, and logical and physical elements within the system being studied. One way to begin is by including all data items from data flow diagrams.

A larger collection of project information is called a repository. CASE tools permit the analyst to create a repository that may include information about data flow, stores, record structures, and elements; about procedural logic screen and report design; about data relationships; about project requirements and final system deliverables; and about project management information.

"You're really doing very well. Snowden says you've given him all sorts of new ideas for running the new department. That's saying quite a lot, when you consider that he has a lot of his own ideas. By now I hope you've had a chance to speak with everyone that you would like to: certainly Snowden himself, Tom Ketcham, Daniel Hill, and Mr. Hyatt.

"Mr. Hyatt is an elusive soul, isn't he? I guess I didn't meet him until well into my third year. I hope you get to find out about him much sooner. Oh, but when you do get to see him, he cuts quite a figure, doesn't he? And those crazy airplanes. I've almost been conked on the head by one in the parking lot. But how can you get angry, when it's The Boss who's flying it? He's also got a secret—or should I say private—oriental garden off his office suite. No, you'll never see it on the building plans. You have to get to know him very well before he'll show you that, but I would wager it's the only one like it in Tennessee and maybe in the whole U.S. He fell in love with the wonderful gardens he saw in Southeast Asia as a young man. It goes deeper than that, however. Mr. Hyatt knows the value of contemplation and meditation. If he has an opinion, you can be sure it has been well thought through."

HYPERCASE QUESTIONS

1. Briefly list the data elements that you have found on three different reports produced at MRE.
2. Based upon your interviews with Snowden Evans and others, list the data elements that you believe you should add to the Management Unit's project reporting systems to better capture important data on project status, project deadlines, and budget estimates.
3. Create a data dictionary entry for a new data store, a new data flow, and a new data process that you are suggesting based on your response to question 2.
4. Suggest a list of new data elements that might be helpful to Jimmie Hyatt but are clearly not being made available to him currently.

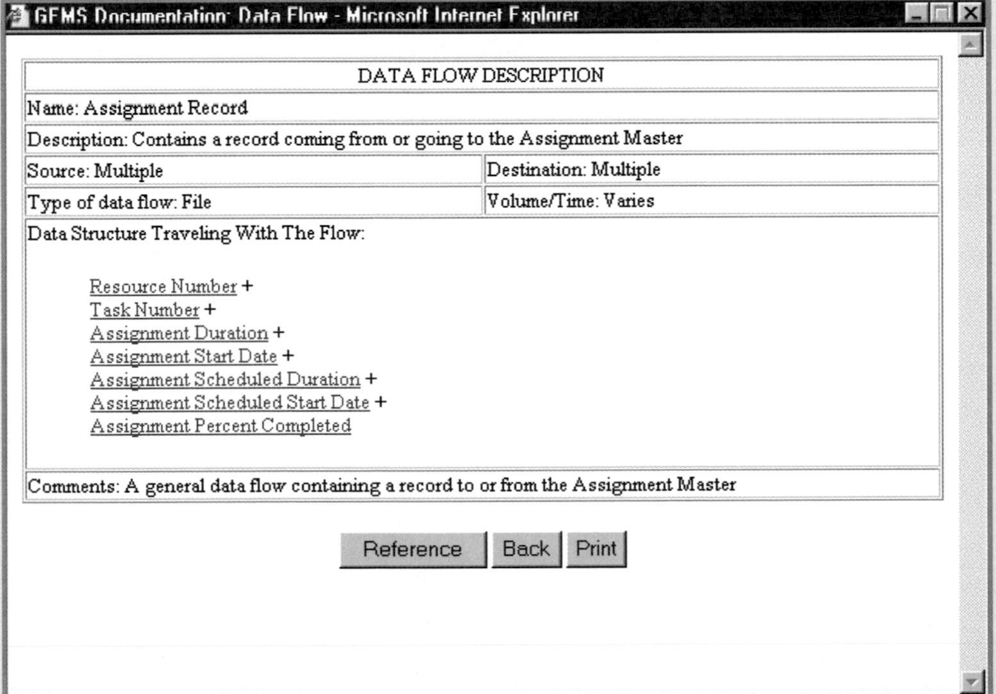

GFMS Documentation: Data Flow - Microsoft Internet Explorer
DATA FLOW DESCRIPTION
Name: Assignment Record
Description: Contains a record coming from or going to the Assignment Master
Source: Multiple
Destination: Multiple
Type of data flow: File
Volume/Time: Varies
Data Structure Traveling With The Flow:
Resource Number +
Task Number +
Assignment Duration +
Assignment Start Date +
Assignment Scheduled Duration +
Assignment Scheduled Start Date +
Assignment Percent Completed
Comments: A general data flow containing a record to or from the Assignment Master
Reference
Back
Print

FIGURE 10.HCI
In HyperCase you can look at the data dictionary kept at MRE.

Each entry in the data dictionary contains the item name, an English description, aliases, related data elements, the range, the length, encoding, and necessary editing information. The data dictionary is useful in all phases of analysis, design, and ultimately documentation, because it is the authoritative source on how a data element is used and defined in the system. Many large systems feature computerized data dictionaries that cross-reference all programs in the database using a particular data element.

KEYWORDS AND PHRASES

- base element
- binary formats
- data dictionary
- data element
- data structure
- derived element
- packed decimal
- physical data structures
- repeating groups
- repeating item
- repository
- structural record
- system deliverables
- zoned decimal

REVIEW QUESTIONS

1. Define the term *data dictionary*. Define *metadata*.
2. What are four reasons for compiling a complete data dictionary?
3. What information is contained in the data repository?
4. What is a structural record?
5. List the eight specific categories that each entry in the data dictionary should contain. Briefly give the definition of each category.
6. What are the basic differences among data dictionary entries prepared for data stores, data structures, and data elements?
7. Why are structural records used?
8. What is the difference between logical and physical data structures?
9. Describe the difference between base and derived elements.
10. How do the data dictionary entries relate to levels within a set of data flow diagrams?
11. List the four steps to take in compiling a data dictionary.
12. Why shouldn't compiling the data dictionary be viewed as an end in itself?
13. What are the main benefits of using a data dictionary?

PROBLEMS

1. Based on Figure 9.EX1 in Chapter 9, Joe, one of your systems analysis team members made the following entry for the data dictionary used by Marilyn's Tours:

   DATA ELEMENT = TOURIST* * * * PAYMENT
   ALIAS = TOURIST PAY
   CHARACTERS = 12–24
   RANGE = $5.00–$1,000
   VARIABLES = $5.00, $10.00, $15.00 up to $1,000, and anything in between in dollars and cents.
   TO CALCULATE = TOTAL COST OF ALL TOURS, ANY APPLICABLE N.Y. STATE TAX, minus any RESERVATION DEPOSITS made.

   a. Is this truly a data element? Why or why not?
   b. Rewrite the data dictionary entry for TOURIST PAYMENT, reclassifying it if necessary. Use the proper form for the classification you choose.

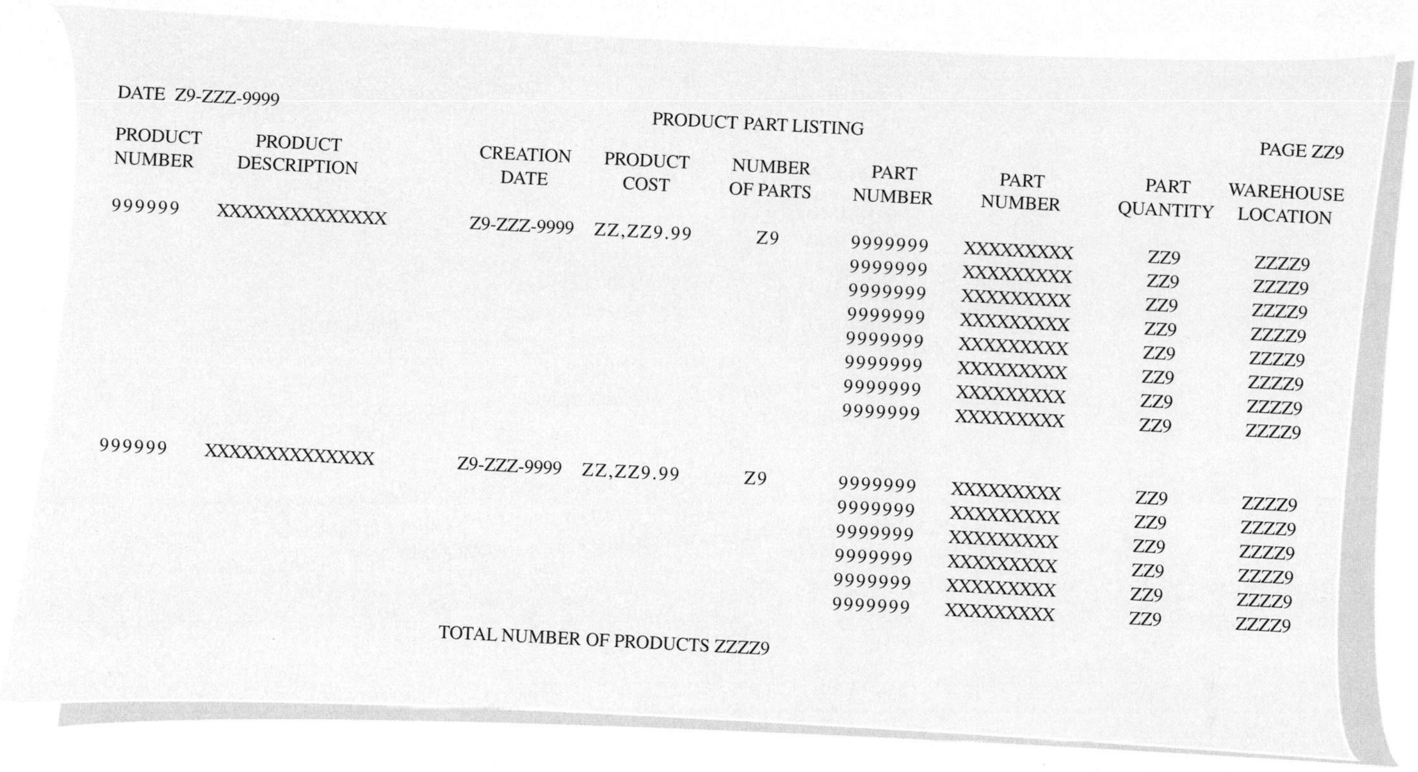

DATE Z9-ZZZ-9999

PRODUCT PART LISTING

PAGE ZZ9

PRODUCT NUMBER	PRODUCT DESCRIPTION	CREATION DATE	PRODUCT COST	NUMBER OF PARTS	PART NUMBER	PART NUMBER	PART QUANTITY	WAREHOUSE LOCATION
999999	XXXXXXXXXXXXXX	Z9-ZZZ-9999	ZZ,ZZ9.99	Z9	9999999	XXXXXXXXX	ZZ9	ZZZZ9
					9999999	XXXXXXXXX	ZZ9	ZZZZ9
					9999999	XXXXXXXXX	ZZ9	ZZZZ9
					9999999	XXXXXXXXX	ZZ9	ZZZZ9
					9999999	XXXXXXXXX	ZZ9	ZZZZ9
					9999999	XXXXXXXXX	ZZ9	ZZZZ9
					9999999	XXXXXXXXX	ZZ9	ZZZZ9
					9999999	XXXXXXXXX	ZZ9	ZZZZ9
999999	XXXXXXXXXXXXXX	Z9-ZZZ-9999	ZZ,ZZ9.99	Z9	9999999	XXXXXXXXX	ZZ9	ZZZZ9
					9999999	XXXXXXXXX	ZZ9	ZZZZ9
					9999999	XXXXXXXXX	ZZ9	ZZZZ9
					9999999	XXXXXXXXX	ZZ9	ZZZZ9
					9999999	XXXXXXXXX	ZZ9	ZZZZ9
					9999999	XXXXXXXXX	ZZ9	ZZZZ9

TOTAL NUMBER OF PRODUCTS ZZZZ9

FIGURE 10.EX1
A prototype of the Product-Part Listing.

2. Pamela, the systems analyst, has made significant progress in understanding the data movement at Bonton's clothing store. To share what she has done with other members of her team as well as the head of franchising in New York, she is composing a data dictionary.
   a. Write an entry in Pamela's data dictionary for one of the data flows that you depicted in your data flow diagram in problem 1 in Chapter 9. Be as complete as possible.
   b. Write an entry in Pamela's data dictionary for one of the data stores that you depicted in your data flow diagram in problem 1 in Chapter 9. Be as complete as possible.
3. Cecile, the manager of the bookstore that your systems analysis team has been working with to build a computerized inventory system, thinks that one of your team members is making a nuisance of himself by asking her extremely detailed questions about data items used in the system. For example, he asks, "Cecile, how much space, in characters, does listing of an ISBN number take?"
   a. What are the problems created by going directly to the manager with questions concerning data dictionary entries? Use a paragraph to list the problems you can see with your team member's approach.
   b. In a paragraph, explain to your team member how he can better gather information for the data dictionary.
4. The Motion Manufacturing Company assembles bicycles, tricycles, scooters, rollerblades, and other outdoor sports equipment. Each outdoor product is built using many parts, which vary from product to product. Interviews with the head parts clerk have resulted in a list of elements for the Product-Part Listing, showing which parts are used in the manufacture of each product. A prototype of the Product-Part Listing is illustrated in Figure 10.EX1. Create a data structure dictionary entry for the Product-Part Listing. The head parts clerk has informed us that there are never more than 50 different parts for each product.

FIGURE 10.EX2
A VDT screen showing cruise availability.

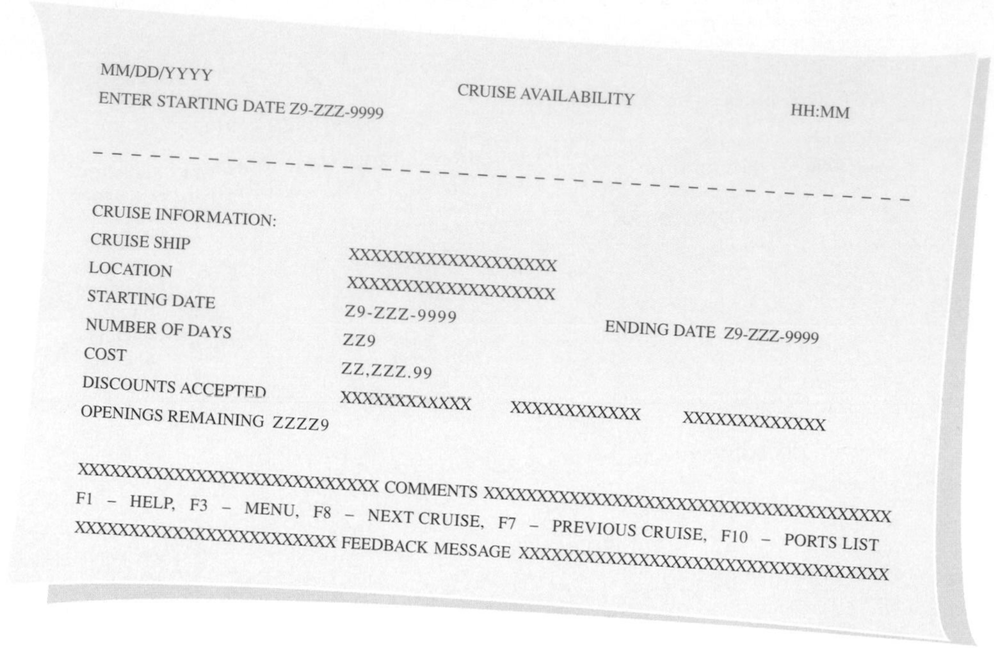

```
MM/DD/YYYY
ENTER STARTING DATE Z9-ZZZ-9999          CRUISE AVAILABILITY                    HH:MM
- - - - - - - - - - - - - - - - - - - - - - - - - - - - - - - - - - - - - - - - - - - -

CRUISE INFORMATION:
CRUISE SHIP              XXXXXXXXXXXXXXXXXXXX
LOCATION                 XXXXXXXXXXXXXXXXXXXX
STARTING DATE            Z9-ZZZ-9999
NUMBER OF DAYS           ZZ9                          ENDING DATE Z9-ZZZ-9999
COST                     ZZ,ZZZ.99
DISCOUNTS ACCEPTED       XXXXXXXXXXXX    XXXXXXXXXXXX    XXXXXXXXXXXX
OPENINGS REMAINING ZZZZ9

XXXXXXXXXXXXXXXXXXXXXXXXXXXXX COMMENTS XXXXXXXXXXXXXXXXXXXXXXXXXXXXXXXXXXXXXXXXX
F1 - HELP, F3 - MENU, F8 - NEXT CRUISE, F7 - PREVIOUS CRUISE, F10 - PORTS LIST
XXXXXXXXXXXXXXXXXXXXXXXXXX FEEDBACK MESSAGE XXXXXXXXXXXXXXXXXXXXXXXXXXXXXXXXXXXXXX
```

5. Analyze the elements found on the Product-Part Listing and create the data structure for the Product Master File and the Part Master File data stores.
6. Which of the elements on the Product-Part Listing are derived elements?
7. The Caribbean Cruise Company arranges cruise vacations of varying lengths at several locations. When customers call to check on the availability of a cruise, a Cruise Availability Inquiry, illustrated in Figure 10.EX2, is used to supply them with information. Create the data dictionary structure for the Cruise Availability Inquiry.
8. List the master files that would be necessary to implement the Cruise Availability Inquiry.
9. The following ports of call are available for the Caribbean Cruise Company:

Kingston	Port-au-Prince	Nassau
Montego Bay	St. Thomas	Freeport
Santo Domingo	Hamilton	Point-à-Pitre
San Juan	Port of Spain	St. Lucia

   Create the PORT OF CALL element. Examine the data to determine the length and format of the element.

GROUP PROJECTS

1. Meet with your group and use a CASE tool or a manual procedure to develop data dictionary entries for a process, data flow, data store, and data structure based on the data flow diagrams you completed for Maverick Transport in the Chapter 9 group exercises. As a group, agree on any assumptions necessary to make complete entries for each data element.
2. Your group should develop a list of methods to help you make complete data dictionary entries for this exercise as well as for future projects. For example, study existing reports, base them on new or existing data flow diagrams, and so on.

SELECTED BIBLIOGRAPHY

Colter, M. "A Comparative Examination of Systems Analyst's Techniques." *MIS Quarterly*, Vol. 8, No. 1, June 1984, pp. 51–66.

Davis, G. B., and M. H. Olson. *Management Information Systems, Conceptual Foundations, Structure, and Development*, 2d ed. New York: McGraw-Hill, 1985.

Gane, C., and T. Sarson. *Structured Systems Analysis and Design Tools and Techniques*. Englewood Cliffs, NJ: Prentice Hall, 1979.

Gore, M., and J. Stubbe. *Elements of Systems Analysis*, 3d ed. Dubuque, IA: William C. Brown, 1983.

Leeson, M. *Systems Analysis and Design*. Chicago: Science Research Associates, Inc., 1985.

Lucas, H. *Information Systems Concepts for Management*, 3d ed. New York: McGraw-Hill, 1986.

Martin, J. *Strategic Data-Planning Methodologies*. Englewood Cliffs, NJ: Prentice Hall, 1982.

McFadden, F. R., and J. A. Hoffer. *Data Base Management*. Menlo Park, CA: Benjamin/Cummings, 1985.

Schmidt, A. *Working with Visible Analyst Workbench for Windows*. Upper Saddle River, NJ: Prentice Hall, 1996.

Semprevivo, P. C. *Systems Analysis and Design: Definition, Process, and Design*. Chicago: Science Research Associates, Inc., 1982.

Senn, J. A. *Analysis and Design of Information Systems*. New York: McGraw-Hill, 1984.

Sprague, R. H., and E. D. Carlson. *Building Effective Decision Support Systems*. Englewood Cliffs, NJ: Prentice Hall, 1982.

10

Allen Schmidt, Julie E. Kendall, and Kenneth E. Kendall

DEFINING WHAT YOU MEAN

"We can use the data flow diagrams we completed to create data dictionary entries for all data flow and data stores," Chip says to Anna at their next meeting. Each of these components has a Composition entry in the repository. The records created for the Microcomputer System are thus linked directly to the data flow diagram components that describe data.

Anna and Chip meet to divide the work of creating records and elements. "I'll develop the data dictionary for the software portion of the system," Anna says.

"Good thing I enjoy doing the hardware," Chip kids her good-naturedly.

Records, or data structures, are created first. They may contain elements, the basic building blocks of the data structure, and they may also contain other records within them called structural records. Visible Analyst also maintains relationships among graph components, records, and elements that may be used for analysis and reporting.

Using information from interviews and the prototype screens, Anna started to create the Software records. Because the output of a system will determine what data need to be both stored and obtained via data entry screens, the starting point was the output data flow SOFTWARE INSTALLATION LIST. This prototype identifies some of the elements that should be stored within the Software Master file:

SOFTWARE INVENTORY NUMBER
VERSION NUMBER
NUMBER OF DISKETTES
CAMPUS LOCATION
TITLE
DISKETTE SIZE
HARDWARE INVENTORY NUMBER
ROOM LOCATION

Other output prototype reports and screens were also examined. Additional elements were obtained from the ADD SOFTWARE prototype screen.

The final element list is shown in Figure E10.1. These elements were arranged into a logical sequence for the SOFTWARE MASTER file. The following standards for arranging elements within a record were used:

1. The major key element that uniquely identified the record. An example is the SOFTWARE INVENTORY NUMBER.
2. Descriptive information, such as TITLE, VERSION NUMBER, and PUBLISHER.
3. Information that is periodically updated, such as NUMBER OF COPIES.
4. Any repeating elements, such as HARDWARE INVENTORY NUMBER, denoting the machines on which the software has been installed.

Next, the SOFTWARE MASTER file record was created using the Visible Analyst Repository. The description screen for creating a record is shown in Figure E10.2. (*Note*: This screen may differ from the data structure screen in your copy of Visible Analyst. To view the screen that is in the same format illustrated in Figure E10.2, click the **Options** menu and then click so there is a check in front of **Classical User Interface**.) Notice the entry area for an Alias, or a different name for the record, used

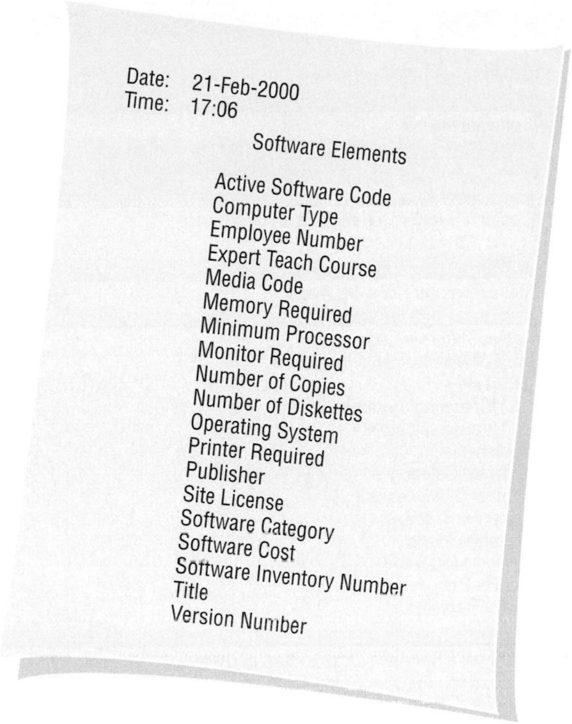

Date: 21-Feb-2000
Time: 17:06

Software Elements

Active Software Code
Computer Type
Employee Number
Expert Teach Course
Media Code
Memory Required
Minimum Processor
Monitor Required
Number of Copies
Number of Diskettes
Operating System
Printer Required
Publisher
Site License
Software Category
Software Cost
Software Inventory Number
Title
Version Number

FIGURE E10.1
Software element list.

by a different user group. Because each user may refer to the same record by a different name, all such names should be documented, resulting in enriched communication among users.

Each element or structural record needs to be defined as part of the whole record, and it is entered in the Composition area. If the element or structural record is a repeating group, the name is enclosed within curly brackets ({ }) and the number of times it repeats is placed in front of the name. If the data are keys, a code is put in brackets ([ ]) in front of the name. The symbol [pk] represents a primary key. The symbol [akn] represents an alternate key, where n is 1, 2, 3 and so on and defines each different key or group of fields that, when combined, make a secondary key. When a group of fields makes up a secondary key, that key is called a concatenated key.

Examine the SOFTWARE MASTER file. It contains a primary key of SOFTWARE INVENTORY NUMBER and a concatenated secondary key of TITLE, OPERATING SYSTEM and VERSION NUMBER.

Visible Analyst allows you to easily describe each structural record or element composing the larger record. Anna places the cursor in each name in the Composition area and presses the Jump button. Further record and element screens are displayed and detailed information is entered.

"This is great!" Anna thinks to herself. "It's so easy to enter the details, and by using this method I won't accidentally forget to describe an element."

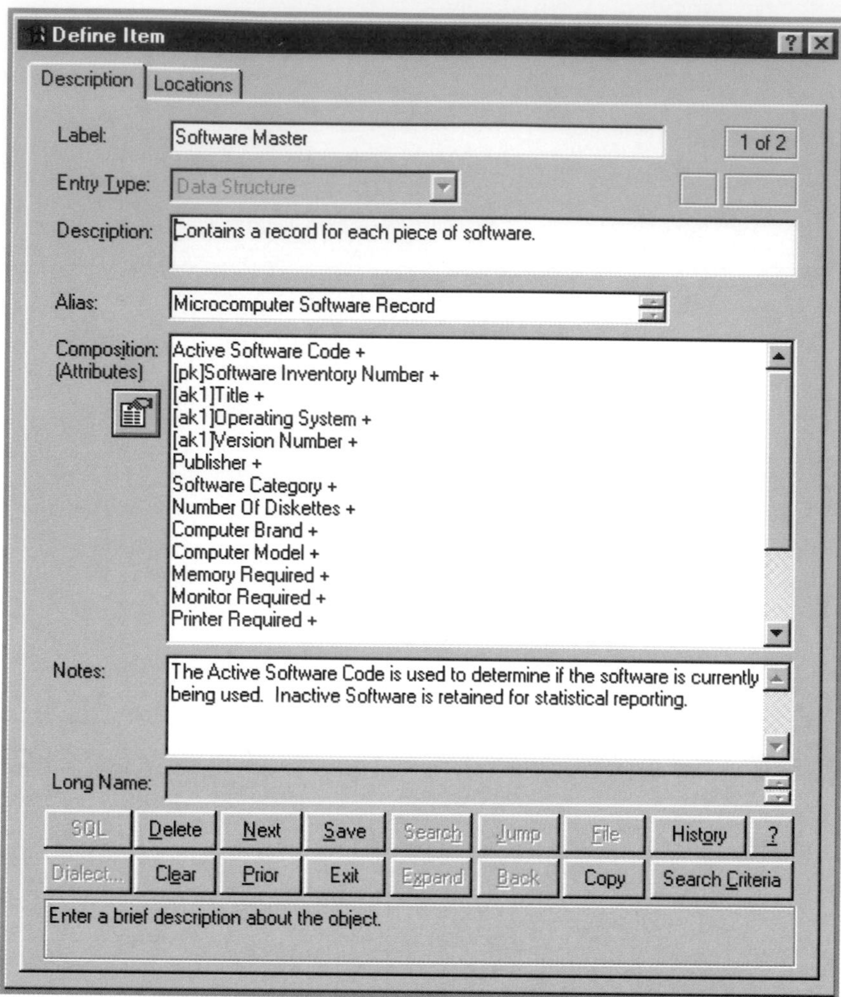

FIGURE E10.2
Record description screen, SOFTWARE MASTER file.

Chip is also impressed with the simplicity of creating the data dictionary. Following a process similar to Anna's, he creates a record description for the MICROCOMPUTER MASTER file. It contains a table of five internal boards and two structural records, PERIPHERAL EQUIPMENT and MAINTENANCE INFORMATION, illustrated in Figure E10.3. The Composition area for entering Element or Record names is a scroll region, meaning that more lines may be keyed than will fit in the screen area. As entries are added to the bottom of the region, top entries scroll out of the area.

As elements are added to the record, Chip decides to describe each in detail. The Element description screen for the HARDWARE INVENTORY NUMBER is shown in Figure E10.4. Observe the areas for entering element attributes. Several aliases may be included along with a definition. A **Notes** area contains any other useful information about the element. Chip and Anna employ this area to enter further edit criteria and other useful notation. The description for the HARDWARE INVENTORY NUMBER details how this HARDWARE INVENTORY NUMBER is used to keep physical track of the machines.

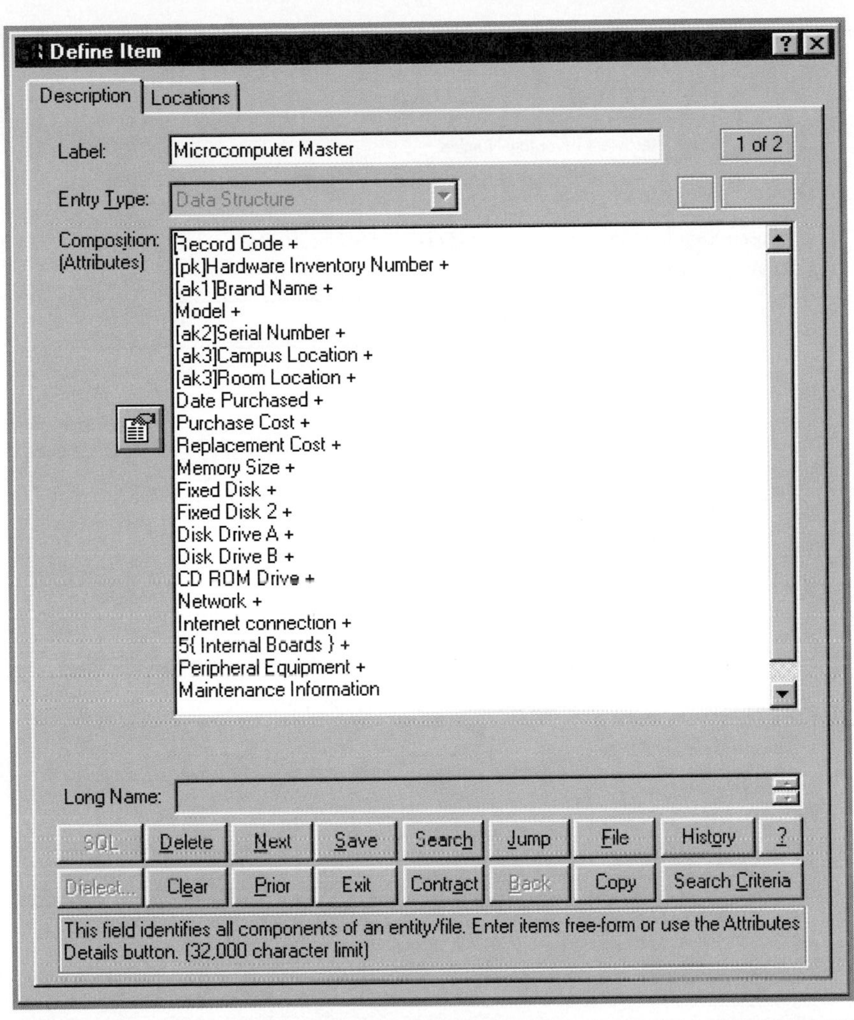

FIGURE E10.3
MICROCOMPUTER MASTER record description screen.

Clicking on the Physical Characteristics tab displays a second screen for the HARDWARE INVENTORY NUMBER, illustrated in Figure E10.5. It contains an area showing within which structures the element is contained as well as an area for the type of data, the length, and the picture used to describe how the data are formatted. Each such picture is a coded entry, similar to those used in programming languages. Examples of some of the codes are as follows:

- 9 Represents numeric data: Only numbers may be entered when prototyping.
- A Alphabetic: Only alphabetic characters may be entered.
- X Alphanumeric: Any characters may be entered.
- Z Zero suppression: Replace leading zeros with spaces.
- $ Dollar sign: Replace leading zeros with a dollar sign.

Chip is careful to include complete entries for these areas, including any default values and whether the entry may be null or not.

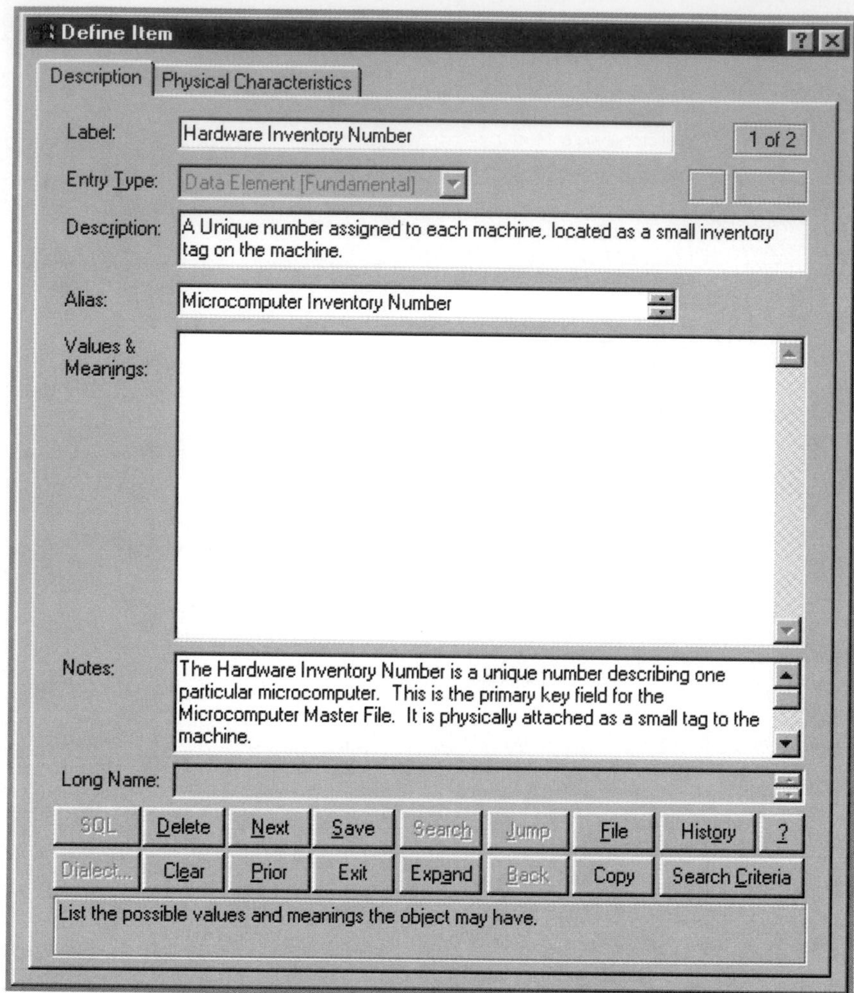

FIGURE E10.4
Elcment description screen, HARDWARE INVENTORY NUMBER.

Anna and Chip repeat this process for all elements found on each record. This effort is time-consuming but worthwhile. After the first few records are created, it becomes easier to create the remaining record structures. Visible Analyst has a search feature that provides lists of the elements contained within the design.

"I think that we've designed a complete set of elements," Chip says at a checkpoint meeting.

"Yes," replies Anna. "There are reports that will show us the details of the data structures and help us to spot duplications and omissions. Let's put Visible Analyst to work producing record layouts for us."

The Reports feature was used to print record layouts for all master files. Figure E10.6 is the Web page preview output for the MICROCOMPUTER MASTER FILE. Information from both screens is included, along with the data flow diagrams within which the data store is contained and the input and output data flow to and from the data store.

Define Item

Description | Physical Characteristics

Label: Hardware Inventory Number — 2 of 2

Entry Type: Data Element [Fundamental]

Locations:
Assoc Entity --> Hardware Software Relational Table
Assoc Entity --> Microcomputer Board Relational Table
Assoc Entity --> Monitor
Attrib Entity --> Component Boards
Attrib Entity --> Microcomputer Maintenance
Data Flow --> Changed Microcomputer
Data Flow --> Completed Installation Listing
Data Flow --> Component Boards
Data Flow --> Confirmed Changes
Data Flow --> Confirmed Microcomputer Record

Physical Characteristics
Type: Char
Length: 8 — Picture: 9(8)
Allow Null: No — Display:
Default: — Owner:
Min: — Max: — Avg:

Long Name:

SQL | Delete | Next | Save | Search | Jump | File | History | ?
Dialect... | Clear | Prior | Exit | Expand | Back | Copy | Search Criteria

A repository object label can be up to 128 characters long, and the first character must be a letter.

FIGURE E10.5
HARDWARE INVENTORY NUMBER, element characteristics screen.

RECORD AND ELEMENT ANALYSIS

"Now let's really put the power of Visible Analyst to use," Anna says. "Let's see how well we've really designed our data."

"What do you mean?" Chip asks.

"I've been studying the analysis features contained within Visible Analyst, and there's a wealth of options for checking our design for consistency and correctness," Anna replies. "The first step is to use the Reports feature to produce a summary report of the elements we've added. Then we can examine the list for duplications and redundancy."

Figure E10.7 is an example of a portion of the element summary report displayed using Microsoft Internet Explorer. Analysts would examine the contents carefully and look for redundancy, or elements defined more than once. These redundancies may be easy to spot because the list is sorted by element name. The elements HARDWARE INVENTORY NUMBER and HARDWARE NUMBER and the elements SOFTWARE INVENTORY NUMBER and SOFTWARE NUM appear to be duplicate elements. Other duplicates, such as ROOM LOCATION and LOCATION, are harder to spot.

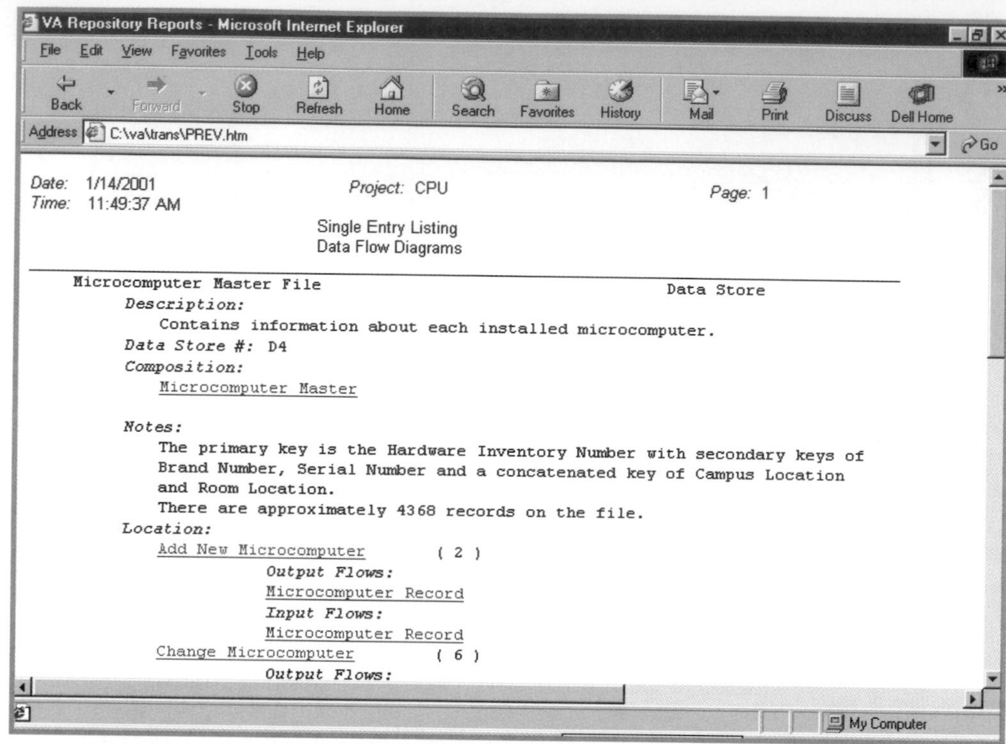

VA Repository Reports - Microsoft Internet Explorer

Address C:\va\trans\PREV.htm

Date: 1/14/2001
Time: 11:49:37 AM

Project: CPU

Page: 1

Single Entry Listing
Data Flow Diagrams

```
Microcomputer Master File                                   Data Store
     Description:
        Contains information about each installed microcomputer.
     Data Store #: D4
     Composition:
        Microcomputer Master

     Notes:
        The primary key is the Hardware Inventory Number with secondary keys of
        Brand Number, Serial Number and a concatenated key of Campus Location
        and Room Location.
        There are approximately 4368 records on the file.
     Location:
        Add New Microcomputer        ( 2 )
                 Output Flows:
                 Microcomputer Record
                 Input Flows:
                 Microcomputer Record
        Change Microcomputer         ( 6 )
                 Output Flows:
```

FIGURE E10.6

MICROCOMPUTER MASTER, data store Web page preview.

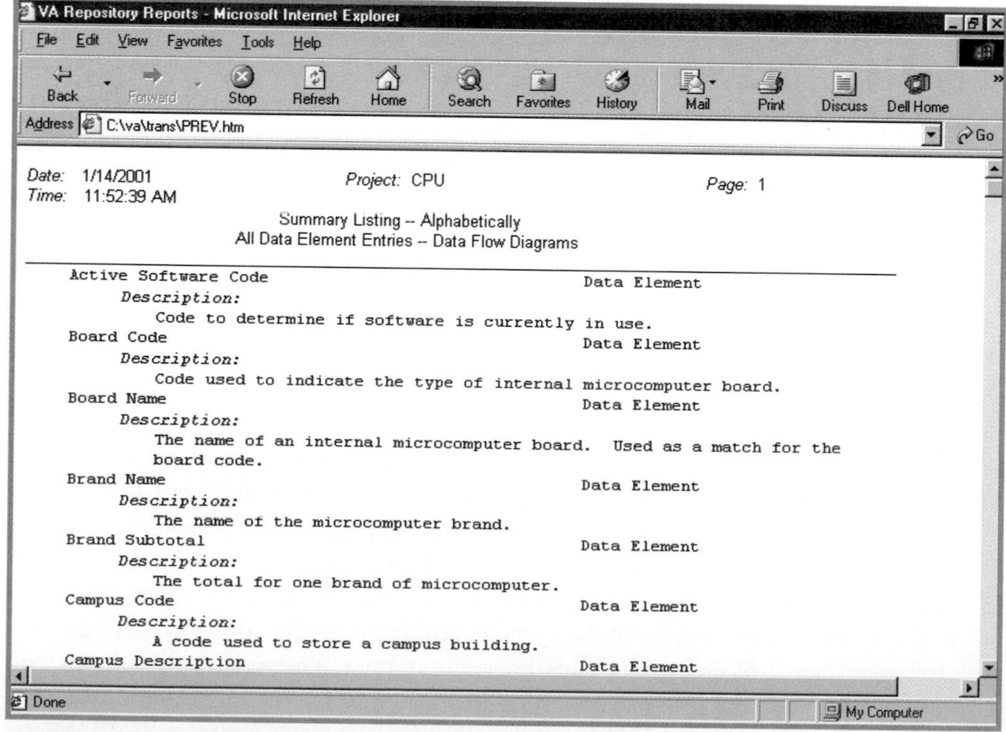

VA Repository Reports - Microsoft Internet Explorer

Address C:\va\trans\PREV.htm

Date: 1/14/2001
Time: 11:52:39 AM

Project: CPU

Page: 1

Summary Listing -- Alphabetically
All Data Element Entries -- Data Flow Diagrams

```
Active Software Code                                  Data Element
     Description:
        Code to determine if software is currently in use.
Board Code                                            Data Element
     Description:
        Code used to indicate the type of internal microcomputer board.
Board Name                                            Data Element
     Description:
        The name of an internal microcomputer board.  Used as a match for the
        board code.
Brand Name                                            Data Element
     Description:
        The name of the microcomputer brand.
Brand Subtotal                                        Data Element
     Description:
        The total for one brand of microcomputer.
Campus Code                                           Data Element
     Description:
        A code used to store a campus building.
Campus Description                                    Data Element
```

FIGURE E10.7

Element summary preview.

"Next we should use the **No Location References** option, which shows all the elements that are not included on any record," says Anna.

Figure E10.8 shows a portion of the report preview produced for elements that are not included anywhere.

"This is terrific!" exclaims Chip. "This **No Location References** shows design work that needs to be completed. We should produce this report for all the design components."

The elements were either added to other structures or deleted as duplicates. Producing the No Location References report a second time revealed no further isolated elements.

"Well, I guess that wraps up the data portion of the system design," Chip says.

"Guess again," replies Anna. "We've only begun to analyze. The **Report Query** feature will provide us with a lot of design information, both for analysis and for documentation."

The analysts select a report called Def Entities without Composition as their first choice. The report shows entries that are a data store or data structure and do have a Composition entry. The output shows that there are no records in error. The next report query is Elements without Pictures, and it shows all elements that do not have pictures defined for them. A portion of this report preview is illustrated in Figure E10.9. A last report that Chip and Anna created is called Undefined Elements, indicating all elements that have not been defined, that is, they exist in the Repository as a name only, but with no physical characteristics. A portion of the report displayed in a Web page is illustrated in Figure E10.10.

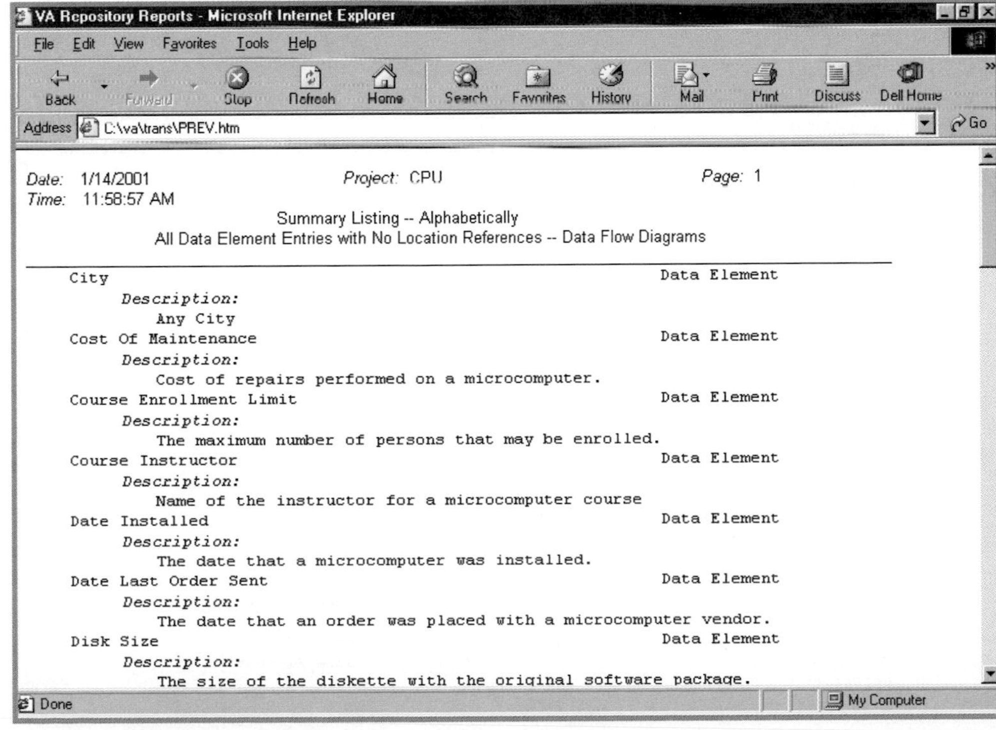

FIGURE E10.8

No Location References Web page preview.

10

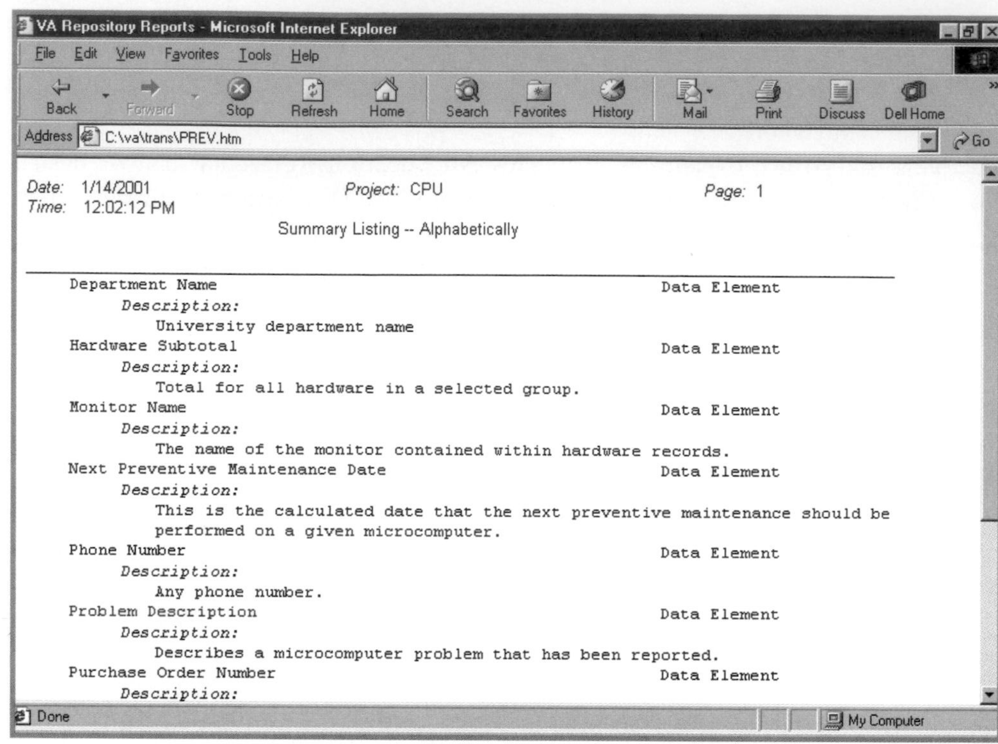

FIGURE E10.9
Elements Without Pictures preview.

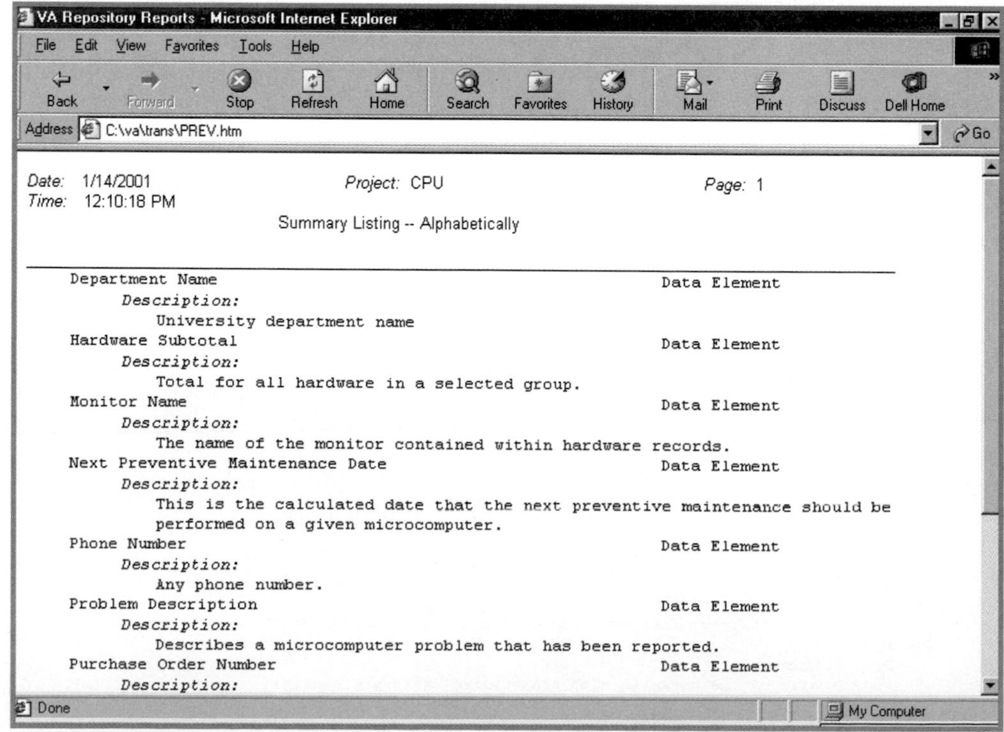

FIGURE E10.10
Undefined Elements preview.

"I'm really impressed with this analysis," Chip says. "Since correcting the errors in our design, I've come to realize how easy it is to feel confident that the design has been completed when there are discrepancies and omissions still needing our attention."

"We're not finished yet. There are some useful matrices that will provide documentation for any changes that may be made in the future. Let's produce the Data Elements versus Data Structures matrix, which shows records and their elements," Anna suggests.

The Report feature has the ability to produce reports as well as matrices in a grid representation. The first matrix produced is the Data Elements versus Data Structures, illustrated in Figure E10.11. It shows all elements and the data structures within which

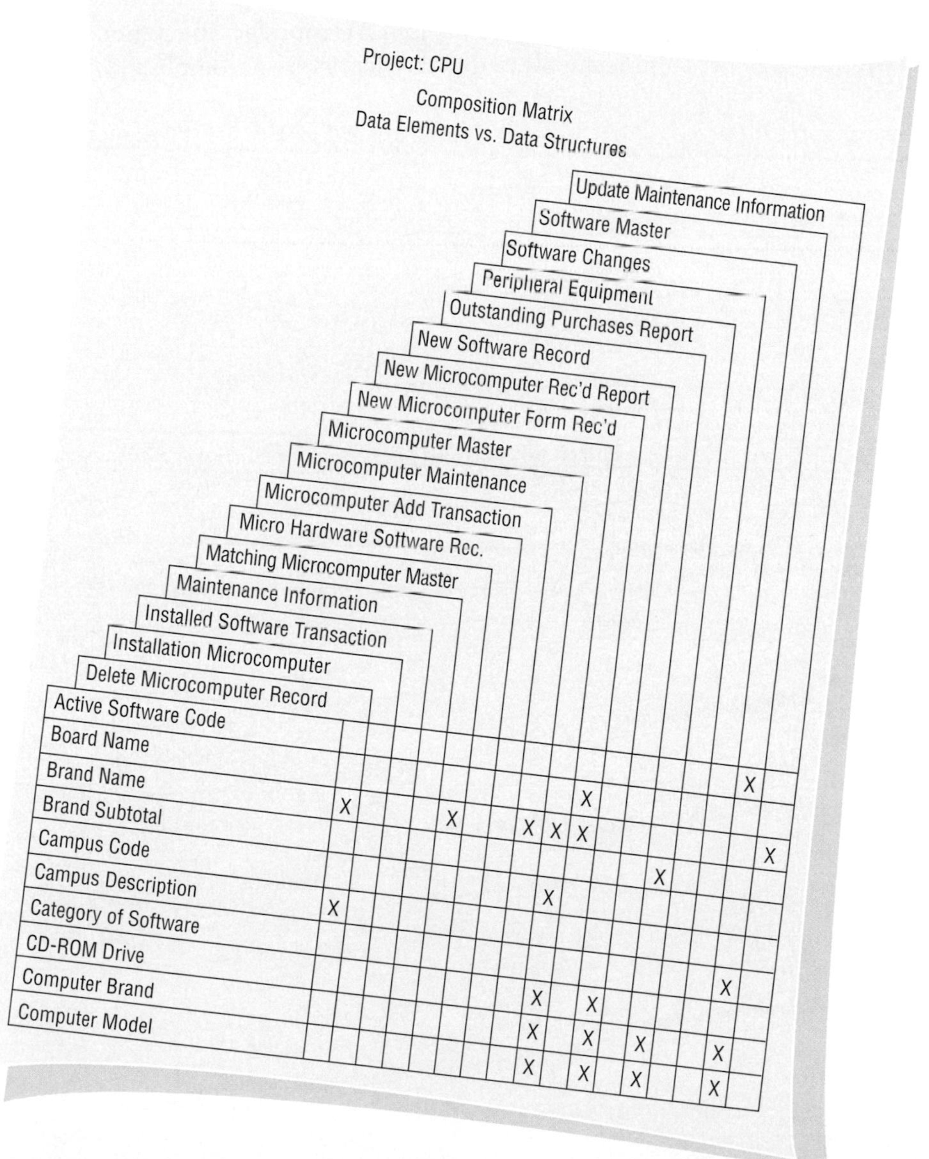

FIGURE E10.11
Data Elements vs. Data Structures Matrix report.

they are contained. This matrix is used to access the effect of changing an element by showing which corresponding data structures must be changed.

The next matrix created is the Diagram Location Matrix, showing all data stores and the diagrams on which they are located. This information is useful if a change needs to be made to the data store, because it will indicate where programs and documentation need to be changed. A portion of this matrix is illustrated in Figure E10.12.

A final grid is the Composition Matrix, showing all Data Elements and the Data Stores within which they are contained. A portion of this matrix is illustrated in Figure E10.13. This matrix gives Chip and Anna a picture of which elements may be stored redundantly, that is, in several data stores rather than one.

"There are many other reports and matrices that would be useful for us to produce," Anna says. "Some of these should be used later for documentation and tracking any proposed changes. I'm really pleased with what we've accomplished."

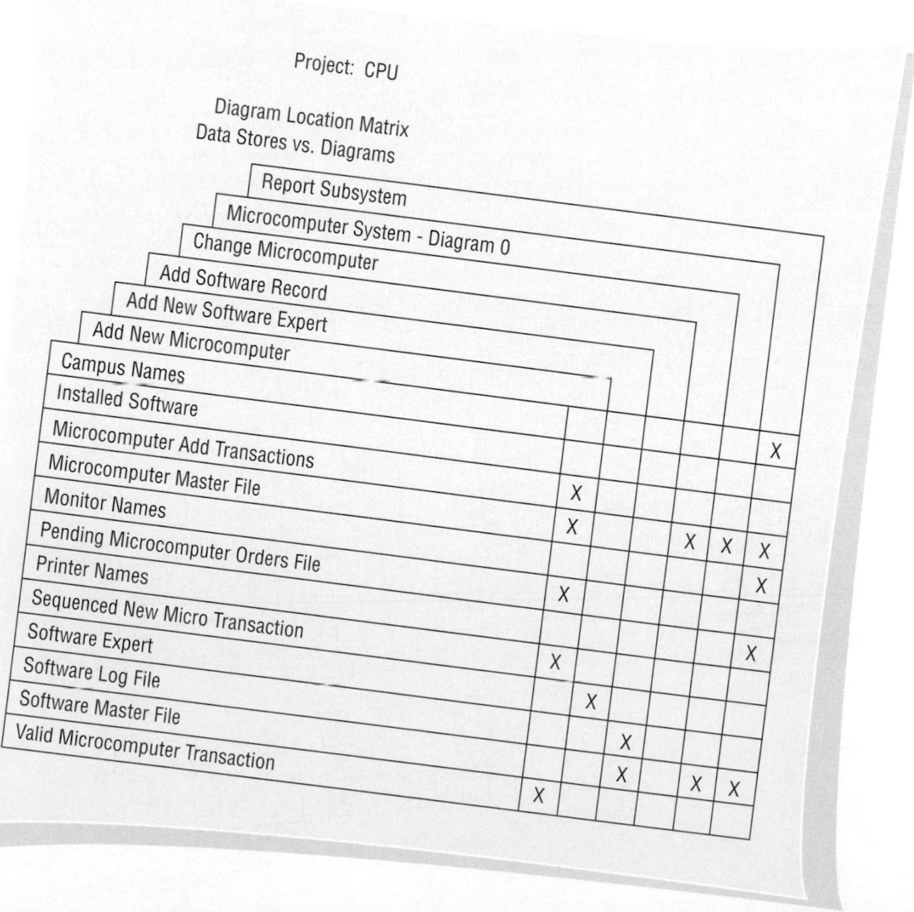

Project: CPU

Diagram Location Matrix
Data Stores vs. Diagrams

	Add New Microcomputer	Add New Software Expert	Add Software Record	Change Microcomputer	Microcomputer System - Diagram 0	Report Subsystem
Campus Names						X
Installed Software						
Microcomputer Add Transactions	X					
Microcomputer Master File	X			X	X	X
Monitor Names						X
Pending Microcomputer Orders File	X					
Printer Names						X
Sequenced New Micro Transaction	X					
Software Expert		X				
Software Log File			X			
Software Master File			X		X	X
Valid Microcomputer Transaction	X					

FIGURE E10.12
Diagram Location Matrix report.

Project: CPU

Composition Matrix
Data Elements vs. Data Stores

Data Element	Microcomputer Master File	Software Master File
Active Software Code		X
Board Code		
Board Name		
Brand Name	X	
Brand Subtotal		
Campus Location	X	
Category of Software		X
CD-ROM Drive	X	
Computer Brand		X
Computer Model		X
Cost		
Cost of Repairs	X	
Date Ordered		
Date Purchased	X	
Disk Drive A	X	
Disk Drive B	X	
Diskette Size		X
Fixed Disk	X	
Fixed Disk 2	X	
Hardware Inventory Number	X	X
Internal Boards	X	
Internet Connection	X	

FIGURE E10.13
Composition Matrix report.

EXERCISES*

E-1. Use Visible Analyst to view the Microcomputer Master File Data Store. Jump to the data structure and browse the elements and structural records.

E-2. Print the Software Master File record using the Report feature.

E-3. Use the Jump button to move to the Software Record Structure. Delete the following elements:
ACTIVE SOFTWARE CODE
INSTALLATION MICRO
SOFTWARE EXPERT

E-4. Modify the Software Changes record, supplying changes to the Software Master record. The modifications are as follows:

a. Add a [pk], for primary key, in front of the Software Inventory Number.
b. Add the following elements: Computer Brand, Computer Model, Memory Required, Monitor Required, Printer Required, Diskette Size, Site License, and Number of Copies.

E-5. Modify the Microcomputer Add Transaction record, which contains new microcomputer records to be placed on the Microcomputer Master data store.

a. Insert the Brand Name and Model above the Serial Number.
b. Place the Campus Location and Room Location after the Serial Number.
c. Add the following elements at the bottom of the list: Fixed Disk, Fixed Disk 2, Disk Drive A, and Zip Disk Size.
d. Delete the Internal Boards element, which will be determined after the microcomputer installation.

E-6. Modify the Installed Software Transaction, which is used to update the Software Master and to produce the Software Installation Listing. Delete the Title and Version Number, because they may be obtained from the Software Master and are redundant keying. Add the Hardware Inventory Number, specifying the installation microcomputer. Delete the Campus Location and Room Location, because they are elements of the installation microcomputer.

E-7. View the alias entry for the Software Master Table.

E-8. Modify the INSTALLED SOFTWARE data store. Add the Composition record INSTALLED SOFTWARE TRANSACTION. The index elements are SOFTWARE INVENTORY NUMBER and HARDWARE INVENTORY NUMBER.

E-9. Define the data store SOFTWARE LOG FILE. This file is used to store information on the new software records plus the date, time, and user ID of the person entering the record. Index elements are SOFTWARE INVENTORY NUMBER, TITLE, VERSION (a concatenated key), and SOFTWARE CATEGORY.

E-10. Define the data store PENDING MICROCOMPUTER ORDERS FILE. This file is created when a purchase order is made for ordering new microcomputers, and it is updated by the Microcomputer system. Place a comment in the **Notes** area stating that the average number of records is 100. Index elements are PURCHASE ORDER NUMBER and a concatenated key consisting of BRAND NAME and MODEL.

E-11. View the entry for the Software Record data flow. Press Jump with the cursor in the Composition area and examine the Software Master record. Press Back to return to the data flow description screen.

E-12. Modify the SOFTWARE UPGRADE INFORMATION data flow. The **Composition** record is SOFTWARE UPGRADE INFORMATION.

E-13. Modify the SOFTWARE CROSS-REFERENCE REPORT data flow. The **Composition** record is SOFTWARE CROSS-REFERENCE REPORT.

E-14. Modify the data flow entity for INSTALL UPDATE. This flow updates the Microcomputer Master record with installation information. Its data structure is INSTALL UPDATE RECORD. Include a comment that it processes about 50 records per month in updating the Microcomputer Master File.

E-15. Use the INSTALL UPDATE data flow to jump to (and create) the INSTALL UPDATE RECORD. Provide a definition based on information supplied in the previous problem. Enter the following elements:

HARDWARE INVENTORY NUMBER (primary key)
CAMPUS LOCATION
ROOM
INTERNAL BOARDS (occurs 5 times)
FIXED DISK 2
MOUSE
PRINTER
MAINTENANCE INTERVAL
DATE INSTALLED

E-16. Create the data flow description for the SOFTWARE INSTALLATION LIST. This flow contains information on specific software packages and the machines on which the software should be installed. The Composition should include the SOFTWARE INSTALLATION LISTING, a data structure.

E-17. Use the SOFTWARE INSTALL LIST to jump to (and therefore create) the SOFTWARE INSTALLATION LISTING. The elements on the listing are as follows:

Name
SOFTWARE INVENTORY NUMBER
TITLE
VERSION NUMBER
NUMBER OF DISKETTES
DISKETTE SIZE
HARDWARE INVENTORY NUMBER
CAMPUS LOCATION
ROOM LOCATION

E-18. Modify and print the element HARDWARE SUBTOTAL. Change the Type to Numeric, the Length to 6,2, and the Picture to Z, ZZZ, ZZ9.99.

E-19. Modify and print the MONITOR NAME element, the result of a table lookup using a monitor code. The Type to should be Character, the Length to 30, and the Picture to X(30).

E-20. Modify and print the DEPARTMENT NAME element. Create an Alias of Staff Department Name. In the Notes area, enter the following comment: Table of codes: Department Table. The Type to should be Character, the Length to 25, and the Picture to X(25).

E-21. Create the following element descriptions. Use the values supplied in the table. Create any alternate names and definitions based on your understanding of the element.

Name	Purchase Order Number	Problem Description
Type	Character	Character
Length	7	70
Picture	9999999	X(70)

Name	Total Microcomputer Cost	Next Preventive Maintenance Date
Type	Numeric	Date
Length	7,2	8
Picture	Z, ZZZ, ZZ9.99	Z9/99/9999
Notes		The Next Preventive Maintenance Date is calculated by adding the Maintenance Interval to the Last Preventive Maintenance Date.

Name	Phone Number	Repair Status
Type	Character	Character
Length	7	1
Picture	999-9999	X
Notes		Table of codes: Repair Table
Default		C

E-22. Use the Repository Reports feature to produce the following reports and matrices, either by printing the reports or by previewing them using your Web browser. The selection criteria from the Repository Reports dialogue box are listed, separated with a slash (/). Explain in a paragraph where the information produced may be effectively used.

a. Data Flow/Cross-Reference Listing/Data Element/Entire Project
b. Data Flow/Cross-Reference Listing/Data Structure/Entire Project
c. Record Contains Element (One Level) matrix
d. Data Flow/Single-Entry Listing/Software Master—Normalized
e. Data Flow/Diagram Location Matrix/Data Stores versus Diagrams
f. Data Flow/Composition Matrix/Data Elements versus Data Flows
g. Data Flow/Composition Matrix/Data Elements versus Data Structures
h. Data Flow/Composition Matrix/Data Element versus Data Stores

E-23. Use the Report Query feature to produce the following reports. Explain in a sentence what information the report is providing you with.

a. The Undefined Elements report
b. The Elements without Pictures report
c. The Coded Elements report
d. The Any Item with Components report

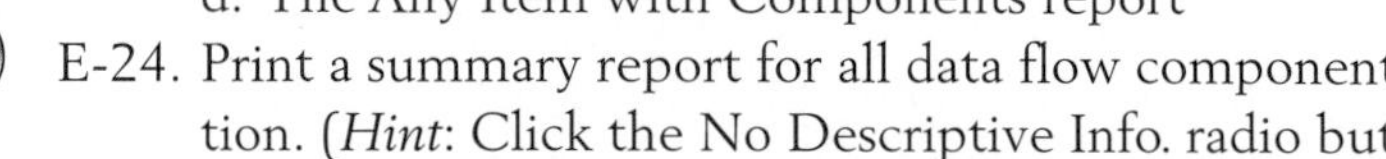

E-24. Print a summary report for all data flow components that do not have a description. (*Hint*: Click the No Descriptive Info. radio button.)

E-25. Print a summary report for all data flow components that are not on a diagram. (*Hint*: Click the No Location References radio button.)

E-26. Print a detailed report for all elements. Include only the physical information and the values and meanings. (*Hint*: Click the Fields button and then the Invert button and select the fields that you want printed.) Why would this report be useful to the analyst?

*Exercises preceded by a CD-ROM icon require the program Visible Analyst or another CASE tool. A CD-ROM containing Visible Analyst examples is provided free of charge to any professor adopting this book. The examples on the disk may be imported into Visible Analyst and then used by students.

DESCRIBING PROCESS SPECIFICATIONS AND STRUCTURED DECISIONS

11

METHODS AVAILABLE

The systems analyst approaching process specifications and structured decisions has many options for documenting and analyzing them. In Chapters 9 and 10 you noted processes such as VERIFY AND COMPUTE FEES, but you did not explain the logic necessary to execute these tasks. The methods available for documenting and analyzing the logic of decisions include structured English, decision tables, and decision trees. It is important to be able to recognize logic and structured decisions that occur in a business and how they are distinguishable from semistructured decisions. Then it is critical to recognize that structured decisions lend themselves particularly well to analysis with systematic methods that promote completeness, accuracy, and communication.

Decision analysis focuses on the logic of the decisions that are made, or need to be made, within the organization to carry out the objectives of the firm. This chapter covers the methods of structured English, decision tables, and decision trees that are used to analyze decisions and describe process logic, and in doing so it complements the material on data flow diagrams from Chapter 9 and on the data dictionary from Chapter 10.

OVERVIEW OF PROCESS SPECIFICATIONS

To determine the information requirements of a decision analysis strategy, the systems analyst must first determine the organization's objectives, using a top-down approach. The systems analyst must understand the principles of organizations (as covered in Chapter 2) and have a working knowledge of data-gathering techniques (as presented in Chapters 4 through 8). The top-down approach is critical because all decisions in the organization should be related, at least indirectly, to the broad objectives of the entire organization.

Process specifications—sometimes called *minispecs*, because they are a small portion of the total project specifications—are created for primitive processes on a data flow diagram as well as for some higher-level processes that explode to a child diagram. These specifications explain the decision-making logic and formulas that will transform process input data into output. Each derived element must have process logic to show how it is produced from the base elements or other previously created derived elements that are input to the primitive process.

The three goals of producing process specifications are as follows:

1. To reduce the ambiguity of the process. This goal compels the analyst to learn details about how the process works. Any vague areas should be noted, written down, and consolidated for all process specifications. These observations form a basis and provide the questions for a follow-up interview with the user community.
2. To obtain a precise description of what is accomplished, which is usually included in a packet of specifications for the programmer.
3. To validate the system design. This goal includes ensuring that a process has all the input data flow necessary for producing the output. In addition, all input and output must be represented on the data flow diagram.

You will find many situations where process specifications are not created. Sometimes the process is very simple or the computer code already exists. This eventuality would be noted in the process description, and no further design would be required. Categories of processes that generally *do not* require specifications are as follows:

1. Processes that represent physical input or output, such as read and write. These processes usually require only simple logic.
2. Processes that represent simple data validation, which is usually fairly easy to accomplish. The edit criteria are included in the data dictionary and incorporated into the computer source code. Process specifications may be produced for complex editing.
3. Processes that use prewritten code. These processes are generally included in a system as subprograms and functions.

Subprograms are computer programs that are written, tested, and stored on the computer system. They usually perform a general system function, such as validating a date or a check digit. These general-purpose subprograms are written and documented only once but form a series of building blocks that may be used within many systems throughout the organization. Thus these subprograms appear as processes on many data flow diagrams. Functions are similar to subprograms but are coded differently.

PROCESS SPECIFICATION FORMAT

Process specifications link the process to the data flow diagram and hence the data dictionary, as illustrated in Figure 11.1. Each process specification should be entered on a separate form or into a CASE tool screen such as the one used for Visible Analyst and shown in the CPU case at the end of this chapter. Enter the following information:

1. The process number, which must match the process ID on the data flow diagram. This specification allows an analyst to work on or review any process and to locate the data flow diagram containing the process easily.
2. The process name, which again must be the same as the name displayed within the process symbol on the data flow diagram.
3. A brief description of what the process accomplishes.
4. A list of data input flow, using the names found on the data flow diagram. Data names used in the formula or logic should match those in the data dictionary to ensure consistency and good communication.
5. The output data flow, also using data flow diagram and data dictionary names.

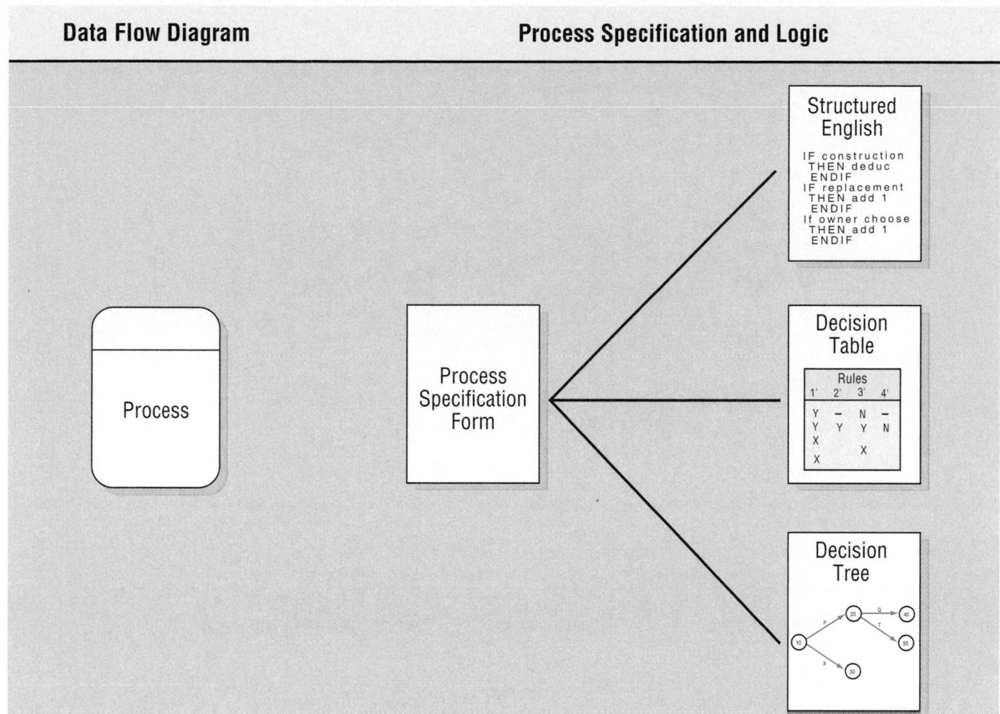

FIGURE 11.1
How process specifications relate to the data flow diagram.

6. An indication of the type of process: batch, online, or manual. All online processes require screen designs, and all manual processes should have well-defined procedures for employees performing the process tasks.
7. If the process uses prewritten code, include the name of the subprogram or function containing that code.
8. A description of the process logic that states policy and business rules in everyday language, not computer language pseudocode. Business rules are the procedures or perhaps a set of conditions or formulas that allow a corporation to run its business. Common business rule formats include the following:
   - Definitions of business terms.
   - Business conditions and actions.
   - Data integrity constraints.
   - Mathematical and functional derivations.
   - Logical inferences.
   - Processing sequences.
   - Relationships among facts about the business.
9. If there is not enough room on the form for a complete structured English description or if there is a decision table or tree depicting the logic, include the corresponding table or tree name.
10. List any unresolved issues, incomplete portions of logic, or other concerns. These issues form the basis of the questions used for follow-up interviews.

The above items should be entered to complete a process specification form, which includes a process number, process name, or both from the data flow diagram as well as the eight other items shown in the World's Trend example (Figure 11.2). Notice that completing this form thoroughly facilitates linking the process to the data flow diagram and the data dictionary. When using an

FIGURE 11.2

An example of a completed Process Specification form for determining whether an item is available.

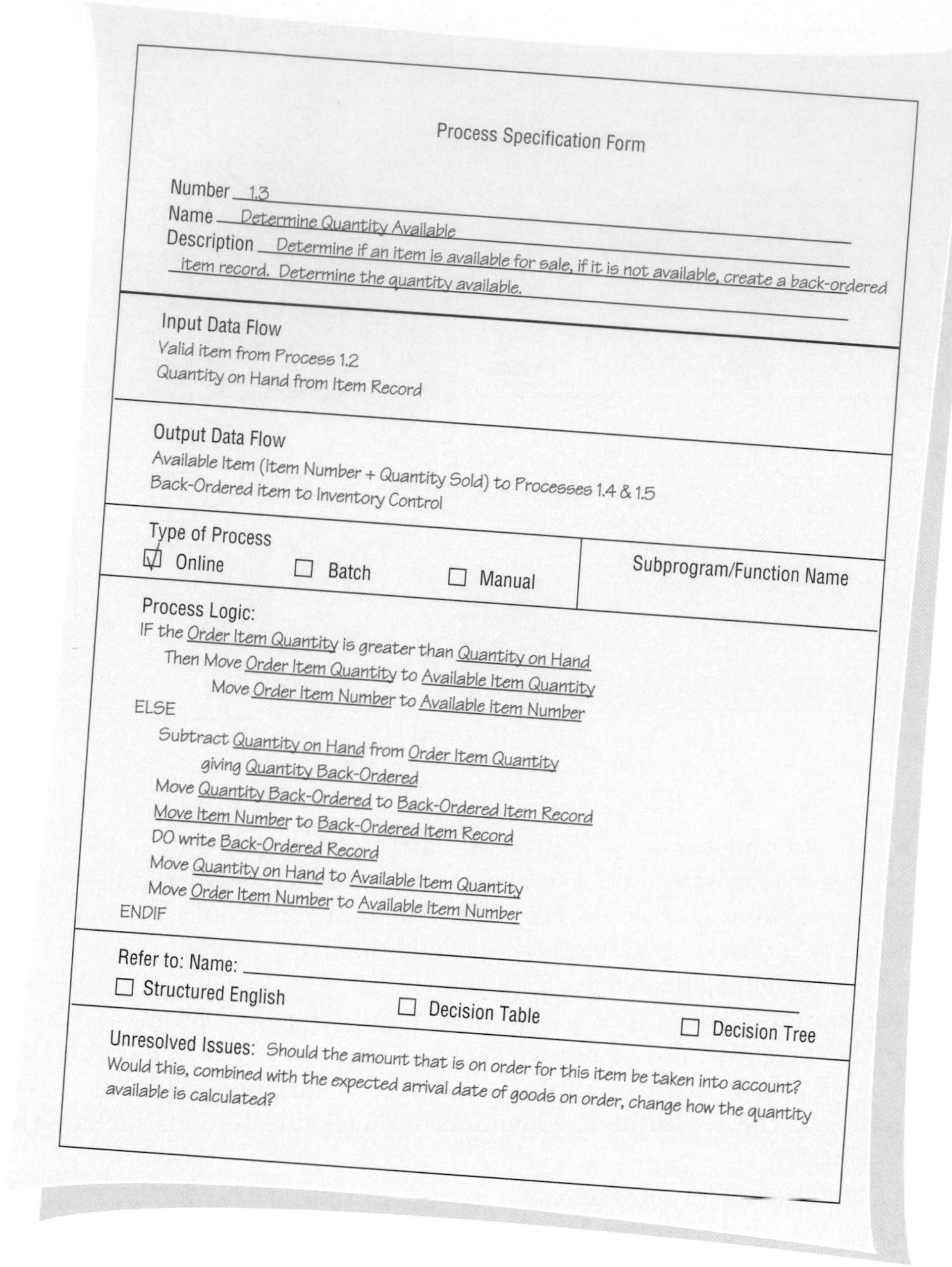

Process Specification Form

Number 1.3
Name Determine Quantity Available
Description Determine if an item is available for sale, if it is not available, create a back-ordered item record. Determine the quantity available.

Input Data Flow
Valid item from Process 1.2
Quantity on Hand from Item Record

Output Data Flow
Available Item (Item Number + Quantity Sold) to Processes 1.4 & 1.5
Back-Ordered item to Inventory Control

Type of Process
☑ Online ☐ Batch ☐ Manual

Subprogram/Function Name

Process Logic:

```
IF the Order Item Quantity is greater than Quantity on Hand
  Then Move Order Item Quantity to Available Item Quantity
       Move Order Item Number to Available Item Number
ELSE
  Subtract Quantity on Hand from Order Item Quantity
       giving Quantity Back-Ordered
  Move Quantity Back-Ordered to Back-Ordered Item Record
  Move Item Number to Back-Ordered Item Record
  DO write Back-Ordered Record
  Move Quantity on Hand to Available Item Quantity
  Move Order Item Number to Available Item Number
ENDIF
```

Refer to: Name:
☐ Structured English ☐ Decision Table ☐ Decision Tree

Unresolved Issues: Should the amount that is on order for this item be taken into account? Would this, combined with the expected arrival date of goods on order, change how the quantity available is calculated?

electronic form, such as the Visible Analyst screens shown in Figure 11.3, the description is entered into a scrolling area. The Expand button allows the analyst to display a larger amount of text, which helps to view the overall logic of the process.

INFORMATION REQUIRED FOR STRUCTURED DECISIONS

Conditions, condition alternatives, actions, and action rules must be known to design systems for structured decisions. The analyst first determines the conditions, that is, an occurrence that might affect the outcome of something else. In the next step the systems analyst defines the condition alternatives as specified by the decision maker; these alternatives can be as simple as "yes" or "no," or they can be more descriptive, such as "less than $50," "between $50 and $100," and "greater than $100."

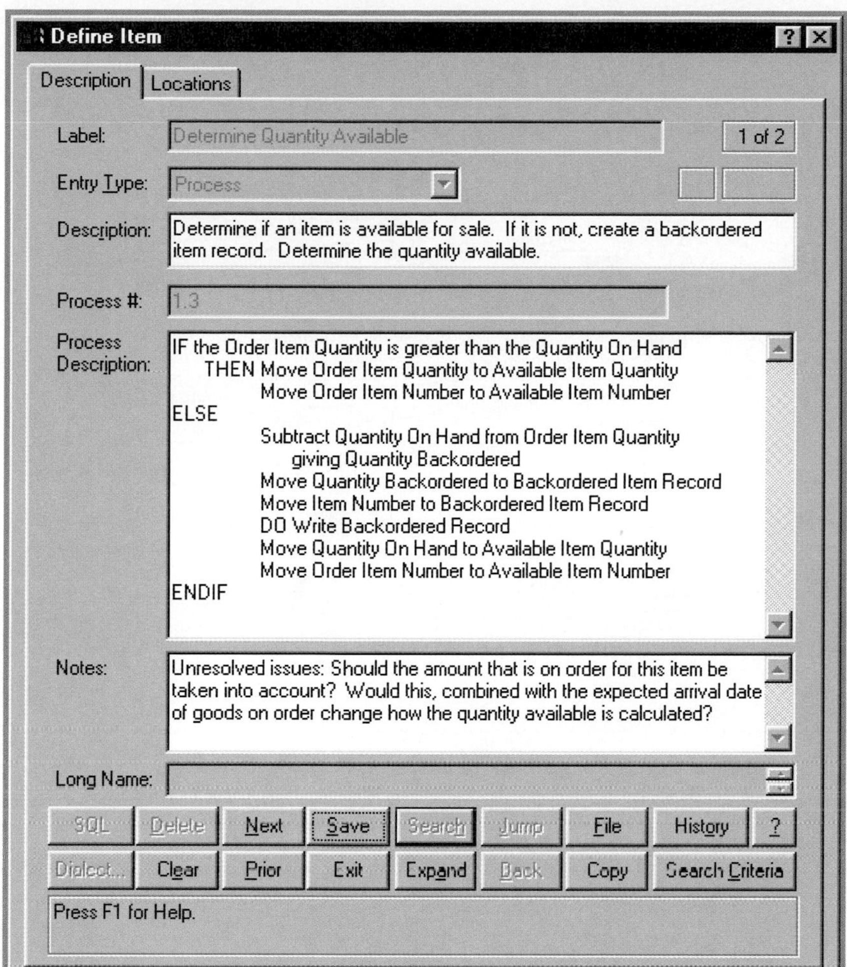
Define Item
Description
Locations
Label:
Determine Quantity Available
1 of 2
Entry Type:
Process
Description:
Determine if an item is available for sale. If it is not, create a backordered item record. Determine the quantity available.
Process #:
1.3
Process Description:
IF the Order Item Quantity is greater than the Quantity On Hand
THEN Move Order Item Quantity to Available Item Quantity
Move Order Item Number to Available Item Number
ELSE
Subtract Quantity On Hand from Order Item Quantity
giving Quantity Backordered
Move Quantity Backordered to Backordered Item Record
Move Item Number to Backordered Item Record
DO Write Backordered Record
Move Quantity On Hand to Available Item Quantity
Move Order Item Number to Available Item Number
ENDIF
Notes:
Unresolved issues: Should the amount that is on order for this item be taken into account? Would this, combined with the expected arrival date of goods on order change how the quantity available is calculated?
Long Name:
SQL
Delete
Next
Save
Search
Jump
File
History
?
Dialect...
Clear
Prior
Exit
Expand
Back
Copy
Search Criteria
Press F1 for Help.

FIGURE 11.3
Visible Analyst can be used to describe process specifications.

Next, actions are identified. Actions can include any instruction that needs to be carried out as a result of one or more of the above conditions. Instructions to somehow manipulate or total numbers, print reports, or even disallow the transaction in question all involve examples of potential actions. They are tied to the conditions by the action rules, which are directions to execute the required actions in order.

Examples of action rules are provided below from a rate document supplied to insurance agents from the Fortress Insurance Corporation:

> Homeowners' insurance, of course, depends on the type of policy and the location of the home, but once that is determined there are other factors that increase or decrease the premium to the homeowner. One factor is the construction. A brick home will save the homeowner 10 percent in the annual premium, and if there is a burglar alarm, 5 percent will be deducted from the premium beyond that. There are choices that homeowners make that will increase the premium, too. If the homeowner wishes to be paid the replacement, rather than depreciated, value, add 10 percent to the base. The homeowner may choose to carry a $100 deductible instead of a $250 deductible, which will increase the premium by 15 percent.

This statement may seem clear at first, but a careful examination reveals ambiguities that need resolution before decision analysis is completed.

In Figure 11.4 this rate document was analyzed to determine the actions and conditions. A box was drawn around each action, and each condition was circled. (Boxes

FIGURE 11.4
Analyzing a rate document to determine actions and conditions.

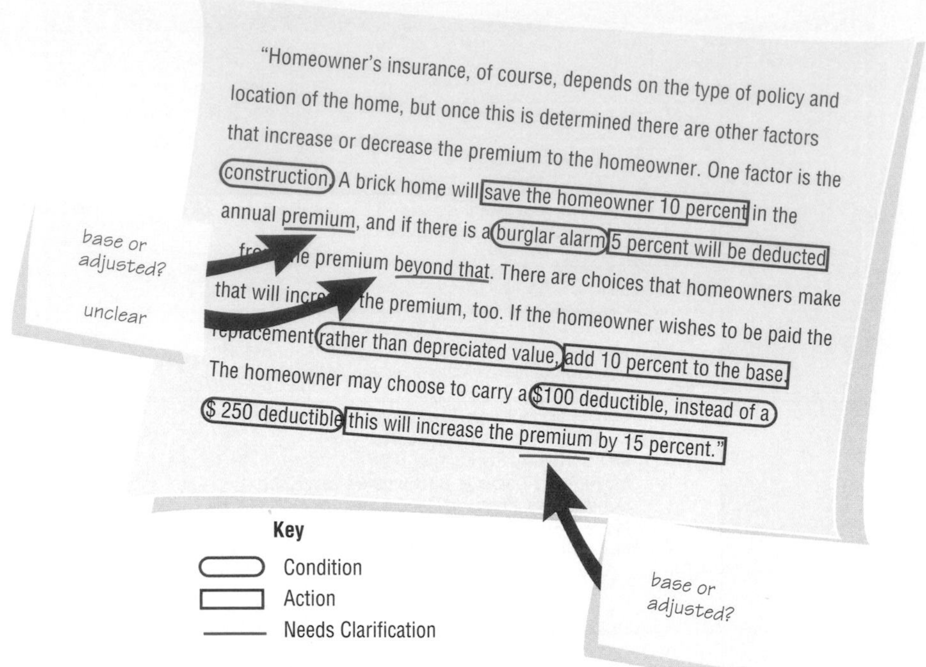
"Homeowner's insurance, of course, depends on the type of policy and location of the home, but once this is determined there are other factors that increase or decrease the premium to the homeowner. One factor is the construction. A brick home will save the homeowner 10 percent in the annual premium, and if there is a burglar alarm 5 percent will be deducted from the premium beyond that. There are choices that homeowners make that will increase the premium, too. If the homeowner wishes to be paid the replacement rather than depreciated value, add 10 percent to the base. The homeowner may choose to carry a $100 deductible, instead of a $ 250 deductible this will increase the premium by 15 percent."
base or adjusted?
unclear
base or adjusted?
Key
Condition
Action
Needs Clarification

and circles are used again later in the decision tree.) Then, questionable terms, ambiguities, unclear adjectives, and instances of "however" and "but" are underlined.

Problems occurred because (1) the "base" is not defined; (2) it isn't clear what the phrase "beyond that" refers to; and (3) when the "premium" is modified, it is not clear if the deduction or increase is applied to the original premium or the adjusted premium, nor is it clear in what order these steps are done. To clear up these details, an interview was conducted and Figure 11.5 was drawn to organize the decision process. Notice that the alternatives are clearly specified, the actions are more specific, "base" is defined, and the action rules are described and ordered.

In the following sections three alternatives for decision analysis of structured decisions are explored. First we discuss structured English, then decision tables, and finally decision trees.

FIGURE 11.5
Organizing the decision process by specifying alternatives and actions, by defining ambiguous terms, and by describing and ordering action rules.

Number	Conditions	Condition Alternatives	Actions	Action Rule
1	Construction	Brick	Deduct 10% of Base* from Subtotal	Do This First
		Other	None	None
2	Home Has Burglar Alarm	Yes	Deduct 5% of Adjusted Subtotal	Do This after Number 4
		No	None	None
3	Replacement Option Is Chosen	Yes	Add 10% of Base* to Subtotal	Do This after Number 1
		No	None	None
4	Deductible	$100 Option	Add 15% of Subtotal to Subtotal	Do This after Number 3
		Standard $250	None	None

*Base is the original premium based on amount the house is insured for and the location of the home.

```
Calculate Base Premium
IF constuction is brick
    THEN deduct 10 percent of base to total
ENDIF
IF replacement option is chosen
    THEN add 10 percent of base to subtotal
ENDIF
If owner chooses $100 deductible
    THEN add 15 percent of subtotal
ENDIF
IF home has burglar alarm
    THEN deduct 5 percent of adjusted subtotal
ENDIF
```

FIGURE 11.6
Using structured English for analyzing the decision process for a simple sequential decision.

STRUCTURED ENGLISH

When the process logic involves formulas or iteration or when structured decisions are not complex, an appropriate technique for analyzing the decision process is the use of structured English. As the name implies, structured English is based on (1) structured logic, or instructions organized into nested and grouped procedures, and (2) simple English statements such as add, multiply, and move.

The previous Fortress Insurance Corporation example provides us with a good use for structured English, and it can be transformed into structured English, as shown in Figure 11.6, by putting the decision rules into their proper sequence and using the convention of IF-THEN-ELSE statements throughout.

Once this example is written in structured English, one can see that it is a rather simple sequential decision. Structured English can be more complex if blocks of instructions are nested within other blocks of instructions, as shown in Figure 11.7.

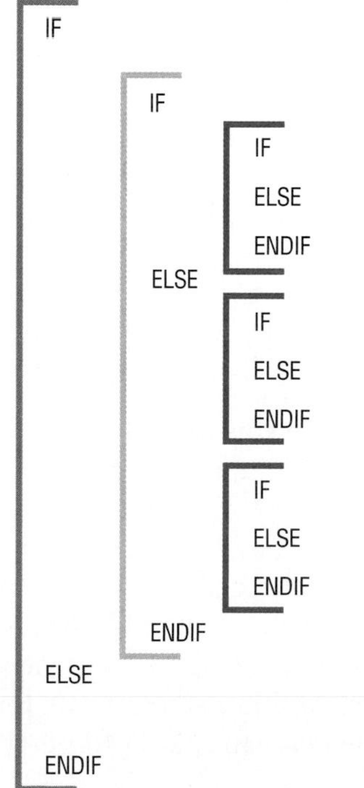

FIGURE 11.7
Proper use of structured English will involve nesting blocks of instructions within others.

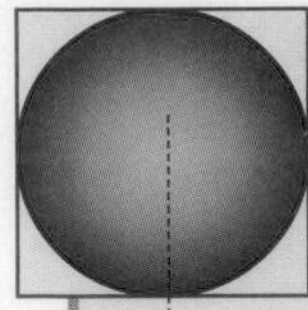

CONSULTING OPPORTUNITY 11.1

KIT CHEN KABOODLE, INC.

"I don't want to get anyone stirred up, but I think we need to sift through our unfilled order policies," says Kit Chen. "I wouldn't want to put a strain on our customers. As you know already, Kit Chen Kaboodle is a Web and mail-order cookware business specializing in 'klassy kitsch for kitchens,' like our latest catalog says. I mean, we've got everything you need to do gourmet cooking and entertaining: nutmeg grinders, potato whisks, egg separators, turkey basters, placemats with cats on 'em, ice cube trays in shamrock shapes, and more.

"Here's how we've been handling unfilled orders. We search our unfilled orders file from Internet as well as mail-order sales once a week. If the order was filled this week, we delete the record, and the rest is gravy. If we haven't written to the customer in four weeks, we send 'em this cute card with a chef peeking into the oven, saying, 'Not ready yet.' (It's a notification that their item is still on back-order.)

"If the back-order date changed to greater than forty-five days from now, we send out a notice. If the merchandise is seasonal (like Halloween treat bags, Christmas cookie cutters, or Valentine's Day cake molds) and the back-order date is thirty days or more, though we send out a notice with a chef glaring at his egg timer.

"If the back-order date changed at all and we haven't sent out a card within two weeks, we send out a card with a chef checking his recipe. If the merchandise is no longer available, we send a notice (complete with chef crying in the corner) and delete the record. We haven't begun to use email in place of mailed cards, but I'd like to.

"Thanks for listening to all this. I think we've got the right ingredients for a good policy; we just need to blend them together and cook up something special."

Because you are the systems analyst whom Kit hired, go through the narrative of how Kit Chen Kaboodle, Inc., handles unfilled orders, drawing boxes around each action Kit mentions and circling each condition brought up. Make notes of any ambiguities you would like to clarify in a later interview.

WRITING STRUCTURED ENGLISH

To write structured English, it is advisable to use the following conventions:

1. Express all logic in terms of sequential structures, decision structures, case structures, or iterations (see Figure 11.8 for examples).
2. Use and capitalize accepted key words such as IF, THEN, ELSE, DO, DO WHILE, DO UNTIL, and PERFORM.
3. Indent blocks of statements to show their hierarchy (nesting) clearly.
4. When words or phrases have been defined in a data dictionary (as in Chapter 10), underline those words or phrases to signify that they have a specialized, reserved meaning.
5. Be careful when using "and" and "or," and avoid confusion when distinguishing between "greater than" and "greater than or equal to" and like relationships. Clarify the logical statements now rather than wait until the program coding stage.

A Structured English Example. The following example demonstrates how a spoken procedure for processing medical claims is transformed into structured English:

> We process all our claims in this manner. First, we determine whether the claimant has ever sent in a claim before; if not, we set up a new record. The claim totals for the year are then updated. Next, we determine if a claimant has policy A or policy B, which differ in deductibles and copayments (the percentage of the claim claimants pay themselves). For both policies, we check to see if the deductible has been met ($100 for plan A and $50 for plan B). If the deductible has not been met, we apply the claim to the deductible. Another step adjusts for the copayment; we subtract the percentage the claimant pays (40 percent for plan A and 60 per-

CONSULTING OPPORTUNITY 11.2

KNEADING STRUCTURE

Kit Chen has risen to the occasion and answered your questions concerning the policy for handling unfilled orders at Kit Chen Kaboodle, Inc. Based on those answers and any assumptions you need to make, pour Kit's narrative (from Consulting Opportunity 11.1) into a new mold by rewriting the recipe for handling unfilled orders in structured English. In a paragraph describe how this process might change if you used email for notification rather than regular mail.

cent for plan B) from the claim. Then we issue a check if there is money coming to the claimant, print a summary of the transaction, and update our accounts. We do this until all claims for that day are processed.

In examining the foregoing statements, one notices some simple sequence structures, particularly at the beginning and end. There are a couple of decision structures, and it is most appropriate to nest them, first by determining which plan (A or B) to use and then by subtracting the correct deductibles and copayments. The last sentence points to an iteration: Either DO UNTIL all the claims are processed or DO WHILE there are claims remaining.

Realizing that it is possible to nest the decision structures according to policy plans, we can write the structured English for the foregoing example (see Figure 11.9). As one begins to work on the structured English, one finds that some logic and relationships that seemed clear at one time are actually ambiguous. For example, do we add the claim to the year-to-date (YTD) claim before or after updating the deductible? Is it possible that an error can occur if some-

Structured English Type	Example
Sequential Structure A block of instructions where no branching occurs	Action #1 Action #2 Action #3
Decision Structure Only IF a condition is true, complete the following statements; otherwise, jump to the ELSE	IF Condition A is True THEN implement Action A ELSE implement Action B ENDIF
Case Structure A special type of decision structure where the cases are mutually exclusive (if one occurs, the others cannot)	IF Case #1 implement Action #1 ELSE IF Case #2 Implement Action #2 ELSE IF Case #3 Implement Action #3 ELSE IF Case #4 Implement Action #4 ELSE print error ENDIF
Iteration Blocks of statements that are repeated until done	DO WHILE there are customers. Action #1 ENDDO

FIGURE 11.8
Examples of logic expressed in a sequential structure, a decision structure, a case structure, and an iteration.

FIGURE 11.9

Structured English for the medical-claim processing system. Underlining signifies that the terms have been defined in the data dictionary.

```
DO WHILE there are claims remaining
   IF claimant has not sent in a claim
      Set up new claimant record
   ELSE continue
 Add claim to YTD Claim
   IF claimant has policy–plan A
      THEN IF deductible of $100.00 has not been met
         THEN subtract deductible–not–met from claim
         Update deductible
      ELSE continue
      ENDIF
      Subtract copayment of 40% of claim from claim
   ELSE IF claimant has policy–plan B.
      THEN IF deductible of $50.00 has not been met
         THEN subtract deductible–not–met from claim
         Update deductible
      ELSE continue
      ENDIF
      Subtract copayment of 60% of claim from claim
   ELSE continue
 ELSE write plan–error–message
 ENDIF
ENDIF
IF claim is greater than zero
   Print check
ENDIF
Print summary for claimant
Update accounts
ENDDO
```

thing other than plan A or B is stored in the claimant's record? We subtract 40 percent of what from the claim? These ambiguities need to be clarified at this point.

Besides the obvious advantage of clarifying the logic and relationships found in human languages, structured English has another important advantage: It is a communication tool. Structured English can be taught to and hence understood by others in the organization, so if communication is important, structured English is a viable alternative for decision analysis.

DATA DICTIONARY AND PROCESS SPECIFICATIONS

All computer programs may be coded using the three basic constructs: sequence, selection (IF . . . THEN . . . ELSE and the case structure), and iteration or looping. The data dictionary indicates which of these constructs must be included in the process specifications.

If the data dictionary for the input and output data flow contains a series of fields without any iteration—{ }—or selection—[ ]—the process specification

```
Shipping Statement =    Order Number +
                        Order Date +
                        Customer Number +
                        Customer Name +
                        Customer Address +
                        {Order Item Lines} (1 to 5) +
                        Number of Items +
                        Merchandise Total +
                        (Tax) +
                        Shipping and Handling +
                        Order Total

Customer Name =         First Name +
                        (Middle Initial) +
                        Last Name

Address =               Street +
                        (Apartment) +
                        City +
                        State +
                        Zip +
                        (Zip Expansion) +
                        (Country)

Order Item Lines =      Item Number +
                        Quantity Ordered +
                        Quantity Backordered +
                        Item Description +
                        Size Description +
                        Color Description +
                        Unit Price +
                        Extended Amount
```

FIGURE 11.10

Data structure for a shipping statement for World's Trend.

will contain a simple sequence of statements, such as MOVE, ADD, and SUBTRACT. Refer to the example of a data dictionary for the SHIPPING STATEMENT, illustrated in Figure 11.10. Notice that the data dictionary for the SHIPPING STATEMENT has the ORDER NUMBER, ORDER DATE, and CUSTOMER NUMBER as simple sequential fields. The corresponding logic, shown in lines 3 through 5 in the corresponding structured English in Figure 11.11, consists of simple move statements.

A data structure with optional elements contained in parentheses or either/or elements contained within brackets will have a corresponding IF . . . THEN . . . ELSE statement in the process specification. Also, if an amount, such as QUANTITY BACK-ORDERED, is greater than zero, the underlying logic will be IF . . . THEN . . . ELSE. Iteration, indicated by braces on a data structure, must have a corresponding DO WHILE, DO UNTIL, or PERFORM UNTIL to control looping on the process specification. The data structure for the Order Items lines allows up to five items in the loop. Lines 8 through 17 show the statements contained within the DO WHILE through the END DO necessary to produce the multiple Order Items.

FIGURE 11.11
Structured English for creating the shipping statement for World's Trend.

Structured English

Format the shipping statement. After each line of the statement has been formatted, write the shipping line.

```
GET Order Record
GET Customer Record
Move Order Number to shipping statement
Move Order Date to shipping statement
Move Customer Number to shipping statement
DO format Customer Name (leave only one space between First/Middle/Last)
DO format Customer Address lines
DO WHILE there are items for the order
        GET Item Record
        DO Format Item Line
        Multiply Unit Price by Quantity Ordered giving Extended Amount
        Move Extended Amount to Order Item line
        Add Extended Amount to Merchandise Total
        IF Quantity Back-Ordered is greater than zero
                Move Quantity Back-Ordered to Order Item line
        ENDIF
ENDDO
Move Merchandise Total to shipping statement
Move 0 to Tax
IF State is equal to CT
        Multiply Merchandise Total by Tax Rate giving Tax
ENDIF
Move Tax to Shipping statement
DO calculate Shipping and Handling
Move Shipping and Handling to shipping statement
Add Merchandise Total, Tax, and Shipping and Handling giving Order Total
Move Order Total to shipping statement
```

DECISION TABLES

A decision table is a table of rows and columns, separated into four quadrants, as shown in Figure 11.12. The upper left quadrant contains the condition(s); the upper right quadrant contains the condition alternatives. The lower half of the table contains the actions to be taken on the left side and the rules for executing the actions on the right. When a decision table is used to determine which action needs to be taken, the logic moves clockwise beginning from the upper left.

Suppose a store wanted to illustrate its policy on noncash customer purchases. The company could do so using a simple decision table as shown in Figure 11.13. Each of the three conditions (sale under $50, pays by check, and uses credit cards) has only two alternatives. The two alternatives are Y (yes, it is true) or N (no, it is not true). Four actions are possible:

1. Ring up the sale.
2. Look up the credit card number in a book before ringing up the sale.
3. Call the supervisor for approval.
4. Call the bank for credit card authorization.

Conditions and Actions	Rules
Conditions	Condition Alternatives
Actions	Action Entries

FIGURE 11.12
The standard format used for presenting a decision table.

The final ingredient that makes the decision table worthwhile is the set of rules for each of the actions. Rules are the combinations of the condition alternatives that precipitate an action. For example, rule 3 says:

IF N (the total sale is NOT under $50.00)

AND

IF Y (the customer paid by check and had two forms of ID)

AND

IF N (the customer did not use a credit card)

THEN

DO X (call the supervisor for approval).

The foregoing example featured a problem with four sets of rules and four possible actions, but that is only a coincidence. The next example demonstrates that decision tables often become large and involved.

DEVELOPING DECISION TABLES

To build decision tables, the analyst needs to determine the maximum size of the table; eliminate any impossible situations, inconsistencies, or redundancies; and simplify the table as much as possible. The following steps provide the analyst with a systematic method for developing decision tables:

1. Determine the number of conditions that may affect the decision. Combine rows that overlap, such as conditions that are mutually exclusive. The number of conditions becomes the number of rows in the top half of the decision table.
2. Determine the number of possible actions that can be taken. That number becomes the number of rows in the lower half of the decision table.
3. Determine the number of condition alternatives for each condition. In the simplest form of decision table, there would be two alternatives (Y or N) for each condition. In an extended-entry table, there may be many alternatives for each condition.

Conditions and Actions	Rules 1	2	3	4
Under $50	Y	Y	N	N
Pays by check with two forms of ID	Y	N	Y	N
Uses credit card	N	Y	N	Y
Ring up sale	X			
Look up credit card in book		X		
Call supervisor for approval			X	
Call bank for credit authorization				X

FIGURE 11.13
Using a decision table for illustrating a store's policy of customer checkout with four sets of rules and four possible actions.

4. Calculate the maximum number of columns in the decision table by multiplying the number of alternatives for each condition. If there were four conditions and two alternatives (Y or N) for each of the conditions, there would be 16 possibilities as follows:

Condition 1:	2 alternatives
Condition 2: ×	2 alternatives
Condition 3: ×	2 alternatives
Condition 4: ×	2 alternatives
	16 possibilities

5. Fill in the condition alternatives. Start with the first condition and divide the number of columns by the number of alternatives for that condition. In the foregoing example, there are 16 columns and two alternatives (Y or N), so 16 divided by 2 is 8. Then choose one of the alternatives, say Y, and write it in the first eight columns. Finish by writing N in the remaining eight columns as follows:

   Condition 1: Y Y Y Y Y Y Y Y N N N N N N N N

   Repeat this step for each condition, using a subset of the table,

   Condition 1: Y Y Y Y Y Y Y Y N N N N N N N N
   Condition 2: Y Y Y Y N N N N
   Condition 3: Y Y N N
   Condition 4: Y N

   and continue the pattern for each condition:

   Condition 1: Y Y Y Y Y Y Y Y N N N N N N N N
   Condition 2: Y Y Y Y N N N N Y Y Y Y N N N N
   Condition 3: Y Y N N Y Y N N Y Y N N Y Y N N
   Condition 4: Y N Y N Y N Y N Y N Y N Y N Y N

6. Complete the table by inserting an X where rules suggest certain actions.
7. Combine rules where it is apparent that an alternative does not make a difference in the outcome. For example,

Condition 1:	Y Y
Condition 2:	Y N
Action 1:	X X

   can be expressed as:

Condition 1:	Y
Condition 2:	—
Action 1:	X

   The dash [—] signifies that Condition 2 can be either Y or N, and the action will still be taken.
8. Check the table for any impossible situations, contradictions, and redundancies. They are discussed in more detail later.
9. Rearrange the conditions and actions (or even rules) if it makes the decision table more understandable.

A Decision Table Example. Figure 11.14 is an illustration of a decision table developed using the steps previously outlined. In this example a company is trying to maintain a meaningful mailing list of customers. The objective is to send out only the catalogs from which customers will buy merchandise.

Conditions and Actions	Rules 1	2	3	4	5	6	7	8
Customer ordered from Fall catalog	Y	Y	Y	Y	N	N	N	N
Customer ordered from Christmas catalog	Y	Y	N	N	Y	Y	N	N
Customer ordered from special catalog	Y	N	Y	N	Y	N	Y	N
Send out this year's Christmas catalog		X		X		X		X
Send out special catalog			X				X	
Send out both catalogs	X				X			

FIGURE 11.14
Constructing a decision table for deciding which catalog to send to customers who order only from selected catalogs.

The company realizes that certain loyal customers order from every catalog and that some people on the mailing list never order. These ordering patterns are easy to observe, but deciding which catalogs to send customers who order only from selected catalogs is more difficult. Once these decisions are made, a decision table is constructed for three conditions (C1: customer ordered from Fall catalog, C2: customer ordered from Christmas catalog, and C3: customer ordered from specialty catalog), each having two alternatives (Y or N). Three actions can be taken (A1: send out this year's Christmas catalog, A2: send out the new specialty catalog, and A3: send out both catalogs). The resulting decision table has six rows (three conditions and three actions) and eight columns (two alternatives × two alternatives × two alternatives).

The decision table is now examined to see if it can be reduced. There are no mutually exclusive conditions, so it is not possible to get by with fewer than three condition rows. No rules allow the combination of actions. It is possible, however, to combine some of the rules as shown in Figure 11.15. For instance, rules 2, 4, 6, and 8 can be combined because they all have two things in common:

1. They instruct us to send out this year's Christmas catalog.
2. The alternative for condition 3 is always N.

It doesn't matter what the alternatives are for the first two conditions, so it is possible to insert dashes [—] in place of the Y or N.

Conditions and Actions	Rules 1	2	3	4	5	6	7	8
Customer ordered from Fall catalog	Y	Y	Y	Y	N	N	N	N
Customer ordered from Christmas catalog	Y	Y	N	N	Y	Y	N	N
Customer ordered from specialty catalog	Y	N	Y	N	Y	N	Y	N
Send out this year's Christmas catalog		X		X		X		X
Send out specialty catalog			X				X	
Send out both catalogs	X				X			

Conditions and Actions	Rules 1'	2'	3'
Customer ordered from Fall catalog	—	—	—
Customer ordered from Christmas catalog	Y	—	N
Customer ordered from specialty catalog	Y	N	Y
Send out this year's Christmas catalog		X	
Send out specialty catalog			X
Send out both catalogs	X		

FIGURE 11.15
Combining rules to simplify the customer-catalog decision table.

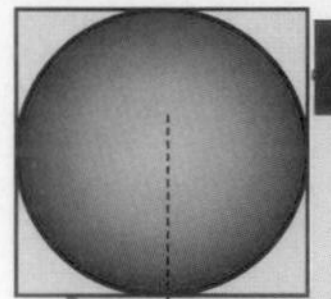

CONSULTING OPPORTUNITY 11.3

SAVING A CENT ON CITRON CAR RENTAL

"We feel lucky to be this popular. I think customers feel we have so many options to offer that they ought to rent an auto from us," says Ricardo Limon, who manages several outlets for Citron Car Rental. "Our slogan is, 'You'll never feel squeezed at Citron.' We have five sizes of cars that we list as A through E.

A Subcompact
B Compact
C Midsize
D Full-size
E Luxury

"Standard transmission is available only for A, B, and C. Automatic transmission is available for all cars."

"If a customer reserves a subcompact (A) and finds on arriving that we don't have one, that customer gets a free upgrade to the next-sized car, in this case a compact (B). Customers also get a free upgrade from their reserved car size if their company has an account with us. There's a discount for membership in any of the frequent flyer clubs run by cooperating airlines, too. When customers step up to the counter, they tell us what size car they reserved, and then we check to see if we have it in the lot ready to go. They usually bring up any discounts, and we ask them if they want insurance and how long they will use the car. Then we calculate their rate and write out a slip for them to sign right there."

Ricardo has asked you to computerize the billing process for Citron so that customers can get their cars quickly and still be billed correctly. Draw a decision table that represents the conditions, condition alternatives, actions, and action rules you gained from Ricardo's narrative that will guide an automated billing process.

Ricardo is thinking about making it possible to reserve a car over the Web. Draw an updated decision table that shows a 10 percent discount for booking a car over the Web.

The remaining rules—1, 3, 5, and 7—cannot be reduced to a single rule because two different actions remain. Instead, rules 1 and 5 may be combined; likewise, rules 3 and 7 can be combined.

CHECKING FOR COMPLETENESS AND ACCURACY

Checking over your decision tables for completeness and accuracy is essential. Four main problems can occur in developing decision tables: incompleteness, impossible situations, contradictions, and redundancy.

Ensuring that all conditions, condition alternatives, actions, and action rules are complete is of utmost importance. Suppose an important condition—if a customer ordered less than $50—had been left out of the catalog store problem discussed earlier. The whole decision table would change because a new condition, a new set of alternatives, new action, and one or more new action rules would have to be added. Suppose the rule is: IF the customer did not order more than $50, THEN do not send any catalogs. A new rule 4 would be added to the decision table, as shown in Figure 11.16.

When building decision tables as outlined in the foregoing steps, it is sometimes possible to set up impossible situations. An example is shown in Figure

FIGURE 11.16
Adding a rule to the customer-catalog decision table changes the entire table.

Conditions and Actions	Rules 1'	2'	3'	4'
Customer ordered from Fall catalog	—	—	—	—
Customer ordered from Christmas catalog	Y	—	N	—
Customer ordered from specialty catalog	Y	N	Y	—
Ordered $50 or more	Y	Y	Y	N
Send out this year's Christmas catalog		X		
Send out specialty catalog			X	
Send out both catalogs	X			
Do not send out any catalog				X

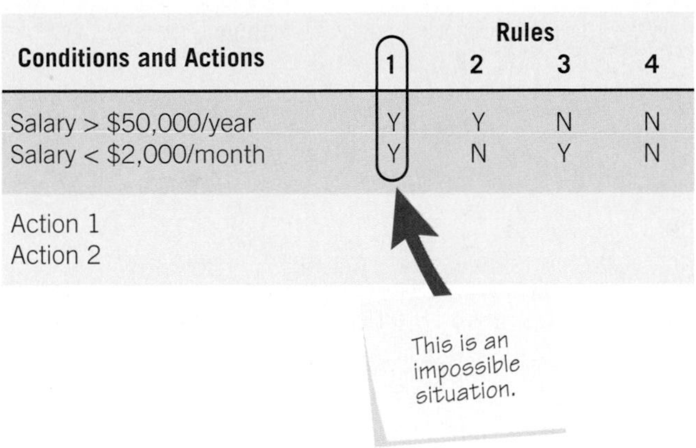

Conditions and Actions	Rules 1	2	3	4
Salary > $50,000/year	Y	Y	N	N
Salary < $2,000/month	Y	N	Y	N
Action 1				
Action 2				

FIGURE 11.17
Checking the decision table for impossible situations is important.

11.17. Rule 1 is not feasible, because a person cannot earn greater than $50,000 per year and less than $2,000 per month at the same time. The other three rules are valid. The problem went unnoticed because the first condition was measured in years and the second condition in months.

Contradictions occur when rules suggest different actions but satisfy the same conditions. The fault could lie with the way the analyst constructed the table or with the information the analyst received. Contradictions often occur if dashes [—] are incorrectly inserted into the table. Redundancy occurs when identical sets of alternatives require the exact same action. Figure 11.18 illustrates a contradiction and a redundancy. The analyst has to determine what is correct and then resolve the contradiction or redundancy.

MORE ADVANCED DECISION TABLES

Decision tables can become very burdensome because they grow rapidly as the number of conditions and alternatives increases. A table with only seven conditions with yes or no alternatives would have 128 columns. To reduce the complexity of unwieldy decision tables, use extended entries or the ELSE rule or to construct multiple tables.

Notice in the following Y or N table that the conditions are mutually exclusive.

C1:	Did not order	Y N N N
C2:	Ordered once	N Y N N
C3:	Ordered twice	N N Y N
C4:	Ordered more than twice	N N N Y

Therefore, the conditions can be written in extended-entry form as follows:

C1: Number of times customer ordered 0 1 2 >2

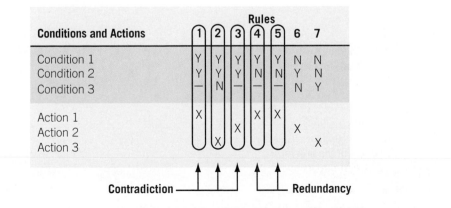

Conditions and Actions	Rules 1	2	3	4	5	6	7
Condition 1	Y	Y	Y	Y	Y	N	N
Condition 2	Y	Y	Y	N	N	Y	N
Condition 3	—	N	—	—	—	N	Y
Action 1	X			X	X		
Action 2			X			X	
Action 3		X					X

Contradiction (Rules 1, 2, 3) — Redundancy (Rules 4, 5)

FIGURE 11.18
Checking the decision table for inadvertent contradictions and redundancy is important.

FIGURE 11.19

Using extended-entry tables reduces the possibility of redundancy and contradiction.

Conditions and Actions	Rules 1	2	3	4	5	6
Cost of the Item		Less Than $10	Between $10 and $50 Inclusive	Greater Than $50	Less Than or Equal to $50	Greater Than $50
Order Quantity	Fewer Than 50 Units	Betweeen 50 and 100 Units Inclusive	Between 50 and 100 Units Inclusive	Between 50 and 100 Units Inclusive	More Than 100 Units	Over 100 Units
Order Immediately			X			
Wait until Regular Order Is Placed	X	X				
Check with Supervisor				X	X	
Send to Purchasing for Bid						X

The number of required columns and rows decreases while the understandability increases. Instead of using four rows for the number of times a customer orders, only one row is needed.

An example of a structured inventory ordering policy is shown in Figure 11.19. The cost of an item can be less than $10, between $10 and $50 inclusive, or greater than $50. In addition, the order quantity can be less than 50 units per order, between 50 and 100 units, or over 100 units. The decision table only has two condition rows, and the alternatives are written in words in the upper right-hand quadrant. By using extended-entry tables, the probability of redundancy and contradiction becomes smaller.

FIGURE 11.20

The ELSE rule can be used to eliminate repetitious rules requiring the same action.

Conditions and Actions	Rules 1	2	3	4	ELSE
Cost of the Item A cost < $10 B $10 ≤ cost ≤ $50 C cost > $50	–	A	B	C	
Order Quantity D quantity < 50 E 50 ≤ quantity ≤ 100 F quantity > 100	D	E	E	F	
Order Immediately			X		
Wait until Regular Order is Placed	X	X			
Send to Purchasing for Bid				X	
Check with Supervisor					X

Another useful technique in building decision tables is to use the ELSE column. This technique is useful in helping to eliminate many repetitious rules requiring the exact same action. It is also useful in preventing errors of omission. Figure 11.20 shows how the automatic inventory ordering policy can take advantage of the ELSE rule.

Decision tables are an important tool in the analysis of structured decisions. One major advantage of using decision tables over other methods is that tables help the analyst ensure completeness. When using decision tables, it is also easy to check for possible errors, such as impossible situations, contradictions, and redundancy. Decision table processors, which take the table as input and provide computer program code as output, are also available.

DECISION TREES

Decision trees are used when complex branching occurs in a structured decision process. Trees are also useful when it is essential to keep a string of decisions in a particular sequence. Although the decision tree derives its name from natural trees, decision trees are most often drawn on their side, with the root of the tree on the left-hand side of the paper; from there, the tree branches out to the right. This orientation allows the analyst to write on the branches to describe conditions and actions.

Unlike the decision tree used in management science, the analyst's tree does not contain probabilities and outcomes, because in systems analysis trees are used mainly for identifying and organizing conditions and actions in a completely structured decision process.

DRAWING DECISION TREES

It is useful to distinguish between conditions and actions when drawing decision trees. This distinction is especially relevant when conditions and actions take place over a period of time and their sequence is important. For this purpose, use a square node to indicate an action and a circle to represent a condition, as shown in Figure 11.21. Using notation makes the decision tree more readable, as does numbering the circles and squares sequentially. Think of a circle as signifying IF, whereas the square means THEN.

When decision tables were discussed in an earlier section, a point-of-sale example was used to determine the purchase approval actions for a department store. Conditions included the amount of the sale (under $50) and whether the

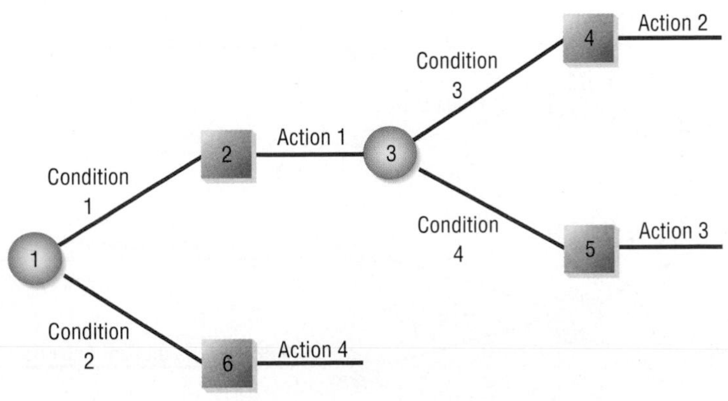

FIGURE 11.21
Conventions for drawing a decision tree where circles can be thought of as meaning IF and squares meaning THEN.

customer paid by check or credit card. The four actions possible were to ring up the sale, look up the credit card in a book, call the supervisor for approval, or call the bank for credit card authorization. Figure 11.22 illustrates how this example can be drawn as a decision tree. In drawing the tree:

1. Identify all conditions and actions and their order and timing (if they are critical).
2. Begin building the tree from left to right while making sure you are complete in listing all possible alternatives before moving over to the right.

This simple tree is symmetrical, and the four actions at the end are unique. A more complex example follows to demonstrate both that the tree does not have to be balanced and that identical actions may appear more than once.

A Complex Decision Tree Example. A systems analyst was hired to help Festival-on-the-Lake, a very popular theatre festival, fill requests for tickets. Here is part of an interview with the box office manager:

> When patrons request seats for a play, we try to fill their request, but often the seat they would like on their first performance date preference is sold out. On our order form, we ask the patron to choose three performance dates in order of preference, to indicate whether they prefer orchestra or balcony, and finally, to select the price of the seat they would like. We have three prices—$25, $20, and $15—and each price is available in both orchestra and balcony.
>
> Because we fill most of our ticket orders by mail, we have to assume a number of things and try to come as close as possible to the desires of the patron.
>
> If orchestra is not available, we assign balcony and vice versa, but if the first night selected is unavailable in the price requested either in orchestra or balcony, we look at the second night. If the second night is unavailable, we look for options on the third. If the third night is unavailable, we select the next lower price and repeat the process. If the patron originally selected the lowest price, we must issue a notice that the ticket request could not be filled.

The foregoing interview suggests the use of a decision tree for two reasons: (1) The process is accomplished in stages ("First we try this, and if it doesn't work, we try this") and (2) the logic is asymmetrical, lending itself to decision trees rather than decision tables.

FIGURE 11.22

Drawing a decision tree to show the noncash purchase approval actions for a department store.

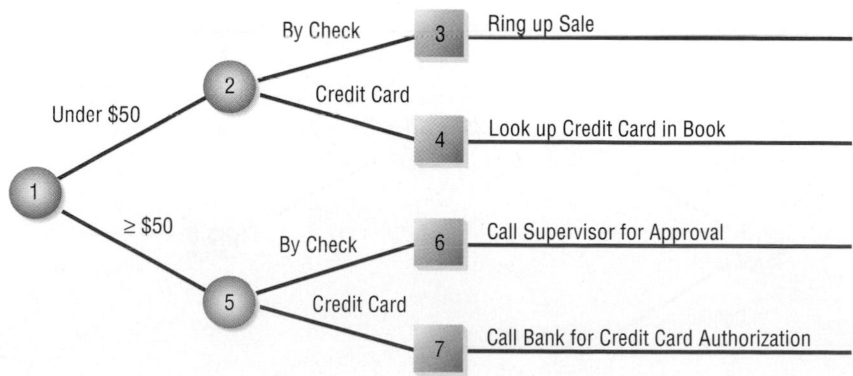

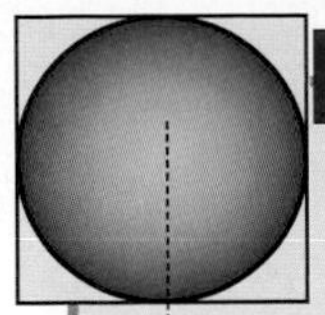

CONSULTING OPPORTUNITY 11.4

A TREE FOR FREE

"I know you've got a plane to catch, but let me try to explain it once again to you, sir," pleads Glen Curtiss, a marketing manager for Premium Airlines. Curtiss has been attempting (unsuccessfully) to explain the airline's new policy for accumulating miles for awards (such as upgrades to first class and free flights) to a member of Premium's "Flying for Prizes" club.

Glen takes another pass at getting the policy off the ground, saying, "You see, sir, the traveler (that's you, Mr. Icarus) will be awarded the miles actually flown. If the actual mileage for the leg was less than 500 miles, the traveler will get 500 miles credit. If the trip was made on a Saturday, the actual mileage will be multiplied by two. If the trip was made on a Tuesday, the multiplication factor is 1.5. If this is the ninth leg traveled during the calendar month, the mileage is doubled no matter what day, and if it is the seventeenth leg traveled, the mileage is tripled. If the traveler booked the flight on the Web, 100 miles are added.

"I hope that clears it up for you, Mr. Icarus. Enjoy your flight, and thanks for flying Premium."

Mr. Icarus, whose desire to board the Premium plane has all but melted away during Glen's long explanation, fades into the sea of people wading through the security lanes, without so much as a peep in reply.

Develop a decision tree for Premium Airlines' new policy for accumulating award miles so that the policy becomes clearer, is easier to grasp visually, and hence is easier to explain.

The steps in building the decision tree are as follows:

1. Identify the conditions:
   first choice of dates
   second choice of dates
   third choice of dates
   first price preference
   second price preference
   third price preference
   orchestra preference
   balcony preference
2. Identify the condition alternatives:
   available
   not available
3. Identify the actions:
   assign seats
   issue tickets
   check first choice of dates
   check second choice of dates
   check third choice of dates
   repeat process for next lower price category
   check location preference
   issue sold out notice
4. Identify action rules (in order):
   start with ideal choice
   assign seats if available
   issue tickets if assigned
   check alternate location (orchestra/balcony)
   look at next choice of dates
   look at next price category
   issue notice if order is unfilled

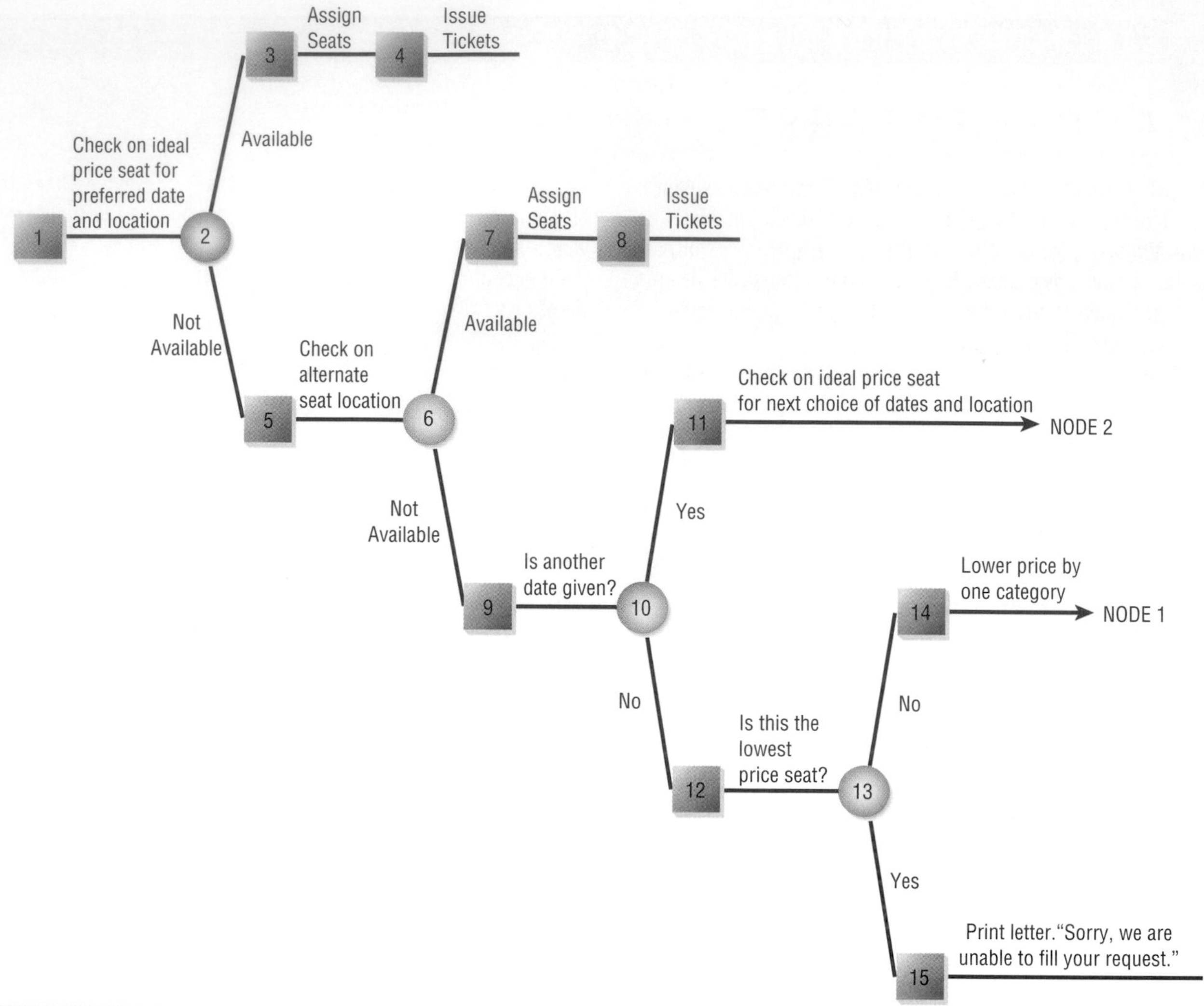

FIGURE 11.23

Building a decision tree for a process accomplished in stages whose logic is asymmetrical.

The decision tree for the theatre festival is found in Figure 11.23. Actions are represented by the square nodes, and conditions are depicted by the circles. Notation of this sort keeps the processes clearer. The circles represent IF, and the squares represent THEN.

In building the tree, focus on the first action that should be taken, place it on the extreme left side, and then build from left to right with conditions or further actions. In the theatre example, choose ideal price and location on first choice of dates. If this ideal choice is available, two actions are taken: (1) The seat is assigned, and (2) tickets are issued. If the ideal choice is not available, the action taken is to check on availability of the alternate location (orchestra or balcony). If that is not available, the second choice of dates is examined, and so forth.

One caution about decision trees bears mentioning. The tree takes up considerable space, and consequently a minimum amount of description on conditions and actions is written on the tree. Instructions such as "Assign Seats" are not spelled out in enough detail at this point. The systems analyst needs to be aware of the implications of this brief statement: In this case, reserve the seats chosen so that they cannot be assigned to more than one patron, and then match the seat numbers with the patron so that they can be mailed. Furthermore, data requirements are not specified on the tree; only the conditions and actions are specified.

The decision tree has three main advantages over a decision table. First, it takes advantage of the sequential structure of decision tree branches so that the order of checking conditions and executing actions is immediately noticeable. Second, conditions and actions of decision trees are found on some branches but not on others, which contrasts with decision tables, where they are all part of the same table. Those conditions and actions that are critical are connected directly to other conditions and actions, whereas those conditions that do not matter are absent. In other words, the tree does not have to be symmetrical. Third, compared with decision tables, decision trees are more readily understood by others in the organization. Consequently, they are more appropriate as a communication tool.

CHOOSING A STRUCTURED DECISION ANALYSIS TECHNIQUE

We have examined the three techniques for analysis of structured decisions: structured English, decision tables, and decision trees. Although they need not be used exclusively, it is customary to choose one analysis technique for a decision rather than employing all three. The following guidelines provide you with a way to choose one of the three techniques for a particular case:

1. Use structured English when
   a. There are many repetitious actions.
      OR
   b. Communication to end users is important.
2. Use decision tables when
   a. Complex combinations of conditions, actions, and rules are found.
      OR
   b. You require a method that effectively avoids impossible situations, redundancies, and contradictions.
3. Use decision trees when
   a. The sequence of conditions and actions is critical.
      OR
   b. When not every condition is relevant to every action (the branches are different).

PHYSICAL AND LOGICAL PROCESS SPECIFICATIONS

The remaining sections in this chapter are advanced topics that may be explored further if you wish. The first topic shows how a data flow diagram can be transformed into process specifications. The second section explains how process specifications can in turn be used to balance (and correct) a data flow diagram.

Each data flow diagram process expands to a child diagram, a structure chart (discussed in Chapter 20), or process specifications (as structured English). If the process is primitive, the specifications show the logic, arithmetic, or algorithm for transforming the input into output. These specifications are a portion of the logical model—the business rules—that would exist regardless of the type of system used to implement the business. Business rules often form the basis for creating procedural language when using code generators.

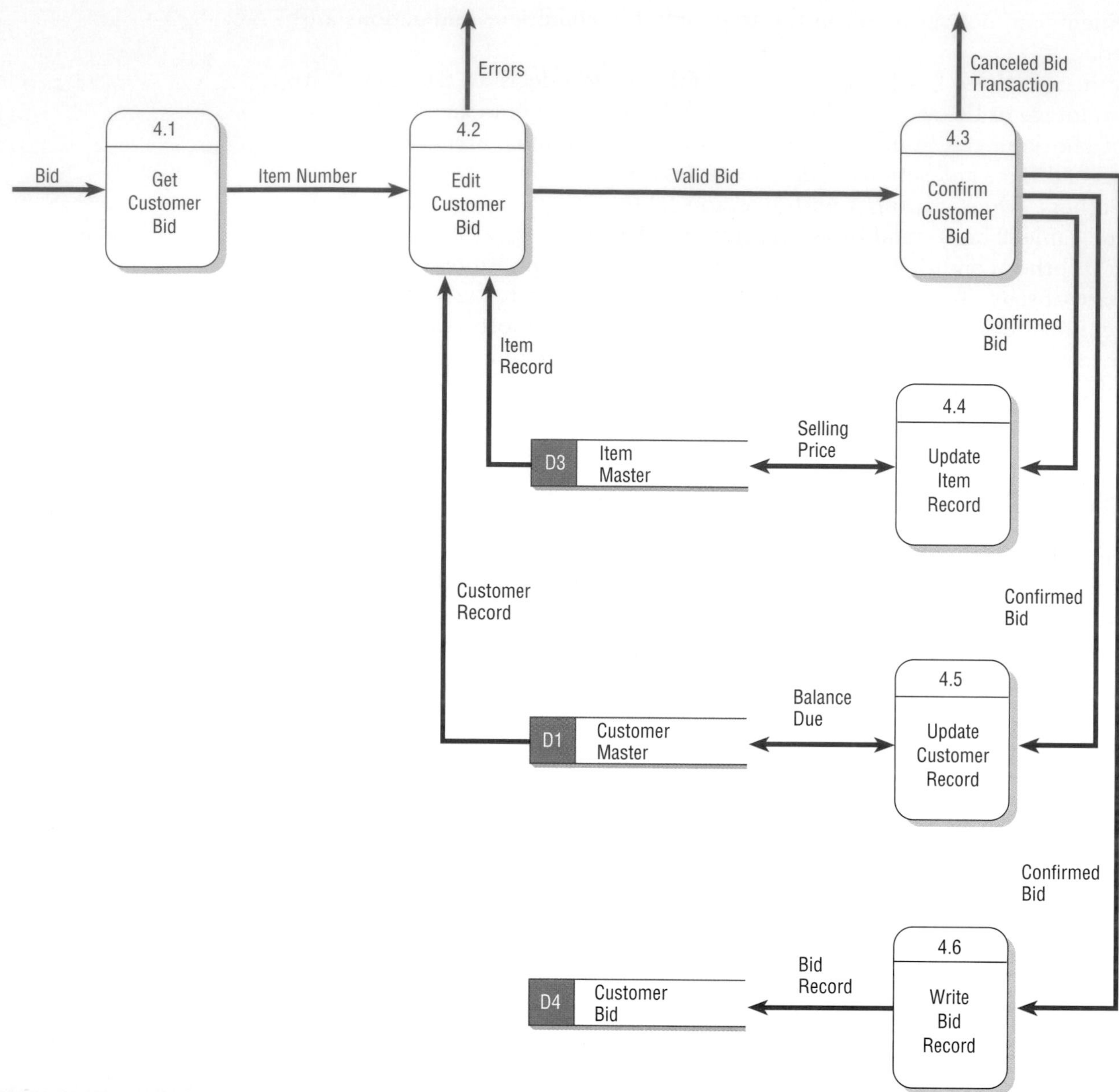

FIGURE 11.24
Data flow diagram explosion of process 4, RECORD CUSTOMER BID.

For example, we observe that an auction house has a computer system to keep track of successful customer bids (process 4) and to produce a payment statement for the person supplying the auctioned item (process 3).

If the process expands to a child diagram or a structure chart, the process specification describes the order and conditions under which the child diagram processes will execute. This control logic is a part of the physical model and would be created after the method of implementation (either batch or online) for the process has been determined. Figure 11.24 shows the auction system Diagram 4, an explosion of process 4, RECORD CUSTOMER BID. Figure 11.25 illustrates the structured English format for process 4. Note that most of the logic involves IF and PERFORM statements, which are typical of controlling program modules.

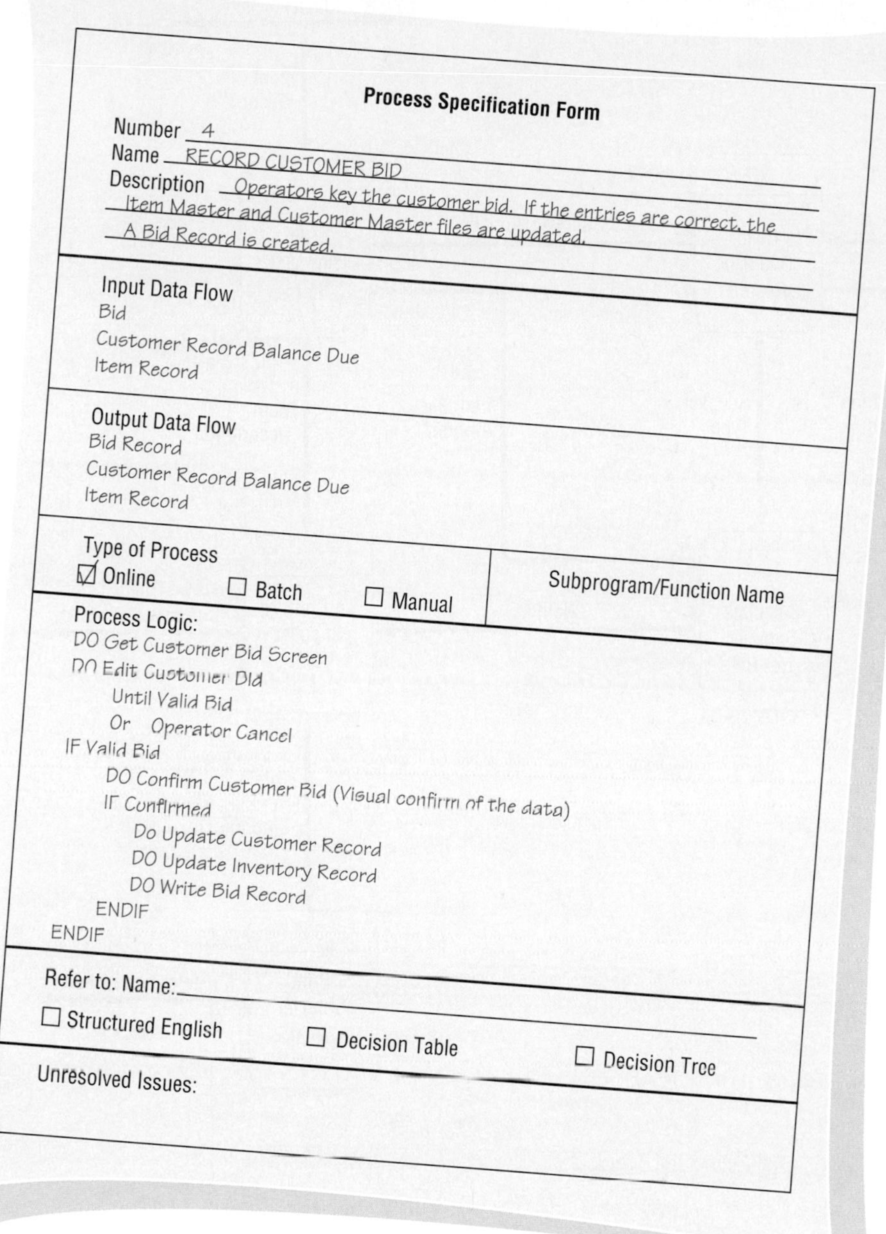
Process Specification Form

Number 4

Name RECORD CUSTOMER BID

Description Operators key the customer bid. If the entries are correct, the Item Master and Customer Master files are updated. A Bid Record is created.

Input Data Flow
Bid
Customer Record Balance Due
Item Record

Output Data Flow
Bid Record
Customer Record Balance Due
Item Record

Type of Process
☑ Online ☐ Batch ☐ Manual

Subprogram/Function Name

Process Logic:

```
DO Get Customer Bid Screen
DO Edit Customer Bid
    Until Valid Bid
    Or  Operator Cancel
IF Valid Bid
    DO Confirm Customer Bid (Visual confirm of the data)
    IF Confirmed
        Do Update Customer Record
        DO Update Inventory Record
        DO Write Bid Record
    ENDIF
ENDIF
```

Refer to: Name:

☐ Structured English ☐ Decision Table ☐ Decision Tree

Unresolved Issues:

FIGURE 11.25
Structured English for a process that explodes to a child diagram.

USING PROCESS SPECIFICATIONS: HORIZONTAL BALANCING

Process specifications, whether on paper or captured using a CASE tool, may be used for generating computer language source code and for analyzing the system design. Computer programs are indicated by partitioning on a data flow diagram. All the individual process specifications for a program are consolidated to become the processing details in a program specification packet.

To attempt to write the specifications for a program without examining each process may lead to omissions and errors. Because process specifications are developed on a small scale, one process at a time, each one may be analyzed for complete and correct logic. When the analysis is finished and corrections are made for all processes within a program, the final program specifications should be complete and accurate.

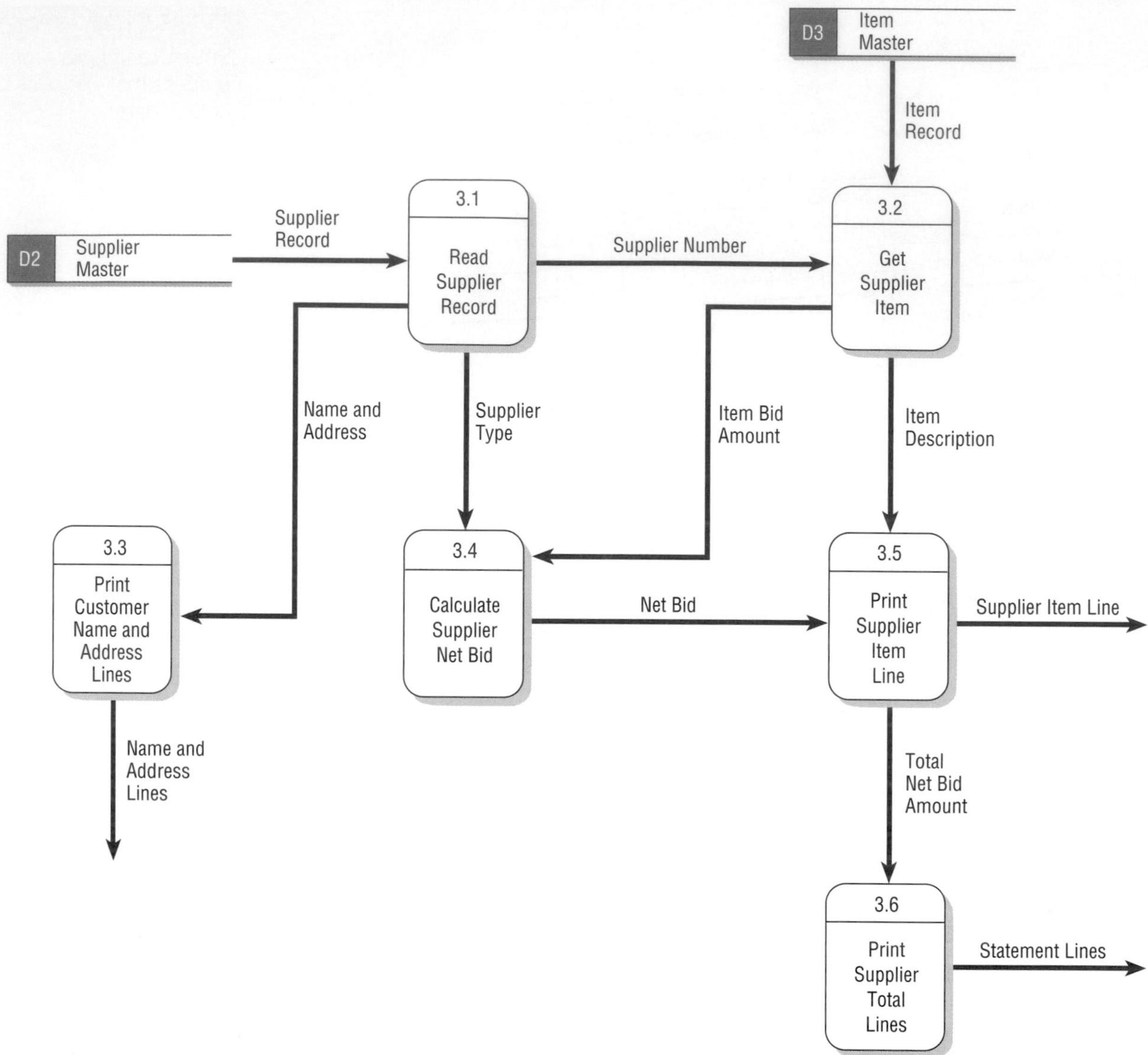

FIGURE 11.26
Diagram 3, Produce Supplier Payment Statement.

Process specifications may be used to analyze the data flow diagram and data dictionary through a method called horizontal balancing. Horizontal balancing dictates that all output data flow elements must be obtained from the input elements and process logic. Base elements on an output data flow must be present on the input flow, and derived elements on an output flow must be either present on an input data flow or created using the process specifications. Unresolved areas should be summarized into a series of interview questions. Be sure to pose these questions during follow-up interviews with key users.

Figure 11.26 illustrates Diagram 3, an explosion of the Auction System process 3, PRODUCE SUPPLIER PAYMENT STATEMENT. Figure 11.27 shows the corresponding data dictionary entries. Figure 11.28 is the structured English for process 3.4, CALCULATE SUPPLIER NET BID, and for process 3.5, PRINT BID LINE. Figure 11.29 is the structured English for process 3.6, PRINT SUPPLIER TOTAL DUE LINE.

Supplier Sales Receipt = Current Date +
Supplier Type +
Supplier Name +
Address +
{Supplier Item Line} +
Auction Surcharge +
Total Amount Paid to Supplier

Supplier Item Line = Item Description +
(Item Bid Amount) +
(Net Bid) +
(Date Sold)

Supplier Record = Supplier Number +
Supplier Type +
Supplier Name +
Address

Item Record = Item Number +
Item Description +
(Item Bid Amount) +
(Amount Paid to Supplier) +
(Date Sold) +
Supplier Number

FIGURE 11.27
Data dictionary entries for producing the Supplier Sales Receipt.

The output from process 3.4 is the NET BID for each item, a derived element. The logic for the process requires as input the SUPPLIER TYPE and ITEM BID AMOUNT, both used in the NET BID calculation. Checking the data flow diagram reveals that both these elements are input to process 3.4. Only the NET BID is shown as output from the process, however. The AMOUNT PAID TO SUPPLIER, included in the structured English figure, is not shown on the data flow diagram, nor is the ITEM MASTER data store. The YEAR-TO-DATE NET BID is not included in the data dictionary for the SUPPLIER RECORD, nor is it shown on the data flow diagram. The data flow diagram and data dictionary must be updated to include these missing components. Figure 11.30 shows Diagram 3 with the necessary corrections.

Examine the output from process 3.5. The SUPPLIER ITEM LINE contains four elements: ITEM DESCRIPTION, ITEM BID AMOUNT, NET BID, and DATE SOLD. The ITEM DESCRIPTION and NET BID are input to process 3.5, but DATE SOLD and ITEM BID AMOUNT, which are base elements, are not on any input flow. They must be added to the data flow diagram. To avoid having three input flows (ITEM DESCRIPTION, ITEM BID AMOUNT, and DATE SOLD) traveling from process 3.2 to 3.3, the entire item record is passed between the two processes. The final process to be examined is 3.6. The structured English requires that SUPPLIER TYPE and TOTAL NET BID AMOUNT be present as input flow. Because only the TOTAL NET BID AMOUNT is present, process 3.6 has missing input.

FIGURE 11.28

Structured English description for processes 3.4 and 3.5.

Structured English: Process 3.4, CALCULATE SUPPLIER NET BID

BEGIN CASE
IF the Supplier Type is a charitable organization
  THEN Commission Rate = 10%
ELSE IF the Supplier Type is a government unit
  THEN Commission Rate = 15%
ELSE IF the Supplier Type is a bankruptcy
  THEN Commission Rate = 18%
ELSE IF the Supplier Type is an estate
  THEN Commission Rate = 20%
ELSE Commission Rate = 25%
END CASE
Multiply Item Bid Amount by Commission Rate giving Commission
Subtract Commission from Item Bid Amount giving Net Bid
Move Net Bid to the Amount Paid in Supplier on the Item Record
Rewrite the Item Record
Add Net Bid to Year-to-Date Net Bid on the Supplier Record
Rewrite the Supplier Record

Structured English: Processes 3.5, PRINT BID LINE

Move Item Description to Supplier Item Line
Move Date Sold to Supplier Item Line
Move Item Bid Amount to Supplier Item Line
Move Net Bid to the Supplier Item Line
Write Supplier Item Line
Add Net Bid to Total Net Bid Amount

FIGURE 11.29

Structured English description for process 3.6.

Structured English: Process 3.6, PRINT SUPPLIER TOTAL DUE LINE

Note: The Auction Surcharge is a one-time cost per auction to cover set-up costs.

BEGIN CASE
IF the Supplier Type is a charitable organization
  THEN Auction Surcharge = $200
ELSE IF the Supplier Type is a governmental unit
  THEN Auction Surcharge = $500
ELSE IF the Supplier Type is a bankruptcy
  THEN Auction Surcharge = $400
ELSE IF the Supplier Type is an estate
  THEN Auction Surcharge = $300
ELSE Auction Surcharge = $500
END CASE
Multiply Item Bid Amount to Statement Line
Write Statement Line
Move Auction Surcharge to Statement Line
Write Statement Line
Subtract Auction Surcharge from Total Net Bid Amount giving Payment Total
Move Payment Total to Statement Line
Write Statement Line

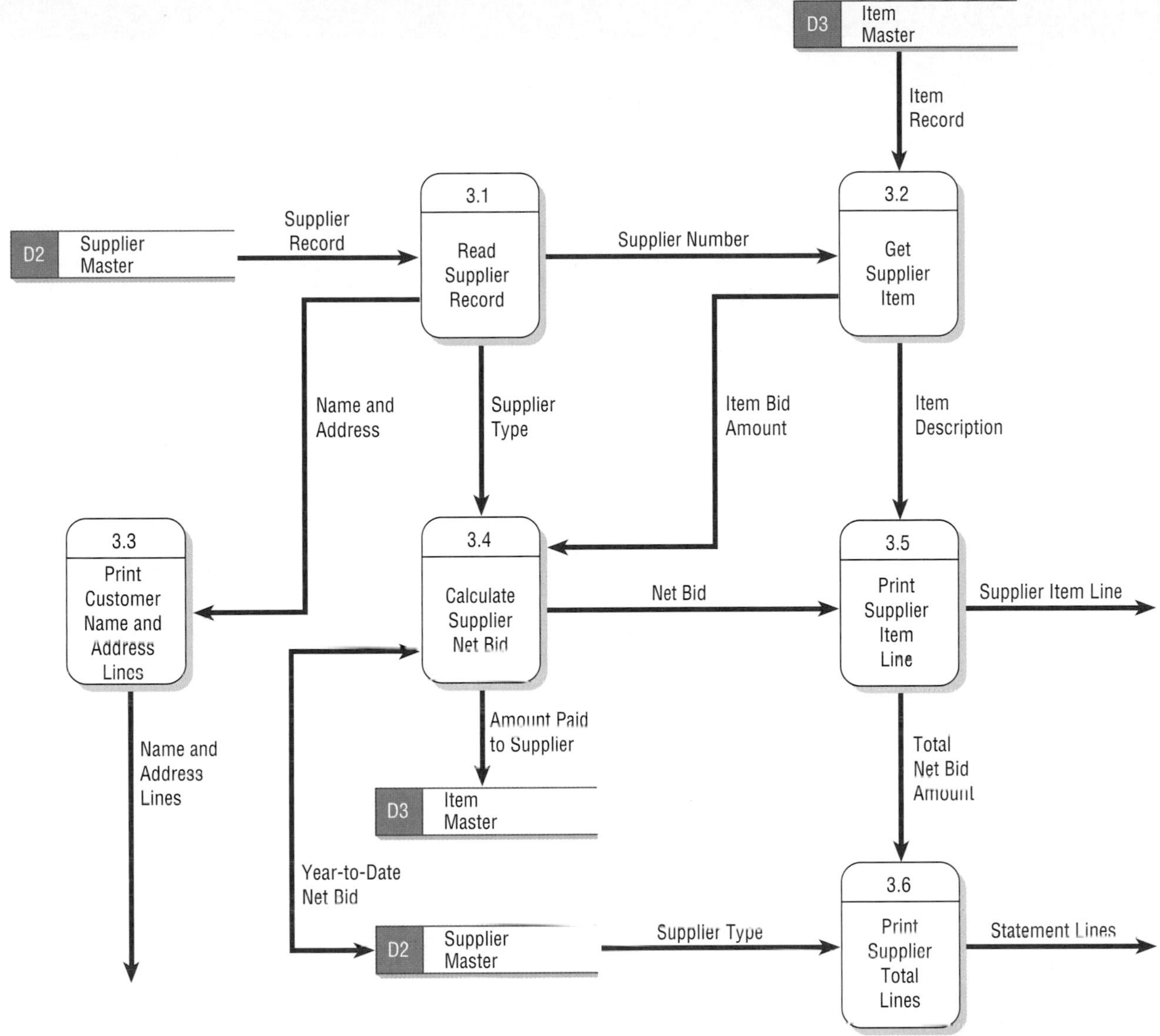

FIGURE 11.30
Corrected data flow diagram, an explosion of process 3, Produce Supplier Payment Statement.

SUMMARY

Once the analyst identifies data flows and begins constructing a data dictionary, it is time to turn to process specification and decision analysis. The three methods for decision analysis and describing process logic discussed in this chapter are structured English, decision tables, and decision trees.

Process specifications (or minispecs) are created for primitive processes on a data flow diagram as well as for some higher-level processes that explode to a child diagram. These specifications explain the decision-making logic and formulas that will transform process input data into output. The three goals of process specification are to reduce the ambiguity of the process, to obtain a precise description of what is accomplished, and to validate the system design.

A large part of a systems analyst's work will involve structured decisions, that is, decisions that can be automated if identified conditions occur. To do that, the analyst needs to define four variables in the decision being examined: conditions, condition alternatives, actions, and action rules.

One way to describe structured decisions is to use the method referred to as structured English, where logic is expressed in sequential structures, decision

HYPERCASE® EXPERIENCE 11

"It's really great that you've been able to spend all of this time with us. One thing's for sure, we can use the help. And clearly, from your conversations with Snowden and others, you must realize we all believe that consultants have a role to play in helping companies change. Well, most of us believe it anyway.

"Sometimes structure is good for a person. Or even a company. As you know, Snowden is keen on any kind of structure. That's why some of the Training people can drive him wild sometimes. They're good at structuring things for their clients, but when it comes to organizing their own work, it's another story. Oh well, let me know if there's any way I can help you."

HYPERCASE® QUESTION

1. Assume you will create the specifications for an automated project tracking system for the Training employees. One of the system's functions will be to allow project members to update or add names, addresses, and phone/fax numbers of new clients. Using structured English, write a procedure for carrying out the process of entering a new client name, address, and phone/fax number. [*Hint*: The procedure should ask for a client name, check to see if the name is already in an existing client file, and let the user either validate and update the current client address and phone/fax (if necessary) or add a new client's address and phone/fax number to the client file.]

structures, case structures, or iterations. Structured English uses accepted key words such as IF, THEN, ELSE, DO, DO WHILE, and DO UNTIL to describe the logic used, and it indents to indicate the hierarchical structure of the decision process.

Decision tables provide another way to examine, describe, and document decisions. Four quadrants (viewed clockwise from the upper left-hand corner) are used to (1) describe the conditions, (2) identify possible decision alternatives (such as Y or N), (3) indicate which actions should be performed, and (4) describe the actions. Decision tables are advantageous because the rules for developing the table itself, as well as the rules for eliminating redundancy, contradictions, and impossible situations, are straightforward and manageable. The use of decision tables promotes completeness and accuracy in analyzing structured decisions.

The third method for decision analysis is the decision tree, consisting of nodes (a square for actions and a circle for conditions) and branches. Decision trees are appropriate when actions must be accomplished in a certain sequence. There is no requirement that the tree be symmetrical, so only those conditions and actions that are critical to the decisions at hand are found on a particular branch.

Each of the decision analysis methods has its own advantages and should be used accordingly. Structured English is useful when many actions are repeated and when communicating with others is important. Decision tables provide a complete analysis of complex situations while limiting the need for change attributable to impossible situations, redundancies, or contradictions. Decision trees are important when proper sequencing of conditions and actions is critical and when each condition is not relevant to each action.

Each data flow diagram process expands to a child diagram, a structure chart, or process specifications (as structured English). If the process is primitive, the specifications show the logic, arithmetic, or algorithm for transforming the input into output. These logical model specifications are part of the business rules (which are often used as the basis for creating procedural language when code generators are used).

If the process expands to a child diagram or a structure chart, the process specification describes the order and conditions under which the child diagram processes will execute. This control logic is part of the physical model.

Process specifications may be used to analyze the data flow diagram and data dictionary through a method called horizontal balancing, which dictates that all output data flow elements must be obtained from the input elements and process logic. Unresolved areas can be posed as questions in follow-up interviews.

KEYWORDS AND PHRASES

action rules
actions
condition alternatives
conditions
contradictions
decision tables
decision trees
horizontal balancing
impossible situations
incompleteness
minispecs
process specifications
redundancy
structured decision
structured English

REVIEW QUESTIONS

1. List three reasons for producing process specifications.
2. Define what is meant by a structured decision.
3. What four elements must be known for the systems analyst to design systems for structured decisions?
4. What are the two building blocks of structured English?
5. List five conventions that should be followed when using structured English.
6. What is the advantage of using structured English to communicate with people in the organization?
7. Which quadrant of the decision table is used for conditions? Which is used for condition alternatives?
8. What is the first step to take in developing a decision table?
9. List the four main problems that can occur in developing decision tables.
10. What is one way to reduce the complexity of unwieldy decision tables?
11. What is one of the major advantages of decision tables over other methods of decision analysis?
12. What are the main uses of decision trees in systems analysis?
13. List the four major steps in building decision trees.
14. What three advantages do decision trees have over decision tables?
15. In which two situations should you use structured English?
16. In which two situations do decision tables work best?
17. In which two situations are decision trees preferable?
18. How do data dictionary structures help in determining the type of structured English statements for a process?
19. What is horizontal balancing? Why is it desirable to balance each process?

PROBLEMS

1. Clyde Clerk is reviewing his firm's expense reimbursement policies with the new salesperson, Trav Farr. "Our reimbursement policies depend on the situation. You see, first we determine if it is a local trip. If it is, we only pay mileage

of 18.5 cents a mile. If the trip was a one-day trip, we pay mileage and then check the times of departure and return. To be reimbursed for breakfast, you must leave by 7:00 A.M., lunch by 11:00 A.M., and have dinner by 5:00 P.M. To receive reimbursement for breakfast, you must return later than 10:00 A.M., lunch later than 2:00 P.M., and have dinner by 7:00 P.M. On a trip lasting more than one day, we allow hotel, taxi, and airfare as well as meal allowances. The same times apply for meal expenses." Write structured English for Clyde's narrative of the reimbursement policies.

2. Draw a decision tree depicting the reimbursement policy given in problem 1.
3. Draw a decision table for the reimbursement policy given in problem 1.
4. A computer supplies firm called True Disk has set up accounts for countless businesses in Dosville. True Disk sends out invoices monthly and will give discounts if payments are made within 10 days. The discounting policy is as follows: If the amount of the order for computer supplies is greater than $1,000, subtract 4 percent for the order; if the amount is between $500 and $1,000, subtract a 2 percent discount; if the amount is less than $500, do not apply any discount. All orders made via the Web automatically receive an extra 5 percent discount. Any special order (computer furniture, for example) is exempt from all discounting.

   Develop a decision table for True Disk discounting decisions, where the condition alternatives are limited to Y and N.
5. Develop an extended-entry decision table for the True Disk company discount policy described in problem 4.
6. Develop a decision tree for the True Disk company discount policy presented in problem 4.
7. Write structured English to solve the True Disk company situation in problem 4.
8. Premium Airlines has recently offered to settle claims for a class-action suit, which was originated for alleged price fixing of tickets. The proposed settlement is stated as follows:

   Initially, Premium Airlines will make available to the settlement class a main fund of $25 million in coupons. If the number of valid claims submitted is 1.25 million or fewer, the value of each claim will be the result obtained by dividing $25 million by the total number of valid claims submitted. For example, if there are 500,000 valid claims, each person submitting a valid claim will receive a coupon with a value of $50.

   The denomination of each coupon distributed will be in a whole dollar amount not to exceed $50. Thus, if there are fewer than 500,000 valid claims, the value of each claim will be divided among two coupons or more. For example, if there are 250,000 valid claims, each person submitting a valid claim will receive two coupons, each having a face value of $50, for a total coupon value of $100.

   If the number of valid claims submitted is between 1.25 million and 1.5 million, Premium Airlines will make available a supplemental fund of coupons, with a potential value of $5 million. The supplemental fund will be made available to the extent necessary to provide one $20 coupon for each valid claim.

   If there are more than 1.5 million valid claims, the total amount of the main fund and the supplemental fund, $30 million, will be divided evenly to produce one coupon for each valid claim. The value of each such coupon will be $30 million divided by the total number of valid claims.

   Draw a decision tree for the Premium Airlines settlement.

9. Write structured English for Premium Airlines in problem 8.
10. "Well, it's sort of hard to describe," says Sharon, a counselor at Less Is More Nutrition Center. "I've never had to really tell anybody about the way we charge clients or anything, but here goes.

    "When clients come into Less Is More, we check to see if they've ever used our service before. Unfortunately for them, I guess, we have a lot of repeat clients who keep bouncing back. Repeat clients get a reduced rate (pardon the pun) of $100 for the first visit if they return within a year of the end of their program.

    "Everyone new pays an initial fee, which is $200 for a physical evaluation. The client may bring in a coupon at this time, and then we deduct $50 from the up-front fee. Half of our clients use our coupons and find out about us from them. We just give our repeaters their $100 off, though; they can't use a coupon, too! Clients who transfer in from one of our centers in another city get $75 off their first payment fee, but the coupon doesn't apply. Customers who pay cash get 10 percent off the $200, but they can't use a coupon with that."

    Create a decision table with Y and N conditions for the client charge system at Less Is More Nutrition Center.
11. Reduce the decision table in Figure 11.EX1 to the minimum number of rules.

GROUP PROJECTS

1. Each group member (or each subgroup) should choose to become an "expert" and prepare to explain how and when to use one of the following structured decision techniques: structured English, decision tables, or decision trees. Each group member or subgroup should then make a case for the usefulness of its assigned decision analysis technique for studying the types of structured decisions made by Maverick Transport on dispatching particular trucks to particular destinations. Each group should make a presentation of its preferred technique.
2. After hearing each presentation, the group should reach a consensus on *which* technique is most appropriate for analyzing the dispatching decisions of Maverick Transport and *why* that technique is best in this instance.

Conditions and Actions	Rules															
	1	2	3	4	5	6	7	8	9	10	11	12	13	14	15	16
Sufficient Quantity on Hand	Y	Y	Y	Y	Y	Y	Y	Y	N	N	N	N	N	N	N	N
Quantity Large Enough for Discount	Y	Y	Y	Y	N	N	N	N	Y	Y	Y	Y	N	N	N	N
Wholesale Customer	Y	Y	N	N	Y	Y	N	N	Y	Y	N	N	Y	Y	N	N
Sales Tax Exemption Filed	Y	N	Y	N	Y	N	Y	N	Y	N	Y	N	Y	N	Y	N
Ship Items and Prepare Invoice	X	X	X	X	X	X	X	X								
Set up Back Order									X	X	X	X	X	X	X	X
Deduct Discount	X	X														
Add Sales Tax		X	X	X		X	X	X								

FIGURE 11.EX1
A decision table for a warehouse.

SELECTED BIBLIOGRAPHY

Adam, E. E., Jr., and R. J. Ebert. *Production and Operations Management*, 3d ed. Englewood Cliffs, NJ: Prentice Hall, 1986.

Anderson, D. R., D. J. Sweeney, and T. A. Williams. *An Introduction to Management Science*, 8th ed. New York: West, 1997.

Awad, E. M. *Systems Analysis and Design*, 2d ed. Homewood, IL: Richard D. Irwin, 1985.

Evans, J. R. *Applied Production and Operations Management*, 4th ed. Minneapolis/St. Paul: West, 1993.

Gane, C., and T. Sarson. *Structured Systems Analysis and Design Tools and Techniques*. Englewood Cliffs, NJ: Prentice Hall, 1979.

ALLEN SCHMIDT, JULIE E. KENDALL, AND KENNETH E. KENDALL

TABLING A DECISION

After doing many follow-up interviews with Dot Matricks, Anna tells Chip, "I've determined the logic needed to update the PENDING MICROCOMPUTER ORDERS data store. Because many micros may be ordered on the same purchase order, as each microcomputer is entered, the matching record is located and one is subtracted from the number of outstanding micros per purchase order."

Anna shows Chip the Process repository screen print (depicted in Figure E11.1). "The name of the corresponding process, UPDATE PENDING MICROCOMPUTER ORDER (process 2.5), links the process specification to the data flow diagram," she explains. Inputs and outputs are listed and should match the data flow into or out of the process. "The VALID MICROCOMPUTER TRANSACTION record is input, and the updated PENDING ORDER is the output flow."

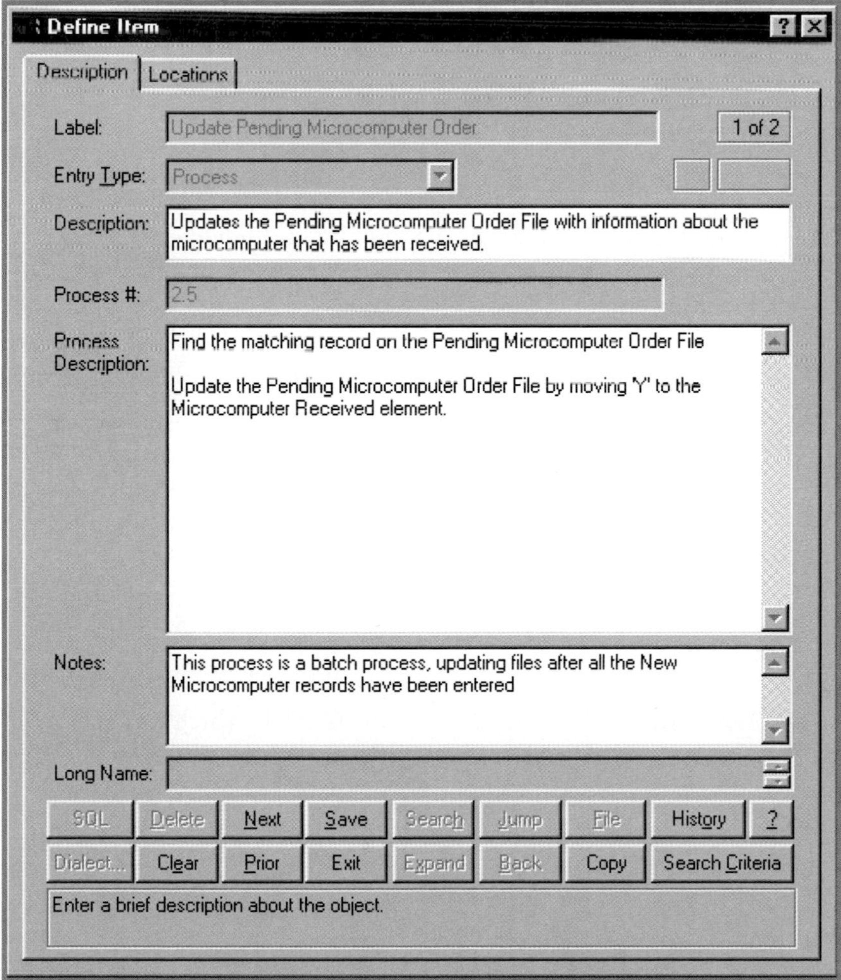

FIGURE E11.1
Process Repository screen, UPDATE PENDING MICROCOMPUTER ORDER.

"That will be useful," Chip says, "even though it took a while to untangle it all."

Anna points out, "The **Process Description** area contains the logic, shown in structured English."

When the logic is complete, Anna further enters a few notes on the nature of the process, notes that it is a batch process, and also adds timing information.

A decision table may be created for control or process logic. Before the decision table is keyed, it is a good idea to create it on paper and optimize the table. This way only the essential conditions and actions will be entered.

"I've been busy too," Chip assures Anna. "I've spoken with Cher Ware several times since you interviewed her. I've finally captured some of the logic for calculating the cost of a software upgrade.

"Cher indicated three different conditions affecting the cost. The site license provides unlimited copies and is used for popular software installed on many micros. An educational discount is provided by many publishers, and a discount for quantity is usually available," he continues.

"First I determined the values for the conditions and the number of combinations," Chip says. He set out the three conditions and their values as follows:

Condition	*Values*	*Number of Values*
SITE LICENSE	Y/N	2
EDUCATIONAL DISCOUNT	Y/N	2
DISCOUNT FOR QUANTITY	Y/N	2

"The total number of combinations is found by multiplying the number of values for each of the conditions, $2 \times 2 \times 2 = 8$. The next step is to decide which conditions should be first." Chip continues, "I reason that a SITE LICENSE would not have a discount for quantity or an additional educational discount, because the actual site license cost already reflects this kind of discount. Therefore, the SITE LICENSE should be the first condition. Each of the two other conditions would not have any particular advantage over the other, so the order is unimportant.

"Because the total number of conditions is eight and the SITE LICENSE condition has two possible values, the repeat factor would be 8/2 or 4." Chip continues by noting that the first row of the decision table would be

Condition	*1*	*2*	*3*	*4*	*5*	*6*	*7*	*8*
SITE LICENSE	Y	Y	Y	Y	N	N	N	N

"The next condition is EDUCATIONAL DISCOUNT, which also has two values. Dividing these two into the previous factor of four yields 4/2 = 2 for the next repeat factor." Chip notes that the decision table now expands to

Condition	*1*	*2*	*3*	*4*	*5*	*6*	*7*	*8*
SITE LICENSE	Y	Y	Y	Y	N	N	N	N
EDUCATIONAL DISCOUNT	Y	Y	N	N	Y	Y	N	N

Chip continues, "The last condition, DISCOUNT FOR QUANTITY, also has two values, and dividing these two into the previous repeat factor of two gives 2/2 = 1, which should always be the repeat factor for the last row of the conditions." He notes that the completed condition entry is

Condition	*1*	*2*	*3*	*4*	*5*	*6*	*7*	*8*
SITE LICENSE	Y	Y	Y	Y	N	N	N	N
EDUCATIONAL DISCOUNT	Y	Y	N	N	Y	Y	N	N
DISCOUNT FOR QUANTITY	Y	N	Y	N	Y	N	Y	N

Chip points out that when the actions are included, the completed decision table is

Condition	*1*	*2*	*3*	*4*	*5*	*6*	*7*	*8*
SITE LICENSE	Y	Y	Y	Y	N	N	N	N
EDUCATIONAL DISCOUNT	Y	Y	N	N	Y	Y	N	N
DISCOUNT FOR QUANTITY	Y	N	Y	N	Y	N	Y	N
Actions								
COST = SITE LICENSE COST	X	X	X	X				
COST = EDUCATIONAL COST × COPIES						X		
COST = DISCOUNT COST × COPIES							X	
COST = UPGRADE COST × COPIES								X
COST = (EDUC COST – DISC) × COPIES					X			

"I have proceeded to reduce some of the redundant actions, specifically those occurring when a SITE LICENSE has been obtained," Chip continues. "Because the actions are the same for SITE LICENSE values of Y, the educational and quantity discounts are meaningless to the condition and don't have to be considered. Rules 1 through 4 may be reduced to one rule." Chip concludes by noting that the final optimized decision table is

Condition	*1*	*2*	*3*	*4*	*5*
SITE LICENSE	Y	N	N	N	N
EDUCATIONAL DISCOUNT	—	Y	Y	N	N
DISCOUNT FOR QUANTITY	—	Y	N	Y	N
Actions					
COST = SITE LICENSE COST	X				
COST = EDUCATIONAL COST × COPIES			X		
COST = DISCOUNT COST × COPIES				X	
COST = UPGRADE COST × COPIES					X
COST = (EDUC COST - DISC) × COPIES		X			

The final decision table, shown in Figure E11.2, contains the optimized decision table. There are three conditions: whether a site license, an educational discount, or a quantity discount is available. The top left quadrant contains the conditions. Directly

Conditions and Actions	1	2	3	4	5
Site License	Y	N	N	N	N
Educational Discount		Y	Y	N	N
Discount for Quantity		Y	N	Y	N
Upgrade cost = Site License Cost	X				
Upgrade cost = Educational Cost * Number of Copies			X		
Upgrade cost = Discount Cost * Number of Copies				X	
Upgrade cost = Cost per Copy * Number of Copies					X
Upgrade cost = (Educational Cost – Discount) *Number of Copies		X			

FIGURE E11.2
Decision Table, UPGRADE COST.

below it are the actions. The condition alternatives are in the upper right quadrant, and the action entries are in the lower right quadrant. The actions show how the UPGRADE COST is determined for each condition, indicated by an **X** in the rule columns.

EXERCISES*

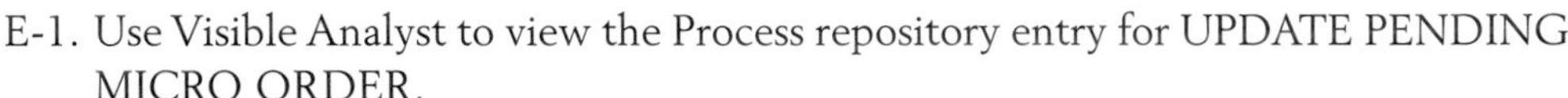
E-1. Use Visible Analyst to view the Process repository entry for UPDATE PENDING MICRO ORDER.

E-2. Modify and print the ACCUMULATIVE HARDWARE SUBTOTALS Process entry. Add the **Process Description** "Accumulate the hardware subtotals. These include the number of machines for each hardware brand."

E-3. Modify and print the CONFIRM MICROCOMPUTER DELETION Process entry. Add the following Description:

Use the Mircrocomputer Record to format the Deletion Confirmation Screen (refer to the Delete Microcomputer Prototype Screen).

Prompt the user to press the **OK** button to confirm the deletion; otherwise, press the **Cancel** button to cancel the deletion.

If the operator presses **OK** to delete the record, delete the record and display a "Record Deleted" message; otherwise, display a "Deletion Canceled" message.

E-4. Create Process specifications for process 6.6, VALIDATE MICROCOMPUTER CHANGES. The Description for the process is as follows:

Validate the changes to the MICROCOMPUTER MASTER. Include a note to use the edit criteria established for each element. Provide the following additional editing criteria:

The ROOM LOCATION must be valid for a particular campus.

The MONITOR must not be a lower grade than the graphics board. An example of this error would be a SVGA (higher resolution) graphics board paired with a VGA (lower resolution) monitor.

There must not be a second fixed disk without the first one.

The LAST PREVENTIVE MAINTENANCE DATE must not be greater than the current date.

The DATE PURCHASED must not be greater than the LAST PREVENTIVE MAINTENANCE DATE or greater than the current date.

The MODEL must conform to the type supported by the BRAND name.

No changes may be made to an inactive record.

E-5. Create process specifications for process 1.4, CREATE SOFTWARE LOG FILE. Use the data flow diagram examples to determine inputs and outputs. Process details are as follows:

Format the SOFTWARE LOG RECORD from the following information:

The confirmed NEW SOFTWARE RECORD elements.

The following system elements: System date, System time, User ID, Network ID.

When the record has been formatted, write to the SOFTWARE LOG FILE.

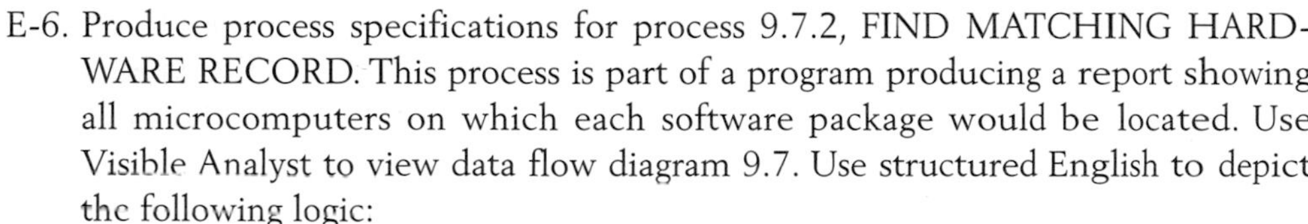

E-6. Produce process specifications for process 9.7.2, FIND MATCHING HARDWARE RECORD. This process is part of a program producing a report showing all microcomputers on which each software package would be located. Use Visible Analyst to view data flow diagram 9.7. Use structured English to depict the following logic:

For each SOFTWARE RECORD, loop while there is a matching hardware inventory number. Within the loop, accomplish the following tasks:

Randomly read the MICROCOMPUTER MASTER file.

If a record is found, format the MATCHING MICROCOMPUTER RECORD information.

If no record is found, format a NO MATCHING error line.

Furthermore, if the found MICROCOMPUTER RECORD is inactive, indicating that it has been removed from service, format an INACTIVE MATCHING MICROCOMPUTER error line.

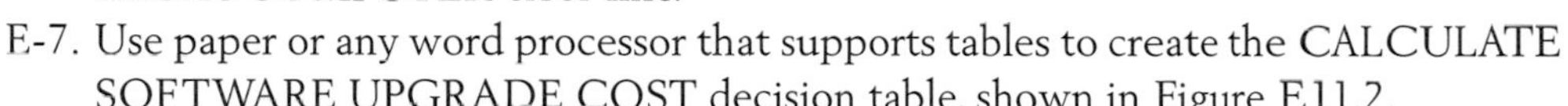

E-7. Use paper or any word processor that supports tables to create the CALCULATE SOFTWARE UPGRADE COST decision table, shown in Figure E11.2.

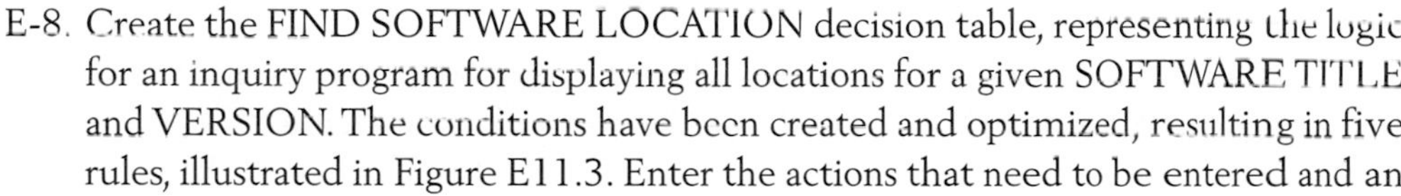

E-8. Create the FIND SOFTWARE LOCATION decision table, representing the logic for an inquiry program for displaying all locations for a given SOFTWARE TITLE and VERSION. The conditions have been created and optimized, resulting in five rules, illustrated in Figure E11.3. Enter the actions that need to be entered and an

Conditions and Actions	1	2	3	4	5
Matching Software Record Found	Y	Y	Y	Y	N
Version of Software Found	Y	Y	Y	N	
Matching Microcomputer Record Found	Y	Y	N		
Campus Code Found in Table	Y	N			
Display 'No Matching Software Record' Error Message					X
Display 'Version Not Available' Error Message				X	
Display 'Machine Not Found' Error Message			X		
Display 'Campus Code Not Found' Error Message		X			
Display Location Information	X				

FIGURE E11.3
Decision Table, FIND SOFTWARE LOCATION.

X in the column related to the conditions. If you are using a word processor, print the final decision table. The conditions and actions are represented by the following logic:

The SOFTWARE MASTER file is located for the specified TITLE. If the matching record is not found, an error message is displayed. Because there may be several versions, the VERSION NUMBER on the record is checked for a match to the version entered. If the requested version is not found, further records are read using the alternate index. If all records are read and the version number is not found, an error message VERSION NOT AVAILABLE is displayed.

Once the correct software has been located, a matching MICROCOMPUTER MASTER record is obtained. If the MICROCOMPUTER MASTER is not found, the error message MACHINE NOT FOUND is displayed. For each matching machine, the CAMPUS TABLE is searched for the CAMPUS LOCATION code. If the code is not found, the message CAMPUS CODE NOT FOUND is displayed.

If no errors occur, the requested information is displayed.

E-9. Create a decision table for a batch update of the MICROCOMPUTER MASTER FILE. There are three types of updates: Add, Delete, and Change.

The MICROCOMPUTER MASTER record must be read. If the transaction is an Add and the master is not found, format and write the new microcomputer master record. Print a valid transaction line on an UPDATE REPORT. For a Change or Delete transaction, print a CHANGE ERROR LINE or a DELETE ERROR LINE if the master record is not found.

If the master record is found, check the active code. If the record is inactive and the transaction is an Add, format and rewrite the new microcomputer master record. Print a valid transaction line on an UPDATE REPORT. For a Change or Delete transaction, print a CHANGE ERROR LINE or a DELETE ERROR LINE.

If the MICROCOMPUTER MASTER RECORD is active and the transaction is an Add, print an ADD ERROR LINE. For a Change transaction, format the changes and rewrite the MICROCOMPUTER MASTER RECORD. Print the VALID TRANSACTION LINE. For a Delete transaction, change the ACTIVE CODE to inactive and rewrite the MICROCOMPUTER MASTER RECORD. Print the VALID TRANSACTION LINE.

*Exercises preceded by a CD-ROM icon require the program Visible Analyst or another CASE tool. A CD-ROM containing Visible Analyst examples is provided free of charge to any professor adopting this book. The examples on the disk may be imported into Visible Analyst and then used by students.

ANALYZING SEMISTRUCTURED DECISION SUPPORT SYSTEMS

12

METHODS AVAILABLE

Three major concerns arise when analyzing semistructured decision support systems. The systems analyst needs to know (1) whether decision makers are primarily analytic or heuristic; (2) how decisions are made in the three problem-solving phases of intelligence, design, and choice; and (3) the multiple-criteria methods that are useful in solving semistructured problems.

Decision support systems (DSS) can function in many ways. They can organize information for decision situations, interact with decision makers, expand the decision makers' horizons, present information for decision makers' understanding, add structure to decisions, and use multiple-criteria decision-making models. Multiple-criteria models include trade-off processes, weighting methods, and sequential elimination methods, which are all well suited to handling the complexity and semistructured nature of many problems supported through DSS. This chapter covers what the systems analyst needs to know about analytic and heuristic decision makers, about the three phases of problem solving supported by DSS, and about the multiple-criteria methods needed to solve semistructured problems.

DECISION SUPPORT SYSTEMS

Decision support systems possess many characteristics that differentiate them from other, more traditional management information systems. End users of DSS, by virtue of the types of problems they address and the learning they undergo, also possess special characteristics that need to be considered.

CHARACTERISTICS OF A DECISION SUPPORT SYSTEM

First and foremost, a decision support system is a way to organize information intended for use in decision making. It involves the use of a database for a specific decision-making purpose. A DSS does not just automate transformations performed on data; neither does it simply provide output in the form of reports. Rather, it supports the decision-making process through the presentation of information that is designed for the decision maker's problem-solving approach and application needs. It neither displaces judgment nor makes a decision for the user.

Decision support systems allow the decision maker to interact with them in a natural manner by virtue of the careful design of the user interface. Useful DSS will challenge and eventually change decision makers. By contrast, management information systems (MIS) provide output but no real impetus for change in the person receiving it. Interacting with DSS will prove to be new and challenging for most decision makers and will provide new perspectives on the decision-making process that are attractive and understandable yet innovative. By furnishing a new way to see problems and opportunities, DSS eventually change the user's decision-making process and, along with it, the user.

Decision support systems are designed to help support decisions involving complex problems that are formulated as semistructured. Because such problems remain resistant to complete computerization, a solution per se is not the goal of DSS; instead they support the decision process that leads to a solution.

Decision support systems may be constructed to support one-time decisions, those that are infrequent, or those that occur routinely. The type of problem or opportunity best addressed through use of DSS is one both that ultimately requires human judgment—either because humans feel it is inappropriate to relinquish their judgment or because the problem cannot be fully automated—and is complex enough to benefit from the time savings to be gained through partial automation. Medical decisions involving differential diagnosis are a good example of such problems.

A decision support system is typically designed for either a particular decision maker or a group of decision makers. This high degree of individualization allows the system designer to customize important system features to adapt to the type of representations (graphs, tables, charts, and so on) and interface that the user understands best. Web sites can also effectively include DSS, specific to the site users. For instance, sites devoted to online testing for graduate schools may offer a DSS that helps students select graduate schools that are appropriate for their needs. Decision support systems are particularly helpful on ecommerce sites selling big ticket items or supporting life changes.

Rather than building a specific system "from scratch," a systems analyst can use a package of interrelated hardware and software called a DSS generator (DSSG). There has been considerable growth in the sales of DSSGs because they can lessen the time and cost associated with building a case-specific DSS. Hardware and software designated as DSS tools aid in the building of both specific DSS and DSS generators.

A decision support system is best conceptualized as a process instead of a product. Figure 12.1 contrasts the product orientation of the traditional MIS with the process-oriented DSS. Notice that the focus of the DSS is on the decision maker's interaction with the system and not on the output generated. As discussed

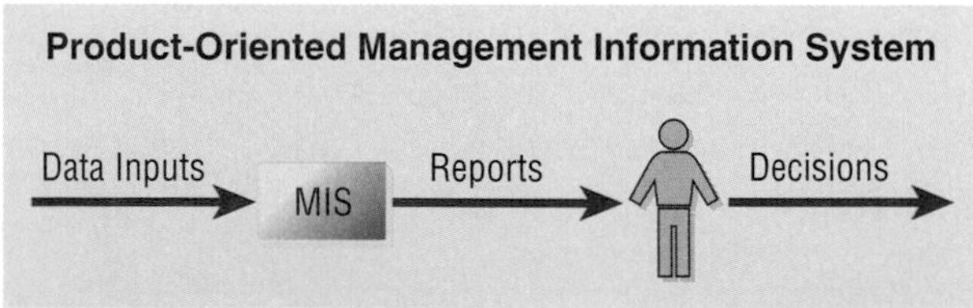

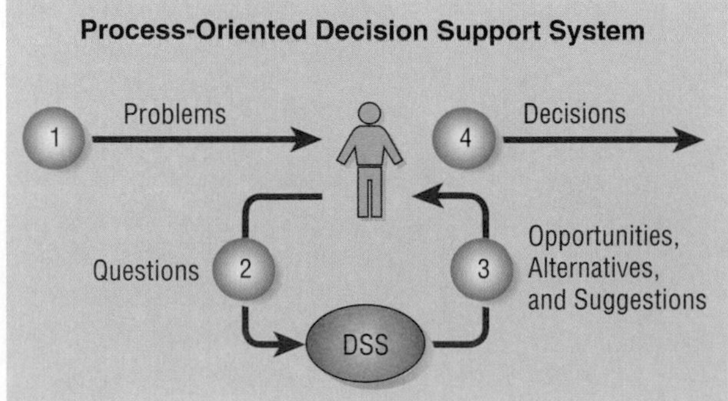

FIGURE 12.1
Decision support systems have a process orientation focusing on the decision maker's interaction with the system.

in Chapter 2, processes are ways to transform inputs into outputs. In the instance of DSS, the process works to transform the user—the decision maker—by changing and improving upon his or her decision-making performance.

What this process orientation means is that DSS are evolutionary in nature. They change as the needs of the end user change, and the decision maker changes by interacting with the DSS process.

DECISION SUPPORT SYSTEM USERS

Decision making in organizations occurs on three main levels: the strategic, the managerial, and the operational, as discussed in Chapter 2. Many of the decisions required on the operational level can be successfully and thoroughly automated. The same can be said for many routine managerial decisions. When problems and opportunities cannot be totally structured and when human judgment and experience are required to make a decision, however, traditional MIS are often considered inadequate for the task. It is in developing the solution of semistructured, complex problems that DSS can be of use. This kind of problem most often occurs on the strategic and managerial levels.

Although the use of a decision support system is not limited to middle- and upper-level managers or chief executive officers, these individuals are most often the primary users. They bring with them differing decision making styles, differing needs, and differing levels of sophistication. The DSS designer therefore needs to take specific decision-making attributes into consideration so that it is possible for the user to interact successfully with the system. If the end user is too busy or too threatened by the prospect of interacting with the DSS, a technical go-between or assistant may be used to interact with the computer. In this way, the decision maker is free to analyze and react to the process, not to the mechanics of it.

Because DSS are designed specifically for a user or a group of users (as in group DSS, commonly abbreviated GDSS), systems analysts need to be acutely aware of how a decision maker's style might influence DSS design. In the following sections, we discuss several concepts pertinent to DSS, including decision making under risk; decision-making style; and how decisions can be classified as structured, unstructured, and semistructured.

DECISION-MAKING CONCEPTS RELEVANT TO DSS

Many concepts help inform us about the relationship between decision making and decision support systems, including theories about certainty, uncertainty, and risk and how they help shape a decision maker's style as either analytic or heuristic. The three problem-solving phases of intelligence, choice, and design as they are supported by decision support systems are also of interest here.

DECISION MAKING UNDER RISK

Classical decision-making theory usually assumes that decisions are made under three sets of conditions: certainty, uncertainty, and risk. Certainty means that we know everything in advance of making our decision. A common management science model that assumes conditions of certainty is linear programming, where the resources, rates of consumption, constraints, and profits are all assumed to be known and correct. Uncertainty implies just the opposite; we know nothing about the probabilities or the consequences of our decisions.

Between the two extremes of certainty and uncertainty lies the huge set of conditions called risk. Decisions made under risk assume that we are somewhat knowledgeable about our alternatives (controllable variables), about what we cannot control but only estimate (environmental variables), and about what the outcomes will be (dependent variables). Not only can we estimate the environmental variables, but we can also estimate the probability that they will occur. This information can be 100 percent accurate, partially accurate, or false, but we still attempt to base the decision on the information we have. Most business decisions are made under risk.

An Example of Uncertainty, Certainty, and Risk. To help understand these three sets of conditions, suppose you had to decide whether to carry an umbrella into work one morning. There is only one circumstance that represents complete certainty; it is raining now. If you open the door, look outside, and it is raining, you know you must use an umbrella or you will get drenched. If the sun is out now, it still might rain in the afternoon.

Risk implies that we have an idea about the alternatives, probabilities, and outcomes. If we listen to the weather forecast on television and the forecaster says there is a 40 percent chance of rain, we can use that information to make our decision about whether to carry an umbrella.

To understand conditions of complete uncertainty, visualize being locked in a room without windows or visitors. Furthermore, your television cable company is having service problems, your radio's batteries are corroded, and your newspaper has been lost en route. In this situation, all your sources of information about the outside weather would be gone.

Is this uncertainty, though? If all your sources of information were taken away, wouldn't you still be able to make a decision under risk? Most everyone would be able to do so, because most people would realize where they were and what time of year it was. In other words, we know from experience what the probability of rain is in our location at this time of year. This example suggests that experience, as well as information, helps us to make decisions.

Some individuals rely more heavily on information, whereas others prefer to base their decisions on experience. The next subsection builds upon this relationship as it discusses decision-making style.

DECISION-MAKING STYLE

The way in which information is gathered, processed, and used, together with the manner in which decisions are communicated and implemented, forms the parameters of decision-making style. Decision makers are often characterized as being either analytic or heuristic.

Analytic Decision Making. An analytic decision maker relies on information that is systematically acquired and systematically evaluated to narrow alternatives and make a choice that is information based. As shown in Figure 12.2, analytic decision makers learn by analyzing a particular situation. They are methodical, using step-

FIGURE 12.2
Decision makers can be characterized as either analytic or heuristic by examining the information they use and how they make their decisions.

Analytic Decision Maker	Heuristic Decision Maker
Learns by analyzing	Learns by acting
Uses step-by-step procedure	Uses trial and error
Values quantitative information and models	Values experience
Builds mathematical models and algorithms	Relies on common sense
Seeks optimal solution	Seeks satisficing solution

by-step procedures to make decisions. Analytic decision makers value quantitative information and the models that generate and use it.

In addition, decision makers with an analytic style use mathematics to model problems and algorithms to solve them. An analytic decision maker seeks optimal—rather than completely satisfying—solutions, being content with answers that satisfy most but not all requirements. Analytic decision makers use decision techniques such as graphing, probability models, and mathematical techniques to ensure a sound decision-making process. These methods, however, require that the information employed is (1) available, (2) reasonable, (3) complete, and (4) accurate.

An example of an analytic decision maker is a manager who, when faced with a problem, employs a computerized management science model to resolve conflicts in an optimal way. To accomplish this goal, the manager determines an objective function and constraints, builds mathematical relationships, gathers information for the model, and solves the problem.

Heuristic Decision Making. A decision maker who uses heuristics is making decisions with the aid of some guidelines (or rules of thumb), although they may not always be applied consistently or systematically. They seek a satisficing, not an optimal, solution. Heuristics are generally experience-based.

Heuristic decision makers learn by acting. They use trial and error to find a solution. Thus, they value experience, relying on common sense to guide them.

An example is a manager deciding heuristically from which company to purchase raw materials. Based on past experience, the manager knows that materials shipped from the East Coast tend to arrive at the plant on time. Formulating that experience as a heuristic, or rule of thumb, the manager decides to try and pick a supplier of raw materials from the East Coast.

The decision making style of managers relates back to the openness and closedness of organizational systems. If information in the business is free-flowing, the opportunities for the use of decision aids and systematic analysis may be greater. If timely information is difficult to acquire, the organization may be encouraging managers (albeit unwillingly) toward a more heuristic style.

Implications for Development of Decision Support Systems. Analytic decision makers are systematic in their evaluation of alternatives in a decision situation. A DSS for an analytic decision maker should include several mathematical and graphical models that allow desired comparisons. It is also important to make supporting evidence for alternatives available through the system.

Heuristic decision makers are somewhat systematic in their approach to decision making, but they use rules of thumb for guidelines instead of carefully reasoning through each alternative each time it is presented. A decision support system supporting a heuristic thinker might present summary information on alternatives rather than detail pros and cons in copious output. One DSS for heuristic decision makers might also display past decisions and their outcomes in support of the preference for experience-based data. Further discussions of DSS output are featured in Chapter 15, which deals with output design.

PROBLEM-SOLVING PHASES

Decision making (or problem solving) is a process and is conceived of in phases rather than in steps. In phases, the occurrence of behavior swells and tapers off, with some overlap between each phase. In stepped behavior, one step occurs independently of the next and is completed before the next step is carried out.

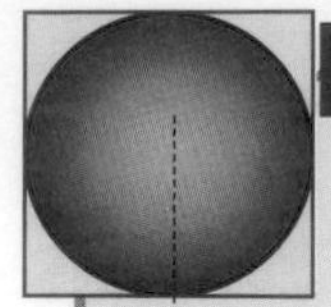

CONSULTING OPPORTUNITY 12.1

WALKING INTO OPEN ARMS

"I helped start Open Arms four years ago, after I earned my M.B.A.," says Maria Ewing-Barton, the 36-year-old administrator of a health-care referral service. "When the County Medical Society realized that the health-care services here in Laramie weren't being fully utilized, a group of us thought about what we could do to fill the gap. We aim to match up patients with appropriate health-care providers. For various reasons (transiency, poverty, newly transferred to the city), patients have no regular contact with a health-care provider. I see anywhere from five to twenty patients a day. Some have chronic illnesses, such as alcoholism; some have a common cold. I assess what level of treatment is relevant and what funding is possible, and then I decide what the next step for the patient will be. For example, should the patient see a private physician today or schedule an appointment with an eye doctor sometime in the future? I called you in to help me strengthen and justify the decision-making process I go through in making referrals, maybe through the creation of a DSS.

"I'd like to have a lot of information at my fingertips so that I can make decisions quickly. Although most cases are not life or death, some are serious. Also, I don't have much time to spend with each patient. I want to be sure I can justify my referral decisions, because some local doctors think I'm usurping their decision-making power. Most of the people in health care here trust my judgment, but the occasional physician is still skeptical of my qualifications. They resent that my training is in management, not health care. As an upshot, I have to be very methodical in defending my referral decisions to them.

"I keep statistics on the patients we've referred, and I read everything I can get my hands on in the medical and health-care administration journals, although they don't bear directly on what we do. I also talk to other administrators to keep current on the services their facilities are offering. What's frustrating is that the services available change so often. It seems as if local hospitals are only concerned with marketing their profitable specialties.

"I've been playing with the data I've accumulated, looking for patterns and trying to create some sort of model for the way Open Arms is working. As you can see, I have an old PC here in my office, and I've spent many evenings entering data to help formulate reasonable models of Open Arms. Although accurate records are important, I want to do more with the data than merely store it.

"Quite frankly, a large part of my decision on whether to refer is based on how I size up a person initially. I know it's important to be objective, but with so little time to make a decision, I usually go with my heart and not my head. Eventually, I would like to rely on the numbers I generate with my PC to match patients with health-care providers. Then I can justify my decisions when necessary. It's a way to make more effective matches, too, that are fairer to everyone."

What type of decision maker is Maria? Which of her characteristics lead to this viewpoint? Describe what you think is Maria's typical decision-making process regarding patient referrals.

The three phases of problem solving are intelligence, design, and choice (Simon, 1965). The decision maker begins with the intelligence phase, with design and choice following in phased succession. Let's examine each phase as it relates to decision-maker behavior.

Intelligence. Intelligence is awareness of a problem or opportunity. In this phase, the decision maker searches the external and internal business environments, checking for decisions that need to be made, problems that need to be solved, or opportunities that need to be examined. Intelligence means an active awareness of changes in environments that call for action.

Intelligence translates to vigilance: continual searching and scanning. The intelligence phase provides the impetus for the other two phases and always precedes them.

Design. In the design phase, the decision maker formulates a problem and analyzes several alternative solutions. The design phase allows the decision maker to generate and analyze alternatives for their potential applicability.

Choice. In the choice phase, the decision maker chooses a solution to the problem or an opportunity identified in the intelligence phase. This choice follows from the foregoing analysis in the design phase and is reinforced by information gained in the choice phase. It also includes implementation of the decision maker's choice. Other authors have added separate phases for implementation and evaluation.

SEMISTRUCTURED DECISIONS

Many people conceive of decisions as existing on a continuum from structured to unstructured. Structured decisions are those where all or most of the variables are known and can be totally programmed. Decisions that are structured are routine and require little human judgment once the variables are programmed. Analysis of structured decisions was covered extensively in Chapter 11.

Unstructured decisions are those that are currently resistant to computerization and depend primarily on intuition. Semistructured decisions are those that are partially programmable but still require human judgment. Decision support systems are most powerful when addressing semistructured decisions, because they support the decision maker in all phases of decision making but do not mandate one final answer.

It is hypothesized that all decisions may contain "deep structure," that is, structure that is present but not yet apparent. If true, eventually all decisions could be treated as semistructured, and the usefulness of DSS would be substantially broader.

An argument used in support of the presence of deep structure is that the game of tic-tac-toe appears to be unstructured to a five-year-old child. While the child is learning to play the game, it isn't clear which move is the best, and the game itself seems suspenseful and exciting. As the child grows up, however, it soon becomes apparent that there is indeed a structure. If certain rules are applied, the outcome can be maximized by guaranteeing at least a tie.

Tic-tac-toe, however, is not like a business situation. First, the rules for tic-tac-toe are simple: Place an X in one of the spaces and wait until your opponent responds. Second, the objective is simple: Tie, or perhaps win, by placing three Xs (or Os) in a row if your opponent makes an error. The only factor that makes the game unstructured is the decision skill of the decision maker. That is, the average child doesn't yet possess the analytical skills or experience needed to win the game.

DIMENSIONS OF SEMISTRUCTURED DECISIONS

Decisions are called semistructured or unstructured for a number of reasons. It is useful to visualize the dimensions of structured decisions as a cube, as illustrated in Figure 12.3. The three dimensions are as follows:

1. The degree of decision-making skill required.
2. The degree of problem complexity.
3. The number of decision criteria considered.

In the tic-tac-toe example, the five-year-old child would not see the game as structured even though there was a single criterion and very simple rules, because the child's decision-making skill has not been sufficiently developed.

Degree of Decision-Making Skill Required. The degree of decision-making skill relates back to the concepts of analytic and heuristic decision makers. Decision-making skill is measured in the analytic and experience-based maturity of the decision maker. A DSS could help upgrade decision-making skill by providing easy-to-comprehend models (for the analytic thinker) or analogies (for the heuristic thinker).

The Degree of Problem Complexity. If a problem is highly complex, it seems semistructured or even unstructured. A DSS can help in this regard by encouraging the decision maker to define the boundary of the system, and he or she would accomplish this goal by clearly defining the problem and limiting the number of variables. In addition, the DSS should provide support in helping the decision maker systemati-

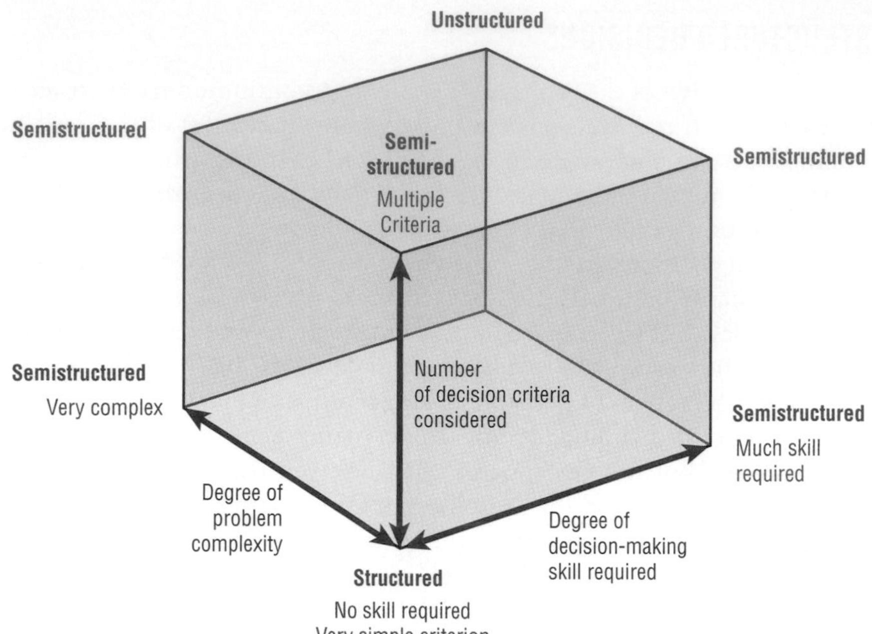

FIGURE 12.3
Visualizing structured decisions as a cube with three dimensions: degree of decision-making skill required, degree of problem complexity, and the number of decision criteria required.

cally eliminate some alternatives. Decision support systems are needed to organize information, track variables, and present the problems, alternatives, and choices in a manner easily understandable to the manager.

Number of Criteria Considered. A manager concerned with a single decision criterion is dealing with a structured problem, but most real-world problems have multiple, conflicting goals and multiple decision criteria. Consequently, they are by nature semistructured. An example of a decision maker dealing with multiple conflicting objectives is a production manager deciding on a production schedule whose objectives include keeping costs to a minimum, maximizing customer satisfaction by meeting the delivery schedule, keeping quality high, and minimizing idle time. The production manager cannot optimize one of these objectives unless other objectives are sacrificed.

A decision involving multiple criteria may include choosing a software package that is powerful, has clear documentation, and is user friendly but inexpensive. Because it is unlikely that any single product fulfills all the requirements, the decision maker must assign importance (rank or weight) to each criterion. Decision support systems are invaluable in helping decision makers handle multiple criteria. A more extensive discussion of multiple-objective and multiple-criteria models follows in a later section of this chapter.

SEMISTRUCTURED DECISIONS IN INTELLIGENCE, DESIGN, AND CHOICE

No matter what the decision is, the decision has to go through each of the three phases (intelligence, design, and choice) in the decision-making process. Semistructured decisions are made in each of the phases of decision making. Sometimes the decision is easy and other times it is very cumbersome, but rarely do we encounter a problem in which every decision-making phase is equally difficult.

Figure 12.4 identifies some of the bottlenecks that occur in each of the three phases of the decision-making process. Bottlenecks in the intelligence phase occur because the decision maker is unable to identify a problem, define it, or set priorities for tackling the problem. During the design phase, bottlenecks can occur because the decision maker cannot generate feasible alternatives, assign values or outcomes to the alternatives, or establish performance criteria to compare alterna-

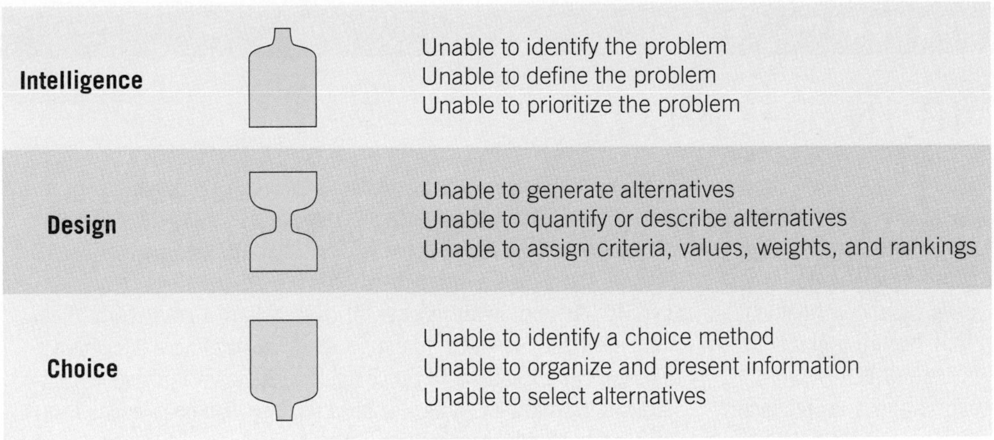

Intelligence		Unable to identify the problem Unable to define the problem Unable to prioritize the problem
Design		Unable to generate alternatives Unable to quantify or describe alternatives Unable to assign criteria, values, weights, and rankings
Choice		Unable to identify a choice method Unable to organize and present information Unable to select alternatives

FIGURE 12.4
Decision support systems can alleviate common bottlenecks in the three phases of decision making.

tives. The choice phase can also present bottlenecks when the decision maker is unable to choose a decision method, organize and present information, or select an alternative. The following discussion focuses on how decision support systems can alleviate bottlenecks and thus assist the decision maker.

Intelligence Phase Decision Support Systems. Decision making is often difficult because the problem itself is difficult to identify. Problems are noticed only if appropriate performance measures are put in place to highlight them. Suppose a clothing store set a goal of selling $30,000 worth of merchandise in a week and always exceeded that goal; no problem would be identified. What if more information, though, revealed that for the square footage leased, the store needs to make a minimum of $35,000 per week to stay in business? Without the additional information, the problem goes unnoticed. An effective DSS, therefore, must contain mechanisms for recognizing problems.

Once a problem is identified, it needs to be defined. A DSS could help the decision maker determine the scope of the problem to minimize the complexity of the decision. A store manager might choose to limit the problem to strategic pricing or strategic advertising. Other possible approaches, such as firing employees or changing the merchandise mix, would not be considered.

The final step in the intelligence phase is to assign a priority to the problem. The problem may be of an immediate nature or it may be a future opportunity that could be pursued if other, more pressing problems are addressed first. For example, an opportunity to expand may look very attractive initially, but it may be delayed because of more pressing problems in the original store. Decision support systems are needed for problem and project selection.

Design Phase Decision Support Systems. In a relationship that parallels the one just discussed, alternatives need to be identified. A decision support system can aid in generating alternatives that might not have occurred to the decision maker. Experience-based DSS can compare the current situation with similar scenarios and guide the manager through the maze of alternatives. Some alternative situations that the clothing manager could investigate include varying the types and amounts of advertising or changing the prices of certain merchandise items to attract more customers.

Next the alternatives need to be quantified or described. A decision maker could retrieve data from a database, collect new data, and manipulate data. A more heuristic decision maker might take an extraorganizational approach, build analogies, and seek opinions or consider various scenarios. The clothing store manager could gather quantitative information on, for example, costs and the effectiveness

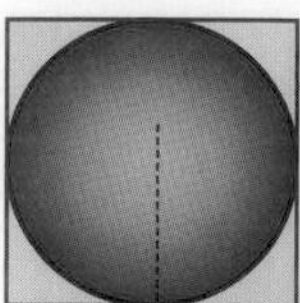

CONSULTING OPPORTUNITY 12.2

THE PROOF IS IN THE PIZZA

"I really enjoy my current position," says 53-year-old Hal Lupino. He is the district manager for the Calliope chain of seven Nebraska pizza palaces, which his father and uncle founded and still run. There are four stores in Lincoln (where the business started), two in Omaha, and one in Grand Island. They have expanded from small take-out places to sit-down restaurants that feature a giant, electronically controlled calliope that plays music while pizza eaters dine. The main attraction of the chain, however, is still a secret family pizza recipe.

"Some people think I'm too old to be a district manager, but I have the best of both worlds in this job," Hal continues. "I've got a lot of decision-making power, with Dad and Uncle Reg getting up in years, but I still talk with lots of customers, too. The stores' profits have been increasing since 1964. We added the last store in Lincoln in 1994.

"'Everyone, inside and out, has been sort of pressuring us to expand again. My son is an MBA student and he recommended you to help me set up a DSS to support location decisions.

"My thinking is that we should make the most of the delivery network we already have set up, because it's so darn expensive to move the ingredients we need. With the rotten luck the farmers are having and that there are more sand dunes than people in western Nebraska, I think expansion there is out. My gut reaction is that competition is too fierce to get anything else going in Omaha.

"I'll let managers and customers be my guide on where to locate next, though. I'm a people person. Why not ask someone a question rather than read a report about it? My father and uncle are different, though. Those guys love numbers. So if I don't give them some numbers to back up my decisions, I'll have a hard time getting them implemented. I've been humoring them by passing out questionnaires on new locations to our customers. Someday I'll have my secretary tally what they wrote. I've been reading their open-ended comments though; they're more interesting to me.

"I keep notes about what I hear from customers and managers on the bulk napkins we buy for the restaurants. They go in here [he indicates an in-basket piled with napkins] when I've read them. Otherwise, I keep 'em in my folder. I don't throw much away, but finding a key comment is a tough task.

"I guess I'm a slowpoke when it comes to making a decision. I go over and over what I've seen in the markets we serve and review all my experience in the business and all the talks with our customers. I keep going back to our last location decision when we opened the new store in Lincoln. If the proof is in the pudding, then Lincoln may just be the spot for the new one, too."

How can a decision support system make Hal a better decision maker? In particular, what features of his style could it support? What features of his style might a decision support system change?

of advertising, the competitor's pricing, and price elasticity. The manager could also gather information from employees and customers. This information could be compiled and summarized for decision-making purposes.

Once alternatives are generated and organized, performance criteria need to be established. Then the decision maker can assign values, risks, weights, or a rank to each alternative. The clothing store manager may set weekly sales volume as one criterion but may also want to include weekly costs as well as intangibles, such as customer satisfaction and employee morale. Spreadsheets are excellent ways to list various alternatives, and graphs provide vivid visuals to limit the number of alternatives being considered.

Decision support systems can represent possible choices to decision makers in ways that may not conform to their typical way of seeing the world. For example, if a problem solver is not accustomed to considering other alternatives, DSS can suggest them. In essence, DSS can expand the field of vision for the decision maker.

The interaction between DSS and the decision maker is strong in this phase. It is entirely feasible to build a system that will work through several what-if scenarios with numerous variables, something that is too complex for lone decision makers to do. In this way, a DSS enhances the design phase by greatly expanding the number of possible actions that can be considered in detail as well as by suggesting new alternatives that the decision maker might otherwise overlook.

Choice Phase Decision Support Systems. Choosing a solution is obviously important, and in semistructured decisions, the choice must be left to the judgment of the decision maker who is supported by the DSS. First, the decision sup-

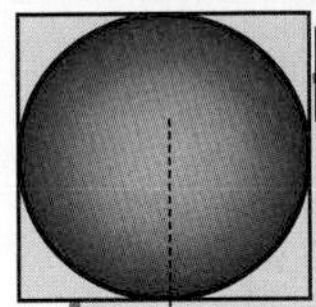

CONSULTING OPPORTUNITY 12.3

THE HISTORY OF SCHEDULING

"I like my work here at the university," Stuart Dent admits. At age 22, he is younger than many of the students and staff he deals with in his job of scheduling classes for the approximately 27,000 students who attend regular semester sessions. "I've been here two years, ever since I graduated in art history," Stu Dent continues.

"The person who had the job before me generated a whole stack of computer reports that showed each department's course schedule, availability of faculty, how often courses needed to be offered, what day they were offered on last, student preferences for course days, and availability of rooms. All that was a jumble, so I called you in to help. I know you teach a DSS course on campus.

"I want to schedule each required course at a time that is satisfactory to both faculty and students. I think the larger picture is important. Courses that occur in a cycle (for example, many graduate-level courses) should be offered when promised. I want to eliminate double-bind situations where a student is unable to take a course because its prerequisite hasn't ever been offered.

"I guess I'm pretty idealistic. I like thinking about historical trends and trying to characterize them. I asked my coworkers about the scheduling problem, but they keep sending me to the mountain of computer reports made by the guy who had my job before me. I'm so sick of their advice that I've 'joined the enemy' and decided to actually do something for the students. I've been interviewing a handful of students from each college on campus.

"I've also reviewed my own scheduling hassles as a student, in case they offer a key. Here are some of the charts I've sketched to summarize information from my experiences and the interviews, but as you can see, it's slow going.

"Although the old computer reports seem virtually useless, I've been careful to inventory them and maintain them. In fact, I labeled them and put them on a shelf here in my office. I know the value of history. The tapes I have of student interviews contain enough 'horror stories' about five-year undergrads being standard because of poor scheduling that they may be potentially damaging to the administration. I've duplicated them, and both sets of tapes are at home.

"I want to make my decisions quickly, because scheduling for each semester is done well in advance and must be done twice a year. The students accept me and respect that I'm trying to do a good job, but Cyril, my boss, likes numbers, and I know he'll never go along with my suggestions for scheduling unless I back up my intuition with some quantitative data. The people I work with think I'm overwhelmed by the volume of data this department generates. They're doing what they always do and ignoring me until I sort everything out, but I'm confident in my ability to make good decisions."

In what ways can a decision support system help Stu? Describe how it can help in the intelligence, design, and choice phases of his decision making.

port system can be of help by reminding the decision maker of what methods of choice are appropriate for the problem. The DSS can include suggestions for analytic techniques (which management science models to rely on) or heuristic techniques (how best to display information when making a choice). For example, our store manager may want to approach the problem on a simple cost-benefit basis, because the approaches being considered do not involve a great many intangibles (unlike the approaches to firing employees or changing store hours). By helping present methods for selection of alternatives, a DSS can support the choice phase.

The decision support system can also help the decision maker organize and present the information. For one decision maker, that might mean defining variables and formulating constraints that make up a management science model. For another manager, that may mean selecting the type of graph (line, bar, or pie chart), the scale, or the range of data to be considered. For our clothing store example, the manager might prefer to see the information displayed on a personal computer's spreadsheet.

Finally, the decision support system needs to be able to handle multiple-criteria decision problems. It would ideally carry out cumbersome manipulations but leave the decision to the personal judgment of the decision maker. Because our store manager preferred information in the form of a spreadsheet, a trade-off or weighting technique might be appropriate. Due to the increasing importance of making decisions based on multiple criteria, the subject is covered in more detail in the next section.

MULTIPLE-CRITERIA DECISION MAKING

Even in problems with complete information, a limited number of variables can be semistructured. These problems have numerous, often conflicting objectives, and they are of much interest to the DSS designer. In modeling these decisions as realistically as possible, researchers have developed numerous approaches to evaluating multiple-objective or multiple-criteria problems.

Multiple-criteria approaches allow decision makers to set their own priorities, and most allow the decision maker to perform sensitivity analysis by asking what-if types of questions. These methods include weighting methods, the conjunctive constraints approach, analytic hierarchy processing, and goal programming. When included in DSS or MIS, multiple-criteria decision-making models give the decision maker a more powerful way to evaluate alternatives in the design phase of decision making.

USING WEIGHTING METHODS

Students encounter a weighting method in every course they take when their grade is computed. Various components of the course are worth a certain percentage, such as 20 percent each for the first and second exams, 20 percent for the term paper, and 40 percent for the final examination. Numerical scores for each component are then multiplied by the percentages, and the final grade is calculated.

A spreadsheet example of a weighting approach is depicted in Figure 12.5. Here a purchase decision is needed regarding three database packages appropriate for a business. Each of the attributes is assigned a value by the decision maker. After examining the software packages, the decision maker assigns a grade to each attribute for each package, in this case, a grade of 1 to 10. The total score is then calculated in the last row of the spreadsheet, and the decision maker chooses the package with the highest total score. The decision maker has the opportunity to change some of the values for attributes or even the scores assigned to each package in case what-if analysis is desired.

FIGURE 12.5
Using a weighting approach to grade software on a spreadsheet.

	A	B	C	D	E
1	%	ATTRIBUTES	DATAQUIX	BIGBASE	FLEXI FILE
2					
3	.20	FAST RESPONSE TIME	10	6	5
4	.05	MANY OUTPUT OPTIONS	6	4	7
5	.25	EASY TO USE	5	8	7
6	.20	EASY ERROR RECOVERY	10	7	6
7	.05	FLEXIBILITY	10	8	10
8	.10	GOOD DOCUMENTATION	8	9	10
9	.05	HOT LINE SUPPORT	8	10	5
10	.10	ABLE TO HANDLE BIG FILES	2	10	5
11					
12					
13					
14					
15		TOTALS	7.45	7.60	6.55
16					
17					
18					
19					
20					

USING SEQUENTIAL ELIMINATION BY LEXICOGRAPHY

Sometimes decision makers feel weighting methods disguise the best features of an alternative by taking a weighted average of all attributes. One method that features the importance of individual attributes is sequential elimination by lexicography. This method is less demanding than weighting because the attributes are simply ranked in order of importance rather than assigned weights. Intra-attribute values are still specified as in weighting.

An example of lexicography is found in Figure 12.6. A matrix of attributes and alternatives is entered into a spreadsheet. The spreadsheet is sorted, first by attribute from top (most important) to bottom (least important). Next, the alternatives are sorted row by row. Only those alternatives that have the highest possible score (in this case 10) for the first attribute are considered further. In other words, only 5, 6, 3, and 4 are potential answers. Then the second attribute is considered, and only 5 and 4 are considered further, as shown in Figure 12.7.

The final result is that alternative 5 is best because the score for the third attribute is higher for alternative 5 than for alternative 4. Once again, the decision maker can perform sensitivity analysis and rerun the sorting procedure to see if the answer changes.

USING SEQUENTIAL ELIMINATION BY CONJUNCTIVE CONSTRAINTS

Another sequential elimination technique is conjunctive constraints. As its name implies, the decision maker sets constraints, or standards, and then proceeds to eliminate all alternatives that do not satisfy the set of all constraints. If the constraints are set too tight, all the alternatives are eliminated, but if they are not tight enough, many alternatives still remain. To use this method, the decision maker is required to employ an interactive approach.

Figure 12.8 shows a spreadsheet approach to sequential elimination by conjunctive constraints. In this example, the decision maker is trying to choose the most appropriate company vehicle. The attributes are listed in column A, the direction of constraints are placed in column B, and the value of the constraints in

	A	B	C	D	E	F	G	H
1	ATTRIBUTE	RANK	CAR 5	CAR 6	CAR 3	CAR 4	CAR 2	CAR 1
2								
3	PRICE	1	10	10	10	10	5	4
4	MILEAGE	2	10	8	5	10	3	2
5	SAFETY	3	5	3	8	3	10	10
6	RESALE	4	7	9	10	5	5	10
7	COMFORT	5	10	3	2	10	2	3
8								
9								
10								
11								
12								
13								
14								
15								
16								
17								
18								
19								
20								

The data are sorted by rank (1 is most important).

FIGURE 12.6
Using sequential elimination by lexicography, which ranks the importance of individual attributes.

FIGURE 12.7

Narrowing alternatives through the use of sequential elimination by lexicography.

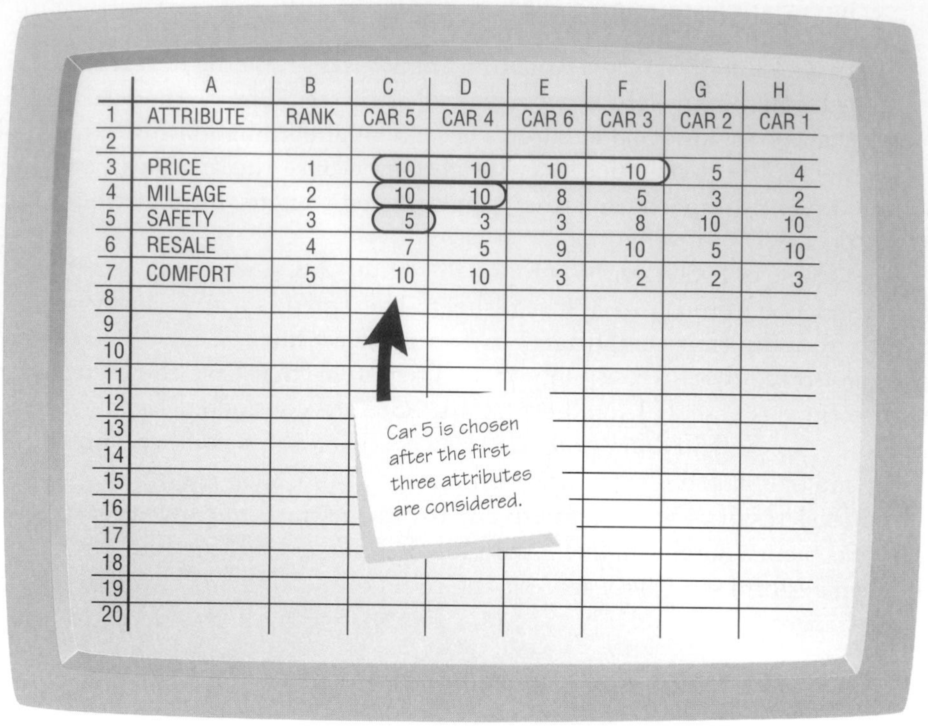

	A	B	C	D	E	F	G	H
1	ATTRIBUTE	RANK	CAR 5	CAR 4	CAR 6	CAR 3	CAR 2	CAR 1
2								
3	PRICE	1	10	10	10	10	5	4
4	MILEAGE	2	10	10	8	5	3	2
5	SAFETY	3	5	3	3	8	10	10
6	RESALE	4	7	5	9	10	5	10
7	COMFORT	5	10	10	3	2	2	3
8								
9								
10								
11								
12								
13								
14								
15								
16								
17								
18								
19								
20								

Car 5 is chosen after the first three attributes are considered.

column C. Columns D, F, and H are used to contain any numerical values, and columns E, G, and I are used to show if the relationship is true or false. At the bottom of the spreadsheet, an "AND" condition is set up so that all the constraints must be met for the relationship at the bottom to be true.

At first, many alternatives may be feasible, so the decision maker then proceeds to change the values in column C until only one alternative remains. In the example shown in Figure 12.9, the decision maker changes the mileage from 18 to 20 miles per gallon, leaving car 3 as the only remaining alternative.

FIGURE 12.8

An example of a spreadsheet approach to using sequential elimination by conjunctive constraints.

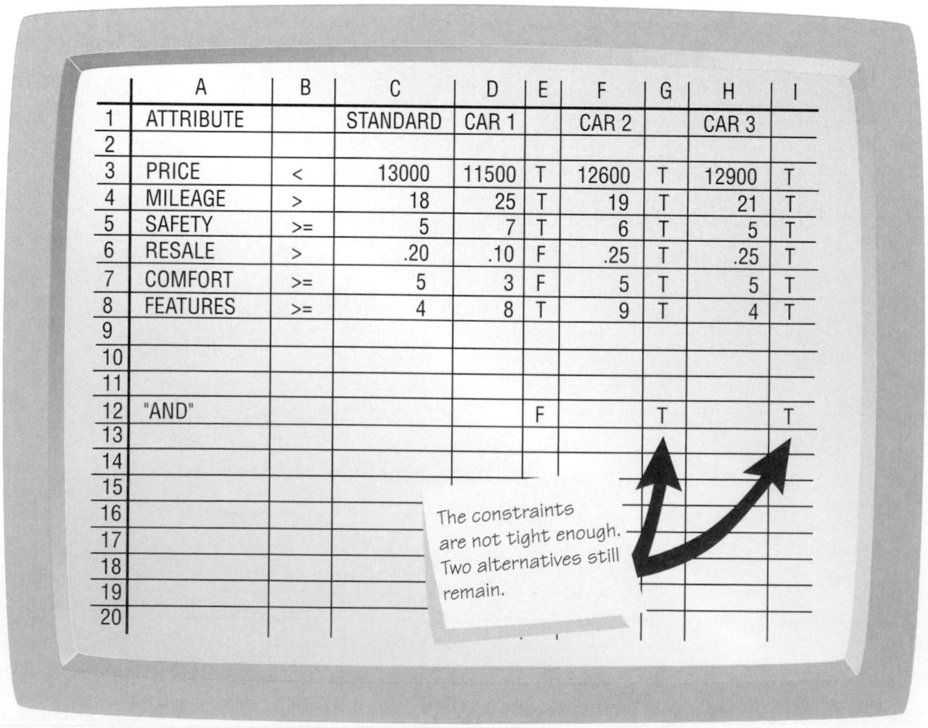

	A	B	C	D	E	F	G	H	I
1	ATTRIBUTE		STANDARD	CAR 1		CAR 2		CAR 3	
2									
3	PRICE	<	13000	11500	T	12600	T	12900	T
4	MILEAGE	>	18	25	T	19	T	21	T
5	SAFETY	>=	5	7	T	6	T	5	T
6	RESALE	>	.20	.10	F	.25	T	.25	T
7	COMFORT	>=	5	3	F	5	T	5	T
8	FEATURES	>=	4	8	T	9	T	4	T
9									
10									
11									
12	"AND"				F		T		T
13									
14									
15									
16									
17									
18									
19									
20									

The constraints are not tight enough. Two alternatives still remain.

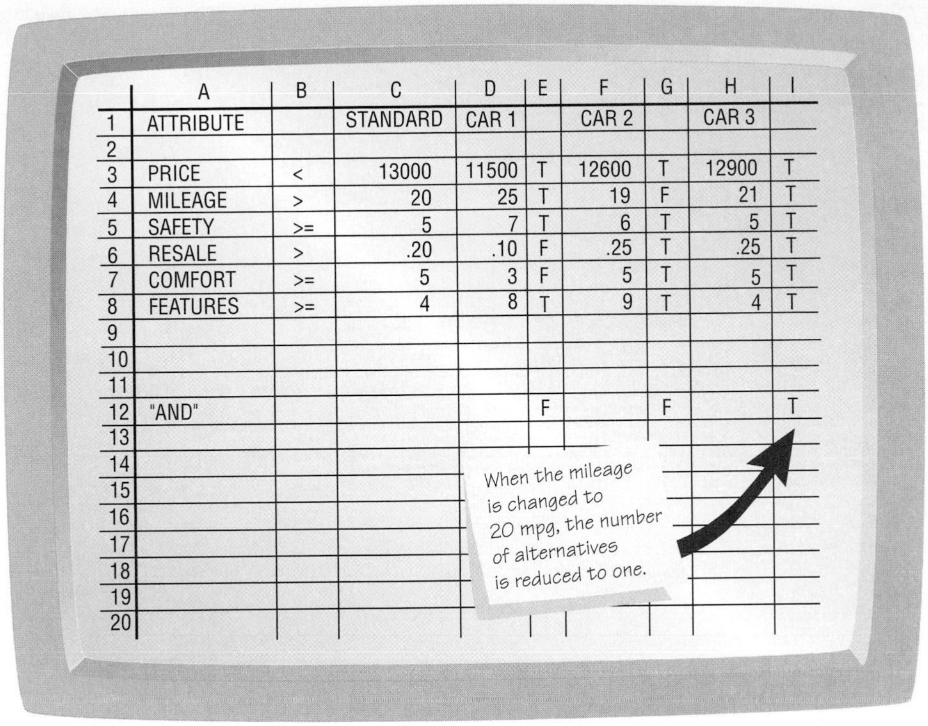

	A	B	C	D	E	F	G	H	I
1	ATTRIBUTE		STANDARD	CAR 1		CAR 2		CAR 3	
2									
3	PRICE	<	13000	11500	T	12600	T	12900	T
4	MILEAGE	>	20	25	T	19	F	21	T
5	SAFETY	>=	5	7	T	6	T	5	T
6	RESALE	>	.20	.10	F	.25	T	.25	T
7	COMFORT	>=	5	3	F	5	T	5	T
8	FEATURES	>=	4	8	T	9	T	4	T
9									
10									
11									
12	"AND"				F		F		T
13									
14									
15									
16									
17									
18									
19									
20									

When the mileage is changed to 20 mpg, the number of alternatives is reduced to one.

FIGURE 12.9 Using sequential elimination by conjunctive constraints allows the decision maker to trade-off attributes by tightening one or more constraints.

The usefulness of this approach becomes apparent when you realize that a person often trades off attributes by interactively relaxing or tightening one or more constraints. That explains why people are willing to spend a little more money than they said initially, even though they rank price as the most important constraint.

USING GOAL PROGRAMMING

A discussion of multiple-criteria decision making is not complete without a discussion of goal programming, which has been widely applied to numerous problems in profit and nonprofit organizations. Goal programming is similar in construct to linear programming, and therefore it has the same assumptions and limitations.

A goal-programming model contains decision variables, deviational variables, priorities, and sometimes weights. Consequently, the decision maker must not only set goals for each of the goal equations in the problem but also choose priorities for minimizing the deviational variables. Goal programming is a valuable technique when the information required is readily available and the decision maker is knowledgeable and confident about goals and priorities. In addition, the decision maker must be skilled at formulating goal equations, which is not a simple task.

USING ANALYTIC HIERARCHY PROCESSING

Another method for solving multiple-criteria problems is analytic hierarchy processing (AHP). This process requires decision makers to make judgments regarding the relative importance of each of the criteria and then to indicate their preference regarding the importance of each of the criteria for each alternative.

Take once again our example about purchasing a car. Figure 12.10 shows a hierarchy tree for the decision. Because the overall goal is choosing the best car (our alternatives), we indicate that there are six branches (our criteria): price, mileage, safety, resale, comfort, and features (our attributes). Underneath each branch, you can see each of our three cars listed. Each one of the cars in our original example has a number associated with it, but that does not have to be the case.

FIGURE 12.10

An AHP screen from Expert Choice.

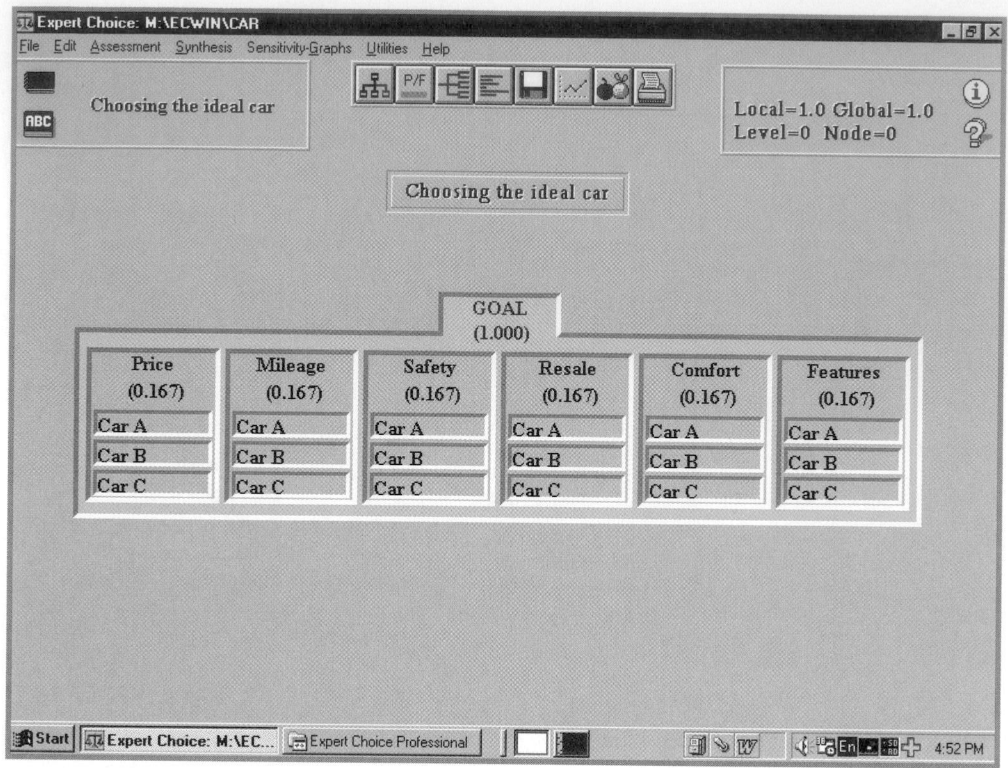

We could have described safety as good (5), above average (6), and much better than average (7). Or, we could have described comfort in words such as poor (3) and adequate (5). One advantage that AHP has over goal programming is that the decision maker does not have to quantify the attributes.

The steps for AHP are simple:

1. Decide which car is preferred over another car and by how much. (In our example, compare A to B, A to C, and B to C.) This step is called pairwise comparison, because two alternatives are compared at a time.
2. Do this step for each of the criteria (in our example, there are six criteria).
3. Rate each of the criteria according to its importance.

To begin our example, step one requires us to compare each car on the first criterion. Suppose we look at safety first (note that we do not know how important safety is at this point). We compare car A to car B, car A to car C, and car B to car C. AHP uses a system as follows to convert verbal judgment into numbers:

Verbal Judgment	*Numerical Preference*
Extremely preferred	9
Very strongly to extremely preferred	8
Very strongly preferred	7
Strongly to very strongly preferred	6
Strongly preferred	5
Moderately to strongly preferred	4
Moderately preferred	3
Equally to moderately preferred	2
Equally preferred	1

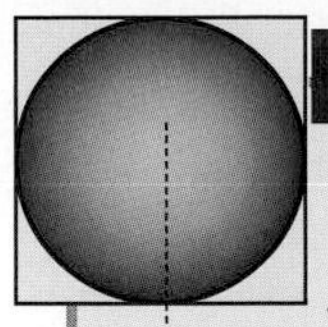

CONSULTING OPPORTUNITY 12.4

SHELVING YOUR CONCERNS

"I never know what to do. Sometimes one supplier's quality is good, but its cost is too high. Then I've got one who isn't that reliable, but when it can deal, its price makes it a steal. I'm just drawn to some suppliers like a magnet. It's rough to know how to decide. I mean, we literally go through tons of steel to make our shelves, and I've got to decide who to buy it from. We're the largest manufacturer of adjustable and nonadjustable shelving for retailers. Sometimes I think we could conquer the world with wall-to-wall shelves," says Abel Gance, director of purchasing for Gaul's Shelving.

"I've been trying to figure out a system for rating our steel suppliers, but most of my data end up sitting on a shelf because we're so busy with other stuff." Abel hands you the following breakdown of how he's been rating the suppliers (see the chart below).

What sort of approach would you recommend for Abel in determining from which supplier to purchase steel? Could you quantify the responses Abel gave you regarding the criteria? What type of multiple-criteria approach would you recommend? Why?

Supplier	Quality of Product	Cost of Items	Reliable Service	Merchandise Out-of-Stock	Return Policy
Bona Parts	Above average	Very high	Always reliable	Sometimes	Excellent
Waterloo	Average	Low	Seldom reliable	Often	Satisfactory
Corsica Corp.	Above average	Average	Often reliable	Never	Poor
Josephine's Supplies	Below average	Low	Often reliable	Never	Satisfactory
Wellington General Stores	Above average	Average	Always reliable	Sometimes	Excellent

So, if our decision maker judges each of the cars on safety, one possible judgment is, "I *moderately prefer* car A to car B, I *very strongly prefer* car A to car C, and I *moderately to strongly prefer* car B to car C." The AHP model would assign preferences to each statement (i.e., a 3 to *moderately prefer*, a 7 to *very strongly prefer*, and a 4 to *moderately to strongly prefer*). The AHP model creates a matrix such as this one:

Alternative	*Car A*	*Car B*	*Car C*
Car A		3	7
Car B			4
Car C			

Then it normalizes and computes a priority vector showing the relative priority of cars A, B, and C. The priority vector for the entries above would be

Alternative	*Priority*
Car A	.656
Car B	.265
Car C	.080

which add up to 1.0 (forgiving the rounding error, of course). It seems to indicate that car A is preferred (with a priority of .656). Once this step is done for each of the six criteria and then each of the criteria themselves are compared on importance, the

FIGURE 12.11
Selected software packages that use either AHP or similar multicriteria techniques.

Product	Company	Web Site
Crystal Ball 2000	Decisioneering	www.decisioneering.com
Criterium Decision Plus	Info Harvest	www.infoharvest.com
Best Choice 3	Logic Technologies	www.logic-gem.com
Expert Choice	ExpertChoice	www.expertchoice.com
AliahTHINK!	Aliah, Inc.	www.aliah.pgh.pa.us

AHP model makes its final calculation and lists the choice, again displaying relative weights that add up to 1.0.

Rather than go through all the calculations, we suggest that you get a copy of Expert Choice or some other AHP-based decision tool and try it out. Some AHP-based tools are found in Figure 12.11. Most of these tools are available on a trial basis from their respective Web sites. You will see that the actual software does much more than help make a decision. These tools are also helpful in performing sensitivity analysis, which will help the decision maker examine what-if scenarios and sleep better at night after making a crucial decision.

One disadvantage of using AHP models comes from the pairwise method used to evaluate alternatives. It is easy sometimes to say that car A is better than B, car B is better than car C, and car C is better than car A, but that doesn't make sense. The principle of transitivity has been violated. There are ways to check for consistency using AHP models, but mistakes can happen.

EXPERT SYSTEMS, NEURAL NETS, AND OTHER DECISION TOOLS

Other decision models available to managers include expert systems and neural nets. Expert systems are rule-based reasoning systems developed around an expert in the field. Gathering expertise is called knowledge acquisition and is the most difficult part of rule set specification. From this point on, you can assume that software tools are widely available in all price categories.

Neural nets are developed by solving a number of problems of one type and letting the software get feedback on the decisions, observing what was involved in successful decisions. This process is referred to as training the neural net. Some widely available neural nets and expert systems are described in Figure 12.12.

Both these models are mentioned in the realm of artificial intelligence. What makes them decision support systems? Typically, it takes a human decision maker to do problem identification, knowledge acquisition, and sensitivity analysis. Rarely are these decisions left solely to the computer. The complexity of the problems solved allows these techniques to be part of the decision support system world.

FIGURE 12.12
Selected expert system and neural net software for the personal computer.

Product	Web Site	Description
Braincel	www.palisade.com	Neural Net add-in for Excel spreadsheets
Exsys CORVID	www.exsys.com	Expert System
Neuralyst	www.palisade.com	Neural Net add-in for Excel spreadsheets

Company	Products	Web Site	Rating System	Sample Web Sites
Macromedia	Macromedia Likeminds	www.macromedia.com	Alphanumeric A–F	www.moviecritic.com
Net Perceptions Inc.	Net Perceptions for Call Centers E-Commerce Analyst	www.netperceptions.com	Numeric 1–5 or 1–13 and implicit (i.e., reading time)	www.amazon.com www.eonline.com

FIGURE 12.13

Selected recommendation systems using numeric and alphanumeric ranking systems.

RECOMMENDATION SYSTEMS

Recommendation systems are software and database systems that allow decision makers to reduce the number of alternatives by ranking or counting or some other method. A restaurant guide, such as *Zagat's*, is an example of a recommendation system. It surveys diners and reports the results both online and in a book, and information for dining in some major cities is available for downloading to wireless handheld devices. A widely used term for the process is *collaborative filtering*.

More sophisticated recommendation systems are being developed all the time. Figure 12.13 lists a number of systems that allow users to rate the alternatives using a numeric system (such as 1 to 7) or an alphanumeric system (A–F, such as grades). Users can get collaborative filtering of books, cars, current films, and so on.

A recommendation system does not depend on numeric weights. This system counts the number of occurrences, such as how many people bookmarked a certain Web site or how many users mentioned an author. A list of some of these systems can be found in Figure 12.14.

GETTING EXTERNAL INFORMATION FROM THE WEB

At times, decision makers want to filter their own information rather than depend on recommendation systems. We can classify this information as news, about the economy, industry, competition, and so on. The Web, however, is dynamic, and it is difficult to predict how executives will get their information in the years ahead. Figure 12.15 lists a sampling of various types of services a decision maker can use to obtain external information about the things like the economy, customers, or trends.

Push technologies (the first group) have enormous potential. Executives can configure one of these products to receive news from the Internet directly on their personal computers, or in some cases on wireless handheld devices, cell phones, or pagers that use wireless application protocol. A version of these push products can also serve as a screen saver, where news crawls across the screen much like a stock ticker. Personalized home pages can be set up to search for specific information. Online newspapers are good for browsing, because the user has control over a broad search. Finally, intelligent agents get to know your personality, learn your behavior, and track the topics they think you want to keep up to date on.

Company	Source of Information	Aggregation Scheme	Recommended as
PHOAKS	Mined from use-net posting	One person one vote (per URL)	Mention of a URL
ReferralWeb	Mined from public data sources	Assemble referral chain to desired person	Mention of a person or a document
Siteseer	Mined from existing bookmark folders	Frequency of mention in overlapping folders	Mention of a URL

FIGURE 12.14

Selected recommendation systems that use methods other than numeric weighting or ranking.

FIGURE 12.15

Selected sources for external information available on the Web.

Type of Service	Product	Web Site
Push technologies	BackWeb Marimba Castanet	www.backweb.com www.marimba.com www.infogate.com
Personalized home pages	My Yahoo! Personal Front Page	www.my.yahoo.com www.msnbc.com
Online Newspapers	CNN Interactive London Times New York Times USA Today	www.cnn.com www.the-times.co.uk www.nytimes.com www.usatoday.com
Intelligent Agents	AgentWare Lite	www.agentwaresystems.com

PROCESS SIMULATION MODELS

Simulations, sometimes called a method of last resort, are making a comeback. New tools for decision makers are called ServiceModel and ProModel, both by the ProModel Corporation. ServiceModel is a full-scale simulation for service system modeling and analysis, and it is used by airports, banks, restaurants, and so on. With this tool, a decision maker can draw an office layout, model employee interaction and machine use, observe animation showing the movement and bottlenecks in the office, and analyze detailed reports and graphs.

Figure 12.16 is a screen from ServiceModel. It depicts bank customers being served in a single-server queue. ProModel does much the same thing in a manufacturing environment.

FIGURE 12.16

Screen from ProModel's ServiceModel software.

Source: Arnie Levin © 2000 The Cartoon Bank. All Rights Reserved

SUMMARY

Decision support systems (DSS) are a special class of information systems that emphasize the process of decision making and changing DSS users through their interaction with the system. They are well suited for addressing semistructured problems where human judgment is still desired or required. Decision support systems do not come up with one solution for users; rather they support the decision-making process by helping the user explore alternatives and consider their ramifications through different modeling techniques.

Users of DSS or group decision support systems (GDSS) come from all three management levels of the organization; semistructured decisions, however, are most often required on the middle managerial and strategic levels. Users of a DSS are eventually changed through the process of interacting with the system. DSS can be added to ecommerce sites to assist decision makers with big ticket items or important life decisions.

The decision-making style of users can be categorized as either analytic or heuristic. Analytic decision makers tend to break problems into quantitative components and use mathematical models to make a decision, whereas heuristic decision makers rely on experience. Decision support systems can be designed with the decision maker's predominant style in mind, so that analytical thinkers are supplied with quantitative models and heuristic decision makers are provided with summary information and memory aids that allow them to recall how they used heuristics in the past.

Semistructured decisions are those where human judgment is still required or considered desirable. Some decisions are considered to be semistructured because the decision maker doesn't possess the decision-making skills to make the decision. Also, if a problem is too complex, it is classified as semistructured. Finally, a problem could be called semistructured if multiple criteria must be addressed. Decision support systems are especially well suited to help solve semistructured problems.

In all problem solving, decision makers go through three phases: intelligence, choice, and design. In the intelligence phase, the decision maker is scanning external

HYPERCASE® EXPERIENCE 12

"So many decisions are made here. You'd be surprised at the types of things even the administrative assistants like me are asked to decide. And at the spur of the moment, not after long hours of analysis. They are not trivial questions, by any means. It seems as if the computers could help us decide most things, if we would just plan for it. It's all these ad hoc decisions we make that could use some support, however. I think Snowden would be all for it. I can certainly see the benefits."

HYPERCASE® QUESTIONS

1. Where might a decision support system fit in at MRE?
2. Who (which MRE employees) would be most likely to benefit from a DSS? Defend your choices.
3. Identify three semistructured decisions that are being made in the Management Systems and Training Unit. Choose one to support with a DSS. Explain your choice.

and internal business environments for potential problems and opportunities. The design phase consists of articulating the problem or opportunity by discovering and creating alternatives, evaluating them, and examining their implications. The choice phase is composed of choosing an alternative from among those that have been considered and determining the reasons and rationales for the adoption of that solution. Decision support systems should be designed to support decisions in all three phases of problem solving.

A complete decision support system should be able to support multiple-criteria decision making. The decision maker using this kind of system has a large repertoire of methods available, including a weighting method, sequential elimination by lexicography, sequential elimination by conjunctive constraints, goal programming, and analytic hierarchy processing.

Decision makers can also use expert systems and neural nets to solve problems. They can also seek advice from recommendation systems, which poll user preferences and arrive at results by either numerical weighting or by occurrence. Executives seek external information, and there are many different ways to obtain that information from the Web. These methods include push technologies, personalized home pages, online newspapers, and intelligent agents. Information to support decisions can even be pushed to handheld devices, cell phones, and pagers. Finally, simulation is being recognized as a tool for analyzing process flows in organizations.

KEYWORDS AND PHRASES

- analytic decision making
- analytic hierarchy processing (AHP)
- choice phase
- collaborative filtering
- decision-making skill
- decision making under certainty
- decision making under risk
- decision making under uncertainty
- decision support systems (DSS)
- decision support systems generator (DSSG)
- deep structure
- design phase
- expert systems
- goal programming
- group decision support systems (GDSS)

heuristic decision making
intelligent agents (re: the Web)
intelligence phase
multiple criteria
neural nets
online newspapers
pairwise comparison
personalized home pages
problem complexity
process simulations
push technologies
recommendation systems
semistructured decision
sequential elimination by conjunctive constraints
sequential elimination by lexicography
weighting methods
wireless application protocol (WAP)

REVIEW QUESTIONS

1. List the functions of decision support systems.
2. Define a decision support system and compare it to a management information system.
3. What is a DSS generator?
4. What type of problem or opportunity is best addressed through the use of a decision support system?
5. Why is a decision support system best conceptualized as a process rather than as a product?
6. How does a decision support system change its user?
7. Define the concept of decision making under certainty.
8. Define the concept of decision making under uncertainty.
9. Define the concept of decision making under risk.
10. Describe the characteristics of an analytic decision maker.
11. Describe the characteristics of a heuristic decision maker.
12. What are some features a decision support system for an analytic decision maker should include?
13. What are some features a decision support system for a heuristic decision maker should include?
14. Define what is meant by the intelligence phase of problem solving. How can a DSS support the intelligence phase?
15. Define what is meant by the design phase of problem solving. How can a DSS support the design phase?
16. Define what is meant by the choice phase of problem solving. How can a decision support system support the choice phase?
17. What is meant by characterizing a decision as structured?
18. What is meant by characterizing a decision as semistructured?
19. Define the concept of deep structure in decisions.
20. What does decision-making skill mean? How can a DSS help in upgrading decision-making skill?
21. What does degree of problem complexity mean? How can a DSS decrease problem complexity?
22. List three multiple-criteria approaches to decision making.
23. Define what is meant by the term *weighting method* as it applies to decision making.
24. Define sequential elimination by lexicography.
25. Define sequential elimination by conjunctive constraints.
26. List the elements that compose a goal-programming model. Why is goal programming of limited use as a standard DSS tool?
27. What does AHP stand for?
28. List the three steps in AHP.

29. Define pairwise comparison.
30. Does AHP have any advantages over goal programming?
31. What is a disadvantage of AHP?
32. What are recommendation systems?
33. What is the difference between the numerical weighting recommendations and the other types of recommendation systems?
34. How can decision makers get external information from the Web?
35. What is the difference between push technologies, personalized home pages, online newspapers, and intelligent agents?
36. Decision support systems on ecommerce Web sites are particularly useful in supporting what types of decisions?
37. How do simulation models help the decision maker?

PROBLEMS

1. While you are working on a large information systems project for a pharmaceutical company, Bob, a newly hired production supervisor, mentions to you that he heard about decision support systems in an M.B.A. class. He would like to have you work on one for him as a sort of subsystem of the larger project. As a production supervisor, Bob reports to a production manager, but he makes routine decisions about production schedules, has input into the purchasing of ingredients, and manages line workers. He thinks that a DSS would be helpful in supporting these functions.
   a. In a paragraph, deny Bob's request for a decision support system. Provide reasons for not pursuing a DSS that would support Bob's decision-making functions.
   b. What changes in Bob's decision-making situation might make a DSS appropriate? Respond in a paragraph.
2. Mary Wren is owner of five Wren's Auto Supply Stores, a chain that sells all manner of auto parts to wholesale and retail customers. She currently uses a small computer rigged up by a friend who was interested in electronics to help her decide what to stock in each of the stores, which are located in five population centers. When her father was alive, he visited each store and based his orders on what managers said was selling well. Mary still visits stores, but she also inputs sales figures into a computer model that displays buying trends for each store. The model also has what-if capabilities so that Mary can experiment with different scenarios, such as falling oil prices, increasing length of original car ownership, and increasing popularity of customizing kits. Mary does not rely solely on the computer to make her ordering decisions; she also uses her own observations about what is selling well to make ordering decisions for each store.
   a. Would you characterize the ordering decisions Mary must make for her five auto supply stores as semistructured? Why or why not? What are some of the variables involved? Respond in two paragraphs.
   b. What features of the information system that Mary is using would support the assertion that it is more of a decision support system than a traditional management information system? List them.
   c. How is Mary's decision process improved upon by the DSS she is using? (Compare it with the way her father made decisions.) Respond in a paragraph.
3. After watching customer patterns for the first few weeks in her new business, Anita knows that most of her customers arrive in the early morning (before 9:30 A.M.) or in the late afternoon (between 3:30 P.M. and 5:00 P.M.). Based on this information, she has decided to close her fabric shop at 5:30 P.M.

a. Is Anita making her decision under conditions of certainty, uncertainty, or risk? Explain in two sentences.
b. Is her information most likely based on information or experience? Explain in two sentences.

4. "If this doesn't work, I can do something else. I just keep experimenting with different glazes for the figurines I sculpt. If they don't look pretty when they come out of the kiln, or if they blow up, or if no one buys them, I can try something else. Live and learn," says Emmy Potts, a sculptor who has a growing business selling custom-made ceramics.
   a. Would you characterize Emmy as an analytic or heuristic decision maker?
   b. Write a paragraph supporting your choice in problem 4a.
   c. Emmy wants a system to keep track of her reactions (and her customers' reactions) to different glazes and ceramics molds for the eventual purpose of ordering supplies and deciding what pieces to make (that is, what pieces will sell even though they are not commissioned). What features would you incorporate into a decision support system for Emmy? List the features and write a sentence giving the reason for each one.
5. Emmy's husband, Buddy, is retiring from his position as a district manager for a chain of hardware stores. For several years, he has been doing market research for the stores. He is systematic in the way he approaches problems and has long used mathematical formulas to project sales and so on. Although her business has been doing well in the last year and a half, Emmy is getting too bogged down in backlogged orders to use the decision support system you built for her. Buddy has agreed to take over the management aspects of her shop so that she can concentrate on artistic aspects.
   a. How would you characterize Buddy as a decision maker? Explain in a paragraph.
   b. What features could you include in a DSS to support Buddy's decision-making style? List them and write a sentence supporting each feature.
   c. Buddy is interested in starting a Web site to showcase Emmy's ceramics. Users could view ceramics and also purchase them. What sort of DSS, if any, would be appropriate for users of the proposed Web site?
6. Louis, a systems analyst, has designed and implemented a decision support system for Scott Weidenfeed, owner of Grass That Grows, a huge suburban lawn care service. Part of the system is designed to flag any unusual problems that are reported by the numerous lawn crews. Some problems are of critical importance (detection of Dutch elm disease, for example), because they will affect much of the community and must be addressed as soon as they are recognized. Resolving them will mean deciding on reassignment of crews from regular lawn care, authorizing the extra expense to the company in terms of chemicals needed, and so on.
   a. Which of the three phases of problem solving is this part of DSS supporting? Explain in a paragraph.
   b. Extend the lawn care example and describe in two paragraphs how the decision support system might support the two phases of problem solving that you did not list in problem 6a. Make any assumption necessary about the company.
7. Howie Johnson is the president of the regional IT manager's association. He is trying to choose a hotel to host the regional convention for the association. About 300 people from five states routinely attend the convention.
   a. Make a list of eight to ten criteria for a good convention hotel.
   b. Choose a multiple-criteria decision method for selecting a hotel.
   c. In a paragraph, defend the method you chose.

8. Jo Pshop wants to determine how to sequence orders that are coming into her specialized invitation shop (invitations for weddings, parties, etc.). Business has been heavy lately, and sequencing of jobs is critical. Her choices are the following:
    1. First come, first served.
    2. Shortest processing time (the easiest jobs are done first).
    3. Due date (do them in the order promised).

   Because her decision is semistructured, you have suggested a trade-off approach. List some criteria and explain how this technique would work.
9. Cary Farr has the job of choosing one of five shipping companies that he will use for his business. The information he has gathered on the shipping companies' attributes is as follows:

	Cost	Damage	Percentage on Time	Complaints	Courteous
All American	2	4	5	3	1
United Truckers	4	3	2	3	4
Rapido Lines	1	1	3	4	3
Carefree Shippers	3	4	5	5	5
We-Do-It-All	4	3	4	3	3

   The numbers are estimates based on the following scale:

1	2	3	4	5
much worse than average		average		much better than average

   Cary gives the order of importance for each criterion as follows:

   Least cost (most important)
   Least damage
   Largest percentage on time
   Fewest complaints
   Most courteous (least important)

   Help Cary by choosing a shipping company. Solve this problem using sequential elimination by lexicography.
10. Solve problem 9 using sequential elimination by lexicography, given the following order of importance:

    Least damage (most important)
    Largest percentage on time
    Fewest complaints
    Least cost
    Most courteous (least important)

11. Assume in problem 9 that weights were assigned for each of the criteria as follows:

Cost	.4
Damage	.2
Percentage on time	.2
Complaints	.1
Courteous	.1
	1.0

    Solve using the weighting method.

12. Solve problem 9 using sequential elimination by conjunctive constraints given the following constraints (remember, 3 = average, 5 = much better than average).

Cost	Must be average or better
Damage	Must be average or better
Percentage on time	Does not matter
Complaints	Must be better than average
Courteous	Must be average or better

13. For problem 9, draw a tree structure for AHP. Label all criteria and alternatives.
14. Using pairwise comparison, show how all five shipping companies would be compared using the courtesy criteria from problem 9. Use the preference ratings (1 to 9) found in the analytic hierarchy processing section of this chapter. Draw the resulting matrix.

SELECTED BIBLIOGRAPHY

Alter, S. *Decision Support Systems: Current Practices and Continuing Challenges*. Reading, MA: Addison-Wesley, 1980.

Bennett, J. L. (ed.). *Building Decision Support Systems*. Reading, MA: Addison-Wesley, 1982.

Davis, G. B., and M. H. Olson. *Management Information Systems, Conceptual Foundations, Structure, and Development*, 2d ed. New York: McGraw-Hill, 1985.

Dhar, V., and R. Stein. *Intelligent Support Methods: The Science of Knowledge Work*. Upper Saddle River, NJ: Prentice Hall, 1997.

Keen, P. W., and M. S. Scott Morton. *Decision Support Systems: An Organizational Perspective*. Reading, MA: Addison-Wesley, 1978.

Kendall, K. E., and B. A. Schuldt. "Case Progression Decision Support System Improves Drug and Criminal Investigator Effectiveness." *Omega*, Vol. 21, No. 3, 1993, pp. 319–28.

———. "Decentralizing Decision Support Systems: A Field Experiment with Drug and Criminal Investigators." *Decision Support Systems*, Vol. 9, 1993, pp. 259–68.

Sauter, V. L. *Decision Support Systems: An Applied Managerial Approach*. New York: John Wiley, 1997.

Simon, H. *The Shape of Automation for Men and Management*. New York: Harper and Row, 1965.

Sprague, R. H., Jr., and E. D. Carlson. *Building Effective Decision Support Systems*. Englewood Cliffs, NJ: Prentice Hall, 1982.

Sprague, R. H., Jr., and H. J. Watson. *Decision Support for Management*. Upper Saddle River, NJ: Prentice Hall, 1996.

Turban, E., and J. E. Aronson. *Decision Support Systems and Intelligent Systems*, 6th ed. Upper Saddle River, NJ: Prentice Hall, 2001.

Watson, H., and R. Sprague (eds.). *Decision Support Systems*. 4th ed. Upper Saddle River, NJ: Prentice Hall, 1996.

EPISODE *CPU CASE*

12

Allen Schmidt, Julie E. Kendall, and Kenneth E. Kendall

AWAITING A WEIGHTY DECISION

"Now that the data flow diagrams, data dictionary, and process logic have been clearly defined, we should spend some time looking at the physical design alternatives for the new microcomputer system," says Chip. "We need to look at which of the several different designs is the best for Dot, Mike, Cher, Paige, and Hy."

"Yes," Anna replies, "and because it is not a problem of economics, we have no cost-benefit figures that would make the choice of the new system obvious. I suggest that we use a weighted method to determine which solution will be the best for the user group."

"That's an excellent idea!" exclaims Chip. "We already have a problem definition with weights assigned by the users. We should list the objectives for each problem and then analyze each alternative, determining the quality of the solution for the users."

Chip and Anna created three alternative solutions:

1. An Internet/intranet solution, which uses Web forms for entering all system input. Updates are performed periodically. This alternative provides forms and screens that are available anywhere, but it has a higher development cost and requires a high degree of security.
2. An online mainframe solution, with terminal emulators used for Web access to the mainframe. This alternative allows centralized control of the data and has a high degree of reliability and security as well as quick response time.
3. A microcomputer solution using a local area network to provide access to a centralized database. This alternative provides a high degree of reliability and quick response time. Each user could use the microcomputer for other tasks.

The microcomputer system has 10 criteria that must be addressed. These objectives are summarized in Figure E12.1.

"Of the 10 system objectives, the highest priority is to provide software and hardware cross-reference information," asserts Chip. "Let's start with this issue and examine how each of the solutions would satisfy it."

"The Internet solution would provide a Web form for entering a software title and version number and then listing each machine and the software installed on it," says Anna. "The online system could also provide a similar screen and, in addition, it could print a report containing complete information for all software. The microcomputer solution could also have reports or inquiries and perhaps generate Web output, depending on the software used."

"Why don't we each determine a performance weight, using a scale from 1 to 10, on how the three solutions compare? Then we can average them together," Chip suggests.

Anna agrees, and the resulting performance weights for the first criterion are as follows:

1. Internet/intranet solution scores a 9, a lower score than the others because it lacks the ability to produce a hard-copy report that the online and microcomputer solutions provided. Any recent changes would be reflected on the screens.
2. Online solution scores a 10.
3. Microcomputer solution also scores a 10.

Objectives	Weight
Provide software/hardware cross-reference.	10
Maintain complete microcomputer information.	9
Automate software installation procedure.	8
Provide software upgrade installation machine information.	7
Provide preventive and other maintenance information.	7
Maintain up-to-date accurate software information.	6
Provide complete cost information for microcomputer inventory.	5
Provide information on the cost to upgrade software.	5
Design a process for performing accurate and efficient physical microcomputer inventory.	3
Maintain and provide training and software expert information.	2

FIGURE E12.1
Computer system objectives.

Anna and Chip continue to analyze the quality of each of the three alternatives for the remaining issues. The results are shown in Figure E12.2, which depicts each of the 10 system objectives and the weight or importance assigned to the objectives by the users. The Internet/intranet solutions seemed to receive lower scores because much of the information is processed centrally and was not needed at remote sites. The Web presented a greater security risk that lowered some of the performances. Finally, although Web pages may be printed, there was a lack of predefined report features, such as group printing, that would make reports easier to read.

Each solution has a brief narrative explaining how the solution will achieve the objective and a performance weight indicating the quality of the solution. The weight for each objective is then multiplied by the performance to determine a score for each solution fragment. The resulting scores are finally summed to provide an overall measure for the solution.

EXERCISES

E-1. Calculate the total score for solution 2, the mainframe online solution.
E-2. Calculate the final score for solution 3, the microcomputer local area network.
E-3. Which solution should be designed and implemented to solve the users' problems? Describe the system in a paragraph. If some of the performances were incorrectly estimated by a small amount, would a different solution be chosen? Explain your reasoning in a paragraph.

Computer System Proposal: System Alternatives

Objectives	Weight	SOLUTION 1 Internet/Intranet	Performance	SOLUTION 2 Mainframe (Internet Telnet access)	Performance	SOLUTION 3 Microcomputer LAN	Performance
Provide software/hardware cross-reference	10	Provide a Web site for viewing software located on particular computers. Enter software title.	9	Software location inquiry screen. Provide a cross-reference report when requested.	10	Software location inquiry screen. Provide a cross-reference report when requested.	10
Maintain complete microcomputer information	9	Update information using a Web page. Update central database.	9	Screens to add, delete, and change microcomputer information, immediate update.	9	Screens to add, delete, and change microcomputer information. Screens to change maintenance data.	10
Automate software installation procedure	8	Create a Web page used to enter software and display a list of machines that should have software installed on them.	8	Use inquiry screens to determine which machines should contain the software. Automatically update master tables.	9	Inquiry screens to determine installation machines. Update files automatically with overrides for exceptions.	10
Provide software upgrade installation machine information	7	Display a Web form to enter the software title and display Web pages listing the machines that need to be upgraded. Modify records using a Web form.	7	Print report of all machines containing software. Note any installation exceptions and use screens to modify software upgrade records.	8	A single screen to automatically upgrade all records. Use a computer in the installation room to immediately update any upgrade exceptions.	10
Provide preventive and other maintenance information	7	Provide Web pages and programs that scan the database and list the machines that require preventive maintenance. Use Web forms to update records.	8	Print reports sorted by location for all machines requiring preventive maintenance. On-line update reflecting work completed.	8	Print reports sorted by location for all machines requiring preventive maintenance. On-site creation of records using local computer.	10
Maintain up to-date accurate software information	6	Use Web forms to update software additions, deletions, and changes.	9	On-line add, delete, and change programs. Automatic deletion of old versions when software is upgraded.	10	Screens for additions, deletions, and changes.	9
Provide complete cost information for microcomputer hardware	5	Display cost information Web pages on a selected machines. Strictly controlled by password.	8	Provide details and summary cost reports. Include cost information on inquiry screens.	9	Provide details and summary cost reports. Include cost and summary information on inquiry screens.	10
Provide information on the cost to upgrade software	5	Create a Web form to enter software title and display a Web page containing the cost of the upgrade.	9	On-line selection of package to be upgraded. Cost figures calculated based on number of upgrades.	10	On-line selection of package to be upgraded. Cost figures calculated based on number of upgrades. User may calculate individual departmental costs.	10
Design a process for performing accurate and efficient physical microcomputer inventory	3	Use a Web form to enter the physical inventory for a particular room. Update a centralized database.	9	Physical inventory listing printed and used as a turn-around document. On-line update of changes.	9	Physical inventory listing printed. On-site entry of changes using local computer.	10
Maintain and provide training and software expert information	2	Use a Web form for adding training and software expert information. Display Web pages of training classes	10	On-line update of software expert and training courses. Inquiry screens to locate experts. Training listing printed.	9	On-line update of software expert and training courses. Inquiry screens provide expert information. Training listing printed. Generate Web site of training classes.	10
Total							

FIGURE E12.2

Three computer system alternatives. Weight is the user-assigned weight, indicating importance of the objective. Performance is the performance weight, indicating the quality of the solution.

PREPARING THE SYSTEMS PROPOSAL

13

METHODS AVAILABLE

The systems proposal is a distillation of all that the systems analyst has learned about the business and about what is needed to improve its performance. To address information requirements adequately, the systems analyst must use systematic methods for acquiring hardware and software, must identify and forecast future costs and benefits, and must perform cost-benefit analysis. All these methods are used in preparing systems proposal material.

Information needs of users drive the selection of computer hardware, data storage media, and any prepackaged software. The hardware and software system that is eventually proposed is the analyst's response to users' information needs. This chapter provides the methods that are needed to project future needs systematically and then to weigh current hardware and software alternatives. Forecasting, guidelines for hardware and software acquisition, and cost-benefit analysis are also considered.

ASCERTAINING HARDWARE AND SOFTWARE NEEDS

In this section, we cover the process of estimating the present and future workloads of a business and the process involved in evaluating the ability of computer hardware and software to handle workloads adequately. Figure 13.1 shows the steps the systems analyst takes in ascertaining hardware and software needs. First, all current computer hardware must be inventoried to discover what is on hand and what is usable. Next, current and future system workloads must be estimated. Then an evaluation of available hardware and software is undertaken.

The systems analyst needs to work along with users to determine what hardware will be needed. Hardware determinations can come only in conjunction with determining information requirements. Knowledge of the organizational structure (as discussed in Chapter 2) can also be helpful in hardware decisions. Only when systems analysts, users, and management have a good grasp of what kinds of tasks must be accomplished can hardware options be considered.

INVENTORYING COMPUTER HARDWARE

Begin by inventorying what computer hardware is already available in the organization. As will become apparent, some of the hardware options involve expanding or recycling current hardware, so it is important to know what is on hand.

FIGURE 13.1
Steps in choosing hardware and software.

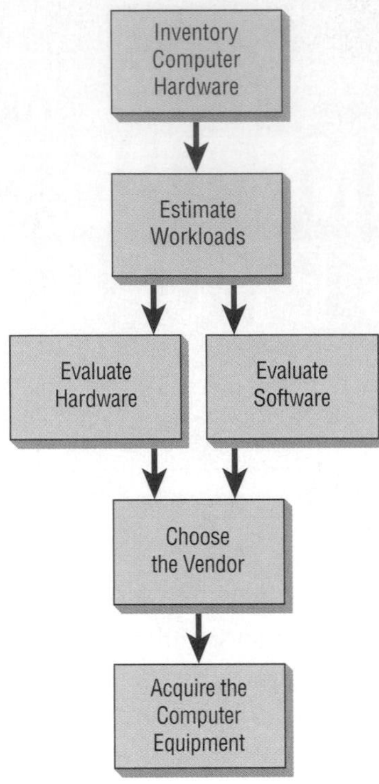

If an updated computer hardware inventory is unavailable, the systems analyst needs to set up one quickly and carry through on it. You need to know the following:

1. The type of equipment: model number, manufacturer.
2. The status of the equipment operation: on order, operating, in storage, in need of repair.
3. The estimated age of the equipment.
4. The projected life of the equipment.
5. The physical location of the equipment.
6. The department or person considered responsible for the equipment.
7. The financial arrangement for the equipment: owned, leased, or rented.

Ascertaining the current hardware available will result in a sounder decision-making process when hardware decisions are finally made, because much of the guesswork about what exists will be eliminated. Through your earlier interviewing, questionnaires, and research of archival data, you will already know the number of people available for data processing as well as their skills and capabilities. Use this information to project how well the staffing needs for new hardware can be met.

ESTIMATING WORKLOADS

The next step in ascertaining hardware needs is to estimate workloads. Thus, systems analysts formulate numbers that represent both current and projected workloads for the system so that any hardware obtained will possess the capability to handle current and future workloads.

If estimates are accomplished properly, the business should not have to replace hardware solely due to unforeseen growth in system use. (Other events, however, such as superior technological innovations, may dictate hardware replacement if the business wants to maintain its competitive edge.)

Task	Existing System	Proposed System
Task	Monthly summary of shipments to distribution warehouses	Same
Method	Manual	Computer
Personnel	Distribution Manager	Computer Operator
Cost/Hour	$20.00	$10.00
When and How	Daily: files shipping receipts for each warehouse Monthly: summarizes daily records using calculator and prepares report	Daily: runs program that totals shipments and writes to disk Monthly: runs program that summarizes and prints reports
Human Time Requirements	Daily: 20 minutes Monthly: 8 hours	Daily: 4 minutes Monthly: 20 minutes
Computer Time Requirements	None	Daily: 4 minutes Monthly: 20 minutes

FIGURE 13.2
Comparisons of workloads between existing and proposed systems.

Out of necessity, workloads are sampled rather than actually put through several computer systems. The guidelines given on sampling in Chapter 4 can be of use here, because in workload sampling, the systems analyst is taking a sample of necessary tasks and the computer resources required to complete them.

Figure 13.2 is a comparison of the times required by an existing and a proposed information system that are supposed to handle a given workload. Notice that the company is currently using a manual system to prepare a monthly summary of shipments to its distribution warehouses, and a computer system is being suggested. The workload comparison looks at the cost per hour of each system, when and how each process is done, how much human time is required, and how much computer time is needed.

EVALUATING COMPUTER HARDWARE

Evaluating computer hardware is the shared responsibility of management, users, and systems analysts. Although vendors will be supplying details about their particular offerings, analysts need to oversee the evaluation process personally because they will have the best interests of the business at heart. In addition, systems analysts may have to educate users and management about the general advantages and disadvantages of hardware before they can capably evaluate it.

Based on the current inventory of computer equipment and adequate estimates of current and forecasted workloads, the next step in the process is to consider the kinds of equipment available that appear to meet projected needs. Information from vendors on possible systems and system configurations becomes more pertinent at this stage and should be reviewed with management and users.

In addition, workloads can be simulated and run on different systems, including those already used in the organization. This process is referred to as benchmarking.

Criteria that the systems analysts and users should use to evaluate performance of different systems hardware include the following:

1. The time required for average transactions (including how long it takes to input data and how long it takes to receive output).
2. The total volume capacity of the system (how much can be processed at the same time before a problem arises).
3. The idle time of the central processing unit.
4. The size of the memory provided.

Some criteria will be shown in formal demonstrations; some cannot be simulated and must be gleaned from manufacturers' specifications. It is important to be clear about the required and desired functions before getting too wrapped up in vendors' claims during demonstrations.

Once functional requirements are known and the current products available are comprehended and compared with what already exists in the organization, decisions are made by the systems analysts in conjunction with users and management about whether obtaining new hardware is necessary. Options can be thought of as existing on a continuum from using only equipment already available in the business all the way to obtaining entirely new equipment. In between are options to make minor or major modifications to the existing computer system.

Computer Size and Use. The rapid advance of technology dictates that the systems analyst research types of computers available at the particular time that the systems proposal is being written. Computer sizes range all the way from the smallest notebook computers to room-sized supercomputers. Each has different attributes that should be considered when deciding how to implement a computer system.

ACQUISITION OF COMPUTER EQUIPMENT

The three main options for acquisition of computer hardware are buying, leasing, or renting it. There are advantages and disadvantages that ought to be weighed for each of the decisions, as shown in Figure 13.3. Some of the more influential factors to consider in deciding which option is best for a particular installation include initial versus long-term costs, whether the business can afford to tie up capital in computer equipment, and whether the business desires full control of and responsibility for the computer equipment.

Buying implies that the business itself will own the equipment. One of the main determinants of whether to buy is the projected life of the system. If the system will be used longer than four to five years (with all other factors held constant), the decision is usually made to buy. Notice in the example in Figure 13.4 that the cost of purchase after six years is dramatically lower than that of leasing or renting. As systems become smaller and distributed systems become increasingly popular, more businesses are deciding to purchase equipment.

Leasing, rather than buying, computer hardware is another possibility. Leasing equipment from the vendor or a third-party leasing company is more practical when the projected life of the system is less than four years. In addition, if significant change in technology is imminent, leasing is a better choice. Leasing also allows the business to put its money elsewhere, where it can be working for the company rather than be tied up in capital equipment. Over a long period, however, leasing is not an economical way to acquire computer power.

	Advantages	Disadvantages
Buying	• Cheaper than leasing or renting over the long run • Ability to change system • Provides tax advantages of accelerated depreciation • Full control	• Initial cost is high • Risk of obsolescence • Risk of being stuck if choice was wrong • Full responsibility
Leasing	• No capital is tied up • No financing is required • Leases are lower than rental payments	• Company doesn't own the system when lease expires • Usually a heavy penalty for terminating the lease • Leases are more expensive than buying
Renting	• No capital is tied up • No financing is required • Easy to change systems • Maintenance and insurance are usually included	• Company doesn't own the computer • Cost is very high because vendor assumes the risk (most expensive option)

FIGURE 13.3
Comparing the advantages and disadvantages of buying, leasing, and renting computer equipment.

Rental of computer hardware is the third main option for computer acquisition. One of the main advantages of renting is that none of the company's capital is tied up and hence no financing is required. Also, renting computer hardware makes it easier to change system hardware. Finally, maintenance and insurance are usually included in rental agreements. Because of the high costs involved and that the company will not own the rented equipment, however, rental should be contemplated only as a short-term move to handle nonrecurring or limited computer needs or technologically volatile times.

Evaluation of Vendor Support for Computer Hardware. Several key areas ought to be evaluated when weighing the support services available to businesses from vendors. Most vendors offer testing of hardware upon delivery and a 90-day warranty covering any factory defects, but you must ascertain what else the vendor has to

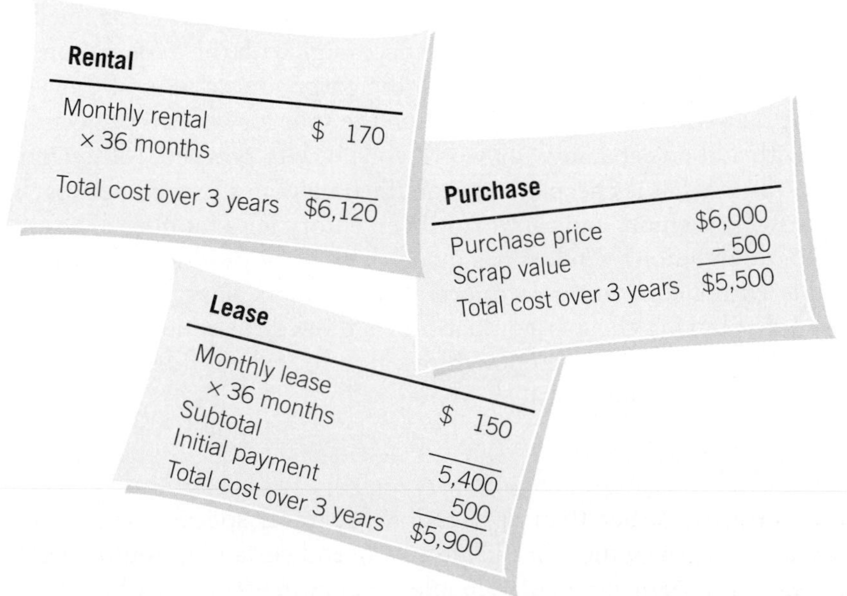

Rental
Monthly rental
× 36 months
$ 170
Total cost over 3 years
$6,120
Purchase
Purchase price
$6,000
Scrap value
– 500
Total cost over 3 years
$5,500
Lease
Monthly lease
× 36 months
$ 150
Subtotal
5,400
Initial payment
500
Total cost over 3 years
$5,900

FIGURE 13.4
Comparison of alternatives for computer acquisition.

FIGURE 13.5
Guidelines for vendor selection.

Vendor Services	Specifics Vendors Typically Offer
Hardware Support	Full line of hardware Quality products Warranty
Software Support	Complete software needs Custom programming Warranty
Installation and Training	Commitment to schedule In-house training Technical assistance
Maintenance	Routine maintenance procedures Specified response time in emergencies Equipment loan while repair is being done

offer. Vendors of comparable quality frequently distinguish themselves from others by the range of support services they offer.

A list of key criteria that ought to be checked when evaluating vendor support is provided in Figure 13.5. Most of the extra vendor support services listed there are negotiated separately from hardware lease or purchase contracts.

Support services include routine and preventive maintenance of hardware, specified response time (within six hours, next working day, etc.) in case of emergency equipment breakdowns, loan of equipment in the event that hardware must be permanently replaced or offsite repair is required, and in-house training or off-site group seminars for users. Remember that it may be more difficult to obtain training for unique hardware that is not widely used by other organizations. Although the possibility of a customized installation may be attractive, the prospects for its long-term support may be diminished. Peruse the support services documents accompanying the purchase or lease of equipment and remember to involve appropriate legal staff before signing contracts for equipment or services.

Unfortunately, evaluating computer hardware is not as straightforward as simply comparing costs and choosing the least expensive option. Some other eventualities commonly brought up by users and management include (1) the possibility of adding on to the system if the need comes up later; (2) the possibility of interfacing with equipment from other vendors if the system needs to grow; (3) the benefits of buying more memory than is projected as necessary, with the expectation that business will eventually "grow into it"; and (4) the corporate stability of the vendor.

Adding on to the existing system is often the spur for systems projects. Installing a system with add-on capability is a worthwhile way to proceed. Although it takes a little extra planning, it is cheaper and more flexible than the third approach, that is, obtaining excess memory and carrying it in inventory for a number of years.

Competition among vendors has made the idea of producing hardware that is compatible with a competitor's important for vendors' survival. Before becoming convinced that buying cheaper compatibles is the way to endow your system with add-on capability, however, do enough research to feel confident that the original vendor is a stable corporate entity.

SOFTWARE EVALUATION

Packaged software, rather than application programs specifically written for an installation, is becoming more readily available and certainly should be given careful consideration. Many hours of valuable programmer time can be saved if pack-

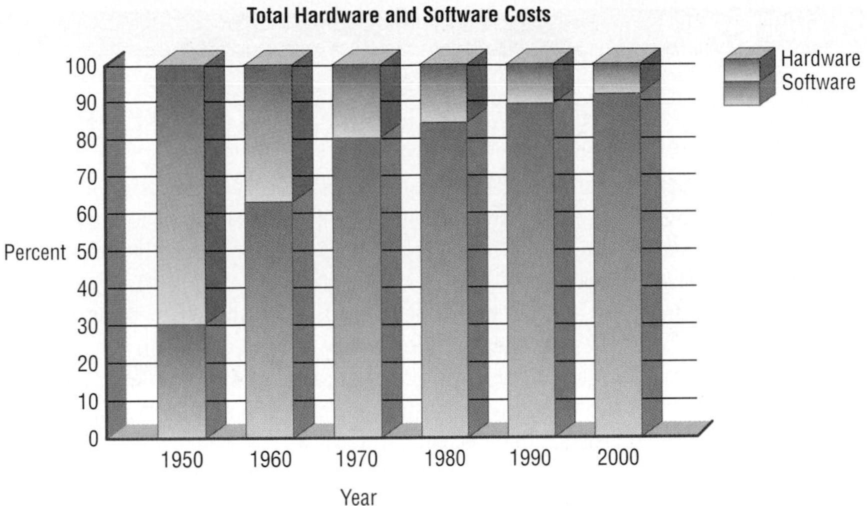

FIGURE 13.6
The cost of software as a percentage of total hardware and software cost is rising.

aged software is deemed suitable for part of or all the system and if extensive customizing is not necessary. The bar chart in Figure 13.6 shows the cost of software (projected as ever-increasing) as part of total hardware and software costs.

Once again, you will be dealing with vendors who may have their own best interests at heart. You must be willing to evaluate software along with users and not be unduly influenced by vendors' sales pitches. Specifically, there are six main categories on which to grade software, as shown in Figure 13.7: performance effectiveness, performance efficiency, ease of use, flexibility, quality of documentation, and manufacturer support.

Software Requirements	Specific Software Features
Performance Effectiveness	Able to perform all required tasks Able to perform all tasks desired Well-designed display screens Adequate capacity
Performance Efficiency	Fast response time Efficient input Efficient output Efficient storage of data Efficient backup
Ease of use	Satisfactory user interface Help menus available ReadMe files for last-minute changes Flexible interface Adequate feedback Good error recovery
Flexibility	Options for input Options for output Usable with other software
Quality of Documentation	Good organization Adequate online tutorial Web site with FAQ
Manufacturer Support	Tech support hot line Newsletter/email Web site with downloadable product updates

FIGURE 13.7
Guidelines for evaluating software.

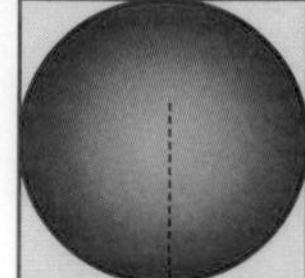

CONSULTING OPPORTUNITY 13.1

VENI, VIDI, VENDI, OR, I CAME, I SAW, I SOLD

"It's really some choice. I mean, no single package seems to have everything we want. Some of them come darn close, though," says Roman, an advertising executive for *Empire Magazine* with whom you have been working on a systems project. Recently, the two of you have decided that packaged software would probably suit the advertising department's needs and stem its general decline.

"The last guy's demo we saw, you know, the one who worked for Data Coliseum, really had a well-rounded pitch. And I like their brochure. Full-color printing, on card stock. Classic," Roman asserts.

"And what about those people from Vesta Systems? They're really fired up. And their package was easy to use with a minimum of ceremony. Besides, they said they would train all 12 of us, on site, at no charge. But look at their advertising. They just take things off their printers."

Roman fiddles in his chair as he continues his ad hoc review of software and software vendors. "That one package from Mars, Inc., really sold me all on its own, though. I mean, it had a built-in calendar. And I like the way the menus for the screen displays could all be chosen by Roman numerals. It was easy to follow. And the vendor isn't going to be hard to move on price. I think they're already in a price war."

"Do you want to know my favorite, though?" Roman asks archly. "It's the one put out by Jupiter, Unlimited. I mean, it has everything, doesn't it? It costs a little extra coin, but it does what we need it to do, and the documentation is heavenly. They don't do any training of course. They think they're above it."

You are already plotting that to answer Roman's burning questions by your March 15 deadline, you need to evaluate the software as well as the vendors, systematically, and then render a decision. Evaluate each vendor and package based on what Roman has said so far (assume you can trust his opinions). What are Roman's apparent biases when evaluating software and vendors? What further information do you need about each company and its software before making a selection?

Evaluate packaged software based on a demonstration with test data from the business considering it and an examination of accompanying documentation. Vendors' descriptions alone will not suffice. Vendors typically certify that software is working when it leaves their supply house, but they will not guarantee that it will be error-free in every instance or that it will not crash when incorrect actions are taken by users. Obviously, they will not guarantee their packaged software if used in conjunction with faulty hardware.

The need for multiple copies or network versions of software (for use at several microcomputer workstations, for instance) means negotiating a multiple-use agreement with the vendor so that copyrights are not infringed through the creation of illegal copies. This negotiation often results in the purchase of one software package at its regular price and the purchase of any additional copies at a reduced price.

It is also possible to negotiate a special vendor services contract covering support for purchased software. This agreement might include extended technical assistance, emergency and preventive maintenance, free or reduced-price updates, additional copies of documentation, and special user training. Many consulting firms, such as JD Edwards and PeopleSoft, now devote the majority of their time to customizing, installing, and supporting expensive, enterprisewide enterprise resource planning software.

IDENTIFYING AND FORECASTING COSTS AND BENEFITS

Costs and benefits of the proposed computer system must always be considered together, because they are interrelated and often interdependent. Although the systems analyst is trying to propose a system that fulfills various information requirements, decisions to continue with the proposed system will be based on a costs and benefits analysis, not on information requirements. In many ways, benefits are measured by costs, as becomes apparent in the next section.

FORECASTING COSTS AND BENEFITS

Systems analysts are required to predict certain key variables before the proposal is submitted to the client. To some degree, a systems analyst will rely on a what-if analysis, such as, "What if labor costs rise only 5 percent per year for the next three years, rather than 10 percent?" The systems analyst should realize, however, that you cannot rely on what-if analysis for everything if the proposal is to be credible, meaningful, and valuable.

The systems analyst has many forecasting models available. The main condition for choosing a model is the availability of historical data. If they are unavailable, the analyst must turn to one of the judgment methods: estimates from the sales force, surveys to estimate customer demand, Delphi studies (a consensus forecast developed independently by a group of experts through a series of iterations), creating scenarios, or drawing historical analogies.

If historical data are available, the next differentiation between classes of techniques involves whether the forecast is conditional or unconditional. Conditional implies that there is an association among variables in the model or that such a causal relationship exists. Common methods in this group include correlation, regression, leading indicators, econometrics, and input/output models.

Unconditional forecasting means the analyst isn't required to find or identify any causal relationships. Consequently, systems analysts find that these methods are low-cost, easy-to-implement alternatives. Included in this group are graphical judgment, moving averages, and analysis of time series data. Because these methods are simple, reliable, and cost-effective, the remainder of the section focuses on them.

Estimation of Trends. Trends can be estimated in a number of different ways. The most widely used techniques are (1) graphical judgment, (2) the method of least squares, and (3) the moving average method. A brief explanation of these techniques is in order.

Graphical judgment The simplest way to identify a trend and forecast future trends is by graphical judgment, which is accomplished by simply looking at the graph and estimating by freehand an extension of a line or curve. An example of graphical judgment is illustrated in Figure 13.8.

The disadvantages of this method are obvious from looking at the graphs in the figure. The extension of the line or curve may depend too much on individual judgment and may not represent the real situation. The graphical judgment method is useful, however, because the ability to perform sensitivity analysis (what-if) has increased with the introduction of electronic spreadsheets.

The method of least squares When a trend line is constructed, the actual data points will fall on either side of that line. The objective in estimating a trend using the least squares method is to find the "best-fitting line" by minimizing the sum of

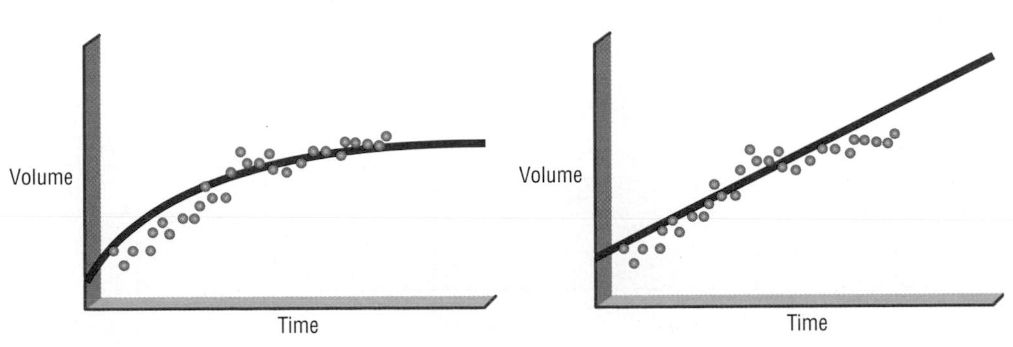

FIGURE 13.8
Different conclusions can easily be drawn from the same set of data.

the deviations from a line. Once the best-fitting line is found, it can be graphed, and the line can then be extended to forecast what will happen.

The best-fitting line, or least square line, is developed from the data points $(X_1, Y_1), (X_2, Y_2), \ldots (X_N, Y_N)$, where the X coordinates signify the time periods and the Y coordinates represent the variable the systems analyst is trying to predict. The equation for the least square line is expressed in the form

$$Y = m * X + b$$

where the variable m represents the slope of the line and b represents the Y intercept, the point at which the line intercepts the Y axis.

We recommend a more computationally efficient method to find the least square equation by calculating the center of gravity of the data by taking $x = X - \overline{X}$ and $y = Y - \overline{Y}$ and then calculating the least square line as

$$y = \left(\frac{\Sigma xy}{\Sigma x^2}\right) * x$$

finally substituting back the $X - \overline{X}$ for x and $Y - \overline{Y}$ for y.

In Excel, you can calculate the trend based on least squares directly by using the TREND function.

Moving averages The method of moving averages is useful because some seasonal, cyclical, or random patterns may be smoothed, leaving the trend pattern. The principle behind moving averages is to calculate the arithmetic mean or data from groups of periods, using the equation

$$\frac{Y_1 + Y_2 + \cdots + Y_N}{N}$$

where N equals the number of periods. Then calculate the next arithmetic mean by discarding the oldest period's data and adding data from the next period

$$\frac{Y_2 + Y_3 + \cdots + Y_{N+1}}{N}$$

and in this manner say the average is moving.

Figure 13.9 shows one type of moving average. Here five years' data are averaged and the resulting figure is indicated. Notice that years 1990 through 1994

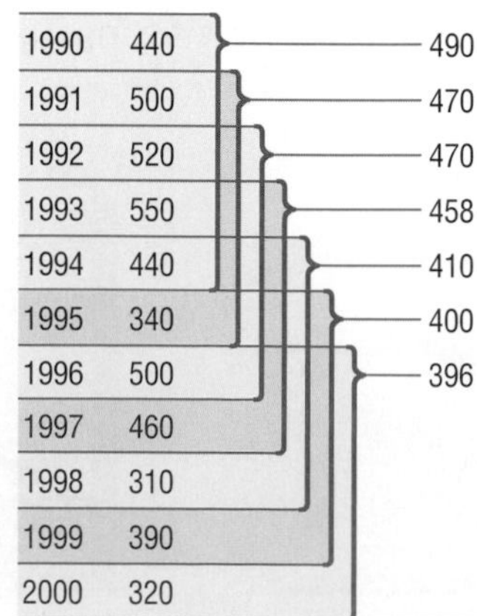

1990 440
1991 500
1992 520
1993 550
1994 440
1995 340
1996 500
1997 460
1998 310
1999 390
2000 320
490
470
470
458
410
400
396

FIGURE 13.9
Calculating a five-year moving average.

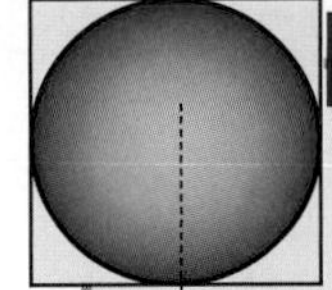

CONSULTING OPPORTUNITY 13.2

THE BIRTH OF A SYSTEM

"Yup, what little there is is all mine. I started in this business because I couldn't keep my hands off the stuff. I loved tinkering with our electronic equipment, taking apart TVs and VCRs. Ask my wife, Carol. Then I started helping friends with their projects, and they thought I was pretty good. When I inherited some money, I opened this little shop selling and repairing TVs and VCRs and renting videotapes," says Roger Corman, owner of a video rental and repair store.

"Right now," Roger continues as he shows you around the small store, "we use a manual system for keeping track of rental videotapes. We make a three-by-five card for each title we own and the name of the person renting the tape is recorded on this card.

"The cards were okay for a while, but now I need your help. I'm afraid that if things keep growing like they have, I'll need a computer soon. I didn't get into this business overnight though, and I want to be sure that the computer will pay for itself. I need proof. I've got to see it in black and white," Roger confides.

"I was interested enough to keep a log of demand for rental tapes for the past 27 months," Roger continues. "Here it is."

Hoping to catch up on some movies you've missed recently, you agree to do a small systems project for Roger in return for a fee and some free videotaped films. Using the methods you have learned so far, forecast the demand for tape rentals and for new titles.

Month	Number of Rentals	Number of Titles
January	1,000	70
February	1,200	100
March	1,400	130
April	1,800	140
May	2,200	150
June	2,000	160
July	1,800	170
August	1,800	180
September	2,500	200
October	2,800	220
November	3,000	280
December	3,500	260
January	4,000	280
February	4,600	300
March	4,800	320
April	5,200	340
May	5,700	360
June	5,000	380
July	4,800	400
August	3,500	260
September	4,000	280
October	4,600	300
November	4,800	320
December	5,200	340
January	5,700	360
February	5,000	380
March	4,800	400

inclusive are averaged to predict 1995, then the years 1991 through 1995 are averaged to predict the amount for 1996, and so on. When the results are graphed, it is easily noticeable that the widely fluctuating data are smoothed.

The moving average method is useful for its smoothing ability, but at the same time it has many disadvantages. Moving averages are more strongly affected by extreme values than the methods of graphical judgment and least squares.

Many worthwhile forecasting packages are available for PCs as well as for mainframes. The analyst should learn forecasting well, as it often provides information valuable in justifying the entire project.

IDENTIFYING BENEFITS AND COSTS

Benefits and costs can be thought of as either tangible or intangible. Both tangible and intangible benefits and costs must be taken into account when systems are considered.

Tangible Benefits. Tangible benefits are advantages measurable in dollars that accrue to the organization through the use of the information system. Examples of tangible benefits are an increase in the speed of processing, access to otherwise inaccessible information, access to information on a more timely basis than was

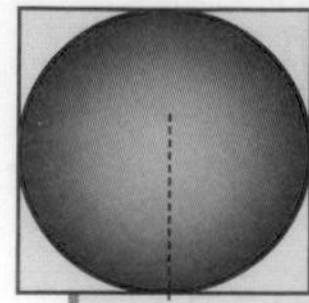

CONSULTING OPPORTUNITY 13.3

THE BIRTH OF A SYSTEM II: THE SEQUEL

"That's a good question. I can think of a million problems we have here. Just off the top of my head, though, here are the big ones," says Roger Corman during his second interview with you. "It's taking us time to keep up records. It's getting to be extremely difficult to sniff out tapes that aren't returned. I like detective flicks, but I'm no Sherlock Holmes. I usually have to track down tapes myself, and that takes me away from my repair service.

"What really gripes me about tapes that aren't returned on time is that they result in lost rentals, because most of my customers choose from the selections in stock that they can see on the shelves. And I don't have any information about the categories of tapes my customers prefer. Part of it is my fault, I guess, because I haven't bothered to set up a reservation system. I mean, there are probably more problems, but those are the ones I handle every day," says Roger.

You analyze the recording process and determine that employees can save two minutes for each tape rented and half a minute for each title in inventory per month. Roger himself gets involved with tracking down late tapes (about one for every 250 tapes rented), and you find he can save half an hour per month for each tape he has to track down if he uses the computer instead.

Assume Roger can take out a small loan at 15 percent to purchase the computer and pay for development costs. The economy is reasonably good with an inflation rate of 5 percent. Your time, as well as his, can be valued at $20 per hour. His employees' time is worth $6 per hour.

You estimate that systems analysis will take 30 hours, system design 40 hours, and development and implementation 30 hours. Roger will be involved in answering questions for about 15 hours, and in addition, you will have to individually train two people and Roger for five hours each.

The microcomputer, printer, and software cost $4,000. After the system is up and running, you have advised Roger that he will need to earmark $20 per month for supplies and maintenance.

After discussing your analysis with him, he says, "I don't like giving you a lot of direction, but would you make a list for me? I'd like to screen all this, so I can project what will happen. Give it your best shot, so I can take a close-up look."

From the foregoing data, prepare a list of tangible and intangible costs and benefits for Roger. You are asked to analyze these costs in Consulting Opportunity 13.4.

possible before, the advantage of the computer's superior calculating power, and decreases in the amount of employee time needed to complete specific tasks. There are still others. Although measurement is not always easy, tangible benefits can actually be measured in terms of dollars, resources, or time saved.

Intangible Benefits. Some benefits that accrue to the organization from the use of the information system are difficult to measure but are important nonetheless. They are known as intangible benefits.

Intangible benefits include improving the decision-making process, enhancing accuracy, becoming more competitive in customer service, maintaining a good business image, and increasing job satisfaction for employees by eliminating tedious tasks. As you can judge from the list given, intangible benefits are extremely important and can have far-reaching implications for the business as it relates to people both outside and within the organization.

Although intangible benefits of an information system are important factors that must be considered when deciding whether to proceed with a system, a system built solely for its intangible benefits will not be successful. You must discuss both tangible and intangible benefits in your proposal, because presenting both will allow decision makers in the business to make a well-informed decision about the proposed system.

Tangible Costs. The concepts of tangible and intangible costs present a conceptual parallel to the tangible and intangible benefits discussed already. Tangible costs are those that can be accurately projected by the systems analyst and the business' accounting personnel.

Included in tangible costs are the cost of equipment such as computers and terminals, the costs of resources, the cost of systems analysts' time, the cost of programmers' time, and other employees' salaries. These costs are usually well established or can be discovered quite easily and are the costs that will require a cash outlay of the business.

Intangible Costs. Intangible costs are difficult to estimate and may not be known. They include losing a competitive edge, losing the reputation for being first with an innovation or the leader in a field, declining company image due to increased customer dissatisfaction, and ineffective decision making due to untimely or inaccessible information. As you can imagine, it is next to impossible to project a dollar amount for intangible costs accurately. To aid decision makers who want to weigh the proposed system and all its implications, you must include intangible costs even though they are not quantifiable.

COMPARING COSTS AND BENEFITS

There are many well-known techniques for comparing the costs and benefits of the proposed system. They include break-even analysis, payback, cash-flow analysis, and present value analysis. All these techniques provide straightforward ways of yielding information to decision makers about the worthiness of the proposed system.

BREAK-EVEN ANALYSIS

By comparing costs alone, the systems analyst can use break-even analysis to determine the break-even capacity of the proposed information system. The point at which the total costs of the current system and the proposed system intersect represents the break-even point, the point where it becomes profitable for the business to get the new information system.

Total costs include the costs that recur during operation of the system plus the developmental costs that occur only once (one-time costs of installing a new system), that is, the tangible costs that were just discussed. Figure 13.10 is an example of break-even analysis on a small store that maintains inventory using a manual

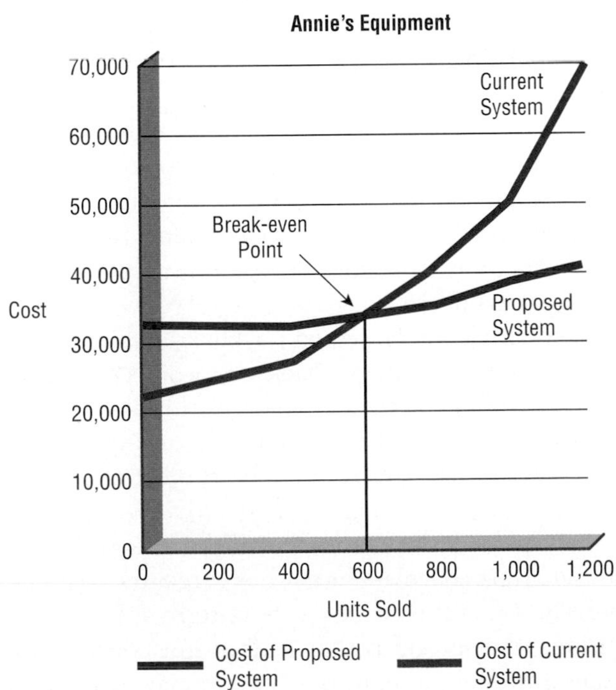

FIGURE 13.10
Break-even analysis for the proposed inventory system.

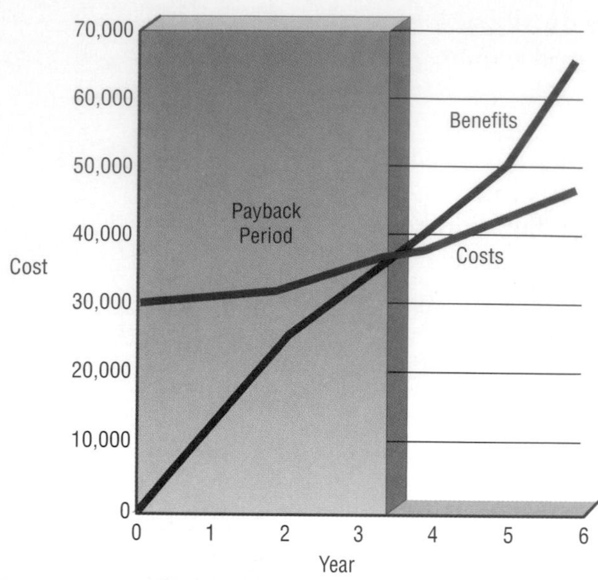

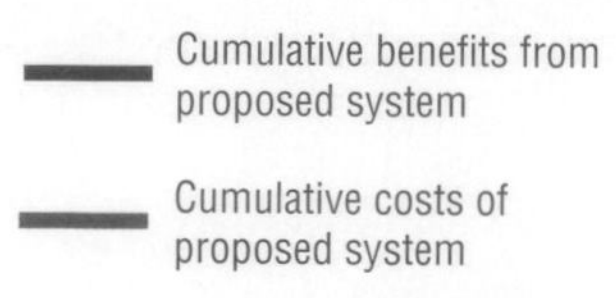

Year	Cost	Cumulative Costs	Benefits	Cumulative Benefits
0	30,000	30,000	0	0
1	1,000	31,000	12,000	12,000
2	2,000	33,000	12,000	24,000
3	2,000	35,000	8,000	32,000
4	3,000	38,000	8,000	40,000
5	4,000	42,000	10,000	50,000
6	4,000	46,000	15,000	65,000

FIGURE 13.11
Payback analysis showing a payback period of three and a half years.

system. As volume rises, the costs of the manual system rise at an increasing rate. A new computer system would cost a substantial sum up front, but the incremental costs for higher volume would be rather small. The graph shows that the computer system would be cost-effective if the business sold about 600 units per week.

Break-even analysis is useful when a business is growing and volume is a key variable in costs. One disadvantage of break-even analysis is that benefits are assumed to remain the same, regardless of which system is in place. From our study of tangible and intangible benefits, we know that is clearly not the case.

PAYBACK

Payback is a simple way to assess whether a business should invest in a proposed information system, one that is based on how long it will take for the benefits of the system to pay back the costs of developing it. Briefly, the payback method determines the number of years of operation that the information system needs to pay back the cost of investing in it. Figure 13.11 illustrates a system with a payback period of three and a half years.

Payback can be determined in one of two ways: either by increasing revenues or by increasing savings. A combination of the two methods can also be used. Because payback is a popular way to assess alternative investments, businesses will typically have a set time period for payback assessments (three years, for example). You can find this information by asking the accounting personnel with whom you are working on the systems project.

If the proposed system has a projected payback of six years in a company that adheres to a three-year maximum payback on projects involving fast-changing technology, the system will be rejected. Payback that is made within the range used by the business but is still longer than typical (for example, four years instead of three) may not be rejected outright but may be subject to scrutiny through other methods.

Although the payback method offers a well-known and simple way to assess the worthiness of the information system, it has three drawbacks that limit its usefulness. One drawback is that it is strictly a short-term approach to investment and replacement decisions, the second is that it does not consider the importance of how repayments are timed, and the third is that the payback method does not consider the total returns from the proposed systems project that may go well beyond the payback year. Other forms of analysis should be used to augment the payback method and overcome some of these flaws.

CASH-FLOW ANALYSIS

Cash-flow analysis examines the direction, size, and pattern of cash flow that is associated with the proposed information system. If you are proposing the replacement of an old information system with a new one and if the new information system will not be generating any additional cash for the business, only cash outlays are associated with the project. If that is the case, the new system cannot be justified on the basis of new revenues generated and must be examined closely for other tangible benefits if it is to be pursued further.

Figure 13.12 shows a small company that is providing a mailing service to other small companies in the city. Revenue projections are that only $5,000 will be generated in the first quarter, but after the second quarter, revenue will grow at a steady rate. Costs will be large in the first two quarters and then level off. Cash-flow analysis is used to determine when a company will begin to make a profit (in this case, it is in the third quarter, with a cash flow of $7,590) and when it will be "out of the red," that is, when revenue has made up for the initial investment (in the first quarter of the second year, when accumulated cash flow changes from a negative amount to a positive $10,720).

The proposed system should have increased revenues along with cash outlays. Then the size of the cash flow must be analyzed along with the patterns of cash flow associated with the purchase of the new system. You must ask when cash outlays and revenues will occur, not only for the initial purchase but also over the life of the information system.

PRESENT VALUE ANALYSIS

Present value analysis helps the systems analyst to present to business decision makers the time value of the investment in the information system as well as the funds flow (as discussed in the previous section). Present value is a way to assess all the economic outlays and revenues of the information system over its economic

	Year 1 Quarter 1	Quarter 2	Quarter 3	Quarter 4	Year 2 Quarter 1
Revenue	5,000	20,000	24,960	31,270	39,020
Costs					
Software					
Development	10,000	5,000			
Personnel	8,000	8,400	8,800	9,260	9,700
Training	3,000	6,000			
Equipment					
Lease	4,000	4,000	4,000	4,000	4,000
Supplies	1,000	2,000	2,370	2,990	3,730
Maintenance	0	2,000	2,200	2,420	2,660
Total Costs	26,000	27,400	17,370	18,670	20,090
Cash Flow	–21,000	–7,400	7,590	12,600	18,930
Cumulative Cash Flow	–21,000	–28,400	–20,810	–8,210	10,720

FIGURE 13.12
Cash-flow analysis for the computerized mail-addressing system.

FIGURE 13.13

Without considering present value, the benefits appear to outweigh the costs.

	Year						
	1	2	3	4	5	6	Total
Costs	40,000	42,000	44,100	46,300	48,600	51,000	272,000
Benefits	25,000	31,200	39,000	48,700	60,800	76,000	280,700

life and to compare costs today with future costs and today's benefits with future benefits.

In Figure 13.13, system costs total $272,000 over six years and benefits total $280,700. Therefore, we might conclude that benefits outweigh the costs. Benefits only started to surpass costs after the fourth year, however, and dollars in the sixth year will not be equivalent to dollars in the first year.

For instance, a dollar investment at 7 percent today will be worth $1.07 at the end of the year and will double in approximately 10 years. The present value, therefore, is the cost or benefit measured in today's dollars and depends on the cost of money. The cost of money is the opportunity cost, or the rate that could be obtained if the money invested in the proposed system were invested in another (relatively safe) project.

The present value of $1.00 at a discount rate of i is calculated by determining the factor

$$\frac{1}{(1 + i)^n}$$

where n is the number of periods. Then the factor is multiplied by the dollar amount, yielding present value as shown in Figure 13.14. In this example, the cost of money—the discount rate—is assumed to be .12 (12 percent) for the entire planning horizon. Multipliers are calculated for each period: $n = 1, n = 2, \ldots, n = 6$. Present values of both costs and benefits are then calculated using these multipliers. When that step is done, the total benefits (measured in today's dollars) are $179,484, less than the costs (also measured in today's dollars). The conclusion to be drawn is that the proposed system is not worthwhile if present value is considered.

Although this example, which used present value factors, is useful in explaining the concept, all electronic spreadsheets have a built-in present value function. The analyst can directly compute present value using this feature.

GUIDELINES FOR ANALYSIS

The use of the methods discussed in the preceding subsections depends on the methods employed and accepted within the organization itself. For general guidelines, however, it is safe to say the following:

1. Use break-even analysis if the project needs to be justified in terms of cost, not benefits, or if benefits do not substantially improve with the proposed system.
2. Use payback when the improved tangible benefits form a convincing argument for the proposed system.
3. Use cash-flow analysis when the project is expensive relative to the size of the company or when the business would be significantly affected by a large drain (even if temporary) on funds.
4. Use present value analysis when the payback period is long or when the cost of borrowing money is high.

	Year						
	1	2	3	4	5	6	Total
Costs	40,000	42,000	44,100	46,300	48,600	51,000	
Multiplier	.89	.80	.71	.64	.57	.51	
Present Value of Costs	35,600	33,600	31,311	29,632	27,702	26,010	183,855
Benefits	25,000	31,200	39,000	48,700	60,800	76,000	
Multiplier	.89	.80	.71	.64	.57	.51	
Present Value of Benefits	22,250	24,960	27,960	31,168	34,656	38,760	179,484

FIGURE 13.14

Taking into account present value, the conclusion is that the costs are greater than the benefits. The discount rate, i, is assumed to be .12 in calculating the multipliers in this table.

Whichever method is chosen, it is important to remember that cost-benefit analysis be approached systematically, in a way that can be explained and justified to management, who will eventually decide whether to commit resources to the systems project. Next, we turn to the importance of comparing many systems alternatives.

EXAMINING ALTERNATIVE SYSTEMS

Through the use of break-even analysis, payback, cash-flow analysis, and present value analysis, it is possible to compare alternatives for the information system. As shown previously, it is important to use multiple analyses to cover the shortcomings of each approach adequately. Although you will consider several alternatives, the proposal itself will recommend only one. Thus, you will have done comparative analyses about which system makes better economic sense before the proposal is written. Those analyses can be included to provide support for the system you are recommending.

Do not think there is only one "correct" system solution to help a business solve its problems and reach its goal. Different businesses call for different system attributes, and system analysts themselves differ about the best way to handle various business problems.

Based on your ascertainment of information requirements, the tangible costs and benefits of the system, and so on, compare the alternatives with which you are working. This process might call for the use of one of the multiple-criteria methods described in Chapter 12. Be open-minded as you compare alternatives so that the system you are recommending can be said to fit within the business as its members experience it.

The key point is that you want to compare and contrast opinions in as fair a manner as possible so that a true choice is offered to organizational decision makers. The closer their initial identification with and acceptance of the proposed system, the greater the likelihood of its continued use and acceptance once the system is in place. Continue including decision makers in the planning, even though you must in some ways expect to play the role of the systems expert now.

SUMMARY

Evaluating hardware and software, identifying and forecasting costs and benefits, and performing cost-benefit analysis are all necessary activities the systems analyst must accomplish in preparing material for the systems proposal. Information requirements help shape what software is purchased or written as well as what hardware is needed to perform required data transformation functions.

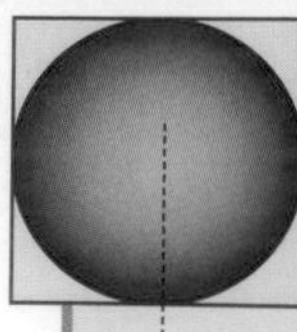

CONSULTING OPPORTUNITY 13.4

THE BIRTH OF A SYSTEM III: MORE GRAPHIC THAN EVER

"The list you made of tangible and intangible costs and benefits really helped," raved Roger. "You had to do some tight editing to make anything out of the melodrama I spun for you the other day. Before we produce a system, however, I'd like you to *analyze* the tangible costs and benefits you wrote for me. Even films have critics," Roger reminds you, as he slips out of view and heads back to his repair bench.

Using one of the methods discussed in this chapter, perform the analysis Roger called for and make a convincing case for the systems solution you preview. Use graphs to get the proper angle on the project and put the debut of your analysis in its best light.

Systems analysts must estimate workloads to characterize current and projected workload capacity necessary for hardware adequately. Sample workloads can then be run on the hardware under consideration.

Although computer equipment changes rapidly, the process used in evaluating hardware need not change too often. By inventorying equipment already on hand and on order, systems analysts will be able to better determine if new, modified, or current computer hardware is to be recommended.

Computer hardware can be acquired through purchase, lease, or rental. Vendors will supply support services such as preventive maintenance and user training that are typically negotiated separately.

Packaged software must also be evaluated by the systems analyst and pertinent users. Much programming time can be saved if such a package is usable without extensive customizing. Software needs to be evaluated on how well it performs desired functions, its ease of use, the adequacy of documentation, and the support services vendors may offer.

Preparing a proposal means identifying all the costs and benefits of a number of alternatives. The systems analyst has a number of methods available to forecast future costs, benefits, volumes of transactions, and economic variables that affect costs and benefits. Costs and benefits can be tangible (quantifiable) or intangible (nonquantifiable and resistant to direct comparison).

A systems analyst has many methods for analyzing costs and benefits. Break-even analysis examines the cost of the existing system versus the cost of the proposed system. The payback method determines the length of time it will take before the new system is profitable. Cash-flow analysis is appropriate when it is critical to know the amount of cash outlays, whereas present value analysis takes into consideration the cost of borrowing money. These tools help the analyst examine the alternatives at hand and make a well-researched recommendation in the systems proposal.

KEYWORDS AND PHRASES

- benchmarking
- break-even analysis
- cash-flow analysis
- estimated workload
- forecasting
- graphical judgment
- intangible benefits
- intangible costs
- method of least squares
- moving averages
- payback
- present value
- tangible benefits
- tangible costs
- vendor support

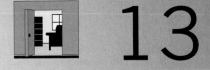

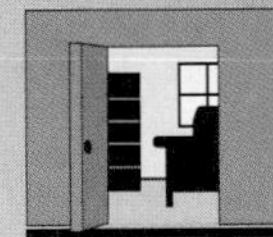

"Sometimes the people who have been here for some time are surprised at how much we have actually grown. Yes, I do admit that it isn't easy to keep track of what each person is up to or even what purchases each department has made in the way of hardware and software. We're working on it, though. Snowden would like to see more accountability for computer purchases. He wants to make sure we know what we have, where it is, why we have it, who's using it, and if it's boosting MRE productivity or, as he so delicately puts it, 'to see whether it's just an expensive toy' that we can live without."

HYPERCASE® QUESTIONS

1. Complete a computer equipment inventory for the Training and Management Systems Unit, describing all of the systems you find. Hint: create an inventory form to simplify your task.
2. Using the software evaluation guidelines given in the text, do a brief evaluation of GEMS, a software package used by the Management Systems employees. In a paragraph, briefly critique this custom-made software by comparing it with off-the-shelf software such as Lotus's "Organizer," or Microsoft's "Project." (Both are mentioned in Chapter 3.)
3. List the intangible costs and benefits of GEMS as reported by employees of MRE.
4. Briefly describe the two alternatives Snowden is considering for the proposed project tracking and reporting system.
5. What organizational and political factors should Snowden consider in proposing his new system at MRE? (In a brief paragraph, discuss three central conflicts.)

REVIEW QUESTIONS

1. List the elements that should be included on a computer hardware inventory form.
2. What is meant by the words *estimated workload*?
3. List four criteria for evaluating system hardware.
4. What are the three main options for the acquisition of computer hardware?
5. Under what conditions is rental of computer hardware appropriate?
6. List four extra support services that are negotiable with vendors of computer hardware.
7. List the six main categories on which to grade software.
8. Why is forecasting a useful tool for the systems analyst?
9. Define unconditional forecasting.
10. What is a disadvantage of graphical judgment?
11. What is the objective in estimating a trend using the least squares method?
12. Why is the method of moving averages a useful one?
13. Define tangible costs and benefits. Give an example of each one.
14. Define intangible costs and benefits. Give an example of each one.
15. List four techniques for comparing the costs and benefits of a proposed system.
16. When is break-even analysis useful?
17. What are the three drawbacks of using the payback method?
18. When is cash-flow analysis used?
19. Define present value analysis.
20. As a general guideline, when should present value analysis be used?

PROBLEMS

1. Delicato, Inc., a manufacturer of precise measuring instruments for scientific purposes, has presented you with a list of attributes that its managers think are probably important in selecting a vendor for computer hardware and software. The criteria are not listed in order of importance.
   1. Low price.
   2. Precisely written software for engineering applications.
   3. Vendor performs routine maintenance on hardware.
   4. Training for Delicato employees.
      a. Critique the list of attributes in a paragraph.
      b. Using its initial input, help Delicato, Inc., draw up a more suitable list of criteria for selecting computer hardware and software vendors.
2. SoftWear Silhouettes is a rapidly growing mail-order house specializing in all-cotton clothing. Management would like to expand sales to the Web with the creation of an ecommerce site. The company has two full-time system analysts and one programmer. Company offices are located in a small, isolated New England town, and the employees who handle the traditional mail-order business have little computer training.
   a. Considering the company's situation, draw up a list of software attributes that SoftWear Silhouettes should emphasize in its choice of software to create a Web site and integrate the mail-order business with business from the Web site.
   b. List the variables that contributed to your response in part a above.
3. Below is 10 years' demand for YarDarts, an outdoor game for the whole family that is part of the 65-game product line of Open Air, Ltd., a manufacturer specializing in outdoor games that can be played in a small area.

Year	Demand
1991	20,900
1992	31,200
1993	28,000
1994	41,200
1995	49,700
1996	46,400
1997	51,200
1998	52,300
1999	49,200
2000	57,600

   a. Graph the demand data for YarDarts.
   b. Forecast the demand for YarDarts for the next five years using the graphical judgment approach.
4. a. Determine the linear trend for YarDarts demand using the least squares method.
   b. Estimate the demand for YarDarts for the next five years using the trend you determined.
5. a. Determine the linear trend for YarDarts using a three-year moving average.
   b. Use least squares on the averages in problem 3a to determine a linear trend.
   c. Estimate demand for YarDarts for the next five years by extending the linear trend found in problem 3b.

6. Do the data for YarDarts appear to have a cyclical variation? Explain.
7. Interglobal Paper Company has asked for your help in comparing its present computer system with a new one its board of directors would like to see implemented. Proposed system and present system costs are as follows:

Year	Proposed System Costs	Present System Costs
Year 1		
Equipment Lease	$20,000	$11,500
Salaries	30,000	50,000
Overhead	4,000	3,000
Development	30,000	—
Year 2		
Equipment Lease	$20,000	$10,500
Salaries	33,000	55,000
Overhead	4,400	3,300
Development	12,000	—
Year 3		
Equipment Lease	$20,000	$10,500
Salaries	36,000	60,000
Overhead	4,900	3,600
Development	—	—
Year 4		
Equipment Lease	$20,000	$10,500
Salaries	39,000	66,000
Overhead	5,500	4,000
Development	—	—

a. Using break-even analysis, determine the year in which Interglobal Paper will break even.
b. Graph the costs and show the break-even point.

8. Below are system benefits for Interglobal Paper Company (from problem 7):

Year	Benefits
1	$55,000
2	75,000
3	80,000
4	85,000

a. Use the costs of Interglobal's proposed system from problem 7 to determine the payback period (use the payback method).
b. Graph the benefits versus the costs and indicate the payback period.

	July	August	September	October	November
REVENUE	35,000	36,000	42,000	48,000	57,000
COSTS					
Office Remodeling	25,000	8,000			
Salaries	11,000	12,100	13,300	14,600	16,000
Training	6,000	6,000			
Equipment Lease	8,000	8,480	9,000	9,540	10,110
Supplies	3,000	3,150	3,300	3,460	3,630

9. Glenn's Electronics, a small company, has set up a computer service. The table on the previous page shows the revenue expected for the first five months of operation, in addition to the costs for office remodeling and so on. Determine the cash flow and accumulated cash flow for the company. When is Glenn's expected to show a profit?
10. Alamo Foods of San Antonio wants to introduce a new computer system for its perishable products warehouse. The costs and benefits are as follows:

Years	Costs	Benefits
1	33,000	21,000
2	34,600	26,200
3	36,300	32,700
4	38,100	40,800
5	40,000	51,000
6	42,000	63,700

a. Given a discount rate of 8 percent (.08), perform present value analysis on the data for Alamo Foods. (*Hint*: Use the formula

$$\frac{1}{(1 + i)^n}$$

to find the multipliers for years 1 to 6.)

b. What is your recommendation for Alamo Foods?

11. a. Suppose the discount rate in problem 10 changes to 13 percent (.13). Perform present value analysis using the new discount rate.

b. What is your recommendation to Alamo Foods now?

c. Explain the difference between problem 10b and problem 11b.

12. Solve problem 7 using an electronic spreadsheet program such as Excel.
13. Use a spreadsheet program to solve problem 9.
14. Solve problem 10 using a function for net present value, such as @NPV (x, range) in Excel.

SELECTED BIBLIOGRAPHY

Alter, S. *Information Systems: A Management Perspective*. Menlo Park, CA: Benjamin-Cummings, 1996.

Lazzaro, V. "Outlining for Conducting and Implementing a Systems Study." In V. Lazarro (ed.), *Systems and Procedures: A Handbook for Business and Industry*, 2d ed. Englewood Cliffs, NJ: Prentice Hall, 1968.

Levine, D. M., P. R. Ramsey, and M. L. Berenson. *Business Statistics for Quality and Productivity*. Upper Saddle River, NJ: Prentice-Hall, 1995.

Lucas, H. *Information Systems Concepts for Management*, 3d ed. New York: McGraw-Hill, 1986.

Meredith, J. R., and T. E. Gibbs. *The Management of Operations*, 2d ed. New York: John Wiley, 1984.

Voich, D., Jr., H. J. Mottice, and W. A. Shrode. *Information Systems for Operations and Management*. Cincinnati: South-Western, 1975.

ALLEN SCHMIDT, JULIE E. KENDALL, AND KENNETH E. KENDALL

PROPOSING TO GO FORTH

"Because we chose to design and implement the new microcomputer system using PCs linked with a local area network, we should work on preparing the systems proposal," Anna begins. She and Chip are meeting to plan the next phase of the design.

"Yes," replies Chip. "We need to make some hardware and software decisions as well as ensure that the users are aware of the benefits the new system will provide."

"We should determine which software will be required to implement the system and the hardware requirements for each user of the system," notes Anna. "Why don't you work on the hardware portion, and I'll investigate software?"

"Sure," Chip replies. "I plan to meet with each of the users again. When I have all of the information, I'll produce a summary report."

Chip proceeds to work with each user to determine what equipment would be required. Some of his findings are as follows:

Mike Crowe has a Pentium 860 desktop computer. This computer is more than adequate for serving the needs of the new system. Additional, necessary equipment is a laptop computer for creating transactions when performing physical inventory and preventive maintenance work.

Dot Matricks has a Pentium 800 desktop computer with mainframe terminal emulation. This computer is adequate for the new system.

Hy Perteks has a Pentium 700 computer on his desk. This computer is adequate for the new system.

Paige Prynter has a Pentium 400 personal computer. Recommend replacing it with a Pentium 800 computer. Add software to run terminal emulation on the new computer.

Cher Ware has an older 400 Pentium computer. Recommend that it be upgraded with a Pentium 800 computer.

Other equipment and supplies: a server microcomputer to manage the network should be a Pentium 900 or better, one fitted with communication boards. A higher-speed laser printer attached directly to the server and smaller inkjet or laser printers attached to each microcomputer should be provided. In addition, cable must be purchased to connect each user to the network.

Meanwhile, Anna is determining the software that would be needed to implement the system. Because each of the users would be receiving software developed by programmers, the major task was to decide what software would be needed for system development and to network the microcomputers. After researching software options, Anna made the following recommendations:

1. Development software to create the system. Three options are available:
   a. Use C++ to write the application software. The advantage of C++ is that it is currently being used by a few members of the programming staff and is object-oriented.
   b. Use a database package and write object-oriented code. Compile the database programs into executable code. Currently, Access is available in the student labs. Other database packages should be evaluated.
   c. Build a client-server solution. Visual Basic, Delphi, and PowerBuilder software are very powerful, and they work with many different databases.

2. Network software is required to establish and make the local area network user friendly with graphical interface screens.

Chip and Anna sit at a work table and examine each other's findings.

"I suppose the next task is to obtain some cost figures for the hardware and software selection," Anna says. "What do you think is our best source of cost information?"

"There are several sources of information," Chip replies. "We could search the Web or examine trade journals for prices. There are many mail-order houses that would have posted prices, often with an Internet discount. We should also call or visit dealers and obtain quotes, especially with educational discounts. The manufacturers may have special programs available. We'll check with the university purchasing officer, too. Once we have all the cost information, we can produce a document as part of the systems proposal."

EXERCISES

E-1. Use microcomputer periodicals in your library to investigate costs for each of the machines and peripheral devices to be purchased. Make a comparison list for each machine.

E-2. Visit a local computer retail store and obtain cost information for each microcomputer listed in this episode. Include printers and high-quality (minimum of 17-inch) monitors. Make a comparison list for each machine.

E-3. Search the Web for Internet stores or computer retailers and obtain cost information for each microcomputer listed in this episode. Include printers and high-quality monitors. Make a comparison list for each machine.

E-4. Scan trade journals and summarize your findings, comparing three different database packages, their features, and costs.

E-5. Investigate the features and prices for microcomputer C++ packages. Make a summary list of your findings.

E-6. Investigate the features and prices for microcomputer database packages. Make a summary list of your findings.

E-7. Investigate the features and prices for PowerBuilder, Visual Basic, and Delphi. Make a summary list of your findings.

E-8. Use the World Wide Web to find out information about the features of three of the software packages mentioned above. Make a summary list of your findings.

E-9. Using the information gathered in the problems above, calculate the total cost for three unique solutions.

WRITING AND PRESENTING THE SYSTEMS PROPOSAL

14

METHODS AVAILABLE

The written proposal serves as a summary of the systems analyst's work in the business up to that point, and as such it is essential that great care is given to writing and presenting it. Through the use of three methods, the analyst can create a successful systems proposal. These methods are effectively organizing the content, writing in a professional style, and orally presenting the proposal in an informative way.

THE SYSTEMS PROPOSAL

ORGANIZING THE SYSTEMS PROPOSAL

Once you have gathered the material to be included in your systems proposal, you need to piece it together in a logical and visually effective way. You need to include 10 main functional sections, use an effective writing style, use figures to supplement your writing, and attend to the visual details of the written proposal.

What to Include in the Systems Proposal. Ten main sections comprise the written systems proposal. Each part has a particular function, and the eventual proposal should be arranged in the following order:

1. Cover letter.
2. Title page of project.
3. Table of contents.
4. Executive summary (including recommendations).
5. Outline of systems study with appropriate documentation.
6. Detailed results of the systems study.
7. Systems alternatives (three or four possible solutions).
8. Systems analysts' recommendations.
9. Proposal summary.
10. Appendices (assorted documentation, summary of phases, correspondence, and so on).

Cover letter. A cover letter to management and the IT task force should accompany the systems proposal. It should list the people who did the study and summarize the objectives of the study. The cover letter can also include information about the prearranged time and place for the oral presentation of the systems proposal. Keep the cover letter concise (one page maximum) and friendly.

Title page. Include on the title page the name of the project, the names of the systems analysis team members, and the date the proposal is submitted. Keep the title page uncluttered in appearance. The proposal title must accurately express the content of the proposal, but it can also exhibit some imagination. The main point is that the title page is important in getting your reader to open the proposal, and it should not be done as an afterthought.

Table of contents. The table of contents can be enormously useful to readers of long proposals. If the proposal is less than 10 pages long, omit the table of contents, because it is superfluous in such a short document.

Executive summary. The executive summary, in 250 to 375 words, provides the who, what, when, where, why, and how of the proposal, just as would the first paragraph in a news story. It goes to the heart of the systems project so that anyone reading it will have an accurate concept of what is going on. It should also include the recommendations of the systems analysts and desired management action, because some people will only have time to read the summary. It should be written last, after the remainder of the proposal is complete.

Outline of systems study. The outline of the systems study provides information about all the methods used in the study and who or what was studied. Any questionnaires, interviews, sampling of archival data, observation, or prototyping used in the systems study should be discussed in this section.

Detailed results of systems study. This section details what the systems analyst has found out about the system through all the methods described in the preceding section. Conclusions about systems problems that have come to the fore through the study—including kinds and rates of errors, current and projected work volume, and how work is being handled by the current system—should be noted here. The material here should raise the problems or suggest opportunities that call forth the alternatives presented in the next section.

Systems alternatives. In the systems alternatives portion of the proposal, the analyst presents two or three alternative solutions that directly address the aforementioned problems. The alternatives you present should include one that recommends keeping the system the same. Each alternative should be explored separately. Describe the costs and benefits of each situation. Because there are usually trade-offs involved in any solution, be sure to include the advantages and disadvantages of each.

Each alternative must clearly indicate what management must do to implement it. The wording should be as clear as possible, such as, "Buy notebook computers for all middle managers," "Purchase packaged software to manage inventory," and "Modify the existing system through funding in-house programming efforts."

Systems analysts' recommendations. After the systems analysis team has weighed the alternatives, it will have a definite professional opinion about which solution is most workable. The systems analysts' recommendations section expresses the *recommended* solution. Include the reasons supporting the team's recommendation so that it is easy to understand why it is being made. The recommendation should flow logically from the preceding analysis of alternative solutions.

Proposal summary. The proposal summary is a brief statement that mirrors the content of the executive summary. It gives the objectives of the study and the recommended solution. It also allows the analyst one more chance to stress the project's importance and feasibility along with the value of the recommendations. Conclude the proposal on a positive note.

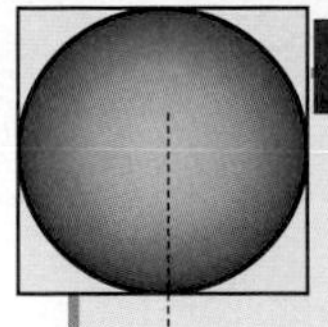

CONSULTING OPPORTUNITY 14.1

MAKING IT REEL

"As you know from working with me the past few weeks, I'm a technical guy," admits Roger as you are about to leave his store after follow-up interviews one afternoon. "I can understand most anything about computers you want to throw at me. Your proposal can be as technical as you like."

Thinking that writing up the proposal might actually be enjoyable for a change, you look happily at Roger, who is relaxing with his feet up on his desk. As you open the door of the store to exit, Roger calls after you casually, "Of course, the proposal isn't really for me. Remember, I want it simple enough so I can explain it to the employees and to my banker."

As a systems analyst, you have been working with Roger Corman on developing a PC based system for tracking the rental of movies and other features on videotape. Based on the assignments you completed in Chapter 13 and what Roger has just told you about the expectations for the proposal, write a one-page executive summary to serve as the first page of the report. It should cogently summarize your recommendations in layman's terms. Use an appealing layout and follow the guidelines for executive summaries.

Appendices. The appendix is the last part of the systems proposal, and it can include any information that the systems analyst feels may be of interest to specific individuals but that is not essential for understanding the systems study and what is being proposed. Appendices might contain pertinent correspondence, a summary of phases completed in the study, detailed graphs for analysis purposes, or even previously done systems studies.

It is difficult to provide useful heuristics about the appropriate length of the systems proposal. Keep in mind that the proposal's size is directly related to the size of the modification or system being proposed. Web Strategy Pro 4.0, for instance, suggests only about 50 pages for an entire ecommerce business plan.

Once the systems proposal is written, carefully select who should receive the report. Personally hand the report to the people you have selected. Your visibility is important for the acceptance and eventual success of the system.

CHOOSING A WRITING STYLE

Although a business style of writing is most often appropriate for writing a systems proposal, your choice of a writing style will ultimately be determined by what you have already witnessed in the organization's own publications. If the people who comprise your target audience favor a certain style, use it to write your proposal.

Present information in a way that is easily comprehensible to them without being condescending. There should be enough detail for management to make informed decisions without being overwhelmed. Keep references to a minimum and do not use footnotes.

USING FIGURES FOR EFFECTIVE COMMUNICATION

The emphasis so far in this chapter has been on considering your audience when composing the systems proposal. Tables and graphs as well as words are important in capturing and communicating the basics of the proposed system.

Integrating figures into your proposal helps demonstrate that you are responsive to the different ways people absorb information. Figures within the report supplement written information and must always be interpreted in words; they should never stand alone.

Effective Use of Tables. Although tables are technically not visual aids, they provide a different way of grouping and presenting analyzed data that the analyst wants to communicate to the proposal reader. Tables are more similar to figures than they are to written text and are therefore discussed here.

Tables use labeled columns and rows to present statistical or alphabetical data in an organized way. Each table must be numbered according to the order in which it appears in the proposal and should be meaningfully titled. Figure 14.1 shows appropriate layout and labeling for a table.

Some guidelines for tables are the following:

1. Print only one table per page, and rather than relegate it to the end of the proposal, integrate it into the body of the proposal.
2. Try to fit the entire table vertically on a single page if possible.
3. Number and title the table at the top of the page. Make the title descriptive and meaningful.
4. Label each row and column. Use more than one line for a title if necessary.
5. Use a boxed table if room permits. Vertically ruled columns will enhance the readability.
6. Use an asterisk if necessary to explain detailed information contained in the table.

Several methods for comparing costs and benefits were presented in Chapter 13. Tabled results of those comparisons should appear in the systems proposal. If a break-even analysis is done, a table illustrating results of the analysis should be

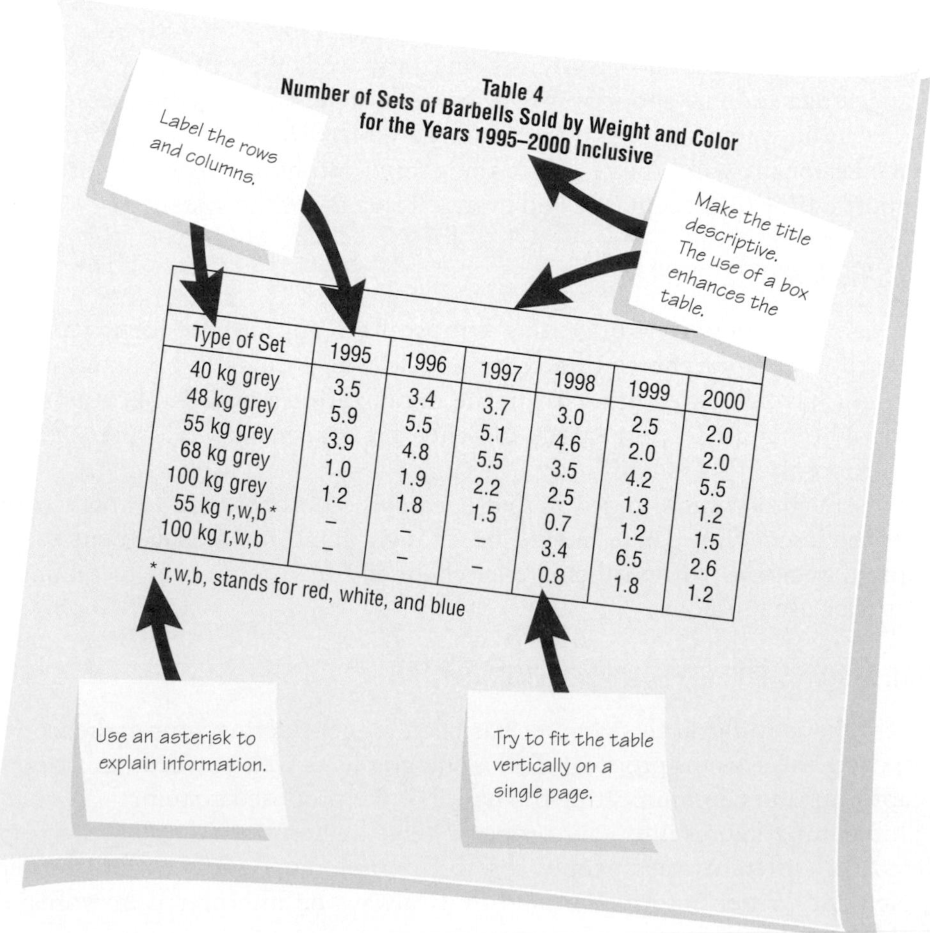

Table 4
Number of Sets of Barbells Sold by Weight and Color for the Years 1995–2000 Inclusive

Type of Set	1995	1996	1997	1998	1999	2000
40 kg grey	3.5	3.4	3.7	3.0	2.5	2.0
48 kg grey	5.9	5.5	5.1	4.6	2.0	2.0
55 kg grey	3.9	4.8	5.5	3.5	4.2	5.5
68 kg grey	1.0	1.9	2.2	2.5	1.3	1.2
100 kg grey	1.2	1.8	1.5	0.7	1.2	1.5
55 kg r,w,b*	–	–	–	3.4	6.5	2.6
100 kg r,w,b	–	–	–	0.8	1.8	1.2

* r,w,b, stands for red, white, and blue

FIGURE 14.1
Guidelines for creating effective tables.

included. Payback can be shown in tables that serve as additional support for graphs. A short table comparing computer systems or options might also be included in the systems proposal.

Effective Use of Graphs. This section covers different kinds of graphs: line graphs, column charts, bar charts, and pie charts. Line and column graphs and bar charts compare variables, whereas pie charts illustrate the composition of 100 percent of an entity.

The guidelines for including effective graphs in a proposal are as follows:

1. Draw only one graph to a page unless you want to make a critical comparison between graphs.
2. Integrate the graph into the body of the proposal.
3. Give the graph a sequential figure number and a meaningful title.
4. Label each axis and any lines, columns, bars, or pieces of the pie on the graph.
5. Include a key to indicate differently colored lines, shaded bars, or crosshatched areas.

An example of how a graph would appear on a page in a systems proposal is shown in Figure 14.2. Our explanation of graphs begins with the simplest type, called a line graph.

Line graphs. Line graphs are used primarily to show change over time. No other type of graph shows a trend more clearly than a line graph. Changes in a single variable or up to five variables can be illustrated in a single line graph.

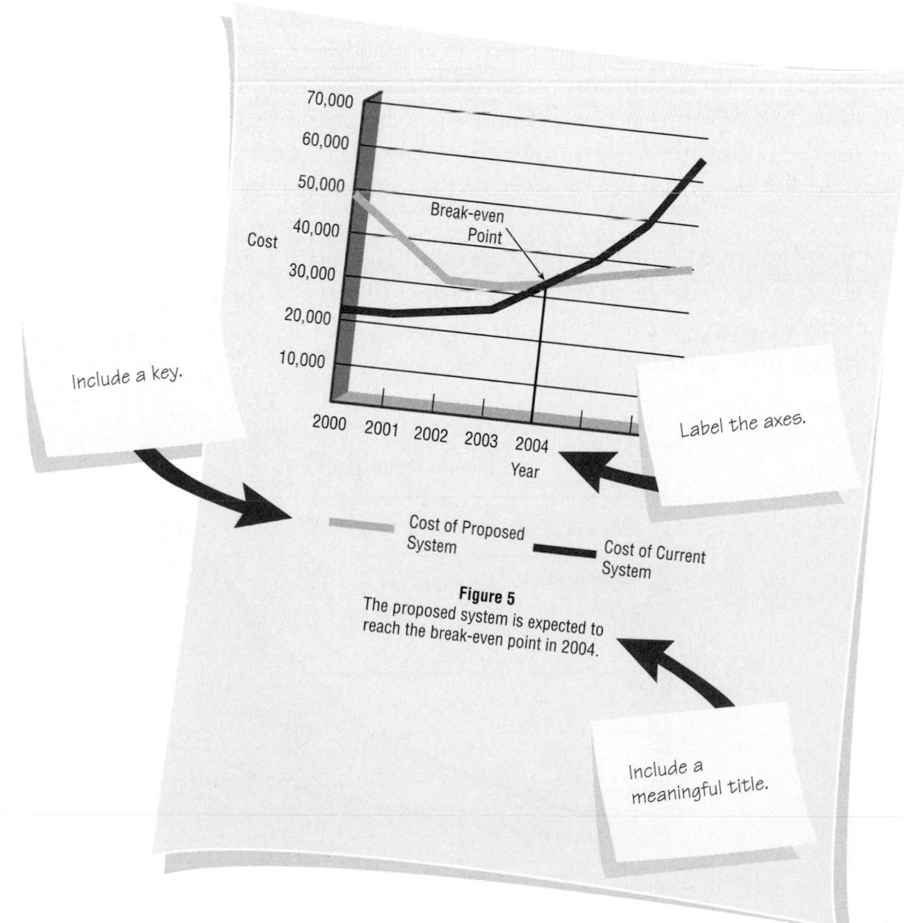

FIGURE 14.2
Guidelines for drawing effective line graphs.

FIGURE 14.3
Depicting each variable with a different kind of line on the line graph.

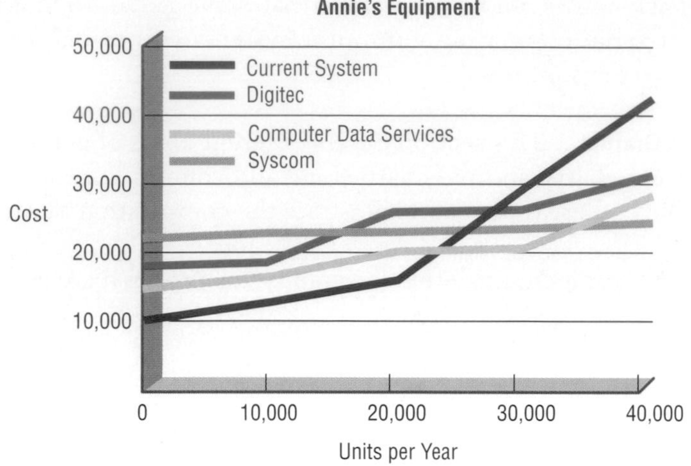

At times, however, a line graph is used to show something other than time on the horizontal axis. This situation occurs when one has to estimate when two or more lines intersect, as shown in Figure 14.3. In this example, the current system is the least expensive until Annie's Equipment grows to approximately 24,000 units per year. Then Computer Data Services offers the least expensive option. Later we find that Syscom becomes the least expensive option if Annie's were to grow to over 28,000 units annually.

A dramatic method of visual comparison, in the same general family as line graphs, is the area chart. Figure 14.4 shows the growth of the videocassette industry over the time period 1995–2000. In this area chart, the total gross receipts consist of both sales (the shaded area) and rentals (the colored area). The area chart is very useful when the difference between two variables expands greatly.

In Chapter 13, the importance of forecasting for justifying the systems project was stressed. Line graphs are excellent ways of showing proposal readers how demand on the computer system may change within a certain number of years or how demand for the products or services of a business may change within a specific time period.

Line graphs are also useful for representing results of payback analysis or break-even analysis to decision makers. Graphic display of the payback period is an excellent way to portray the economic feasibility of the proposed system, as is a graph of break-even results.

FIGURE 14.4
An area chart is a form of line graph that may make more of an impact.

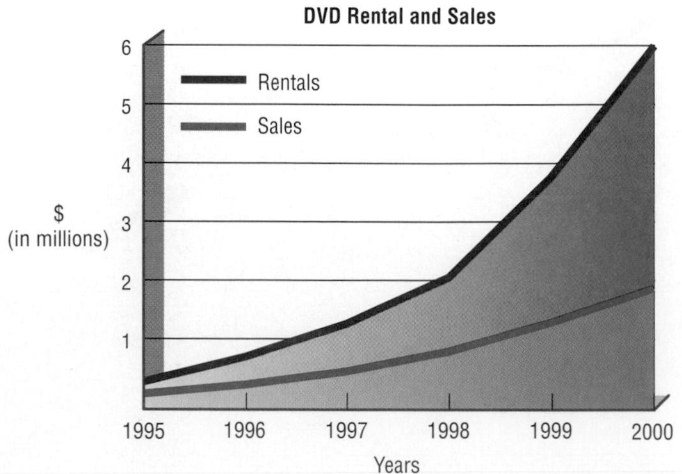

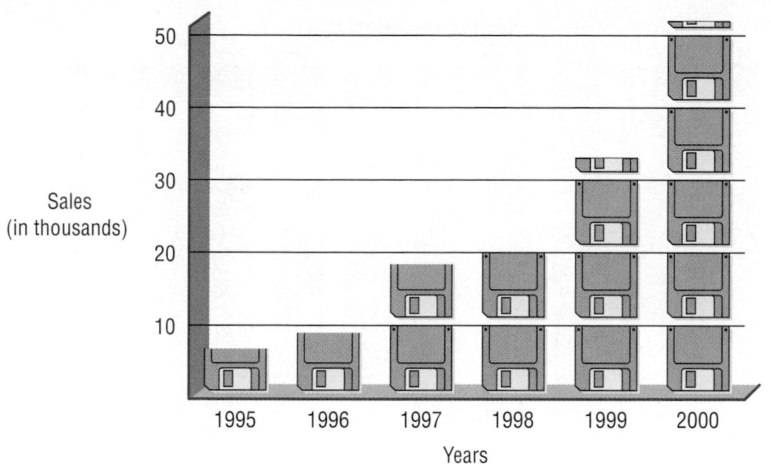

FIGURE 14.5

A familiar icon such as a diskette can be used to create columns on a column chart.

Column charts. Another familiar kind of graph is the column chart. Column charts can depict a comparison between two or more variables over time, but they are used more often to compare different variables at a particular point in time. Although they do not show trends as well as line graphs—nor can one easily estimate value between columns using them—many people find column charts easier to understand than line graphs.

A simple column chart is shown in Figure 14.5. In this example, the columns consist of diskettes. One way to attract the reader, particularly someone who disdains statistics and graphs, is to make the graph more human. A familiar icon, such as a diskette, can be useful in this regard.

Figure 14.6 shows a column chart with more than one variable. In this situation, the columns are drawn in different colors or shades to distinguish between the variables. Notice that there is space between each of the two classes (HQ and Troops A, B, C, D, and E) but no space between the two variables, "current strength" and "minimum required."

There are also special forms of column charts. A 100 percent stacked column chart is shown in Figure 14.7. This type of chart is used to show the relationship between two variables that make up 100 percent of an entity. Here, sporting goods sales are made up of competitive sporting equipment and individual achievement equipment. The chart depicts the competitive sporting equipment as shrinking as a percentage of total sales. (It does not, however, show the actual sales, which may indeed be growing even though the percentage is diminishing.)

FIGURE 14.6

More than one variable can be displayed on a column chart by shading or coloring the column bars.

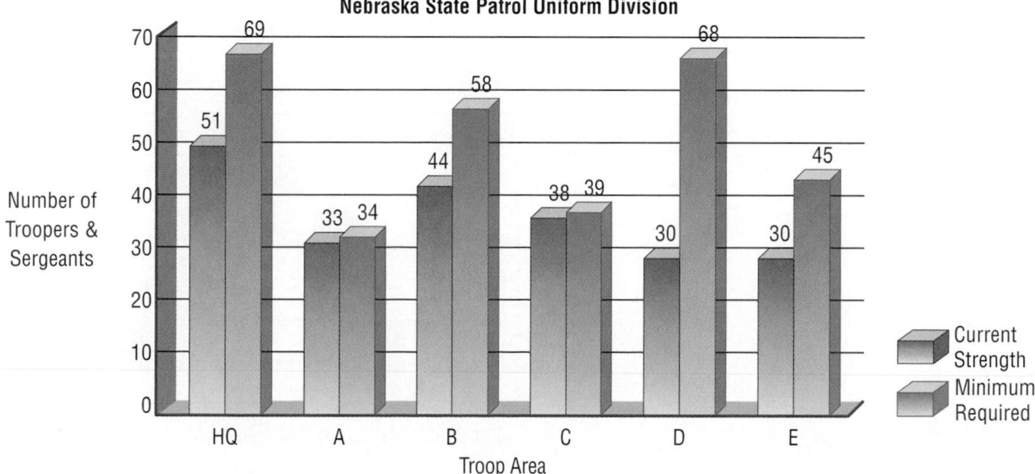

FIGURE 14.7

A 100-percent column chart can be used to show the percentage share over time.

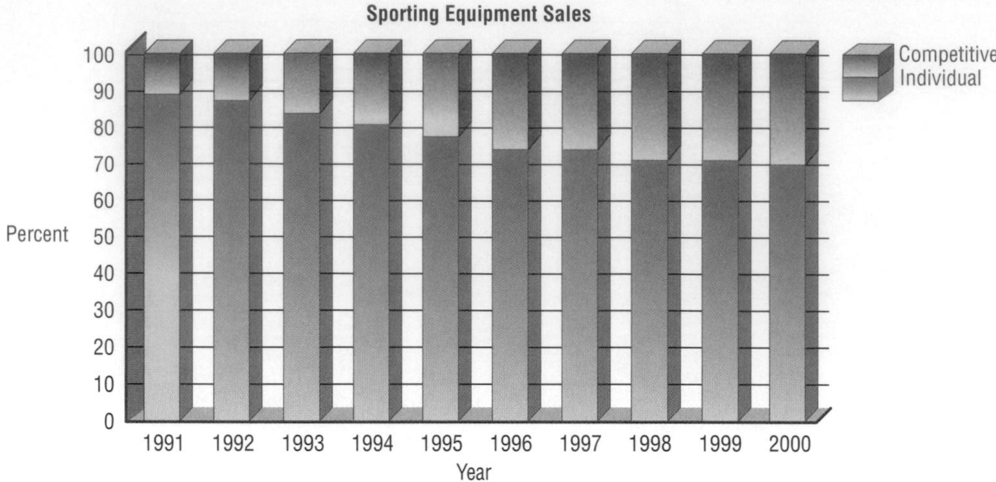

Another special type of column chart is the deviation column chart. This type of chart is useful for emphasizing years that show a loss or pointing out the year in which the company intends to break even. Furthermore, the chart can be drawn to show the deviation from an average. An example of a deviation column chart is shown in Figure 14.8, where the differences in above- and below-average months are emphasized.

Bar charts. Bar charts are similar to column charts, but they are never used to show a relationship over a period of years. Rather, they are used to show one or more variables within certain classes or categories during a specific time period.

The bars themselves may be organized in many different ways. They can be in alphabetical, numerical, geographical, or progressive order, or they can be sorted by magnitude. For instance, in a systems proposal, a bar chart would be useful in comparing the volumes of shipping invoices, customer accounts, and vendor invoices processed by the computer system during July, as shown in Figure 14.9. A bar chart is one of the most widely known forms of graphs and can make a comparison in a straightforward way.

Pie charts. Another commonly used type of graph is the circle or pie chart. It is used to show how 100 percent of a commodity is divided at a particular point in time, as in Figure 14.10.

FIGURE 14.8

A deviation column chart can be more effective in showing which months have above-average transactions.

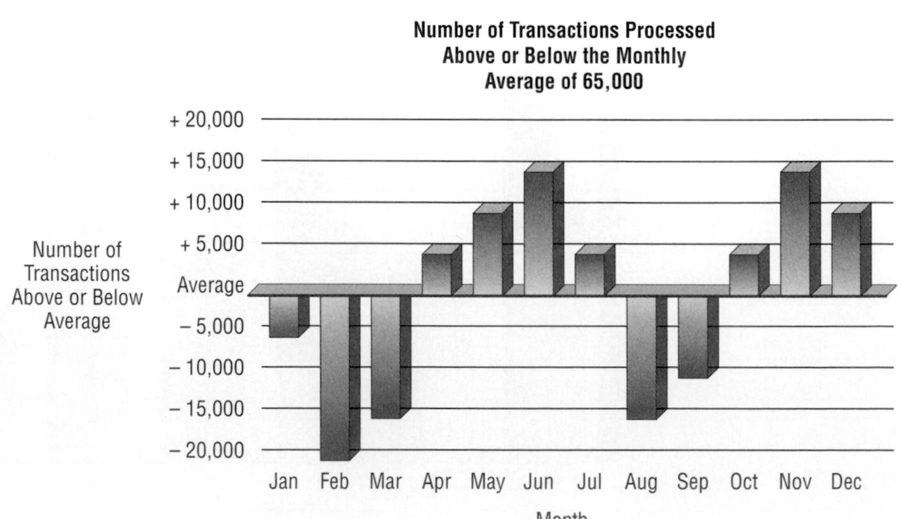

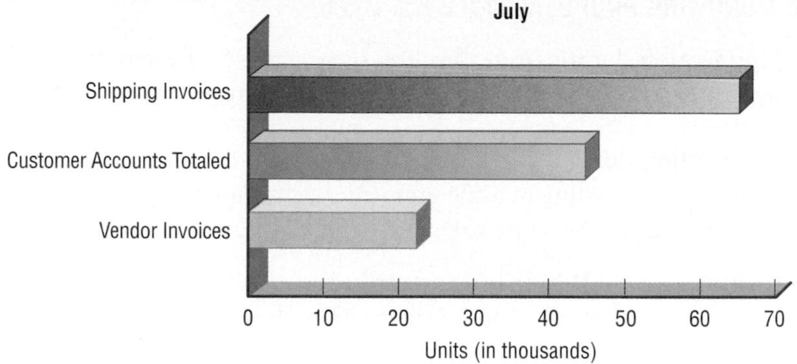

FIGURE 14.9
A bar chart is like a column chart, but it is mainly used to compare different items at a specific point in time.

Pie charts are easier to read than 100 percent stacked column charts or 100 percent subdivided bar charts. Their main disadvantage is that they take up a lot of room on a page.

Guidelines for Using Figures in the Systems Proposal. Figures (tables and graphs) can communicate in a way that is not possible in words alone. When preparing the systems proposal, remember to take advantage of the graphs and tables that you are already using for planning purposes. The following guidelines help enhance the systems proposal through the use of figures:

1. Whenever possible, integrate the figure into the body of the proposal itself. The figure may be placed on the page following its first reference. If your first thought is to relegate the figure to an appendix, it may not be important enough to include at all.
2. Always introduce figures in the text before they appear.
3. Always interpret figures in words; never leave them to stand on their own.
4. Title all figures, label each axis, and provide legends where necessary.
5. Use more than one figure if necessary so that the point you make is clear and figures are uncluttered.

This section has been a cursory glance at using tables and graphs to help you make a point within the systems proposal. For a more detailed, in-depth discussion of the use of figures, see Chapter 15.

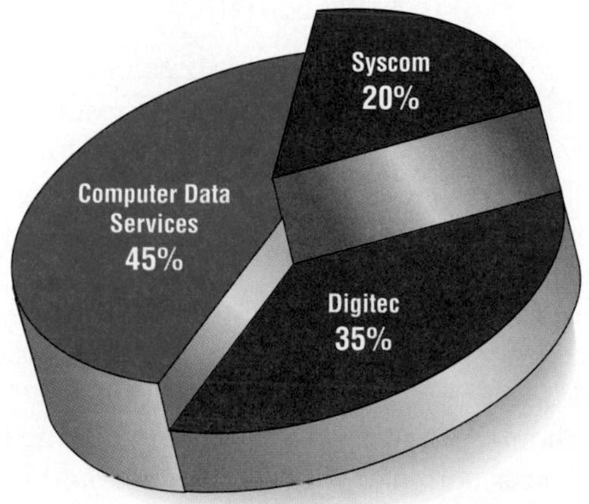

FIGURE 14.10
A visually appealing way to display how 100 percent of an entity is divided up at a particular time is a pie chart.

ADOPTING A UNIFYING PROPOSAL STYLE

The following section demonstrates how to use a variety of formatting elements to improve the appearance and coherence of your written proposal.

Visual and Formatting Considerations. Proposals are persuasive documents. All the best arguments for proceeding in a specific way are brought to the fore when writing the systems proposal. By the same token, the proposal must be visually persuasive.

Use of white space. White space distributed throughout the text of the systems proposal helps to set off ideas, ensuring that they will be noticed. Leave margins of one inch at the top, bottom, left, and right of each page. Leave one-and-a-half inch margins if you intend to bind the left side.

Use of headings and subheadings. The use of headings and subheadings is critical, especially if the proposal is long. Headings set apart each section and point the reader to a main section. If imaginatively written, headings help readers to follow the logic of the writing and to maintain their interest. Subheadings function in much the same way but refer to more specific points. When taken altogether, headings and subheadings should provide a useful and instructive outline of the entire proposal.

References and appendices. Keep references to external support materials to a minimum. If necessary, you may include references in a consistent format at the end of your report.

Proposal readers usually have diverse organizational interests. If the proposal seeks to address technical as well as other users, the addition of appendices that include technical specifications may be wise. A separate, more detailed proposal addressing technical concerns is also an option.

PRESENTING THE SYSTEMS PROPOSAL

As a systems analyst, you should understand your audience and how to organize, support, and deliver the oral presentation.

UNDERSTANDING THE AUDIENCE

Just as the audience for the written proposal helps dictate the writing style, level of detail, and type of figures, the audience for the oral presentation helps the speaker discover how formal to be, what to present, and what types of visual aids to include. It is imperative that you know *who* you will be addressing.

ORGANIZING THE SYSTEMS PROPOSAL PRESENTATION

Page through the data collected from the organization that are summarized in the written proposal. Find four to six main points that capsulize the proposal. In particular, check the executive summary, the recommendation sections, and the proposal summary. If the time allotted for the oral presentation is longer than half an hour, main points can be expanded to nine or more.

Once main points and supporting points are worked out, an introduction and conclusion can be written. Notice that writing the introduction comes last, not first, because the introduction should preview the proposal's four to six main points, which are impossible to determine at the outset.

The introduction should also include a "hook," something that will get the audience intrigued with what is coming next. The hook should be a creative approach to the proposal that directly unites the audience's interests with the new material

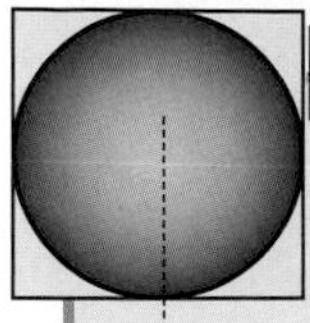

CONSULTING OPPORTUNITY 14.2

SHOULD THIS CHART BE BARRED?

"Gee, I'm glad they hired you guys. I know the Redwings will be better next season because of you. My job'll be a lot easier, too," says Andy Skors, ticket manager for the Kitchener, Ontario, hockey team, the Kitchener Redwings. Andy has been working with your systems analysis team on analyzing the systems requirements for computerizing ticket sales.

Recall that when we last heard from the systems analysis team, consisting of Hy Sticking (your leader), Rip Shinpadd, Fiona Wrink, and you, you were wrestling with whether to expedite the project and to set team productivity goals (in Consulting Opportunity 3.3).

Andy is talking with the team about what to include in the systems proposal to make it as persuasive as possible to the Redwings' management. "I know they're going to like this chart," Andy continues. "It's a little something I drew up after you asked me all those questions on past ticket sales, Rip."

Andy hands the bar chart to Rip, who looks at it and suppresses a slight smile. "As long as we have you here Andy, why don't you explain it for us?"

Like a player fresh out of the penalty box, Andy skates smoothly into his narrative of the graph. "Well, our ticket sales reached an all-time high in 1993. We were real crowd pleasers that year. Could've sold seats on the scoreboard if they let me. Unfortunately, ticket sales were at an all-time low in 1994. I mean, we're talking about a disaster. Tickets moved slower than a glacier. I had to convince the players to give tickets away when they made appearances at the shopping mall. Why, just look at this table, it's terrible.

"I think computerizing the ticket sales will help us pick out who our season supporters are. We've got to figure out who they are and get them back. Get them to stick with us. That would be a good goal to shoot for," Andy concludes.

As Andy's presentation finally winds down, Hy looks as if he thought the 20-minute period would never end. Picking up on his signal, Fiona says, "Thanks for the data, Andy. We'll work on getting them into the report somehow."

As Fiona and Rip head out of the room with Andy, Hy realizes the bench has emptied, so he asks you, the fourth team member, to coach Andy on his bar chart by making a list of the problems you see in it. Hy would also like to sketch some alternative ways to graph the data on ticket sales so that a correct and persuasive graph of ticket sales can be included in the systems proposal.

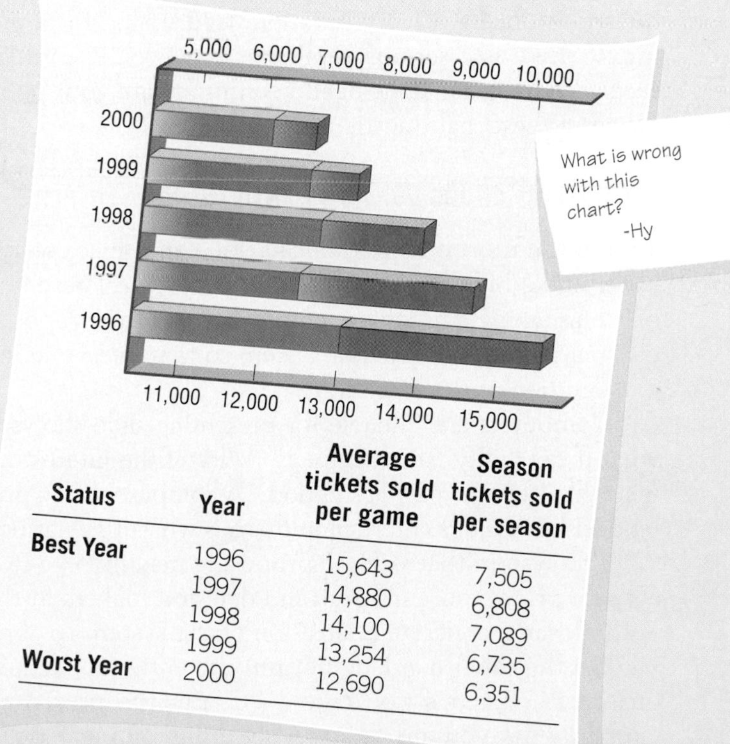

Status	Year	Average tickets sold per game	Season tickets sold per season
Best Year	1996	15,643	7,505
	1997	14,880	6,808
	1998	14,100	7,089
	1999	13,254	6,735
Worst Year	2000	12,690	6,351

FIGURE 14.C1
An incorrectly drawn graph.

Source: © The New Yorker Collection from cartoonbank.com.

"Great graphics, Dave, but the answer is still no."

being presented. For example, an anecdote, an analogy, a quotation, poetry, or even a joke can open a presentation successfully. If humor is used, it should be directly relevant to the topic and should make a point about what is coming up.

Conclusions should mirror introductions. The analyst shouldn't quote verbatim from the introduction, but the main ideas should be reiterated and a closing thought (similar to the creative hook of the introduction) should be given.

Questions can be taken either during or after the presentation. Answering questions during the presentation itself makes for a more informal, relaxed meeting. If there is a serious challenge, however, it could derail the proposal prematurely. To maintain control and communicate your points effectively, it is permissible to request that questions be saved until the end.

TAKING A NOTEBOOK COMPUTER INTO THE PRESENTATION

Two of the most effective visuals you can use are software using a notebook computer and a high-quality computer projector. There are some logistics to be aware of when taking a notebook computer into the proposal meeting. If possible, try to use a large-screen projector system so that everyone in the room is able to see the presentation at the same time.

Computers are valuable for presenting slide shows as well as spreadsheets with what-if capability. An example is, "What if the interest rate drops dramatically? How will that affect the payback period?" A computer and a properly set up spreadsheet can let decision makers entertain numerous what-if scenarios during the meeting itself.

The system that you are proposing may involve PCs. In this instance, a PC can act as a prototype, and users and decision makers in the meeting will be provided with an early, concrete realization of the system to come.

On the other hand, do not put the entire burden of your systems presentation on the PC. Even if you take a PC into the presentation, you will still need to explain what you are doing. Also, things can and do go wrong. You must plan a backup method for presenting material just in case you encounter software or hardware problems that are not solvable on the spot.

Now that audiences, supporting materials, and visuals have been discussed, we turn to oral delivery of the systems proposal. At this stage, the systems presentation departs radically from the written systems proposal.

USING PRESENTATION IN GRAPHICS PACKAGES

One of the most exciting ways to present your systems proposal is to make a slide show using a presentation software package such as Microsoft PowerPoint, which includes easy-to-use, professional-looking templates. Notice that a variety of interesting and engaging clipboard art is available to users. Remember to choose images that help to enliven what you are trying to communicate. Using clip art purchased on CD-ROM greatly expands your choices. The combinations you can create are virtually limitless, because clip art library disks purchased with software can include over 500,000 images, symbols, and fonts.

Using the software presentation package to create your talk permits you to rearrange the order of your presentation, as shown in Figure 14.11. Rearranging becomes important if you want to tailor your presentation to suit different audiences. In addition, you may want to add or delete slides to shorten or expand the time it takes to present your talk, depending on the time constraints for your meeting. A page of thumbnail views of the slides can be printed, copied, and distributed to audience members after your talk to serve as notes of your presentation.

You have the option of taking advantage of multimedia capabilities by including sound in your slide show. Again, use your creativity to employ sound in a memorable way. For example, launching a new project may call for the sound of a rocket blastoff countdown or the sound of the blastoff itself. Music and, in some cases, full-motion video can be used to highlight a presentation as well. Remember to include charts and graphs in your slide show where appropriate. They are easy to prepare, and they make your point nicely, with meaningful graphics, when words just cannot say it all. Many of the presentation software packages allow you to create presentations that run on the Internet, using the World Wide Web.

Some packages, such as Micrografx FlowCharter, allow the user to create living flowcharts and network diagrams that permit people to walk through business

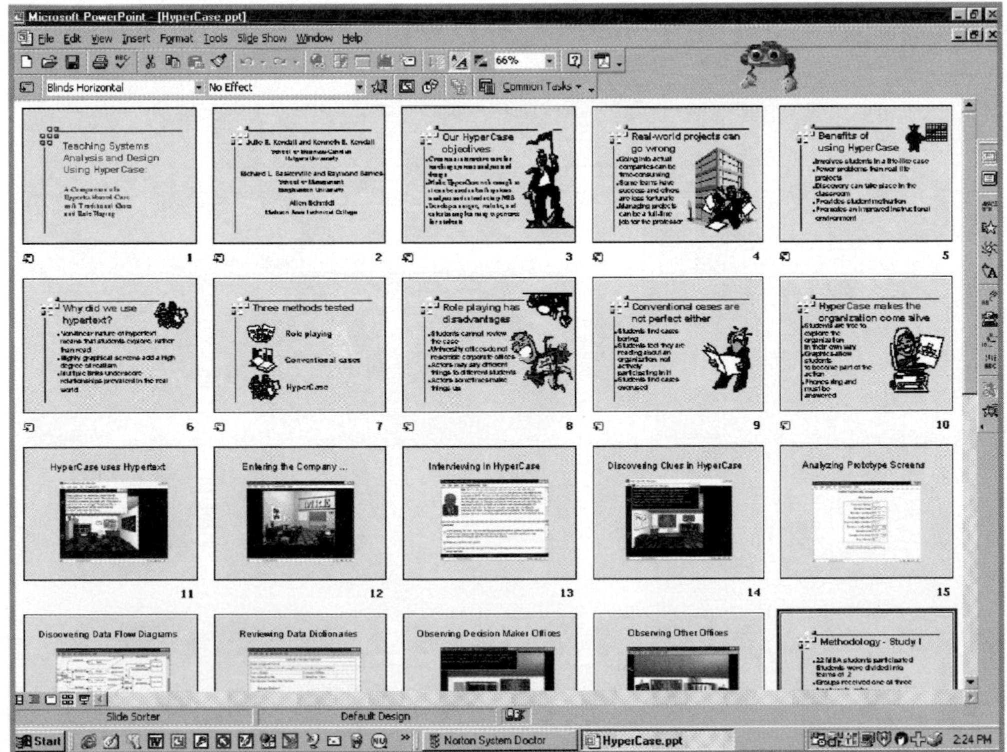

FIGURE 14.11
Rearrange the order of your slide presentation, such as this one displayed in Microsoft Powerpoint, to suit a specific audience or to compress or expand the time it takes to present your talk.

processes dynamically. Figure 14.12 illustrates the CoolSheets feature for developing bar charts, checklists, and timelines, using very colorful and expressive templates. Another software package, Astound, allows the designer to control precisely the timeline and animation sequences within the presentation. It is one of the most powerful presentation packages.

There are a few guidelines to follow when creating a slide show. Though many of the design lessons you will learn in Chapters 15 and 16 will be helpful, some criteria are unique to this situation. When creating a slide show, remember to do the following:

1. Use the templates provided by most packages for a well-designed, consistent effect as well as to save time.
2. Use a combination of graphics and text to communicate. Data can be charted in many different ways for effective display. Fonts and font sizes should be chosen for readability and appropriateness. The recommended font size is 20 to 22 points minimum. San serif fonts (those without small feet at the bottom of the letters) are better for screen displays. Serif fonts are better for printing.
3. Keep a clean look to each slide and guard against clutter. Include no more than five key points per slide.
4. Use color in a meaningful way. Provide a color key for complicated visuals such as charts. Avoid using a white background.
5. Use clip art with text to add humor and to reinforce your points.
6. Use sound to help underscore the points of your presentation. Judicious use of sound will help your audience recall your talk later.
7. Take a multimedia approach to your presentation, integrating slides, video, and sound through hypermedia to communicate in a memorable way.
8. If you are using a laptop computer, plug it in. Most notebook computers will go into "sleep" mode after a period of inactivity, and the screen will be blank.
9. Avoid using too many contrasting transitions (the way one slide moves to the next).

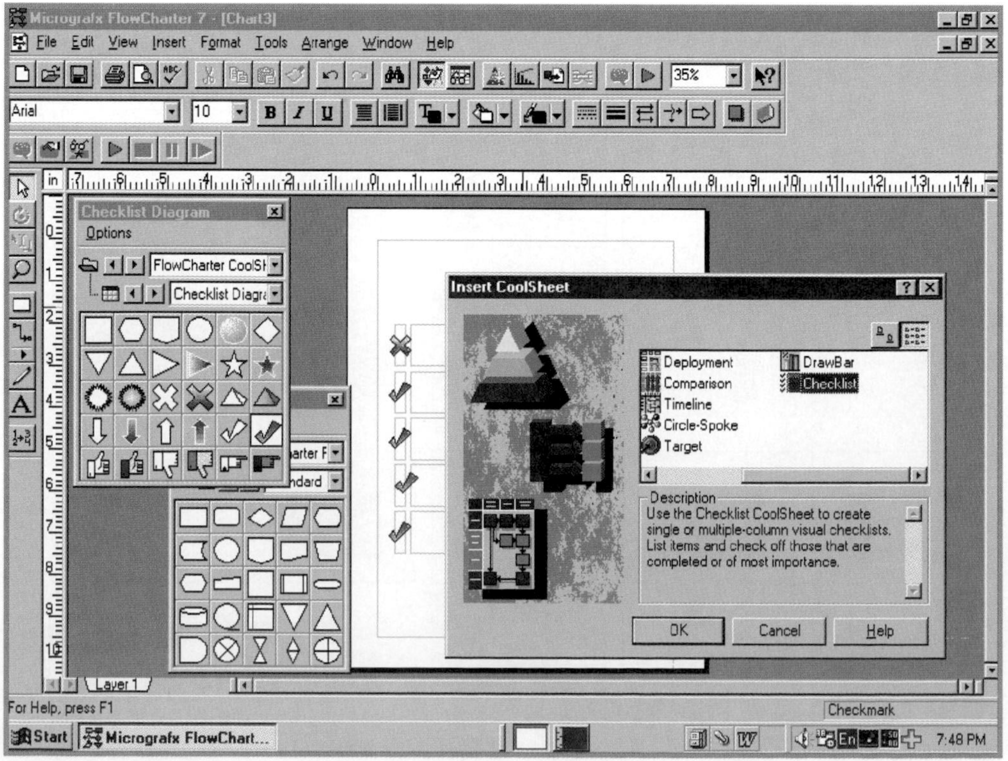

FIGURE 14.12

A graphics package, such as Micrografx FlowCharter, can be used to quickly create commonly used images such as checklists, bar charts, and timelines with very little effort.

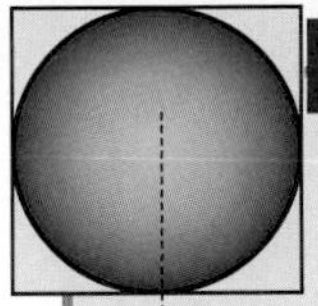

CONSULTING OPPORTUNITY 14.3

FAST FORWARD ON THE SYSTEMS PROPOSAL

Entering Roger Corman's office, you ask him when he would like to schedule an oral presentation of the systems proposal. Corman replies, "Look, your presentation can be pretty informal for me. After talking to you for so many weeks, I feel like we've known each other a long time. If you're ready now, let's grab a cup of coffee and talk over your findings."

An hour later, Roger is enthusiastically discussing the finer points of your proposed system. "You know what would be great? We have all this video stuff here, so why not take advantage of it? It's hard to get time to talk to the employees together. Why don't you make a 15-minute tape of your presentation of the systems proposal and let employees watch it at their convenience? It will help convince them of the benefits of the system," Roger asserts persuasively.

Prepare an outline of your systems proposal talk that lists main points (four to six) along with their supporting points. Also indicate on the outline where visuals such as graphs will be shown. What are some changes to consider when presenting the system to potential users rather than to decision makers? How does a taped presentation differ from a live one?

Using presentation software to create a slide show results in an interesting, consistent, and easy-to-follow presentation with many potential uses. Save the slides you create on disk because they can be presented and used in many different ways. Many speakers like to use a combination approach, perhaps projecting their slide show and then reinforcing it with an audience handout. Some speakers make nonproprietary talks available on the Web following a presentation.

PRINCIPLES OF DELIVERY

Knowing who is in the audience will tell the analyst how formal to make the presentation. If the chief executive officer is included in the meeting, chances are it will be quite formal. If primary users comprise the audience, perhaps a less formal, workshop presentation will be more appropriate.

One of the best ways to gauge the formality of presentations is by observing many different organizational meetings prior to the systems proposal presentation. Expectations are usually based on customs and culture and may dictate that every presenter must use a PowerPoint presentation or provide an outline of his or her remarks.

The rules for delivery are basic:

1. Project loud enough so that the audience can hear you.
2. Look at each person in the audience as you speak.
3. Make visuals large enough so that the audience can see them.
4. Use gestures that are natural to your conversational style.
5. Introduce and conclude your talk confidently.

The very thought of getting up in front of people can make presenters extremely nervous; in fact, the greatest fear of men is said to be public speaking (it's the second greatest fear of women). The following section details four guidelines that can help presenters overcome anxiety.

The First Guideline: Be Yourself. Being yourself means that the persona of the speaker is very important in persuading an audience. Presenters need to develop themselves in all aspects of personality: intellectually, emotionally, spiritually. The ancient Greeks held to the idea that speakers had to earn the right to address the audience through development of a complete self.

The Second Guideline: Be Prepared. Being prepared means that the more thoroughly the analyst knows his or her material, the easier it will be to deliver it. This guideline translates into the necessity of adequate rehearsal time for presenters. Few, if any, speakers are so gifted as to be able to ad lib a talk. Being prepared gives speakers the confidence that they do indeed have something important to tell the audience members and that they will do anything to communicate with them.

The Third Guideline: Speak Naturally. Speaking naturally, not memorizing or reading, may seem antithetical to being prepared, because a speaker reading from a prepared text has little chance of departing from his or her key points. Reading, however, destroys speaker credibility by limiting vital eye contact. Memorizing a talk lessens chances for successfully adapting to a particular audience. In addition, when a speaker loses his or her place in a memorized talk, it is nearly impossible to recover gracefully.

Instead of reading or memorizing, know your five or six key points very well. Know what support materials accompany each main point, but do not write them out. The notes for your oral presentation should be limited to an easily readable listing of main points, perhaps with a symbol to indicate when to show a visual.

The Fourth Guideline: Remember to Breathe. An excellent way to achieve a strong speaking voice is to become aware of the breathing process. Take long, deep breaths immediately before addressing the group. Allow yourself to breathe between sentences and during normal pauses in your speaking. Your voice will remain strong and you will be calm.

SUMMARY

The systems analyst has three main steps to follow for putting together an effective systems proposal: effectively organizing the proposal content, writing the proposal in an appropriate business style, and orally presenting an informative systems proposal. To be effective, the proposal should be written in a clear and understandable manner, and its content should be divided into 10 functional sections.

Visual considerations are important when putting together a proposal that communicates well. Much of what is important in the systems proposal can be enhanced through the correct use of figures, including tables and graphs. Graphs compare two or more variables over time or at a particular point in time. Figures are always accompanied with a written interpretation in the proposal. The graphs and tables used for planning prior to the proposal can be incorporated into it when relevant.

The oral presentation of the system is based on the written proposal and is another way of effectively selling the system. One option for presentation is to create a slide show using presentation software such as PowerPoint. Also, graphics presentation packages and clip art can be used to enhance the visual presentation of the systems proposal. To give a strong oral presentation, the analyst should know who will comprise the audience, the topic (presumably the systems proposal or some part of it), the amount of time allotted for the presentation, and the equipment available (including room setup). All four elements are interrelated, and each needs to be thought through and planned to ensure success.

"I know it's hard to get your feet wet, but you've been here long enough to know that we're all curious about what you've come up with so far. We're especially interested in what you think of us! Are we one big happy family, or is this a zoo? Seriously, Snowden would like it very much if you gave a brief oral presentation of a preliminary proposal for a new automated project reporting system for the Training Group. Who should we include? Well, Mr. Torrey, Dan Hill, Tom Ketcham, and Snowden, of course, will want to be there. Let's see . . . I've got the executive calendar on the screen here. Everyone we need is free a week from Thursday at 3:00. You can bring your whole team along if you want. That room has multimedia capabilities, if you want to get fancy, but keep it to about 15 minutes at the most. Oh, one more thing, I'm sure Mr. Hyatt will want to come. Have fun!"

HYPERCASE® QUESTIONS

1. Prepare an outline of the preliminary proposal for a new automated project reporting system for the Training Group. Include enough detail so that it would be possible to use your outline as speaking notes during a presentation.
2. Use a software package such as Microsoft PowerPoint to create a short (3 to 5 slides) slide show to illustrate the preliminary proposal for the automated project reporting system you outlined in problem 1.
3. Have your teammates role play the parts of Warren Torrey, Dan Hill, Tom Ketcham, and Snowden Evans (the part of Mr. Hyatt is optional). Present your brief preliminary proposal for the new automated project reporting system to them. Use the slide show you have created for problem 2.
4. Write a two-paragraph report based on feedback received on the preliminary proposal during the role playing in problem 3. What questions arose? What changes will you make?

KEYWORDS AND PHRASES

- analysts' recommendations
- bar charts
- column charts
- executive summary
- line graphs
- oral presentation
- pie charts
- principles of delivery
- systems proposal
- visuals

REVIEW QUESTIONS

1. What are the three steps the systems analyst must follow to put together an effective systems proposal?
2. List the 10 main sections of the systems proposal.
3. Which sections of the systems proposal should include the solution the analyst thinks is *most* workable?
4. What relationships does a line graph depict?
5. What relationships does a column chart depict?
6. What relationships does a bar chart depict?
7. What relationships does a pie chart depict?
8. List the five guidelines for using figures effectively in the systems proposal.
9. What purpose do headings and subheadings serve in the written systems proposal?

10. What sort of support material should be included in an oral presentation of the systems proposal to executive audiences?
11. When should the introduction for an oral presentation of the systems proposal be written?
12. How can a notebook computer be used as a visual aid in an oral presentation of the systems proposal?
13. List the seven guidelines for creating a slide show on your computer.
14. List the five rules for effective oral delivery of the systems proposal.

PROBLEMS

1. "I think it's only fair to write up *all* the alternatives you've considered," says Lou Cite, a personnel supervisor for Day-Glow Paints. "After all, you've been working on this systems thing for a while now, and I think my boss and everyone else would be interested to see what you've found out." You are talking with Lou as you prepare to put together the final systems proposal that your team will be presenting to upper management.
   a. In a paragraph, explain to Lou why your proposal will not (and should not) contain all the alternatives that your team has considered.
   b. In a paragraph, discuss the sorts of alternatives that should appear in the final systems proposal.
2. In going over the data you have collected for your proposal for Linder's Machine Parts of Duluth, Minnesota, you find a forecast of demand for parts for the next five years as well as the forecast of the number of companies purchasing parts. You would like to include the data in your systems proposal to help support the need for a new system, and the numbers currently given in this narrative are as follows: "The columns show that demand of 120,000 will increase to 130,000 in year 2, go up 20,000 in year 3, go up 40,000 in year 4, and level off in year 5. Although demand for parts will be going up, the total number of companies who will be buying will be 700 in year 1 and will be reduced by 50 companies each year through the next five years."
   a. Based on the narrative, draw a bar graph to depict demand over the next five years for Linder's Machine Parts.
   b. Based on the narrative, draw a column chart to depict demand over the next five years for Linder's Machine Parts.
   c. Based on the narrative, draw a bar graph to show the decline in the total number of companies ordering machine parts from Linder's.
   d. Based on the narrative, draw a line graph to depict the increase in demand and the decrease in the total number of companies purchasing parts together.
3. "I was thinking of how I'll handle my portion of the presentation to management," says Margaret, a member of your systems analysis team. "Even though some of them told us they 'haven't been keeping up with computers,' I think they need to know the technical aspects of our recommended system inside and out; otherwise they may not accept it. So I'll begin by defining basic terms such as 'byte' and 'program code,' and then I'll turn the meeting into a short tutorial on computing. What do you think?"
   a. In a paragraph, critique Margaret's approach to the systems proposal presentation to the executive audience.
   b. In a paragraph, suggest a different way to approach the executive audience for the systems proposal presentation. Be sure to include types of support—as well as topics—that would be more appropriate than what Margaret has in mind.

SELECTED BIBLIOGRAPHY

Carey, P., and J. Carey. *Microsoft® PowerPoint®97 at a Glance*. Redmond, WA: Microsoft Press, 1997.

Deetz, S. *Transforming Communication, Transforming Business: Building Responsible and Responsive Workplaces*. Cresskill, NJ: Hampton Press, 1995.

Di Salvo, V. S. *Business and Professional Communication*. Columbus, OH: Merrill, 1976.

Himstreet, W. C., and W. M. Baty. *Business Communication Principles and Methods*, 6th ed. Boston: Kent, 1981.

Humes, J. C. *The Sir Winston Method: The Five Secrets of Speaking the Language of Leadership*. New York: William Morrow, 1991.

Lewis, P. V., and W. H. Baker. *Business Report Writing*. Columbus, OH: Grid, 1978.

Stefik, M., G. Foster, D. G. Bobrow, K. Kahn, S. Lanning, and L. Suchman. "Beyond the Chalkboard: Computer Support for Collaboration and Problem Solving in Meetings." *Communications of the ACM*, Vol. 30, No. 1, January 1987, pp. 32–47.

EPISODE *CPU CASE*

14

Allen Schmidt, Julie E. Kendall, and Kenneth E. Kendall

SHOW AND TELL

"That completes the list," Anna says. "I've contacted each user of the system as well as management. The presentation meeting for the proposed new computer system is scheduled for 10 A.M. next Tuesday. We have most of the materials ready, so let's put the finishing touches on the documentation."

Chip looks up from the task of creating an executive summary. "It seems hard to believe that we're almost finished with the analysis stage. I'm a little apprehensive about the meeting. I hope everything goes well."

"I'm sure it will," Anna says reassuringly. "We've been thorough in our analysis, and the proposal seems to be coming along nicely too."

Chip completes the executive summary, shown in Figure E14.1. This document gives an overview of the nature of the system and the recommended solution. After completing the executive summary, he proceeds to create the outline of the study. The outline, shown in Figure E14.2, is another summary, because many of the people reviewing the proposal have been involved in the interviews and prototypes. It provides a concise review of the methods that have been used in analyzing the needs of the microcomputer system.

Meanwhile, Anna was working on polishing the results section of the analysis, called the *problem definition*. Earlier in the analysis, she had taken information from the interviews, survey, and prototypes, and she produced 10 concise points reflecting system needs. These points had been reviewed by the users and modified with minor changes to clarify them. Each user was then requested to rank the issues on relative importance, using a scale from 1 to 10. These final ranks were then averaged to become weights, indicating overall importance. The completed problem definition is shown in Figure E14.3.

Alternatives for the proposed system have already been created and evaluated (refer to Chapter 12). The recommended solution is a PC-based local area network.

"We should include information showing how we are going to evaluate the installed system," Chip says. "How do you think we should create evaluation criteria?"

"Well," Anna replies thoughtfully, "perhaps we should restate the objectives in more concise terms. The original objectives were broadly stated, allowing us to be more flexible and creative in our choice of solutions. The new objectives would be measurable, describing deliverables: the specific screens, reports, and other products of systems development."

Chip and Anna work together to develop the measurable objectives. They start with the first objective and work down the list, point by point. Following are the broadly stated objectives:

1. Provide software/hardware cross-reference.
2. Maintain complete microcomputer information.
3. Automate software installation procedure.
4. Provide information on software upgrade installation by machine.
5. Provide preventive and other maintenance information.
6. Maintain up-to-date, accurate software information.
7. Provide complete cost information for microcomputer hardware.

Computer System Proposal
Executive Summary

The system for managing computer hardware and software is inadequate for the current level of devices and software packages. The designers of the original system, installed in the early 1980s, could not foresee the rapid development of products culminating in the tremendous variety presently available.

Analysts Anna Liszt and Chip Puller have conducted extensive interviews with system users. Persons interviewed include Mike Crowe, Dot Matricks, Hy Perteks, Paige Prynter, and Cher Ware. The problems determined from these interviews include missing information and a lack of cross-referencing hardware and software information. As software and hardware demands escalate in the future, this lack of information will result in an increase in redundant information and a duplication of costly services.

We recommend a computer system with a local area network linking all parties involved in the system. Menu screens will allow users to select options to update hardware and software information. Reports will be periodically produced and a variety of inquiry screens made available.

Additional benefits include improved cost information, automating the preventive maintenance scheduling, ease of updating software versions, and improved techniques for performing physical device inventory.

Programs and data will be secured by requiring users to enter a user identification number and password. The user ID will further control who has access to the various system functions: update, report generation, or inquiry.

Training sessions will be conducted prior to final installation of the computer system. It is recommended that at least two persons should be trained for each system aspect.

The estimated completion date is August 19th. The estimated time for development is 8 person months at an expense of $50,000. Hardware expenditures should not exceed $15,000.

FIGURE E14.1
Executive summary.

8. Provide information on the cost to upgrade software.
9. Design a process for performing accurate and efficient physical microcomputer inventory.
10. Maintain and provide training and software expert information.

"How do you think we can restate item one—'Provide software/hardware cross-reference'—as a measurable objective?" asks Chip.

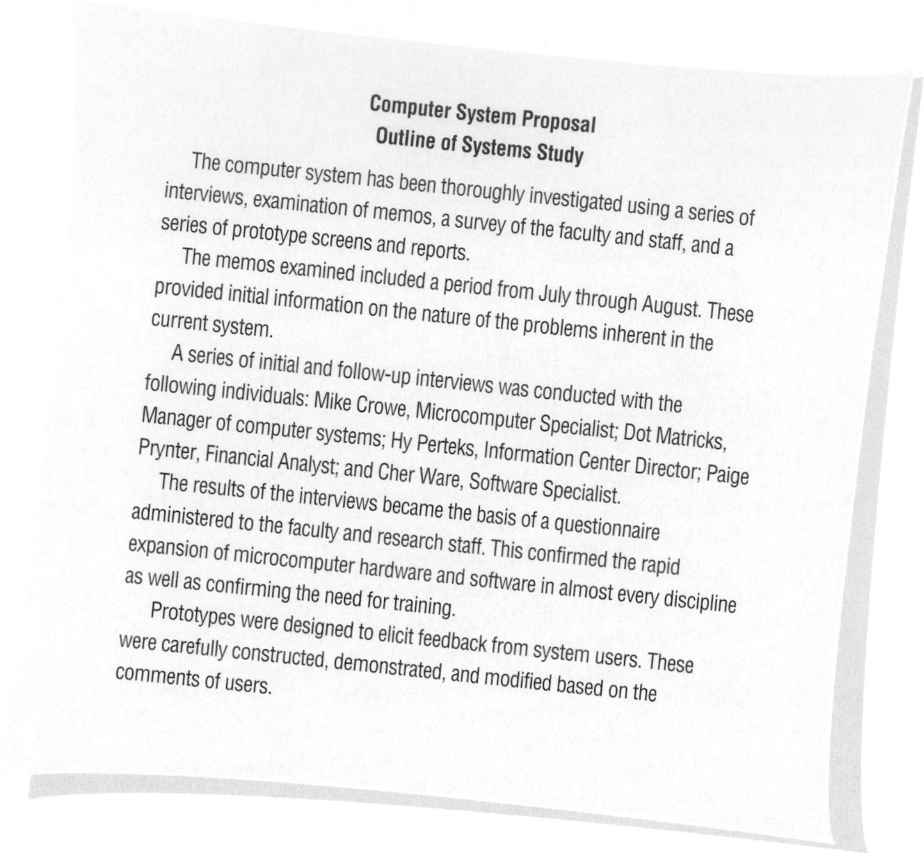

Computer System Proposal
Outline of Systems Study

The computer system has been thoroughly investigated using a series of interviews, examination of memos, a survey of the faculty and staff, and a series of prototype screens and reports.

The memos examined included a period from July through August. These provided initial information on the nature of the problems inherent in the current system.

A series of initial and follow-up interviews was conducted with the following individuals: Mike Crowe, Microcomputer Specialist; Dot Matricks, Manager of computer systems; Hy Perteks, Information Center Director; Paige Prynter, Financial Analyst; and Cher Ware, Software Specialist.

The results of the interviews became the basis of a questionnaire administered to the faculty and research staff. This confirmed the rapid expansion of microcomputer hardware and software in almost every discipline as well as confirming the need for training.

Prototypes were designed to elicit feedback from system users. These were carefully constructed, demonstrated, and modified based on the comments of users.

FIGURE E14.2
Outline of a Systems Study.

"Let's look at the data flow diagrams, data dictionary records, and prototypes to review how we plan to provide the cross-reference information," responds Anna. "How about 'Provide a software inquiry listing the machine and its location for each copy of the software. Produce hardware/software cross-reference report on demand'?"

"Sounds good," says Chip, "The next one—'Maintain complete microcomputer information'—could be stated as 'Add maintenance, preventive maintenance, boards, cost, and peripheral information to the HARDWARE MASTER file. Create separate online programs for adding, deleting, changing, and modifying maintenance information.' "

Chip and Anna work together to restate the third objective—"Automate software installation procedure"—as "Create an online inquiry screen to determine the location of suitable microcomputers. Use the screen to choose locations. Print the location list. Provide automatic update of files after installation."

When the material has been completed and proofed, Chip starts to work on creating a microcomputer presentation of the project.

"This presentation software makes it easy to create a slide show," says Chip with a smile. "I think it will help to control the sequence of the presentation and make it easier for us to relax."

Computer System Proposal
Detailed Results—Problem Definition

The computer system has become inadequate to handle the current volume of microcomputers, their peripheral devices, and software installed upon them. Additional information needs to be added to existing files, and there is a lack of cross-referencing software packages installed upon machines.

The computer Maintenance department has difficulty determining which machines require preventive maintenance, and a physical inventory of microcomputers and the peripheral devices is not regularly performed.

Cost information is not accurately maintained or reported.

Issues—Present Situation

	Issue	Weight
1.	There is a lack of information about which software is installed on any given microcomputer.	10
2.	Incomplete information is maintained for each microcomputer.	9
3.	The present system does not provide microcomputer capacity information used to determine which machine software may be installed upon.	8
4.	There is no method of determining which machines contain software scheduling for an upgrade.	7
5.	Preventive and other maintenance information is not acquired. There is no reliable method for predicting preventive maintenance dates.	7
6.	Software information is missing and out of date, with redundant records for older versions.	6
7.	Cost information is incomplete for microcomputers.	5
8.	Software upgrade cost data is unavailable.	5
9.	The process for performing physical microcomputer inventory is inaccurate and inefficient.	3
10.	There is no method for maintaining training and software expert information.	2

FIGURE E14.3
Problem definition.

"I agree," replies Anna, "and I especially enjoy the pieces of clip art that you have included to liven it up a bit. I'll arrange to have one of the new computer projection units delivered to the conference room. They have an exceptionally bright image."

Tuesday morning finds Chip and Anna in the conference room. Copies of the proposal are neatly placed on the table in front of each seat. The presentation equipment is connected and has been carefully tested to make sure that everything is working properly.

"I feel a few butterflies in my stomach," murmurs Chip. "Even though I've spoken with all these people before, today is different somehow."

"I'm a little nervous myself," answers Anna. "It's normal to feel that way. Don't worry, though. We're well prepared, and we'll take turns presenting different topics and features."

The users arrive and the presentation begins. Anna starts with opening remarks and an introductory summary. Chip takes a deep breath and walks to the front of the room. He gains confidence as he reviews some of the problems and displays the graphs showing the increase in software and hardware over the past ten years. The audience is listening intently, visibly interested in what he is saying.

The presentation lasts about an hour. User feedback is detailed and enthusiastic. Chip and Anna are delighted when the users unanimously approve their recommendation.

EXERCISES*

E-1. Rewrite the last seven broadly stated objectives as measurable objectives.

E-2. Create a cover letter based on the information presented in the CPU case.

E-3. Work in small groups to deliver a presentation of the CPU system to the class based on information mentioned in Chapters 13 and 14 and on previous exercises.

E-4. Research presentation software packages. Use both information from corporate Web sites as well as magazine reviews comparing features. What are their features, advantages, and disadvantages? Summarize the information for three presentation packages.

E-5. Work as a small group to create a slide show using presentation software that explains the features of presentation software. Have other students view the slide show and comment on it.

E-6. Use Access to view the MICROCOMPUTER GROWTH presentation graph (under the Reports tab), illustrated in Figure E14.4. Notice that the number of computers is somewhat leveling off and is not growing as fast as in the past.

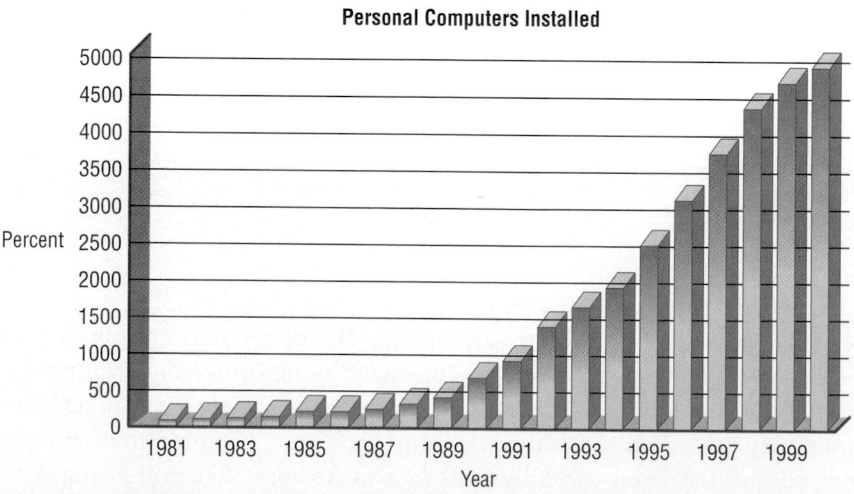

FIGURE E14.4
A graph showing the growth of the number of personal computers installed.

Leveling occurs when an organization provides most of the employees with a computer and then moves to replacing the oldest computers with newer ones instead of adding to the total number of machines.

E-7. Using Access, view the SOFTWARE GROWTH presentation graph, then modify the database table, Software Installed, and print the SOFTWARE GROWTH chart report (under the Reports tab). Add diskette symbols to reflect the following total number of software packages for each of the following years:

Year	Cumulative Software Packages
1997	5,309
1998	5,402
1999	5,480
2000	5,520

E-8. Use Visible Analyst to modify the MICROCOMPUTER CONFIGURATION presentation graph shown in Figure E14.5. Include the notebook computer that Mike Crowe will use to update records from remote locations.

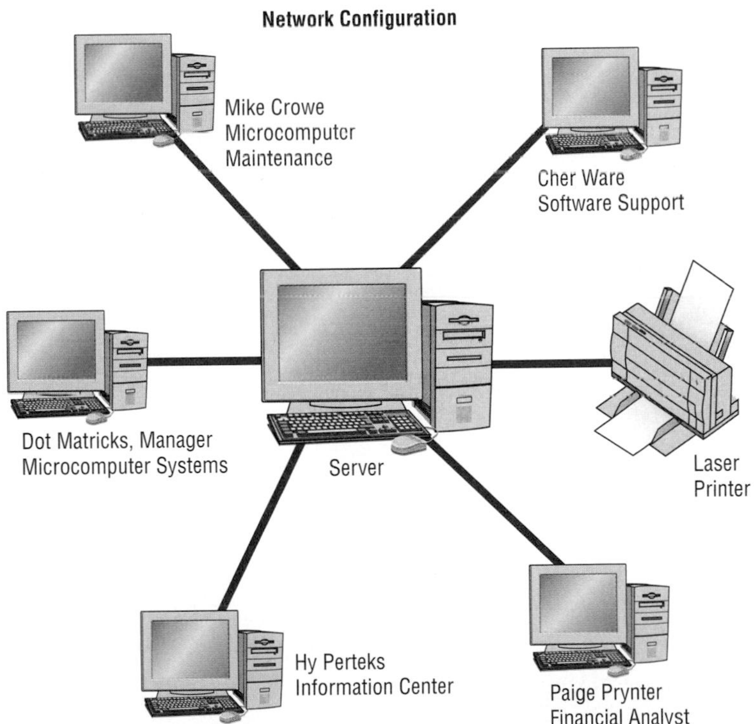

FIGURE E14.5
Computer network configuration.

*Exercises preceded by a CD-ROM icon require the program Visible Analyst or another CASE tool. A CD-ROM containing Visible Analyst examples is provided free of charge to any professor adopting this book. The examples on the disk may be imported into Visible Analyst and then used by students.

DESIGNING EFFECTIVE OUTPUT

15

OUTPUT DESIGN OBJECTIVES

Output is information delivered to users through the information system by way of intranets, extranets, or the World Wide Web. Some data require extensive processing before they become suitable output; other data are stored, and when they are retrieved, they are considered output with little or no processing. Output can take many forms: the traditional hard copy of printed reports and soft copy such as computer screens, microforms, and audio output. Users rely on output to accomplish their tasks, and they often judge the merit of the system solely by its output. To create the most useful output possible, the systems analyst works closely with the user through an interactive process until the result is considered to be satisfactory.

Because useful output is essential to ensuring the use and acceptance of the information system, there are several objectives that the systems analyst tries to attain when designing output. As shown in Figure 15.1, there are six objectives for output:

1. Designing output to serve a specific purpose.
2. Making output meaningful to the user.
3. Delivering the appropriate quantity of output.
4. Providing appropriate output distribution.
5. Providing output on time.
6. Choosing the most effective output method.

DESIGNING OUTPUT TO SERVE THE INTENDED PURPOSE

All output should have a purpose. It is not enough to make a report, screen, or Web page available to users because it is technologically possible to do so. During the information requirements determination phase of analysis, the systems analyst finds out what purposes must be served. Output is then designed based on those purposes.

You will see that you have numerous opportunities to supply output simply because the application permits you to do so. Remember the rule of purposiveness, however. If the output is not functional, it should not be created, because there are costs of time and materials associated with all output from the system.

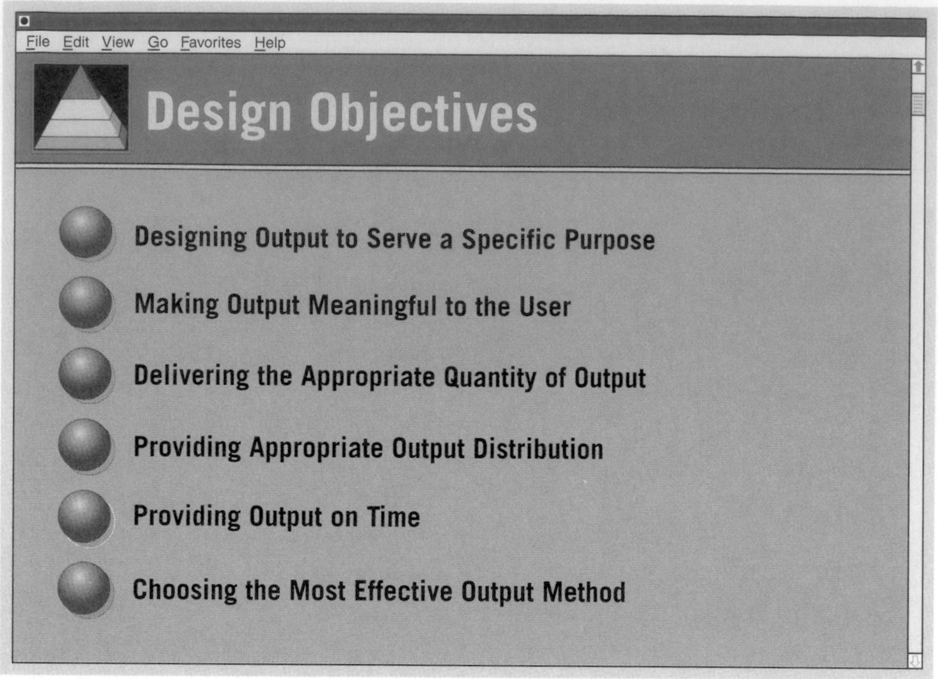

FIGURE 15.1
Six objectives for the design of output.

DESIGNING OUTPUT TO FIT THE USER

With a large information system serving many users for many different purposes, it is often difficult to personalize output. On the basis of interviews, observations, cost considerations, and perhaps prototypes, it will be possible to design output that addresses what many, if not all, users need and prefer.

Generally speaking, it is more practical to create user-specific or user-customizable output when designing for a decision support system or other highly interactive applications such as those mounted on the Web. It is still possible, however, to design output to fit a user's function in the organization, which leads us to the next objective.

DELIVERING THE APPROPRIATE QUANTITY OF OUTPUT

More is not always better, especially where the amount of output is concerned. Part of the task of designing output is deciding what quantity of output is correct for users. You can see that this task is very difficult, because information requirements are in continuing flux.

A useful heuristic is that the system must provide what each person needs to complete his or her work. This answer, however, is still far from a total solution, because it may be appropriate to display a subset of that information at first and then provide a way for the user to access additional information easily. For example, rather than cluttering a screen with an entire year's sales, each of 12 screens or Web pages might provide a month's sales, with subsequent months and summary information available on separate screens or through hyperlinks, which are traceable as the user desires more information.

The problem of information overload is so prevalent as to have become a cliché, but it remains a valid concern. No one is served if excess information is given only to flaunt the capabilities of the system. Always keep the decision makers in mind when deciding about quantity of output. Often they will not need great amounts of output, especially if there is an easy way to access more.

MAKING SURE THE OUTPUT IS WHERE IT IS NEEDED

Output is printed on paper, displayed on screens, piped over speakers, made available on the Web, and stored on microforms. Output is often produced at one location (for example, in the data-processing department) and then distributed to the user.

The increase in online, screen-displayed output that is personally accessible has cut down somewhat on the problem of distribution, but appropriate distribution is still an important objective for the systems analyst. To be used and useful, output must be presented to the right user. No matter how well-designed reports are, if they are not seen by the pertinent decision makers, they have no value.

PROVIDING THE OUTPUT ON TIME

One of the most common complaints of users is that they do not receive information in time to make necessary decisions. The systems analysts' objectives for output are thus compounded. Not only do you have to be conscientious about who is receiving what output, but you must also be concerned about the timing of output distribution.

Although timing isn't everything, it does play a large part in how useful output will be to decision makers. By this phase in the systems development life cycle, you have learned what output is necessary—and at what time—to drive each stage of the organization's processes. Many reports are required on a daily basis, some only monthly, others annually, and others only by exception. Using well-publicized Web-based output can alleviate some problems with the timing of output distribution as well. Accurate timing of output can be critical to business operations.

CHOOSING THE RIGHT OUTPUT METHOD

As mentioned earlier, output can take many forms, including printed paper reports, information on screens, audio with digitized sounds that simulate the human voice, microforms, and Web documents. Choosing the right output method for each user is another objective in designing output.

For many people, the term *output* still conjures up the vision of stacks of paper computer printouts, but this situation is changing rapidly. With the movement to online systems, much output now appears on display screens, and users have the option of printing it out with their own printer. The analyst needs to recognize the trade-offs involved in choosing an output method. Costs differ; for the user, there are also differences in the accessibility, flexibility, life span, distribution, storage and retrieval possibilities, transportability, and overall impact of the data. The choice of output methods is not trivial, nor is it usually a foregone conclusion.

RELATING OUTPUT CONTENT TO OUTPUT METHOD

The content of output from information systems must be considered as interrelated to the output method. Whenever you design output, you need to think of how function influences form and how the intended purpose will influence the output method that you choose.

Output should be thought of in a general way so that any information put out by the computer system that is useful to people in some way can be considered output. It is possible to conceptualize output as either external (going outside the business), such as information that appears on the Web, or internal (staying within the business), such as material available on an intranet.

External output is familiar to you through utility bills, advertisements, paychecks, annual reports, and myriad other communications that organizations have with their customers, vendors, suppliers, industry, and competitors. Some of this output, such as utility bills, is designed by the systems analyst to serve double duty as a turnaround document. Figure 15.2 is a gas bill that is a turnaround document for the gas company's data processing. In other words, the output for one stage of processing becomes the input for the next. When the customer returns the designated portion of the document, it is optically scanned and used as computer input.

External output differs from internal output not only in its distribution but often in its design and appearance as well. Many external documents must include instructions to the recipient if they are to be used correctly. In addition, many external outputs are placed on preprinted forms or Web sites bearing the company logo and corporate colors.

Internal outputs include various reports to decision makers. They range all the way from short summary reports to lengthy, detailed reports. An example of a summary report is a report summarizing monthly sales totals. A detailed report might give weekly sales by salesperson.

Other kinds of internal reports include historical reports that recount an occurrence and exception reports that are output only at the time an exception (a deviation from the expected) occurs. Examples of exception reports are a listing of all employees with no absences for the year, a listing of all salespeople who did *not* meet their monthly sales quota, or a report on consumer complaints made in the last six months.

OUTPUT TECHNOLOGIES

Producing different types of output requires different technologies. For printed computer output, the options include impact and nonimpact printers. For screen output, the options include attached or stand-alone cathode-ray tubes or liquid crystal displays. Audio output can be amplified over a loudspeaker or listened to through small speakers on a PC. Electronic output is created with special software tools. As you can see, the choices are numerous. Figure 15.3 is a comparison of output methods.

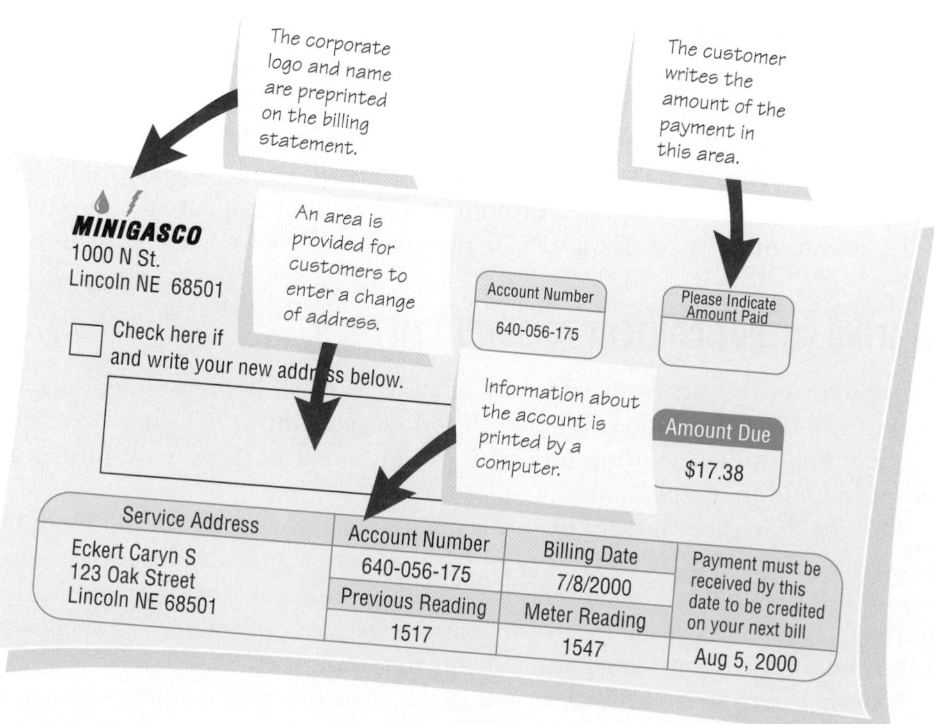

FIGURE 15.2
A turnaround document for Minigasco's data processing.

Output Method	Advantages	Disadvantages
Printer	• Affordable for most organizations • Flexible in types of output, location, and capabilities • Handles large volumes of output • Reaches many inexpensively • Highly reliable with little down time	• May be noisy • Compatibility problems with computer software • May require special, expensive supplies • Still requires some operator intervention • Depending on model, may be slow
Display Screen	• Interactive • Works in online, real-time transmission through widely dispersed network • Quiet • Takes advantage of computer capabilities for movement within databases and files • Good for frequently accessed, ephemeral messages	• Requires cabling and setup space • Still may require printed documentation • Can be expensive if required for many users
Audio Output	• Good for individual user • Good for transient messages • Good where worker needs hands free • Good if output is highly repetitive	• Is expensive to develop • Needs dedicated room where output will not interfere with other tasks • Has limited application • Is not yet perfected
DVD, CD-ROM, and CD-RW	• Has large capacity • Allows multimedia output • Has speedy retrieval • Is less vulnerable to damage	• Is expensive to develop • Is more difficult to update • Is more difficult to use on a network
Electronic Output (email, faxes, and Web pages)	• Reduces paper • Can be updated very easily • Eliminates "telephone tag" • Can be "broadcast" • Can be made interactive	• Has generally lower resolution • Is not conducive to formatting (email) • Is difficult to convey context of messages (email) • Web sites need diligent maintenance

FIGURE 15.3
A comparison of output methods.

Printers. Because printed reports are still the most common kind of output, it is logical to assume that in any large organization printers are ubiquitous. Although other types of output are gaining popularity, for the foreseeable future it is likely that businesses will still desire printed output or will want to design output that will look good if customers, suppliers, or vendors print it out using their own software and hardware.

The trend in printers for mainframe computer systems as well as for personal computers is toward increased flexibility. This trend translates into expanding the options for the location of the printing site itself, accommodating different numbers of characters per page, including numerous type styles and type fonts, changing the position of print on the page, including more graphics capability (including the use of color), producing quieter printing, reducing the number of preprinted forms in inventory, simplifying operator tasks, and reducing the amount of overall operator intervention. Even when printers are dedicated to one particular use, vendors are stressing flexibility to facilitate that use.

Together with users, the systems analyst must determine the purpose for the printer. Once that is established, three key factors of printers to keep in mind:

1. Reliability.
2. Compatibility with software and hardware.
3. Manufacturer support.

Screens as Output. Screens are an increasingly popular output technology. Once used mostly for data entry, screens are also becoming a feasible technology for many other uses as their size and price decrease and as their compatibility with other system components increases.

Screens have distinct advantages over printers because of their quietness and potential for interactive user participation. In the latter regard, screen output (depending, of course, on systems design) can afford flexibility in allowing the user to change output information in real time either through deletion, addition, or modification. Screens also permit the review of stored output through access to and the display of items from a relevant database, permitting individual decision makers to move away from storing redundant printouts.

Video, Audio, and Animation. Many of the tools and application packages you will be working with facilitate the inclusion of video in the output options. Video is a complex form of output since it combines the strength and potential emotional impact of audio (including sound effects, voice and music) with a stimulating visual channel. Some familiar applications are those that are Web-based. Examine Figure 15.4 to see a Web page that provides a series of six brief video clips of an actual event, the Decision Sciences Institute's Knowledge Bowl. Video output is useful here, since the event was held to commemorate an important anniversary in the organizations' history and it can be added to the archives. Although not all of the people who were interested in the event were able to attend it, they can now share in the event through streaming video.

Video can make a good impression if the user has a fast cable modem, DSL (digital subscriber line), or T1 line. However, video can be disappointing to a user who is using a standard dial-up modem over the phone lines.

There are many uses for including video output in your user's displays. A moving, talking picture is an excellent way to supplement static information, such as that found in a user's manual posted on the Web. Additionally, it is a wonderful way to dramatically show progress on a project. For example, an architectural firm may want to post a video of the completion of a building that they have designed for a client. Video clips make useful output for:

1. Supplementing static, printed output.
2. Distance collaboration that is connecting people who do not often get to see each other. For example, this can be helpful for virtual project team members who must work together, but who do not typically meet face to face.
3. Showing "how to" perform an action, such as demonstrating how a form should be filled out (i.e. income tax forms), how software should be installed, or how a product should be assembled.
4. Providing brief training episodes that are job specific in order to emphasize a new or unfamiliar skill.
5. Shifting the time of an actual event by recording it for later output.
6. Preserving an important occasion for addition to an organization's archives.

In a way, audio output can be thought of as the exact opposite of printed output. Audio output is transient, whereas the printed word is permanent. Audio output is usually output for the benefit of one user, whereas printed output is often widely distributed.

Audio output is interpreted by the human ear as speech, although it is actually produced by discrete digital sounds that are then put together in such a way as to be perceived as continuous words. Telephone companies were among the first businesses to produce systems using audio output for customers.

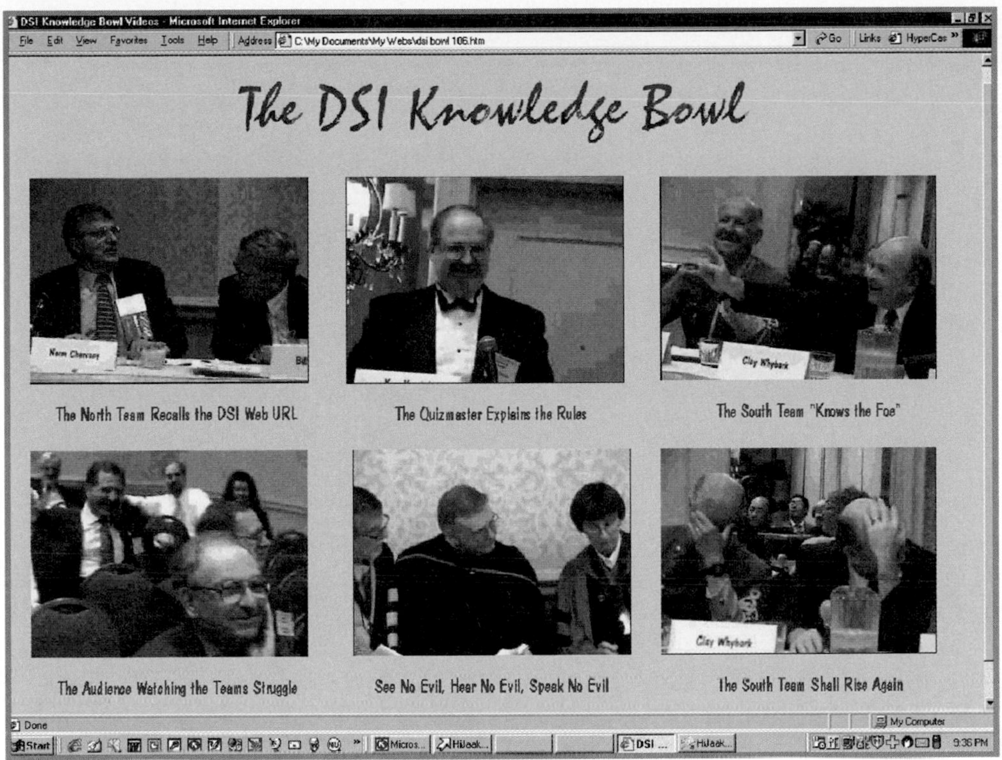

FIGURE 15.4
Streaming video can be used effectively for telling a story or sharing an event. This Web page chronicles an event called the DSI Knowledge Bowl (www.thekendalls.org/dsi-bowl).

Sound can also enhance a presentation. Public domain music and sound effects are readily available. Presentation packages like Microsoft PowerPoint allow users to insert sound, music, and even videos. Sound files come in various formats, but one of the most common for PCs is the .WAV files that can be played in Microsoft Windows. These .WAV files can be obtained from various bulletin boards and are also available on CD-ROM.

Audio output is being used to "staff" toll-free catalog numbers 24 hours a day, seven days a week. By using a digital phone, consumers can call the number and, in response to instructions via audio output, enter the item number, quantity, price, and their credit card number. Stores are capturing sales that would otherwise be missed, because hiring actual employees might be too expensive to justify offering a 24-hour number.

Animation is another form of output that can be used to enhance a Web site or presentation. Animation is the presentation of different images in a series, one at a time. The basic unit of animation is the image, and it is composed of four elements:

1. Elemental symbols.
2. Spatial orientation.
3. Transition effects.
4. Alteration effects.

Elemental symbols can be abstract or real, and they can take on different colors, forms and textures. Spatial orientation helps the user grasp whether symbols are closely related to one another. Transitional effects are either gradual or abrupt. Alteration effects include changing the color, size, texture, and can also include transforming the image through morphing.

If animation is used to support decision making, experiments have shown that the use of realistic, rather than abstract images results in a better quality of decisions. Experimental subjects who viewed gradual, rather than abrupt animated transitions made better decisions.

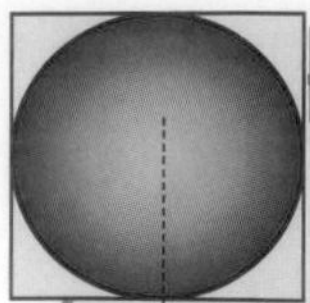

CONSULTING OPPORTUNITY 15.1

YOUR CAGE OR MINE?

"Why can't they get this right? It's driving me to distraction. The zoo in Colombia is writing to me about a tiger that has been on loan from our place since 1993. They should be writing to Tulsa," trumpets Ella Fant, waving a letter in the air. Ella is general curator in charge of the animal breeding program at the Gotham Zoo.

She is talking with members of the zoo's five-person committee about the proposals before them. The committee meets every month to decide which animals to loan to other zoos and which animals to get on loan so as to breed them. The committee is composed of Ella Fant, the general curator; Ty Garr, the zoo's director; two zoo employees, Annie Malle and Mona Key; and a layperson, Rex Lyon, who is in business in the community.

Ty paces in front of the group and continues the meeting, saying, "We have the possibility of loaning out two of our gold lion tarmarins, and we have the opportunity to play matchmaker for two lesser pandas. Because three of you are new to the committee, I'll briefly discuss your responsibilities. As you know, Ella and I would pounce on any chance to lure animals in for the breeding program. Your duties are to assess the zoo's financial resources and to look at our zoo's immediate demands. You also must consider the season and our shipping capability as well as that of the zoos we're considering. The other zoos charge us nothing for the loan of their animals for the breeding program. We pay the shipping for any animal being loaned to us and then maintain them, and that gets expensive."

"We are linked, via the Internet, to a database of selected species with 164 other zoos," says Ella as she picks up the story from Ty. "My office has a computer equipped with a monitor. I can access the records of all captive animals in the system, including those from the two zoos we are negotiating with right now."

As the committee members work they begin asking questions. "I need to read some information, get some meat to sink my teeth into, before I'm ready to decide whether the loan of the lesser pandas is a good idea. Where are the data on the animals we're considering?" growls Rex.

Annie replies, "We have to go to Ella's office to get to it. Mostly, the other employees who need to know just use her computer. We've been waiting for other users for our computer so that we could justify getting more high-quality printers. Right now the one we're using is just for drafts, and the output doesn't copy very well."

Mona gets into the swing of the discussion and says, "Some information on the current state of the budget would be divine, too. I'll go bananas with new expenditures until we at least have a summary of what we're spending. I bet it's a bunch."

Ty answers, "We don't mean to monkey around, but frankly we feel trapped. Costs of reproducing all the financial data seem high to us. We'd rather put our money into reproducing rare and endangered species! Paperwork multiplies on its own."

The group laughs nervously together, but there is an air of expectancy in the room. The consensus is that the committee members need more internal information about the zoo's financial status and the prospective loan animals.

Ella, aware that the group cannot be tamed in the way the previous one was, says, "The old committee preferred to get their information informally, through chattering with us. Let's spend this first meeting discovering what kinds of documents you think you need to do your work as a committee. Financial data are on a stand-alone PC that our financial director uses. It's his baby, of course."

What are some of the problems related to output that the committee is experiencing? What suggestions do you have for improving output to the committee? How can the budget constraints of the zoos be met while still allowing the committee to receive the output it needs to function? Comment on the adequacy of the output technology that is currently in use at the zoo. Suggest alternatives or modifications to output and output technology that would enhance what is being done. (*Hint*: Consider ways in which the committee can leverage its use of the Internet—say more use of the Web—to get the output that it needs and that it needs to share.) Analyze both internal and external output requirements.

There are many software packages that facilitate the insertion of animated images into Web sites. Figure 15.5 is an example that shows how Macromedia Flash is used to animate an image of a honey bee flying between its hive and a flower. Flash is often used in the promotion of feature-length films on Web sites, as well as for promotion of products and services on other commercial Web sites. Another popular animation package for PCs is Ulead GIF Animator by Ulead Systems. Animation Factory provides an animated GIF library, in which many of the images can be used free of charge on personal Web pages.

CD-ROMs and DVDs. With the demand for multimedia output growing, the display of material on CD-ROMs has become increasingly widespread. Once used almost exclusively for reference works because of their large storage capacity, CD-

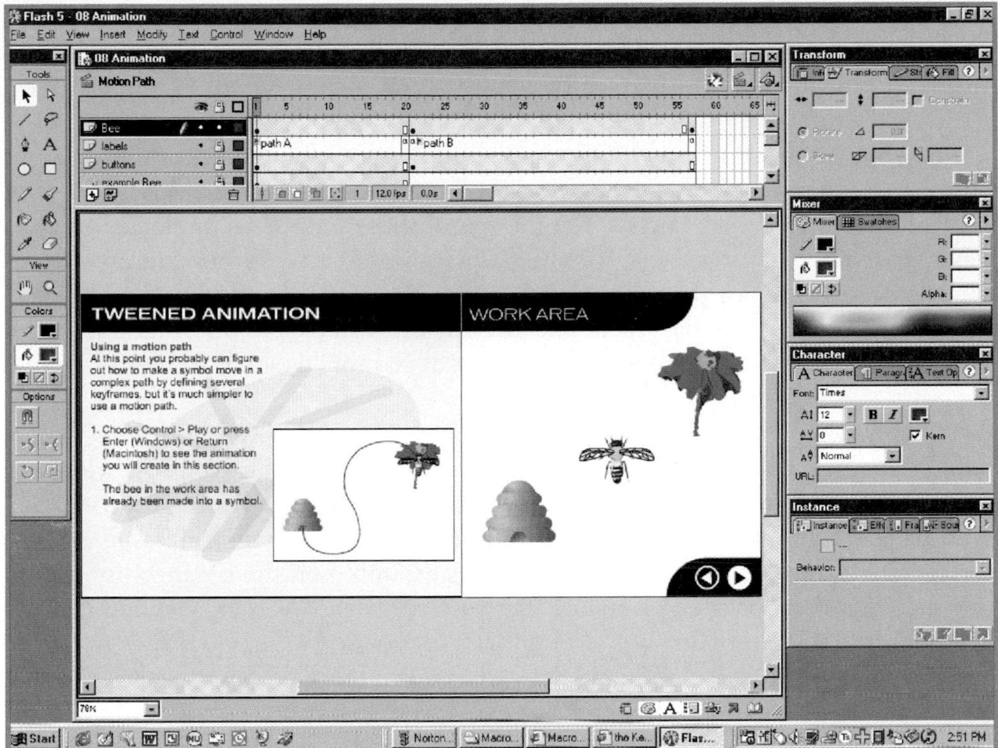

FIGURE 15.5
One of the best animation tools is Macromedia Flash. In this example, animation will be used to move the bee from the beehive to the flower.

ROMs are being used to output any information that is voluminous and somewhat stable in content.

Retrieval of CD-ROM output is faster than older methods, such as accessing paper and microforms. In addition, CD-ROMs are less vulnerable to damage from human handling than other output. CD-ROMs can include full-color text and graphics as well as music and full motion video, so as an output medium they provide a designer maximum creativity.

The DVD (digital versatile disk) is expected to replace the CD-ROM soon. A DVD has more capacity, and a DVD drive can read CD-ROMs as well as DVDs. Not only will DVDs be used for output, but they will also be used for storage, as in the DVD-RAM (random access memory).

Electronic Output. Many of the new systems you design will have the capability of sending electronic output in the form of email, faxes, and bulletin board messages that can be sent from one computer to another without the need for hard copy. Many of the advances in electronic output are paving the way for what is often called the information superhighway, the Internet, or more simply the Net. Part of the Internet is the World Wide Web (or, simply, the Web). Web sites feature Web pages that are newer forms of electronic output that are accessible to Internet users who have special software to view the hypertext-based Web pages. This special software is called a browser. In an upcoming section we provide some key definitions of design terms that will be helpful in designing Web sites for the World Wide Web. Although we leave it to you to trace out the far-reaching social implications of these shifts, in a very practical way, as an analyst, you are already involved with them.

Electronic mail (email) is an exchange of messages between computers that you can set up and run internally within the organization through an intranet or that can be set up through communication companies or online services such as America Online (AOL). By designing email systems, you can support communication throughout the organization, help reduce paper waste, cut down on the tiresome game of telephone tag, help users broadcast messages to many others, and provide a means for updating output very easily. A useful and flexible email system can form the basis of support for workgroups. In Chapter 21, we cover groupware in detail.

Two newer groups of technologies that allow users to pull information from the Web and also allow organizations to send information to them periodically are being designed for organizations. The designs of these output technologies are new ways for analysts to facilitate communication within, between, and among organizations and individuals. These output technologies are called pull and push technologies, reflecting the way users and organizations look for information on the Web and either "pull it" in downloads or have it sent or "pushed" to them.

Pull Technology. An important output technology made possible by the Web is pull technology. If you have tried to pull information from the Web by clicking on links, you have used the most basic type of pull technology. Figure 15.6 shows a Web page for an international IS research organization. When each issue is complete, *OASIS*, the organization's newsletter, is mounted on the organization's Web site, and members of the association can pull it off the Web by viewing it as an Adobe Acrobat document.

This type of pull technology has several advantages compared with sending output as a simple paper newsletter. For example, whenever the newsletter is complete, it can be mounted on the Web; there is no delay in delivery. In addition, if the user has a color printer, color copies can be obtained, whereas reproducing the newsletter in color on paper for all members is prohibitively expensive for this nonprofit organization.

In the future, you may be asked to program evolutionary agents (intelligent agent software) to help organizational members find what they need on the Web. These agents will relieve some of the users' typical burden of searching the Web, because the agents will observe and understand users' behavior as they interact

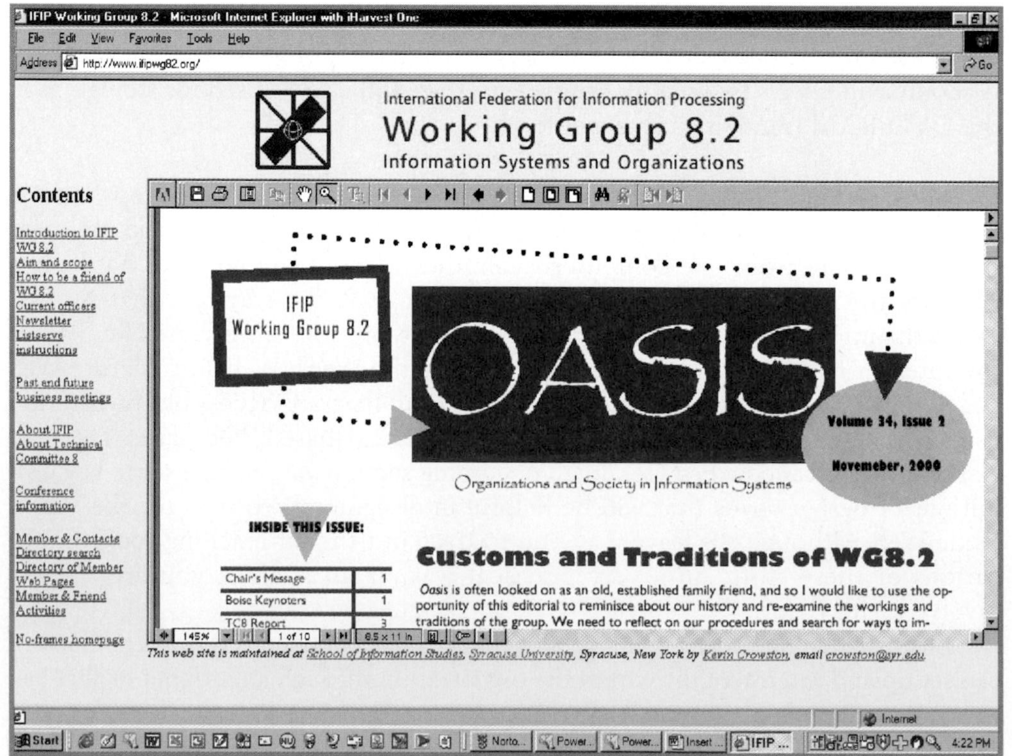

FIGURE 15.6
Pull technology refers to a user pulling information from a Web site. In this example, the newsletter OASIS can be accessed from the IFIP WG 8.2 Web site (www.ifipwg82.org).

with a variety of material on the Web. Evolutionary agents are then programmed to seek out the information users want. In addition, searches on the Web will be more efficient and more effective for users, because satisfying results from initial searches can be obtained with the use of an evolutionary agent, one that observes the behavior of the user and customizes the Web searches accordingly.

Push Technology. Another type of output analysts design is Web and wireless content delivered via push technology. Push technology can be used for external communication to push (electronically send) solicited or unsolicited information to a customer or client. It can also be used within the organization to call immediate attention of an employee or a decision maker who is facing a critical deadline to critical items. The term *push technology* can be described as any content sent to users at specified times, from basic Web casting all the way to selective content delivery using sophisticated evolutionary filtering agents.

Many traditional as well as Internet-based businesses are experimenting with push technology. The output in Figure 15.7 is an example of push technology used to distribute an electronic newsletter about information technology, called *Woody's Windows Watch,* to subscribers via email. Notice that the page design, including a variety of fonts, strategic uses of color, and engaging writing, intentionally emulates the look of a Web page even though it is delivered via email. This particular newsletter provides the full text of the material on email and embeds links to URLs that send the user to other useful Web sites and relevant products.

Push technology can also get the information to the person who needs it, which is one output design objective that analysts try to meet. Broadcasting information to all employees is less expensive than printing out information and then distributing it to a select few. Although in this instance managers do not need to be concerned whether or not a particular employee should get a report, the analyst needs to guard against flooding employees with meaningless pushed information just because the technology exists to do so.

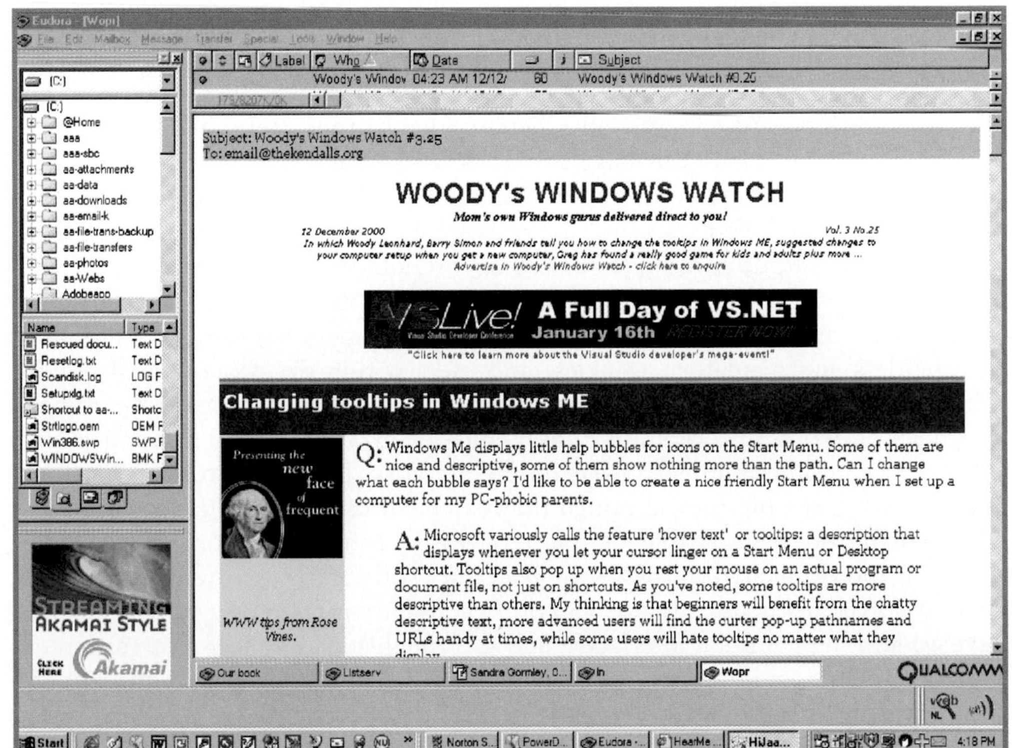

FIGURE 15.7
Push technology is used to distribute information to users. In this example, the newsletter Woody's Window Watch is distributed automatically via email but appears to be much like a Web page.

When working with push technologies, you will be struck by their flexibility compared with paper output. When data are delivered over an Intranet to a PC, the user is able to take it and customize it in many ways. For instance, an employee may decide to look at a single product or may want to generate a graph of sales over time.

Many organizations are experimenting with push technology. National Semiconductor added its own channel to a Webcaster that includes three types of product-related information. Wheat First Securities used a different Webcaster to deliver information to its brokers. MCI's network operations group uses another commercial Webcaster to send information about outage alerts to the 7,000 employees who run its long-distance network.

FACTORS TO CONSIDER WHEN CHOOSING OUTPUT TECHNOLOGY

As you can determine from this brief discussion of output technology, there are several factors to consider when choosing it. Although the technology changes rapidly, certain usage factors remain fairly constant in relation to technological breakthroughs. These factors, some of which present trade-offs, must be considered. They include the following:

1. Who will use (see) the output (requisite quality)?
2. How many people need the output?
3. Where is the output needed (distribution/logistics)?
4. What is the purpose of the output?
5. What is the speed with which output is needed?
6. How frequently will the output be accessed?
7. How long will (or must) the output be stored?
8. Under what special regulations is the output produced, stored, and distributed?
9. What are the initial and ongoing costs of maintenance and supplies?
10. What are the environmental requirements (noise absorption, controlled temperature, space for equipment, and cabling) for output technologies?

Examining each factor separately will allow you to see the interrelationships and how they may be traded off for one another in a particular system.

Who will use (see) the output? Discovering who will use the output is important because job requirements help dictate what output method is appropriate. For example, when district managers must be away from their desks for extended periods, they need printed output that can travel with them or technology that can access appropriate Web sites and databases as they visit the managers in their region. Screen output or interactive Web documents are excellent for people such as truck dispatchers who are deskbound for long periods.

External recipients of output (clients and customers, vendors and suppliers, shareholders and regulatory agencies) and users within the business will require different output. Clients, vendors, and suppliers can be part of several extranets, which are networks of computers built by the organization, providing applications, processing, and information to users on the network. These networks are not necessarily part of the Internet, although they can be. If users have difficulty accessing electronic output because they do not have the equipment, it is sometimes worthwhile for the company to provide the necessary software or hardware to keep the client involved. Alternatively, you can provide quality printed output for clients or develop another common interface (such as a touch-tone phone), whereas screen, Web documents, audio, or print output all might be viable options for internal use.

Examine the Web site shown in Figure 15.8 for an ecommerce company called MerchantsBay.TV. The Web designer is clearly attuned to the intended users of the

FIGURE 15.8

When designing a Web site it is important to choose a metaphor that can be used throughout the Web site. This example from Merchants Bay (www.merchantsbay.TV) employs a nautical theme.

wholesale gift site. The ecommerce company's Web site is powered by a patented negotiating algorithm in which users submit bids (for one item or 400) on an array of merchandise, dubbed "good stuff" by the company president. The company's strategy is based on the president's personal experience with flea markets and the observation that people are powerfully attracted to bargaining for a deal.

The Web site has intentionally invoked a cluttered feel, similar to what one gets walking through a flea market. The site is intended for customers who would frequent flea markets in person: they are known to be collectors, gregarious, and curious by nature. The Web site is a profusion of colors, a variety of sale signs in a mixture of lettering, and even a video that provides new layers of color and action. Colloquial language is used throughout the site. Notice that company's catchphrase is "Purveyor of Good Stuff." The Web designer has successfully carried out a nautical metaphor throughout the site. For instance, the user is invited to "Search the Bay" for merchandise. In addition, the company's logo includes a wave and a sun on the horizon, and an icon of a ship's steering wheel is placed above a column that invites the user to "navigate" for products, services, and customer service.

To complete a transaction on the site, a customer has an opportunity to accept the "Captain's Price" as posted or to submit a bid. If the bid submitted is too low according to the stored negotiation algorithm, a natural language response is returned in a pop-up window stating: "Thanks for your offer, mate. You don't like to part with your money if you don't have to, heh? Yet hey, I like ya, mate. Please try again by offering a better price or by ordering a larger quantity." In this way, the bid is rejected in a friendly, humorous way, and bidders are even given two hints on how to improve the chances that their next bids will be successful. It is clear that the Web designer had a solid profile of the intended customer in mind when designing the site.

How many people need the output? Choice of output technology is also influenced by how many users need the output. If many people need output, Web-based documents or printed copies are probably justified. If only one user needs the output, a screen, microform, or even audio may be more suitable.

"You are entitled to one call, one fax, or one e-mail."

If many users in the business need different output at different times for short periods and they need it quickly, Web documents or screens connected to online terminals that are able to access database contents are a viable option.

Where is the output needed (distribution/logistics)? Another factor influencing the choice of output technology is the physical destination of the output. Information that will remain close to its point of origin, that will be used by a few users within the business, and that may be stored or referred to frequently can safely be printed or mounted on an intranet. An abundance of information that must be transmitted to users at great distances in branch operations may be better distributed electronically, via the Web or extranets, with the recipient deciding whether to customize and print output, display it on screens, or store it. One example of such information is documentation for an application.

Sometimes federal or state regulations dictate that a printed form remain on file at a particular location for a specified period of time. In those instances, it is the responsibility of the systems analyst to see that the regulation is observed for any new or modified output that is designed.

What is the purpose of the output? The purpose of the output is another factor to consider when choosing output technology. If the output is intended to be a report whose purpose is to attract shareholders to the business by allowing them to peruse corporate finances at their leisure, well-designed, printed output such as an annual report is desirable. A variety of media may also be used so that the annual report is available on the Web as well as in printed form. If the purpose of the output is to provide 15-minute updates on stock market quotations and if the material is highly encoded and changeable, screen displays, Web pages, or even audio presentations are preferable.

What is the speed with which output is needed? As we go through the three levels of strategic, middle, and operations management within the organization, we find that decision makers at the lowest level of operations management need output rapidly so that they can quickly adjust to events such as a stopped assembly line, raw materials that have not arrived on time, or a worker who is absent unexpectedly. Online, onscreen output may be useful here.

As we ascend the management levels, we observe that strategic managers are much less likely than others to need output rapidly. They are more in need of output for a specific time period, which helps in forecasting business cycles and trends.

In addition, particular businesses are more likely than others to need rapid output of the sort that is often provided on screen through online systems. A case in point is an intensive care unit; here a nurse continually surveys a heart attack victim's pulse using a monitor connected to a video display terminal. In contrast, a doctor reads the printout from an electrocardiogram during a healthy patient's routine physical examination. The first is literally a life-or-death situation; the second presumably is not. In the intensive care unit, reaction time to output is everything. Therefore, real-time output is needed. In the second situation, there is likely a time lapse between the completion of the test and the reading of the results, with no criticality involved. Fortunately, most technological decisions on output are not life or death, but this example does point up the burden on the systems analyst to make reasonable choices.

How frequently will the output be accessed? The more frequently output is accessed, the more important is its display on screens connected to local area networks or the Web. Infrequently accessed output that is needed by only a few users is well-suited to microforms such as microfiche or microfilm. For example, university librarians at U.S. schools make observations about the frequency of use when deciding to microfilm daily newspapers from foreign countries.

Output that is accessed frequently is a good candidate for incorporation into Web-based or other online systems or networks with display on screens. Adopting this type of technology allows users easy access and alleviates physical wear and tear that cause frequently handled printed output to deteriorate.

How long will (or must) the output be stored? As just noted, output printed on paper deteriorates rapidly with age. Output preserved on microfiche and microfilms is not as prone to succumb to environmental disturbances such as light, humidity, and human handling. Therefore, to store output for long periods, it may be necessary to microfilm it.

The business in question may be subject to governmental regulations on local, state, or federal levels, regulations that dictate how long output must be kept on file. As long as the corporation is willing to maintain it and it is nonproprietary, archival information, it can be maintained in Web documents as part of the organization's Web site. Organizations themselves also enact policies about how long output must be retained. For example, some universities require professors to retain student work (output) for a full semester after a final grade for the student has been recorded. Although student papers may take a lot of storage space, the short retention period does not justify the cost of microfilming them.

Under what special regulations is the output produced, stored and distributed? The appropriate format for some output is actually regulated by the government. For example, an employee's W-2 form must be printed; its final form cannot be a screen or microform output. Each business exists within a different complex of regulations under which it produces output. To that extent, appropriate technology for some functions may be dictated by law.

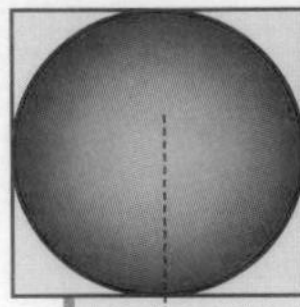

CONSULTING OPPORTUNITY 15.2

A RIGHT WAY, A WRONG WAY, AND A SUBWAY

"So far so good. Sure, there have been some complaints, but any new subway will have those. The 'free ride' gimmick has helped attract some people who never would have ridden otherwise. I think there are more people than ever before interested in riding the subway," says Bart Rayl. "What we need is an accurate fix on what ridership has been so far so we can make some adjustments on our fare decisions and scheduling of trains."

Rayl is an operations manager for S.W.I.F.T., the newly built subway for Western Ipswich and Fremont Transport that serves a major northeastern city in the United States. He is speaking with Benton Turnstile, who reports to him as operations supervisor of S.W.I.F.T. The subway system is in its first month of operation, offering limited lines. Marketing people have been giving away free rides on the subway to increase public awareness of S.W.I.F.T.

"I think that's a good idea," says Turnstile. "It's not just a token effort. We'll show them we're really on the right track. I'll get back to you with ridership information soon," he says.

A month later, Rayl and Turnstile meet to compare the projected ridership with the new data. Turnstile proudly presents a two-inch-high stack of computer printouts to Rayl. Rayl looks a little surprised but proceeds to go through it with Turnstile. "What all is in here?" Rayl asks, fingering the top page of the stack hesitantly.

"Well," says Turnstile, training his eyes on the printout, "it's a list of all the tickets that were sold from the computerized machines. It tells us how many tickets were bought and what kinds of tickets were bought. The guys from Systems That Think, Inc., told me this report would be the most helpful for us, just like it was for the operations people in Buffalo and Pittsburgh," says Turnstile, turning quickly to the next page.

"Maybe, but remember those subway systems began with really limited service. We're bigger. And what about the sales from the three manned ticket booths in the Main Street Terminal?" asks Rayl.

"The clerks in the booth can get information summarizing ticket sales onscreen any time they want it, but it's not included here. Remember that we projected that only 10 percent of our sales would be from the booths anyway. Let's go with our original idea and add that to the printout," suggests Turnstile.

Rayl replies, "But I've been observing riders. Half of them seem to be afraid of the computerized ticket machines. Others start using them, get frustrated reading the directions, or don't know what to do with the ticket that comes out, and they wind up at the ticket booth blowing off steam. Furthermore, they can't understand the routine information posted on the kiosks, which is all in graphics. They wind up asking clerks what train goes where." Rayl pushes the printout holding the ticket sales to one side of the conference table and says, "I don't have much confidence in this report. I feel as if we're sitting here trying to operate the most sophisticated subway system in the United States by peering down a tunnel instead of at the information as we should be. I think we need to think seriously about capturing journey information on magnetically stripped cards like New York City is doing. Every time you insert the card to take a ride, the information is stored."

What are some of the specific problems with the output that the systems consultants and Benton Turnstile gave to Bart Rayl? Evaluate the media that are being used for output as well as the timing of its distribution. Comment on the external output that users of the computerized ticket machines are apparently receiving. Suggest some changes in output to help Rayl get the information he needs to make decisions on fares and scheduling of trains and to help users of the subway system get the information they need. What are some decisions facing organizations like the New York Transit Authority if they collect and store input concerning an individual's destinations each time a trip is taken? What changes would S.W.I.F.T. have to make to its output and its tickets if it adopted this technology?

Much of this regulation, however, is industry-dependent. For example, a regional blood system is required by federal law to keep a medical history of a blood donor—as well as his or her name—on file. The exact output form is not specified, but the content is strictly spelled out. Other governmental regulations may require the printing and storing of output on standardized forms. The analyst must be aware of these regulations and make sure the business is in compliance.

What are the initial and ongoing costs of maintenance and supplies? The initial costs of purchasing or leasing equipment must be considered as yet another factor that enters into the choice of output technology. Most vendors will help you estimate the initial purchase or lease costs of computer hardware, including the cost of printers and monitors, the cost of access to online service providers (Internet access), or the costs of building intranets and extranets. Many vendors, however, do not provide information about how much it costs to keep a printer working (paper, toner, photoelectric unit, repairs, and maintenance). Therefore, it falls to the analyst to research the costs of operating different output technologies or of maintaining a corporate Web site over time.

Beyond that, the systems analyst may need to gather cost information about other, less-used alternatives, such as audio output and microforms. Both may cost substantially more than traditional output methods, but their costs can decrease with use so that in the long run, the user would be better off financially with an audio system. As you have seen, original and maintenance costs are far from the only factors influencing the choice of output method.

What are the environmental requirements (noise absorption, controlled temperature, space for equipment, and cabling) for output technologies? Printers require a dry, cool environment to operate properly. Monitors for screen-displayed output require space for setup and viewing. Audio and video output require a relatively quiet place that allows the user to make sense of digitized sounds.

Output technologies themselves actually create environmental disturbances. Audio and video output require a quiet environment if they are to be heard, and they should not be audible to employees (or customers) who are not using it. Thus, the analyst should not specify audio output for a work situation where many employees are engaged in a variety of tasks unrelated to the output. For example, grocery store managers using microprocessor-based audio output attached to scanners at checkstands have found that cashiers working side by side with several checkers often turn down the volume on their speakers (which name aloud each item scanned and its price) to avoid irritation and errors. Unfortunately, this action defeats the audio double-check.

Conversely, some output technologies are prized for their unobtrusiveness. Libraries that emphasize silence in the workplace make extensive use of display screens for Web documents and for the display of other networked database information, and printers might be scarce.

REALIZING HOW OUTPUT BIAS AFFECTS USERS

Whatever form it takes, output is not just a neutral product that is subsequently analyzed and acted upon by decision makers. Output affects users in many different ways. The significance of this fact for the systems analyst is that great thought and care must be put into designing the output so as to avoid biasing it.

RECOGNIZING BIAS IN THE WAY OUTPUT IS USED

It is a common error to assume that once the systems analyst has signed off on a system project, his or her impact is ended. Actually, the analyst's influence is long-lasting. Much of the information on which organizational members base their decisions is determined by what analysts perceive is important to the business.

Bias is present in everything that humans create. This statement is not to judge bias as bad but to make the point that it is inseparable from what we (and consequently our systems) produce. The concerns of systems analysts are to avoid unnecessarily biasing output and to make users aware of the possible biases in the output they receive.

Presentations of output are unintentionally biased in three main ways:

1. How information is sorted.
2. Setting of acceptable limits.
3. Choice of graphics.

Each source of bias is discussed separately in the following subsections.

Introducing Bias When Information Is Sorted. Bias is introduced to output when the analyst makes choices about how information is sorted for a report. Common sorts include alphabetical, chronological, and cost.

Information presented alphabetically may overemphasize the items that begin with the letters A and B, because users tend to pay more attention to information presented first. For example, if past suppliers are listed alphabetically, companies such as "Aardvark Printers," "Advent Supplies," and "Barkley Office Equipment" are shown to the purchasing manager first.

Introducing Bias by Setting Limits. A second major source of bias in output is the predefinition of limits for particular values being reported. Many reports are generated on an exception basis only, which means that when limits on values are set beforehand, only exceptions to those values will be output. Exception reports make the decision maker aware of deviations from satisfactory values.

For example, limits that are set too low for exception reports can bias the user's perception. For example, an insurance company that generates exception reports on all accounts one week overdue has set too low a limit on overdue payments. The decision maker receiving the output will be overwhelmed with "exceptions" that are not really cause for concern. The one-week overdue exception report leads to the user's misperception that there are a great many overdue accounts. A more appropriate limit for generating an exception report would be accounts 30 days overdue.

Users can also be biased by output that is printed in an exception report when too high a limit is to be met before information is included. For example, a quality assurance manager for a large manufacturer receives an exception report if more than 10 lots of rubber hoses out of 100 that are produced are defective according to specifications such as tensile strength. Typically, the manager can expect 3 lots to be defective. The limit is too high to be of any tangible value, however, because the problem could have been detected earlier as the defective rate began to rise. For the information to be useful, limits on allowable defects must be tightened. The resultant bias of setting too high a limit is that a false sense of security is generated. In fact, there is little, if any, timely output when allowances are set at unrealistically high limits.

Introducing Bias through Graphics. Output is subject to a third type of presentation bias, which is brought about by the analyst's choice of graphics for output display. Bias can occur in the selection of the graphic size, its color, the scale used, and even the type of graphic.

Graphic size must be proportional so that the user is not biased as to the importance of the variables that are presented. For example, Figure 15.9 shows a column chart comparing the number of no-shows for hotel bookings in 1999 with no-shows for hotel bookings in 2000. Notice that the vertical axis is broken, and it appears that the number of no-shows for 2000 is twice as much as the number of no-shows in 1993, although the number of no-shows has actually gone up only slightly.

Choice of graphic color is also important because it, too, may unduly bias the user. The analyst needs to be aware that any colored output naturally claims more user attention than black-and-white printed or displayed output. Certain colors may be assigned meanings within business, and those should not be changed by the analyst without good cause. For instance, red might be used to indicate a problem, such as a budget that is "in the red," as part of an exception report. Some colors are not easily legible (see Chapter 16 for specifics on screen color), so they should be avoided. Otherwise, users will be biased against even reading information or

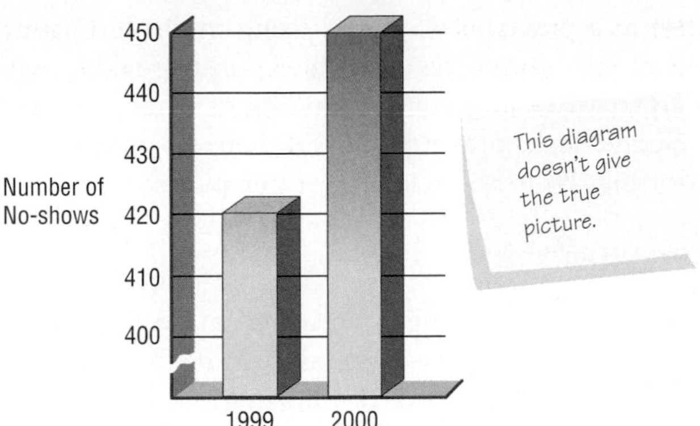

FIGURE 15.9
A misleading graph will most likely bias the user.

related graphics. Notice that Web graphics are included in Web pages. Care must be taken to view color combinations with many different browsers, because not all browsers can handle all colors, and strange combinations occur when millions of colors are viewed as only 256 colors.

AVOIDING BIAS IN THE DESIGN OF OUTPUT

Systems analysts can use specific strategies to avoid biasing the output they design:

1. Be aware of the sources of bias.
2. Create an interactive design of output that includes users and a variety of differently configured systems during the testing of Web document appearance.
3. Work with users so that they are informed of the output's biases and can recognize the implications of customizing their displays.
4. Create output that is flexible and that allows users to modify limits and ranges.
5. Train users to rely on multiple output for conducting "reality tests" on system output.

All these except the first focus on the relationship between the systems analyst and the user as it involves output. Systems analysts must recognize the potential impact of output and be aware of the possible ways in which output is unintentionally biased.

Interactive design of output means that output cannot be successfully designed in a vacuum. Rather, the systems analyst must actively solicit user feedback regarding output. The design process will require several iterations before users feel output is useful. Unavoidably, users will incorporate their own biases into output, but they will also be clearer about how to interpret output that they helped design or that they customized.

The third way to avoid unintentional bias is to work with users so that they are informed of the output's biases through training sessions as well as in the documentation. Output (especially screens) can include information about the way the information has been generated, thus pinpointing potential biases included in the report.

Biased output can also be avoided (or at least made explicit) if users are able to modify the limits and ranges set for output. This approach is especially suited to online, interactive systems. The drawback is that there may be a lack of consistency in output that could lead to problems if future comparisons are necessary. Flexibility in output may also lead to communication problems among users who, for example, may be using different ranges of values to spot problems.

A fifth way to help avoid unintentional bias in output is to train users to conduct "reality tests" on the output they receive. Such tests include comparing output with their own experience and expectations (asking whether what they are reading make sense in a practical way) and using multiple output from different sources (daily and monthly reports, observation and speaking with others in the business, and so on) to arrive at an understanding of what is actually happening in the organization. Such checks prevent an overreliance on any one source of output and help correct for bias by introducing other perspectives.

DESIGNING PRINTED OUTPUT

Using the information gained through the information requirements determination phase and having decided to use printed output, the systems analyst is ready to begin its physical design. The source of information to be included in reports is the data dictionary, the compilation of which was covered in Chapter 10. Recall that the data dictionary includes names of data elements as well as the required field length of each entry. Reports fall into three categories: detailed, exception, and summary. Detailed reports print a report line for every record on the master file. They are used for mailing to customers, sending student grade reports, printing catalogs, and so on. Inquiry screens have replaced many detailed reports. Exception reports print a line for all records that match a set of conditions, such as which books are overdue at a library or which students are on the honor roll. They are usually used to help operational managers and clerical staff to run a business. Summary reports print one line for a group of records and are used to make decisions, such as which items are not selling and which are hot selling.

GUIDELINES FOR PRINTED REPORT DESIGN

Figure 15.10 is an output report that is intended for divisional managers of a food wholesaler that supplies a number of franchise grocery stores. We focus on different aspects of the report as we cover the tools, conventions, and functional and stylistic design attributes of printed output reports.

Report Design Conventions. Conventions to follow when designing a form include the type of data (alphabetic, special, or numeric) that will appear in each position, showing the size of the form being prepared, and showing the way to indicate a continuation of data on consecutive layout forms. Most form design software that analysts' now use features standard conventions for designing forms on-screen. In addition, they feature familiar drag and drop interfaces that allow you both to select attributes such as an address "block" with a mouse click and then to drop it on the screen where you want to position it on your form. You will be using "WYSIWYG" or "What You See Is What You Get," so it makes the design of forms a very visual exercise.

Constant information is information that remains the same whenever the report is printed. The title of the report and all of the column headings are written in as constant information. *Variable information* is information that can vary each time the report is printed out. In our example, the sales figures in thousands of dollars will change; hence, it is indicated as variable information.

Your software design tool will also have conventions to permit you to easily signify whether you will be using alphabetic or numeric characters in a field. In addition, form design software permits you to easily mark information that must be repeated in the same position in a column.

Paper Quality, Type, and Size. Output can be printed on innumerable kinds of paper. The overriding constraint is usually cost. Paper that is treated in any special way—either preprinted, inked in color, multipart with carbon interleaves, or carbonless transfer—is more expensive than plain paper.

Franchise Store Information
Ranked by Earnings in Dollars
For the Month Ending MM/DD/YYYY

Page 2

F NO	Store Names	DIV	Dist	Rank	Sales Dollar 1,000's	Gross Profit 1,000's	%	Other Income 1,000's	%	Allacated Expenses 1,000's	%	Earnings Dollars	%
C 5112	Front Royal, VA	20	23	51	126	5	3.93	2	1.8	5	4.0	2,144	1.7
S 4311	Rockville, MD	40	41	52	144	6	4.27	0	0.3	4	3.1	2,062	1.4
R 3021	Middleburg, VA	20	22	53	95	4	4.29	2	1.9	4	4.0	2,057	2.2
S 5021	Culpeper, VA	20	26	54	219	8	3.78	3	1.5	10	4.4	2,005	0.9
R 2820	Waldorf, MD	40	42	55	72	3	4.69	1	1.2	2	3.3	1,903	2.6
C 4424	Fairfax-Lee Hgwy	20	22	56	131	5	4.16	2	1.3	5	4.0	1,869	1.4
C 4423	Baileys X-Roads	20	22	57	98	5	4.70	2	1.7	5	4.6	1,727	1.8
S 3821	Herndon, VA	20	23	58	221	7	3.35	4	1.7	9	4.2	1,703	0.8
C 7126	Frederick, MD	30	32	59	125	5	4.04	2	1.6	5	4.3	1,615	1.3
S 8029	Centreville, VA	20	27	60	175	7	3.73	3	1.9	8	4.7	1,593	0.9
R 5029	Minnieville, VA	20	34	61	34	2	5.28	1	3.3	1	4.0	1,572	4.7
S 7520	Mount Vernon	20	24	62	90	5	5.22	2	1.7	5	5.2	1,558	1.7
C 4712	D.C. M Street	40	44	63	235	10	4.35	4	1.8	13	5.5	1,489	0.6
S 4716	Annandale	20	25	64	126	6	4.52	0	0.1	4	3.5	1,457	1.2
S 7922	Vienna, VA	20	25	65	177	9	4.86	2	1.2	9	5.3	1,447	0.8
R 4491	Great Falls	20	24	66	86	4	4.39	2	1.9	4	4.7	1,364	1.6
R 3926	Harper's Ferry	30	33	67	68	3	4.80	0	0.3	2	3.1	1,325	1.9
C 2422	Falls Church	20	27	68	144	6	4.06	2	1.4	7	4.6	1,322	.9
R 3024	Clifton, VA	20	23	69	53	3	5.17	1	1.6	2	4.3	1,273	2.4
C 4511	Silver Spring, MD	20	42	70	121	5	4.06	1	1.2	5	4.3	1,237	1.0
R 5120	Olney, MD	30	31	71	43	2	4.60	1	.2.	2	4.0	1,217	2.8
C 4527	D.C Connecticut Ave	40	45	72	110	5	4.28	0	0.2	4	3.4	1,200	1.1
C 4526	Pennsylvania Ave	40	42	73	134	6	4.55	0	0.2	5	4 0	1,073	0.8
S 2923	Manassas	20	25	74	198	7	3.54	0	0.1	6	3.1	1,057	0.5
	City Stores				6,025	255	4.23	67	1.1	190	3.2	69,987	1.2
	Suburhan Stores				3,402	171	5.03	54	1.6	133	3.9	35,020	1.0
	Rural Stores				2,018	92	4.56	27	1.3	47	2.3	43,223	2.1
					=======	====	=====	===	====	====	====	========	====
	Total (All Stores in Region)				11,445	518	4.52	148	1.3	370	3.2	148,230	1.3

FIGURE 15.10
A printed output report for divisional managers of a food wholesaler.

The kind of paper used has a predictable effect on the user, however, so investing in special papers may be justified. For example, using bond to print letters of appreciation to employees would be a way of showing that the letters have special content and differ in importance from everyday internal memos. Other examples are the use of security paper for checks and check envelopes as well as documents that must bear official, inalterable seals or even holograms, such as passports and birth certificates.

Paper also differs as to its weight, depending on the rag (cotton) content. More cotton means better quality, durability, and a higher price, but a business may still want particular correspondence to be printed on cotton bond to present a more distinguished image. Common lengths for reports in the United States,—whether they appear on preprinted forms or computer printouts—are $3\frac{1}{2}$, $3\frac{2}{3}$, $5\frac{1}{2}$, 6, $7\frac{1}{2}$, and 11 inches.

Special Output Forms. The variety of special output forms is seemingly endless, because virtually any color ink or paper can be used. Many options on positioning of headings and logotypes are open. The systems analyst lays out the content of the report onscreen using a template, in the same manner as on the computer-prepared report.

Preprinted forms are used for many purposes. For example, they may be sent to customers as turnaround documents. Preprinted forms can easily convey a distinctive corporate image through the use of corporate colors and design. Using innovative shapes, colors, and layouts is also a dramatic way of drawing users' attention to the report contained on the preprinted form.

The chief drawback of preprinted forms is their cost. They are extremely expensive in comparison to computer-generated report forms, sometimes costing

"Hold on a second, my printer's going nuts."

three times as much. Also, more forms must be kept in inventory. The expense may be justified, however, especially in instances where creating interest is of competitive importance. The advent of Web-based forms has alleviated some of the inventory costs, because when the output form is created as a Web document, it is not stocked in inventory per se; rather it is maintained on the company's Web site.

Design Consideration. In designing the printed report, the systems analyst incorporates both functional and stylistic or aesthetic considerations so that the report supplies the user with necessary information in a readable format. Because function and form reinforce each other, one should not be emphasized at the expense of the other.

Functional attributes. The functional attributes of a printed report include (1) the heading or title of the report, (2) the page number, (3) the date of preparation, (4) the column headings, (5) the grouping of related data items together, and (6) the use of control breaks. Each of these serves a distinctive purpose for the user.

The heading or the title of the report immediately orients users to what it is they are reading. The title should be descriptive, yet concise. It is often redundant to include the word *report* in the title.

Each page should be numbered so that the user has an easy point of reference when discussing output with others or when locating important figures. Also, if pages of output become separated, page numbers are invaluable in reconstructing the document.

Include the date when the report was prepared on each printout. Sometimes that helps users estimate the value of the output. Often, the more timely the output, the more valuable it is.

Column heading serve to further orient the user as to the report contents. Each data item must have a heading. Headings should be short and descriptive. It is permissible to use abbreviations if they are common to users and meaningful to them. Check that abbreviations are not used elsewhere in the organization with a meaning different from that intended in the report.

Data items that are related to one another should be grouped together on the report. Grouping facilitates understanding for the user, and in many instances it fulfills their expectations of where items should appear. In our sample worksheet,

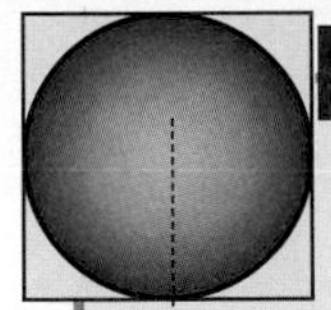

CONSULTING OPPORTUNITY 15.3

IS YOUR WORK A GRIND?

"I want everything I can get my hands on, and the tighter the information is packed, the better. Forget that stuff you hear about information overload. It's not in my vocabulary. I want it all, and not in a bunch of pretty looking, half-page reports either. I want it all together, packed on one sheet that I can take into a meeting in case I need to look something up. And I need it every week," proclaims Stephen Links, vice president of a large, family-owned sausage company.

During an interview, Links has been grilling Paul Plishka, who is part of the systems analysis team that is busy designing an information system for Links Meats. Although Paul is hesitant about what Links has told him, he proceeds to design a printed report that includes all the important items the team has settled on during the analysis phase.

When a prototype of the new report, designed to his specifications, is handed to Stephen, however, there appears to be a change of heart. Links says in no uncertain terms that he can't find what he needs.

"This stuff looks terrible. It looks like scraps. My kindergartner makes better reports in crayon. Look at it. It's all ground up together. I can't find anything. Where's the summary of the number of pork items sold in each outlet? Where is the total volume of items sold for *all* outlets? How about the information on our own shop downtown?" says Links, slicing at the report.

The report clearly needs to be redesigned. Design a report (or reports) that better suits Stephen Links. What tack can the analyst take in suggesting more reports with a less-crowded format? Comment on the difficulty of implementing user suggestions that go against your design training. What are the trade-offs involved (as far as information overload goes) in generating numerous reports as opposed to generating one large report containing all the information Stephen wanted? Devise a heuristic concerning the display of report information on one report in contrast to the generation of numerous reports. Consider advocating a Web-based solution that would permit hyperlinks to all the information Stephen desires. How feasible is that?

store name and store number are grouped together because divisional managers use them interchangeably in referring to stores.

Use control breaks (breaks in data where summaries occur) to help readability. Separate these breaks from the rest of the data with additional lines of space. For example, if 200 franchise grocery stores are grouped by division, the report might feature a control break at the end of each division.

Stylistic/aesthetic attributes. There are several stylistic or aesthetic considerations for the systems analyst to observe when designing a printed report. If printed output is unappealing and difficult to read, it will not be used effectively or may not be used at all. The upshot is uninformed decision makers and a waste of computer time as well as other organizational resources.

Printed reports should be well organized, reflecting the way that the eye sees. In this culture, that means that the report should read from top to bottom and left to right. As mentioned before, related data items should be grouped together. The aesthetics of Web site and Web page design are covered in an upcoming subsection.

Additional blank spaces between columns also contribute to the readability of a report. Users should be able to locate key figures easily on a page; adding blank spaces will facilitate that.

As with the systems proposal report discussed in Chapter 14, printed output reports require ample margins on the right and left as well as on the top and bottom. Margins serve to focus the user's attention on the material centered on the page and to make reading easier. Because the point of creating output is to make it usable, the importance of organization, readability, and ease of use cannot be overstated.

Other stylistic elements that can be used to distinguish printed output are the use of color coding, organizational logos, or preprinted forms that feature these items. Graphics are also a possibility for printed output; they are covered in the subsections on designing output for decision support systems. Show users a prototype of the output report (either on the Web, if that is where it is to reside, or in printed form) so that they can make changes. The mock-up should look as realistic as possible. In addition, some organizations print reports that are saved as report files that may be viewed online. The report viewing software uses scrolling, next page, and previous page buttons to help the user view the report.

STEPS IN DESIGNING OUTPUT REPORTS WITH A COMPUTER-AIDED SOFTWARE TOOL

The following is a step-by-step guide for preparing screen-based prototypes of output reports:

1. Determine the need for the report.
2. Determine the users.
3. Determine the data items to be included.
4. Estimate the overall size of the report.
5. Title the report.
6. Number the pages of the report.
7. Include the preparation date on the report.
8. Label each column of data appropriately.
9. Define variable data, indicating on the screen whether each space or field is to be used for an alphabetic, special, or numeric character.
10. Indicate the positioning of blank lines used to help organize information.
11. Review prototype reports (use screens for Web-based documents and use hard copy output for printed reports) with users and programmers for feasibility, usefulness, readability, understandability, and aesthetic appeal.

Your expertise as a systems analyst enters in when you are required to design printed reports that bring content and form together in a meaningful way. Printed output is the embodiment of your belief about what users need to know to make decisions. There is a trade-off between wanting decision makers to have all available information and designing a useful report that does not overwhelm them with extraneous details.

DESIGNING SCREEN OUTPUT

Chapter 16 covers designing screens for input, and the same guidelines also apply here for designing screen output, although the contents will change. Notice that screen output differs from printed output in a number of ways. It is ephemeral (that is, a screen display is not "permanent" in the same way that printouts are), it can be more specifically targeted to the user, it is available on a more flexible schedule, it is *not* portable in the same way, and sometimes it can be changed through direct interaction.

In addition, users must be instructed on which keys to press when they want to continue reading further screens, when they want to know how to end the display, and when they want to know how to interact with the display (if possible). Access to screen displays may be controlled through the use of a password, whereas distribution of printed output is controlled by other means.

GUIDELINES FOR SCREEN DESIGN

Four guidelines facilitate the design of screens:

1. Keep the screen simple.
2. Keep the screen presentation consistent.
3. Facilitate user movement among screens.
4. Create an attractive screen.

Voter Registration Records
New Rochelle Office

Dist	Number	Street	Last Name	First	Party	Last
69	11413	Bonny Meadow Lane	Petrie	Laura	D	11/2000
69	11413	Bonny Meadow Lane	Petrie	Robert	D	11/1998
69	11414	Bonny Meadow Lane	Helper	Jerry	D	11/1998
69	11414	Bonny Meadow Lane	Helper	Millicent	D	06/1999
69	11415	Bonny Meadow Lane	Brady	Alan	R	11/1998
69	11416	Bonny Meadow Lane	Sorel	Buddy	D	11/1996
69	11416	Bonny Meadow Lane	Sorel	Pickles	D	11/1996
69	11417	Bonny Meadow Lane	Coolie	Melvin	R	06/2000
69	11418	Bonny Meadow Lane	Rogers	Sally	D	11/1999
69	11341	Elm Street	Cleaver	June	R	06/2000
69	11341	Elm Street	Cleaver	Ward	R	06/2000

Press the PGDN key to continue

F1 = Help   F2 = Return to the Main Menu   F3 = Change Party

F4 = Change Name   F5 = Add a Voter   F6 = Change Last Voted

FIGURE 15.11

Poorly designed screen output that repeats redundant information.

Two examples of output screens are shown in Figures 15.11 and 15.12. Notice the difference between the readability of the two screens. On the first, the district number is listed in every instance for the voting register. The readability of the screen is improved when the district number is listed only once, thus eliminating repeating information. This section concentrates on how this kind of graphical and tabular output is created for onscreen display. Chapter 16 covers specific information on how to achieve the four guidelines.

Voter Registration Records
New Rochelle Office
District Number 69

Street	Number	Last Name	First	Party	Last
Bonny Meadow Lane	11413	Petrie	Laura	D	11/2000
			Robert	D	11/1998
	11414	Helper	Jerry	D	11/1998
			Millicent	D	06/1999
	11415	Brady	Alan	R	11/1998
	11416	Sorel	Buddy	D	11/1996
			Pickles	D	11/1996
	11417	Coolie	Melvin	R	06/2000
	11418	Rogers	Sally	D	11/1999
Elm Street	11341	Cleaver	June	R	06/2000
			Ward	R	06/2000

Press the PGDN key to continue

F1 = Help   F2 = Return to the Main Menu   F3 = Change Party

F4 = Change Name   F5 = Add a Voter   F6 = Change Last Voted

FIGURE 15.12

An improved screen report that eliminates redundant information.

A Screen Design Example. Just as form design software tools are used to plan for printed output, the same tools can be used to plan screens and communicate that detail to programmers. The same conventions of notation are followed for screen layout as for printed output layout. Differences in screen design include the necessity of putting instructions on screen for changing screens, moving between screens, and terminating the display of output.

Figure 15.13 is an example of screen layout that is being prepared for outputting tabular reports for use by the shipping department of a stuffed animal manufacturer called New Zoo. The shipping department is currently using a manual system to keep track of shipments, but with the growth in outlets it has experienced in the past year, this approach is no longer practical. The shipping department would like a running summary of shipping activity so that any problems can be handled quickly, especially during the heavy shipping season in September.

Notice that the screen is generally divided into three parts. The top gives the title or heading of the screen report, "NEW ZOO ORDER STATUS." Just as the columns were all given headings on the paper printout, so too are they given headings on the screen. Note also that the screen reads from top to bottom and from left to right.

Continuing our analysis of this example, the columns are written in exactly as they are to appear on the screen. RETAILER, ORDER #, ORDER DATE, and ORDER STATUS are all columns that will appear on the screen. In the RETAILER column, notice that the output is designated as alphabetic characters, whereas ORDER # is shown as numeric, with the use of 9s. ORDER STATUS is also noted as alphabetic characters.

Notice that the two bottom lines of the screen are used for instructions to the user. When the user positions the cursor over any ORDER # and hits ENTER, he or she will automatically bring up another screen that gives more detailed information about that particular retailer's shipment.

When the screen is in a preliminary design phase, before it is finalized, it is wise to show a prototype screen to users and get their feedback about changes or

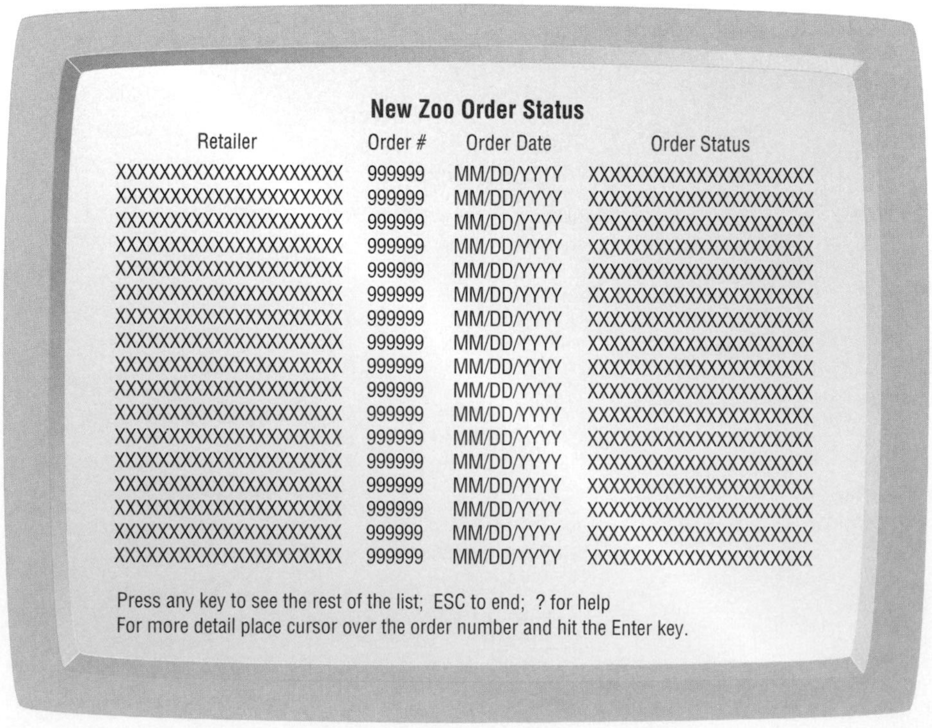

FIGURE 15.13
An electronic form used for planning a screen layout.

New Zoo Order Status

Retailer	Order #	Order Date	Order Status
XXXXXXXXXXXXXXXXXXXXX	999999	MM/DD/YYYY	XXXXXXXXXXXXXXXXXXXXX
XXXXXXXXXXXXXXXXXXXXX	999999	MM/DD/YYYY	XXXXXXXXXXXXXXXXXXXXX
XXXXXXXXXXXXXXXXXXXXX	999999	MM/DD/YYYY	XXXXXXXXXXXXXXXXXXXXX
XXXXXXXXXXXXXXXXXXXXX	999999	MM/DD/YYYY	XXXXXXXXXXXXXXXXXXXXX
XXXXXXXXXXXXXXXXXXXXX	999999	MM/DD/YYYY	XXXXXXXXXXXXXXXXXXXXX
XXXXXXXXXXXXXXXXXXXXX	999999	MM/DD/YYYY	XXXXXXXXXXXXXXXXXXXXX
XXXXXXXXXXXXXXXXXXXXX	999999	MM/DD/YYYY	XXXXXXXXXXXXXXXXXXXXX
XXXXXXXXXXXXXXXXXXXXX	999999	MM/DD/YYYY	XXXXXXXXXXXXXXXXXXXXX
XXXXXXXXXXXXXXXXXXXXX	999999	MM/DD/YYYY	XXXXXXXXXXXXXXXXXXXXX
XXXXXXXXXXXXXXXXXXXXX	999999	MM/DD/YYYY	XXXXXXXXXXXXXXXXXXXXX
XXXXXXXXXXXXXXXXXXXXX	999999	MM/DD/YYYY	XXXXXXXXXXXXXXXXXXXXX
XXXXXXXXXXXXXXXXXXXXX	999999	MM/DD/YYYY	XXXXXXXXXXXXXXXXXXXXX
XXXXXXXXXXXXXXXXXXXXX	999999	MM/DD/YYYY	XXXXXXXXXXXXXXXXXXXXX
XXXXXXXXXXXXXXXXXXXXX	999999	MM/DD/YYYY	XXXXXXXXXXXXXXXXXXXXX
XXXXXXXXXXXXXXXXXXXXX	999999	MM/DD/YYYY	XXXXXXXXXXXXXXXXXXXXX
XXXXXXXXXXXXXXXXXXXXX	999999	MM/DD/YYYY	XXXXXXXXXXXXXXXXXXXXX
XXXXXXXXXXXXXXXXXXXXX	999999	MM/DD/YYYY	XXXXXXXXXXXXXXXXXXXXX

Press any key to see the rest of the list; ESC to end; ? for help
For more detail place cursor over the order number and hit the Enter key.

improvements that they would like to see. This interactive process continues until users feel satisfied that the output provides what they need to see in a usable format. Just as with printed output, good screen layouts cannot be created in isolation. Systems analysts need the feedback of users to design worthwhile screens. Once approved by users, the screen layout can be finalized.

The screen produced from the screen design display is pictured in Figure 15.14. Notice that the screen is uncluttered, yet it gives a basic summary of the shipping status. The display orients users as to what they are looking at with the use of a heading. Instructions at the bottom of the screen provide users with several options, including continuing the present display, ending the display, getting help, or getting more detail.

Output screens within an application should display information consistently from screen to screen. Figure 15.15 shows the screen that results when the user positions the cursor over the ORDER # for a particular retailer. The new screen presents more details on Bear Bizarre. In the body of the screen, the user can see the retailer's order number, complete address, the order date, and the status. In addition, a detailed breakdown of the shipment and a detailed status of each part of the shipment are given. A contact name and phone number are supplied, along with the account balance, credit rating, and shipment history. Notice that the bottom portion of the screen advises the user of options, including more details, ending the display, or getting help.

Rather than crowding all retailer information onto one screen, the analyst has made it possible for the user to bring up a particular retailer if a problem or question arises. If, for example, the summary screen indicates that an order was only "partially shipped," the user can check further on the order by calling up a detailed retailer screen and then following up with appropriate action.

So far, we have been discussing general tabular output for information system screen displays. Although there is much overlap, some aspects of interaction with displays as well as how the decision maker uses them change as we turn to designing tabular output specifically for decision support systems.

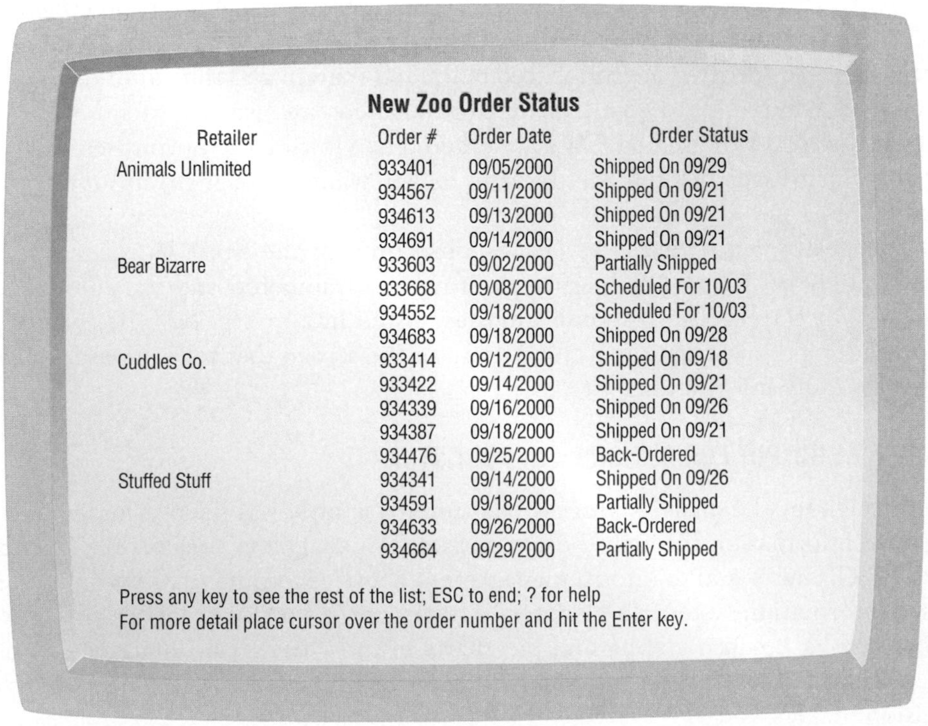

New Zoo Order Status

Retailer	Order #	Order Date	Order Status
Animals Unlimited	933401	09/05/2000	Shipped On 09/29
	934567	09/11/2000	Shipped On 09/21
	934613	09/13/2000	Shipped On 09/21
	934691	09/14/2000	Shipped On 09/21
Bear Bizarre	933603	09/02/2000	Partially Shipped
	933668	09/08/2000	Scheduled For 10/03
	934552	09/18/2000	Scheduled For 10/03
	934683	09/18/2000	Shipped On 09/28
Cuddles Co.	933414	09/12/2000	Shipped On 09/18
	933422	09/14/2000	Shipped On 09/21
	934339	09/16/2000	Shipped On 09/26
	934387	09/18/2000	Shipped On 09/21
	934476	09/25/2000	Back-Ordered
Stuffed Stuff	934341	09/14/2000	Shipped On 09/26
	934591	09/18/2000	Partially Shipped
	934633	09/26/2000	Back-Ordered
	934664	09/29/2000	Partially Shipped

Press any key to see the rest of the list; ESC to end; ? for help
For more detail place cursor over the order number and hit the Enter key.

FIGURE 15.14
The resulting screen shows how the information actually appears.

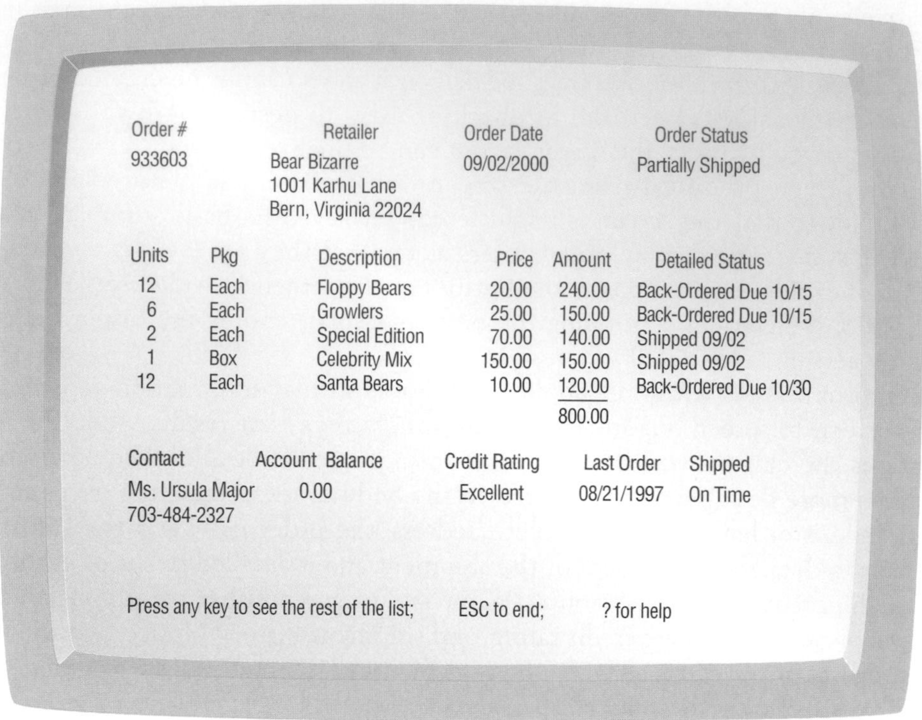

Order #
Retailer
Order Date
Order Status
933603
Bear Bizarre
1001 Karhu Lane
Bern, Virginia 22024
09/02/2000
Partially Shipped
Units
Pkg
Description
Price
Amount
Detailed Status
12 Each Floppy Bears 20.00 240.00 Back-Ordered Due 10/15
6 Each Growlers 25.00 150.00 Back-Ordered Due 10/15
2 Each Special Edition 70.00 140.00 Shipped 09/02
1 Box Celebrity Mix 150.00 150.00 Shipped 09/02
12 Each Santa Bears 10.00 120.00 Back-Ordered Due 10/30
800.00
Contact
Account Balance
Credit Rating
Last Order
Shipped
Ms. Ursula Major
703-484-2327
0.00
Excellent
08/21/1997
On Time
Press any key to see the rest of the list;
ESC to end;
? for help

FIGURE 15.15

If users want more details regarding the shipping status, they can call up a separate screen.

TABULAR OUTPUT FOR DECISION SUPPORT SYSTEMS

In Chapter 12, information systems that were designed to support decision makers in semistructured decision systems, called decision support systems (DSS), were introduced. A DSS helps the decision maker to make a decision through interaction with the system. It often uses screen output with a mouse or keyboard interface to accomplish that interaction. Output designed for decision support systems is thus characterized by modifiability and flexibility, which are not as important in output from a traditional MIS.

Figure 15.16 shows a screen for the Nebraska State Patrol Workforce Planning DSS (based on an earlier work with Merrill Warkentin). The screen is from a Lotus spreadsheet, so the rows are numbered in the left margin and the columns are indicated by the letters Q to V in the fourth row. There are also two menu command rows at the top of the screen. Data are entered by viewing the output screen, placing the cursor over the cell the decision maker wants to change, and typing in a new value.

The performance measures for response time of the State Patrol troops are shown in Figure 15.17. Once again, the rows are numbered and the columns are indicated by letters. The response times are listed in a matrix by type of call and troop. Notice the instruction at the bottom of the screen that tells the user to page down for more information.

GRAPHICAL OUTPUT FOR DECISION SUPPORT SYSTEMS

All throughout the analysis process, the systems analyst has been using graphs: to organize information through the use of data flow diagrams, to schedule activities with PERT charts, and to inform management about recommendations on the proposed information system. In Chapter 14, we covered the use of line graphs as well as bar graphs, column graphs, and pie charts in the written and oral systems proposal. Refer to that chapter for details on the effects of different kinds of graphs on decision makers.

040: CMD MENU

Help Update Results Print Graph Quit
Incidents Variables Parameters Decision Quit

	O	P	Q	R	S	T	U	V
21	Number of Calls (by Type) recorded by the Incident Reporting System							
22	May 2000							
23								
24		============		TROOP	===============			State
25	Type of Call	HQ	A	B	C	D	E	Totals
26	===============	------	------	------	------	------	------	------
27	Accidents	79	67	73	52	53	45	369
28	Criminal complaints	47	47	49	41	37	43	264
29	Motorist assists	36	13	6	32	34	22	143
30	Traffic violations	29	18	22	20	12	12	113
31	Relays	7	8	11	9	18	4	57
32	REDDI calls	44	16	12	49	34	22	177
33	Other hway emergencies	25	18	7	19	22	51	142
34	Other requests	47	4	19	38	42	4	154
35	HISTORICAL	------	------	------	------	------	------	------
36	Total number of calls	314	191	199	260	252	203	1419
37	# Troopers & Sergeants	51	33	44	38	30	30	226
38	Average Response Time	27.0	20.6	26.4	20.4	45.2	30.0	28.5
39								
40	PgDn for additional Incident Reporting data							

FIGURE 15.16

A tabular screen report for the Nebraska State Patrol Workforce Planning Decision Support System.

We now turn briefly to designing graphical output for decision makers using decision support systems. Extensive growth in the use of graphical output for decision support systems is expected in the next few years. There are already numerous graphics packages on the software market. Graphical output can aid decision makers by showing them a trend or concept quickly, with analysis to follow. Note, however, that graphical output is not considered to be a replacement for tabular output, but serves to supplement it.

020: READY

	O	P	Q	R	S	T	U	V
1	Response Times (by troop area) from the Incident Reporting System							
2	May 2000							
3								
4		============		TROOP	===============			
5	Average Response to	HQ	A	B	C	D	E	State
6	===============	------	------	------	------	------	------	------
7	Accidents	32.6	13.0	36.5	16.3	38.8	18.0	26.6
8	Criminal complaints	19.6	14.2	17.3	12.9	32.7	40.7	22.4
9	Motorist assists	19.5	12.8	10.8	17.1	19.8	35.3	20.5
10	Traffic violations	24.7	18.9	11.3	30.7	47.7	37.1	26.0
11	Relays	52.7	35.3	66.4	19.3	62.8	42.0	50.1
12	REDDI calls	13.3	25.4	18.5	24.4	72.6	25.0	30.7
13	Other hway emergencies	20.2	61.8	18.1	33.1	45.9	31.0	35.0
14	Other requests	44.7	22.7	18.4	19.9	54.1	0.0	36.2
15	---------------							------
16	Average Response Time	27.0	20.6	26.4	20.4	45.2	30.0	28.5
17								
18								
19								
20	PgDn for additional Incident Reporting data							

FIGURE 15.17

A DSS screen showing the weighted average response time of the Nebraska State Patrol troops.

FIGURE 15.18

A bar chart display for on-screen inspection of troop time response.

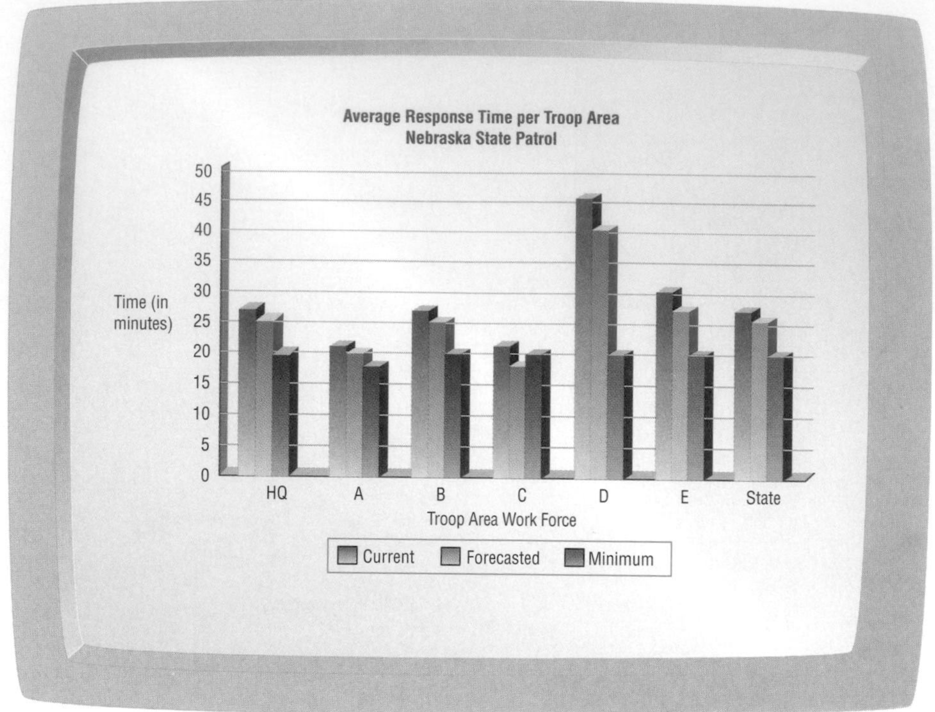

As with the presentation of tabular output, graphical output needs to be accurate and easy to understand and use if it is to be effective in communicating information to users. As discussed earlier in this chapter, decision makers using the graphs need to know the assumptions (biases) under which the graphs are being constructed so that they can adjust to or compensate for them.

In designing graphical output, the systems analyst must determine (1) the purpose of the graph, (2) the kind of data that need to be displayed, (3) its audience, and (4) the effects on the audience of different kinds of graphical output. In the instance of a decision support system, the purposes of graphical displays are to support any of the three phases of problem solving: intelligence, design, or choice. The data to display will vary, and the user is often a lone decision maker.

The Nebraska State Patrol Workforce Planning DSS also allows the decision maker to graph response times, as in Figure 15.18. Here current response times, forecasted response times, and minimum requirements are graphed as differently shaded bars.

Graphical output can be enormously useful to decision makers if they are trained in how to interpret it and when to use it. Without the training, it is doubtful that graphs mean much. That is in contrast to the assumed ability of the decision maker to use tabular output, which is the most common kind of report.

In a good decision support system, the output should be able to call up a variety of user views as well as a library of possible commands. In addition, the user needs to be able to call up a workspace and move data from one model or database to another workspace and back to others. All these functions can be accomplished by using Windows, as is discussed in Chapter 18.

DESIGNING A WEB SITE

You can use some of the design principles from designing screens when you design a Web site. Remember, though, that the key word here is *site*. The first documents displayed on the Internet using the http protocol were called home pages, but it became apparent very quickly that companies, universities, governments, and

Web Terms	Meaning
bookmark	A stored address of a Web page. (In Microsoft Internet Explorer, bookmarks are called "favorites.") You can jump to a page by clicking on its bookmark.
browser	Software that allows you to read Web pages and copy, save, and print these pages. It also allows you to navigate the Web by following links, going backward and forward, and jumping to favorite Web pages you have bookmarked. Popular browsers are Netscape Communicator and Microsoft Internet Explorer.
FAQ	Stands for "Frequently Asked Questions." Web sites often have a page devoted to these so that the company sales force or tech support is not inundated with the same questions over and over again and users can have 24-hour access to answers.
FTP	"File Transfer Protocol" is currently the most common way to move files between computer systems.
GIF	Stands for "Graphic Interchange Format." A popular compressed image format best suited for artwork.
Java	An object-oriented language that allows dynamic applications to be run on the Internet. Nonprogrammers can use software packages such as Symantec's *Visual Café for Java.*
JPEG	"Joint Photographic Expert Group" developed and gave the acronym of its name to title this popular compressed image format best suited for photographs. The quality or the image can be adjusted by the designer.
HTML	"Hypertext Markup Language" is the language behind the appearance of documents on the Web. It is actually a set of conventions that mark the portions of a document, telling a browser what distinctive format should appear on each portion of a page.
http://	"Hypertext Transfer Protocol" is used to move Web pages between computers, such as from a Web site on a computer in another country to your personal computer.
hyperlink	In a hypertext system, words, phrases, or images that are underlined or emphasized in some way (often with a different color). When the user clicks on one of them with the mouse, another document is displayed. HTML has features that allow authors to insert hyperlinks in their documents, and hyperlinks can point to a local page or another URL. Often links will change color to indicate that the user has already clicked on them before.
plug-ins	Additional software (often developed by a third party) that can be used with another program; for example, RealNetworks' Real Player or Macromedia Flash are used as plug-ins in Web browsers to play CD-quality streaming audio or video and view vector-based animation while you are visiting the Web site.
URL	"Uniform Resource Locator" is the address of a document or program on the Internet. Familiar extensions are .com for commercial, .edu for educational institution, .gov for government, and .org for organization.
VRML	"Virtual Reality Markup Language" is a language similar to HTML that allows users to browse in three dimensions.
Webmaster	The person responsible for maintaining the Web site.
www	Stands for "World Wide Web." A global hypertext system that uses the Internet. Now we refer to it as just "the Web."

FIGURE 15.19
Web vocabulary terms.

people were not going to be displaying just one page. The term *Web site* replaced the words *home page*, indicating that the array of pages would have to be organized, coordinated, designed, developed, and maintained in an orderly process.

Printing is a highly controlled medium, and the analyst has a very good idea of what the output will look like. GUI and character based screens (CHUI, character-based user interface) screens are also highly controlled. The Web is a very uncontrolled environment. In the case of an intranet, a Web site used within a corporation, the analyst can control some of the ways that a Web page is browsed. When creating Internet Web sites, however, the sites display differently depending on whether the computer used is a Macintosh or an IBM compatible. Different browsers display the images differently, and screen resolution has a large impact on the look and feel of a Web site. The standard low resolution is 640 × 480 pixels, but many computers are set at 800 × 600 or 1024 × 768 pixels. Web TV is 540 pixels, and the issue is further complicated by the use of handheld devices that are used to browse the Web. The complexity deepens when you realize that each person may set a browser to use different fonts and may disable the use of Javascript, cookies, and other Web programming elements. Clearly, the analyst has many decisions to make when designing a Web site.

In addition to the general design elements discussed earlier in this chapter, there are specific guidelines appropriate for the design of professional quality Web sites. Web terms are defined in Figure 15.19. The following subsections address these guidelines.

GENERAL GUIDELINES FOR DESIGNING WEB SITES

There are many tools as well as examples that can guide you in designing Web sites.

Use Professional Tools. Use software called a Web editor such as Macromedia Dreamweaver, Microsoft's FrontPage, or Adobe Page Mill. These tools are definitely worth the price. You will be more creative and you'll get the Web site finished much faster than by using other products. There is no need to work directly with HTML (hypertext markup language) anymore.

Study other Web Sites. Look at Web sites you think are engaging. Analyze what design elements are being used and see how they are functioning, then try to emulate what you see by creating prototype pages. (If you are really ambitious, you can usually print out the HTML and try to do similar things on your page.)

Use the Resources That the Web Has to Offer. Look at Web sites that give hints on design. One such site is www.clever.net.

Examine the Web Sites of Professional Designers. Some design houses are listed in Figure 15.20, along with some of the often-visited and praised Web sites they developed. As you look at these pages ask yourself, "What works? What doesn't work? In what ways can users interact with the site? For example, does the site have hot links to email addresses, interactive forms to fill in, consumer surveys, games, quizzes, chat rooms and so on?"

Use the Tools You've Learned. Figure 15.21 provides a form that has been used successfully by Web designers to evaluate Web pages systematically. You might want to use copies of the form to help you compare and contrast the many Web sites you will visit as you go about learning Web page design.

FIGURE 15.20
Selected Web site designer houses.

Design House	Its Web Address	Sites It Designed
Archetype	www.archetypedesignco.uk	www.ctdu.org.uk
Organic	www.organic.com	www.avis.com www.macys.com www.unibanco.com
Modem media	modemmedia.com	www.att.com/traveler/ www.kraftfoods.com/ www.artmuseum.net

FIGURE 15.21
A Web site evaluation form.

Web Site Critique

Date Visited: __ / __ / __   Analyst's Name ____________
Time Visited: ________

URL Visited ____________

DESIGN	**Needs Improvement**				**Excellent**
Overall Appearance	1	2	3	4	5
Use of Graphics	1	2	3	4	5
Use of Color	1	2	3	4	5
Use of Sound/Video (multimedia)	1	2	3	4	5
Use of New Technology and Products	1	2	3	4	5
CONTENT & INTERACTIVITY					
Content	1	2	3	4	5
Navigability	1	2	3	4	5
Site Management and Communications	1	2	3	4	5

SCORE   /40

COMMENTS:

Consult the Books. Something that can add to your expertise in this new field is to read a book or books about Web design. Note that the following books do not cover much on HTML anymore. Some books on Web site design are the following:

Flanders, V. and Willis, M., *Web Pages That Suck: Learn Good Design by Looking at Bad Design*, Alameda, CA, SYBEX, 1998.

Horton, W. K., et al. *The Web-Page Design Cookbook: All the Ingredients You Need to Create 5-Star Web Pages*. New York: John Wiley, 1996.

Pring, R. *www.type: Effective Typographic Design for the World Wide Web*. New York: Watson-Guptill, 1999.

Siegel, D. *Creating Killer Web Sites*. New York: Hayden, 1997.

Weinman, L. *Designing Web Graphics 2: How to Prepare Images and Media for the Web*, 2d ed.: Indianapolis, IN, NRP, 1997.

Look at Some Poor Examples of Web Pages, Too. Critique poor Web pages and remember to avoid those mistakes. Examine the Web site found at: www.webpagesthatsuck.com

Despite its counter-culture name, this is a wonderful site that provides links to many poorly designed sites, and points out the errors that designers have made on them. However, the site also provides links to material that take the reader through creating a Web site, improving site navigation, learning JavaScript and much more. The authors are humorous, and vigilant at identifying Web sites both good and bad, and they provide a wealth of useful information.

Create Templates of Your Own. If you adopt a standard-looking page for most of the pages you create, you'll get the Web site up and running quickly and it will consistently look good. Web sites may be made using cascading style sheets that allow the designer to specify the color, font size, font type and many other attributes only

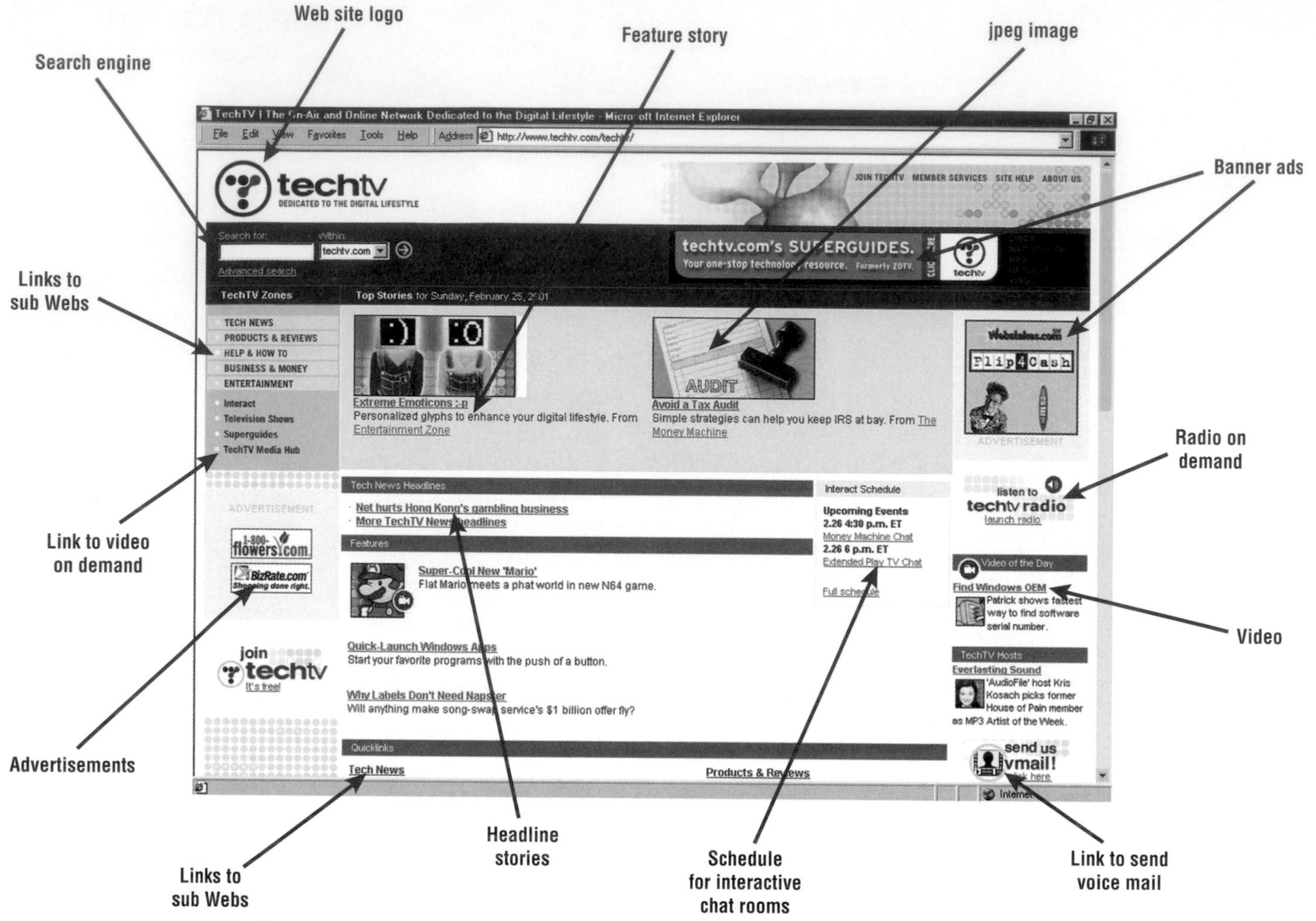

FIGURE 15.22
The techtv Web site makes the most of hyperlinks, video clips, and banner ads.

once. These attributes are stored in a style sheet file and are then applied to many Web pages. If a designer changes a specification in the style sheet file, all the Web pages using that style sheet will be updated to reflect the new style.

Use Plug-Ins, Audio, and Video Very Sparingly. It is wonderful to have features that the professional pages have, but remember that everyone looking at your site doesn't have every new plug-in. Don't discourage visitors to your page.

Plan Ahead. Good Web sites are well thought out. Pay attention to the following:

1. Structure.
2. Content.
3. Text.
4. Graphics.
5. Presentation.
6. Navigation.
7. Promotion.

Each of these items is described in more detail below.

Structure Planning a Web site is one of the most important steps in developing a professional Web site. Web pages should be designed according to content. Each page within the overall Web structure should have a distinct message or other related information. The home page should link to all other pages. Sometimes it is useful to examine professional pages to analyze them for content and features. Figure 15.22 is a screen capture from the techtv Web site. The purpose for the site and the Web medium work exceptionally well together. In this outstanding site, notice that there is great attention to detail. There are words, graphics, jpeg images, and icons. In addition, there are many kinds of links: to radio, video, voice mail, subwebs, chat rooms, a search engine, and many other features. JavaScript is used to play headline stories, and run video clips.

The planning of a Web site takes a considerable amount of thought. Webmasters must consider loading times and limit the amount of material squeezed on to each page. Providing a main page that links to more detailed pages is the best strategy. Links include pointers to other pages in the Web site; pointers to images, composed of graphics interchange format (GIF) or Joint Photographic Experts Group (JPEG) files, on the local Web site; pointers to return to the Web site home page, and links to other Web sites. To keep track of the links, the Webmaster draws a site map, using tree-like diagrams to show the variety of links that sprout forth from each page.

A Webmaster can benefit from using one of the many Web site diagramming and mapping tools available. Microsoft Visio has a web charting option built into the software. Although helpful for development, these tools become even more important when maintaining a Web site. Given the dynamic nature of the Web, sites that may be linked to your site may move at any time, requiring you or your Webmaster to update the links.

In Figure 15.23, a map of a section of the authors' Web site is shown in the Visio window. In this example, we explored the Web site down to all the existing levels. This example shows links to HTML pages, documents, images (GIF or JPEG files), and mail-tos (a way to send email to a designated person). The links can be either internal or external. If a link is broken, a red "X" appears and the analyst can investigate further. This Visio file can be printed out in sections and posted on the wall to get an overall picture of the Web site.

Find a book that discusses Web site structure. One such book is *Large-Scale Web Sites* by Darrell Sano, which was published in 1996 by Netscape Communications Corporation.

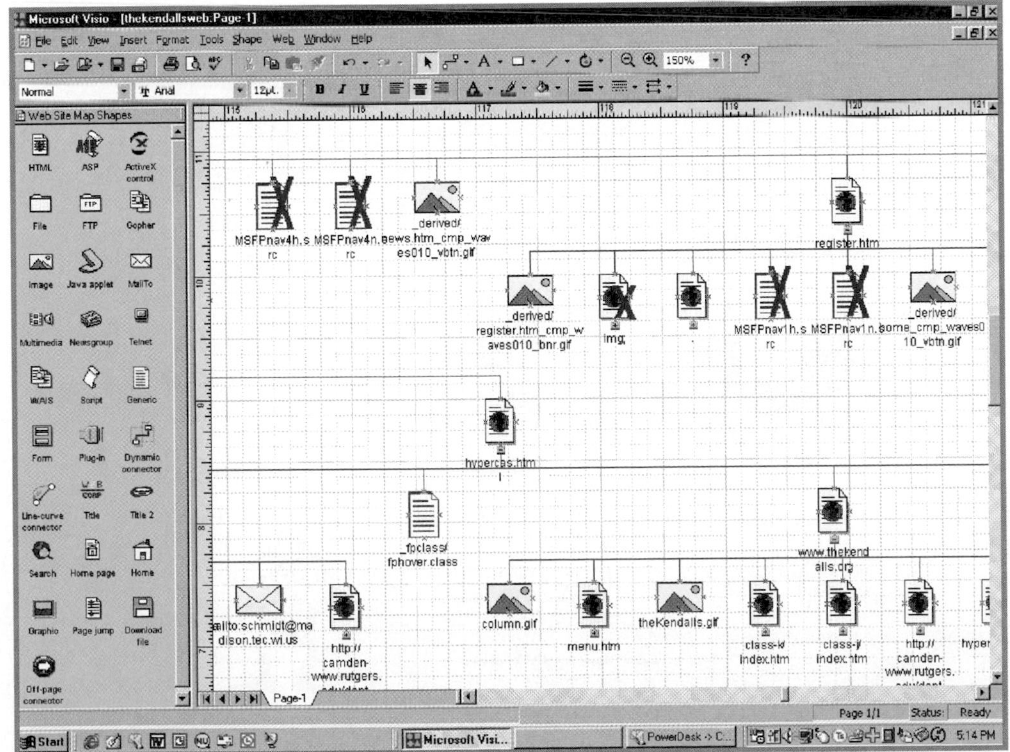

FIGURE 15.23

A Web site can be evaluated for broken links by using a package such as Microsoft Visio.

Content Content is critical. Without anything to say, your Web site will fail. A 12-year-old friend of ours, wise beyond his age, confided, "I could make my own Web site, but what's the point? I have nothing to say!" Exciting animation, movies, and sounds are fun, but you have to include appropriate content to keep the reader interested.

Web pages should be designed according to content. Each page within the overall Web structure should have a distinct message or other related information. The home page should link to all other pages.

Provide something important to Web site viewers. Supply some timely advice, important information, a free offer, or any activity that you can provide that is interactive and moves users away from a browsing mode and into an interactive one.

Use a metaphor or images that provide a metaphor for your site. You can use a theme, such as a storefront, with additional pages having various metaphors related to the storefront, such as a deli. Avoid the overuse of cartoons, and don't be repetitive. An example of the use of metaphor can be found in the Web site www.javaranch.com, which is used as a resource for those learning and using the Java programming language. Refer to Figure 15.24. Notice the use of ranch terms throughout. The Big Moose Saloon is a discussion area, the Cattle Drive gives actual experience writing Java code, and so on.

Remember to adjust the content for the intended audience. If you're selling products or services, you want to write in a style that is distinctly different from the tone you take when writing technical support manuals. Web sites intended for a global audience should contain links that will display the site in the native language. Statistics show that people are three times more likely to do business in their native tongue than in a foreign language.

Every Web site should include a FAQ page. By having answers readily available, 24 hours a day, you will save valuable employee time and also save user time. FAQ pages also demonstrate to users of your site that you are in concert with them and have a good idea of what they would like to know.

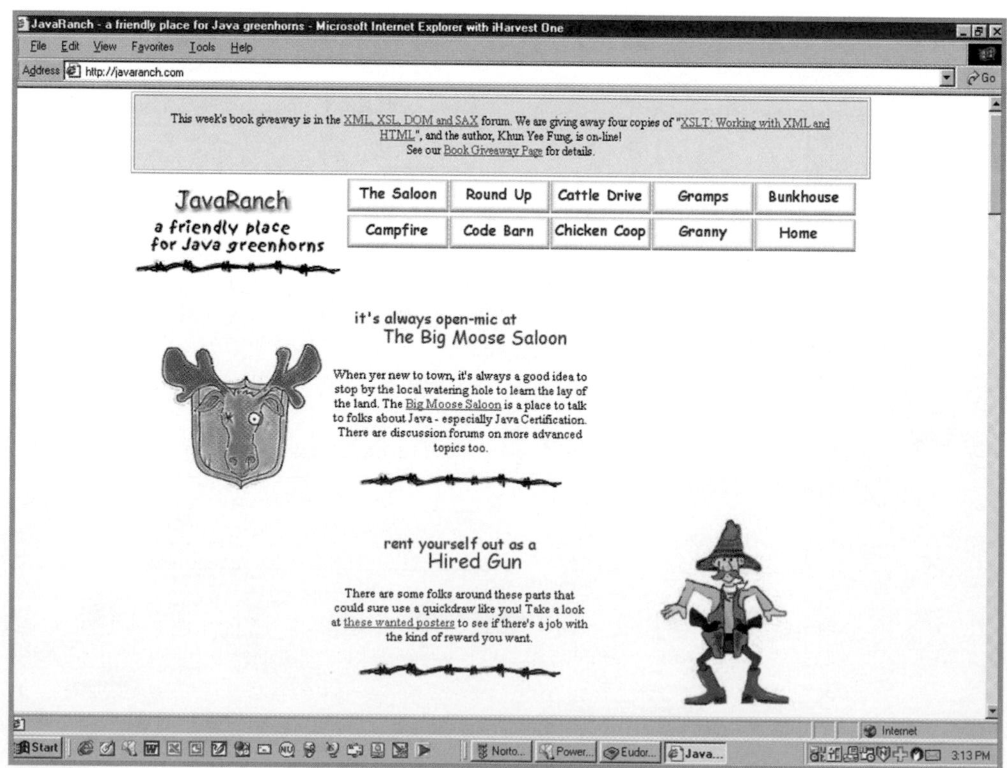

FIGURE 15.24

A good Web site will use a main metaphor as an organizing principle. This example is from www.javaranch.com.

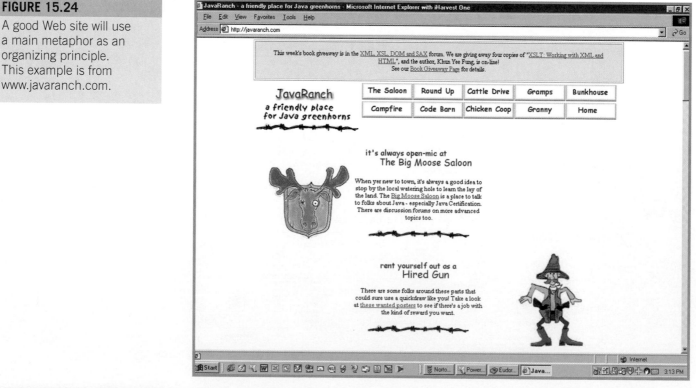

Text Each Web page should have a title. Place meaningful words in the first sentence appearing on your Web page. Let people know that they have indeed navigated to the right Web site. Write text in a way that people can find Web pages when using word searches. Clear writing is especially important for the headings, subheadings, and the first paragraph. It helps to repeat the same key words a few times in your many pages.

Graphics The following list provides details about creating effective graphics for Web sites.

1. Use one of the two most commonly used image formats, JPEG or GIF. JPEGs are best for photographs, and GIFs are best for artwork images. GIFs are limited to 256 colors but may include a transparent background, pixels that allow the background to show through the GIF image. GIF images may also be interlaced, meaning that the Web browser will show the image in successive stages, presenting a clearer image with each stage.
2. Create a few professional-looking graphics for use on your pages. They create a much better impression than many poor-quality graphics.
3. Keep the background simple and make sure users can read (and print out) the text clearly. When using a background pattern, make sure that you can see the text clearly on top of it. Many viewers will be using the page at 256 colors, and as a result, it may be hard to read the text.
4. Examine your Web site on a variety of monitors and low-end graphics boards. Scenes and text that look great on a high-end video display may not look good to others with poorer-quality equipment. Also the amount of material that displays on a 13-inch monitor and a 21-inch monitor differs, depending on the resolution.
5. Save JPEG images, a type of compressed file format for graphics, at the highest quality possible but yet keeping to a reasonable file size. Because JPEG is a compression, the files can be saved as low-, medium-, or high-quality images.
6. Use horizontal rules (lines, images, or icons) to break apart sections of your Web pages in a meaningful way. In addition, use section headings in a larger font, a different color, and sometimes a different font to introduce sections.
7. Use colorful bullets for lists and consider using hot buttons for links to other pages or sites.
8. Keep graphics images small and reuse bullet or navigational buttons such as BACK, TOP, EMAIL, and NEXT. These images are stored in a cache, an area on the browsing computer's hard drive. Once an image has been received, it will be taken from the cache whenever it is used again. Using cached images improves the speed with which a browser can load a Web page.
9. Although JPEG images are rectangular, the image itself doesn't have to be. The trick is to make the image a transparent GIF or even place it on a more interesting patterned background.
10. Use a package like Microsoft Image Composer (included with FrontPage) to tilt your image or to add a shadow below the image.
11. Follow the three clicks rule: a user should be able to move to the desired page in three clicks of the mouse.

Presentation The following list gives added details about how to design engaging entry screens for Web sites.

1. Provide an entry screen (also called a home page) that introduces the visitor to the Web site. The entry screen must be designed to load quickly. A useful rule of thumb is to design a page that will load in 14 seconds while assuming that a user has a 28,000- or 56,000-baud modem. (Although you may be designing the page on a workstation at the university, a visitor to your Web site may be

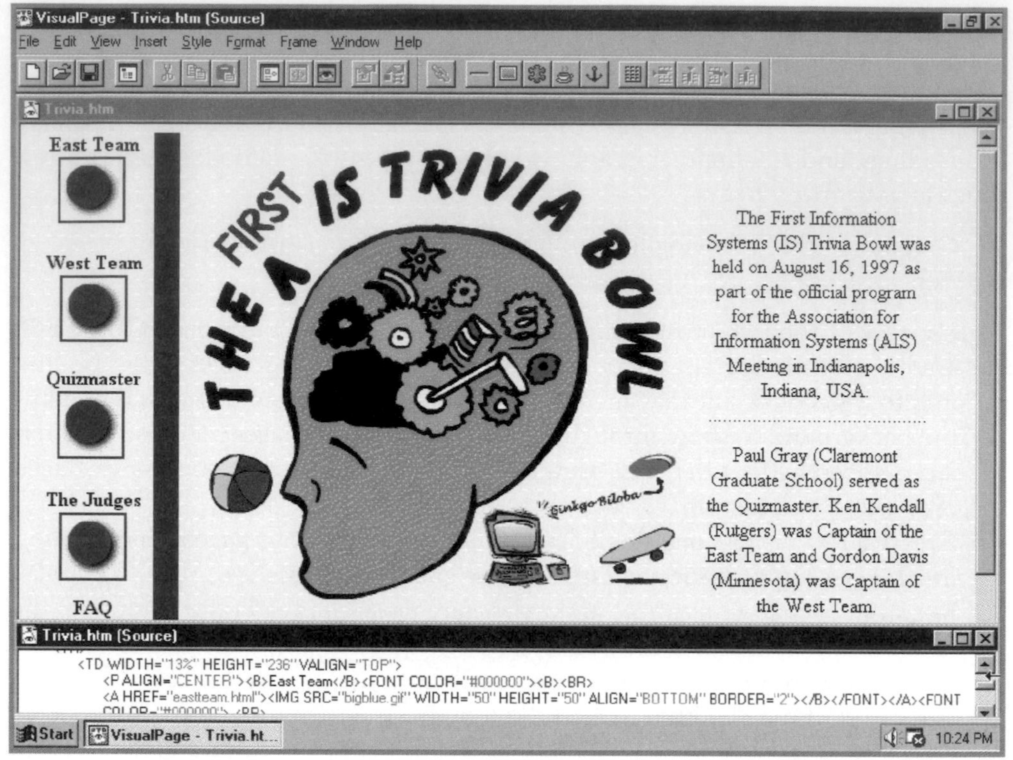

FIGURE 15.25
Using a visual HTML editor (in this example Visual Page), a Web site designer can see what a page looks like in a browser and the HTML Code (see bottom of screen) at the same time.

accessing it from home.) This entry page should be 100 kilobytes or less, including all graphics. The entry screen should contain a number of choices, much like a menu. An easy way to accomplish that is to design a set of buttons and position them on the left side or the top of the screen. These buttons can be linked to other pages on the same Web site or even linked to different Web sites. A text menu should be included in a smaller font at the bottom of the page. An example of this is shown in Figure 15.25, an entry screen that contains a large image and some content, but then directs the visitor to journey elsewhere in the site. This page was constructed using Symantec's VisualPage, which allows designers to see HTML code (at the bottom of the screen) at the same time they see what the page would look like in a browser.

2. Keep the number of graphics to a reasonable minimum. It takes additional download time to transfer a graphics intensive site.
3. Use large and colorful fonts for headings.
4. Use interesting images and buttons for links. A row of images combined into a single image is called an image map, and Web designers like to refer to hot spots that act as links to other pages. A single larger image will load faster than several smaller ones. Tool-tip help (a small amount of text) should display when the cursor is moved over these hot spots. Tables are useful and are commonly used to create Web pages and Web forms. It is important to ensure that text does not bump up against a table or against the browser's left border. As a result, most tables include cell padding to keep the text slightly indented.

Frames are very effective and widely used, but frames make it a bit more difficult for visitors to bookmark your site and print the information they want. Although the pertinent information inside of the frame prints out, other information does not. Frames can disrupt users' work and cause extra steps to be taken; hence, many programmers and users find them to be very controversial. The advantage of using frames is that a menu and title can be continuously displayed while the user scrolls through a page or moves to other pages.

5. Use the same graphics image on several Web pages. Consistency will be improved, and the pages will load more quickly because the computer stores the image in a cache and doesn't have to load again.
6. Avoid overusing animation, sound, and other elements that might make the site too jumpy or cluttered. Do not put everything possible on a page. The space should be functional and interesting. Avoid busy animation or blinking text that is not serving a specific function.
7. Provide an area on the left side of the Web page for buttons (or icons) that link to other pages in the Web site. Alternatively, you can place a menu bar at the top of the page and a text menu at the bottom of the page. A visitor will expect to see navigational links in one of these places. Wherever you place them, make sure that they are located consistently from one page to another. Some menus change the color or the current page menu item to indicate to the user which page is currently displaying or, in other words, which page they are on.
8. It is important to design your Web page so that it does not scroll horizontally. Such scrolling makes it difficult to read text and is slow. Browsers are designed to scroll vertically. In general, a Web page should not be more than two and a half screens long.
9. Make sure that your initial page on your Web site loads quickly. If you need to include large graphical images, put them in a gallery or on pages that are not the home page. Keep the initial page relatively short, keep the images small and simple, and keep photos to a minimum. You can always link to other local pages.
10. Add one or two Java applets if it makes sense to do so. Many Web sites offer free Java applets. Simply copy and use them.

Navigation Is it fun for you to follow links on the Web? The answer most likely is: it depends. When you discover a Web site that loads easily, has meaningful links, and allows you to easily return to the places you want to go back to, then chances are you think it is fun. Fun is not just play; it can be an important part of work, too. Recent research shows that fun can have a powerful effect on making computer training effective. If, on the other hand, you (1) can't decide which button or hot spot to push and (2) are afraid to choose the wrong one because you might get into the wrong page that takes a long time to load, navigation is more painful than fun. An example is visiting a software company's page to find out information about the features of the latest version of a product. You have choices such as products, download, FAQ, and tech support. Which button will lead to the answers you're looking for?

If a page is lengthy, provide links to information lower on the page and links back to the top. Links can act as a short table of contents. Do not cram too much information onto one long lengthy page. Neither should you create pages that have only a solitary image on them that says "Click here."

Promotion Promote your site. Don't assume that search engines will find you right away. Submit and resubmit your site every few months to various search engines. Include key words, called metatags, that search engines will use to link search requests to your site. You can also purchase software to make this process easier. Use email to promote your site, but unless you have a specific reason for doing so, others will consider it junk email or spam. Track site usage by using hit counters and techniques that provide statistics on the number of users by day, hour, and so on.

Encourage your readers to bookmark your Web site. If you link to and suggest that they go to affiliated Web sites that feature the "best movie review page in the world" or to the "get music for free" Web site, don't assume they'll be coming back to your site in the near future. You will encourage them to revisit if they bookmark your site (bookmarks are called "favorites" on Microsoft Explorer).

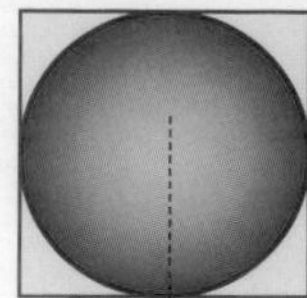

CONSULTING OPPORTUNITY 15.4

A FIELD DAY

"The thing of it is, I get impatient," says Seymour Fields, owner of a chain of 15 highly successful florist shops/indoor floral markets called "Fields" that are located in three Midwestern cities. "See this thing here?" He taps his PC and screen irritatedly. "We do all the payroll and all the accounting with these things, but I don't use it like I should. I actually feel a little guilty about it. See?" he says, as he makes a streak on the screen with his finger. "It's even got dust on it. I'm a practical person, though. If it's sitting here, taking up space, I want to use it. Or smell it, or at least enjoy looking at it, like flowers, right? Or weed it out, that's what I say. The one time I tried something with it, it was a real disaster. Well look, I can show you if I still remember how." Seymour proceeds to try to boot a program, but can't seem to get it working.

Clay Potts, a systems analyst, has been working on a systems project for the entire Fields chain. Part of the original proposal was to provide Seymour and his vice presidents with a group decision support system that would help them devise a strategy to determine which European markets to visit to purchase fresh flowers, which outlets to ship particular kinds of flowers to, and how much general merchandise such as planters, vases, note cards, and knickknacks to stock in each outlet.

Seymour continues, "I can tell you what we disliked about the program I worked with. There were too many darn layers, too much foliage, or whatever you call it, to go through. Even with a screen in front of me, it was like paging through a thick report. What do you call that?"

"Menus?" Potts suggests helpfully. "The main point is that you didn't like having to go through lots of information to get to the display you needed."

Seymour Fields looks happily at Potts and says, "You've got it. I want to see more fields on each screen."

How should Potts design screen output so that Fields and his group can get what they want on each screen while observing the guidelines for good screen design? Remember that the group members are busy and that they are infrequent computer users. Design a hyperlinked page that would work well in a DSS for the vice presidents. What should be included in the first display, and what should be stored in hyperlinks? List elements for each and explain why you have decided upon this strategy.

Make clear who is maintaining the Web site (the Webmaster) by including his or her name, title, and email address on the page. After building several Web sites ourselves and talking frequently to those who design pages and do their upkeep, we cannot stress enough the importance of almost perpetual *maintenance* of the Web site. Maintenance is a newly added responsibility in the corporate world, one that takes some time and patience. Your efforts toward maintenance will be rewarded, however. Remember that the Web is dynamic and that sites appear and disappear at alarming rates. The Webmaster is maintaining not only corporate information, but also—and maybe even more importantly—maintaining the hyperlinks to other documents and pages. Once your users have your site bookmarked, your maintenance efforts must be an essential part of the plan to keep them coming back.

SUMMARY

Output is any useful information or data delivered by the information system or decision support system to the user. Output can take virtually any form, including print, screen, audio, microforms, CD-ROM or DVD, electronic, and Web-based documents.

The systems analyst has six main objectives in designing output. They are to design output to serve the intended purpose, to fit the user, to deliver the right quantity of output, to deliver it to the right place, to provide output on time, and to choose the right output method.

It is important that the analyst realize that output content is related to output method. Output of different technologies affects users in different ways. Output technologies also differ in their speed, cost, portability, flexibility, and storage and retrieval possibilities. All these factors must be considered when deciding among print, onscreen, audio, electronic, or Web-based output, or a combination of these output methods.

The presentation of output can bias users in their interpretation of it. Analysts must be aware of the sources of bias, must interact with users to design and customize output, must inform users of the possibilities of bias in output, must create flexible and modifiable output, and must train users to use multiple output to help verify the accuracy of any particular report.

Printed reports are designed with the use of computer-aided software design tools that feature form design templates and drag and drop interfaces. The data dictionary serves as the source for necessary data on each report. Mock-ups or prototypes of reports and screens are shown to users before the report or design of the Web page is completed, and any necessary changes are then made. The systems analyst uses newly designed screens to communicate physical design to the programmer.

Display screens, which are an especially important form of output for decision support systems as well as the Web, are also designed using onscreen templates. Once again, aesthetics and usefulness are important when creating a well-designed screen. It is important to produce prototypes of screens and Web documents that allow users to make changes where desired.

Onscreen graphical output is becoming increasingly popular, especially for decision support systems. The systems analyst must consider the effect of graphs on users, the kind of data to be displayed, the purpose of the graphs, and the intended audience. It is essential that decision makers receive training in how to interpret graphs if graphs are to be useful to them.

KEYWORDS AND PHRASES

- audio output
- bookmark
- browser
- CD-ROM
- constant information
- display screen
- DVD
- electronic bulletin boards
- electronic output
- email
- external output
- FAQ
- hyperlink
- hypertext
- internal output
- Java
- output bias
- output design
- plug-ins
- URL (uniform resource locator)
- variable information
- Webmaster
- Web pages
- Web site
- World Wide Web (www)

REVIEW QUESTIONS

1. List six objectives the analyst pursues in designing system output.
2. Contrast external output with internal output produced by the system.
3. What are three situations that point to printers as the best choice for output technology?
4. Give two instances that indicate that onscreen output is the best solution for the choice of output technology.
5. List potential electronic output methods.
6. What are the drawbacks of electronic and Web-based output?
7. List 10 factors that must be considered when choosing output technology.
8. What output type is best if frequent updates are a necessity?
9. What kind of output is desirable if many readers will be reading, storing, and reviewing output over a period of years?

15

"I'd say the reception you received (or should I say your team received) for your proposal presentation was quite warm. How did you like meeting Mr. Hyatt? What? He didn't come? Oh [*laughing*], he's his own man. Anyway, don't worry about that too much. The reports I got from Snowden were encouraging. In fact, now he wants to see some preliminary designs from you all. Can you have something on his desk (or screen) in two weeks? He'll be in Singapore on business next week, but then when he recovers from the jet lag, he'll be looking for those designs. Thanks."

HYPERCASE® QUESTIONS

1. Consider the reports from the Training Unit. What are Snowden's complaints about these reports. Explain in a paragraph.
2. Using either a layout paper form or a CASE tool, design a prototype output screen based on the Training Unit's reports that will summarize the following information for Snowden:
   Number of accepted projects in the Training Unit.
   Number of projects currently being reevaluated.
   Training subjects areas for which a consultant is being requested.
3. Design an additional output screen that you think will support Snowden in the kind of decision making he does frequently.
4. Show your screens to three classmates. Get written feedback from them about how to improve the output screens you have designed.
5. Redesign the screens to capture the improvements suggested by your classmates. In a paragraph, explain how you have addressed each of their concerns.

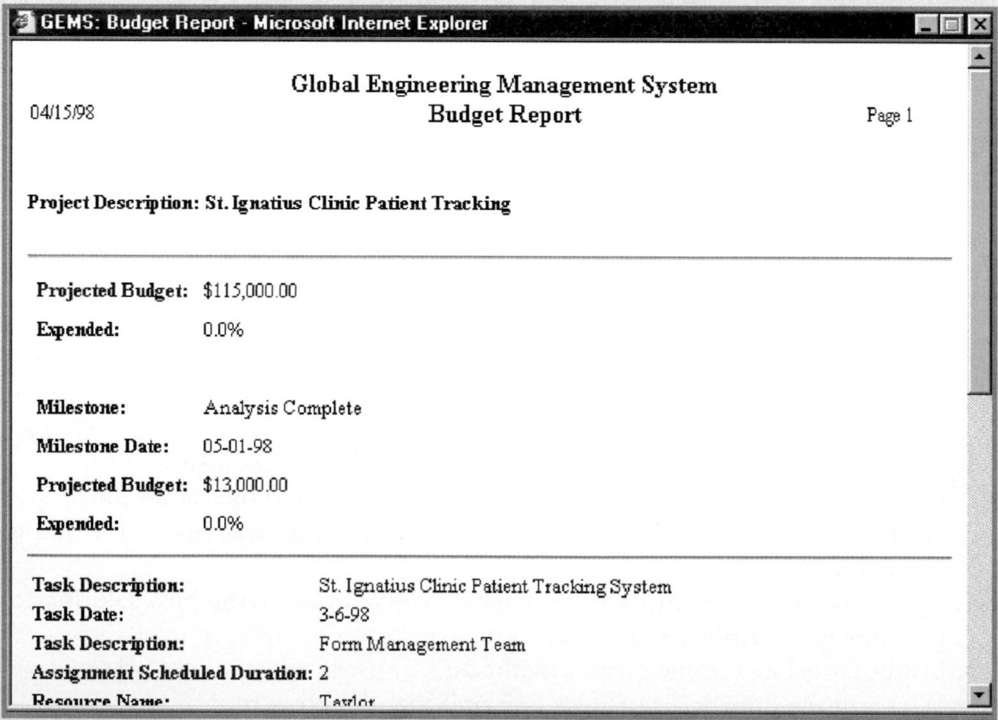

GEMS: Budget Report - Microsoft Internet Explorer

Global Engineering Management System
Budget Report

04/15/98 Page 1

Project Description: St. Ignatius Clinic Patient Tracking

Projected Budget: $115,000.00
Expended: 0.0%

Milestone: Analysis Complete
Milestone Date: 05-01-98
Projected Budget: $13,000.00
Expended: 0.0%

Task Description: St. Ignatius Clinic Patient Tracking System
Task Date: 3-6-98
Task Description: Form Management Team
Assignment Scheduled Duration: 2
Resource Name: Taylor

FIGURE 15.HC1
You have the ability to view and critique output screens in HyperCase.

10. What are two of the drawbacks to audio output?
11. List three main ways in which presentations of output are unintentionally biased.
12. What are five ways the analyst can avoid biasing output?
13. What is the difference between constant and variable information presented on a report?
14. Why is it important to show users a prototype output report or screen?
15. List six functional elements of printed reports.
16. List five stylistic or aesthetic elements of printed reports.
17. In what ways do screens, printed output, and Web-based documents differ?
18. List four guidelines to facilitate the design of good screen output.
19. What differentiates output for a DSS from that of a more traditional MIS?
20. What are the four primary considerations the analyst has when designing graphical output for decision support systems?
21. What is the difference between an Internet and an intranet?
22. List 10 guidelines for creating good Web sites.
23. List 11 guidelines for using graphics in designing Web sites.
24. What are 11 ideas for ways to improve the presentation of corporate Web sites that you design?
25. Give two hints for improving navigation of your Web sites for users.
26. In what ways can you encourage companies to promote their Web sites that you have developed?

PROBLEMS

1. "I'm sure they won't mind if we start sending them the report on these oversized computer sheets. All this time we've been condensing it, retyping it, and sending it to our biggest accounts, but we just can't now. We're so understaffed, we don't have the time," says Otto Breth. "I'll just write a comment on here telling them how to respond to this report, and then we can send it out."
   a. What potential problems do you see in casually changing external output? List them.
   b. Discuss in a paragraph how internal and external output can differ in appearance and function.
2. "I don't need to see it very often, but when I do, I have to be able to get at it quickly. I think we lost the last contract because the information I needed was buried in a stack of paper on someone's desk somewhere," says Luke Alover, an architect describing the company's problems to one of the analysts assigned to the new systems project. "What I need is instant information about how much a building of that square footage cost the last time we bid it; what the basic materials such as steel, glass, and concrete now cost from our three top suppliers; who our likely competition on this type of building might be; and who comprises the committee that will be making the final decision on who gets the bid. Right now, though, it's in a hundred reports somewhere. I have to look all over for it."
   a. Given the limited details you have here, write a paragraph to suggest an output method for Luke's use that will solve some of his current problems. In a second paragraph, explain your reasons for choosing the output method you did. (*Hint*: Be sure to relate output method to output content in your answer.)
   b. Luke's current thinking is that no paper record of the output discussed above need be kept. In a paragraph, discuss what factors should be weighed before screen output is used to the exclusion of printed reports.

c. Make a list of five to seven questions concerning the output's function in the organization that you would ask Luke and others before deciding to do away with any printed reports currently being used.

3. Here are several situations calling for decisions about output content, output methodology, distribution, and so on. For each situation, note the appropriate output decision.
   a. A large, well-regarded supplier of key raw materials to your company's production process requires a year-end summary report of totals purchased from it. Output method?
   b. Internal "brainstorming" memos are circulated through the staff regarding plans for a company picnic and fundraiser. Output method?
   c. A summary report of the financial situation is needed by a key decision maker who will use it when presenting a proposal to potential external backers. Output method?
   d. A listing of the current night's hotel room reservations is needed for front desk personnel. Output method?
   e. A listing of the current night's hotel room reservations is needed by the local police. Output method?
   f. A real-time count of people passing through the gates of Wallaby World (an Australian theme park) will be used by parking lot patrols. Output method?
   g. An inventory system must register an item each time it has been scanned by a wand. Output method?
   h. A summary report of merit pay increases allotted to each of 120 employees will be used by 22 supervisors during a joint supervisors' meeting and subsequently when explaining merit pay increases to the supervisors' own departmental employees. Output method?
   i. Competitive information is needed by three strategic planners in the organization, but it is industrially sensitive if widely distributed. Output method?

4. "I think I see now where that guy was coming from, but he had me going for a minute there," says Miss deLimit. She is discussing a prototype of screen output, one designed by the systems analyst, that she has just seen. "I mean, I never considered it a problem before if even as much as 20 percent of the total class size couldn't be fit into a class," she says. "We know our classes are in demand, and because we can't hire more faculty to cover the areas we need, the adjustment has to come in the student demand. He's got it highlighted as a problem if only 5 percent of the students who want a class can't get in, but that's okay. Now that I know what he means, I'll just ignore it when the computer beeps."
   a. In a sentence or two, describe the problem Miss deLimit is experiencing with the screen output.
   b. Is her solution to "ignore the beeps" a reasonable one given that output is in the prototype stage?
   c. In a paragraph, explain how the screen output for this particular problem can be changed so that it better reflects the rules of the system Miss deLimit is using.

5. Here is a log sheet for a patient information system used by nurses at a convalescent home to record patient visitors and activities during their shifts. Design a printed report using form design software that provides a summary for the charge nurse of each shift and a report for the activities coordinator at the end of a week. Be sure to use proper conventions to indicate constant data, variable data, and so on. These reports will be used to determine staffing patterns and future activities offerings.

Date	Patient	Visitors	Relationship	Activities
2/14	Clarke	2	Mother, father	Walked about halls, attended chapel, meals in cafeteria
	Coffey	6	Coworkers	Played games, party in room
	Martine	0	—	Meals in room
	Laury	4	Husband and friends	Games in sunroom, watched TV
	Finney	2	Parents	Conversation, meals in cafeteria
	Cartwright	1	Sister	Conversation, crafts room
	Goldstein	2	Sister, brother	Conversation, games out of room, whirlpool

6. Design a screen for problem 5 using form design software. Make any assumptions about system capability necessary and follow screen design conventions for onscreen instructions. (*Hint*: You can use more than one screen if you wish.)
   a. In a paragraph, discuss why you designed each report as you did in problems 5 and 6. What are the major differences in your approach to each one? Can the printed reports be successfully transplanted to the screen without changes? Why or why not?
   b. Some of the nurses are interested in a Web-based system that patients' families can access from home with a password. Design an output screen for the Web. In a paragraph, describe how your report had to be altered so that it could be viewed by one patient's family.
7. Clancy Corporation manufactures uniforms for police departments worldwide. Its uniforms are chosen by many groups because of their low cost and simple yet dignified design. You are helping to design a DSS for Clancy Corp., and it has asked for tabular output that will help it in making various decisions about what designers to use, about where to market its uniforms, and about what changes to make to uniforms to keep them looking up to date. Here are some of the data the company would like to see in tables

Style Name	Example Buyer	Designers
Full military	(NYPD)	Claudio, Rialtto, Melvin Mine
Half military	(LAPD)	Rialtto, Calvetti, Duran, Melvin Mine
Formal dress	Australian Armed Forces	Claudio, Dundee, Melvin Mine
Casual dress	("Miami Vice")	Johnson, Melvin Mine

includes uniform style names, example of a buyer group for each style, and which designers design which uniform styles. Prepare an example of onscreen tabular output that incorporates the foregoing data about Clancy's. Follow proper conventions for onscreen tabular output. Use codes and a key where appropriate.

8. Clancy's is also interested in graphical output for its DSS. It wants to see a graphical comparison of how many of each style of uniform are being sold each year.
   a. Choose an appropriate graph style and design an onscreen graph that incorporates the following data:
      1996 full military (57 percent of all sales).
      1997 full military (59 percent of all sales).
      1998 casual dress (62 percent of all sales).
      1999 casual dress (55 percent of all sales).
      2000 casual dress (40 percent) and half military (22 percent).

      Be sure to follow proper screen design conventions. Use codes and a key if necessary.
   b. Chose a second method of graphing that might allow the decision makers at Clancy's to see a trend in the purchase of particular uniform styles over time. Draw an onscreen graph for use as part of the output for Clancy's DSS. Be sure to follow proper screen design conventions. Use codes and a key if necessary.
   c. In a paragraph, discuss the differences in the two onscreen graphs you have chosen. Defend your choices.
9. Browse the Internet to view well-designed and poorly designed Web sites. Comment on what makes the sites good or bad, using the critique form given earlier in the chapter to compare and contrast them.
10. Propose a Web site for Clancy's, the uniform company described in problems 7 and 8. Sketch by hand or use form design software to create a prototype of a Clancy's home page. Indicate hyperlinks and include a sketch of one hyperlink document. Remember to include graphics, icons, and even sound or other media if appropriate. In a paragraph, describe who the intended users of the Web site are and state why it makes sense for Clancy's to have a Web presence.

GROUP PROJECTS

1. Brainstorm with your team members about what types of output are most appropriate for a variety of employees of Maverick Transport. Include a list of environments or decision-making situations and type of output. In a paragraph, discuss why the group suggested particular options for output.
2. Have each group member design an output screen or form for the output situations you listed in problem 1. (Use either a CASE tool or paper layout form to complete each screen or form.)
3. Share each output screen or form among your team members. Using the feedback gathered, improve the screens or forms you have designed.
4. Design a Web site, either on paper or using software with which you are familiar, for Maverick Transport. Although you may sketch documents or graphics for hyperlinks on paper, create a prototype home page for Maverick, indicating hyperlinks where appropriate. Obtain feedback from other groups in your class and modify your design accordingly. In a paragraph, discuss how designing a Web site is different from designing screens for other online systems.

SELECTED BIBLIOGRAPHY

Davenport, T. H. "Saving IT's Soul: Human-Centered Information Management." *Harvard Business Review*, March–April 1994, pp. 119–31.

Davis, G. B., and H. M. Olson. *Management Information Systems, Conceptual Foundations, Structure, and Development*, 2d ed. New York: McGraw-Hill, 1985.

Fahey, M. J., and J. Brown. *Web Publishers Design Guide*. Scottsdale, AZ: Coriolis Group, 1995.

Horton, W. K. et al. *The Web-Page Design Cookbook: All the Ingredients You Need to Create 5-Star Web Pages*. New York: John Wiley, 1996.

Jarvenpaa, S. L., and G. W. Dickson. "Myth vs. Facts about Graphics in Decision Making." *Spectrum*, Vol. 3, No 1, February 1986, pp. 1–3.

Laudon, K. C., and J. P. Laudon. *Management Information Systems*, 4th ed. Upper Saddle River, NJ: Prentice-Hall, 1996.

McCombie, K. "Connecting Your Enterprise LAN to the Internet." *Internet World*, June 1994.

Merholz, P. "10 Hottest Web Designers and Design Houses." *The Net*, Vol. 2, Issue 1, No. 6, 1996, p. 46.

Pfaffenberger, B., and D. Wall. *Que's Computer and Internet Dictionary*, 6th ed. Indianapolis: Que Corporation, 1995.

Pring, R. *www.type: Effective Typographic Design for the Worldwide Web*. New York: Watson-Guptill, 1999.

Quarterman, J. S. "What Can Businesses Get Out of Internet." *Computerworld*, February 22, 1993.

Senn, J. A. *Analysis and Design of Information Systems*, 2d ed. New York: McGraw-Hill, 1987.

Siegel, D. *Creating Killer Web Sites*. New York: Hayden, 1997.

Weinman, L. *Designing Web Graphics 2: How to Prepare Images and Media for the Web*, 2d ed., New Riders Publication, 1997.

Whitten, J. L., L. D. Bentley, and V. M. Barlow. *Systems Analysis and Design*, 3d ed. Barr Ridge, IL: Irwin, 1994.

15

Allen Schmidt, Julie E. Kendall, and Kenneth E. Kendall

REPORTING ON OUTPUTS

"Let's create output specifications and then work backward through the data flow to determine the corresponding in-put data," says Anna during her next meeting with Chip.

"Of course," Chip agrees.

Output was separated into two categories: reports and screens. Reports were further defined as external reports such as the User Software Notification or internal reports such as the Hardware Inventory Listing. Each report was further classified as a detailed, exception, or summary report.

Based on conversations with Paige Prynter, the analysts think the HARDWARE INVESTMENT REPORT has the highest priority. It is needed as soon as possible because the budget process will soon reach a critical phase and there are many requests for new hardware as well as upgrades for existing equipment.

The process used for creating the HARDWARE INVESTMENT REPORT is similar to the process for creating all reports. Chip examines the data flow diagrams for the new system and locates the data flow labeled HARDWARE INVESTMENT REPORT. Double clicking on the data flow line brings up the repository entry for this report, illustrated in Figure E15.1, HARDWARE INVESTMENT REPORT. It contains a definition giving information about the type of report and an Alias. The **Composition** area provides a list of all the elements on the report. The **Notes** area provides additional information necessary for creating the report.

"I'm really glad we took the time to document the prototype reports and screens when creating the data flow diagrams," remarks Chip. "I can easily identify the elements required to produce the report."

Chip places the cursor in a composition element and presses the **Jump** key to display the details for each element.

"This is great," exclaims Chip. "It was a good idea to define all the elements as we learned about them."

Chip then proceeds to create a sample report using Access. After the first draft, Chip uses the Print Preview feature to preview the report.

"Hmmm," murmers Chip, "Some of the fields need rearranging, and the horizontal spacing needs some work."

The report design is modified and reviewed again. By the third try, the report was in its final form. The next step is crucial: Chip asks Paige to review the report and make any changes she likes. Chip asks, "Are there any additional columns or other data missing that would make for a more useful report? Are all the data on the report necessary?"

Paige studies the output for a few minutes and remarks, "Subtotals for each BRAND, including the NUMBER OF MACHINES and grand totals, are necessary. We receive requests for different types of machines, and knowing how many of each machine may help determine what is purchased."

Chip returns to his microcomputer and makes the necessary changes. The final HARDWARE INVESTMENT REPORT sample is shown in Figure E15.2. This version is again reviewed by Paige, and she signs off on the layout as complete.

The logic for this summary report is outlined in a process specification. The MICROCOMPUTER MASTER file is sorted by MODEL within BRAND. Records are

Define Item

Description | Locations

Label: Hardware Investment Report — 1 of 2

Entry Type: Data Flow

Description: A summary report listing the total number of machines and amount invested by brand and model for all microcomputers.

Alias: Microcomputer Investment Report

Composition: (Attributes)
Brand Name +
Model +
Number of Machines +
Total Invested +
Brand Subtotal +
Grand Total

Notes: Print a line for each unique brand and model, including the number of machines for this model and the total amount invested for this model.

Long Name:

SQL | Delete | Next | Save | Search | Jump | File | History | ?

Dialect... | Clear | Prior | Exit | Expand | Back | Copy | Search Criteria

Enter a brief description about the object.

FIGURE E15.1
HARDWARE INVESTMENT REPORT data flow screen.

read from the MICROCOMPUTER MASTER file, and totals for each BRAND and MODEL are accumulated. When either BRAND or MODEL changes, a report line is printed. When a change in BRAND occurs, BRAND SUBTOTALS are printed. GRAND TOTALS are printed after all records are processed.

Anna spends some time speaking with Cher Ware about her report needs. Several printed reports were outlined when Cher asks the question, "Will I get reports on the screen, ones that I can quickly view, that have the latest information?"

The discussion that followed resulted in the creation of several screen reports.

"How would you like to view the software categories?" asks Anna. "Would you like to see all the software on one large scrolling screen?"

"Well, I would like to have some way of finding one category and then displaying all the software available for that category," replies Cher. "It would also be useful to be able to move to subsequent and previous categories."

Anna works with the Visible Analyst repository for the SOFTWARE BY CATEGORY data flow, shown in Figure E15.3. She enters the contents of the screen

1/12/01 Hardware Investment Report Page 1 of 1

Brand Name	Model		Number of Machines	Total Invested
Xxxxxxxxxxxx	Xxxxxxxxxxxxxxxxxx		3	$29,997.00
		Brand Subtotal	3	$29,997.00
Xxxxxxxxxxxxxx	Xxxxxxxxxxxxxxxx		4	$39,996.00
Xxxxxxxxxxxxxx	Xxxxxxxxxxxxxxxxxx		2	$19,998.00
		Brand Subtotal	6	$59,994.00
Xxxxxxxxxxxxxxxx	Xxxxxxxxxxxxxxxx		3	$29,997.00
Xxxxxxxxxxxxxxxx	Xxxxxxxxxxxxxxxxxxxx		8	$79,992.00
		Brand Subtotal	11	$109,989.00
		Grand Total	20	$199,980.00

FIGURE E15.2
HARDWARE INVESTMENT REPORT sample output.

and makes some notes about what additional things are required for a successful program.

Anna creates the SOFTWARE BY CATEGORY screen by creating an Access form, shown in Figure E15.4. There is a button for finding records as well as buttons to move to the previous and next categories. In the lower area of the screen is an area to display multiple software packages for the category. The Operating System field is stored as a code on the corresponding database table and is converted to the code description on the screen.

Anna shows both Chip and Cher the completed screen. "I'm impressed," exclaims Cher. "That's exactly what I need!"

At that moment, Hy Perteks saunters in. "What's going on?" he asks. After viewing the screen, he remarks, "I've been involved in the intranet project under way. Is their any chance of getting some information posted to a Web page?"

"What do you have in mind?" inquires Chip.

"Well, I have been giving it some thought," replies Hy. "I envision that it would be useful for the faculty and staff to be able to look up information about the software courses we are planning to offer. Later we could add an intranet form for them to enroll in the courses."

"I've heard a lot about the intranet and have created some prototypes for it," remarks Chip. "That would be a fun project to work on! We could include a link to the page from our Technology Support menus."

"Count me in on it," replies Anna. "I've been creating some Web pages myself. What would you like on the page?"

"I would like to create a main page that lists the courses, followed by other pages that list the level, such as beginning or intermediate for the course and the dates that the courses start," replies Hy.

Define Item

Description | Locations

Label: Software by Category 1 of 2

Entry Type: Data Flow

Description: List on a screen all the software by software category.

Alias:

Composition: (Attributes) Software Category Description +
{Title} +
{Version} +
{Operating System Name} +
{Publisher} +
{Site License} +
{Number of Copies}

Notes: Include a Find Record button, allowing the user to search for a word in the category.

Include Next Record and Previous Record buttons to allow scrolling

Long Name:

SQL | Delete | Next | Save | Search | Jump | File | History | ?

Dialect... | Clear | Prior | Exit | Expand | Back | Copy | Search Criteria

Enter a brief description about the object.

FIGURE E15.3
SOFTWARE BY CATEGORY data flow screen.

Chip and Anna set to work on the Web page. The fields were identified and grouped onto the TRAINING CLASSES OFFERED data flow, illustrated in Figure E15.5. Note that the Web address is included as an Alias. Anna creates the final intranet Web page, illustrated in Figure E15.6. Chip and Hy review the page.

"I like the menus on the top of the page and the submenu that displays below it for specific features," remarks Chip.

"The calendar makes it very useful for the staff to view the currently scheduled courses by date and buttons to change the month and year," comments Hy.

"Yes, and I think allowing the staff to change how the data are displayed is also very good. Many staff members like to view courses offered at their campus," remarks Chip.

"It would add some pizzazz if we include an image for the mascot," adds Hy, "and the university motto."

"I'll get right on it," replies Anna. "These are really good suggestions."

The final intranet screen is finished and approved by Hy.

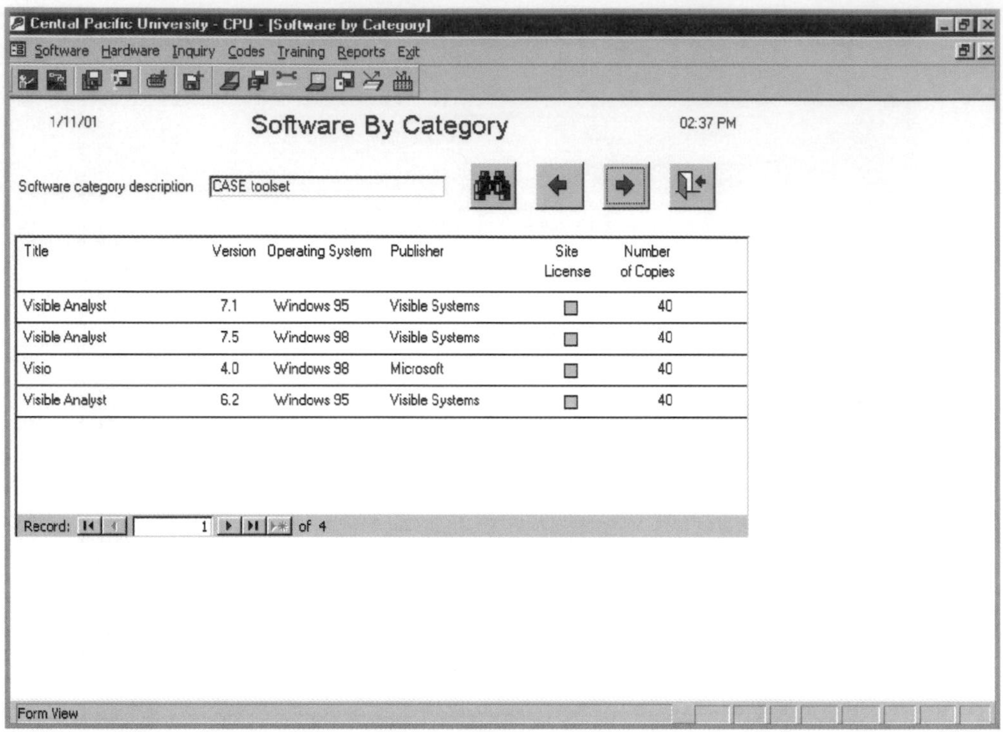

FIGURE E15.4
SOFTWARE BY CATEGORY, Access screen.

"I'll put out an email to all the faculty and staff on the listserv," remarks Hy. "Thanks for including my email address. It should help to facilitate registering for courses and answering any questions. I think we are really making progress!"

The following exercises may be done by designing the report or screen using Printer or Screen layout forms, or they may be created using any word processor with which you are familiar. The fields and other related information for the reports are contained in Visible Analyst data flow repository entries. The names for the data flow are listed for each exercise.

Corresponding reports and screens (called forms in Access) have been created. All the information is present in the Access database; you only have to modify the existing reports and screens to produce the final versions. Modifications are made by clicking on the desired report or screen and then clicking the Design button. The following modifications may be made. The Page Header contains column headings. The Detail area contains the print fields for the report. Refer to Figure E15.7, which shows the design layout for the OUTSTANDING MICROCOMPUTER PURCHASE ORDERS report.

Click in a field to select it. Click on several fields while holding the shift key to select them.
Drag a selected field (or fields) to move them.
Click on one of the small boxes surrounding the field to change the field size.
Select several fields and click **Format** and one of the following:
    Align, to align all fields with the top, left, and so forth field.
    Size, to make fields equal to the widest, tallest, and so forth field.

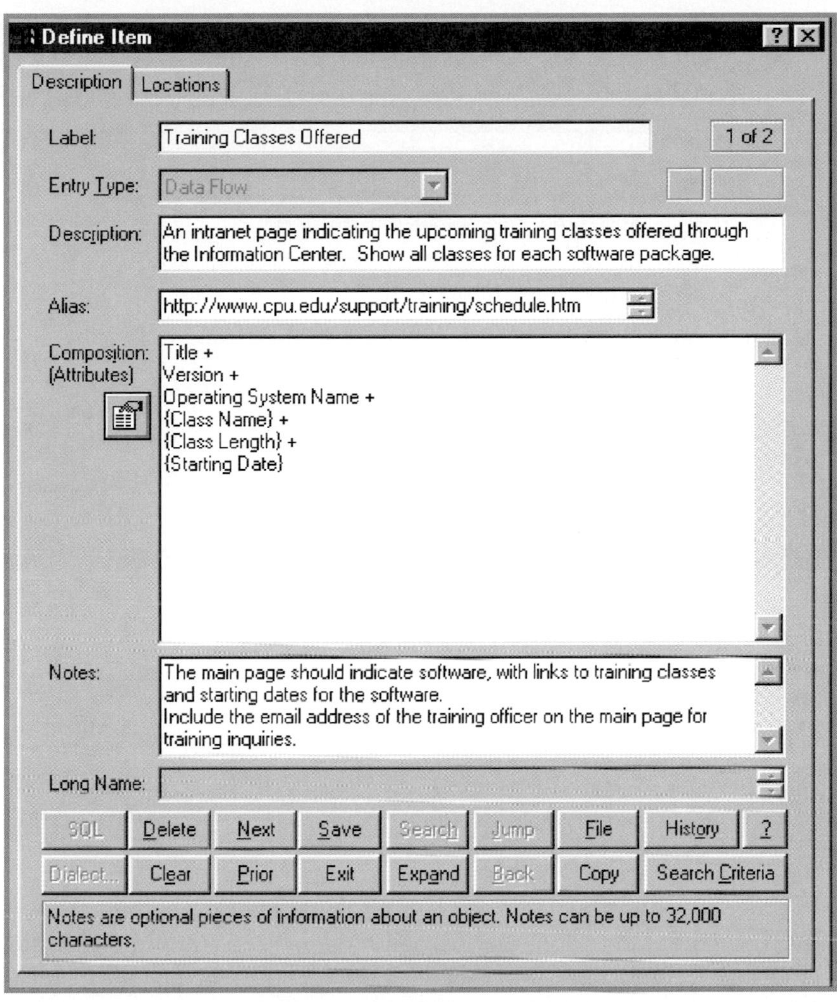
Define Item
Description
Locations
Label: Training Classes Offered
1 of 2
Entry Type: Data Flow
Description: An intranet page indicating the upcoming training classes offered through the Information Center. Show all classes for each software package.
Alias: http://www.cpu.edu/support/training/schedule.htm
Composition: (Attributes)
Title +
Version +
Operating System Name +
{Class Name} +
{Class Length} +
{Starting Date}
Notes: The main page should indicate software, with links to training classes and starting dates for the software.
Include the email address of the training officer on the main page for training inquiries.
Long Name:
SQL Delete Next Save Search Jump File History ?
Dialect... Clear Prior Exit Expand Back Copy Search Criteria
Notes are optional pieces of information about an object. Notes can be up to 32,000 characters.

FIGURE E15.5
TRAINING CLASSES OFFERED data flow screen.

Horizontal Spacing, to make horizontal spacing equal or to increase or decrease the spacing.

Vertical Spacing, to make vertical spacing equal or to increase or decrease the spacing.

EXERCISES*

E-1. Use Access to view the HARDWARE INVESTMENT report. If you are familiar with Access, use the File/Export . . . menu option to save the report as a Web page. When the Export dialogue box opens, click in the Save As Type drop-down list and select HTML Documents.

E-2. Chip, Dot, and Mike participated in several brainstorming sessions resulting in the outlining of several reports. Design (or modify using Access) the HARDWARE MASTER REPORT. This report is large, and you will have to be careful to include all the data in the report area. You may want to have several detail lines for each record. Print the completed report.

15

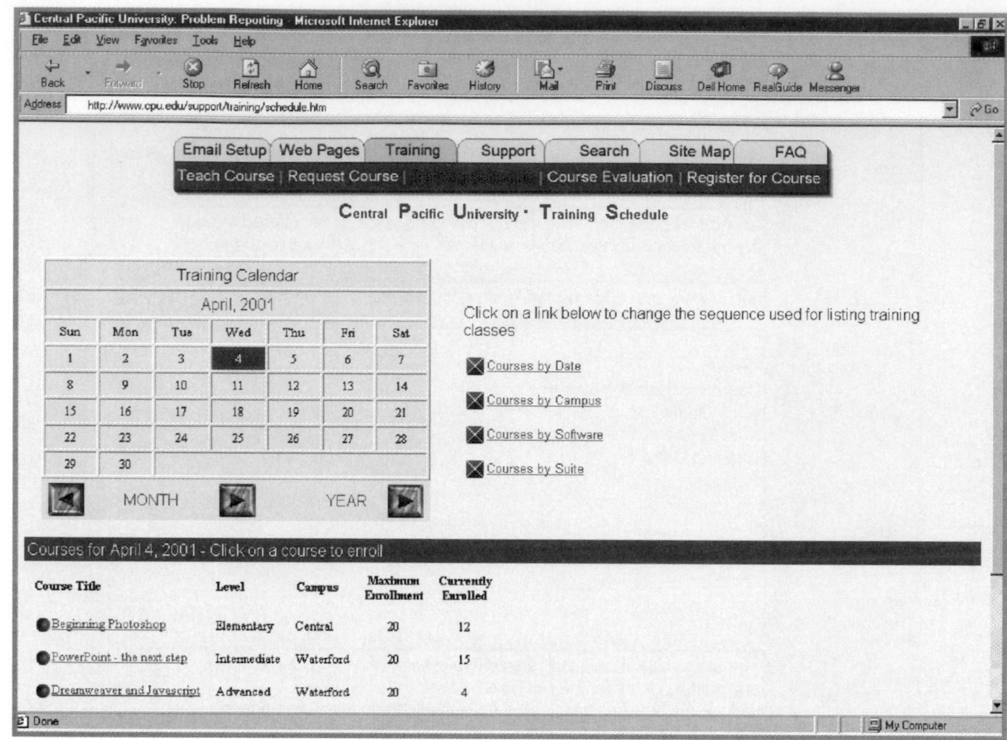

FIGURE E15.6
An intranet Web page for Central Pacific University.

E-3. After meeting with Cher Ware and Hy Perteks to discuss reporting needs, Anna has identified the fields for the partially completed the NEW SOFTWARE INSTALLED report. Design (or modify) the report to include the elements found in the data flow repository entry. Is the report a summary or detailed report? In a paragraph, outline the logic that you think the report-producing program must use.

E-4. Both Dot and Mike need to know when new computers have been received. Create the NEW MICROCOMPUTER RECEIVED REPORT. The data flow MICROCOMPUTER RECEIVED REPORT contains the necessary elements.

E-5. Design the SOFTWARE MASTER REPORT containing pertinent information that helps Cher and Hy to locate the various copies of any software package easily. The elements necessary to produce the report are located on the SOFTWARE MASTER REPORT data flow.

The TITLE, VERSION, OPERATING SYSTEM NAME, PUBLISHER, CATEGORY, and the FIRST and LAST NAME of the software expert should be group printed. Totals are to be included for each Title/Operating System/Version combination. Print the completed report design.

E-6. Design the HARDWARE INVENTORY LISTING, showing the software available in at each room within each campus. The Campus field should be the CAMPUS DESCRIPTION, not the code representing the campus.

E-7. Design the INSTALLED MICROCOMPUTER REPORT, showing personal computers that have been installed in each room. Use the CAMPUS DESCRIPTION and group print by CAMPUS DESCRIPTION and ROOM LOCATION. The Installed boards is a repeating group, with up to five entries per computer.

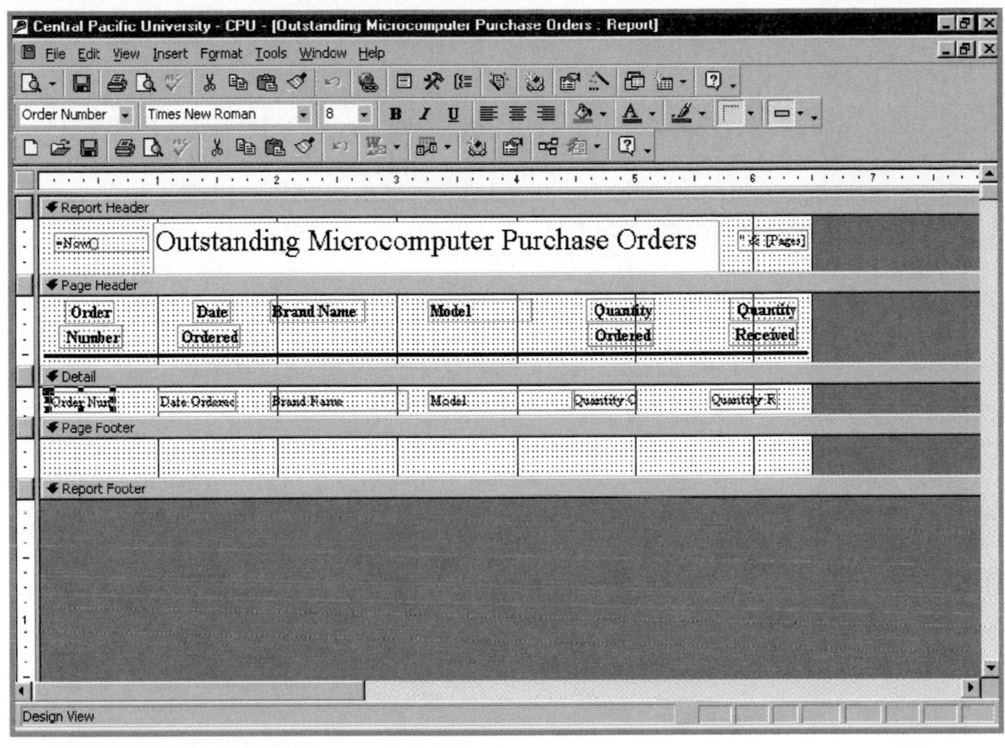

FIGURE E15.7
Modifications can be made to the reports and screens residing in an Access database.

E-8. Use Access to view the SOFTWARE BY CATEGORY screen report. Press the **Find** button and locate CASE toolset. Press the **Next** and **Previous** buttons to view next and previous Software Categories.

E-9. Design the SOFTWARE BY MACHINE screen report. Refer to the data flow repository entry for elements.

E-10. Design the MICROCOMPUTER PROBLEM REPORT. This report shows all personal computers that have a large number of repairs or a large repair cost. Refer to the repository description for the data flow for the elements or modify the Access report.

E-11. Design or modify the INSTALLATION REPORT. Refer to the repository entry for the data flow for the elements. This report shows which computers have been recently received and are available for installation.

E-12. Design the NEW MICROCOMPUTER RECEIVED report. Refer to the repository description for the data flow for the elements or modify the Access report. This summary report shows the number of PCs of each brand and model. These computers need to be unpacked and have component boards and other hardware installed before they may be installed in rooms.

E-13. Design or modify the PREVENTIVE MAINTENANCE REPORT. Refer to the repository entry for the data flow for the elements. This report shows which computers need to have preventive maintenance performed on them.

E-14. Design the SOFTWARE CROSS REFERENCE report. Refer to the repository description for the data flow for the elements or modify the Access report. This report shows the computer upon which each software package is installed. The

TITLE, VERSION, OPERATING SYSTEM MEANING, and PUBLISHER are group printed. The detail lines under the group contain data showing the machine and the installation campus and room.

E-15. Design or modify the OUTSTANDING MICROCOMPUTER PURCHASE ORDERS report. Refer to the repository entry for the data flow for the elements. This report would be produced for all PURCHASE ORDER records that have a purchase order code of M101, representing computers, with the further condition that the QUANTITY ORDERED on the record must be greater than the QUANTITY RECEIVED. In a paragraph, state whether this report is a summary, exception, or detailed report. Explain.

E-16. Design the SOFTWARE INVESTMENT report. Refer to the repository description for the data flow for the elements or modify the Access report.

*Exercises preceded by a CD-ROM icon require the program Visible Analyst or another CASE tool. A CD-ROM containing Visible Analyst examples is provided free of charge to any professor adopting this book. The examples on the disk may be imported into Visible Analyst and then used by students.

DESIGNING EFFECTIVE INPUT

16

INPUT DESIGN OBJECTIVES

The quality of system input determines the quality of system output. It is vital that input forms, screens, and interactive Web documents be designed with this critical relationship in mind. By insisting on well-designed input, the systems analyst is acknowledging that poor input calls into question the trustworthiness of the entire system.

Well-designed input forms, screens, and interactive Web fill-in forms should meet the objectives of effectiveness, accuracy, ease of use, consistency, simplicity, and attractiveness, as depicted in Figure 16.1. All these objectives are attainable through the use of basic design principles, the knowledge of what is needed as input for the system, and an understanding of how users respond to different elements of forms and screens.

Effectiveness means that input forms, input screens, and fill-in forms on the Web all serve specific purposes in the information system, whereas accuracy refers to design that ensures proper completion. Ease of use means that forms and screens are straightforward and require no extra time to decipher. Consistency in this case means that all input forms, whether they are input screens or fill-in forms on the Web, group data similarly from one application to the next, whereas simplicity refers to keeping those same designs purposely uncluttered in a manner that focuses the user's attention. Attractiveness implies that users will enjoy using, or even be drawn to using, input forms, input screens, and fill-in Web forms through their appealing design.

GOOD FORM DESIGN

Although an in-house forms specialist may be available, the systems analyst should be capable of designing a complete and useful form. It is also important to be able to recognize poorly designed, overlapping, or unnecessary forms that are wasting the organization's resources and therefore should be eliminated. Some types of analysis, such as workflow analysis, focus on how work (often filled-in forms) moves through the organization.

Forms are important instruments for steering the course of work. By definition, they are preprinted or duplicated papers that require people to fill in responses in a standardized way. Forms elicit and capture information required by organizational members that will often be input to the computer. Through this

FIGURE 16.1
Six objectives for the design of input.

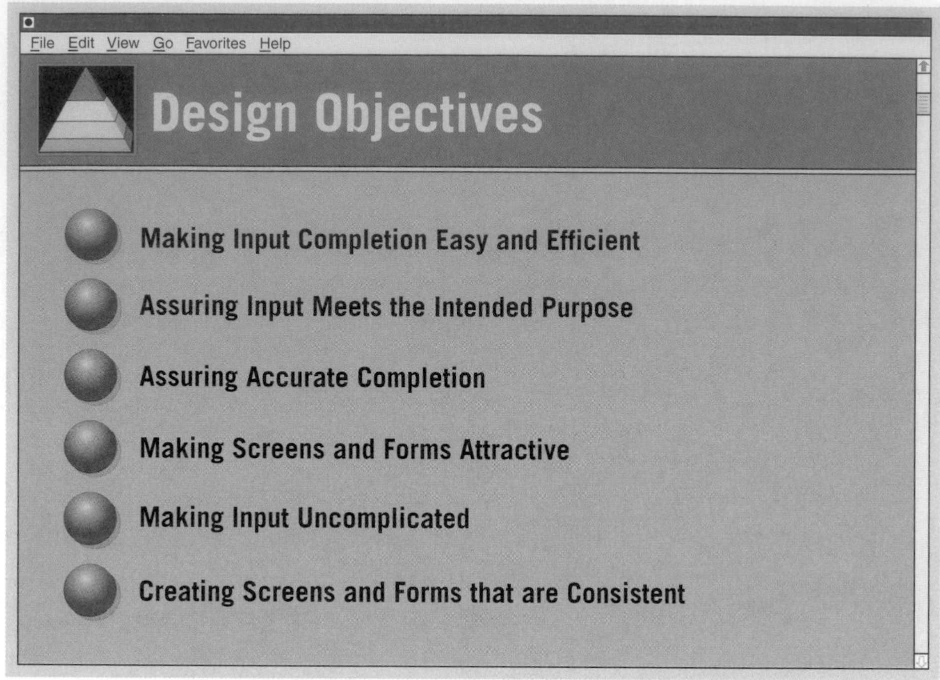

process, forms often serve as source documents for data entry personnel or for input to ecommerce applications.

To design useful forms, four guidelines for form design should be observed:

1. Make forms easy to fill in.
2. Ensure that forms meet the purpose for which they are designed.
3. Design forms to ensure accurate completion.
4. Keep forms attractive.

There are a number of ways to achieve each guideline for form design. Each of the four guidelines is considered separately in the following subsections.

MAKING FORMS EASY TO FILL IN

To reduce error, speed completion, and facilitate the entry of data, it is essential that forms be easy to fill in. This guideline is based on more than well-placed empathy for the user. The cost of the forms is minimal compared with the cost of the time employees spend filling them in manually and then entering data into the information system. Sometimes it is possible to eliminate the process of transcripting data that are entered on a form into the system by using electronic submission. That method often features data keyed in by users themselves, who visit Web sites set up for informational or ecommerce transactions.

Form Flow. Designing a form with proper flow can minimize the time and effort expended by employees in form completion. Forms should flow from left to right and top to bottom, as shown in the police incident report depicted in Figure 16.2.

The flow of the incident report works because it is based on the way people in the Western culture read a page. The incident report is designed so that the attending officer first fills in the date, then the time, and then continues on to the bottom of the form, which elicits suggestions on further handling of the situation.

Illogical flow takes extra time and is frustrating. A form that requires people to go directly to the bottom of the form and then skip back up to the top for completion exhibits poor flow.

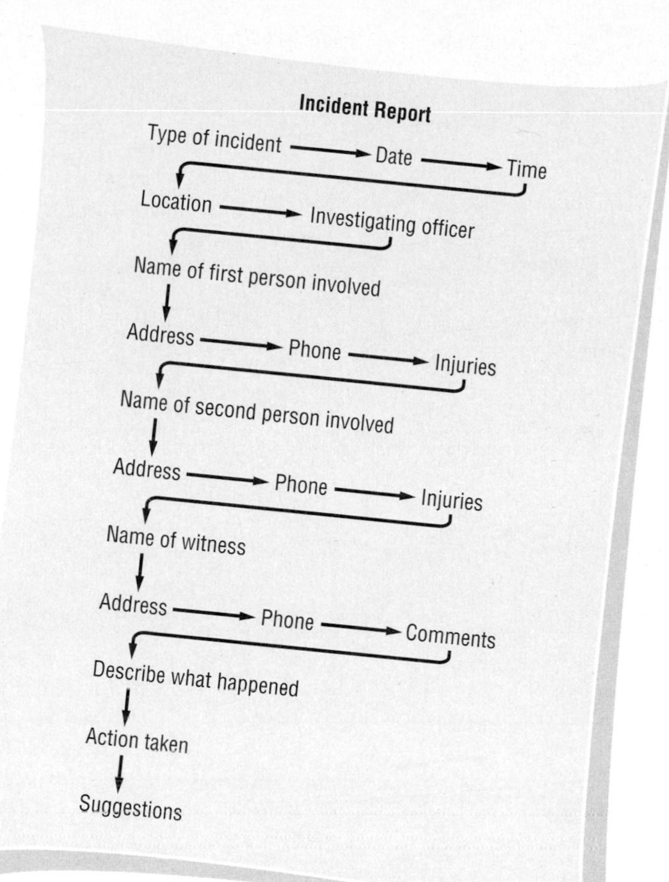

FIGURE 16.2
Good form flow makes the form easy to use.

Seven Sections of a Form. A second technique that makes it easy for people to fill out forms correctly is logical grouping of information. The seven main sections of a strong form are the following:

1. Heading.
2. Identification and access.
3. Instructions.
4. Body.
5. Signature and verification.
6. Totals.
7. Comments.

Ideally, these sections should appear on a page grouped as they are in Figure 16.3. Notice that the seven sections cover the basic information required on most forms. The top quarter of the form is devoted to three sections: the heading, the identification and access section, and the instructions section.

The heading section usually includes the name and address of the business originating the form. The identification and access section includes codes that may be used to file the report and gain access to it at a later date. (In Chapter 17, we discuss in detail how to access specially keyed information in a database.) This information is very important when an organization is required to keep the document for a specified number of years. The instructions section tells how the form should be filled out and where it should be routed when complete.

The middle of the form is its body, which composes approximately half of the form. This part of the form requires the most detail and development from the

FIGURE 16.3
Seven sections found in well-designed forms.

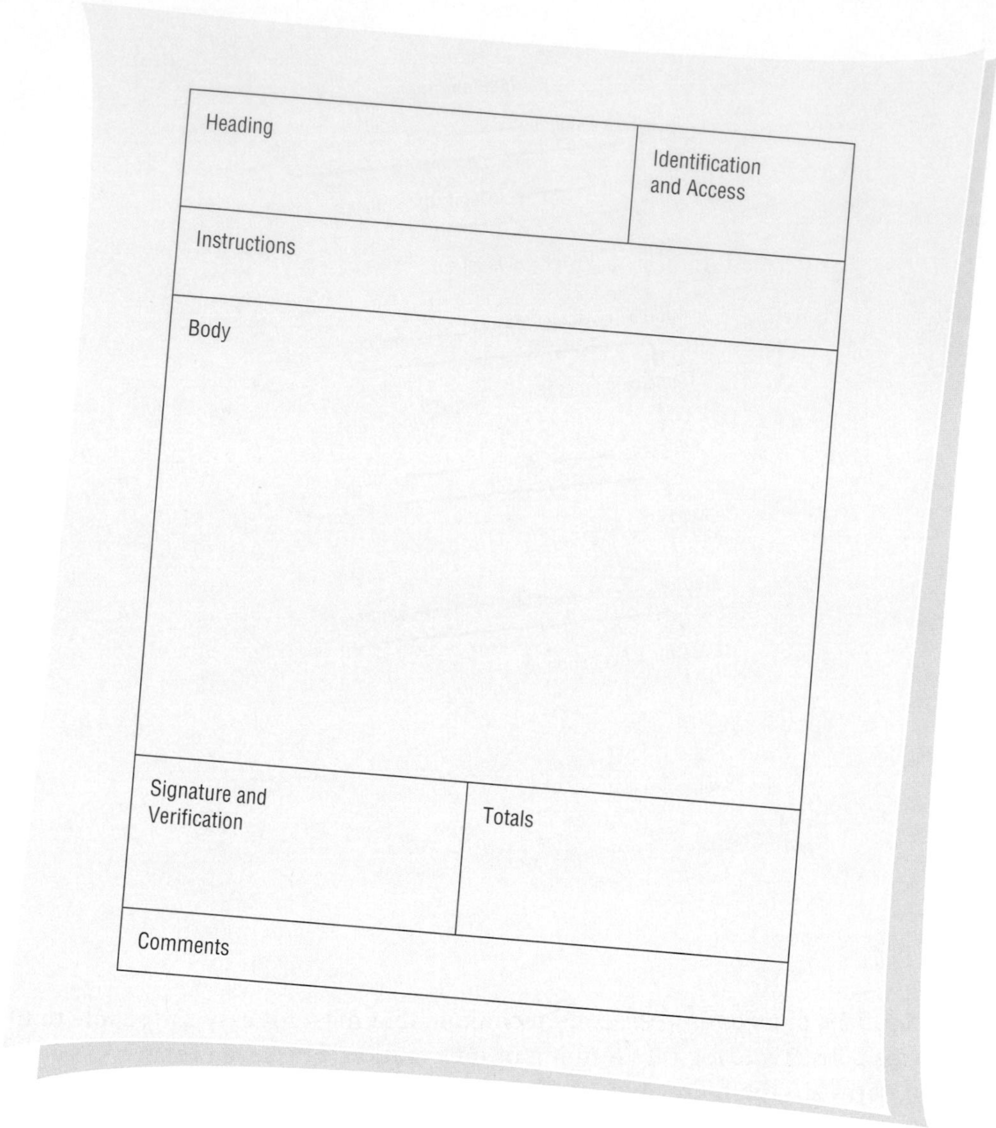

person completing it. The body is the part of the form most likely to contain explicit, variable data. For example, on a parts requisition form, this section might include data such as the firm ordering the part, part number, quantity ordered, and price.

The bottom quarter of the form is composed of three sections: signature and verification, totals, and comments. By requiring a signature in this part of the form, the designer is echoing the design of other familiar documents, such as letters. Requiring ending totals and a summary of comments is a logical way to provide closure for the person filling out the form.

Captioning. Clear captioning is another technique that can make easy work of filling out a form. Captions tell the person completing the form what to put on a blank line, space, or box. Several options for captioning are shown in Figure 16.4. Two types of line captions, two types of check-off captions, and examples of a boxed caption and table caption are shown.

The advantage of putting the caption below the line is that there is more room on the line itself for data. The disadvantage is that it is sometimes unclear which line is associated with the caption: the line above or below the caption. The person filling out the form may realize that the wrong line has been used and will subsequently have to complete a new form.

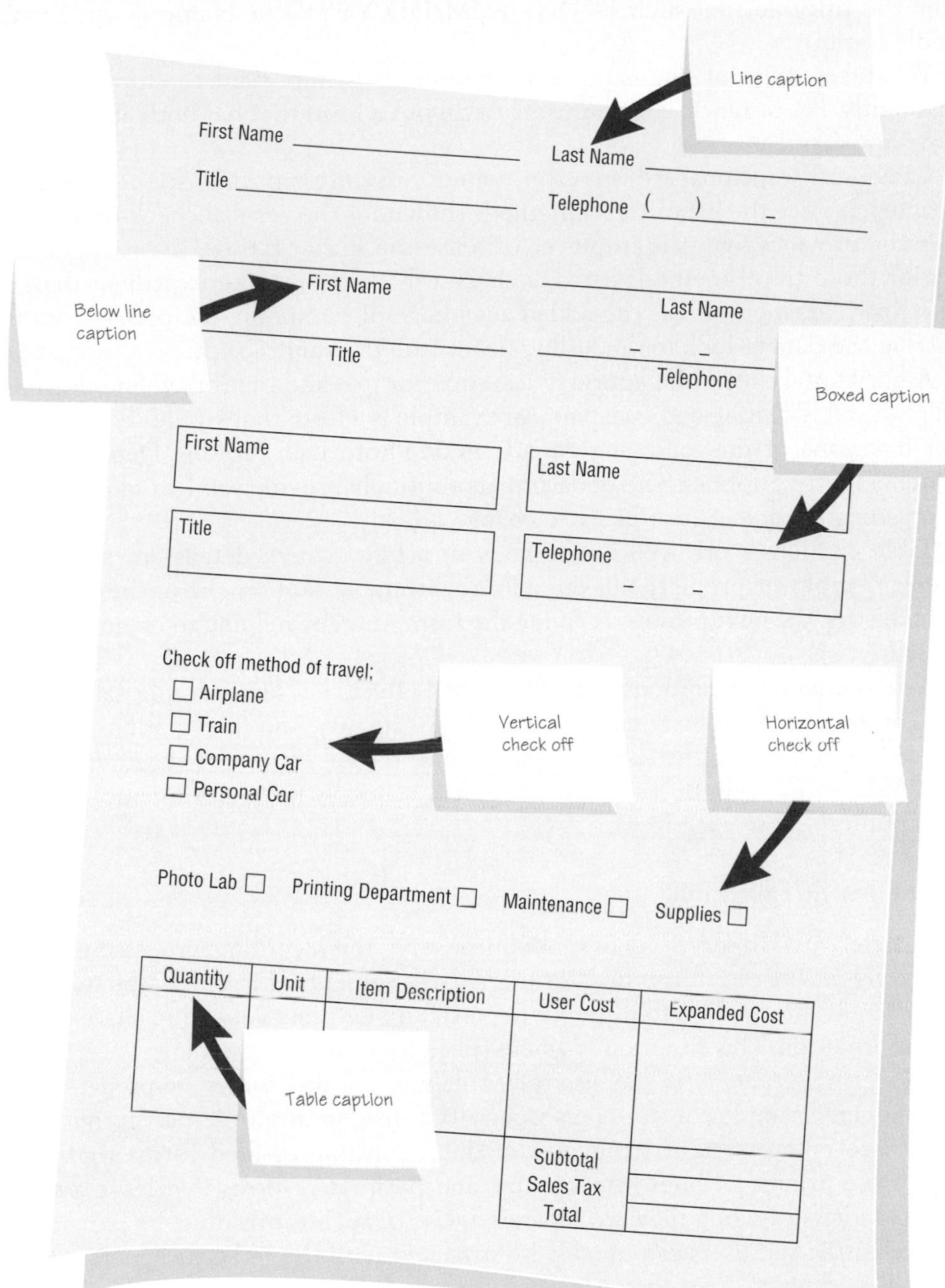

FIGURE 16.4
Major captioning alternatives.

Line captions can be to the left of blanks and on the same line, or they can be printed below the line where data will be entered.

Another way to caption is to provide a box for data instead of a line. Captions can be placed inside, above, or below the box. Boxes on forms help people enter data in the correct place, and they also make reading the form easier for the form's recipient. The caption should use a small type point size so that it does not dominate the entry area. Small vertical "tick marks" may be included in the box if the data is intended for entry into a computer system. If there is not enough room on a record for the data, the person filling out the form, rather than the data-entry operator, has the freedom to determine how the data should be abbreviated. Captions may also include small clarification notes to help the user correctly enter the information, such as Date (MM/DD/YYYY) or Name (Last, First, Middle Initial).

Whatever styles of line caption are chosen, it is important to employ them consistently. For instance, it is confusing to fill out a form that has both above- and below-line captions.

Check-off captions are superior when response options are necessarily restricted. Notice the list of travel methods shown for the vertical check-off example in the previous figure. If employee expenses for business travel are reimbursed only for those travel methods listed, a check-off system is more expedient than a blank line. This method has the added advantage of reminding the person who is verifying the data to look for an airline ticket stub or other receipt.

A horizontal check-off caption is also superior to a line caption when information required is routine and constant. An example is a form that would request services from one of the following departments: Photo Lab, Printing Department, Maintenance, or Supplies. The departments routinely provide services to others in the organization and are not likely to change quickly.

Table captions work well in the body of a form where details are required. When an employee properly fills out a form with table captions, he or she is creating a table for the next person receiving the form, thereby helping to organize data coherently.

A combination of captions can also be used effectively. For example, table captions can be used to specify categories such as quantity, and line captions can be used to indicate where the subtotal, sales tax, and total should be typed. Because different captions serve different purposes, it is generally necessary to employ several caption styles in each form.

MEETING THE INTENDED PURPOSE

Forms are created to serve one or more purposes in the recording, processing, storing, and retrieving of information for businesses. Sometimes it is desirable to provide different information to different departments or users yet still to share some basic information. This situation is where specialty forms are useful.

The term *specialty form* can also refer solely to the way forms are prepared by the stationer. Examples of stationers' specialty forms are multiple-part forms that are used to create instant triplicates of data, continuous-feed forms that run through the printer without intervention, and perforated forms that leave a stub behind as a record when they are separated. Use of such forms must be judicious; they are costly, and users can quickly be strangled with the red tape generated by meaningless multiple-part forms.

ENSURING ACCURATE COMPLETION

Error rates typically associated with collecting data will drop sharply when forms are designed to ensure accurate completion. Design is important in making people do the right thing with the form, whether it is the first or four-hundredth time they are using it. When service employees such as gas meter readers or inventory takers use handheld devices to scan or otherwise key in data at the appropriate site, the extra step of transcription during data entry is avoided. Handheld devices use wireless transmission or are plugged back into larger computer systems to which they can upload the data that the service worker has stored. No further transcription of what has occurred in the field is necessary.

The Bakerloo Brothers employee expense voucher, shown in Figure 16.5, goes a long way toward securing accurate form completion. Many of the form

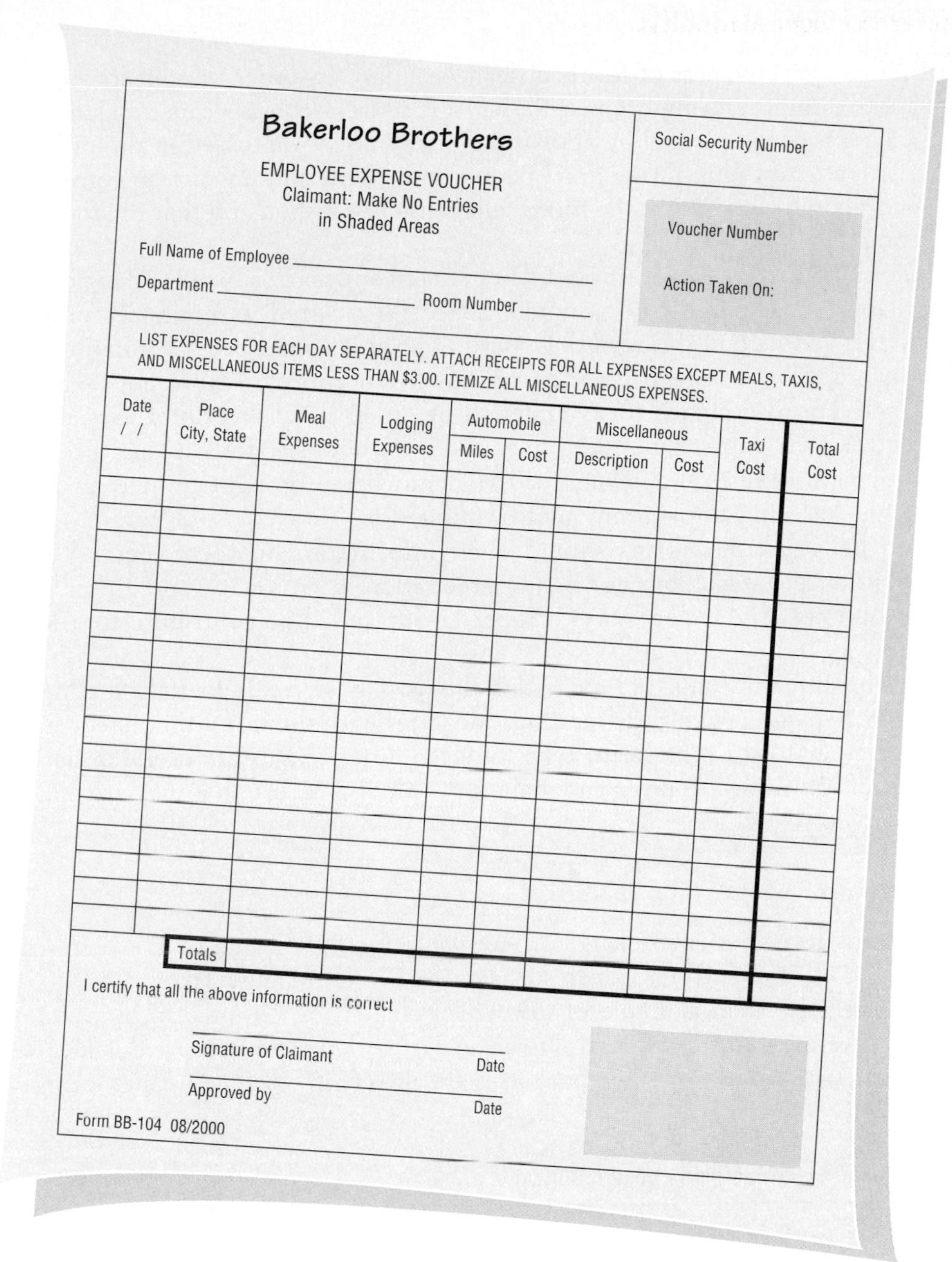

Bakerloo Brothers

EMPLOYEE EXPENSE VOUCHER
Claimant: Make No Entries
in Shaded Areas

Social Security Number

Voucher Number

Action Taken On:

Full Name of Employee ______________

Department ______________ Room Number ______

LIST EXPENSES FOR EACH DAY SEPARATELY. ATTACH RECEIPTS FOR ALL EXPENSES EXCEPT MEALS, TAXIS, AND MISCELLANEOUS ITEMS LESS THAN $3.00. ITEMIZE ALL MISCELLANEOUS EXPENSES.

Date / /	Place City, State	Meal Expenses	Lodging Expenses	Automobile		Miscellaneous		Taxi Cost	Total Cost
				Miles	Cost	Description	Cost		
	Totals								

I certify that all the above information is correct

Signature of Claimant Date

Approved by Date

Form BB-104 08/2000

FIGURE 16.5
A form that encourages accurate completion.

design techniques we have discussed are used in this sample expense voucher. The form design implements the correct flow: top to bottom and left to right. It also observes the idea of seven main sections or information categories. In addition, the employee expense voucher uses a combination of clear captions and instructions.

Because Bakerloo Brothers employees are reimbursed only for actual expenses, getting a correct total expenditure is essential. The form design provides an internal double check with column totals and row totals expected to sum to the same number. If the row and column totals don't sum to the same number, the employee filling out the form knows there is a problem and can correct it on the spot. An error is prevented, and the employee can be reimbursed the amount due; both outcomes are attributable to a suitable form design.

KEEPING FORMS ATTRACTIVE

Although attractiveness of forms is dealt with last, its order of appearance is not meant to diminish its importance. Rather, it is addressed last because making forms appealing is accomplished by applying the techniques discussed in the preceding subsections. Aesthetic forms draw people into them and encourage completion. Hence, people who fill out the forms will be more satisfied and that the forms will be completed.

Forms should look uncluttered. They should appear organized and logical after they are filled in. Providing enough space for printed or manually prepared responses will help in this regard. Computer printer entries require a minimum of 1/6-inch spacing between lines, and handwritten entries require approximately 1/4 inch. Forms designed for completion by either hand or printer should allow about 1/3-inch intervals between lines. A useful heuristic for gauging the appropriate length of lines is allowing for five handwritten or eight printed characters per inch. Allowing ample room invites completion.

To be attractive, forms should elicit information in the expected order: Convention dictates asking for name, street address, city, state, and zip or postal code (and country, if necessary). Proper layout and flow contribute to a form's attractiveness.

Using different fonts for type within the same form can help make it appealing to fill in. Separating categories and subcategories with thick and thin lines can also encourage interest in the form. Type fonts and line weights are useful design elements for capturing attention and making people feel secure that they are filling in the form correctly.

COMPUTER-ASSISTED FORM DESIGN

Numerous form design packages are available for PCs. Some of the features of paper and electronic forms design software are given in Figure 16.6.

Figure 16.7 is an example of the screen created by OmniForm by ScanSoft. This software is enormously useful to an analyst seeking to automate quickly existing business processes where paper forms are already in existence. Paper forms can

FIGURE 16.6
Software for electronic form design has many dynamic features.

Features of Electronic Form Design Software

- Gives the ability to design paper forms, electronic forms, or Web-based forms using one, integrated package
- Allows forms design using forms templates
- Enables forms design by cutting and pasting familiar shapes and objects
- Facilitates electronic form completion through use of a companion data-entry software package
- Permits customization of electronic form completion with the capability to customize menus, toolbars, keyboards, and macros
- Supports integration with popular databases
- Enables the sending and broadcasting of electronic forms
- Permits sequential routing of forms
- Assists tracking of routed forms
- Encourages automatic delivery and processing (push technology for forms)
- Allows the development of roles databases (that show relationships between people and types of information)
- Establishes security protection for electronic forms
- Takes scanned paper forms and permits publishing them to the Web
- Creates electronic fields automatically from scanned paper forms
- Permits form fill-in on the Web
- Allows calculations to be accomplished automatically

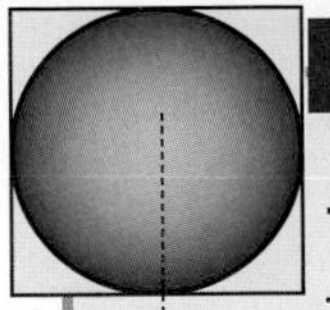

CONSULTING OPPORTUNITY 16.1

THIS FORM MAY BE HAZARDOUS TO YOUR HEALTH

Figure 16.C1 is a printed medical history form that Dr. Mike Robe, a family practitioner, has his receptionist give to all new patients. All patients must fill it out before they see the doctor.

The receptionist is getting back many incomplete or confusing responses, which makes it difficult for Dr. Robe to review the forms and understand why the new patient is there. In addition, the poor responses make it time-consuming for the receptionist to enter new patients into the files.

Redesign the form on $8\frac{1}{2}'' \times 11''$ paper so that pertinent new patient data can be collected in a logical and inoffensive way. Make sure the form is self-explanatory to new patients. It should also be easy for Dr. Robe to read and easy for the receptionist to enter into the patient database, which is sorted by patient name and Social Security number. The office uses PCs connected by a LAN. How would you redesign the form so that it can be electronically submitted by the receptionist? Which office procedures would you have to change?

Medical History Form

Name ____________ Employer ____________ Age ______

Address ____________ Zip ______ Phone ________ Office ________

Insurer ________ Is this [ ] your policy [ ] your spouse's policy

Blue Cross [ ] State Physician's Service [ ] Other [ ] (state) ________

- -

Have you ever had surgery? Yes___ No___ If so, when? ______

Describe the surgery____________________

Have you ever been hospitalized? Yes___ No___ If so, when?______

Why?____________________

Complete the following.

	I have had	Family history
Diabetes	☐	☐
Heart trouble	☐	☐
Cancer	☐	☐
Seizure	☐	☐
Fainting	☐	☐

What have you been immunized for?

Family: ____________ ____________ ____________
Spouse or next of kin / Relationship / Address

Date of last exam ___/___ Who referred you? ____________

Why are you seeing the doctor today?

Are you currently having pain? ______ Constant ______ Sporadic ______

How long does it last? ________ Please give us your soc. sec. # ________

IMPORTANT! We need your correct insurance carrier number ____________

FIGURE 16.C1
Your help in improving this form is greatly appreciated.

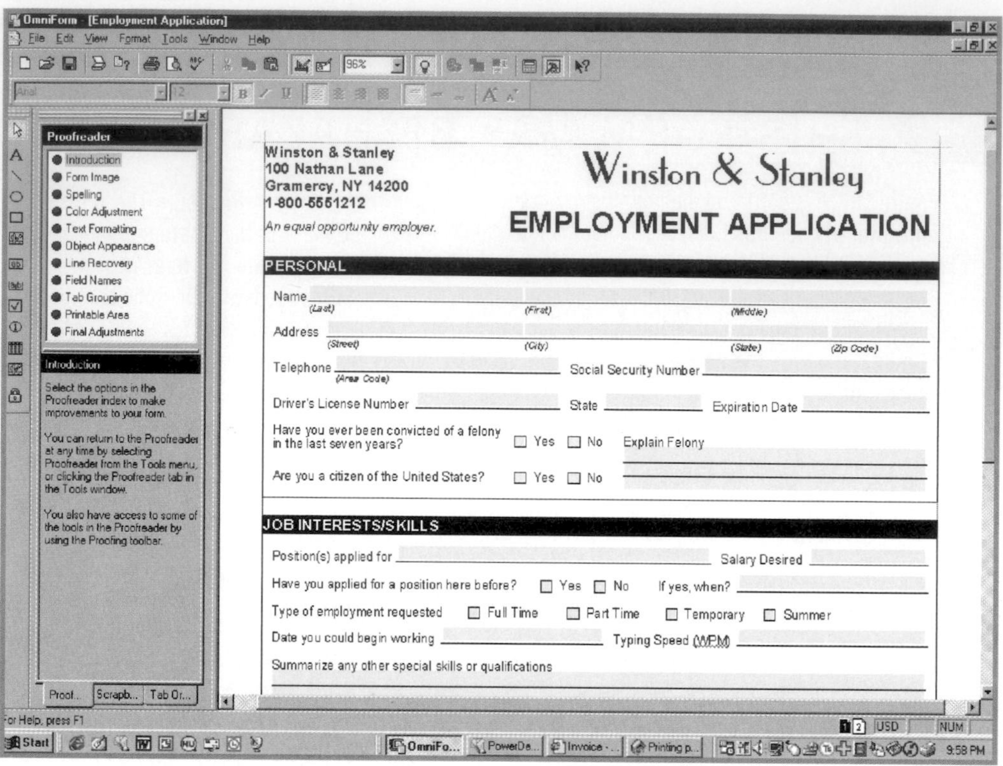

FIGURE 16.7

OmniForm from ScanSoft allows the user to take an existing form, scan it into the computer and define fields so the form can be easily filled out on a PC.

be scanned in and then published to the Web. The analyst can use a set of tools to set up fields, check boxes, lines and boxes, and many other features.

Figure 16.8 shows the scanning process. The bottom of the split screen shows the form as it was scanned in, and the top portion of the screen shows an enlarged view of some of the fields automatically identified by the software. After scanning in a form, the analyst used a wizard to proofread, enhance, identify fields, and change the tab order so that the form could be used electronically.

The functionality of the form is extended because OmniForm automatically creates field names for fields in forms that are scanned. Notice that when it is activated in this mode, the software displays the created fields in green. On the left hand side of the screen is a description of the field creation feature in this software. This description can dramatically speed up the automating of standard processes where time is limited and the desire for innovation may also be limited.

Once a form is scanned, it can be easily modified and published to the Web. ScanSoft currently offers a form hosting service called eOmniForm.com, where you can store up to 10,000 filled records on the Web site. This service is an advantage not only in B2C ecommerce applications, but also in B2B applications. Furthermore, employees can have easy access to company forms without extra administrative intervention.

Electronic forms can have intelligence. OmniForm also enables calculations to be done automatically, so items can be totaled and sales tax calculated. It can also check the field and validate that the data are entered properly. An example is checking that a date is entered as 99/99/9999.

CONTROLLING BUSINESS FORMS

Controlling business forms is an important task. Businesses often have a forms specialist who controls forms, but sometimes this job falls to the systems analyst, who sets up and implements forms control.

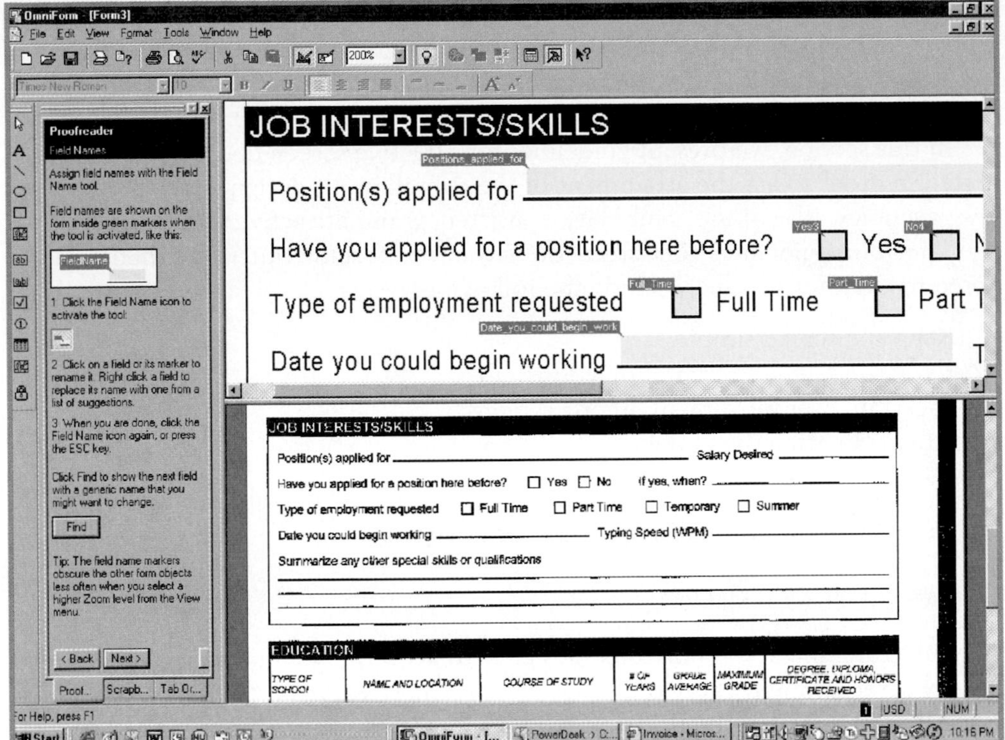

FIGURE 16.8

An example from OmniForm by ScanSoft of the scanning process, where fields are automatically generated by the software.

The basic duties for controlling forms include making sure that each form in use fulfills its specific purpose and that the specified purpose is integral to organizational functioning, preventing duplication of the information that is collected and of the forms that collect it, designing effective forms, deciding on how to get forms reproduced in the most economical way, and establishing stock control and inventory procedures that make forms available (when needed) at the lowest possible cost. A unique form number and revision date (month/year) should be included on each form, regardless of whether it is completed and submitted manually or electronically.

Even if a form specialist is available, control of paperwork is still an area that must be double-checked by the systems analyst. Improvement and change in information systems are highly dependent on managerial information activities, many of which derive from data originally captured on forms.

GOOD SCREEN AND WEB FORMS DESIGN

Much of what we have already said about good form design is transferable to screen design and eventually to the good design of Web sites and their pages. Once again, the user must remain foremost in the analyst's thoughts during the design of display screens.

There are differences, however, and systems analysts should strive to realize the unique qualities of screen displays rather than to adopt blindly the conventions of paper forms. One big difference is the constant presence of a cursor (a block of light or other pointer) on the screen, which orients the user to the current data-entry position. As data are entered on screen, the cursor moves one character ahead, pointing the way.

Another major difference among electronic, Web, and static forms is that designers can include context-sensitive help in any electronic fill-in form. This practice can reduce the need for instructions being displayed for each line, thus reducing the clutter of the form and cutting down on calls to Technical Support. Using a Web-based approach also permits the designer to take advantage of hyperlinks, thus ensuring that the forms are filled out correctly by providing users with hyperlinked examples of correctly completed forms to the Web form.

In this section, we present guidelines for effective screen design. They are presented in order to aid the attainment of the overall input design goals of effectiveness, accuracy, ease of use, simplicity, consistency, and attractiveness.

The four guidelines for screen design are important but not exhaustive. As noted in Chapter 15, they include the following:

1. Keep the screen simple.
2. Keep the screen presentation consistent.
3. Facilitate user movement among screens.
4. Create an attractive screen.

In the next subsections, we develop each of these guidelines, and we present many design techniques for observing the four guidelines.

KEEPING THE SCREEN SIMPLE

The first guideline for good screen design is to keep the screen display simple. The display screen should show only that which is necessary for the particular action being undertaken. For the occasional user, 50 percent of the screen should contain useful information. When designing for the regular user, the screen may be up to 90 percent of the available space.

Three Screen Sections. Figure 16.9 shows a division of the screen into three sections; this method is useful because it simplifies interactions with screens. The top of the screen features a "heading" section, part of which is written into the software to describe to the user where he or she is in the package. The rest of the heading might consist of the file name created by the user.

FIGURE 16.9
Three sections of a VDT screen.

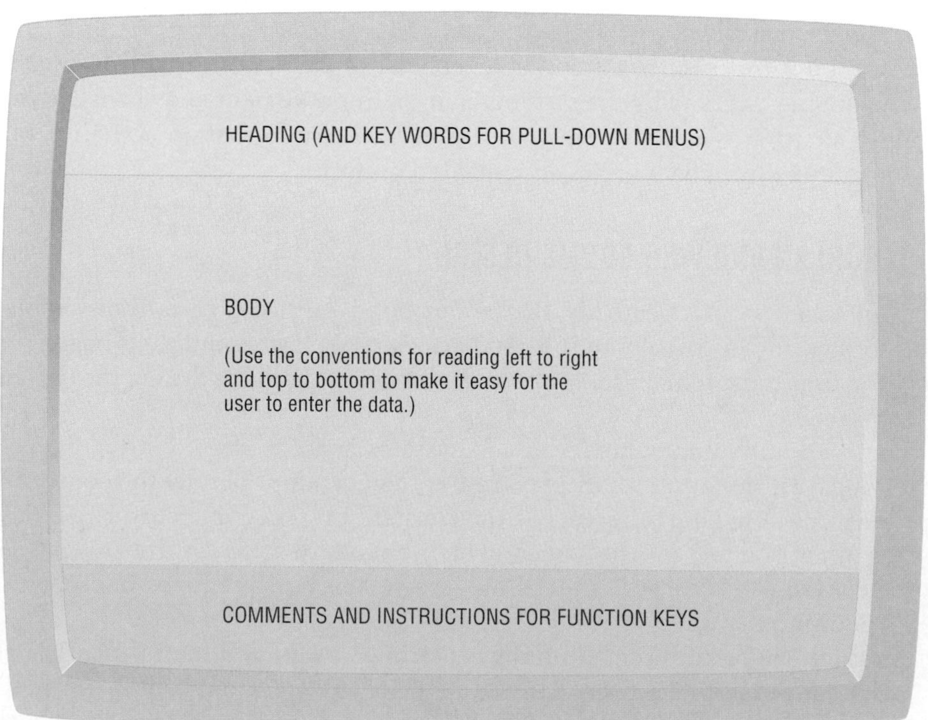
HEADING (AND KEY WORDS FOR PULL-DOWN MENUS)
BODY
(Use the conventions for reading left to right and top to bottom to make it easy for the user to enter the data.)
COMMENTS AND INSTRUCTIONS FOR FUNCTION KEYS

The middle section is called the "body" of the screen. The body can be used for data entry and is organized from left to right and top to bottom, because people in Western cultures move their eyes on a page in this way. Captions and instructions should be supplied in this section to help the user enter the pertinent data in the right place.

Field definitions showing how much data are allowable in each field of the screen's body should also be supplied to the user. That can be accomplished through several different design features: color; inverse video to highlight the precise length of each field; onscreen boxes within which data must fit; brackets, braces, or ampersands to denote the beginnings and endings of fields; or instructions to the user to type directly over a display of distinctive symbols (such as periods) that will then be automatically replaced with data. Sounds such as "beeps" can be included on Web and other interactive fill-in forms to warn users that they are entering numerics rather than alphabetics. In addition, the computer can "beep" or make a sound when the user has reached the end of a field but doesn't notice and so continues typing without entering anything in.

Context-sensitive help can also be made available by having the user click the right mouse button within the body section of the screen.

The third section of the screen is the "Comments and Instructions" section. This section may display a short menu of commands that remind the user of basics such as how to change screens or functions, save the file, or terminate entry. Inclusion of such basics can make inexperienced users feel infinitely more secure about their ability to operate the computer without causing a fatal error. Instructions in the third section could also list acceptable code choices the user needs for completing data entry or ways to get help.

Figure 16.10 shows a display screen of a check register program used by a women's clothing store. The basic check register screen is simple, following the three basic sections suggested. The heading contains the title "Check Register"; the body asks for specific variable data—"Check Number," "Date," and so on—to

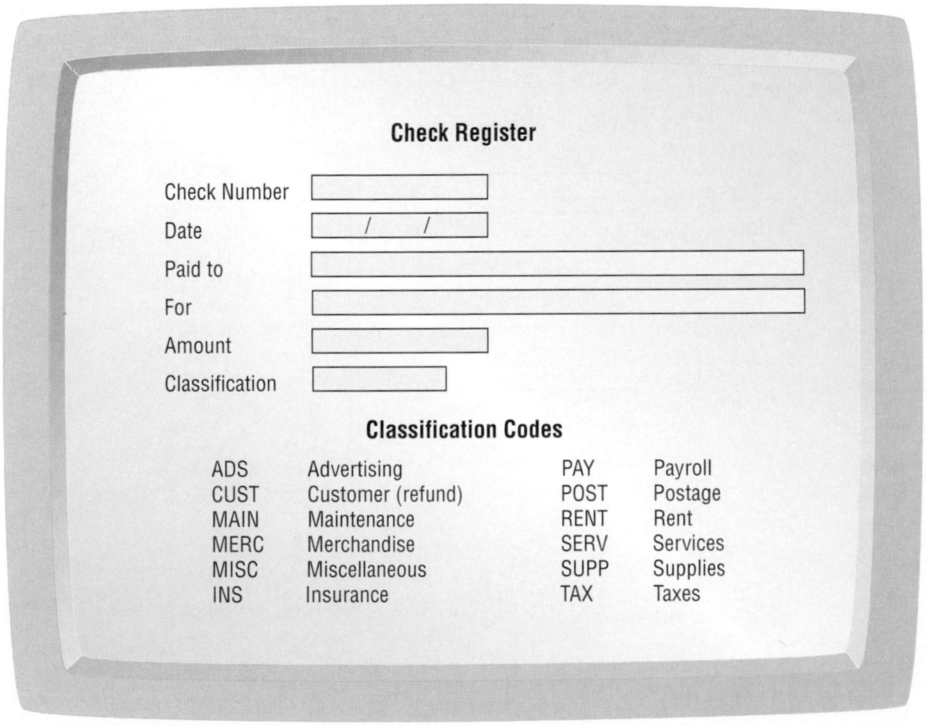

FIGURE 16.10
A screen with classification codes at the bottom.

be entered by the user; and the comments section provides a list of "Classification Codes" to aid the operator in correctly classifying the expense for later analysis.

Using Windows or Hyperlinks. Another way to keep the screen display simple is to list a few basic commands that, when used, will overlay windows to fill the current screen partially or totally with new information. Users can minimize or maximize the size of windows as needed. In this way, users start with a simple, well-designed screen whose complexity they can customize and control through the use of multiple windows. Hyperlinks on a Web-based fill-in form serves a similar purpose.

Figure 16.11 shows an alternative screen display of the check register program along with the window of classification codes the user has asked for. The basic check register screen is simple; it is designed in a format that is familiar through other check forms. By typing a "?" the user is able to access the necessary classification codes.

Making this window available facilitates quick and correct entry, because the user need not remember the infrequently used codes or leave the screen to check a hardcopy list for them before completing data entry. In addition, the bookkeeping software can be programmed to accept only the classification codes listed on the screen, automatically producing the window if the code entered was incorrect and preventing errors that might have occurred if all classification codes were accepted by the system.

Windows and hyperlinks have almost limitless applications. For instance, they may allow users to stop data entry and jump to or pull down another menu choice, to get visual help showing how the data-entry process should proceed, to calculate a value onscreen and then return to data entry to enter the sum calculated, to set an alarm clock for an appointment reminder, or to perform a number of other possible actions.

Because multiple windows are possible, any or all of the preceding functions could appear in varying sizes or as hyperlinked graphics, images, icons, or words on a screen at the same time. Clicking the right mouse button might bring up more window options. Entering another command or pressing a predefined key might

FIGURE 16.11

A window can keep the screen simple.

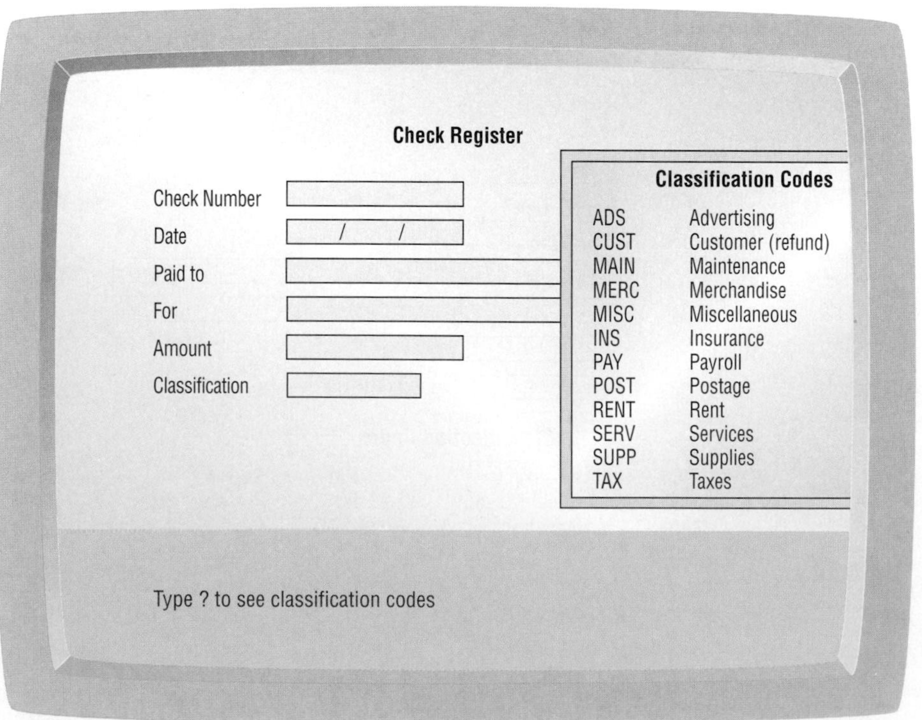

allow the operator to clear the windowed information from the screen and return to the original display.

One of the main disadvantages of windows is that they allow the user to overcomplicate a simple screen. By including too many windowing operations, the designer may be inviting the creation of a chaotic-looking screen that subsequently results in a user who is lost and frustrated by all the clutter. Windows and hyperlinks must be used judiciously so that they will support the purpose of the user, not hinder it.

KEEPING THE SCREEN CONSISTENT

The second guideline for good screen design is to keep the screen display consistent. If users are working from paper forms, screens should follow what is shown on paper. Screens can be kept consistent by locating information in the same area each time a new screen is accessed. Also, information that logically belongs together should be consistently grouped together: Name and address go together, not name and zip code. Although the screen should have a natural movement from one region to another, information should not overlap from one group into another. You would not want name and address in one area and zip code in another.

The check register depicted earlier helps users input data correctly because it requests data that are consistent with familiar paper checks. For example, "Date," "Paid to," and "Amount" are all expected to be found on a check.

When the user accesses the next screen so as to input the next check, "Check Number," "Date," and "Paid to" are all in the same place as on the previous screen. Keeping the data-entry fields in the same place for every screen helps maintain consistency. Recalling and using the three sections of a screen presented previously will also help designers keep screens consistent. Consistency also means using the same terms and acronyms consistently on several screens. For example, do not use "FAQ" on one screen and "Questions" on another.

FACILITATING MOVEMENT

The third guideline for good screen design is to make it easy to move from one screen to another. The three click rule says that users should be able to get to the screens they need within three mouse or keyboard clicks. Web-based forms facilitate movement with the use of hyperlinks to other relevant screens. Another common method for movement is to have users feel as if they are physically moving to a new screen. There are at least three ways this illusion of physical movement among screens is developed.

Scrolling. An example of effective scrolling is shown in Figure 16.12, which depicts employee records with name, address, and city displayed on the first screen. This example is from an Excel spreadsheet. The screen designer used the command "freeze pane" to ensure that row 1 and columns A and B will always be displayed no matter which way the user scrolls, as shown in Figure 16.13.

Another method mimicking physical movement from screen to screen is to employ keys already assigned on the computer keyboard. PC keyboards have keys labeled "Pg Up" for page up and "Pg Dn" for page down, which in effect take the user to a new page (screen display). These keys make it extremely easy for inexperienced users to change screens, although they do take up additional keyboard space and force users to take their hands off the alphabetic keyboard to use the keypad. These keys are commonly employed on Web sites as well, with the inclusion of commands such as "Go to Top of Page" or a "Back" command that allows the user to return to the immediately previous screen. Standard mainframe function keys for scrolling are F8, next page; F7, previous page; F11, scroll right; and F10, scroll left.

	Last Name	First	Address	City
1	Last Name	First	Address	City
2	Adams	Susan	4392 68th Street	Haddonfield
3	Bernard	David	5143 Hill Street	Cherry Hill
4	Clayton	Murray	242 Pawnee	Cherry Hill
5	Creighton	Stanley	6340 75th Street	Haddonfield
6	Dempsey	Janet	78 Fox Hollow Court	Haddonfield
7	Dodsworth	Helen	1134 Harrison Ave.	Cherry Hill
8	Gershwin	Ira	254 Decker Street	Voorhees
9	Hernandez	Jorge	7022 Wright Street	Haddonfield
10	Jester	Carol	2550 Deleware Avenue	Voorhees
11	Koni	Karl	4773 Hennipin Ave.	Haddonfield
12	Morgan	Janice	2002 Elmwood Ave. Apt 301	Cherry Hill
13	Newton	Issac	344 Memorial Drive	Haddonfield
14	Otto	Ronald	3114 Seneca Road	Voorhees
15	Simpson	Sheila	56 Armin Place	Cherry Hill
16	Vargo	Bradley	9550 Hayes Blvd.	Haddonfield
17	Wilcox	Mary	2680 Shadywood Road	Cherry Hill
18	Wu	Yi	203 Lincoln Avenue	Voorhees
19	Yarrow	Barbara	1020 Rogers Street	Cherry Hill

FIGURE 16.12
The first screen shows columns A to D. To see more, the user must scroll. This example was done in Microsoft Excel.

GUI screens should not scroll if at all possible. Sometimes there is a smaller window displaying repeating information, such as the courses for a student on a student screen, that may scroll if there is more information than will fit in that display area. On the other hand, Web pages often involve scrolling.

Calling Up More Detail. Another general approach to movement between screens allows users to call up another screen quickly by using cursor positioning along with a specific command. For example, Figure 16.14 exhibits a screen devoted to

	Last Name	First	Phone	Dept	Ext	Hired
1	Last Name	First	Phone	Dept	Ext	Hired
2	Adams	Susan	(856) 897-5454	Operations	3371	May-94
3	Bernard	David	(856) 534-8975	Marketing	3389	Jun-94
4	Clayton	Murray	(856) 534-8765	Operations	3372	Jan-97
5	Creighton	Stanley	(856) 897-5871	Acccounting	3364	May-00
6	Dempsey	Janet	(856) 897-4418	Research	3351	Dec-99
7	Dodsworth	Helen	(856) 534-0812	Operations	3375	Apr-88
8	Gershwin	Ira	(856) 226-6593	Operations	3376	Jun-94
9	Hernandez	Jorge	(856) 897-6884	Acccounting	3362	Aug-98
10	Jester	Carol	(856) 897-7731	Acccounting	3369	Feb-97
11	Koni	Karl	(856) 897-0091	Research	3353	Aug-00
12	Morgan	Janice	(856) 534-9125	Marketing	3386	Mar-00
13	Newton	Issac	(856) 897-2660	Marketing	3382	May-98
14	Otto	Ronald	(856) 897-7214	Shipping	3391	Jul-89
15	Simpson	Sheila	(856) 534-3009	Information Systems	3304	Oct-92
16	Vargo	Bradley	(856) 897-5131	Advertising	3321	Sep-96
17	Wilcox	Mary	(856) 534-1778	Information Systems	3309	Jun-94
18	Wu	Yi	(856) 897-2810	Security	3000	Nov-99
19	Yarrow	Barbara	(856) 534-2077	Operations	3343	Jan-98

FIGURE 16.13
After scrolling, the user still sees columns A and B and also sees columns F to I. This example was done in Microsoft Excel.

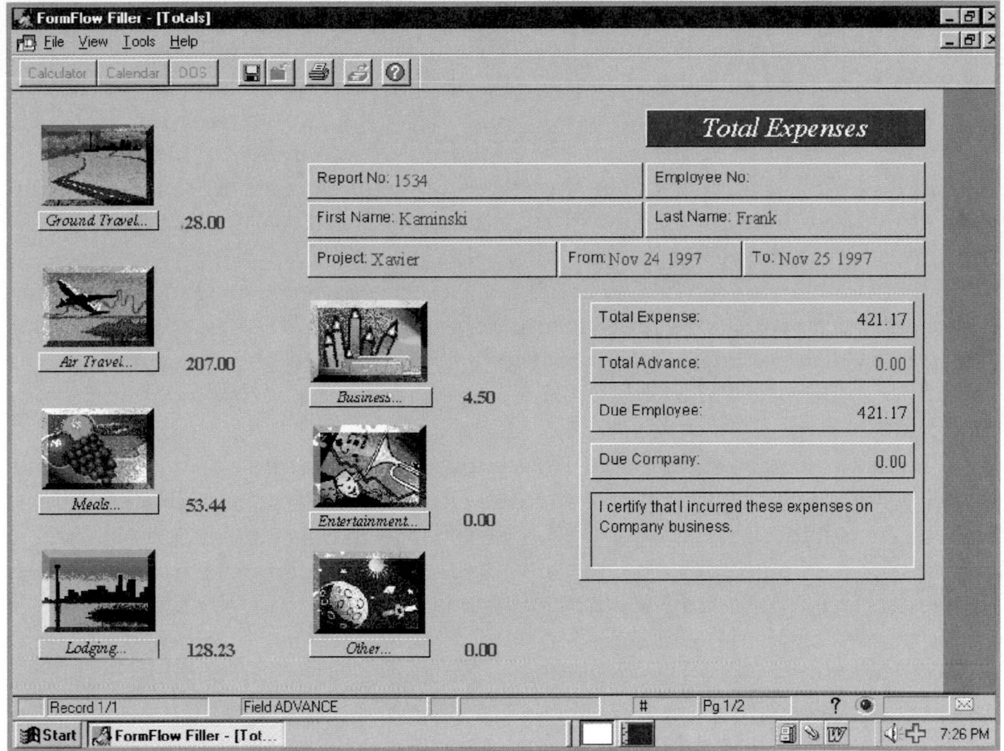

FIGURE 16.14
An overview screen of employee travel expenses designed using JetForm's FormFlow Filler.

employee travel expenses records. For the budget manager to call up a specific employee meal expense, he or she just positions the cursor over the button labeled "Meal . . ." and clicks the mouse.

That command brings up the next screen of detailed employee meal expenses, as shown in Figure 16.15. The screens could be effectively hyperlinked as well. The manager sees a breakdown of employee meal expenses.

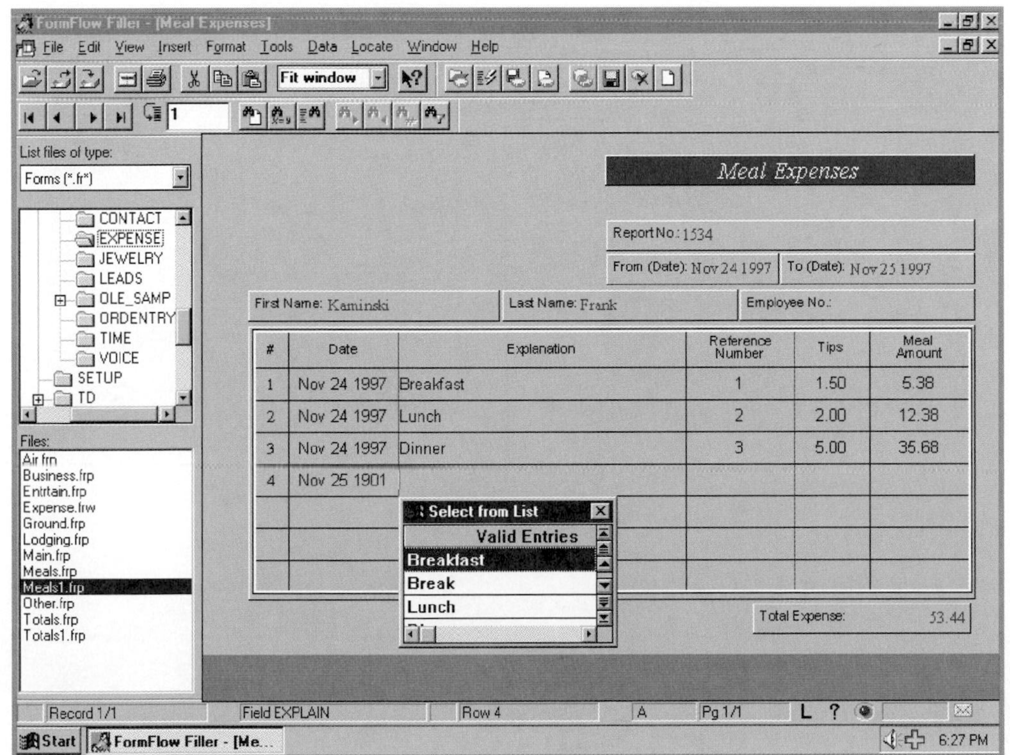

FIGURE 16.15
Calling up more detail on an employee's meal expenses (screen designed using FormFlow Filler by JetForm).

Onscreen Dialog. Displaying prompts facilitates a special kind of user movement between screens. Prompts are extremely useful in applications such as telemarketing. The initial screen used by Summerfest volunteers at public station DETV for their Summerfest fund drive is shown in Figure 16.16. Volunteers who answer the phone are facing this first screen when the phone rings. The left side of the screen shows the dialogue (prompts) that the volunteer should use. The right-hand side of the screen shows default (most likely) responses for each question.

The phone answerer begins a call by asking if the person is calling to contribute. If the answer is yes, the default response (the Enter key, for instance) is pressed, and the volunteer continues the dialogue with the second question. The volunteer types the caller's last name into the database that contains active contributors.

The next question should confirm what the caller has said about contributing. If the person never before contributed to DETV, a window is pulled up on the screen (or a hyperlink can be used for Web-based fill-in forms), as shown in Figure 16.17. The prompt reminds the volunteer to ask for and enter the caller's address and then to inquire about the existence of an apartment number.

Alternatively, if the caller responds that he or she has contributed before, a different window is called up, as shown in Figure 16.18, or a hyperlink can be followed. This screen provides the caller's last-known address and then prompts the volunteer to ask if it is still correct. The dialogue proceeds until the contribution amount and method of payment are finalized. All throughout the contribution process, the volunteer has used split-screen dialogue prompts and defaults to move from screen to screen to procure and enter the correct data.

An interesting method of creating the feel of physically moving from screen to screen is direct manipulation (a GUI discussed in detail in Chapter 18). An example is designing a cursor to look like a human hand, in other words, an

FIGURE 16.16

First screen for public broadcasting station DETV's Summerfest drive.

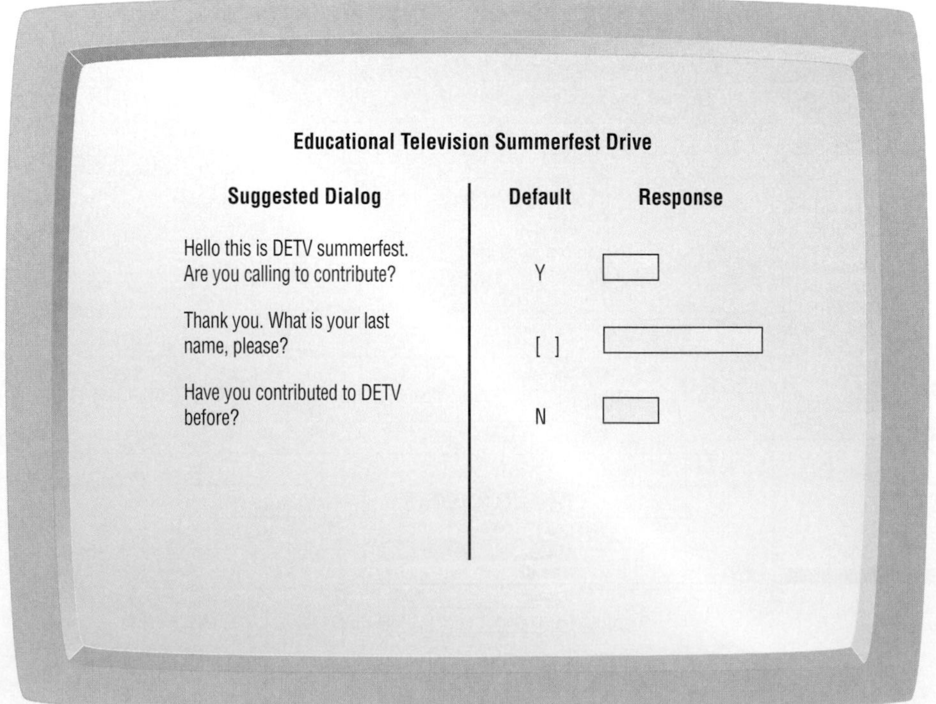
Educational Television Summerfest Drive
Suggested Dialog
Default
Response
Hello this is DETV summerfest. Are you calling to contribute?
Y
Thank you. What is your last name, please?
[ ]
Have you contributed to DETV before?
N

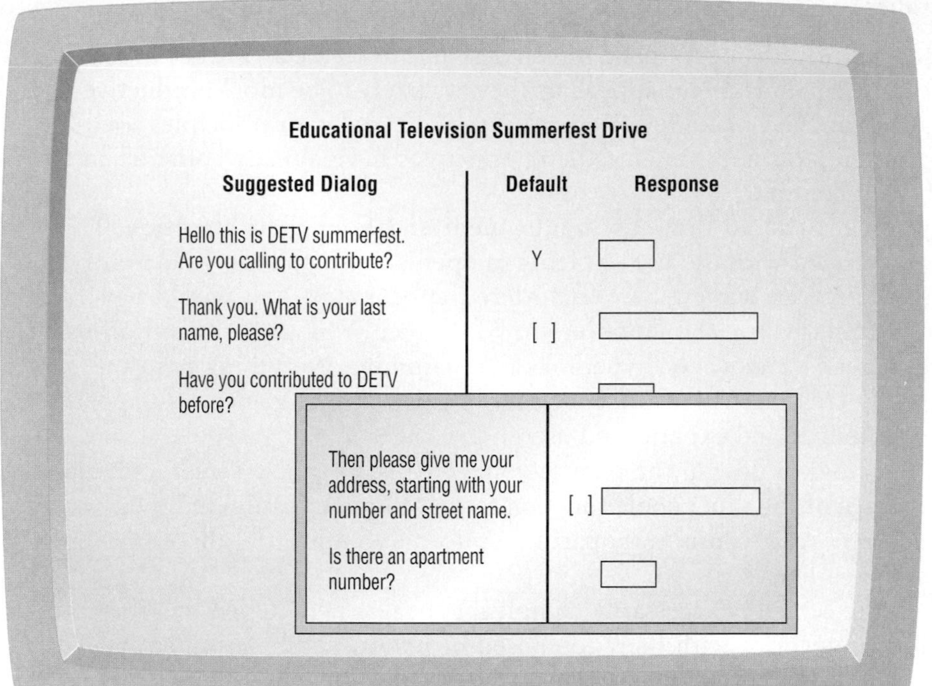
Educational Television Summerfest Drive
Suggested Dialog
Default
Response
Hello this is DETV summerfest.
Are you calling to contribute?
Y
Thank you. What is your last
name, please?
[ ]
Have you contributed to DETV
before?
Then please give me your
address, starting with your
number and street name.
[ ]
Is there an apartment
number?

FIGURE 16.17
Window displayed if the caller has not contributed to DETV before.

onscreen icon. This onscreen hand would be used to push forward to a new screen or back to a previous one by use of a mouse in a way that is analogous to pushing a piece of paper upward or downward on a desk. This method also works on Web-based fill-in forms. Remember that you have standard features such as a "Back" button and a "Home" button that can be invoked when designing a Web form.

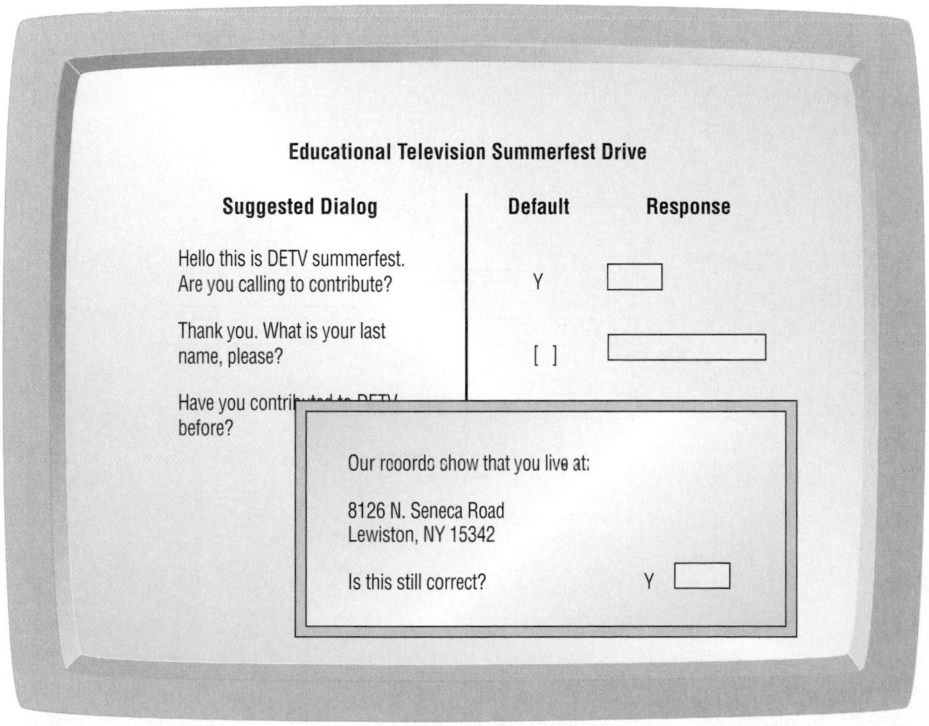

Educational Television Summerfest Drive

Suggested Dialog | Default | Response

Hello this is DETV summerfest.
Are you calling to contribute? Y

Thank you. What is your last
name, please? [ ]

Have you contributed to DETV
before?

Our records show that you live at:

8126 N. Seneca Road
Lewiston, NY 15342

Is this still correct? Y

FIGURE 16.18
Window displayed if the caller has contributed to DETV before.

DESIGNING AN ATTRACTIVE SCREEN

The fourth guideline for good screen design is to create an attractive screen for the user. If users find screens appealing, they are likely to be more productive, need less supervision, and make fewer errors. Some of the design principles used for forms apply here, too, and some aesthetic principles have already come up in a slightly different context.

Screens should draw users into them and hold their attention. This goal is accomplished with the use of plenty of open area surrounding data-entry fields so that the screen achieves an uncluttered appearance. You would never crowd a form; similarly, you should never crowd a screen. You are far better off using multiple screen windows or hyperlinks than jamming everything onto one screen or page. By creating screens that are easy to grasp at first glance, you appeal to both inexperienced and experienced users.

Use logical flow in the plan to your screens. Organize screen material to take advantage of the way people function so that they can easily find their way around the screen. Also, consistently partition information into the three smaller sections detailed earlier.

If the screen is necessarily complex, appeal is heightened by separating information categories with lines composed of periods, dashes, ampersands, exclamation points, or boxes. The check register screen shown in Figure 16.19 uses boxes to define data-entry fields. Note that the boxes are left-justified, which results in an orderly look for the screen.

Thickness of separation lines between subcategories can also be varied to add further distinctions. Variety helps the user to see quickly both the purpose of the screen and what data items are required.

With the advent of GUIs, it is possible to make input screens very attractive. By using color or shaded boxes and creating three-dimensional boxes and arrows, you can make forms user-friendly and fun to use. Figure 16.20 shows an example of an order entry screen that is effective. These features are also available for Web-based fill-in forms.

FIGURE 16.19

A check register screen that shows balanced design.

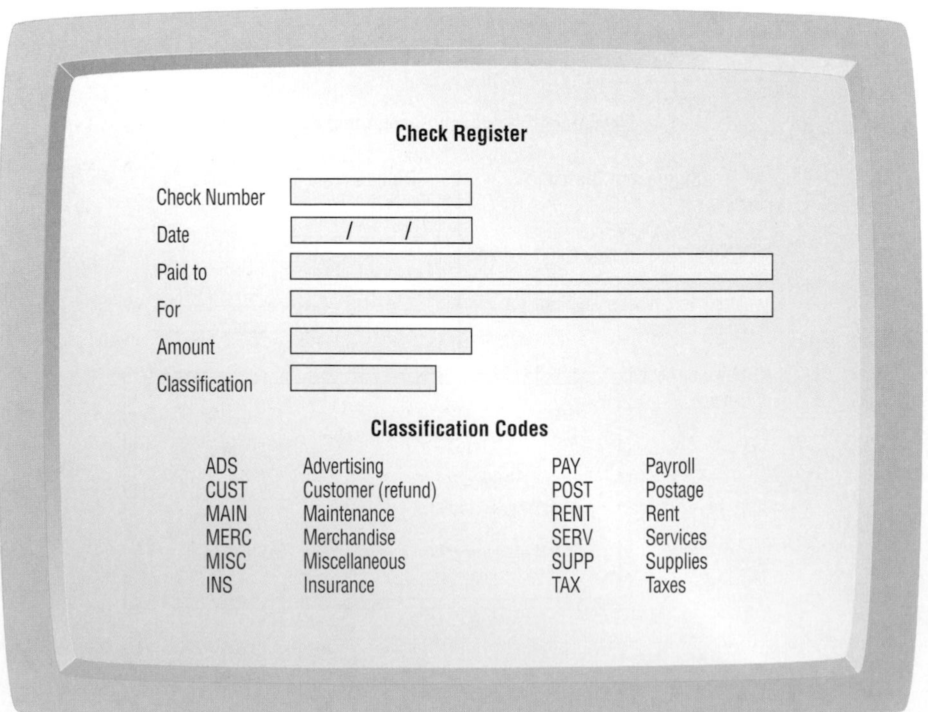

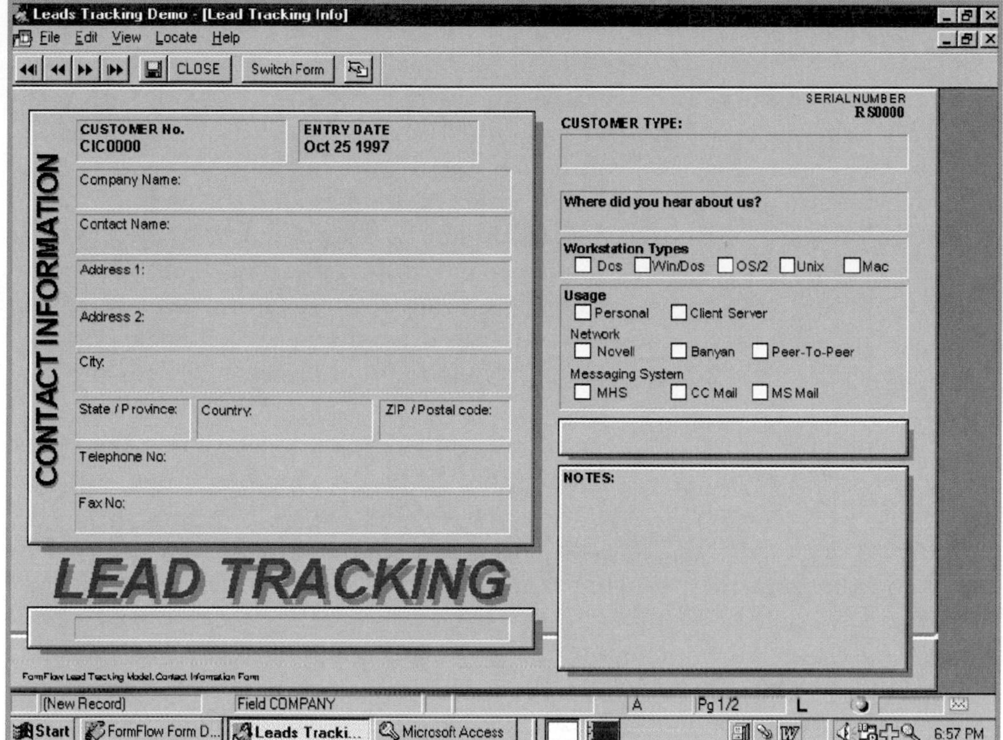

FIGURE 16.20
You can design an attractive data entry screen with a 3-dimensional effect using JetForm's FormFlow.

Inverse Video and Blinking Cursors. Other techniques can also effectively enhance the attractiveness of screens, but only if they are used sparingly. They include inverse video, blinking cursor or fields, and type fonts in various styles and sizes.

When you are considering the use of these techniques, simplicity is still the watchword. Design the basic screen that will include basic information first. Then, if greater differentiation is still needed, the basics can be embellished. Fortunately, additional enhancements may not be as costly for screen design as they are for forms.

Inverse video swaps the foreground color for the background color. It is an excellent way to highlight an important field, but employing it can cause users to be so overwhelmed with the brightness of the inverse that they ignore other fields.

Blinking video displays, when used for a cursor, are one way to alert an inexperienced user to its location. Blinking can also be used for error control by calling attention to a field that was skipped. Some users, however, find blinking video annoying, imagining that the computer is impatiently awaiting the next entry.

Using Different Type Fonts. State-of-the-art computer systems and software allow type fonts of different styles and sizes. Type fonts are another way to make screens attractive to users. Different styles enhance differentiation among categories. For instance, thick, sans serif type styles can be used to denote main categories and to give screens a modern look. Larger type can indicate captions for data-entry fields. Thinner type with serifs can be used to designate subcategories on the same screen and can provide a more conservative look. Fonts should not have a size of less than 8 points. A font of 10 points is more readable.

When contemplating the use of different type styles and sizes, ask yourself if they truly assist the user in understanding and liking the screen. If they draw undue attention to the art of screen design or if they serve as a distraction, leave them out. Be aware that not all Web pages are viewed identically by different browsers. Test your forms with a variety of combinations to see if the resultant color combinations will be pleasing or distressing to the majority of users.

DIFFERENCES IN MAINFRAME AND PC SCREEN DESIGN

Mainframe and PC design have much in common, but there are some critical differences between them. A mainframe computer is designed to work with many terminals, whereas a PC is a self-contained processing unit and terminal. To increase mainframe efficiency, screens are sent as a whole rather than as a series of individual keystrokes. Conversely, a PC is designed to respond to any keystroke.

When using a PC, the user may be prompted to press a certain key, and the program responds by displaying information. For example, most spreadsheet programs allow the user to press a slash (/) to bring up a menu of options. A microcomputer program may pause for the user to examine the screen, which displays the message "Press any key to continue." When the user presses a key, the program continues.

A mainframe computer operates differently. The data entered on the screen are stored at the terminal and not transmitted to the mainframe until the user presses one of several keys designated as transmission keys. These keys (on an IBM mainframe) are called AID keys, for *A*ttention *ID*entifier, and they include the Enter key, the Clear key, function keys 1 to 24 (PF1–24), and three *P*rogram *A*ttention keys called PA1 through PA3. When one of these keys is pressed, the screen data, along with whichever of the keys has been pressed, are transmitted to the mainframe. By using AID keys, the mainframe does not have to respond to each terminal keystroke, and each user thus benefits from improved response time. (Response time is the time that elapses between the moment the user presses an AID key and the appearance of a new screen generated by the computer.)

Another technique used in mainframe systems is to transmit only the data fields on a screen that have been altered (that is, those fields in which the user has entered new data or changed existing data). Clearly defining the screen fields ensures that only the minimal number of characters are transmitted from each terminal to the central computer, and response time again is thereby improved.

Attributes. The data fields on a screen are defined using a field attribute character. The attribute character occupies one screen position, immediately preceding the data field, and it always appears as a blank (invisible) character on the screen. Figure 16.21 illustrates a screen used to enter new item information with the attribute characters indicated by small rectangles. Notice that these rectangles are located before, and sometimes after, each screen data field.

Attribute characters control the characteristics of the screen field to the right of them and include the following qualities:

1. Protection.
2. Intensity.
3. Shift and extended attributes.

These qualities are described in the following subsections.

Protection Protection determines whether the user may enter data into the screen field or not. There are three types of protection: unprotected, which allows any users to enter data; protected, used for output display information such as an operator message; and auto skip, used for captions, titles, and other screen features that do not change. The cursor will only be placed in unprotected areas of the screen.

On the ADD NEW ITEM screen, there is a protected attribute in front of the date, the time, the operator message (indicating the function key assignments), and the feedback message line. As the program processes input data, different information will be sent to these fields, but the user will not be allowed to key any data into them.

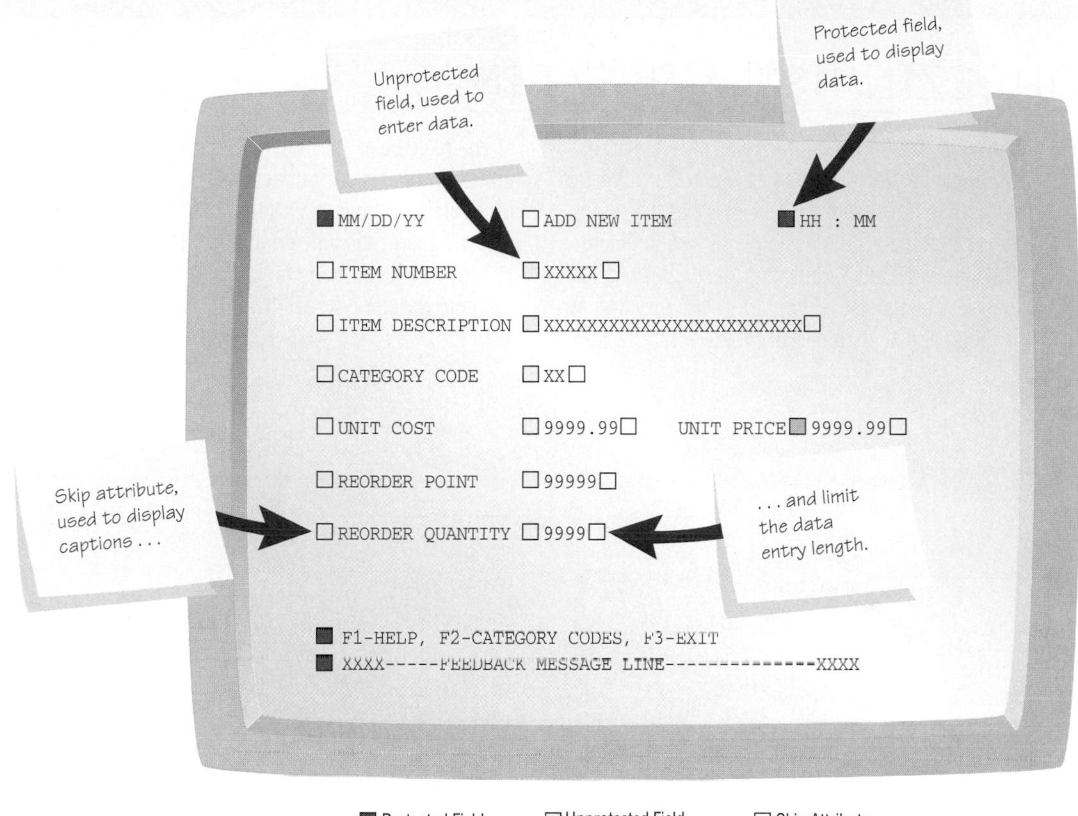

FIGURE 16.21
A screen showing attribute characters.

The title ADD NEW ITEM and the captions—ITEM NUMBER, ITEM DESCRIPTION, CATEGORY CODE, and so on—all have the skip attribute, because these displays do not change on the ADD NEW ITEM screen.

The Xs representing where the ITEM NUMBER, ITEM DESCRIPTION, and CATEGORY CODE—as well as the 9s representing areas where numeric data is to be entered—are all designated as unprotected. After each unprotected screen field is another auto skip attribute character used to control the number of characters the user may enter for each screen field. As each character is entered, the cursor advances to the right. When the auto skip attribute character is reached, the cursor "jumps" to the next unprotected screen field. When the last entry field is completed, the cursor wraps around to the first screen field.

Intensity Intensity is how bright a screen field will appear. Three choices are available: normal, generally used both for displaying captions, titles, date, time, and other fixed information and for entering data; bright, or high intensity, often used to display error messages and to highlight an error in a field; and invisible, usually reserved for passwords. A combination of protected or unprotected and normal or bright intensities provides different screen colors for data and captions: usually white, green, blue, and red.

Shift and extended attributes The shift attribute limits data fields to numeric or alphanumeric entries when data are keyed. If a screen field uses the numeric shift attribute, data entered as 123 in an 8-character field will be both right-justified in the field and padded with zeros on the left after the Enter key has been pressed, resulting in 00000123. This attribute is frequently used for numeric data-entry fields. Extended attributes may be included, which provide for reverse video, underline, extra color, and blinking.

CONSULTING OPPORTUNITY 16.2

SQUEEZIN' ISN'T PLEASIN'

The Audiology Department in a large veteran's hospital is using a PC and monitor so that audiology technicians can enter data directly into the patient records system. After talking with Earl Lobes, one of the technicians, you determine that the screen design is a major problem.

"We used a form at one time, and that was decent," said Mr. Lobes. "The screen doesn't make sense, though. I guess they had to squeeze everything on there, and that ruined it."

You have been asked to redesign the screen (Figure 16.C2) to capture the same information, but simplify it, and by doing so, reduce the errors that have been plaguing the technicians. You realize that squeezing isn't the only problem with the screen.

Explain your reasons for changing the screen as you did. You may use more than one screen if you think it is necessary.

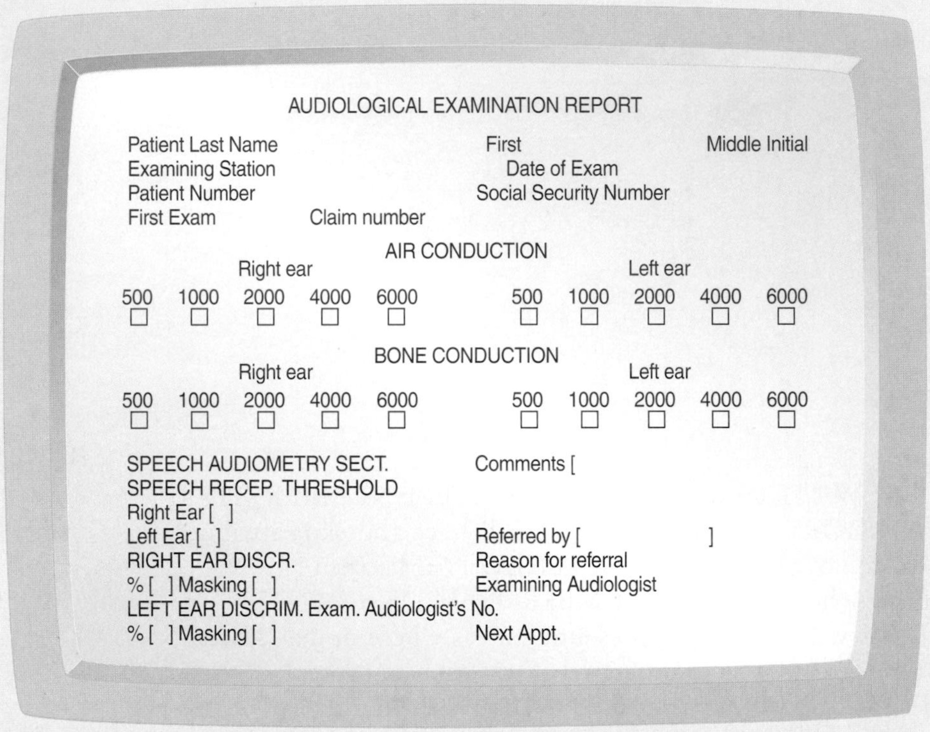

FIGURE 16.C2
This screen can be designed to be more user-friendly.

Attribute Character Considerations. When designing mainframe terminal screens for entering data, the attribute character must be taken into consideration. At least one space must be reserved between captions and data-entry areas for the attribute character. Dates are usually entered without any slashes, spaces, or hyphens separating the month, day, and year. The reason for not entering a slash, for example, is that either the operator would have to key the slash or the program would have to display the slash. Having the operator key the slash slows the data-entry process and may lead to the inconsistent entry of dates. Some users might enter dates with slashes, whereas other users might forget to enter the slashes.

If the program displays the slashes and the entire date entry field is unprotected, the operator may accidentally overtype the slashes. If the slashes are placed on the screen as protected entry fields, there must be an attribute byte before and after each of the slashes to protect them. The field would have the following attribute character format: **uMMs/uDDs/uYYYYs**, where **u** represents the unprotected attribute character necessary for entering a month, day, or year and **s** represents the auto skip attribute for jumping the cursor over the slashes. The data would appear to be spread out across the screen and would display as MM/DD/YYYY.

This consideration is also taken into account when designing entry fields for other fields that have editing characters in them, such as a standard U.S. telephone number: (nnn)nnn-nnnn. Social Security numbers are another case in point. Finally, including a fixed screen location for the placement of the decimal point in amount fields is a common example of the need for attribute characters.

Screen Code Generation. Screens may be designed using a number of CASE tools. Visible Analyst screens may be used to generate COBOL and other language code. A commonly used IBM mainframe screen programming language is CICS (Customer Information Control System, sometimes pronounced "kicks"). CICS code makes extensive use of attribute character control and is one of the most efficient means for sending and receiving mainframe screens. Many of the powerful CASE code generators create CICS code. Products will take the CICS code and transform it into a GUI screen in a process called wrapping. CICS is also used in many large client/server systems to extract the data from the server.

Figure 16.22 is an example of a mainframe CICS screen for adding customer payments. A CUSTOMER NUMBER is entered, and customer information (NAME, STREET, BALANCE DUE, and so on) is displayed using data from the CUSTOMER MASTER file. An INVOICE NUMBER, CHECK NUMBER, and PAYMENT AMOUNT are entered on the lower portion of the screen. Figure 16.23 is an example of the corresponding Windows screen that is generated using wrapping software. Notice that all entry or protected display fields are surrounded by rectangles, the standard entry field for a graphical user interface screen. Captions do not have a rectangle, because they do not change as various payments are entered. Protection, color, and so on are all translated onto the graphical screen.

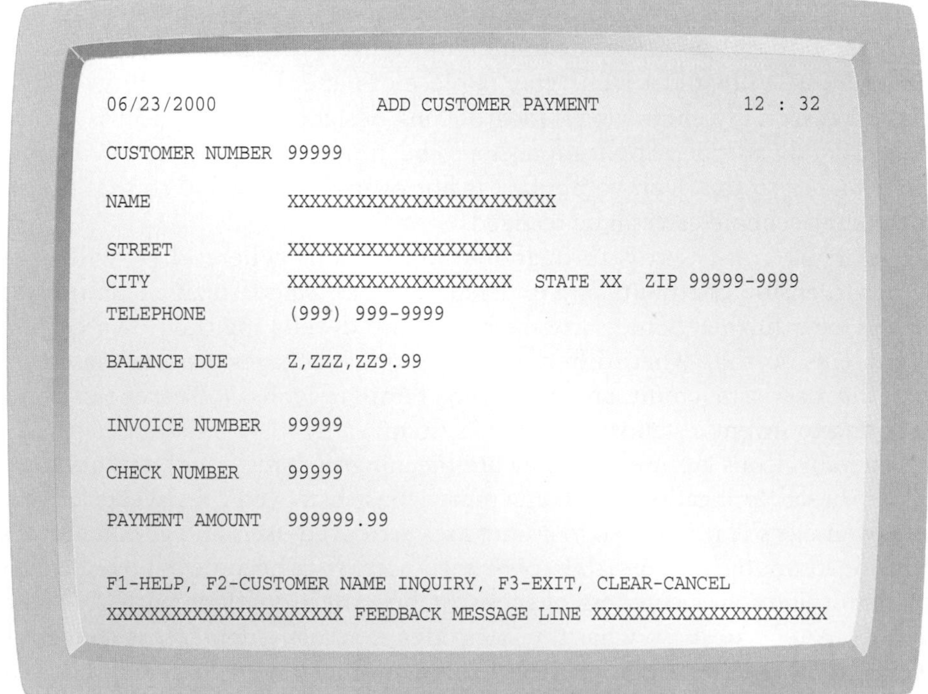

FIGURE 16.22
Mainframe screen for adding customer payments.

FIGURE 16.23

Windows screen that corresponds to the mainframe screen.

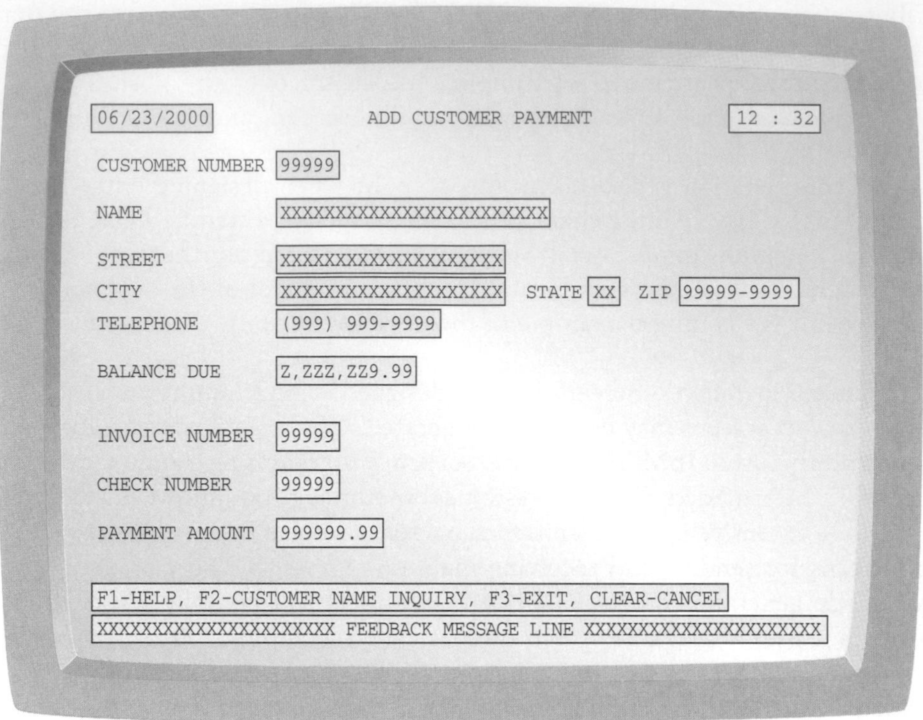

USING ICONS IN SCREEN DESIGN

Icons are pictorial, onscreen representations symbolizing computer actions that users may select using a mouse, keyboard, lightpen, or joystick. Icons serve functions similar to those of words and may replace them in many menus, because their meaning is more quickly grasped than words. Icons designed for the spreadsheet for Excel shown in Figure 16.24.

There are some guidelines for the design of effective icons. Shapes should be readily recognizable so that the user is not required to master a new vocabulary. Numerous icons are already known to most users. Use of standard icons can quickly tap into this reservoir of common meaning. A user may point to a file cabinet, "pull out" a file folder icon, "grab" a piece of paper icon, and "throw it" in the wastebasket icon. By employing standard icons, designers and users all save time.

Icons for a particular application should be limited to approximately 20 recognizable shapes so that icon vocabulary is not overwhelming and so that a worthwhile coding scheme can still be realized.

Use icons consistently throughout applications where they will appear together to ensure continuity and understandability. Standardizing icon usage can be taken even further. Some software houses are developing their own corporate icon systems so that when different application packages are purchased from them, the user can count on employing familiar icons. Researchers are also attempting to invent a standardized icon system.

Generally, icons are useful if they are meaningful. New environments for PCs and Web-based applications use many more icons than even a few years ago. Their chief advantage so far has been to attract inexperienced users and get them excited about the computer's potential. Experienced users may become annoyed with the pseudosimplicity and cuteness of icons or become impatient with the way in which icons tend to mask what the computer is actually doing. It is conceivable, however, that experienced users prefer command language (discussed in Chapter 18) more out of habit than out of any real dislike for icons.

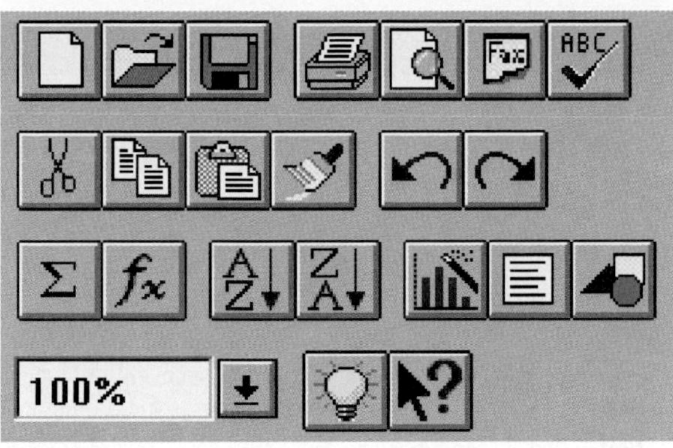

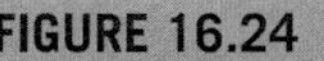

FIGURE 16.24
Icons from Microsoft Excel.

GRAPHICAL USER INTERFACE DESIGN

A graphical user interface (GUI, pronounced "gü-é") uses a Windows, Macintosh, or other graphics screen for entering and displaying data. Although these screens have traditional data entry and display fields, several additional features are also included in the screen design. Figure 16.25 is a Microsoft Access input screen showing a variety of GUI controls.

Text Boxes. A rectangle represents a text box, as mentioned previously, and is used to outline data-entry and display fields. Care must be made to ensure that the text box is large enough to accommodate all the characters that must be entered. Each text box should have a caption or label to the left, identifying what is to be entered or what is displayed in the box. Care should be used to select the proper alignment of data within the box. Character data should be aligned on the left, and numeric data should be aligned on the right. The border should be the same for all text boxes on the screen, usually a simple line, a sunken look, or a raised appearance. Rectangles are also aligned, often on the left within columns and with the top border within rows.

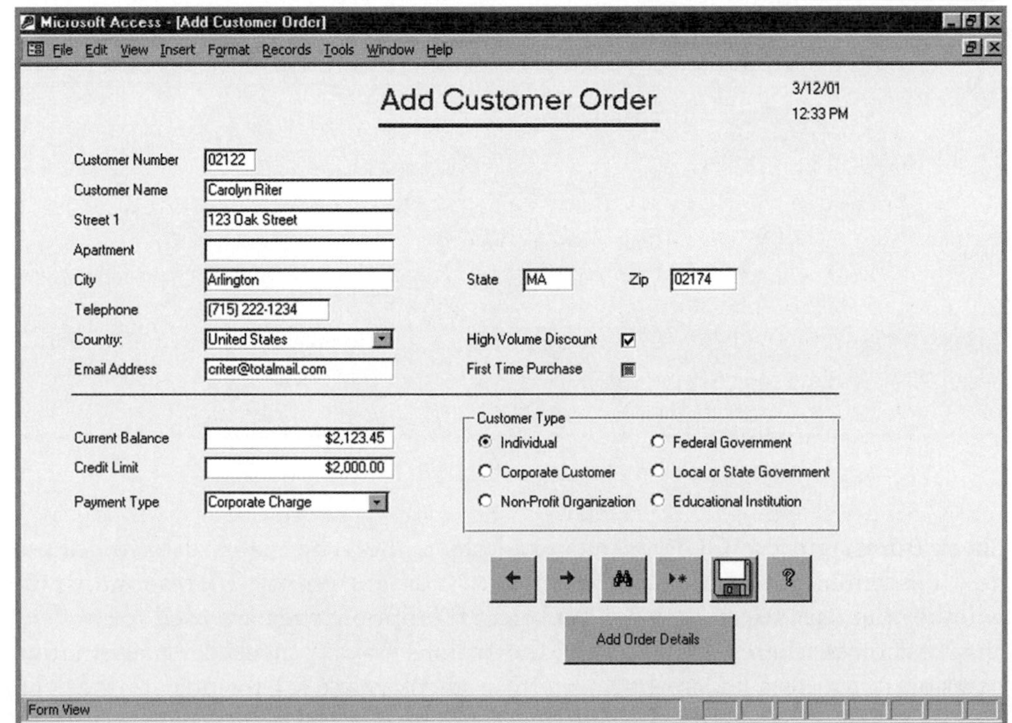

FIGURE 16.25
The designer has many GUI components which allows flexibility in designing input screens for the Web or other software packages. This example is from Microsoft Access.

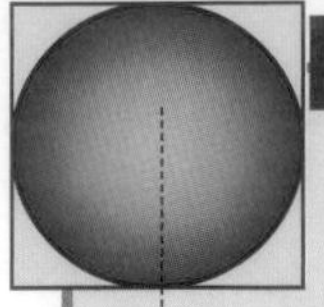

CONSULTING OPPORTUNITY 16.3

WHAT'S THAT THING SUPPOSED TO BE?

Art Istik flips off his monitor with a loud click. "I've just about had it," he says, turning impatiently to his colleague. Looking at Art with mock sympathy, Sim Ball says, "New system is too much for you, isn't it?" Art replies, "No, it's not, but I'll tell you what's really wrong. It's these silly pictures."

Art clicks on his newly installed microcomputer, rebooting a database management program that appears on his monitor. The first screen shows icons shaped like a Sherlock Holmes cap, some kind of tree, a pair of socks, an apple, a door, and a rabbit. Sim, leaning over Art's shoulder, takes one look at the screen and laughs uncontrollably.

Art says sarcastically, "I knew you would be able to help." Sim manages to stop laughing long enough to point to the pair-of-socks icon and demands, "What's that thing supposed to be?"

Art replies, "I have no idea. All I know is that this database management package is from some West Coast company called 'Organic Outputs.' The software is called 'DATAPIX: The icon-based database' by a guy named Drew Ikahn. Maybe we ought to call him up. His idea of a good screen is way out. No way can I learn all these crazy pictures." Sim returns to his desk, saying, "Yeah, but at least it's entertaining."

As a last resort, Art turns to the DATAPIX user's manual, which provides translations for the unconventional icons.

Based on Art Istik's and Sim Ball's comments (and laughter), describe what you feel is amiss with the DATAPIX icons (see Figure 16.C3). To what do Art and Sim seem to attribute some of the problems with their database management program? Applying some of the information you learned about using icons effectively, redraw the DATAPIX icons to improve them. What icons have become universally recognizable with the widespread use of Windows? Draw three of them and write the meaning for each of them next to their name. In a paragraph, discuss the importance of standardizing icons. Add a paragraph voicing your opinion about whether it is possible or desirable to create one universal icon "dictionary" for use with all applications.

Description	Icon	Meaning
Apple		**Create** a file (as in Adam and Eve)
Door		**Enter** data (as you would a door)
Sherlock's Cap		**Find** (after Holmes' famous investigative powers)
Pair of Socks		**Sort** (as in laundry)
Tree		**Print** (a gentle reminder that printing destroys the trees)
Rabbit		**Copy** a file (for making multiple copies)

FIGURE 16.C3
These DATAPIX icons can be improved.

Check Boxes. In the GUI controls example, a check box is used to indicate a new customer. Check boxes contain an X or are empty, corresponding to whether the user selected or did not select the option; they are used for nonexclusive choices where one or more of the options may be checked. An alternative notation is to use a square button with a check mark (√) to indicate that the option has been selected. Note that check box text, or label, is usually placed to the right of the box. Do not state the check box text in a negative way. For example, do not use the text "Not Current Customer." If there is more than one check box, the labels should have some natural order to them, perhaps alphabetic or most commonly checked appearing first in a list. If there are more than 10 check boxes, put them in a group surrounded with a border or separated with either a line or some white space.

Option Buttons. A circle, called an option button or a radio button, is used to select exclusive choices. Either one or the other option can be chosen, but not both. An example of this device is shown by the buttons located under the caption SELECT CUSTOMER TYPE. The customer is either a REGULAR CUSTOMER or a HIGH-VOLUME CUSTOMER, but not both. The selected button has a darkened center. Choices are again listed to the right of the button, usually in some sequence. If there is a commonly selected option, it is usually selected as a default when the screen first displays. Often there is a rectangle, called an option group, surrounding the radio buttons. If there are more than six option buttons, consider using a list box or a drop-down list. If there are only two radio buttons stating "Yes" or "No," use a check box. Leave some white space between radio buttons and check boxes.

List and Drop-Down List Boxes. A list box displays several options that may be selected with the mouse. A drop-down list box is used when there is little room available on the screen. (In the figure, a drop-down list box is used for selecting the method of payment.) A single rectangle with an arrow points down toward a line located on the right side of the rectangle. Selecting this arrow causes a list box to be displayed. In the ADD CUSTOMER ORDER example, the method of payment may be chosen from this list. The scroll bars on the right side of the list box are used if there are more choices than will display within the box. Once a choice has been made, it is displayed in the drop-down selection rectangle and the list box disappears. As illustrated, the previous credit card choice in this example was MasterCard. Care must be taken that the selection text will completely display within the size of the box. Sometimes a list box will allow the users to select more than one choice (used on Web forms), but a drop-down list box allows only one choice. If there is a commonly selected choice, it is usually displayed in the drop-down list by default.

Sliders and Spin Buttons. Sliders and spin buttons are used to change data that have a continuous range of values. Moving the slider in one direction (either left/right or up/down) increases and decreases the values. Figure 16.26 illustrates the use of sliders to change the amount of red, green, and blue when selecting a new color. Spin buttons are also used to change a continuous value and are shown to the right of the sliders. The advantage of using spin buttons is that the user has much greater control when choosing values. A text entry area for entering the numeric values often accompanies spin buttons.

Image Maps. Image map fields are used to select values within an image. The user clicks on a point within an image and the corresponding x and y coordinates are sent to the program. The color image illustrated on the left in Color Picker allows the user to click on any color to select it. Arrowhead indicators on the top and left indicate the position of the selected point. The corresponding numerical values are placed in the text boxes on the right, and the sliders adjust to match the values. Image maps are used when creating Web pages containing maps with instructions to click in a certain area to show a detailed map of the region.

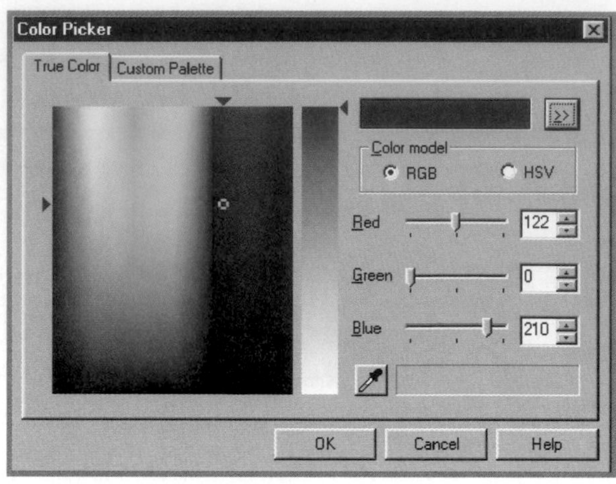

FIGURE 16.26

Sliders and spin buttons are two additional GUI components the analyst can use to design input screens.

Text Areas. A text area is used for entering a larger amount of text. These areas include a number of rows, columns, and scroll bars that allow the user to enter and view text that is greater than the size of the box area. There are two ways to handle the text within the area. One is to avoid the use of word wrap, forcing the user to press the Enter key to move to the next line. The text will scroll to the right if it exceeds the width of the text area. The other option is to allow word wrap within the text area.

Message Boxes. Message boxes are used to display warning and other messages in a dialogue box, often overlapping the screen. These message boxes have different formats, with unique buttons that display. Each should appear within a rectangular window and should clearly spell out the message. An "i" within a circle represents "information" and alerts the user with a message. The information symbol in Figure 16.27 alerts the user that the customer is already on file. An exclamation point indicates a warning message, and a stop sign is used for a critical error or other action message. Some have an **OK** button, and others have **OK** or **Cancel**.

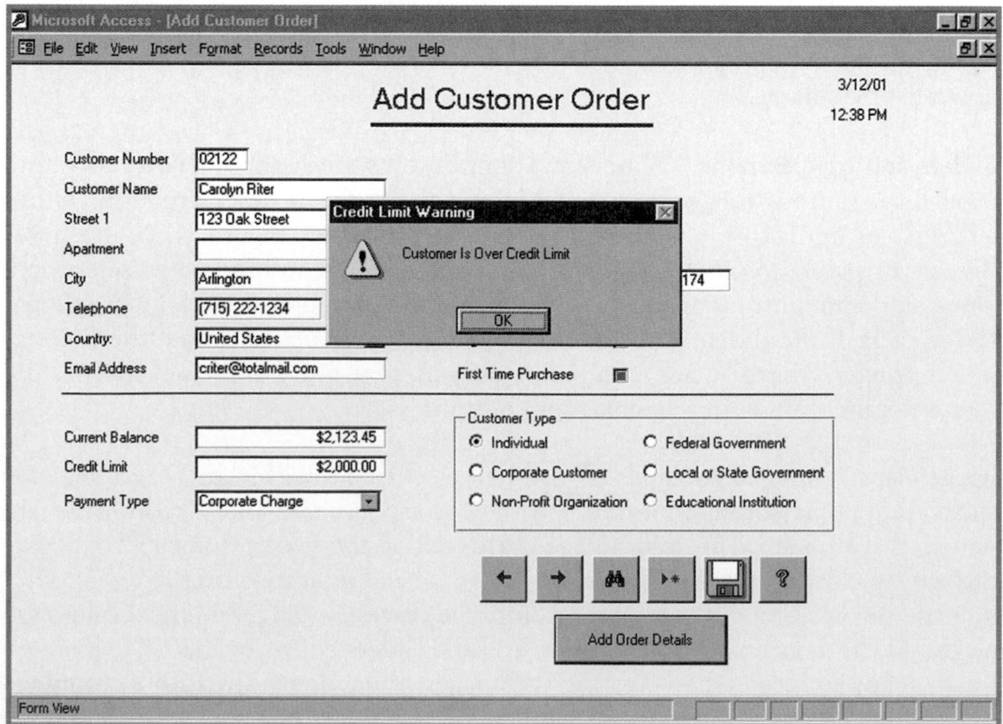

FIGURE 16.27

The designer must also develop effective message boxes to provide feedback to users. This example is from Microsoft Access.

Command Buttons. A command button performs an action when the user selects it with the mouse. CALCULATE TOTAL, ADD ORDER, and OK are examples of command buttons. The text is centered inside the button, which has a rectangular shape. If there is a default action, such as the OK button, the text is generally surrounded with a dashed line. The button may also be shaded to indicate that it is the default. The user may press the Enter key to select the default button.

TAB CONTROL DIALOG BOXES

Tab control dialog boxes are another part of graphical user interfaces and another way to get users organized and into system material efficiently. Figure 16.28 provides an example of a tab control dialog box. Guidelines for designing the control dialog boxes include the following:

1. Create a separate tab for each unique feature (for example, a tab to select color and another to select text, background, grid, or other font characteristics).
2. Place the most commonly used tabs in front and display them first.
3. Consider including three basic buttons in your design: OK, Cancel, and Help.

Microsoft Office introduced a new type of dialog box when they introduced Office 2000. The dialog box now has the look and feel or a Web page, as can be seen in Figure 16.29. On the left side are buttons called places. These buttons are hyperlinked to items a user would want to access more frequently than others. The default places are "History," which pulls up a list of the most recent files worked on; "My Documents," which is the default location for saving files; "The Desktop"; "Favorites," which are the Web sites the user bookmarked in his or her Web browser; and "Web Folders," which are the Web sites the user constructed. These places can be customized using special software such as the WOPR Placebar Customizer so that users can construct their own shortcut buttons.

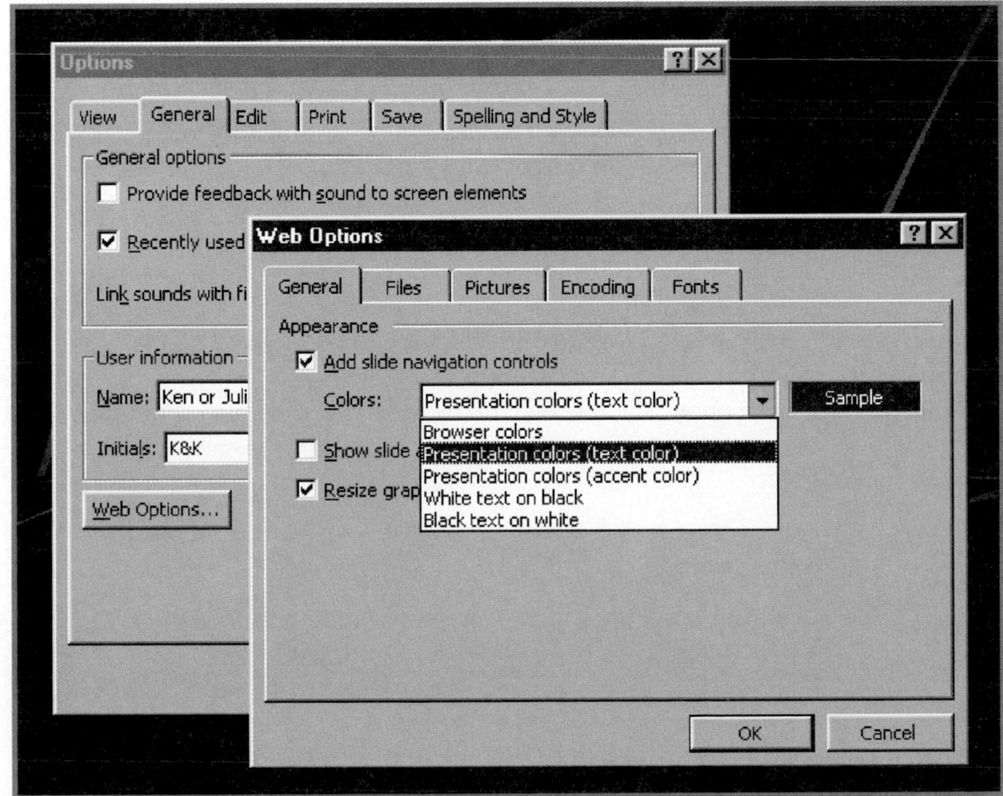

FIGURE 16.28
A dialog tab control box from Microsoft PowerPoint showing drop-down boxes and check boxes.

FIGURE 16.29

A new dialog box with the look and feel of a Web page was introduced by Microsoft in Office 2000.

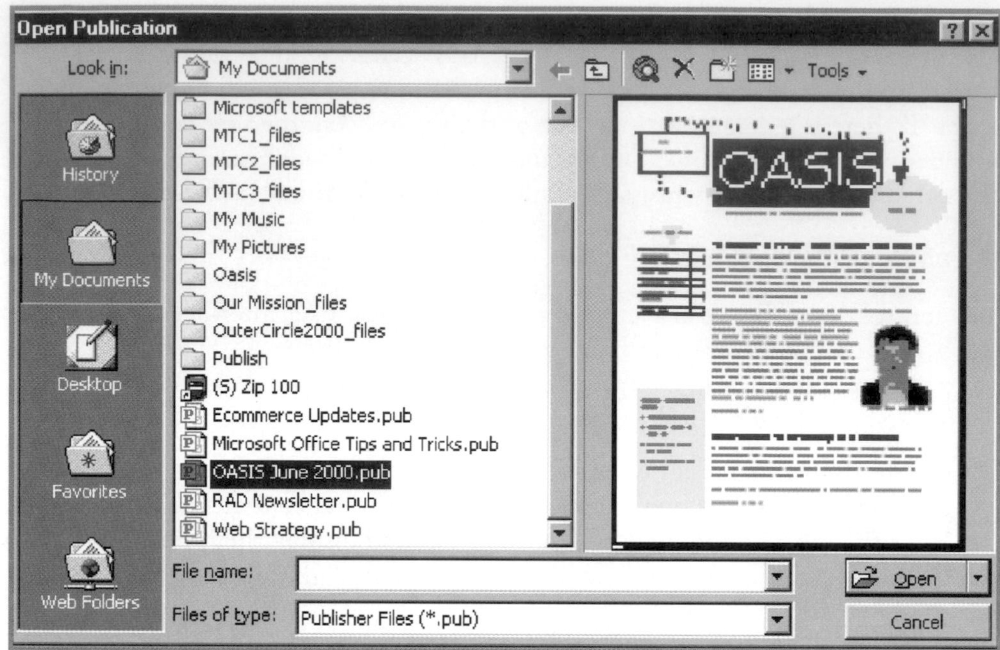

In the center of the dialogue box is the current directory; any files or folders in the current directory are displayed in this center box. The box to the right is called the viewing area. By clicking on an icon, a user can see details about the files, properties of a single file, or a preview of the current file. In this example, the document "OASIS June 2000.pub" shows a preview of a Microsoft Publisher document.

This dialogue box also has a pull-down box for easy navigation and icons that allow the user to create new folders and navigate or search the Web. It also has standard buttons to open a file or to cancel the operation. This new Web-style dialogue box is more functional than others, because it has a variety of different elements that are laid out systematically.

USING COLOR IN SCREEN DESIGN

Color is an appealing and proven way to facilitate computer input. Appropriate use of color on display screens allows you to contrast foreground and background, highlight important fields on forms, feature errors, specially code input, and call attention to many other special attributes.

Highly contrasting colors should be used for screen foreground and background. This use of color helps users grasp what is presented quickly without straining. Also, background color will affect perception of foreground color. For example, dark green may look like a different color if taken off a white background and placed on a yellow one. Specifically, the top five most legible combinations of foreground lettering on background are (starting with the most legible combination) the following:

1. Black on yellow.
2. Green on white.
3. Blue on white.
4. White on blue.
5. Yellow on black.

The least legible are red on green and blue on red. As can be gathered from these possible foreground and background combinations, bright colors should be used for foregrounds, with less bright colors for the background. Strongly contrasting colors should be assigned first to fields that must be differentiated; then other colors can be assigned.

Use color to highlight important fields on screens. Fields that are important can be colored differently from the rest. Alternatively, important fields and data can be programmed in a color that is brighter than those that are less important. Fields that are used often should be colored differently from other fields. When creating Web-based forms, hyperlinks are usually color-coded to show users that a hypertext path can be taken. Hyperlinks can be made to change color after a user has clicked on them once with the mouse. This color coding prevents users from pursuing previously-used hyperlinks and thus helps organize their search and save them valuable time.

Like colors can mean similar situations so that the red end of the spectrum (red, pink, fuchsia) might, for example, indicate dangerous or error situations. As is shown in Figure 16.30, red may mean that entering data at the time of its display will permanently alter a file.

On this screen the column depicting June's sales data for all car salespeople is red on a white background, which means it is ready to be updated. The programmer has also highlighted headings for the rows and columns in blue so that the user can easily distinguish them from other material.

Color can assist in the special coding of input and should be used in addition to a well-formatted screen. Color coding is of great help in assisting users with searching and counting tasks. Alphanumeric blocks, groups, or columns are usefully distinguished through the use of color. If displays are high density, it is wise to use a double coding strategy, one employing both color and shape or color and patterns. Many applications now permit users to customize colors.

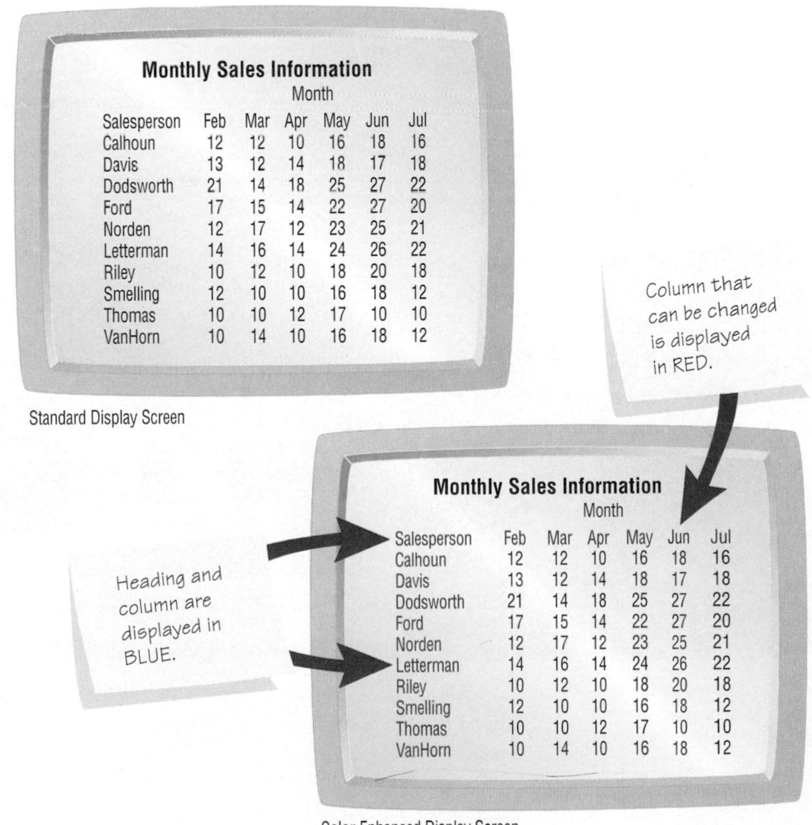

FIGURE 16.30
Color enhancement of a computer display screen.

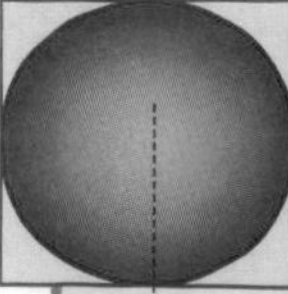

CONSULTING OPPORTUNITY 16.4

IT'S ONLY SKIN DEEP

When contemplating upgrading the design of the ecommerce Web site for Marathon Vitamin Shops, Bill Berry, the owner, realized that his customers were diverse.

"We've worked hard to attract many different types of customers. As far as the store goes, we are succeeding. People with many different interests come in. I've met sports enthusiasts who want high-energy vitamins to boost their power. Other customers want to lose weight with the help of vitamin supplements or herbal remedies. Some of our customers are health conscious and believe that a vitamin a day keeps the doctor away. Some even cling to the hippie lifestyle they cultivated in the seventies and want only organic supplements. By the way the store is set up, you can see that we're trying to segment the space so that each kind of consumer feels welcome. It's hard to translate that to the Web though."

Bill turns to one of his employees, Jin Singh, and asks her, "Is there anything we can do to transform the online catalog so that it attracts different customers? And what about being responsive to the different people who visit the site?"

Jin, who just happens to be an Internet Web cast enthusiast, says, "I have just the thing," as she turns to her computer and brings up her Windows Media Player. "Personally, I like to get into a frame of mind that matches the music or videos I am experiencing on the Web."

Jin shows Bill examples of some "skins" on the screen. You can see a variety of "skins" for the Microsoft Windows Media Player displayed in Figure 16.C4.

Jin continues, "Skins allow me to customize the appearance of my Media Player. When I play oldies, I choose a rusty skin. When I am playing something new age, I opt for a skin that has a rainbow of colors, and so on."

Peering at the screen, Bill exclaims, "I think you're on to something. What did you call those things again?"

Jin laughs and explains, "They're called skins, but they're just fun overlays that customers can add to whatever it is they're viewing. I can envision that eventually the Web site can take on an entirely new appearance depending on customer preferences for a particular kind of skin."

Based on your assessment of the different types of customers Marathon would like to attract to its Web site, design, draw, and describe a series of skins that would be appropriate for the company's purposes. Explain how the inclusion of user-controlled skins on a Web site can further the analyst's design objectives of attractiveness and ease of use for input.

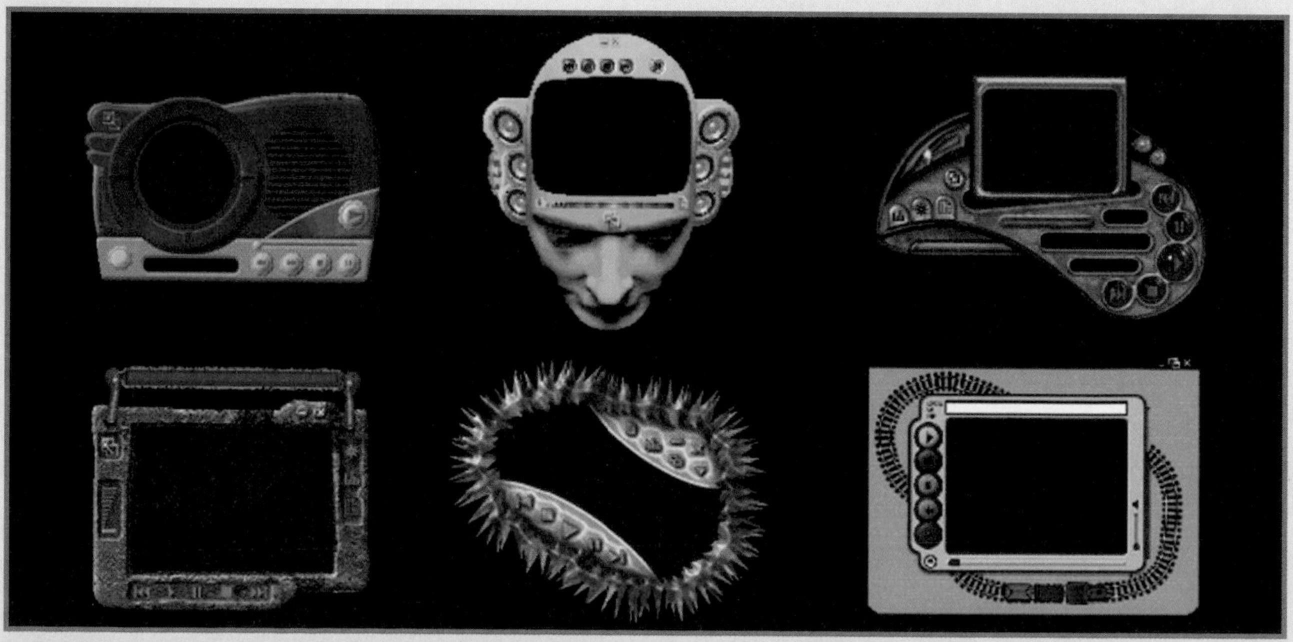

FIGURE 16.C4
Six skins from Microsoft's Windows Media Player allow users to customize their player to fit their mood.

As with any enhancement, screen designers need to question the added value of using color. Use of color can be overdone; a useful heuristic is no more than four colors for new users and only up to seven for experienced ones. Irrelevant colors distract users and detract from their performance. In numerous instances, however, color has been shown to facilitate use in very specific ways. Color should be considered an important way to contrast foreground and background, highlight important fields and data, point out errors, and allow special coding of input.

INTRANET AND INTERNET PAGE DESIGN

In Chapter 15, the rudiments of designing Web sites were discussed. There are more hints about designing a good Internet or intranet fill-in form that should be noted now that you have learned some of the elementary aspects of input form and screen design. Figure 16.31 shows a fill-in form order screen that shows many elements of good design for the Web. Guidelines include the following:

1. Use a variety of text boxes, push buttons, drop-down menus, check boxes, and radio buttons to serve specific functions and to create interest in the form. For example, a text box can be used to capture pieces of information about a Website visitor.
2. Provide clear instructions, because Web users may not be familiar with the terminology or the computer, and users are likely to be international rather than local. If the site is intended for international users, provide options for instructions in a variety of languages, if possible.
3. Include radio buttons when users must choose one answer in a bipolar, close-ended question, the responses to which could be "Yes" or "No," "Agree" or "Disagree," or "True" or "False," but not both.
4. Employ check boxes to allow users to show whether a test condition is either true or untrue. For example, "This is the first time I have visited this site."

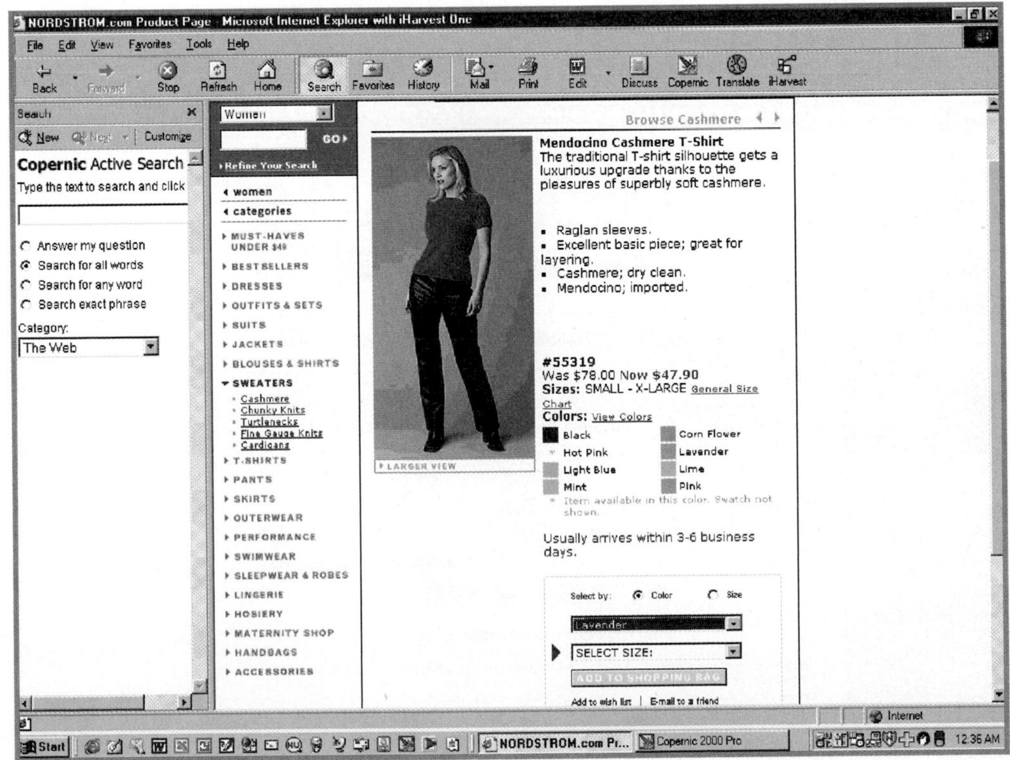

FIGURE 16.31
The order screen from the Nordstrom Web site (www.nordstrom.com) is a good example of how to design an input form that is clear, easy to use, and functional.

5. Demonstrate a logical entry sequence for fill-in forms, especially because the users may have to scroll down to a region of the page that is not visible at the top.
6. Prepare two basic buttons on every Web fill-in form: "Submit" and "Clear Form."
7. Create a feedback screen that refuses submission of a form unless mandatory fields are filled in correctly. The returned form screen can provide detailed comments to the user in a different color. Red is appropriate here. For example, a user may be required to fill in a "Country" in the country field or indicate a credit card number if that type of payment has been checked off.
8. Provide a scrolling text box at times when you are uncertain about how much space users will need to respond to a question or about what language, structure, or form users will use to enter data.
9. If the form is lengthy and the users must scroll excessively, divide the form into several simpler forms on separate pages.

Ecommerce applications involve more than just good design of Web sites. Customers need to feel confident that they are buying the correct amount, that they are getting the right price, and that the total cost of an Internet purchase including shipping charges is what they expect. The most common way to establish this confidence is to use the metaphor of a shopping cart. Figure 16.32 shows the contents of a shopping cart for a customer making a purchase on the Merchants Bay Web site. An important feature of the shopping cart is that the customer can edit the quantity of the item ordered or can remove the item entirely. In this example, the quantity ordered, unit price, and totals are clearly identified and the buttons at the bottom allow the customer to continue shopping or proceed to the check out.

A good ecommerce Web site will instill customer confidence in the company in many ways. Notice on the Merchants Bay Web site that many links are provided. These links allow customers to read more about MerchantsBay.TV, learn how to

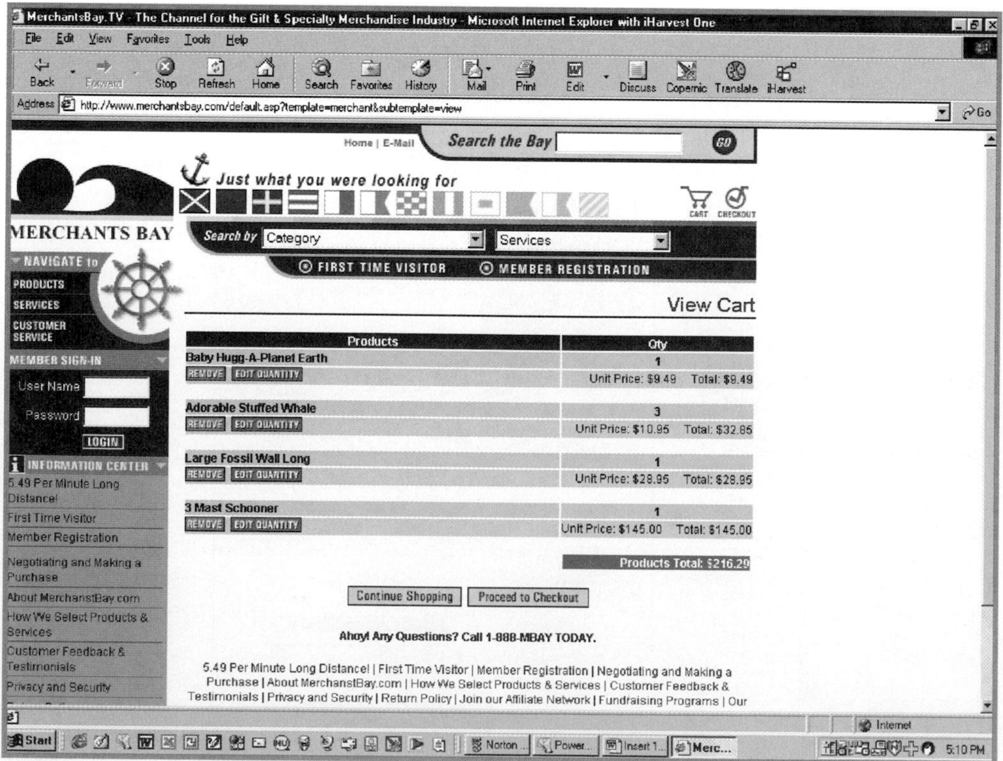

FIGURE 16.32

The Merchants Bay (www.merchantsbay.TV) Web site is a good example of a shopping cart. Customers should be able to modify or remove an item from the shopping cart easily.

negotiate with the Web site to obtain a better price, and find out about how the company selects products and services to sell on the Web site. In addition, customers can read the testimonials of other satisfied customers or read the corporate statement on privacy and security. Good Web sites also allow the customers to give feedback and contact customer service for help with orders. Ecommerce applications add more demands to the analyst who must design Web sites to meet several business objectives, including setting forth the corporate mission and values regarding confidentiality, privacy, and product returns; the efficient processing of transactions; and building good customer relationships.

SUMMARY

This chapter has covered elements of input design for forms, screens, and Web fill-in forms. Well-designed input should meet the goals of effectiveness, accuracy, ease of use, simplicity, consistency, and attractiveness. Knowledge of many different design elements will allow the systems analyst to reach these goals.

The four guidelines for well-designed input forms are the following:

1. Forms must be easy to fill out.
2. Forms must meet the purpose for which they are designed.
3. Forms must be designed to ensure accurate completion.
4. Forms must be attractive.

Design of useful forms, screens, and Web fill-in forms overlaps in many important ways, but there are some distinctions. Screens display a cursor that continually orients the user. Screens often provide assistance with input, whereas with the exception of preprinted instructions, it may be difficult to get additional assistance with a form. Web-based documents have additional capabilities such as embedded hyperlinks, context-sensitive help functions, and feedback forms to correct input before final submission. Skins can be added as an option to personalize a Web site.

The four guidelines for well-designed screens are as follows:

1. Screens must be kept simple.
2. Screens must be consistent from screen to screen.
3. Screen design must facilitate movement between screens.
4. Screens must be attractive.

Many different design elements allow the systems analyst to meet these guidelines.

The proper flow of paper forms, screens, and fill-in forms on the Web is important. Forms should group information logically into seven categories, and screens should be divided into three main sections. Captions on forms and screens can be varied, as can type fonts and the weights of lines dividing subcategories of information. Multiple-part forms are another way to ensure that forms meet their intended purposes. Designers can use windows, prompts, dialogue boxes, and defaults onscreen to ensure the effectiveness of design. There are many similarities, but some critical differences, between screen design for mainframe systems and screen design for microcomputers. To increase efficiency, mainframe screens are sent as a whole rather than as a series of individual keystrokes.

Data fields on a mainframe screen are defined using a field attribute character, which controls the qualities of protection, intensity, shift, and extended attributes. The attribute character must be considered when designing mainframe terminal screens.

Screens can be designed using a number of CASE tools. A commonly used IBM mainframe screen programming language is CICS (Customer Information Control System). Many of the powerful CASE code generators generate CICS code. Icons, color, and graphical user interfaces can also be used to enhance user understanding of input screens.

Web fill-in forms should be constructed with the following nine guidelines in mind as well as those in Chapter 15:

1. Use a variety of text boxes, push buttons, drop-down menus, check boxes, and radio buttons.
2. Provide clear instructions.
3. Include radio buttons when users must choose one answer in a bipolar.
4. Employ check boxes to allow users to show whether a test condition is either true or untrue.
5. Demonstrate a logical entry sequence for fill-in forms.
6. Prepare two basic buttons on every Web fill-in form: "Submit" and "Clear Form."
7. Create a feedback screen that highlights errors in an appropriate color and refuses submission of the form until mandatory fields are correctly filled in.
8. Provide a scrolling text box at times when you are uncertain about how much space users will need to respond to a question.
9. If the form is lengthy and the users must scroll extensively, divide the form into several simpler forms on separate pages.

KEYWORDS AND PHRASES

attribute characters
blinking cursor
box captions
check box
CICS
command button
control of business forms
cursor
drop-down list box
facilitating screen movement
form flow
horizontal check-off captions
image maps
Internet/intranet fill-in forms
inverse video
line captions
list box
message boxes
onscreen color
onscreen dialogue
onscreen icons
option buttons
prompt
radio buttons
response time
screen color combinations
scrolling on screen
seven sections of a form
skins
sliders
specialty forms
spin buttons
tab control dialog box
table caption
text boxes
three screen sections
vertical check-off captions

REVIEW QUESTIONS

1. What are the design objectives for paper input forms, input screens, or Web-based fill-in forms?
2. List the four guidelines for good form design.
3. What is proper form flow?
4. What are the seven sections of a good form?
5. List four types of captioning for use on forms.
6. What is a specialty form? What are some disadvantages of using specialty forms?

"Isn't Spring the most beautiful season here? The architect really captured the essence of the landscape, didn't he? I mean, you can't go anywhere in the building without seeing another beautiful vista through those huge windows. When Snowden came back, he looked at your output screens. The good news is that he thinks they'll work. The project is blossoming, just like the flowers and trees. When Snowden returns from Finland, would you have some input screens ready to demonstrate? He doesn't want things to slow down just because he's out of the country. By the way, the Singapore trip was very successful. Maybe MRE will be worldwide someday."

HYPERCASE® QUESTIONS

1. Using either a paper layout form or software such as JetForm's FormFlow, design a prototype paper form that captures client information for the Training Unit.
2. Test your form on three classmates by having each of them fill it out. Ask them for a written critique of the form.
3. Redesign your input form to reflect your classmates' comments.
4. Using either a paper layout form or a CASE tool, design a prototype screen form that captures client information for the Training Unit.
5. Test your input screen on three classmates by having each of them try it out. Ask them for a written critique of the screen's design.
6. Redesign the input screen based on the comments you receive. In a paragraph, explain how you have addressed each comment.

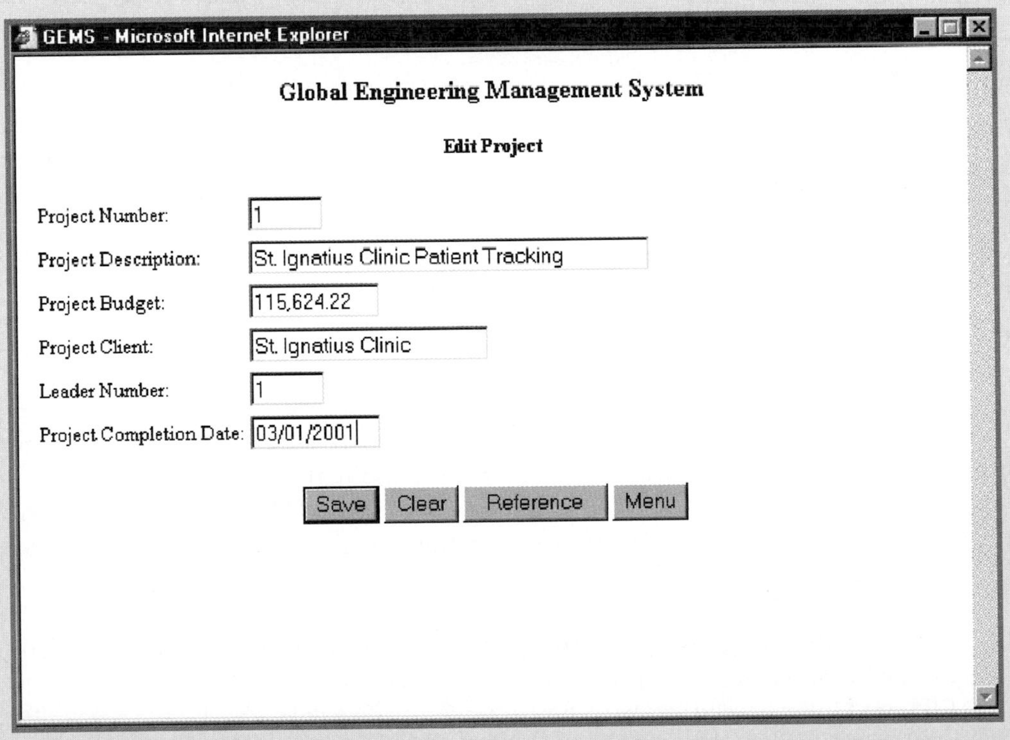

FIGURE 16.HCI
Take a look at some of the input screens in HyperCase. You may want to redesign some of the electronic forms.

7. List the guidelines for the spacing of handwritten forms and printed forms.
8. What are the basic duties involved in controlling forms?
9. List the four guidelines for good screen design.
10. What are the three sections useful for simplifying a screen?
11. What are the advantages of using onscreen windows?
12. What are the disadvantages of using onscreen windows?
13. List two ways screens can be kept consistent.
14. Give three ways to facilitate movement between screens.
15. What are the differences between mainframe and PC screen design?
16. What is an attribute character? Where is it used?
17. What screen characteristics are stored in an attribute character?
18. List four graphical interface design elements. Alongside each one, describe when it would be appropriate to incorporate each of them in a screen design or on a Web-based fill-in form.
19. Define what is meant by onscreen icons. When are icons generally useful for screen design? For the design of Web-based fill-in forms?
20. List the five most legible foreground and background color combinations for display use.
21. Define what is meant by the term "skins" when used in Web design.
22. What three buttons should be included with a tab control dialogue box?
23. What are four situations where color may be useful for screen and Web-based fill-in form design?
24. List nine design guidelines for a Web-based fill-in form.

PROBLEMS

1. Here are captions used for a state census form:

   Name

   ....

   Occupation

   ....

   Address

   ....

   Zip code

   ....

   Number of people in household?

   ....

   Age of head of household

   ....

   a. Redo the *captions* so that the state census bureau can capture the same information requested on the old form without confusing respondents.
   b. Redesign the form so that it exhibits proper flow. (*Hint*: Make sure to provide an access and identification section so that the information can be stored in the state's computers.)
   c. Redesign the form so it can be filled in by citizens who visit the state's Web site. What changes were necessary in moving from a paper form to one that will be submitted electronically?
2. Elkhorn College needs to keep better track of the books checked out from its Buck Memorial Library.
   a. Design and draw a form on $4\frac{1}{2}'' \times 5\frac{1}{2}''$ paper to use for checking out library books. Label the seven sections of a form that you included.
   b. Design and draw a representation of a screen to accomplish the same thing. Label the three sections of a screen that you included.

3. Refer to Figure 16.24, which shows the icons from Excel, the spreadsheet program from Microsoft, and try to explain what each icon means. Propose new icons if the actual ones are confusing.
4. Take a look at Figure 16.EX1. These icons are from Freelance Graphics, a presentation package from Lotus Development Corporation. Try to guess what each icon means. Does the lightbulb have the same meaning as the lightbulb in the Microsoft application in problem 3? Explain. Suggest other icons that are better.
5. Speedy Spuds is a fast-food restaurant offering all manner of potatoes. The manager has a 30-second rule for serving customers. Servers say they could achieve that rule if the form they must fill out and give to the kitchen crew were simplified. The information from the completed form is keyed into the computer system at the end of the day, when the data-entry person needs to enter the kind of potato purchased, additional toppings purchased, the quantity, and the price charged. The current form is difficult for servers to scan and fill out quickly.
   a. Design and draw a form (you choose the size, but be sensible) that lists possible potatoes and toppings in a manner that is easy for servers and kitchen crew to scan and can also be used as input for the inventory/reorder system that is on the extranet connecting Speedy Spuds and Idaho potato growers. (*Hint*: Remember to observe *all* the guidelines for good form design.)
   b. Design and draw a representation of a display screen that can be used by the servers and clerks to fill in the information captured on the form.
   c. Design a display screen based on the screen you designed in part b. This time, it should function as a screen that shows a kitchen crew member what to prepare for each Spuds order. List three changes to the existing screen that you made to adapt it to function as an output screen.
6. Sherry's Meats, a regional meat wholesaler and retailer, needs to collect up-to-date information on how much of each meat product it has in each store. It will then use that information to schedule deliveries from its central warehouse. Currently, customers entering the store fill out a detailed form specifying their

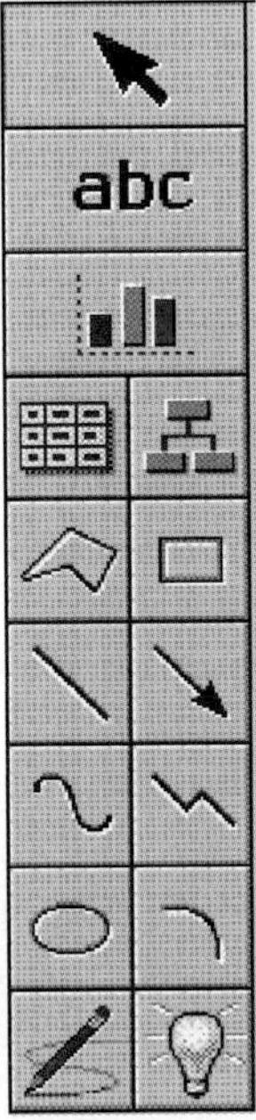

FIGURE 16.EX1
Icons from Freelance Graphics (Lotus Corporation).

individual orders. The form lists over 150 items; it includes meat and meat products available in different amounts. At the end of the day, between 250 and 400 customer orders are tabulated and deducted from the store's inventory. Then the office worker in each store phones in an order for the next day. Store employees have a difficult time tabulating sales because of the mistakes customers make in filling out their forms.

a. It is not possible to have the solitary office worker in each store fill out the numerous customer order forms. Change the form ($3\frac{1}{2}'' \times 6''$ *either* horizontal *or* vertical) and draw it so that it is easier for customers to fill out correctly and for office workers to tabulate.
b. Design and draw a specialty form of the same size that will meet the needs of Sherry's customers, office workers, and warehouse workers.
c. Design and draw two different forms of the same size to meet the purposes in part b, because Sherry's carries both poultry and beef products. (*Hint*: Think about ways to make forms easy to distinguish visually.)
d. Design a fill-in form for onscreen display. When a customer submits an order, it is entered into Sherry's inventory system by any person who is serving customers at the counter. This information will be captured and sent to the central warehouse computer to help control inventory.
e. In a paragraph, describe the drawbacks of having lots of different people at different locations enter data. In a paragraph, list steps you can take as the designer to ensure that the fill-in form is designed to ensure accuracy of entry.

7. R. George's, a fashionable clothing store that also has a mail-order business, would like to keep track of the customers coming into the store so as to expand its mailing list.
   a. Design and draw a simple form that can be printed on $3'' \times 5''$ cards and given to in-store customers to fill out. (*Hint*: The form must be aesthetically appealing to encourage R. George's upscale clientele to complete it.)
   b. Design and draw a representation of a display screen that captures in-store customer information from the cards in part a.
   c. Design and draw an onscreen tab control dialogue box that can be used with the screen in part b, one that allows a comparison between in-store customer information and a listing of customers who hold R. George's credit cards.
   d. Design and draw a second onscreen tab control dialogue box that compares in-store customers with mail-order customers.
   e. The owner is having you help set up an ecommerce site. Design an onscreen form to capture visitor information to the Web site. In a paragraph, explain how it will differ from the printed form.
8. Zero Corp. likes what you have achieved for it in forms control, but it would like to set up an in-house forms specialist to continue your work. Compose a concise job description for an in-house forms specialist that lists four of his or her most important job responsibilities. Remember to include Web-based forms for informational and transactional purposes on their ecommerce site as part of this employee's responsibilities.
9. Figure 16.EX2 depicts the icons from the personal information manager (PIM) discussed in Chapter 3, called Organizer. See if you can guess what each icon means. Why do you think that good icon design is important? Explain in a paragraph why you think (or do not think) these icons conform to the principles of good design.

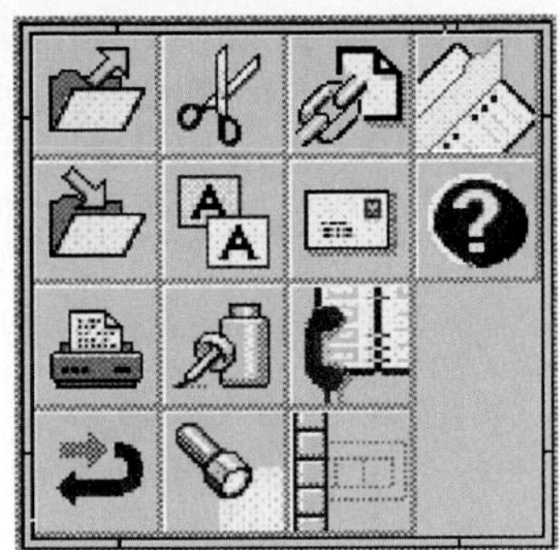

FIGURE 16.EX2
Icons from Lotus Corporation's Organizer.

10. Design a system of onscreen icons with readily recognizable shapes that allows account executives in brokerage houses to determine at a glance what actions (if any) need to be taken on a client's account. (*Hint*: Use color coding as well as icons to facilitate quick identification of extreme conditions.)
    a. Design and draw icons that correspond to the following:
        i. Transaction completed on the same day.
        ii. Account needs updating.
        iii. Client has requested information.
        iv. Account in error.
        v. Account inactive for two months.
        vi. Account closed.
    b. Recently, an up-and-coming discount brokerage house expressed an interest in developing its own portfolio management software that clients could use at home on their PCs to make trades, get real-time stock quotes, and so on. Design two input screens that make data entry easy for the client. The first screen should allow users to enter stock symbols for the stocks they want to track on a daily basis. The second screen should allow the client to use an icon-based system to design a customized report showing stock price trends in a variety of graphs or text.
    c. Suggest two other input screens that should be included in this new portfolio management software.
11. My Belle Cosmetics is a large business that has sales well ahead of any other regional cosmetics firm. As an organization, it is very sensitive to color, because it introduces new color lines in its products every fall and spring. The company has recently begun using technology to electronically show in-store customers how they appear in different shades of cosmetics without requiring them to actually apply the cosmetics.
    a. Design and draw a representation of a display screen that can be used by sales clerks at a counter to try many shades of lipstick and makeup on an individual customer very quickly and with a high degree of accuracy. Input from customers should be their hair color, the color of their favorite clothing, and their typical environmental lighting (fluorescent, incandescent, outdoor, and so on).
    b. Design and draw a representation of a display screen that is equivalent to the one in part a but that vividly demonstrates to decision makers in My Belle how color improves the understandability of the screen.
    c. One of the affiliates My Belle has on the Web is a large department store chain. In a paragraph, describe how the screen in part a can be altered so that an individual can use it and My Belle can put it on the department store's ecommerce site to attract customers.

12. The Home Finders Reality Corporation specializes in locating homes for prospective buyers. Home information is stored in a database and is to be displayed on an inquiry screen. Design a GUI interface, Web-based screen to enter the following data fields, which are used to select and display homes matching the criteria. Keep in mind the features available for a GUI screen. The screen elements (which are not in any particular sequence) are as follows:
    a. Minimum size (in square feet).
    b. Maximum size (optional, in square feet).
    c. Minimum number of bedrooms.
    d. Minimum number of bathrooms.
    e. Garage size (optional, number of cars).
    f. School district (a limited number of school districts are available for each area).
    g. Swimming pool (yes/no, optional).
    h. Setting (either city, suburban, or rural).
    i. Fireplace (yes/no, optional).
    j. Energy efficient (yes/no).

    In addition,

    k. Describe the hyperlinks necessary to achieve this type of interaction.
13. Use index cards to design a tab control dialogue box, one that changes the following database screen settings. Use one card for each tab. Be sure to group each tab by function.
    a. Change the background color.
    b. Change the font.
    c. Change the object border to a raised look.
    d. Set the foreground color.
    e. Change the object border to a sunken look.
    f. Set the border color.
    g. Change the font size.
    h. Set the text to bold.
    i. Change the object border to a flat look.
    j. Set the text to underline.
    k. Change the object background color.
14. Design a Web entry page for the Home Finders Reality Corporation screen created in problem 12.
15. The five-year-old TowerWood hotel chain needs help designing its Web site. The company maintains properties in all the large U.S. tourist communities such as Orlando, Florida (near Disney World); Maui, Hawaii; Anaheim, California (near Disneyland); Las Vegas, Nevada; and New Orleans, Louisiana. Their properties feature a variety of rooms in all these locations.
    a. In a paragraph, discuss how the company can use skins on its Web site to attract different types of clientele, including families with small children, young couples on their honeymoon, retired couples who want to travel on a budget, and business travelers who need business services.
    b. Design and draw a series of skins that would appeal to the different types of hotel clientele listed in part a. (*Hint*: Use a graphics package or drawing program to help design the skins.)
    c. Add a group of potential Web site users for the TowerWood hotel chain who were *not* mentioned in part a and design and draw additional skins for them. Then create a table that matches each client group with a particular skin you designed.

GROUP PROJECTS

1. Maverick Transport is considering updating its mainframe screens. With your team, brainstorm about what should appear on input screens of mainframe terminal operators who are entering delivery load data as loads are approved. Fields will include date of delivery, contents, weight, special requirements (for example, whether contents are perishable), and so on.
2. Each team member should design an appropriate input screen using either a CASE tool or paper and pencil. Share your results with your team members.
3. Make a list of other input screens that Maverick Transport should develop. Remember to include dispatcher screens as well as screens to be accessed by customers and drivers. Indicate which should be PC screens, displays on wireless handheld devices, or mainframe screens.
4. Design a Web-based screen that will allow Maverick Transport customers to track the progress of a shipment. Brainstorm with team members for a list of elements or perform an interview with a local trucking company to find out its requirements. List what hyperlinks will be essential. How will you control access so that only customers can track their own shipments?

SELECTED BIBLIOGRAPHY

Dahlboom, B., and L. Mathiassen. *Computers in Context.* Cambridge, MA: NCC Blackwell, 1993.

Gibbs, M. "Forms Design and Control." In V. Lazarro (ed.), *Systems and Procedures: A Handbook for Business and Industry,* 2d ed. Englewood Cliffs, NJ: Prentice Hall, 1968.

Ives, B. "Graphical User Interfaces for Business Information Systems." *MIS Quarterly* (Special Issue), December 1982, pp. 15–48.

Reisner, P. "Human Factors Studies of Data Base Query Languages: A Survey and Assessment." *Computing Surveys,* Vol. 4, No. 1, 1981.

16

ALLEN SCHMIDT, JULIE E. KENDALL, AND KENNETH E. KENDALL

FORMING SCREENS AND SCREENING FORMS

Pooling information from the output design and reviewing their progress, Chip and Anna proceeded to the next stage, the design of input. "Forms and screens must be designed to capture input information easily and accurately," remarks Anna.

Chip replies, "Special attention should be placed on creating input screens that are easy to use and require minimal operator entry."

Chip loads Visible Analyst and examines Diagram 0. "Perhaps the first form that we should create is the NEW MICROCOMPUTER RECORD, flowing from the SHIPPING/RECEIVING DEPARTMENT into process 2, ADD NEW MICROCOMPUTER." Chip double-clicks on the data flow representing the form to bring up its repository record. Its **Composition** area contains a data structure called NEW MICROCOMPUTER FORM RECORD. "I decided to create a separate structure for the form, because the elements are used both on the form and on the matching screen," muses Chip. He clicks in the area and presses the **Jump** button. The elements contained on the form are included in the structure's **Composition** area, illustrated in Figure E16.1. Notice that the **Notes** area contains information about which fields should be implemented as drop-down lists and check boxes.

Chip goes to work at designing the form. It is zoned to group logically related elements, which are arranged in a manner that would allow the user to complete the form easily. Because a prototype of the data-entry screen was previously approved, the task of designing the form is considerably simplified.

Chip schedules a meeting with Dot to review the form. She looks thoughtfully at the document for a few minutes and remarks, "It looks very good. I can see you've considered our viewpoint when designing the form. The only change I would recommend is to separate the initial information we have when the microcomputer is received from the data supplied as we make decisions on which printer and monitor to attach."

Chip revises the form with the suggested changes and obtains final approval from Dot. The completed form is shown in Figure E16.2. Notice the zoning and the use of tick marks indicating the number of characters to be keyed. These help the user decide how to abbreviate data that would not fit within the file or database field length.

With the form complete, Chip starts working on modifying the screen used to enter the form data. The screen design matched the form, with fields on the screen in the same order and rough placement as on the form. The ADD NEW MICROCOMPUTER entry screen is shown in Figure E16.3.

One of the considerations of the entry screen is ease of entering data, and another is accuracy. Still another consideration is the availability of help. New employees would not be familiar with the operation of the system or with what is required for a particular field entry. To achieve these goals, Chip included pull-down lists for the MONITOR, PRINTER, INTERNET CONNECTION, and MICROCOMPUTER BOARDS. "I like the way these pull-down lists work," he remarks to Anna. "The users can easily select the codes that should be stored in the database."

"Why have the users select codes," replies Anna. "There must be a way for them to select descriptive code meanings, such as the name of the printer, and have the computer store the codes."

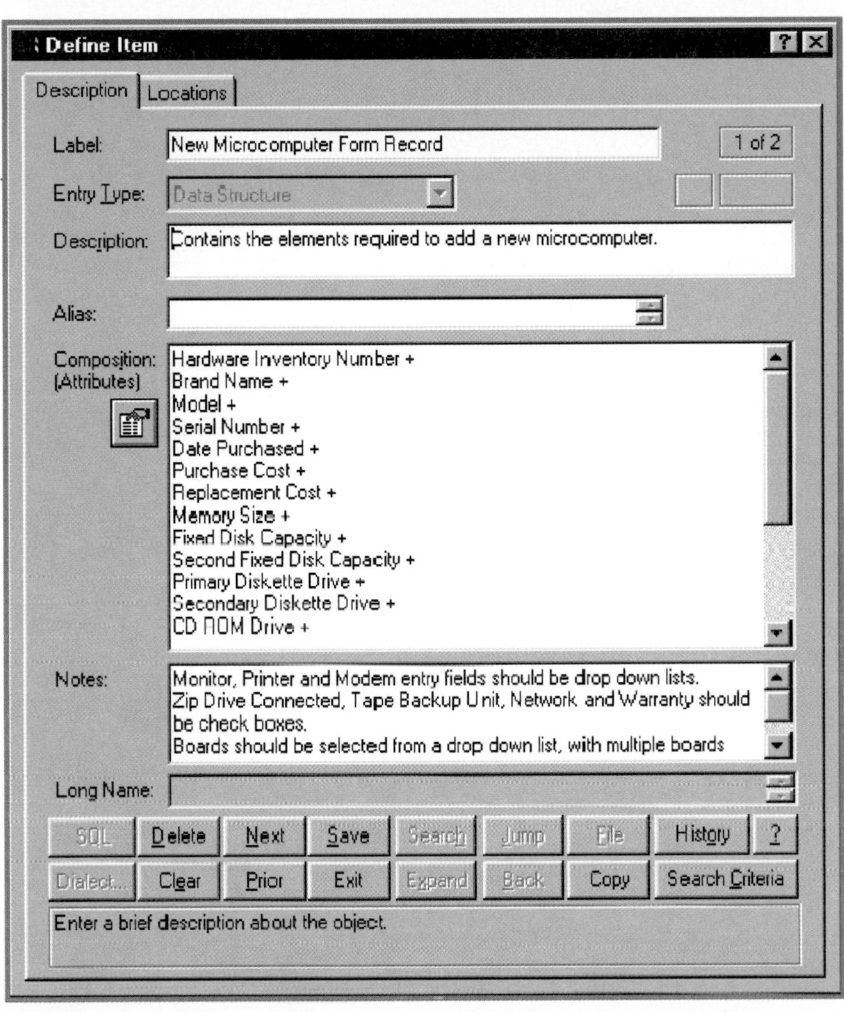

FIGURE E16.1
NEW MICROCOMPUTER FORM RECORD data structure screen showing elements.

"That's an excellent idea," exclaims Chip. A short time later, the modifications have been implemented.

Anna reviews the screen and remarks, "This looks terrific! I like the grouping of the check boxes and descriptive information contained in the drop-down lists."

"Watch this screen in action," replies Chip. "I've added a button for the users to press when they have entered all the data and made all the selections. They can also print the completed form."

"What about help?" asks Anna.

"I've thought about that also," answers Chip. "As the cursor moves from field to field, the status line on the bottom of the screen displays one line of help appropriate for that field. I can also add tool-tip help, a small box of help options that appears when the mouse cursor remains over one entry area for a short amount of time." Notice that the pull-down lists have meaningful names in the data areas. There is room for three microcomputer boards in the BOARD CODE entry area. Scroll bars provide additional entry areas if needed. Help is displayed in the status line on the bottom of

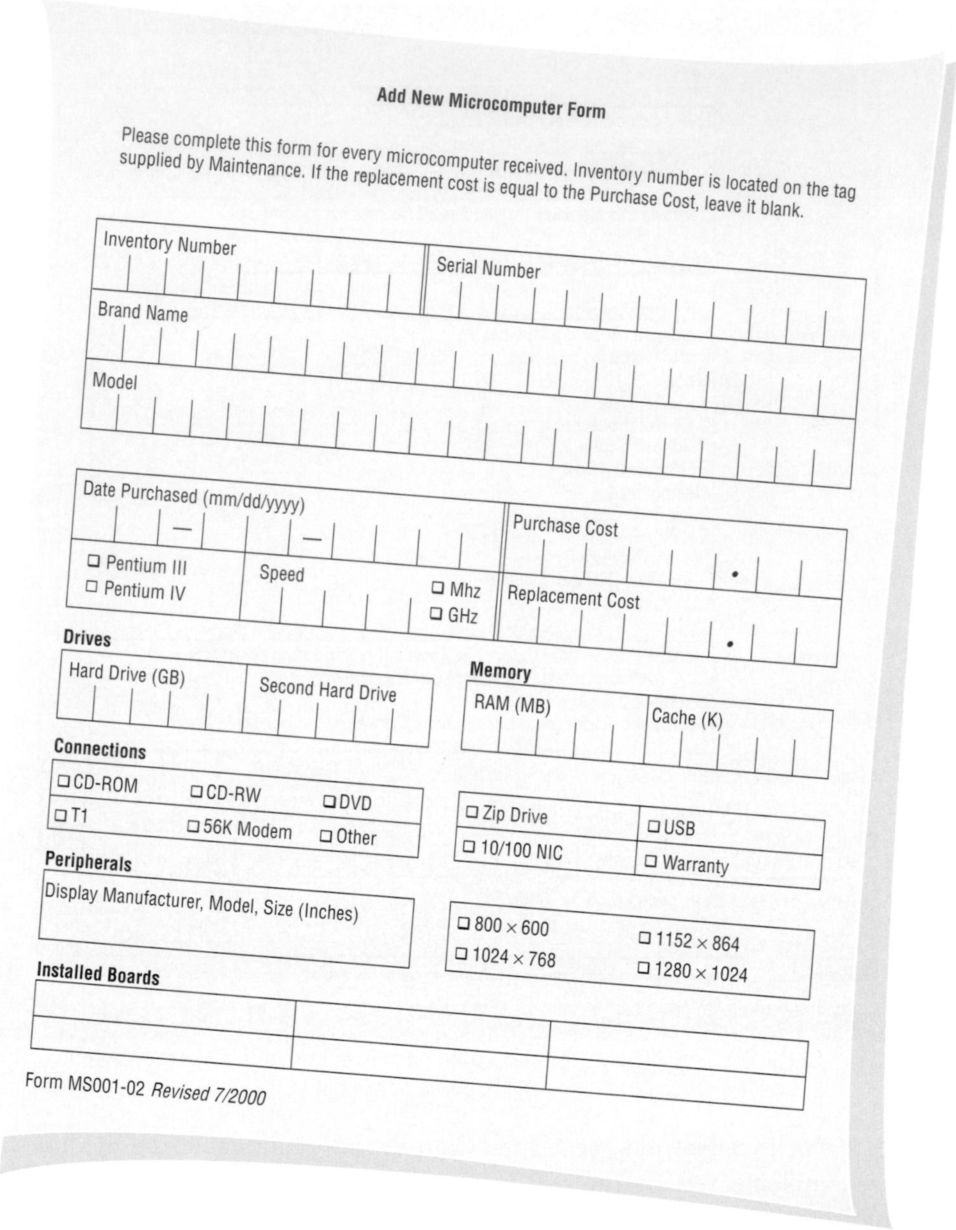

Add New Microcomputer Form

Please complete this form for every microcomputer received. Inventory number is located on the tag supplied by Maintenance. If the replacement cost is equal to the Purchase Cost, leave it blank.

Inventory Number

Serial Number

Brand Name

Model

Date Purchased (mm/dd/yyyy)

— —

Purchase Cost

.

☐ Pentium III
☐ Pentium IV

Speed

☐ Mhz
☐ GHz

Replacement Cost

.

Drives

Hard Drive (GB)

Second Hard Drive

Memory

RAM (MB)

Cache (K)

Connections

☐ CD-ROM ☐ CD-RW ☐ DVD
☐ T1 ☐ 56K Modem ☐ Other

☐ Zip Drive ☐ USB
☐ 10/100 NIC ☐ Warranty

Peripherals

Display Manufacturer, Model, Size (Inches)

☐ 800 × 600 ☐ 1152 × 864
☐ 1024 × 768 ☐ 1280 × 1024

Installed Boards

Form MS001-02 *Revised 7/2000*

FIGURE E16.2
ADD NEW MICROCOMPUTER form.

the screen. An example of selecting monitor types from a pull-down list is illustrated in Figure E16.4.

Dot reviews the completed screen and enters some test data. "I'm really impressed!" she exclaims. "It is much smoother than I ever expected. When can we expect the rest of the system?" Chip smiles with appreciation and remarks that great progress is being made. "I do hope that the rest of the system is as clear to use and easy to operate!" Dot says appreciatively.

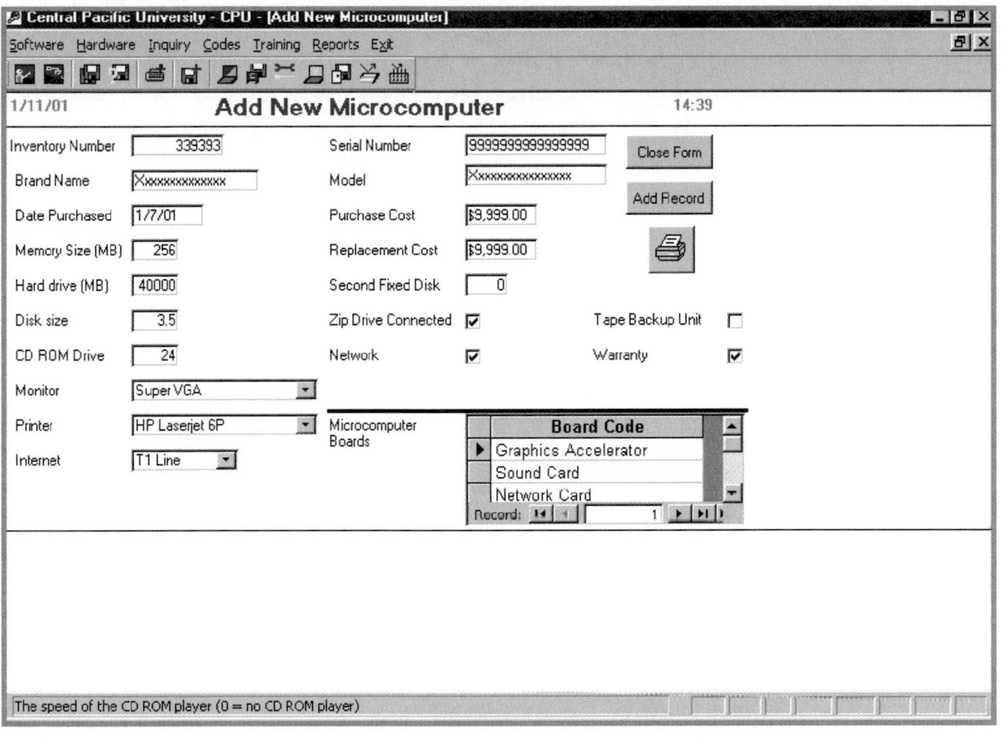

FIGURE E16.3
The ADD NEW MICROCOMPUTER Microsoft Access screen.

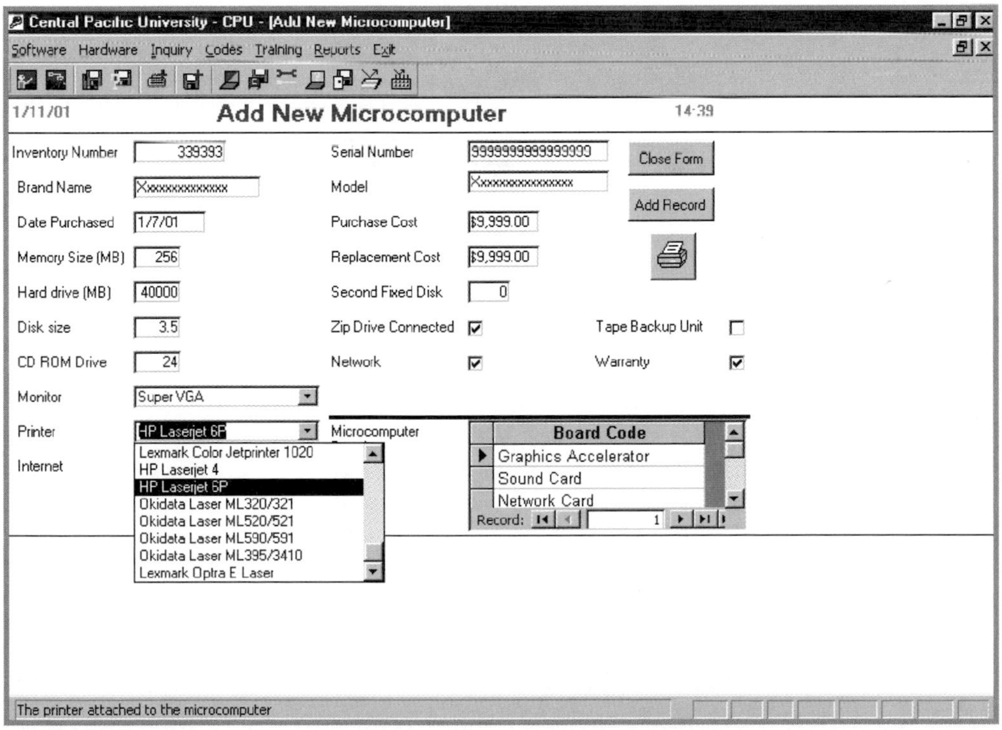

FIGURE E16.4
Pull-down lists on the ADD NEW MICROCOMPUTER Microsoft Access screen.

16

Meanwhile, Anna is meeting with Hy Perteks, who is desperately seeking help. "I'm swamped with requests for help on software packages! Is there any way to design a portion of the system for maintaining information on the available software experts?" asks Hy. "I have names written on scraps of paper and I keep misplacing them. Often I find out who these experts are only after someone else finds them first."

Anna asks some questions about what information would be required and how Hy would like to maintain and display the records. Hy replies, "There is so much expertise available, but the only way I have of locating the person's information is by using their name as an index. And, I confess, I'm awful at remembering the correct spelling of the first name, let alone the last name." Anna assures him that there will be an easy-to-use system available soon.

Back at her desk, Anna thinks about the problem. "The ADD screen would be easy to create, but what about the CHANGE screen?" She wonders, "How can I?" and then thinks, "Ah ha!" as she snaps her fingers. The design becomes clear. There would be a screen with two distinct regions on it. The first region would contain the last and first name of the Software Expert. Included with the screen is a **Find** button as well as buttons for scrolling back and forth through records. If the users made a mistake entering data, there is an **Undo** button, and there is also a button to save the changes. The completed screen is illustrated in Figure E16.5.

Anna tests the **Find** button. "Check this out," says Anna, glancing over at Chip. Clicking the **Find** button displays a **Find** dialogue box. Anna selects **Any Part of Field** for a **Match** choice. She enters "rock" in the **Find What** area and clicks the **Find First** button. Refer to the screen shown in Figure E16.6.

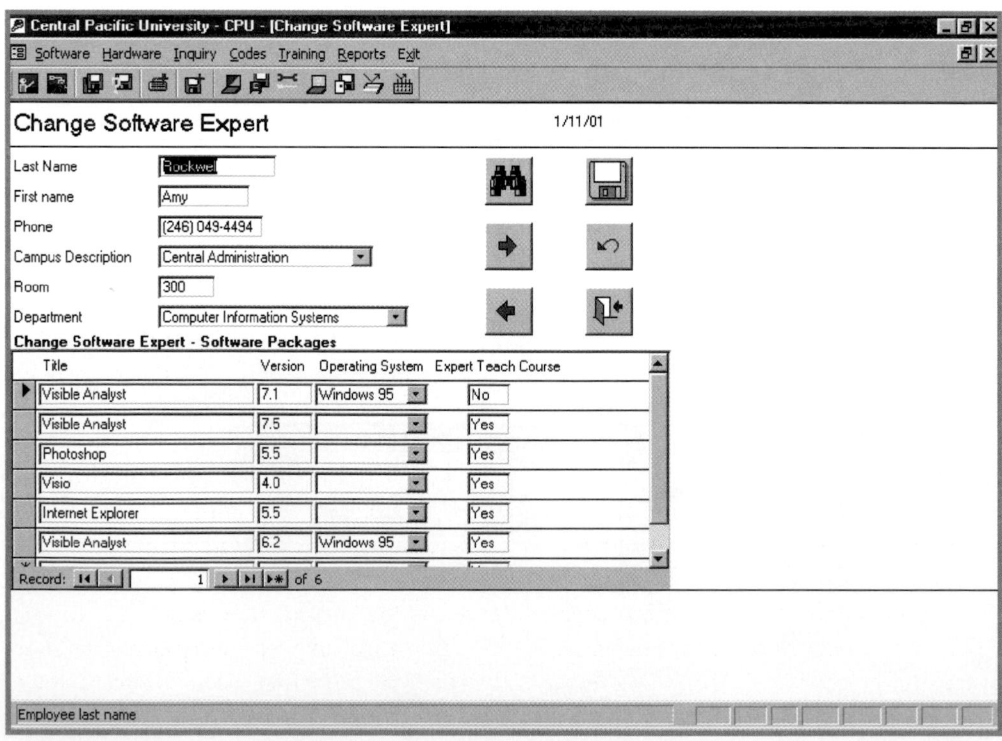

FIGURE E16.5
The CHANGE SOFTWARE EXPERT Microsoft Access screen.

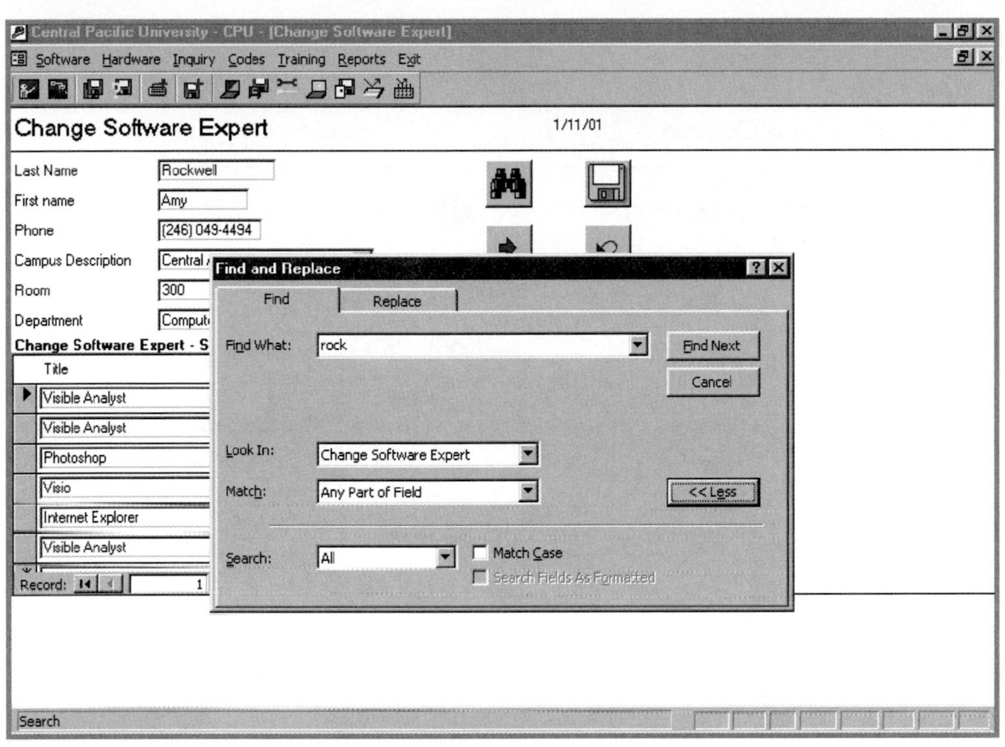

FIGURE E16.6

The Find and Replace Dialog box used to search for a name in the Change Software Expert screen. Microsoft Access was used to design this screen.

"Great-looking screen," grins Chip. "I want to be here when you show it to Hy."

The problem of deleting Software Course records for software that is no longer in use required a different approach. Anna reasoned that it would be easy if she used the **Find** feature to locate a record and then used a **Find Next** button to locate the next record that matched the criteria. There would also be buttons that allowed her to move to the next or previous records.

After the record is located, the DELETE SOFTWARE COURSE program would display pertinent information. All codes on the file, such as COURSE LEVEL and OPERATING SYSTEM, would be replaced with the full code meaning. None of the data would be able to be modified at this time. The operator would have the opportunity to review the record and then choose to either delete or not delete the record. When the delete button is pressed, a dialogue box is displayed asking the users if they really want to delete the record. They may choose to cancel the delete at that time. The final screen is shown in Figure E16.7.

Hy is delighted with the prototype screens. As he tests each of them, he remarks, "You don't know how easy it's going to be for me to answer help requests. These screens are fabulous!" He pauses for a long moment and then asks, "I have a lot of requests about providing periodically scheduled training courses. Do you think we could work on a system to register for courses?"

Anna purses her lips for a moment and remarks, "Did you ever hear of a project having scope creep, always adding little things and the project never ends? The

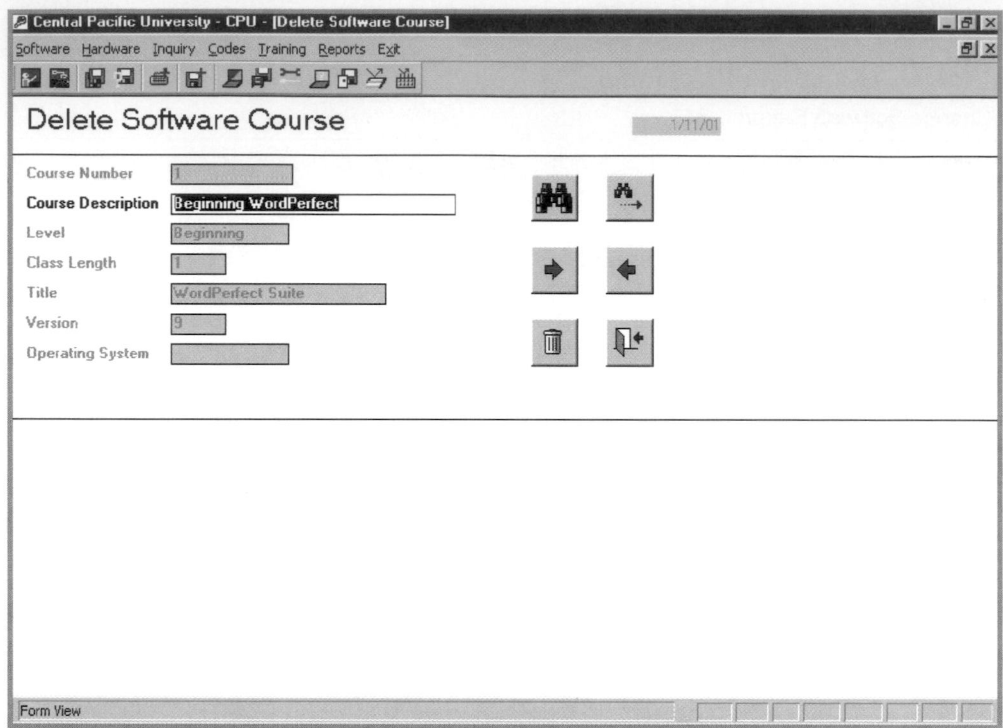

FIGURE E16.7
The DELETE SOFTWARE COURSE screen designed in Microsoft Access.

University does, however, have an intranet initiative going, and it is looking for volunteers. Perhaps we can design a Web page for registering courses."

"That's great!" replies Hy. "That's more than I ever hoped for."

Anna starts to design the Web page, including the users' first and last names as well as their Internet addresses and office phones. Additional areas are used to enter the campus where they are located, the software they use, and their class level. Chip reviews the form and remarks, "Rather than have them enter the Campus and Software, why not have them select them from a pull-down list? And what about allowing them to select convenient times for training?"

"Good idea," replies Anna. "And I think the level of training should be a radio button." The completed intranet Web page is illustrated in Figure E16.8. Notice that there are buttons to submit the query or reset it to spaces and pull-down list default values. There is also a link to email questions to the Training Officer.

Hy is thrilled. "This form is better than I ever imagined. I think we are really providing effective training registration, and I know that my phone will not be ringing as much. I've got another great idea!"

The following exercises may be done by designing the report or screen using Printer or Screen layout forms, or they may be created using any word processor with which you are familiar. The fields and other related information for the reports are contained in Visible Analyst data flow repository entries. The names for the data flow are listed for each exercise.

Corresponding reports and screens (called forms in Access) have been created. All the information is present in the Access database; you only have to modify the existing

Central Pacific University: Class Registration Form - Microsoft Internet Explorer

File Edit View Favorites Tools Help

Back Forward Stop Refresh Home Search Favorites History Mail Print Discuss Dell Home RealGuide Messenger

Address http://www.cpu.edu/support/training/register.htm Go

Email Setup | Web Pages | Training | Support | Search | Site Map | FAQ

Teach Course | Request Course | Training Schedule | Course Evaluation | Registration Course

Central Pacific University · Training Course Registration Form

Complete this form and press **Submit** to register for a training class. Click here to view the training calendar. Classes are offered to all faculty and staff on a first come, first served basis. Please include your email address and you will receive a registration confirmation. You will receive notification of the time and location shortly before the start of the course.

First Name: Last Name:

Email Address: Office Phone:

Campus: Central Software: Visible Analyst

Select the level of training — Training Level: Beginning, Intermediate, Advanced, Special Topics

Check all available days — Monday, Tuesday, Wednesday, Thursday, Friday

Check all available times — Morning, Noon Hour, Afternoon, Evening

Questions or comments:

Submit Reset

Email Training Staff

Modified on 04/01/2001 by *Chip Fuller*

Done My Computer

FIGURE E16.8

An intranet Web form for Training Registration on the CPU Web site.

reports and screens to produce the final versions. Modifications are made by clicking on the desired report or screen and then clicking the **Design** button. The following modifications may be made. The Page Header contains column headings. The Detail area contains the print fields for the report. Refer to Figure E15.7.

Click in a field to select it. Click on several fields while holding the Shift key to select them.
Drag a selected field (or fields) to move them.
Click on one of the small boxes surrounding the field to change the field size.
Select several fields and click **Format** and either:
- Align, to align all fields with the top, left, and so on, field.
- Size, to make fields equal to the widest, tallest, and so on, field.
- Horizontal spacing, to make horizontal spacing equal, or to increase or decrease the spacing.
- Vertical spacing, to make vertical spacing equal, or to increase or decrease the spacing.

EXERCISES*

E-1. Cher Ware has remarked several times that a good form would make the task of adding new software much easier. It would also provide permanent paper documentation for software additions.

Design a form to add software to the SOFTWARE MASTER file. Open Diagram 0 in Visible Analyst and double click on the SOFTWARE RECEIVED

FORM data flow to view the Repository entry for the data flow. Click on the NEW SOFTWARE RECORD in the **Composition** area and press the **Jump** button to view the data structure containing the elements required on the form. **Jump** to each element to determine the length of the screen field. You may also use the **Repository Reports** and **Single Entry Listing** to print a list of elements for the form.

E-2. Design the ADD SOFTWARE RECORD screen, either on paper or by modifying the Access screen. Use the fields created in Exercise E-1. The Visible Analyst data structure name is NEW SOFTWARE RECORD.

E-3. Hy Perteks would like a form to fill in as he learns about new Software Experts. Use the Visible Analyst ADD SOFTWARE EXPERT Data Structure to determine the fields required for the form.

E-4. Create the ADD SOFTWARE EXPERT screen, either on paper, using a word processor, or by modifying the Access form. Test the ADD SOFTWARE EXPERT screen, using the drop-down lists and observing the status bar on the bottom of the screen.

E-5. Design or modify the Access form for the DELETE SOFTWARE EXPERT screen. Which fields are drop-down lists? Use the Visible Analyst DELETE SOFTWARE EXPERT data structure.

E-6. Design or modify the Access form for the DELETE MICROCOMPUTER RECORD screen. The Visible Analyst structure is called DELETE MICROCOMPUTER RECORD.

E-7. Cher Ware and Anna spent the better part of a morning working out the details on the software portion of the system. Plagued by the problem of providing consistent software upgrades for all machines, Cher would like an easy method of upgrading. A few older versions of software may also be retained for special needs.

Part of the solution is to produce a report, sorted by location, of all machines containing the software to be upgraded. As the new software is installed, a check mark is placed on the report after each machine.

Design the UPGRADE SOFTWARE screen design. Add a **Find** button to locate the title and to provide a field that can be used to enter the new VERSION NUMBER. The update program will display a line for each machine containing the old version of the installed software. These lines are sorted by CAMPUS LOCATION and ROOM LOCATION.

Columns are CAMPUS LOCATION, ROOM LOCATION, INVENTORY NUMBER, BRAND NAME, MODEL, UPGRADE, and RETAIN OLD VERSION. The UPGRADE column contains a check box that is to be checked if the software is to be upgraded. The RETAIN OLD VERSION is also a check box, unchecked by default. The users would check the box for a specific machine that must retain the old and new versions of the software.

Look in the Visible Analyst SOFTWARE UPGRADE data structure for the elements contained on the screen.

E-8. Explain why the UPGRADE SOFTWARE screen would display machines rather than have Cher enter the machine IDs. In a paragraph, discuss why the screen displays records in a CAMPUS/ROOM sequence.

E-9. Design the CHANGE SOFTWARE screen. This screen allows Cher Ware to modify data that have been entered incorrectly as well as information that routinely changes, such as SOFTWARE EXPERT and NUMBER OF COPIES. The

SOFTWARE INVENTORY NUMBER is the primary key and may not be changed. The other SOFTWARE MASTER fields that should be included on the screen are found in the Visible Analyst SOFTWARE CHANGES data structure. Use these fields to design the screen. A limited screen, CHANGE SOFTWARE RECORD, has been created in Access. Use the Access **Field List** to add fields to the screen. Include the following buttons: Find, Find Next, Previous Record, Next Record, Save Record, and Cancel Changes.

E-10. Hy Perteks is concerned that old courses for obsolete versions of software are cluttering the disks. Create and print the Delete Software Course screen.

Entry fields are the SOFTWARE TITLE, OPERATING SYSTEM, and VERSION NUMBER. The program displays a line for each course taught for the software version. The first column contains an entry field with a D (for Delete) presented as a default. Placing a space in the field will prevent the record from being deleted. The other columns for each line are COURSE TITLE, LEVEL, and CLASS LENGTH. Add a meaningful operator message.

E-11. Design the UPDATE MAINTENANCE INFORMATION screen. It contains entry fields that allow Mike Crowe to change maintenance information as computers are repaired or as routine maintenance is performed on them. The Visible Analyst data structure is UPDATE MAINTENANCE INFORMATION.

*Exercises preceded by a CD-ROM icon require the program Visible Analyst or another CASE tool. A CD-ROM containing Visible Analyst examples is provided free of charge to any professor adopting this book. The examples on the disk may be imported into Visible Analyst and then used by students.

DESIGNING DATABASES

17

DESIGN OBJECTIVES

Data storage is considered by some to be the heart of an information system. The general objectives in the design of data storage organization are shown in Figure 17.1.

First, the data have to be available when the user wants to use them. Second, the data must be accurate and consistent (it must possess integrity). Beyond this requirement, the objectives of database design include efficient storage of data as well as efficient updating and retrieval. Finally, it is necessary that information retrieval be purposeful. The information obtained from the stored data must be in a form useful for managing, planning, controlling, or decision making.

CONVENTIONAL FILES AND DATABASES

There are two approaches to the storage of data in a computer-based system. The first method is to store the data in individual files, each unique to a particular application. Figure 17.2 illustrates an organization with a number of information systems using conventional, separate files: SALES-FILE, which contains historical sales information; CURRENT-ACTIVITY, which is updated often; and PERSONNEL-FILE, which contains addresses, titles, and so on.

Notice that the NUM and NAME exist in each file. Besides the extra effort needed to input the NAME three times, a name change (because of a change in marital status, for instance) would require updating three separate files.

The second approach to the storage of data in a computer-based system involves building a database. A database is a formally defined and centrally controlled store of data intended for use in many different applications. Figure 17.3 shows that different users in different departments within the organization can share the same database. A user may select a portion of the database, as shown in data set 1, or certain rows, as shown in data set 2. Data set 3 contains selected columns, and data set 4 uses some rows to calculate totals. These data sets are overlapping, indicating that data that were redundant in conventional files are only stored once in the database.

CONVENTIONAL FILES

Conventional files will remain a practical way to store data for some (but not all) applications. A file can be designed and built quite rapidly, and any concerns about data availability and security are minimized. When file designs are carefully

FIGURE 17.1

Five objectives for the design of data storage.

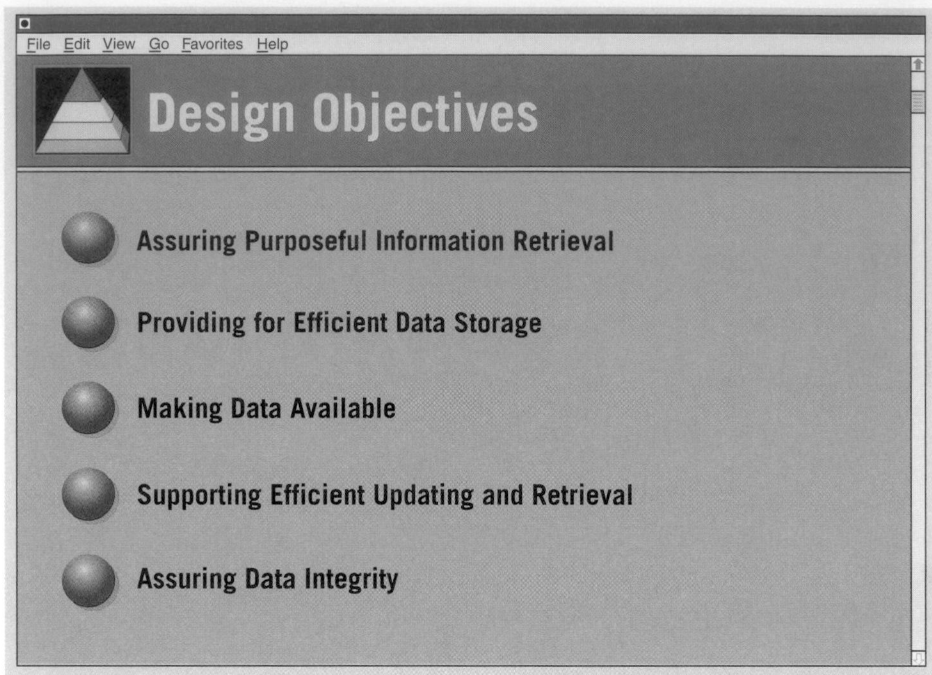

File Edit View Go Favorites Help
Design Objectives
Assuring Purposeful Information Retrieval
Providing for Efficient Data Storage
Making Data Available
Supporting Efficient Updating and Retrieval
Assuring Data Integrity

FIGURE 17.2

The use of separate files means that the same data often are stored in more than one place.

SALES-FILE

NUM	NAME	1998	1999	2000

CURRENT-ACTIVITY

NUM	NAME	AREA	2000

PERSONNEL-FILE

NUM	DEPT	NAME	ADDRESS	PHONE	HIRED	SALARY	TITLE

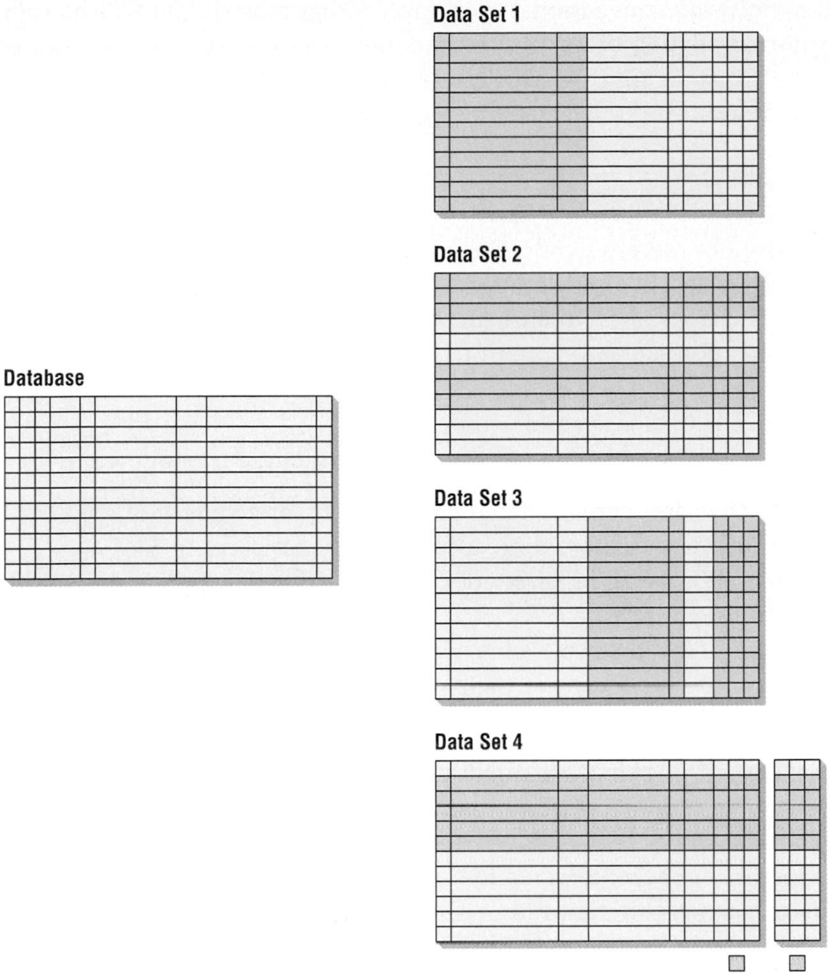

FIGURE 17.3
The database approach allows different users to share the same database yet access different data sets.

thought out, all the necessary information can be included and the risk of unintentionally omitting data will be low. If the user is personally involved with organizing the file, there will be few problems with understanding how to access the data.

An analyst rarely encounters a situation in which there are no existing applications. Chances are that separate files are already in service. If development time is an important consideration, the systems analyst cannot take on the problem of redoing data storage in order to achieve a database approach. The obvious solution given time constraints is to limit the scope of the project by designing another, separate file for the new application.

Processing speed is another advantage of using files. It is possible to choose the optimal technique for the file processing of a single application, but it is impossible to come up with an optimal design for many different tasks. Consequently, if the systems analyst is faced with designing a system for one specific application when processing efficiency is of the utmost concern, the best approach may be to design an individual file for that purpose.

The use of individual files has many consequences. One major problem is the lack of potential for files to evolve. Files are often designed only with immediate needs in mind. When it becomes important to query the system for a combination of some of the attributes, these attributes may be contained in separate files or may not even exist. Redesigning files often implies that programs that access the files must be rewritten accordingly, which translates into expensive programmers time for file and program development and maintenance.

A system using conventional files implies that stored data will be redundant. Furthermore, updating files is more time-consuming. Data integrity is a concern, because a change in one file will also require modification of the same data in other files. Seldom-used files may be neglected when it is time for updating.

DATABASES

Databases are not merely a collection of files. Rather, a database is a central source of data meant to be shared by many users for a variety of applications. The heart of a database is the database management system (DBMS), which allows the creation, modification, and updating of the database; the retrieval of data; and the generation of reports. The person who ensures that the database meets its objectives is called the database administrator.

The effectiveness objectives of the database include the following:

1. Ensuring that data can be shared among users for a variety of applications.
2. Maintaining data that are both accurate and consistent.
3. Ensuring that all data required for current and future applications will be readily available.
4. Allowing the database to evolve and the needs of the users grow.
5. Allowing users to construct their personal view of the data without concern for the way the data are physically stored.

The foregoing list of objectives provides us with a reminder of the advantages and disadvantages of the database approach. First, the sharing of the data means that data need to be stored only once. That in turn helps achieve data integrity, because changes to data are accomplished more easily and reliably if the data appear once rather than in many different files.

When a user needs particular data, a well-designed database anticipates the need for such data (or perhaps it has already been used for another application). Consequently, the data have a better chance of being available in a database than in a conventional file system. A well-designed database can also be more flexible than separate files; that is, a database can evolve as the needs of users and applications change.

Finally, the database approach has the advantage of allowing users to have their own view of the data. Users need not be concerned with the actual structure of the database or its physical storage.

The first disadvantage of the database approach is that all the data are stored in one place. Therefore, data are more vulnerable to catastrophes and require complete backup. There is a risk that the database administrator becomes the only one privileged or skilled enough to go near the data. The bureaucratic procedures required to modify or even update the database can seem insurmountable.

Other disadvantages come about when attempting to achieve two efficiency objectives for the management of the data resource:

1. Keeping the time required to insert, update, delete, and retrieve data to a tolerable amount.
2. Keeping the cost of storing the data to a reasonable amount.

Remember that a database cannot be optimized for retrieving data for a specific application, because it may be shared by many users for various applications. Furthermore, additional software for the DBMS is required, and occasionally a larger computer is required.

The database approach is a concept that is becoming increasingly important. The use of relational databases (covered later in this chapter) on networked PCs

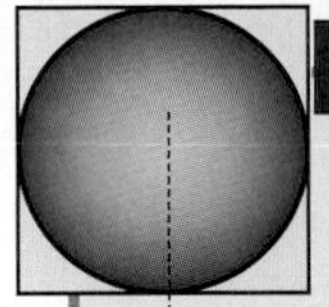

CONSULTING OPPORTUNITY 17.1

HITCH YOUR CLEANING CART TO A STAR

The Marc Schnieder Janitorial Supply Company has asked for your assistance in cleaning up its data storage. As soon as you begin asking Marc Schnieder detailed questions about his database, his face gets flushed. "We don't really have a database as you describe it," he says with some embarrassment. "I've always wanted to clean up our records, but I couldn't find a capable person to head the effort."

After talking with Mr. Schnieder, you walk down the hall to the closet-sized office of Stan Lessink, the chief programmer. Stan fills you in on the historical development of the current information system. "The Marc Schnieder Janitorial Supply Company is a rags-to-riches story," Stan remarks. "Mr. Schnieder's first job was as a janitor in a bowling alley. He saved enough money to buy some products and started selling them to other alleys. Soon he decided to expand the janitorial supply business. He found out that as his business grew, he had more product lines and types of customers. Salespeople in the company are assigned to different major product lines (stores, offices, and so on), some are in-house sales, and some specialize in heavy equipment, such as floor strippers and waxers. Records were kept in separate files."

"The problem is that we have no way to compare the profits of each division. We would like to set up incentive programs for salespeople and provide better balance in allocating salespeople to each product line," states Mr. Schnieder.

When you talk with Stan, however, he adds, "Each division has its own incentive system. Commissions vary. I don't see how we can have a common system. Besides, I can get our reports out quickly because our files are set up the way we want them. We have never issued a paycheck late."

Describe how you would go about analyzing the data storage needs of the Marc Schnieder Janitorial Supply Company. Would you trash the old system or just polish it up a bit? Discuss the implications of your decision.

means that the concept is becoming understandable to many users. Many users are extracting parts of the central database from mainframes and downloading them onto PCs. These smaller databases are then used to generate reports or answer queries specific to the end user.

Relational databases for PCs have improved dramatically over the last few years. The competition among database software vendors is keen. Most standard PC databases are extremely flexible, a feature that aids in the design of reports and labels. They allow the end user to read in databases from other software programs. All have good query capability. Some are more user-friendly than others.

One major technological change has been the design of database software that takes advantage of the GUI. With the advent of programs such as Microsoft Access, users can drag and drop fields between two or more tables. Developing relational databases with these tools has been made relatively easy.

DATA CONCEPTS

It is important to understand how data are represented before considering the use of files or the database approach. In this section, critical definitions are covered, including the abstraction of data from the real world to the storage of data in files.

REALITY, DATA, AND METADATA

The real world itself will be referred to as reality. Data collected about people, places, or events in reality will eventually be stored in the file or database. To understand the form and structure of the data, information about the data itself is required. The information that describes data is referred to as metadata.

The relationship between reality, data, and metadata is pictured in Figure 17.4. Within the realm of reality are entities and attributes, within the realm of actual data are record occurrences and data item occurrences, and within the realm of metadata are record definitions and data item definitions. The meanings of these terms are discussed in the following subsections.

FIGURE 17.4
Reality, data, and metadata.

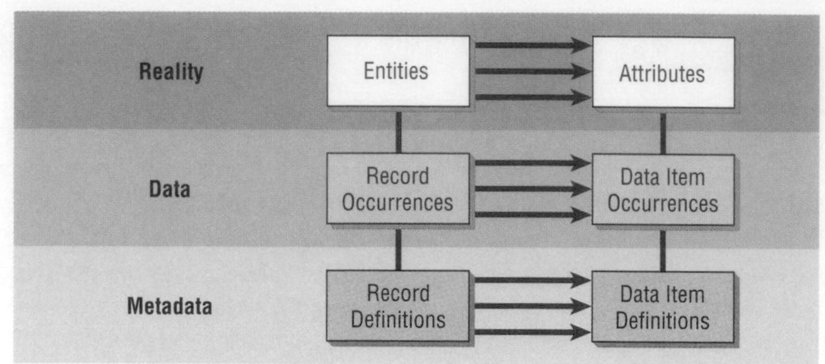

Entities. Any object or event about which someone chooses to collect data is an entity. An entity may be a person, place, or thing (for example, a salesperson, a city, or a product). Any entity can also be an event or unit of time such as a machine breakdown, a sale, or a month or year. In addition to the entities discussed in Chapter 2 is an additional minor entity called an *entity subtype*. Its symbol is a smaller rectangle within the entity rectangle. An entity subtype is a special one-to-one relationship used to represent additional attributes (fields) of another entity that may not be present on every record of the first entity. Entity subtypes eliminate the situation where an entity may have null fields stored on database tables. An example is the primary entity of a customer. Preferred customers may have special fields containing discount information, and this information would be in an entity subtype. Another example is students who have internships. The STUDENT MASTER should not have to contain information about internships for each student, because perhaps only a small number of students have internships.

Relationships. Relationships are associations between entities (sometimes they are referred to as data associations). Figure 17.5 is an entity-relationship (E-R) diagram that shows various types of relationships.

The first type of relationship is a one-to-one relationship (designated as 1:1). The diagram shows that there is only one PRODUCT PACKAGE for each PRODUCT. The second one-to-one relationship shows that each EMPLOYEE has a unique OFFICE. Notice that all these entities can be described further (a PRODUCT PRICE would not be an entity, nor would a phone extension).

Another type of relationship is a one-to-many (1:M) or a many-to-one association. As shown in the figure, a PHYSICIAN in a health maintenance organization is assigned many PATIENTS, but a PATIENT is assigned only one PHYSICIAN. Another example shows that an EMPLOYEE is a member of only one DEPARTMENT, but each DEPARTMENT has many EMPLOYEE(s).

Finally, a many-to-many relationship (designated as M:N) describes the possibility that entities may have many associations in either direction. For example, a STUDENT can have many COURSES, while at the same time a COURSE may have many STUDENT(s) enrolled in it. The second example shows that a SALESPERSON can call on many cities and a CITY can be a sales area for many SALESPERSON(s).

The standard symbols for crow's foot notation, the official explanation of the symbols, and what they actually mean are all given in Figure 17.6. Notice that the symbol for an entity is a rectangle. An entity is defined as a class of a person, place, or thing. The rectangle with the diamond inside stands for an associative entity, which is used to join two entities. A rectangle with an oval in it stands for an attributive entity, which is used for repeating groups.

The other notations necessary to draw E-R diagrams are the connections, of which there are five different types. In the lower portion of Figure 17.6, the meaning

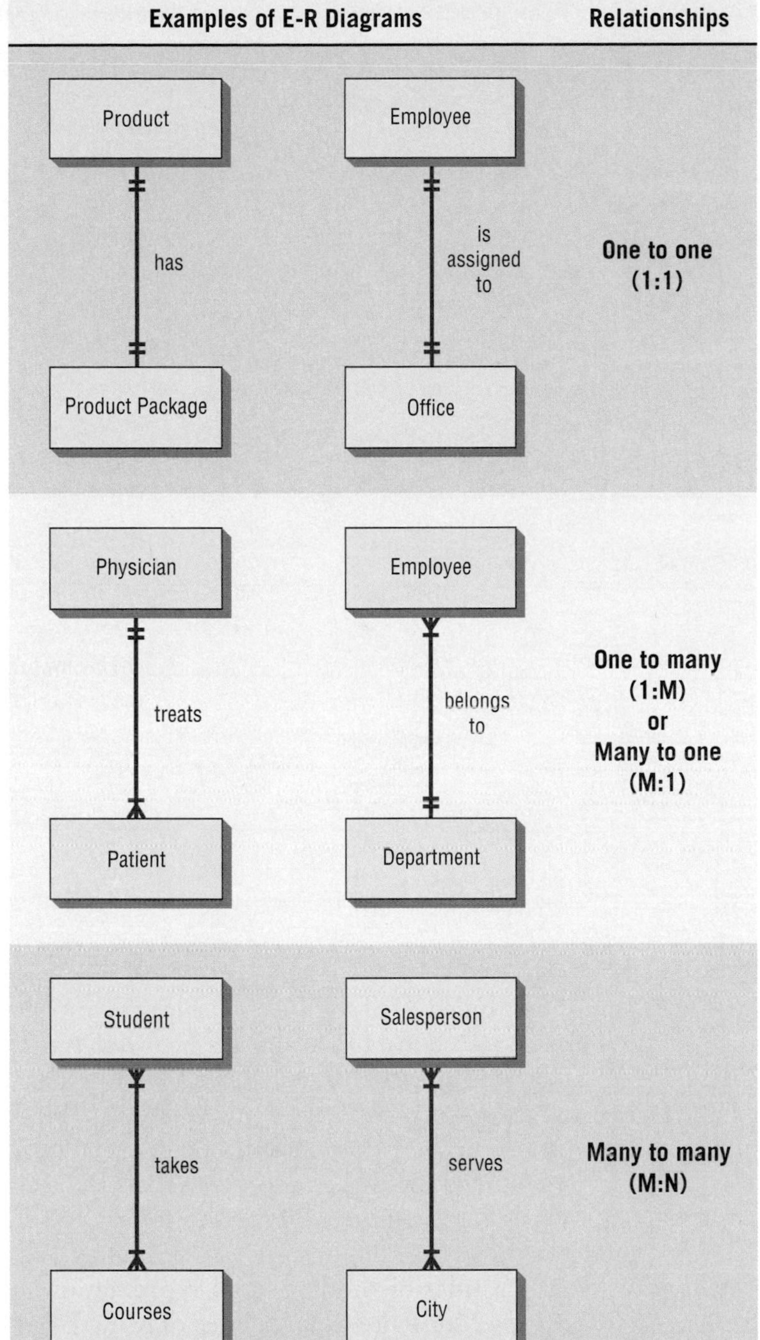

FIGURE 17.5
Entity-relationship diagrams can show one-to-one, one-to-many, many-to-one, or many-to-many associations.

of the notation is explained. When a straight line connects two plain entities and the end of the lines are marked with two short marks (||), a one-to-one relationship exists. Following that you will notice a crow's foot with a short mark (|); when this notation links entities, it indicates a relationship of one-to-one or one-to-many (to one or more). Entities linked with a straight line plus a short line (|) and a zero (but which looks more like a circle or **O**) are depicting a relationship of one-to-zero or one-to-one (only zero or one). A fourth type of link for relating entities is drawn with a straight line marked on the end with a zero (**O**) followed by a crow's foot. This type shows a zero-to-zero, zero-to-one, or zero-to-more relationship. Finally, a straight line with a crow's foot at the end linking entities depicts a relationship to more than (greater than) one. An entity may have a relationship connecting it to

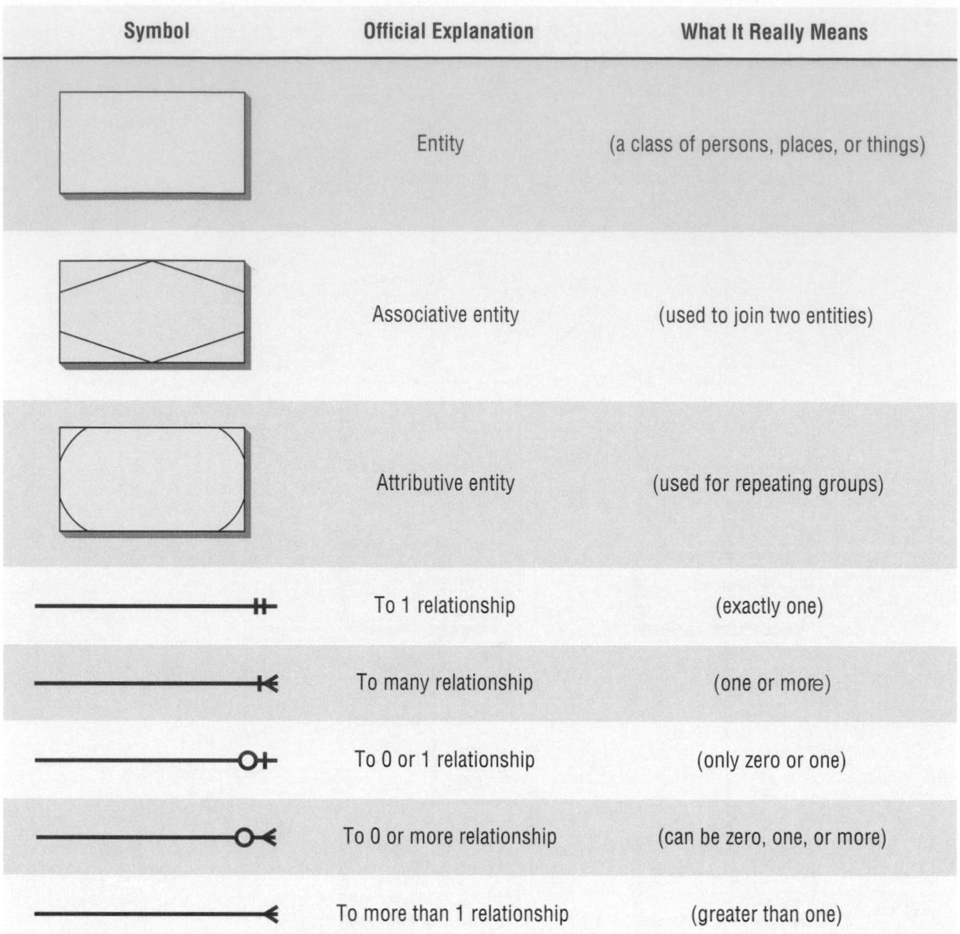

Symbol	Official Explanation	What It Really Means
	Entity	(a class of persons, places, or things)
	Associative entity	(used to join two entities)
	Attributive entity	(used for repeating groups)
	To 1 relationship	(exactly one)
	To many relationship	(one or more)
	To 0 or 1 relationship	(only zero or one)
	To 0 or more relationship	(can be zero, one, or more)
	To more than 1 relationship	(greater than one)

FIGURE 17.6
The entity-relationship symbols and their meaning.

itself. This type of relationship is called a self-join relationship; the implication is that there must be a way to link one record in a file to another record in the same file. An example of a self-join relationship may be found in the HyperCase® simulations found throughout these chapters. A task may have a precedent task (that is, one that must be completed before starting the current task). In this situation, one record (the current task) points to another record (the precedent task) on the same file.

The relationships in words can be written along the top or the side of each connecting line. In practice, you see the relationship in one direction, although you can write relationships on both sides of the line, each representing the point of view of the entity. (See Chapter 2 for more details about drawing E-R diagrams.)

An Entity-Relationship Example. An entity-relationship diagram containing many entities, many different types of relations, and numerous attributes is featured in Figure 17.7. In this E-R diagram, we are concerned about a billing system and in particular with the prescription part of the system. (For simplicity, we assume that office visits are handled differently and are outside the scope of this system.)

The entities are therefore PRESCRIPTIONS, PHYSICIANS, PATIENTS, and the INSURANCE CARRIER. The entity TREATMENTS is not important for the billing system, but it is part of the E-R diagram because it is used to bridge the gap between PRESCRIPTION and PATIENT. We therefore drew it as an associative entity in the figure.

Here, a PHYSICIAN treats many PATIENT(s) (1:M), who each subscribe to an individual INSURANCE CARRIER. Of course, the PATIENT is only one of many patients that subscribe to that particular INSURANCE CARRIER (M:1).

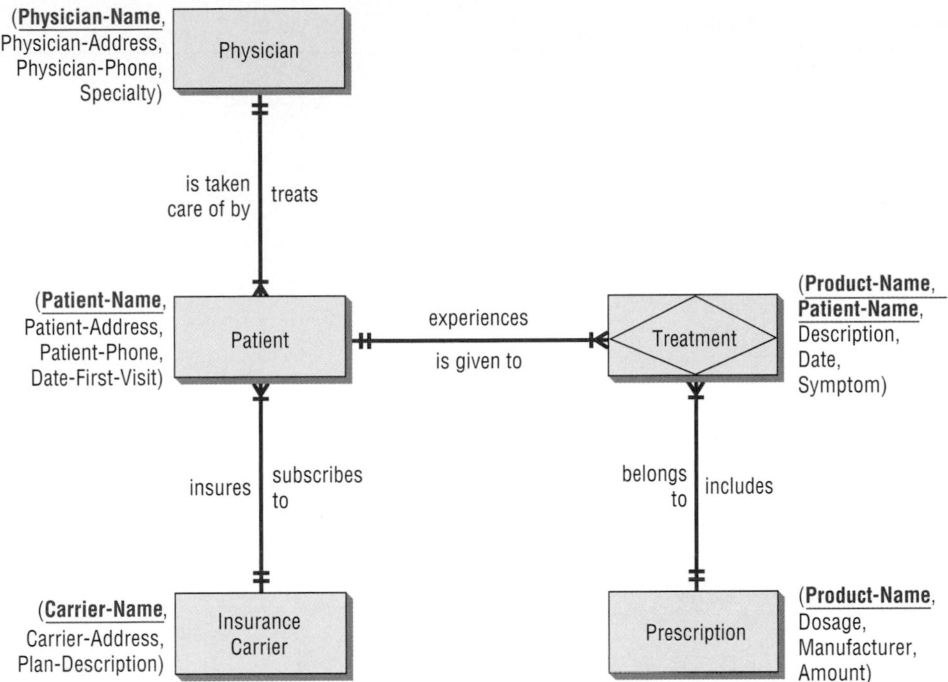

FIGURE 17.7
The entity-relationship diagram for patient treatment. Attributes can be listed alongside the entities. In each case, the key is underlined.

To complete the PHYSICIAN's records, the physician needs to keep information about the treatments a PATIENT has. Many PATIENT(s) experience many TREATMENT(s), making it a many-to-many (M:N) relationship. TREATMENT is represented as an associative entity because it is not important in our billing system by itself. TREATMENT(s) can include the taking of PRESCRIPTION(s), and likewise that is a M:N relationship because many treatments may call for combinations of pharmaceuticals and many drugs may work for many treatments.

Some detail is then filled in for the attributes. The attributes are listed next to each of the entities, and the key is underlined. For example, the entity PRESCRIPTION has a <u>PRODUCT-NAME</u>, DOSAGE, MANUFACTURER, and AMOUNT. Ideally, it would be beneficial to design a database in this fashion, using entity-relationship diagrams and then filling in the details concerning attributes. This top-down approach is desirable, but it is sometimes very difficult to achieve.

Attributes. An attribute is some characteristic of an entity. There can be many attributes for each entity. For example, a patient (entity) can have many attributes, such as last name, first name, street address, city, state, and so on. The date of the patient's last visit as well as the prescription details are also attributes. When the data dictionary was constructed in Chapter 10, the smallest particular described was called a data element. When files and databases are discussed, these data elements are generally referred to as data items. Data items are in fact the smallest units in a file or database. The term *data item* is also used interchangeably with the word *attribute*.

Data items can have values. These values can be of fixed or variable length; they can be alphabetic, numeric, special characters, or alphanumeric. Examples of data items and their values can be found in Figure 17.8.

Sometimes a data item is also referred to as a field. A field, however, represents something physical, not logical. Therefore, many data items can be packed into a field; the field can be read and converted to a number of data items. A common example of this is to store the date in a single field as MM/DD/YYYY. To sort the file in order by date, three separate data items are extracted from the field and sorted first by YYYY, then by MM, and finally by DD.

FIGURE 17.8
Typical values assigned to data items may be numbers, alphabetic characters, special characters, and combinations of all three.

Entity	Data Item	Value
Salesperson	Salesperson Number	87254
	Salesperson Name	Kaytell
	Company Name	Music Unlimited
	Address	45 Arpeum Circle
	Sales	$20,765
Package	Width	2
	Height	16
	Length	16
	Weight	3
	Mailing Address	765 Dulcinea Drive
	Return Address	P.O. Box 341 Spring Valley, MN
Order	Product(s)	B521
	Description(s)	"My Fair Lady" compact disc
	Quantity Ordered	1
	Last Name of Person Who Placed the Order	Kiley
	First Initial	R.
	Street Address	765 Dulcinea Drive
	City	La Mancha
	State	CA
	Zip Code	93407
	Credit Card Number	65-8798-87
	Date Order Was Placed	05/01/2000
	Amount	$6.99
	Status	Back Ordered

Records. A record is a collection of data items that have something in common with the entity described. Figure 17.9 is an illustration of a record with many related data items. The record shown is for an order placed with a mail-order company. The <u>ORDER-#</u>, LAST NAME, INITIAL, STREET ADDRESS, CITY, STATE, and CREDIT CARD are all attributes. Most records are of fixed length, so there is no need to determine the length of the record each time.

Under certain circumstances (for instance, when space is at a premium), variable-length records are used. A variable-length record is used as an alternative to reserving a large amount of space for the longest possible record, such as the maximum number of visits a patient has made to a physician. Each visit would contain many data items that would be part of the patient's full record (or file folder in a manual system). Later in this chapter, normalization of a relation is discussed. Normalization is a process that eliminates repeating groups found in variable-length records.

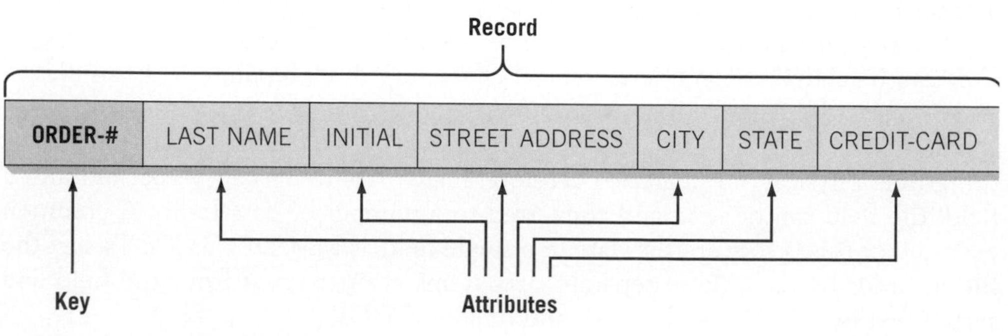

FIGURE 17.9
A record has a primary key and may have many attributes.

Keys. A key is one of the data items in a record that is used to identify a record. When a key uniquely identifies a record, it is called a primary key. For example, ORDER-# can be a primary key because only one number is assigned to each customer order. In this way, the primary key identifies the real-world entity (customer order).

A key is called a secondary key if it cannot uniquely identify a record. Secondary keys can be used to select a group of records that belong to a set (for example, orders that come from the state of Virginia).

When it is not possible to identify a record uniquely by using one of the data items found in a record, a key can be constructed by choosing two or more data items and combining them. This key is called a concatenated key. When a data item is used as a key in a record, the description is underlined. Therefore, in the ORDER RECORD (ORDER-#, LAST NAME, INITIAL, STREET ADDRESS, CITY, STATE, CREDIT CARD), the key is ORDER-#. If an attribute is a key in another file, it should be underlined with a dashed line.

Metadata. Metadata are data about the data in the file or database. Metadata describe the name given and the length assigned each data item. Metadata also describe the length and composition of each of the records.

Figure 17.10 is an example of metadata for a database for some generic software. The length of each data item is indicated according to a convention, where 5.2

Data Item	Value		
Salesperson Number	N	5	
Salesperson Name	A	20	
Company Name	A	26	
Address	A	36	
Sales	N	9.2	
Width	N	2	
Height	N	2	
Length	N	2	
Weight	N	2	
Mailing Address	A	36	
Return Address	A	36	
Product(s)	A	4	
Description(s)	A	30	
Quantity Ordered	N	2	
Last Name of Person Who Placed the Order	A	24	
First Initial	A	1	
Street Address	A	28	
City	A	12	
State	A	2	
Zip Code	N	9	
Credit Card Number	N	10	
Date Order Was Placed	D	8	MM/DD/YYYY
Amount	$	7.2	
Status	A	22	

Fields	
N	Numeric
A	Alphanumeric or text
D	Date MM/DD/YYYY
$	Currency
M	Memo

7.2 means that the field takes up 7 digits, two of which are right of the decimal.

Special formats for fields may be specified.

FIGURE 17.10
Metadata includes a description of what the value of each data item looks like.

means that 5 spaces are reserved for the number, two of which are to the right of the decimal point. The letter N signifies "numeric," and the A stands for "alphanumeric." The D stands for "date" and is automatically in the form MM/DD/YYYY. Some programs, such as Microsoft Access, use plain English for metadata, so words such as *text*, *currency*, and *number* are used. Microsoft Access provides a default of 50 characters as the field length for names, which is fine when working with small systems. If, however, you are working with a large database for a bank or a utility company, for example, you do not want to devote that much space to that field. Otherwise, the database would become quite large and filled with wasted space. That is when you can use metadata to plan ahead and design a more efficient database.

FILE ORGANIZATION

A file contains groups of records used to provide information for operations, planning, management, and decision making. The types of files used are discussed first, followed by a description of the many ways conventional files can be organized.

File Types. Files can be used for storing data for an indefinite period of time, or they can be used to store data temporarily for a specific purpose. Master files and table files are used to store data for a long period. The temporary files are usually called transaction files, work files, or report files.

Master files Master files contain records for a group of entities. The attributes may be updated often, but the records themselves are relatively permanent. These files tend to have large records containing all the information about a data entity. Each record usually contains a primary key and several secondary keys. Master files are contained as tables with a database or as indexed or indexed-sequential files. Although the analyst is free to arrange the data elements within a master file in any order, a standard arrangement is usually to place the primary key field first, followed by descriptive elements, and finally by elements that reflect the business and change frequently with business activities. This procedure allows analysts or other operational people to identify records easily when a file is listed with a print utility.

Descriptive information is data that do not change with business events, such as an item description, customer name, address, or employee department. These elements are usually changed by maintenance programs using direct access methods. Usually, these elements contain alternate keys or indexes, and the data are in display format.

Business information elements are those that periodically change with business events, such as year-to-year gross pay, grade point average, customer account balance, and the customer date-of-last-purchase. These elements are changed by update programs, which usually read both files and sequentially match records for efficiency. The record elements are modified only when the data are in error by correcting programs using random update methods. Often the dollar amount fields are in a compressed data format called packed decimal to save room on files and to speed up program execution time.

If the master file is stored using conventional file methods, an expansion area is reserved at the end of each record. This area provides room for adding new fields to the record as business needs change. If the file is part of a database structure, expansion area is not required. Examples of a master file include patient records, customer records, a personnel file, and a parts inventory file.

Table files A table file contains data used to calculate more data or performance measures. One example is a table of postage rates used to determine the shipping costs of a package. Another example is a tax table. Table files usually are read only by a program.

Transaction files A transaction file is used to enter changes that update the master file and produce reports. Suppose a newspaper subscriber master file needs to be updated; the transaction file would contain the subscriber number, a transaction code such as E for extending the subscription, C for canceling the subscription, or A for address change. Then only information relevant to the updating needs to be entered, that is, the length of renewal if E, and the address if A. No additional information would be needed if the subscription were canceled. The rest of the information already exists in the master file. As a result, transaction files are usually kept to a minimum length. Transaction files may contain several different types of records, such as the three used for updating the newspaper subscription master, with a code on the transaction file indicating the type of transaction.

Work files A program can sometimes run more efficiently if a work file is used. A common example of a work file is a file that is resorted so that records may be accessed more quickly.

Report files When it is necessary to run a program but when no printer is available (or a printer is busy printing other jobs), a report file is used. Sending the output to a file rather than a printer is called spooling. Later, when the device is ready, the document can be printed. Report files are very useful, because users can take files to other computer systems and output to specialty devices such as plotters, laser printers, microfiche units, and even computerized typesetting machines.

Sequential Organization. When records are physically in order in a file, the file is said to be a sequential file. When a sequential file is updated, it is necessary to go through the entire file. Because records cannot be inserted in the middle of the file, a sequential file is usually copied over during the updating process.

Figure 17.11 illustrates a file of current orders for a mail-order company that sells compact discs. The file contains 12 records and is stored sequentially according to the ORDER-#. If we want to look up order 13432, we would start at the beginning and read through the file until we arrived at order 13432.

Sequential master files are used when the hardware requires it (remember that a magnetic tape is a sequential device) or when the normal access requires that most of the records be accessed. In other words, when we need to read or update only a few records, it is inefficient to use a sequential structure, but when many records need to be read or modified, sequential organization would make sense. Sequential organization is normally used for all types of files except master files.

	ORDER-#	LAST NAME	I	STREET-ADDRESS	CITY	ST	CREDIT-CARD
1	10784	MacRae	G	2314 Curly Circle	Lincoln	NE	45-4654-76
2	10796	Jones	S	34 Dream Lane	Oklahoma City	OK	45-9876-74
3	11821	Preston	R	1008 Madison Ave.	River City	IA	34-7642-64
4	11845	Channing	C	454 Harmonia St.	New York	NY	34-0876-87
5	11872	Kiley	R	765 Dulcinea Drive	La Mancha	CA	65-8798-87
6	11976	Verdon	G	7564 K Street	Chicago	IL	67-8453-18
7	11998	Rivera	C	4342 West Street	Chicago	IL	12-2312-54
8	12765	Orbach	J	1345 Michigan Ave.	Chicago	IL	23-4545-65
9	12769	Steele	T	3498 Burton Lane	Finnian	NJ	65-7687-09
10	12965	Crawford	M	1986 Barnum Cir.	London	NH	23-0098-23
11	13432	Cullum	J	354 River Road	Shenandoah	VT	45-8734-33
12	13542	Mostel	Z	65 Fiddler Street	Anatevka	ND	34-6723-98

FIGURE 17.11
A sequential file sorted by ORDER-#.

Linked Lists. When files are stored on direct-access devices such as a disk, the options are expanded. Records can be sorted logically, rather than physically, using linked lists. Linked lists are achieved by using a set of pointers to direct you to the next logical record located anywhere in the file.

Figure 17.12 shows the compact disc ordering file with an additional attribute used to store the pointer. Because the file is already stored in sequential order according to ORDER-#, the pointer is used to point to records in logical (alphabetical) order by LAST NAME. This example shows an obvious advantage of using linked lists: Files can be sorted logically in many different ways by using a variety of pointers.

Hashed File Organization. Direct-access devices also permit access to a given record by going directly to its address. Because it is not feasible to reserve a physical address for each possible record, a method called hashing is used. Hashing is the process of calculating an address from the record key.

Suppose there were 500 employees in an organization and we wanted to use Social Security number as a key. It would be inefficient to reserve 999,999,999 addresses, one for each Social Security number. Therefore, we could take the Social Security number and use it to derive the address of the record.

There are many hashing techniques. A common one is to divide the original number by a prime number that approximates the storage locations and then to use the remainder as the address, as follows: Begin with the Social Security number 053-46-8942. Then divide by 509, yielding 105047. Note that 105047 multiplied by 509 does not equal the original number; it equals 53468923 instead. The difference between the original number, 53468942, and the dividend, 53468923, is the remainder, and it equals 19. The storage location of the record for an employee whose Social Security number is 053-46-8942 would thus be 19.

A problem arises, however, when a person with a different Social Security number (say, 472-38-4086) has the same remainder. When this occurs, the second person's record has to be placed in a special overflow area.

Indexed Organization. An index is different from a pointer, because it is stored in a file that is separate from the data file. Figure 17.13 shows that four separate index files are generated on the ORDER FILE for compact discs.

The numerous index files have practical applications. If we want to list the details of the order placed by "Cullum," we can look up "Cullum" in the LAST-NAME INDEX file and go directly to record 11. The CITY INDEX could be used

FIGURE 17.12
A linked list uses pointers to designate the logical order of the records.

ORDER FILE START = 4

	ORDER-#	LAST NAME	I	STREET-ADDRESS	CITY	ST	CREDIT-CARD	POINTER
1	10784	MacRae	G	2314 Curly Circle	Lincoln	NE	45-4654-76	6
2	10796	Jones	S	34 Dream Lane	Oklahoma City	OK	45-9876-74	4
3	11821	Preston	R	1008 Madison Ave.	River City	IA	34-7642-64	9
4	11845	Channing	C	454 Harmonia St.	New York	NY	34-0876-87	1
5	11872	Kiley	R	765 Dulcinea Drive	La Mancha	CA	65-8798-87	5
6	11976	Verdon	G	7564 K Street	Chicago	IL	67-8453-18	END
7	11998	Rivera	C	4342 West Street	Chicago	IL	12-2312-54	10
8	12765	Orbach	J	1345 Michigan Ave.	Chicago	IL	23-4545-65	8
9	12769	Steele	T	3498 Burton Lane	Finnian	NJ	65-7687-09	11
10	12965	Crawford	M	1986 Barnum Cir.	London	NH	23-0098-23	2
11	13432	Cullum	J	354 River Road	Shenandoah	VT	45-8734-33	3
12	13542	Mostel	Z	65 Fiddler Street	Anatevka	ND	34-6723-98	7

LAST-NAME INDEX

Channing	4
Crawford	10
Cullum	11
Jones	2
Kiley	5
MacRae	1
Mostel	12
Orbach	8
Preston	3
Rivera	7
Steele	9
Verdon	6

CITY INDEX

Anatevka	12
Chicago	6 7 8
Finnian	9
La Mancha	5
Lincoln	1
London	10
New York	4
Oklahoma City	2
River City	3
Shenandoah	11

CREDIT-CARD INDEX

12-2312-54	7
23-0098-23	10
23-4545-65	8
34-0876-87	4
34-6723-98	12
34-7642-64	3
45-4654-76	1
45-8734-33	11
45-9876-74	2
65-7687-09	9
65-8798-87	5
67-8453-18	6

STATUS INDEX

Back Ordered	5 8 9 12
In Process	3 4 10 11
Shipped 5/12	1 6 7
Shipped 5/14	2

ORDER FILE

	ORDER-#	LAST NAME	I	STREET-ADDRESS	CITY	ST	CREDIT-CARD	STATUS
1	10784	MacRae	G	2314 Curly Circle	Lincoln	NE	45-4654-76	Shipped 5/12
2	10796	Jones	S	34 Dream Lane	Oklahoma City	OK	45-9876-74	Shipped 5/14
3	11821	Preston	R	1008 Madison Ave.	River City	IA	34-7642-64	In process
4	11845	Channing	C	454 Harmonia St.	New York	NY	34-0876-87	In process
5	11872	Kiley	R	765 Dulcinea Drive	La Mancha	CA	65-8798-87	Back Ordered
6	11976	Verdon	G	7564 K Street	Chicago	IL	67-8453-18	Shipped 5/12
7	11998	Rivera	C	4342 West Street	Chicago	IL	12-2312-54	Shipped 5/12
8	12765	Orbach	J	1345 Michigan Ave.	Chicago	IL	23-4545-65	Back Ordered
9	12769	Steele	T	3498 Burton Lane	Finnian	NJ	65-7687-09	Back Ordered
10	12965	Crawford	M	1986 Barnum Cir.	London	NH	23-0098-23	In process
11	13432	Cullum	J	354 River Road	Shenandoah	VT	45-8734-33	In process
12	13542	Mostel	Z	65 Fiddler Street	Anatevka	ND	34-6723-98	Back Ordered

FIGURE 17.13

Inverted lists can have many indices.

for a mailing list, the STATUS INDEX could be used to write notices to people with back-ordered items, and so on.

When the number of index files increases, it is possible to use index files for storing much of the data. The amount of information in the index files can be quite large. In this case, the structure is referred to as an inverted list.

Inverted lists are most appropriate when used to find information on a specific combination of keys. For example, a query such as "Who in Chicago still has an item that is back-ordered?" could be answered efficiently. Inverted lists, however, are difficult to maintain. Hence, inverted lists are more appropriate for a file that is rarely updated.

Indexed-Sequential Organization. A widely used method of file organization is called indexed-sequential organization, or indexed-sequential access method (ISAM). In an ISAM file, the records are arranged in blocks. The records within blocks are stored in order physically, but the blocks of records may be in any order. Therefore, an index is needed to locate the blocks of records.

A newer organizational format used for mainframe computers is the virtual storage access method (VSAM), a more modern and efficient method for handling indexed-sequential files. Indexed-sequential files allow programs to read records directly (that is, randomly) without reading other records in the file. Records written using an indexed-sequential method are placed in sequence within the file. When organized in this way, records may be deleted or rewritten without reading other records as well. When a record is rewritten, it is physically placed on the disk in the same location from which the original record was obtained.

DATABASE ORGANIZATION

Databases can be organized in several ways. Here we will consider three most common approaches.

Logical and Physical Views of Data. A database, unlike a file, is intended to be shared by many users. It is clear that the users all see the data in different ways. We refer to the way a user pictures and describes the data as a user view. The problem, however, is that different users have different user views. These views are examined by the systems analyst, and an overall logical model of the database needs to be developed. Finally, the logical model of the database must be transformed into

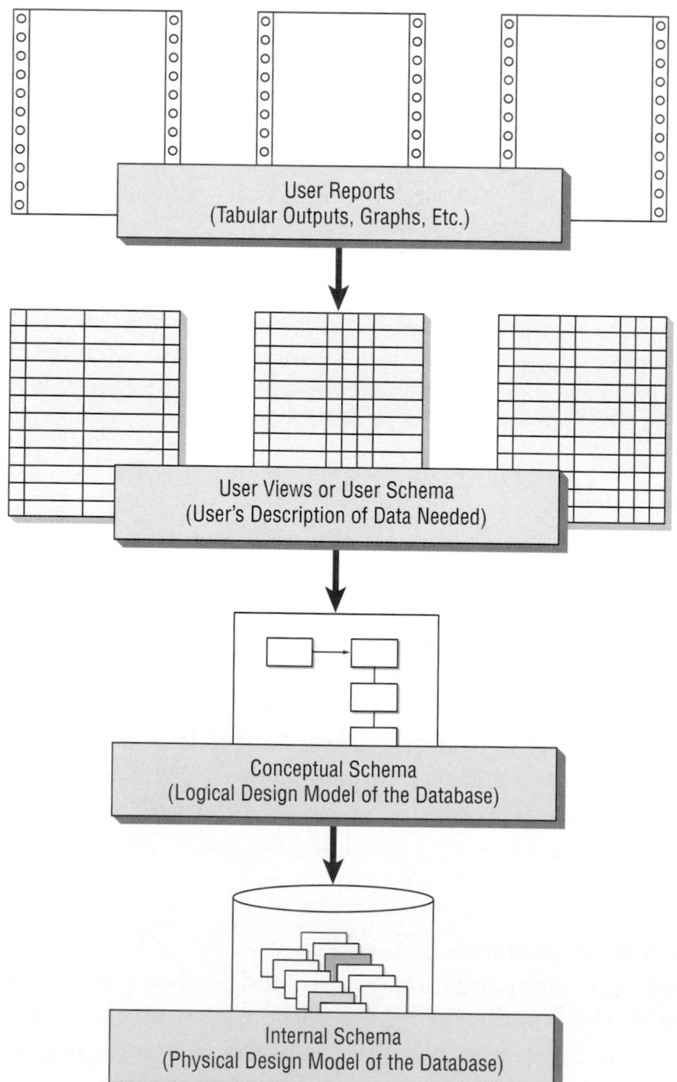

FIGURE 17.14
Database design includes synthesizing user reports, user views, and logical and physical designs.

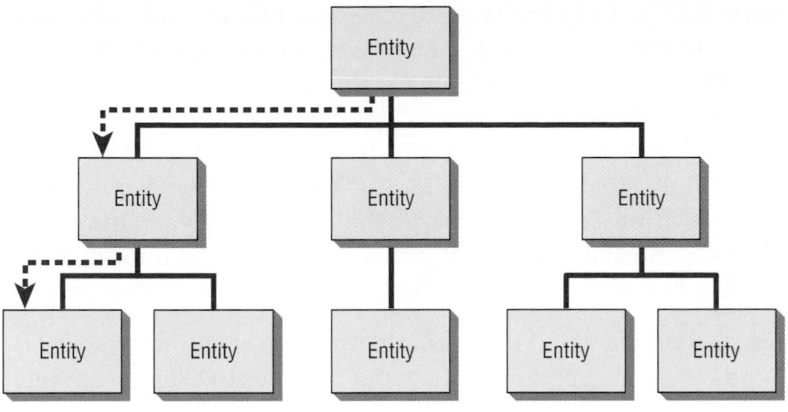

FIGURE 17.15
Retrieving data in a hierarchical data structure is accomplished by following the branches.

a corresponding physical database design. Physical design is involved with how data are stored and related as well as how they are accessed.

In database literature, the views are referred to as *schema*. Figure 17.14 shows how the user reports and user views (user schema) are related to the logical model (conceptual schema) and physical design (internal schema).

There are three main types of logically structured databases: hierarchical, network, and relational.

Hierarchical data structures Hierarchical data structures imply that an entity can have no more than one owning entity. Therefore, it is a structure made up of many one-to-many or one-to-one associations. Other associations such as many-to-one or many-to-many are not allowed.

Hierarchical structures are sometimes called trees because the subordinates connected to the owning entities resemble the branches of a tree, but they are usually drawn upside down as shown in Figure 17.15.

Sometimes it is very easy to retrieve information from a hierarchical database. As an example, consider the hierarchical structure in Figure 17.16. In this example,

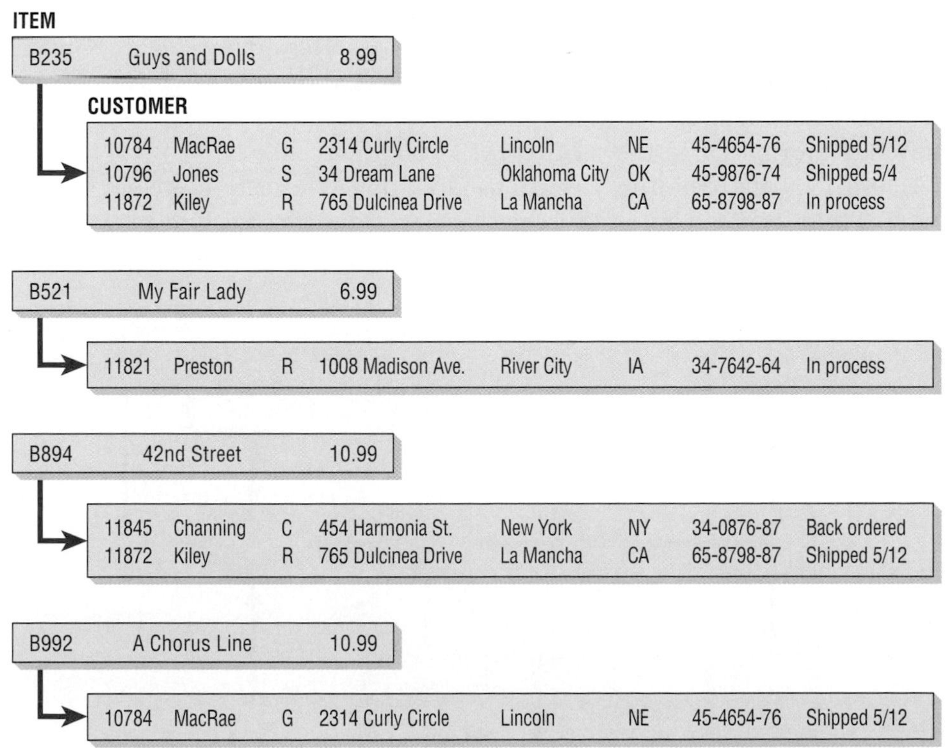

FIGURE 17.16
In a hierarchical data structure, the CUSTOMERs are represented as the subordinates to the ITEM.

each compact disc (ITEM) has one or more subordinates (CUSTOMERS). If we wanted to see who ordered the CD *42nd Street*, we would go to the owning entity B894 only and look at each subordinate (in this case, 11845 and 11872) to find the names "Channing" and "Kiley."

Sometimes, however, operations may become difficult. For example, if we found an error in the credit card number of "G. MacRae," we would have to search through each owning entity to ensure that we find every occurrence of "G. MacRae." Notice also that we cannot add a new customer until a specific item is chosen, which is one of the disadvantages of the hierarchical structure.

Network data structures A network structure allows any entity to have any number of subordinates or superiors. A network structure is shown in Figure 17.17. Entities are connected using network links, which are data items common to both of the connected entities. Some of the problems inherent in hierarchical structures can be alleviated using the network structure, but the network structure is more complex.

An example of the compact disc ordering database using a network structure is shown in Figure 17.18. The entities (ITEM-DESCRIPTION and ORDER-DETAILS) are connected by network links (STATUS-LINK). Updating a record (such as correcting a person's credit card number) is easier than in the hierarchical structure, because the order record (10784 for "MacRae") appears only once. It is also possible to insert records for customers who have not yet placed orders (for instance, if they wish to be on a catalog mailing list). The appropriate ITEM-DESCRIPTION can be added at a later date, when the order is placed.

Relational data structures A relational structure consists of one or more two-dimensional tables, which are referred to as relations. The rows of the table represent the records, and the columns contain attributes.

The compact disc ordering database is depicted as a relational structure in Figure 17.19. Here, three tables are needed to (1) describe the items and keep track of the current price of compact discs (ITEM-PRICE), (2) describe the details of the order (ORDER), and (3) identify the status of the order (ITEM-STATUS).

To determine the price of an item, we need to know the item number to be able to find it in the relation ITEM-PRICE. To update "G. MacRae's" credit card number, we can search the ORDER relation for MacRae and correct it only once, even though he ordered many compact discs. To find out the status of part of an order, however, we must know the ITEM-# and ORDER-#, and then we must locate that information in the relation ITEM-STATUS.

Maintaining the tables in a relational structure is usually quite simple when compared with maintaining a hierarchical or network structure. One of the primary advantages of relational structures is that ad hoc queries are handled efficiently.

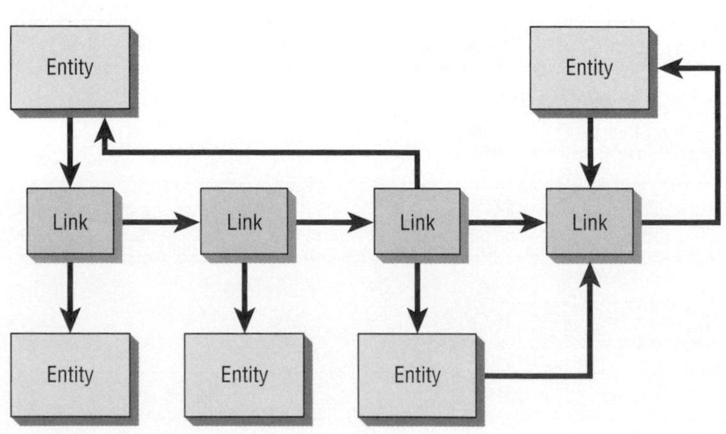

FIGURE 17.17

Network structures allow the entity to have any number of subordinates or superiors, and entities are connected by common links.

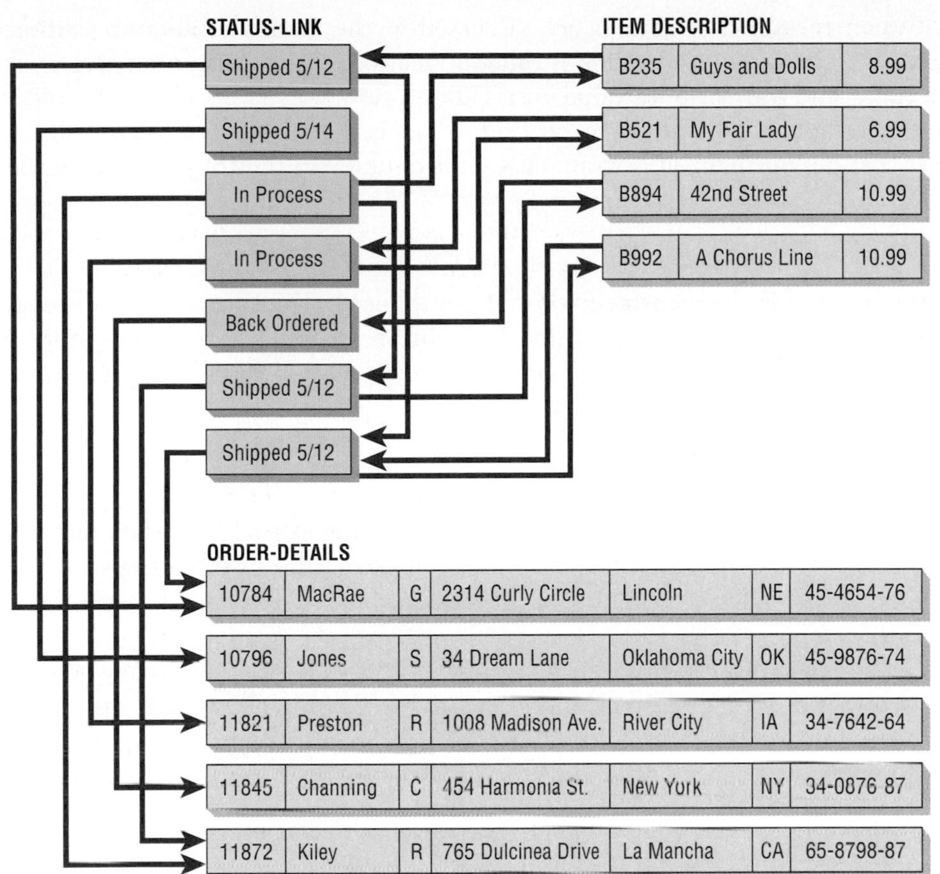

FIGURE 17.18
The STATUS-LINK is used in this example to link the ITEM-DESCRIPTION to the ORDER-DETAILS in this network data structure.

ITEM-PRICE

ITEM-#	TITLE	PRICE
B235	Guys and Dolls	8.99
B521	My Fair Lady	6.99
B894	42nd Street	10.99
B992	A Chorus Line	10.99

ORDER

ORDER-#	LAST NAME	I	STREET-ADDRESS	CITY	ST	CHARGE-ACCT
10784	MacRae	G	2314 Curly Circle	Lincoln	NE	45-4654-76
10796	Jones	S	34 Dream Lane	Oklahoma City	OK	44-9876-74
11821	Preston	R	1008 Madison Ave.	River City	IA	34-7642-64
11845	Channing	C	454 Harmonia St.	New York	NY	34-0876-87
11872	Kiley	R	765 Dulcinea Drive	La Mancha	CA	65-8798-87

ITEM-PRICE

ITEM-#	ORDER-#	STATUS
B235	10784	Shipped 5/12
B235	19796	Shipped 5/14
B235	11872	In Process
B521	11821	In Process
B894	11845	Back Ordered
B894	11872	Shipped 5/12
B992	10784	Shipped 5/12

FIGURE 17.19
In a relational data structure, data are stored in many tables.

When relational structures are discussed in the database literature, different terminology is often used. A file is called a relation, a record is usually referred to as a tuple, and the attribute value set is called a domain.

For relational structures to be useful and manageable, the relational tables must first be "normalized." Normalization is detailed in the following section.

NORMALIZATION

Normalization is the transformation of complex user views and data stores to a set of smaller, stable data structures. In addition to being simpler and more stable, normalized data structures are more easily maintained than other data structures.

THE THREE STEPS OF NORMALIZATION

Beginning with either a user view or a data store developed for a data dictionary (see Chapter 10), the analyst normalizes a data structure in three steps, as shown in Figure 17.20. Each step involves an important procedure, one that simplifies the data structure.

The relation derived from the user view or data store will most likely be unnormalized. The first stage of the process includes removing all repeating groups and identifying the primary key. To do so, the relation needs to be broken up into two or more relations. At this point, the relations may already be of the third normal form, but more likely more steps will be needed to transform the relations to the third normal form.

FIGURE 17.20
Normalization of a relation is accomplished in three major steps.

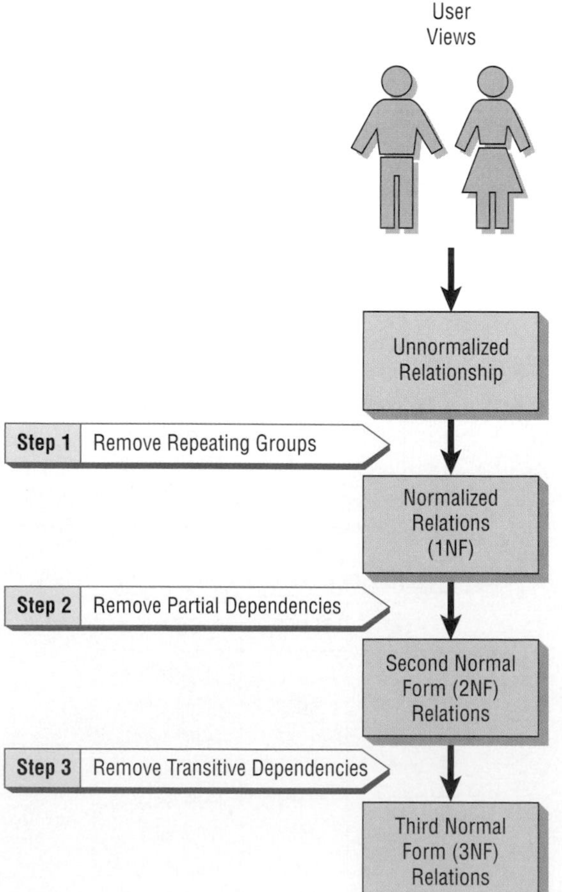

The second step ensures that all nonkey attributes are fully dependent on the primary key. All partial dependencies are removed and placed in another relation.

The third step removes any transitive dependencies. A transitive dependency is one in which nonkey attributes are dependent on other nonkey attributes.

A NORMALIZATION EXAMPLE

Figure 17.21 is a user view for the Al S. Well Hydraulic Equipment Company. The report shows (1) the SALESPERSON-NUMBER, (2) the SALESPERSON-NAME, and (3) the SALES-AREA. The body of the report shows the (4) CUSTOMER-NUMBER and (5) CUSTOMER-NAME. Next is (6) the WAREHOUSE-NUMBER that will service the customer, followed by (7) the WAREHOUSE-LOCATION, which is the city in which the company is located. The final information contained in the user view is (8) the SALES-AMOUNT. The rows (one for each customer) on the user view show that items 4 through 8 form a repeating group.

If the analyst was using a data flow/data dictionary approach, the same information in the user view would appear in a data structure. Figure 17.22 shows how the data structure would appear at the data dictionary stage of analysis. The repeating group is also indicated in the data structure by an asterisk (*) and indentation.

Before proceeding, note the data associations of the data elements in Figure 17.23. This type of illustration is called a bubble diagram or data model diagram. Each entity is enclosed in an ellipse, and arrows are used to show the relationships. Although it is possible to draw these relationships with an E-R diagram, it is sometimes easier to use the simpler bubble diagram to model the data.

In this example, there is only one SALESPERSON-NUMBER assigned to each SALESPERSON-NAME, and that person will cover only one SALES-AREA, but each SALES-AREA may be assigned to many salespeople: hence, the double arrow notation from SALES-AREA to SALESPERSON-NUMBER. For each SALESPERSON-NUMBER, there may be many CUSTOMER-NUMBER(s).

Furthermore, there would be a one-to-one correspondence between CUSTOMER-NUMBER and CUSTOMER-NAME; the same is true for

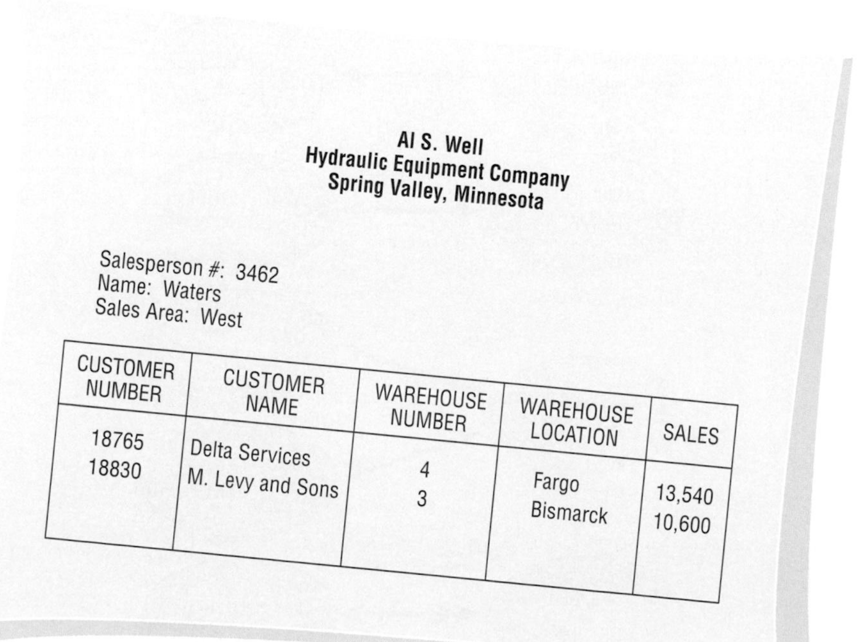

Al S. Well
Hydraulic Equipment Company
Spring Valley, Minnesota

Salesperson #: 3462
Name: Waters
Sales Area: West

CUSTOMER NUMBER	CUSTOMER NAME	WAREHOUSE NUMBER	WAREHOUSE LOCATION	SALES
18765	Delta Services	4	Fargo	13,540
18830	M. Levy and Sons	3	Bismarck	10,600

FIGURE 17.21
A user report for the Al S. Well Hydraulic Equipment Company.

FIGURE 17.22
The analyst would find a data structure (from a data dictionary) useful in developing a database.

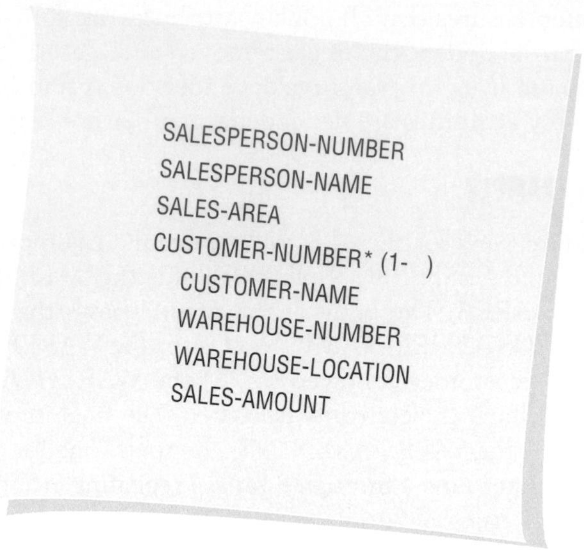

SALESPERSON-NUMBER
SALESPERSON-NAME
SALES-AREA
CUSTOMER-NUMBER* (1-)
CUSTOMER-NAME
WAREHOUSE-NUMBER
WAREHOUSE-LOCATION
SALES-AMOUNT

WAREHOUSE-NUMBER and WAREHOUSE-LOCATION. CUSTOMER-NUMBER will have only one WAREHOUSE-NUMBER and WAREHOUSE-LOCATION, but each WAREHOUSE-NUMBER or WAREHOUSE-LOCATION may service many CUSTOMER-NUMBER(s). Finally, to determine the SALES-AMOUNT for one salesperson's calls to a particular company, it is necessary to know both the SALESPERSON-NUMBER and the CUSTOMER-NUMBER.

FIGURE 17.23
Drawing data model diagrams for data associations sometimes helps analysts appreciate the complexity of data storage.

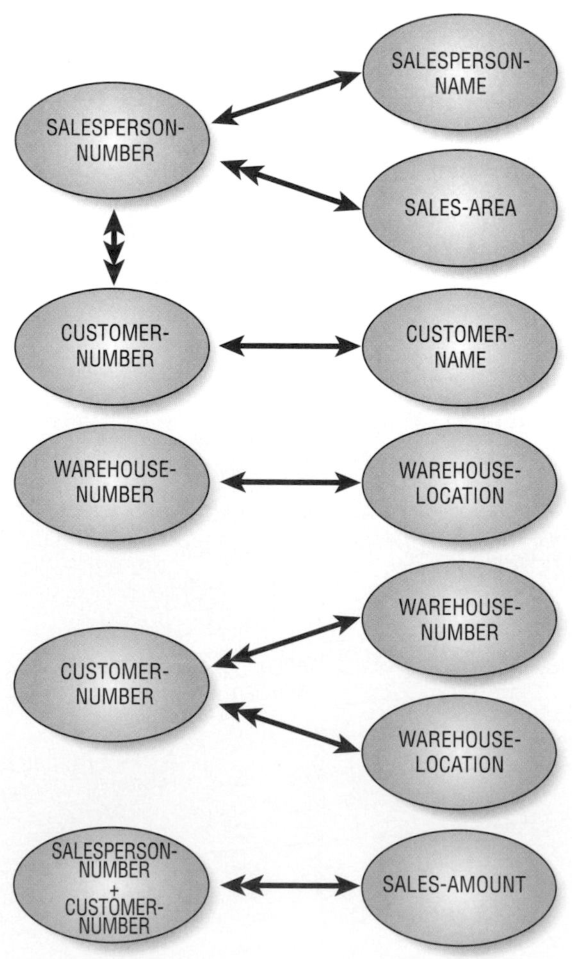

SALESPERSON NUMBER	SALESPERSON NAME	SALES AREA	CUSTOMER NUMBER	CUSTOMER NAME	WAREHOUSE NUMBER	WAREHOUSE LOCATION	SALES AMOUNT
3462	Waters	West	18765	Delta Systems	4	Fargo	13540
			18830	A. Levy and Sons	3	Bismarck	10600
			19242	Ranier Company	3	Bismarck	9700
3593	Dryne	East	18841	R. W. Flood Inc.	2	Superior	11560
			18899	Seward Systems	2	Superior	2590
			19565	Stodola's Inc.	1	Plymouth	8800
etc.							

FIGURE 17.24

If the data were listed in an unnormalized table, there could be repeating groups.

The main objective of the normalization process is to simplify all the complex data items that are often found in users views. For example, if the analyst were to take the user view discussed above and attempt to make a relational table out of it, the table would look like Figure 17.24. Because this relation is based on our initial user view, we refer to it as SALES-REPORT.

SALES-REPORT is an unnormalized relation, because it has repeating groups. It is also important to observe that a single attribute such as SALESPERSON-NUMBER cannot serve as the key. The reason is clear when one examines the relationships between SALESPERSON-NUMBER and the other attributes in Figure 17.25. Although there is a one-to-one correspondence between SALESPERSON-NUMBER and two attributes (SALESPERSON-NAME and SALES-AREA), there is a one-to-many relationship between SALESPERSON-NUMBER and the other five attributes (CUSTOMER-NUMBER, CUSTOMER-NAME, WAREHOUSE-NUMBER, WAREHOUSE-LOCATION, and SALES-AMOUNT).

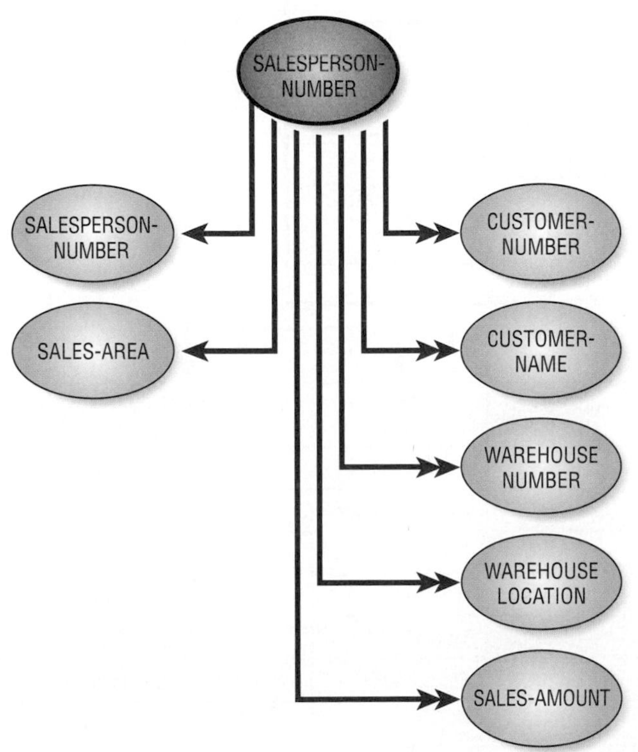

FIGURE 17.25

A data model diagram shows that in the unnormalized relation, the SALESPERSON-NUMBER has a one-to-many association with some attributes.

SALES-REPORT can be expressed in the following shorthand notation:

SALES REPORT (SALESPERSON-NUMBER, SALESPERSON-NAME, SALES-AREA, (CUSTOMER-NUMBER, CUSTOMER-NAME, WAREHOUSE-NUMBER, WAREHOUSE-LOCATION, SALES-AMOUNT))

where the inner set of parentheses represent the repeated group.

First Normal Form (1NF). The first step in normalizing a relation is to remove the repeating groups. In our example, the unnormalized relation SALES-REPORT will be broken into two separate relations. These new relations will be named SALESPERSON and SALESPERSON-CUSTOMER.

Figure 17.26 shows how the original, unnormalized relation SALES-REPORT is normalized by separating the relation into two new relations. Notice that the relation SALESPERSON contains the primary key SALESPERSON-NUMBER and all the attributes that were not repeating (SALESPERSON-NAME and SALES-AREA).

The second relation, SALESPERSON-CUSTOMER, contains the primary key from the relation SALESPERSON (the primary key of SALESPERSON is SALESPERSON-NUMBER) as well as all the attributes that were part of the repeating group (CUSTOMER-NUMBER, CUSTOMER-NAME,

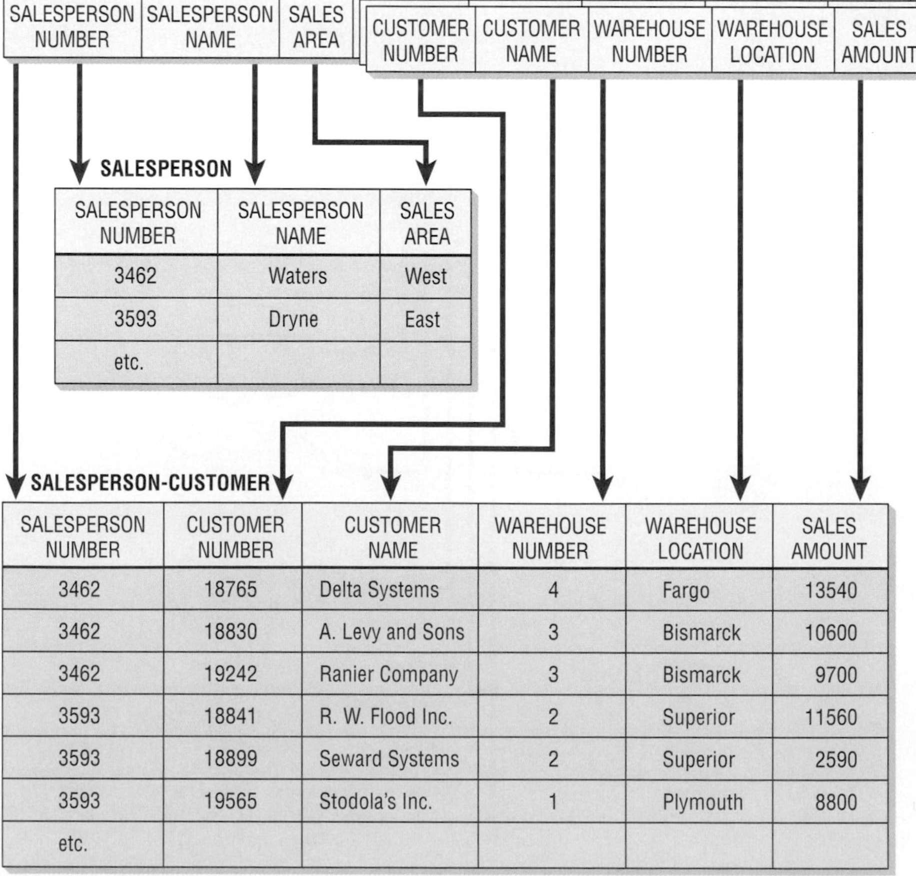

SALESPERSON

SALESPERSON NUMBER	SALESPERSON NAME	SALES AREA
3462	Waters	West
3593	Dryne	East
etc.		

SALESPERSON-CUSTOMER

SALESPERSON NUMBER	CUSTOMER NUMBER	CUSTOMER NAME	WAREHOUSE NUMBER	WAREHOUSE LOCATION	SALES AMOUNT
3462	18765	Delta Systems	4	Fargo	13540
3462	18830	A. Levy and Sons	3	Bismarck	10600
3462	19242	Ranier Company	3	Bismarck	9700
3593	18841	R. W. Flood Inc.	2	Superior	11560
3593	18899	Seward Systems	2	Superior	2590
3593	19565	Stodola's Inc.	1	Plymouth	8800
etc.					

FIGURE 17.26
The original unnormalized relation SALES-REPORT is separated into two relations, SALESPERSON (3NF) and SALESPERSON-CUSTOMER (1NF).

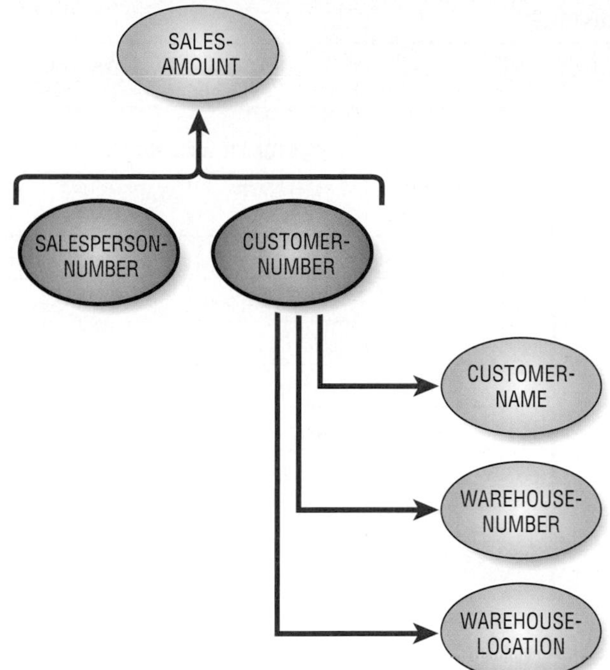

FIGURE 17.27

A data model diagram shows that three attributes are dependent on CUSTOMER-NUMBER, so the relation is not yet normalized. Both SALESPERSON-NUMBER and CUSTOMER-NUMBER are required to look up SALES-AMOUNT.

WAREHOUSE-NUMBER, WAREHOUSE-LOCATION, and SALES-AMOUNT). Knowing the SALESPERSON-NUMBER, however, does not automatically mean that you will know the CUSTOMER-NAME, SALES-AMOUNT, WAREHOUSE-LOCATION, and so on. In this relation, one must use a concatenated key (both SALESPERSON-NUMBER and CUSTOMER-NUMBER) to access the rest of the information. It is possible to write the relations in shorthand notation as follows:

SALESPERSON (SALESPERSON-NUMBER, SALESPERSON-NAME, SALES-AREA)

and

SALESPERSON-CUSTOMER (SALESPERSON-NUMBER, CUSTOMER-NUMBER, CUSTOMER-NAME, WAREHOUSE-NUMBER, WAREHOUSE-LOCATION, SALES-AMOUNT)

The relation SALESPERSON-CUSTOMER is a first normal relation, but it is not in its ideal form. Problems arise because some of the attributes are not functionally dependent on the primary key (that is, SALESPERSON-NUMBER, CUSTOMER-NUMBER). In other words, some of the nonkey attributes are dependent only on CUSTOMER NUMBER and not on the concatenated key. The data model diagram in Figure 17.27 shows that SALES-AMOUNT is dependent on both SALESPERSON-NUMBER and CUSTOMER-NUMBER, but the other three attributes are dependent only on CUSTOMER-NUMBER.

Second Normal Form (2NF). In the second normal form, all the attributes will be functionally dependent on the primary key. Therefore, the next step is to remove all the partially dependent attributes and place them in another relation. Figure 17.28

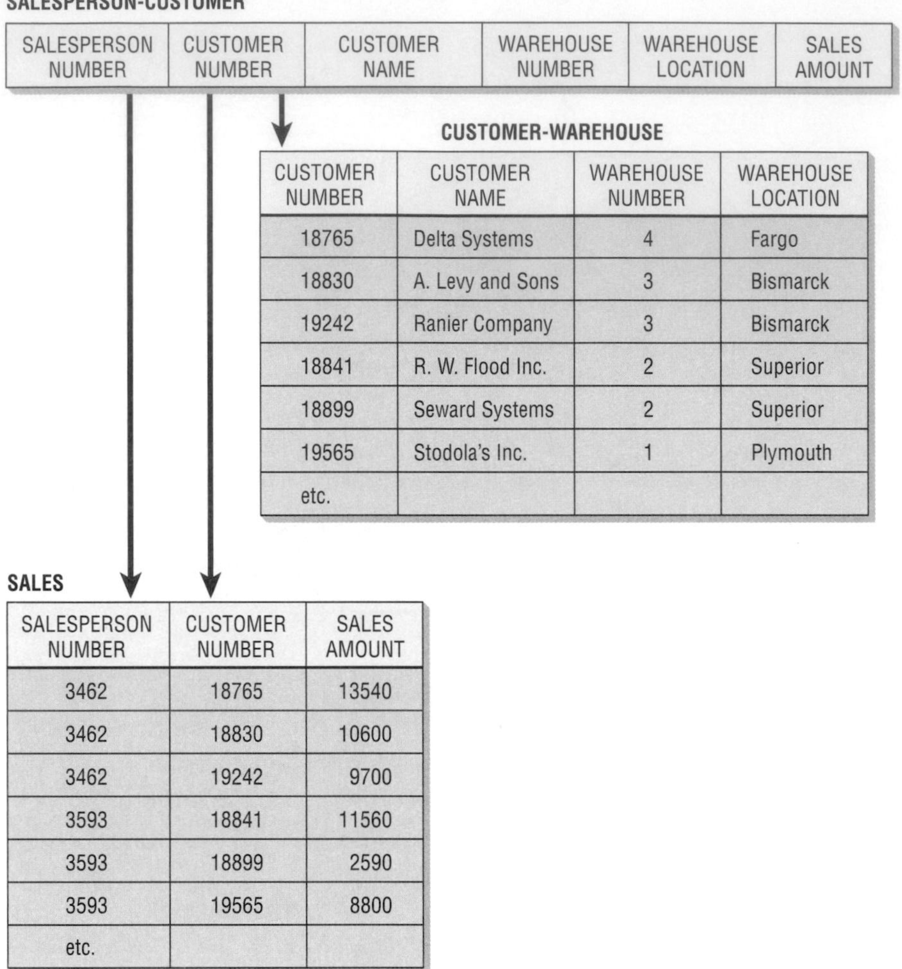

SALESPERSON-CUSTOMER

SALESPERSON NUMBER	CUSTOMER NUMBER	CUSTOMER NAME	WAREHOUSE NUMBER	WAREHOUSE LOCATION	SALES AMOUNT

CUSTOMER-WAREHOUSE

CUSTOMER NUMBER	CUSTOMER NAME	WAREHOUSE NUMBER	WAREHOUSE LOCATION
18765	Delta Systems	4	Fargo
18830	A. Levy and Sons	3	Bismarck
19242	Ranier Company	3	Bismarck
18841	R. W. Flood Inc.	2	Superior
18899	Seward Systems	2	Superior
19565	Stodola's Inc.	1	Plymouth
etc.			

SALES

SALESPERSON NUMBER	CUSTOMER NUMBER	SALES AMOUNT
3462	18765	13540
3462	18830	10600
3462	19242	9700
3593	18841	11560
3593	18899	2590
3593	19565	8800
etc.		

FIGURE 17.28
The relation SALESPERSON-CUSTOMER is separated into a relation called CUSTOMER-WAREHOUSE (2NF) and a relation called SALES (1NF).

shows how the relation SALESPERSON-CUSTOMER is split into two new relations: SALES and CUSTOMER-WAREHOUSE. These relations can also be expressed as follows:

SALES (<u>SALESPERSON-NUMBER</u>, <u>CUSTOMER-NUMBER</u>,
SALES-AMOUNT)

and

CUSTOMER-WAREHOUSE (<u>CUSTOMER-NUMBER</u>,
CUSTOMER-NAME,
WAREHOUSE-NUMBER,
WAREHOUSE-LOCATION)

The relation CUSTOMER-WAREHOUSE is in the second normal form. It can still be simplified further because there are additional dependencies within the relation. Some of the nonkey attributes are dependent not only on the primary key, but also on a nonkey attribute. This dependency is referred to as a transitive dependency.

Figure 17.29 shows the dependencies within the relation CUSTOMER-WAREHOUSE. For the relation to be a second normal form, all the attributes must be dependent on the primary key <u>CUSTOMER-NUMBER</u>, as shown in the diagram. WAREHOUSE-LOCATION, however, is obviously dependent on WAREHOUSE-NUMBER also. To simplify this relation, another step is required.

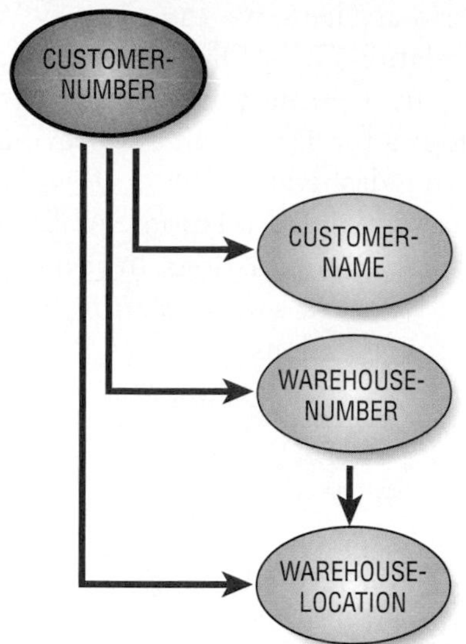

FIGURE 17.29
A data model diagram shows that a transitive dependency exists between WAREHOUSE-NUMBER and WAREHOUSE-LOCATION.

Third Normal Form (3NF). A normalized relation is third normal if all the nonkey attributes are fully functionally dependent on the primary key and there are no transitive (nonkey) dependencies. In a manner similar to the previous steps, it is possible to break apart the relation CUSTOMER-WAREHOUSE into two relations, as shown in Figure 17.30.

The two new relations are called CUSTOMER and WAREHOUSE and can be written as follows:

CUSTOMER (CUSTOMER-NUMBER, CUSTOMER-NAME, WAREHOUSE-NUMBER)

and

WAREHOUSE (WAREHOUSE-NUMBER, WAREHOUSE-LOCATION)

The primary key for the relation CUSTOMER is CUSTOMER-NUMBER, and the primary key for the relation WAREHOUSE is WAREHOUSE-NUMBER.

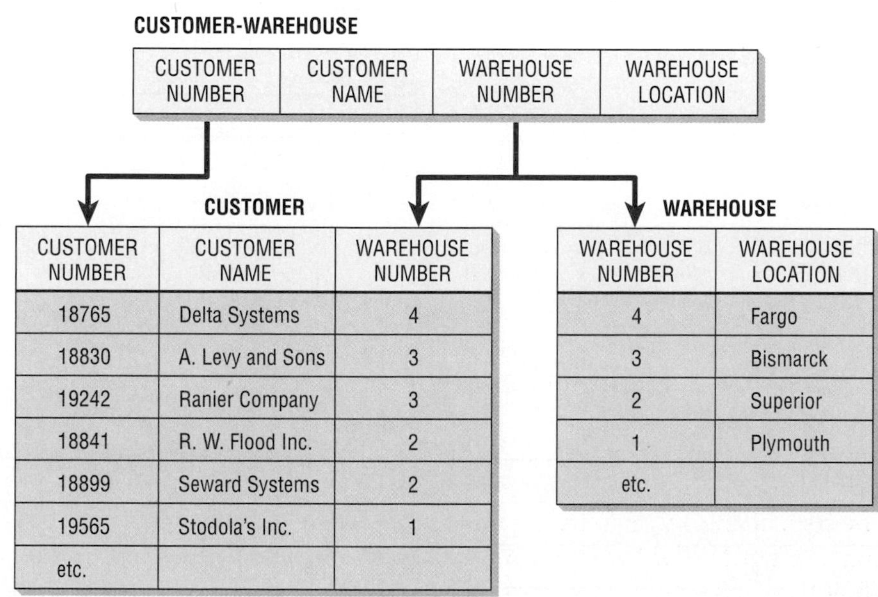

CUSTOMER-WAREHOUSE

CUSTOMER NUMBER	CUSTOMER NAME	WAREHOUSE NUMBER	WAREHOUSE LOCATION

CUSTOMER

CUSTOMER NUMBER	CUSTOMER NAME	WAREHOUSE NUMBER
18765	Delta Systems	4
18830	A. Levy and Sons	3
19242	Ranier Company	3
18841	R. W. Flood Inc.	2
18899	Seward Systems	2
19565	Stodola's Inc.	1
etc.		

WAREHOUSE

WAREHOUSE NUMBER	WAREHOUSE LOCATION
4	Fargo
3	Bismarck
2	Superior
1	Plymouth
etc.	

FIGURE 17.30
The relation CUSTOMER-WAREHOUSE is separated into two relations called CUSTOMER (1NF) and WAREHOUSE (1NF).

In addition to these primary keys, we can identify WAREHOUSE-NUMBER to be a foreign key in the relation CUSTOMER. A foreign key is any attribute that is nonkey in one relation but a primary key in another relation. We designated WAREHOUSE-NUMBER as a foreign key in the previous notation and in the figures by underscoring it with a dashed line:_ _ _ _ _ _ _ _ _ _ _ _.

Finally, the original, unnormalized relation SALES-REPORT has been transformed into four third normal (3NF) relations. In reviewing the relations shown in Figure 17.31, one can see that the single relation SALES-REPORT was transformed into the following four relations:

SALESPERSON (SALESPERSON-NUMBER, SALESPERSON-NAME, SALES-AREA)
SALES (SALESPERSON-NUMBER, CUSTOMER-NUMBER, SALES-AMOUNT)
CUSTOMER (CUSTOMER-NUMBER, CUSTOMER-NAME, WAREHOUSE-NUMBER)

and

WAREHOUSE (WAREHOUSE-NUMBER, WAREHOUSE-LOCATION)

The third normal form is adequate for most database design problems. The simplification gained from transforming an unnormalized relation into a set of 3NF relations is a tremendous benefit when it comes time to insert, delete, and update information in the database.

An entity-relationship diagram for the database is shown in Figure 17.32. One SALESPERSON serves many CUSTOMER(s) who generate SALES and

FIGURE 17.31
The complete database consists of four 1NF relations called SALESPERSON, SALES, CUSTOMER, and WAREHOUSE.

SALESPERSON

SALESPERSON NUMBER	SALESPERSON NAME	SALES AREA
3462	Waters	West
3593	Dryne	East
etc.		

SALES

SALESPERSON NUMBER	CUSTOMER NUMBER	SALES AMOUNT
3462	18765	13540
3462	18830	10600
3462	19242	9700
3593	18841	11560
3593	18899	2590
3593	19565	8800
etc.		

CUSTOMER

CUSTOMER NUMBER	CUSTOMER NAME	WAREHOUSE NUMBER
18765	Delta Systems	4
18830	A. Levy and Sons	3
19242	Ranier Company	3
18841	R. W. Flood Inc.	2
18899	Seward Systems	2
19565	Stodola's Inc.	1
etc.		

WAREHOUSE

WAREHOUSE NUMBER	WAREHOUSE LOCATION
4	Fargo
3	Bismarck
2	Superior
1	Plymouth
etc.	

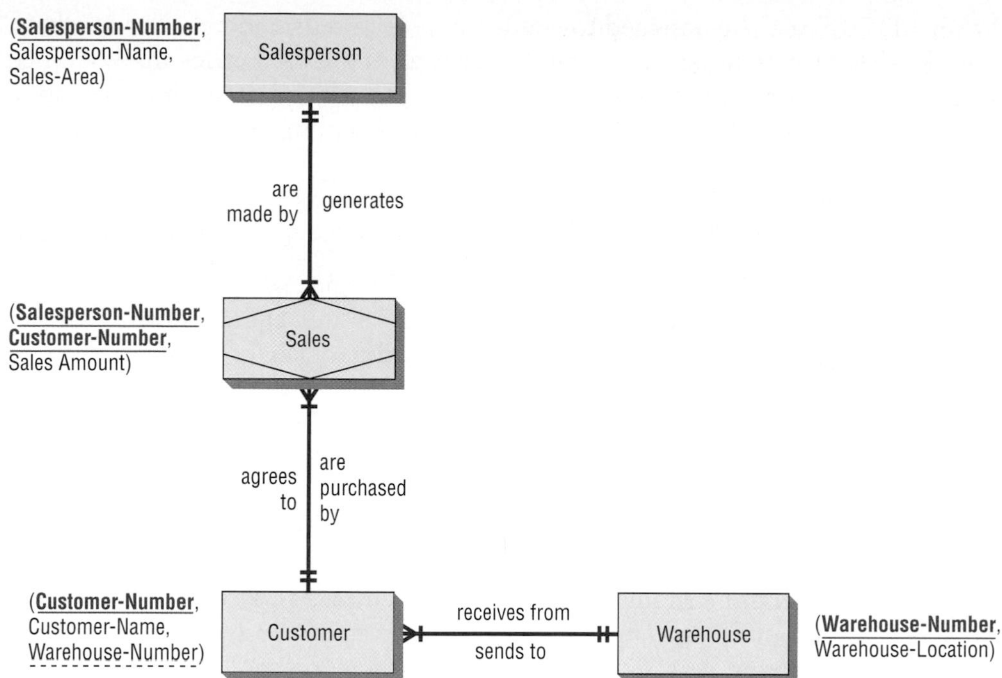

FIGURE 17.32
An entity-relationship diagram for the Al S. Well Hydraulic Company database.

receive their items from one WAREHOUSE (the closest WAREHOUSE to their location). Take the time to notice how the entities and attributes relate to the database.

USING THE ENTITY-RELATIONSHIP DIAGRAM TO DETERMINE RECORD KEYS

The entity-relationship diagram may be used to determine the keys required for a record or a database relation. The first step is to construct the entity-relationship diagram and label a unique (primary) key for each data entity. Figure 17.33 shows an entity-relationship diagram for a customer order system. There are three data entities: CUSTOMER, with a prime key of CUSTOMER-NUMBER; ORDER, with a primary key of ORDER-NUMBER; and ITEM, with ITEM-NUMBER as the prime key. One CUSTOMER may place many orders, but each ORDER can be placed by one CUSTOMER only, so the relationship is one-to-many. Each ORDER may contain many ITEMS, and each ITEM may be contained within many ORDERS, so the ORDER-ITEM relationship is many-to-many.

A foreign key, however, is a data field on a given file that is the primary key of a different master file. For example, a DEPARTMENT-NUMBER indicating a student's major may exist on the STUDENT MASTER file. DEPARTMENT-NUMBER could also be the unique key for the DEPARTMENT MASTER file.

ONE-TO-MANY RELATIONSHIP

A database file cannot contain a repeating group or table, but a traditional indexed-sequential file may have one. The file on the "many" end may have foreign keys stored in a table on the file at the "one" end. For example, the CUSTOMER

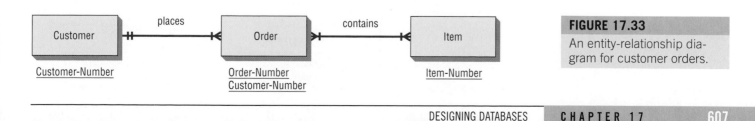

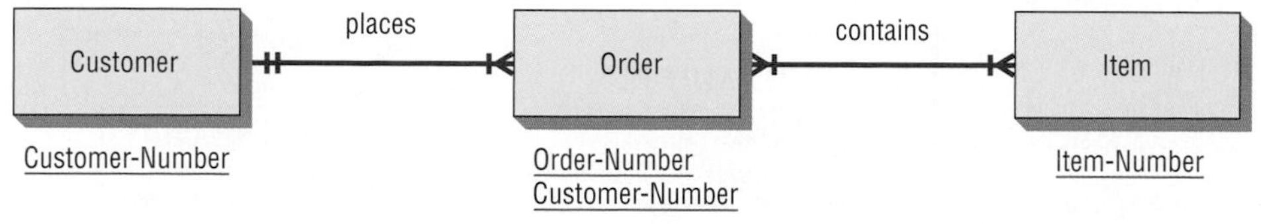

FIGURE 17.33
An entity-relationship diagram for customer orders.

MASTER FILE may be designed to contain a table of outstanding order numbers. The disadvantage of using such a table is that all the table entries may be filled with order numbers, and the analyst is then faced with the decision of having to expand the table (which is expensive and time-consuming) or lose some of the data in the table.

MANY-TO-MANY RELATIONSHIP

When the relationship is many-to-many, three files are necessary: one for each data entity and one for the relationship. The ORDERS and ITEMS entities in our example have a many-to-many relationship. The primary key of each data entity is stored as a foreign key of the relational file. The relational file may simply contain the primary keys for each data entity or may contain additional data, such as the grade received for a course or the quantity of an item ordered. Refer to the file layout illustrated in Figure 17.34. The ORDER ITEM FILE contains information about which order contains which items and provides a link between the ORDER FILE and the ITEM MASTER FILE.

The relationship file should be indexed on each foreign key—one for each of the files in the relationship—and may have a primary key consisting of a combination of the two foreign keys. To find many records from a second file given the first file, directly read the relational file for the desired key. Locate the matching record in the second "many" file. Continue to loop through the relational file until the desired key is no longer found. For example, to find records in the ITEM MASTER for a specific record in the ORDER MASTER FILE, directly read the ORDER-ITEM FILE using the ORDER-NUMBER as the index. Records are logically sequenced based on the data in the index, so all records for the same ORDER-NUMBER are grouped together. For each ORDER ITEM record that matches the desired ORDER-NUMBER, directly read the ITEM MASTER FILE using the ITEM-NUMBER as an index.

The logic is the same for the reverse situation, such as finding all the orders for a back-ordered item that has been received. Use the desired ITEM-NUMBER to read the ORDER-ITEM FILE directly. The ORDER ITEM INDEX is set to the ITEM-NUMBER. For all matching ORDER ITEM records, use the ORDER-NUMBER to read the ORDER FILE directly. Finally, read the CUSTOMER MASTER FILE directly to obtain the CUSTOMER-NAME and ADDRESS using the CUSTOMER-NUMBER on the ORDER FILE.

With indexed-sequential files, one of the files may contain a table of keys for the other file, eliminating the need for a relational file. Refer to the example illustrated in Figure 17.35. The ORDER-KEY FILE has a table of order item informa-

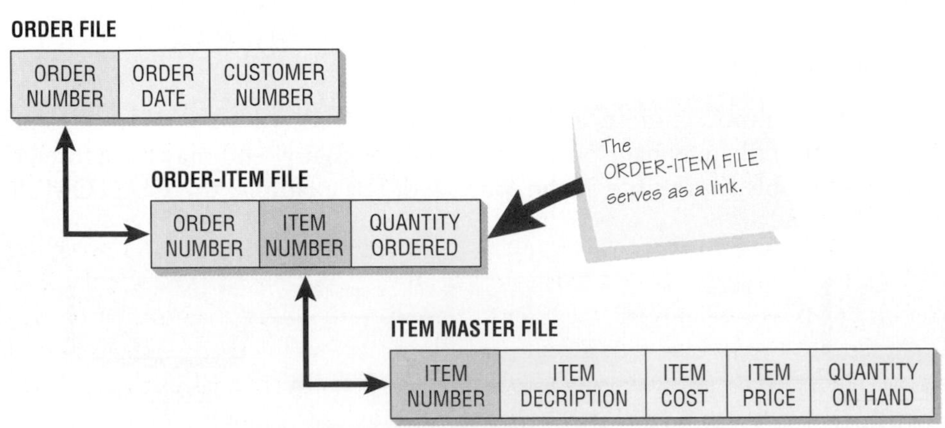

FIGURE 17.34
When the relationship is many-to-many, three files are necessary.

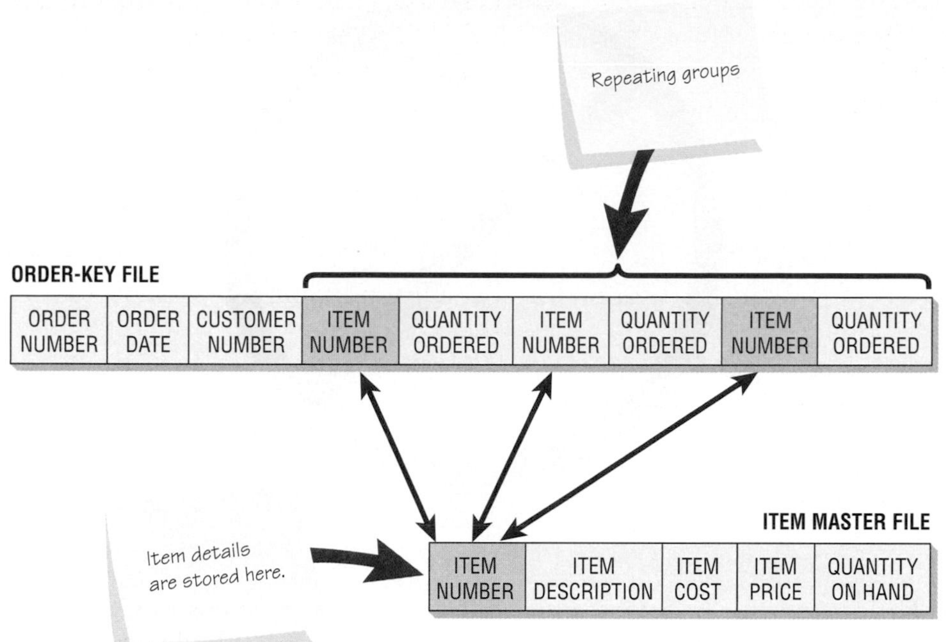

FIGURE 17.35
With indexed-sequential files, one of the files may contain a table of keys.

tion. Each element in the table contains an ITEM-NUMBER and the QUANTITY-ORDERED. The remaining information about all items is found in the ITEM MASTER FILE, which is organized sequentially by ITEM-NUMBER. The disadvantage of this method is that it is difficult and time-consuming to find the orders that match a specific item number. Many conventional file systems use tables on master file records.

GUIDELINES FOR FILE/DATABASE RELATION DESIGN

The following guidelines should be taken into account when designing master files or database relations:

1. Each separate data entity should create a master file. Do not combine two distinct entities on one file. For example, items are purchased from vendors. The ITEM MASTER FILE should contain only item information, and the VENDOR MASTER FILE should contain only vendor information. Figure 17.36 illustrates the data dictionary for a poorly designed ITEM MASTER FILE, which includes information about the vendor. There may be many items purchased from one vendor, and the VENDOR NAME and ADDRESS in this arrangement would be stored on many records. If a vendor changes its address, multiple item records would have to be modified. If some were modified and others were not, the system would produce inconsistent results, such as purchase orders for one item going to the old address and other purchase orders being sent to the new address. Figure 17.37 illustrates the data dictionary for the improved ITEM MASTER FILE and VENDOR MASTER FILE.
2. A specific data field should exist only on one master file. For example, the CUSTOMER NAME should exist only on the CUSTOMER MASTER FILE, not on the ORDER FILE or any other master file. The exceptions to this guideline are the key or index fields, which may be on as many files as necessary. If a report or screen needs information from many files, the indexes should provide the linkage for obtaining the required records.

FIGURE 17.36
Example of a poorly designed item master file.

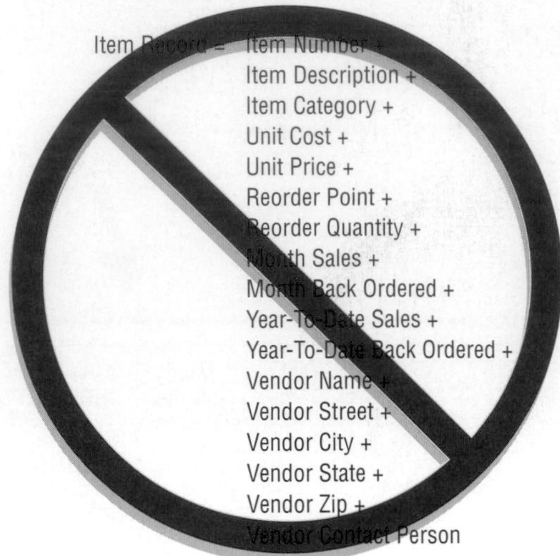

3. Each master file or database relation should have programs to **C**reate, **R**ead, **U**pdate, and **D**elete (abbreviated CRUD) the records. Ideally, only one program should add new records and only one program should delete specified records. Data records may be changed—for instance, to change an address or correct an incorrect customer balance—by one program or several, depending on the data. These file maintenance programs are responsible for infrequent, unpredictable changes to the data.

FIGURE 17.37
The improved item and vendor master files.

Item Record = Item Number +
Item Description +
Item Category +
Unit Cost +
Unit Price +
Reorder Point +
Reorder Quantity +
Month Sales +
Month Back Ordered +
Year-To-Date Sales +
Year-To-Date Back Ordered +
Vendor Number

Vendor Record = Vendor Number +
Vendor Name +
Vendor Street +
Vendor City +
Vendor State +
Vendor Zip +
Vendor Contact Person

Usually, many update programs are responsible for changing data fields in the course of normal business activities. For example, a customer master file may have a CURRENT BALANCE field that is increased by the ORDER TOTAL within the order processing program and decreased by a PAYMENT AMOUNT or an AMOUNT RETURNED (using output by two additional programs).

INTEGRITY CONSTRAINTS

Integrity constraints are rules that govern changing and deleting records and that help keep the data in the database accurate. Three types of integrity constraints apply to a database:

1. Entity integrity constraint.
2. Referential integrity.
3. Domain integrity.

Entity integrity constraints are rules that govern the composition of primary keys. The primary key cannot have a null value, and if the primary key is a composite key, none of the component fields in the key can contain a null value. Some databases allow you to define a unique constraint or a *unique key*. This unique key identifies only one record, which is not a primary key. The difference between a unique key and a primary key is that a unique key may contain a null value.

Referential integrity governs the nature of records in a one-to-many relationship. The table that is connected to the **one** end of the relationship is called the parent. The table connected to the **many** end of the relationship is called the child table. Referential integrity means that all foreign keys in the many table (the child table) must have a matching record in the parent table. Hence, you cannot add a record in the child (many) table without a matching record in the parent table.

A second implication is that you cannot change a primary key that has matching child table records. If you could change the parent record, the result would be a child record that would have a different parent record or an orphan record, or a child record without a parent record. Examples are a grade record for a student number that would not be on the STUDENT MASTER file and an order record for a CUSTOMER NUMBER that did not exist. The last implication of referential integrity is that you cannot delete a parent record that has child records. That would also lead to the orphan records mentioned earlier.

Referential integrity is implemented in two different ways. One way is to have a restricted database, in which the system can update or delete a parent record only if there are no matching child records. A cascaded database will delete or update all child records when a parent record is deleted or changed (the parent triggers the changes). A restricted relationship is better when deleting records. You would not want to delete a customer record and have all the outstanding invoices deleted as well! The cascaded approach is better when changing records. If the primary key of a student record is changed, all the course records for that student would have their foreign keys (the CUSTOMER NUMBER on the COURSE MASTER) changed as well.

Domain integrity rules are used to validate the data, such as table, limit, range, and other validation checks. They are further explained in Chapter 19. The domain integrity rules are usually stored in the database structure in one of two forms. Check constraints are defined at the table level and can refer to one or more fields in the table. An example is that the "Date of Purchase" is always less than or equal to the current date. Rules are defined at the database level as separate objects and can be used with a number of fields. An example is a value that is greater than zero, used to validate a number of elements.

MAKING USE OF THE DATABASE

There are several steps you must take in sequential order to assure that the database will be useful for presenting data.

FIGURE 17.38

Data are retrieved and presented in eight distinct steps.

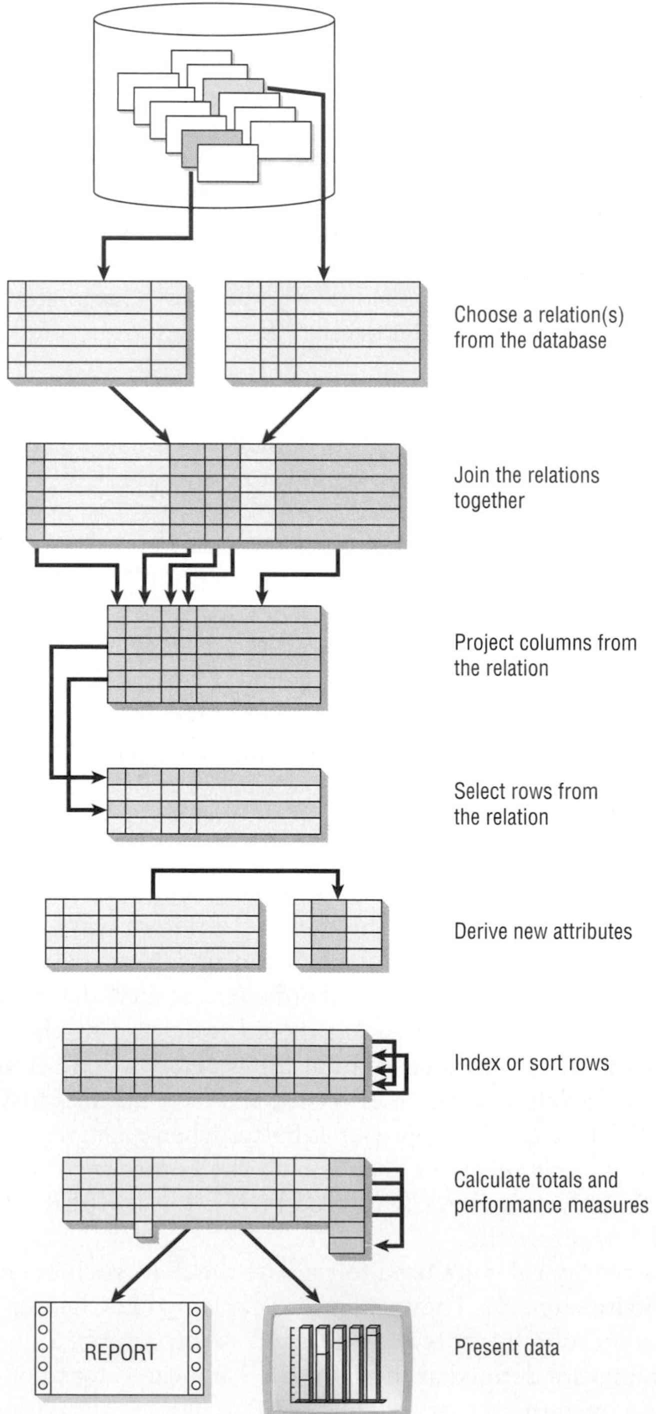

STEPS IN RETRIEVING AND PRESENTING DATA

There are eight steps in the retrieval and presentation of data:

1. Choose a relation from the database.
2. Join two relations together.
3. Project columns from the relation.
4. Select rows from the relation.
5. Derive new attributes.
6. Index or sort rows.
7. Calculate totals and performance measures.
8. Present data.

The first and last steps are mandatory, but the six steps in between are optional, depending on how data are to be used. Figure 17.38 is a visual guide to the steps, which are described in the following section.

Choose a Relation from the Database. The first and obvious step is to choose a relation from the database. A good way to accomplish this step is to keep a directory of user views as a memory aid. Even if the user wants an ad hoc query, it is useful to have similar views available.

CUSTOMER

CUSTOMER NUMBER	CUSTOMER NAME	WAREHOUSE NUMBER
18765	Delta Systems	4
18830	A. Levy and Sons	3
19242	Ranier Company	3
18841	R. W. Flood Inc.	2
18899	Seward Systems	2
19565	Stodola's Inc.	1
etc.		

WAREHOUSE

WAREHOUSE NUMBER	WAREHOUSE LOCATION
4	Fargo
3	Bismarck
2	Superior
1	Plymouth
etc.	

Join

CUSTOMER-WAREHOUSE-LOCATION

CUSTOMER NUMBER	CUSTOMER NAME	WAREHOUSE NUMBER	WAREHOUSE LOCATION
18765	Delta Systems	4	Fargo
18830	A. Levy and Sons	3	Bismarck
19242	Ranier Company	3	Bismarck
18841	R. W. Flood Inc.	2	Superior
18899	Seward Systems	2	Superior
19565	Stodola's Inc.	1	Plymouth
etc.			

FIGURE 17.39
The operation join takes two relations and puts them together to form a single relation.

FIGURE 17.40
Relations can be joined subject to certain conditions.

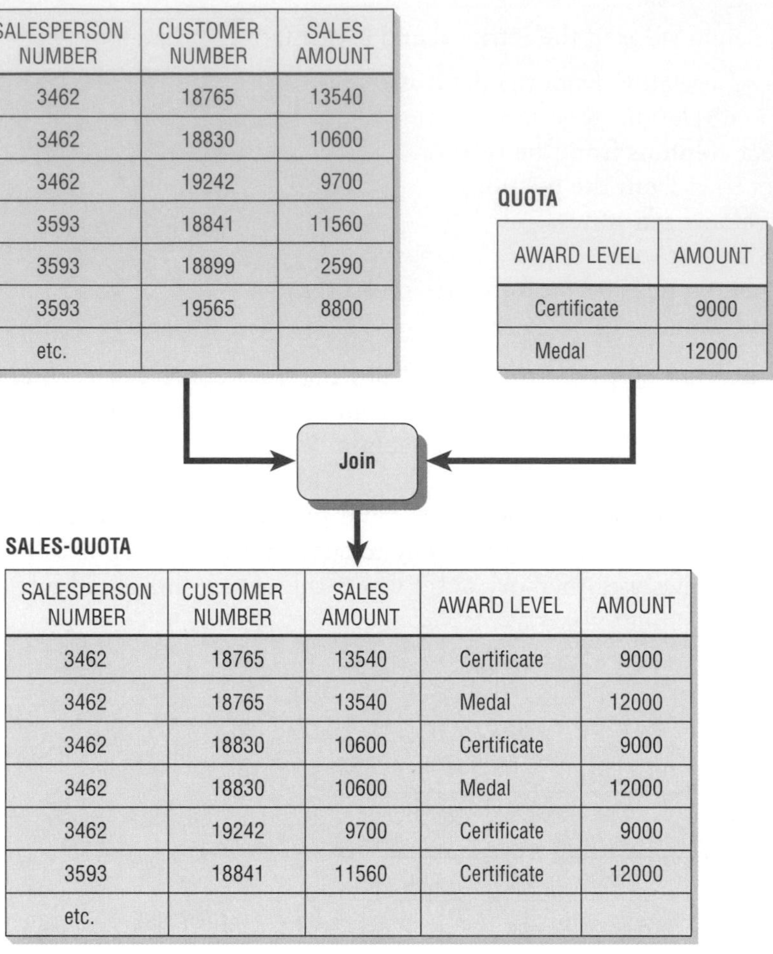

SALES

SALESPERSON NUMBER	CUSTOMER NUMBER	SALES AMOUNT
3462	18765	13540
3462	18830	10600
3462	19242	9700
3593	18841	11560
3593	18899	2590
3593	19565	8800
etc.		

QUOTA

AWARD LEVEL	AMOUNT
Certificate	9000
Medal	12000

SALES-QUOTA

SALESPERSON NUMBER	CUSTOMER NUMBER	SALES AMOUNT	AWARD LEVEL	AMOUNT
3462	18765	13540	Certificate	9000
3462	18765	13540	Medal	12000
3462	18830	10600	Certificate	9000
3462	18830	10600	Medal	12000
3462	19242	9700	Certificate	9000
3593	18841	11560	Certificate	12000
etc.				

Join Two Relations Together. The operation join is intended to take two relations and put them together to make a larger relation. For two relations to be joined, they must have a common attribute. For instance, take two relations from our example:

CUSTOMER (CUSTOMER-NUMBER, CUSTOMER-NAME, WAREHOUSE-NUMBER)

and

WAREHOUSE (WAREHOUSE-NUMBER, WAREHOUSE-LOCATION)

Suppose we join these relations over WAREHOUSE-NUMBER to get a new relation, CUSTOMER-WAREHOUSE-LOCATION. The joining of these relations is illustrated in Figure 17.39. Also note that the new relation is not 3NF.

The operation join may also go one step further; that is, it may combine files for rows that have an attribute that meets a certain condition. Figure 17.40 shows an example in which two relations, SALES and QUOTA, are joined by satisfying the condition that a salesperson has met or exceeded predetermined quotas.

CUSTOMER-WAREHOUSE-LOCATION

CUSTOMER NUMBER	CUSTOMER NAME	WAREHOUSE NUMBER	WAREHOUSE LOCATION
18765	Delta Systems	4	Fargo
18830	A. Levy and Sons	3	Bismarck
19242	Ranier Company	3	Bismarck
18841	R. W. Flood Inc.	2	Superior
18899	Seward Systems	2	Superior
19565	Stodola's Inc.	1	Plymouth
etc.			

Projection

CUSTOMER-LOCATION

CUSTOMER NUMBER	WAREHOUSE LOCATION
18765	Fargo
18830	Bismarck
19242	Bismarck
18841	Superior
18899	Superior
19565	Plymouth
etc.	

FIGURE 17.41
Projection creates a smaller relation by choosing only relevant attributes (columns) from the relation.

Join is an important operation because it can take many 3NF relations and combine them to make a more useful relation. Together with the operations below it, join is a powerful operation.

Project Columns from the Relation. Projection is the process of building a smaller relation by choosing only relevant attributes from an existing relation. In other words, projection is the extraction of certain columns from a relational table.

An example of projection is featured in Figure 17.41. The relation

CUSTOMER-WAREHOUSE-LOCATION (<u>CUSTOMER-NUMBER</u>,
CUSTOMER-NAME,
WAREHOUSE-NUMBER),
WAREHOUSE-LOCATION

is projected over CUSTOMER-NUMBER and WAREHOUSE-LOCATION, and during the projection process, duplicate records are removed.

Select Rows from the Relation. The operation referred to as selection is similar to projection, but instead of extracting columns it extracts rows. Selection creates a new (smaller) relation by extracting records that contain an attribute meeting a certain condition.

FIGURE 17.42

Selection extracts the relevant records (rows) from the relation.

PERSONNEL

NUMBER	EMPLOYEE NAME	DEPARTMENT	S/H	GROSS
72845	Waters	Outside Sales	S	48960
72888	Dryne	Outside Sales	S	37200
73712	Fawcett	Distribution	H	23500
80345	Well, Jr.	Marketing	S	65000
84672	Piper	Maintenance	H	20560
89760	Acquia	Accounting	H	18755
etc.				

Selection

SALARIED-EMPLOYEES

NUMBER	EMPLOYEE NAME	DEPARTMENT	S/H	GROSS
72845	Waters	Outside Sales	S	48960
72888	Dryne	Outside Sales	S	37200
80345	Well, Jr.	Marketing	S	65000
etc.				

Figure 17.42 illustrates how the selection operation works. Selection is performed on the relation PERSONNEL to extract salaried employees only. There is no need to remove duplicate records here as there was in the previous illustration of projection.

Selection may also be performed for a more complex set of conditions, such as select all the employees who are salaried *and* who make more than $40,000 annually, or select employees who are hourly *and* make more than $15.00 per hour. Selection is an important operation for ad hoc queries.

Derive New Attributes. The fifth step involves the manipulation of the existing data plus some additional parameters (if necessary) to derive new data. New columns are created for the resulting relation. An example of derivation of new attributes can be found in Figure 17.43. Here, two new attributes are determined: (1) GIRTH (by multiplying the sum of width and height by 2 and adding it to length) and (2) SHIPPING-WEIGHT (which depends on the girth).

Index or Sort Rows. People require data to be organized in a certain order so that they can either locate items in a list more easily or group and subtotal items more easily. Two options for ordering data are available: indexing and sorting.

Indexing is the logical ordering of rows in a relation according to some key. As discussed in the previous section, the logical pointer takes up space, and list-

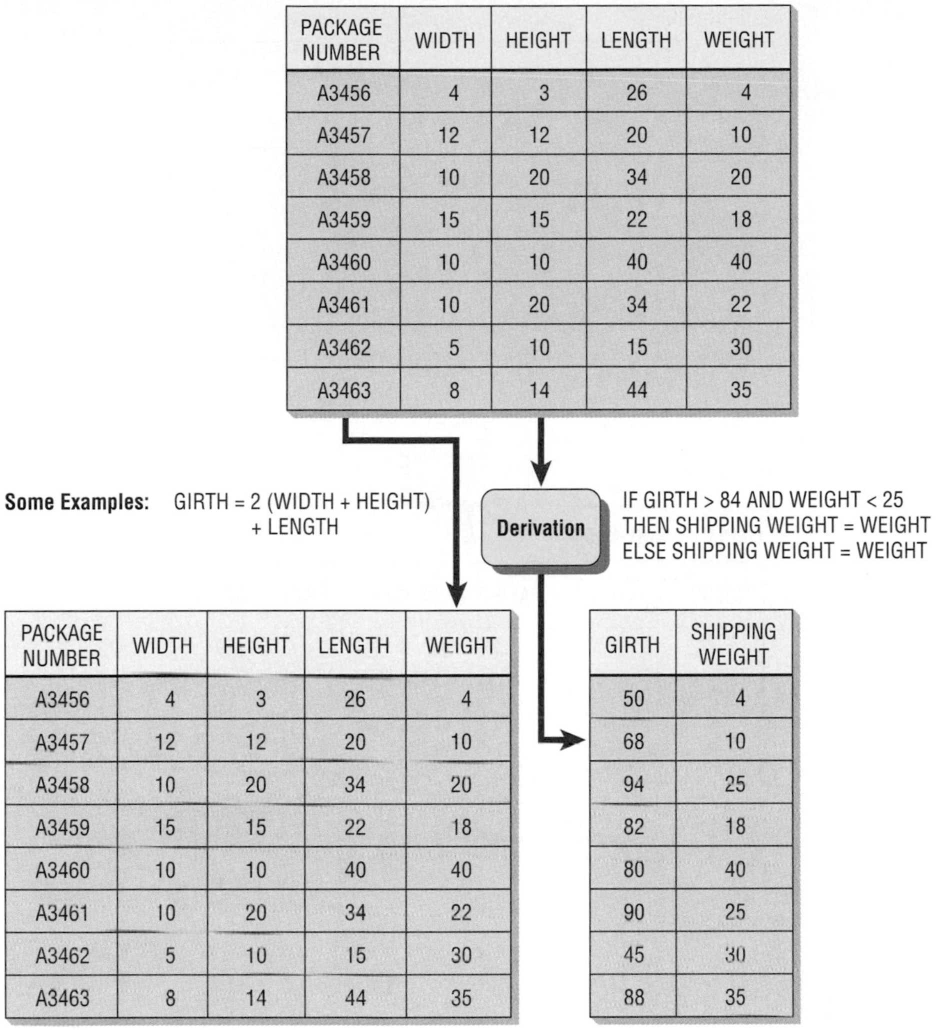

PACKAGE NUMBER	WIDTH	HEIGHT	LENGTH	WEIGHT
A3456	4	3	26	4
A3457	12	12	20	10
A3458	10	20	34	20
A3459	15	15	22	18
A3460	10	10	40	40
A3461	10	20	34	22
A3462	5	10	15	30
A3463	8	14	44	35

Some Examples: GIRTH = 2 (WIDTH + HEIGHT) + LENGTH

Derivation

IF GIRTH > 84 AND WEIGHT < 25
THEN SHIPPING WEIGHT = WEIGHT
ELSE SHIPPING WEIGHT = WEIGHT

PACKAGE NUMBER	WIDTH	HEIGHT	LENGTH	WEIGHT
A3456	4	3	26	4
A3457	12	12	20	10
A3458	10	20	34	20
A3459	15	15	22	18
A3460	10	10	40	40
A3461	10	20	34	22
A3462	5	10	15	30
A3463	8	14	44	35

GIRTH	SHIPPING WEIGHT
50	4
68	10
94	25
82	18
80	40
90	25
45	30
88	35

FIGURE 17.43
Derivation creates new attributes (columns) in the relation by manipulating the data contained in existing attributes.

ing the relation by using an index is slower than if the relation were in the proper physical order. The index, however, takes up far less space than a duplicate file.

Sorting is the physical ordering of a relation. The result of physical sorting is a sequential file as discussed earlier in the chapter. Figure 17.44 illustrates indexing and sorting of the relation PERSONNEL by employee name in alphabetical order.

Calculate Totals and Performance Measures. Once the appropriate subset of data is defined and the rows of the relation are ordered in the required manner, totals and performance measures can be calculated. Figure 17.45 shows how calculation is performed.

Present Data to the User. The final step in the retrieval of data is presentation. Presentation of the data abstracted from the database can take many forms. Sometimes the data will be presented in tabular form, sometimes in graphs, and other times as a single-word answer on a screen. Output design, as covered in Chapter 15, provides a more detailed look at presentation objectives, forms, and methods.

FIGURE 17.44
The operation sort orders the records (rows) in the relation so that records can be displayed in order and can be grouped for subtotals. Here the PERSONNEL relation is sorted alphabetically according to EMPLOYEE NAME.

PERSONNEL

NUMBER	EMPLOYEE NAME	DEPARTMENT	S/H	GROSS
72845	Waters	Outside Sales	S	48960
72888	Dryne	Outside Sales	S	37200
73712	Fawcett	Distribution	H	23500
80345	Well, Jr.	Marketing	S	65000
84672	Piper	Maintenance	H	20560
89760	Acquia	Accounting	H	18755

Sort

EMPLOYEES-BY-NAME

NUMBER	EMPLOYEE NAME	DEPARTMENT	S/H	GROSS
89760	Acquia	Accounting	H	18755
72888	Dryne	Outside Sales	S	37200
73712	Fawcett	Distribution	H	23500
84672	Piper	Maintenance	H	20560
72845	Waters	Outside Sales	S	48960
80345	Well, Jr.	Marketing	S	65000

DENORMALIZATION

One of the main reasons for normalization was to organize data so as to reduce redundant data. If you are not required to store the same data over and over again, you can save a great deal of space. Such organization allowed the analyst to reduce the amount of storage needed, something that was very important when storage was expensive.

We learned in the last section that to use normalized data we had to progress through a series of steps that involved joining, sorting, and summarizing. When speed of querying the database (that is, asking a question and requiring a rapid response) is critical, it may be important to store data in other ways.

Denormalization is the process of taking the logical data model and transforming it into a physical model that is efficient for the most often needed tasks. These tasks can include report generation, but they can also mean more efficient queries. Complex queries such as online analytic processing (OLAP) as well as data mining and knowledge data discovery (KDD) processes can also make use of databases that are denormalized.

Denormalization can be accomplished in a number of different ways. Figure 17.48 depicts some of these approaches. First, we can take a many-to-many relationship, like that of SALESPERSON and CUSTOMER which share the associative entity SALES. By combining the attributes from SALESPERSON and SALES we can avoid one of the join processes. This may result in a considerable amount of data duplication, but it helps the queries about sales patterns to be more efficient.

Another reason for denormalization is to avoid repeated reference to a look-up table. It may be more efficient to repeat the same information—for example, the

SHIPPING-WEIGHT

PACKAGE NUMBER	WIDTH	HEIGHT	LENGTH	WEIGHT	GIRTH	SHIPPING WEIGHT
A3456	4	3	26	4	50	4
A3457	12	12	20	10	68	10
A3458	10	20	34	20	94	25
A3459	15	15	22	18	82	18
A3460	10	10	40	40	80	40
A3461	10	20	34	22	90	25
A3462	5	10	15	30	45	30
A3463	8	14	44	35	88	35

FIGURE 17.45
Calculation provides subtotals, totals, and other performance measures.

city, state, and zip code—even though this information can usually be stored as a zip code only. Hence, in the sales example, CUSTOMER and WAREHOUSE may be combined.

Finally, we look at one-to-one relationships because they are very likely to be combined for practical reasons. If we learn that many of the queries regarding orders also are interested in how the order was shipped, it would make sense to combine, or denormalize. Hence, in the example, some of the details can appear in both ORDER-DETAILS and SHIPPING-DETAILS when we go through denormalization.

DATA WAREHOUSES

Data warehouses differ from traditional databases. The purpose of a data warehouse is to organize information for quick and effective queries. In effect, they store denormalized information, but they go one step farther. They organize data around subjects. Most often, a data warehouse is more than one database processed so that information is represented in uniform ways. Therefore, the information stored in data warehouses comes from different sources, usually databases that were set up for different purposes.

The data warehouse concept is unique. Differences between data warehouses and traditional databases include the following:

1. In a data warehouse, data are organized around major subjects rather than individual transactions.
2. Data in a data warehouse are typically stored as summarized data rather than the detailed, raw data found in a transaction-oriented database.

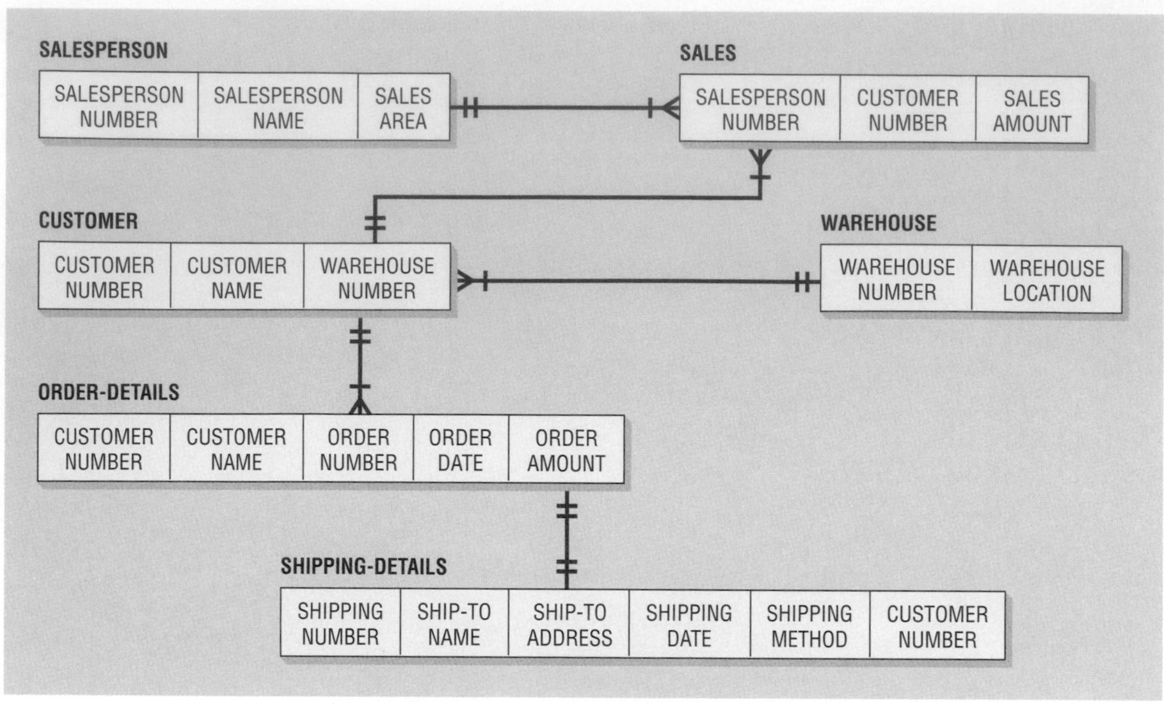

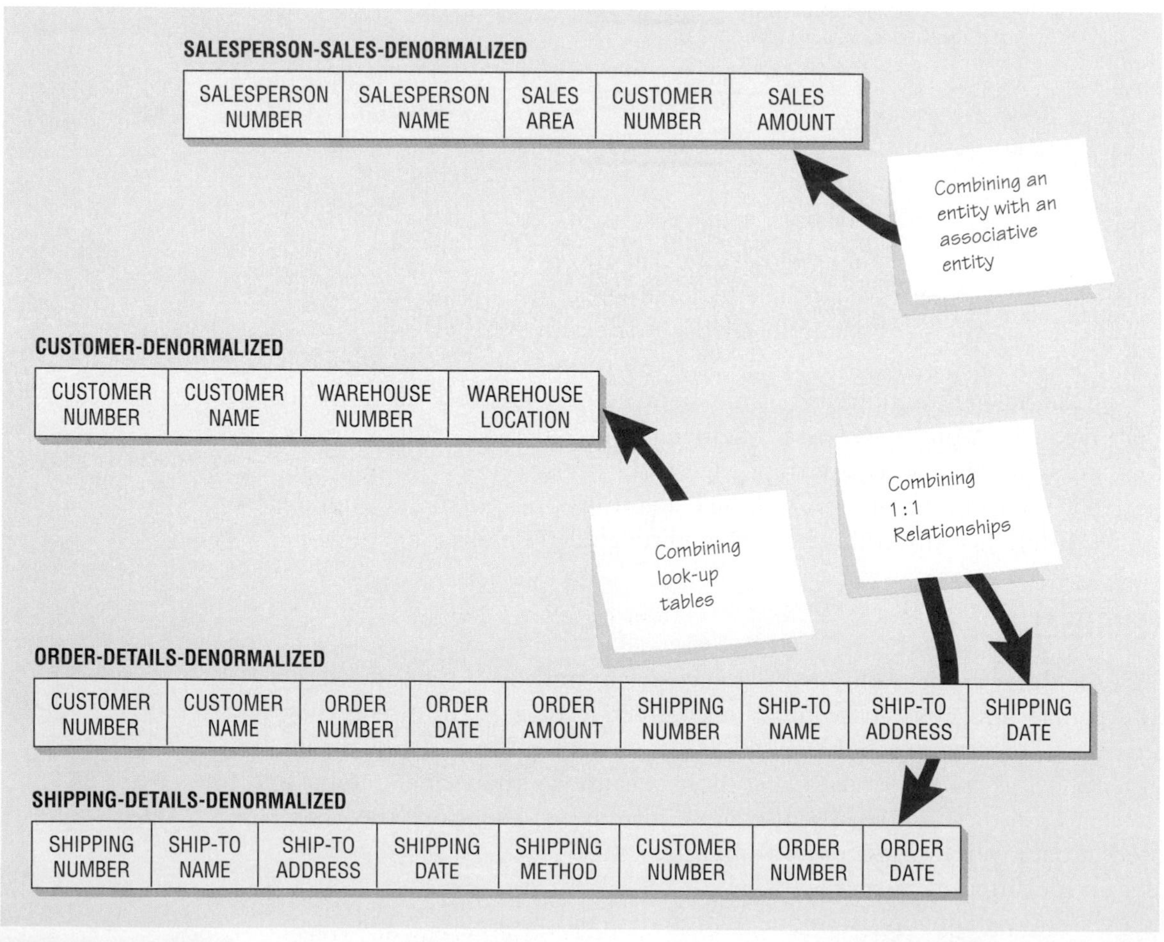

FIGURE 17.46
Three examples of denormalization in order to make access more efficient.

3. Data in a data warehouse cover a much longer time frame than data in traditional transaction-oriented databases because queries usually concern longer-term decision making rather than daily transaction details.
4. Most data warehouses are organized for fast queries, whereas the more traditional databases are normalized and structured in such a way as to provide efficient storage of information.
5. Data warehouses are usually optimized for answering complex queries, known as OLAP, from managers and analysts rather than simple, repeatedly asked queries.
6. Data warehouses allow easy access via data-mining software (called siftware) that searches for patterns and are able to identify relationships not imagined by human decision makers.
7. Data warehouses include not just one but multiple databases that have been processed so that the warehouse's data are defined uniformly. These databases are referred to as "clean" data.
8. Data warehouses usually include data from outside sources (such as an industry report, the company's Security and Exchange Commission filing, or even information about competitors' products) as well as data generated for internal use.

Building a data warehouse is a monumental task. The analyst needs to gather data from a variety of sources and translate that data into a common form. For example, one database may store information about gender as Male and Female, another may store it as M and F, and a third may store it as 1 and 0. The analyst needs to set a standard and convert all the data to the same format.

Once the data are "clean," the analyst has to decide how to summarize the data. Once summarized, the detail is lost, so an analyst has to predict the type of queries that might be asked.

Then, the analyst needs to design the data warehouse by logically organizing, and perhaps even physically clustering, the data by subject, requiring much analysis and design. The analyst needs to know a substantial amount about the business.

Typical data warehouses tend to be from 50 gigabytes to tens of terabytes in size. Because they are large, they are also expensive. Most data warehouses cost millions of dollars.

ONLINE ANALYTIC PROCESSING (OLAP)

First introduced in 1993 by E. F. Codd, OLAP was meant to answer decision makers' complex questions. Codd concluded that a decision maker had to look at data in a number of different ways. Therefore, the database itself had to be multidimensional. Many people picture OLAP as a Rubik's cube of data. You can look at the data from all different sides and can also manipulate the data by twisting or turning it so that it makes sense.

This OLAP approach validated the concept of data warehouses. It now made sense for data to be organized in ways that allowed efficient queries. Of course, OLAP involves the processing of data through manipulation, summarization, and calculation, so more than a data warehouse is involved.

DATA MINING

Data mining can identify patterns that a human is unable to detect. Either the decision maker cannot see a pattern or perhaps the decision maker is not able to think about asking whether that pattern exists. Data-mining algorithms search data warehouses for patterns using algorithms.

Data mining is known by another name, knowledge data discovery (KDD). Some think that KDD differs from data mining because KDD is meant to assist the decision makers find patterns rather than turning control over to an algorithm to find them. The decision aids available are called "siftware"; they include statistical analysis, decision trees, neural networks, intelligent agents, fuzzy logic, and data visualization.

The types of patterns decision makers try to identify include associations, sequences, clustering, and trends. Associations are patterns that occur together at the same time. For example, a person who buys cereal usually buys milk to go with the cereal. Sequences, on the other hand, are patterns of actions that take place over a period of time. For example, if a family buys a house this year, they will most likely buy durables (a refrigerator or washer and dryer) next year. Clustering is the pattern that develops among a group of people. For example, customers who live in a particular zip code may tend to buy a particular car. Finally, trends are patterns that are noticed over a period of time. For example, consumers may move from buying generic goods to premium products.

Data mining is discussed further in Chapter 18, along with queries in general.

PUBLISHING DATABASES TO THE WEB

Web-based databases are all about sharing information. For example, you can set up a sales and inventory management system that lets employees remotely view, edit, and update orders. You can develop task management applications for the Web that permit employees to assign tasks or track projects.

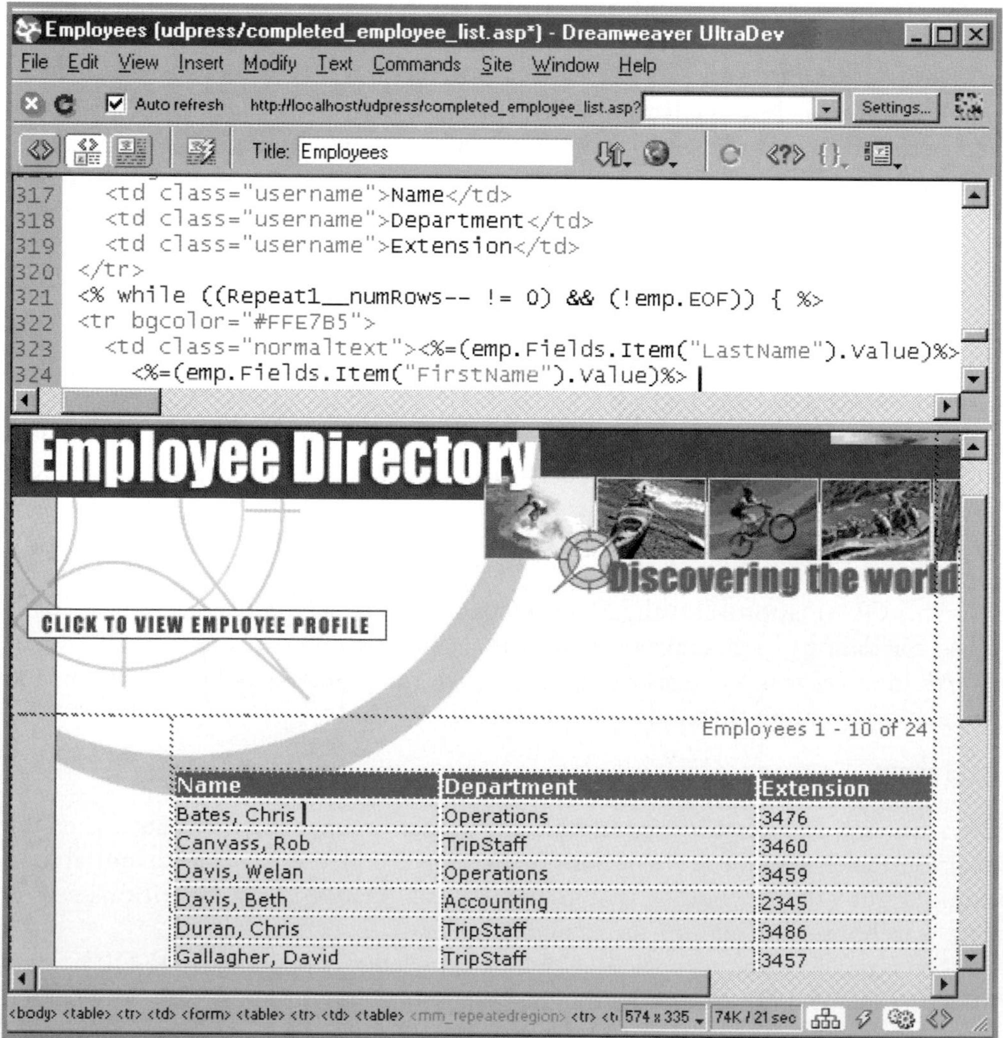

FIGURE 17.47 Macromedia Dreamweaver UltraDev 4 layout screen used to publish an employee directory database to the Web.

It is also possible to set up an online storefront that allows customers to browse a product database and make a purchase. It is even possible to personalize Web applications that use dynamic pages with customized content based on interests that the customer enters.

The example presented in Figure 17.47 illustrates how to set up an online employee directory. Any person visiting the Web site can search for an employee to locate a correct phone extension. The example shows a split screen with the source code on the top and a visual representation of the Web page on the bottom of the screen. The example was generated using Macromedia Dreamweaver UltraDev 4.

Macromedia Dreamweaver UltraDev is a premium software package for the development of ecommerce sites. UltraDev is geared toward publishing databases to the Web and supports extensible markup language, discussed below.

UltraDev is visually oriented rather than oriented toward source code editing. Thus, the developer can draw cells directly on a page, drag cells to other locations, or group them to create a nested table. UltraDev also allows the developer to work with live data, pages and logic which can continue to be laid out while working with server data. UltraDev allows developers to rapidly create database-driven Web applications for multiple server platforms.

Figure 17.48 shows how to set up a form that allows a user to insert a record into the database. Of course, you also need to set up security restrictions so that users first enter a password, and be approved, before they could modify the database. UltraDev can be used to create active server pages (ASP) applications, ColdFusion (CF) applications, or JavaServer Pages (JSP) applications.

XML

Extensible markup language (XML) is a supported markup language, similar to HTML, widely used primarily for business data exchange. Whereas HTML is adequate for expressing how each word, image, and photo is displayed on a Web page, XML describes how content or data are rendered. The XML document itself contains

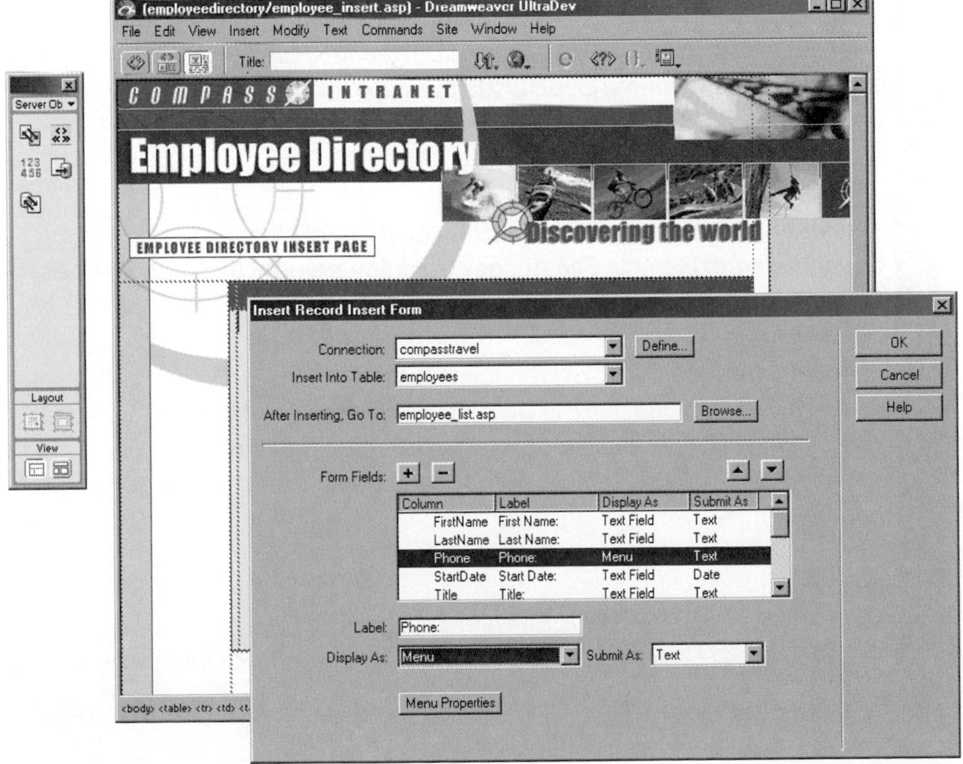

FIGURE 17.48
Macromedia Dreamweaver UltraDev 4 INSERT RECORD insertion form used to allow users to modify the database records remotely.

only data, so applications are necessary to decide how the data are presented or displayed.

As in HTML, XML uses tags or codes. The tags describe the data they contain, so it is therefore possible to search, sort, and manipulate an XML document. XML documents are very portable. It is an open technology, so a user needs no proprietary software to read an XML document. An example of simple XML code is shown in Figure 17.49, a window from Macromedia Dreamweaver UltraDev.

XML provides the mechanism to take raw data and translate them into a universal language that can be read by anyone with the appropriate translator tools. Many companies are using XML to improve electronic data exchange and further enhance the Web experience for customers. XML makes it possible for data to be delivered over the Internet, shared, and presented in the best possible way.

Many software packages now available allow Web database publishing by making use of this nonproprietary standard, XML. One example for personal computers is FileMaker Pro, by FileMaker, Inc. It is easy to integrate XML with FileMaker Pro through forms and style sheets. FileMaker Pro receives data from a person using a standard HTML form and sends a response in XML that includes the data and the details about where a style sheet for the data can be found. Supplying a different style sheet or JavaScript program can show each user a different view of

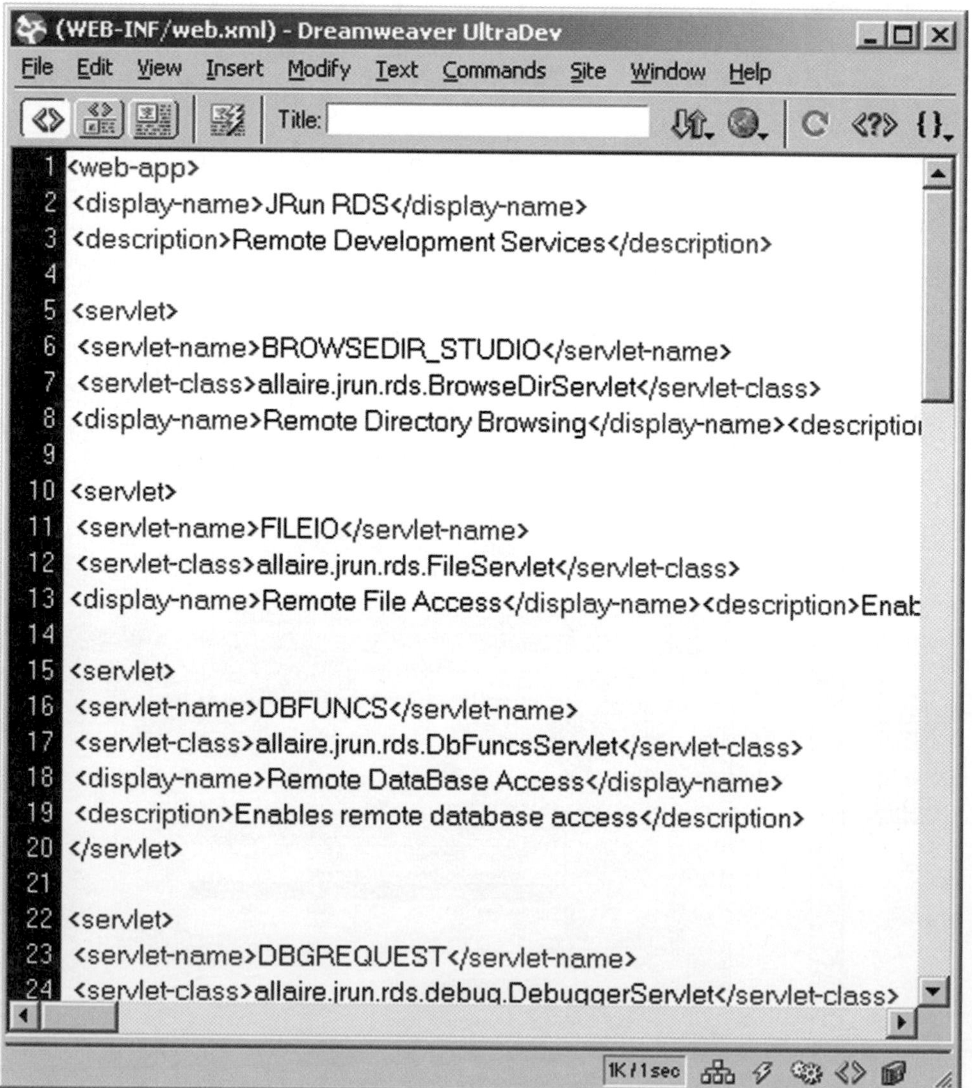
(WEB-INF/web.xml) - Dreamweaver UltraDev
File Edit View Insert Modify Text Commands Site Window Help
Title:
<web-app>
<display-name>JRun RDS</display-name>
<description>Remote Development Services</description>
<servlet>
<servlet-name>BROWSEDIR_STUDIO</servlet-name>
<servlet-class>allaire.jrun.rds.BrowseDirServlet</servlet-class>
<display-name>Remote Directory Browsing</display-name><descriptio
<servlet>
<servlet-name>FILEIO</servlet-name>
<servlet-class>allaire.jrun.rds.FileServlet</servlet-class>
<display-name>Remote File Access</display-name><description>Enab
<servlet>
<servlet-name>DBFUNCS</servlet-name>
<servlet-class>allaire.jrun.rds.DbFuncsServlet</servlet-class>
<display-name>Remote DataBase Access</display-name>
<description>Enables remote database access</description>
</servlet>
<servlet>
<servlet-name>DBGREQUEST</servlet-name>
<servlet-class>allaire.jrun.rds.debug.DebuggerServlet</servlet-class>
1K / 1 sec

FIGURE 17.49
An example of XML coding used for software development using Macromedia Dreamweaver UltraDev 4.

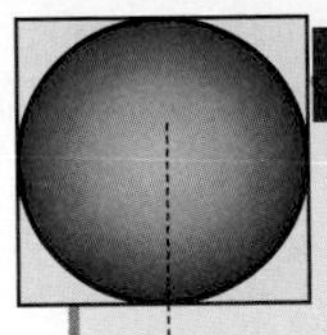

CONSULTING OPPORTUNITY 17.2

STORING MINERALS FOR HEALTH, DATA FOR MINING

One of Marathon Vitamin Shops' employees, Esther See, approached the owner, Bill Berry about an observation she had. "I've noticed that our customers have different habits. Some come in regularly, and others are less predictable," Esther says. "When I see a regular customer, I pride myself on knowing what the customer will buy and maybe even suggest other vitamins they might like. I think I generate more sales that way. The customer is happier, too."

Esther continues, "I wish I could be better at helping out some of the customers who come in less frequently, though."

"That's a very nurturing attitude, Esther, and it helps out our store as well," Bill replies. "I know that we can benefit in other ways by getting a better handle on customer patterns. For instance, we can be sure that we have an item in stock."

Esther nods in agreement and adds, "It's not just the type of vitamin I'm talking about. Some customers prefer one brand over another. I don't know if it depends on their income level or the interests they have in leisure activities. Sports, for example."

"I see, Miss See," Bill chuckles at his own joke, "but do you have anything in mind?"

"Yes, Mr. Berry," she says more formally. "We should organize the data we have about our customers using a data warehouse concept. We can merge the data we have with data from other sources. Then we can look for patterns in our data. Maybe we can identify existing patterns and predict new trends."

Think about how you would organize a data warehouse for Marathon Vitamin Shops. What other databases would you like to merge into the data warehouse? What sort of patterns should Bill Berry be looking for? Identify these patterns by type (associations, sequences, clustering, or trends).

the same data. In this way, users who prefer one presentation style to another receive what they actually want.

Moreover.com is a service that, for a price, allows a Web site developer to display any number of news feeds precisely tailored to the customer's needs. Moreover.com's database is built on XML. It filters information into over 300 special categories from which developers can choose. Moreover.com offers delivery of the feeds in a number of different browser formats, including five XML formats.

SUMMARY

How to store data is often an important decision in the design of an information system. There are two approaches to storing data. The first approach is to store data in individual files, one file for each application. The second approach is to develop a database that can be shared by many users for a variety of applications as the need arises. Dramatic improvements have been made in the design of database software to take advantage of the graphical user interface.

The conventional file approach may at times be a more efficient approach, because the file can be application-specific. On the other hand, the database approach may be more appropriate because the same data need to be entered, stored, and updated only once.

An understanding of data storage requires a grasp of three realms: reality, data, and metadata. An entity is any object or event for which we are willing to collect and store data. Attributes are the actual characteristics of these entities. Data items can have values and can be organized into records that can be accessed by a key. Metadata describe the data and can contain restrictions about the value of a data item (such as numeric only).

Examples of conventional files include master files, table files, transaction files, work files, and report files. They can have a sequential organization, linked lists, hashed file organization, indexed organization, or indexed-sequential organization.

HYPERCASE® EXPERIENCE

17 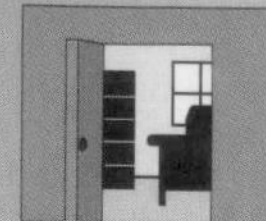

"I hear very good things about your team from the people in Management Systems. You even got some hard-earned praise from Training people. You know, Tom Ketcham isn't easy to please these days. Even *he* is seeing some possibilities. I think you'll pull us together yet . . . unless we all go off in different directions again. I'm just teasing you. I told you to think about whether we are a family, a zoo, or a war zone. Now's the time to start designing systems for us that fit us. You've been here long enough now to form those opinions. I hope they're favorable. I think our famous Southern hospitality should help influence you, don't you? I was so busy persuading you that we're worth the effort that I almost forgot to tell you: Tom and Snowden have agreed to think about moving toward a database of some sort. Would you have this ready in the next two weeks? Tom is at a conference in Minneapolis, but when he returns you should have some database ideas worked up for Snowden and him to discuss it. Keep at it."

HYPERCASE® QUESTIONS

1. Assume your team members have used the Training Unit Client Characteristics Report to design a database table to store the relevant information contained on this report, with the following result:

   Table name: CLIENT TABLE

COLUMN NAME	DESCRIPTION
CLIENT ID (primary key)	Mnemonic made up by users, such as STHSP for State Hospital
CLIENT NAME	The actual, full client name
ADDRESS	The client's address
CONTACTS	The names of contact persons
PHONE NUMBER	The phone numbers of contact persons
CLASS	The type of institution (Veteran's Administration hospital, clinic, other, etc.)
STAFF-SIZE	Size of client staff (number)
TRAINING LEVEL	Minimum required expertise level of the staff (as defined by the class)
EQUIP-QTY	The number of medical machines that the client has
EQUIP TYPE	The type of medical machines (e.g., X-ray, MRI, CAT)
EQUIP MODEL-YR	The model and year of each medical machine

2. Apply normalization to the table your team has developed to remove repeating groups. Display your results.
3. Remove transitive dependencies from your table and show your resulting database table.

A more modern and efficient way to handle indexed-sequential files is the VSAM. Databases can have hierarchical, network, or relational structures.

Normalization is the process that takes user views and transforms them into less complex structures called normalized relations. There are three steps in the normalization process. First, all repeating groups are removed. Second, all partial dependencies are removed. Finally, the transitive dependencies are taken out. After these three steps are completed, the result is the creation of numerous relations that are of third normal form (3NF).

The entity-relationship diagram may be used to determine the keys required for a record or a database relation. The three guidelines to follow when designing master files or database relations are that (1) each separate data entity should create a master file (do not combine two distinct entities on one file); (2) a specific data field should exist only on one master file; and (3) each master file or database relation should have programs to create, read, update, and delete.

The process of retrieving data may involve as many as eight steps: (1) a relation or relations are chosen and (2) joined, (3) projection and (4) selection are performed on the relation to extract the relevant rows and columns, (5) new attributes may be derived, (6) rows are sorted or indexed, (7) totals and performance measures are calculated, and finally (8) the results are presented to the user.

Denormalization is a process that takes the logical data model and transforms it into a physical model that is efficient for tasks that are most needed. Data warehouses differ from traditional databases in many ways; one is that they store denormalized data, which is organized around subjects. Data warehouses allow easy access via data-mining software called siftware that searches for patterns and identifies relationships not imagined by human decision makers.

Extensible markup language (usually just called XML) is a nonproprietary standard language that serves as a mechanism to take raw data and translates it into a universal language that can be read by anyone with the appropriate translation tools. It is used primarily for business data exchange.

KEYWORDS AND PHRASES

- attribute
- bubble diagram
- "clean" data
- concatenated key
- conventional file
- CRUD (Create, Read, Update, and Delete)
- database
- database administrator
- database management system (DBMS)
- data elements
- data item
- data mining
- data model diagram
- data storage
- data warehouse
- denormalization
- domain integrity
- entity
- entity integrity constraint
- entity-relationship (ER) diagram
- entity subtype
- first normal form (1NF)
- hashed file organization
- hierarchical data structure
- indexed organization
- indexed-sequential access method (ISAM)
- indexed-sequential organization
- key
- linked list
- logical view
- master file
- network data structure
- normalization
- online analytical processing (OLAP)
- partial dependencies
- patterns
  - associations
  - sequences
  - clustering
  - trends
- physical view
- primary key
- reality, data, and metadata
- record
- referential integrity
- relational data structure
- relationship
- repeating groups
- report file
- retrieval
- secondary key
- second normal form (2NF)
- siftware
- special characters

steps in information retrieval
- choose
- join
- project
- select
- derive
- index or sort
- calculate
- present

table file
third normal form (3NF)
transaction file
transitive dependencies
unnormalized relation
virtual storage access method (VSAM)
work file
XML

REVIEW QUESTIONS

1. What are the advantages of organizing data storage as separate files?
2. What are the advantages of organizing data storage using a database approach?
3. What are the effectiveness measures of database design?
4. What are the efficiency measures of database design? Why do they conflict with each other?
5. List some examples of entities and their attributes.
6. Define the term *metadata*. What is the purpose of metadata?
7. List file types of commonly used conventional files. Which of these are temporary files?
8. What is a linked list?
9. What often occurs when a hashed file organization is used?
10. What is an inverted list? When is it valuable to have an inverted list?
11. Name the three main types of database organization.
12. Define the term *normalization*.
13. What is removed when a relation is converted to the first normal form?
14. What is removed when a relation is converted from 1NF to 2NF?
15. What is removed when a relation is converted from 2NF to 3NF?
16. List the three entity constraints. In a sentence, describe the meaning of each entity constraint.
17. List the eight steps for retrieving, presorting, and presenting data.
18. What does join do? What is projection? What is selection?
19. State the differences between "sort" and "index."
20. List two ways of storing a many-to-many relationship and the differences between the two methods.
21. Define denormalization.
22. Explain the differences between traditional databases and data warehouses.
23. Define what siftware does when used in data mining.
24. Explain how XML works to facilitate business data exchange.

PROBLEMS

1. Given the following file of renters:

Record Number	Last Name	Apartment Number	Rent	Lease Expires
41	Warkentin	102	550	4/30
42	Buffington	204	600	4/30
43	Schuldt	103	550	4/30
44	Tang	209	600	5/31
45	Cho	203	550	5/31
46	Yoo	203	550	6/30
47	Pyle	101	500	6/30

a. Develop a linked list by apartment number in ascending order.
b. Develop a linked list according to last name in ascending order.

2. Develop an inverted file. Develop an index for each of the attributes.
3. The following is an example of a grade report for two students at the University of Southern New Jersey:

USNJ Grade Report
Spring Semester 2000

Name: I. M. Smarte — *Major: MIS*
Student: 053-6929-24 — *Status: Senior*

Course Number	Course Title	Professor	Professor's Department	Grade
MIS 403	Systems Analysis	Diggs, T.	MIS	A
MIS 411	Conceptual Foundations	Barre, G.	MIS	A
MIS 420	Human Factors in IS	Barre, G.	MIS	B
CIS 412	Database Design	Menzel, I.	CIS	A
DESC 353	Management Models	Murney, J.	MIS	A

USNJ Grade Report
Spring Semester 2000

Name: E.Z. Grayed — *Major: MIS*
Student: 472-6124-59 — *Status: Senior*

Course Number	Course Title	Professor	Professor's Department	Grade
MIS 403	Systems Analysis	Diggs, T.	MIS	B
MIS 411	Conceptual Foundations	Barre, G.	MIS	A

Draw a hierarchical data structure for this user view. Focus on each individual course and follow the approach used in Figure 17.15.

4. Draw a network data structure for the user view in problem 3. Use grade as a link and follow the approach in Figure 17.17.
5. Draw a data model diagram with associations for the user view in problem 3.
6. Convert the user view in problem 3 to a 3NF relation. Show each step along the way.
7. Draw an entity-relationship diagram for the following situation: Many students play many different sports. One person, called the head coach, assumes the role of coaching all these sports. Each of the entities have a number and a name. (Make any assumptions necessary to complete a reasonable diagram. List your assumptions.)
8. The entity-relationship diagram you drew in problem 7 represents the data entities that are needed to implement a system for tracking students and the sports teams that they play. List the files that are needed to implement the system, along with primary, secondary, and foreign keys that are required to link the files.
9. Draw an entity-relationship diagram for the following situation: A commercial bakery makes many different products. These products include breads, desserts, specialty cakes, and many other baked goods. Ingredients such as flour, spices, and milk are purchased from vendors. Sometimes an ingredient is purchased from a single vendor, and other times an ingredient is purchased from many vendors. The bakery has commercial customers, such as schools and restaurants, that regularly place orders for baked goods. Each baked good has a specialist that oversees the setup of the bake operation and inspects the finished product.

10. List the files and keys that are needed to implement the commercial bakery system.
11. Draw an E-R diagram for the ordering system in Figure 17.34.
12. Draw a data flow diagram for placing an order. Base your data flow diagram on the E-R diagram.

GROUP PROJECTS

1. Gregg Baker orders tickets for two concerts over the Web. His orders are processed, exact seat locations are assigned, and the tickets are mailed separately. One of the sets of tickets gets lost in the mail. When he calls the service number, he does not remember the date or the seat numbers, but the ticket agency was able to locate his tickets quickly because the agency denormalized the relationship. Describe the ticket ordering system by listing the data elements that are kept on the order form and the shipping form. What information did Gregg give the ticket agency to retrieve the information?

SELECTED BIBLIOGRAPHY

Avison, D. E. *Information Systems Development: A Database Approach*, 2d ed. London: Blackwell Scientific, 1992.

Avison, D. E., and G. Fitzgerald. *Information Systems Development: Methodologies, Techniques, and Tools*, 2d ed. New York: McGraw Hill, 1995.

Codd, E. F. "Twelve Rules for On-Line Analytic Processing." *Computerworld*, April 13, 1995.

Dietel, H. M., P. J. Dietel, and T. R. Nieto. *E-Business and e-Commerce: How to Program*. Upper Saddle River, NJ: Prentice Hall, 2001.

Everest, G. C. *Database Management: Objectives, System Functions, and Administration*. New York: McGraw-Hill, 1985.

Gane, C., and T. Sarson. *Structured Systems Analysis: Tools and Techniques*. Englewood Cliffs, NJ: Prentice Hall, 1979.

Gray, P. "Data Warehousing: Three Major Applications and Their Significance." In K. E. Kendall, ed., *Emerging Information Technologies, Improving Decision, Cooperation, and Infrastructure*. Thousand Oaks, CA: Sage Publications, 1999.

McFadden, F., and J. A. Hoffer. *Modern Database Management*, 4th ed. Redwood City, CA: Benjamin-Cummings, 1994.

Sanders, G. L. *Data Modeling*. New York: Boyd and Fraser, 1995.

Website for XML http://www.xml.com last accessed February 9, 2001.

ALLEN SCHMIDT, JULIE E. KENDALL, AND KENNETH E. KENDALL

BACK TO DATA BASICS

After numerous interviews, prototypes, data flow diagrams, and data dictionary entries had been completed, Anna and Chip both start work on the entity-relationship model. "I'll be responsible for creating the Microsoft Access table relationships," Anna promises. Chip volunteers to complete an entity-relationship diagram. "Let's compare the two diagrams for accuracy and consistency when we're done," Anna suggests, and so they did.

Figure E17.1 shows the entity-relationship diagram for the microcomputer system. Visible Analyst calls each of the rectangles an *entity*. An entity is distinct from an external entity, which represents a source or recipient of information on a data flow diagram. The entity represents a file of information stored within the system, one corresponding to a data store on the data flow diagram. Each of the diamond rectangles represents a relationship between the data entities. A rectangle with an oval in it represents an entity that cannot exist without the connecting entity, such as a table of codes.

"I've created the entity-relationship diagram, starting with the simplest portions of the system," Chip tells Anna. "The first data entities created are SOFTWARE and HARDWARE. The relationship is that software is installed on the hardware. Next I determined the cardinality of the relationship. Because one software package could be installed on many microcomputers, this relationship is one-to-many. Each microcom-

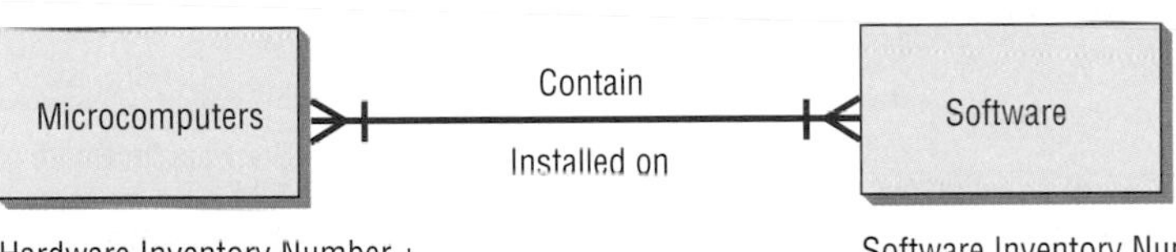

Microcomputers	Software
Hardware Inventory Number +	Software Inventory Number +
Brand Name +	Title +
Model +	Operating System Name +
Serial Number +	Version Number +
Date Purchased +	Publisher +
Purchase Cost +	Software Category Description +
Replacement Cost +	Number of Diskettes +
Memory Size +	Computer Brand +
Fixed Disk Capacity +	Computer Model +
Second Fixed Disk Capacity +	Memory Required +
Primary Diskette Drive +	Monitor Name +
Secondary Diskette Drive +	Printer Description +
CD-ROM Drive +	Diskette Size +
Zip Drive Connected +	Site License +
Tape Backup Unit +	Number of Copies +
Network +	Expert Last Name +
Internet Connection Name +	Expert First Name +
Monitor Name +	Office Phone
Printer Description +	
Warranty +	
Campus Description +	
Room Location	
{Board Code} +	
{Software Inventory Number}	

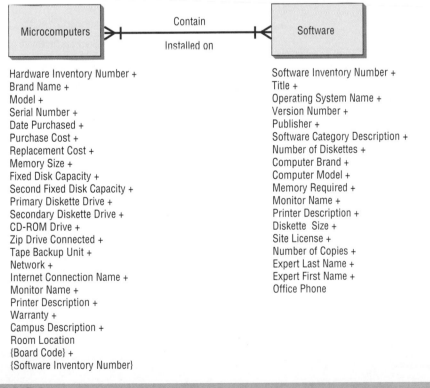

FIGURE E17.1
Unnormalized entity-relationship diagram for the computer system.

puter may also have many different software packages installed on it so that it also provides a one-to-many relationship. Because there is a one-to-many relationship for each of the data entities, the full relationship between them becomes many-to-many." The unnormalized entity-relationship diagram is shown in Figure E17.1, along with the data dictionary entries for the tables.

Chip continues by saying, "This first view is far from normalized. Notice that the BOARD CODE and SOFTWARE INVENTORY NUMBER are repeating elements on the HARDWARE entity. I will have to create several entities for each of them." A bit later Chip reviews his work with Anna. The BOARD CODES have become a separate entity, linked by a relational entity, and the SOFTWARE INVENTORY NUMBER has been removed and placed in a relational entity. Refer to the entity-relationship diagram illustrated in Figure E17.2. "This places the data in the first normal form," remarks Chip. "Also, there are no elements that are dependent on only a part of the key, so the data are also in the second normal form. There are, however, elements that are not part of the entity that is represented on the diagram, and they will have to be removed. For example, look at the MONITOR NAME and PRINTER DESCRIPTION. These elements are not a part of the microcomputer but are connected to it. They should have their own entity. That makes it easier to change the spelling of, say, a printer. Rather than having to change the spelling of the printer on many of the MICROCOMPUTER records, it would only have to be changed once."

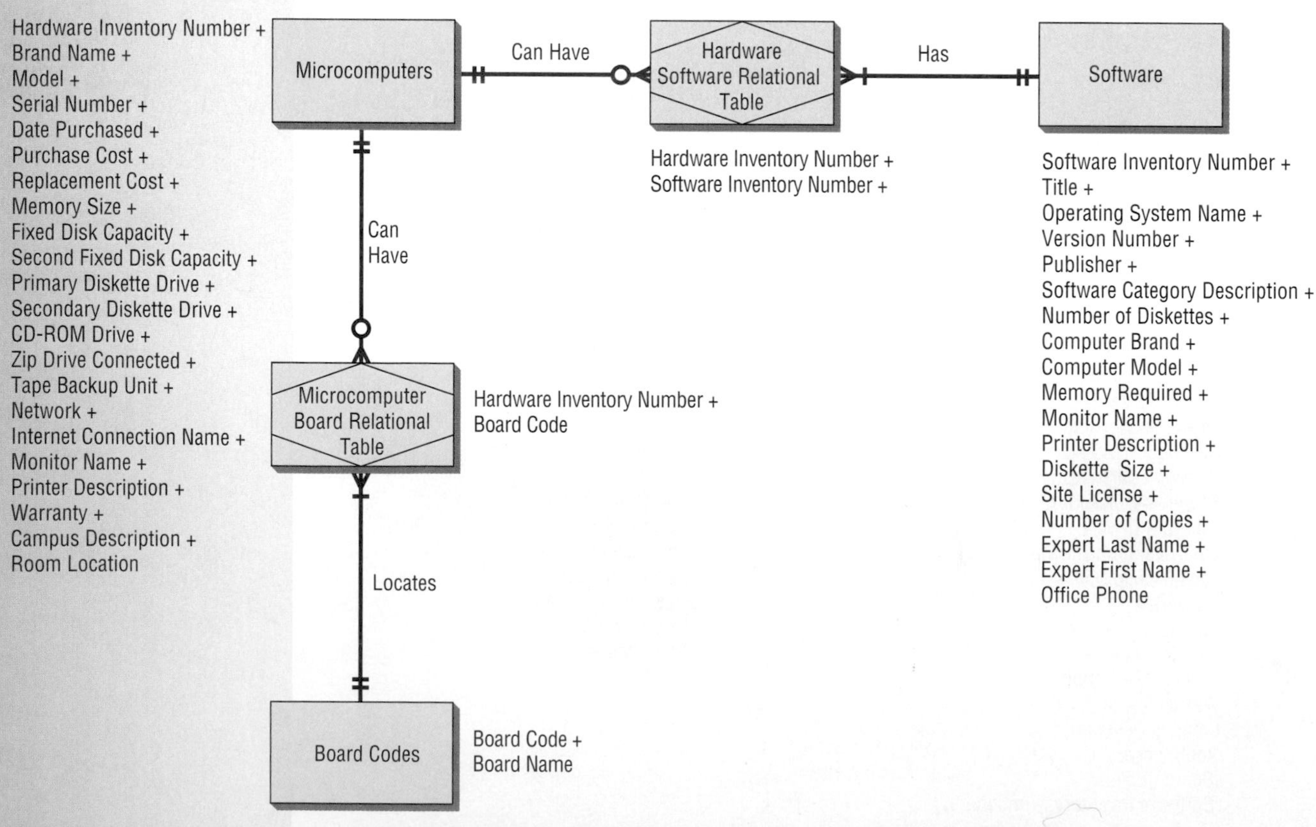

FIGURE E17.2
The computer system's entity-relationship diagram in the first normal form.

Anna agrees, remarking, "That's really a good assessment of the situation. It will make implementing the Microsoft Access tables easier."

Chip continues to work on the entity-relationship diagram. After a few hours he exclaims, "I think that it's done. Would you take a look at the final version?" The final version is shown in Figure E17.3. All the entities and relationships have been described in the repository. Figure E17.4 shows the first repository screen for the entity MICROCOMPUTERS. Notice that the Composition area contains the final elements for keeping track of Microcomputers and that the keys are indicated using the notation [Pk] for the primary key and [Akn] for an alternate key, where n represents any unique number. Foreign keys are lower in the list and are indicated using the notation [Fk]. The second screen is illustrated in Figure E17.5 and contains information about the relationships that the MICROCOMPUTERS entity has with other entities, including the cardinality. This screen also contains the locations of the entity.

The PROVIDES SOLUTIONS FOR relationship, linking SOFTWARE EXPERT and SOFTWARE, is shown in Figure E17.6, and it has a definition as well as the entities that are linked at either end of the relationship. Cardinality is included for each end of the relationship.

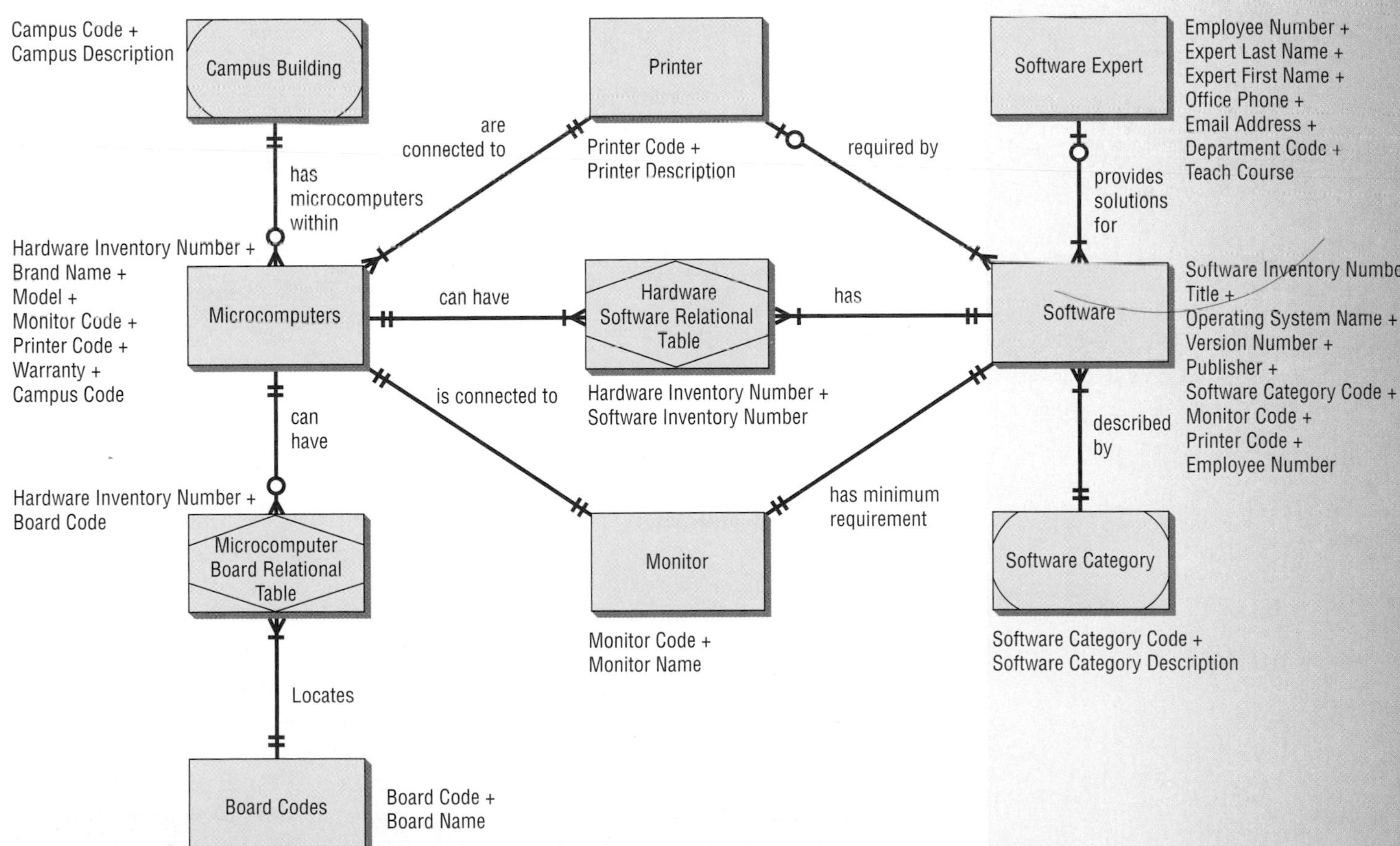

FIGURE E17.3

Final entity-relationship diagram for the computer system.

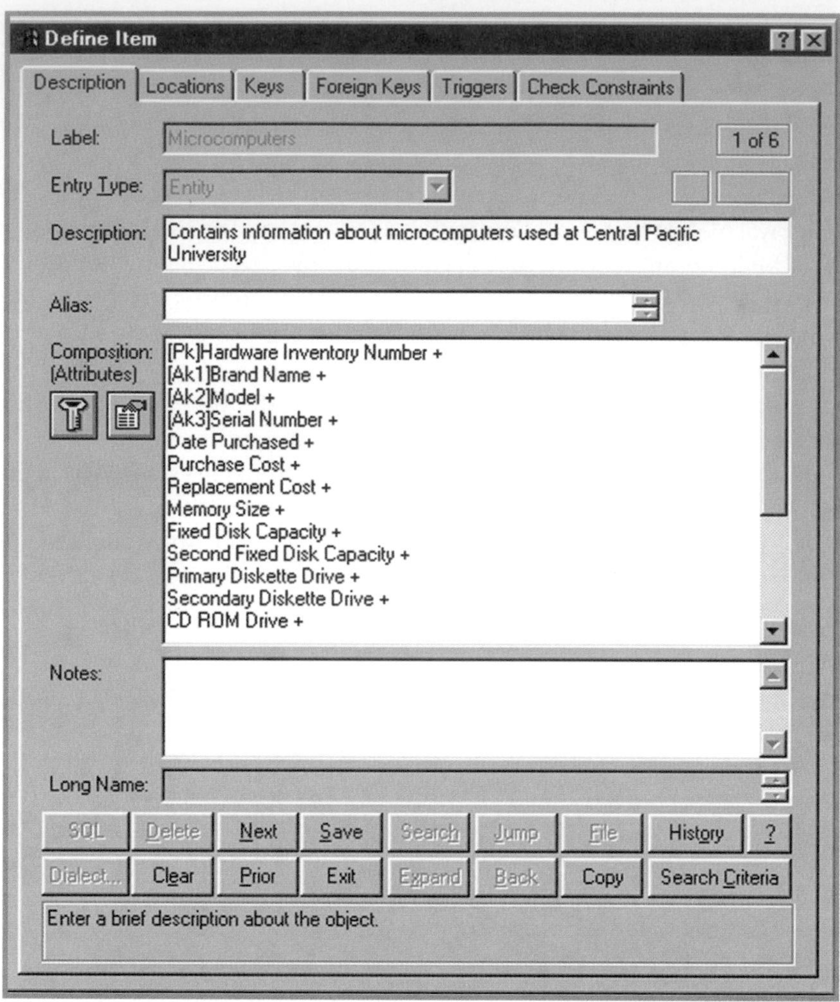

FIGURE E17.4

Entity description screen for MICROCOMPUTERS, the first repository screen.

Anna reviews the final version and exclaims, "It looks great! You are right in moving the PRINTER and MONITOR to their own entities. I see that the CAMPUS BUILDING has been moved to its own entity. Good idea, as the building is not a part of the microcomputer. Also, the SOFTWARE EXPERT is definitely not a part of the software entity. How about the SOFTWARE CATEGORY?"

"I moved the SOFTWARE CATEGORY into its own entity to save room on the master files when they are constructed," Chip answers. "It is really a table of codes, and we can store a small code, rather than a lengthy description. Why don't you double check the various keys on the diagram? Each related entity, on the many end, should have a foreign key that matches the primary key of the entity on the one end."

Anna examines the diagram for a while and remarks, "It looks good to me, but why don't you run some of the Visible Analyst analysis reports for the entity-relationship diagram?"

"That's an excellent idea!" exclaims Chip. The first step in the diagram is to run the Visible Analyst syntax check for the final entity-relationship diagram. This option

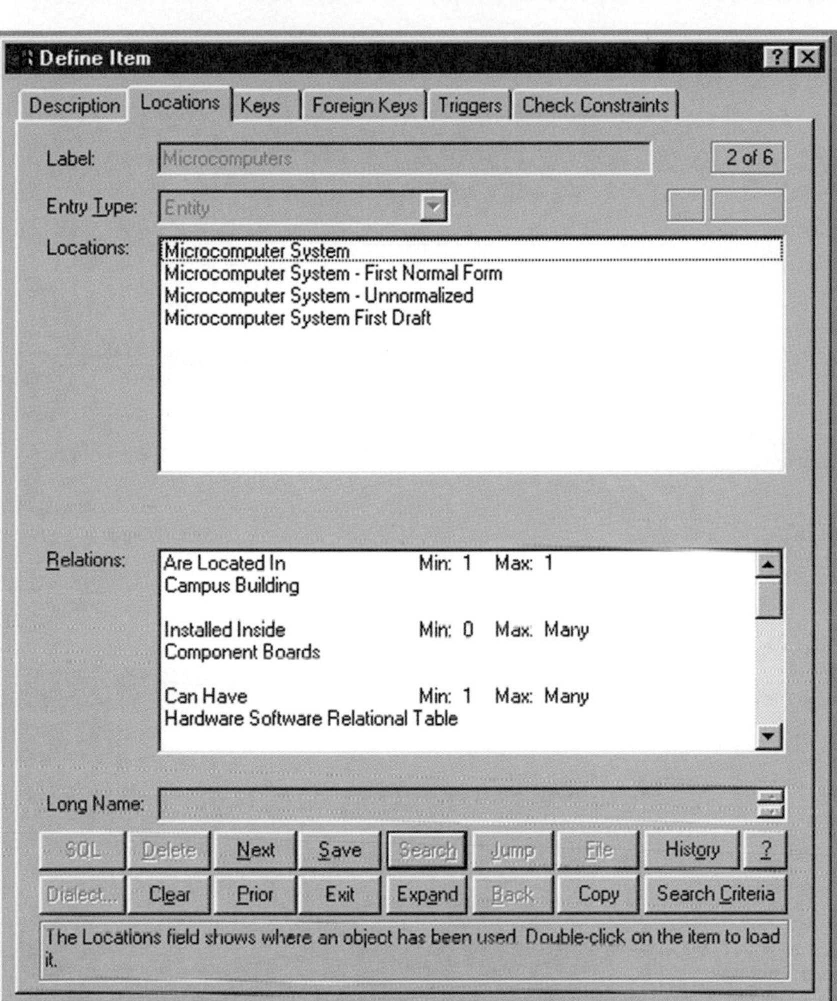

FIGURE E17.5
The second repository screen for the MICROCOMPUTERS entity.

checks for entities and relationships that have not been named, and it produces no errors when Chip runs it. Next, Chip runs the Normalization analysis option, which displays a series of warning error messages, illustrated in Figure E17.7. These messages inform Chip that some of the relationships have a one-to-one cardinality, which is acceptable. The other warning messages let Chip know that the SOFTWARE CATEGORY and CAMPUS BUILDING do not have identifying relationships. Hence, the record keys have not been defined in the repository.

The next analysis that Chip runs is Key Analysis. This option performs syntax and normalization analysis as well as reporting problems with primary and foreign keys. The analysis report, which shows that the SOFTWARE CATEGORY has no primary key and that the entity PRINTER has no identifying relationship, is illustrated in Figure E17.8 The final analysis report that Chip runs is the Model Balancing report, which is shown in Figure E17.9. This report produces some of the errors detected by previous reports, but, in addition, it shows all elements that are contained in the Composition area of the entities and are not used by a data flow diagram process.

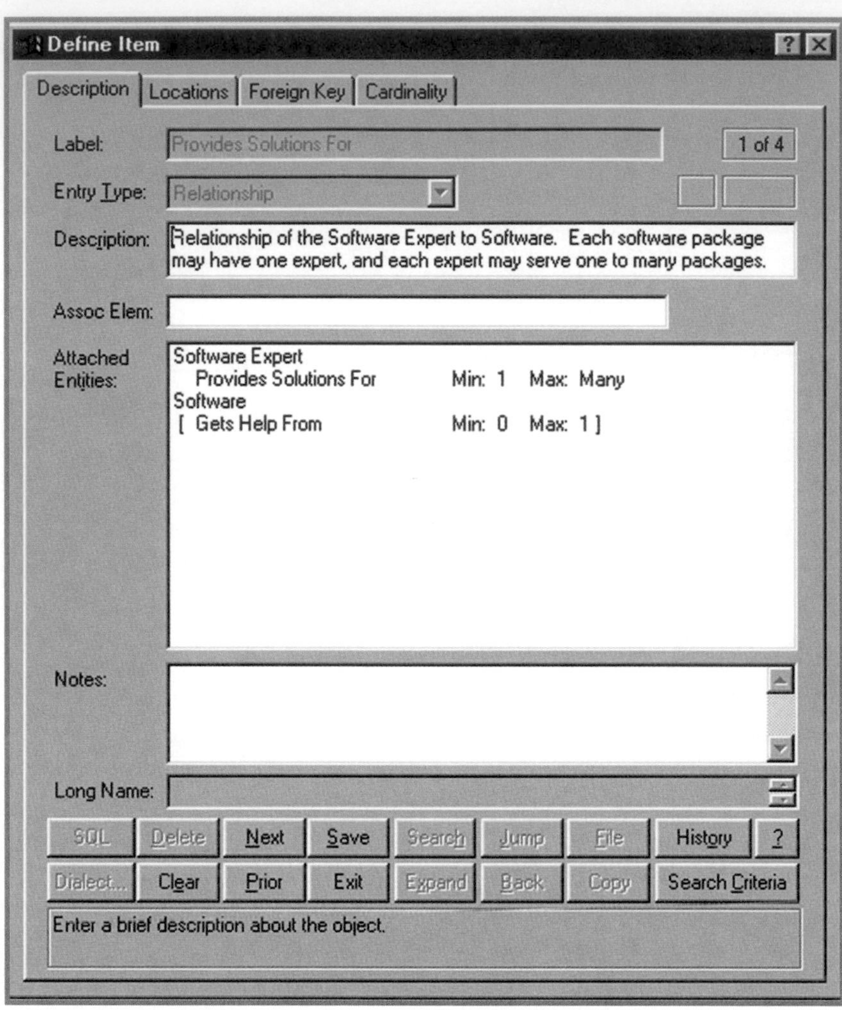

FIGURE E17.6

A relationship screen for the item PROVIDES SOLUTIONS FOR.

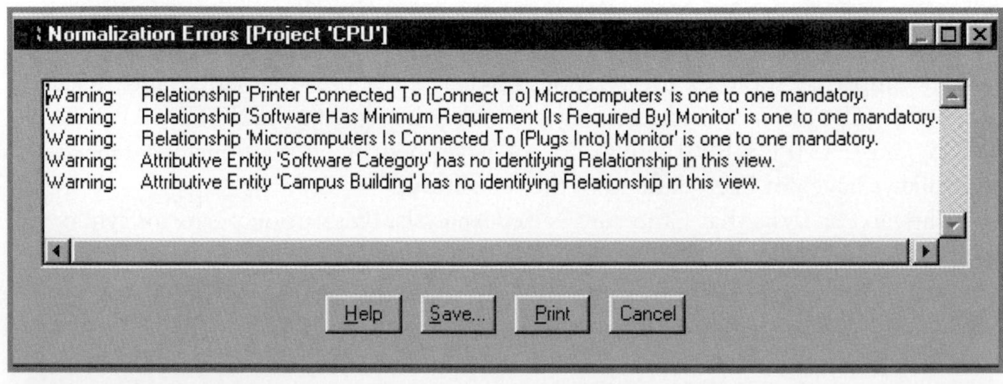

FIGURE E17.7

The output screen NORMALIZATION ERRORS, showing warning messages.

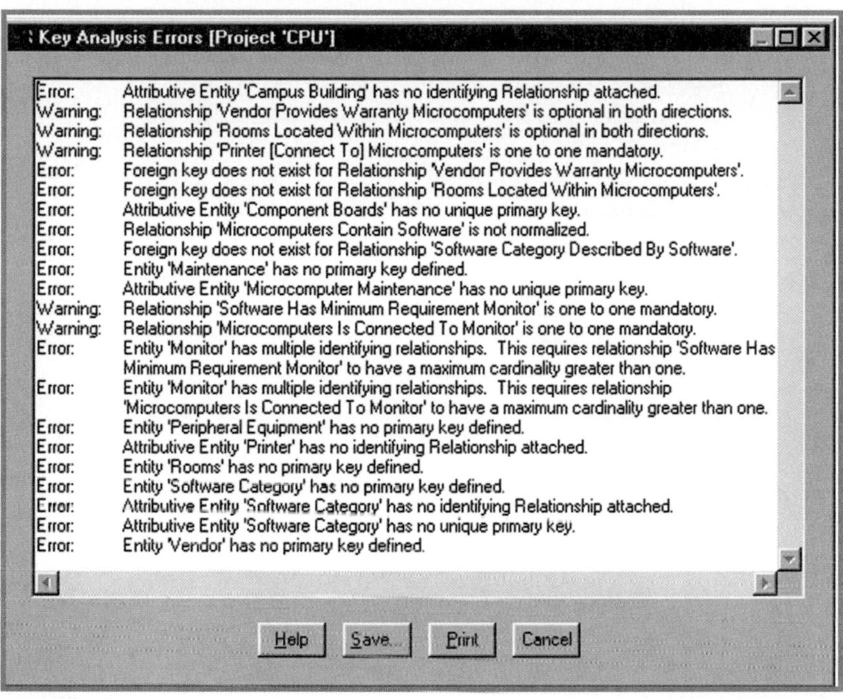

FIGURE E17.8
The KEY ANALYSIS ERRORS output screen.

"Now that we've done the analysis, watch this feature in action," remarks Chip. "I'm going to run Key Synchronization."

"I thought we were finished with the reports," replies Anna.

"Key Synchronization is more than a report," answers Chip. "It will define all the foreign keys for the project."

"This I've got to see!" exclaims Anna, as she slides a chair next to Chip.

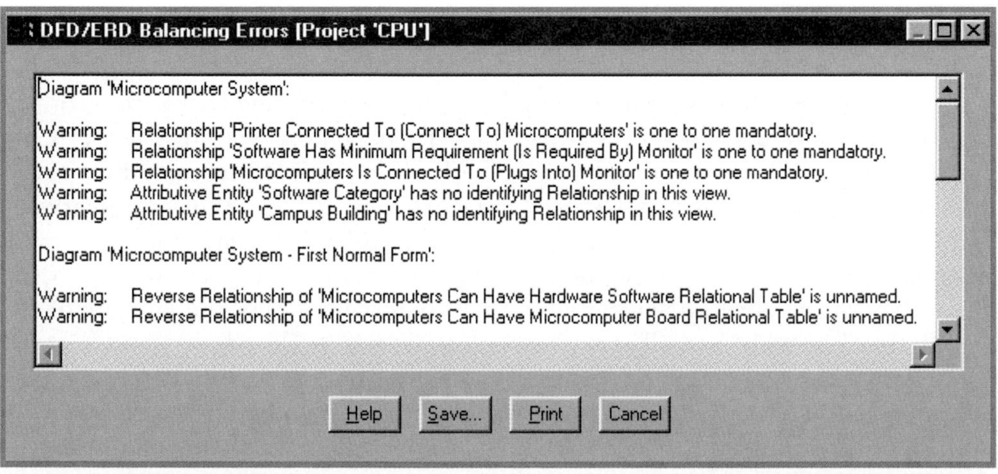

FIGURE E17.9
A BALANCING ERRORS screen.

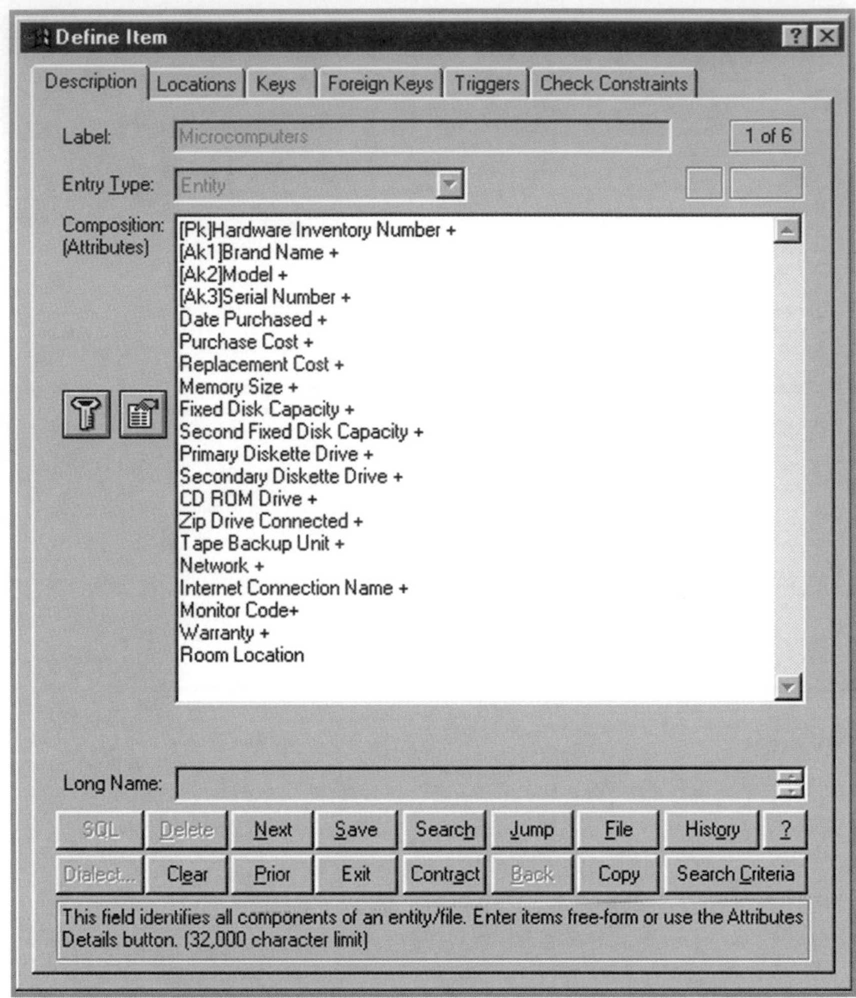

FIGURE E17.10
A repository entry, MICROCOMPUTERS.

"First, we have to draw the entity-relationship diagram and define the primary keys, which we've done," explains Chip. "Take a look at this repository entry." Chip opens the entity-relationship diagram and double clicks on the **Microcomputers** entity, displaying its repository entry, illustrated in Figure E17.10. Notice that a primary key (the [pk] notation in front of the Hardware Inventory Number element in the composition area) and several alternate keys ([Ak1], [Ak2], and [Ak3]) are defined. Chip clicks Repository and Key Synchronization. A report informing the analysts of the changes made to tables as well as any remaining errors is produced.

"Let's take a look at that repository entry again," muses Chip. "Look, Key Synchronization has added the required foreign keys." These keys are indicated with the [FK] notation in front of the new keys. The modified **Microcomputers** entity is shown in Figure E17.11.

After spending time examining the diagram as well as the repository entries and analysis reports, both Anna and Chip are satisfied that the relationships between the

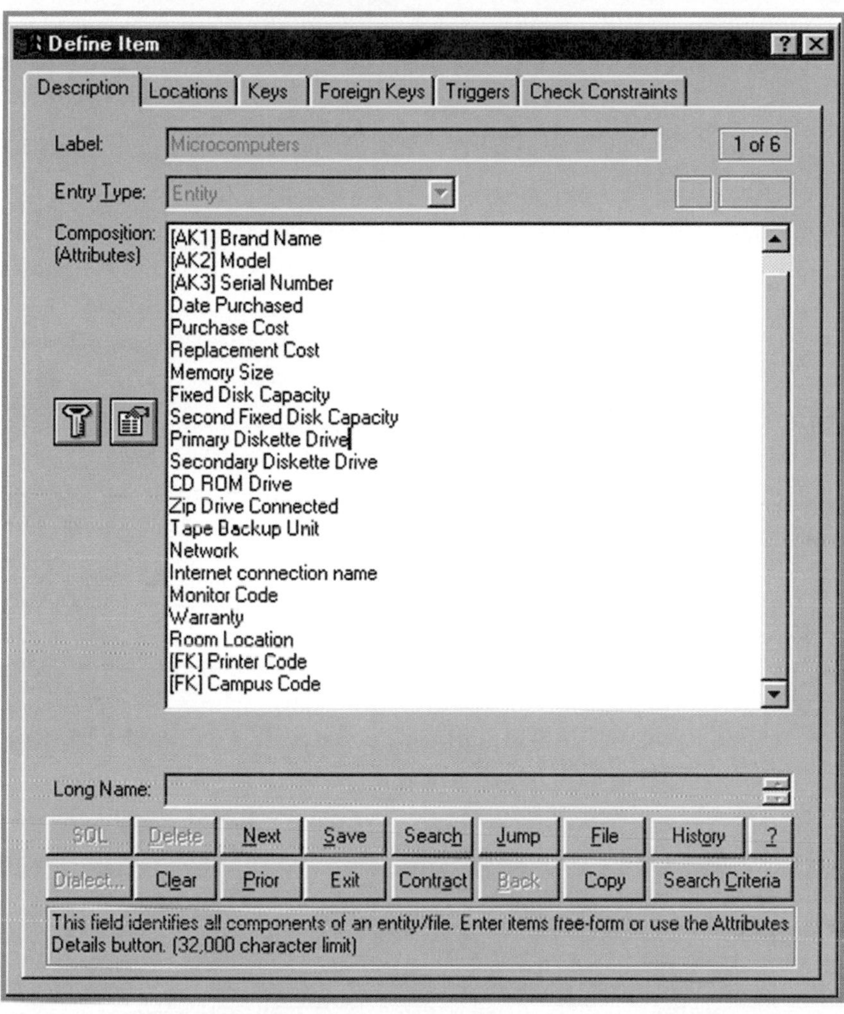

FIGURE E17.11
The repository entry MICROCOMPUTERS that was modified to add two foreign keys.

data have been accurately portrayed. Next they decided how to design the files or database from the diagrams. A critical choice had to be made: whether to use files or a database. "I think a series of indexed files and COBOL language is best," Chip says.

Anna says, "The system should be implemented using Microsoft Access, because a database structure would easily accommodate the many relationships."

The HARDWARE/SOFTWARE relationship is analyzed first. Because there is a many-to-many relationship between these two data entities, it may be implemented in one of two forms:

1. Use three indexed files, which is the database solution.
   a. A HARDWARE MASTER file.
   b. A SOFTWARE MASTER file.
   c. A hardware/software relationship file, which would contain the key fields for the HARDWARE and SOFTWARE master file for all software installed on all machines.

2. Although the preceding database solution could also be used when designing a COBOL-indexed file structure, a better alternative would be to use a repeating group on one of the files that references the other master file. This solution is easier for accomplishing file maintenance and producing information.
   a. A HARDWARE MASTER file.
   b. A SOFTWARE MASTER file with a repeating group or table of HARDWARE INVENTORY NUMBERS, the key field of the MICROCOMPUTER MASTER on which the software would be installed. This table could also be included on the HARDWARE MASTER file, which references the SOFTWARE INVENTORY NUMBER for each software package installed on the machine.

"I finally see your way of thinking," Chip says. After a lengthy discussion, they decide to go with the database solution. It will be easier for the programmers to design, code, implement, and maintain the system than other methods. Furthermore, the file structures will be easy to modify as the system evolves. "I guess it's my turn to work on the relationships," says Anna as she takes a copy of the entity-relationship diagram. "I'll modify the Microsoft Access tables from the prototyping sessions."

Anna starts by setting up the primary keys for each of the tables. When the tables are in their final form, she creates the relationships between them. The Microsoft Access Relationships diagram is illustrated in Figure E17.12. Rectangles

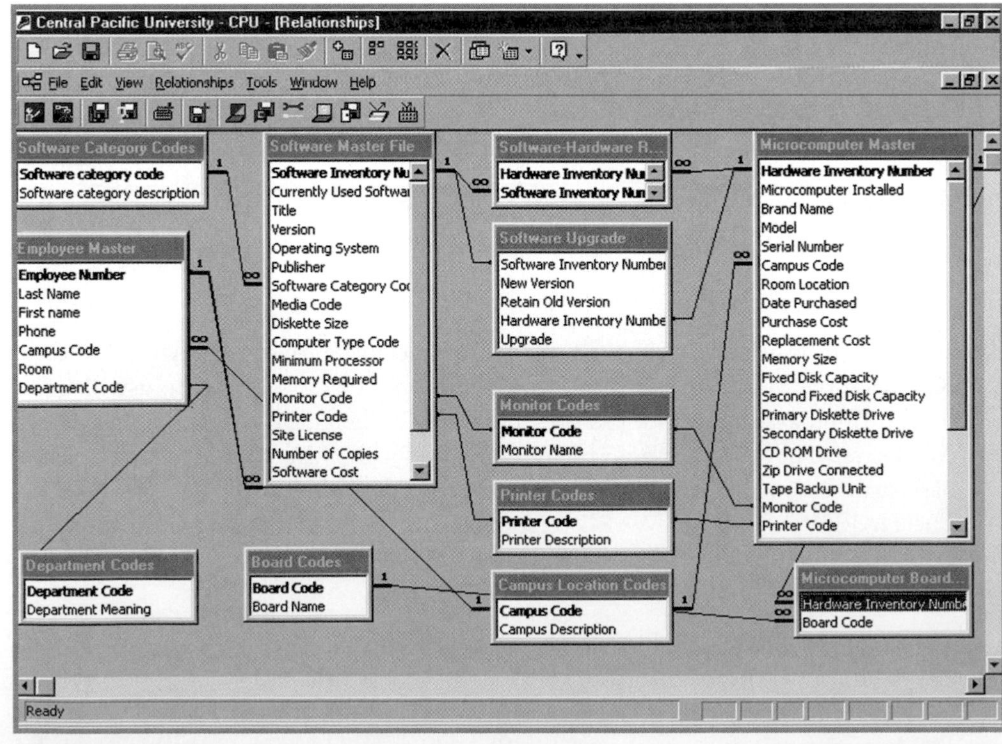

FIGURE E17.12
A Microsoft Access RELATIONSHIPS diagram.

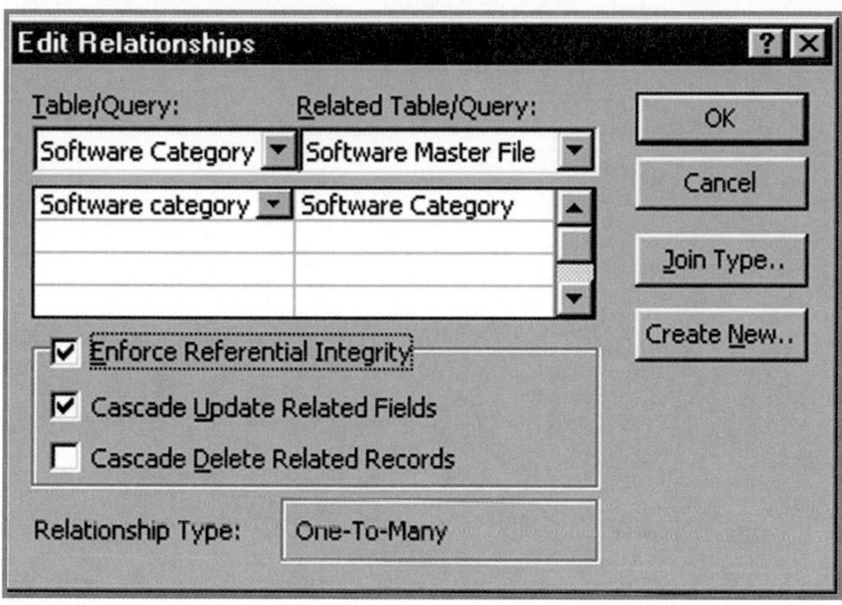

FIGURE E17.13
An example that shows setting referential integrity for the relationship between SOFTWARE CATEGORY CODES and SOFTWARE MASTER.

on the diagram represent the database tables and correspond to the various entity types found on the entity-relationships diagram. Notice that the cardinality is represented by 1 and infinity symbols. The primary key fields are listed as the first field of each rectangle; they are also displayed in boldface type. Foreign keys are shown attached to the other end of the relationship line, if the foreign key is visible in the table rectangle. Keys are dragged from one table to another to establish a relationship, and a dialog box appears to determine properties of the relationship. For example, the property called Enforce Referential Integrity means that you cannot create a record in the many end of the table without first creating it in the one end (containing the primary key). Figure E17.13 illustrates setting referential integrity for the relationship between Software Category Codes and the Software Master. Note that **Cascade Update Related Fields** is checked. If you change a Software Category code value, the same code will be updated on the Software Master. **Cascade Delete Related Records**, however, is not checked. You would not want the system to delete a Software Category code and have all the related Software Master records also deleted.

"I would like to produce some documentation for the system," remarks Anna. "It would help when we need to modify the design as well as the Microsoft Access objects and code." There are several matrices found in the Repository Reports feature that are useful to produce. The first is the Entities versus Data Stores Matrix, which is illustrated in Figure E17.14. It shows the entities found on all the entity-relationship diagrams and data stores that contain similar elements. This matrix is useful for mapping the entity-relationship diagram into a data flow diagram.

The Composition Matrix is produced next. It provides a cross-reference grid of elements and the entities within which they are contained. This matrix is useful for

Date: 4/19/2001
Time: 11:57 AM

	DD1 MICROCOMPUTER MASTER	DD2 SOFTWARE MASTER
Board Codes		
Campus Building		
Component Board	X	X
Hardware Software Relational Table	X	X
Microcomputer Board Relational Table	X	X
Microcomputer Maintenance	X	X
Microcomputers	X	X
Monitor	X	X
Peripheral Equipment	X	X
Printer		
Rooms		
Software	X	X
Software Category		
Software Expert		
Software Hardware Relation	X	X
Vendor		

FIGURE E17.14
An entities vs. data stores matrix.

determining what entities will need to be modified (or what tables in Microsoft Access) if the element size or any other characteristic of the element changes. A portion of this matrix is illustrated in Figure E17.15. A last matrix that is useful in assessing changes to the entire system is the Diagram Location Matrix. This shows the entities and the diagrams within which they are contained. Figure E17.16 is an example of this matrix. Notice that it also shows how more entities are added as the diagram becomes more normalized, a process resulting from pulling elements off entities into new ones.

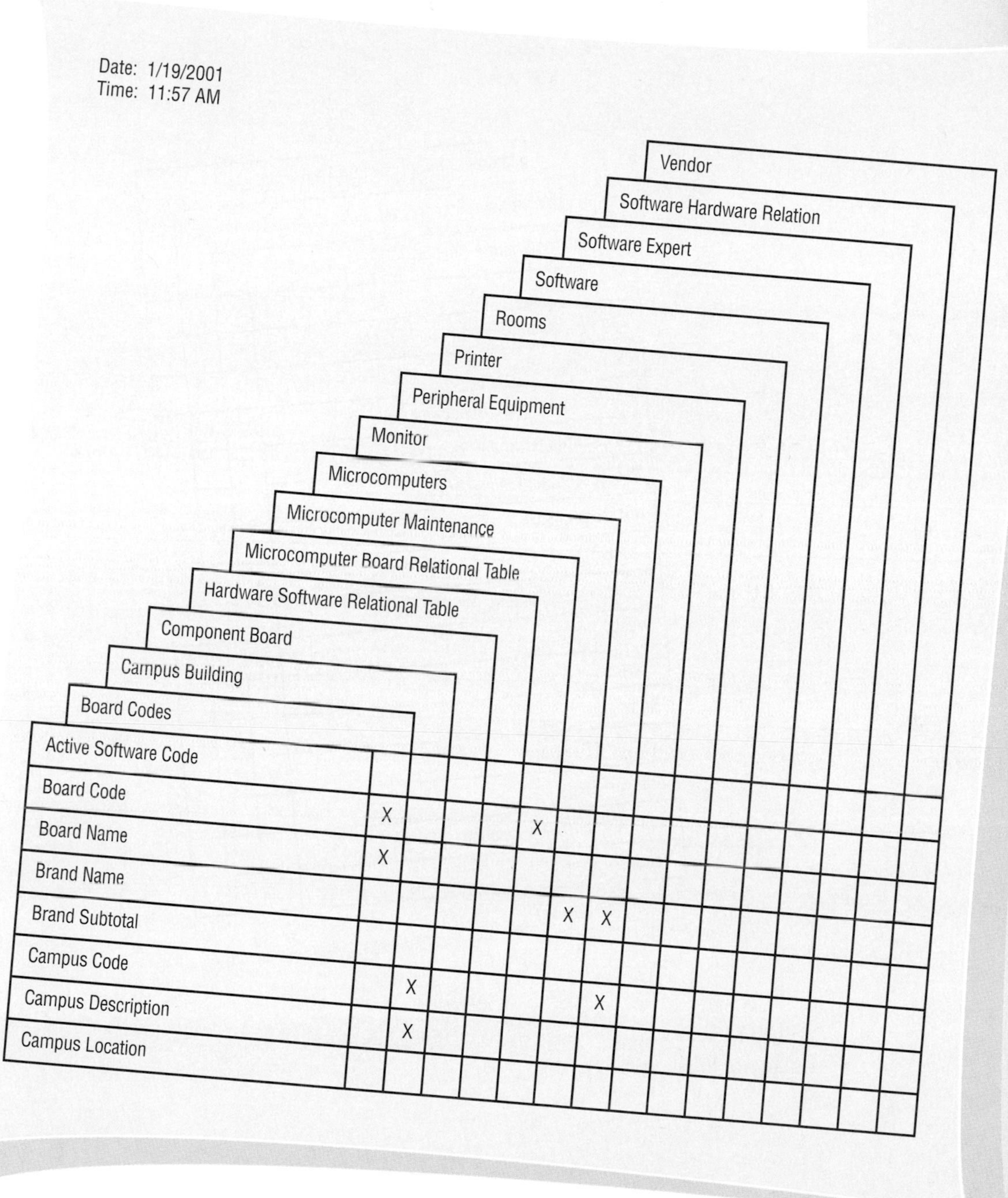

Date: 1/19/2001
Time: 11:57 AM
Vendor
Software Hardware Relation
Software Expert
Software
Rooms
Printer
Peripheral Equipment
Monitor
Microcomputers
Microcomputer Maintenance
Microcomputer Board Relational Table
Hardware Software Relational Table
Component Board
Campus Building
Board Codes
Active Software Code
Board Code
Board Name
Brand Name
Brand Subtotal
Campus Code
Campus Description
Campus Location

FIGURE E17.15

A data elements vs. entities composition matrix.

17

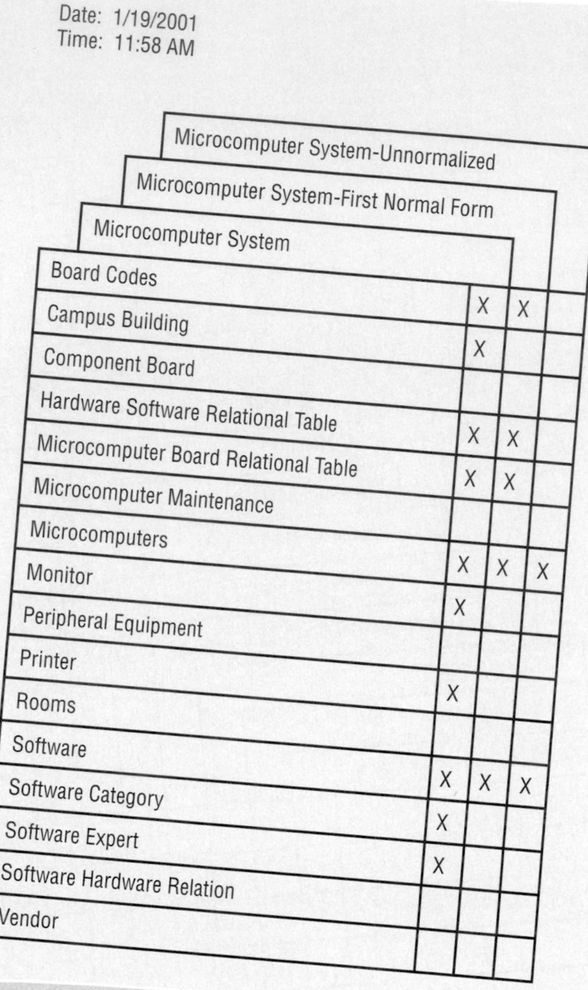
Date: 1/19/2001
Time: 11:58 AM

	Microcomputer System	Microcomputer System-First Normal Form	Microcomputer System-Unnormalized
Board Codes	X	X	
Campus Building	X		
Component Board			
Hardware Software Relational Table	X	X	
Microcomputer Board Relational Table	X	X	
Microcomputer Maintenance			
Microcomputers	X	X	X
Monitor	X		
Peripheral Equipment			
Printer	X		
Rooms			
Software	X	X	X
Software Category	X		
Software Expert	X		
Software Hardware Relation			
Vendor			

FIGURE E17.16
A diagram location matrix.

EXERCISES*

E-1. Use Visible Analyst to view the unnormalized and first normal form entity-relationship diagrams for the Microcomputer System.

E-2. Use Visible Analyst to view the entity-relationship diagram for the Microcomputer System.

E-3. Add the VENDOR entity to the diagram. The vendor warrants the microcomputers, and the relationship between VENDOR and MICROCOMPUTERS is that one VENDOR can warrant many MICROCOMPUTERS.

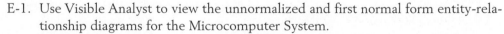

E-4. Add the MAINTENANCE entity to the diagram. Maintenance repairs are performed on microcomputers, and the relationship between MAINTENANCE and MICROCOMPUTERS is such that one MICROCOMPUTER may have many MAINTENANCE records.

E-5. Describe the SOFTWARE CATEGORY entity in the repository. Include the elements found on the entity-relationship diagram below SOFTWARE CATEGORY in the **Composition** area.

E-6. Describe the MAINTENANCE entity in the repository. The elements are as follows:

a. MAINTENANCE ORDER NUMBER.
b. HARDWARE INVENTORY NUMBER.
c. MAINTENANCE DATE.
d. TYPE OF MAINTENANCE.
e. COST OF MAINTENANCE.
f. MAINTENANCE COVERED BY WARRANTY.

E-7. Describe the VENDOR entity. The elements are as follows:

a. VENDOR NUMBER.
b. VENDOR NAME.
c. STREET.
d. CITY.
e. STATE.
f. ZIP CODE.
g. TELEPHONE NUMBER.
h. DATE LAST ORDER SENT.
i. TOTAL AMOUNT PURCHASED FROM VENDOR.
j. TOTAL NUMBER OF ORDERS SENT TO VENDOR.

E-8. Produce the following reports using Visible Analyst:

a. Open the Microcomputer System entity-relationship diagram; then syntax check the diagram (Diagram/Analyze/Syntax Check).
b. Run the Normalization analysis report for the Microcomputer System entity-relationship diagram (Diagram/Analyze/Normalization).
c. The Key Analysis Report.
d. The Key Synchronization Report. What has changed in the composition field for each of the entities?
e. The Model Balancing Report.
f. The Entities versus Data Stores Matrix.
g. The Composition Matrix.
h. The Diagram Location Matrix.

E-9. Explain in a paragraph why you think a table would be acceptable on a record in an indexed file system written in COBOL.

E-10. Explain in a paragraph the relationship between a foreign key and a primary key and why it is necessary to have them on separate entities when there is a relationship between the entities.

*Exercises preceded by a CD-ROM icon require the program Visible Analyst or another CASE tool. A CD-ROM containing Visible Analyst examples is provided free of charge to any professor adopting this book. The examples on the disk may be imported into Visible Analyst and then used by students.

DESIGNING USER INTERFACES

18

USER INTERFACE OBJECTIVES

The interface *is* the system for most users. However well or poorly designed, it stands as the representation of the system and, by reflection, your competence as a systems analyst.

Your goal must be to design interfaces that help users and businesses get the information they need in and out of the system by addressing the following objectives:

1. Matching the user interface to the task.
2. Making the user interface efficient.
3. Providing appropriate feedback to users.
4. Generating usable queries.
5. Improving productivity of knowledge workers.

With these goals in mind, we move to more detailed discussions of how each of the objectives, as shown in Figure 18.1, can be met.

TYPES OF USER INTERFACE

In this section, several different kinds of user interface are described, including natural-language interfaces, question-and-answer interfaces, menus, form-fill interfaces, command-language interfaces, graphical user interfaces (GUIs), and a variety of Web interfaces for use on the Internet. The user interface has two main components: presentation language, which is the computer-to-human part of the transaction, and action language, which characterizes the human-to-computer portion. Together, both concepts cover the form and content of the term *user interface*.

NATURAL-LANGUAGE INTERFACES

Natural-language interfaces are perhaps the dream and ideal of inexperienced users, because they permit users to interact with the computer in their everyday or "natural" language. No special skills are required of the user, who interfaces with the computer using natural language.

The screen depicted in Figure 18.2 lists three natural-language questions from three different applications. Notice that interaction with each seems very easy. For instance, the first sentence—which says, "List all the salespeople who met their quotas this month"—seems straightforward.

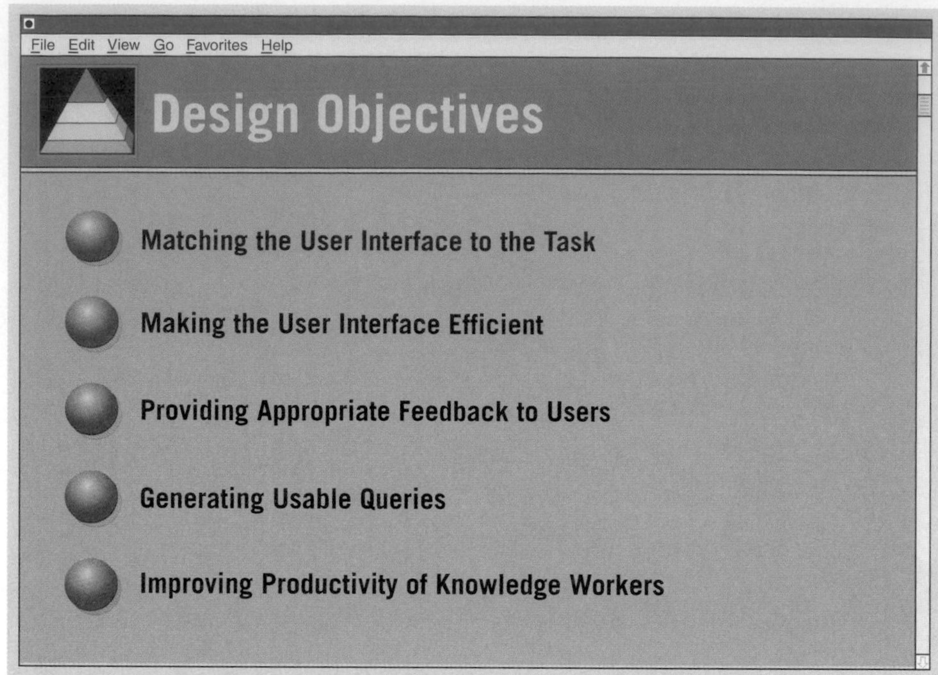

FIGURE 18.1
The four user interface objectives.

The subtleties and irregularities residing in the ambiguities of English produce an extremely exacting and complex programming problem. Attempts at natural-language interfacing for particular applications where any other type of interface is nonfeasible (say in the case of a user who is disabled) are meeting with some success, however, although these interfaces are typically expensive. Implementation problems and extraordinary demand on computing resources have so far kept natural-language interfaces to a minimum. The demand exists, though, and so many programmers and researchers are working diligently on natural-language

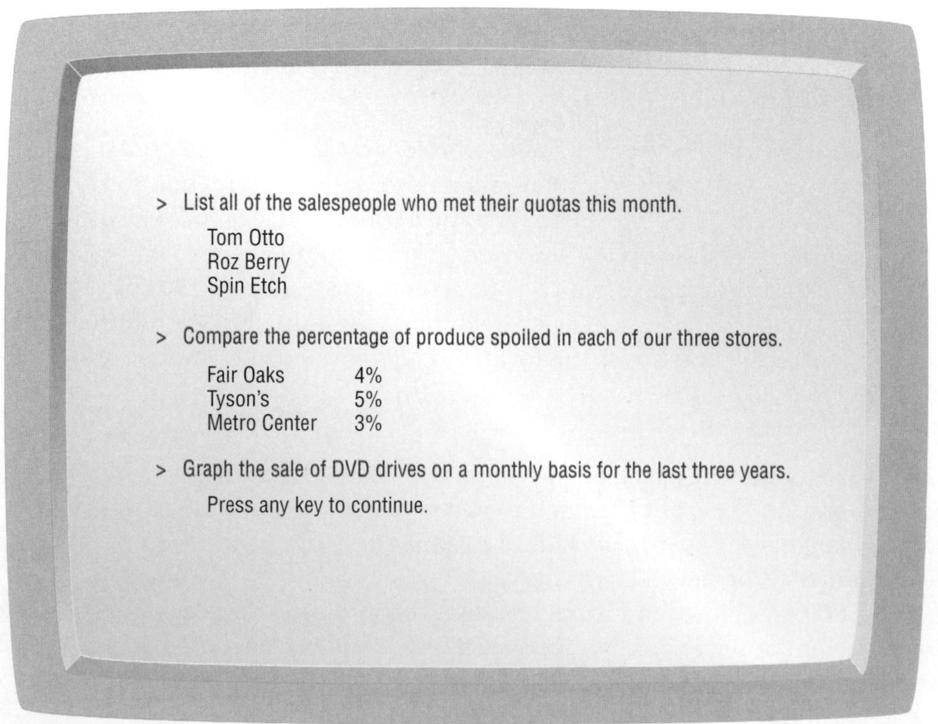

> List all of the salespeople who met their quotas this month.
Tom Otto
Roz Berry
Spin Etch
> Compare the percentage of produce spoiled in each of our three stores.
Fair Oaks 4%
Tyson's 5%
Metro Center 3%
> Graph the sale of DVD drives on a monthly basis for the last three years.
Press any key to continue.

FIGURE 18.2
Natural-language interfaces.

interfaces. It is a growth area, and it therefore merits your continued monitoring. Some Web sites, such as Ask Jeeves (www.askjeeves.com), use a natural interface for users to enter their search query. When the query is entered, Ask Jeeves responds with a list of queries that match the question entered by the user.

QUESTION-AND-ANSWER INTERFACES

In a question-and-answer interface, the computer displays a question to the user on the screen. To interact, the user enters an answer (via a keyboard stroke or a mouse click), and the computer then acts on that input information in a preprogrammed manner, typically by moving to the next question.

Many management science applications use a question-and-answer interface. A goal programming example is shown in Figure 18.3. As with most question-and-answer interfaces, the computer system directs the questioning sequence. The user responds to what is asked. For instance, in responding to the phrase "Now enter the number of constraints," the user types in an appropriate number.

Another type of question-and-answer interface, called a dialog box, is shown in Figure 18.4. A dialog box acts as a question-and-answer interface within another application, in this case a PERT chart for a systems analysis project for the Bakerloo Brothers. Notice that the rectangle for "Yes" is highlighted, indicating that it is the most likely answer for this situation. The main interface for this application need not necessarily be question and answer. Rather, by incorporating a dialog box, the programmer has included an easy-to-use interface within a more complicated one.

Programmers attempt to phrase questions for display in a question-and-answer interface in a concise and understandable manner, but they also need to anticipate the kinds of answers that both the user will input and the system will accept. Greater latitude for user response translates directly into an increase in the complexity of the programming required.

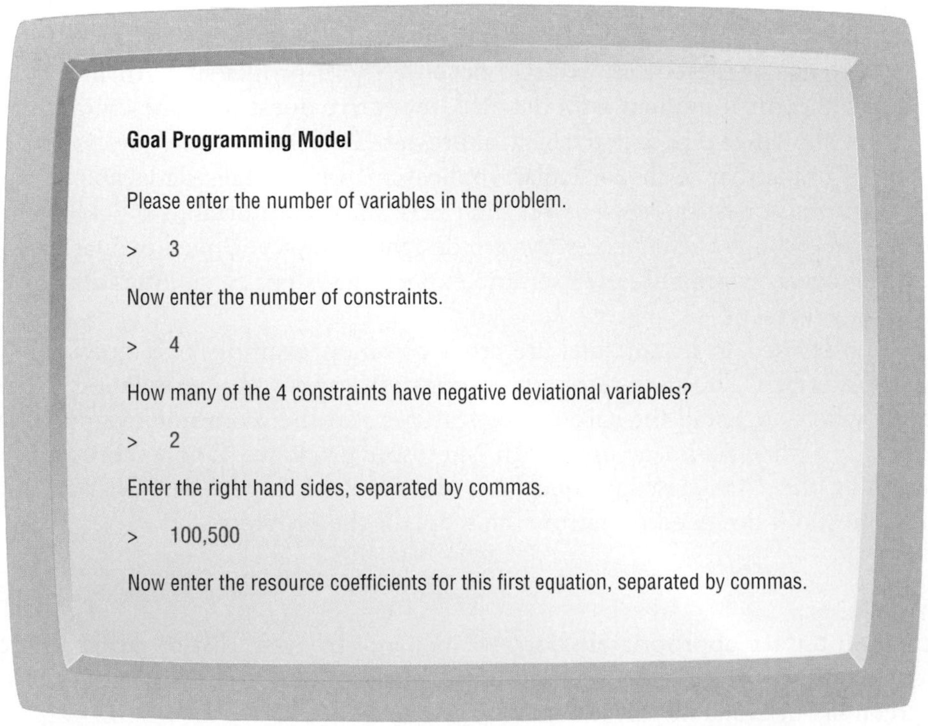

Goal Programming Model

Please enter the number of variables in the problem.

> 3

Now enter the number of constraints.

> 4

How many of the 4 constraints have negative deviational variables?

> 2

Enter the right hand sides, separated by commas.

> 100,500

Now enter the resource coefficients for this first equation, separated by commas.

FIGURE 18.3
Question-and-answer interface.

FIGURE 18.4

A dialog box: one type of question-and-answer interface.

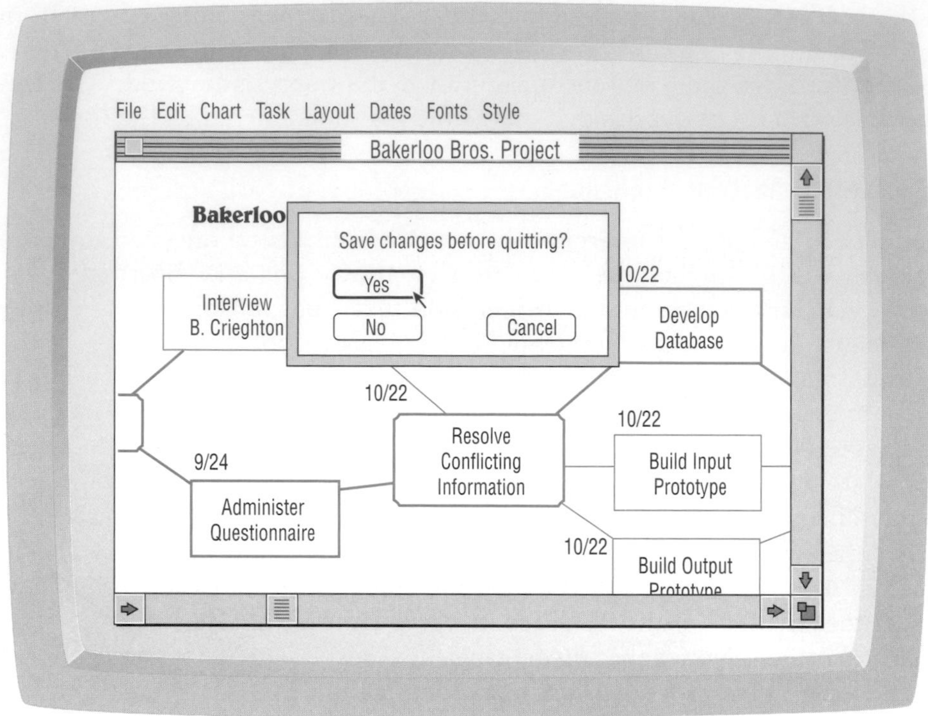

When interfaces are designed, a decision is made on how much flexibility to allow the user in responding to questions. Users require instruction about how much flexibility they are being afforded. For instance, users must know if typing "Y" or clicking on a "Y" in a highlighted box is an acceptable replacement for typing "Yes" as a response.

It is possible to include additional help or prompting to remind the user of acceptable responses, and many programmers do. Help is important because if users expend too much additional effort in searching for responses or remembering how to respond, they could be dissatisfied and not want to use the system.

Experience suggests that as users become more proficient with the system, they may become impatient with detailed, repetitive questions. They might prefer an option allowing responses to short, abbreviated versions of the questions.

Users unfamiliar with particular applications or not knowledgeable about a topic may find question-and-answer interfaces most comfortable, quickly gaining confidence through their success. System designers, however, need to guard against lulling an overly confident new user into expecting system capabilities above what are actually present.

Wizards used to install software are a common example of a question-and-answer interface. The user responds to questions about the installation process, such as where to install the software or features. Another common example is the use of the Office Assistant used with Microsoft products. When the user needs help, the Office Assistant asks questions and responds to the answers with additional questions designed to narrow the scope of the problem.

MENUS

A menu interface appropriately borrows its name from the list of dishes that can be selected in a restaurant. Similarly, a menu interface provides the user with an onscreen list of available selections.

In responding to the menu, a user is limited to the options displayed. The user need not know the system but *does* need to know what task should be accomplished. For example, with a typical word-processing menu, users can choose "edit," "copy," or "print" options. To utilize the menu best, however, users must know which task they desire to perform.

Menus as an interface are not hardware-dependent. Variations abound. Menus can be set up to use keyboard entry, lightpen, or mouse. Selections can be identified with a number, letter, or key word, or users can click on a selection with a mouse.

An example of a menu-driven program is shown in Figure 18.5. The user has a menu of seven different actions that may be taken on personnel files. To choose selection "5) Add an employee to this file," the user has only to type a 5. This keystroke takes the user to a new screen, which is ready for entry of new employee data.

Consistency is important in designing a menu interface. To access a menu selection, a user may be required to press the Enter key, or the computer may turn directly to the desired program when only a single key must be hit. For example, if the menu offers only the numbers 0 through 9, it is possible to use only one keystroke without hitting Enter. If more numbers are required, for example, Enter is needed to distinguish whether the user intends to enter 1, 10, or 100.

Menus can also be put aside until the user wants to employ them. Figure 18.6 shows how a pull-down menu is used while constructing a PERT diagram for a systems analysis project being completed for the Bakerloo Brothers. The user puts the pointer on the word "Dates" and pulls it down. Then the user puts the arrow on "Calendar," selecting the option to display the project on a conventional monthly calendar.

Menus can be nested within one another to lead a user through options in a program. Nested menus have some advantages. They allow the screen to appear less cluttered, which is consistent with good screen design. Nested menus also

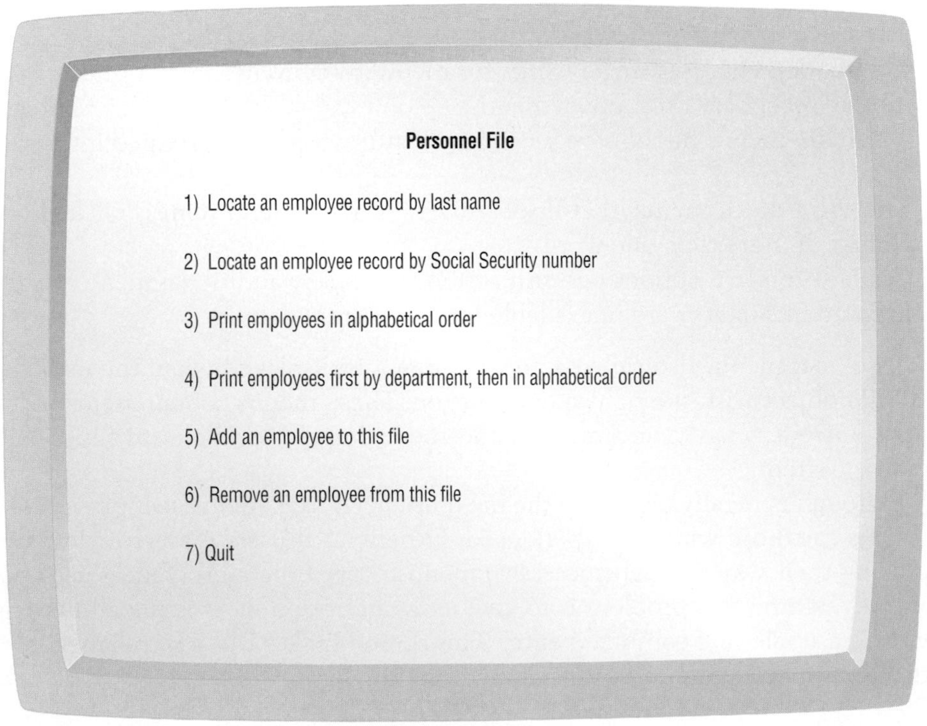

Personnel File
1) Locate an employee record by last name
2) Locate an employee record by Social Security number
3) Print employees in alphabetical order
4) Print employees first by department, then in alphabetical order
5) Add an employee to this file
6) Remove an employee from this file
7) Quit

FIGURE 18.5

A menu using numbers to select an action.

FIGURE 18.6
A pull-down menu is there when the user needs it.

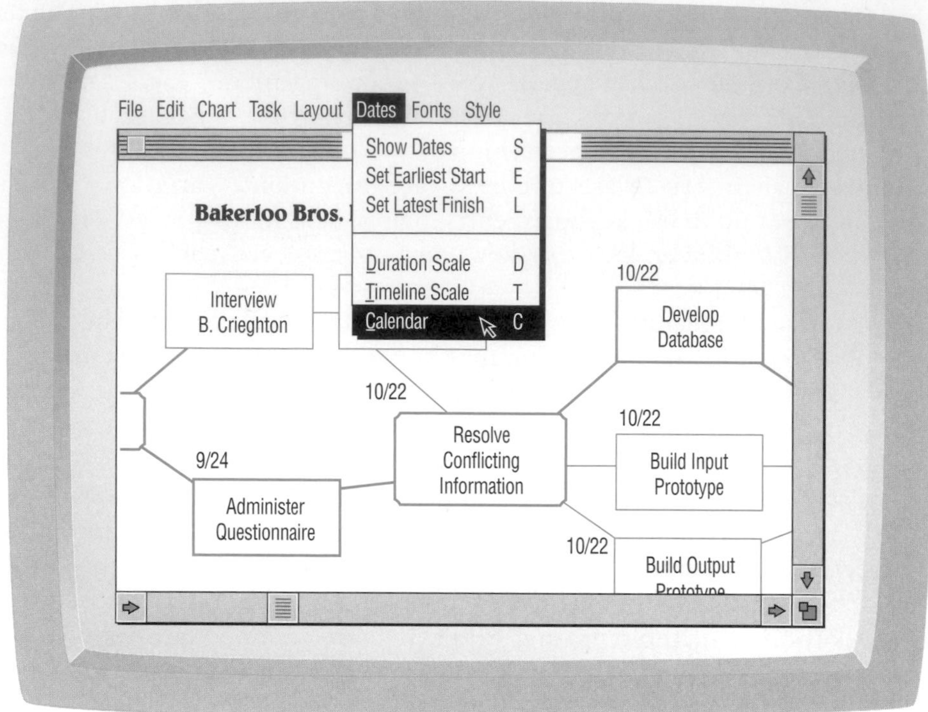

allow users to avoid seeing menu options in which they have no interest, so irrelevant user information is reduced. In addition, nested menus can move users quickly through the program.

When mainframe terminals are used, the menu is widely used to control the system. Pull-down menus and icons are generally used when the operator is using a PC.

GUI menus are used to control personal computer software and have the following guidelines:

1. The main menu bar is always on the screen.
2. The main menu uses single words for menu items. Main menu options always display secondary drop-down menus.
3. The main menu should have secondary features zoned or grouped into similar features.
4. The drop-down menus that display when a main menu item is clicked often consist of more than one word.
5. These secondary options perform actions or display additional menu items.
6. Menu items in grey are unavailable for the current activity.

An object menu, also called a pop-up menu, is displayed when the user clicks on a GUI object with the right mouse button. These menus contain items specific for the current activity, and most of the menu items are duplicate functions of main menu items.

Although generally a boon to the inexperienced user, menus can present some problems for those who are experienced. Experienced users may grow impatient at picking their way through successive menus every time a program is used. One way to overcome this problem is to give users an option of entering all necessary menus in a single-line command entry. This option makes the system amenable to both experienced and inexperienced users.

Source: © 2000 The New Yorker Collection from Cartoonbank.com. All Rights Reserved.

FORM-FILL INTERFACES (INPUT/OUTPUT FORMS)

Form-fill interfaces consist of onscreen forms or Web-based forms displaying fields containing data items or parameters that need to be communicated to the user. The form often is a facsimile of the paper form already familiar to the user. This interface technique is also known as a form-based method and input/output forms.

Figure 18.7 shows a form fill interface from FormFlow 99 by JetForm. A pull-down menu for Part No. automatically enters a Description for the item and Unit Price for Part No. When the user tabs to the quantity field and enters the number of items being purchased, the software automatically calculates the Extended Price by multiplying quantity by unit price.

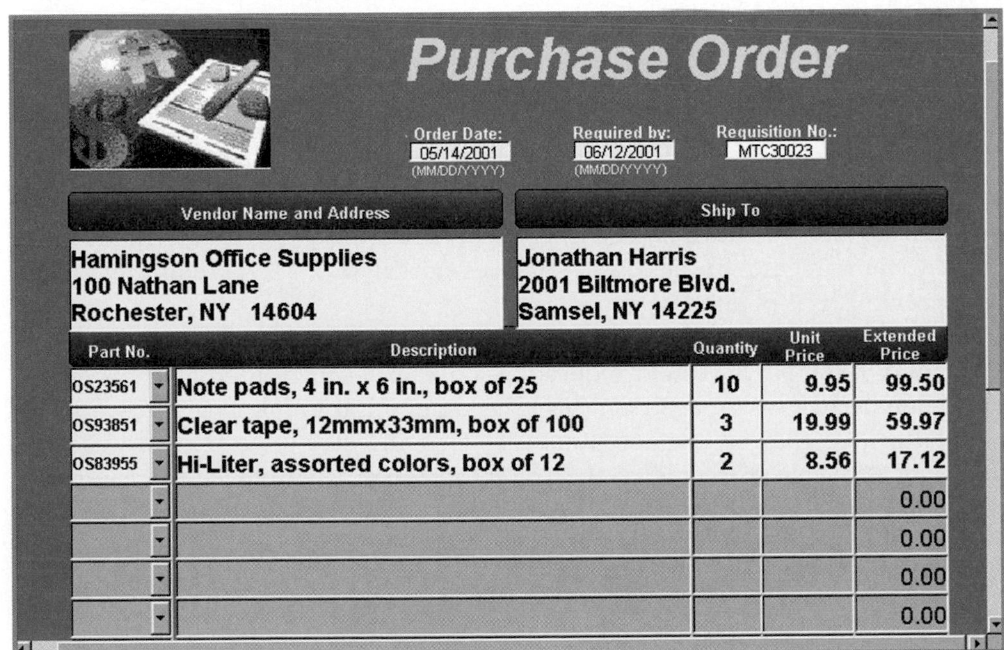

Purchase Order

Order Date: 05/14/2001 (MM/DD/YYYY)
Required by: 06/12/2001 (MM/DD/YYYY)
Requisition No.: MTC30023

Vendor Name and Address
Hamingson Office Supplies
100 Nathan Lane
Rochester, NY 14604

Ship To
Jonathan Harris
2001 Biltmore Blvd.
Samsel, NY 14225

Part No.	Description	Quantity	Unit Price	Extended Price
OS23561	Note pads, 4 in. x 6 in., box of 25	10	9.95	99.50
OS93851	Clear tape, 12mmx33mm, box of 100	3	19.99	59.97
OS83955	Hi-Liter, assorted colors, box of 12	2	8.56	17.12
				0.00
				0.00
				0.00
				0.00

FIGURE 18.7

An example of the form-fill interface from FormFlow by JetForm.

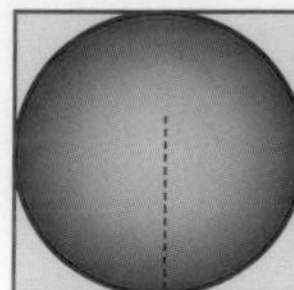

CONSULTING OPPORTUNITY 18.1

I'D RATHER DO IT MYSELF

"I can get Mickey to download any data I need from the Web or our server to my PC," DeWitt Miwaye, an upper-level manager for Yumtime Foods (a Midwest food wholesaler) tells you. "Getting data is no problem. What I don't want are a lot of reports. I'd rather play with the data myself."

Miwaye goes on to tell you that as an executive, he doesn't use his PC as often as he'd like, maybe only three times a month, but he has some very specific ideas about what he'd like to do with it.

"I'd like to be able to make some comparisons myself. I could compare the turnover rate for all 12 of our warehouses. I'd also like to see how effectively the capacity of each of our warehouses is being used. Sometimes I'd like to be able to graph the comparisons or see a chart of them over time."

In three paragraphs, compare three different types of interfaces that Miwaye could use. Then recommend one interface for his use that takes into account his infrequent use of the PC, his enjoyment in working with raw data, and his desire to see data displayed in a variety of ways.

Onscreen forms are set up to show what information should be input and where. Blank fields requiring information can be highlighted with inverse or flashing characters. The cursor is moved by the user from field to field by a single stroke of the arrow key, for instance. This arrangement allows movement one field backward or one field forward by hitting the arrow key. Web-based forms afford the opportunity to include hyperlinks to examples of correctly filled-out forms or to further help and examples.

Form input for screens can be simplified by supplying default values for fields and then allowing users to modify default information if necessary. For example, a database management system designed to show a form for inputting checks may supply the next sequential check number as a default when a new check form is exhibited. If checks are missing, the user changes the check number to reflect the actual check being input.

A powerful use of input/output forms is shown in Figure 18.8. In this example, the programmer has set up a form for users in the State Motor Vehicle Division. The form is more helpful than the typical input/output form, however, because it allows users who possess only a minimum of data to fill in the form partially. The computer will then take the data supplied (last name begins with McK, no valid license) and search the appropriate database to find information that completes the records already begun. The computer fills in the form with two names, McKinley and McKinnon. Notice that the computer does not pull up the record for McNeil, even though he fits the requirement of no valid driver's license, because McNeil does not fit the "McK" requirement the user previously stipulated.

Input for onscreen fields can be alphanumerically restricted so that users can enter only numbers in a field requesting a Social Security number or they can input only letters where a person's name is required. If numbers are input where only letters are allowed, the computer may alert the user that the field was filled out incorrectly. No matter what type of enforcement is employed to ensure that the form is filled out properly, users should be considered.

The chief advantage of the input/output form interface is that the printed version of the filled-in form provides excellent documentation. It shows field labels as well as the context for entries. Web-based documents can be sent directly to billing if a transaction is involved, or they can go directly to a consumer database if a survey is being submitted. Web-based forms push the responsibility for accuracy to

FIGURE 18.8

Input/output interface for locating records in a state motor vehicle database.

State Motor Vehicle Division

Name (Last)	Moving Violation	Parking Citation	Valid License	Insurance Confirmation
Mck.			No	

This query searches the database . . .

VEHICLE DATABASE

NAME (LAST)	MOVING VIOLATION	PARKING VIOLATION	VALID LICENSE	INSURANCE CONFIRMATION
McKenzie	Yes	No	Yes	Yes
McKibben	No	No	Yes	Yes
McKinley	Yes	No	No	Yes
McKinnon	No	No	No	Yes
McMaster	Yes	No	Yes	Yes
McMichael	Yes	Yes	Yes	Yes
McNeil	No	No	No	Yes

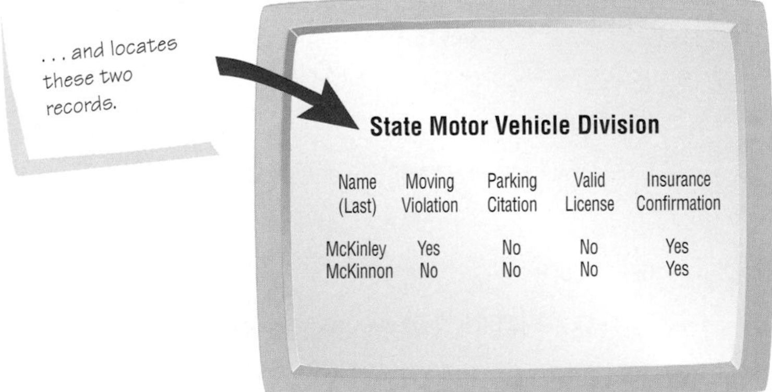

the user and make the form available for completion and submission on a 24-hour, 7 day a week, worldwide basis.

There are few disadvantages to input/output forms. The main drawback is similar to the "experienced user" problem discussed for question-and-answer interfaces and menus. Experienced users may become impatient with input/output forms and may want ways to enter data that are more efficient.

COMMAND-LANGUAGE INTERFACES

A command-language interface allows the user to control the application with a series of keystrokes, commands, phrases, or some sequence of these three methods. It is a popular interface that is more refined than those previously discussed.

Two application examples of command language are shown in Figure 18.9. The first example shows the user who asks to use a file containing data on all salespeople and then asks the computer to display all last names, then first names, for all salespeople whose current sales (CURSALES) are greater than their quota. In the second example, the user asks to use a file called GROCER. Then the user directs the computer to calculate the spoilage (SPOILS) by subtracting produce sold from produce bought. After that is done, the user asks to go back to the top of the file and to print out (LIST) the file.

Contrast the comprehension difficulty of the command language with that of the natural-language interface. The command language has no inherent meaning for the user, and that fact makes it quite dissimilar to the other interfaces discussed so far.

Command languages manipulate the computer as a tool by allowing the user to control the dialog. Therefore, command language affords the user more flexibility and control. When the user gives a command to the computer using command language, it is executed by the system immediately. Then the user may proceed to give it another command.

Command languages require memorization of syntax rules that may prove to be obstacles for inexperienced users (anyone can learn them with practice, however). Other interfaces resemble human exchanges more closely and are therefore more easily understood by users unaware of how a computer functions. Experienced users tend to prefer command languages, possibly because of the faster completion time they allow.

FIGURE 18.9
Command-language interfaces.

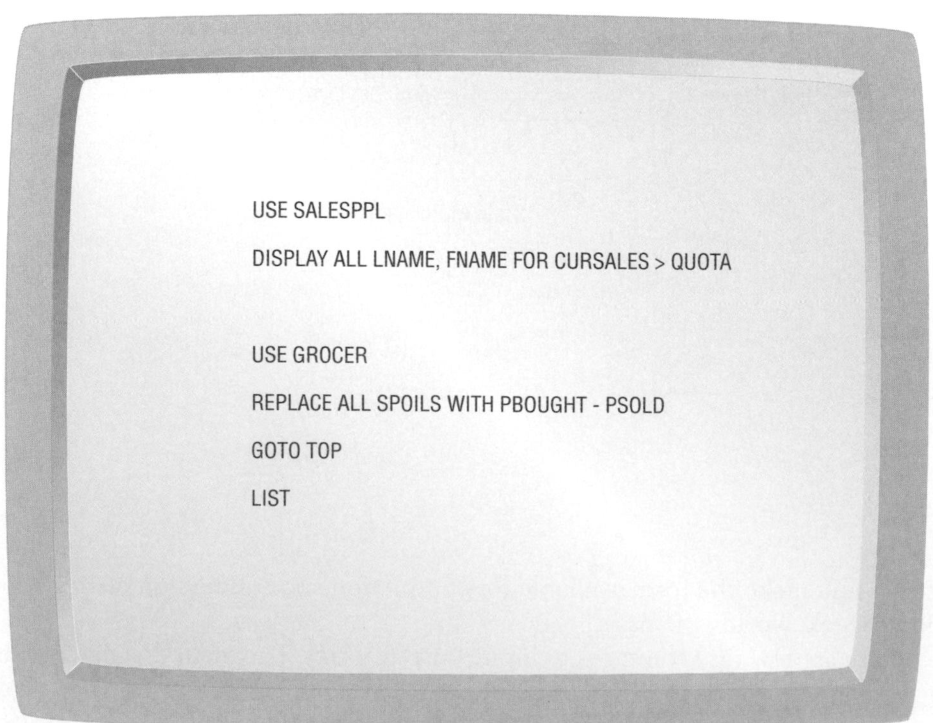

USE SALESPPL
DISPLAY ALL LNAME, FNAME FOR CURSALES > QUOTA
USE GROCER
REPLACE ALL SPOILS WITH PBOUGHT - PSOLD
GOTO TOP
LIST

GRAPHICAL USER INTERFACES

Graphical user interfaces (GUIs) allow direct manipulation of the graphical representation on the screen, which can be accomplished with keyboard input, a joystick, or a mouse. Direct manipulation requires more system sophistication than the interfaces discussed previously.

An example of the drag-and-drop feature of a GUI (sometimes simply called a drag-and-drop interface) is shown in Figure 18.10. In this screen from Microsoft Visio Professional, the symbols arrayed on the left are called templates. To draw with this interface, users drag and drop a master shape into their drawing. By creating entity-relationship diagrams or other specialized drawings, this kind of interface makes the work of the analyst much easier. Visio comes with many more stencils for drawing data flow diagrams, program flowcharts, network diagrams, and object-oriented diagrams, and it, as shown in the figure, has templates for designing office spaces.

The key to GUI is the constant feedback on task accomplishment that it provides. Continuous feedback on the manipulated object means that changes or reversals in operations can be made quickly, without incurring error messages. The concept of feedback for users is discussed thoroughly in an upcoming section, "Feedback for Users."

The creation of GUIs poses a challenge, because an appropriate model of reality or an acceptable conceptual model of the representation must be invented. To do so requires combining several skills in a way that stretches the capabilities of most systems analysts and programmers. Designing GUIs for use on intranets, extranets, and, even more pressingly, on the Web requires even more careful planning (see Chapter 16 on Web-site design). By and large, the users of Web sites are unknown to the developer, so a design must be clear-cut. Choice of icons, language, and hyperlinks becomes an entire set of decisions and assumptions about what kinds of users we are hoping to attract. In fact, the Web is creating change so

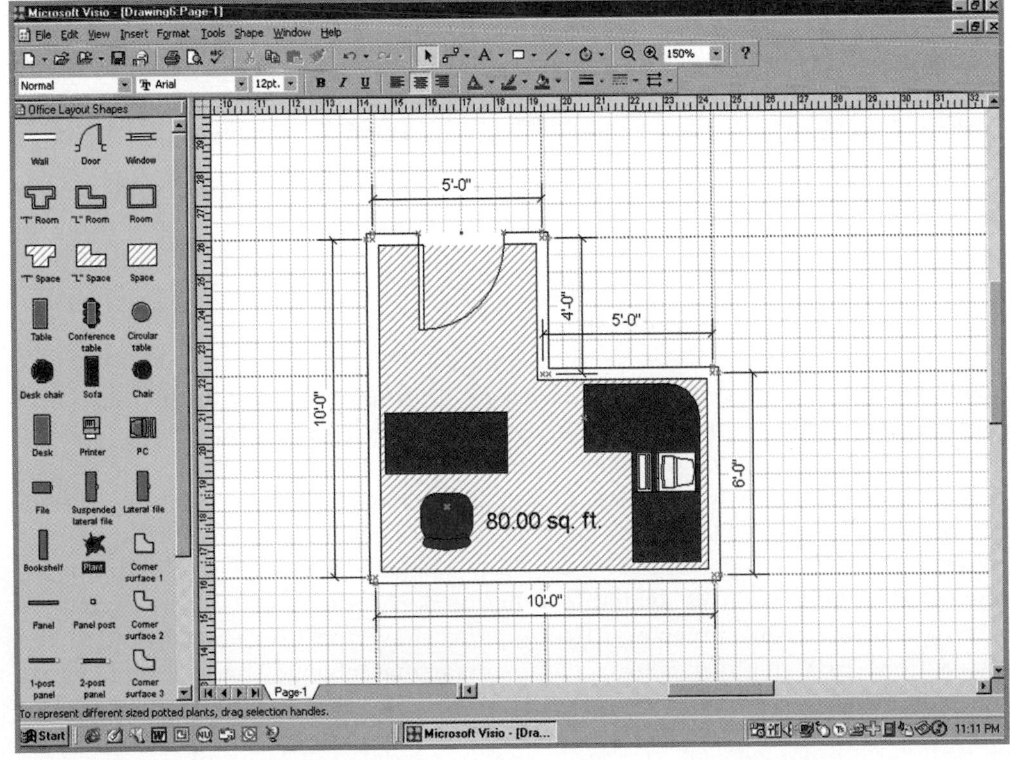

FIGURE 18.10
One common feature of GUI interfaces is "drag-and-drop." This office layout diagram is being created by dragging template symbols onto the screen using Microsoft Visio Professional.

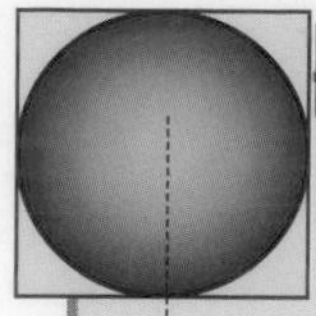

CONSULTING OPPORTUNITY 18.2

DON'T SLOW ME DOWN

"I've seen 'em all," Carrie Moore tells you. "I was here when they got their first computer. I guess I've sort of made a career of this," she says cheerfully, pointing to the large stack of medical insurance claim forms she has been entering into the computer system. As a systems analyst, you are interviewing Carrie, a data-entry operator for HealthPlus (a large, medical insurance company), about changes being contemplated in the computer system.

"I'm really fast compared with the others," she states as she nods toward the six other operators in the room. "I know, because we have little contests all the time to see who's the fastest, with the fewest errors. See that chart on the wall? That shows how much we enter and how quickly. The gold stars show who's the best each week."

"I don't really mind if you change computers. Like I say, 'I've seen 'em all.' " She resumes typing on her keyboard as she continues the interview. "Whatever you do, though, don't slow me down. One of the things I'm most proud of is that I can still beat the other operators. They're good too, though," Carrie adds.

Based on this partial interview with Carrie Moore, what type of user interface will you design for her and the other operators? Assume the new system will still require massive amounts of data entry from a variety of medical insurance forms sent in by claimants.

Compare and contrast interfaces such as natural language, question and answer, menus, input/output forms, and Web-based form-fill documents. Then choose and defend one alternative. What qualities possessed by Carrie and the other operators—and the data they will be entering—shaped your choice? Make a list of them. Is there more than one feasible choice? Why or why not? Respond in a paragraph.

rapidly, that Strom (1997, p. 25) quotes one software developer as saying, "Application developers who aren't using the Web as their primary end-user interface will be in the same position as scribes after the invention of the printing press." This thought is definitely something to reflect on as you go about conceptualizing new projects.

DIALOG AND DESKTOPS

GUIDELINES FOR DIALOG DESIGN

Dialog is the communication between the computer and a person. Well-designed dialog makes it easier for people to use a computer and leads to less frustration with the computer system. There are several key points for designing good dialog. Some of them were mentioned in Chapter 16. They include the following:

1. Meaningful communication, so that the computer understands what people are entering and people understand what the computer is presenting or requesting.
2. Minimal user action.
3. Standard operation and consistency.

Communication. The system should present information clearly to the user. Clear presentation means having an appropriate title on each screen, minimizing the use of abbreviations, and providing clear user feedback. Inquiry programs should display code meanings as well as data in an edited format, such as displaying slashes between the month, day, and year in a date field or commas and decimal points in an amount field. User instructions should be supplied (usually on one line only) regarding details, such as available function key assignments. Displaying the instruction line using inverse video helps to draw the user's attention to the instructions. In a graphical interface, the cursor may change shape depending on the work being performed.

Users with less skill require a greater amount of communication. Internet Web sites must display more text and instructions to guide the user through the site.

Intranet sites may have less dialog, because there is a measure of control over how well trained users are. Internet graphics should have pop-up text descriptions when images are used as hyperlinks, because there may be uncertainty in interpreting their meaning, especially if the site is used internationally. Status-line information for GUI screens is another way of providing instructions for users.

Easy-to-use help screens should be provided. Many PC help screens have additional topics that may be directly selected using highlighted text displayed on the first help screen. These hyperlinks are usually in a different color, which makes them stand out in contrast to the rest of the help text. They are usually selected using a mouse. Many of the newer GUIs also incorporate tool-tip help, displaying a small help message identifying the function of a command button when the cursor is placed over it.

The other side of communication is that the computer should "understand what the user has entered." Hence, all data entered on the screen should be edited for validity.

Minimal User Action. Keying is often the slowest part of a computer system, and good dialog will minimize the number of keystrokes required. You can accomplish this goal in a number of different ways:

1. Keying codes instead of whole words on entry screens. Codes are also keyed when using a command-language interface. An example is entering a two-letter state postal abbreviation. On a GUI screen, the codes may be entered by selecting them from a pull-down list of available codes. To enhance communication, the pull-down lists should contain the names of the codes, even though codes are stored on the files.
2. Only entering data that are not already stored on files. For example, when changing or deleting item records, only the item number should be entered. The computer responds by displaying descriptive information that is currently stored on the item file. When entering an order for customers, the customer number is entered and the name and address is displayed, allowing the operator to sight-verify that the customer number has been entered correctly.
3. Supplying the editing characters (for example, slashes as date field separators) by means of software for personal computers. Users should not have to enter formatting characters such as leading zeros, commas, or a decimal point when entering a dollar amount; nor should they have to enter slashes or hyphens when entering a date. (On a mainframe terminal, these characters are omitted.)
4. Using default values for fields on entry screens. Defaults are used when a user enters the same value in a screen field for the majority of the records being processed. The values are displayed, and the user may press the Enter key to accept the default or overtype the default value with a new one. For example, an inventory control clerk may use a screen to generate purchase orders for items that are low in stock. The screen would display the reorder quantity from the item record as a default purchase quantity. Under typical circumstances, the user would simply press Enter to accept this quantity, but if the item were a popular sale item, the operator might choose to enter a higher amount to increase the quantity purchased.

   If the software must be flexible enough to accommodate different user situations, it should be designed so that the preset defaults may be changed. The new defaults should display each time the software is invoked. An example is changing the screen colors, which should not have to be manually reset each time the screen is used.

GUIs may contain check boxes and radio buttons that are selected when a dialog box opens. Provide context sensitive menus that appear when an object is clicked with the right mouse button. These menus contain options specific for the object under the mouse.

5. Designing an inquiry (or change or delete) program so that the user needs to enter only the first few characters of a name or item description. The program displays a list of all matching names, and when the operator chooses one, the matching record is displayed.
6. Providing keystrokes for selecting pull-down menu options. Often, these options are selected using a mouse, followed by some keying. Users must move their hands from the keyboard to the mouse and back to the keyboard. As users become familiar with the system, keystrokes often provide a faster method for manipulating the pull-down menus, because both hands remain on the keyboard. On a PC, keystrokes usually involve pressing a function key or the Alt key followed by a letter. Figure 18.11 is an example of nested pull-down menus with shortcut keys from Microsoft Visio Professional. Notice that the user, who is creating a structure chart, can get into a series of ever more specific menus. At first, the user highlights the word "Stencils"; then the next menu unfolds and the user chooses "Network Diagram." After that, the user chooses "Network Devices" on the next level of menu to use this symbol on their chart.

Any combination of these six approaches can help the analyst decrease the number of keystrokes required by the user, thereby speeding up data entry and minimizing errors.

Standard Operation and Consistency. The system should be consistent throughout its set of different screens and in the mechanisms for controlling the operation of the screens throughout different applications. Consistency makes it easier for the

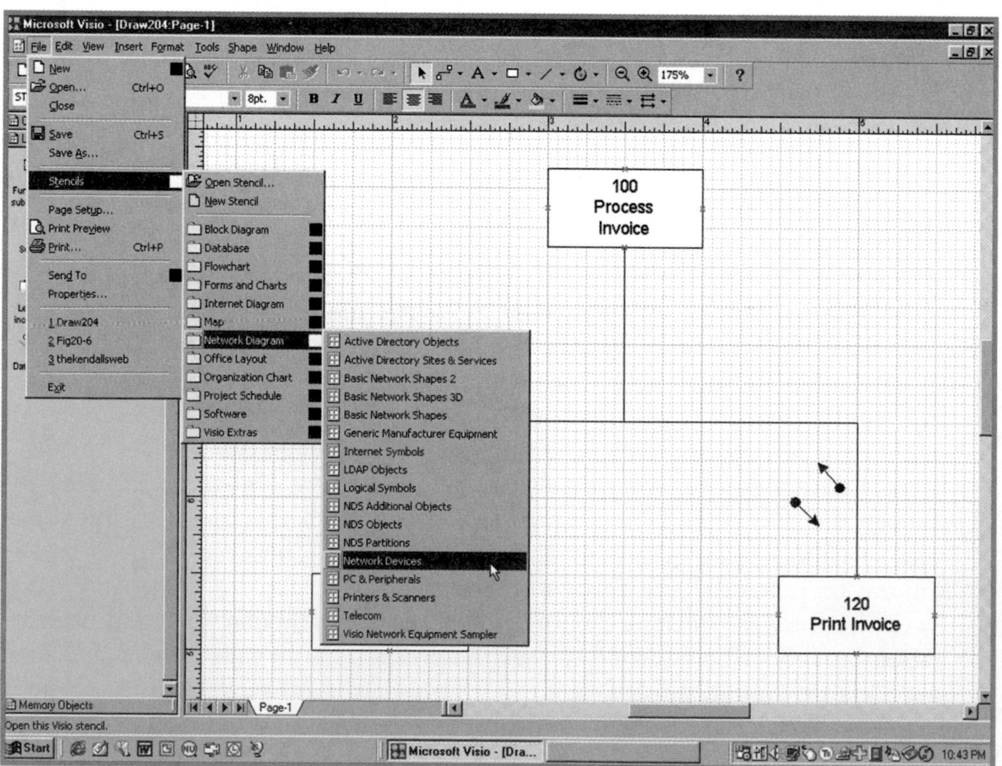

FIGURE 18.11
Example of nested pull-down menus with shortcut keys from Microsoft Visio Professional.

users to learn how to use new portions of the system once they are familiar with one component. You can achieve consistency by the following:

1. Locating titles, date, time, and operator and feedback messages in the same places on all screens.
2. Exiting each program by the same key or menu option. It would be a poor design that used function key 4 (F4) to exit the ADD CUSTOMER program and function key 6 (F6) to exit the CHANGE CUSTOMER program.
3. Canceling a transaction in a consistent way, usually by using a function key (usually F12) on a mainframe and the Escape key on a PC.
4. Obtaining help in a standardized way. The industry standard for help is function key 1 (F1), and most microcomputer software developers are adopting this convention.
5. Standardizing the color used for all screens. Error messages are typically displayed in red. Remember to keep the background screen color the same for all applications.
6. Standardizing the use of icons for similar operations when using a graphical user interface. For example, a small piece of paper with a bent upper corner often represents a document.
7. Consistent use of terminology within a screen or Web site.
8. Providing a consistent way to navigate through the dialog. For example, find a consistent way to add records or to work with a Web site, such as using the same buttons for "Back" and "Next."
9. Consistent font alignment, size, and color on a Web page.

An example of good GUI design is the tab control dialog box shown in Figure 18.12. Currently, the user is choosing HP Laser Jet print options, and he or she is in the "Paper" tab, but he or she also has the choice of six other tabs, including "Fonts" and "Graphics." This screen shows selections that a user has made by clicking on the left or right arrows on the horizontal sliding bar that runs beneath the envelope sizes,

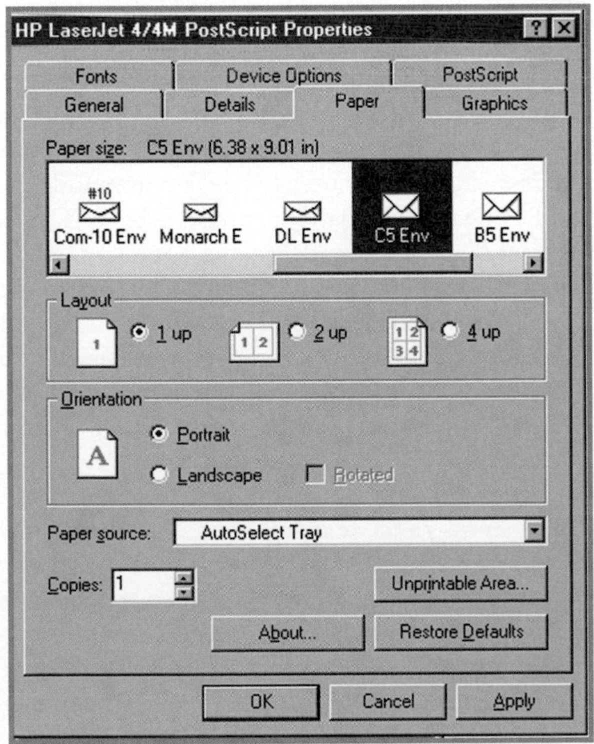

FIGURE 18.12

This tab control dialog box has seven tabs. The chosen tab "Paper" appears as if it is in front of the other tabs.

"Com-10 Env, Monarch E, DL Env, C5 Env," and so on. The darkened area indicates that the user has chosen to print a C5 envelope. Notice that the designer of this interface has used option buttons for both "Layout" and "Orientation." The user has clicked on a choice of "Portrait" for orientation. A drop-down menu is also used to select the paper source. In this instance, the user has chosen "AutoSelect Tray." The designer has also used push buttons at the very bottom of the screen that allow users to enter "OK," "Cancel," or "Apply" in regard to the options they have just chosen.

OTHER USER INTERFACES

Other user interfaces, although less common than those discussed previously, are growing in popularity. These interfaces include pointing devices such as the stylus, touch-sensitive screens, and speech recognition and synthesis. Each of these interfaces has its own special attributes that uniquely suit it to particular applications.

The stylus (a pointed stick that resembles a pen) is becoming popular because of new handwriting recognition software and personal digital assistants (PDAs). The Palm device has been a success because it does a limited number of things very well and sells for a low price. It includes a calendar, address book, to-do list, and memo pad. Data entry is also facilitated with a docking cradle so that you can synchronize data with your PC.

Touch-sensitive screens allow a user to use a finger (or object) to activate the screen when it comes close to the screen surface. Coming close to the screen breaks a grid of light beams within or just over the screen's surface. Touch-sensitive screens are useful in public information displays such as maps of cities and their sights, which are posted in hotel lobbies. They can also be used to explain dioramas in museums and to locate camping facilities in state parks. Touch-sensitive screens require no special expertise from users, and the screen is self-contained, requiring no special input device that might be broken or stolen.

Voice recognition has long been the dream of scientists and science fiction writers alike. It is intuitively appealing, because it seems to approximate human communication. With voice recognition, the user speaks to the computer, and the system is able to recognize an individual's vocal signals, convert them, and store the input. Voice recognition inventory systems are already in operation.

An advantage of voice recognition systems is that their use can speed data entry enormously, while freeing the user's hands for other tasks. Speech input adds still another dimension to the PC. It is now possible to add equipment and software that allows a personal computer user to speak commands such as "Open File" or "Save File" to avoid using the keyboard or mouse. The obvious advantages of this technology are increased accuracy and greater speed than what conventional mouse movements afford as well as the avoidance of repetitive stress injuries such as carpal tunnel syndrome, which can debilitate the wrist and hand.

The two main developments in speech recognition are (1) continuous speech systems, which allow for the input of regular text in word processors, and (2) speaker independence so that any number of people can enter commands or words at a given workstation.

Lernout and Hauspie's Dragon Naturally Speaking products include dictation systems, command systems, and text to speech systems. Dragon Naturally Speaking was the first large vocabulary continuous speech and recognition product for the PC. It is now available in a network version so that speech recognition can be shared throughout the organization. (The current version has an active vocabulary starting at 62,000 words and expanding up to 230,000 words.) The vocabulary is not merely a spelling list, but includes speaker-independent acoustic and language usage information, which means more accurate recognition.

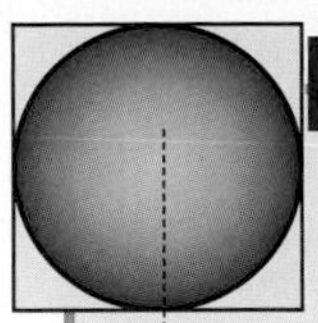

CONSULTING OPPORTUNITY 18.3

THAT'S NOT A LIGHTBULB

From your preliminary analysis, it appears that many reductions in error will be realized if sales clerks at Bright's Electric (which sells electrical parts, bulbs, and fixtures to wholesale customers) adopt an online system. The new system would allow sales clerks to withdraw a part from inventory (and thereby update inventory), return a part to inventory, check on the inventory status, and check on whether a part is back-ordered. Currently, to update inventory, sales clerks fill out a three-part form by hand. The customer gets one, inventory keeps one, and at the end of the day the originals are deposited in the front office.

The next morning, the first thing the lone office worker does is enter the data from the forms into the computer. Errors occur when she enters the wrong part numbers or quantities. Additional time is consumed when inventory workers hunt for a part they think might be in stock but that is not. Updated inventory sheets are available to the sales clerks around noon, but by that time they have already taken from inventory twice the number of parts that will be taken out after noon. Clearly, a well-designed online system would help reduce these errors and would also help with inventory control.

The owner, Mr. Bright, has entertained the idea of an online system and dropped it several times over the last five years. The chief reason is that the sales clerks, who would be the heaviest users of the system, do not think the systems analysts they've talked to can fulfill their needs.

M. T. Sockette, the sales clerk who has been with Bright's the longest, is the most vocal, telling you, "We know the parts, we know our customers. What we could do with a computer here would be great. The guys they've brought in here to get it going, though. . . . I mean, they say things like, 'You can step right up and type one 60-watt General Electric lightbulb into the computer.' "

"To us, that's not a lightbulb, it's a GE60WSB. All of us know the part numbers here. We pride ourselves on it. Typing in all that junk will take all day."

After talking to Mr. Bright, you decide to implement an online system. You have talked to M. T. and the others and reassured them that the system will use the part numbers they're familiar with and will save them time. Although skeptical, they have been persuaded by you to give it a try.

What type of user interface will you design for the sales clerks? Before you come to your solution, do a careful analysis in three paragraphs that compares and contrasts various user interfaces—natural language, question-and-answer, menus, input/output forms, command language, and Web-based form-fill documents—for their suitability as at Bright's. Then choose one interface and explain in a paragraph why you find this one the most appropriate based on what you know about Bright's sales clerks and their current system. Draw a prototype of a screen that will be part of your solution. Describe in a paragraph how you will test it with the sales clerks.

A user can say a command to the computer and it will be executed. In the example shown in Figure 18.13, the user corrects a word by pulling down a menu of alternative words that sound the same.

When evaluating the interfaces you have chosen, there are some standards to keep in mind:

1. The necessary training period for users should be acceptably short.
2. Users early in their training should be able to enter commands without thinking about them or without referring to a help menu or manual. Keeping interfaces consistent throughout applications can help in this regard.
3. The interface should be "seamless" so that errors are few and those that do occur are not occurring because of poor design.
4. The time that users and the system need to bounce back from errors should be short.
5. Infrequent users should be able to relearn the system quickly.

Many different interfaces are available, and it is important to realize that an effective interface goes a long way toward successfully involving users. Users should want to use the system. In the next section, we discuss the importance of providing feedback for users to support and sustain their involvement with the system.

FIGURE 18.13

Using software such as Dragon Naturally Speaking by Lernout & Hauspie, a user can speak commands to their computer. In this example, the user corrects a word by pulling up a menu of alternative words that sound the same.

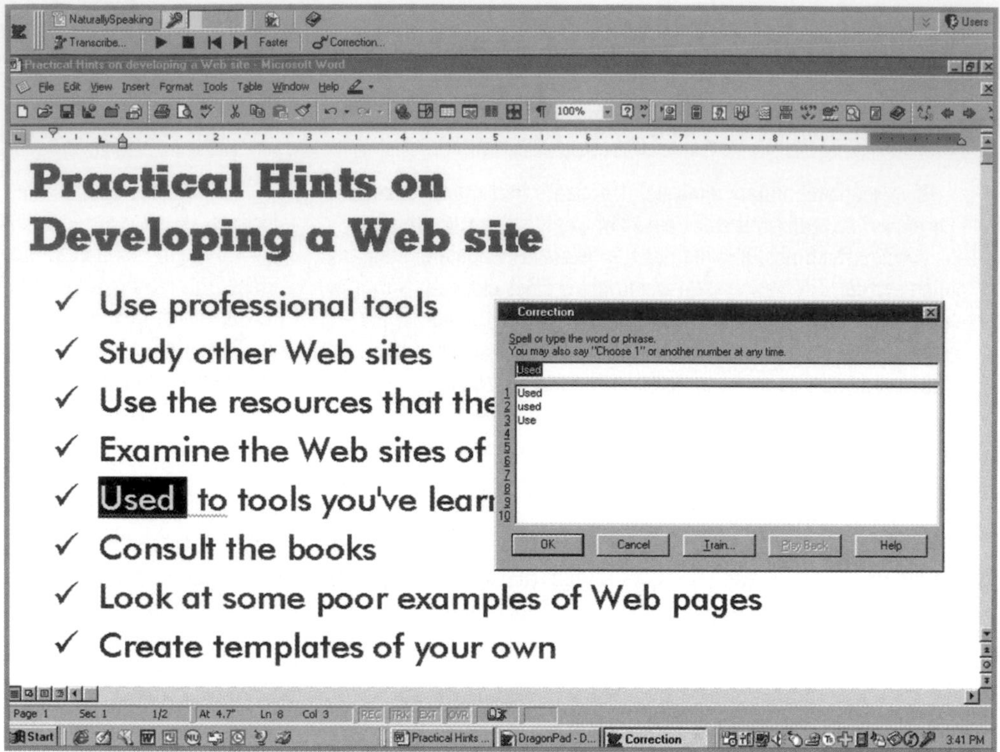

FEEDBACK FOR USERS

All systems require feedback, so as to monitor and change behavior, as discussed in Chapter 2. Feedback usually compares current behavior with predetermined goals and gives back information describing the gap between actual and intended performance.

Because humans themselves are complex systems, they require feedback from others to meet psychological needs. Feedback also increases human confidence. How much feedback is required is an individual characteristic.

When users interface with machines, they still need feedback about how their work is progressing. As designers of user interfaces, systems analysts need to be aware of the human need for feedback and build it into the system. In addition to text messages icons can often be used. For example, displaying an hourglass while the system is processing encourages the user to wait for some time rather than repeatedly hitting keys to try and invoke another screen or response.

Feedback to the user from the system is necessary in seven distinct situations, as shown in Figure 18.14. Feedback that is ill-timed or too plentiful is not helpful, because we can process only a limited amount of information. Each of the seven situations where feedback is appropriate is explained in the upcoming subsections. Web sites should display a status message or some other way of notifying the user that the site is responding and that input is either correct or in need of further information.

FIGURE 18.14

Feedback is used in many ways.

Feedback Is Needed to Tell the User That:

- The computer has accepted the input
- The input is in the correct form
- The input is not in the correct form
- There will be a delay in the processing
- The request has been completed
- The computer is unable to complete the request
- More detailed feedback is available (and how to get it)

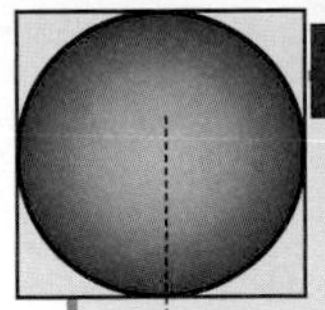

CONSULTING OPPORTUNITY 18.4

WAITING TO BE FED

"Yeah, we were sold a package all right. This one right here. Don't get me wrong, it gets the work done. We just don't know when."

You are talking with Owen Itt, who is telling you about the sales unit's recent purchase of new software for its PCs that allows input of sales data for each of its 16 salespeople, provides output comparison data for them, and projects future sales based on past sales records.

"We've had some odd experiences with this program, though," Owen continues. "It seems slow or something. For instance, we're never sure when it's done. I type in a command to get a file and nothing happens. About half a minute later, if I'm lucky, the screen I want might come up, but I'm never sure. If I ask it to save sales data, I just get a whirring sound. If it works, I'm returned to where I was before. If it doesn't save data, I'm still returned to where I was before. It's confusing, and I never know what to do. There's nothing on the screen that tells me what to do next. See the manual that came with it? It's dog-eared because we have to keep thumbing through trying to figure out what to do next."

Based on what you've heard in the interview, take this opportunity to supplement the program by designing some onscreen feedback for Owen and his sales team. The feedback should address all Owen's concerns, while following the guidelines for giving feedback to users and the guidelines for good screen design. Draw a prototype of the screen or screens you think are necessary to address the problems he lists above.

TYPES OF FEEDBACK

Acknowledging Acceptance of Input. The first situation in which users need feedback is to learn that the computer has accepted the input. For example, when a user enters a name on a line, the computer provides feedback to the user by advancing the cursor one character at a time when the letters are entered correctly.

Recognizing That Input Is in the Correct Form. Users need feedback to tell them that the input is in the correct form. For example, a user inputs a command, and the onscreen computer feedback is "READY" as the program progresses to a new point. A poor example of feedback that tells the user that input is in the correct form is the message, "INPUT OK," because that message takes extra space, is cryptic, and does nothing to encourage input of more data.

Notifying That Input Is Not in the Correct Form. Feedback is necessary to warn users that input is not in the correct form. When data are incorrect, one way to inform the user is to generate a window that briefly describes the problem with the input and tells how the user can correct it, as shown in Figure 18.15.

Notice that the message concerning an error in inputting the length of subscription is polite and concise but not cryptic so that even inexperienced users will be able to understand it. The subscription length entered is wrong, but the feedback given does not dwell on the user's mistake. Rather, it offers options (13, 26, or 52 weeks) so that the error can be corrected easily. On a GUI screen, feedback is often in the form of a message box with an **OK** button on it. Web messages are usually sent on a new page with the message on the side of the field containing the error. The new Web page may have a link for additional help.

So far, we have discussed visual feedback in text or iconic form, but many systems have audio feedback capabilities as well. When a user inputs data in the incorrect form, as shown previously for subscription length, the system might beep instead of providing a window. But audio feedback alone is not descriptive, so it is not as helpful to users as onscreen directions. Use audio feedback sparingly, perhaps to denote urgent situations. The same advice also applies to the design of Web sites, which may be viewed in an open office, where sounds carry and a coworker's workstation speakers are within earshot of several other people.

FIGURE 18.15

Feedback informs the user that input was not in the correct form and lists options.

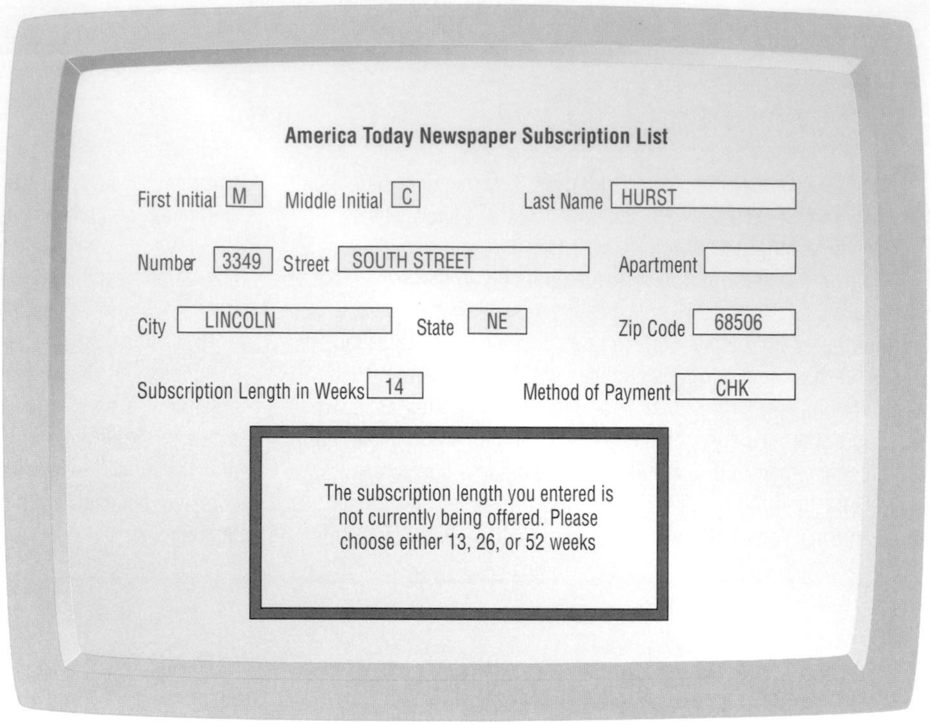

America Today Newspaper Subscription List

First Initial M Middle Initial C Last Name HURST

Number 3349 Street SOUTH STREET Apartment

City LINCOLN State NE Zip Code 68506

Subscription Length in Weeks 14 Method of Payment CHK

The subscription length you entered is not currently being offered. Please choose either 13, 26, or 52 weeks

Explaining a Delay in Processing. One of the most important kinds of feedback informs the user that there will be a delay in processing his or her request. Delays longer than 10 seconds or so require feedback so that the user knows the system is still working.

Figure 18.16 shows a screen providing feedback in a window for a user who has just requested a printout of the newspaper's subscription list. The screen displays a sentence reassuring the user that the request is being processed as well as a sign in the upper-right-hand corner instructing the user to "WAIT" until the current command has been executed. The screen also provides a way to stop the operation if necessary.

One PC system uses an icon for feeding back information on processing delays. The icon is in the familiar shape of a wristwatch and serves as visual reassurance that the system is working. Sometimes during delays, while new software is being installed, a short tutorial on the new application is run, which is meant to be more of a distraction than feedback about installation. Often, a list of files that are being copied and a status bar are used to reassure the user that the system is functioning properly. Web browsers usually display the Web pages that are being loaded and the time remaining.

Timing feedback of this sort is critical. Too slow a system response could cause the user to input commands that impede or disrupt processing, and too quick a response may make users feel as if the pace of their work is being propelled by the system.

Acknowledging That a Request Is Completed. Users need to know when their request has been completed and new requests may be input. Figure 18.17 shows a screen where the user has requested preparation of an index by name. When the indexing is finished, the system replies with "DONE." The user then knows that it is safe to proceed.

Often a specific feedback message is displayed when an action has been completed by a user, such as "EMPLOYEE RECORD HAS BEEN ADDED," "CUSTOMER RECORD HAS BEEN CHANGED," or "ITEM NUMBER 12345 HAS BEEN DELETED."

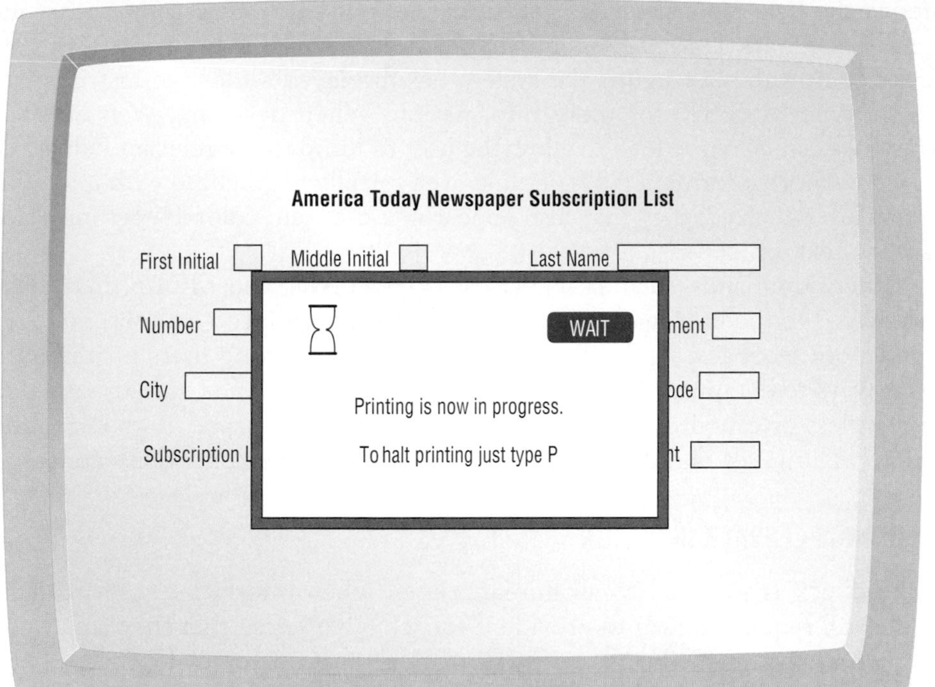

Printing is now in progress.
To halt printing just type P

FIGURE 18.16
Feedback tells the user that there will be a delay while printing is being done.

Notifying That a Request Was Not Completed. Feedback is also needed to let the user know that the computer is unable to complete a request. Notice that in the last figure the system has displayed the message "UNABLE TO PROCESS REQUEST. CHECK REQUEST AGAIN" in response to the user's request to locate subscribers by zip code. The user can then go back and check to see if the request has been input correctly rather than continue to enter commands that cannot be executed.

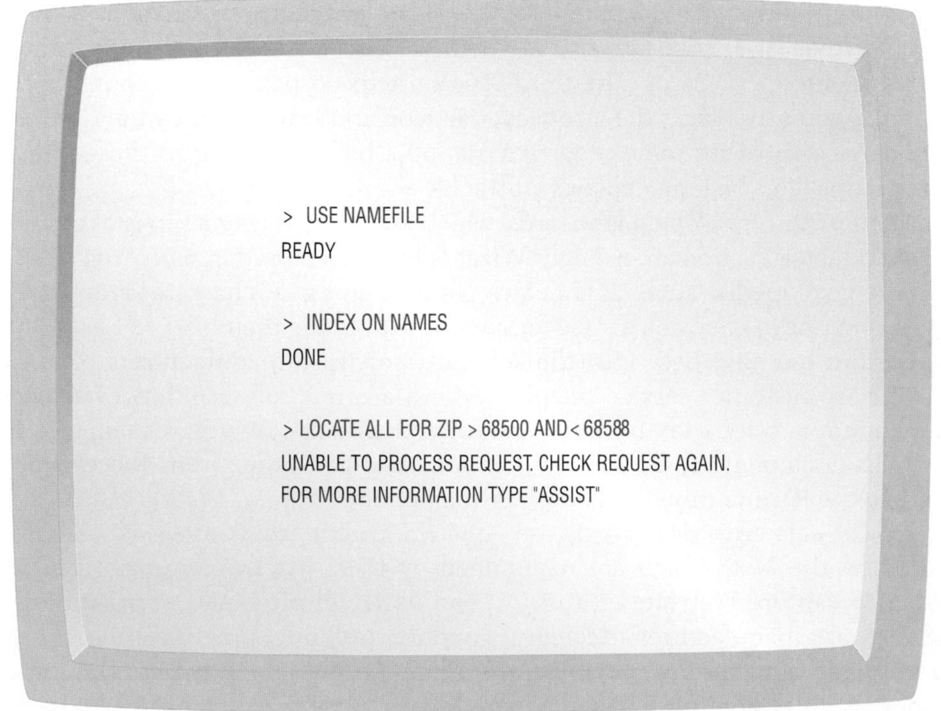

> USE NAMEFILE
READY
> INDEX ON NAMES
DONE
> LOCATE ALL FOR ZIP > 68500 AND < 68588
UNABLE TO PROCESS REQUEST. CHECK REQUEST AGAIN.
FOR MORE INFORMATION TYPE "ASSIST"

FIGURE 18.17
Feedback tells when input is accepted, whether processing is completed or not, and how to ask for further assistance.

Offering the User More Detailed Feedback. Users need to be reassured that more detailed feedback is available, and they should be shown how they can get it. Notice that in the last figure, the system has displayed a message instructing the user to type "ASSIST" for more information. When designing Web interfaces, hyperlinks can be embedded to allow the user to jump to the relevant help screens or to view more information. Hyperlinks are typically highlighted with underlining or they are italicized; they may also appear in a different color. Hyperlinks can be graphics, text, or icons.

Other commands, such as INSTRUCT, EXPLAIN, and MORE, may also be employed. Or, the user may type a question mark or point to an appropriate icon to get more feedback. Using the command HELP or having users hit the Escape key as ways to obtain further information has been questioned, because users may feel helpless or caught in a trap from which they must escape. Both conventions are in use, however, and their familiarity to users may overcome these concerns.

INCLUDING FEEDBACK IN DESIGN

Feedback is essential to all humans. Even when interfacing with machines, people still require it. Plan to provide feedback for users so that they are aware of whether their input is being accepted, whether input is or is not in the correct form, whether processing is going on, whether requests can or cannot be processed, and whether more detailed information is available and how to get it.

It is well worth the systems analyst's time to provide user feedback. If used correctly, feedback can both be a powerful reinforcer of users' learning processes and serve to improve their performance with the system and increase their motivation to produce.

A Variety of Help Options. Feedback on personal computers has developed over the years. "Help" originally started as a response to the user who pressed a function key such as F1; the GUI alternative is the pull-down help menu. This approach, however, was cumbersome, because end users had to navigate through a table of contents or search using an index. Next came context-sensitive help. End users could simply click on the right mouse button, and topics or explanations about the current screen or area of the screen would be revealed. Some software manufacturers call these cue cards. The third type of help on personal computers occurs when the end user places the arrow over an icon and leaves it there for a couple of seconds. At this point, some programs pop up a balloon similar to those found in comic strips. This balloon explains a little bit about the icon.

The fourth type of help is wizards, which ask the end user a series of questions and then take an action accordingly. Wizards have been used in narrowing a search in an encyclopedia such as Encarta, in designing a chart in Freelance or PowerPoint, or in choosing a style for a word-processing memo.

Besides building help into the software, software manufacturers offer help lines (most customer service telephone lines are not toll-free, however). Some manufacturers offer a fax-back system. An end user can request a catalog of various help documents to be sent by fax and then can order from the catalog by entering the item number with a touch-tone phone.

Finally, software forums exist on America Online; they also exist as bulletin boards on the Net, which are maintained by the software company itself that wishes to capture Web site visitors. An end user will often get information from the software manufacturer's technical support staff, but sometimes the end user can also get valuable information from other users of the product. This type of support is, of course, unofficial, and the information thus obtained may be true, may be partially true, or may even lead the user astray. The principles regarding the use of software forums are the same for those mentioned in Chapter 20, where FOLKLORE and recommendation systems are discussed. Read this section before you accept what is said on bulletin boards. Beware!

Besides informal help on software, vendor Web sites are extremely useful for updating drivers, viewers, and the software itself. Most computer magazines have some sort of "driver watch" or "bug report" that monitors the bulletin boards and Web sites for useful programs that can be downloaded. Programs such as Oil Change will forage vendor Web sites for the latest updates, inform the user of them, assist with the downloads, and actually upgrade user applications.

SPECIAL DESIGN CONSIDERATIONS FOR ECOMMERCE

Many of the user interface design principles you have learned concerning feedback also extend to designing ecommerce Web sites. A few extra considerations shown in this section, however, can give your Web interface designs improved functionality. They include learning to incorporate methods for eliciting feedback on the Web site from ecommerce customers and four ways to provide one-click navigation on ecommerce sites that will ensure that customers can easily navigate the site, and that they can readily return to it.

SOLICITING FEEDBACK FROM ECOMMERCE WEB SITE CUSTOMERS

Not only do you need to give users feedback about what is happening with an order, but you need to elicit feedback as well. Most ecommerce Web sites have a feedback button. There are two standard ways to design what users will experience when they click on the feedback button.

The first way is to launch the user's email program with the email address of the company's contact automatically entered into the "To:" field of the message. This method prevents typing errors and facilitates ease in contacting the organization. The user does not need to leave the site to communicate with it. These messages, however, raise expectations that they will be answered just as regular mail or phone calls are. Research indicates that 60 percent of organizations with this type of email contact feature on their sites do not have anyone assigned to reply to the email messages received. Thus, the business is losing valuable feedback, allowing customers to harbor the impression that they are communicating and engendering ill will when no response is received. If you design this type of feedback opportunity, you also need to design procedures for the organization to use in replying to email from the Web site.

The second type of design for garnering feedback from customers using an ecommerce Web site is to take users to a blank message template when they click on a button labeled "Feedback." Even a familiar tool such as Microsoft FrontPage permits you to create and insert a feedback form onto your site easily. This form might begin with a header that states "Company X Feedback" and then proceeds to read, "You can use the form below to send suggestions, comments, and questions about the X site to our Customer Service team."

Fields can include First Name, Last Name, Email Address, Regarding (a subject field that supplies a drop-down menu of the company's product or service selections, asking the user to "Please make a selection") an "Enter your Message here:" section (a free-form space where users can write in their message), and the standard "Submit" and "Clear" buttons at the bottom of the form. Using this type of form permits the analyst to have the user data already formatted correctly for storage in a database. Consequently, it makes the data entered into a feedback form easier to analyze in the aggregate.

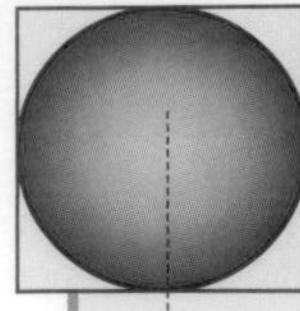

CONSULTING OPPORTUNITY 18.5

WHEN YOU RUN A MARATHON IT HELPS TO KNOW WHERE YOU'RE GOING

Marathon Vitamin Shops was successful in getting its Web site up and running. The Web developers put the company's entire catalog online and included a choice of skins so that each type of customer would enjoy using the Web site. (See Consulting Opportunities 1.1 and 16.4 for more details.)

The analysts are meeting with owner Bill Berry and some employees to evaluate customer feedback as well as give their own reactions to the new Web site. They are meeting in a large conference room, where they have a computer with Internet access and a projector. As they sit down at the table, the entry screen for the Web site is projected at the front of the room. "The Web site has attracted lots of attention, but we want to give the customers even more so that they keep coming back," says Bill, gesturing to the screen.

He continues, "It's not like we're closing our retail stores or anything. In fact, it's just the opposite. When customers notice we're on the Web, they're eager to locate the store in their community. They want to be able to walk into a store and talk to a trained expert rather than buying everything over the Internet. We need to tell people how to get there."

"We think we can improve the site by adding special enhancements and features," says Al Falfa, a member of the systems team who originally developed and implemented the ecommerce Web site.

"Yes," says Ginger Rute, one of the other members of the systems development team, as she nods in agreement. "Blockbuster and Borders use a mapping facility from MapQuest, and Home Depot uses maps from Vicinity, the parent company of MapBlast!"

Vita Minn, another member of the original systems development team, speaks up enthusiastically, saying, "We know of a couple good message board services and chat rooms we can build into our Web site. We think they can improve the stickiness of the site, making people stay on the site longer and also making them want to return."

"That's a great idea," says Jin Singh, one of the technologically savvy Marathon employees. "We can let customers talk with one another, tell each other about a product they liked and so on."

Vita continues by moving to the computer keyboard and saying, "Let me show you the sites at www.planetgov.com and www.worldviewer.com." As she types in the first URL, the group sees the site projected. "They use chat systems from ichat and Multicity.com, respectively," she continues.

"Customers also need to search for more information about a product or manufacturer," Al adds. "Let's make it easier for them. Let's look at www.Cincinnati.com for an example. They use Atomz to search for information."

After listening intently, Bill speaks up. "Medical information could also be useful, too," he says. "I've notice that www.medpool.com has medical news from NewsEdge. I've seen people on the treadmills at my health and fitness center watching the financial channels while they exercise."

"While we're at it, why don't we add news and financial information to the Website"? Ginger asks. "I notice that www.nmm.com has market news from a company called Moreover.com."

Think about the conversation between the systems development team and the people from Marathon Vitamins. Some of the enhancement suggestions involved taking advantage of free services; others required payment ranging from $1,000 to $5,000 annually. Although some were good ideas, others may not be practical or feasible. Perhaps some of the ideas just do not make sense for the company.

For each of the following, review what you know about the mission and business activities of Marathon Vitamin Shops. Then make a recommendation regarding each option the analysts and clients have made and defend it:

- Mapping software linked to the Web site.
- Chat rooms and message boards.
- Search engines.
- Medical information.
- News feeds and financial markets information.

Thus, the analyst does more than just design a response to individual email. The analyst helps the organization capture, store, process, and analyze valuable customer information in a manner that makes it more likely that the company will be capable of spotting important trends in customer response, rather than simply reacting to individual queries.

EASY NAVIGATION FOR ECOMMERCE WEB SITES

Many authors speak of what is known as "intuitive navigation" for ecommerce Web sites. Users need to know how to navigate the site without having to learn a new interface and without having to explore every inch of the Web site before they can find what they want. The standard for this type of navigational approach is called "one click."

There are four ways to design easy, one-click navigation for an ecommerce site: (1) creating a rollover menu, (2) building a collection of hierarchical links so that the home page becomes an outline of the key topic headings associated with the Web site, (3) placing a site map on the home page and emphasizing the link to it (there as well as on every other page on the site), and (4) placing a navigational bar on every inside page (usually at the top or to the right of the page) that repeats the categories used on the entry screen.

A rollover menu can be created with a Java applet or with Java script and HTML layers, if you do not want to make users run a Java applet. The rollover menu appears when the customer using the Web site pauses the pointer over a link.

Creating an outline of the content of the site through the presentation of a table of contents on the home page is another way to speed navigation of the site. This design, however, imposes severe constraints on the designer's creativity, and sometimes simply presenting a list of topics does not adequately convey the strategic mission of the organization to the user.

Designing and then prominently displaying the link to a site map is a third way to improve navigational efficiency. Remember to include the link to the site map on the home page and on every other page as well.

Finally, you can design navigation bars that are consistently displayed on the home page as well as at the top and to the right of all other pages that comprise the site. Once you have established (during the information requirements phase) the most useful and most used categories (usually categories such as "Our Company" "Our Products," "Buy Now," "Contact Us," "Site Map," and "Search"), remember to include them on all pages.

The main priority in navigation is, however, that whatever you do, you must make it extremely easy for users to return to a previous start and make it somewhat easy to return to the place where they entered the client's site. Your main concern is keeping customers on the Web site. The longer customers are on the site, the greater the chance that they will purchase one of the products offered there. So, make sure that if users navigate to a link in your Web site, they can easily find their way back. Doing these things will ensure the "stickiness" of the Web site. Do not create any barriers to the customer who wants to return to the client's Web site.

DESIGNING QUERIES

When users want to ask questions of the database, they are said to query it. Six different types of queries are among the most common.

QUERY TYPES

The questions we pose concerning data from our database are referred to as queries. There are six basic query types. Each query involves three items: an entity, an attribute, and a value. In each case, two of these are given, and the intent of the query is to find the remaining item.

Query Type 1. In the first type of query, the entity and one of the entity's attributes are given. The purpose of the query is to find the value. The query can be expressed as follows:

What is the value of a specified attribute for a particular entity?

Sometimes it is more convenient to use notation to help formulate the query. This query can be written as

$$V \longleftarrow (E, A)$$

where V stands for the value, E for entity, and A for attribute and the variables in parenthesis are given.

The question

What did employee number 73712 make in year 2000?

can be stated more specifically as

What is the value of the attribute YEAR-2000 for the entity EMPLOYEE-NUMBER 73712?

Basic query type 1 is illustrated in Figure 18.18. The record containing employee number 73712 was found, and the answer to the query is $27,100.

Query Type 2. The intent of query type 2 is to find an entity or entities when an attribute and value are given. Query type 2 can be stated as follows:

What entity has a specified value for a particular attribute?

Because values can also be numeric, it is possible to search for a value equal to, greater than, less than, not equal to, greater than or equal to, and so on. An example of this type of query is as follows:

What employee(s) earned more than $50,000 in 2000?

or, more specifically,

What entities (EMPLOYEE-NUMBER) have the value > 50,000 for the attribute YEAR-2000?

The notation for query type 2 is

$$E \longleftarrow (V, A)$$

Figure 18.19 illustrates query type 2. In this case, three employees made more than $50,000, so the response turned out to be a listing of the employee numbers for the three employees.

Query Type 3. The purpose of query type 3 is to determine which attribute(s) fits the description provided when the entity and value are given. Query type 3 can be stated as follows:

What attribute(s) has a specified value for a particular entity?

This query is useful when there are many similar attributes that have the same property. The example below has similar attributes (specific years) that contain the annual salaries for the employees of the company:

What years did employee #72845 make over $50,000?

or, more precisely,

What attributes {YEAR-1997, YEAR-1998, YEAR-1999, YEAR-2000} have a value > 50,000 for the entity EMPLOYEE-NUMBER = 72845?

where the optional list in braces { } is the set of eligible attributes.

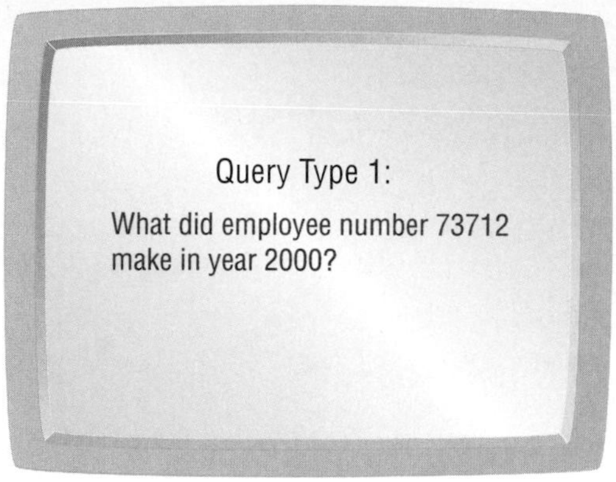

Query Type 1:

What did employee number 73712 make in year 2000?

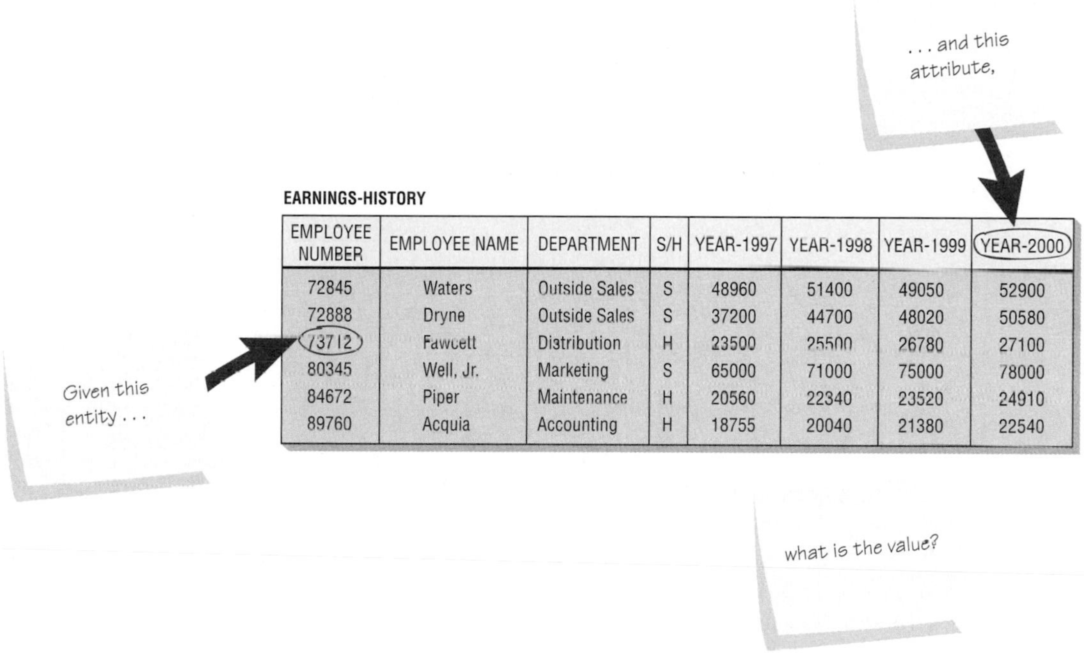

. . . and this attribute,

EARNINGS-HISTORY

EMPLOYEE NUMBER	EMPLOYEE NAME	DEPARTMENT	S/H	YEAR-1997	YEAR-1998	YEAR-1999	YEAR-2000
72845	Waters	Outside Sales	S	48960	51400	49050	52900
72888	Dryne	Outside Sales	S	37200	44700	48020	50580
73712	Fawcett	Distribution	H	23500	25500	26780	27100
80345	Well, Jr.	Marketing	S	65000	71000	75000	78000
84672	Piper	Maintenance	H	20560	22340	23520	24910
89760	Acquia	Accounting	H	18755	20040	21380	22540

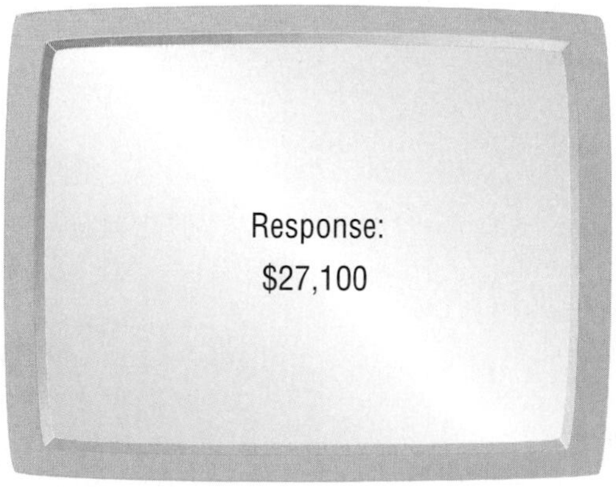

Given this entity . . .

what is the value?

Response:

$27,100

FIGURE 18.18

Query type 1 finds the value of an attribute for a given entity.

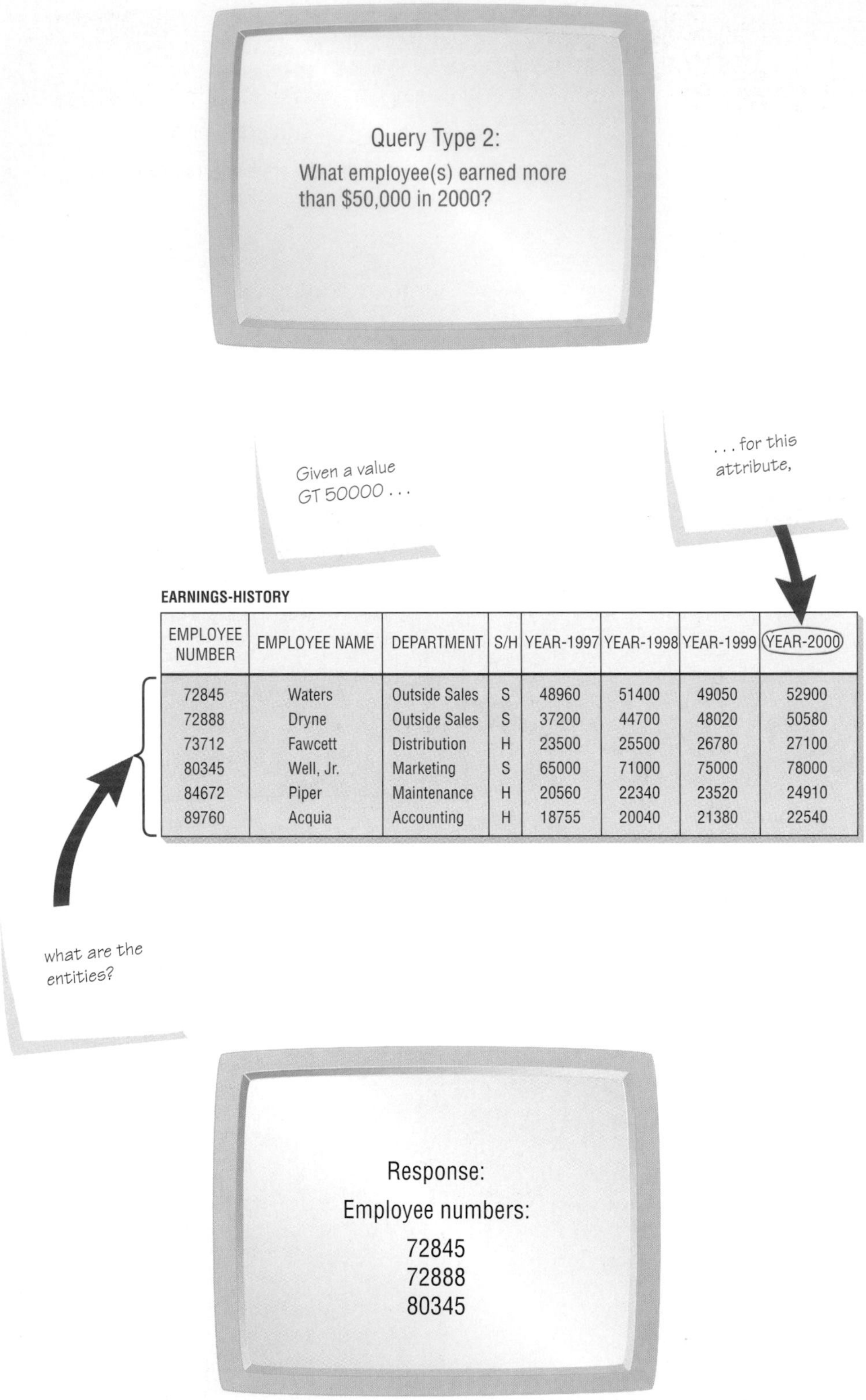

FIGURE 18.19

Query type 2 finds the entities that satisfy a specific value for a particular attribute.

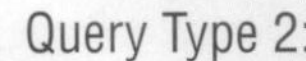
Query Type 2:
What employee(s) earned more than $50,000 in 2000?
Given a value GT 50000 . . .
. . . for this attribute,
what are the entities?

EARNINGS-HISTORY

EMPLOYEE NUMBER	EMPLOYEE NAME	DEPARTMENT	S/H	YEAR-1997	YEAR-1998	YEAR-1999	YEAR-2000
72845	Waters	Outside Sales	S	48960	51400	49050	52900
72888	Dryne	Outside Sales	S	37200	44700	48020	50580
73712	Fawcett	Distribution	H	23500	25500	26780	27100
80345	Well, Jr.	Marketing	S	65000	71000	75000	78000
84672	Piper	Maintenance	H	20560	22340	23520	24910
89760	Acquia	Accounting	H	18755	20040	21380	22540

Response:
Employee numbers:
72845
72888
80345

The notation for query type 3 is

$$A \longleftarrow (V, E)$$

An illustration of query type 3 is given in Figure 18.20. In this example, Waters (#72845) made over $50,000 for two years. These years are listed in the response. Query type 3 is rarer than either type 1 or type 2 due to the requirement of having similar attributes exhibiting the same properties.

Query Type 4. Query type 4 is similar to query type 1. The difference is that the values of all attributes are desired. Query 4 can be expressed as follows:

List all the values for all the attributes for a particular entity.

Examples of query 4 include the following:

List all the details in the earnings history file for employee number 72888.
List all the information regarding the inventory status on part HV-5678.

The notation for query type 4 is

$$\text{all } V \longleftarrow (E, \text{all } A)$$

An illustration of query type 4 is shown in Figure 18.21. The response for this query was simply the record for the employee named Dryne (#72888).

Query Type 5. The fifth type of query is another global query, but it is similar in form to query type 2. Query type 5 can be stated as follows:

List all entities that have a specified value for all attributes.

An example of query type 5 is the following:

List all the employees whose earnings exceeded $50,000 in any of the years available.

The notation for query type 5 is

$$\text{all } E \longleftarrow (V, \text{all } A)$$

An example of query type 5 is shown in Figure 18.22.

Query Type 6. The sixth query type is similar to query type 3. The difference is that query type 6 requests a listing of the attributes for all entities rather than one particular entity. Query type 6 can be stated as follows:

List all the attributes that have a specified value for all entities.

The following is an example of query type 6:

List all the years for which earnings exceeded $20,000 for all employees in the company.

The notation for query type 6 is

$$\text{all } A \longleftarrow (V, \text{all } E)$$

An illustration of query type 6 can be found in Figure 18.23. As with query type 3, query type 6 is not used as much as other types.

FIGURE 18.20

Query type 3 finds the attributes when the entity and value are given.

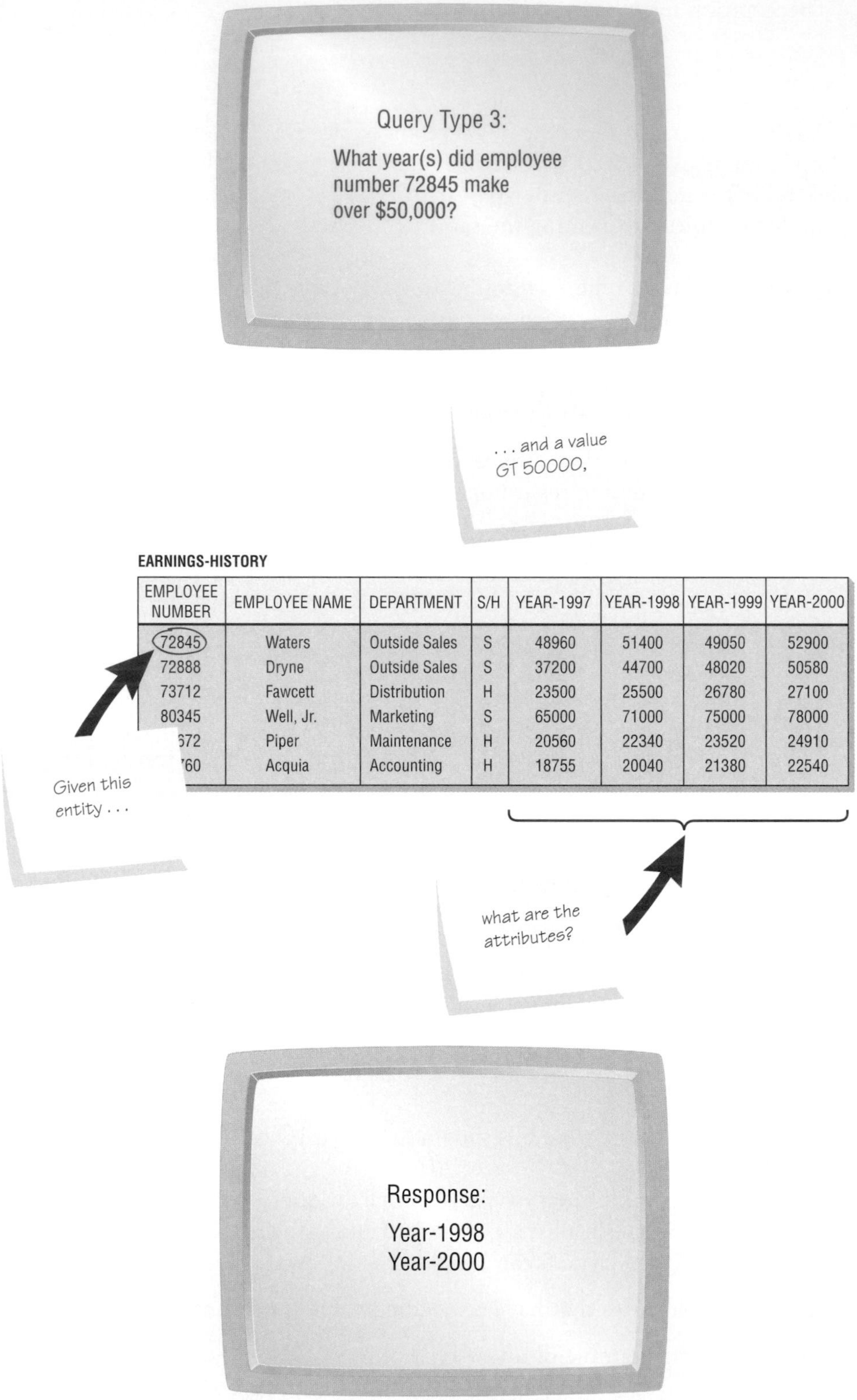

Query Type 3:

What year(s) did employee number 72845 make over $50,000?

. . . and a value GT 50000,

EARNINGS-HISTORY

EMPLOYEE NUMBER	EMPLOYEE NAME	DEPARTMENT	S/H	YEAR-1997	YEAR-1998	YEAR-1999	YEAR-2000
72845	Waters	Outside Sales	S	48960	51400	49050	52900
72888	Dryne	Outside Sales	S	37200	44700	48020	50580
73712	Fawcett	Distribution	H	23500	25500	26780	27100
80345	Well, Jr.	Marketing	S	65000	71000	75000	78000
672	Piper	Maintenance	H	20560	22340	23520	24910
760	Acquia	Accounting	H	18755	20040	21380	22540

Given this entity . . .

what are the attributes?

Response:

Year-1998
Year-2000

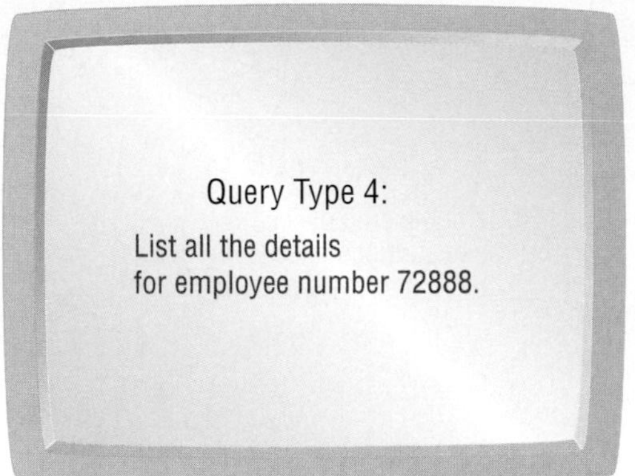

FIGURE 18.21

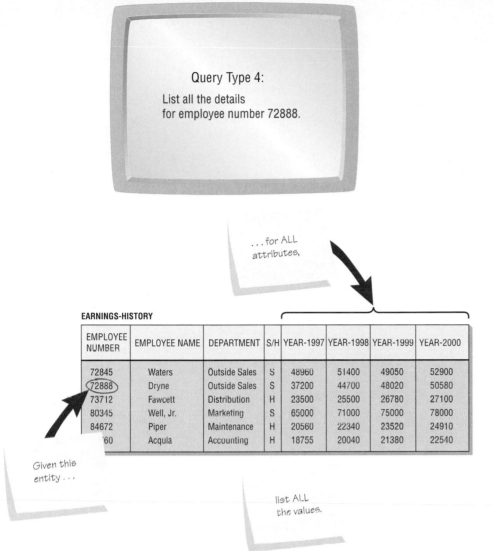

Query type 4 lists all the values for all attributes of a particular entity.

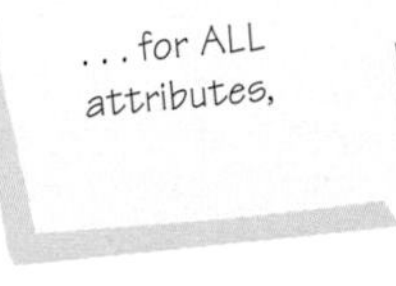

EARNINGS-HISTORY

EMPLOYEE NUMBER	EMPLOYEE NAME	DEPARTMENT	S/H	YEAR-1997	YEAR-1998	YEAR-1999	YEAR-2000
72845	Waters	Outside Sales	S	48960	51400	49050	52900
72888	Dryne	Outside Sales	S	37200	44700	48020	50580
73712	Fawcett	Distribution	H	23500	25500	26780	27100
80345	Well, Jr.	Marketing	S	65000	71000	75000	78000
84672	Piper	Maintenance	H	20560	22340	23520	24910
[illegible]60	Acqula	Accounting	H	18755	20040	21380	22540

Given this entity . . .

list ALL the values.

Response:

72888 Dryne Outside Sales S 37200 44700 48020 50580

FIGURE 18.22

Query type 5 lists all of the entities that satisfy a value for all attributes described.

Query Type 5:
List all the employees whose earnings exceeded $50,000 in any of the years available.

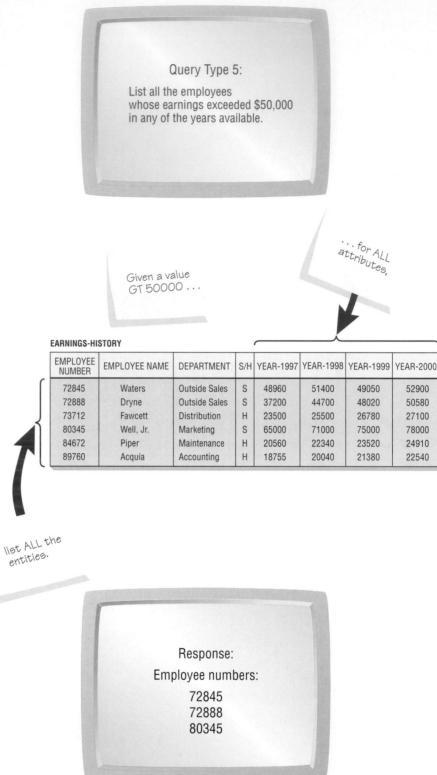

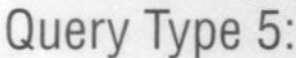

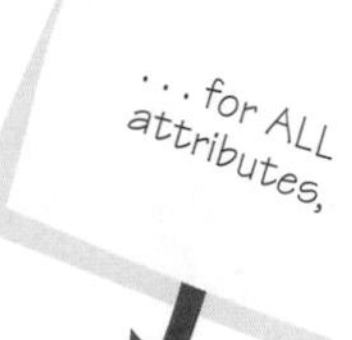

. . . for ALL attributes,

Given a value GT 50000 . . .

EARNINGS-HISTORY

EMPLOYEE NUMBER	EMPLOYEE NAME	DEPARTMENT	S/H	YEAR-1997	YEAR-1998	YEAR-1999	YEAR-2000
72845	Waters	Outside Sales	S	48960	51400	49050	52900
72888	Dryne	Outside Sales	S	37200	44700	48020	50580
73712	Fawcett	Distribution	H	23500	25500	26780	27100
80345	Well, Jr.	Marketing	S	65000	71000	75000	78000
84672	Piper	Maintenance	H	20560	22340	23520	24910
89760	Acquia	Accounting	H	18755	20040	21380	22540

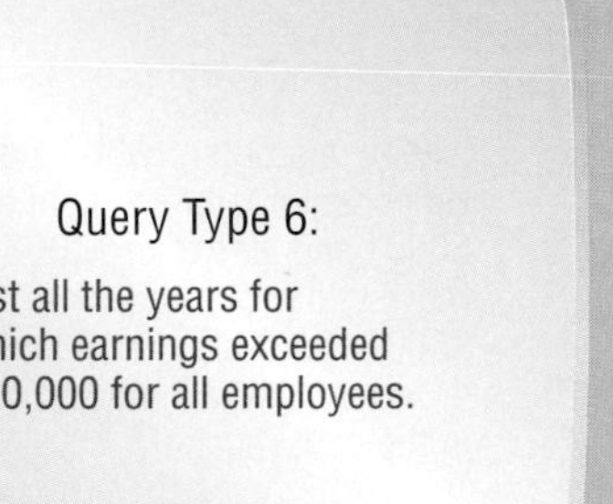

Query Type 6:
List all the years for which earnings exceeded $20,000 for all employees.

FIGURE 18.23

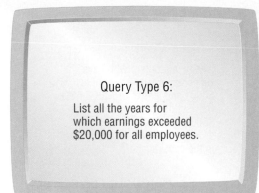

Query type 6 lists all of the attributes that have a specified value for all entities.

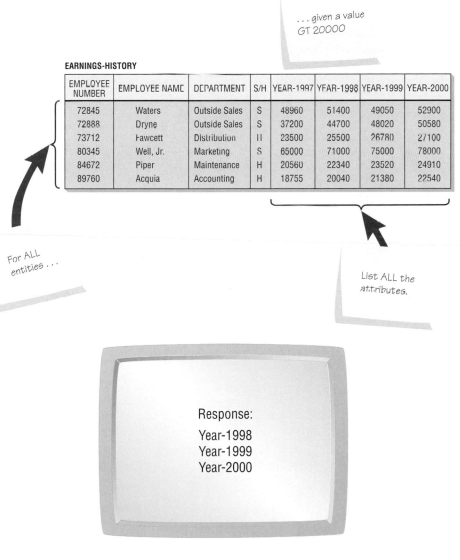

. . . given a value GT 20000

EARNINGS-HISTORY

EMPLOYEE NUMBER	EMPLOYEE NAME	DEPARTMENT	S/H	YEAR-1997	YEAR-1998	YEAR-1999	YEAR-2000
72845	Waters	Outside Sales	S	48960	51400	49050	52900
72888	Dryne	Outside Sales	S	37200	44700	48020	50580
73712	Fawcett	Distribution	H	23500	25500	26780	27100
80345	Well, Jr.	Marketing	S	65000	71000	75000	78000
84672	Piper	Maintenance	H	20560	22340	23520	24910
89760	Acquia	Accounting	H	18755	20040	21380	22540

For ALL entities . . .

List ALL the attributes.

Response:

Year-1998
Year-1999
Year-2000

FIGURE 18.24

Complex queries can be created using Boolean logic.

Query:

List all customers who have zip codes greater or equal to 60001 and less than 70000 AND (have either ordered more than $500 from catalogs OR have ordered at least 5 times in the past year).

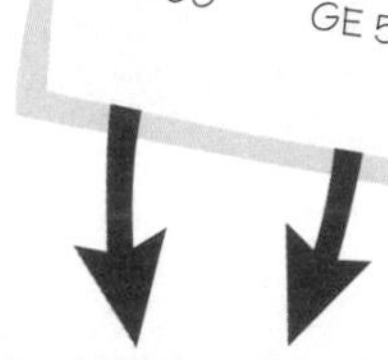

GE 60001 and LT 70000

ORDER-TIMES

ORDER-#	LAST NAME	I	STREET ADDRESS	CITY	ST	ZIP	AMOUNT	TIMES
10784	MacRae	G	2314 Curly Circle	Lincoln	NE	68506	322	8
10796	Jones	S	34 Dream Lane	Oklahoma City	OK	73118	47	2
11821	Preston	R	1008 Madison Ave.	River City	IA	52101	36	1
11845	Channing	C	454 Harmonia St.	New York	NY	10453	98	4
11872	Kiley	R	765 Dulcinea Drive	La Mancha	CA	93407	125	7
11976	Verdon	G	7564 K Street	Chicago	IL	60637	187	5
11998	Rivera	C	4342 West Street	Chicago	IL	60625	559	10
12765	Orbach	J	1345 Michigan Ave.	Chicago	IL	60616	58	3
12769	Steele	T	3498 Burton Lane	Finnian	NJ	07860	323	6
12965	Crawford	M	1986 Barnum Cir.	London	NH	03570	145	2
13432	Cullum	J	354 River Road	Shenandoah	VT	05201	237	4
13542	Mostel	Z	65 Fiddler Street	Anatevka	ND	58501	38	1

OR

AND

List ALL entities that satisfy the conditions.

Response:

10784
11976
11998

Type	Level	Symbol
Arithmetic Operators	1 2 3	* * * / + –
Comparative Operators	4	GT LT EQ NE GE LE
Boolean Operators	5 6	AND OR

FIGURE 18.25

Arithmetic, comparative, and Boolean operators are processed in a hierarchical order of precedence unless parentheses are used.

Building More Complex Queries. The preceding six query types are only building blocks for more complex queries. Expressions referred to as Boolean expressions can be formed for queries. One example of a Boolean expression is the following:

> List all the customers who have zip codes greater than or equal to 60001 and less than 70000 and who have ordered more than $500 from our catalogs or have ordered at least five times in the past year.

One difficulty with this statement is determining which operator (for example, AND) belongs with which condition; it is also difficult to determine the sequence in which the parts of the expression should be carried out. The following may help to clarify this problem:

> LIST ALL CUSTOMERS HAVING (ZIP-CODE GE 60001 AND ZIP-CODE LT 70000) AND (AMOUNT-ORDERED GT 500 OR TIMES-ORDERED GE 5)

Now some of the confusion is eliminated. The first improvement is that the operators are expressed more clearly as GE, GT, and LT than as English phrases, such as "at least." Second, the attributes are given distinct names, such as AMOUNT-ORDERED and TIMES-ORDERED. In the earlier sentence, these attributes were both referred to as "have ordered." Finally, parentheses are used to indicate the order in which the logic is to be performed. Whatever is in parentheses is done first. Figure 18.24 shows how a more complex query is performed on a relation.

Operators are generally performed in a predetermined order of precedence. Arithmetic operators are usually performed first (exponentiation, then either multiplication or division, and then addition or subtraction). Next, comparative operators are performed. These operators are GT (greater than), LT (less than), and others. Finally, the Boolean operators are performed (first AND and then OR). Within the same level, the order generally goes from left to right. The precedence is summarized in Figure 18.25.

QUERY METHODS

Two popular query methods are query by example and Structured Language Query.

Query by Example. Query by example (called QBE) is a simple yet powerful method for implementing queries in database systems, such as Microsoft Access. The database fields are selected and displayed in a grid, and requested query values are either entered in the field area or below the field. The query should be able to select both rows from the table that match conditions as well as specific columns (fields). Complex conditions may be set to select records, and the user may easily specify the columns to be sorted. Figure 18.26 is an example of query by example using Microsoft Access. The query design screen is divided into two portions. The top portion contains

FIGURE 18.26

Query by example using Microsoft Access.

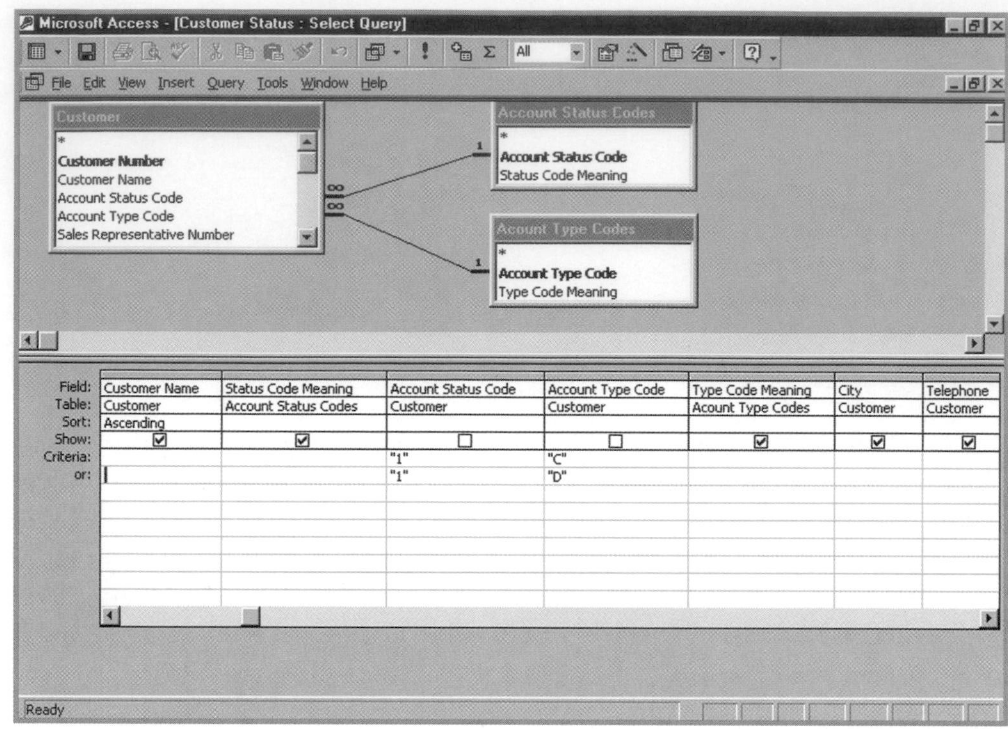

the tables selected for the query and their relationships, and the bottom portion contains the query selection grid. Fields from the database tables are dragged to the grid.

The first two rows contain the field and the table in which the field is located. The next row contains sorting information. In this example, the results will be sorted by CUSTOMER NAME. A check mark in the Show box (fourth row down) indicates that the field is to be displayed in the results. Notice that the CUSTOMER NUMBER, CUSTOMER NAME, and STATUS CODE MEANING are selected for the resulting display (other fields are displayed as well, but they do not

FIGURE 18.27

A query by example for CUSTOMER STATUS yields these results.

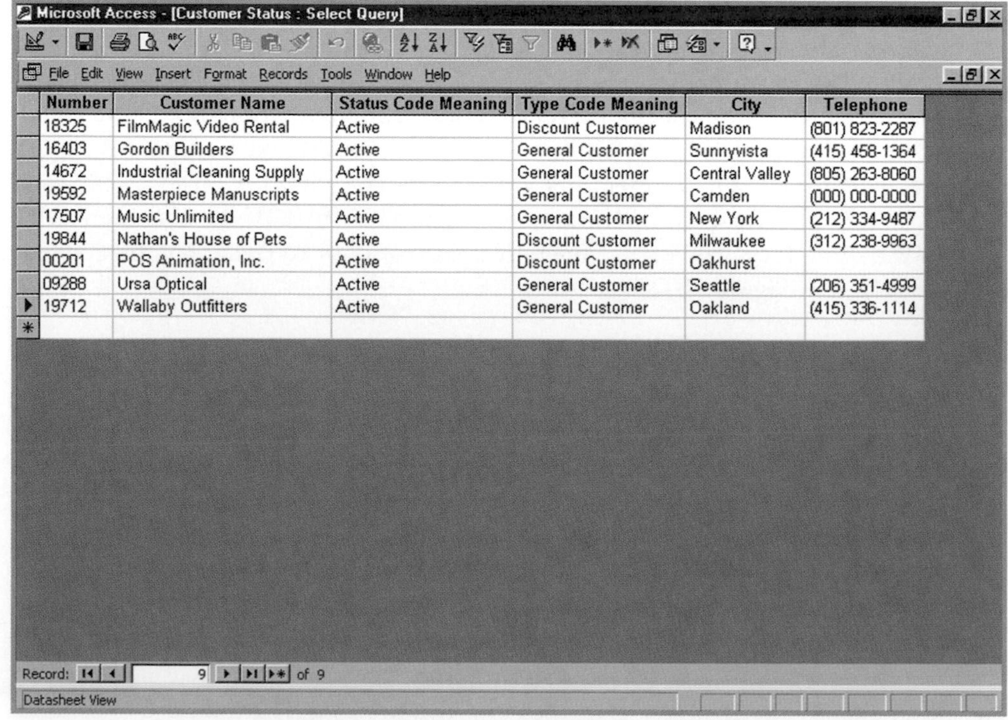

Microsoft Access - [Customer Status : Select Query]

Number	Customer Name	Status Code Meaning	Type Code Meaning	City	Telephone
18325	FilmMagic Video Rental	Active	Discount Customer	Madison	(801) 823-2287
16403	Gordon Builders	Active	General Customer	Sunnyvista	(415) 458-1364
14672	Industrial Cleaning Supply	Active	General Customer	Central Valley	(805) 263-8060
19592	Masterpiece Manuscripts	Active	General Customer	Camden	(000) 000-0000
17507	Music Unlimited	Active	General Customer	New York	(212) 334-9487
19844	Nathan's House of Pets	Active	Discount Customer	Milwaukee	(312) 238-9963
00201	POS Animation, Inc.	Active	Discount Customer	Oakhurst	
09288	Ursa Optical	Active	General Customer	Seattle	(206) 351-4999
19712	Wallaby Outfitters	Active	General Customer	Oakland	(415) 336-1114

Record: 9 of 9

Datasheet View

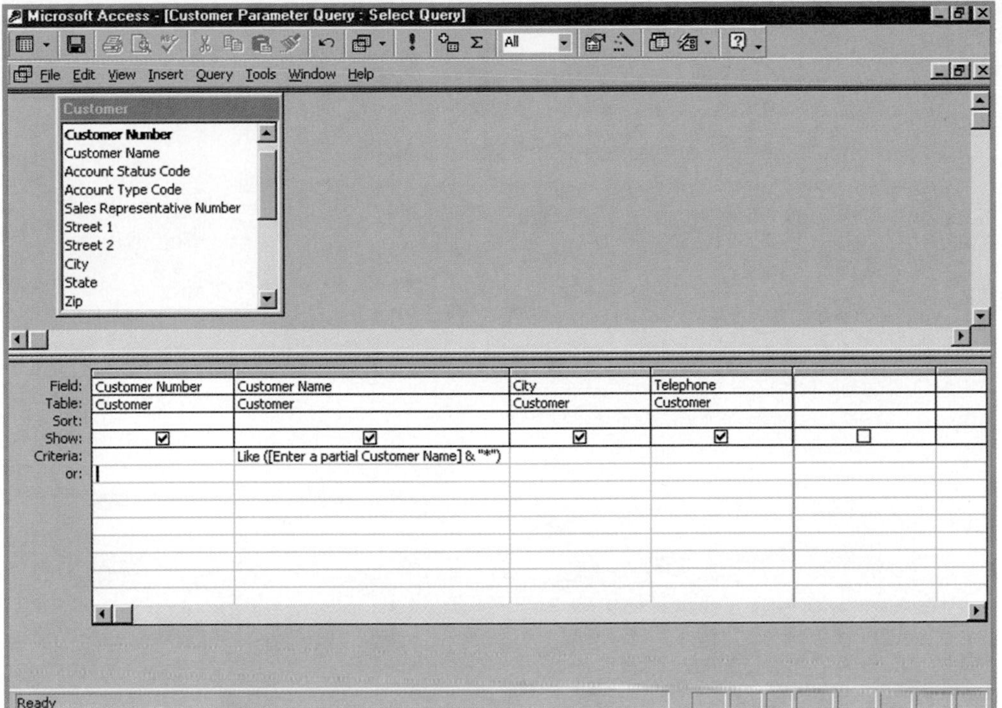

FIGURE 18.28
A parameter query design screen.

show in the screen). Notice that the ACCOUNT STATUS CODE and ACCOUNT TYPE CODE are not checked and therefore will not be in the final results. In the Criteria rows, there is a 1 in the ACCOUNT STATUS CODE (indicating an active record) and a C and D (selecting a General Customer or a Discount Customer) in the ACCOUNT TYPE CODE columns. Two conditions in the same row indicate an AND condition, and two conditions in different rows represent the OR conditions. This query specifies that the user should select both an Active Customer and either a General or Discount Customer.

The results of a query are displayed in a table, illustrated in Figure 18.27. Notice that the ACCOUNT STATUS CODE and ACCOUNT TYPE CODE do not display. They are not checked and are included in the query for selection purposes only. Instead, the code meanings are displayed, which are more useful to the user. The customer names are sequenced alphabetically.

One of the problems encountered when designing queries is that either the user must modify the query parameters or the same conditions are selected each time the query is executed. A solution to this problem is to use a parameter query. This type of query allows the user to enter the conditions in a dialog box each time the query is run. Figure 18.28 illustrates a parameter query. Notice that the criteria has the message "Enter a partial Customer Name" included inside brackets. Preceding the message is the word "Like," and following the message is an ampersand, indicating that an exact match is not required. When the query is executed, a dialog box opens with the query message on the top. Refer to Figure 18.29. The

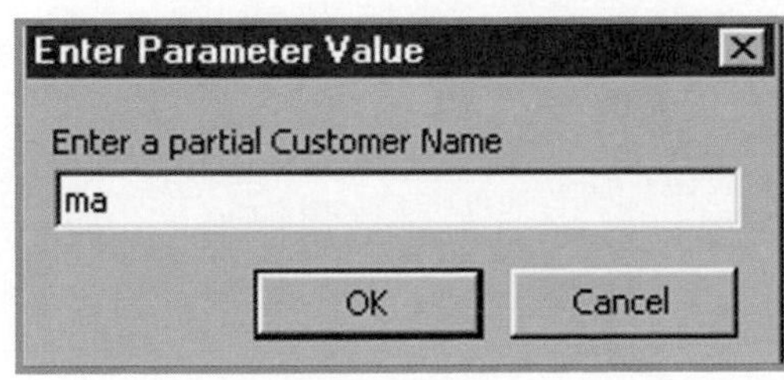

FIGURE 18.29
A parameter query entry dialog box.

FIGURE 18.30
The results of the parameter query for CUSTOMER NAME show businesses that begin with "Ma."

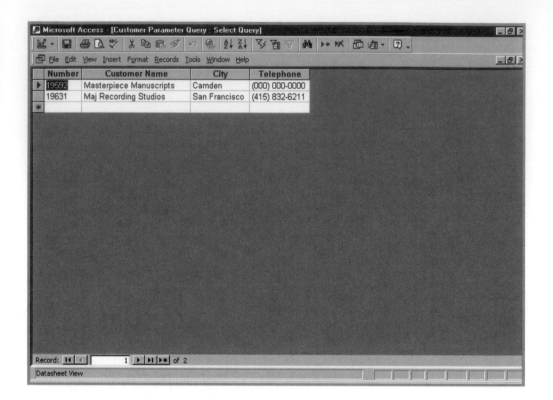

value "ma" is entered and used to select the name. The results are displayed in Figure 18.30. Notice that only customers whose names begin with the letters "ma" are selected and displayed.

Structured Query Language. Structured query language (SQL) is another popular way to implement queries. It uses a series of words and commands to select the rows and columns that should be displayed in the resulting table. Figure 18.31 illustrates the SQL code that is equivalent to the parameter query shown above. The SELECT DISTINCTROW key word determines which rows are to be selected. The WHERE key word specifies the condition that the CUSTOMER NAME should be used to select the data entered in the LIKE parameter.

Figure 18.32 is an example of the SQL code used to produce the results that match query by example. Notice both that there are additional key words that include the relationship between the tables (INNER JOIN) and that the AND and OR logic is included in parenthesis to permit the correct evaluation of the conditions. The ORDER BY parameter indicates the sorting sequence of the resulting table.

FIGURE 18.31
Structured Query Language (SQL) for the CUSTOMER NAME parameter query.

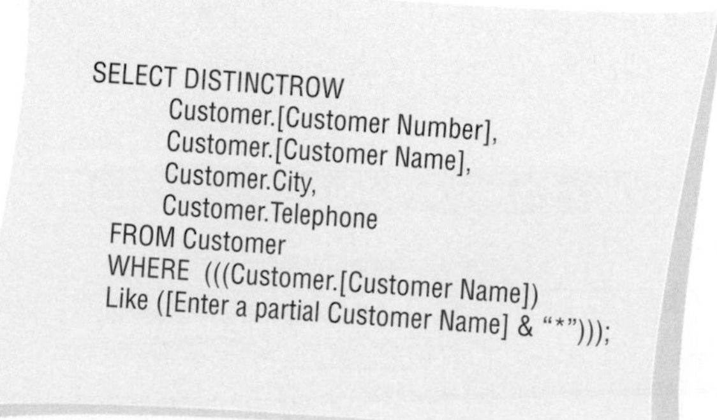

```
SELECT DISTINCTROW
        Customer.[Customer Number],
        Customer.[Customer Name],
        Customer.City,
        Customer.Telephone
FROM Customer
WHERE (((Customer.[Customer Name])
Like ([Enter a partial Customer Name] & "*")));
```

```
SELECT DISTINCTROW
        Customer.[Customer Number],
        Customer.[Customer Name],
        [Account Status Codes].[Status Code Meaning],
        [Account Type Codes].[Type Code Meaning],
        Customer.City,
        Customer.Telephone
FROM [Account Type Codes]
INNER JOIN ([Account Status Codes]
INNER JOIN Customer
ON [Account Status Codes].[Account Status Code]
      = Customer.[Account Status Code])
ON [Account Type Codes].[Account Type Code]
      = Customer.[Account Type Code]
WHERE (((Customer.[Account Status Code])="1")
      AND ((Customer.[Account Type Code])="C"))
      OR
      (((Customer.[Account Status Code])="1")
      AND ((Customer.[Account Type Code])="D"))
ORDER BY Customer.[Customer Name];
```

FIGURE 18.32

Structured Query Language commands for the CUSTOMER STATUS query.

SEARCHING THE WEB

It is impossible to discuss queries without talking about searching the Web or Internet. There are many search engines, Internet directories, and subject trees that all work in much the same fashion. Typically, a spider or robot will be sent out across the Web to discover Web sites that fit certain criteria or key words. These spiders and 'bots (short for "robots") bring URLs (addresses) back and store them in a database.

Search engines refer to the database when answering a query. Some popular search engines are the following:

AltaVista	www.altavista.digital.com
Excite	www.excite.com
Four11	www.four11.com
Google	www.google.com
HotBot	www.hotbot.com
Infoseek	www.infoseek.com
Yahoo!	www.yahoo.com

It is impossible to tell which search engine is best because (1) these companies bring out updated versions with new features often and (2) the interface of one search engine might be more suitable for you than another.

In addition to the search engines listed above are metasearch engines such as Copernic and Symantec's Internet Fast Find, offline browsers such as Traveling Software's WebEx, and programs that manage your bookmarks such as Starfish's Internet Utilities. These all can be helpful, and their features will likely be incorporated into new versions of the latest Web browsers.

GUIDELINES FOR SEARCHING THE WEB

You can improve your chances of finding what you want by following some of these strategies:

1. Decide whether you really want to search or surf. If you know what information you want, use a search engine, such as Infoseek, that will find specific sites. If you want to browse, use a Web directory service such as Yahoo!

2. Think of your key terms before you sit down at the computer. It is usually better to plan than react.
3. Construct your search questions logically. Are you searching for "decision" AND "support" rather than "decision" OR "support" (you will get very different results)? Do you want to find all sites that contain "decision," "support," and "systems," or are you looking for a phrase "decision support systems"? You must let the search engine know your intentions. What happens when you enter "DSS"? (You get an avalanche of information about direct satellite systems and little about decision support systems.)
4. Use a search engine that saves your searches.
5. Use a search engine that informs you of changes in the Web sites you select.
6. Remember that the search engine business is very competitive. Check on the search engines periodically. You will find that some search engines that did not possess a feature in an earlier version have subsequently brought out an improved update. This new version could easily surpass the previous leader.

DATA MINING

The concept of data mining came from the desire to use the database for a more selective targeting of customers. Early approaches to direct mail included using zip code information as a way to determine what a family's income might be (assuming a family has to generate sufficient income to afford to live in the prestigious Beverly Hills zip code 90210 or some other affluent neighborhood). It was a way (not perfect, of course) to limit the number of catalogs sent out.

Data mining takes this concept one step further. Assuming past behavior is a good predictor for future purchases, a large amount of data is gathered on a particular person from, for example, credit card purchases. The company can tell what stores we shop in, what we have purchased, how much we paid for an item, and when and how frequently we travel. Data are also entered, stored, and used for a variety of purposes when we fill out warranties, apply for a driver's license, respond to a free offer, or apply for a membership card at a video rental store. Moreover, companies share these data and often make money on the sale of it as well. Figure 18.33 illustrates the data mining concept.

American Express has been a leader in data mining for marketing purposes. American Express will send you discount coupons for new stores or entertainment when it sends you a credit card bill after it determines you have shopped in similar stores or attended similar events. General Motors offers a MasterCard that allows customers to build up bonus points toward the purchase of a new car and then attempts to send out information about new vehicles at the most likely time that a consumer would be interested in purchasing a new car. The process of data mining includes using powerful supercomputers that process very large databases—or data warehouses—using techniques such as neural nets (see Chapter 12).

The data mining approach is not without problems, however. First, the costs may be too high to justify data mining, something that may only be discovered after huge setup costs have been accrued. Second, data mining has to be coordinated so that various departments or subsidiaries do not all try to reach the customer at the same time. In addition, customers may think their privacy has been invaded and resent the offers that are coming their way. Finally, customers may think profiles created solely on the basis of their credit card purchases present a highly distorted image of who they are.

Several years ago, some disturbing civil rights issues arose when it was discovered that police authorities in England were observing citizens in their neighborhoods, making inferences about the people from their observed behavior, and stor-

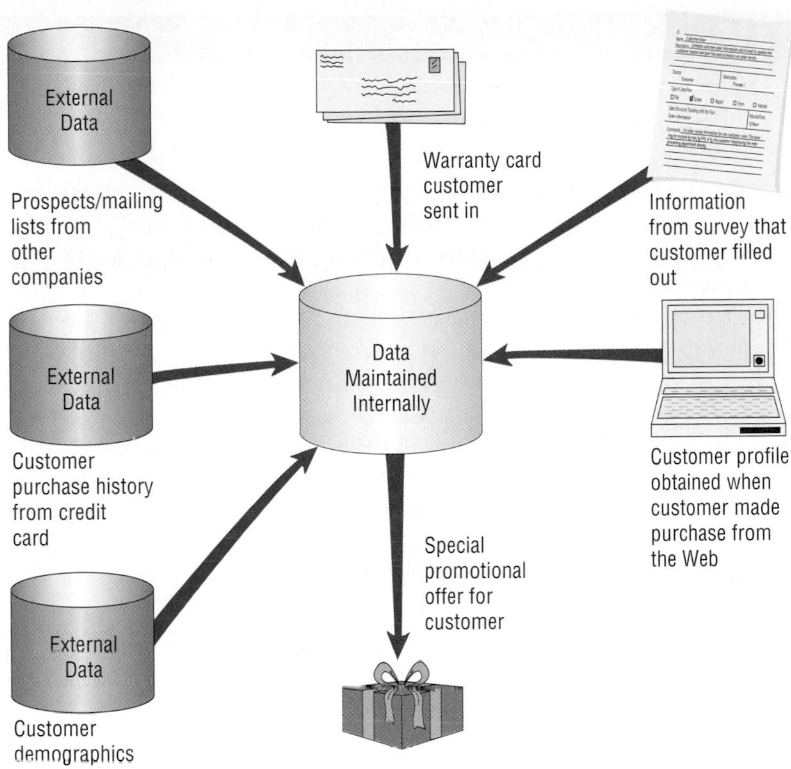

FIGURE 18.33
Data mining collects personal information about customers in an effort to be more specific in interpreting and anticipating their preferences.

ing those inferences in secret databases. Other police had access to those records, and the inferences were just that, because they were never seen or subject to review by the "suspected" citizen and they were never validated by other types of data. Erroneous profiles were created, stored, acted upon, and not deleted. Even in a democracy, data mining has applications beyond benign marketing attempts to reach us with the latest product.

Analysts should take responsibility for considering the ethical aspects of any data mining projects that are proposed. Questions about the length of the time profile material is kept, the confidentiality of it, the privacy safeguards included, and the uses to which inferences are put should all be asked and considered with the client. The opportunities for abuse are apparent and must be guarded against. For consumers, data mining is another push technology, and if consumers do not want be pushed, the data mining efforts will backfire.

PRODUCTIVITY AND ERGONOMICS DESIGN

Users might avoid using a system or become dissatisfied with it simply because the systems analyst has not bothered to visualize what it would be like to sit at a workstation and enter data into the system day in and day out. The powerful influence of the workstation on user involvement with the system must be considered. Continued use as well as user productivity, comfort, and satisfaction with the system are all related to how ergonomically well-designed the workspace is. (*Ergonomics* refers to the application of biological and engineering data to the problems related to people and the machines they use in work.)

Systems analysts can skillfully use their knowledge about the influence of well-designed workstations to support the behavior they want from users, that is, the productive use of the system. An example from a typical university computer room serves to underscore how workstation design determines the use of computers.

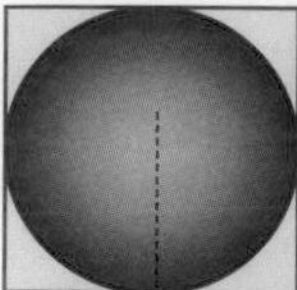

CONSULTING OPPORTUNITY 18.6

LOSING PROSPECTS

"Market share can be a real problem," says Ryan Taylor, Director of Marketing Systems for a large East Coast health insurer. "One of the greatest challenges we face is how to identify good leads for our salespeople. With over 50% market share we must eliminate the names of most of the prospects we buy before populating our marketing database. It is critical that we get it right because our marketing database is a critical part of our company's arsenal of strategic information tools."

Ryan explains to Chandler, one of your systems analysis team members, "A marketing database, or MDB for short, is a powerful, relational database that is the heart of marketing systems. Our marketing database is used to provide information for all marketing systems. They include productivity tools, such as our Sales Force Automation and our Mass Mailing Systems, which are designed to aid our salespeople in managing the sales cycle. They also include analytical tools, like our geographic information systems (GIS) or graphical query language tools (GQL), which are designed to provide decision support.

The primary function of a marketing database, though, is to track information on our customers and prospects. We currently track geographic information, demographic information, and psychographic information, or, as I like to say it, where they live, who they are, and how they think.

"The simplest marketing databases can be made up of just three files: Prospect Profile >>> Customer Profile >>> Purchase and Payment History.

"Once you have designed your marketing database, the next challenge is deciding how to populate it. We currently purchase our prospect information from a list vendor. Because our company's marketing strategy is based on mass marketing, we buy every business in our area. Because of this volume, we pay less than a dime for each prospect. If, however, a company is practicing product differentiation, their prospect base will likely be more defined. This company would likely pay a premium for more detailed data that have been carefully validated," explains Ryan.

"We face a real challenge. If I had a dollar for every time a rep complained to me about the address on a prospect being wrong, I could retire and move to Florida," Ryan quips. "I'm expected to identify which prospects are bad. That's not too hard if you only have a thousand of them, but what do you do when you have over a quarter of a million?"

Ryan continues, "Because we use these data frequently for large mailings, it is very important for us to ensure that the names and addresses on that file are as accurate as possible. For example, they should conform to postal standards and should not be duplicates.

"We achieve this through a technique called data hygiene. How's that for a geeky term? Data hygiene is usually accomplished with specialized software, which is used to determine the validity of an address. This software matches the database address to its own internal database of valid streets and number ranges in a given city or zip code."

Ryan resumes, "One of the other data challenges faced by marketers is eliminating duplicate records in the marketing database. There are two types of duplicates we look for: internal duplicates, which are the existence of multiple records of the same customer or prospect, and external duplicates, which represent our inability to eliminate customers from our prospect data.

"Internal duplicates create reporting problems and increase mailing costs. External duplicates are even worse; they are both costly and embarrassing," Ryan explains. "One of the most embarrassing things for a sales representative is to make a prospecting call only to find out that the business is already our customer. The customer is generally left feeling like only a number in one of our computers. It creates a poor impression while wasting valuable time and resources."

In two paragraphs, describe some techniques Ryan could use to help identify internal and external duplicates in his company's marketing database. Describe how you would build a marketing database to minimize duplicates (use a paragraph). Are there operational methods that might cut down on this problem? List them. Who else in the organization could help with this process? Provide a brief list. In a paragraph, recommend methods to Chandler and your other systems analysis team members that can be used to help enlist and secure the assistance of other relevant organizational members.

Universities often allow free use of PCs or workstations, but one or two workstations will usually have time restrictions posted. These workstations can be used by individuals who only have to make minor changes to a computer program. The concept is similar to a grocery store express lane, but instead of "eight items or fewer," the posted sign in the computer room about the restricted workstation states, "Five minutes or less." Policing this limit, however, is difficult. This stage is where the design takes over. Rather than monitoring usage of the workstation and cutting off users after five minutes, designers solve the problem by making users stand, rather than sit. Because no chairs are provided for the express workstations, there is rarely a problem with the five-minute rule. Comfortable seating is just one of the variables in getting users to use the system.

Sometimes what exists in the user workspace will not be under your control as a systems analyst. If the opportunity arises, however, you should be able to recommend ergonomically sound workstation designs. The user's office or workstation should be viewed as an improvable context that must be adapted to the individual if you want the system to be used. The workspace variables important to consider include: room color and lighting; screen displays, user keyboards, mice; and computer desks and user seating.

COMPUTER ROOM COLOR AND LIGHTING

Computer rooms should be painted in muted colors with flat paints that neither assault the senses of their inhabitants nor give off a glare. Some colors and decoration schemes produce negative effects on users. Painting every wall in a light, bright color (brighter than the brightest possible computer displays) should be avoided, because it increases glare on computer displays. Decorating computer rooms in several different colors should also be avoided. Too many wall colors distract users from their tasks and can be confusing.

Rooms where computers or terminals are used should not be lit by fluorescent light, which is too bright for computer display use and has an almost imperceptible, yet continuous, flicker. It is better to provide incandescent lamps that produce more natural light or lamps that use halogen bulbs (although these can become very hot). Rooms with computer displays should be about half as bright as rooms in typical offices, where people process paperwork.

Do not allow displays to be placed where direct window light can hit the screen, because that creates tremendous glare. Ideally, users should be furnished with adjustable task lights at their workstations.

SCREEN DISPLAYS, KEYBOARDS, AND MICE

The design of efficient and effective screen displays was covered in Chapter 17, along with the design of user input and the use of screen color and icons. Some remaining ergonomic specifications and health considerations, however, also figure into the use of screen displays (also called monitors).

Researchers have found that the following guidelines will enhance the comfort of users who are viewing a display. In general, flexibility and adjustability for individual preferences are the watchwords. The more control given users over the features of their displays the better.

1. Displays should be within comfortable sightlines. Set displays at an angle 10 to 15 degrees from the vertical, away from the user. The user should look slightly downward; the screen should be 5 to 35 degrees from the horizontal.
2. Displays should be perceived as continuous images, which means no glare, no flickering of images, or no bleeding of characters into one another.
3. Displays should provide sharp contrast, with adjustable brightness and crisp image resolution.

Because the main form of data entry for most users is using a keyboard in front of a display, it becomes extremely important that displays are properly configured. Ideally, the refresh rate should be at least 85 Hz. A subtle flicker will cause eyestrain. In addition to the refresh rate, the colors, brightness, and screen geometry must be set up properly.

Excellent tools are available to adjust monitors. For example, DisplayMate for Windows from Sonera Technologies (www.displaymate.com) uses a number of test patterns to optimize your display. You can use DisplayMate test screens to adjust the size of the screen, using the pincushion, trapezoid, hourglass, and tilt

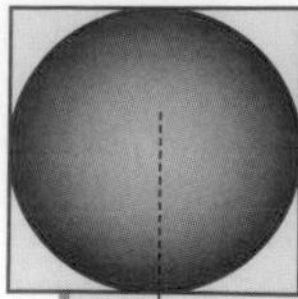

CONSULTING OPPORTUNITY 18.7

HEY, LOOK ME OVER (REPRISE)

You have been called back to take another look at Merman's Costumes. Here is part of the database created for Annie Oaklea of Merman's (with whom you last worked in Consulting Opportunities 9.2 and 10.1). The database contains information, such as the cost of the rental, date checked out, date due back, and the number of days the costume has been rented since the beginning of the year (YTD DAYS OUT) (see Figure 18.C1).

Analyzing Annie's typical day in the costume rental business, you realize there are several requests she must make of the database so that she can make decisions on when to replace frequently used costumes, or even when to buy more costumes of a particular type. She also needs to remember to keep in the good graces of customers she has previously turned down for a particular costume rental, she needs to know when to recall an overdue costume, and so on.

Formulate several queries that will help her get the information she needs from the database. (*Hint*: Make any assumptions necessary about the types of information she needs to make decisions and use as many of the different query types discussed in this chapter as you can.) In a paragraph, describe how Annie's queries would be different if she were working with a Web-based or hyperlinked system.

COSTUME-RENTAL

COSTUME NUMBER	DESCRIPTION	SUIT NUMBER	COLOR	COST OF	DATE CHECKED OUT	DUE DATE	YTD DAYS OUT	TYPE OF COSTUME	REQUESTS TURNED DOWN
0003	Lady MacBeth F, SM	01	Blue	15.00	10/15	11/30	150	Standard	2
1342	Bear F, MED	01	Dk. Brown	12.50	10/24	11/09	26	Standard	0
1344	Bear F, MED	02	Dk. Brown	12.50	10/24	11/09	115	Standard	0
1347	Bear F, LG	01	Black	12.50	10/24	11/09	22	Standard	0
1348	Bear F, LG	02	Black	12.50	11/01	11/08	10	Standard	0
1400	Goldilocks F, MED	01	Light Blue	7.00	10/24	11/09	140	Standard	0
1402	Goldilocks F, MED	02	Light Blue	7.00	10/28	11/09	10	Standard	0
1852	Hamlet M, MED	01	Dark Green	15.00	11/02	11/23	115	Standard	3
1853	Ophelia F, SM	01	Light Blue	15.00	11/02	11/23	22	Standard	0
4715	Prince M, LG	01	White/purple	10.00	11/04	11/21	145	Standard	5
4730	Frog M, SM	01	Green	7.00	11/04	11/21	175	Standard	2
7822	Jester M, MED	01	Multi	7.50	11/10	12/08	12	Standard	0
7824	Jester M, MED	02	Multi	7.50	11/09	11/15	10	Standard	0
7823	Executioner M, LG	01	Black	7.00	11/19	12/05	21	Standard	0
8645	Mr. Spock N, LG	01	Orange	18.00	09/07	09/12	150	Trendy	4
9000	Pantomime F, LG	01	Red	7.00	08/25	09/15	56	Standard	0
9001	Pantomime M, MED	01	Blue	7.00	08/25	09/15	72	Standard	0
9121	Juggler M, MED	01	Multi	7.00	11/05	11/19	14	Standard	0
9156	Napoleon M, SM	01	Blue/white	15.00	10/26	11/23	56	Standard	1

FIGURE 18.C1

A portion of the database from Merman's costume rental shop.

controls on your monitor so that your lines are indeed straight and your circles round. You can also eliminate distracting moiré patterns and calibrate the color and intensity. All these adjustments make for a better visual working environment.

Concerns have been voiced about possible health risks for users, some of whom have complained of physical symptoms ranging from headaches, eyestrain, and backaches to problem pregnancies and the early onset of cataracts. Studies of the effects of the extremely low-frequency radiation of cathode-ray tubes (CRTs) have been controversial to date. CRTs use the same technology as television tubes. A stream of electrons are usually beamed from three guns (or one, as in Trinitron tubes) at a layer of phosphors. The glowing phosphors are what we see.

There are alternatives to CRTs. Liquid crystal displays (LCDs) have a thin layer of material that can either block or transmit light. Because they are thin and light, they are used in notebooks. Expect LCDs to show up on desktops, replacing large CRTs. Another alternative, the field emission display (FED), has a layer of phosphors similar to a CRT but has hundreds of tiny electron transmitters behind each pixel rather than three large electronic guns. Use of LCDs and FEDs should cut down on some of the health concerns traditionally associated with CRTs.

User keyboards for computer interface should be as flexible and adjustable as possible. Detachability and light weight allow changes in sitting position for users, who need a break from doing data entry in a continuous pose. Remote mice, wireless mice, egg-shaped mice, and foot pedal controls are examples of designs developed to reduce health problems associated with overuse of hand and wrist muscles.

WORKSPACE FURNISHINGS

Workspace furnishings include the desks or stands on which computers and keyboards rest. Many times computers are put into existing furniture setups. Although an old typing table might work in a pinch, because it is 26 inches high, putting a keyboard on an already existing desktop is not a good idea. The standard desktop height of 29 inches is too high for continuous keyboard use, because that height causes back, neck, and arm strain. If users have both a desk and a computer table, the desk should be 29 inches high and the computer table should be 26 inches high. Both can be joined into an L-shaped configuration for ease of access.

As we saw in the example at the beginning of this section, user seating is important enough to influence whether systems are used and with what satisfaction. Getting the correct chair can also increase user comfort and productivity. Well-designed seating often looks as if it had been sculpted around a person. Users' chairs should be armless with firm upper- and lower-back support (preferably adjustable for the individual), and a firm seat cushion should provide back support. Chairs with five casters provide excellent balance and mobility.

Seating height should be adjustable for the individual, but clearly it should preserve the prescribed relationship between the user's gaze and the display as well as preserve the relationship among the user's posture, keyboard, and mouse (or other input device). To get users to use the system, well-designed workstations are essential.

SUMMARY

We have focused on system users, their interface with the computer, their need for feedback, designing ecommerce Web site feedback and navigation, the design of database queries, and the design of user workstations in this chapter. The success of the systems you design depends on user involvement and acceptance. Therefore, thinking about users in systematic and empathic ways is of utmost importance and is not a peripheral issue for systems analysts.

A variety of user interfaces and input devices were covered in this chapter. Some interfaces are particularly well-suited to inexperienced users, such as natural language, question and answer, menus, form-fill and Web-based form-fill, graphical user interfaces (especially on Web pages), the mouse, lightpens, the stylus, touch-sensitive screens, and voice recognition systems. Command language is better suited to experienced users. Some authors believe that the Web itself is the single most important interface for the future work of applications designers.

Combinations of interfaces can be extremely effective. For example, using pull-down menus with graphical user interfaces or employing nested menus within question-and-answer interfaces yields interesting combinations. Each interface

18

"I have no problem with using a mouse, or any other rodent you throw my way. Really, though, whatever Snowden needs is what I try to do. Everyone is different, however. I've seen people here go out of their way to avoid using a computer altogether. Other people would prefer not to talk with a human. In fact, they would be as happy as a puppy chewing on a new bedroom slipper if they could use command language to interact. I have a hunch they would prefer not talking to people at all, but that's just an impression. Most of the folks we have here are open to new things. Otherwise, they wouldn't be here at MRE. We do pride ourselves on our creativity. I have you signed up for a meeting with people from the Training group, including Tom Ketchem, Melissa Smith, and Kathy Blandford. You can invite anyone else you think should be included. Snowden may sit in as well, if he has time. That's why he asked me to relay the message, I guess. They'll be very curious to see what kind of interface you are suggesting for them on the new project reporting system."

HYPERCASE® QUESTIONS

1. Write a short proposal describing what type of user interface would be appropriate for the users of the project reporting system who are in the Training group. Include reasons for your decision.
2. Design a user interface using a CASE tool, such as Visible Analyst, a software package such as Microsoft Access, or paper layout forms. What are the key features that address the needs of the people in the Training group?
3. Demonstrate your interface to a group of students who can role play as members of the Training group. Ask for reactions.
4. Redesign the interface based on the feedback you have received. Write a paragraph to say how your new design addresses any comments you have received.

FIGURE 18 HC.1

In HyperCase, you can see how users process information in order to create a more effective user interface.

poses a different level of challenge for programmers, with natural language being the most difficult to program. The Web has posed new challenges for designers, because the user is not known. Web design takes advantage of hyperlinks to allow users to take numerous paths as they interact with the Web site.

Users' need for feedback from the system was also stressed. System feedback is necessary to let users know if their input is being accepted, if input is or is not in the correct form, if processing is going on, if requests can or cannot be processed, and if more detailed information is available and how to get it. Feedback is most often visual, with text, graphics, or icons being used. Audio feedback can also be effective.

Special considerations apply to designing ecommerce Web sites. Build improved functionality into the application by eliciting customer feedback through automatic email feedback buttons or by including blank feedback forms on the Web site.

In addition, four important navigation design strategies improve the stickiness of ecommerce Web sites: rollover menus, hierarchical displays of links on the entry screen, site maps, and navigation bars that provide one-click navigation that makes navigating the site and returning to the site as easy as possible for the customer.

Queries are designed to allow users to extract meaningful data from the database. There are six basic types of queries, and they can be combined using Boolean logic to form more complex queries.

Some of the principles about data queries you learned can be used in Web searches. Internet search tools are called search engines. Users can be more effective if searches are carefully planned and logically structured.

Data mining involves using a database for the more selective targeting of customers. Assuming past behavior is a good predictor for future purchases, companies collect data about a person from past credit card purchases, driver's license applications, warranty cards, and so on. Data mining can be powerful, but it may be costly and it needs to be coordinated. In addition, it may infringe on consumer privacy or even a person's civil rights.

Finally, we considered both how users' workspaces influence their willingness to use the system and how workstations can be improved through the implementation of relevant ergonomic principles. There are specific productivity and comfort guidelines on the construction and positioning of displays, keyboards, computer stands, and users seating, but generally they all should be flexible enough to permit adjustment for individual use.

KEYWORDS AND PHRASES

- Boolean operators
- command-language interfaces
- continuous speech systems
- cue cards
- data mining
- data warehousing
- dialog box
- ergonomics
- ergonomic specifications
- feedback
- feedback for users
- form-fill (input/output form) interfaces
- graphical user interface (GUI)
- intuitive navigation
- menus
- natural-language interfaces
- navigation bar
- nested menus
- one-click navigation
- pull-down menus
- queries
- question-and-answer interfaces
- rollover menu
- search engine
- site map
- speaker independence
- speech recognition and synthesis
- stickiness
- structured query language (SQL)

stylus
templates
touch-sensitive screen
Web form-fill interfaces
Web searching
wizards

REVIEW QUESTIONS

1. What are the five objectives for designing user interfaces?
2. Define natural-language interfaces. What is their major drawback?
3. Explain what is meant by question-and-answer interfaces. To what kind of users are they best suited?
4. Describe how users use onscreen menus.
5. What is a nested menu? What are its advantages?
6. Define onscreen input/output forms. What is their chief advantage?
7. What are the advantages of Web-based fill-in forms?
8. What are the drawbacks of Web-based form-fill interfaces?
9. Explain what command-language interfaces are. To what types of users are they best suited?
10. Define graphical user interfaces. What is the key difficulty they present for programmers?
11. For what type of user is a GUI particularly effective?
12. What are the three guidelines for designing good screen dialog?
13. What are the roles of icons, graphics, and color in providing feedback?
14. List six ways for achieving the goal of minimal operator action when designing a user interface.
15. List five standards that can aid in evaluating user interfaces.
16. What are the seven situations that require feedback for users?
17. What is an acceptable way of telling the user that input was accepted?
18. When a user is informed that his or her input is not in the correct form, what additional feedback should be given at the same time?
19. Why is it unacceptable to notify the user that input is not correct solely through the use of beeping or buzzing?
20. When a request is not completed, what feedback should be provided to the user?
21. Describe two types of ecommerce Web site design for eliciting feedback from Web site customers.
22. List four practical ways that an analyst can improve the ease of user navigation and the stickiness of an ecommerce Web site.
23. What are hypertext links? Where should they be used?
24. List in shorthand notation the six basic query types.
25. List six guidelines for searching the Web.
26. What is the purpose of data mining?
27. What sort of information is "mined for" in data mining?
28. Describe four problems with data mining.
29. Give three guidelines for setting up computer displays for user productivity and comfort.
30. What is the preferred height for computer desks? Why is the height important?

PROBLEMS

1. Design a nested menus interface for a check-in and check-out hotel reservation system. Use numbers to select a menu item. Show how each menu would look on a standard PC display.
2. Design a form-fill interface for the inventory control of a musical computer disc and tape wholesale company that could be used on a PC display screen.

3. Design a Web-based form-fill interface to accomplish the same task as in problem 2 above.
   a. What difficulties did you encounter? Discuss them in a paragraph.
   b. Of the two designs you did, which would you say is better suited to the task? Why? List three reasons for your choice.
4. Design a command-language interface that a travel agent would use to book seats for an airline.
   a. Show what it would look like on a standard display screen.
   b. Make a list of commands needed to book an airline seat and write down what each command means.
5. Design a graphical user interface for an executive desktop. Use icons for file cabinets, wastebasket, telephone, and so on. Show how they would appear on the computer display.
6. Design a screen that provides appropriate feedback for a user whose command cannot be executed.
7. Design a screen for a payroll software package that displays information telling the user how to get more detailed feedback.
8. Design a Web-based screen that displays an acceptable way to tell users that their input was accepted.
9. Design a feedback form for customers using an ecommerce Web site.
10. Write six different queries for the file in problem 1 in Chapter 17.
11. Write six different queries for the 3NF relation in problem 6 in Chapter 17.
12. Design a search that will find potential competitors of a company such as World's Trend on the Web. Assume you are the customer.
13. Search for World's Trend's potential competitors on the Web. (Remember that you won't find World's Trend itself on the Web. It is a fictional company.) Make a list of those you've found.
14. Design a data mining project for World's Trend. What information is needed? Suggest some projects you can undertake with data mining to improve marketing efforts at World's Trend.
15. Taking the data from your search for competitors in problem 12 above, make a brief list of Web site innovations they are using that could be of use to World's Trend in its data mining project (see problem 13 above).
16. Draw up a set of ethical guidelines or create a policy to help analysts evaluate the appropriate uses for data mining (you may also have to define what uses are not appropriate to do so). In a paragraph, discuss some precautions that should be taken by the analyst to protect the consumer privacy that may be at stake in data mining projects.

GROUP PROJECTS

1. With your group members, create a pull-down menu for an employment agency that matches professional candidates to position openings. Include a list of keystrokes that would directly invoke the menu options using the ALT-X format. The menu has the following options:

Add employee	Add employer	Add position
Change employee	Change employer	Change position
Delete employee	Delete employer	Delete position
Employee inquiry	Match employee to opening	
Position inquiry	Print open positions report	
Employer inquiry	Print successful matches report	

2. In a paragraph, describe the problems your group faced in creating this menu.
3. The drag-and-drop feature is used in GUIs and allows the user to move sentences around in a word-processing package. As a group, suggest how drag-and-drop can be used to its fullest potential in the following applications:
   a. Project management software (Chapter 3).
   b. Relational database program (Chapter 17).
   c. Screen or forms designer (Chapter 16).
   d. Presentation package (Chapter 14).
   e. Spreadsheet program (Chapter 12).
   f. CASE tool for drawing data flow diagrams (Chapter 9).
   g. Fax program (Chapter 15).
   h. File management program (Chapter 18).
   i. Personal information manager (PIM) (Chapter 3).
   j. Illustration in a drawing package (Chapter 14).
   k. CASE tool for developing data dictionaries (Chapter 10).
   l. Decision tree drawing program (Chapter 11).
   m. Web site for collecting consumer opinions on new products (Chapter 15).
   n. Organizing bookmarks for Web sites.

   For each solution your group designs, draw the screen and show movement by using an arrow.
4. Ask all the members of your group to request a search based on their hobbies or leisure activities. If there are four people in your group, there will be four unique searches to perform. Now go ahead and do all the searches. Compare your results. Does the person who is involved with the hobby have an advantage over the people who know less about the hobby? Explain.

SELECTED BIBLIOGRAPHY

Benbesat, I., and R. G. Schroeder. "An Experimental Investigation of Some MIS Design Variables." *MI Systems Quarterly*, Vol. 2, No. 2, 1978, pp. 43–54.

Bennett, J. L. *Building Decision Support Systems*. Reading, MA: Addison-Wesley, 1983.

Bort, J. "Navigation: An Art for E-Com Sites," *Microtimes.com*, Issue #201, December 1999, http://microtimes.com/201/ecombort201a.html, last accessed on February 9, 2001.

Davis, G. B., and M. H. Olson. *Management Information Systems: Conceptual Foundations, Structure, and Development*. New York: McGraw-Hill, 1985.

Dietsch, D. "Ergo, Ergonomics." *A+*, Vol. 3, Issue 6, 1985.

Gane, C., and T. Sarson. *Structured Systems Analysis: Tools and Techniques*. Englewood Cliffs, NJ: Prentice Hall, 1979.

Kleiner, A. (ed.). "The Health Hazards of Computers." *Whole Earth Review*, No. 48, 1985, pp. 80–93.

Large, P. *The Micro Revolution Revisited*. London: Franses Pinter (Publishers), 1984.

Laudon, K. C., and J. P. Laudon. *Management Information Systems*, 4th ed. New York: Macmillan, 1996.

Newman, W. M., and M. G. Lamming. *Interactive System Design*. Reading, MA: Addison-Wesley, 1995.

Sano, D. *Designing Large-Scale Web Sites: A Visual Design Methodology*. New York: John Wiley, 1996.

Strom, D. "Net Management via Web: Another Pretty Interface?" *PC Week*, Vol. 14, No. 32A, 1997, p. 25.

ALLEN SCHMIDT, JULIE E. KENDALL, AND KENNETH E. KENDALL

UP TO THE USERS

"Let's take our prototypes and some new screens, reports, and forms to create the final user interface," Anna says to Chip. "It's about time, isn't it?" replies Chip. He was all too aware of the importance of designing a good interface.

After talking, they set up the following screen dialog guidelines:

1. Well-designed screens should:
   - Communicate actions and intentions clearly to users.
   - Show options available to operators. Examples are:
     - MAKE CORRECTIONS OR PRESS ESC TO CANCEL
     - ENTER HARDWARE INVENTORY NUMBER OR PRESS F10 TO EXIT
     - PRESS ENTER KEY
     - PRESS ENTER TO CONFIRM DELETE, ESC TO CANCEL
     - Buttons that say OK or CANCEL
   - Standardize use of any abbreviations.
   - Avoid the use of codes, substituting the code meaning.
   - Provide help screens for complicated portions of the dialog.
   - Provide tool-tip help for toolbar icons.
2. Feedback should be provided to the users. Feedback includes:
   - Titles to show the current screen.
   - Actions successfully completed messages, such as:
     - RECORD HAS BEEN ADDED
     - RECORD HAS BEEN CHANGED
   - Error messages. Examples are:
     - INVALID DATE
     - CHECKDIGIT IS INVALID
     - SOFTWARE IS NOT ON FILE
   - An invalid data dialog box, with an OK button on a graphical user interface screen.
   - Processing delay messages similar to:
     - PLEASE WAIT—REPORT IS BEING PRODUCED
   - An hourglass turning upside down on a graphical user interface.
3. There should be consistency in the design, including:
   - Location of the OPERATOR MESSAGE line, line 23 on every screen.
   - Location of the FEEDBACK MESSAGE line, chosen as line 24.
   - Date, time, system name, and screen reference number should appear in heading lines.
   - Consistent exit of all screens, such as through the use of the same function key.
   - Standard use of keys, such as PgDn and PgUp, to display a next or previous screen within a multiple-screen display.
   - A consistent method of canceling an operation, such as through the use of the Escape key (Esc).
   - Standardized use of color and high-intensity display, such as all error messages appearing in red.
   - Standardized use of icons on a GUI screen.
   - Standardized pull-down menus on a GUI screen.

4. Minimum operator actions should be required to use the system. Some examples are:
   - The use of **Y** and **N** as Yes and No replies. The use of the plus and minus signs on the number pad as a substitute for **Y** and **N**.
   - When changing or deleting records, only the record key need be specified. The system would obtain the record and display pertinent information.
   - When names are required as key entries, only the first few letters of the name need be entered. The program should find all matching record key names and present them for selection by the operator.
   - Data-entry screens should allow the entry of codes.
   - All numeric entries may omit leading zeros, commas, or a decimal point.
   - As each data field is completed, the cursor should advance to the next entry field.
   - After each option is completed, the same screen, with blank entry areas, should be redisplayed until the Exit key is pressed.
   - When an option is exited, the previous menu is to be displayed.
   - Drop-down list boxes should be used whenever possible on GUI screens.
   - Check boxes and radio buttons should be used to make selections whenever possible.
   - Default buttons should be outlined so the user can press the Enter key to press them.
5. Data entering the system should be validated. Guidelines are:
   - Specific fields should be verified according to edit criteria.
   - As errors are detected, operators should be given a chance either to correct the error or to cancel the transaction.
   - When no errors have been detected in a transaction, the screen should be presented to the operator for visual confirmation. The operator should have the opportunity either to accept the screen or to make corrections to the data entered.

Because the system is to be written using Microsoft Access, Chip and Anna need to decide whether to use a menu screen or pull-down menus to control the system. The advantage of using a menu is that all the options may be written as independent objects and linked under the menu screen. Such a procedure also allows the flexibility of easy expansion as the system evolves. After Anna and Chip conferred with the users, a consensus was obtained that a menu screen was indeed the best choice. Hy Perteks and Mike Crowe were strongly in favor of the menu screen, but others wanted to use pull-down menus. Dot was strongly in favor of the pull-down menu-controlled system since it was easy for new employees to learn. She also wanted a button bar for the most commonly used features.

Upon examining the many screens and reports (over 30 in all), Chip and Anna decided to split the menu into several functions. "How do we divide these various functions into a set of menus?" asked Chip.

"Why don't we use a decomposition diagram to organize the functions into a hierarchy," replied Anna. Chip and Anna began working on the diagram. The menu interactions will be represented in a hierarchical structure, with options shown as rectangles and the overall menu represented by the rectangle on the top. Each secondary menu will be shown beneath the primary menu, with screen programs at the lowest level. The main menu will have six main choices—Update Software, Update Hardware, Inquiry, Modify Codes, Training, and Reports—as illustrated in Figure E18.1. Each of these options is further subdivided into smaller menus or individual functions. The Inquiry menu is subdivided into two smaller menus, Software Options and Hardware Options, as well as options for running the Software Expert Inquiry and the Printer Location Inquiry.

The rectangles on the functional decomposition diagram are implemented using a series of pull-down menu lists, which are shown in Figure E18.2. Notice that the Inquiry menu has functions corresponding to the rectangles on the decomposition diagram illustrated in the first figure in this episode. A row of buttons for common functions is included below the menus. The menu functions are included as a set of buttons in the main area of the screen, and these buttons may be clicked to run corresponding programs. It was decided that the Add Microcomputer, Add Software Package, and Change Microcomputer programs would be run directly from the main menu. Clicking the other buttons causes selection dialog boxes to display, with choices for selecting programs. Figure E18.3 shows the dialog box for the **Reports . . .** option. All the reports are listed, with **Print Preview**, **Print**, and **Close Form** buttons for selecting actions.

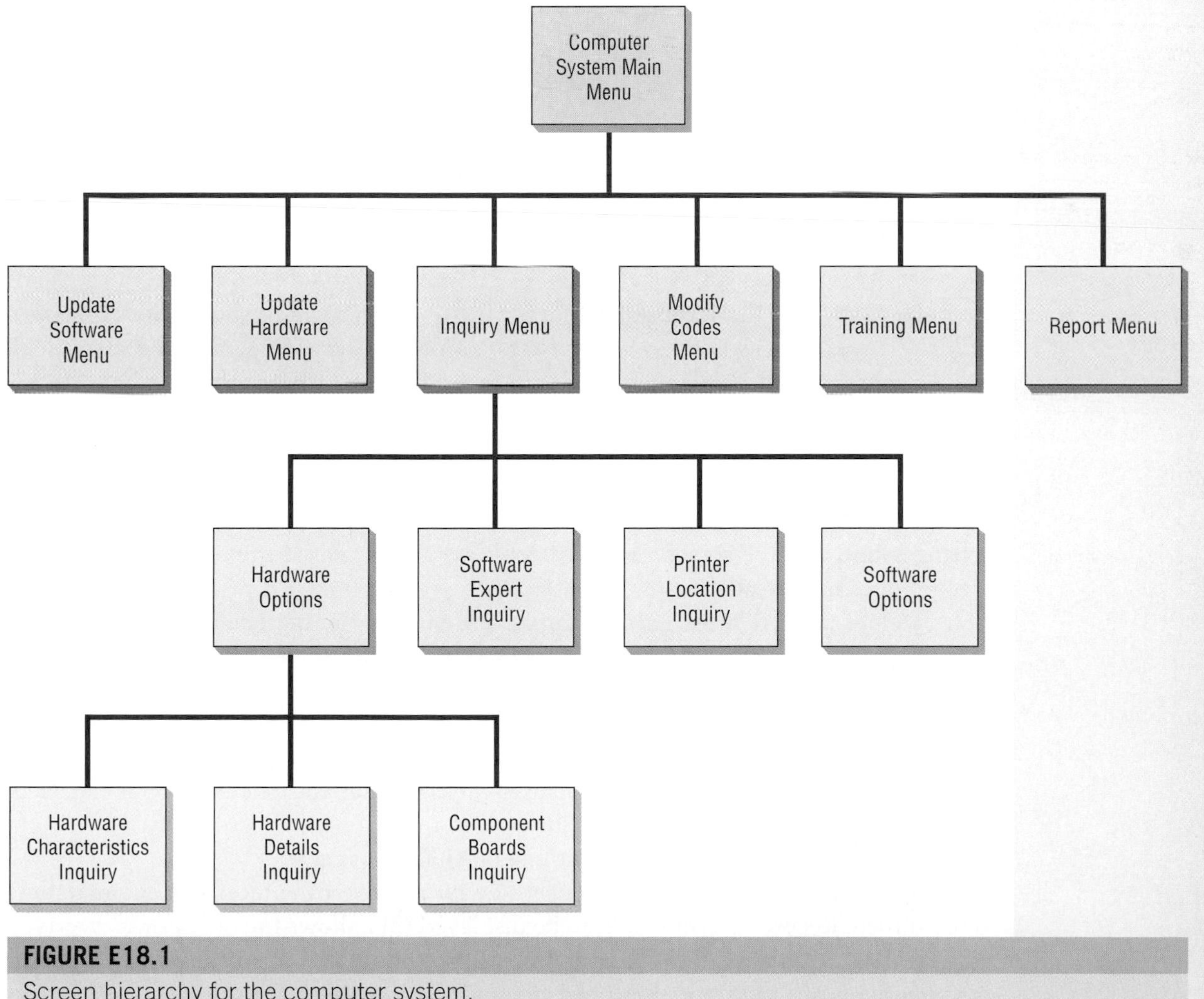

FIGURE E18.1
Screen hierarchy for the computer system.

18

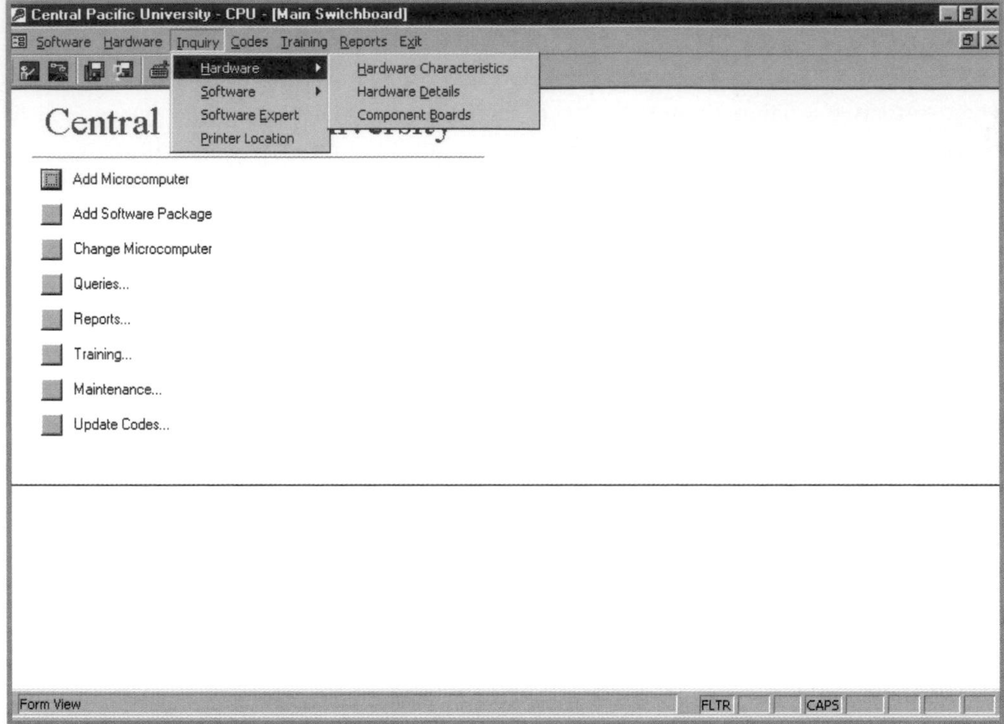

FIGURE E18.2
The main menu for the computer system.

"Here's what I think the guidelines for the update programs should be," Anna tells Chip. "The key focus is on accuracy, with comprehensive editing for each data field. Add programs will display an entry screen and allow either hardware or software records to be created. After all entries are complete, a user should double-check the data and press the Add Software Record button. Any data that are already on the system should be implemented using drop-down lists. [Refer to Figure E18.4.] There are also buttons to undo changes, move to different records, print the record, save the changes, and exit the screen. A record could be added only if the primary key for the record does not already exist.

"Delete screens must have a simple, primary key entry, such as the COURSE DESCRIPTION in the DELETE SOFTWARE COURSE screen," Anna continues. The DELETE SOFTWARE COURSE screen is illustrated in Figure E18.5. "The use of a **Find** button (the binoculars) helps to locate the desired record. The corresponding record is read and the information is displayed. Users press the **Delete** button and are prompted to confirm the delete. If the user clicks **Cancel**, the delete action is canceled. How does all that sound?" she asks Chip.

"So far, so good," he replies. "Anything on change screens?"

"Yes. Change screens have a primary key for the record entered and the matching record read. Record information is to be displayed that allows the operator to overtype the data with changes. All changes are to be validated with full editing. When all change fields are valid, the user must press a button to save the changes. Is that clear enough for the user?" Anna asks.

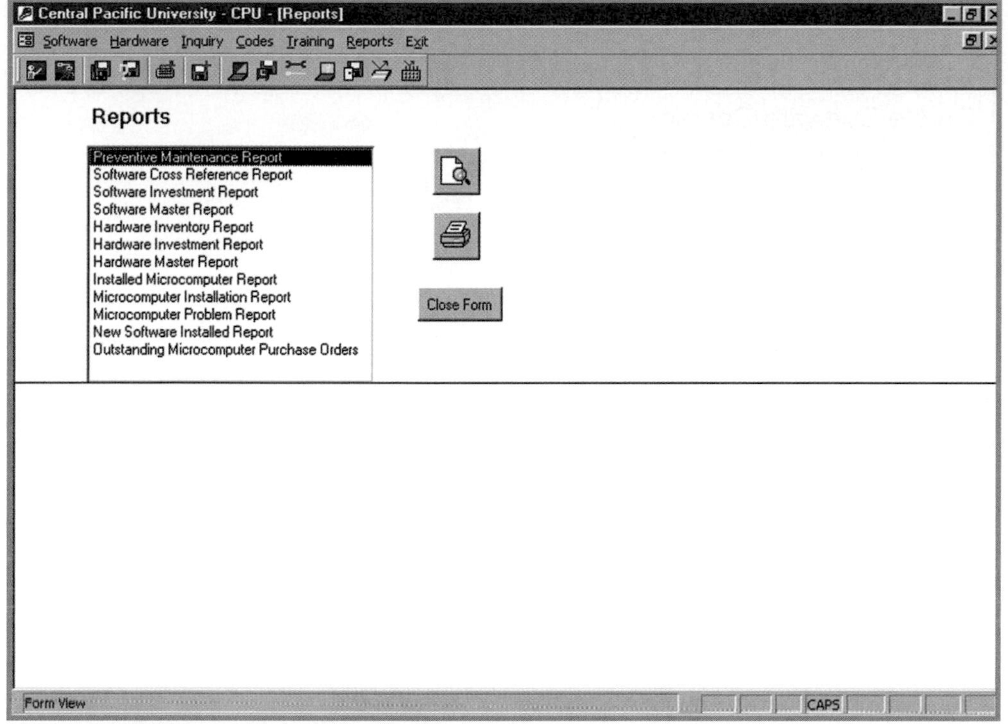

FIGURE E18.3
A dialog box for the computer system report menu.

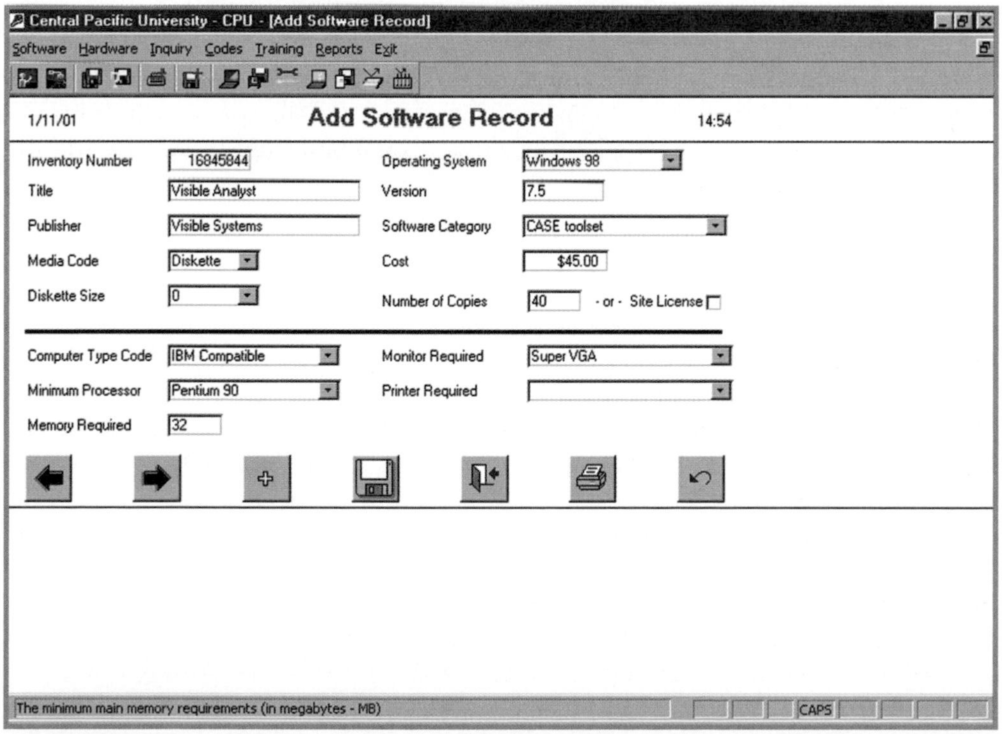

Central Pacific University - CPU - [Add Software Record]
Software Hardware Inquiry Codes Training Reports Exit

1/11/01 **Add Software Record** 14:54

Inventory Number 16845844
Title Visible Analyst
Publisher Visible Systems
Media Code Diskette
Diskette Size 0
Operating System Windows 98
Version 7.5
Software Category CASE toolset
Cost $45.00
Number of Copies 40 - or - Site License

Computer Type Code IBM Compatible
Minimum Processor Pentium 90
Memory Required 32
Monitor Required Super VGA
Printer Required

The minimum main memory requirements (in megabytes - MB)
CAPS

FIGURE E18.4
ADD SOFTWARE RECORD screen showing command buttons and drop-down lists.

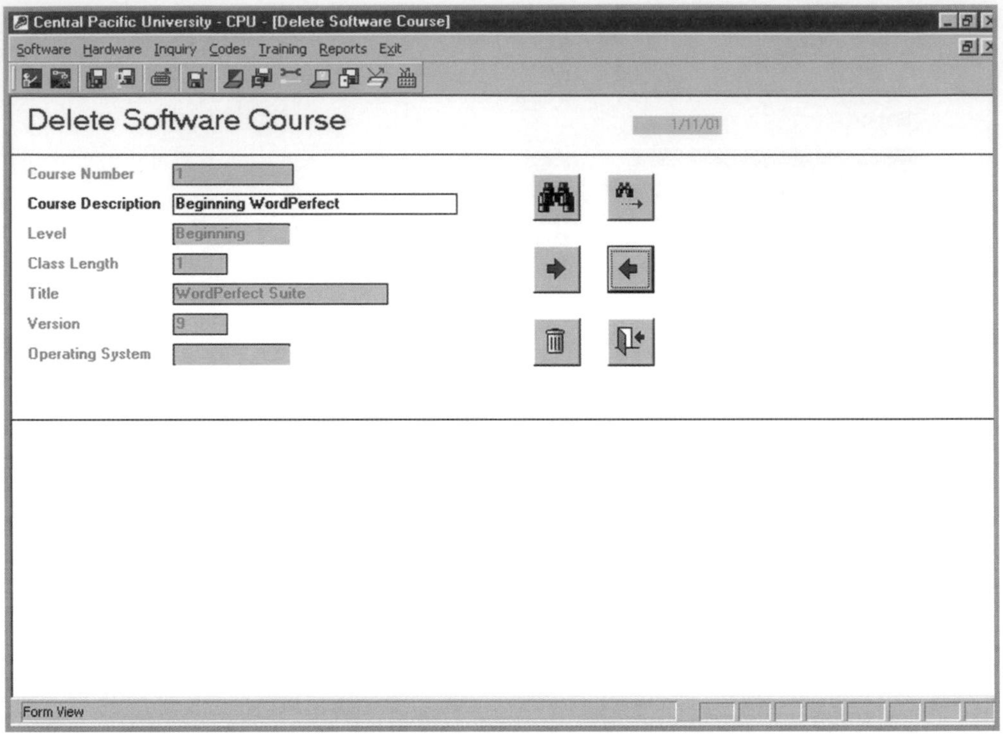

Central Pacific University - CPU - [Delete Software Course]
Software Hardware Inquiry Codes Training Reports Exit

Delete Software Course 1/11/01

Course Number	1
Course Description	Beginning WordPerfect
Level	Beginning
Class Length	1
Title	WordPerfect Suite
Version	9
Operating System	

Form View

FIGURE E18.5
DELETE SOFTWARE COURSE screen.

"I think it's very good," Chip acknowledges.

Chip is responsible for the inquiry portion of the system. The focus on these programs is speed. A short entry is obtained from the user, and the corresponding records are read. Information is formatted for maximum communication and displayed. "I've met with various users," he tells Anna. "Here's a list of inquiry programs." Each of the inquiry screens is designed, along with the database files needed and possible errors that could occur.

"The first screen I designed was the HARDWARE INQUIRY," Chip continues. "I used the description of the screen that we had put into the Visible Analyst repository after the prototypes had been created." The repository screen is shown in Figure E18.6. Notice that the **Notes** area contains information about how the screen should operate. An INVENTORY NUMBER or partial INVENTORY NUMBER should be entered. The first matching record is read (in the case of a partial INVENTORY NUMBER), and the user can scroll to the next record or to previous records.

"I produced a rough layout and met with Dot to obtain feedback on the design," Chip says. "After pointing out some minor corrections, she mentioned that the maintenance details should be included, providing complete information for each microcomputer."

The program logic is to use the HARDWARE INVENTORY NUMBER as the entry field of a Parameter Value dialog box (illustrated in Figure E18.7), with the starting number of 3 entered. The record is located in the database. If it is not found, a message is displayed. Once the record is located, the matching board records are read. Board records contain a code for the type of board, and the BOARD CODE TABLE is searched for the

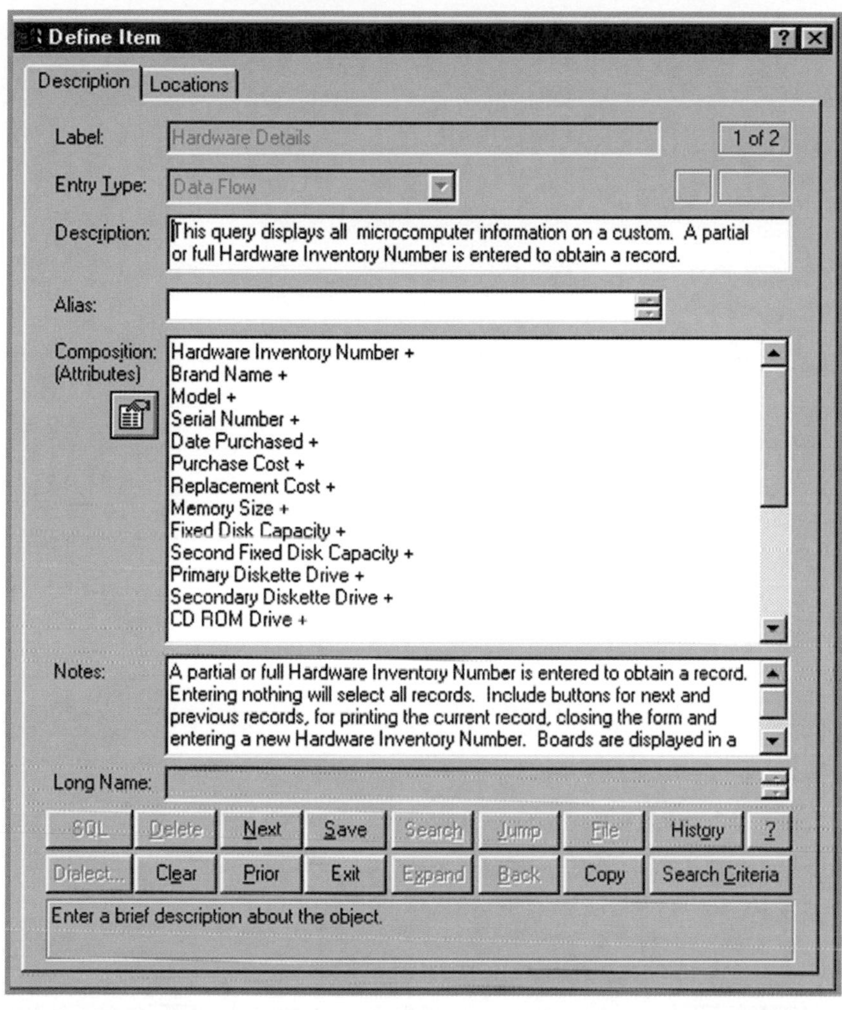

FIGURE E18.6
HARDWARE DETAILS repository screen using Visible Analyst software.

matching code. The meaning of the code is formatted on the screen. The resulting screen is displayed in Figure E18.8. Notice that there are buttons for printing the current record on the form and closing the form. The **New Inventory Number** button redisplays the Parameter Entry dialog box, and it allows the user to choose a new record.

"I selected the SOFTWARE LOCATION inquiry as the next screen to develop," Chip tells Anna. "After talking at length with Cher, I produced the details and documented them in the Visible Analyst repository [illustrated in Figure E18.9]. The entry field is a partial software TITLE, entered in a Parameter Value dialog box. The first record matching the partial title is displayed, and because there are different operating systems and versions of the software, the user can click buttons to advance to the next (or previous) record. Five columns of information are displayed: HARDWARE INVENTORY NUMBER, BRAND NAME, MODEL, CAMPUS, and ROOM. Cher can quickly locate a machine containing the desired software. She seems to be happy with this idea so far," Chip adds.

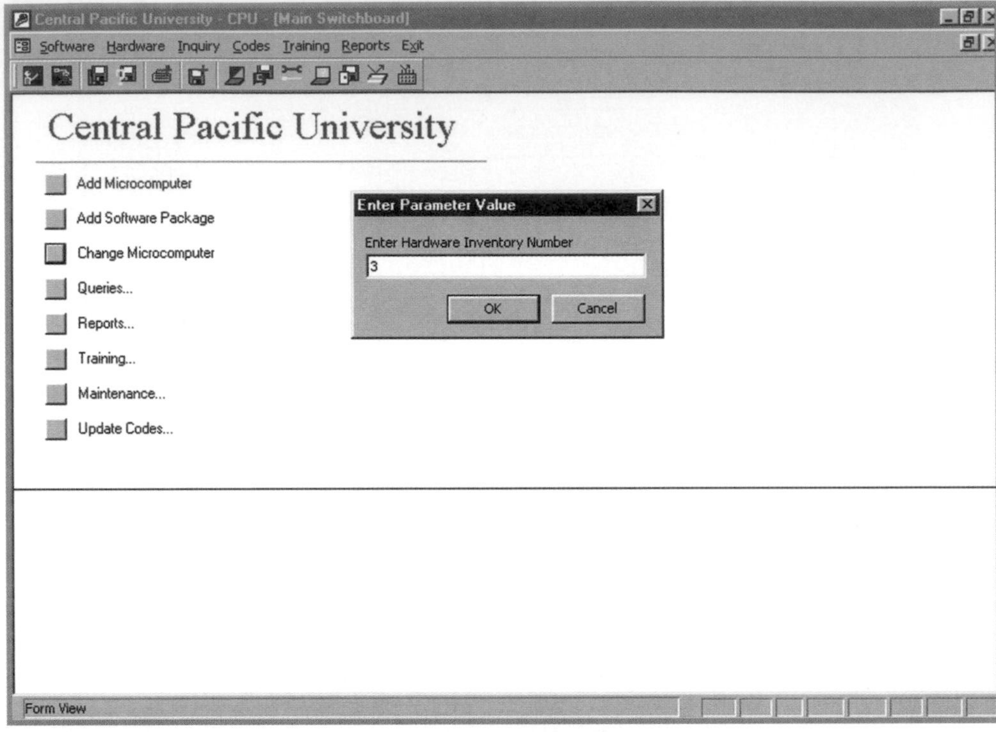

FIGURE E18.7

HARDWARE INVENTORY NUMBER parameter value dialog box.

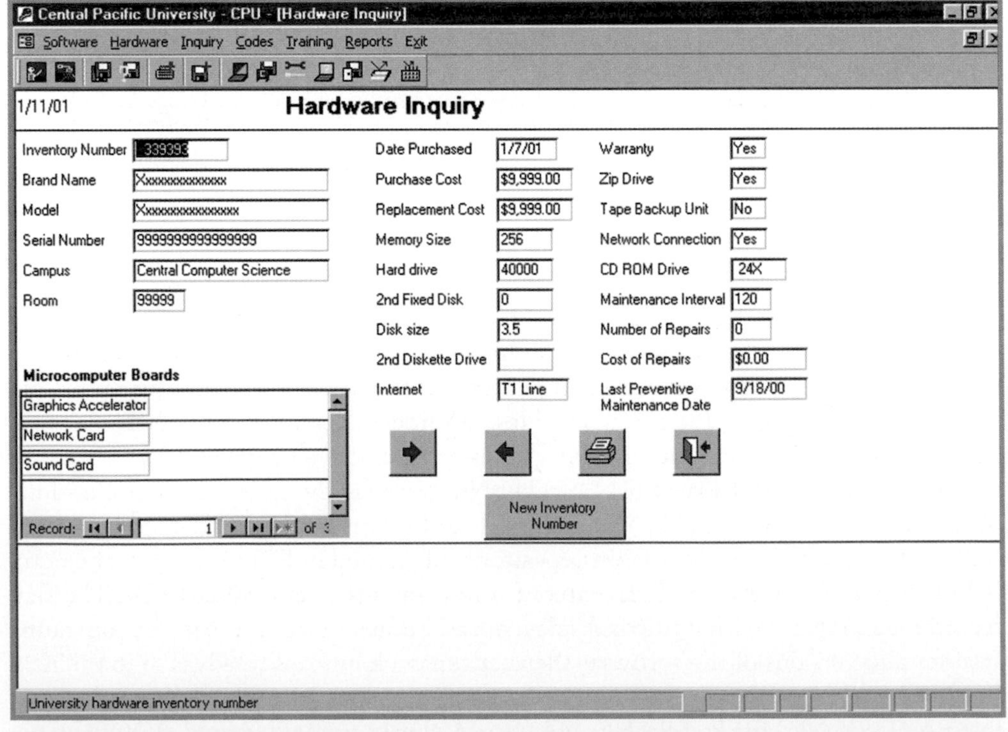

FIGURE E18.8

HARDWARE INVENTORY NUMBER inquiry screen as represented in Microsoft Access.

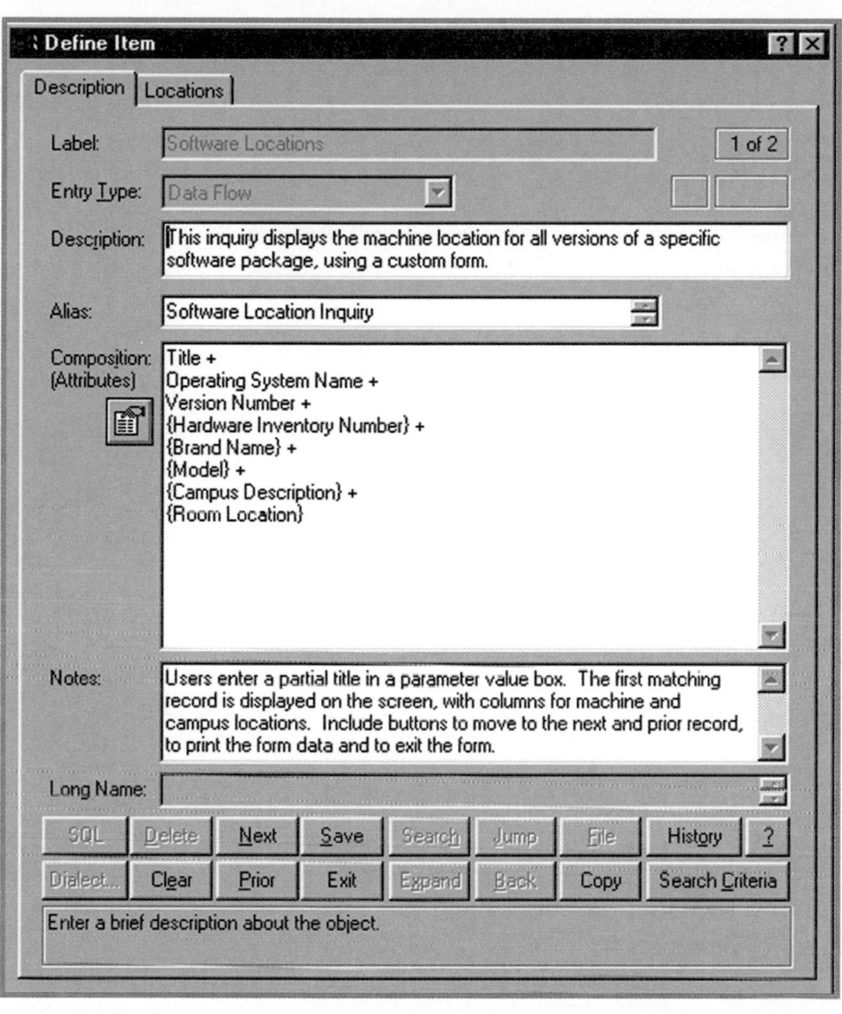

FIGURE E18.9
Visible Analyst repository screen for the SOFTWARE LOCATION inquiry.

The Parameter Value dialog box is illustrated in Figure E18.10, and the SOFTWARE LOCATION inquiry screen is shown in Figure E18.11. The program locates the SOFTWARE MASTER file using the alternate key TITLE. If the matching record is not found, an error message is displayed. Because there may be several versions, the **Next Record** button may be pressed until the correct OPERATING SYSTEM and VERSION NUMBER are obtained.

Once the correct software has been obtained, the relational file is used to find the matching SOFTWARE INVENTORY NUMBER. This relational file contains the SOFTWARE INVENTORY NUMBER and the matching HARDWARE INVENTORY NUMBER, which is used to locate the matching record in the MICROCOMPUTER MASTER file. For each matching machine, the CAMPUS table is used to locate the CAMPUS LOCATION code and to display the matching CAMPUS DESCRIPTION. The area for displaying the machines containing the software is a scroll region, because it may contain more machines than will fit on a single screen.

"I think we've got a good start on designing our user interfaces," Anna comments. Chip nods in agreement.

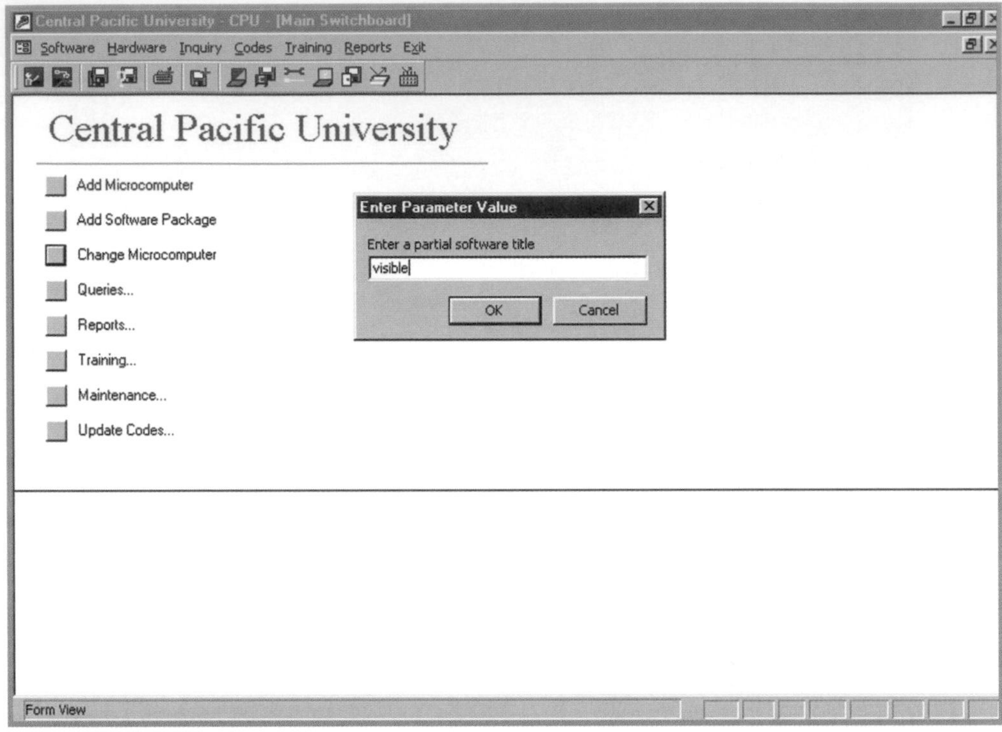

FIGURE E18.10
SOFTWARE LOCATION parameter value entry dialog box.

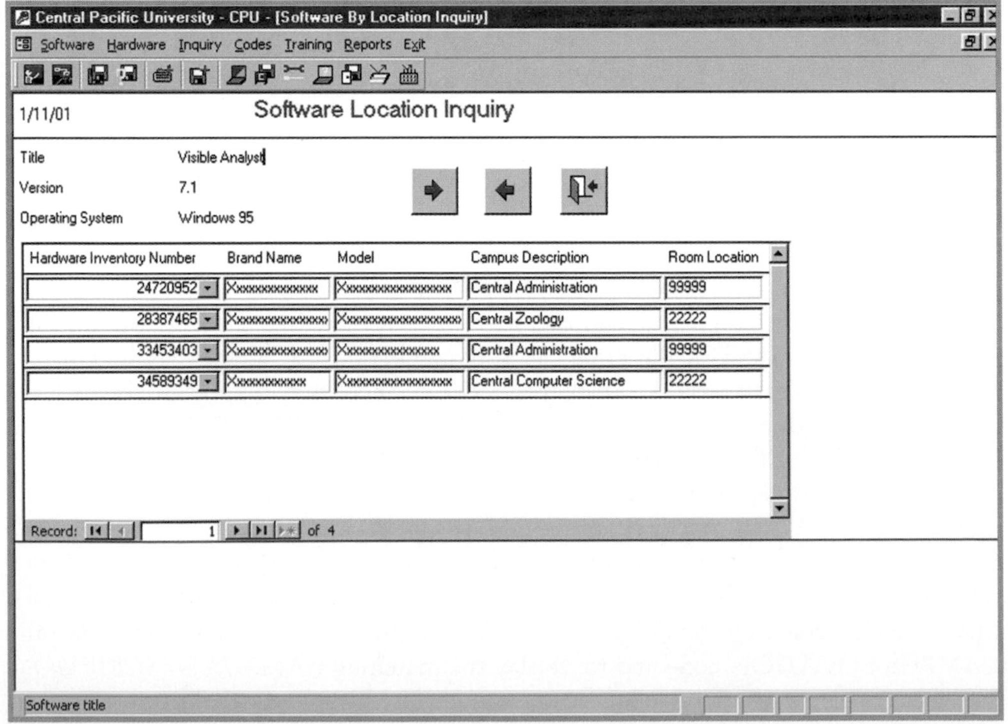

FIGURE E18.11
SOFTWARE LOCATION INQUIRY screen.

EXERCISES*

E-1. Use Microsoft Access to view the menu options for the Microcomputer System.

E-2. Examine the HARDWARE INQUIRY shown in Figures E18.5 and E18.6. Explain the inquiry **type** using the value, entity, and attribute (V, E, A) notation.

E-3. In a paragraph, explain why a data-entry screen should emphasize accuracy, whereas an inquiry screen emphasizes how fast results may be displayed.

E-4. Modify and print the hierarchy chart representing the Hardware Update menu. Add rectangles to represent the following menu options:
CHANGE MICROCOMPUTER
DELETE MICROCOMPUTER RECORD
UPDATE INSTALLED MICROCOMPUTER

E-5. Use the Functional Decomposition feature of Visible Analyst to draw a hierarchy chart representing the options found on the UPDATE SOFTWARE menu. Start with the top rectangle representing the UPDATE SOFTWARE menu.
ADD SOFTWARE PACKAGE
CHANGE SOFTWARE RECORD
DELETE SOFTWARE RECORD
UPGRADE SOFTWARE PACKAGE

E-6. Chip and Anna realize that the menu that has been designed is for the users involved in the installation and maintenance of microcomputer hardware and software. This menu would not be suitable for general faculty and staff members, because they should not have the ability to update the records. Design a menu, either on paper or using software with which you are familiar, that would provide the general user with the ability to perform inquiries and reports.

E-7. Discuss in a paragraph why the users would need to move to another screen (by pressing the Next Record button) to display the correct record for the SOFTWARE LOCATION inquiry.

E-8. Design the SOFTWARE DETAILS inquiry screen. The entry field is SOFTWARE INVENTORY NUMBER, and all Software information, with the exception of Expert and Machines Installed On, should be displayed. Refer to the Visible Analyst **Software Details** data flow repository entry.

E-9. When scheduling classrooms for student use, Cher Ware needs to know all the software packages in a given room. She would like to enter the CAMPUS LOCATION and the ROOM on an inquiry screen. The display fields would be TITLE, VERSION, SITE LICENSE, and NUMBER OF COPIES.

Design the **Software by Room** inquiry, which is described as a data flow in the Visible Analyst repository.

E-10. Mike Crowe needs to know which component boards are installed in each machine. Use Visible Analyst to view the data flow entry for **Component Boards** and to design the COMPONENT BOARD inquiry. The input field is the HARDWARE INVENTORY NUMBER. Output fields are BRAND NAME, MODEL, and a scroll region for BOARDS. The logic is to randomly read the MICROCOMPUTER MASTER using the HARDWARE INVENTORY NUMBER. If the record is not found, display an error message to that effect. Find the matching BOARD records. Write the notation using value, entity, and attribute (V, E, A) for the type of inquiry.

18

E-11. Every so often, Hy Perteks receives a request for help concerning a given software package. Staff members and students need to perform advanced options or transfer data to and from different packages, and they are having difficulties. Hy would like to enter the software TITLE and VERSION NUMBER. The resulting display would show the SOFTWARE EXPERT NAME and his or her CAMPUS LOCATION and ROOM NUMBER. Design the screen for the LOCATE SOFTWARE EXPERT inquiry. Describe the logic and files needed to produce the inquiry. Write the notation for this inquiry using value, entity, and attribute (V, E, A). The details for this inquiry are included in the Visible Analyst **Software Expert** data flow repository entry.

E-12. In a follow-up interview with Cher Ware, it was determined that she needs to know what machines are available to install any software package, given the package's graphics requirements. Produce an inquiry that would allow Cher to enter the monitor code and, optionally, a graphics board and campus location for the software. Four columns should be displayed:

HARDWARE INVENTORY NUMBER
CAMPUS LOCATION
ROOM LOCATION
GRAPHICS BOARD

Refer to the Visible Analyst **Monitor Required** data flow. Write a paragraph describing the logic involved in obtaining the results. Include the type of inquiry using value, entity, and attribute (V, E, A) notation.

E-13. Both Cher and Hy expressed an interest in finding machines of a specified brand connected to different printers. Sometimes the engineering students need a plotter, whereas other situations demand a laser or PostScript printer.

Design an inquiry that would have the PRINTER and BRAND NAME of the microcomputer as input fields. Output would be two columns: CAMPUS LOCATION (full name, not a code) and ROOM LOCATION. Refer to the Visible Analyst **Printer Location** data flow.

Briefly describe the logic used in producing the output. Would this inquiry need a scroll region to display all the information? Why or why not? Use a paragraph to describe the type of inquiry using value, entity, and attribute (V, E, A) notation.

E-14. Hy receives a number of requests for training classes. He would like to plan training and place the upcoming classes on the intranet so that faculty would have an adequate amount of lead time to schedule a class. Design the SOFTWARE TRAINING CLASSES inquiry. The details may be found in the Visible Analyst data flow repository entry called **Software Training Classes**.

*Exercises preceded by a CD-ROM icon require the program Visible Analyst or another CASE tool. A CD-ROM containing Visible Analyst examples is provided free of charge to any professor adopting this book. The examples on the disk may be imported into Visible Analyst and then used by students.

DESIGNING ACCURATE DATA-ENTRY PROCEDURES

19

DATA-ENTRY OBJECTIVES

Making sure that data are entered into the system accurately is of utmost importance. It is by now axiomatic that the quality of data input determines the quality of information output. The systems analyst can support accurate data entry through the achievement of four broad objectives, as shown in Figure 19.1. They are creating meaningful coding for data; designing efficient data capture approaches; assuring complete and effective data capture; and assuring data quality through validation.

The quality of data is a measurement of how consistently correct the data are within certain preset limits. Effectively coded data facilitate accurate data entry by cutting down on the sheer quantity of data and thus the time required to enter the information.

When data are being entered efficiently, data entry is meeting predetermined performance measures that give the relationship between the time spent on entry and the number of data items entered. Efficient data entry also means that data to be input are quickly and easily decipherable by data-entry operators or information systems. Effective coding, effective and efficient data capture and entry, and ensuring data quality through validation procedures are all data-entry objectives covered in this chapter.

EFFECTIVE CODING

One of the ways that data can be entered more accurately and efficiently is through the knowledgeable employment of various codes. The process of putting ambiguous or cumbersome data into short, easily entered digits or letters is called coding (not to be confused with program coding).

Coding aids the systems analyst in reaching the objective of efficiency, because data that are coded require less time to enter, and thus reduces the number of items entered. Coding can also help in the appropriate sorting of data at a later point in the data transformation process. In addition, coded data can save valuable memory and storage space. In sum, coding is a way of being eloquent but succinct in capturing data. Besides providing accuracy and efficiency, codes should have a purpose. Specific types of codes allow us to treat data in a particular manner. Purposes for coding include the following:

1. Keeping track of something.
2. Classifying information.

FIGURE 19.1

The analyst needs to work toward four data-entry objectives to assure accurate input to the system.

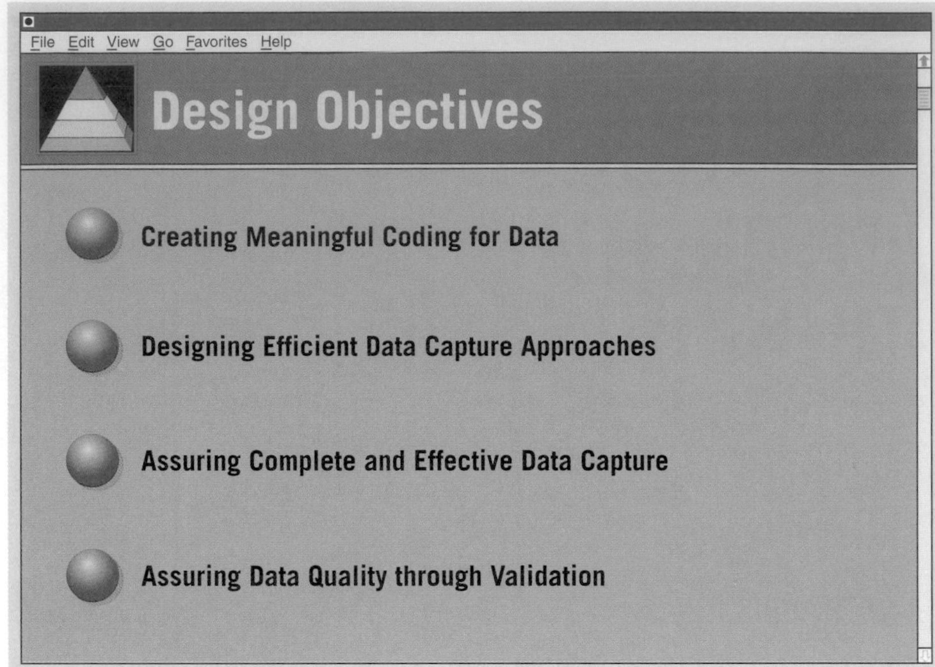
File Edit View Go Favorites Help
Design Objectives
Creating Meaningful Coding for Data
Designing Efficient Data Capture Approaches
Assuring Complete and Effective Data Capture
Assuring Data Quality through Validation

3. Concealing information.
4. Revealing information.
5. Requesting appropriate action.

Each of these purposes for coding is discussed in the following subsections, along with some example of codes.

KEEPING TRACK OF SOMETHING

Sometimes we want merely to identify a person, place, or thing just to keep track of it. For example, a shop that manufactures custom-made upholstered furniture needs to assign a job number to a project. The salesperson needs to know the name and address of the customer, but the job shop manager or the workers who assemble the furniture need not know who the customer is. Consequently, an arbitrary number is assigned to the job. The number can be either random or sequential, as described in the following subsection.

Simple Sequence Codes. The simple sequence code is a number that is assigned to something if it needs to be numbered. It therefore has no relation to the data themselves. Figure 19.2 shows how a furniture manufacturer's orders are assigned an order number. With this easy reference number, the company can keep track of the order in process. It is more efficient to enter job "5676" than "that brown and black rocking chair with the leather seat for Arthur Hook, Jr."

Using a sequence code rather than a random number has some advantages. First, it eliminates the possibility of assigning the same number. Second, it gives users an approximation of when the order was received.

Sequence codes should be used when order-of-processing requires knowledge of the sequence in which items enter the system or the order in which events unfold. An example is found in the situation of a bank running a special promotion

Order #	Product	Customer
5676	Rocking Chair/with Leather	Arthur Hook, Jr.
5677	Dining Room Chair/Upholstered	Millie Monice
5678	Love Seat/Upholstered	J. & D. Pare
5679	Child's Rocking Chair/Decals	Lucinda Morely

FIGURE 19.2
Using a simple sequence code to indicate the sequence in which orders enter a custom furniture shop.

that makes it important to know when a person applied for a special, low-interest home loan, because (all other things being equal) the special mortgage loans will be granted on a first-come, first-served basis. In this case, assigning a correct sequence code to each applicant is important.

Alphabetic Derivation Codes. At times it is undesirable to use sequence codes. The most obvious instance is when you do *not* wish to have someone read the code to figure out how many numbers have been assigned. Another situation in which sequence codes may not be useful is when a more complex code is desirable to avoid a costly mistake. One possible error would be to add a payment to account 223 when you meant to add it to account 224, because you entered an incorrect digit.

The alphabetic derivation code is a commonly used approach in identifying an account number. The example in Figure 19.3 comes from a mailing label for a magazine. The code becomes the account number. The first five digits come from the first five digits of the subscriber's zip code, the next three are the first three consonants in the subscriber's name, the next four numbers are from the street address, and the last three make up the code for the magazine. The main purpose of this code is to identify an account.

A secondary purpose is to print mailing labels. When designing this code, the zip code is the first part of the account number. The subscriber records are usually updated only once a year, but the primary purpose of the records is to print mailing labels once a month or once per week. Having the zip code as the first part of a primary key field means that the records do not have to be sorted by zip code for bulk mailing, because records on a file are stored in primary key sequence. Notice that the expiration date is not part of the account number, because that number can change more frequently than the other data.

One disadvantage of an alphabetic derivation code occurs when the alphabetic portion is small (for example, a name of Po) or when the name contains fewer consonants than the code requires. A name of Roe has only one consonant and would have to be derived as RXX or derived using some other scheme. Another disadvantage is that some of the data may change. Changing one's address or name would change the primary key for the file.

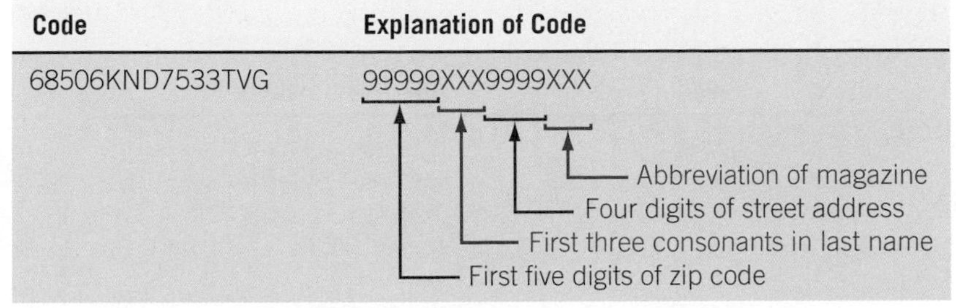

FIGURE 19.3
Identifying the account of a magazine subscriber with an alphabetic derivation code.

CLASSIFYING INFORMATION

Coding affords the ability to distinguish among classes of items. Classifications are necessary for many purposes, such as reflecting what parts of a medical insurance plan an employee carries or showing which student has completed the core requirements of his or her coursework.

To be useful, classes must be mutually exclusive. For example, if a student is in class F, meaning freshman, having completed 0 to 36 credit hours, he or she should not also be classifiable as a sophomore (S). Overlapping classes would be F = 0–36 credit hours, S = 32–64 credit hours, and so on. Data are unclear and not as readily interpretable when coding classes are not mutually exclusive.

Classification Codes. Classification codes are used to distinguish one group of data with special characteristics from another. Classification codes can consist of either a single letter or a number. They are a shorthand way of describing a person, place, thing, or happening.

Classification codes are listed in manuals or posted so that users can locate them easily. Many times, users become so familiar with frequently used codes that they memorize them. A user classifies an item and then enters its code directly into the terminal of an online system or onto a source document of a batch system.

An example of classification coding is the way you may wish to group tax-deductible items for the purpose of completing your income taxes. Figure 19.4 shows how codes are developed for items such as interest, medical payments, contributions, and so on. The coding system is simple: Take the first letter of each of the categories. Contributions are C, interest payments are I, and supplies are S.

All goes well until we get to other categories (such as computer items, insurance payments, and subscriptions) that begin with the same letters we used previously. Figure 19.5 demonstrates what happens in this case. The coding was stretched so that we could use the P for "comPuter," the N for "iNsurance," and the B for "suBscriptions." Obviously, this situation is far from perfect. One way to avoid the confusion of this type is to allow for codes longer than one letter, discussed later in this chapter under the subheading of mnemonic codes. Pull-down menus in a GUI system often use classification codes as a shortcut for running menu features, such as Alt-F for the File menu.

Block Sequence Codes. Earlier we discussed sequence codes. The block sequence code is an extension of the sequence code. Figure 19.6 shows how a business assigns numbers to microcomputer software. Main categories of software are browser packages, database packages, word-processing packages, and presentation packages. These were assigned sequential numbers in the following "blocks" or ranges: browser 100–199, database 200–299, and so forth. The advantage of the block sequence code is that the data are grouped according to common characteristics, while still taking advantage of the simplicity of assigning the next available number (within the block, of course) to the next item needing identification.

FIGURE 19.4
Grouping tax-deductible items through the use of a one-letter classification code.

Code	Tax-Deductible Item
I	**I**nterest Payments
M	**M**edical Payments
T	**T**axes
C	**C**ontributions
D	**D**ues
S	**S**upplies

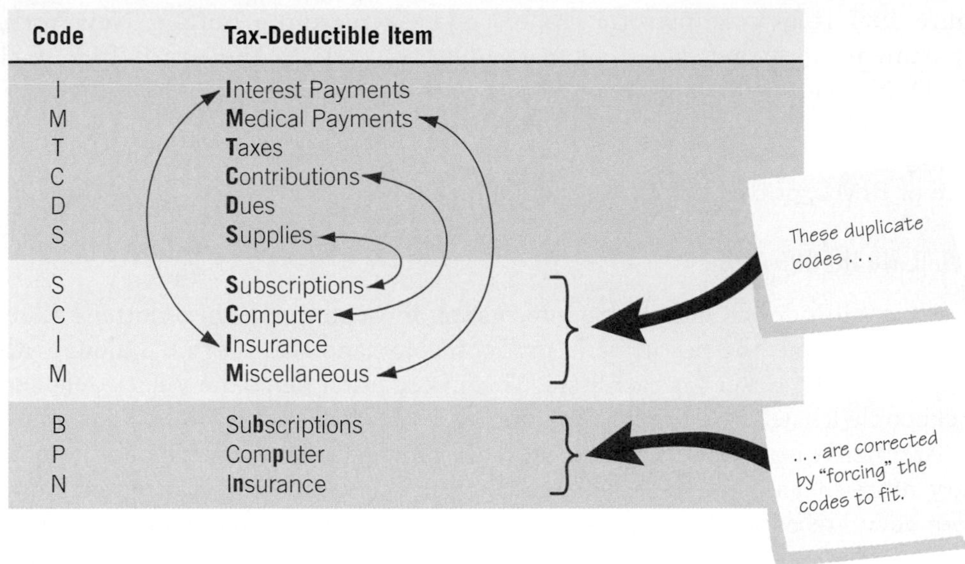

FIGURE 19.5
Problems in using a one-letter classification code occur when categories share the same letter.

CONCEALING INFORMATION

Codes may be used to conceal or disguise information we do not wish others to know. There are many reasons why a business may want to do that. For example, a corporation may not want information in a personnel file to be accessed by data-entry workers. A store may want its salespeople to know the wholesale price to show them how low a price they can negotiate, but they may encode it on price tickets to prevent customers from finding that out. A restaurant may want to capture information about the service without letting the customer know the name of the server. Concealing information and security have become very important in the last few years. Corporations have started to allow vendors and customers to access their databases directly, and handling business transactions over the Internet has made it necessary to develop tight encryption schemes. The following subsection describes an example of concealing information through codes.

Cipher Codes. Perhaps the simplest coding method is the direct substitution of one letter for another, one number for another, or one letter for a number. A popular type of puzzle called a cryptogram is an example of letter substitution.

Code	Name of Software Package	Type
100	Netscape	Browser
101	Internet Explorer	
102	Lynx	
.	.	
.	.	
.	.	
200	Access	Database
201	Paradox	
202	Oracle	
.	.	
.	.	
.	.	
300	Microsoft Word	Word Processing
301	WordPerfect	
.	.	
.	.	
.	.	
400	Astound	Presentation
401	Micrografx Designer	
402	PowerPoint	

FIGURE 19.6
Using a block sequence code to group similar software packages.

Figure 19.7 is an example of a cipher code taken from a Buffalo, New York, department store that coded all markdown prices with the words BLEACH MIND. No one really remembered why whose words were chosen, but all the employees knew them by heart, and so the cipher code was successful. Notice in this figure that an item with a retail price of $25.00 would have a markdown price of BIMC, or $18.75 when decoded letter by letter.

REVEALING INFORMATION

Sometimes it is desirable to reveal information through a code. In a clothing store, information about the department, product, color, and size is printed along with the price on the ticket for each item. This information helps the salespeople and stockpeople locate the place for the merchandise.

Another reason for revealing information through codes is to make the data entry more meaningful. A familiar part number, name, or description supports more accurate data entry. The examples of codes in the following subsection explain how these concepts can be realized.

Significant-Digit Subset Codes. When it is possible to describe a product by virtue of its membership in many subgroups, we can use a significant-digit subset code to help describe it. The clothing-store price ticket example in Figure 19.8 is an example of an effective significant-digit code.

To the casual observer or customer, the item description appears to be one long number. To one of the salespeople, however, the number is made up of a few smaller numbers, each one having a meaning of its own. The first three digits represent the department, the next three the product, the next two the color, and the last two the size.

Significant-digit subset codes may consist of either information that actually describes the product (for example, the number 10 means size 10) or numbers that are arbitrarily assigned (for instance, 202 is assigned to mean the maternity department). In this case, the advantage of using a significant-digit subset code is that it makes it possible to locate items that belong to a certain group or class. For example, if the store's manager decided to mark down all winter merchandise for an upcoming sale, salespeople could locate all items belonging to departments 310 through 449, the block of codes used to designate "winter" in general.

Another advantage of using significant-digit subset codes is that inquiries may be performed on a portion of the code. Using the example illustrated in Figure 19.8, a salesperson might look for other matching red items, for other items in size

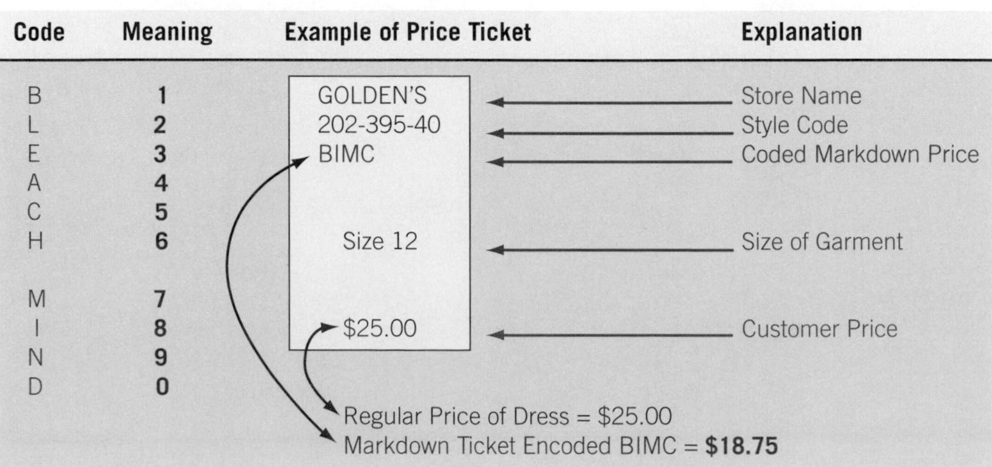

Code
Meaning
Example of Price Ticket
Explanation
B 1
L 2
E 3
A 4
C 5
H 6
M 7
I 8
N 9
D 0
GOLDEN'S
202-395-40
BIMC
Size 12
$25.00
Store Name
Style Code
Coded Markdown Price
Size of Garment
Customer Price
Regular Price of Dress = $25.00
Markdown Ticket Encoded BIMC = $18.75

FIGURE 19.7
Encoding markdown prices with a cipher code is a way of concealing price information from customers.

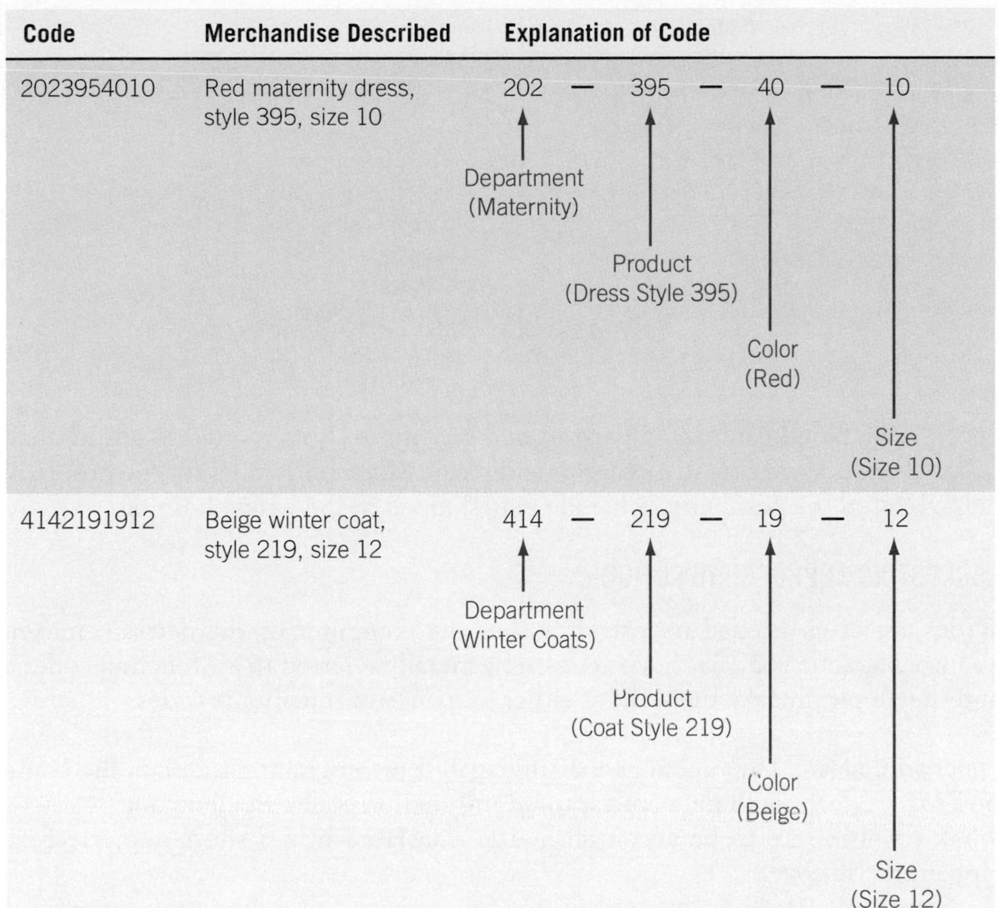

Code
Merchandise Described
Explanation of Code
2023954010
Red maternity dress, style 395, size 10
202 — 395 — 40 — 10
Department (Maternity)
Product (Dress Style 395)
Color (Red)
Size (Size 10)
4142191912
Beige winter coat, style 219, size 12
414 — 219 — 19 — 12
Department (Winter Coats)
Product (Coat Style 219)
Color (Beige)
Size (Size 12)

FIGURE 19.8

Using a significant-digit subset code helps employees locate items belonging to a particular department.

10, for other maternity items, or for similar dresses with the same style. This code is also very useful for marketing a product. Internet businesses often recommend products that they think a customer might like. For example, if a customer purchased a certain type of music, a Web site might recommend other music in the genre. When a customer purchases a certain type of book, a Web site might recommend other titles that have similar content or style.

Mnemonic Codes. A mnemonic (pronounced nî-môn′-ĭk) is a memory aid. Any code that helps either the data-entry person remember how to enter the data or the end user remember how to use the information can be considered a mnemonic. Using a combination of letters and symbols affords a strikingly clear way to code a product so that the code is easily seen and understood.

The city hospital codes formerly used by the Buffalo regional blood center were mnemonic, as shown in Figure 19.9. The simple codes were invented precisely

Code	City Hospitals
BGH	**B**uffalo **G**eneral **H**ospital
ROS	**Ros**well Park Memorial Institute
KEN	**Ken**more Mercy
DEA	**Dea**coness Hospital
SIS	**Sis**ters of Charity
STF	**Saint F**rancis Hospital
STJ	**Saint J**oseph's Hospital
OLV	**O**ur **L**ady of **V**ictory Hospital

FIGURE 19.9

Mnemonic codes function as memory aids by using a meaningful combination of letters and numbers.

FIGURE 19.10
Function codes compactly capture functions that the computer must perform.

Code	Function
1	Delivered
2	Sold
3	Spoiled
4	Lost or Stolen
5	Returned
6	Transferred Out
7	Transferred In
8	Journal Entry (Add)
9	Journal Entry (Subtract)

because the blood-bank administrators and systems analysts wanted to ensure that hospital codes were easy to memorize and recall. Mnemonic codes for the hospitals helped lessen the possibility of blood being shipped to the wrong hospital.

REQUESTING APPROPRIATE ACTION

Codes are often needed to instruct either the computer or the decision maker about what action to take. Such codes are generally referred to as "function codes," and they typically take the form of either sequence or mnemonic codes.

Function Codes. The functions that the analyst or programmer desires the computer to perform with data are captured in function codes. Spelling out precisely what activities are to be accomplished is translated into a short numerical or alphanumeric code.

Figure 19.10 shows examples of a function code for updating inventory. Suppose you managed a dairy department; if a case of yogurt spoiled, you would use the code 3 to indicate this event. Of course, data required for input vary depending on what function is needed. For example, appending or updating a record would require only the record key and function code, whereas adding a new record would require all data elements to be input, including the function code.

GENERAL GUIDELINES FOR CODING

In the previous subsections, we examined the purposes for using different types of codes when entering and storing data. Next, we examine a few heuristics for establishing a coding system. These rules are highlighted in Figure 19.11.

Be Concise. Codes should be concise. Overly long codes mean more keystrokes and consequently more errors. Long codes also mean that storing the information in a database will require more memory.

FIGURE 19.11
There are eight general guidelines for establishing a coding system.

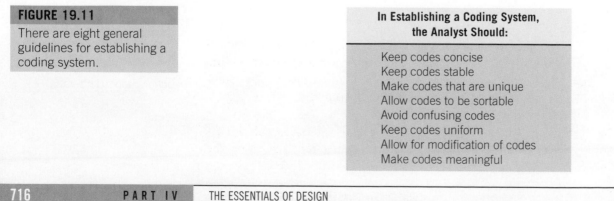

In Establishing a Coding System, the Analyst Should:
Keep codes concise
Keep codes stable
Make codes that are unique
Allow codes to be sortable
Avoid confusing codes
Keep codes uniform
Allow for modification of codes
Make codes meaningful

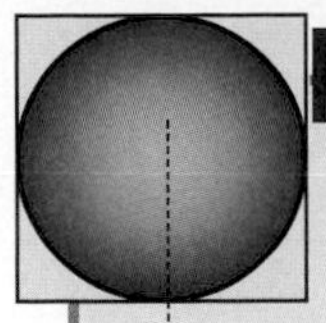

CONSULTING OPPORTUNITY 19.1

IT'S A WILDERNESS IN HERE

"I can't stand this. I've been looking for this hat for the last 45 minutes," complains Davey, as he swings a coonskin cap by its tail above his head. He is one of the new warehouse workers for Crockett's, a large catalog sales firm. "The catalog slip calls it a Coo m5–9w/tl. Good thing you told me Coo stands for coonskin. Then, of course, I thought about caps and looked over here. I found it here in this bin labeled BOYS/CAP. Wouldn't it be easier if the catalog matched the bins? To me, this invoice says, 'Cookware, metallic, 5–9-piece set with Teflon.' I've been stranded in the cookware sets the whole time."

Daniel, Davey's coworker, barely listens as he hurriedly pulls items out of bins to fill another order. "You'll get used to it. They've got to have it this way so that the computers can understand the bill later. Mostly, I look at the catalog page number on the invoice, then I look it up in the book and sort of translate it to back here . . . unless I remember it from finding it before," Daniel explains.

Davey persists, saying, "Computers are smart, though, and we have to fill so many orders. We should tell the people up in billing the names we've got on our bins."

Daniel replies cynically, "Oh, sure. They're dying to know what we think." Then he continues in a quieter tone. "You know, we used to have it like that, but when they got all the new computers and went to 24-hour phone orders, it all changed. They said the operators had to know more about what they were selling, so they changed their codes to be more like a story."

Davey, surprised at Daniel's revelation, asks, "What's the story for the one I was working on?"

Inspecting the code on the cap's invoice, Daniel replies, "The one you were working on was 'Coo m5–9w/tl.' After looking it up real fast on her computer the operator can tell the customer, 'It's a coonskin (Coo) cap for boys (*m* for male) ages 5–9 with a real tail (w/tl).' We can't see the forest for the trees because of their codes, but you know Crockett's. They've got to make the sale."

How important is it that the warehouse bins and invoices are encoded inconsistently? Respond in a paragraph. What are some of the problems created when a code appears to be mnemonic but employees are never given an appropriate "key" to decode it? Discuss your response in two paragraphs. What changes would you make to invoice/warehouse coding for Crockett's? Document your changes, identify the type of code you would use, and use the code in an example of a product that Crockett's might sell. Remember to decipher it as well.

Short codes are easier to remember and easier to enter than long codes. If codes must be long, they should be broken up into subcodes. For example, 5678923453127 could be broken up with hyphens as follows: 5678-923-453-127. This approach is much more manageable and takes advantage of the way people are known to process information in short chunks. Sometimes codes are made longer than necessary for a reason. Credit card numbers are often long to prevent people from guessing a credit card number. Visa and MasterCard use 16-digit numbers, which would accommodate nine trillion customers. Because the numbers are not assigned sequentially, chances of guessing a credit card number are very slight.

Keep the Codes Stable. Stability means that the identification code for a customer should not change each time new data are received. Earlier, we presented an alphabetic derivation code for a magazine subscription list. The expiration date was not part of the subscriber identification code because it was likely to change.

Don't change the code abbreviations in a mnemonic system. Once you have chosen the code abbreviations, do not try to revise them, because that makes it extremely difficult for data-entry personnel to adapt.

Ensure That Codes Are Unique. For codes to work, they must be unique. Make a note of all codes used in the system to ensure that you are not assigning the same code number or name to the same items. Code numbers and names are an essential part of the entries in data dictionaries, discussed in Chapter 10.

Allow Codes to Be Sortable. If you are going to manipulate the data usefully, the codes must be sortable. For example, if you were to perform a text search on the months of the year in ascending order, the "J" months would be out of order

FIGURE 19.12

Plan ahead in order to be able to do something useful with data that have been entered. In this example, the person creating the codes did not realize the data would have to be sorted.

Incorrect Sorting Using MMM-DD-YYYY	Incorrect Sorting Using MM-DD-YYYY	Incorrect Sorting (Year 2000 Problem) YY-MM-DD	Correct Sorting Using YYYY-MM-DD
Dec-25-1998	06-04-1998	00-06-11	1997-06-12
Dec-31-1997	06-11-2000	97-06-12	1997-12-31
Jul-04-1999	06-12-1997	97-12-31	1998-06-04
Jun-04-1998	07-04-1999	98-06-04	1998-10-24
Jun-11-2000	10-24-1998	98-10-24	1998-12-25
Jun-12-1997	12-25-1998	98-12-25	1999-07-04
Oct-24-1998	12-31-1997	99-07-04	2000-06-11

(January, July, and then June). Dictionaries are sorted in this way, one letter at a time from right to left. So, if you sorted MMMDDYYYY where the MMM stood for the abbreviation for the month, DD for the day, and YYYY for the year, the result would be in error.

Figure 19.12 shows what would happen if a text search were performed on different forms of the date. The third column shows a problem that is part of the Year 2000 (Y2K) crisis that caused some alarm and even made the cover of *Time* magazine.

One of the lessons learned is to make sure that you can do what you intend to do with the codes you create. Numerical codes are much easier to sort than alphanumerics; therefore, consider converting to numerics wherever practical.

Avoid Confusing Codes. Try to avoid using coding characters that look or sound alike. The characters O (the letter oh) and 0 (the number zero) are easily confused, as are the letter I and the number 1 and the letter Z and the number 2. Therefore, codes such as B1C and 280Z are unsatisfactory.

One example of a potentially confusing code is the Canadian Postal Code, as shown in Figure 19.13. The code format is X9X 9X9, where X stands for a letter and 9 stands for a number. One advantage to using letters in the code is to allow more data in a six-digit code (there are 26 letters, but only 10 numbers). Because the code is used on a regular basis by Canadians, the code makes perfectly good sense to them. To foreigners sending mail to Canada, however, it may be difficult to tell if the second-to-last symbol is a Z or a 2.

Keep the Codes Uniform. Codes need to follow readily perceived forms most of the time. Codes used together, such as BUF-234 and KU-3456, are poor because the first contains three letters and three numbers, whereas the second has only two letters followed by four numbers.

FIGURE 19.13

Combining look-alike characters in codes can result in errors.

Code Format for Canadian Postal Code X9X 9X9

Handwritten Code	Actual Code	City, Province	Problem
L8S 4M4	L8S 4M4	Hamilton, Ontario	S looks like a 5
T3A ZE5	T3A 2E5	Calgary, Alberta	2 looks like a Z 5 looks like an S
LOS 1JO	LOS 1JO	Niagara-on-the-Lake, Ontario	Zero and Oh look alike S looks like a 5 1 looks like an I

When you are required to add dates, try to avoid using the codes MMDDYYYY in one application, YYYYDDMM in a second, and MMDD19YY in a third. It is important to keep codes uniform among as well as within programs.

In the past, uniformity meant that all codes be kept the same length. With the introduction of online systems, the length is not as important as it once was. With online systems, the Enter key is hit after data entry is verified by the operator as correct, so it doesn't make much difference if the code is three characters or four characters long.

Allow for Modification of Codes. Adaptability is a key feature of a good code. The analyst must keep in mind that the system will evolve over time, and the coding system should be able to encompass change. The number of customers should grow, customers will change names, and suppliers will modify the way they number their products. The analyst needs to be able to forecast the predictable and anticipate a wide range of future needs when designing codes.

Make Codes Meaningful. Unless the analyst wants to hide information intentionally, codes should be meaningful. Effective codes not only contain information, but they also make sense to the people using them. Meaningful codes are easy to understand, work with, and recall. The job of data entry becomes more interesting when working with meaningful codes instead of just entering a series of meaningless numbers.

Using Codes. Codes are used in a number of ways. In validation programs, input data is checked against a list of codes to ensure that only valid codes have been entered. In report and inquiry programs, a code stored on a file is transformed into the meaning of the code. Reports and screens should not display or print the actual code. If they did, the user would have to memorize code meanings or look them up in a manual. Codes are used in GUI programs to create drop-down lists.

EFFECTIVE AND EFFICIENT DATA CAPTURE

To ensure the quality of data entered into the system, it is important to capture data effectively. Data capture has received increasingly more attention as the point in information processing at which excellent productivity gains can be made. Great progress in improving data capture has been made since the 1970s, as we have moved from a multiple-step, slow, and error-prone system such as keypunching to sophisticated, optical character recognition (OCR), bar codes, point-of-sale terminals and scanning special characters in magazines and catalogs to access a Web site directly.

DECIDING WHAT TO CAPTURE

The decision about what to capture precedes user interaction with the system. Indeed, it is vital in making the eventual interface worthwhile, for the adage "Garbage in, garbage out" is still true.

Decisions about what data to capture for system input are made among systems analysts and systems users. Much of what will be captured is specific to the particular business. Capturing data, inputting it, storing it, and retrieving it are all costly endeavors, mostly due to the labor costs involved. With all these factors in mind, determining what to capture becomes an important decision.

There are two types of data to enter: data that *change* or *vary* with every transaction and data that concisely *differentiate* the particular item being processed from all other items.

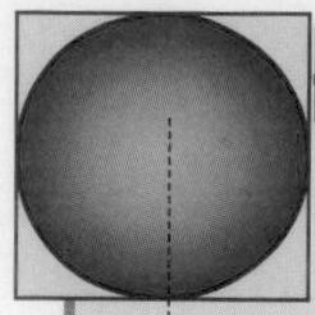

CONSULTING OPPORTUNITY 19.2

CATCHING A SUMMER CODE

Vicky takes her fingers off her keyboard and bends over her workstation to verify the letters on the invoices stacked in front of her. "What on earth?" Vicky asks aloud as she further scrutinizes the letters that encode cities where orders are to be shipped.

Shelly Overseer, her supervisor, who usually sits a couple of workstations away, is passing by and sees Vickie's consternation. "What's the matter? Did the sales rep forget to write in the city code again?"

Vicky swings around in her chair to face Shelly. "No, there are codes here, but they're weird. We usually use a three-letter code, right? Like CIN for Cincinnati, SEA for Seattle, MIN for Minneapolis, BUF for Buffalo. They're all *five*-letter codes here, though."

"Look," Vicky says, lifting the invoice to show Shelly. "CINNC, SEATT, MINNE. It'll take me all day to enter these. No kidding, it's really slowing me down. Maybe there's a mistake. Can't I just use the standard?"

Shelly backs away from Vicky's workstation as if the problem were contagious. Excusing herself apologetically, Shelly says, "It's the part-timers. They are learning sales now, and management was worried that they'd get messed up on their cities. I think it has something to do with mixing up Newark and New Orleans on the last orders. So, a committee decided to make the cities more recognizable by having them add two letters. Those kids can't learn everything we know overnight, even though they try. It's just until August 19, though, when the part-timers go back to school."

As Vicky glumly turns back to her keyboard, Shelly puts her hand sympathetically on Vicky's shoulder and says, "I know it's a strain and it's making you feel miserable, but don't worry. You'll get over it. It's just a summer code."

What general guidelines of coding has management overlooked in its decision to use a summer code for cities? Make a list of them. What is the effect on full-time data-entry personnel of changing codes for the ease of temporary help? Respond in two paragraphs. What future impact could the temporary change in codes have on sorting and retrieving data entered during the summer period? Take two paragraphs to discuss these implications. What changes can you suggest so that the part-timers don't get mixed up on codes in the short term? In a memo to the supervisor of this work group, make a list of five to seven changes in the data capture or data-entry procedures that can be made to accommodate short-term hires without disrupting normal business. In a paragraph, indicate how can this goal be accomplished without marring the productivity of data-entry personnel.

An example of changeable data is the quantity of supplies purchased each time an advertising firm places an order with the office supply wholesaler. Because quantities change depending on the number of employees at the advertising firm and on how many accounts they are servicing, quantity data must be entered each time an order is placed.

An example of differentiation data is the inclusion on a patient record of the patient's Social Security number and the first three letters of his or her last name. In this way, the patient is uniquely differentiated from other patients in the same system.

LETTING THE COMPUTER DO THE REST

When considering what data to capture for each transaction and what data to leave to the system to enter, the systems analyst must take advantage of what computers do best. In the preceding example of the advertising agency ordering office supplies, it is not necessary for the operator entering the stationery order to reenter each item description each time an order is received. The computer can store and access this information easily.

Computers can automatically handle repetitive tasks, such as (1) recording the time of the transaction, (2) calculating new values from input, and (3) storing and retrieving data on demand. By employing the best features of computers, efficient data capture design avoids needless data entry, which in turn alleviates much human error and boredom.

Software can be written to indicate the date of data entry automatically so that the person inputting data does not have to bother with that for every transaction processed. Alternatively, the computer program can be written to ask the user to enter today's date. Once entered, the system proceeds to use that date on all transactions processed in that data-entry session.

Part of a display screen for hotel reservations and guest check-ins is shown in Figure 19.14. Notice that when a reservation is made initially, the guest's name and credit card number are entered. When the guest checks in, the desk clerk calls up the record without having to entirely reenter name or number. The system also automatically records date and time, saving further data entry.

As you can see, it is not necessary to enter data for each transaction that can be easily retrieved by the computer from storage. A prime example is that of the online computer library center (OCLC) used by thousands of libraries in the United States. OCLC was built upon the idea that each item bought by a library should only have to be cataloged once for all time. Once an item is entered, cataloging information goes into the huge OCLC database and is shared with participating libraries. In this case, implementation of the simple concept of entering data only once has saved enormous data-entry time.

The calculating power of the computer should also be taken into account when deciding what *not* to reenter. Computers are adept at long calculations, using data already entered.

For example, the person doing data entry may enter the flight numbers and account number of an air trip taken by a customer belonging to a frequent-flyer incentive program. The computer then calculates the number of miles accrued for each flight, adds it to the miles already in the customer's account, and updates the total miles accrued to the account. The computer may also flag an account that, by virtue of the large number of miles flown, is now eligible for a prize. Although all this information may print out on the customer's updated account sheet, the only new data entered were the flight numbers of the flights flown.

Although users will have suggestions about what data are unnecessary, these decisions should not be made at the time of entry. Rather, the decision is best made by the systems analyst in conjunction with users at an earlier stage in the process.

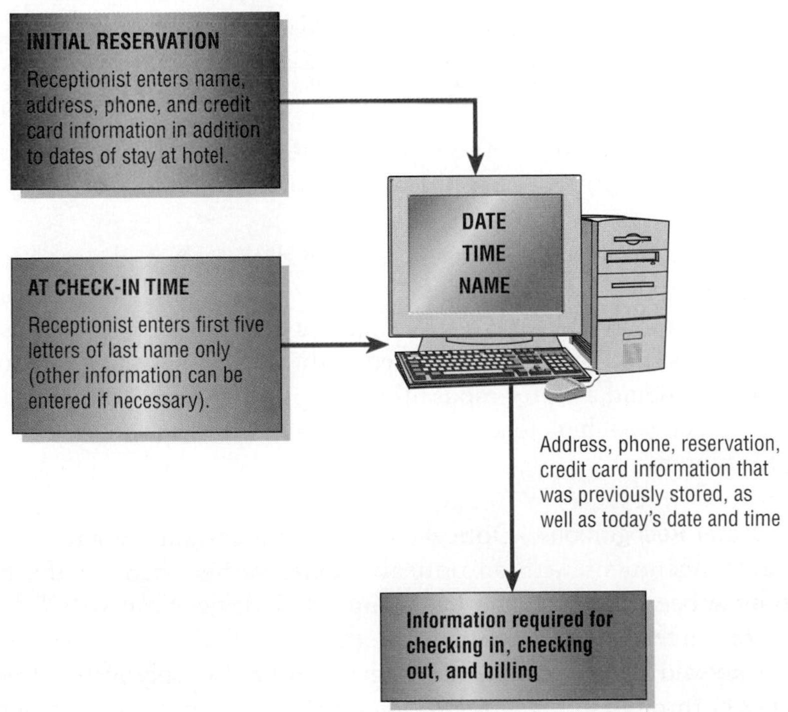
INITIAL RESERVATION
Receptionist enters name, address, phone, and credit card information in addition to dates of stay at hotel.
AT CHECK-IN TIME
Receptionist enters first five letters of last name only (other information can be entered if necessary).
DATE
TIME
NAME
Address, phone, reservation, credit card information that was previously stored, as well as today's date and time
Information required for checking in, checking out, and billing

FIGURE 19.14
Automatically recording date and time in a hotel reservation system eliminates unnecessary data entry.

Through your prior analyses, you will possess the larger picture of what is being input to the system and for what purpose. When analyses are complete, unnecessary input becomes apparent, as do redundancies.

AVOIDING BOTTLENECKS AND EXTRA STEPS

A bottleneck in data entry is an apt allusion to the physical appearance of a bottle. Data are poured rapidly into the wide mouth of the system only to be slowed in its "neck" because of an artificially created instance of insufficient processing for the volume or detail of the data being entered. One way a bottleneck can be avoided is by ensuring that there is enough capacity to handle the data that are being entered.

Ways to avoid extra steps are determined not only at the time of analysis, but also when users begin to interact with the system. The fewer steps involved in inputting data, the fewer chances there are for the introduction of errors. So, beyond the obvious consideration of saved labor, avoiding extra steps is also a way to preserve the quality of data. Once again, use of an online, real-time system that captures customer data without necessitating the completion of a form is an excellent example of saving steps in data entry.

STARTING WITH A GOOD FORM

Effective data capture is achievable only if prior thought is given to what the source document should contain. The data-entry operator inputs data from the source document (usually some kind of form); this document is the source of a large amount of all system data. Online systems (or special data-entry methods such as bar codes) may circumvent the need for a source document, but often some kind of paper form, such as a receipt, is created anyway.

With effective forms, it is not necessary to reenter information that the computer has already stored or data such as time or date of entry that the computer can determine automatically. Chapter 15 discussed in detail how a form or source document should be designed to maximize its usefulness for capturing data and to minimize the time users need to spend entering data from it.

CHOOSING A DATA-ENTRY METHOD

Several efficient data-entry methods are available, and choosing one of them is shaped by many factors, including the need for speed, accuracy, and operator training; the cost of the data-entry method (whether it is materials- or labor-intensive); and the methods currently in use in the organization.

Keyboards. Keyboarding is the oldest method of data entry, and certainly it is the one with which organizational members are the most familiar. Some improvements have been made over the years to standardize keyboards, such as built-in numerical keypads that can be programmed with macros to reduce the number of keystrokes required and a better mouse (one shaped to the contours of the hand). Software permits researchers to analyze how many keystrokes are needed to enter certain types of data.

Optical Character Recognition. Optical character recognition lets a user read input from a source document with an optical scanner rather than off the magnetic media we have been discussing so far. Using OCR devices can speed data input from 50 to 75 percent over some keying methods.

What is needed is a source document that can be optically scanned when it is filled out, either through special block printing or by hand, as shown in Figure 19.15.

ABCDEFGHIJKLMN
OPQRSTUVWXYZ ,.
$/*-1234567890

FIGURE 19.15
Optical character recognition (OCR) of special-character source documents facilitates data entry.

How the source document is filled in depends on the requirements of the particular hardware and software. Corporations have recently started incorporating the use of devices such as a Cue Cat, a handheld scanning device used by consumers. Sliding the Cue Cat over a specially coded Web site in magazines and catalogs will automatically insert the Web address into a browser.

The increased speed of OCR comes through not having to encode or key in data from source documents. It eliminates many of the time-consuming and error-fraught steps of other input devices. In doing so, OCR demands few employee skills and commensurately less training, resulting in fewer errors and less time spent by employees in redundant efforts. It also decentralizes responsibility for quality data directly to the unit that is generating it. OCR, which has become available to all, has one additional, highly practical use: the transformation of faxes into documents that can be edited.

Other Methods of Data Entry. Other methods of data entry are also becoming more widely employed. Most of these methods reduce labor costs by requiring few operator skills or little training, they move data entry closer to the source of data, and they eliminate the need for a source document. In doing so, they have become fast and highly reliable data-entry methods. The data-entry methods discussed in the following sections include magnetic ink character recognition, mark-sense forms, punch-out forms, bar codes, and data strips.

Magnetic ink character recognition Magnetic ink characters are found on the bottom of bank checks and some credit card bills. This method is akin to OCR in that special characters are read, but its use is limited. Data entry through magnetic ink character recognition (MICR) is done through a machine that reads and interprets a single line of material encoded with ink that is made up of magnetic particles.

Some advantages of using MICR are (1) it is a reliable and high-speed method that is not susceptible to accepting stray marks (because they are not encoded magnetically); (2) if it is required on all withdrawal checks, it serves as a security measure against bad checks; and (3) data-entry personnel can see the numbers making up the code if it is necessary to verify it.

Mark-sense forms Mark-sense forms allow data entry through the use of a scanner that senses where marks have been made by lead pencil on special forms. A common usage is for scoring answer sheets for survey questionnaires, as shown in Figure 19.16. Little training of entry personnel is necessary, and a high volume of forms can be processed quickly.

One drawback of mark-sense forms is that although the readers can determine whether a mark has been made, they cannot interpret the mark in the way

FIGURE 19.16

A mark-sense form for data entry that is filled out in pencil and read by a scanner speeds data entry.

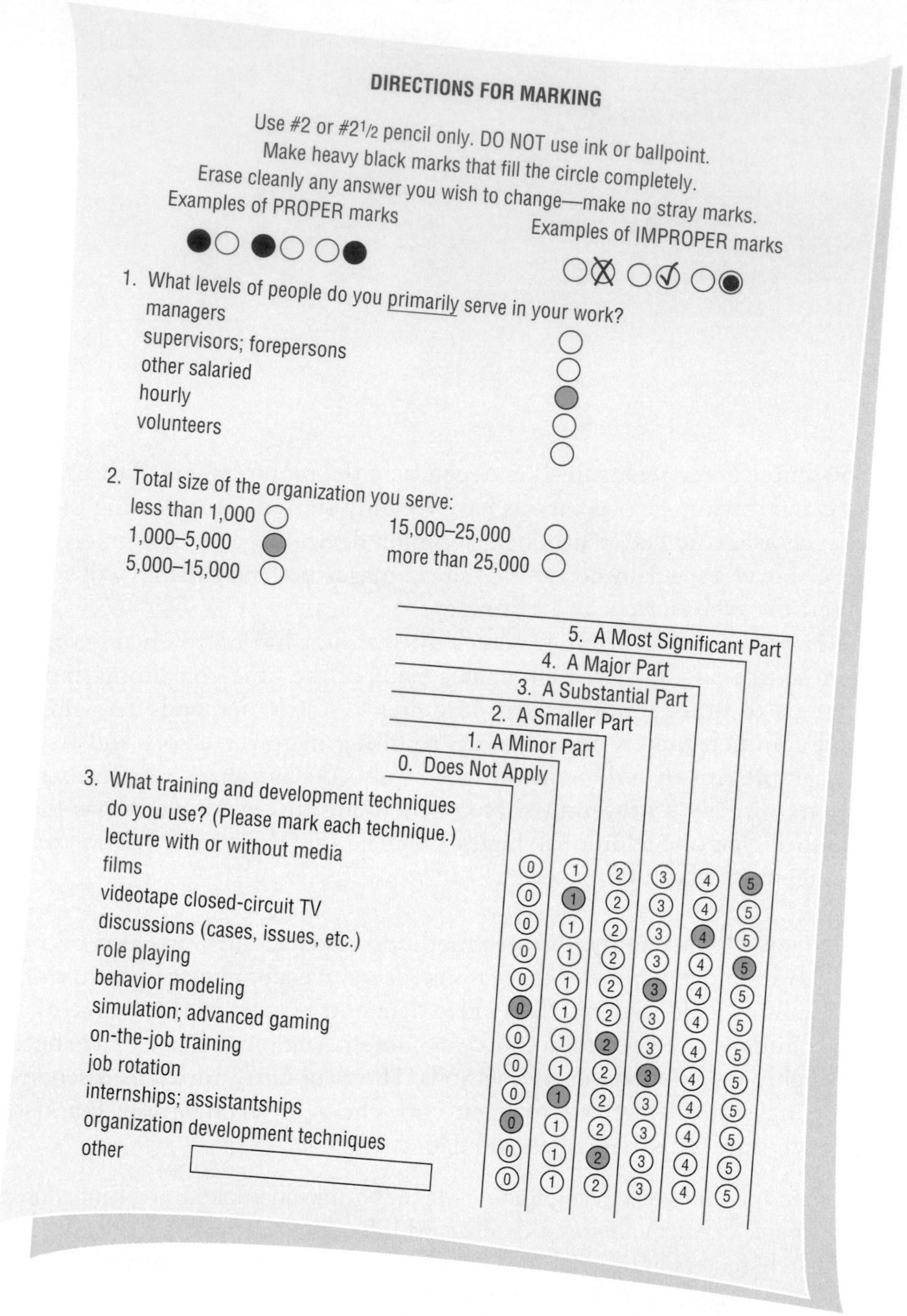

DIRECTIONS FOR MARKING

Use #2 or #2½ pencil only. DO NOT use ink or ballpoint.
Make heavy black marks that fill the circle completely.
Erase cleanly any answer you wish to change—make no stray marks.

Examples of PROPER marks

Examples of IMPROPER marks

1. What levels of people do you primarily serve in your work?
   - managers
   - supervisors; forepersons
   - other salaried
   - hourly
   - volunteers

2. Total size of the organization you serve:
   - less than 1,000
   - 1,000–5,000
   - 5,000–15,000
   - 15,000–25,000
   - more than 25,000

5. A Most Significant Part
4. A Major Part
3. A Substantial Part
2. A Smaller Part
1. A Minor Part
0. Does Not Apply

3. What training and development techniques do you use? (Please mark each technique.)
   - lecture with or without media
   - films
   - videotape closed-circuit TV
   - discussions (cases, issues, etc.)
   - role playing
   - behavior modeling
   - simulation; advanced gaming
   - on-the-job training
   - job rotation
   - internships; assistantships
   - organization development techniques
   - other

that optical character readers do. Stray marks on forms can thus be entered as incorrect data. In addition, choices are limited to the answers provided on the mark-sense form, forms have difficulty in capturing alphanumeric data because of the space required for a complete set of letters and numbers, and it is easy for those filling out mark-sense forms to get confused and put a mark in an incorrect position.

Bar codes. Bar codes typically appear on product labels, but they also appear on patient identification bracelets in hospitals and in almost any context in which a person or object needs to be checked into and out of any kind of inventory system. Bar codes can be thought of as "metacodes," or codes encoding

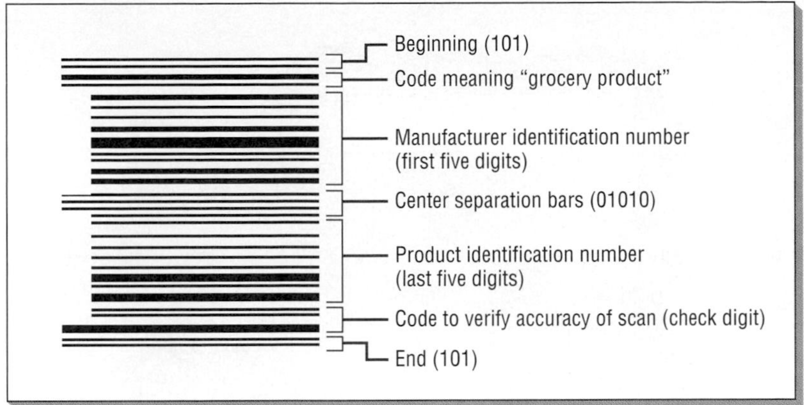

FIGURE 19.17
Bar coding, as shown in this label for a grocery product, affords highly accurate data entry. Used with the permission of the Uniform Code Council, Dayton, Ohio.

codes, because they appear as a series of narrow and wide bands on a label which encodes numbers or letters. These symbols in turn have access to product data stored in computer memory. A beam of light from a scanner or lightpen is drawn across the bands on the label either to confirm or record data about the product being scanned.

A bar-coded label, such as the one shown in Figure 19.17, includes coding for a particular grocery product: the manufacturer identification number, the product identification number, a code to verify the scan's accuracy, and codes to mark the beginning and end of the scan.

Bar coding affords an extraordinarily high degree of accuracy for data entry. It saves labor costs for retailers because each item does not have to be individually price-marked. In addition, bar coding allows the automatic capturing of data that can be used for reordering, more accurate inventory tracking, and the forecasting of future needs. Sale prices or other changes in the meaning of the bar codes are entered into the central processor, thus saving the trouble of marking down numerous items.

One new use of bar coding is the tracking of an individual's credit card purchases for the purpose of building a consumer profile that can then in turn be used to refine marketing to that individual or type of consumer. New input devices are constantly being developed. Of course, it has been possible to transfer photographic images for some time now using systems such as the Kodak Photo CD process, but using a digital camera eliminates the middle step of users needing to digitize their own photographs.

Using Intelligent Terminals. Intelligent terminals can be considered a step above dumb terminals and a step below intelligent workstations and portable microcomputers in their capabilities. In many instances, intelligent terminals eliminate the need for a source document.

The biggest advantage of using intelligent terminals is that, through the use of a microprocessor, they are able to relieve the central processing unit (CPU) of many of the burdens of editing, controlling, transforming, and storing data, processes that the dumb terminals require. Dumb terminals rely on the CPU for all data manipulation, including editing and updating.

The configuration for intelligent terminals is a microprocessor, display screen, and a keyboard. The intelligent terminal has access to the CPU through a network and can be either online or deferred online. In an online intelligent terminal, all the steps in entry, processing, verifying, and output are done immediately with the customer present. The closer data entry is to the source of the data, the more

Source: Gahan Wilson © 1993 The Cartoon Bank. All Rights Reserved.

"Garbage in, garbage out!"

accurate it is likely to be. A well-known example of an online intelligent terminal is the airline ticketing system.

Deferred online intelligent terminals allow data to be entered and verified immediately, but processing is batched and done later (which is less expensive). Electronic cash registers combine these attributes, with both input and output capabilities at point-of-sale terminals.

ENSURING DATA QUALITY THROUGH INPUT VALIDATION

So far, we have discussed ensuring the effective capturing of data onto source documents and the data's efficient entry into the system through various input devices. Although these conditions are necessary for ensuring quality data, they alone are not sufficient.

Errors cannot be ruled out entirely, and the critical importance of catching errors during input, *prior* to processing and storage, cannot be overemphasized. The snarl of problems created by incorrect input can be a nightmare, not the least of which is that many problems take a long time to surface. The systems analyst must assume that errors in data *will* occur and must work with users to design input validation tests to prevent erroneous data from being processed and stored, because initial errors that go undiscovered for long periods are expensive and time-consuming to correct.

You cannot imagine everything that will go awry with input, but you must cover the kinds of errors that give rise to the largest percentage of problems. A summary of potential problems that must be considered when validating input are given in Figure 19.18.

This Type of Validation	Can Prevent These Problems
Validating Input Transactions	Submitting the wrong data Data submitted by an unauthorized person Asking the system to perform an unacceptable function
Validating Input Data	Missing data Incorrect field length Data have unacceptable composition Data are out of range Data are invalid Data do not match with stored data

FIGURE 19.18
Validating input is important to ensure that most potential problems with data are eliminated early.

VALIDATING INPUT TRANSACTIONS

Validating input transactions is largely done through software, which is the programmer's responsibility, but it is important that the systems analyst know what common problems might invalidate a transaction. Businesses committed to quality will include validity checks as part of their routine software.

Three main problems can occur with input transactions: submitting the wrong data to the system, the submitting of data by an unauthorized person, or asking the system to perform an unacceptable function.

Submitting the Wrong Data. An example of submitting the wrong data to the system is the attempt to input a patient's Social Security number into a hospital's payroll system. This error is usually an accidental one, but it should be flagged before data are processed.

Submitting of Data by an Unauthorized Person. The system should also be able to discover if otherwise correct data are submitted by an unauthorized person. For instance, only the supervising pharmacist should be able to enter inventory totals for controlled substances in the pharmacy. Invalidation of transactions submitted by an unauthorized individual applies to privacy and security concerns surrounding payroll systems and employee evaluation records that determine pay levels, promotions, or discipline; files containing trade secrets; and files holding classified information, such as national defense data.

Asking the System to Perform an Unacceptable Function. The third error that invalidates input transactions is asking the system to perform an unacceptable function. For instance, it would be logical for a human resources manager to update the existing record of a current employee, but it would be invalid to ask the system to create a new file rather than merely to update an existing record.

VALIDATING INPUT DATA

It is essential that the input data themselves, along with the transactions requested, are valid. Several tests can be incorporated into software to ensure this validity. We consider eight possible ways to validate input.

Test for Missing Data. The first kind of validity test examines data to see if there are any missing items. For some situations, *all* data items must be present. For example, a Social Security file for paying out retirement or disability benefits would be invalid if it did not include the payee's Social Security number.

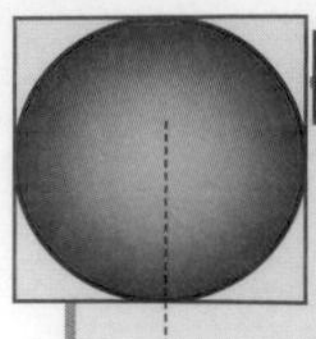

CONSULTING OPPORTUNITY 19.3

TO ENTER OR NOT TO ENTER: THAT IS THE QUESTION

"I've just taken on the presidency of Elsinore Industries," says Rose N. Krantz. "We're actually part of a small cottage industry that manufactures toy villages for children seven years old and up. Our tiny hamlets consist of various kits that will build what children want from interlocking plastic cubes, essentials such as city hall, the police station, the gas station, and a hot dog stand. Each kit has a unique part number from 200 to 800, but not every number is used. The wholesale price varies from $54.95 for the city hall to $1.79 for a hot dog stand.

"I've been melancholy over what I've found out since signing on at Elsinore. 'Something is rotten' here, to quote a famous playwright. In fact, the invoicing system was so out of control that I've been working around the clock with our bookkeeper, Gilda Stern," Krantz soliloquizes.

"I would like you to help straighten things out," Rose continues. "We ship to 12 distribution warehouses around the country. Each invoice we write out includes the warehouse number 1 through 12, its street address, and the zip code. We also put on each invoice the date we fill the order, code numbers for the hamlet kits they order, a description of each kit, the price per item, and the quantity of each kit ordered. Of course, we also include the subtotals of kit charges, shipping charges, and the total that the warehouse owes us. No sales tax is added, because they resell what we send them to toy stores in all 50 states. I want you to help us design a computerized order entry system that will be part of the invoicing system for Elsinore Industries."

For your design of a data-entry system for Elsinore, take into consideration all the objectives for data entry discussed throughout this chapter. Draw any screens necessary to illustrate your design. How can you make the order entry system efficient? Respond in a paragraph. Specify what data can be stored and retrieved and what data must be entered anew for each order. How can unnecessary work be avoided? Write a paragraph to explain why the system you propose is more efficient than the old one. How can data accuracy be ensured? List three strategies that will work with the type of data that are being entered for Elsinore Industries.

In addition, the record should include both the key data that distinguish one record from all others and the function code telling the computer what to do with the data. The systems analyst needs to interact with users to determine what data items are essential and to find out whether exceptional cases ever occur that would allow data to be considered valid even if some data items were missing. For example, a second address line containing an apartment number or a person's middle initial may not be a required entry.

Test for Correct Field Length. A second kind of validity test checks input to ensure it is of the correct length for the field. For example, if the Omaha, Nebraska, weather station reports into the national weather service computer but mistakenly provides a two-letter city code (OM) instead of the national three-letter city code (OMA), the input data might be deemed invalid and hence would not be processed.

Test for Class or Composition. The test for class or composition validity test checks to see that data fields that are supposed to be exclusively composed of numbers do not include letters, and vice versa. For example, a credit card account number for American Express should not include any letters. Using a composition test, the program should not accept an American Express account number that includes both letters and numbers.

Test for Range or Reasonableness. Validity tests for range or reasonableness are really a common-sense measure of input that answers the question of whether data fall within an acceptable range or whether they are reasonable within predetermined parameters. For instance, if a user was trying to verify a proposed shipment date, the range test would neither permit a shipping date on the thirty-second day of October nor accept shipment in the thirteenth month, the respective ranges being 1 to 31 days and 1 to 12 months.

A reasonableness test ascertains whether the item makes sense for the transaction. For example, when adding a new employee to the payroll, entering an age of 120 years would not be reasonable. Reasonableness tests are used for data that are continuous, that is, data that have a smooth range of values. These tests can include a lower limit, an upper limit, or both a lower and an upper limit.

Test for Invalid Values. Checking input for invalid values works if there are only a few valid values. This test is not feasible for situations in which values are neither restricted nor predictable. This kind of test is useful for checking responses where data are divided into a limited number of classes. For example, a brokerage firm divides accounts into three classes only: class 1 = active account, class 2 = inactive account, and class 3 = closed account. If data are assigned to any other class through an error, the values are invalid. Value checks are usually performed for discrete data, which is data that have only certain values. If there are many values, they are usually stored in a table of codes file. Having the values in a file provides an easy way to add or change values.

Cross-Reference Checks. Cross-reference checks are used when one element has a relationship with another one. To perform a cross-reference check, each field must be correct in itself. For example, the price for which an item is sold should be greater than the cost paid for the item. Price must be entered, numeric, and greater than zero. The same criterion is used to validate cost. When both price and cost are valid, they may be compared.

A geographical check is another type of cross-reference check. In the United States, the state abbreviation may be used to ensure that a telephone area code is valid for that state and that the first two digits of the zip code are valid for the state.

Test for Comparison with Stored Data. The next test for validity of input data that we consider is one comparing it with data that the computer has already stored. For example, a newly entered part number can be compared with the complete parts inventory to ensure that the number exists and is being entered correctly.

Setting Up Self-Validating Codes (Check Digits). Another method for ensuring the accuracy of data, particularly identification numbers, is to use a check digit in the code itself. This procedure involves beginning with an original numerical code, performing some mathematics to arrive at a derived check digit, and then adding the check digit to the original code. The mathematical process involves multiplying each of the digits in the original code by some predetermined weights, summing these results, and then dividing this sum by a modulus number. The modulus number is needed because the sum usually is a large number, and we need to reduce the result to a single digit. Finally, the remainder is subtracted from the modulus number, giving us the check digit.

Figure 19.19 shows how a five-digit part number for a radiator hose (54823) is converted to a six-digit number containing a check digit. In this example, the weights chosen were the "1-3-1" system; in other words, the weights alternate between 1 and 3. After the digits 5, 4, 8, 2, and 3 were multiplied by 1, 3, 1, 3, and 1, they became 5, 12, 8, 6, and 3. These new digits sum to 34. Next, 34 is divided by the chosen modulus number, 10, with the result of 3 and a remainder of 4. The remainder, 4, is subtracted from the modulus number, 10, giving a check digit of 6. The digit 6 is now tacked onto the end of the original number, giving the official product code for the radiator hose (548236).

FIGURE 19.19

Steps in converting a five-digit part number to a six-digit number containing a check digit.

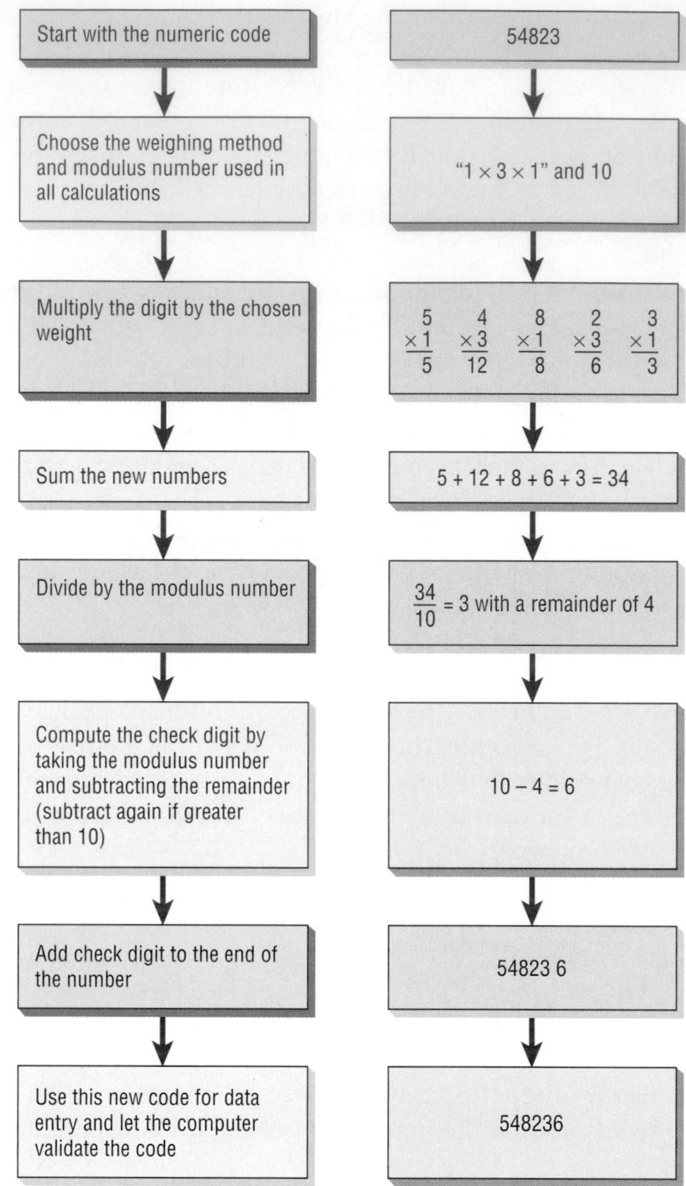

Using check digits The check digit system works in the following way. Suppose we had the part number 53411. This number has to be typed into the system, and while that is being done, different types of errors can occur. One possible error is the single digit miskey; for example, the clerk types in 54411 instead of 53411. Only the digit in the thousand place is incorrect, but this error may result in the wrong part being shipped.

A second type of error is transposed digits. It commonly occurs that the intended number 53411 gets typed in as number 54311 instead, just because two keys are pressed in reverse order. Transposition errors are also difficult for humans to detect.

These errors are avoidable through the use of a check digit because each of these numbers—the correct one and the error—would have a different check digit number, as shown in Figure 19.20. Now if part number 53411 was modified to 534118 (including the check digit 8) and either of the two errors just described occurred, the mistake would be caught. If the second digit was miskeyed as a 4, the computer would not accept 544118 as a valid number, because the check digit for 54411 would be 5, not 8. Similarly, if the second and third digits were transposed,

Status	Original Code	Check Digit	New Code
Correct	5 3 4 1 1	8	534118
Single digit miskey	5 **4** 4 1 1	5	544115
Transpose	5 **4** **3** 1 1	6	543116

FIGURE 19.20
Avoiding common data-entry errors through the use of a check digit.

as in 543118, the computer would also reject the number because the check digit for 54311 would be 6, not 8.

The systems analyst chooses the weights and the modulus number, but once chosen, they must not change. Some examples of weighting methods and modulus numbers can be found in Figure 19.21.

The check digit system is not foolproof. It is conceivable that two part numbers (732463 and 732413, for example) may have the same check digit. A single miskey in the second digit from the right would not be detected in that case.

The check digit system also has a cost. The added space taken up by the check digit must be considered as well as the added computation involved in calculating and verifying the check digit. The check digit approach is useful when the original codes are five or more digits, when codes are simple numerics with no meaning, and when the cost of making miskey and transposition errors is high.

The seven tests for checking on validity of input can go a long way toward protecting the system from the entry and storage of erroneous data. Always assume errors in input are more likely than not to occur. It is your responsibility to understand which errors will invalidate data and how to use the computer to guard against those errors and thus to limit their intrusion into system data.

Check Digit Method	Calculations for Check Digit to Be Added to the Original Number 29645	
Modulus 10 "2-1-2"	2 × 2 = 4; 9 × 1 = 9; 6 × 2 = 12; 4 × 1 = 4; 5 × 2 = 10 4 + 9 + 12 + 4 + 10 = 39/10 = 3 remainder Check digit equals Code with check digit is 296451.	10 (9) 1
Modulus 10 "3-1-3"	2 × 3 = 6; 9 × 1 = 9; 6 × 3 = 18; 4 × 1 = 4; 5 × 3 = 15 6 + 9 + 18 + 4 + 15 = 52/10 = 5 remainder Check digit equals Code with check digit is 296458.	10 (2) 8
Modulus 11 "Arithmetic"	2 × 6 = 12; 9 × 5 = 45; 6 × 4 = 24; 4 × 3 = 12; 5 × 2 = 10 12 + 45 + 24 + 12 + 10 = 103/11 = 9 remainder Check digit equals Code with check digit is 296457.	11 (4) 7
Modulus 10 "Geometric"	2 × 32 = 64; 9 × 16 = 144; 6 × 8 = 48; 4 × 4 = 16; 5 × 2 = 10 64 + 144 + 48 + 16 + 10 = 282/11 = 25 remainder Check digit equals Code with check digit is 296454.	11 (7) 4

FIGURE 19.21
Examples of weighting methods and modulus numbers.

CONSULTING OPPORTUNITY 19.4

DO YOU VALIDATE PARKING?

"What are we going to do, Mercedes?" Edsel asks wearily. Together, Mercedes and Edsel are reviewing the latest billing printout for their firm, Denton and Denton Parking Garages. They have been purchasing batch billing services from a small, local computer services company since they acquired three parking garages in a medium-sized metropolitan area. Denton and Denton Parking Garages rents daily, monthly, and yearly parking places to corporations and individuals.

Mercedes replies, "I'm not sure what our next move is, but the billing is all wrong. Maybe we should try to talk to the computer people."

"They said they could figure out how to compute these charges from looking at what the old owners did by hand before, and they said they didn't want to run the old and new systems in parallel," Edsel remarks, shaking his head. "That isn't right, though. At least I can't figure it out. Maybe you can."

Mercedes accepts the notion of chasing the suspect output and starts looking at the report in detail. "Well, for one thing, they don't realize we get cars from all over in here. Wherever we've got a car with plates that aren't in-state, it seems as if the computer stops figuring. Look, our plates start with a number and then a letter, right? Well this one from New York begins with three letters. The computer can't handle it," she says.

Edsel catches on and starts to think about the business as he looks at the printout. "Yeah, and look here. This person doesn't have a yearly account number, just a monthly one, so no bill came out," he says. "We've got monthlies, too, and the computer doesn't know it?"

"And look at this. It still made daily charges for the three days in November when we told them right out there weren't any vacancies for daily customers. It isn't reasonable," Mercedes asserts.

Edsel continues paging through the printout, but Mercedes stops him, saying, "Don't look any further. I'm calling the computer people so we can get this mess straightened out."

How would you characterize the problems being encountered with the current garage billing system? Use a paragraph to formulate a response. What are some tests for validity of data that could be included in the software for a revised billing system for the parking garages? List them. What could the programmer/analysts for the computer services company have done differently so that the customer was not faced with correcting the poor-quality output? Use three paragraphs to do a critical analysis of what was done and what *should* have been done.

ACCURACY, CODES, AND THE GRAPHICAL USER INTERFACE

In systems that use a graphical user interface (GUI), codes are often stored either as a function or as a separate table in the database. There is a trade-off on creating too many tables, because the software must find matching records from each table, which may lead to slow access. If the codes are relatively stable and rarely change, they may be stored as a database function. If the codes change frequently, they are stored on a table so that they may be easily updated.

Figure 19.22 shows how a drop-down list is used to select the codes for adding or changing a record in the CUSTOMER table. Notice that the code is stored, but the drop-down list displays both the code and the code meaning. This method helps to ensure accuracy, because the user does not have to guess at the meaning of the code and there is no chance of typing an invalid code.

ACCURACY ADVANTAGES IN ECOMMERCE ENVIRONMENTS

One of the many bonuses of ecommerce transactions is increased accuracy of data, due to four main reasons:

1. Customers generally key or enter data themselves.
2. Data entered by customers are stored for later use.
3. Data entered at the point of sale are reused throughout the entire order fulfillment process.
4. Information is used as feedback to customers.

An analyst needs to be aware of the advantages that have resulted from ecommerce and the electronic capture and use of information.

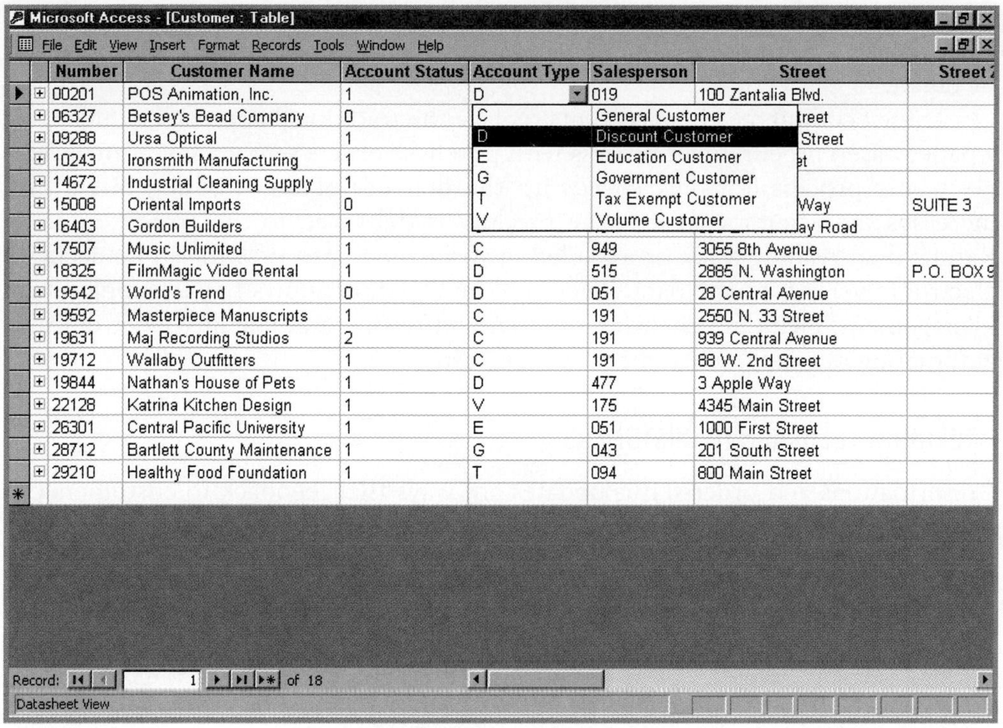

Microsoft Access - [Customer : Table]

Number	Customer Name	Account Status	Account Type	Salesperson	Street	Street 2
00201	POS Animation, Inc.	1	D	019	100 Zantalia Blvd.	
06327	Betsey's Bead Company	0			treet	
09288	Ursa Optical	1			Street	
10243	Ironsmith Manufacturing	1			et	
14672	Industrial Cleaning Supply	1				
15008	Oriental Imports	0			Way	SUITE 3
16403	Gordon Builders	1			ay Road	
17507	Music Unlimited	1	C	949	3055 8th Avenue	
18325	FilmMagic Video Rental	1	D	515	2885 N. Washington	P.O. BOX 9
19542	World's Trend	0	D	051	28 Central Avenue	
19592	Masterpiece Manuscripts	1	C	191	2550 N. 33 Street	
19631	Maj Recording Studios	2	C	191	939 Central Avenue	
19712	Wallaby Outfitters	1	C	191	88 W. 2nd Street	
19844	Nathan's House of Pets	1	D	477	3 Apple Way	
22128	Katrina Kitchen Design	1	V	175	4345 Main Street	
26301	Central Pacific University	1	E	051	1000 First Street	
28712	Bartlett County Maintenance	1	G	043	201 South Street	
29210	Healthy Food Foundation	1	T	094	800 Main Street	

C	General Customer
D	Discount Customer
E	Education Customer
G	Government Customer
T	Tax Exempt Customer
V	Volume Customer

Record: 1 of 18

Datasheet View

FIGURE 19.22
A table of codes used in drop-down list. This list is used to select a code for adding or changing an item in a record.

CUSTOMERS KEYING THEIR OWN DATA

First, customers know their own information better than anyone else. They know how to spell their street address, they know whether they live on a drive or a street, they know their own area code. If this information is transmitted by phone, it is easier to make a mistake spelling the address; if it is entered by using a faxed paper form, mistakes can occur if the fax transmission is difficult to read. If users enter their own information, however, accuracy increases.

STORING DATA FOR LATER USE

After customers enter information, it may be stored on their own personal computers. If they return to that ecommerce site and fill out the same form to complete a second transaction, they will witness the advantage of storing this information. As they begin to type their name, drop-down menus will prompt them with their full name even though only a couple of characters were entered. By clicking on this prompt, the full name is entered and no further typing is necessary for this field. This "autocomplete" feature can suggest matches for credit card and password information as well, and this information is encrypted so that Web sites cannot read the information stored on the user's computer.

Companies that want to store information to enable faster and more accurate transactions do so in files called cookies. Specific credit card and other personal information can only be accessed by the company that placed the cookie on the user's computer.

USING DATA THROUGH THE ORDER FULFILLMENT PROCESS

When companies capture information from a customer order, they can use and reuse that information throughout the entire order fulfillment process. Hence, the information gathered to complete an order can also be used to send an invoice to a customer, obtain the product from the warehouse, ship the product, send feedback to the customer, and restock the product by notifying the manufacturer. It can also be used again to send a paper catalog to the customer or send a special offer by email.

These ecommerce enhancements replace the traditional approach, which used a paper-based procurement process with purchase orders sent via fax or mail. This electronic process not only speeds up the delivery of the product, but it also increases the accuracy so that the product is delivered to the correct address. Rather than reading a fax or a mailed-in form, a shipper uses the more accurate electronic version of the data. Electronic information allows better supply chain management, including checking product and resource availability electronically, automating planning, scheduling, and forecasting.

PROVIDING FEEDBACK TO CUSTOMERS

Confirmations and order status updates are ways that feedback to customers can be enhanced. If a customer receives confirmation of a mistake in an order just placed, the order can be corrected immediately. For example, suppose a customer mistakenly submits an order for two copies of a DVD rather than one. After submitting the order, the customer receives an email confirming the order. The customer notices the mistake, immediately contacts the company, and has the order corrected, thereby avoiding having to return the extra copy of the DVD. Accuracy is improved by better feedback.

SUMMARY

Ensuring the quality of the data input to the information system is critical to ensuring quality output. The quality of data entered can be improved through the attainment of the three major data-entry objectives: effective coding, effective and efficient data capture, and the validation of data.

One of the best ways to speed data entry is through effective use of coding, which puts data into short sequences of digits and/or letters. Both simple sequence codes and alphabetic derivation codes can be used to follow the progress of a given item through a system. Classification codes and block sequence codes are useful for distinguishing classes of items from each other. Codes such as the cipher code are also useful because they can conceal information that is sensitive or is restricted to personnel within the business.

Revealing information is also a worthwhile use of codes, because it can enable business employees to locate items in stock and can also make data entry more meaningful. Significant-digit subset codes use subgroups of digits to describe a product. Mnemonic codes also reveal information by serving as memory aids that can help a data-entry operator enter data correctly or help the end user in using information. Codes that are useful for informing computers or people about what functions to perform or what actions to take are called function codes; they circumvent having to spell out in detail what actions are necessary.

Another part of assuring effective data entry is attention to the input devices being used. A well-designed, effective form that serves as a source document (where needed) is the first step. Data can be input through many different methods, each with varying speed and reliability. Keyboards have been redesigned for efficiency and improved ergonomics. Optical character recognition (OCR) allows the reading of input data through the use of special software, which eliminates some steps and also requires fewer employee skills.

Other data-entry methods include magnetic ink character recognition (MICR), which is used by banks to encode customer account numbers, and mark-

"Sometimes I think I'm the luckiest person on Earth. Even though I've been here five years, I still enjoy the people I meet and what I do. Yes, I know Snowden's demanding. You've experienced some of that, haven't you? He, for one, loves codes. I, for another, think they are a pain. I always forget them or try to make up new ones or something. Some of the physicians, though, think they're great. It must be all those Latin abbreviations they studied in med school. I hear that your most pressing assignment this week has to do with actually getting the information into the project reporting system. The Training Unit wants your ideas, and it wants them fast. Good luck with it. Oh, and when Snowden gets back from Thailand, I'm certain he'll want to take a peek at what your team has been up to."

HYPERCASE® QUESTIONS

1. Using a CASE tool, a software package such as Microsoft Access, or a paper layout form, design a data-entry procedure for the proposed project reporting system for the Training Unit. Assume we are particularly concerned about the consulting physicians staff, who don't want to spend a great deal of time keying in large amounts of data when using the system.
2. Test your data-entry procedure on three teammates. Ask for feedback concerning the appropriateness of the procedure, given the type of users the system will have.
3. Redesign the data-entry procedure to include the feedback you have received. Explain in a paragraph how your changes reflect the comments you were given.

sense forms, which are used for high-volume data entry. Bar codes (applied to products or human identification and then scanned) also speed data entry and improve data accuracy and reliability. New input technologies such as digital cameras expand the ease of use and the range of functions available. Intelligent terminals are input devices (often microprocessor-based) with a screen display and keyboard, and they can be networked to the CPU. They permit transactions to be entered and completed in real time.

Along with appropriate coding, data capture, and input devices, accurate data entry can be enhanced through the use of input validation. The systems analyst must assume that errors in data *will* occur and must work with users to design input validation tests to prevent erroneous data from being processed and stored, because initial errors that go undiscovered for long periods are expensive and time-consuming to correct.

Input transactions should be checked to ensure that the transaction requested is acceptable, authorized, and correct. Input data can be validated through the inclusion in the software of several types of tests that check for missing data, length of data item, range and reasonableness of data, and invalid values for data. Input data can also be compared with stored data for validation purposes. Once numerical data are input, they can be checked and corrected automatically through the use of check digits.

Ecommerce environments afford the opportunity for increasing accuracy of data. Customers can enter their own data, store data for later use, use the same stored data throughout the order fulfillment process, and receive feedback regarding order confirmations and updates.

KEYWORDS AND PHRASES

- alphabetic derivation code
- autocomplete feature
- bar codes
- block sequence code
- bottlenecks
- changeable
- check digits
- cipher code
- classification code
- coding
- cookies
- cross-reference test
- differentiated
- function code
- intelligent terminals
- keyboarding
- magnetic ink character recognition (MICR)
- mark sense
- mnemonic code
- optical character recognition (OCR)
- redundancy in input data
- self-validating codes
- significant-digit subset code
- simple sequence code
- supply chain management
- test for class or composition
- test for comparison with stored data
- test for correct field length
- test for invalid values
- test for missing data
- test for range or reasonableness
- validating input

REVIEW QUESTIONS

1. What are the four primary objectives of data entry?
2. List the five general purposes for coding data.
3. Define the term *simple sequence code*.
4. When is an alphabetic derivation code useful?
5. Explain what is accomplished with a classification code.
6. Define the term *block sequence code*.
7. What is the simplest type of code for concealing information?
8. What are the benefits of using a significant-digit subset code?
9. What is the purpose of using a mnemonic code for data?
10. Define the term *function code*.
11. List the eight general guidelines for proper coding.
12. What are changeable data?
13. What are differentiation data?
14. What is one specific way to reduce the redundancy of data being entered?
15. Define the term *bottleneck* as it applies to data entry.
16. What three repetitive functions of data entry can be done more efficiently by the computer than by the data-entry operator?
17. List six data-entry methods.
18. List the three main problems that can occur with input transactions.
19. What are the eight tests for validating input data?
20. Which test checks to see whether data fields are correctly filled in with either numbers or letters?
21. Which test would not permit a user to input a date such as October 32?
22. Which test ensures data accuracy by the incorporation of a number in the code itself?
23. List four improvements to data accuracy that transactions conducted over ecommerce Web sites can offer.

PROBLEMS

1. A small, private university specializing in graduate programs wants to keep track of when a particular student actually enrolls. Suggest a kind of code for this purpose and give an example of its use in the university that demonstrates its appropriateness.
2. The military has been using a simple sequence code to keep track of new recruits. There have been some upsetting mixups, however, between recruit files because of similar-looking recruit numbers.
   a. In a paragraph, suggest a different coding scheme that will help uniquely identify each recruit and explain how it will prevent mixups.
   b. The military is concerned that confidential information in its coding of new recruits (such as IQ score and rating of physical condition upon entering the service) *not* be revealed to clerks who do not have appropriate clearance, but it wants this information to be encoded on a recruit's identification number so that those conducting basic training are immediately aware of the type of recruit they are training. Suggest a type of code (or combination of codes) that can accomplish this task and give an example.
3. A code used by an ice cream store to order its products is 12DRM215-220. This code is deciphered in this manner: 12 stands for the count of items in the box, DRM stands for DREAMCICLES (a particular kind of ice cream novelty), and 215-220 indicates the entire class of low-fat products carried by the distributor.
   a. What kind of code is used? Describe the purpose behind each part (12, DRM, 215-220) of the code.
   b. Construct a coded entry using the same format and logic for an ice cream novelty called Pigeon Bars, which come in a six-count package and are *not* low-fat.
   c. Construct a coded entry using the same format and logic for an ice cream novelty called Airwhips, which come in a 24-count package and are low-fat.
4. The data-entry operators at Michael Mulheren Construction have been making errors when entering the codes for residential siding products, which are as follows: U = stUcco, A = Aluminum, R = bRick, M = Masonite, EZ = EZ color-lok enameled masonite, N = Natural wood siding, AI = pAInted finish, SH = SHake SHingles. Only one code per address is permitted.
   a. List the possible problems with the coding system that could be contributing to erroneous entries. (*Hint*: Are the classes mutually exclusive?)
   b. Devise a mnemonic code that will help the operators understand what they are entering and subsequently help their accuracy.
   c. How would you redesign the classes for siding materials? Respond in a paragraph.
5. The following is a code for one product in an extensive cosmetic line: L02002Z621289. L means that it is a lipstick, 0 means it was introduced without matching nail polish, 2002 is a sequence code indicating in what order it was produced, Z is a classification code indicating that the product is hypoallergenic, and 621289 is the number of the plant (there are 15 plants) where the product is produced.
   a. Critique the code by listing the features that might lead to inaccurate data entry.
   b. Designer Brian d'Arcy James owns the cosmetic firm that uses this coding scheme. Always interested in new design, Brian is willing to look at a more

elegant code that encodes the *same* information in a better way. Redesign the coding scheme and provide a key for your work.

c. Write a sentence for each change you have suggested, indicating what data-entry problem (from part a) the change will eliminate.

6. The d'Arcy James cosmetic firm requires its salespeople to use notebooks to enter orders from retail department stores (their biggest customers). This information is then relayed to warehouses, and orders are shipped on a first-come, first-served basis. Unfortunately, the stores are aware of this policy and are extremely competitive about which one of them will offer a new d'Arcy James product first. Many retailers have taken the low road and persuaded salespeople to falsify their order dates on sales forms by making them earlier than they actually were.
   a. This problem is creating havoc at the warehouse. Disciplining any of the personnel involved is not feasible. How can the warehouse computer be used to certify when orders are actually placed? Explain in a paragraph.
   b. Salespeople are complaining that they have to ignore their true job of selling so that they can key in order data. List the data items relating to sales of cosmetics to retailers that should be stored in and retrieved from the central computer rather than keyed in for every order.
   c. Describe in a paragraph or two how bar coding might help solve the problem in part b.
7. List the best data-entry method and your reason for choosing it for each of the five situations listed below:
   a. Turnaround document for a utility company that wants notification of change in the customer address.
   b. Data retrieval allowed only if there is positive machine identification of the party requesting data.
   c. Not enough trained personnel available to interpret long written responses; many forms submitted that capture answers to multiple-choice examinations; high reliability necessary; fast turnaround not required.
   d. Warehouse set up for a discount compact disc operation; bins are labeled with price information, but individual discs are not; and few skilled operators available to enter price data.
   e. Poison control center that maintains a large database of poisons and antidotes; needs a way to enter data on poison taken; also enter weight, age, and general physical condition of the victim when a person calls the center's toll-free number for emergency advice.
   f. Online purchase of a CD by a consumer with a credit card.
8. Ben Coleman, one of your systems analysis team members surprises you by asserting that when a system uses a test for correct field length, it is redundant also to include a test for range or reasonableness. In a paragraph, give an example that demonstrates that Ben is mistaken on this one.
9. Several retailers have gotten together and begun issuing a "state" credit card that is good only in stores within their state. As a courtesy, sales clerks are permitted to transcribe the 15-digit account number by hand (after getting it from the accounting office) if the customer is not carrying his or her card. The only problem with accounts that retailers have noticed so far is that sometimes erroneous account numbers are accepted into the computer system, resulting in a bill being issued to a nonexistent account.
   a. What sort of validity test would clear up the problem? How? Respond in a paragraph.
   b. Suggest an alternative data-entry method that might alleviate this problem altogether.

10. The following are part numbers:

    38902
    38933
    39402
    35693
    35405
    39204

    Develop a check digit for the preceding part numbers using 1-3-1-3-1 multipliers and modulus 11. Use the method presented in this chapter. Why do some numbers have the same check digit?
11. Develop a check digit system for the preceding part numbers using 5-4-3-2-1 multipliers and modulus.
12. Why would a check digit system such as 1-1-1-1-1 not work as well as other methods? What errors would it miss?

GROUP PROJECTS

1. Along with your group members, read Consulting Opportunity 19.3, "To Enter or Not to Enter: That Is the Question," presented earlier in this chapter. Design an appropriate data-entry system for Elsinore Industries. Your group's design should emphasize efficiency and accuracy. In addition, distinguish between data that are changeable and data that differentiate an item being entered from all others. Draw prototypes of any screen necessary to explain what you are recommending.
2. Divide your group into analysts and Elsinore Industries employees to role play. The analysts should present the new data-entry system, complete with prototype screens. Ask for feedback on the design from Elsinore employees.
3. Write a brief paragraph describing how to improve the original data-entry design based on the comments received.

SELECTED BIBLIOGRAPHY

Davis, G. B., and M. H. Olson. *Management Information Systems, Conceptual Foundations, Structure, and Development*, 2d ed. New York: McGraw-Hill, 1985.

Galitz, W. O. *Human Factors in Office Automation*. Atlanta: Life Office Automation Management Association, 1980.

Lamming, M. G., P. Brown, K. Carter, M. Eldridge, M. Flynn, G. Louie, P. Robinson, and A. Sellan. "The Design of a Human Memory Prosthesis." *Computer Journal*, Vol. 37, 1994, pp. 153–63.

Miller, G. A. "The Magical Number Seven, Plus or Minus Two: Some Limits on Our Capability for Processing Information." *Psychological Review*, Vol. 63, No. 2, March 1956, pp. 81–97.

Newman, W. N., and M. G. Lamming. *Interactive System Design*. Reading, MA: Addison-Wesley, 1995.

Owsowitz, S., and A. Sweetland. "Factors Affecting Coding Errors." *Rand Memorandum RM-4346-PR*. Santa Monica, CA: Rand Corporation, 1965.

Robey, D., and W. Taggart. "Human Processing in Information and Decision Support Systems." *MIS Quarterly*, Vol. 6, No. 2, June 1982, pp. 61–73.

19

ALLEN SCHMIDT, JULIE E. KENDALL, AND KENNETH E. KENDALL

ENTERING NATURALLY

Tuesday afternoon finds Anna and Chip having their weekly analysis and design review session. Chip waves toward a large stack of documents that are neatly organized on a large table. "I can't believe that we're almost finished with the design of this system," he remarks. "It's been a long process, but I'll bet we've obtained enough user feedback to ensure a high-quality system. All that's left is the design of the data-entry procedures, and we'll be ready to start packaging the specs for the programmers."

"Yes," replies Anna, "the end is in sight. Let's start by examining the design of the input portion of the system." The data flow diagram shown in Figure E19.1 represents a portion of two Add programs. The top diagram depicts adding new microcomputers using a batch process.

New Microcomputer Number
1.1 Key Micro-computer Record
Keyed Microcomputer
1.2 Verify Micro-computer Record
Microcomputer Transaction
D8 Microcomputer Add Transactions
Errors

Errors
New Software
2.1 Key Software Record
Keyed Software
2.2 Validate Software Record
Valid Software
2.3 Confirm New Software
Software Record
Confirmed Software
D5 Software Master File

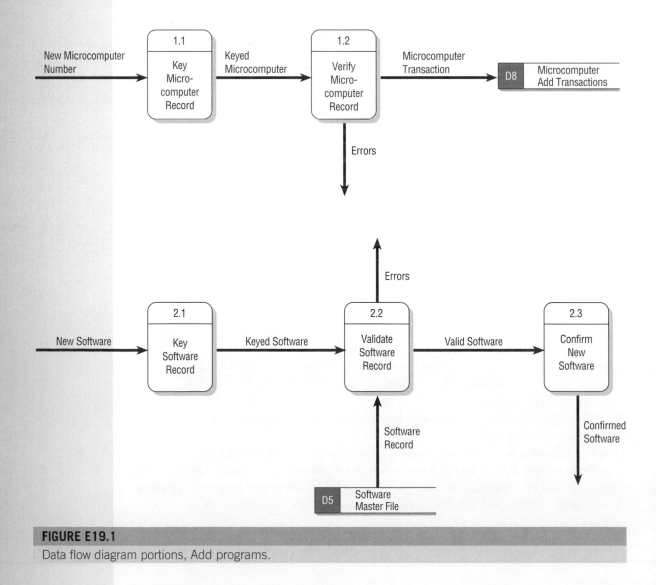

FIGURE E19.1
Data flow diagram portions, Add programs.

"Because this portion is to be performed by data entry, we should use key-to-diskette data entry," remarks Anna. "The microcomputers are usually ordered for an entire lab, so they often arrive in shipments of 20 or more. We can take advantage of the keying speed of the data-entry operators. The new microcomputer information is on easy-to-enter forms that will be keyed by one operator and then rekeyed by a different person to verify the data. The results will be stored on a transaction file for use by the edit and update programs."

"The Add Software process will have a different design," notes Chip. "Because software does not arrive in a batch, the add program is online. Instead of a key-and-verify operation, the operator will have to search-verify each transaction. After all data fields have been edited for accuracy, a message will appear on the bottom of the screen. It will prompt the operators to check the data on the screen for accuracy against the form and click the **Save Record** button if correct. The operators will have a chance to make changes if the data are keyed incorrectly. This method may not be keyed as fast as data entry, but the errors are corrected by the person seeing the data, which speeds up the actual updating of the files."

Regardless of the method for adding information, every data field must be edited for accuracy. Chip notes, "In the long run, it's better to have complete editing for accuracy in the programs rather than to find that erroneous data have been stored on master files and printed on reports."

The strategy for field editing is to check the data in the following order:

1. Syntax—whether the data are numeric or alphabetic—and the length of the data. An example is the HARDWARE INVENTORY NUMBER, which must be eight characters in length and numeric.
2. The contents of the field, including range, limit, and values for the data. When validating the DATE PURCHASED, the month must be from 1 to 12. This check should occur only after the month has been verified as numeric.
3. Cross-reference checks between two or more data elements. For those who are to check the day portion of the DATE PURCHASED, a table of the number of days possible for each month will be used for an upper limit. This table could not be used if the month number was not between 1 and 12. Check digits are another example of a cross-reference edit.
4. External edits, such as reading a file to verify if the record to be added already exists in the file. Reading records is slower than editing, which is performed in main memory, and it should occur only after the data successfully pass all other edits.

Edit criteria have been entered on the Visible Analyst Element Repository screen as the elements were added to the design. These elements include simple editing criteria and table checking. The **Notes** area may be used to enter editing criteria. The HARDWARE INVENTORY NUMBER entry shown in Figure E19.2 includes a reference for using the modulus-11 method of verifying the check digit portion of the number. Furthermore, when adding a new microcomputer, the MICROCOMPUTER MASTER file must be read to ensure that a record does not already exist with the same HARDWARE INVENTORY NUMBER. Figure E19.3 shows the details for the HARDWARE INVENTORY NUMBER. Notice that the Picture is 9(8); here the 9s

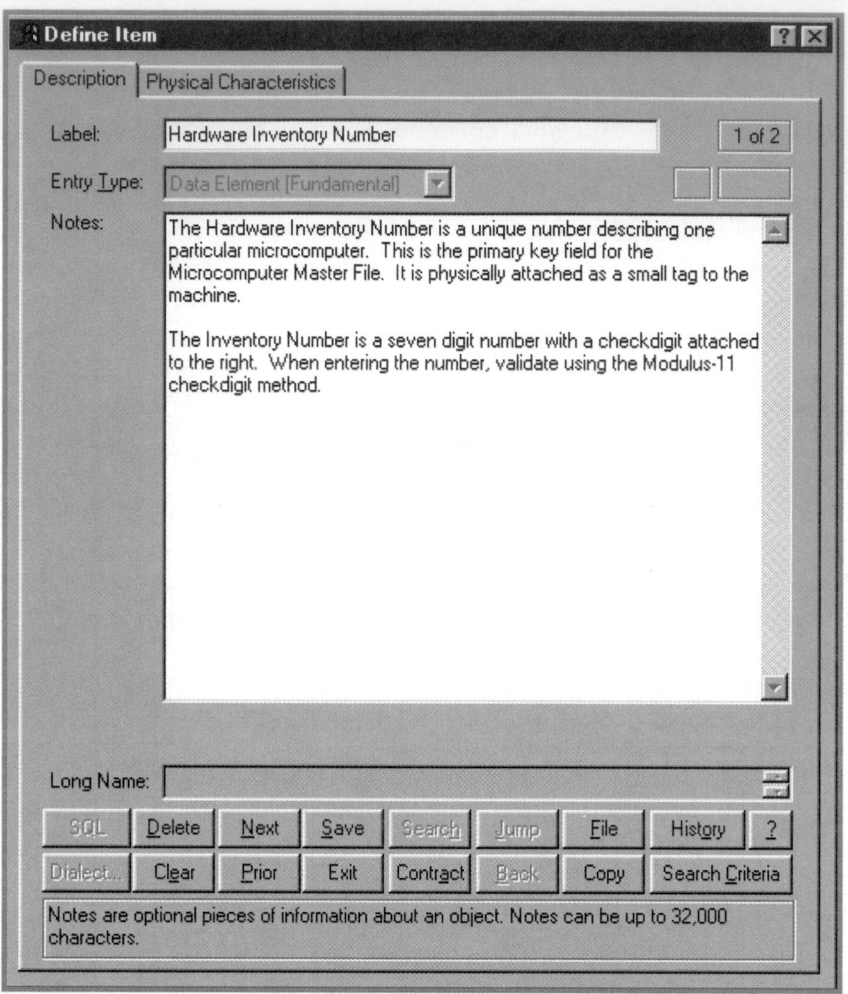

FIGURE E19.2
Element repository screen, HARDWARE INVENTORY NUMBER, showing edit criteria.

indicate that the data should be numeric, and the number in parentheses indicates that the length is 8.

"I think these reports will be useful," Chip tells Anna. "The first report involves creating a final list of all the elements found both on the MICROCOMPUTER MASTER file and the structural records contained within the master. The report is produced using the Report feature and shows the elements, their length, pictures, and edit criteria in the **Notes** area." (This report is shown in Figure E19.4.) This report is used to create the edit criteria table, which became part of the program specifications. The table is shown in Figure E19.5.

Several of the elements have **Notes** areas referring to tables as well as entries for the codes in the Values and Meanings area. An example is the INTERNAL BOARDS element. "I'll produce a list of all tables we'll require," Anna offers. This time, the **Report Query** feature is used to produce the necessary information. A list of all ele-

Define Item

Description | Physical Characteristics

Label: Hardware Inventory Number 2 of 2

Entry Type: Data Element [Fundamental]

Locations:
Assoc Entity --> Hardware Software Relational Table
Assoc Entity --> Microcomputer Board Relational Table
Assoc Entity --> Monitor
Attrib Entity --> Component Boards
Attrib Entity --> Microcomputer Maintenance
Data Flow --> Changed Microcomputer
Data Flow --> Completed Installation Listing
Data Flow --> Component Boards
Data Flow --> Confirmed Changes
Data Flow --> Confirmed Microcomputer Record

Physical Characteristics
Type: Char
Length: 8 Picture: 9(8)
Allow Null: No Display:
Default: Owner:
Min: Max: Avg:

Long Name:

SQL | Delete | Next | Save | Search | Jump | File | History | ?
Dialect... | Clear | Prior | Exit | Contract | Back | Copy | Search Criteria

A repository object label can be up to 128 characters long, and the first character must be a letter.

FIGURE E19.3
Element characteristics in the repository, HARDWARE INVENTORY NUMBER.

ments containing **Notes** starting with "Table of codes" was printed, as shown in Figure E19.6. Included on the list is the Picture and Length, showing the syntax of the code. With this list, tables are created.

Each table of codes is defined using Microsoft Access tables. Chip and Anna each spend time working on the tables. A mnemonic code is chosen for BOARDS and MONITORS, because these two would be easy for maintenance personnel to work with. Mnemonic codes are also used to represent the SOFTWARE CATEGORY, because these will be easy for users to remember.

"There are a wide variety of printers available," remarks Chip. "I think a significant-digit coding scheme would be the best here. The first digit represents the type of printer . . . laser and so on. The next two digits are for manufacturer, and the last two a sequence number representing different model numbers."

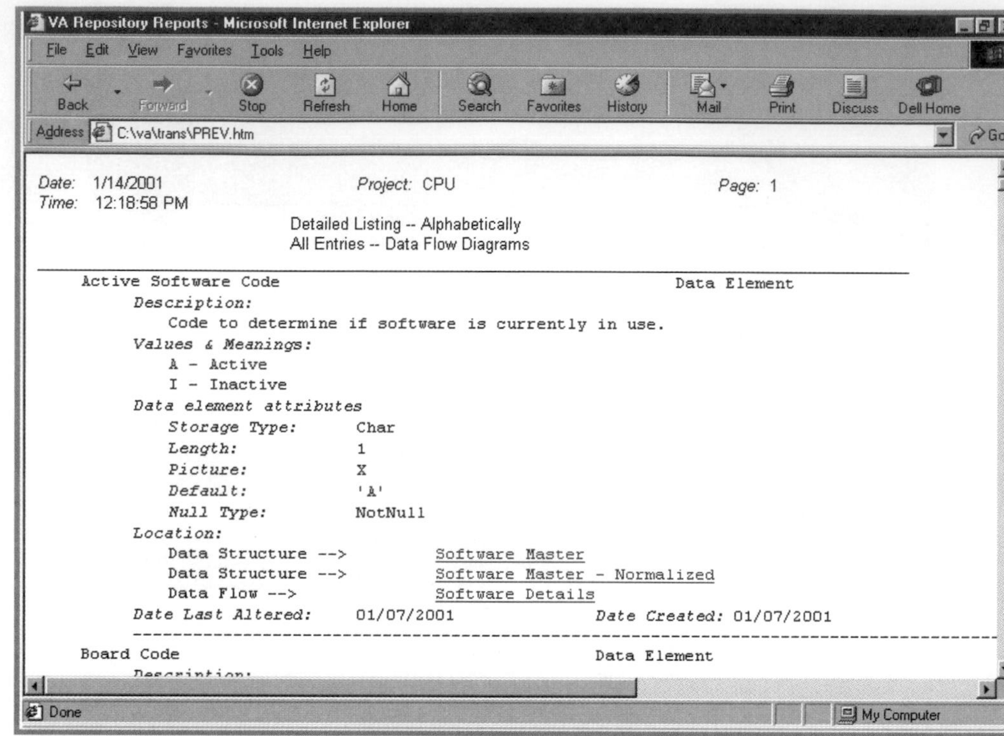

FIGURE E19.4
MICROCOMPUTER MASTER FILE ELEMENTS repository report.

Anna agrees. "That's good, Chip. That strategy can also be used for the campus buildings: the first digit for the campus location, and the remaining two digits representing individual buildings within the campus."

Chip designs the codes used for the BOARD TABLE. The Microsoft Access screen is shown in Figure E19.7. Two columns are used to define codes. The left column contains the code, and the right column contains the meaning of the code. These entries may be modified, and new entries may be added, providing flexibility in the final system.

"Here's the SOFTWARE CATEGORY table that I created," says Anna. (The Microsoft Access table is illustrated in Figure E19.8.) "This table may be easily updated as new software is developed and acquired by the university."

"That's a valuable component of the system," Chip comments. "It provides consistency for all codes and their meanings."

Chip and Anna finish their work the next morning at about 11:30. They glance around the room happily, frequently reexamining the final design. The months of analysis, design work, consultation with the users, and careful adherence to standards are finally complete.

"I feel really good about this project," says Anna.

Chip agrees, "I'm proud of the quality we put in."

Element Name	Numeric	Alphabetic	Length	Date	Limit	Range	Table	Check Digit	Cross Refer	File Check
Brand Name			>0				X			
Campus Location	X						X			
CD ROM Drive						3				
Cost of Repairs	X									
Date Purchased	X		8	X						
Disk Drive A										
Disk Drive B									4	
Fixed Disk	X				5					
Fixed Disk 2	X				5				6	
Hardware Inventory Number	X		8		1			X		X
Internal Boards							X			
Internet Connection						11				
Last Prevent Maintenance Date	X		8	X						
Maintenance Interval	X					7				
Memory Size	X				8					
Model			>0							
Monitor							X			
Network		X	1			9				
Number of Repairs	X									
Printer							X			
Purchase Cost	X				1					
Record Code		X	1				10			
Replacement Cost	X									
Room Location			>0							
Serial Number			>0							
Tape Backup Unit		X	1			9				
Warranty		X	1			9				
Zip Drive Connected		X	1			9				

Legend of codes:
1 Limit: Must be greater than 0
2 Date must be not greater than current date
3 Values are CD-ROM, CD-RW
4 Disk drive A must exist
5 Value is either 0 or greater than 18
6 First fixed disk must exist
7 Range: 7 to 250
8 Limit: Greater than 255
9 Values are Y or N
10 Values are A or I
11 Values are 0, 1, or M

FIGURE E19.5
MICROCOMPUTER MASTER edit table.

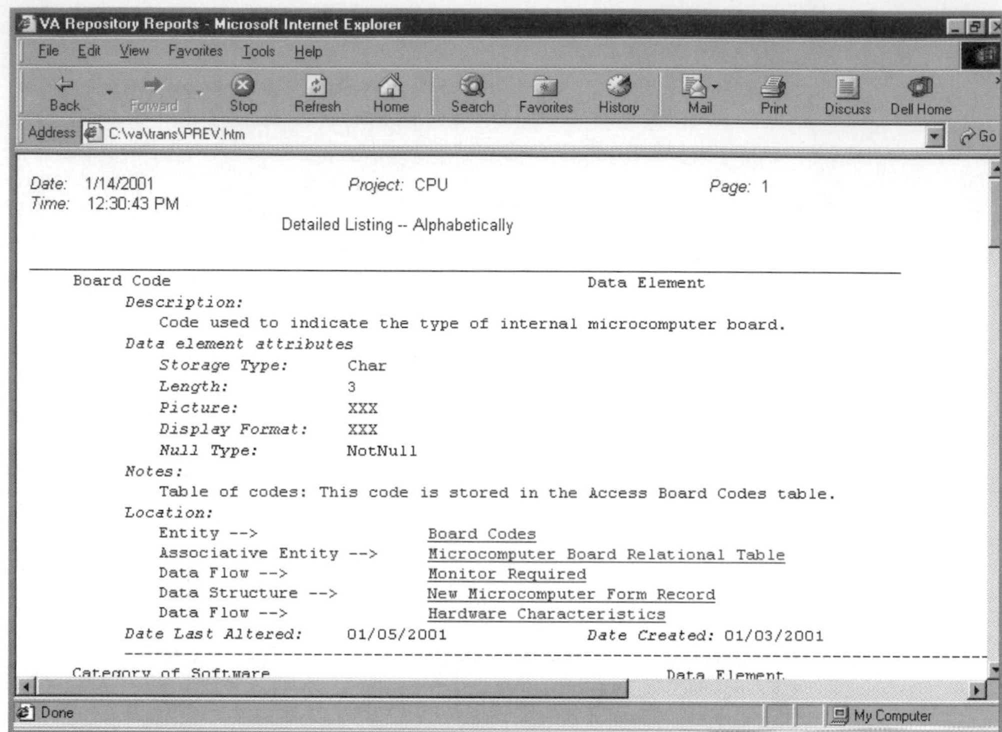

FIGURE E19.6
Visible Analyst report listing all coded elements.

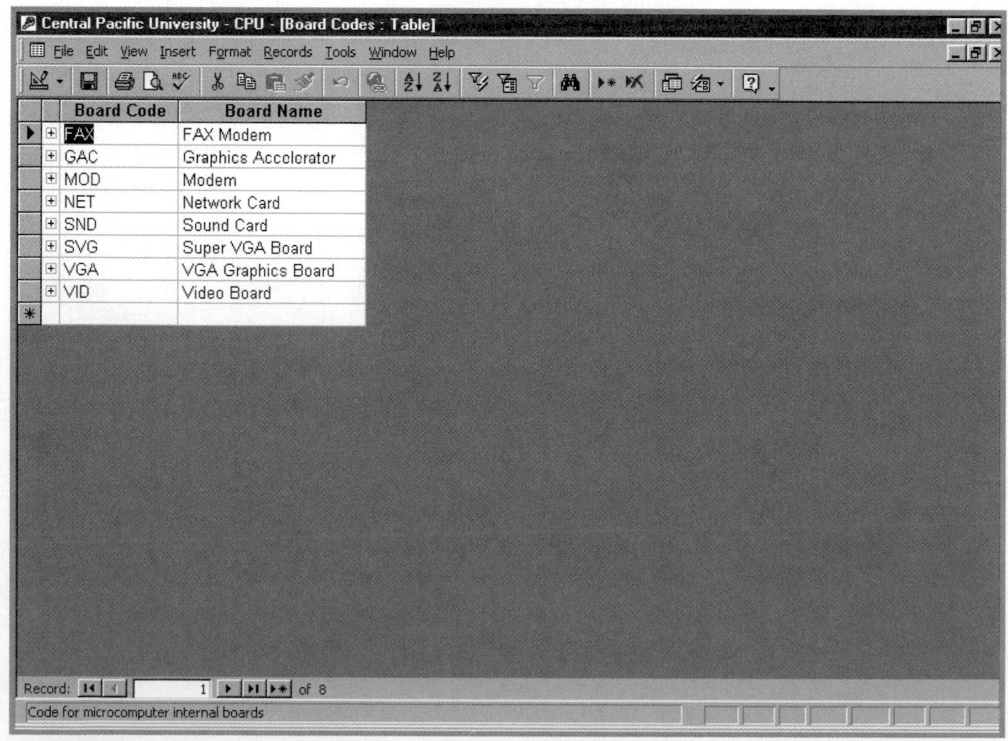

Board Code	Board Name
FAX	FAX Modem
GAC	Graphics Accelerator
MOD	Modem
NET	Network Card
SND	Sound Card
SVG	Super VGA Board
VGA	VGA Graphics Board
VID	Video Board

FIGURE E19.7
BOARD TABLE defined using Microsoft Access.

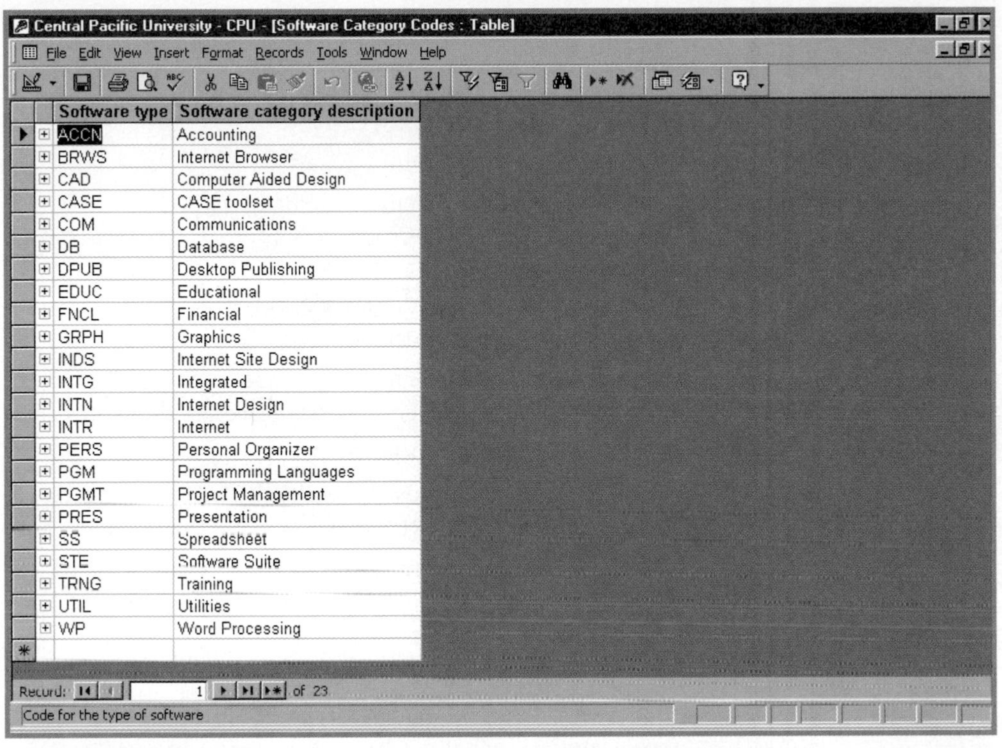

Software type	Software category description
ACCN	Accounting
BRWS	Internet Browser
CAD	Computer Aided Design
CASE	CASE toolset
COM	Communications
DB	Database
DPUB	Desktop Publishing
EDUC	Educational
FNCL	Financial
GRPH	Graphics
INDS	Internet Site Design
INTG	Integrated
INTN	Internet Design
INTR	Internet
PERS	Personal Organizer
PGM	Programming Languages
PGMT	Project Management
PRES	Presentation
SS	Spreadsheet
STE	Software Suite
TRNG	Training
UTIL	Utilities
WP	Word Processing

FIGURE E19.8
SOFTWARE CATEGORY code table.

EXERCISES*

E-1. Explain the difference between verifying input data in a batch data-entry situation and in an online program.

E-2. Modify and print the following elements with edit criteria in the **Notes** (or Values and Meanings for specific codes) area.

Element	*Edit Criteria*
a. SOFTWARE CATEGORY	Table of codes: Software Category Code
b. COURSE TRAINING LEVEL CODE	B - Beginning; I - Intermediate; A - Advanced
c. INTERNET CONNECTION NAME	0 - No Internet; M - Modem; 1 - T1 Line
d. SOFTWARE MEDIA CODE	D - Diskette; C - CD ROM
e. ZIP DRIVE CONNECTED	Y - Yes; N - No

Element	*Edit Criteria*
f. OPERATING SYSTEM	W - Windows 3.X; M - Macintosh; O - OS/2; 9 - Windows 95; 8 - Windows 98; N - Windows NT; 6 – DOS 6.X; 0 - Windows 2000; E - Millennium; U - Unix

E-3. Modify and print the following elements with edit criteria placed in the **Notes** area:

a. Element: SOFTWARE INVENTORY NUMBER
Notes: A modulus-11 check digit must be verified when entering the number. The ADD SOFTWARE program creates the check digit.
The ADD SOFTWARE program should also check the SOFTWARE MASTER file to ensure that a record with the same inventory number does not already exist.

b. Element: DATE PURCHASED
Notes: Verify that the DATE PURCHASED is less than or equal to the current date.

c. Element: QUANTITY RECEIVED
Notes: Verify that the QUANTITY RECEIVED is less than or equal to the QUANTITY ORDERED.

d. Element: SOFTWARE UPGRADE VERSION
Notes: Ensure that the upgrade version is greater than the current version.

e. Element: FIXED DISK 2
Notes: FIXED DRIVE 2 may exist only if there is anentry for FIXED DISK.

E-4. View and print the Coded Elements Report Query.

E-5. Use Microsoft Access to view the SOFTWARE CATEGORY CODES table. What is wrong with the design of these codes?

E-6. Use Microsoft Access to modify and print the BOARD CODES table. Add the following codes.

PCM PCMCIA Fax Modem
XJK XJACK Ethernet Modem
FAX FAX BOARD

E-7. Use Microsoft Access to modify and print the PRINTER CODES table. The format of this significant digit code is as follows:

TMMSS where

T is the type of printer
M is the manufacturer
S represents a sequence number, a higher number indicating an improved model

Values for the type of printer are:

0	Dot matrix
1	Daisy wheel
2	Ink jet
3	Thermal
4	Laser
5	PostScript
6	Plotter

Values for the manufacturer are as follows:

01	IBM
02	Epson
03	Hewlett-Packard
04	Panasonic
05	Star Micronics
06	Okidata
07	Kodak
08	Texas Instruments

Add the following codes:

Code	*Meaning*
20301	Hewlett-Packard DeskJet PLUS
50801	Texas Instruments MicroLaser
40201	Epson Laser EPL-6000
40305	Hewlett-Packard LaserJet V
40401	Panasonic KX-p4420

E-8. Use Microsoft Access to modify and print the MONITOR CODES table using the mnemonic form. Add the following entries:

Code	*Meaning*
EGAM	VGA GRAPHICS MONITOR
MONO	MONOCHROME MONITOR-GREEN
ORAN	MONOCHROME MONITOR-ORANGE
SVGA	SUPER VGA MONITOR
PLSM	PLASMA MONITOR
AMTX	ACTIVE COLOR MATRIX

E-9. After speaking with Dot Matricks and Mike Crowe, it has become apparent that the campus codes must be sortable for installing hardware and software as well as for creating inventory sheets. Use Microsoft Access to modify and print the

CAMPUS LOCATION CODES table. The first digit represents the campus location. Values are as follows:

1	Central Campus
2	Waterford Campus
3	Hillside Campus

The next three digits represent buildings within the campus, with the following building codes:

001	Administration	010	Environmental Studies
002	Admissions	011	Geology
003	Agricultural	012	Law
004	Astronomy	013	Library
005	Business	014	Mathematics
006	Chemical Engineering	015	Medicine
007	Computer Science	016	Physics
008	Education	017	Psychology
009	Engineering	018	Zoology

Use a combination (your choice) of campus and building codes to build the final table of codes. Include the meaning of the code.

*Exercises preceded by a CD-ROM icon require the program Visible Analyst or another CASE tool. A CD-ROM containing Visible Analyst examples is free of charge to any professor adopting this book. The examples on the disk may be imported into Visible Analyst and then used by students.

QUALITY ASSURANCE THROUGH SOFTWARE ENGINEERING

20

APPROACHES TO QUALITY

Quality has long been a concern of businesses, as it should be for systems analysts in the analysis and design of information systems. It is too risky to undertake the entire analysis and design process without using a quality assurance approach. The three approaches to quality assurance through software engineering are (1) securing total quality assurance by designing systems and software with a top-down, modular approach; (2) documenting software with appropriate tools; and (3) testing, maintaining, and auditing software.

Two thoughts guide quality assurance. The first is that the user of the information system is the single most important factor in establishing and evaluating its quality. The second is that it is far less costly to correct problems in their early stages than it is to wait until a problem is articulated through user complaints or crises.

We already have learned about the huge investment of labor and other business resources that are required to launch a system successfully. Using quality assurance throughout the process is a way to minimize risks, and it helps ensure that the resulting system is what is needed and wanted and will demonstrably improve some aspect of business performance. This chapter provides the analyst with three major approaches to quality.

THE TOTAL QUALITY MANAGEMENT APPROACH

Total quality management (TQM) is essential throughout all the systems development steps. According to Dean and Evans (1994), the primary elements of TQM are meaningful only when occurring in an organizational context that supports a comprehensive quality effort. It is within this context that the elements of customer focus, strategic planning and leadership, continuous improvement, empowerment, and teamwork are united to change employees' behavior and, ultimately, the organization's course. Notice that the concept of quality has broadened over the years to reflect an organizational, rather than an exclusively production, approach. Instead of conceiving of quality as controlling the number of defective products produced, quality is now thought of as an evolutionary process toward perfection that is otherwise referred to as total quality management.

Systems analysts must be aware of the factors that are driving the interest in quality. It is important to realize that the increasing commitment of businesses to TQM fits extraordinarily well into the overall objectives for systems analysis and design.

RESPONSIBILITY FOR TOTAL QUALITY MANAGEMENT

Practically speaking, a large portion of the responsibility for the quality of information systems rests with systems users and management. Two things must happen for TQM to become a reality with systems projects. First, the full organizational support of management must exist, which is a departure from merely endorsing the newest management gimmick. Such support means establishing a context for management people to consider seriously how quality of information systems and information itself affects their work.

Early commitment to quality from the analyst and the business is necessary to achieve the goal of quality. This commitment results in exerting an evenly paced effort toward quality throughout the systems development life cycle, and it stands in stark contrast to having to pour huge amounts of effort into ironing out problems at the end of the project.

Organizational support for quality in management information systems can be achieved by providing on-the-job time for IS quality circles, which consist of six to eight organizational peers specifically charged with considering both how to improve information systems and how to implement improvements.

Through work in IS quality circles or through other mechanisms already in place, management and users must develop guidelines for quality standards of information systems. Preferably, standards will be reshaped every time a new system or major modification is to be formally proposed by the systems analysis team.

Hammering out quality standards is not easy, but it is possible and it has been done. Part of the systems analyst's job is encouraging users to crystallize their expectations about information systems and their interactions with them.

Departmental quality standards must then be communicated through feedback to the systems analysis team. The team is often surprised at what has developed. Expectations typically are less complex than what experienced analysts know could be done with a system. In addition, human issues that have been overlooked or underrated by the analyst team may be designated as extremely pressing in users' quality standards. Getting users involved in spelling out quality standards for IS will help the analyst avoid expensive mistakes in unwanted or unnecessary systems development.

STRUCTURED WALKTHROUGH

One of the strongest quality management actions the systems analysis team can take is to do structured walkthroughs routinely. Structured walkthroughs are a way of using peer reviewers to monitor the system's programming and overall development, point out problems, and allow the programmer or analyst responsible for that portion of the system to make suitable changes.

Structured walkthroughs involve at least four people: the person responsible for the part of the system or subsystem being reviewed (a programmer or analyst), a walkthrough coordinator, a programmer or analyst peer, and a peer who takes notes about suggestions.

Each person attending a walkthrough has a special role to play. The coordinator is there to ensure that the others adhere to any roles assigned to them and to ensure that any activities scheduled are accomplished. The program author or analyst is there to listen, not to defend his or her thinking, rationalize a problem, or argue. The programmer or analyst peer is present to point out errors or potential problems, not to specify how the problems should be remedied. The note-taker records what is said so that the others present can interact without encumbrance.

Structured walkthroughs can be done whenever a portion of coding, a subsystem, or a system is finished. Just be sure that the subsystem under review is comprehensible outside of its larger context. Structured walkthroughs fit well within a total quality management approach when accomplished throughout the systems development life cycle. The time they take should be short—half an hour to an hour at most—which means that they must be well coordinated. Figure 20.1 shows a form that is useful in organizing the structured walkthrough as well as in reporting its results. Because walkthroughs take time, do not overuse them.

Use structured walkthroughs as a way to obtain (and then act on) valuable feedback from a perspective that you lack. As with all the quality assurance measures, the point of walkthroughs is to evaluate the product systematically on an ongoing basis rather than to wait until completion of the system.

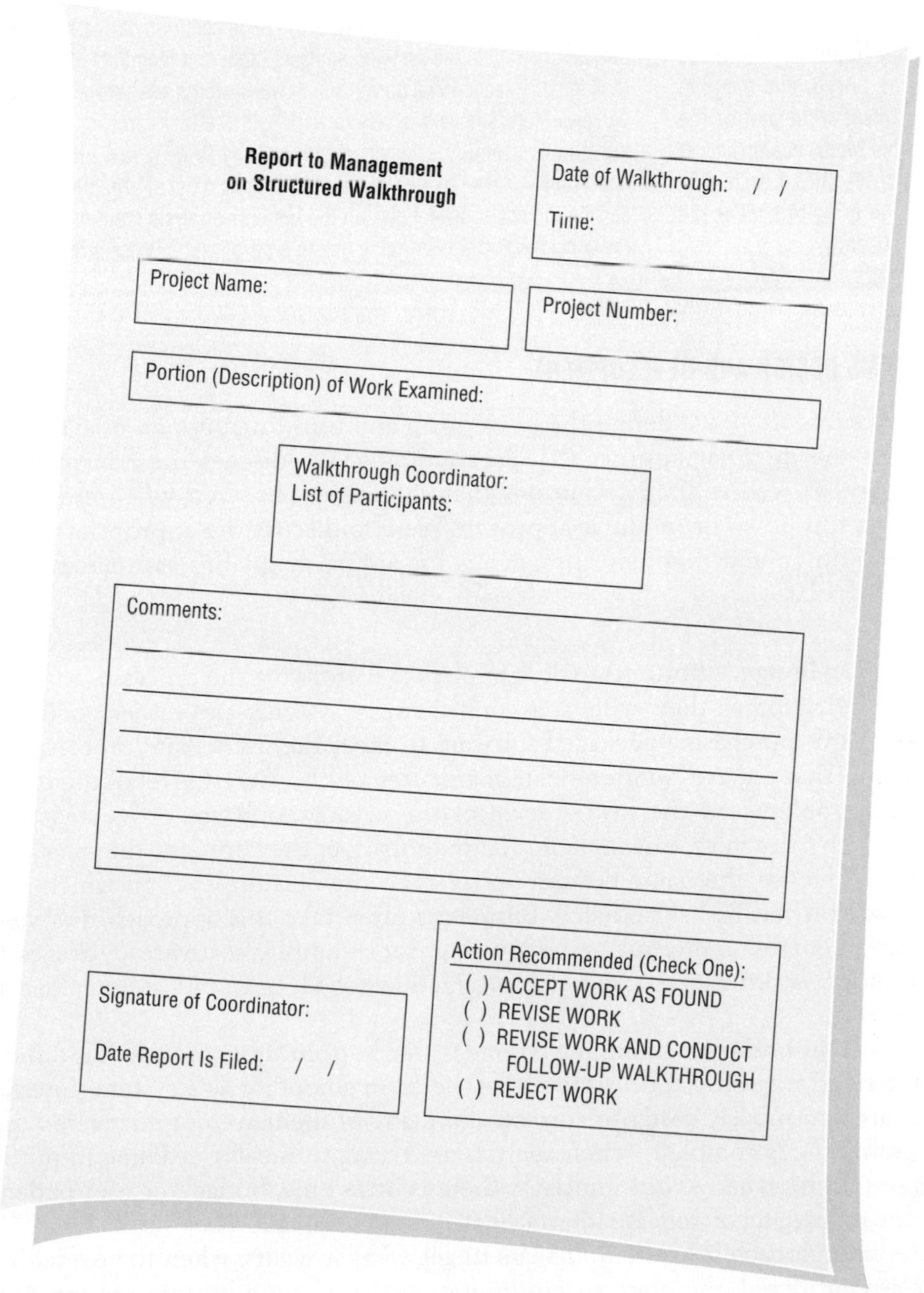

Report to Management on Structured Walkthrough

Date of Walkthrough: / /

Time:

Project Name:

Project Number:

Portion (Description) of Work Examined:

Walkthrough Coordinator:
List of Participants:

Comments:

Signature of Coordinator:

Date Report Is Filed: / /

Action Recommended (Check One):
( ) ACCEPT WORK AS FOUND
( ) REVISE WORK
( ) REVISE WORK AND CONDUCT FOLLOW-UP WALKTHROUGH
( ) REJECT WORK

FIGURE 20.1

A form to document structured walkthroughs; walkthroughs can be done whenever a portion of coding, a system, or a subsystem is complete.

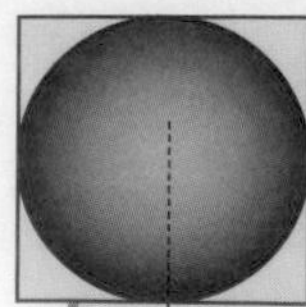

CONSULTING OPPORTUNITY 20.1

THE QUALITY OF MIS IS NOT STRAINED

"Merle, come here and take a look at these end-of-the-week reports," Portia pleads. As one of the managers on the six-person IS task force/quality assurance committee, Portia has been examining for her marketing department the system output that has been produced by the prototype. The systems analysis team has asked her to review the output.

Merle Chant walks over to Portia's desk and takes a look at the prospectus she's holding. "Why, what's wrong?" he asks. "It looks okay to me. I think you're taking this task force deal too much to heart. We're supposed to get our other work done as well, you know." Merle turns to leave and returns to his desk slightly perturbed at being interrupted.

"Merle, have a little mercy. It is really silly to put up with these reports the way they are. I can't find anything I need, and then I'm supposed to tell everyone else in the department what part of the report to read. I, for one, am disappointed. This report is slipshod. It doesn't make any sense to me. It's a rehash of the output we're getting now. Actually, it looks worse. I am going to bring this up at the next task force meeting," Portia proclaims insistently.

Merle turns to face her, saying, "Quality is their responsibility, Portia. If the system isn't giving us good reports, they'll fix it when it's all together. All you're doing is making waves. You're acting as if they actually value our input. I wouldn't give them the time of day, let alone do their work for them. They're so smart, let them figure out what we need."

Portia looks at Merle blankly, then starts getting a little angry. "We've been on the task force for four weeks," she says. "You've sat in on four meetings. We're the ones that know the business. The whole idea of TQM is to tell them what we need, what we're satisfied with. If we don't tell them what we need, then we can't complain. I'm bringing it up the next time we meet."

How effective do you think Merle will be in communicating his standards of quality to the systems analysis team and members of the MIS task force? Respond in a paragraph. If the systems analysts are able to perceive Merle's unwillingness to work with the task force on developing quality standards, what would you say to convince him of the importance of user involvement in TQM? Make a list of arguments supporting the use of TQM. How can the systems analysis team respond to the concerns Portia is bringing up? In a paragraph, devise a response.

SYSTEMS DESIGN AND DEVELOPMENT

In this subsection, we define the bottom-up and top-down design of systems as well as the modular approach to programming. We discuss the advantages of each one as well as the precautions that should be observed when employing either a top-down or modular approach. We also discuss the appropriateness of the top-down and modular approaches for aiding in quality assurance of systems projects.

Bottom-Up Design. Bottom-up design refers to identifying the processes that need computerization as they arise, analyzing them as systems and either coding the processes or purchasing packaged software to meet the immediate problem. The problems that require computerization are often on the lowest level of the organization. Problems on the lowest level of the organization are often structured and thereby the most amenable to computerization; they are also the most cost-effective. Hence, the name bottom-up refers to the bottom level on which computerization was first introduced. Businesses often take this approach to systems development by going out and acquiring, for example, software packages for accounting, a different package for production scheduling, and another one for marketing.

When in-house programming is done with a bottom-up approach, it is difficult to interface the subsystems so that they perform smoothly as a system. Interface bugs are enormously costly to correct, and many of them are not uncovered until programming is complete, when analysts are trying to meet a deadline in putting the system together. At this juncture, there is little time, budget, or user patience for the debugging of delicate interfaces that have been ignored.

Although each subsystem appears to get what it wants, when the overall system is considered, there are severe limitations to taking a bottom-up approach.

One is that there is a duplication of effort in purchasing software and even in entering data. Another is that much worthless data are entered into the system. A third, and perhaps the most serious drawback of the bottom-up approach, is that overall organizational objectives are not considered and hence cannot be met.

Top-Down Design. It is easy to visualize the top-down approach; it means looking at the large picture of the system and then exploding it into smaller parts or subsystems, as shown in Figure 20.2. Top-down design allows the systems analyst to ascertain overall organizational objectives first as well as to ascertain how they are best met in an overall system. Then the analyst moves to divide that system into subsystems and their requirements.

Top-down design is compatible with the general systems thinking that was discussed in Chapter 2. When systems analysts employ a top-down approach, they are thinking about the interrelationships and interdependencies of subsystems as they fit into the existing organization. The top-down approach also provides desirable emphasis on synergy or the interfaces that systems and their subsystems require, which is lacking in the bottom-up approach.

The advantages of using a top-down approach to systems design include avoiding the chaos of attempting to design a system "all at once." As we have seen, planning and implementing management information systems is incredibly complex. Attempting to get all subsystems in place and running at once is agreeing to fail.

A second advantage of taking a top-down approach to design is that it enables separate systems analysis teams to work in parallel on different but necessary subsystems, which can save a great deal of time. The use of teams for subsystems design is particularly well suited to a total quality assurance approach.

A third advantage is that it avoids a major problem associated with a bottom-up approach. Using a top-down approach prevents systems analysts from getting so mired in detail that they lose sight of what the system is supposed to do.

There are some pitfalls of top-down design that the systems analyst needs to know. The first is the danger that the system will be divided into the "wrong" subsystems. Attention must be paid to overlapping needs and to the sharing of

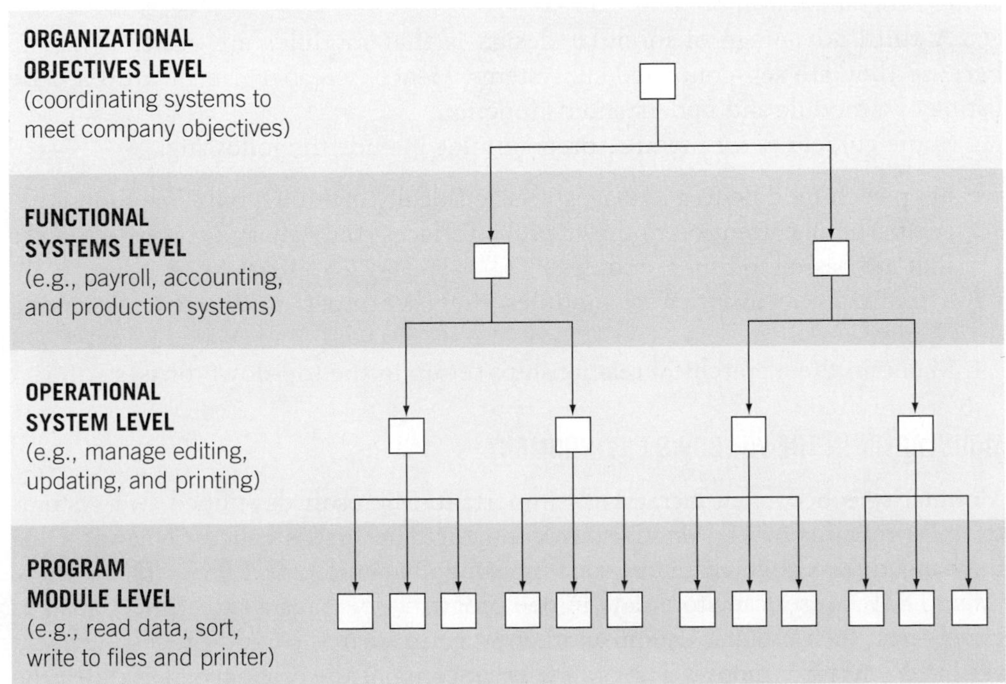

FIGURE 20.2
Using the top-down approach to first ascertain overall organizational objectives.

resources so that the partitioning of subsystems makes sense for the total systems picture. Furthermore, it is important that each subsystem address the correct problem.

A second danger is that once subsystem divisions are made, their interfaces may be neglected or ignored. Responsibility for interfaces needs to be detailed.

A third caution that accompanies the use of top-down design is that subsystems must be reintegrated eventually. Mechanisms for reintegration need to be put in place at the beginning. One suggestion is regular information trading between subsystem teams; another is using tools that permit flexibility if changes to interrelated subsystems are required.

Total quality management and the top-down approach to design can go hand in hand. The top-down approach provides the systems group with a ready-made division of users into task forces for subsystems. Task forces set up in this manner can then serve a dual function as quality circles for the management information system. The necessary structure for quality assurance is then in place, as is proper motivation for getting the subsystem to accomplish the departmental goals that are important to the users involved.

MODULAR DEVELOPMENT

Once the top-down design approach is taken, the modular approach is useful in programming. This approach involves breaking the programming into logical, manageable portions, or modules. This kind of programming works well with top-down design because it emphasizes the interfaces between modules and does not neglect them until later in systems development. Ideally, each individual module should be functionally cohesive so that it is charged with accomplishing only one function.

Modular program design has three main advantages. First, modules are easier to write and debug because they are virtually self-contained. Tracing an error in a module is less complicated, because a problem in one module should not cause problems in others.

A second advantage of modular design is that modules are easier to maintain. Modifications usually will be limited to a few modules and will not spread over an entire program.

A third advantage of modular design is that modules are easier to grasp, because they are self-contained subsystems. Hence, a reader can pick up a code listing of a module and understand its function.

Some guidelines for modular programming include the following:

1. Keep each module to a manageable size (ideally including only one function).
2. Pay particular attention to the critical interfaces (the data and control variables that are passed to other modules).
3. Minimize the number of modules the user must modify when making changes.
4. Maintain the hierarchical relationships set up in the top-down phases.

MODULARITY IN THE WINDOWS ENVIRONMENT

Modularity is becoming increasingly important. Microsoft developed two systems to link programs in its Windows environment. The first is called Dynamic Data Exchange (DDE), which shares code by using Dynamic Link Library (DLL) files. Using DDE, a user can store data in one program—perhaps a spreadsheet such as Excel—and then use that data in another program such as a word-processing package like Word for Windows. The program that contains the original data is called the server, and the program that uses the data is called the client (another term for *client* host). The DDE link can be set up so that whenever the client's word-processing file is opened, the data are automatically updated and any changes made to the server spreadsheet file since the word-processing file was last opened will be reflected. (See Chapter 21 for an extended discussion of the client/server model.)

One of the most commonly used DLL files is COMMDLG.DLL, which contains Windows' File Open, File Save, Search, and Print dialogue boxes. One advantage of using this file is that programs will have the same look and feel as other Windows programs. It also speeds development, because programmers do not have to write the code contained in common DLL files.

There is a big disadvantage in using a file such as COMMDLG.DLL, however: it is limited in features. Perhaps a systems designer thinks that it is important for the user to have the ability to create a subdirectory, or to move or rename files, when saving a file. Maybe it is important to search for a file when you try to open a file. In such cases, it becomes necessary to design your own File Open and File Save commands. The main disadvantage to using common DLL files is that these programs tend to use the least common denominator rather than take advantage of potentially powerful features; therefore, they appeal to the average user rather than the power user.

A second approach to linking programs in Windows is called Object Linking and Embedding (OLE). This method of connecting programs is superior to DDE because it ties in application data and graphics. Whereas DDE uses a cut-and-paste approach to linking data and does not retain formatting, OLE retains all the properties of the originally created data. This object-oriented approach (see Chapter 22 for a discussion of object-oriented principles) allows the end user to remain in the client application and still edit the original data in the server application. With OLE, when an end user clicks on the embedded object, a toolbar pops up to allow visual editing.

USING STRUCTURE CHARTS TO DESIGN SYSTEMS

The recommended tool for designing a modular, top-down system is called a structure chart. A structure chart is simply a diagram consisting of rectangular boxes, which represent the modules, and connecting arrows.

Figure 20.3 shows three modules labeled 1, 1.1, and 1.2. As noted earlier with data-flow diagrams, the number to the right of the decimal point in 1.1 and 1.2 signifies that these modules are subsets of module 1.

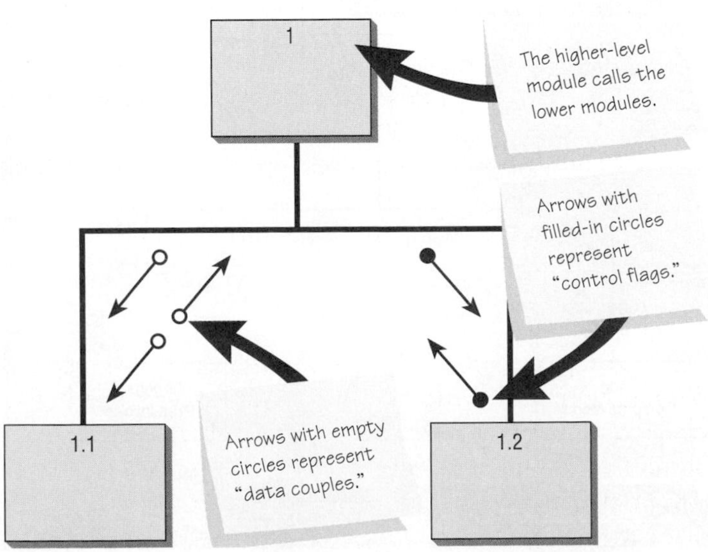
1
The higher-level module calls the lower modules.
Arrows with filled-in circles represent "control flags."
1.1
Arrows with empty circles represent "data couples."
1.2

FIGURE 20.3
A structure diagram encourages top-down design using modules.

Structure charts created using the program Microsoft Visio can be seen in Figure 20.4. An alternative notation for the three modules is shown. They are labeled 000, 100, and 200 and are connected using right-angle lines. Higher-level modules are numbered by 100s or 1,000s, and lower-level modules are numbered by 10s or 100s. This numbering allows programmers to insert modules using a number between the adjacent module numbers. For example, a module inserted between modules 110 and 120 would receive number 115. If two modules were inserted, the numbers might be 114 and 117. These numbering schemes vary, depending on the organizational standards used.

Off to the sides of the connecting lines, two types of arrows are drawn. The arrows with the empty circles are called *data couples*, and the arrows with the filled-in circles are called *control flags* or *switches*. A switch is the same as a control flag except that it is limited to two values: either "Yes" or "No." These arrows indicate that something is passed either down to the lower module or back up to the upper one.

Ideally, the analyst should keep this coupling to a minimum. The fewer data couples and control flags one has in the system, the easier it is to change the system. When these modules are actually programmed, it is important to pass the least number of data couples between modules.

Even more important is that numerous control flags should be avoided. Control is designed to be passed from lower-level modules to those higher in the structure. On rare occasions, however, it will be necessary to pass control downward in the structure. Control flags govern which portion of a module is to be executed and are associated with IF . . . THEN . . . ELSE . . . and other similar types of statements. When control is passed downward, a low-level module is allowed to make a decision, and the result is a module that performs two different tasks. This result violates the ideal of a functional module: it should perform only one task.

Figure 20.5 illustrates a portion of a structure chart for adding new employees. The program reads an EMPLOYEE TRANSACTION FILE and verifies that each record in the file contains only acceptable data. Separate reports are printed for

FIGURE 20.4

An alternative style for structure charts, drawn using Microsoft Visio.

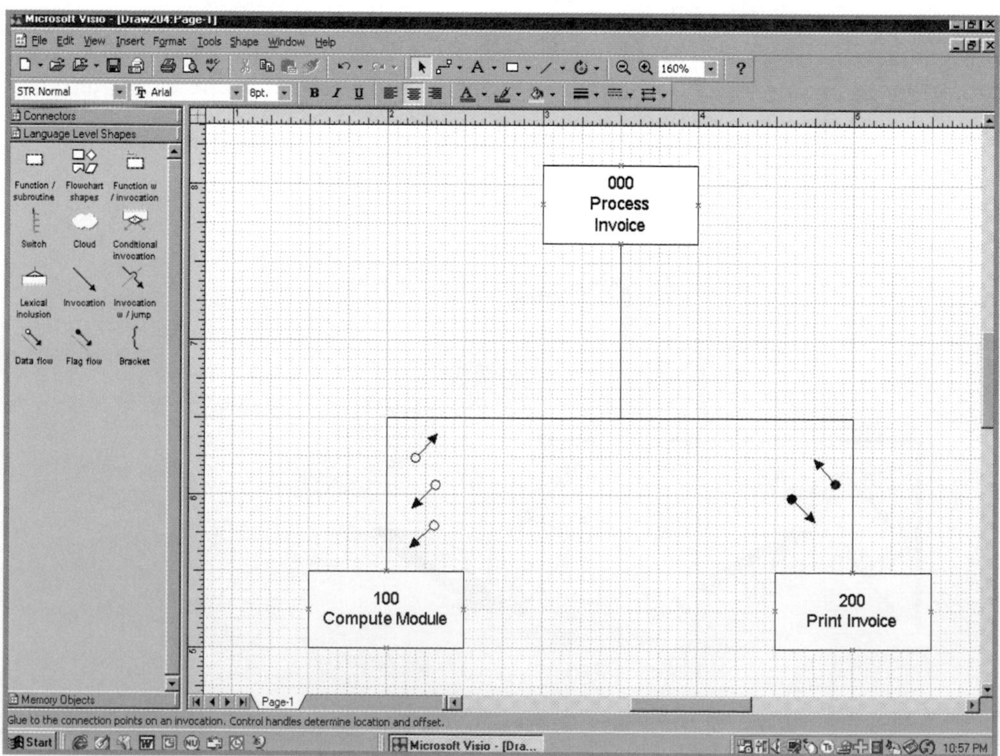

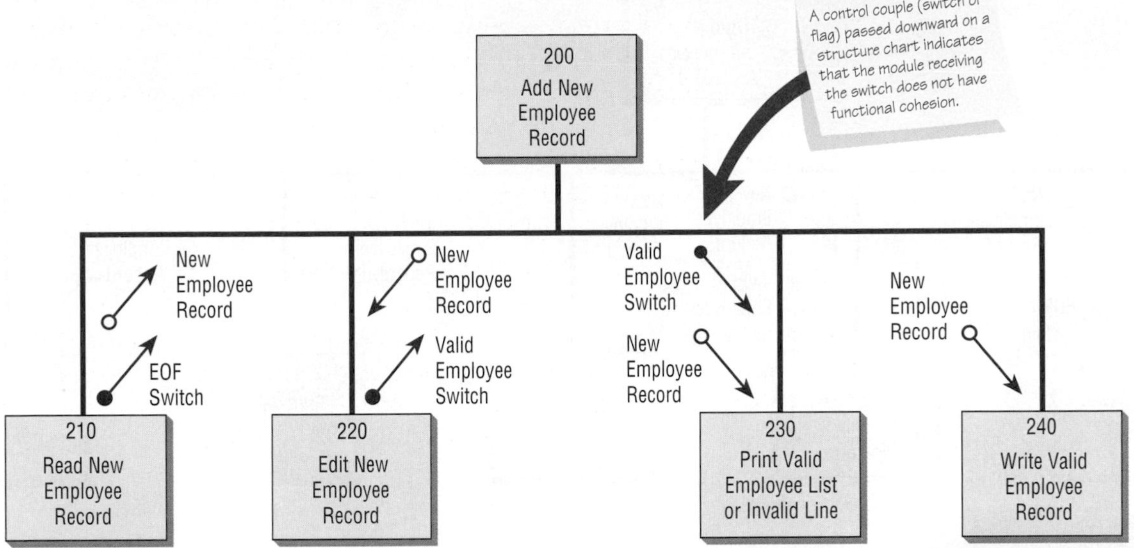

FIGURE 20.5
This structure chart illustrates control moving downward and also shows nonfunctional modules.

both valid and invalid records, providing an audit trail of all transactions. The report containing invalid records is sent to the users for error correction. Records that are valid are placed on a valid transaction file, which is passed to a separate program for updating the Employee master file. Module 200, ADD NEW EMPLOYEE RECORD, represents the logic of adding one record. Because module 230 is used to print both reports, a control flag must be sent down to tell the module which report to print. The logic of module 230 is thus entirely controlled by an IF statement, which is illustrated in Figure 20.6.

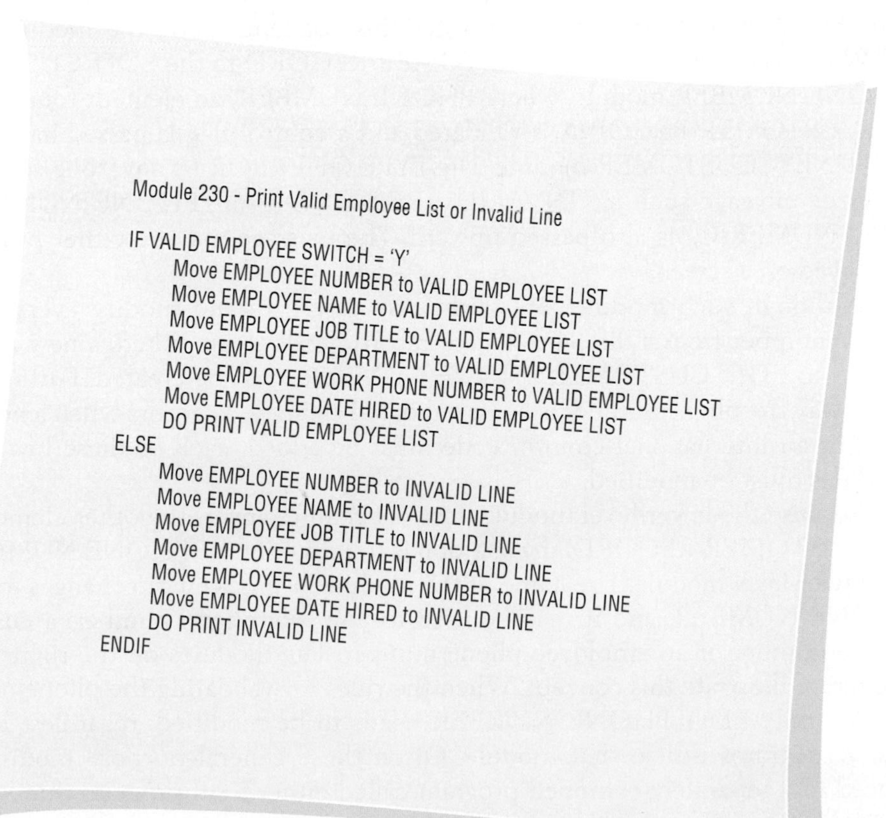

```
Module 230 - Print Valid Employee List or Invalid Line

IF VALID EMPLOYEE SWITCH = 'Y'
    Move EMPLOYEE NUMBER to VALID EMPLOYEE LIST
    Move EMPLOYEE NAME to VALID EMPLOYEE LIST
    Move EMPLOYEE JOB TITLE to VALID EMPLOYEE LIST
    Move EMPLOYEE DEPARTMENT to VALID EMPLOYEE LIST
    Move EMPLOYEE WORK PHONE NUMBER to VALID EMPLOYEE LIST
    Move EMPLOYEE DATE HIRED to VALID EMPLOYEE LIST
    DO PRINT VALID EMPLOYEE LIST
ELSE
    Move EMPLOYEE NUMBER to INVALID LINE
    Move EMPLOYEE NAME to INVALID LINE
    Move EMPLOYEE JOB TITLE to INVALID LINE
    Move EMPLOYEE DEPARTMENT to INVALID LINE
    Move EMPLOYEE WORK PHONE NUMBER to INVALID LINE
    Move EMPLOYEE DATE HIRED to INVALID LINE
    DO PRINT INVALID LINE
ENDIF
```

FIGURE 20.6
Pseudocode for module 230 illustrating the effect of passing a switch downward.

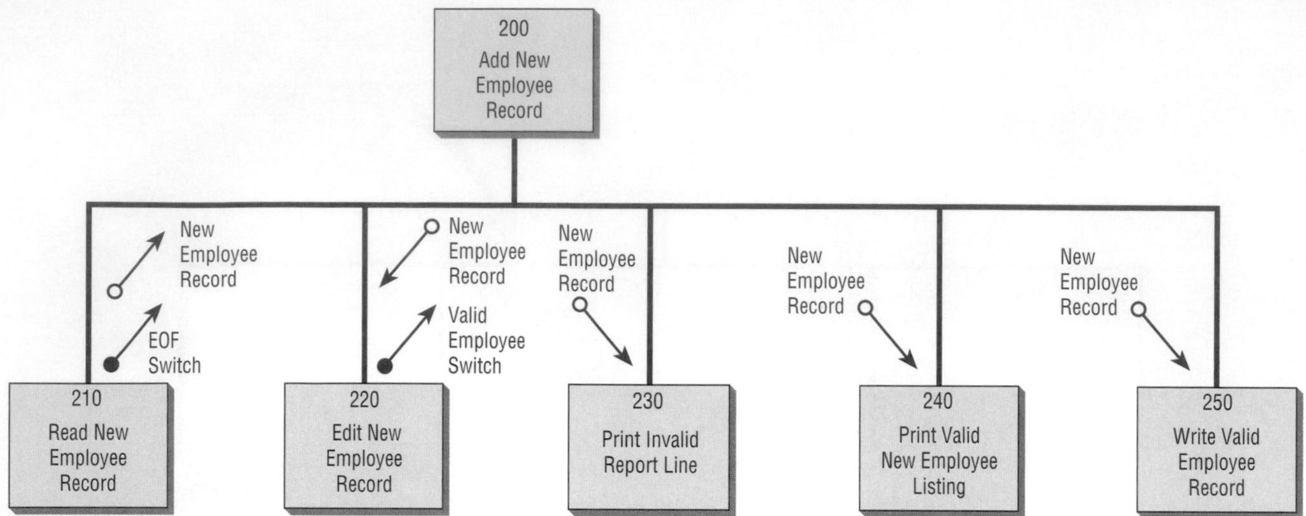

FIGURE 20.7

An improved structure chart showing control flowing upward.

Figure 20.7 shows the correct way to design the structure underneath module 200, ADD NEW EMPLOYEE RECORD. Here, each print function has been placed in a separate module, and control flags are only passed up the structure to the higher-level module.

The data that are passed through data couples must also be examined. It is best to pass only the data required to accomplish the function of the module. This approach is called data coupling. Passing excessive data is called stamp coupling, and although it is relatively harmless, it reduces the possibility of creating a reusable module. Figure 20.8 illustrates this concept. Here, the module EDIT NEW CUSTOMER passes the CUSTOMER RECORD to the EDIT CUSTOMER PHONE NUMBER module, where PHONE NUMBER, an element found within the CUSTOMER RECORD, is validated and a control flag is passed back to the EDIT NEW CUSTOMER module. The TYPE OF ERROR (if any), one containing an error message such as "INVALID AREA CODE" or "PHONE NUMBER IS NOT NUMERIC," is also passed upward. The message may be either printed or displayed on a screen.

Although such modules are fairly easy to create and modify every time a phone number from a different source record needs to be edited, a new module, similar to EDIT CUSTOMER PHONE NUMBER must be created. Furthermore, if the way the phone number is being validated changes, as occurs when a new area code or an international country code must be added, each of these lower-level modules must be modified.

Because the lower-level module does not require any of the other elements on the CUSTOMER RECORD, the solution is to pass only the PHONE NUMBER to the lower-level module. The name of the module in this scenario changes to EDIT PHONE NUMBER, and it may be used to edit *any* phone number: a customer phone number or an employee phone number. The modules on the right side of the figure illustrate this concept. When the rules for validating the phone number change, only EDIT PHONE NUMBER needs to be modified, regardless of how many programs utilize that module. Often these general-purpose modules are placed in a separately compiled program called either a subprogram, function, or procedure, depending on the computer language.

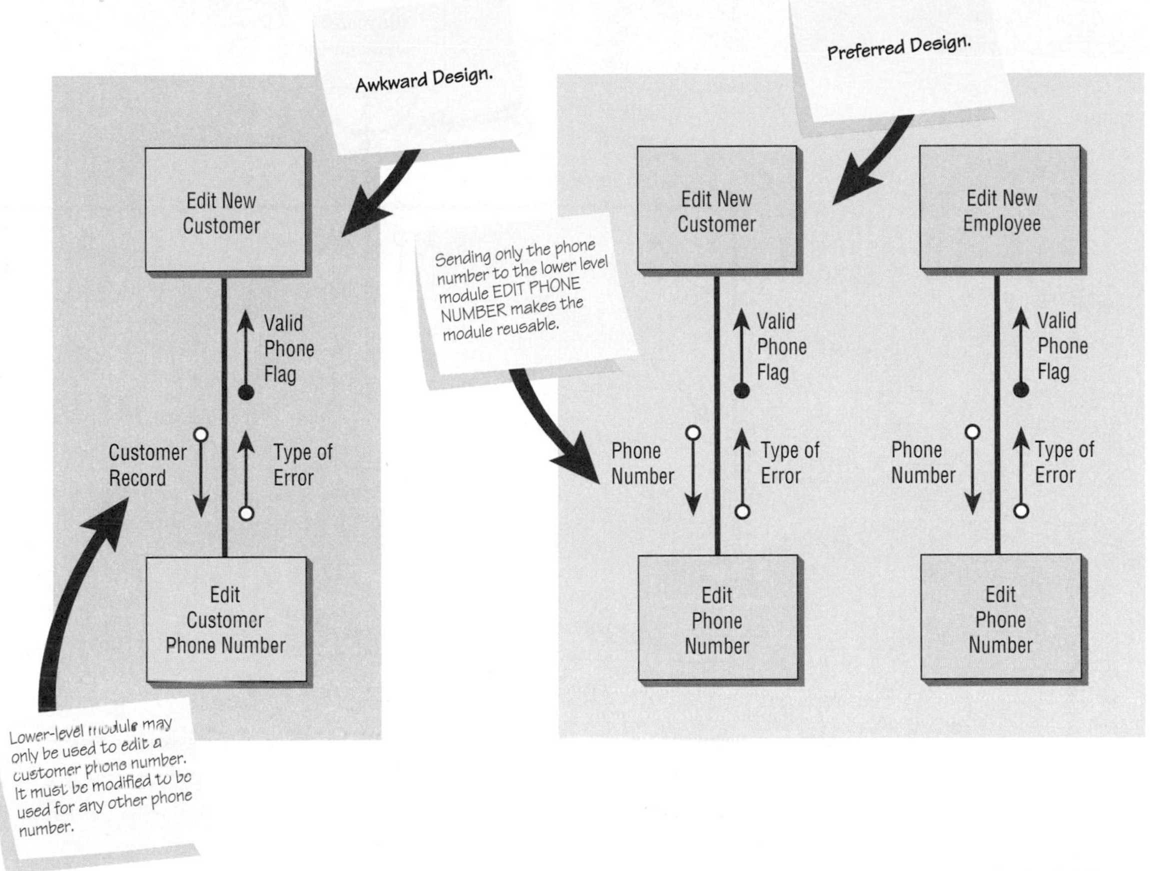

FIGURE 20.8
Creating reusable modules.

Another symbol used in structure charts is the loop, as shown in Figure 20.9. This symbol indicates that some procedures found in modules 2.1 and 2.2 are to be repeated until finished. In this example of organizing part numbers, the processes READ PART NUMBER and EDIT PART NUMBER are repeated until there are no more numbers. Then the numbers are sorted in module 2.3. In this example, the data couples are "raw input," "part numbers," and "sorted part numbers." Notice that some of these data couples occur twice, once going up to the main module 2 and once going down to the submodule. The control flags are "E-O-F flag" and "Edit flag."

Still another symbol used in structure charts is the small diamond. The diamond appears on the bottom of one of the rectangles, as shown in Figure 20.10, and signifies that only some of the modules below the diamond will be performed. Notice that the diamond doesn't indicate which modules will be selected, nor does the loop indicate which modules will be repeated. They are meant to be general, not specific.

Drawing a Structure Chart. Obviously, structure charts are meant to be drawn from the top down, but where does one start to find the processes that are to become the modules? Most likely, the best place to find this information is in the data flow diagram (see Chapter 9).

FIGURE 20.9
Iterations are depicted on structure charts by drawing a loop.

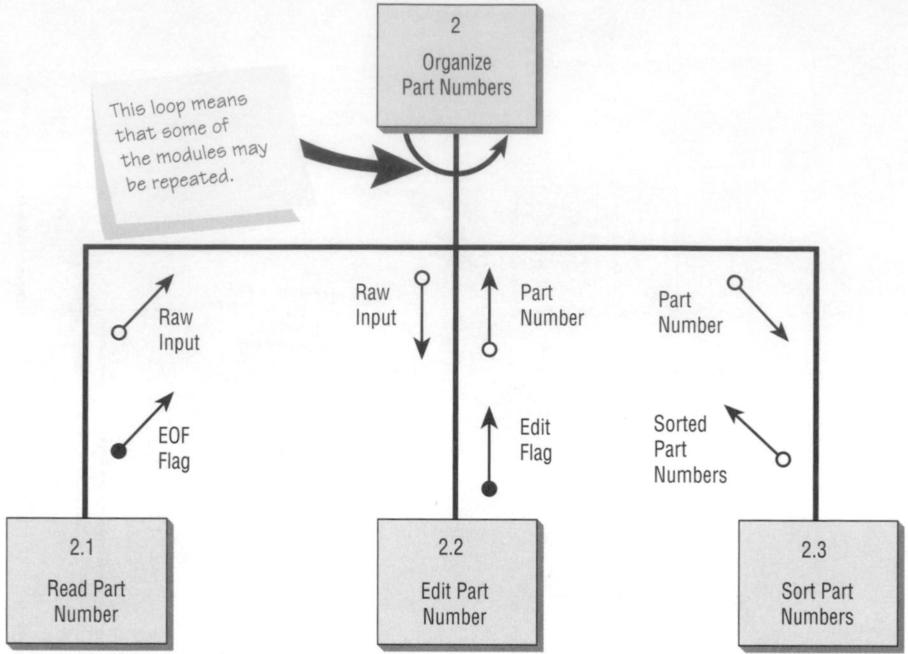

Figure 20.11 is a data flow diagram of a payroll system. The entities are the EMPLOYEES and the PAYROLL MANAGER. Four processes are drawn in the data-flow diagram:

1. Verify pay request.
2. Calculate pay.
3. Update payroll records.
4. Prepare checks and summary.

Notice that there are two data stores and many data flows. More important, observe that the data flow diagram is laid out in a linear fashion. Now the data flow diagram will be used to draw the hierarchical structure chart.

FIGURE 20.10
The small diamond in a structure chart indicates that certain modules are to be performed only when a specified condition exists.

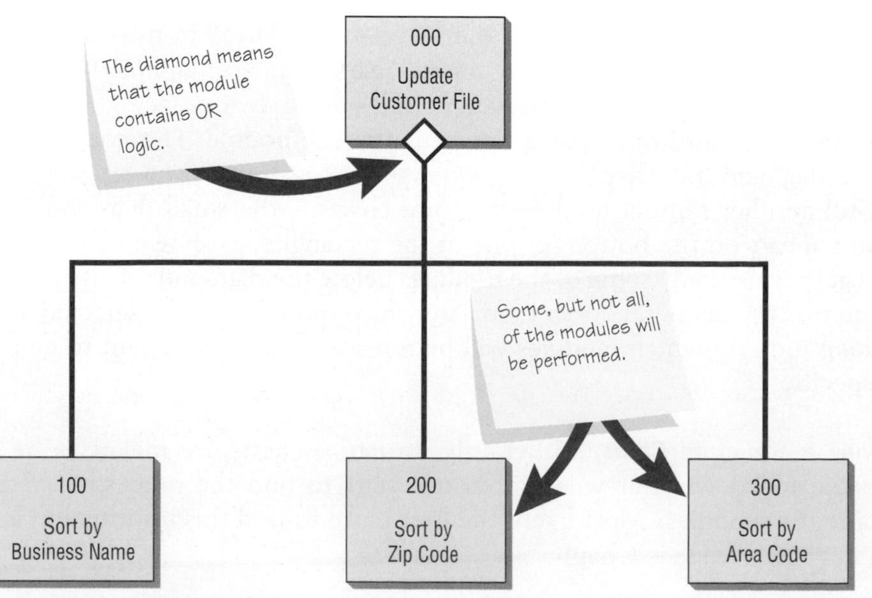

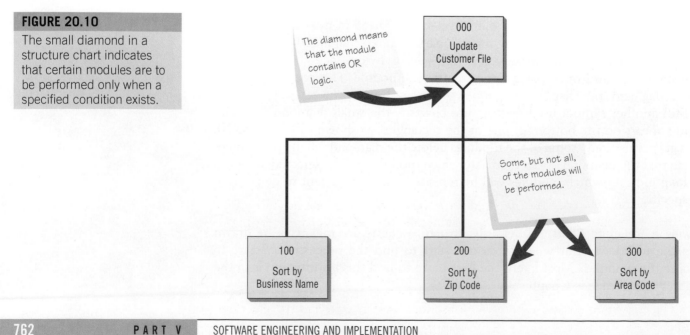

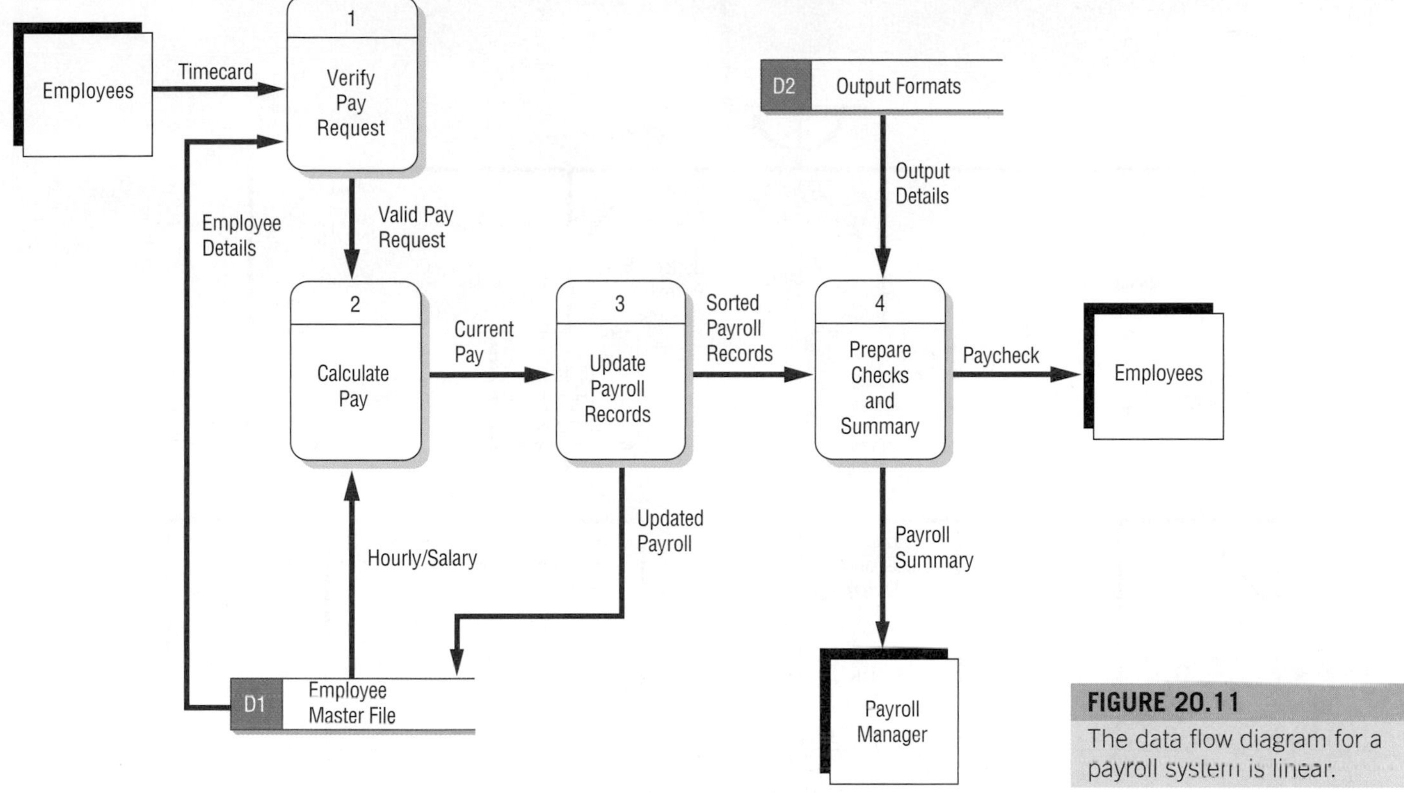

FIGURE 20.11
The data flow diagram for a payroll system is linear.

The structure chart is shown in Figure 20.12. The main process at the top of the chart, called PERFORM PAYROLL, represents the module that controls everything underneath. The modules on the second level bear the same names and perform the same processes shown in the data flow diagram. Because the data flow diagram is intended to be a logical representation of the system, it is not unusual that the modules derived from the diagram are the same.

The modules on the second level will control the operations of the modules on the third level. These third-level modules accomplish separate functions "READ TIMECARD and get EMPLOYEE DETAILS" and "EDIT and VERIFY TIMECARD." Each of these modules contains only one function, which is the ideal case.

The data flows on a data flow diagram turn into the data couples found on a structure chart. Notice that the TIMECARD, VALID PAY REQUEST, and other data are present on both diagrams.

In this example, the data couples and control flags are kept to a minimum. The hierarchical structure makes the structure chart appear to be an inverted tree, but the symmetry is only a coincidence. This example is said to be transform-centered, because all the transactions follow the same path.

When all of the transactions do *not* follow the same path, the structure chart is said to be transaction-centered. A simple example of a transaction-centered structure chart is shown in Figure 20.13. The two diagrams accomplish the same thing, but the top diagram is awkward because the same decision is made in two places. The bottom design is preferred because it can reduce the number of decisions (or, alternatively, reduce the number of control flags that have to be passed between modules).

A data flow diagram for a transaction-centered system is shown in Figure 20.14. Here, the major metropolitan newspaper company publishes a variety of newspapers such as the *Morning Star*, the *Daily Planet*, and the special *Weekend Nova*. Depending on the transaction the subscriber wants, a different set of actions

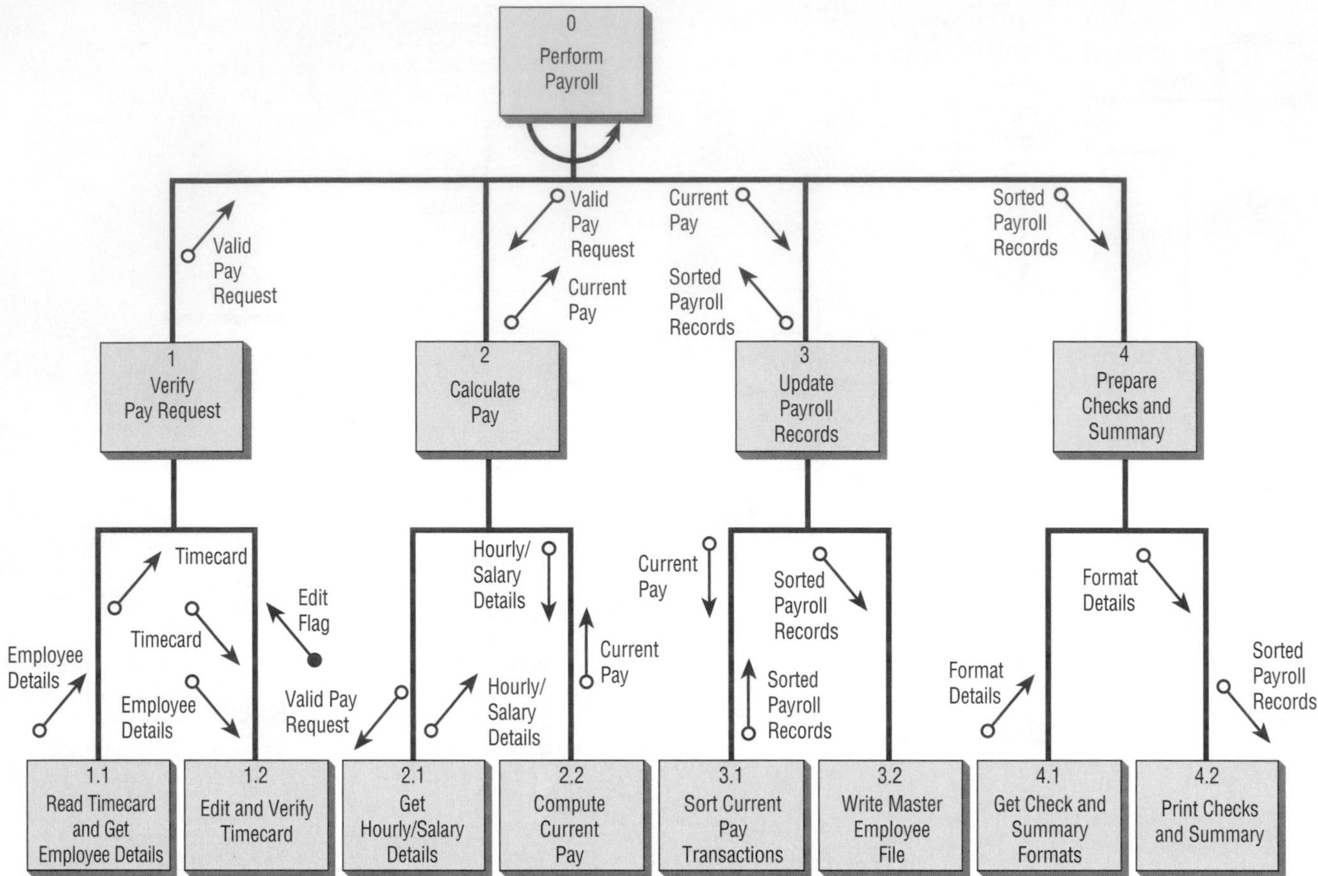

FIGURE 20.12

The structure chart for the payroll example shows hierarchy: data couples and control flags are kept to a minimum.

will be taken. This sort of transaction-centered system is often drawn as a second-level data-flow diagram.

The resulting structure chart can be seen in Figure 20.15. It is important that the decision be made at a high level in the structure chart. One module simply determines the transaction type. Following this step, one of the actions is performed. Notice that the module entitled "UPDATE AND PRINT SUBSCRIBERS" is common to all transactions.

This particular structure chart served the main objectives of drawing structure charts,

1. To encourage a top-down design.
2. To support the concept of modules and identify the appropriate modules.
3. To identify and limit as much as possible the data couples and control flags that pass between modules.

When transforming a data flow diagram into a structure chart, there are several additional considerations to keep in mind. The data flow diagram will indicate the sequence of the modules in a structure chart. If one process provides input to another process, the corresponding modules must be performed in the same sequence. Figure 20.16 is a data flow diagram for preparing a student report card. Notice that process 1, READ GRADE RECORD, provides input to process 2, READ COURSE RECORD, and to process 3, READ STUDENT RECORD. The structure chart created for this diagram is illustrated in Figure 20.17. Notice that module 110, READ GRADE RECORD, must be executed first. Processes 2 and 3

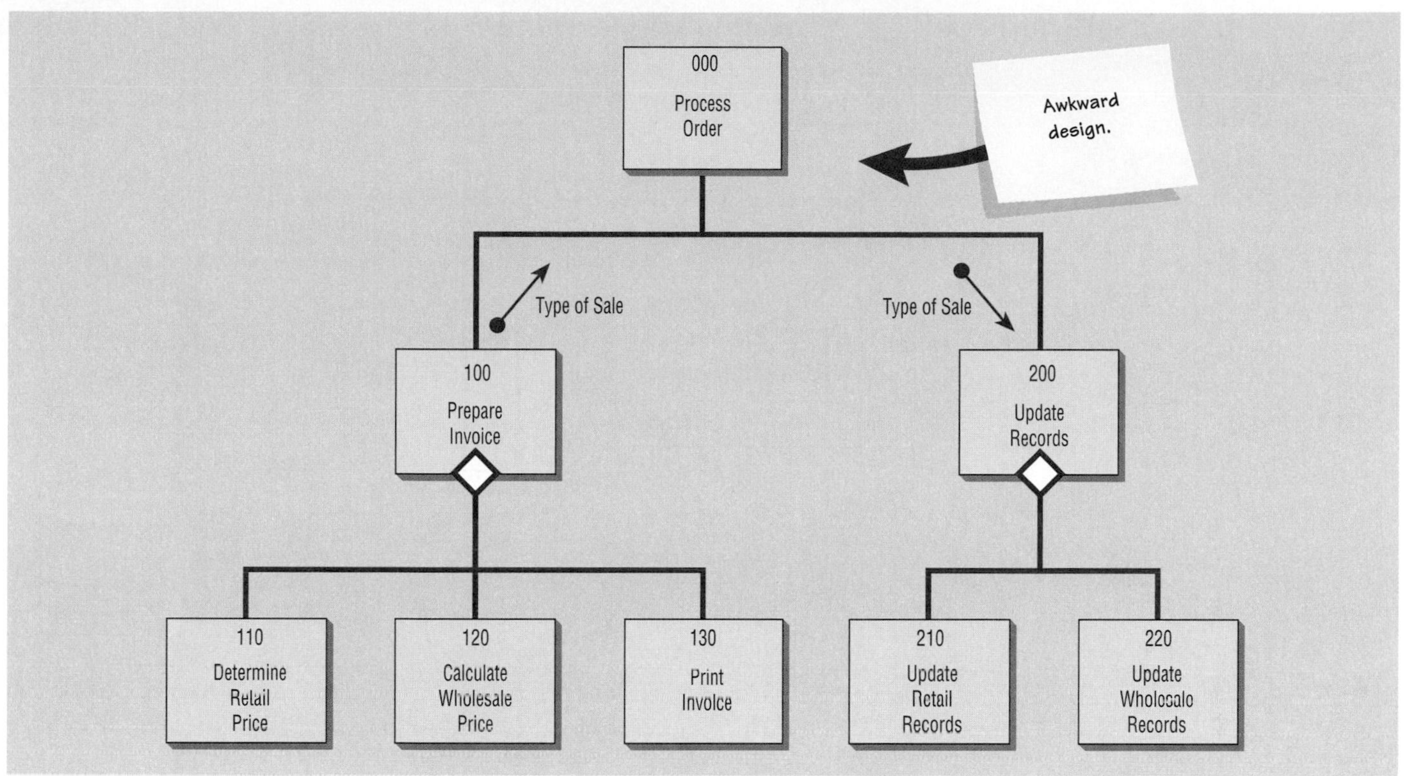

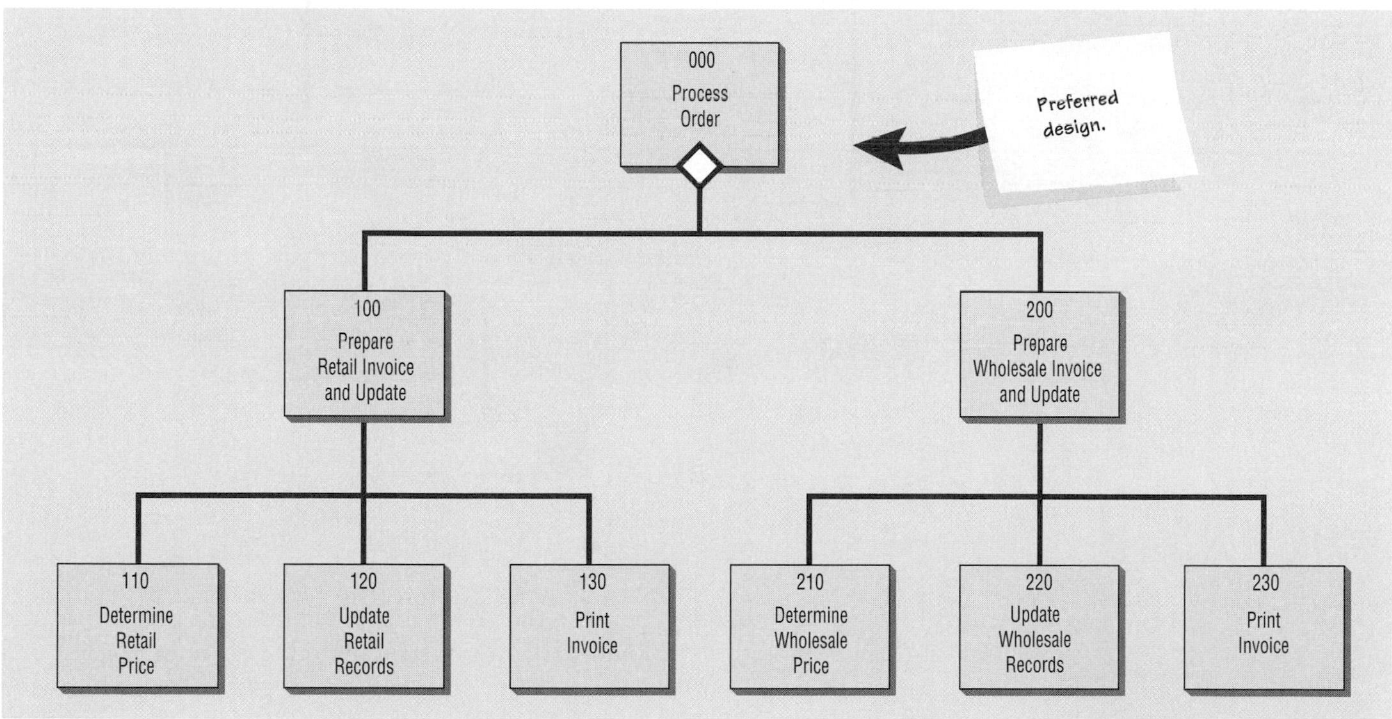

FIGURE 20.13

To prevent redundant decisions (or the passing of a control flag such as type of sale), the structure chart can be redrawn to make the decision at the top.

FIGURE 20.14

The data flow diagram for newspaper subscriber transaction processing branches out depending on the transaction type.

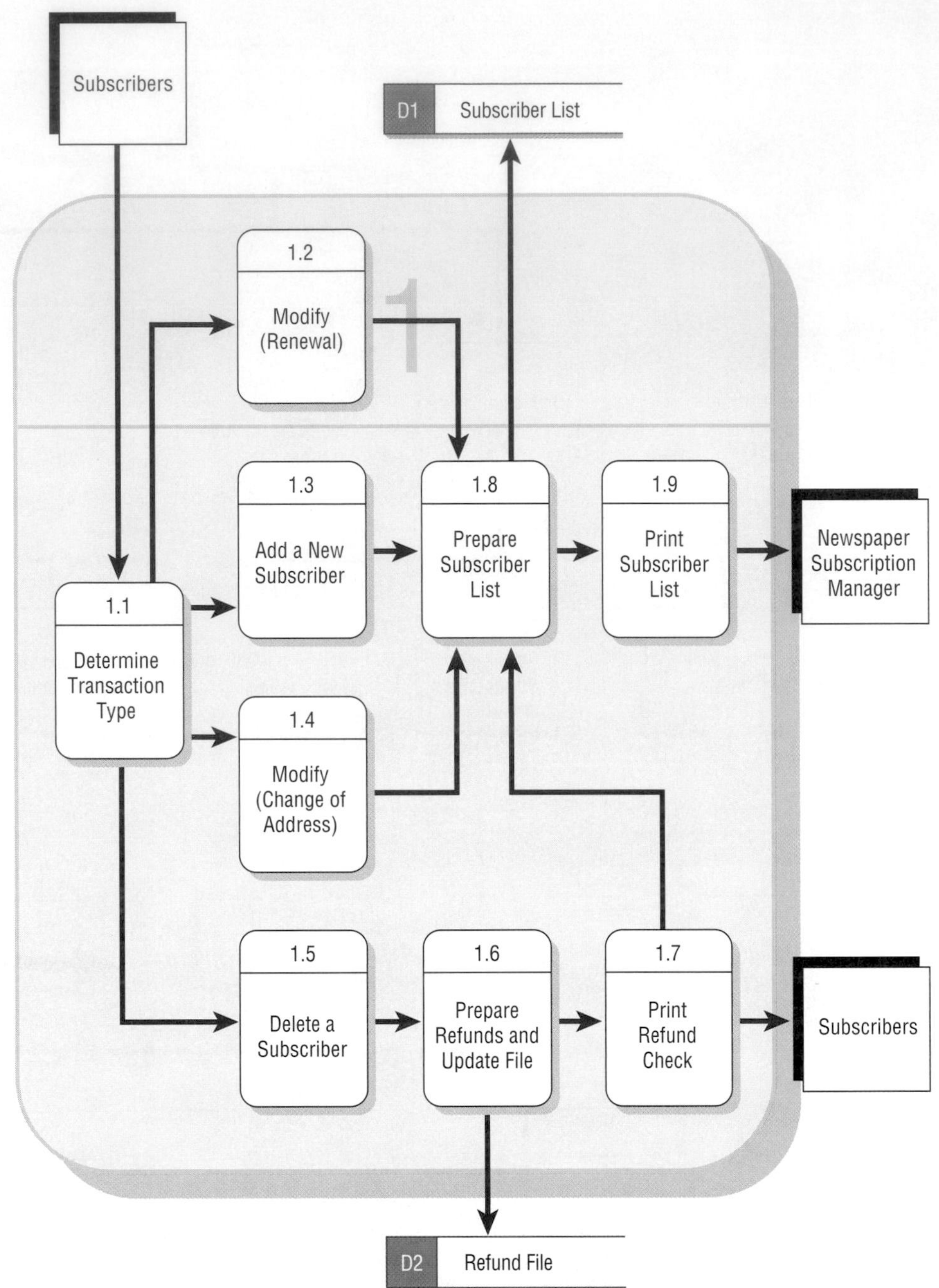

must be executed next, but because they do not provide input to each other, the order of these modules (120 and 130) in the structure chart is unimportant and may be reversed without any effect on the final results. Processes 1 and 2 provide input to process 4, CALCULATE GRADE POINT AVERAGE (also known as module 140). Process 5, PRINT STUDENT REPORT CARD (module 150), receives data flow from all the other processes and must be the last module to be performed.

If a process explodes to a child data-flow diagram, the module corresponding to the parent process will have subordinate modules that correspond to the processes found on the child diagram. Process 5, PRINT STUDENT REPORT CARD, has four input data flows and one output and is therefore a good candidate

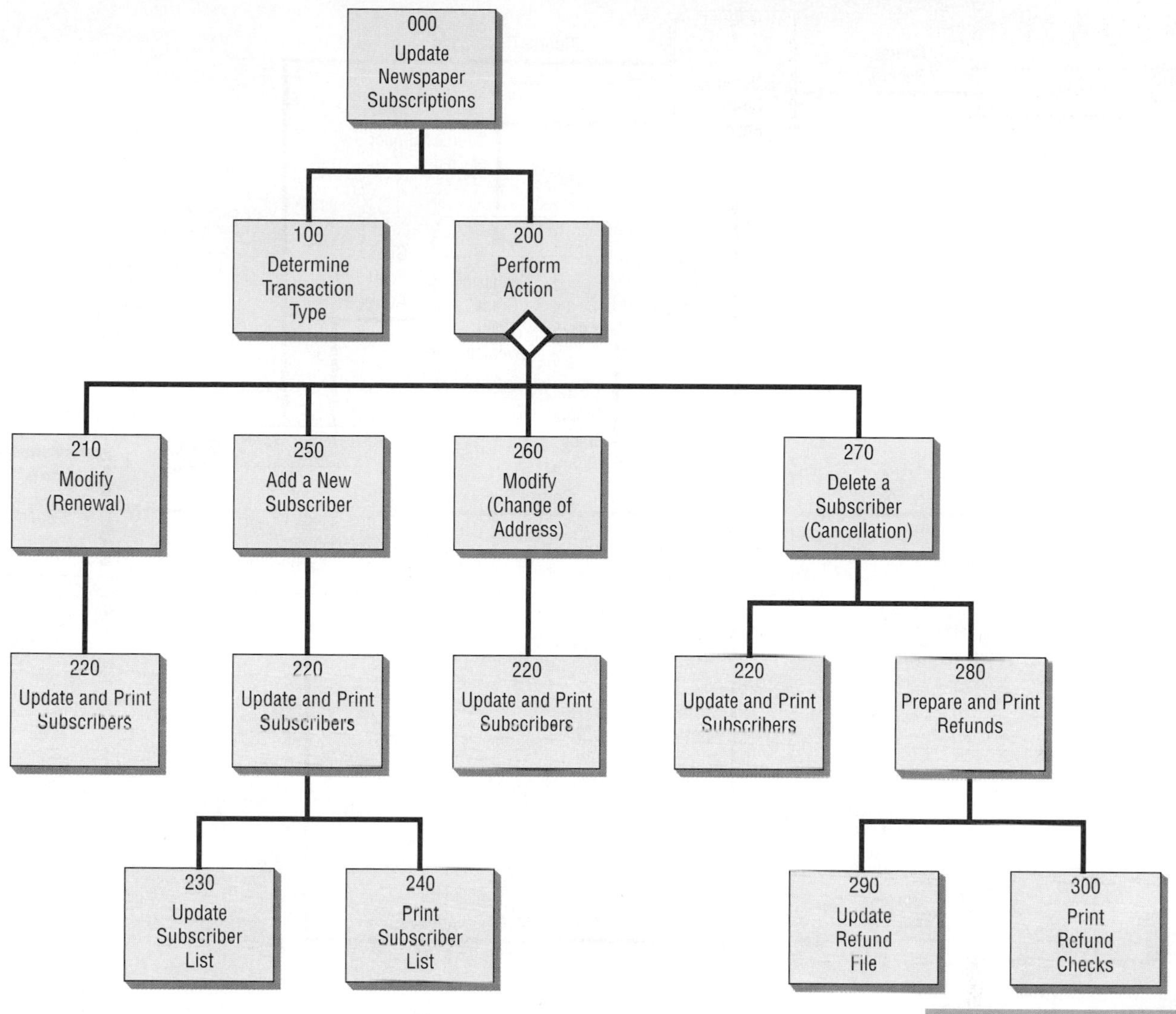

FIGURE 20.15
A structure chart for a transaction centered system can branch out from a diamond and then later share the same module.

for a child diagram. Figure 20.18 illustrates Diagram 5, the details of process 5. The processes on Diagram 5 translate to the modules subordinate to module 150, PRINT STUDENT REPORT CARD.

TYPES OF MODULES

Structure chart modules fall into one of three general categories: control, transformational (sometimes called worker), and functional or specialized. When producing a structure chart that is easy to develop and modify, one should take care not to mix the different types of modules.

Control modules are usually found near the top of the structure chart and contain the logic for performing the lower-level modules. The control modules may or may not be represented on the data flow diagram. The types of statements that are usually in control modules are IF, PERFORM, and DO. Detailed statements such as ADD and MOVE are usually kept to a minimum. Control logic is usually the most difficult to design; therefore, control modules should not be very large in size. If a control module has more than nine subordinate modules, new control modules should be created that are subordinate to the

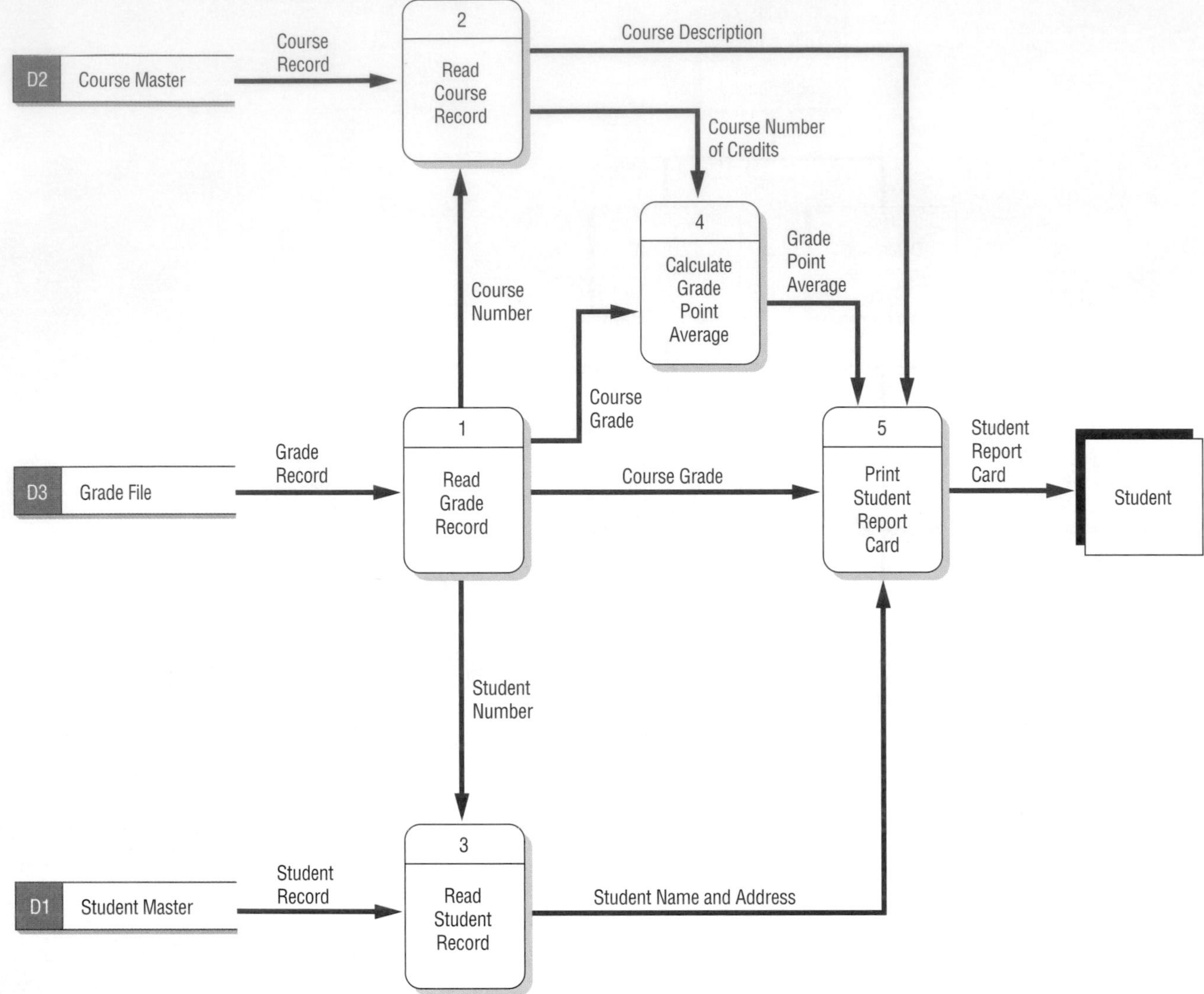

FIGURE 20.16
Data flow diagram for printing a student report card.

original control module. The logic of a control module may be determined from a decision tree or decision table. A decision table with too many rules is split into several decision tables, with the first table performing the second table. Each decision table would create a resulting control module. (See Chapter 11 for more on decision trees and tables.)

Transformational modules are those created from a data flow diagram. They usually perform only one task, although several secondary tasks may be associated with the primary task. For example, a module named "PRINT CUSTOMER TOTAL LINE" may format the total line, print the line, add to the final totals, and finally set the customer totals to zero in preparation for accumulating the amounts of the next customer. Transformational modules usually have mixed statements, a few IF and PERFORM or DO statements, and many detailed statements such as MOVE and ADD. These modules are lower in the structure than control modules.

Functional or specialized modules are the lowest in the structure, with a rare subordinate module beneath them. They perform only one task, such as formatting, reading, calculating, and writing. Some of these modules are found on a data

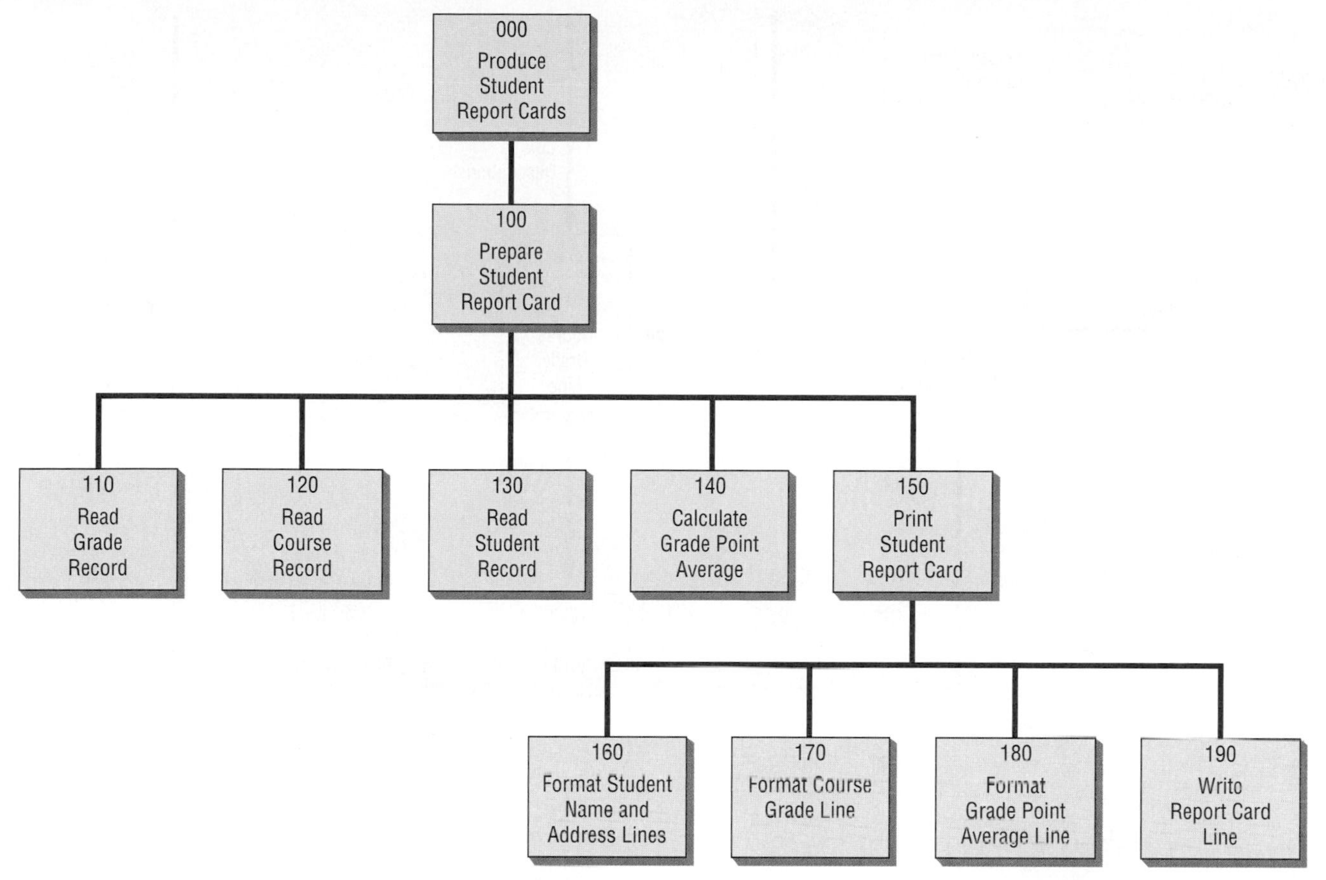

FIGURE 20.17
A structure chart for producing student report cards.

flow diagram, but others may have to be added, such as reading a record or printing an error line.

Figure 20.19 represents the structure chart for adding reservations for hotel guests. Modules 000, ADD GUEST RESERVATIONS, and 100, ADD ROOM RESERVATION, are control modules, representing the entire program (module 000), and they provide the control necessary for making one room reservation (module 100). Module 110, DISPLAY RESERVATION SCREEN, is a functional module responsible for displaying the initial reservation screen. Modules 120, GET VALID ROOM RESERVATION, and 160, CONFIRM ROOM RESERVATION, are lower-level control modules.

Module 120, GET VALID ROOM RESERVATION, is performed iteratively until the reservation data are valid or until the reservation operator cancels the transaction. This type of GET VALID . . . module relieves the 100 module of a fair amount of complex code. The modules subordinate to GET VALID ROOM RESERVATION are functional modules responsible for receiving the reservation screen, editing or validating the room reservation, and displaying an error screen if the input data are not valid. Because these modules are in a loop, control remains in this portion of the structure until the screen data are valid.

Module 160, CONFIRM ROOM RESERVATION, is also performed iteratively and allows the operator to sight-verify that the correct information has been entered. In this situation, the operator will inspect the screen and press a specified

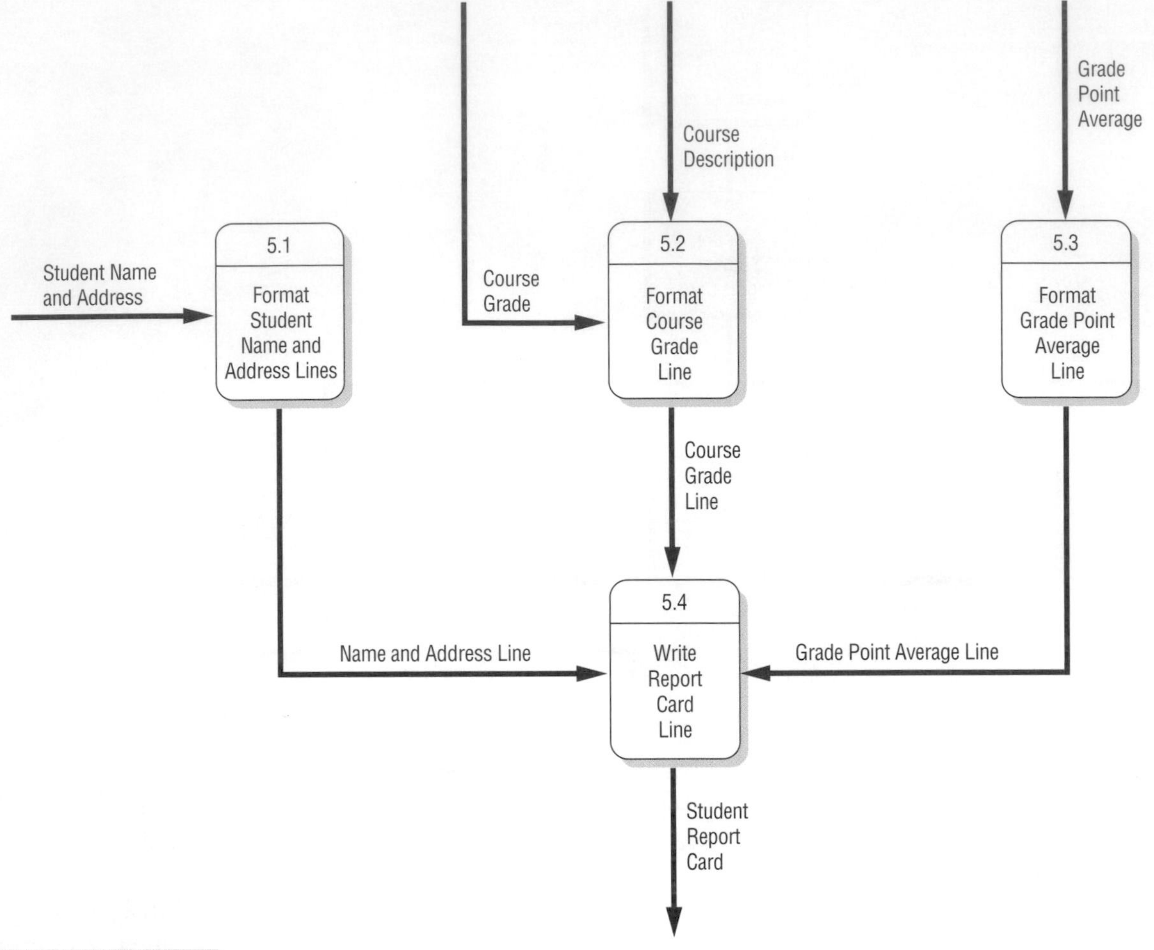

FIGURE 20.18
Child diagram for process 5, PRINT STUDENT REPORT CARD.

key (such as the Enter key if the data are correct or a different key) to modify or cancel the transaction. Again, the program will remain in these modules, looping until the operator accepts or cancels the reservation.

Module 190 is a transformational module that formats the RESERVATION RECORD and performs module 200 to write the RESERVATION RECORD. Modules 130, 140, 150, 170, 180, and 200 are functional modules, performing only one task: accepting a screen, displaying a screen, or editing or writing a record. These modules are the easiest to code, debug, and maintain.

MODULE SUBORDINATION

A subordinate module is one lower on the structure chart called by another module higher in the structure. Each subordinate module should represent a task that is a part of the function of the higher-level module. Allowing the lower-level module to perform a task not required by the calling module is called improper subordination. In such a case, the lower module should be moved higher in the structure.

Figure 20.20 illustrates this concept using a structure chart for changing a customer master file. Examine module 120, READ CUSTOMER MASTER. It has the task of using the CUSTOMER NUMBER from the CHANGE TRANS-

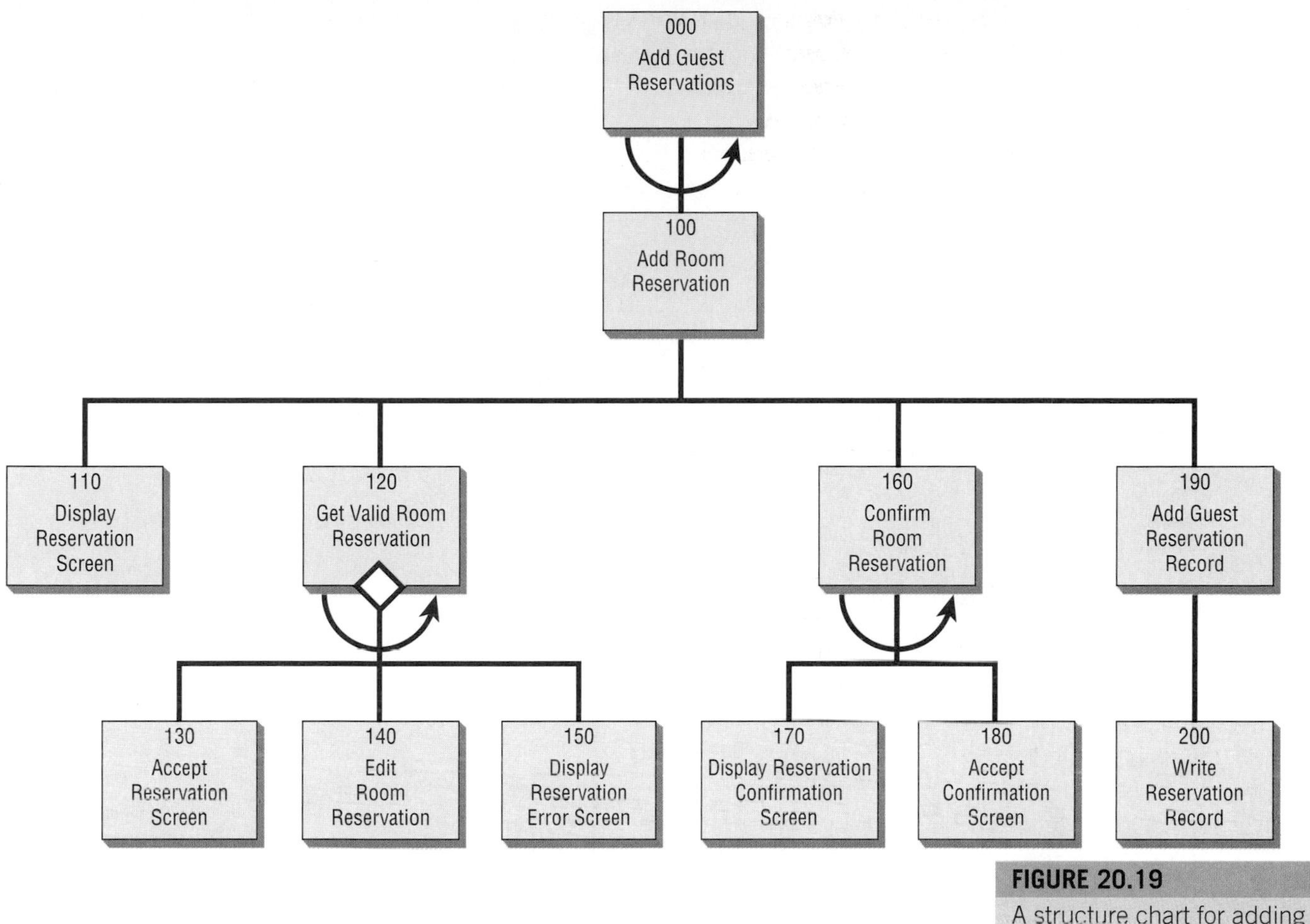

FIGURE 20.19
A structure chart for adding hotel guest reservations on-line.

ACTION RECORD to obtain the matching CUSTOMER RECORD directly. If the record is not found, an error line is printed. Otherwise, the CUSTOMER MASTER is changed, and the record is rewritten. This module should be a functional module, simply reading a record, but instead it has three subordinate modules. The question must be asked, "Does an error line have to be printed to accomplish reading the CUSTOMER MASTER?" Furthermore, "Does the new CUSTOMER MASTER RECORD have to be formatted and rewritten so as to read the CUSTOMER MASTER?" Because the answer to both questions is "No," modules 130, 140, and 150 should not be subordinate to READ CUSTOMER MASTER.

Figure 20.21 shows the corrected structure chart. Control statements are moved out of the READ CUSTOMER MASTER record and into the primary control module, CHANGE CUSTOMER RECORD. READ CUSTOMER MASTER becomes a functional module (module 120).

Even when a structure chart accomplishes all the purposes for which it was drawn, the structure chart cannot stand alone as the sole design and documentation technique. First, it doesn't show the order in which the modules should be executed (a data flow diagram will accomplish that). Second, it doesn't show enough detail (Nassi-Shneiderman charts and pseudocode will accomplish that). The remainder of this chapter discusses these more detailed design and documentation techniques for software development using the newspaper subscription problem presented earlier, which we now view in more detail.

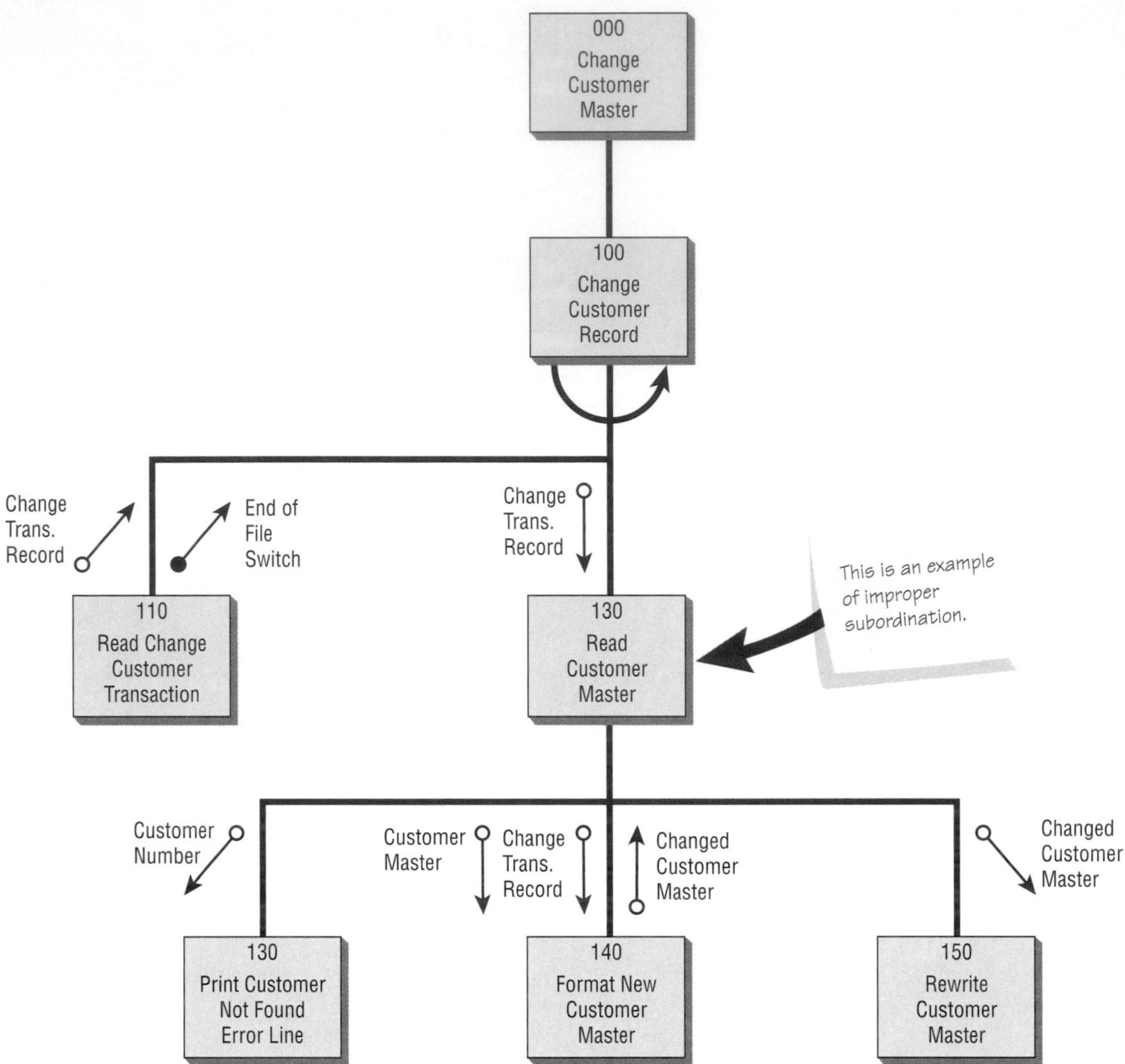

FIGURE 20.20
A structure chart illustrating the principle of improper subordination.

SOFTWARE ENGINEERING AND DOCUMENTATION

Planning and control are essential elements of every successful system. In developing software for the system, the systems analyst should know that planning takes place in the design before programming is even begun. We need techniques to help us set program objectives, so that our programs are complete. We also need design techniques to help us break apart the programming effort into manageable modules.

It is not satisfactory, however, to try to get by with just the planning stages. After programs are completed, they must be maintained, and maintenance efforts typically outweigh the effort expended on the original design and programming.

The techniques described in the upcoming section are meant not only to be used initially in the design of software, but also in its maintenance. Because most systems are not considered disposable, they will need to be maintained. The total quality assurance effort requires that programs be documented properly.

Software and procedures are documented so that they are encoded into a format that can be easily accessed. Access to procedures is necessary for new people learning the system and as a reminder to those who use the program infrequently. Documentation allows users, programmers, and analysts to "see" the system, its software, and procedures without having to interact with it.

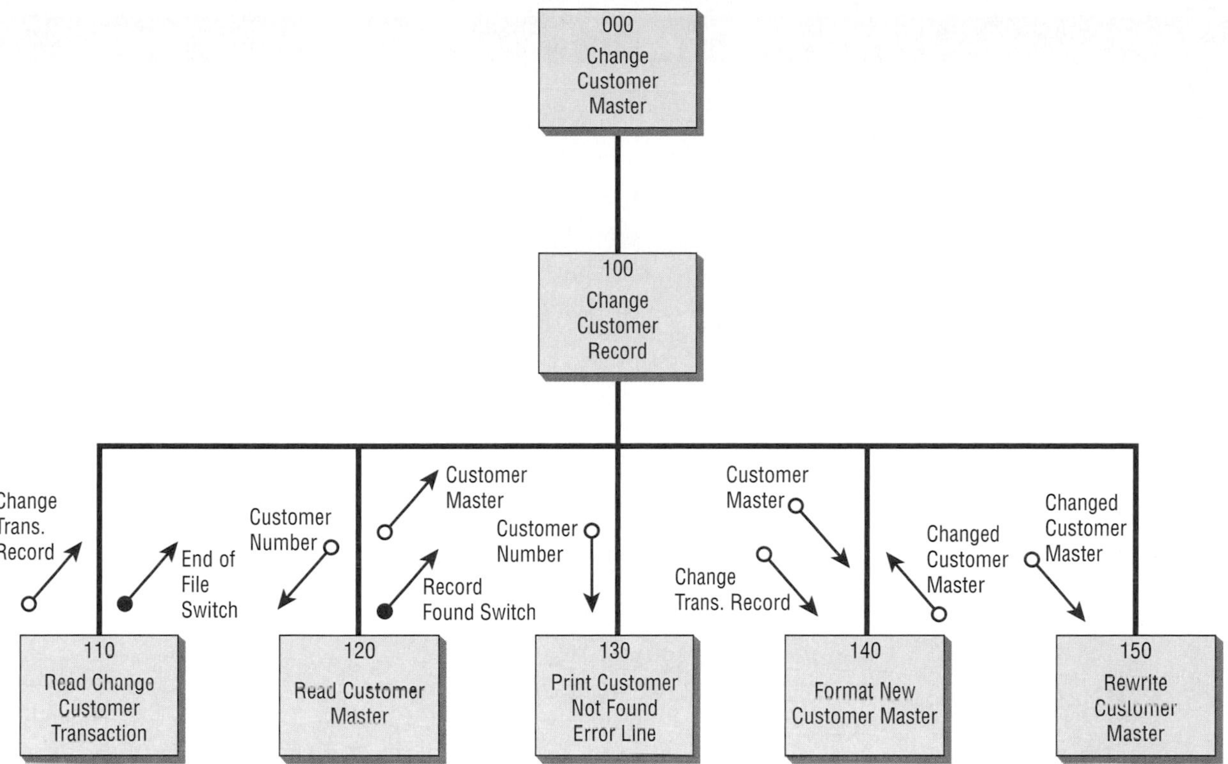

FIGURE 20.21
A corrected structure chart showing proper subordination.

Some documentation provides an overview of the system itself, whereas procedural documentation details what must be done to run software on the system and program documentation details the program code that is used.

Turnover of information service personnel has traditionally been high in comparison with other departments, so chances are that the people who conceived of and installed the original system will not be the same ones who maintain it. Consistent, well-updated documentation will shorten the number of hours required for new people to learn the system before performing maintenance.

There are many reasons why systems and programs are undocumented or underdocumented. Some of the problems reside with the systems and programs themselves, others with systems analysts and programmers.

Some legacy systems were written before the business standardized its documentation techniques, but they are still in use (without documentation). Many other systems have tolerated major and minor modifications and patches over the years, but their documentation has not been modified to reflect them. Some systems featuring specialized programs were purchased for their important applications despite their lack of accompanying documentation.

Systems analysts may fail to document systems properly because they do not have the time or are not rewarded for time spent documenting. Some analysts do not document because they dread doing so or think it is not their real work. Furthermore, many analysts are reticent about documenting systems that are not their own, perhaps fearing reprisals if they include incorrect material about someone else's system. Documentation accomplished by means of a CASE tool during the analysis phases can address many of these problems.

There is no single standard design and documentation technique in use today. In this section, we discuss several different techniques that are currently in use. Each technique has its own advantages and disadvantages, because each one has unique properties.

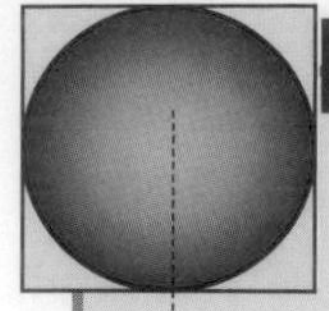

CONSULTING OPPORTUNITY 20.2

GETTING A LEG UP

"I think it's magnificent," proclaims Glen. He's been taking a bird's-eye view of the decision tree you drew of Premium Airline's 'Flying for Prizes' policies. "It will help me straighten up and fly right when I'm explaining our program to everyone. You've put us on course, all right. Now if you can just get the computers to cooperate! No, I'm just kidding you. You know, though, that something else has come up since you've talked with the ticket agents. We've got new instructions on handling fliers' accounts.

"They say that we must update each 'Flying for Prizes' account on a daily basis, and we have to do this for each flyer, for each leg they travel. An example of a leg is from Omaha to Chicago; another one is from Chicago to New York. If the miles flown in an account go over a multiple of 10,000, we must issue an award certificate. Then, each day, we need to print a summary of total miles flown (not for each person, just an aggregate of everybody's miles together). I hope you can computerize this information for us. They've got pretty lofty expectations, but I guess it's good to aim high. Let us know if it's just pie in the sky, though."

Premium Airlines has asked you to help it with its Flying for Prizes system. Draw a Nassi-Shneiderman chart that depicts the functions Glen has just described to you. Use the logic details given in Consulting Opportunity 11.4 to draw your diagram.

NASSI-SHNEIDERMAN CHARTS

One structured approach for design and documentation is the Nassi-Shneiderman (N-S) chart. The main advantage of the N-S chart is that it adopts the philosophy of structured programming. Second, it uses a limited number of symbols so that the flowchart takes up less space and can be read by someone unfamiliar with symbols used in other types of flowcharts. Figure 20.22 shows the three basic symbols that are used in N-S charts.

The first symbol is a box, used to represent any process in the program. The second symbol is a column-splitting triangle that represents a decision. The most basic form of a decision—"true" or "false"—is modeled here, but any form of a decision including several condition alternatives can be depicted using this symbol. The third symbol is the box-within-a-box symbol that is used to show that an iteration takes place. The box-within-a-box also appears as an indentation on the whole chart.

FIGURE 20.22
The three basic symbols used to draw Nassi-Shneiderman charts.

Action	Symbol
Process	
Decision	
Iteration	

In structured programming, a top-down approach is used. The analyst would begin by drawing the major loops first and then indent to complete the inner loops later.

A newspaper subscription updating system is depicted in the N-S chart in Figure 20.23. The example shows that a process is repeated for every newspaper on a daily basis. Within each newspaper iteration, a number of operations, such as clearing newspaper totals, printing the date, and printing the newspaper name, are performed. Then another loop is encountered; this loop is performed for every subscriber update. A search for the subscriber record is performed and—depending on whether the transaction is a (1) renewal, (2) new subscription, (3) cancellation, or (4) change of address—a different action is taken. Other tasks (such as updating totals and printing information) are performed, and the iterations are continued while there are subscribers and newspapers to update.

Nassi-Shneiderman charts must be complete and comprehensive to be understood. This requirement is a disadvantage when compared to other methods, because the fear of being incomplete might prevent analysts and programmers from even starting an N-S chart. If changes to the system will be made regularly, N-S charts may not be appropriate. Because they must be entirely redrawn to accommodate change, N-S charts are not easily modified.

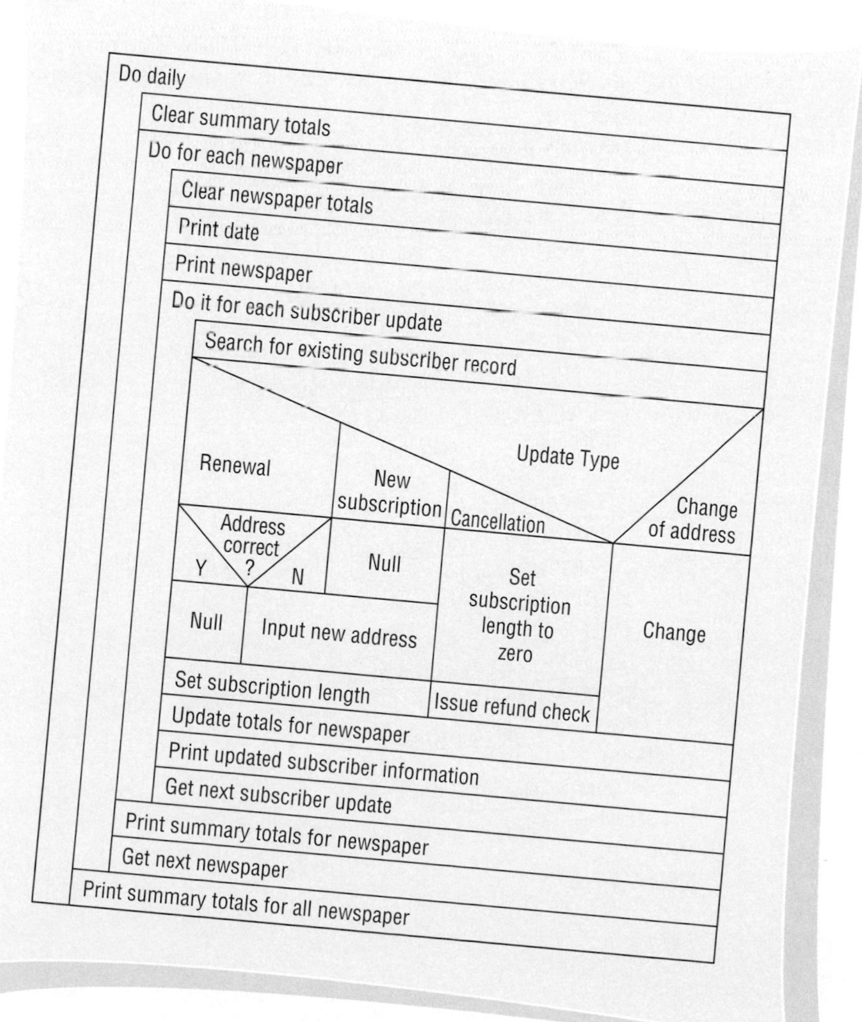

FIGURE 20.23
Using a Nassi-Shneiderman chart to depict a daily subscription updating service for newspapers.

On the other hand, the benefits of using N-S charts are many. They provide analysts with a tool for aiding in the program design and development process because they are compatible with structured programming. The N-S chart is easy to read because no knowledge of complex symbols is required. It does not take up precious space either. In summary, the N-S chart can be a very valuable tool for the analyst, and it is supported by some CASE tools.

PSEUDOCODE

In Chapter 11, we introduced the concept of structured English as a technique of analyzing decisions. Pseudocode is similar to structured English because it is not a particular type of programming code, but it can be used as an intermediate step for developing program code. Pseudocode for the newspaper example is given in Figure 20.24.

FIGURE 20.24

Using pseudocode to depict a daily subscription update service for newspapers.

```
Open Files
Summary.total = 0
Read the first newspaper.name
DO WHILE there are more newspaper.name(s)
  PRINT date
  PRINT newspaper.name
  Newspaper.total = 0
  Read first subscriber.record
  DO WHILE there are more subscriber.record(s)
    IF Action = Renewal
      THEN subs.length = subs.length + num.weeks
      IF address < > cur.address
        THEN PERFORM Address.change
      ELSE continue
    ELSE IF Action = New
        THEN PERFORM Address.change
        subs.length - num.weeks
    ELSE IF Action = Cancellation
        subs.length = 0
        PERFORM Refund
    ELSE IF Action = Address.change
        PERFORM Address.change
    ELSE PERFORM Action.error
    ENDIF
    Newspaper.total = Newspaper.total + 1
    PERFORM Print.subscriber
    Read another subscriber.record
  ENDDO
  PERFORM Print.newspaper
  Get next newspaper.name
ENDDO
Close Files
```

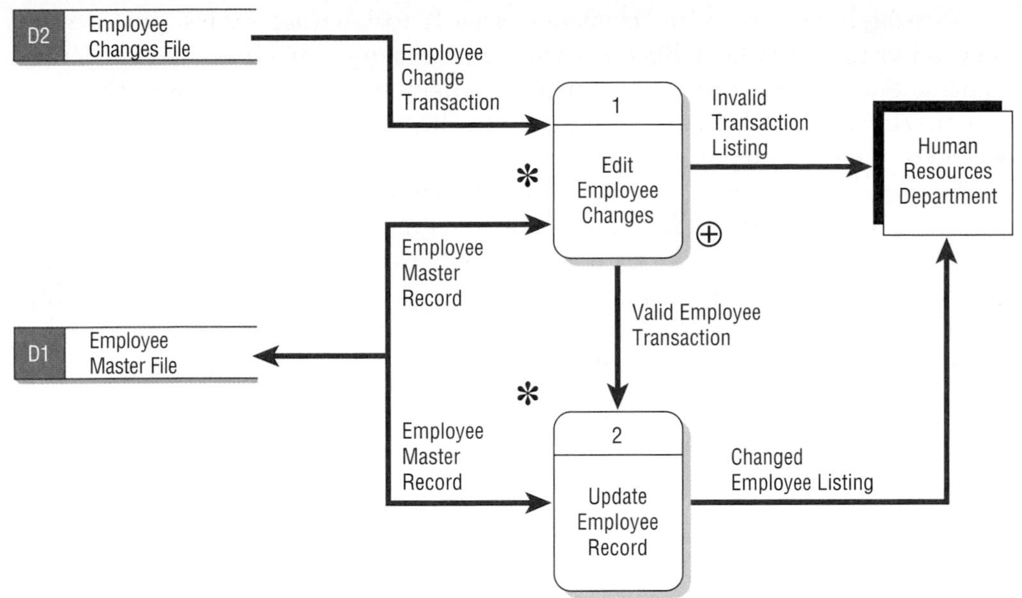

FIGURE 20.25
Use of "and" (*) and "or" (_) connectors to determine program logic.

The use of pseudocode is common in the industry, but lack of standardization will prevent it from being accepted by everyone. Because pseudocode is so close to program code, it is naturally favored by programmers and consequently is not as favored by business analysts. Pseudocode is frequently used to represent the logic of each module on a structure chart.

The data flow diagram may be used to write the pseudocode logic. When used at a program level rather than at a system level, the data flow diagram may incorporate several additional symbols. The asterisk (*), meaning "and," is used to indicate that both named data flows must be present. Refer to the portion of a data flow diagram that is illustrated in Figure 20.25. If the input data flows are from different records, the presence of the "and" connector signifies that the process receiving the flow must perform some sort of file matching, either a sequential match, reading all the records from both files, or an indexed read of a second file using a key field obtained from the first file.

The plus sign enclosed in a circle (⊕) represents an exclusive "or" and indicates that one or the other data flow is present at any given time. Use of this symbol implies that the process receiving or producing the data flow must have a corresponding IF . . . THEN . . . ELSE statement. In the figure, process 1, EDIT EMPLOYEE CHANGES, must have an IF statement to determine if the change transaction is valid or not. Process 2, UPDATE EMPLOYEE RECORD, must have both the VALID EMPLOYEE TRANSACTION and the EMPLOYEE MASTER RECORD, and it must match the records somehow to produce the output.

PROCEDURE MANUALS

Procedure manuals are common organizational documents that most people have seen. They are the English component of documentation, although they may also contain program codes, flowcharts, and so on. Manuals are intended to communicate to those who use them. They may contain background comments, steps required to accomplish different transactions, instructions on how to recover from problems, and what to do next if something isn't working (troubleshooting). Many manuals are now available online, with hypertext capability that facilitates use.

A straightforward, standardized approach to creating user support documentation is desirable. A business will often make a person or even an entire department responsible for producing and maintaining manuals and other documents. It is essential that manuals are thought of as current, rather than historical, documents. To be useful, user documentation must be kept up to date. Use of the Web has revolutionized the speed with which assistance can be obtained by users. Many software developers are moving user support—complete with FAQ, help desks, technical support, and fax-back service—to the Web. In addition, many software vendors include "read me" files with downloads or shipments of new software. These files serve a variety of purposes: they document changes, patches, or fixes for newly discovered bugs in the application that have occurred too late in its development to be included in the accompanying user's manual.

Manuals should not contain overblown rhetoric touting the benefits of the application. Remember that the manual is not an advertisement.

A good printed manual will still be used repeatedly as a reference. As such, it needs to be organized in a logical way, with careful thought given to the circumstances that would call forth the use of the manual. Online manuals using a hypertext system assist users in accessing items that otherwise might be difficult to find.

Key sections of a manual should include an introduction, how to use the software, what to do if things go wrong, a technical reference section, an index, and information on how to contact the manufacturer. Online manuals on Web sites should also include information on downloading updates and a FAQ page. The biggest complaints with procedure manuals are that (1) they are poorly organized, (2) it is hard to find needed information in them, (3) the specific case in question does not appear in the manual, and (4) the manual is not written in plain English.

In addition to the manual's organization and clarity, careful thought should be given to the kinds of people who will be using the manual or accessing the Web site. In an upcoming section on testing, we discuss the importance of having users "test" systems manuals and prototype Web sites before they are finalized.

THE FOLKLORE METHOD

FOLKLORE is a system documentation technique that was created to supplement some of the techniques just covered. Even with the plethora of techniques available, many systems are inadequately documented or not documented at all. FOLKLORE gathers information that is often shared among users but is seldom written down.

FOLKLORE is a systematic technique, based on traditional methods used in gathering folklore about people and legends. This approach to systems documentation requires the analyst to interview users, investigate existing documentation in files, and observe the processing of information. The objective is to gather information corresponding to one of four categories: customs, tales, sayings, and art forms. Figure 20.26 suggests how each category relates to the documentation of information systems.

When documenting customs, the analyst (or other folklorist) tries to capture in writing what users are currently doing to get all programs to run without problems. An example of a custom is: "Usually, we take two days to update the monthly records because the task is quite large. We run commercial accounts on day one and save the others for the next day."

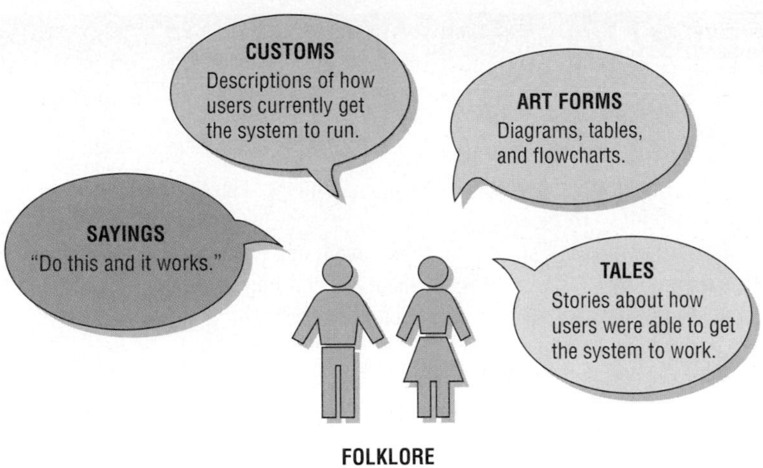
CUSTOMS
Descriptions of how users currently get the system to run.
ART FORMS
Diagrams, tables, and flowcharts.
SAYINGS
"Do this and it works."
TALES
Stories about how users were able to get the system to work.
FOLKLORE

FIGURE 20.26
Customs, tales, sayings, and art forms used in the FOLKLORE method of documentation apply to information systems.

Tales are stories that users tell regarding how the systems worked. The accuracy of the tale, of course, depends on the user's memory and is at best an opinion about how the program worked. The following is an example of a tale:

> The problem occurred again in 1995. This time, the LIB409 job (monthly update) was run with only the "type 6" records in it. Because of this mistake, there were no financial records in the LIBFIN file. When we tried to read the empty file, it was immediately closed, and the totals were consequently reported as zero. We were able to correct this problem by adding a "type 7" record and rerunning the job.

Tales normally have a beginning, a middle, and an end. In this instance, we have a story about a problem (the beginning), a description of the effects (the middle), and the solution (the end).

Sayings are brief statements representing generalizations or advice. We have many sayings in everyday life, such as "April showers bring May flowers" or "A stitch in time saves nine." In systems documentation, we have many sayings, such as "Write-protect the original before you try to back it up," "Omit this section of code and the program will bomb," or "Always back up frequently." Users like to give advice, and the analyst should try to capture this advice and include it in the FOLKLORE documentation.

Gathering art forms is another important activity of traditional folklorists, and the systems analyst should understand its importance, too. Flowcharts, diagrams, and tables that users draw sometimes may be better or more useful than flowcharts drawn by the original system author. Analysts will often find such art posted on bulletin boards, or they may ask the users to clean out their files and retrieve any useful diagrams.

The FOLKLORE approach works because it can help fill the knowledge gap created when a program author leaves. Contributors to the FOLKLORE document do not have to document the entire system, only the parts they know about. Finally, it is fun for users to contribute, taking some of the burden from analysts. Recently, FOLKLORE was automated in Finland to aid in the documentation of a new system. Notice that the class of recommendation systems that was discussed in conjunction with decision support systems in Chapter 12 is very close to the FOKLORE conceptualization. These systems expand the idea of FOLKORE to include all kinds of recommendations, such as ratings of restaurants and movies. Through low or no-cost email "on the Web," some initial barriers to gathering and sharing informal information have been overcome.

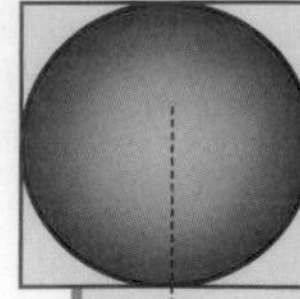

CONSULTING OPPORTUNITY 20.3

WRITE IS RIGHT

"It's so easy to understand. I say if everybody uses Nassi-Shneiderman charts, we won't have trouble, you know, with things not being standardized," said Al Gorithm, a new programmer who will be working with your systems analysis team. Al is speaking to an informal meeting among three members of the systems analysis team, a six-person MIS task force from the advertising department, and two programmers who were all working to develop an information system for advertising personnel.

Philip, an advertising account executive and one of the members of the MIS task force, looks up in surprise. "What is this method called?" The two programmers reply at the same time, "Nassi-Shneiderman." Philip looks unimpressed and says, "That doesn't say anything to me."

Neeva Phail, one of the systems analysts, begins explaining. "It probably won't matter one way or the other what we use, if . . ."

Flo Chart, another systems analyst, breaks in saying, "I hate looking at Nassi-Shneiderman. Pseudocode is much better." She looks hopefully at the programmers. "I'm sure we can agree on a better technique."

David, an older advertising executive, seems slightly upset, stating, "I learned about flowcharting from the first systems analysts we had years ago. Don't you people do that anymore? I think they work best."

What was at first a friendly meeting suddenly seems to have reached an impasse. The participants are looking at each other warily. As a systems analyst who has worked on many different projects with many different kinds of people, you realize that the group is looking to you to make some reasonable suggestions.

Based on what you know about the various documentation techniques, what technique or techniques would you propose to the members of the group? How will the technique(s) you proposed overcome some of the concerns they have voiced? What process will you use to decide on appropriate techniques?

The danger of relying on FOLKLORE is that the information gathered from users may be correct, partially correct, or even incorrect. Unless someone takes the time to redo program documentation entirely, however, the description of customs, tales, sayings, and art forms may be the only written information about how a set of programs work.

CHOOSING A DESIGN AND DOCUMENTATION TECHNIQUE

The techniques discussed in this chapter are extremely valuable as design tools, memory aids, productivity tools, and as a means of reducing dependencies on key staff members. The systems analyst, however, is faced with a difficult decision regarding which method to adopt. The following is a set of guidelines to help the analyst use the appropriate technique.

Choose a technique that:

1. Is compatible with existing documentation.
2. Is understood by others in the organization.
3. Allows you to return to working on the system after you have been away from it for a period of time.
4. Is suitable for the size of the system on which you are working.
5. Allows for a structured design approach if that is considered to be more important than other factors.
6. Allows for easy modification.

CODE GENERATION AND DESIGN REENGINEERING

Code generation is the process of using software—often a lower or integrated CASE product—to create all or a part of a computer program. Many different code generators exist for both almost every popular computer language and every platform from the microcomputer to midrange and mainframe computers.

Full code generators require a formal methodology for entering all data, business rules, screen designs, and so on. Otherwise, much of the computer program code must be entered into the CASE toolset. Partial code generators generate only specific pieces of code that may be incorporated into a program being constructed by programmers. An example of partial code generation is a CASE tool's ability to generate record layouts and screen design layouts. With add-on products (such as a Microfocus interface module), Excelerator can generate COBOL paragraph names from a structure chart as well as mainframe screen design code.

TESTING, MAINTENANCE, AND AUDITING

Once the analyst has designed and coded the system, testing, maintenance and auditing of it are prime considerations.

THE TESTING PROCESS

All the system's newly written or modified application programs—as well as new procedural manuals, new hardware, and all system interfaces—must be tested thoroughly. Haphazard, trial-and-error testing will not suffice.

Testing is done throughout systems development, not just at the end. It is meant to turn up heretofore unknown problems, not to demonstrate the perfection of programs, manuals, or equipment.

Although testing is tedious, it is an essential series of steps that helps ensure the quality of the eventual system. It is far less disruptive to test beforehand than to have a poorly tested system fail after installation. Testing is accomplished on subsystems or program modules as work progresses. Testing is done on many different levels at various intervals. Before the system is put into production, all programs must be desk-checked, checked with test data, and checked to see if the modules work together with one another as planned.

The system as a working whole must also be tested. Included here are testing the interfaces between subsystems, the correctness of output, and the usefulness and understandability of system documentation and output. Programmers, analysts, operators, and users all play different roles in the various aspects of testing, as shown in Figure 20.27. Testing of hardware is typically provided as a service by vendors of equipment who will run their own tests on equipment when it is delivered onsite.

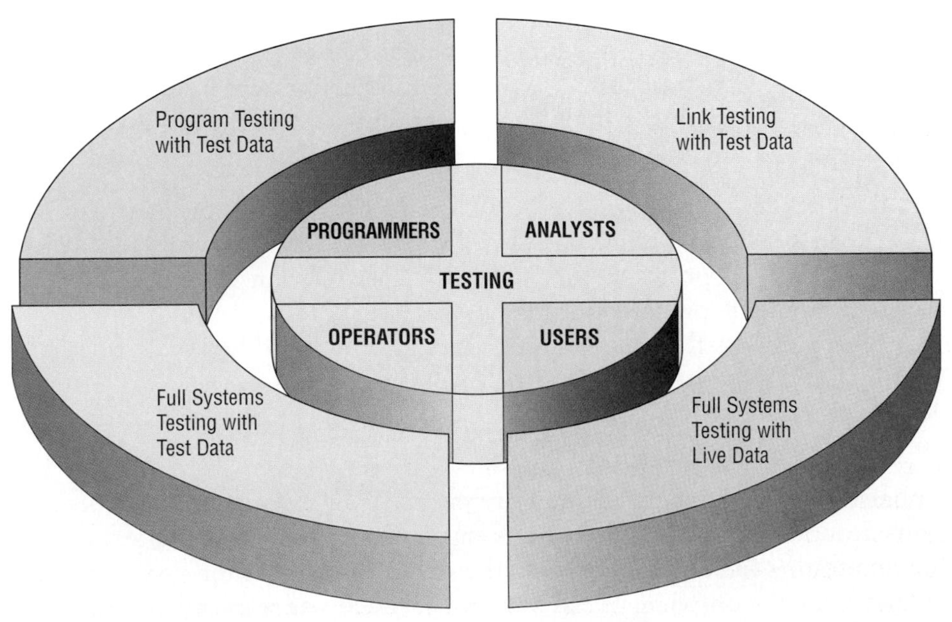

FIGURE 20.27
Programmers, analysts, operators, and users all play different roles in testing software and systems.

Program Testing with Test Data. Much of the responsibility for program testing resides with the original author(s) of each program. The systems analyst serves as an advisor and coordinator for program testing. In this capacity, the analyst works to ensure that correct testing techniques are implemented by programmers but probably does not personally carry out this level of checking.

At this stage, programmers must first desk check their programs to verify the way the system will work. In desk checking, the programmer follows each step in the program on paper to check whether the routine works as it is written.

Next, programmers must create both valid and invalid test data. These data are then run to see if base routines work and also to catch errors. If output from main modules is satisfactory, you can add more test data so as to check other modules. Created test data should test possible minimum and maximum values as well as all possible variations in format and codes. File output from test data must be carefully verified. It should never be assumed that data contained in a file are correct just because a file was created and accessed.

Throughout this process, the systems analyst checks output for errors, advising the programmer of any needed corrections. The analyst will usually not recommend or create test data for program testing but might point out to the programmer omissions of data types to be added in later tests.

Link Testing with Test Data. When programs pass desk checking and checking with test data, they must go through link testing, which is also referred to as string testing. Link testing checks to see if programs that are interdependent actually work together as planned.

A small amount of test data, usually designed by the systems analyst to test system specifications as well as programs, is used for link testing. It may take several passes through the system to test all combinations, because it is immensely difficult to unravel problems if you try to test everything all at once.

The analyst creates special test data that cover a variety of processing situations for link testing. First, typical test data are processed to see if the system can handle normal transactions, those that would make up the bulk of its load. If the system works with normal transactions, variations are added, including invalid data used to ensure that the system can properly detect errors.

Full Systems Testing with Test Data. When link tests are satisfactorily concluded, the system as a complete entity must be tested. At this stage, operators and end users become actively involved in testing. Test data, created by the systems analysis team for the express purpose of testing system objectives, are used.

As can be expected, there are a number of factors to consider when systems testing with test data:

1. Examining whether operators have adequate documentation in procedure manuals (hard copy or online) to afford correct and efficient operation.
2. Checking whether procedure manuals are clear enough in communicating how data should be prepared for input.
3. Ascertaining if work flows necessitated by the new or modified system actually "flow."
4. Determining if output is correct and whether users understand that this output is, in all likelihood, how output will look in its final form.

Remember to schedule adequate time for system testing. Unfortunately, this step often gets dropped if system installation is lagging behind the target date.

Systems testing includes reaffirming the quality standards for system performance that were set up when initial system specifications were made. Everyone

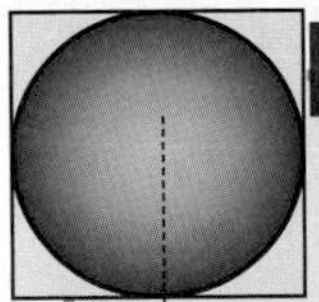

CONSULTING OPPORTUNITY 20.4

CRAMMING FOR YOUR SYSTEMS TEST

"We're strapped for time. Just look at this projection," says Lou Scuntroll, the newest member of your systems analysis team, showing you the PERT diagram that the team has been using to project when the new system would be up and running. "We can't possibly make the July target date for testing with live data. We're running three weeks behind because of that slow equipment shipment."

As one of the systems analysts who has seen deadlines come and go on other projects, you try to remain calm and to size up the situation carefully before you speak. Slowly, you question Lou about the possibility of delaying testing.

Lou replies, "If we try to push the testing off until the first weeks of August, there are two key people from accounting who are going to be out on vacation." Lou is visibly upset at the possibility of missing the deadline.

Stan Dards, another junior member of your systems analysis team, enters Lou's office. "You two look terrible. Things are going okay, aren't they? I'm not reassigned to program a payroll application, am I?"

Lou looks up, obviously neither appreciating Stan's sense of humor nor what seems like his single-minded self-concern. "Good thing you came in when you did. We've got some big decisions to make about scheduling." Lou holds up the PERT diagram for Stan's inspection. "Notice the July test date. Notice that there is no way we can make it. Any bright ideas?"

Stan contemplates the chart momentarily, then states, "Something's got to go. Let's see here . . . maybe move testing of the accounting module to—"

Lou interrupts, saying bluntly, "Nope, already thought of that, but Stanford and Binet from accounting are out of town in August. Maybe we can skip that portion of the testing. They've been really cooperative. I don't think they'd object if we just 'do it for real' and test as we actually go into production."

"I think that's a good idea, Lou," Stan agrees, trying to make up for his earlier jokes. "We haven't had any real trouble with that, and the programmers sure are confident. That way we could stay on schedule with everything else. I vote for *not* testing the accounting portion, but just sort of winging it when it starts up."

As the most senior member of the team present, what can you do to convince Lou and Stan about the importance of testing the accounting module with live data? What can systems analysts do in planning their time to allow adequate time for testing with test and live data? What are some of the possible problems the team members may encounter if they do not test the system completely with live data before putting the system into production? Realistically, are there steps in the system analysis and design process that can be collapsed to bring a delayed project in on time?

involved should once again agree on how to determine whether the system is doing what it is supposed to do. This step will include measures of error, timeliness, ease of use, proper ordering of transactions, acceptable down time, and understandable procedure manuals.

Full Systems Testing with Live Data. When systems tests using test data prove satisfactory, it is a good idea to try the new system with several passes on what is called "live data," data that have been successfully processed through the existing system. This step allows an accurate comparison of the new system's output with what you know to be correctly processed output as well as a good idea for how actual data will be handled. Obviously, this step is not possible when creating entirely new outputs. As with test data, only small amounts of live data are used in this kind of system testing.

Testing is an important period for assessing how end users and operators actually interact with the system. Although much thought is given to user-system interaction (see Chapter 18), you can never fully predict the wide range of differences in the way users will actually interact with the system. It is not enough to interview users about how they are interacting with the system; you must observe them firsthand.

Items to watch for are ease of learning the system; the adjustment to ergonomic factors; and user reaction to system feedback, including what happens when an error message is received and what happens when the user is informed that the system is executing his or her commands. Be particularly sensitive to how users react to system response time and to the language of responses. Also listen to what users say about the system as they encounter it. Any real problems need to be addressed before the system is put into production, not just glossed over as adjustments to the system that users and operators "ought" to make on their own.

As mentioned earlier, procedure manuals also need to be tested. Although manuals can be proofread by support staff and checked for technical accuracy by the systems analysis team, the only real way to test them is to have users and operators try them, preferably during full systems testing with live data. Have them use accurate but not final versions of the manuals.

It is difficult to communicate procedures accurately. Difficulty is compounded when a systems analyst or programmer has been working with a system for a long time. Recalling your own early experiences with intimidating or frustrating new systems can help focus your thinking on producing readable and useful documentation. Remember that manuals need to be organized in different ways for users who will interact with the system in countless ways. Too much information will be just as much a deterrent to system use as too little. Use of Web-based documents can help in this regard. Users can jump to topics of interest and download and print what they want to keep. Consider user suggestions, and when possible, incorporate them into the final versions of Web pages, printed manuals, and other forms of documentation.

MAINTENANCE PRACTICES

Your objective as a systems analyst should be to install or modify systems that have a reasonably useful life. You want to create a system whose design is comprehensive and farsighted enough to serve current and projected user needs for several years to come. Part of your expertise should be used to project what those needs might be and then build flexibility and adaptability into the system. The better the system design, the easier it will be to maintain and the less money the business will have to spend on maintenance.

Reducing maintenance costs is a major concern, because software maintenance alone can devour upwards of 50 percent of the total data-processing budget for a business. Excessive maintenance costs reflect directly back on the system's designer, because approximately 70 percent of software errors have been attributed to inappropriate software design. From a systems perspective, it makes sense that detecting and correcting software design errors early on is less costly than letting errors remain unnoticed until maintenance is necessary.

Maintenance is performed most often to improve the existing software rather than to respond to a crisis or system failure. As users' requirements change, software and documentation should be changed as part of the maintenance work. In addition, programs might be recoded to improve on the efficiency of the original program. Over half of all maintenance is composed of such enhancement work.

Maintenance is also done to update software in response to the changing organization. This work is not as substantial as enhancing the software, but it must be done. Emergency and adaptive maintenance comprises less than half of all system maintenance.

Part of the systems analyst's job is to ensure that there are adequate channels and procedures in place to permit feedback about—and subsequent response to—maintenance needs. Users must be able to communicate problems and suggestions easily to those who will be maintaining the system. It is very discouraging (sometimes so much so that the system will fall into disuse) if the system is not properly maintained. Solutions are to provide users email access to technical support as well as to allow them to download product updates or patches from the Web.

The systems analyst also needs to set up a classification scheme to allow users to designate the perceived importance of the maintenance being suggested or requested. Classifying requests enables maintenance programmers to understand how users themselves estimate the importance of their requests. This viewpoint, along with other factors, can then be taken into account when scheduling maintenance.

AUDITING

Auditing is yet another way of ensuring the quality of the information contained in the system. Broadly defined, auditing refers to having an expert who is not involved in setting up or using a system examine information in order to ascertain its reliability. Whether or not information is found to be reliable, the finding on its reliability is communicated to others for the purpose of making the system's information more useful to them.

For information systems, there are generally two kinds of auditors: internal and external. Whether both are necessary for the system you design depends on what kind of system it is. Internal auditors work for the same organization that owns the information system, whereas external (also called "independent") auditors are hired from the outside.

External auditors are used when the information system processes data that influences a company's financial statements. External auditors audit the system to ensure the fairness of the financial statements being produced. They may also be brought in if there is something out of the ordinary occurring that involves company employees, such as suspected computer fraud or embezzlement.

Internal auditors study the controls used in the information system to make sure that they are adequate and that they are doing what they are purported to be doing. They also test the adequacy of security controls. Although they work for the same organization, internal auditors do not report to the people responsible for the system they are auditing. The work of internal auditors is often more in-depth than that of external auditors.

SUMMARY

The systems analyst uses three broad approaches to total quality management (TQM) for analyzing and designing information systems: designing systems and software with a top-down, modular approach; designing and documenting systems and software using systematic methods; and testing systems and software so that they can be easily maintained and audited.

Users are critically important for establishing and evaluating the quality of several dimensions of management information systems and decision support systems. They can be involved in the entire evolution of systems through the establishment of MIS task forces or quality circles.

TQM can be successfully implemented by taking a top-down approach to design. This approach refers to looking at overall organizational objectives first and then decomposing them into manageable subsystem requirements. Modular development makes programming, debugging, and maintenance easier to accomplish. Programming in modules is well suited to taking a top-down approach.

Two systems link programs in the Windows environment. One is DDE (Dynamic Data Exchange), which shares code by using Dynamic Link Library (DLL) files. Using DDE, a user can store data in one program and then use it in another. A second approach to linking programs in Windows is OLE (Object Linking and Embedding). Because of its object-oriented approach, this linking method is superior to DDE for linking application data and graphics.

20 

This is a fascinating place to work. I'm sure you agree now that you've had a chance to observe us. Sometimes I think it must be fun to be an outsider . . . don't you feel like an anthropologist discovering a new culture? I remember when I first came here. Everything was so new, so strange. Why, even the language was different. It wasn't a 'customer'; it was a 'client.' We didn't have 'departments'; we have 'units.' It's not an employee cafeteria; it's the 'canteen.' That goes for the way we work, too. We all have our different ways to approach things. I think I'm getting the hang of what Snowden expects, but every once in a while I make a mistake, too. For instance, if I can give him work on disk, he'd just as soon see it that way than get a printed report. That's why I have two computers on my desk, too! I always see you taking so many notes . . . I guess it makes sense, though. You're supposed to document what *we* do with our systems and information as well as what your team is doing, aren't you?"

HYPERCASE® QUESTIONS

1. Use the FOLKLORE method to complete the documentation of the Management Information Systems Unit GEMS system. Be sure to include customs, tales, sayings, and art forms.
2. In two paragraphs, suggest a PC-based approach for capturing the elements of FOLKLORE so that it is not necessary to use a paper-based log. Make sure that your suggested solution can accommodate graphics as well as text.
3. Design input and output screens for FOLKLORE that facilitate easy entry and provide prompting so that recall of FOLKLORE elements is immediate.

FIGURE 20.HC1

In HyperCase, use FOLKLORE to document art forms that users have created or collected to make sense of their systems.

A recommended tool for designing a top-down, modular system is called a structure chart. Two types of arrows are used to indicate the kinds of parameters that are passed between the modules. The first is called a data couple, and the second is called a control flag. Structure chart modules fall into one of three categories: control, transformational (sometimes called worker), and functional or specialized.

Part of TQM is to see that programs and systems are properly designed, documented, and maintained. Some of the structured techniques that can aid the systems analyst are Nassi-Shneiderman charts, pseudocode, procedure manuals, and FOLKLORE. Pseudocode is frequently used to represent the logic of each module or structure chart. Pseudocode can be used for structured walkthroughs. Systems analysts must choose a technique both that fits in well with what was previously used in the organization and that allows flexibility and easy modification.

Code generation is the process of using software to create all or part of a computer program. Many code generators are now available commercially. Reengineering and reverse engineering refer to using software to analyze existing program code and to create CASE design elements from the code. CASE design may then be modified and used to generate new computer program code.

Testing of specific programs, subsystems, and total systems is essential to quality. Testing is done to turn up any existing problems with programs and their interfaces before the system is actually used. Testing is usually done in a bottom-up fashion, with program codes being desk checked first. Following several intermediate test steps, testing of the full system with live data (actual data that have been successfully processed with the old system) is accomplished. This testing provides an opportunity to work out any problems that arise before the system is put into production.

System maintenance is an important consideration. Well-designed software can help reduce maintenance costs. Systems analysts need to set up channels for user feedback on maintenance needs, because systems that are not maintained will fall into disuse. Web sites can help in this regard by providing access to product updates and email exchanges with technical staff.

Both internal and external auditors are used to determine the reliability of the system's information. They communicate their audit findings to others so as to improve the usefulness of the system's information.

KEYWORDS AND PHRASES

- bottom-up design
- code generation
- code generators
- control flags (switches)
- data couples
- desk check
- Dynamic Data Exchange (DDE)
- Dynamic Link Library (DLL)
- FOLKLORE method
- full systems testing with live data
- full systems testing with test data
- improper subordination
- internal auditor
- IS quality circles
- link testing with test data (string testing)
- modular development
- Nassi-Shneiderman charts
- Object Linking and Embedding (OLE)
- program testing with test data
- pseudocode
- reengineering (reverse engineering)
- software documentation
- software maintenance
- stamp coupling
- structure charts
- structured walkthrough
- top-down design
- total quality management (TQM)
- transaction centered
- transform centered

REVIEW QUESTIONS

1. What are the three broad approaches available to the systems analyst for attaining quality in newly developed systems?
2. Who or what is the most important factor in establishing and evaluating the quality of information systems or decision support systems? Why?
3. Define the total quality management (TQM) approach as it applies to the analysis and design of information systems.
4. What is an IS quality circle?
5. Define what is meant by doing a structured walkthrough. Who should be involved? When should structured walkthroughs be done?
6. List the disadvantages to taking a bottom-up approach to design.
7. List the advantages of taking a top-down approach to design.
8. What are the three main disadvantages of taking a top-down approach to design?
9. Define modular development.
10. List four guidelines for correct modular programming.
11. How do structure charts help the analyst?
12. Name the two types of arrows used in structure charts.
13. Why do we want to keep the number of arrows to a minimum when using structure charts?
14. What are the two main types of structure charts?
15. Why should control flags be passed upward in a structure chart?
16. List two ways that the data flow diagram helps to build a structure chart.
17. List the three categories of modules. Why are they used in structure charts?
18. How can a Web site help in maintaining the system and its documentation?
19. Give two reasons that support the necessity of well-developed system and software documentation.
20. Give the two chief advantages of Nassi-Shneiderman charts.
21. Define pseudocode.
22. List the four biggest complaints users voice about procedure manuals.
23. In what four categories does the FOLKLORE documentation method collect information?
24. List six guidelines for choosing a design and documentation technique.
25. What are the advantages of using code generation and design reengineering for building systems?
26. Whose primary responsibility is it to test computer programs?
27. What is the difference between test data and live data?
28. What are the two types of systems auditors?

PROBLEMS

1. One of your systems analysis team members has been discouraging user input on quality standards, arguing that because you are the experts, you are really the only ones who know what constitutes a quality system. In a paragraph, explain to your team member why getting user input is critical to system quality. Use an example.
2. Draw a structure chart for the credit-reporting system in Figure 20.EX1.
3. Construct a structure chart for the payroll system in Figure 9.7 in Chapter 9.
4. Draw a Nassi-Shneiderman (N-S) chart to update your checkbook. Start with the balance the last time you balanced it. Subtract checks, electronic withdrawals, and any service charges. Add deposits and interest.

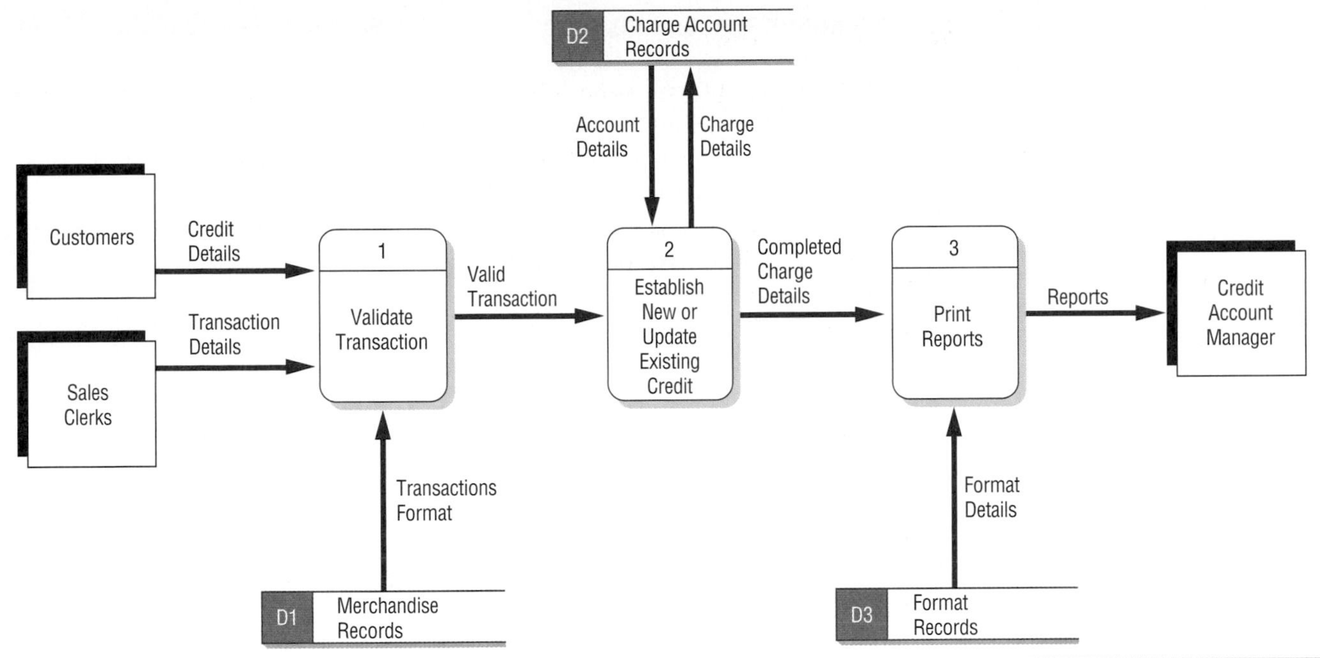

FIGURE 20.EX1
A data flow diagram for a credit reporting system.

5. Write pseudocode for problem 5.
6. Bestmonth Lumber Company is setting up a monthly payroll report. The report has the following information:

Heading	(left)	department number
		run date
	(center)	company name
		title of report
	(right)	page number
Body		last name
		first name
		monthly gross
		federal taxes
		state taxes
		FICA taxes
		other deductions
		monthly net

7. Draw an N-S chart for problem 7.
8. Write pseudocode for the Citron Car rental policy provided in Consulting Opportunity 11.3 in Chapter 11.
9. Write a detailed table of contents for a procedure manual that explains to users how to log onto your school's computer network as well as explains network policies: who is an authorized user and so on. Make sure that the manual is written with the user in mind.
10. Your systems analysis team is close to completing a system for Meecham Feeds. Roger is quite confident that the programs that he has written for Meecham's inventory system will perform as necessary, because they are similar to programs he has done before. Your team has been very busy and would ideally like to begin full systems testing as soon as possible.

    Two of your junior team members have proposed the following:

    a. Skip desk checking of the programs (because similar programs were checked in other installations; Roger has agreed).

b. Do link testing with large amounts of data to prove that the system will work.
c. Do full systems testing with large amounts of live data to show that the system is working.

Respond to each of the three steps in their proposed test schedule. Use a paragraph to explain your response.

11. Propose a revised testing plan for the Meecham Feeds problem (problem 10). Break down your plan into a sequence of detailed steps.

GROUP PROJECTS

1. Divide your group into two subgroups. One subgroup should interview the members of the other group about their experiences encountered in registering for a class. Questions should be designed to elicit information on customs, tales, sayings, and art forms that will help document the registration process at your school.
2. Reunite your group to develop a Web page for a short excerpt for a FOLKLORE manual that documents the process of registering for a class, one based on the folklore passed on in the interviews in problem 1. Remember to include examples of customs, tales, sayings, and art forms.

SELECTED BIBLIOGRAPHY

Dean, J. W., Jr., and J. R. Evans. *Total Quality*. Minneapolis/St. Paul: West, 1994.

Deming, W. E. *Management for Quality and Productivity*. Cambridge, MA: MIT Center for Advanced Engineering Study, 1981.

Katzan, H., Jr. *Systems Design and Documentation: An Introduction to the HIPO Method*. New York: Van Nostrand Reinhold, 1976.

Kendall, J. E., and P. Kerola. "A Foundation for the Use of Hypertext Based Documentation Techniques." *Journal of End User Computing*, Vol. 6, No. 1 Winter 1994, pp. 4–14.

Kendall, K. E., and R. Losee. "Information System FOLKLORE: A New Technique for System Documentation." *Information and Management*, Vol. 10, No. 2, 1986, pp. 103–11.

Kendall, K. E., and S. Yoo. "Pseudocode-Box Diagrams: An Approach to More Understandable, Productive, and Adaptable Software Design and Coding." *International Journal on Policy and Information*, Vol. 12, No. 1, June 1988, pp. 39–51.

Lee, S. M., and M. J. Schniederjans. *Operations Management*. Boston: Houghton-Mifflin, 1994.

ALLEN SCHMIDT, JULIE E. KENDALL, AND KENNETH E. KENDALL

20 CHARTING THE STRUCTURE

"Here they are, as promised," Chip and Anna say triumphantly as they hand over their specifications to Mack Roe, the project programmer.

"Thanks," Mack says. "I've got a lot of work ahead."

Mack starts by creating a structure chart for each program and then for each module design. The PRODUCE SOFTWARE CROSS-REFERENCE REPORT structure chart is shown in Figure E20.1. The "C" in front of each module number refers to the

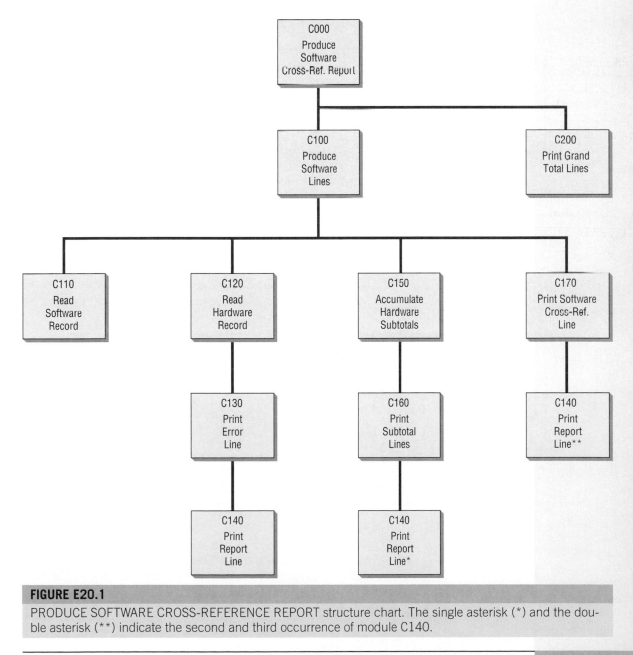

C000 Produce Software Cross-Ref. Report

C100 Produce Software Lines

C200 Print Grand Total Lines

C110 Read Software Record

C120 Read Hardware Record

C150 Accumulate Hardware Subtotals

C170 Print Software Cross-Ref. Line

C130 Print Error Line

C160 Print Subtotal Lines

C140 Print Report Line**

C140 Print Report Line

C140 Print Report Line*

FIGURE E20.1

PRODUCE SOFTWARE CROSS-REFERENCE REPORT structure chart. The single asterisk (*) and the double asterisk (**) indicate the second and third occurrence of module C140.

20

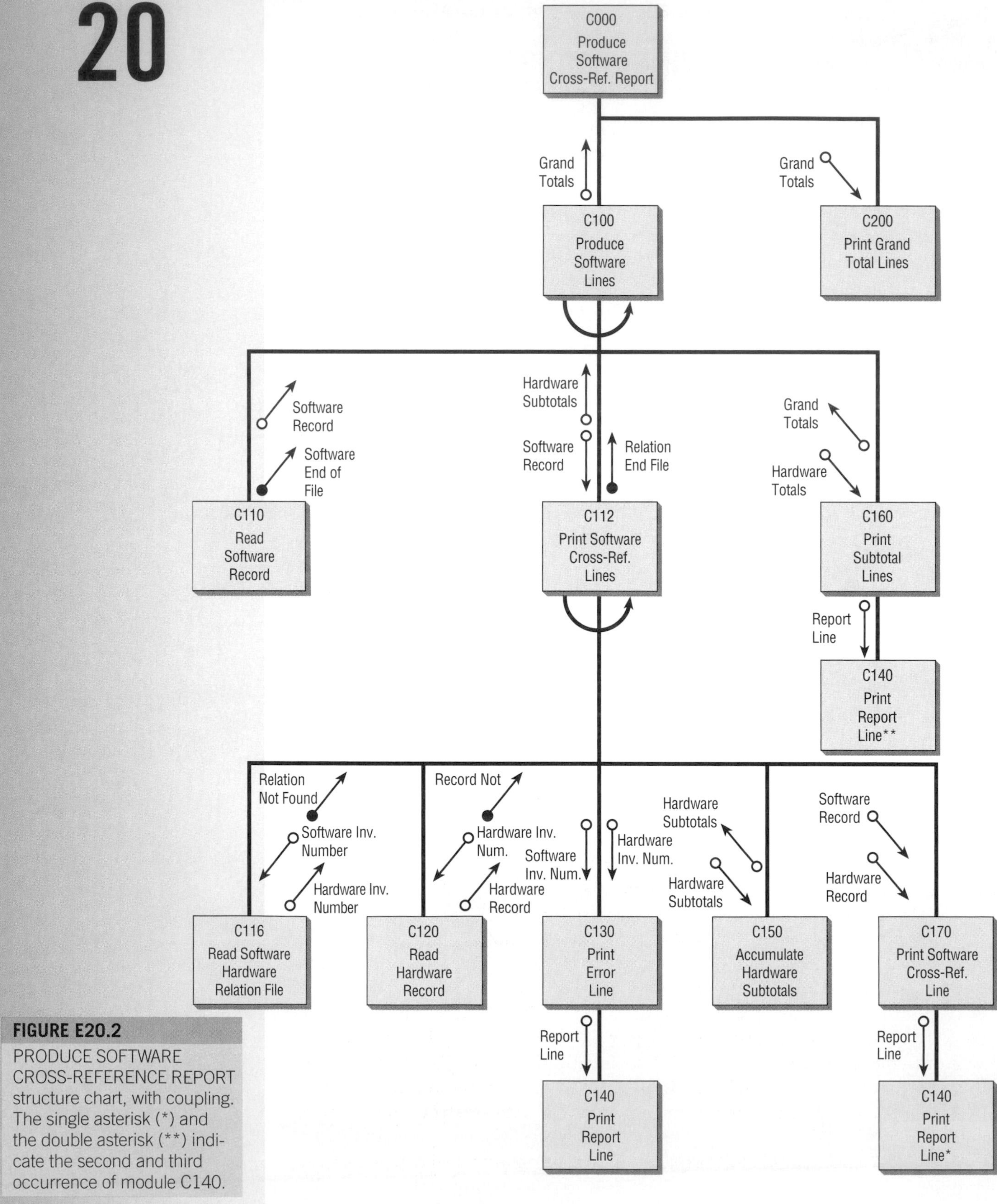

FIGURE E20.2
PRODUCE SOFTWARE CROSS-REFERENCE REPORT structure chart, with coupling. The single asterisk (*) and the double asterisk (**) indicate the second and third occurrence of module C140.

CROSS-REFERENCE REPORT (a letter is required by Visible Analyst as the first character in a module name). This draft is the first one Mack uses in a structured walk-through with Dee Ziner, a senior programmer.

Dee Ziner has several important suggestions for improving the structure. She says, "Module C130, PRINT ERROR LINE, is improperly subordinate to the calling module C120, READ HARDWARE RECORD. The question may be asked, 'Must the program print an error line to accomplish reading a HARDWARE RECORD?' Because the answer is 'No,' the module should be placed at the same level as C120, READ HARDWARE RECORD."

She continues discussing the situation with Mack, saying, "The same is true concerning module C160, PRINT SUBTOTAL LINES. That is not a function of accumulating hardware subtotals and should not be called from module C150, ACCUMULATE HARDWARE SUBTOTALS." Dee continues the walkthrough by asking the question, "May one SOFTWARE RECORD be located on many machines?" Mack responds that that is true, and another controlling module, PRINT SOFTWARE CROSS-REFERENCE LINES, was included in the structure chart.

Mack proceeds to incorporate the changes to the structure chart. When the correct hierarchy is established, the coupling is added. Careful attention is given to pass minimal data and to only pass control *up* the structure chart. The final version is illustrated in Figure E20.2. Module C116 is new, using the SOFTWARE/HARDWARE RELATIONAL file to link one SOFTWARE RECORD to many HARDWARE RECORDS. The SOFTWARE INVENTORY NUMBER is passed down to the module, and the relational file is randomly read. The HARDWARE INVENTORY NUMBER and control switch RELATION NOT FOUND are passed up the structure.

The final structure chart has a functional shape to it. A few control modules at the top of the structure, several worker modules in the middle, and a few specialist modules at the bottom provide a general fan-out, fan-in shape. The module names are all in verb-adjective-noun form, describing what has been accomplished after the module has finished executing. For example, module C150 has the verb ACCUMULATE, describing the work accomplished by the module. SUBTOTALS, a noun, are being accumulated, and HARDWARE describes which subtotals are accumulated.

Each of the modules on the structure chart were described in the repository. Figure E20.3 illustrates the screen describing the function for module C100, PRODUCE SOFTWARE LINES. Notice that the module description contains pseudocode depicting the logic of the module. Because PRODUCE SOFTWARE LINES is a control module, its logic should consist of looping and decision making, with minimal statements concerning processing details such as ADD or READ.

Each data and control couple on the structure chart may also be described in the repository. Figure E20.4 illustrates the repository screen for the SOFTWARE RECORD data couple. Notice that it has the entry SOFTWARE MASTER FILE in the **Related to** area. This area provides a link to the repository entry that contains the details of elements contained in the SOFTWARE RECORD. The **Notes** area contains information about how the SOFTWARE RECORD is used on the structure chart.

"Well, I guess that we're about done creating diagrams for the programmers," remarks Chip.

"Creating diagrams, yes," counters Anna, "but there's a bit more that we can give them."

20

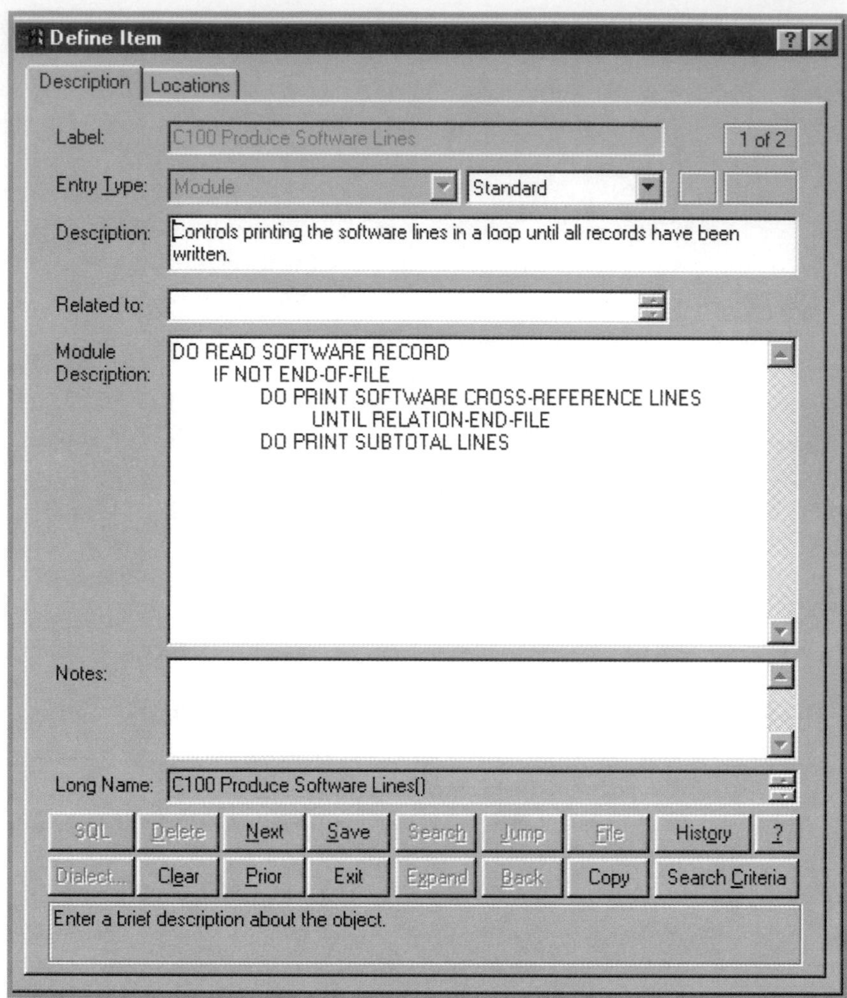

FIGURE E20.3
PRODUCE SOFTWARE LINES module repository screen.

"What do you mean?" asks Chip, with a puzzled look on his face.

"Let's use Visible Analyst to generate the database tables for Microsoft Access," exclaims Anna. "I thought we would start with one of the major entities, such as the SOFTWARE MASTER and use the code generation feature of Visible Analyst."

Anna and Chip proceed to work with Visible Analyst to ensure that all the elements have been defined for the SOFTWARE MASTER. Anna clicks **Repository** and **Generate Database Schema**. They select the MICROCOMPUTER SYSTEM diagram and give the schema the same name. The entire schema for the computer system is generated. A portion of the generated code is illustrated in Figure E.20.5.

"I'm going to copy a portion of the schema to work with just the SOFTWARE MASTER," Anna remarks to Chip. She copies the generated SQL for the SOFTWARE MASTER FILE. The next step is to create a blank query in Microsoft Access. Anna runs Microsoft Access and creates a new empty query. She clicks on the SQL button and pastes the SOFTWARE MASTER FILE into the SQL window.

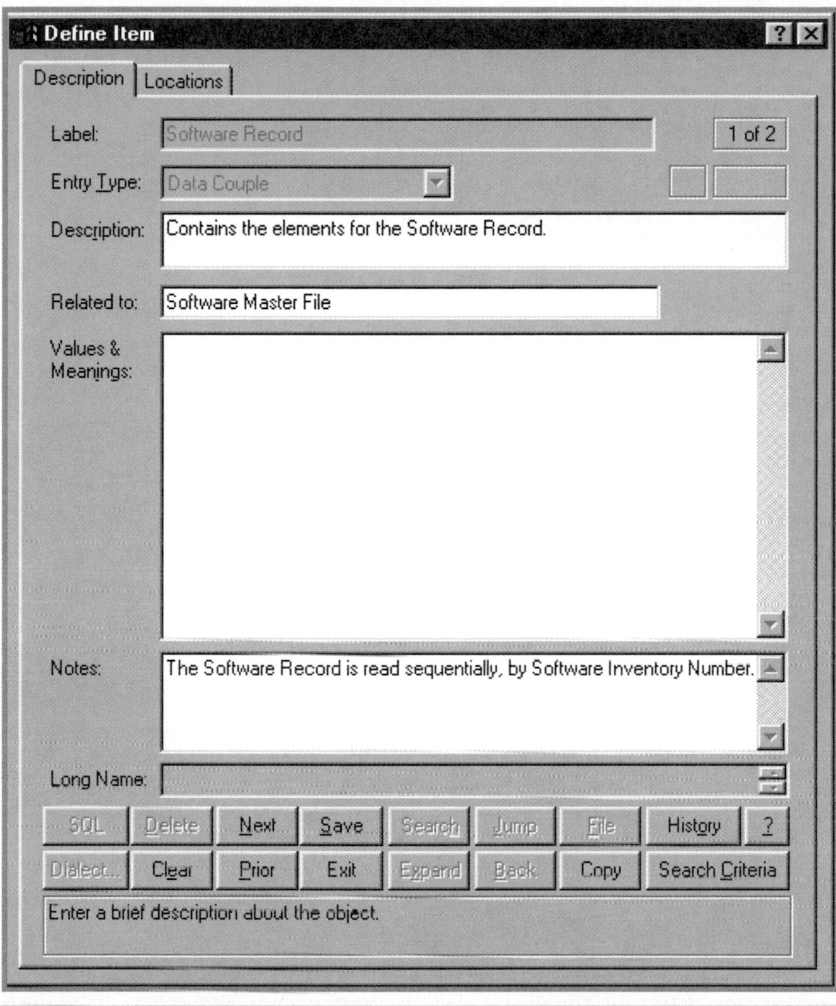

FIGURE E20.4
Repository screen for the SOFTWARE RECORD data couple.

"I need to change the name of the table to SOFTWARE and to change the query type to MAKE TABLE," continues Anna. She gives the new table the name SOFTWARE. Anna clicks the **Run Query** button and closes the query.

"What happened?" asks Chip with a puzzled look. "I don't see any output."

Anna clicks on the Tables button. "Take a look at our new table!" Anna exclaims. She clicks the SOFTWARE table and design view. "Here's our structure from Visible Analyst, implemented in Microsoft Access."

"Now that is cool," beams Chip.

The analysts continue to generate tables until the design is complete.

"I think that we can leave the rest to the programming staff," remarks Anna. "We better start developing the test plans for each program."

Test plans contain details about how to determine if the programs are working correctly, and they are sent to Mack and Dee, who will create the actual test data. Invalid and valid data are included on each test file used in batch programs. The same is true for interactive systems, except that the test data are written on forms imitating the

EPISODE CPU CASE 20

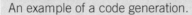

Access 2000 SQL Schema [Project 'CPU']

```
CREATE TABLE Microcomputer_System.Board_Codes
(
   Board_Code  CHAR(3) NOT NULL,
   Board_Name  CHAR(20) NOT NULL
);

CREATE TABLE Microcomputer_System.Campus_Building
(
   Campus_Code         NUMBER NOT NULL,
   Campus_Description  CHAR(30) NOT NULL
);

CREATE TABLE Microcomputer_System.Hardware_Software_Relational_Table
(
   Hardware_Inventory_Number  CHAR(8) NOT NULL,
   Software_Inventory_Number  CHAR(8) NOT NULL
);

CREATE TABLE Microcomputer_System.Microcomputer_Board_Relational_Table
(
   Hardware_Inventory_Number  CHAR(8) NOT NULL,
   Board_Code                 CHAR(3) NOT NULL
);

CREATE TABLE Microcomputer_System.Microcomputers
(
   Hardware_Inventory_Number   CHAR(8) NOT NULL,
   Brand_Name                  CHAR(10) NOT NULL,
   Model                       CHAR(12) NOT NULL,
   Serial_Number               CHAR(12) NOT NULL,
```

Help | Save... | Print | Catalog | Cancel — Select: Errors / Schema

FIGURE E20.5

An example of a code generation.

screen design. After Mack has finished testing his programs and is satisfied that they are working correctly, he challenges Dee to find any errors in the programs. In turn, Dee has Mack test her programs in a round of friendly competition. They both realize that programmers may not always catch their own errors, because they are intimately familiar with their own programs and may not recognize subtle errors in logic.

EXERCISES*

E-1. View the PRODUCE SOFTWARE CROSS-REF REP structure chart. Double click on the modules to view the repository entries for some of the modules.

E-2. Modify the PRODUCE HARDWARE INVESTMENT RPT structure chart. Add the function PRINT INVESTMENT LINE in the empty rectangle provided. Subordinate to this module is PRINT HEADING LINES and WRITE REPORT LINE. Describe each function in the repository.

E-3. Modify the CHANGE MICROCOMPUTER FILE structure chart. Include the Loop symbol and add the following modules subordinate to 160, CHANGE MICROCOMPUTER RECORD (also see below):

A. DISPLAY CHANGE SCREEN
B. ACCEPT MICROCOMPUTER CHANGES
C. VALIDATE CHANGES
D. DISPLAY ERROR MESSAGE
E. CONFIRM CHANGES

The following modules should be subordinate to 220, PUT MICROCOMPUTER RECORD:

A. FORMAT MICROCOMPUTER RECORD
B. REWRITE MICROCOMPUTER RECORD

E-4. Modify the ADD SOFTWARE RECORDS structure chart by adding a looping symbol and coupling for the connections. The following coupling should be placed on the connection line above each module (also see below):

A.	Module:	DISPLAY ADD SOFTWARE SCREEN
	Passed Up:	ADD SOFTWARE SCREEN
B.	Module:	ACCEPT ADD SOFTWARE SCREEN
	Passed Up:	EXIT INDICATOR (Control)
		ADD SOFTWARE SCREEN DATA
C.	Module:	VALIDATE ADD SOFTWARE DATA
	Passed Down:	ADD SOFTWARE SCREEN DATA
	Passed Up:	CANCEL TRANSACTION (Control)
		VALID ADD SOFTWARE DATA
D.	Module:	READ SOFTWARE RECORD
	Passed Down:	SOFTWARE INVENTORY NUMBER
	Passed Up:	RECORD FOUND (Control)
E.	Module:	VALIDATE HARDWARE REQUIREMENTS
	Passed Down:	ADD SOFTWARE SCREEN DATA
	Passed Up:	VALID DATA (Control)
		ERROR MESSAGE
F.	Module:	DISPLAY ERROR MESSAGE
	Passed Down:	ERROR MESSAGE
G.	Module:	PUT NEW SOFTWARE RECORD
	Passed Down:	VALID ADD SOFTWARE DATA
H.	Module:	FORMAT SOFTWARE RECORD
	Passed Down:	VALID ADD SOFTWARE DATA
	Passed Up:	FORMATTED SOFTWARE RECORD I
	Module:	WRITE SOFTWARE RECORD
	Passed Down:	FORMATTED SOFTWARE RECORD

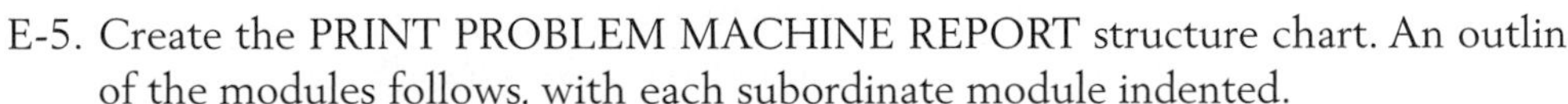

E-5. Create the PRINT PROBLEM MACHINE REPORT structure chart. An outline of the modules follows, with each subordinate module indented.

```
PRINT PROBLEM MACHINE REPORT
    PRINT PROBLEM MACHINE LINES
        READ MACHINE RECORD
        DETERMINE PROBLEM MACHINE
        PRINT PROBLEM MACHINE LINE
            PRINT HEADING LINES
            WRITE REPORT LINE
    PRINT FINAL REPORT LINES
        WRITE REPORT LINE
```

E-6. Create the CHANGE SOFTWARE RECORD structure chart. Modules of the program are shown with subordinate modules indented.

CHANGE SOFTWARE FILE
  CHANGE SOFTWARE RECORDS
    GET SOFTWARE RECORD
      DISPLAY SOFTWARE ID SCREEN
      ACCEPT SOFTWARE ID SCREEN
      FIND SOFTWARE RECORD
      DISPLAY ERROR LINE
    OBTAIN SOFTWARE CHANGES
      DISPLAY CHANGE SCREEN
      ACCEPT SOFTWARE CHANGES
      VALIDATE CHANGES
      DISPLAY ERROR LINE
    PUT SOFTWARE RECORD
      FORMAT SOFTWARE RECORD
      REWRITE SOFTWARE RECORD

E-7. Create the SOFTWARE DETAILS INQUIRY structure chart. Modules are listed with subordinate modules indented.

INQUIRE SOFTWARE DETAILS
  INQUIRE SOFTWARE RECORD
    GET SOFTWARE RECORD
      DISPLAY SOFTWARE ID SCREEN
      ACCEPT SOFTWARE ID SCREEN
      FIND SOFTWARE RECORD
      DISPLAY ERROR LINE
    DISPLAY INQUIRY SCREEN
      FORMAT SOFTWARE INQUIRY SCREEN
      DISPLAY SOFTWARE INQUIRY SCREEN

E-8. View the ADD MICROCOMPUTER system flowchart.

E-9. Modify the ADD SOFTWARE system flow. Add the following program rectangles below the INSTALL SOFTWARE manual process. Include input and output files and reports specified for each program.

Program:	UPDATE SOFTWARE RELATIONAL FILE
Input:	UPDATE SOFTWARE INSTALLATION LIST, document
	UPDATE INSTALLED SOFTWARE SCREEN, display
Output:	SOFTWARE RELATIONAL FILE, disk
	INSTALLED SOFTWARE TRANSACTION, disk
Program:	PRINT USER NOTIFICATION REPORT
Input:	INSTALLED SOFTWARE TRANSACTION, disk
Output:	USER NOTIFICATION REPORT, report

E-10. Create the ADD STAFF system flowchart. There are two programs: ADD STAFF and PRINT NEW STAFF LIST. Input to the ADD STAFF program is a NEW STAFF listing and an ADD NEW STAFF entry screen. The STAFF MASTER file is updated and a NEW STAFF LOG FILE is produced. The NEW

STAFF LOG FILE is input to the PRINT NEW STAFF LIST program, producing the report NEW STAFF LIST.

E-11. Design test data on paper to test the ADD MICROCOMPUTER program. Use Microsoft Access to test the screen. Note any discrepancies.

E-12. Design test data and predicted results for the ADD SOFTWARE program. Use Microsoft Access to test the screen and note whether the results conformed to your predictions.

E-13. Design test data and predicted results for the ADD TRAINING CLASS program. Use Microsoft Access to test the screen and note whether the results conformed to your predictions.

E-14. Design test data on paper to test the CHANGE SOFTWARE EXPERT program. Use Microsoft Access to test the screen. Note any discrepancies.

*Exercises preceded by a CD-ROM icon require the program Visible Analyst or another CASE tool. A CD-ROM containing Visible Analyst examples is free of charge to any professor adopting this book. The examples on the disk may be imported into Visible Analyst and then used by students.

SUCCESSFULLY IMPLEMENTING THE INFORMATION SYSTEM

21

IMPLEMENTATION APPROACHES

The process of ensuring that the information system is operational and then allowing users to take over its operation for use and evaluation is called implementation. The systems analyst has several approaches to implementation that should be considered as the changeover to the new system is being prepared. They include shifting more computer power to users through distributed processing, training users, converting from the old system, and evaluating the new one.

The first approach to implementation concerns the movement of computer power to individual users by setting up and shifting computer power and responsibility to groups throughout the business with the help of distributed computing.

The second approach to implementation is using different strategies for training users and personnel, including taking them on their own level, using a variety of training techniques, and making sure that each user understands any new role that he or she must take on because of the new information system.

Another approach to implementation is choosing a conversion strategy. The systems analyst needs to weigh the situation and propose a conversion plan that is appropriate for the particular organization and information system.

The fourth approach to implementation involves evaluating the new or modified information system. The analyst needs to formulate performance measures on which to evaluate the system. Evaluations come from users, management, and analysts themselves.

IMPLEMENTING DISTRIBUTED SYSTEMS

If the reliability of a telecommunications network is high, it is possible to have distributed systems for businesses, a setup that can be conceived of as an application of telecommunications. The concept of distributed systems is used in many different ways. Here it will be taken in a broad sense so that it includes workstations that can communicate with each other and data processors as well as different hierarchical architectural configurations of data processors that communicate with each other and that have differing data-storage capabilities.

The information architecture model that will likely dominate networking in the next few years is that of the client/server. In this model, the processing functions are delegated either to "clients" (users) or to "servers," depending on which machines are most suitable for executing the work. In this type of architecture, the client portion of a network application will run on the client system, with the server part of the application running on the file server. With a client/server model, users interact with limited parts of the application, including the user interface, data input, database queries, and report generation. Controlling user access to centralized databases, retrieving or processing data, and other functions (such as managing peripheral devices) are handled by the server.

CLIENT/SERVER TECHNOLOGY

The client/server (C/S) model, client/server computing, client/server technology, and client/server architecture all refer to a design model that can be thought of as applications running on a local area network (LAN). In very basic terms, you can picture the client requesting—and the server executing or in some way fulfilling—the request. The computers on the network are programmed to perform work efficiently by dividing up processing tasks among clients and servers. Figure 21.1 shows how a client/server model might be configured with a LAN. Note that several "clients" are depicted as user workstations.

When you think of client/server, you should think of a system that accentuates the users as the center of the work, with their interaction with data being the key concept. Although there are two elements working—the client and the server—it is the intent of the C/S model that users view it as one system. Indeed, the hope is that users are unaware of how the client/server network is performing its distributed processing, because it should have the look and feel of a unified system. In a peer-to-peer network, PCs can act as either the server or the client, depending on the requirements of the application.

Clients as Part of the C/S Model Using a LAN. When you see the term *client*, you might be tempted to think of people or users; for example, we speak of "clients of our consulting practice." In the C/S model, however, the term *client* refers not to people but to networked machines that are typical points of entry to the client/server system that is used by humans. Therefore, "clients" could be networked desktop computers, a workstation or laptop computers, or any way in which the user can enter the system.

Using a graphical user interface (GUI), individuals typically interface directly only with the client part. Client workstations use smaller programs that reside in the client to do "front-end" processing (as opposed to the "back-end" processing, mentioned below), including communicating with the user. If an application is called a client-based application, the application resides in a client computer and cannot be accessed by other users on the network. Note that client-based applications require separate installation on each workstation if the LAN has not purchased a site license.

File server *File server* is the term used to denote a computer on a LAN that stores on its hard disk the application programs and data files for all the clients on the network. "Server-based" applications are types of client processing capabilities that permit the user to request network applications (programs stored on a network server rather than on a user's computer) from the server. "Back-end" processing (such as a physical search of a database, for instance) will usually take place on a server. If a file server crashes, client-based applications would not be affected.

Designing a client/server network is a way to allocate resources in a LAN so as to distribute computing power among the computers in the network. Note, however, that it still makes sense to share some resources, which can be centralized in a file server. Client/server networks are proving to be a good, solid way to embody workgroup applications.

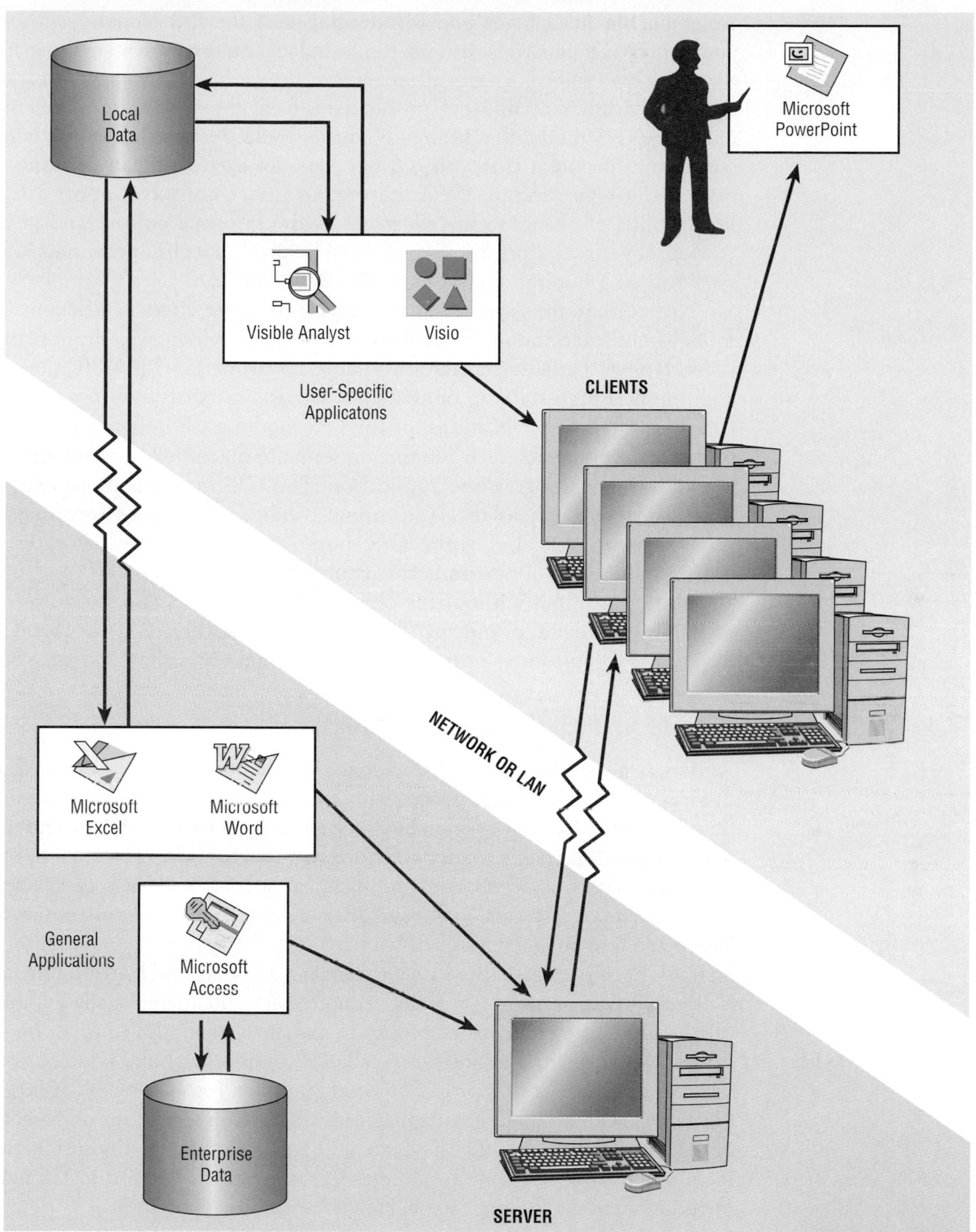

FIGURE 21.1
A client/server system configuration.

Print server A print server on a LAN is accessible to all workstations. In contrast to a file server, a print server is a PC dedicated to receiving and (temporarily) storing files to be printed. The specialized software that the print server uses first enables it to store print jobs and then helps it to manage the distribution of printing tasks to printers hooked into the network.

See the section on designing Web sites in Chapter 15 for more information about Web applications. Note that although it sounds as if they are the same as print servers and file servers, Web servers are software, not a combination of software and hardware like the print servers and file servers discussed before.

Weighing the Advantages and Disadvantages of the C/S Model. Although many companies jumped right in and requested client/server systems, we have found from the experience of early adopters that they are not always the best solution to an organization's computing problems. Often, the systems designer is asked to endorse a C/S model that is already in the works. Just as with any other corporate computing proposal that you did not have an active part in creating, you must review the plan carefully. Will the organization's culture support a C/S model? What kinds of changes must be made in the informal culture and in the formal work procedures before a C/S model can be used to its full potential? What should your role as a systems analyst be in this situation?

Notice that although lower processing costs are cited as a benefit of the C/S model—and even though there is some anecdotal evidence to support this claim—there is very little actual data available to prove it. Certainly there are well-documented high start-up or switch-over costs associated with a movement to a C/S architecture. Applications for the C/S model must be written as two separate software components, each running on separate machines, but they must appear as if they are operating as one application. The C/S model is more expensive than other options, such as the X terminal model, which uses terminals to access remote computers. Using the C/S model, however, affords greater computer power and greater opportunity to customize applications.

Notice also that without the organizational support and structure required to realize the potential of putting "decision-making authority at user level and therefore closer to customers," this benefit is meaningless.

TYPES OF DISTRIBUTED SYSTEMS NETWORKS

Although networks can be characterized by their shape or topology, they are also discussed in terms of their geographic coverage and the kinds of services they offer. Standard types of networks include a wide-area network (WAN) and a local area network (LAN). Local area networks are standard for linking local computers and/or terminals within a department, building, or several buildings of an organization. Wide-area networks can serve users over several miles or across entire continents. There are four main types of distributed systems networks: hierarchical, star, ring, and bus. Each requires different hardware and software and has different capabilities.

Networking is now technically, economically, and operationally feasible for small offices as well, and it provides a solution that analysts should consider for small businesses. Network kits are sometimes called "networks in a box," because everything to set up a small network is provided for in one price (usually below $500). An example of one such kit includes adapters for PCs, an eight-port hub, and a good amount of cabling. Sometimes, worksheets to help you plan the network installation are also included in these bundles. Several vendors are offering easy-to-install network kits, including Lynksys, 3Com, Dayna, Hewlett-Packard, and NetGear.

Networking provides advantages for small businesses when one considers the possibility for the sharing of software and the improvement of group work. Also, small businesses may be able to invest in a better quality printer, CD-ROM drives, modems, or other peripherals because they need fewer of them when users can share resources over the network. Most network kits exclude network operating system software, and this possibility allows the business flexibility in using whichever one fits the requirements of the small business.

Hierarchical Networks. In a basic hierarchical configuration, the host—a mainframe computer—controls all the other nodes that include minicomputers and microcomputers. Notice that computers on the same level do not communicate with each other. With this arrangement, large-scale computing problems are handled by the mainframe and lesser computing demands are handled on the correct level by either minicomputers or microcomputers.

Star Networks. Another popular configuration for distributed computing is the star network. A mainframe, PC, or workstation is designated the central node. As such, it communicates with the lesser nodes, but they cannot communicate directly with one another. A need for the microcomputers to communicate with one another would be met by one microcomputer sending data to the central node, which in turn would relay the data to a second microcomputer.

Ring Networks. Ring networks are another possibility for distributed computing. There is no central computer for a ring. Rather, its shape reminds us that all the nodes are of equal computing power. With the use of a ring network, all microcomputers can communicate directly with one another, passing along all the messages they read to their correct destinations on the ring.

Bus Configurations. Another type of network for distributed processing is the bus configuration. Bus configurations work well in close quarters, such as in a suite of offices where several different devices can be hooked together using a central cable. A bus configuration allows a great deal of change by permitting users to add or remove devices quite easily. In a bus configuration, the single, central cable serves as the only path.

NETWORK MODELING

Because networking has become so important, the systems designer needs to consider network design. Whether a systems designer gets involved with decisions about token rings or Ethernet networks—or whether he or she worries about hardware such as routers and bridges that must be in place when networks meet—the systems designer must always consider the logical design of networks. Network modeling enters here.

CASE tools such as Visible Analyst are not sufficient to help the systems designer with network modeling. It is possible to use some of a CASE tool's drawing capabilities, but forcing the data modeling tools to do network modeling just does not work. Therefore, we suggest using some sort of symbols (like the ones in Figure 21.2) to model the network. It is useful to have distinct symbols to distinguish among hubs, external networks, and workstations. It is also useful to adopt a convention for illustrating multiple networks and workstations.

Usually, a top-down approach is appropriate. The first step is to draw a network decomposition diagram that provides an overview of the system. Next, draw a hub connectivity diagram. Finally, explode the hub connectivity diagram to show the various workstations and how they are to be connected.

Drawing a Network Decomposition Diagram. We can illustrate drawing a network decomposition model by referring once again to the World's Trend Catalog Division example from earlier chapters. Start by drawing a circle at the top and labeling it "World's Trend Network." Now draw a number of circles on the level below, as shown in Figure 21.3. These circles represent hubs for the marketing division and each of the three order-entry and distribution centers (the U.S. Division, the Canadian Division, and the Mexican Division).

We can extend this drawing further by drawing another level. This time, we can add the workstations. For example, the Marketing Division has two workstations connected to it, whereas the U.S. Division has 33 workstations on its LAN

FIGURE 21.2
Use special symbols when drawing network decomposition and hub connectivity diagrams.

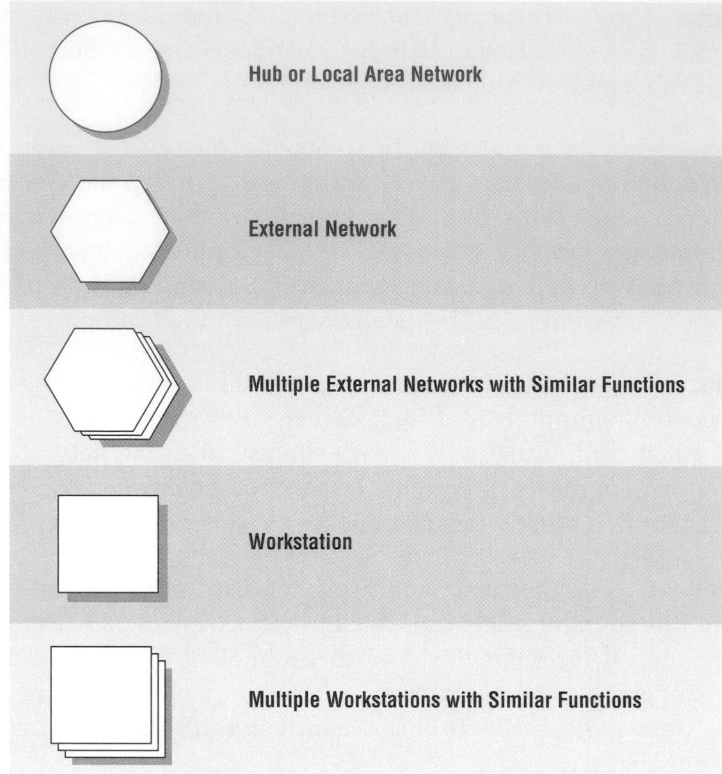

FIGURE 21.3
A network decomposition diagram for World's Trend.

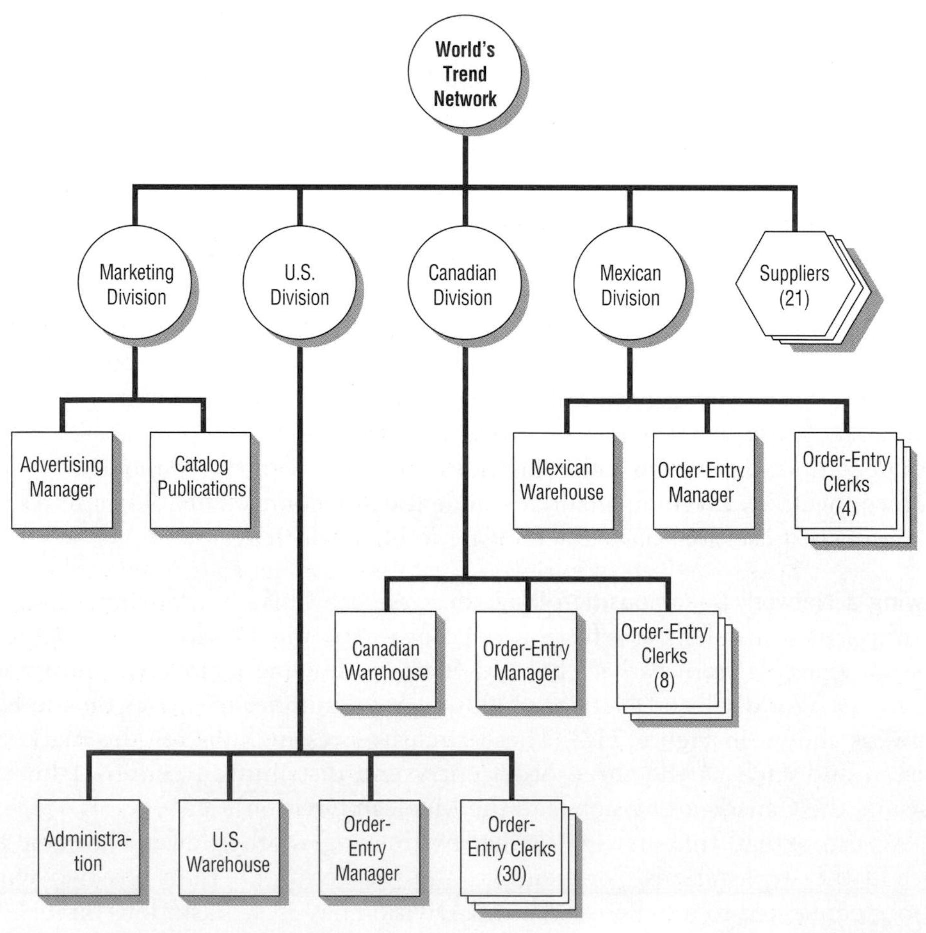

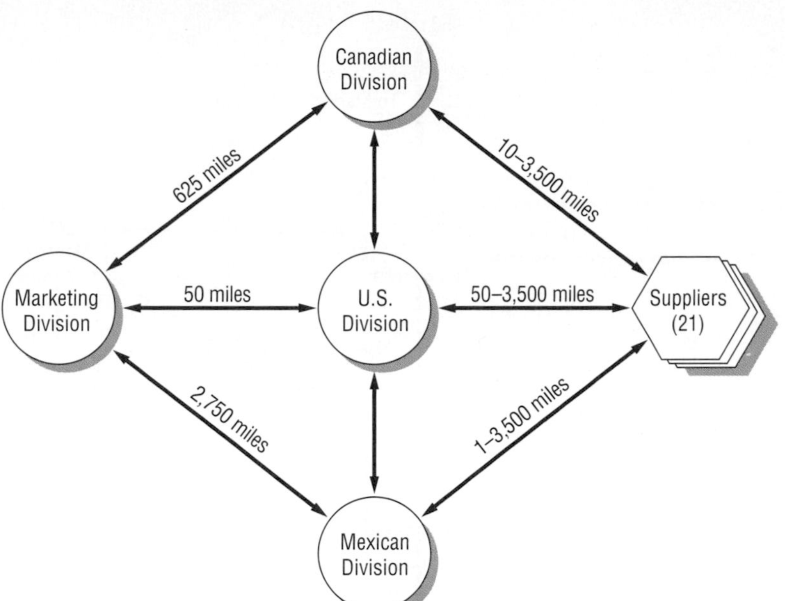

FIGURE 21.4
A hub connectivity diagram for World's Trend.

(Administration, the U.S. Warehouse, the Order-Entry Manager, and 30 Order-Entry Clerks). This network is simplified for the purpose of providing a readily understandable example.

Creating a Hub Connectivity Diagram. The hub connectivity diagram is useful for showing how the major hubs are connected. At World's Trend (see Figure 21.4), there are four major hubs that are all connected to one another. In addition, there are external hubs (suppliers) that need to be notified when inventory drops below a certain point, and so on. Each of the three country divisions are connected to the 21 suppliers; the Marketing Division, however, does not need to be connected to suppliers.

To produce an effective hub connectivity diagram, start by drawing all the hubs. Then experiment (perhaps sketching it first on a sheet of paper) to see which links are necessary. Once that is done, you can redraw the diagram so that it is attractive and communicates well to users.

Exploding the Hub Connectivity Diagram into a Workstation Connectivity Diagram. The purpose of network modeling is to show the connectivity of workstations in some detail. To do so, we explode the hub connectivity diagram. Figure 21.5 shows each of the 33 workstations for the U.S. Division and how they are to be connected.

Draw the diagrams for this level by examining the third level of the network decomposition diagram. Group items such as "Order-Entry Manager" and "Order-Entry Clerks" together, because you already recognize that they must be connected. Use a special symbol to show multiple workstations and indicate in parentheses the number of similar workstations. In our example, there are 30 order-entry clerks.

On the perimeter of the diagram, place workstations that must be connected to other hubs. In this way, it will be easier to represent these connections using arrows. Draw the external connections in a different color or use thicker arrows. External connections are usually long distance. For example, Administration is connected to the Marketing Division, which is 50 miles away, and also to the Canadian and Mexican Divisions. The warehouse needs to communicate directly with the Canadian and Mexican warehouses in case it is possible to obtain the merchandise from another warehouse. Order-Entry Managers and Clerks do not have to be connected to anyone outside their LAN.

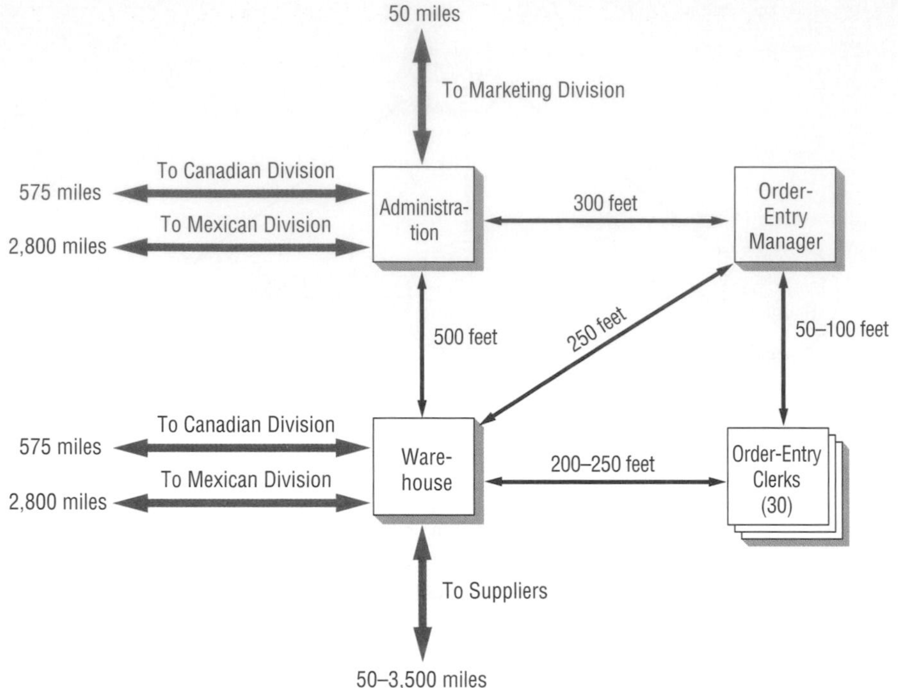

FIGURE 21.5
A workstation connectivity diagram for World's Trend.

Hub connectivity diagrams can be exploded to many levels. If doing so makes sense in your particular application, go ahead and draw it that way. There is no limit to the possible number of explosions.

Hub connectivity and workstation connectivity diagrams can be drawn using software packages as well. Although it may be difficult if you restrict yourself to using CASE tool software, it is relatively simple if you use flexible drag-and-drop software, such as Microsoft Visio. (See Figure 21.6.)

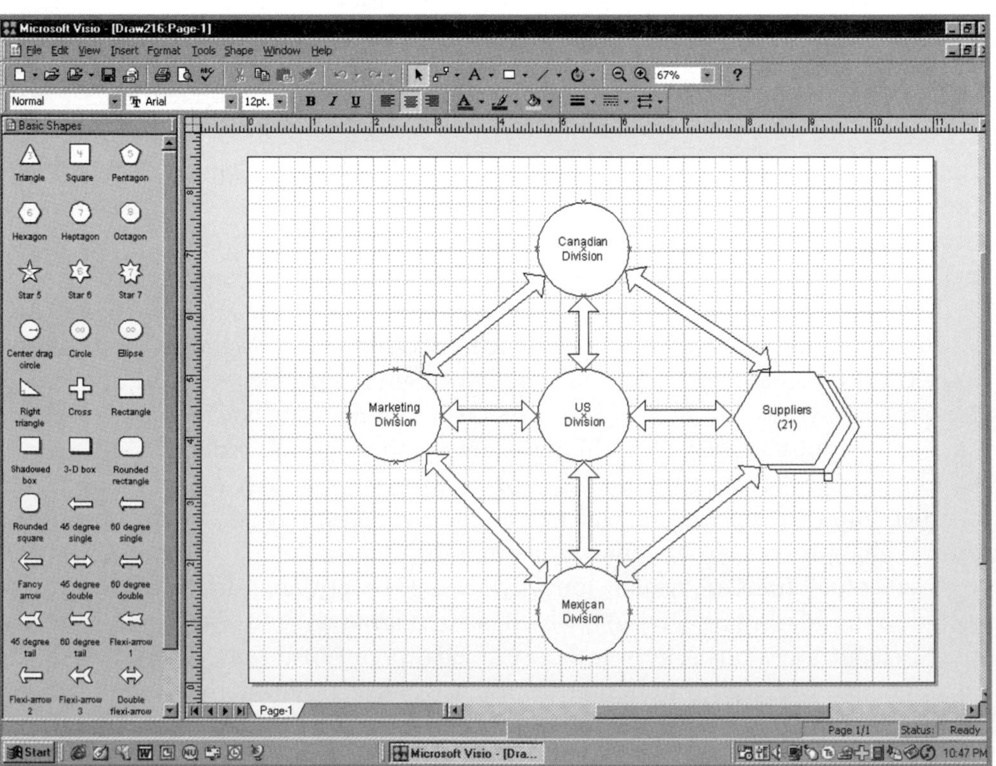

FIGURE 21.6
Analysts can draw hub connectivity diagrams using software like Microsoft Visio Professional.

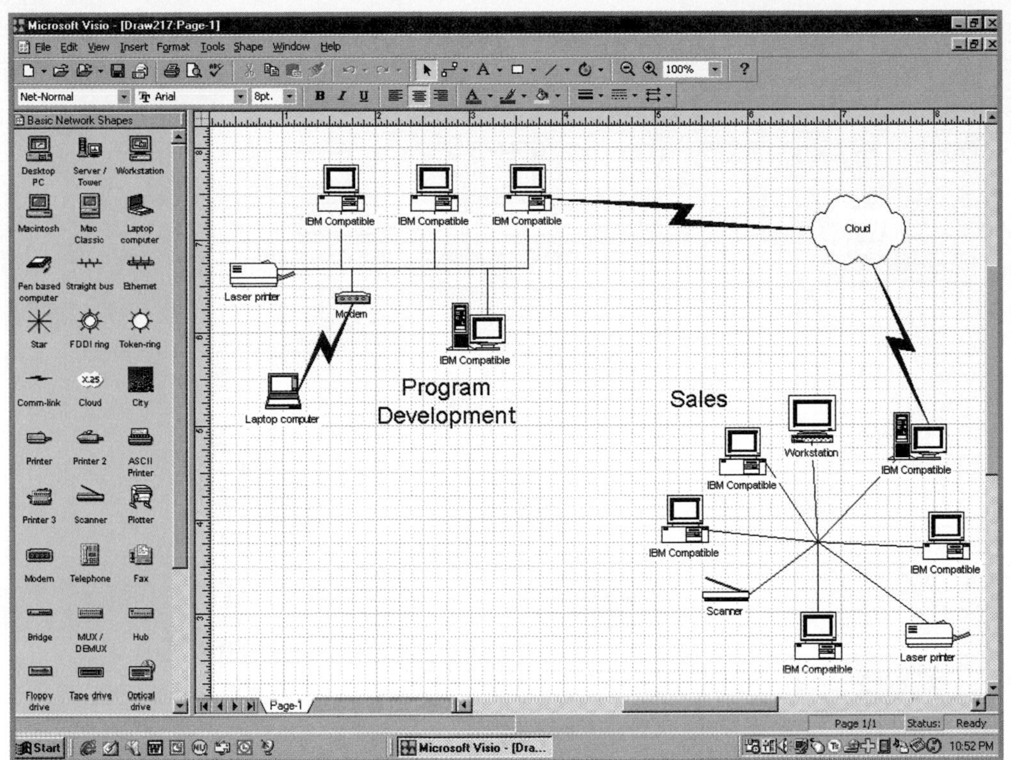

FIGURE 21.7

A more detailed diagram of two networks, drawn by selecting symbols from a template and dragging them to the piece of paper. This diagram was drawn using Visio Professional by Microsoft.

Visio even contains stencils that depict certain pieces of equipment if you wish to go into that much detail. Using specific stencils, an analyst could drag each symbol from the template to the piece of paper. Figure 21.7 shows how two networks are set up for a software manufacturer. The top part of the figure illustrates the program development sector's network of the organization, and the bottom part shows the sales department's network.

Additional detail is shown by zooming in on the program development network. In Figure 21.8, we can see that there are three PCs hooked up to a bus network that has a laser printer, a scanner, a modem for connection to laptop computers, and a server. Packages such as Visio are extremely useful to the systems analyst because they save time and enhance communication by using standard symbols.

GROUPWARE

Writing applications either for an entire organization or a solitary decision maker has also begun to change dramatically. Because much of the organization's work is actually accomplished in groups or teams, there is now a powerful movement afoot to develop special software called "groupware" that supports people who work together in the organization. Groupware takes advantage of the potential synergies and power available from networked PCs in LANs, WANs, or the Web. Groupware products can help group members to schedule and attend meetings, share data, create and analyze documents, communicate in unstructured ways with one another via email, hold group conferences, do image management on the departmental level, and manage and monitor workflow.

We are still far from an integrated approach to groupware, which means that often developers will create a product that supports some combination of, but not all, the foregoing functions. At the heart of most groupware is a useful, easy-to-use, flexible email system. If asked to develop workgroup applications, you should first assess whether there is currently a usable email system in place upon which to build other functions.

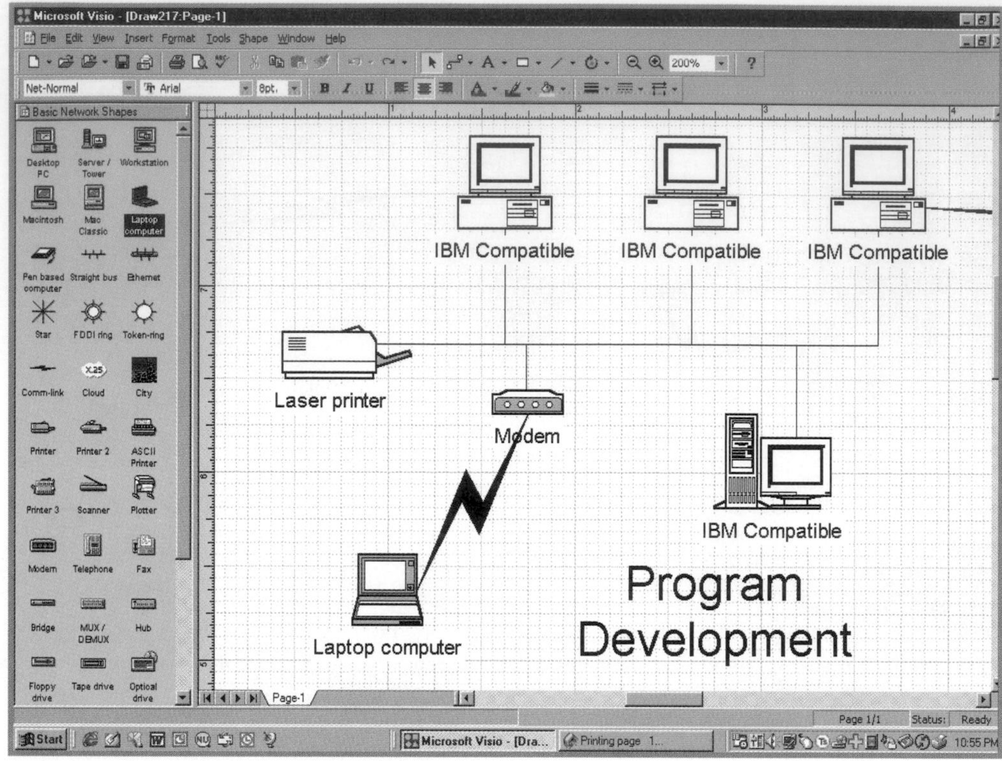

FIGURE 21.8

A closer view of the network shows what type of equipment is connected to the program development group.

You can see from the wide range of possible capabilities that there is no standard definition for what comprises groupware. In fact, each software company seems to be putting forward a different vision of what kind of support groups need to function best.

The product that has defined groupware for many users is Lotus Notes, which Lotus corporate officers aptly refer to as "the Network GUI." Notes features a document-oriented database that has a wide range of options and far fewer restrictions than traditional databases. Group users working with Notes will find that its fields can handle numerous fonts, text colors, scanned images, hypertext links to other documents in Notes, and even push buttons that are actually Notes macros. Using a simple, straightforward, no-frills interface, Notes helps novices get up and running quickly. All Notes databases use an identical interface, so that once a group member has learned one database, it is easy to use a new one.

In addition, Notes has been called "mail-integrated" and provides flexible, wide-ranging mail capabilities that support the way group members work. It also lets them use mail in some novel way (for example, to send Notes messages to a database).

Many organizations experience difficulty in getting networked applications to run smoothly with adequate backups and security. In other words, many network applications purporting to be groupware do not deliver what has been promised because of both initial faulty development and subsequent poor implementation and backup. Lotus Notes has successfully addressed this pitfall by using a breakthrough replication technology that enables users to keep many copies of a database (whether they are on LANs or remote workstations) and to synchronize them by using background dial-up phone connections from time to time. Lotus Notes has been criticized as a very expensive, monolithic option for supporting group work, but it has attracted a large following.

Although most software companies agree to the importance of supporting group work, they differ in their visions of what groups and individuals need for support. For example, Novell currently offers groupware features such as email, messaging, and calendaring in a networked environment. In the future, WordPerfect envisions supporting users by enabling them to pull together a report containing objects gathered from anywhere on the network, regardless of what systems were used to create them. Thus, a report could have text from the annual report, bar charts from last week's sales meeting, and scanned images of hand-drawn sketches that portray the latest departmental brainstorming session. In such a case, the computer would search out the various pieces, and the software would coordinate them for use in one document.

Another approach to developing groupware has been taken by Microsoft, which is giving rudimentary workgroup capability to products such as Windows for Workgroups and Windows NT. New graphical user interfaces include an "object store" similar to the one currently available in Lotus Notes so that Microsoft operating systems will have built-in workgroup capabilities. In the short term, Microsoft is working with smaller developers and consultants to develop workgroup application solutions. This bottom-up approach seems to be showing successful results as well. Different groupware offers different forms of support.

Advantages of Distributed Systems. Distributed systems allow the storage of data where they are not "in the way" of any online real-time transactions. For example, response time on inquiries might be improved if not all records need to be searched through before a response is made. In addition, not all data are needed by all users all the time, so they can be stored in less expensive media at a different site and only accessed when needed.

Use of distributed systems can also lower equipment costs, because not all parts of the system need to be able to perform all functions. Some capabilities, such as processing and storage, can be shared.

Distributed systems can also help lower costs by permitting flexibility in the choice of manufacturer, because the whole focus of networks is on communicating between nodes and manufacturers make compatible components. This compatibility allows the user to shop for price as well as for function. Furthermore, distributed systems can be less expensive initially than large systems because it is feasible to plan for expansion without actually having to buy hardware at the time the system is implemented. Notice that developing corporate intranets is a proactive way to network organizational members, a way that can also serve as a means for cutting down on problematical aspects of the Internet (such as aimless Web surfing during corporate time or possible security breakdowns caused by lack of firewalls) while at the same time supporting group work with useful applications. Extranets formed with suppliers and other important partners are also excellent ways of demonstrating that a business is outward looking and accessible. Advantages of distributed systems are given in Figure 21.9.

Advantages of Distributed Systems

- Allow data storage out of the way of online, real-time transactions
- Allow less expensive media for data storage when all data are not needed all the time by all users
- Lower equipment cost because not all system parts need to perform all functions
- Lower equipment cost by permitting flexibility in choice of manufacturer
- Less expensive than large systems initially because expansion can be planned for without actually purchasing hardware

FIGURE 21.9
There are five main advantages to creating distributed systems.

FIGURE 21.10

There are four chief disadvantages to creating distributed systems.

Disdvantages of Distributed Systems
• Difficulty in achieving a reliable system • Security concerns increase commensurately when more individuals have access to the system • Analysts must emphasize the network and the interactions it provides and deemphasize the power of subsystems • Choosing the wrong level of computing to support (i.e., individual instead of department, department instead of branch)

Disadvantages of Distributed Systems. Distributed systems pose some unique problems that centralized computer systems do not. The analyst needs to weigh these problems against the advantages just presented and to raise them with the concerned business as well.

The first problem is that of network reliability. To make a network an asset rather than a liability, it must be possible to transmit, receive, process, and store data reliably. If there are too many problems with system reliability, the system will be abandoned.

Distributing greater computing power to individuals increases the threat to security because of widespread access. The need for secret passwords, secure computer rooms, and adequate security training of personnel are all concerns that multiply when distributed systems are implemented.

Systems analysts creating distributed systems need to focus on the network itself or on the synergistic aspect of distributed systems. Their power resides in their ability to interact as user work groups share data. If the relationship between subsystems is ignored or deemphasized, you are creating more problems than you are solving. Disadvantages of distributed systems are listed in Figure 21.10.

TRAINING USERS

Systems analysts engage in an educational process with users that is called training. Throughout the systems development life cycle, the user has been involved so that by now the analyst should possess an accurate assessment of the users who must be trained. As we have seen, information centers retain trainers of their own.

In the implementation of large projects, the analyst will often be managing the training rather than be personally involved in it. One of the most prized assets the analyst can bring to any training situation is the ability to see the system from the user's viewpoint. The analyst must never forget what it is like to face a new system. Those recollections can help analysts empathize with users and facilitate their training.

TRAINING STRATEGIES

Training strategies are determined by who is being trained and who will train them. The analyst will want to ensure that anyone whose work is affected by the new information system is properly trained by the appropriate trainer.

Who to Train. All people who will have secondary or primary use of the system must be trained. They include everyone from data-entry personnel to those who will use output to make decisions without personally using a computer. The amount of training a system requires thus depends on how much someone's job will change because of the new system.

You must ensure that users of different skill levels and job interests are separated. It is certain trouble to include novices in the same training sessions as experts, because novices are quickly lost and experts are rapidly bored with basics. Both groups are then lost.

People Who Train Users. For a large project, many different trainers may be used depending on how many users must be trained and who they are. Possible training sources include the following:

1. Vendors.
2. Systems analysts.
3. External paid trainers.
4. In-house trainers.
5. Other system users.

This list gives just a few of the options the analyst has in planning for and providing training.

Large vendors often provide offsite, one- or two-day training sessions on their equipment as part of the service benefits offered when corporations purchase expensive software. These sessions include both lectures and hands-on training in a focused environment.

Because systems analysts know the organization's people and the system, they can often provide good training. The use of analysts for training purposes depends on their availability, because they also are expected to oversee the complete implementation process.

External paid trainers are sometimes brought into the organization to help with training. They may have broad experience in teaching people how to use a variety of computers, but they may not give the hands-on training that is needed for some users. In addition, they may not be able to custom-tailor their presentations enough to make them meaningful to users.

Full-time, in-house trainers are usually familiar with personnel and can tailor materials to their needs. One of the drawbacks of in-house trainers is that they may possess expertise in areas other than information systems and may therefore lack the depth of technical expertise that users require.

It is also possible to have any of these trainers train a small group of people from each functional area that will be using the new information system. They in turn can then be used to train the remaining users. This approach can work well if the original trainees still have access to materials and trainers as resources when they themselves are providing training. Otherwise, it might degenerate into a trial-and-error situation rather than a structured one.

GUIDELINES FOR TRAINING

The analyst has four major guidelines for setting up training. They are (1) establishing measurable objectives, (2) using appropriate training methods, (3) selecting suitable training sites, and (4) employing understandable training materials.

Training Objectives. Who is being trained in large part dictates the training objectives. Training objectives for each group must be spelled out clearly. Well-defined objectives are of enormous help in letting trainees know what is expected of them. In addition, objectives allow evaluation of training when it is complete. For example, operators must know such basics as turning on the machine, what to do when common errors occur, basic troubleshooting, and how to end an entry.

Training Methods. Each user and operator will need slightly different training. To some extent, their jobs determine what they need to know, and their personalities, experience, and background determine how they learn best. Some users learn best by seeing, others by hearing, and still others by doing. Because it is usually not possible to customize training for an individual, a combination of methods is often the best way to proceed. That way, most users are reached through one method or another.

Methods for those who learn best by seeing include demonstrations of equipment and exposure to training manuals. Those who learn best by hearing will benefit from lectures about procedures, discussions, and question-and-answer sessions among trainers and trainees. Those who learn best by doing need hands-on experience with new equipment. For jobs such as that of computer operator, hands-on experience is essential, whereas a quality assurance manager for a production line may only need to see output, learn how to interpret it, and know when it is scheduled to arrive.

Training Sites. Training takes place in many different locations, some of which are more conducive to learning than others. Large computer vendors provide special offsite locations where operable equipment is maintained free of charge. Their trainers offer hands-on experience as well as seminars in a setting that allows users to concentrate on learning the new system. One of the disadvantages of offsite training is that users are away from the organizational context within which they must eventually exist.

Onsite training within the users' organization is also possible with several different kinds of trainers. The advantage is that users see the equipment placed as it will be when it is fully operational. A serious disadvantage is that trainees often feel guilty about not fulfilling their regular job duties if they remain onsite for training. Thus, full concentration on training may not be possible.

Offsite training sites are also available for a fee through consultants and vendors. Training sites can be set up in places with rented meeting space, such as a hotel, or may even be permanent facilities maintained by the trainers. These arrangements allow workers to be free from regular job demands, but they may not provide equipment for hands-on training.

Training Materials. In planning for the training of users, systems analysts must realize the importance of well-prepared training materials. These materials include training manuals; training cases, in which users are assigned to work through a case that incorporates most of the commonly encountered interactions with the system; and prototypes and mock-ups of output. Users of larger systems will sometimes be able to train on elaborate Web-based simulations or software that is identical to what is being written or purchased. Most packaged software provides online tutorials that illustrate basic functions, and vendors may maintain Web sites that feature pages devoted to FAQ, the answers to

FIGURE 21.11

Appropriate training objectives, methods, sites, and materials are contingent on many factors.

Elements	Relevant Factors
Training Objectives	Depend on requirements of user's job
Training Methods	Depend on user's job, personality, background, and experience; use combination of lecture, demonstration, hands on, and study
Training Sites	Depend on training objectives, cost, availability; free vendor sites with operable equipment; in-house installation; rented facilities
Training Materials	Depend on user's needs; operating manuals, cases, prototypes of equipments and output; online tutorials

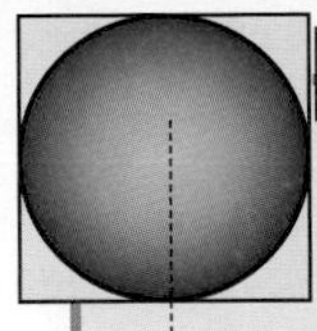

CONSULTING OPPORTUNITY 21.1

YOU CAN LEAD A FISH TO WATER . . . BUT YOU CAN'T MAKE IT DRINK

Sam Monroe, Belle Uga, Wally Ide, and you make up a four-member systems analysis team that is developing an information system to help managers monitor and control water temperature, the number of fish released, and other factors at a large commercial fish hatchery. (They were last seen in Consulting Opportunity 8.3, when they asked you, as their fourth member, to help solve a problem involving the timely delivery of a system prototype.)

With your input, the team successfully turned the tide of the earlier dilemma, and the project has continued. Now you are discussing the training that you have begun to undertake for managers and other systems users. Due to some scheduling difficulties, you have decided to cut down on the number of different training sessions offered, which has resulted in primary and secondary users being in the same training sessions in some instances.

Lauric Hook, one of the operators who is being trained, has been in the same training "tank" with Wade Boot, one of the managers with whom you have been working. Both Laurie and Wade have come to the team privately with different concerns.

Wade told you, "I'm mad that I have to type in my own data in the sessions. The Mississippi will freeze solid before I ever do that on my job. I've got to know *when* to expect output and how to interpret it when it comes. I'm not spending time in training sessions if I can't get that."

Laurie, who shares training sessions with Wade, also complained to your group. "We should be getting more hands-on training. All we hear is a bunch of lectures. It's like school. Not only that, but the managers in the group like to spin these 'fish stories' about what happened to them with the old system. It's boring. I want to know how to operate the thing. It's bait and switch, if you ask me. I'm not learning what you said I would, and besides, with all those bosses in there, I feel like a fish out of water."

What problems are occurring with the training sessions? How can they be addressed, given the scheduling constraints mentioned? What basic advice on setting up training sessions did your team ignore?

which can be downloaded and printed. Changes to manuals can also be gleaned from many vendors' Web sites.

Because the user's understanding of the system depends on them, training materials must be clearly written. Training materials should also be well indexed, written for the correct audience with a minimum of jargon, and available to everyone who needs them. A summary of considerations for training objectives, methods, sites, and materials is provided in Figure 21.11.

CONVERSION

A third approach to implementation is physically converting the old information system to the new or modified one. There are many conversion strategies available to analysts and also a contingency approach that takes into account several organizational variables in deciding which conversion strategy to use. There is no single best way to proceed with conversion. The importance of adequate planning and scheduling of conversion (which often takes many weeks), file backup, and adequate security cannot be overemphasized.

CONVERSION STRATEGIES

The five strategies for converting from the old system to the new are given in Figure 21.12:

1. Direct changeover.
2. Parallel conversion.
3. Phased or gradual conversion.
4. Modular prototype conversion.
5. Distributed conversion.

Each of the five conversion approaches is described separately in the upcoming subsections.

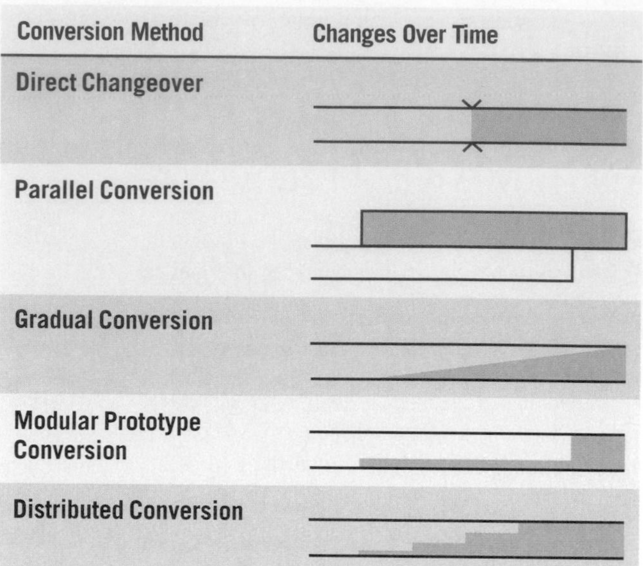

FIGURE 21.12
Five conversion strategies for information systems.

Direct Changeover. Conversion by direct changeover means that on a specified date, the old system is dropped and the new system is put into use. Direct changeover can only be successful if extensive testing is done beforehand, and it works best when some delays in processing can be tolerated. Sometimes, direct changeover is done in response to a government mandate. An advantage of direct changeover is that users have no possibility of using the old system rather than the new one. Adaptation is a necessity.

Direct changeover is considered a risky approach to conversion, and its disadvantages are numerous. For instance, long delays might ensue if errors occur, because there is no alternate way to accomplish processing. In addition, users may resent being forced into using an unfamiliar system without recourse. Finally, there is no adequate way to compare new results with old.

Parallel Conversion. Parallel conversion refers to running the old system and the new system at the same time, in parallel. It is the most frequently used conversion approach, but its popularity may be in decline because it works best when a computerized system replaces a manual one. Both systems are run simultaneously for a specified period of time, and the reliability of results is examined. When the same results can be gained over time, the new system is put into use and the old one is stopped.

The advantages of running both systems in parallel include the possibility of checking new data against old data to catch any errors in processing in the new system. Parallel processing also offers a feeling of security to users, who are not forced to make an abrupt change to the new system.

There are many disadvantages to parallel conversion. They include the cost of running two systems at the same time and the burden on employees of virtually doubling their workload during conversion. Another disadvantage is that unless the system being replaced is a manual one, it is difficult to make comparisons between outputs of the new system and the old one. Supposedly, the new system was created to improve on the old one. Therefore, outputs from the systems should differ. Finally, it is understandable that employees who are faced with a choice between two systems will continue to use the old one because of their familiarity with it.

Gradual Conversion. Gradual or phased conversion attempts to combine the best features of the earlier two plans, without incurring all the risks. In this plan, the volume of transactions handled by the new system is gradually increased as the system is phased in. The advantages of this approach include allowing users to get involved with the system gradually and the possibility of detecting and recovering from errors without a lot of down time. Disadvantages of gradual conversion include taking too long to get the new system in place and its inappropriateness for conversion of small, uncomplicated systems.

Modular Prototype Conversion. Modular prototype conversion uses the building of modular, operational prototypes (as discussed in Chapter 8) to change from old systems to new in a gradual manner. As each module is modified and accepted, it is put into use. One advantage is that each module is thoroughly tested before being used. Another advantage is that users are familiar with each module as it becomes operational.

Prototyping is often not feasible, which automatically rules out this approach for many conversions. Another disadvantage is that special attention must be paid to interfaces so that the modules being built actually work as a system.

Distributed Conversion. Distributed conversion refers to a situation in which many installations of the same system are contemplated, as is the case in banking or in franchises such as restaurants or clothing stores. One entire conversion is done (with any of the four approaches considered already) at one site. When that conversion is successfully completed, other conversions are done for other sites.

An advantage of distributed conversion is that problems can be detected and contained rather than inflicted simultaneously on all sites. A disadvantage is that even when one conversion is successful, each site will have its own peculiarities to work through, and they must be handled accordingly.

A contingency approach to deciding on a conversion strategy is recommended; that is, the analyst considers many factors (including the wishes of clients) in choosing a conversion strategy. Obviously, no particular conversion approach is equally suitable for every system implementation.

SECURITY CONCERNS FOR TRADITIONAL AND WEB-BASED SYSTEMS

Security of computer facilities, stored data, and the information generated is part of a successful conversion. Recognition of the need for security is a natural outgrowth of the belief that information is a key organizational resource, as discussed in Chapter 1. With increasingly complex transactions and many innovative exchanges, the Web has brought heightened security concerns to the IS professional's world.

It is useful to think of security of systems, data, and information on an imaginary continuum from totally secure to totally open. Although there is no such thing as a totally secure system, the actions analysts and users take are meant to move systems toward the secure end of the continuum by lessening the system's vulnerability. It should be noted that as more people in the organization gain greater computer power, gain access to the Web, or hook up with intranets and extranets, security becomes increasingly difficult and complex. Sometimes, organizations will hire a security consultant to work with the systems analyst when security is crucial to successful operations.

Security is the responsibility of all those who come into contact with the system and is only as good as the most lax behavior or policy in the organization. Security has three interrelated aspects: physical, logical, and behavioral. All three must work together if the quality of security is to remain high.

Physical Security. Physical security refers to securing the computer facility, its equipment, and software through physical means. It can include controlling access to the computer room by means of machine-readable badges or a human sign-in/sign-out system, using closed-circuit television cameras to monitor computer areas, and backing up data frequently and storing backups in a fireproof, waterproof area.

In addition, small computer equipment should be secured so that a typical user cannot move it, and it should be guaranteed uninterrupted power. Alarms that notify appropriate people of fire, flood, or unauthorized human intrusion must be in working order at all times.

Decisions about physical security should be made when the analyst is planning for computer facilities and equipment purchase. Obviously, physical security can be much tighter if anticipated in advance of actual installation and if computer rooms are specially equipped for security when they are constructed rather than outfitted as an afterthought.

Logical Security. Logical security refers to logical controls within software itself. The logical controls familiar to most users are passwords or authorization codes of some sort. When used, they permit the user with the correct password to enter the system or a particular part of a database.

Passwords, however, are treated cavalierly in many organizations. Employees have been overheard yelling a password across crowded offices, taping passwords to their terminals, and sharing personal passwords with authorized employees who have forgotten their own.

Special encryption software has been developed to protect commercial transactions on the Web, and business transactions are proliferating. Internet fraud is also up sharply, however, with few authorities trained in catching Internet criminals and a "Wild West, or Last Frontier" mentality clearly at play in many instances when authorities have been able to apprehend Web criminals.

One way for networks to cut down on the risk of exposure to security challenges from the outside world is to build what is called a firewall or a firewall system. Briefly, a firewall constructs a barricade between an internal organization's network and an external (inter)network, such as the Internet. The internal network is assumed to be trustworthy and secure, whereas the Internet is not. Firewalls are intended to prevent communication into or out of the network that has not been authorized and that is not wanted. A firewall system is not a perfect remedy for organization and Internet security; it is an additional layer of security that is now widely accepted. There is still no fully integrated way to address security problems with internal and external networks, but they do deserve analysts' attention when planning any new or improved systems.

Logical and physical controls are important but clearly not enough to provide adequate security. Behavioral changes are also necessary.

Behavioral Security. The behavioral expectations of an organization are encoded in its policy manuals and even on signs posted on bulletin boards, as we saw in Chapter 4. The behavior that organization members internalize, however, is also critical to the success of security efforts. (Note that one reason firewalls are not attack-proof is because many attacks to information systems come from within the organization.)

Security can begin with the screening of employees who will eventually have access to computers, data, and information to ensure that their interests are consistent with the organization's interests and that they fully understand the importance of carrying through on security procedures. Policies regarding security must be written, distributed, and updated so that employees are fully aware of expectations and responsibilities. It is typically that the systems analyst will first have contact with the behavioral aspects of security. Some organizations have written rules or policies prohibiting employees from surfing the Web during work hours or even prohibiting Web surfing at all, if company equipment is involved. Other corporations use software locks to limit access to Web sites that are judged to be objectionable in the workplace, such as game, gambling, or pornographic sites.

Part of the behavioral facet of security is monitoring behavior at irregular intervals to ascertain that proper procedures are being followed and to correct any behaviors that may have eroded with time. Having the system log the number of unsuccessful sign-on attempts of users is one way to monitor whether unauthorized users are attempting to sign on to the system. Periodic and frequent inventorying of equipment and software is desirable. In addition, unusually long sessions or atypical after-hours access to the system should be examined.

Employees should clearly understand what is expected of them, what is prohibited, and the extent of their rights and responsibilities. You should disclose all monitoring that is being done or that is being contemplated, and you should supply the rationale behind it. Such disclosure should include the use of video cameras, and software monitoring.

Output generated by the system must be recognized for its potential to put the organization at risk in some circumstances. Controls for output include screens that can only be accessed via password, the classification of information (that is, to whom it can be distributed and when), and secure storage of printed and magnetically stored documents.

In some cases, provision for shredding documents that are classified or proprietary must be made. Shredding or pulverization services can be contracted from an outside firm that, for a fee, will shred magnetic media, printer cartridges, and paper. A large corporation may shred upwards of 76,000 pounds of output in a variety of media annually.

SPECIAL SECURITY CONSIDERATIONS FOR ECOMMERCE

It is well known that intruders can violate the integrity of any computer system. As an analyst, you need to take a series of precautions to protect the computer network from both internal and external Web security threats. A number of actions and products can help you:

1. Virus protection software.
2. Email filtering products (such as Symantec's Mail) that provide policy-based email and email attachment scanning and filtering to protect companies against both incoming and outgoing email. Incoming scanning protects against spam (unsolicited email such as advertising) attacks, and outgoing scanning protects against the loss of proprietary information.
3. URL filtering products (such as Symantec's I-Gear) that provide employees with access to the Web by user, by groups of users, by computers, by the time, or by the day of the week.
4. Firewalls, gateways, and virtual private networks that prevent hackers from gaining backdoor access to a corporate network.
5. Intrusion detection products (such as Intruder Alert or Net Prowler) that continually monitor usage, provide messages and reports, and suggest actions to take.
6. Vulnerability management products (such as NetRecon and Symantec Expert) that assess the potential risks in a system and discover and report vulnerabilities. Some products correlate the vulnerabilities to make it easier to find the root cause of the security breach. Risk cannot be eliminated, but this software can help manage the risk by balancing security risk to the financial bottom line.

7. Security technologies such as secure socket layering (SSL) for authentication.
8. Encryption technologies such as secure electronic translation (SET).
9. Public key infrastructure (PKI) use and obtaining a digital certificate from a company such as Verisign. Use of digital certificates ensures that the reported sender of the message is really the company that sent the message.

PRIVACY CONSIDERATIONS FOR ECOMMERCE

The other side of security is privacy. To make your Web site more secure, you must ask the user or customer to give up some privacy.

As a Web site designer, you will recognize that the company for which you design exercises a great deal of power over the data their customers are providing to them. The same tenets of ethical and legal behavior apply to Web site design as to the design of any traditional application that accepts personal data from customers. The Web, however, allows the data to be collected faster and allows different data to be collected (such as the browsing habits of the customer). In general, information technology makes it possible to store more data in data warehouses, process that data, and distribute the data more widely.

Every company for which you design an ecommerce application should adopt a privacy policy. Here are some guidelines:

1. Start with a corporate policy on privacy. Make sure it is prominently displayed on the Web site so that all customers can access the policy whenever they complete a transaction.
2. Only ask for information the application requires to complete the transaction at hand. Is it necessary to the transaction to ask a person's age or gender?
3. Make it optional for customers to fill out personal information on the Web site. Some customers do not mind receiving targeted messages, but you should always give customers an opportunity to maintain the confidentiality of their personal data by not responding.
4. Use sources that allow you to obtain anonymous information about classes of customers. For example, Engage is a company that offers audience profiling technology and technology solutions for management of advertisements, their targeting, and their delivery. It does so by maintaining a dynamic database of consumer profiles without linking them to individuals, thereby respecting customers' rights to privacy.
5. Be ethical. Avoid the latest cheap trick that permits your client to gather information about the customer in highly suspect ways. Tricks like screen scraping (capturing remotely what is on a customer's screen) and email cookie grabbing are clear violations to privacy and may prove to be illegal as well.

A coordinated policy of security and privacy is essential. It is essential to establish these policies and adhere to them when implementing an ecommerce application.

OTHER CONVERSION CONSIDERATIONS

Conversion also entails other details for the analyst, which include the following:

1. Ordering equipment (up to three months ahead of planned conversion).
2. Ordering any necessary materials that are externally supplied to the information system, such as toner cartridges, paper, preprinted forms, and magnetic media.
3. Appointing a manager to supervise or personally supervising the preparation of the installation site.
4. Planning, scheduling, and supervising programmers and data-entry personnel who must convert all relevant files and databases.

For many implementations, your chief role will be accurately estimating the time needed for each activity, appointing people to manage each subproject, and coordinating their work. For smaller projects, you will do much of the conversion work on your own. Many of the project management techniques learned in Chapter 3, such as Gantt charts, PERT, and successfully communicating with team members, are useful for planning and controlling implementation.

ORGANIZATIONAL METAPHORS AND THEIR RELATIONSHIP TO SUCCESSFUL SYSTEMS

Be aware of organizational metaphors when you attempt to implement a system you have just developed. Our recent exploratory research has suggested that the success or failure of a system might have something to do with the metaphors used by organizational members.

When people in the organization describe the company as a zoo, you can infer that the atmosphere is chaotic; if it is described as a machine, everything is working in an orderly fashion. When the predominant metaphor is war, journey, or jungle, the environment is chaotic, like the zoo. The war and journey metaphors are oriented toward an organization goal, however, whereas the zoo and jungle metaphors are not.

In addition to the machine, metaphors such as society, family, and the game all signify order and rules. Although the machine and game metaphors are goal oriented, the society and zoo metaphors do not stress the company's goal but instead allow the individuals in the corporation to set their own standards and rewards. Another metaphor, the organism, appears balanced between order and chaos, corporate and individual goals.

Our research suggests that the success or failure of a system may have something to do with the predominant metaphor. Figure 21.13 shows that a traditional MIS will tend to succeed when the predominant metaphor is society, machine, or family, but it might not succeed if the metaphor is war or jungle (two chaotic metaphors). Notice, however, that competitive systems will most likely succeed if the metaphor is war.

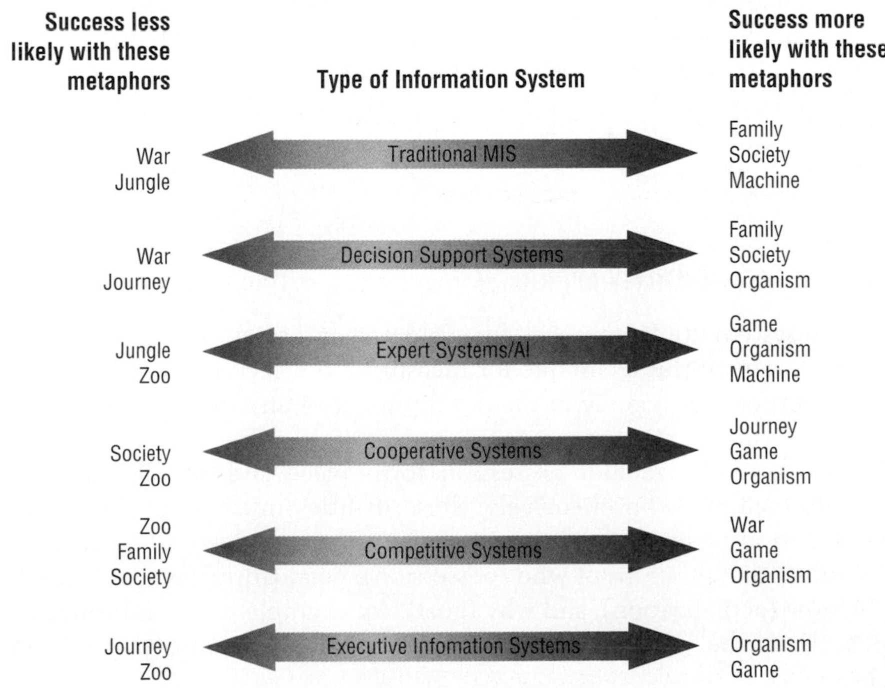

FIGURE 21.13
Organizational metaphors may contribute to the success or failure of an information system.

Positive metaphors appear to be the game, organism, and machine. Negative metaphors appear to be the jungle and the zoo. The others (journey, war, society, and family) show mixed success depending on the type of information system being developed. More research needs to be done in this area. In the meantime, the systems analyst should be aware that metaphors communicated in interviews could be meaningful and may even be a contributing factor toward the success of the information system implementation.

EVALUATION

Throughout the systems development life cycle, the analyst, management, and users have been evaluating the evolving information systems and networks to give feedback for their eventual improvement. Evaluation is also called for following system implementation.

EVALUATION TECHNIQUES

In recognition that the ongoing evaluation of information systems and networks is important, many evaluation techniques have been devised. These techniques include cost-benefit analysis (as discussed in Chapter 13), models that attempt to estimate the value of a decision based on the effects of revised information using information theory, simulation or Bayesian statistics, user evaluations that emphasize implementation problems and user involvement, and information system utility approaches that examine the properties of information.

Each type of evaluation serves a different purpose and has inherent drawbacks. Cost-benefit analysis may be difficult to apply, because information systems provide information about objectives for the first time, making it impossible to compare performance before and after implementation of the system or distributed network. The revised decision evaluation approach presents difficulty, because all variables involved with the design, development, and implementation of the information system cannot be calculated or quantified. The user involvement approach yields some insight for new projects by providing a checklist of potentially dysfunctional behavior by various organizational members, but it stresses implementation over other aspects of IS design. The information system utility approach to evaluation can be more comprehensive than the others if it is expanded and systematically applied.

THE INFORMATION SYSTEM UTILITY APPROACH

The information system utility approach for evaluating information systems can be a comprehensive and fruitful technique for measuring the success of a developed system. It also can serve as a guide in the development of any future projects the analyst might undertake.

Utilities of information include possession, form, place, and time. To evaluate the information system comprehensively, these utilities must be expanded to include actualization utility and goal utility. Then the utilities can be seen to address adequately the questions of who (possession), what (form), where (place), when (time), how (actualization), and why (goal). An example of this information utility approach can be seen in the evaluation of a blood inventory system in Figure 21.14.

Information Systems Modules	Form Utility	Time Utility	Place Utility	Possession Utility	Actualization Utility	Goal Utility
Inventory Lists Good	Good. Acronyms used were the same as shipping codes. As systems grew, too much information was presented; this overload called for summary information.	Good. Reports were received at least one hour before scheduled shipments on a daily basis.	Good. Inventory lists were printed at the regional blood center. Lists were delivered to hospitals with the current shipments.	Good. The same people who originally kept manual records received these reports.	Good. Implementation was easy because hospitals found the inventory lists to be extremely useful.	Good. Information about the location of particular units was made available.
Management Summary Reports Good	Good. Summary report was designed to exact format specifications of manual summary reports developed by the blood administrator for city hospitals.	Good. Same as listings.	Good. Summary reports were printed at the center where they were needed.	Good. Blood administrators who originally kept manual reports for tion.	Good. Blood administrators participated in the design of the reports.	Good. Summary reports helped reduce outdating and prevent shortages.
Short-Term Forecasting Good	Good. A forecast was issued for each blood type.	Good. Forecasts were updated daily.	Good. Printed at blood center.	Good. Administrators concerned with distribution and collections received the report.	Good. Output design could have been more participative.	Good. Shortages were prevented by calling in more donors.
Heuristic Allocation Failure	Poor. The people who allocated blood mistrusted the mysterious numbers produced by the computer.	Good. Reports were provided one hour before allocation decisions were made.	Good. Printed at blood center.	Fair. Administrators responsible for daily blood allocation received the original.	Poor. Too many people were involved with the level of blood inventories to be able to participate in the design of the system. Hospitals were not prepared to participate with the computer-based rationing of blood units.	Poor. Although consultants were concerned about reducing shipping costs, this was not an immediate goal of the blood region. Shipping costs were passed on to patients.
Decentralized Performance Measurement Good	Good. Peer review committee understood and accepted the reports.	Good. Monthly performance measures were available within one day after the month ended.	Good. Reports were mailed to the peer review committee.	Good. Performance reports were given to peer review committee and the regional blood administrator at the same time.	Fair. Trial period with peer committee was satisfactory, but the regional blood administrator was unwilling to participate. Consulting team was unable to convince blood administrator that outdating should be kept low.	Good. Goals were consistent with regional objectives. The blood administrator did not consider minimizing blood outdating to be a priority goal, even though it is national blood policy.
Bloodmobile Scheduling Partially Successful	Good. Tabular form included sufficient detail, whereas computer-produced graph provided summary information as well.	Good. Feedback within one day.	Good. Computer terminal was in the same building so assistant administrator could make changes independently.	Good. Assistant administrator used system for one year. When assistant administrator left, improving scheduling was not considered a priority item.	Fair. Trial period was successful, but when assistant administrator left, no one else had the knowledge to run the system. Perhaps the origination was not ready for the refinement in this module.	Fair. Goals of this module were overshadowed by more immediate goals.
Policy Planning Model Partially Successful	Good. The form was the same as the bloodmobile scheduling output.	Good. Immediate feedback.	Not applicable.	Results were published in blood management and health care journals.	Not applicable.	Good. Goals were long-range strategic planning alternatives.

FIGURE 21.14

Evaluating a blood inventory information and decision support system using the information system utility approach.

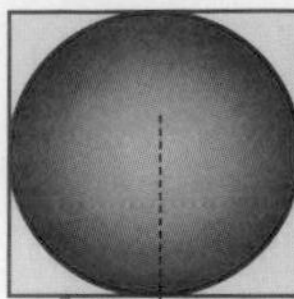

CONSULTING OPPORTUNITY 21.2

THE SWEET SMELL OF SUCCESS

Recall that in Consulting Opportunity 3.1, "The Sweetest Sound I've Ever Sipped," you met Felix Straw. Devise a systems solution that will address the problems discussed there. (*Hint*: The technology is important, but so are the way that people can use it.) Your solution should stress collaboration, flexibility, adaptability, and access. Use network diagramming to illustrate your solution. In a few paragraphs, write a rationale for why your solution should be chosen.

Possession Utility. Possession utility answers the question of who should receive output or, in other words, who should be responsible for making decisions. Information has no value in the hands of someone who lacks the power to make improvements in the system or someone who lacks the ability to use the information productively.

Form Utility. Form utility answers the question of what kind of output is distributed to the decision maker. The documents must be useful for a particular decision maker in terms of the document's format and the jargon used. Acronyms and column headings must be meaningful to the user. Furthermore, information itself must be in an appropriate form. For example, the user should not have to divide one number by another to obtain a ratio. Instead, a ratio should be calculated and prominently displayed. At the other extreme is the presentation of too much irrelevant data. Information overload certainly decreases the value of an information system.

Place Utility. Place utility answers the question of where the information is distributed. Information must be delivered to the location where the decision is made. More detailed reports or previous management reports should be filed or stored to facilitate future access.

Time Utility. Time utility answers the question of when information is delivered. Information must arrive before a decision is made. Late information has no utility. At the other extreme is the delivery of information too far in advance of the decision. Reports may become inaccurate or may be forgotten if delivered prematurely.

Actualization Utility. Actualization utility involves how the information is introduced and used by the decision maker. First, the information system has value if it possesses the ability to be implemented. Second, actualization utility implies that an information system has value if it is maintained after its designers depart or if a one-time use of the information system obtains satisfactory and long-lasting results.

Goal Utility. Goal utility answers the "why" of information systems by asking whether the output has value in helping the organization obtain its objectives. The goal of the information system must not only be in line with the goals of decision makers, but it must also reflect their priorities.

EVALUATING THE SYSTEM

An information system can be evaluated as successful if it possesses all six utilities. If the system module is judged as "poor" in providing one of the utilities, the entire module will be destined to failure. A partial or "fair" attainment of a utility will

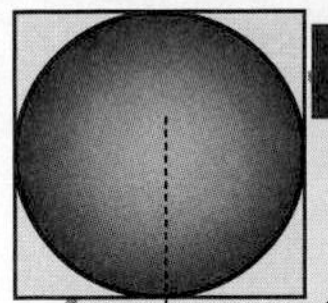

CONSULTING OPPORTUNITY 21.3

MOPPING UP WITH THE NEW SYSTEM

"I don't know what happened. When the new system was installed, the systems analysts made a clean getaway, as far as I can tell," says Marc Schnieder, waxing philosophic. Recall that he is owner of the Marc Schnieder Janitorial Supply Company. (You last met Marc in Consulting Opportunity 17.1, where you helped him with his data-storage needs. In the interim, he has had a new information system installed.)

"The systems analysis team asked us some questions about how we liked the new system," Marc supplies eagerly. "We didn't really know how to tell them that the output wasn't as spotless as we'd like. I mean, it's confusing. It isn't getting to the right people at the right time or anything. We never really did get into the nitty-gritty about the finished system with that consulting team. I feel as if we had to hire your group just to mop up after what they left."

After further discussions with Stan Lessink, the company's chief programmer, you realize that the team that did the initial installation had no evaluation mechanism. Suggest a suitable framework for evaluating the kinds of concerns that Mr. Schnieder raised about the system. What are the problems that can occur when a system is not evaluated systematically?

result in a partially successful module. If the information system module is judged as "good" in providing every utility, the module is a success.

The information system utility approach of "who, what, when, where, why, and how" used to evaluate the regional blood inventory management information system resulted in the subjective judgments concerning the utility of the information system summarized in the table. As you can see, four of the modules were rated as "good" in each category of utility, and consequently these modules were considered successful. Two were evaluated as partially successful, and one module was judged a failure after the trial period. Explanations of each judgment made for the seven modules are also provided.

The information system utility approach is a workable and straightforward framework for evaluating large-scale information systems projects and ongoing efforts. It also can be usefully employed as a checklist to monitor the progress of systems under development. Furthermore, evaluation following implementation allows the analyst to acquire ideas about how to proceed with future systems projects.

EVALUATING CORPORATE WEB SITES

Evaluating the corporate Web site that you are developing or are maintaining is an important part of any successful implementation effort. Analysts can use the information system utility approach previously described to assess the aesthetic qualities, content, and delivery of the site. As an analyst or Webmaster, you should go one step farther and analyze Web traffic.

A visitor to your Web site can generate a large amount of useful information for you to analyze. This information can be gathered automatically by capturing information about the source, including the previous Web site the user visited and the key words used to find the site; the information can also be obtained through using cookies (files left on a visitor's computer about when the visitor last visited the site).

A leading Web activity monitoring package is Webtrends. Figure 21.15 is a sample report showing the most downloaded files on the Web site by day of the week. The graph displays the top five downloaded files, and the table at the bottom is a sorted list of all downloads.

An analyst or Webmaster can gain valuable information by using a service such as Webtrends. (Although some services are free, the pay services usually provide

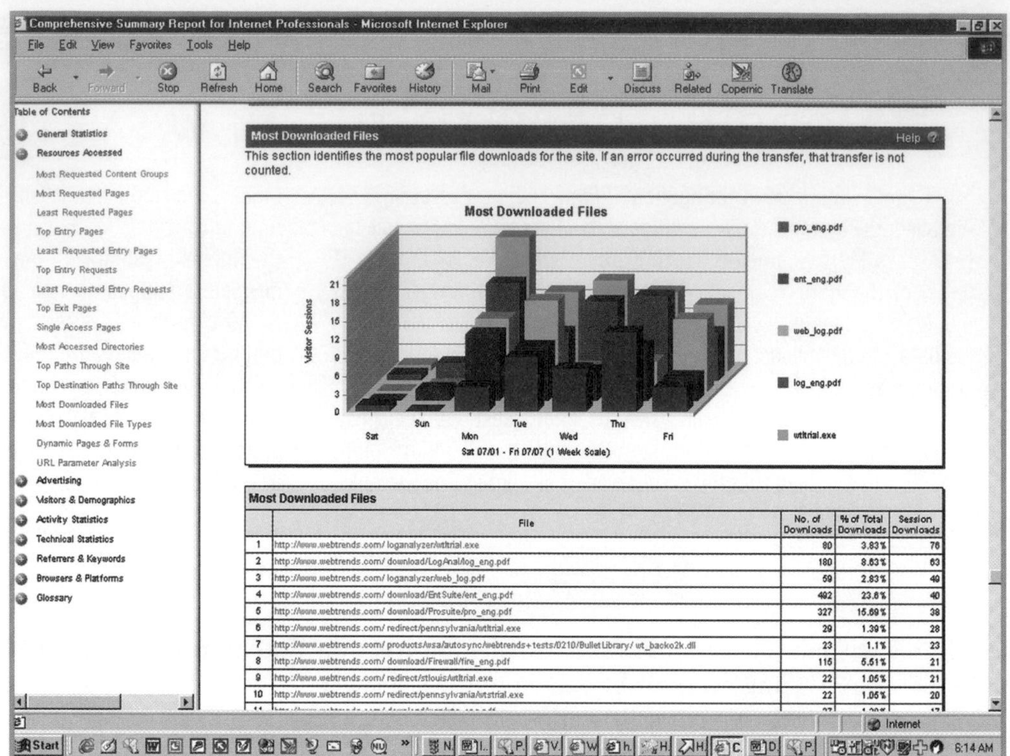

FIGURE 21.15
A sample report from Webtrends Corporation showing the most downloadable files on the corporate Web site.

the detail needed to evaluate the site in depth. The cost is an ongoing budget item for maintaining the Web site.) Information to help you evaluate your client's site and make improvements is plentiful and easy to obtain. The seven essential items are described next.

1. **Know how often your client's Web site is visited.** The number of hits a Web site had in the last few days, the number of visitor sessions, and the number of pages visited are a few of the general things you need to know. To evaluate a site properly, you should get more detailed information about how much traffic the client's site is experiencing.
2. **Learn details about specific pages on the site.** Detailed information about pages accessed helps in evaluating both the content and the ability to navigate the site properly. It is often telling when the last page visited is a page containing the prices of your products. In addition, by knowing the top exit page, you can get statistics on the most requested pages, most requested topics, top paths a visitor takes through the client's Web site, or even the most downloadable files. If the Web site is a commercial one, shopping cart reports can show how many visitors were converted into buyers and how many abandoned their carts or failed to complete the checkout process.
3. **Find out more about the Web site's visitors.** Visitor demographics and information such as the number of visits by a particular visitor in a period of time, whether the visitor is a new or a returning one, and who the top visitors are, are valuable when evaluating a Web site. It is possible to get summary data about the geographic region or even the city most represented by visitors to the site.

   One service, Commerce Trends, is available from Webtrends Corporation. Figure 21.16 shows a page that compares selected statistics about visitors over

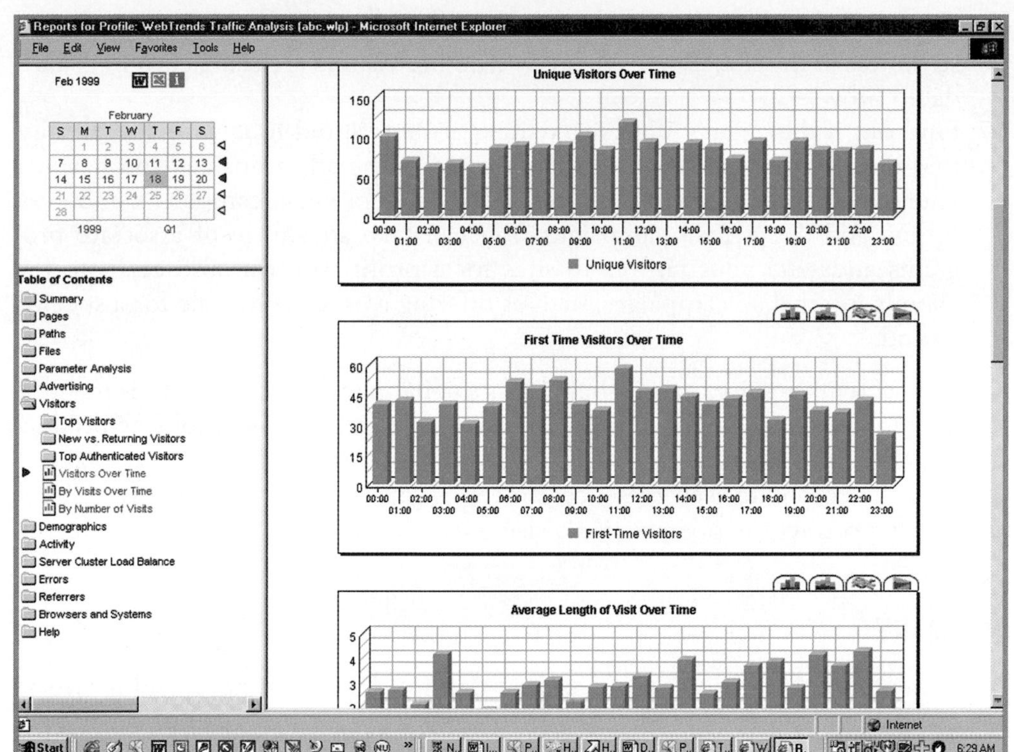

FIGURE 21.16

A report comparing statistics on visitors generated by Commerce Trends (from Webtrends corporation)

time. The top graph shows unique visitors, the middle graph shows first-time visits, and the bottom graph displays the average length of visits over time. This information allows an analyst to evaluate the Web site in terms of the ability to attract new visitors and keep them once they have visited the site. Notice that the calendar at the top left corner can be used to change the view from daily to weekly or monthly.

4. **Discover if visitors can properly fill out the forms you designed.** Once you attract and keep visitors, you need to know whether they are filling out forms properly and whether they understand the client's site in general. If the error rate is high, redesign the form and see what happens. Analysis of the statistics will reveal whether bad form design was to blame for errors in response.
5. **Find out who is referring Web site visitors to the client's site.** Find out which sites are responsible for referring visitors to the client's Web site. Get statistics on the top referring site, the top search engines leading to the site, and even the key words visitors used to locate your client's Web site.

   To increase your company's Web presence, site promotion is critical. Many available services such as NetAnnounce Premier and NetMechanic assist in optimizing your home page and automatically create and install meta tags (HTML code that search engines use to classify a Web site). These services then submit the site to top search engines such as Yahoo!, Alta Vista, Excite, Google, and HotBot.

   After promoting a site, you can use Web traffic analysis to track whether the site promotion really made a difference. Because the majority of users searching for a Web site do not look beyond the first page of search results, promotion of your client's Web site is essential.
6. **Determine what browsers visitors are using.** By knowing what browsers are being used, you can add browser-specific features that can improve the look

and feel of the site and encourage visitors to stay longer, thereby improving the stickiness of the site. It helps to know whether visitors are using current or outdated browsers.

7. **Find out if the client's Web site visitors are interested in advertising.** Finally, find out if visitors to the site are interested in the ad campaigns you have on your site. Get reports on the success of ad banners, email campaigns, and even offline and direct mail campaigns. You can take advantage of associate programs and send your visitors to sites for a profit. You can also evaluate the client's internal ad campaigns, such as offering a product for sale for a specific period.

Web activity services can be helpful in evaluating whether the site is meeting its stated objectives in terms of traffic, advertising effectiveness, employee productivity, and return on investment. It is one of the ways an analyst can evaluate whether the corporate Web presence is meeting management goals and whether it accurately portrays the organization's vision.

SUMMARY

Implementation is the process of ensuring that information systems and networks are operational and then involving well-trained users in their operation. In large systems projects, the primary role of the analyst is overseeing implementation by correctly estimating the time needed and then supervising the installation of equipment for information systems (which may be set up with a client/server approach over a local area network), training users, and converting files and databases to the new system.

Distributed systems take advantage of telecommunications technology and database management to interconnect people manipulating some of the same data in meaningful but different ways. As hardware and software are evaluated, the systems analyst also needs to consider the costs and benefits of employing a distributed system to fulfill user requirements.

One of the most popular ways to approach distributed systems is through the use of a client/server (C/S) model. Standard types of organizational networks include the local area network (LAN) and the wide-area network (WAN). Using a top-down approach, analysts can use five symbols to help draw network decomposition and hub connectivity diagrams. Specialized software, called groupware, is written specifically to support groups or teams of workers with functional applications. Its purpose is to help group members to work together through networks.

Training users and personnel to interact with the information system is an important part of implementation, because users must usually be able to run the system without the intervention of the analyst. The analyst needs to consider who needs to be trained, who will train them, the objectives of training, the methods of instruction to be used, the training sites, and the training materials.

Conversion is also part of the implementation process. The analyst has several strategies for changing from the old information system to the new. The five conversion strategies are direct changeover, parallel conversion, phased or gradual conversion, modular prototype conversion, and distributed conversion. Taking a contingency approach to conversion strategies can help the analyst in choosing an appropriate strategy, one that suits different system and organizational variables.

Security of data and systems has taken on increased importance for analysts who are designing more ecommerce applications. Security has several facets—

"As you know, Snowden is determined to implement some kind of automated tracking for the Training people. Even after having you and your team here at MRE for all of this time, though, it isn't clear to me how that will ever come about. You've probably noticed by now that people like Tom Ketcham are pretty set in their ways, but so is Snowden, and he definitely has the upper hand. I'm not telling you anything you don't know already, am I? I think when Snowden comes back from Poland, you should be ready to show him how we can implement an automated tracking system for the Training group, but it really has to be acceptable to the new users. After all, they're the ones who have to live with it. I'll pencil you in for a meeting with Snowden two weeks from today."

HYPERCASE® QUESTIONS

1. Develop an *implementation plan* that would be useful to the Training group in changing to an automated project tracking system. Use a paragraph to explain your approach. Be sure that what you are doing also meets Snowden's expectations.
2. In two paragraphs, discuss what *conversion* approach is appropriate for adopting a new automated project tracking system for the Training group.
3. Provide an outline of steps you would take to train the users in the Training group so that they could use their new system. In a paragraph, discuss any obstacles you see to training the users in the Training group and also list how you would overcome these problems.

physical, logical, and behavioral—that must all work together. Analysts can take a series of precautions such as virus protection software, email filtering, URL filters, firewalls, gateways, virtual private networks, intrusion detection products, secure socket layering, secure electronic translation, and public key infrastructure to improve privacy, confidentiality, and the security of systems, networks, data, individuals, and organizations.

Recent exploratory research suggests that systems analysts can improve the chances that newly implemented systems will be accepted if they develop systems with predominant organizational metaphors in mind. Nine main metaphors in use are society, the family, machine, organism, journey, game, war, jungle, and zoo. For example, traditional MIS are more likely to succeed when metaphors such as the family, society, or machine are used, and they are less likely to succeed with organizational metaphors such as war and jungle.

After implementation, the new information system and the approach taken (perhaps client/server technology) should be evaluated. Many different evaluation approaches are available, including cost-benefit analysis, the revised decision evaluation approach, and user involvement evaluations.

The information system utility framework is a direct way to evaluate a new system based on the six utilities of possession, form, place, time, actualization, and goal. These utilities correspond to, and answer the questions of who, what, where, when, how, and why, so as to evaluate the utilities of the information system. Utilities can also serve as a checklist for systems under development.

KEYWORDS AND PHRASES

- audience profiling
- behavioral security
- bus configuration
- client/server model
- corporate privacy policy
- direct changeover
- distributed conversion
- distributed processing
- email filtering products
- encryption software
- file server
- firewall or firewall system
- gateways
- groupware
- gradual or phased conversion
- hierarchical network
- hits
- hub connectivity
- information system utility
  - approach to evaluation
  - possession utility
  - form utility
  - place utility
  - time utility
  - actualization utility
  - goal utility
- local area network (LAN)
- logical security
- modular prototype conversion
- network decomposition
- network modeling
- organizational metaphors
  - zoo
  - machine
  - war
  - journey
  - jungle
  - society
  - family
  - game
  - organism
- page views
- parallel conversion
- physical security
- print server
- Public Key Infrastructure (PKI)
- referring site
- ring network
- Secure Electronic Translation (SET)
- Secure Socket Layering (SSL)
- star network
- training materials
- training methods
- training objectives
- training sites
- unique visitors
- URL filtering products
- Virtual Private Networks (VPN)
- virus protection software
- Web activity monitoring
- Web site promotion
- Web traffic analysis
- wide-area network (WAN)

REVIEW QUESTIONS

1. List the four approaches to implementation.
2. Describe what is meant by distributed system.
3. What is a hierarchical network?
4. Draw a star network and label the nodes appropriately.
5. How does a ring network differ from a star network?
6. What is a bus configuration for distributed processing?
7. What is the client/server model?
8. Describe how a "client" is different from a user.
9. What is a peer-to-peer network? How does it differ from other client/server networks?
10. What is a file server?
11. What are the advantages of using a client/server approach?
12. What are the disadvantages of using a client/server approach?
13. What is the purpose of groupware?
14. Who should be trained to use the new or modified information system?
15. List the five possible sources of training for users of information systems.

16. Why is it important to have well-defined training objectives?
17. Some users learn best by seeing, others by hearing, and still others by doing. Give an example of how each kind of learning can be incorporated into a training session.
18. State an advantage and a disadvantage of onsite training sessions.
19. List the attributes of well-executed training materials for users.
20. List the five conversion strategies for converting old information systems to new ones.
21. Define the terms physical, logical, and behavioral security and give an example of each one that illustrates the differences among them.
22. Define what encryption software means.
23. What is a firewall or firewall system?
24. List five of the several measures an analyst can take to improve the security, privacy, and confidentiality of data, systems, networks, individuals, and organizations that use ecommerce Web applications.
25. List five guidelines for designing a corporate privacy policy for ecommerce applications.
26. List the nine organizational metaphors and the hypothesized success of each type of system given their presence.
27. List and describe the utilities of information systems that can be used to evaluate the information system.
28. What are seven essential items that the analysis should include in performing a Web site traffic analysis?

PROBLEMS

1. Draw a local area network or some other configuration of distributed processing using the client/server approach to solve some of the data-sharing problems that Bakerloo Brothers construction company is having. It wants to be able to allow teams of architects to work on blueprints at headquarters, let the construction supervisor enter last-minute changes to plans under construction from the field, and permit clients to view plans from almost anywhere. Currently, the company has a LAN for the architects who are in one city (Philadelphia) that lets them share some drawing tools and any updates that team members make with architects in other cities (New York, Terre Haute, Milwaukee, Lincoln, and Vancouver). The supervisor uses a laptop, cannot make any changes, and is not connected to a database. Clients view plans on screen displays, but sales representatives are not able to enter modifications to show them "what if" a wall was moved or a roof line altered. (*Hint*: List the problems that the company is encountering, analyze the symptoms, think of a solution, and then start drawing.) More than one network may be necessary, and not all problems will be amenable to a systems solution.
2. Cramtrack, the regional commuter train system, is trying to train users of its newly installed computer system. For the users to get the proper training, the systems analysts involved with the project sent a memo to the heads of the four departments that include both primary and secondary users. The memo said in part, "Only people who feel as if they require training need to make reservations for offsite training; all others should learn the system as they work with it on the job." Only 3 of a possible 42 users signed up. The analysts were satisfied that the memo effectively screened people who needed training from those who did not.
   a. In a paragraph, explain how the systems analysts got off the track in their approach to training.

b. Outline the steps you would take to ensure that the right people at Cramtrack are trained.
c. Suggest in a paragraph how the Web might be used to assist in training for Cramtrack.

3. A beautiful, full-color brochure arrived on Bill Cornwell's desk describing the Benny Company's offsite training program and facilities in glowing terms; it showed happy users at terminals and professional-looking trainers leaning over them with concerned looks. Bill ran excitedly into Roseann's office and told her, "We've got to use these people. This place looks terrific!" Roseann was not persuaded by the brochure but didn't know what to say in defense of the onsite training for users that she had already authorized.
   a. In a few sentences, help Roseann argue the usefulness of onsite training with in-house trainers in contrast to offsite training with externally hired trainers.
   b. If Bill does decide on Benny Company training, what should he do to verify that this company is indeed the right place to train the company's information system users? Make a list of actions he should take.
4. "Just a little longer . . . I want to be sure this is working correctly before I change over," says Buffy, the owner of three bathroom accessories boutiques called Tub 'n Stuff. Her accountant, who helped her set up a new accounting information system, is getting desperate to persuade Buffy to change over completely to the new system. Buffy has insisted on running the old and new systems in parallel for an entire year.
   a. Briefly describe the general problems involved in using a parallel conversion strategy for implementing a new information system.
   b. In a paragraph, try to convince the owner of Tub 'n Stuff that a year of running a system in parallel is long enough. Suggest a way to end Tub 'n Stuff's dual systems that will provide enough reassurance to Buffy. (Assume the new system is reliable.)
5. Draft a plan to perform Web traffic analysis for the ecommerce application developed for Marathon Vitamin Shops. (See Consulting Opportunities 1.1, 16.4, and 18.7 for more information about the organization, their products, and their goals.) Your plan should take the form of a written report to the owner of the chain, Bill Berry. Be sure to indicate what statistics you will monitor and why they are important for Marathon Vitamin Shops to know.
6. FilmMagic, a chain of video rental stores introduced in Chapter 9 and revisited in Consulting Opportunity 22.3, is experimenting with adding a new Web-based service to its store (similar to www.netflix.com) that would, for a monthly fee, permit customers to choose a list of DVDs, have them sent to their home, and return them in prepaid mailers when they had finished viewing them. Based on what you know about FilmMagic, write a corporate privacy policy that would work well on their newly proposed Web site. Create a prototype screen (either with a graphics package or on a word processor) that includes appropriate language, fonts, and icons to show how your policy will appear as a page on FilmMagic's Web site.
7. Ayman's Office Supplies Company recently had a new information system installed to help its managers with inventory. In speaking with managers, you notice that they seemed disgruntled with the system output, which is a series of screens that show current inventory, customer and supplier addresses, and so on. All screens need to be accessed through both several special commands and the use of a secret password. The managers had several opinions about the system but had no systematic way to evaluate it.
   a. Devise a checklist or form that helps Ayman's managers evaluate the utilities of an information system.

b. Suggest a second way to evaluate the information system. Compare it with what you did in part a.

8. Visit a number of ISPs such as Verio, AT&T, and others and investigate what sort of Web traffic analysis features they offer to Webmasters whose Web sites they host. Make a list of reports and statistics they offer, and write this list up as the evaluation portion of an ecommerce application systems development proposal.

GROUP PROJECT

1. Visit six different Web sites. Choose one Web site from each of the categories below:

   a. A portal, such as Yahoo! or Excite.
   b. A news page such as ABC News or the New York Times.
   c. A software company.
   d. A university Web site.
   e. An official Web site for a sports team or a theatre company.
   f. A Web site from a continent other than the one on which you live.

   Evaluate each using an information utility approach.

   Prepare a table similar to Figure 21.14 with your answers. There will be one row for each of the six Web sites. Indicate the URL of the Web site. When you think you need Web traffic analysis to evaluate one of the utilities, state so in the appropriate cell within the table.

2. With your group members, prepare for your final exam by completing the crossword puzzle in Figure 21.EX1. The clues are found in Figure 21.EX2. Notice that there are hints regarding where to learn more about the material.

SELECTED BIBLIOGRAPHY

Baskerville, R. L. "An Analytical Survey of Information Systems Security Design Methods: Implications for Information Systems Development." *Computing Surveys*, 1994.

Carlyle, R. E. "Squeezing the Middle." *Datamation*, Vol. 32, No. 10, 1986, pp. 26–28.

Derfler, F. J., Jr., and L. Freed. *How Networks Work*. Emeryville, CA: Ziff-Davis Press, 1993.

FitzGerald, J., and T. S. Eason. *Fundamentals of Data Communication*. New York: John Wiley, 1978.

Ginzberg, M. J. "Key Recurrent Issues in the MIS Implementation Process." *MIS Quarterly*, Vol. 5, No. 2, 1981, pp. 47–59.

Gore, M., and J. Stubbe. *Elements of Systems Analysis*. Dubuque, IA: William C. Brown, 1983.

Jessup, L. M., and J. S. Valacich. *Group Support Systems*. New York: Macmillan, 1993.

Kendall, J. E. "Using Metaphors for Knowledge Elicitation during Expert Systems Development." *Proceedings of the First International Meeting of the Decision Sciences Institute*. Brussels, June 1991, pp. 153–55.

———. "Using Metaphors to Enhance Intelligence in Information Systems: Rationale for an Alternative to Rule-Based Intelligence." In F. Vogt (ed.), *Information Processing 92, Volume III: Intelligent Systems and Personal Computers*. Amsterdam: Elsevier-North Holland, 1992, pp. 213–19.

Kendall, K. E. "Evaluation of a Regional Blood Distribution Information System." *International Journal of Physical Distribution and Materials Management*, Vol. 10, No. 7, 1980.

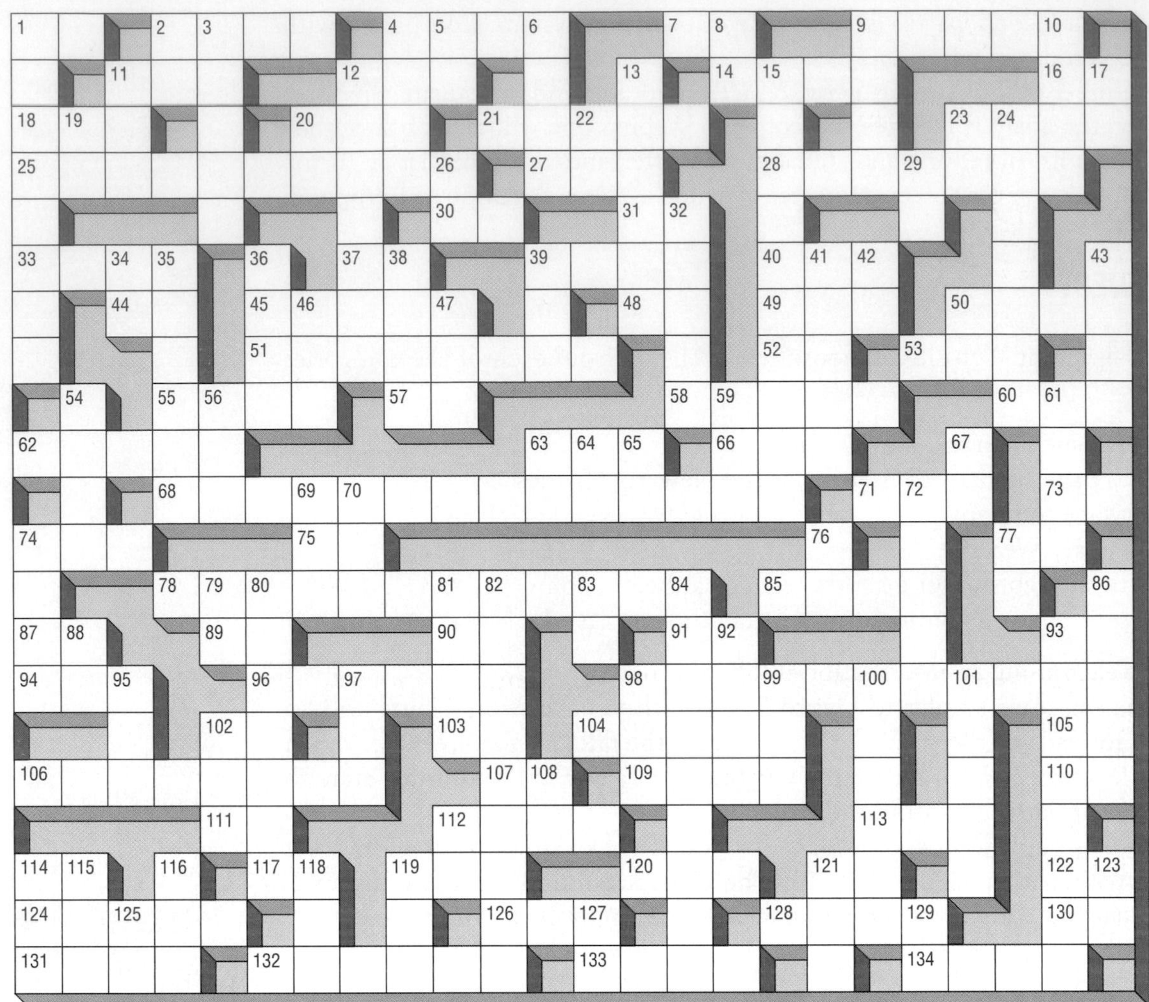

FIGURE 21.EX1
A comprehensive crossword puzzle for systems analysts.

Kendall, J. E., and K. E. Kendall. "Metaphors and Methodologies: Living Beyond the Systems Machine." *MIS Quarterly*, Vol. 17, No. 2, June 1993, pp. 149–71.

———. "Metaphors and Their Meaning for Information Systems Development." *European Journal of Information Systems*, Vol. 3, No. 1, 1994, pp. 37–47.

Labriola, D. "Remote Possibilities." *PC Magazine*, June 14, 1994.

Laudon, K. C., and J. Laudon. *Management Information Systems: Organization and Technology*, 4th ed. Upper Saddle River, NJ: Prentice Hall, 1996.

O'Hara, M. T., and R. T. Watson. "Automation, Business Process Reengineering and Client Server Technology: A Three-Stage Model of Organizational Change." In V. Grover and W. J. Kittinger (eds.), *Business Process Change: Concepts, Methods, and Technologies*. Harrisburg, PA: Idea Group Publishing, 1995.

Oppliger, R. "Internet Security: Firewalls and Beyond." *Communications of the ACM*, Vol. 40, No. 5, May 1997, pp. 92–102.

Pfaffenberger, B., and D. Wall. *Que's Computer and Internet Dictionary*, 6th ed. Indianapolis: Que Corporation, 1995.

Rigney, S. "Network in a Box." *PC Magazine*, Vol. 16, No. 16, September 1997, pp. 167–95.

Shaffer, G. "Coping with Change." *PC Magazine*, June 14, 1994, pp. 167–95.

Swanson, E. B. *Information System Implementation*. Homewood, IL: Irwin, 1988.

Zmud, R. W., and J. F. Cox. "The Implementation Process: A Change Approach." *MIS Quarterly*, Vol. 3, No. 2, 1979, pp. 35–44.

ACROSS

1 Always appears with ENNDO (Ch. 11)
2 File operation (Ch.17)
4 Type of file organization; Abbr. (Ch. 17)
7 Systems Analyst: Abbr.
9 Opposite of Output (Ch. 16)
11 Important person
12 Executive Information System: Abbr. (Ch. 1)
14 Systems Development Life Cycle: Abbr. (Ch. 1)
16 Expert System (Ch. 1)
18 ______ /CAM system
20 Like a beer
21 File operation (Ch. 17)
23 Opposite of East
25 Data gathering technique: Plural (Ch. 5)
27 Type of bar code (Ch. 19)
28 Overview in a DFD (Ch. 9)
30 Information Technology: Abbr.
31 Operating system: Abbr.
33 International Conference on Information Systems: Abbr.
37 Type of diagram: Abbr. (Chs. 2 and 17)
39 Project management technique (Ch. 3)
40 Big Blue
44 Foot: Abbr.
45 Not a renter (Ch. 13)
48 District Attorney: Abbr.
49 Warnier-______ diagrams (Ch.20)
50 ______ Utility (Ch. 21)
51 Type of diagram: 2 wds. (Ch. 9)
52 Doesn't apply: Abbr.
53 Check digit method: Abbr. (Ch. 19)
55 Follow rules
57 Research and development: Abbr.
58 Educate users (Ch. 21)
60 Shape in a bubble diagram (Ch. 17)
62 Sylvester Stallone character
63 CAD/______ system
66 Vase
68 DFD symbol: 2 wds. (Ch. 9)
71 Quality approach: Abbr.
73 Symbol for gold in the periodic table of elements
74 Female deer
75 Mnemonic postal code for Pennsylvania (Ch. 19)
77 Mnemonic postal code for Illinois (Ch. 19)
78 Its symbol is a diamond (Ch. 17)
85 Method for documenting structured decisions (Ch. 11)
87 Direction of a data flow (Ch. 9)
89 ______ UNTIL or ______ WHILE (Ch. 11)
90 ______ Box: Abbr.
91 Code for arsenic in the periodic table of elements
93 Mnemonic postal code for Michigan (Ch. 19)
94 Noah's boat
96 A service on the Internet (Ch. 15)
98 Rapid development method (Ch 8)
102 Spanish for yes
103 Mnemonic postal code for New Mexico (Ch. 19)
104 Type of scale (Ch. 6)
105 ______ LSE ______ F in structured English (Ch. 11)
106 Type of data flow diagram (Ch. 9)
107 Good ______ gold
109 River that flows North
110 Mnemonic postal code for Montana (Ch. 19)
111 Mnemonic postal code for Massachusetts (Ch. 19)
112 Data ______ (Ch. 9)
113 The CPU case takes place in a computer ______
114 Mnemonic postal code for Washington (Ch. 19)
117 Los Angeles: Abbr.
119 Type of interface (Ch. 18)
120 Type of chart (Ch. 14)
121 To be or not to ______
122 Type of diagram: Abbr. (Ch. 20)
124 Electronic communication (Ch. 15)
126 Metaphor for a chaotic organization (Ch. 21)
128 Type of automated tool used by analysts (Ch. 1)
130 Information Center: Abbr. (Ch. 21)
131 Distortion of data (Chs. 4 and 15)
132 Part of the population (Ch. 4)
133 Type of type (Ch. 15)
134 Compact ______ (Ch. 15)

DOWN

1 ______ maker (Ch. 12)
2 Code for silicon in the periodic table of elements
3 Puccini wrote many of these
4 User ______ of data (Ch. 17)
5 Social security: Abbr.
6 Type of interface (Ch. 18)
8 An adverb
9 Symbol used in Windows (Ch. 18)
10 Validating computer code (Ch. 19)
11 Type of display (Ch. 15)
12 Data ______, part of a data dictionary (Ch. 10)
13 Contains data (Ch. 17)
15 Data ______ (Ch. 10)
17 Street: Abbr.
19 Indefinite article
20 Artificial Intelligence: Abbr. (Ch. 1)
22 Data processing: Abbr. (Ch. 1)
23 Us
24 Describe a DFD process in greater detail (Ch. 9)
26 code for silicon in the periodic table of elements
29 Manufacturer of a CASE tool: Abbr. (Ch. 1)
32 Beginning
34 A word used in structured English (Ch. 11)
35 Method for structured observation (Ch. 7)
36 Cipher (Ch. 19)
38 Opposite of front
39 Expert ______ (Ch. 15)
41 Human computer
42 Mister: Abbr.
43 Control ______ (Ch. 20)
46 Custom or direction
47 Rural free delivery: Abbr
50 CompuServe command
54 ______ effect (Ch. 6)
56 Type of caption (Ch. 16)
59 Groove or depression
61 ______ programming (Ch. 12)
63 Code for cerium in the periodic table of elements
64 Systems ______ alyst (Ch. 1)
65 Mnemonic postal code for Montana (Ch. 19)
67 Before noon
69 U. S. agency that protects the environment: Abbr.
70 Rodent, but not a mouse
72 Operation performed on a database (Ch. 18)
74 ______ store (Ch. 10)
76 Boolean operator (Ch. 18)
77 Information Analyst: Abbr.
79 Edward: Abbr.
80 Type of data flow diagram (Ch. 9)
81 Type of system (Ch. 2)
82 Transformation of user views and data stores (Ch. 17)
83 Laser printer manufacturer
84 To break apart data flow diagrams (Ch. 9)
86 Check ______ (Ch. 19)
88 Operations Research: Abbr.
92 Earth
93 Type of code (Ch. 19)
95 Data item used to identify a record (Ch. 17)
97 Buddy or friend
98 Skillet
99 Digit
100 Decision ______: Plural (Ch. 11)
101 Type of question (Ch. 5)
102 Society for Information Management: Abbr.
108 ______ what!
112 Kung ______
114 World Wide ______
115 ______ Pro, a word processing package
116 Type of system: Abbr. (Ch. 1)
118 Also known as: Abbr
119 Rift
121 Type of chart (Ch. 14)
123 Mnemonic postal code for South Carolina (Ch. 19)
125 American Airlines: Abbr.
127 King ______ Hearts
129 Edition: Abbr.

FIGURE 21.EX2
Clues for the crossword puzzle.

EPISODE CPU CASE

21

Allen Schmidt, Julie E. Kendall, and Kenneth E. Kendall

SEMPER REDUNDATE

Mack Roe walks to Anna's desk where Chip is standing and says, "The last program has been tested and incorporated into the system test. The results indicate that the system is finally complete. Every program and subsystem is working as planned. The whole system checks out. Testing has been thorough and exacting, with all the problems and program bugs satisfactorily resolved. I've reviewed the deliverables, and each one has been developed into programs. I'll leave you two to install it and then celebrate."

"That's fantastic!" Anna replies as Mack leaves. "We've been anticipating this moment for a long time. We now have the task of installing the system. I've checked with Mike Crowe, and all the hardware has arrived and has been installed. The computers have been connected in a star configuration, and the network software has been installed. Why don't we make a list of the tasks to be completed?"

"Sure," answers Chip. "We'll need to train the users on the operation of the system. It would be good to provide some general training, followed by specific training for each user. We might want to train several people—the user and a backup person—for each specific operation."

"That's all right with me," responds Anna, "but I don't think we should have a backup person for Paige Prynter. Somehow I don't think she would be fond of the idea."

"Speaking of backup," says Chip, "what about creating backups of master and other system files? We should design an automated procedure for creating these copies."

"Yes," replies Anna. "We also need to be concerned with system security. Who may access the data, and who has clearance to update various database elements?"

"I agree," remarks Chip. "Another consideration is converting the production files from the old system to the new format. We don't want to rekey all the records from the hardware and software master files."

"Why don't we have one of the programmers write a one-time program that will convert each file from the old format to the new?" suggests Anna. "The indexes could be automatically updated and additional fields initialized to spaces or zeros."

The programmers complete the file conversion programs within a short time. The new files are created and painstakingly verified for accuracy. This effort is rewarded with new master files that contain all the necessary records loaded with correct information.

Training is scheduled to start in the Information Center. Hy Perteks is more than willing to reserve a block of time for installing the software and providing the training sessions. Chip and Anna alternated in providing instruction, each for portions of the system that they had created.

With the training sessions concluded, the last task is the conversion of the old system to the new. The phased method is selected as the best approach. First, the computer hardware programs are installed. Records are updated with information for the additional elements included in the system design.

Next, the software update programs are installed. Again, updates to master file records are entered. When the records contain complete information, the inquiry screens are installed. Last, report and menu programs are added to the system.

21

"The installation is a great success," exults Chip. "Everything is working correctly, without a bug in the system. I guess we should knock on wood. Have you heard any comments from the users?"

"Yes," replies Anna. "They are happy and relieved to have their new system. Mike Crowe has already started to use the preventive maintenance feature and has his students help tackling one lab room at a time. Cher and Dot were running through the various screens and several times commented on how easy it is to perform tasks. I paid a visit to Paige Prynter, and she asked me what she should do with all her free time."

The analysts smile at each other. Chip says, "It has been a really great project to work on."

"It certainly has," answers Anna. "The best system we've ever created here at CPU."

"I've learned a lot about the university in my short time here, too. It's a great place to work," Chip muses.

"And as long as you remember our motto, you should do fine," Anna replies. "Semper redundate," she says to Chip.

"Yeah, I see it on all the letterhead. I must admit, though, that I never took Latin in school. What does the motto actually mean?" Chip asks.

"Always backup!" Anna says securely.

EXERCISES

E-1. Use a paragraph to speculate on why the star network configuration was used. Does it matter that users are in several different rooms?

E-2. Describe procedures that should be designed to create automatic backup files. In your paragraph, be sure to consider the pros and cons of these procedures.

E-3. List security measures that should be taken to prevent unauthorized persons from using the computer system.

E-4. Explain in a paragraph why a phased conversion would be used to install the computer system.

OBJECT-ORIENTED SYSTEMS ANALYSIS AND DESIGN AND UML

22

The challenge of developing new information systems for ecommerce, wireless, and handheld applications in dynamic economic, legal, social, and physical environments calls for new analysis and design methods. Object-oriented (O-O) analysis and design can offer an approach that facilitates logical, rapid, yet thorough methods for creating new systems responsive to a changing business landscape. O-O techniques are thought to work well in situations where complicated information systems are undergoing continuous maintenance, adaptation, and redesign.

O-O languages have new structures that are thought to improve program maintenance and make large parts of programs reusable. The consequent recycling of program parts should reduce the costs of development in computer-based systems. It has already proved very effective in the development of GUIs and databases. Because O-O languages have different constructs, analysts need to create specifications for these computer systems that maximize the effective use of these constructs. This constraint has led to a number of new O-O systems analysis and design techniques as well as the agreed-upon standard for O-O modeling called the Unified Modeling Language (UML).

In this chapter, we take a conceptual as well as practical view of O-O analysis and design. The first part of the chapter explains the O-O approach by using the methods originally developed by Peter Coad and Ed Yourdon. This approach is favored by many analysts as a straightforward way to begin thinking about systems from an O-O perspective. This logical approach is laid out in layers. Many analysts believe that depicting systems in layers makes it easier to conceptualize the O-O perspective.

In the latter part of the chapter, we introduce UML, the industry standard for modeling O-O systems. The UML toolset includes diagrams that allow you to visualize the construction of an O-O system. Whereas the Coad and Yourdon OOA-OOD method focuses on a five-layer approach, UML breaks down objects and their relationships differently.

UML is a powerful tool that can greatly improve the quality of your systems analysis and design and thereby help create higher-quality information systems. By using UML in an iterative cycle of systems analysis, you can achieve a greater understanding between the business and IT teams regarding the system requirements and the processes that need to occur within the system to meet those requirements. Each iteration takes a successively more detailed look at the design of the system until the things and relationships within the system are clearly and

by Michael E. Anderson, Richard L. Baskerville, Julie E. Kendall, Kenneth E. Kendall, and Allen Schmidt precisely defined within the UML documents. When your analysis and design are complete, you should have an accurate and detailed set of specifications for the classes and processes within the system that will help you avoid the expense of recoding because of poor initial planning.

THE OBJECT-ORIENTED IDEA

Because O-O analysis and design is strongly related to O-O programming, we should briefly explore this O-O programming context *before* proceeding to O-O analysis and design. Six basic ideas characterize O-O programming: (1) objects, (2) classes, (3) messages, (4) encapsulation, (5) inheritance, and (6) polymorphism.

OBJECTS

An object is a computer representation of some real-world thing or event. Figure 22.1 shows how a computer might represent your car. For example, if you own a Jeep Wrangler, the computer would store the name of the model (Jeep Wrangler), the vehicle identification number (VIN) (#51Y62BG826341Y), and the motor type (6-Cyl). Objects can have both attributes (such as the model, VIN, and motor type) and behaviors (such as "lights go on" and "lights go off").

CLASSES

A class is a category of similar objects. Objects are grouped into classes. Figure 22.2 shows how a group of objects representing automobiles might be formed into a class called "Automobile." A class defines the set of shared attributes and behaviors found in each object in the class. For example, every automobile will have attributes for Make/Model, VIN, and Engine. The programmer must define the classes in the program. When the program runs, objects can be created from the established class. The term *instantiate* is used when an object is created from a class. For example, a program could instantiate the Jeep Wrangler as an object from the class Automobile.

MESSAGES

Information can be sent by one object to another. In Figure 22.3, an object (Julie) of the class "Operator" is sending a message to an object (Jeep) of the class "Automobile." The message is "Start Engine." These messages are not free-form in any sense; rather, the classes Operator and Automobile have been carefully programmed to send and receive a Start Engine message. The Operator class has been programmed to transmit a Start Engine message under certain conditions. The Automobile class has been programmed to react to a Start Engine message in some way.

FIGURE 22.1
An example of the attributes of an object from the class "Automobile."

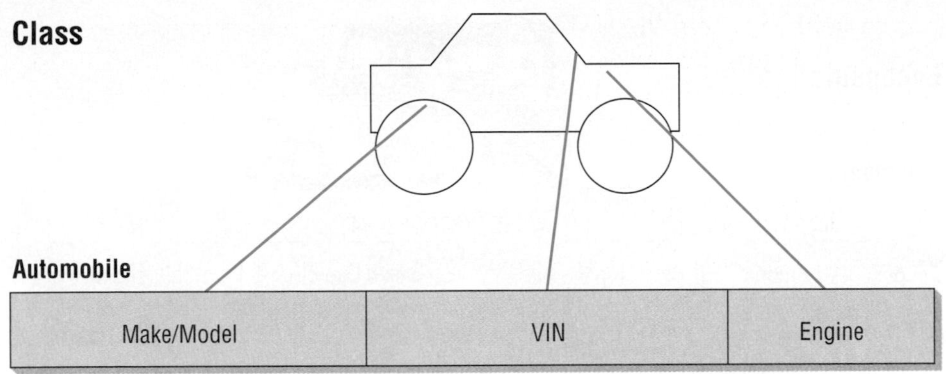

FIGURE 22.2
Shared attributes of objects grouped into the class "Automobile."

ENCAPSULATION

Usually, the information about an object is encapsulated by its behavior. Thus, an object maintains data about the real-world things it represents in a true sense. An object must usually be "asked" or "told" to change its own data with a message; that is, it does not wait for such data from outside processes to change the nature of an object. In Figure 22.4, the Ken object (class Mechanic) sends a Pull Engine message to the Jeep object (class Automobile). The Automobile class reacts to this message with a behavior (also called a "method" or "procedure") that changes the Engine attribute to None. This method is named Pull Engine. We see that the Jeep object reacts to the message by changing one of its attributes to None.

It may seem trivial whether an attribute of an object is changed by directly altering its data or by sending a message to the object to trigger internal behavior that changes that data. This difference, however, is an extremely important characteristic of O-O programs. Encapsulated data can be protected in such a way that only the object itself can make such changes through its own behavior. This construct makes it easier to build objects that are very reliable and consistent because they have complete control over their own attributes. It also makes program maintenance and change much easier. For example, the Mechanic class is isolated completely from the internal details of the Automobile class. The Automobile class can be totally reprogrammed without changing anything in the Mechanic class as long

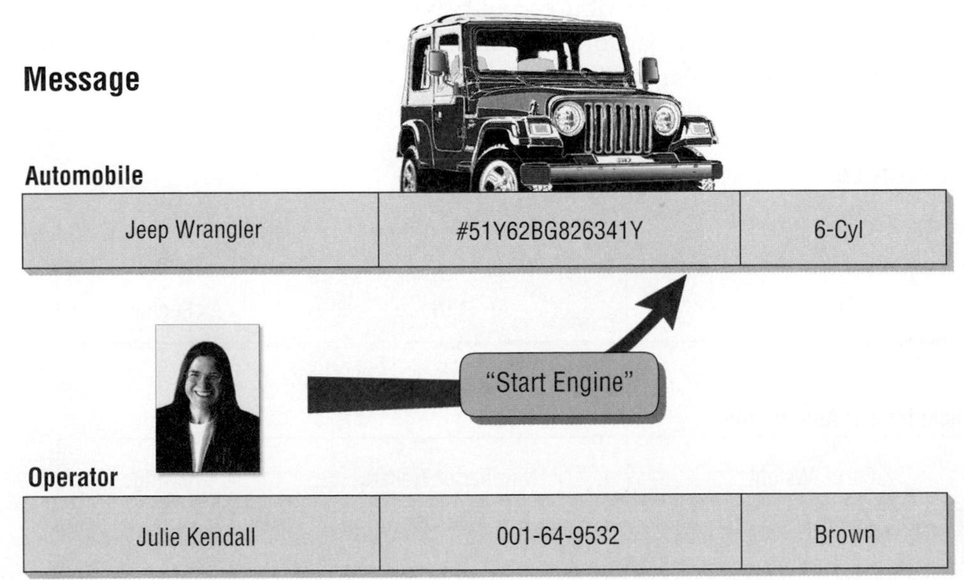

FIGURE 22.3
Example of the information that can be sent by an object of one class to an object of another class.

FIGURE 22.4
Example illustrating how a message from one object causes another object from a different class to change one of its attributes.

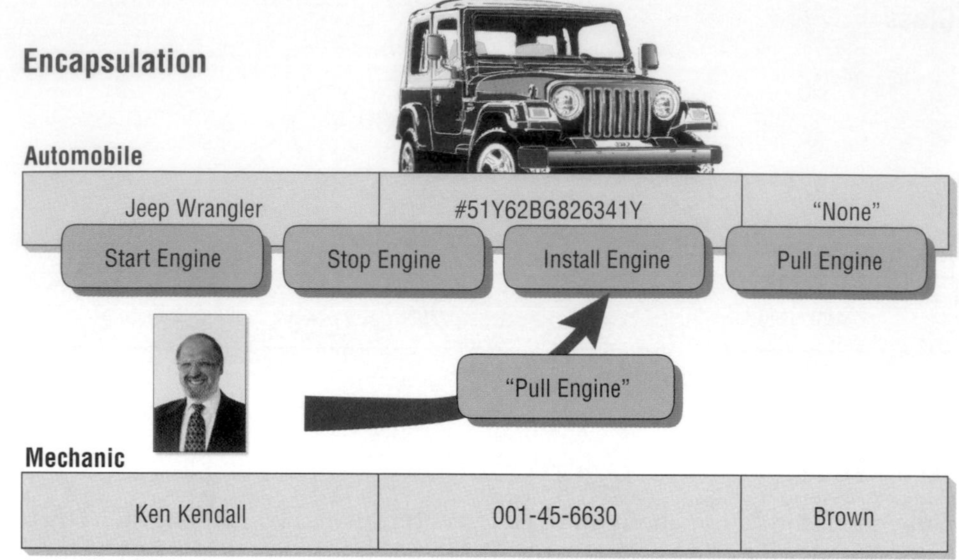

as the Automobile class continues to receive a Pull Engine message properly. This isolation makes it much easier to change one part of a program without causing problems to cascade out into other parts of the program.

INHERITANCE

Classes can have "children"; that is, one class can be created out of another class. The original—or parent class—is known as a "base class." The child class is called a "derived class." A derived class can be created in such a way that it will inherit all the attributes and behaviors of the base class. In Figure 22.5, a derived class (Truck) is created such that it inherits all the attributes of the base class Automobile. A derived class may have *additional* attributes and behaviors as well. For example, the class Truck not only has attributes for Make/Model, VIN, and Engine, but it also has attributes for Cargo Weight, Trailers, and Refridge. Automobile objects do not have these new attributes. Inheritance reduces programming labor by reusing old objects easily. The programmer only needs to declare that the Truck class inherits from the Automobile class and then to provide any additional details about new attributes or behaviors (shown in the solid-line box in the figure). All the old attributes and behaviors of the Automobile class are automatically and implicitly part of the Truck class (shown in the dashed box) and require no new programming at all.

FIGURE 22.5
Example of inheritance of parent class attributes by a child class.

Inheritance

Automobile

Make/Model	VIN	Engine

Truck: Inherit Automobile

Cargo Weight	Number of Trailers	Refridge
Make/Model	VIN	Engine

Polymorphism

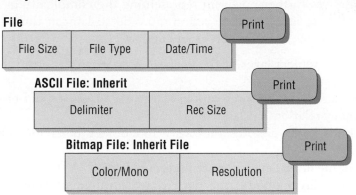

File

File Size	File Type	Date/Time	Print

ASCII File: Inherit

Delimiter	Rec Size	Print

Bitmap File: Inherit File

Color/Mono	Resolution	Print

FIGURE 22.6 An example of polymorphism among related classes.

POLYMORPHISM

The term *polymorphism* regards alternative behaviors among related derived classes. When several classes inherit both attributes and behaviors, there can be cases where the behavior of a derived class might be different from that of its base class or its sibling-derived classes. Hence, a message may have different effects depending on exactly what class of object receives the message. In Figure 22.6, we see three classes: File, ASCII File, and Bitmap File. Both ASCII File and Bitmap File inherit all the attributes of File *except* the Print behavior. A message to activate the Print behavior of an object of the generic parent File class might cause the File Size, File Type, and Date/Time attributes to be printed. The same message sent to an ASCII object might cause the text in the file to be sent to the printer. The same message sent to a bitmap object might cause a graphic display program to execute.

OBJECT-ORIENTED ANALYSIS

The Coad and Yourdon approach to O-O analysis is based on a five-layer model. These layers consist of the (1) class and object layer, (2) structure layer, (3) service layer, (4) attribute layer, and (5) subject layer. We can visualize the entire process of analysis and design as being the development and assembly of these five layers into one laminated design package. Figure 22.7 illustrates how the five layers

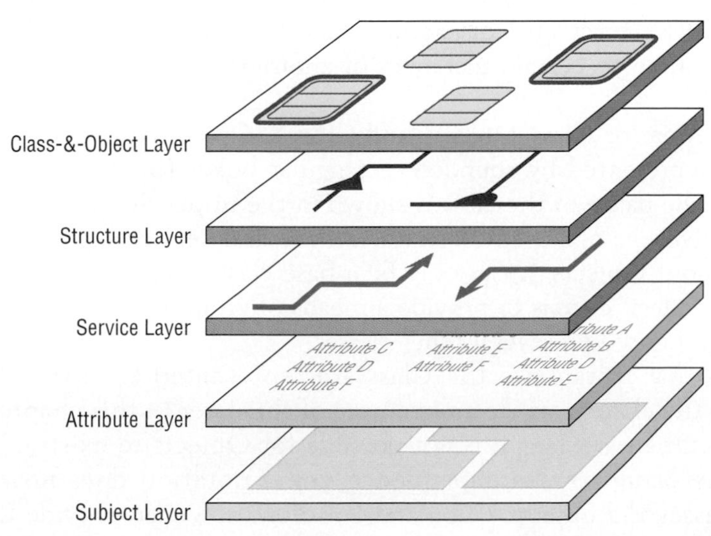

FIGURE 22.7 The five layers of object-oriented analysis.

interlock. These layers add a three-dimensional structure to the analysis and design notation that gives further power in representing the complexity in flexible systems. Each of these layers is discussed in more detail later in this chapter as we consider the activities of O-O analysis and design.

1. The **Class & Object layer** of the analysis and design denotes the classes and objects.
2. The **Structure layer** captures various structures of classes and object, such as one-to-many relationships and inheritance.
3. The **Service layer** denotes messages and object behaviors (services and methods).
4. The **Attribute layer** details the attributes of classes.
5. The **Subject layer** divides the design into implementation units or team assignments.

ANALYZING CLASSES AND OBJECTS

Coad and Yourdon distinguish Class, Object, and Class-&-Object in the following ways:

1. *Object*: An abstraction of something in a problem domain, reflecting the capabilities of a system to keep information about it, interact with it, or both. An object represents an encapsulation of attribute values and their exclusive services. A synonym is "an instance."
2. *Class*: A description of one or more objects with a uniform set of attributes and services, including a description of how to create new objects in the class.
3. *Class-&-Object*: A term referring to both the class and the objects that are instantiated in the class.

Five general types of objects can be discovered during analysis. Objects often represent *tangible things* such as vehicles, devices, and books. Sometimes objects represent *roles* enacted by persons or organizations; roles include objects such as customer, owner, or department. Objects may also be derived from *incidents* or events such as flight, accident, or meeting; incidents typically happen at a specific time. Other objects may denote *interactions* such as a sale or a marriage; interactions have a transaction or contract quality. Objects may also detail *specifications*. Specifications have standards or a definition quality and generally imply that other objects will represent instances of tangible things; for example, a class of object such as "insurance policy type" may have instances like "whole life," "term life," or "homeowners." Such a class of objects specifies qualities common to certain instances of another class of objects called "insurance policy."

The notation for Class, Objects, and Class-&-Object is shown in Figure 22.8. Classes are represented by rounded rectangular boxes (bubtangles) divided into three parts. The name of the class is shown in the upper division of the box. The other two divisions are used for the attribute and service layers. When a class appears without Objects, it can only be a base class, because the only reason for such an "objectless" class is to provide a means of grouping attributes and services that will be inherited by several other classes.

Objects that instantiate the Class are represented by a shaded box surrounded by the class (you see an example of this later in this chapter). Because Objects instantiate a Class, it is not possible for Objects to exist independently of their class. Due to this dependence, some notation does not distinguish between classes and objects. Coad and Yourdon, however, provide the Class-&-

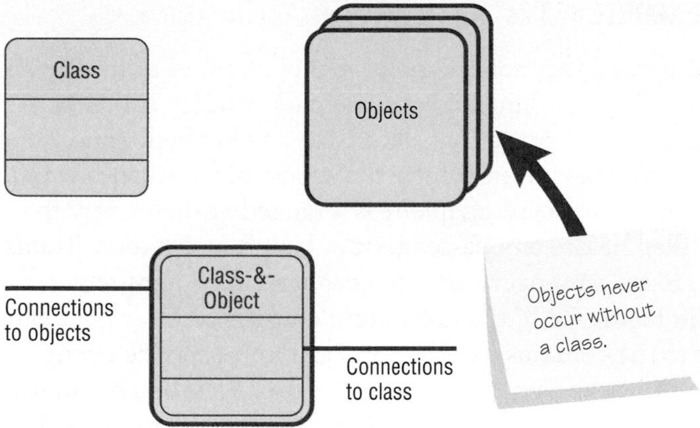

FIGURE 22.8
Notation representing the concepts Class, Objects, and Class-&-Object.

Object notation so as to graphically distinguish between structures and messages that are intended for the Class (such as a "create a new instance object" message) from structures and messages intended for the Object (such as "pull engine").

In general, techniques for discovering objects are the same as those discussed in earlier chapters for discovering processes and data entities. There are, however, certain criteria that we can use to help determine whether a new class of objects is justified:

1. There is a need to remember the object; that is, the object can be described in a definite sense, and its attributes are relevant to the problem.
2. There is a need for certain behaviors of the object; that is, even though an object has no attributes, there are services that it must provide or object states that must be recalled.
3. Usually, an object will have multiple attributes. Objects that have only one or two attributes suggest overanalyzed designs.
4. Usually, a class will have more than one object instantiation unless it is a base class.
5. Usually, attributes will always have a meaningful value for each object in a class. Objects that produce a null value for an attribute or for which an attribute is not applicable usually imply a generalization, specialization structure (described later in this chapter).
6. Usually, services will always behave in the same way for every object in a class. Services that vary dramatically for some objects in a class or that return without action for some objects also suggest a generalization–specification structure.
7. Objects should implement requirements that are derived from the problem setting, not the solution technology. The analysis portion of the O-O project should not become dependent on a particular implementation technology such as a specific computer system or a specific programming language. Objects that address such technical details should not appear until very late in the design stage. Technology-dependent objects suggest that the analysis process if faulty.
8. Objects should not duplicate attributes and services that could be derived from other objects in the system. For example, an object that stores the age of an employee is superfluous when a separate employee object exists that maintains a date-of-birth attribute. The age object can be eliminated by an age service that is a component of the employee object.

KAYJAY WORLD EXAMPLE 1

Kayjay World is a small vacation theme park that operates a circular, six-train monorail system connecting a parking lot, a theme park, a hotel, and a concert hall/restaurant complex. There are four stations, each with a ticket booth and a boarding queue. Passengers obtain a ticket for one of the three possible destination stations and enter the boarding queue. The boarding queue is arranged in such a way that every seat on a train will be filled before any passenger can be left at a station. Trains consist of an engine and one to six passenger cars, each car carrying 50 passengers.

As shown in Figure 22.9, the rail system consists of 15 safety segments of track and a segment (S16) contains the barn. Each train in service occupies one of these segments. A train may not enter a segment that is occupied by another train. Each station counts as one segment, and each link between stations is divided into two or three segments. The barn is capable of storing all six trains with two access segments: one exit segment (S14) leading to the barn and yard and one entrance segment (S15) departing from the barn and yard. There is only one junction switch, joining segment S10 to either segment S11 or S14. Segment S14 leads to the barn and thus is joined to the main circuit with the only system junction switch. The segment from the barn (S15) merges with the main track in a fixed junction and does not require a switch. When a train is on S13 and another is on S15—and both are thus competing for segment S1—the train on S15 is given priority.

The capacity manager initializes the system by ordering one train to leave the barn and thus be placed in service. At least one train remains in service until the capacity manager shuts down the system, ordering the last train out of service. However, when the excess capacity of every train in service falls below 50 percent and at least one train falls below 25 percent, another train leaves the barn and is placed in service. If the excess capacity of every train exceeds 50 percent, one train is removed from service and sent to the barn. (This train accepts no new passengers beginning with station one and thus the remaining passengers disembark at station three.) A yard manager declares out-of-service trains in the barn and yard either "operable" or "inoperable." Operable trains are rotated back into service on a first-in, first-out basis. The yard manager switches inoperable trains out of the automatic system by removing them from the service queue. The inoperable trains then enter a manually switched maintenance yard. Although each train usually has four cars, if an engine fails during high season, the yard master may detach one or more of the inoperable train's cars, take an operable train into the service yard, and add one or more of the orphaned passenger cars to the

FIGURE 22.9
Diagram of the Kayjay World monorail system.

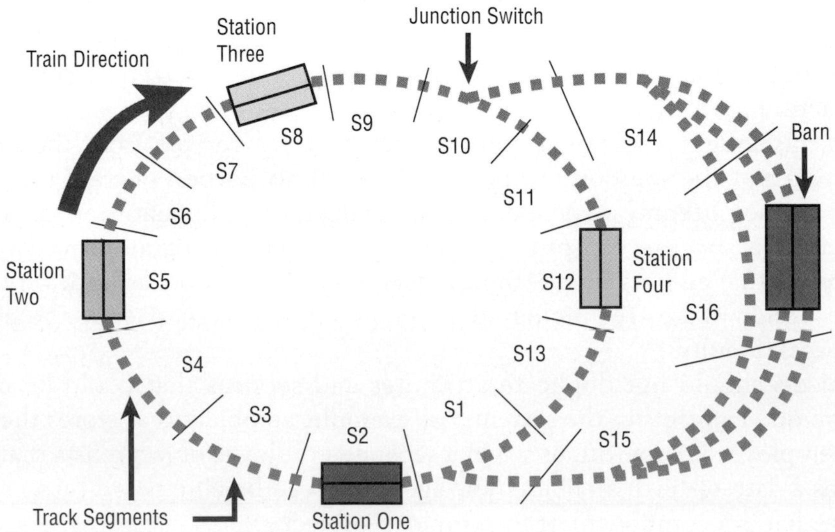

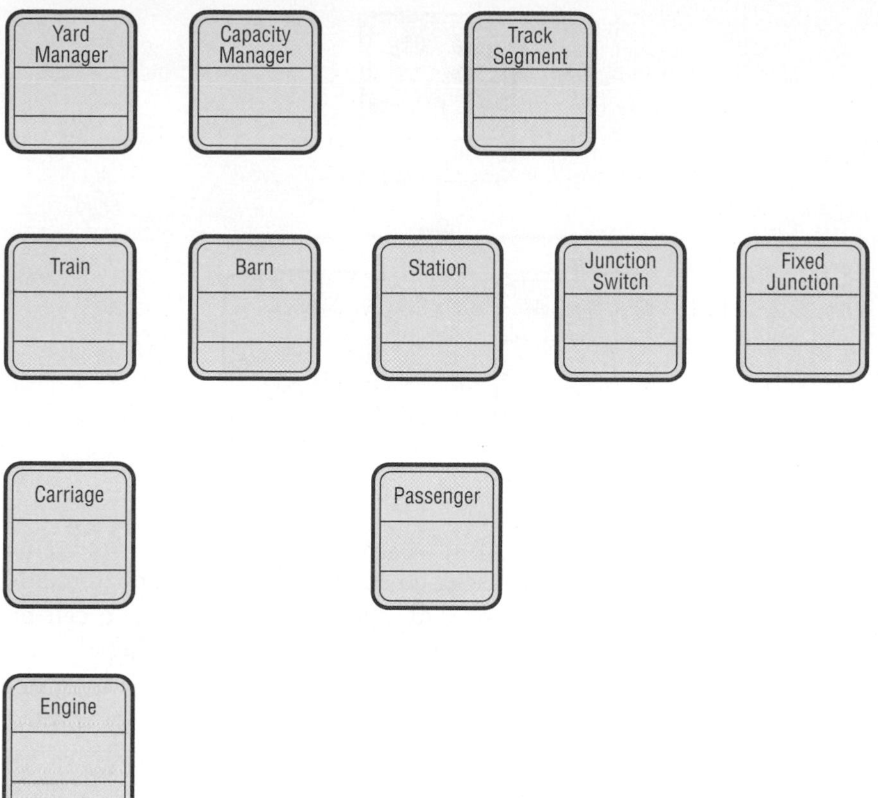

FIGURE 22.10
Initial outline of the Class-&-Object layer of Kayjay World's monorail system. The Class and Object layer is shown in dark blue.

operable train. An engine can pull six full cars, so the monorail can lose two engines and still operate at full capacity (1,200 passengers). This event, however, raises the minimum capacity of the system from 200 to 300 passengers.

Based on the preceding problem description, we can sketch out a preliminary class and object analysis. At this early stage in this analysis, all the objects are Class-&-Objects. As is typical of an early analysis, base classes will soon emerge from further analysis. Figure 22.10 is the initial outline of the Class-&-Object layer of the O-O design package.

ANALYZING STRUCTURES

There are two basic types of structures that might be imposed on classes and objects, the generalization–specialization structure (known as "Gen–Spec") and the whole–part structure.

1. Gen–Spec structures: Inheritance is created with Gen–Spec structures. These relationships between classes are sometimes called classification, subtype, or ISA (pronounced "ĭs-uh") relationships. Gen–Spec structures are denoted by a semicircle with its rounded edge toward the generalized class. These structures always connect class-to-class. They are usually of a hierarchical form. Figure 22.11 shows a Gen–Spec structure between Class-&-Objects in which the classes Bus and Motorcycle inherit all the properties of the class Automobile.
2. Whole–Part structures: Whole–Part structures denote collections of different objects that compose another whole object. Such relationships between objects are sometimes called assemblies, aggregations, or HASA (pronounced "hăs-uh") relationships. Whole–Part structures are noted by a triangle pointing toward the "whole" object. These structures always connect object-to-object. Figure 22.12 shows a Whole–Part structure in which vehicle objects are shown to be composed of two other objects: Motor and Chassis.

FIGURE 22.11
A Gen–Spec structure involving thc rclationship of the classes Motorcycle and Bus to the class Automobile.

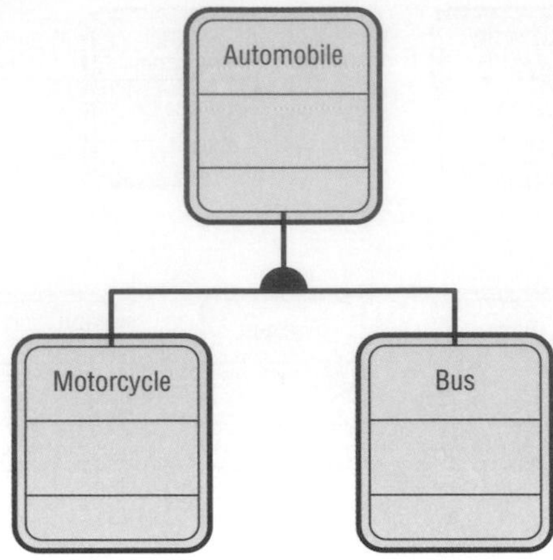

Whole–Part structures also have cardinality, as represented by one-to-many or many-to-many. This concept was discussed in the section on entity-relationship models in Chapter 17. The notation "0,m" specifies that a Vehicle can have no motor (0) or one or more motors (m). The notation "0,1" specifies that a Motor can be part of no vehicle (0) or one vehicle (1), but never more than one. The "1,m" specifies that a Vehicle can have one (or more) Chassis, but never fewer than one. The "1" specifies that a Chassis is always related to one and only one Vehicle.

KAYJAY WORLD EXAMPLE 2

We continue our work based on the problem description for Kayjay World given earlier by imposing a preliminary structure analysis—or structure layer—onto the Class-&-Object diagram. This structure analysis is shown in Figure 22.13. Note that the Barn, Station, Junction Switch, and Fixed Junction classes share similar functions with Track Segments. We can thus design a Gen–Spec structure between Track Segment and these other classes. This way, Station and the other classes inherit all the attributes and services of a Track Segment.

We can also design several Whole–Part structures. For instance, a Station object may have Passenger objects. Similarly, a Train object may have Carriage objects and Passenger objects. We can also decide that the Yard Manager will have all the Train assemblies and that the Capacity Manager will have all the Track Segment objects.

The relationship between Carriage and Engine is a tricky one. A train engine is just a special kind of train carriage, a kind of carriage that has a motor and very few seats. Also, it seems that the Barn object can have Train objects as part of it. This

FIGURE 22.12
A Whole–Part structure illustrating the relationships of the objects Motor and Chassis to the object Vehicle.

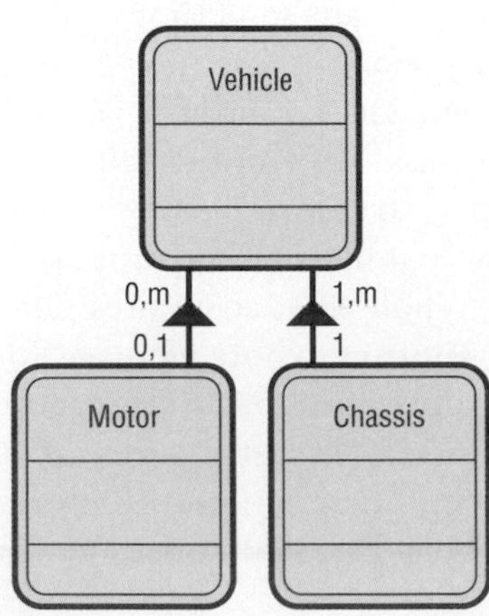

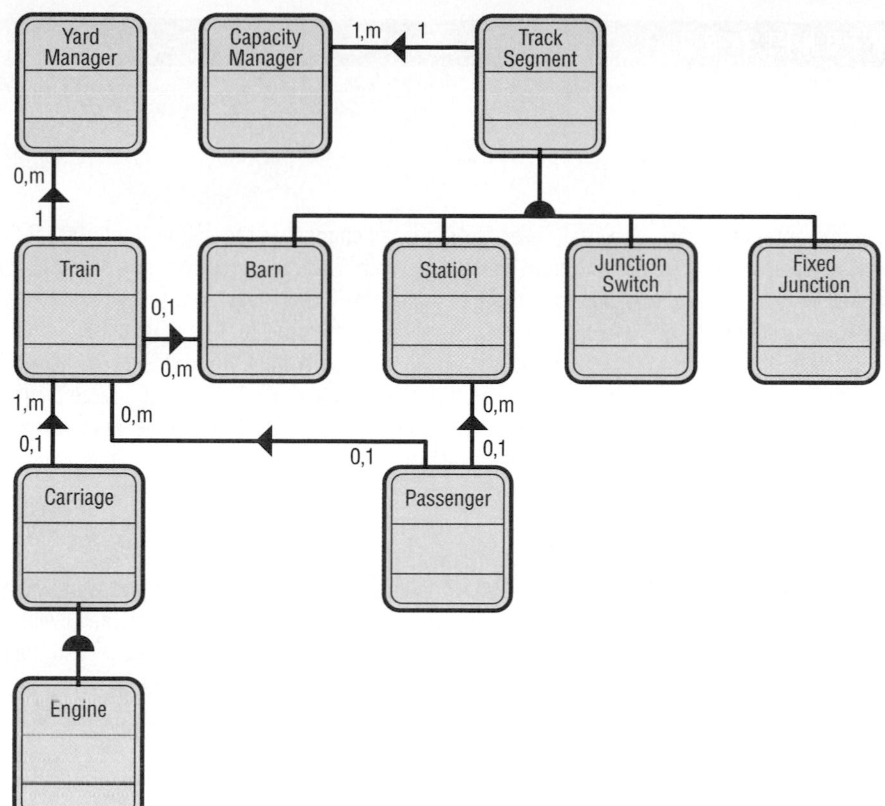

FIGURE 22.13
Preliminary structure analysis of Kayjay World's monorail system. The Structure layer is shown in red.

situation may create problems at implementation time because Train objects can then be part of both the Yard Manager and the Barn, but we will see how this seeming conflict can be sorted out during the design stage of the project.

ANALYZING ATTRIBUTES

The names of the attributes of a class are written in the center section of the class box in the design package. In Figure 22.14, the attributes Oname and Oaddress have been layered over the Owner object and the attributes Model and Color have been layered over the Vehicle object. The basic idea of an attribute is unchanged from our earlier discussion of this topic in Chapter 17. Three new related ideas, however, are germane to our object-oriented perspective. First, attributes are always more prone to change than classes. If a structure or a set of classes seems to be getting cluttered because an object is changing from class to class, perhaps the Class-&-Object in question should simply become a set of attributes in another class. Second, attributes should be kept as high as possible in Gen–Spec structures. This constraint reduces programming and maintenance because a change made in one Gen object will be automatically inherited by all the Spec objects. Third, associations or relationships between objects (other than structures) should be detailed as instance connections rather than as foreign keys.

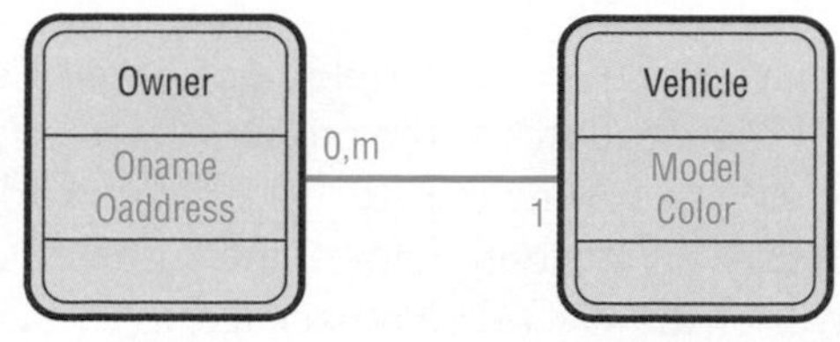

FIGURE 22.14
Depiction of two classes and their respective attributes. The Attribute layer is shown in cyan.

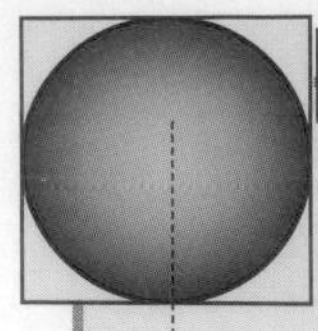

CONSULTING OPPORTUNITY 22.1

REEL MAGIC

"They want the core of the customer service representative's user interface to be radically reprogrammed again!" says Bradley Vargo, the Information Systems Development Director at C-Shore Mutual Funds. "Only eight months ago, we completed a two-year development project of the CSR System, the Customer Service Representative System. During that entire project, we endured a parade of moving requirements. Every month, those guys in the Marketing Department would invent some competitive new customer service feature, and within a week, the CSR group would be down here with vast changes to the CSR System specification. I thought we'd never finish that project! Now it looks as if we will have to start a new reprogramming project on a system less than a year old. We had forecast this system for a seven-year lifespan! Now I think it may be going into eternal reconstruction."

Bradley is talking with Rachael Ciupek, the senior application systems analyst responsible for the CSR system, and Bridget Ciupek, her sister and the programmer who wrote most of the user interface. "Calm down, Bradley," says Rachael. "It is not the fault of the kids in Marketing or CSR. The nature of our business has been affected by fast-paced competition. Marketing doesn't invent these changes out of boredom. They are often responding to new, computer-based customer services offered by our competition. We have to stay ahead or at least keep up, or we'll all be looking for new jobs!"

"Bradley, Rachael, I think you better know that the situation may be worse than you think," Bridget chips in. "The programmers have actually been making small changes in the CSR user interface for the past eight months anyway. The CSR users have been calling us directly and begging for help. They usually want just a small change to one isolated part of the system, but that has created a high labor drain because we have to recertify the entire system. You know how the effects of a small change can ripple throughout a large program. We've billed the time to program maintenance on the grounds that we thought we were just fine-tuning the completed system. Although the changes have been gradual, in eight months we've pretty much rewritten about a quarter of the CSR user interface code already. The work has not been falling off. It's still pretty steady."

"So what you're telling me," says Bradley, "is that we have system needs in this area that have been changing constantly while we tried to write specifications, tried to write program code, and tried to make a fixed solution work against a fluid problem. How can we afford to write programs if they will only last a few months without needing expensive maintenance?"

How can Bradley manage a systems development process that no longer has fixed or constant business processes as part of its goal set? Is there a way for Rachael to manage a specification and control maintenance costs when programmers are constantly asked to tinker with isolated parts of a large program? Keep in mind that an important goal is to provide good support for the users' needs and the organization's business strategies.

Instance Connections. The concept of primary and foreign keys was discussed earlier in Chapter 17. Rather than clutter up the design package with such implementation details, primary key attributes are not specified. Consequently, references between objects, such as associations or relationships, are denoted by a single line between objects with the same cardinality notation used in Whole–Part structures. Notice that instance connections always occur between objects, not classes. For example, an instance connection exists between owner and vehicle objects. The cardinality notes tell us that an owner may be related to zero, one, or more vehicles, but a vehicle must always be related to only one owner.

Preliminary Specification Template. With the introduction of attributes, we need additional analysis details to support the layered diagram package. At this stage of the analysis, these details only regard descriptions of the attributes and constraints on their values.

KAYJAY WORLD EXAMPLE 3

The analysis for the Kayjay World project would next have to spend some time collecting information about the attributes of the monorail system's objects and the relationships between those objects, adding the attribute details and instance connections to the preliminary analysis. The attribute names and instance connections form the preliminary attribute layer of the O-O design package, as shown in Figure 22.15. We

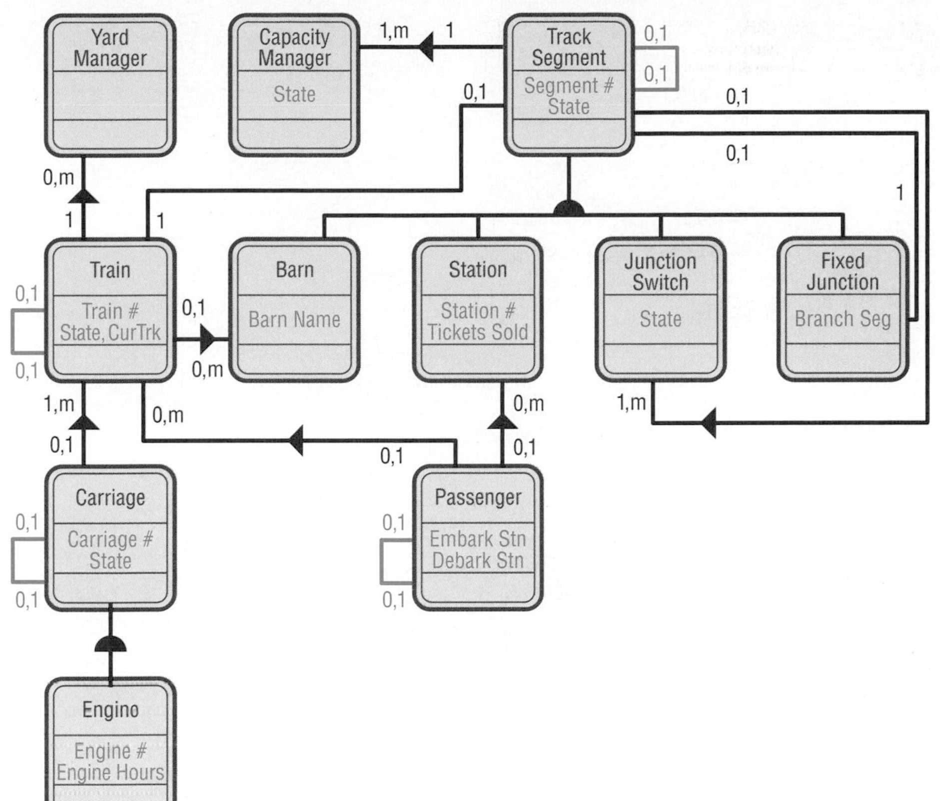

FIGURE 22.15

Preliminary Attribute layer of the O-O design for Kayjay World. The Attribute layer is shown in cyan.

first spot the relationship between a Train and the Track Segment that it occupies. Also, carriages may be linked to other carriages when they form a train in a relationship called a self-association, represented with a line looping back to the same object. We continue by analyzing how passengers queue to buy tickets and board trains so that each passenger is related to the preceding and the following passengers in the queue. We also note that Track Segments are related to other Track Segments because one segment follows a preceding segment and precedes a following segment. Fixed Junctions are branch segments related to a Track Segment joining the main line. There are many alternative ways to design branch track lines like the line going to the yard and the barn. One good way is to ensure that a Junction Switch has a branch line and that the line to the yard and barn belongs to the Junction Switch in a Whole–Part structure.

Now we can expand on the diagram of the attribute layer by creating a corresponding textual specification, as shown in Figure 22.16. This preliminary specification template contains more details about the attributes of the objects. It is called a template because it becomes the basic outline for the full specification that will be developed as the project continues.

ANALYZING SERVICES

Services—also called methods or procedures—become part of objects in much the same way as attributes. Because services frequently involve changes in the state of an object, they are most commonly analyzed and designed using state diagrams. Consequently, service analysis consists of three activities: object state analysis, service specification, and message specification.

Object State Analysis. We can discover state changes most easily by finding those attributes in each object that affect the object's behavior. As we examine each attribute, we ask, "Will the object behavior change when this attribute's value is

Specification: Track Segment
Attribute: Segment Number
Attribute: State (Occupied/Free)
Attribute: Next Segment
Attribute: Previous Segment

Specification: Station
Attribute: Station Number
Attribute: Tickets Sold

Specification: Junction Switch
Attribute: State (Normal/Branch); Normal means a connection to the Next Segment, Branch means a connection to Branch Segment

Specification: Fixed Junction
Attribute: Branch Segment; The merging segment number

Specification: Barn
Attribute: Barn Name

Specification: Train
Attribute: Train Number
Attribute: Service State (In/Out/Inop)

Specification: Carriage
Attribute: Carriage Number
Attribute: State (Operable/Inoperable)

Specification: Engine
Attribute: Engine Number
Attribute: Engine Hours

Specification: Passenger
Attribute: Embarking Station
Attribute: Destination Station

Specification: Capacity Manager
Attribute: State (Operating/Closed)

FIGURE 22.16
Preliminary specifications template for the Kayjay World monorail system.

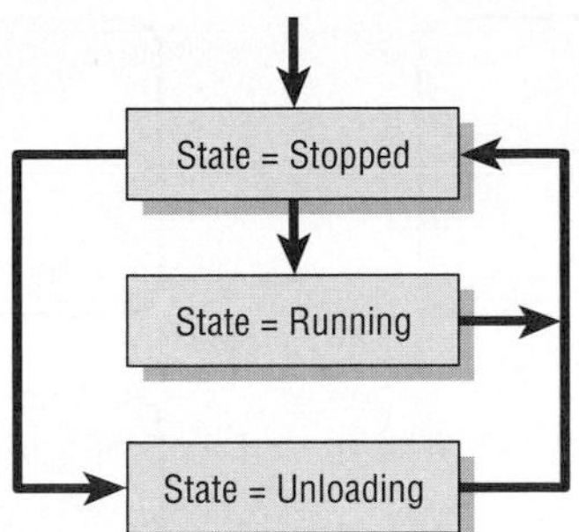

FIGURE 22.17
State diagram of a state variable for train carriages in the Kayjay World monorail system.

changed?" Where no attributes change the object behavior—yet when we know the object will behave differently under certain conditions—we should probably add a "state" attribute. For example, if we are analyzing a train carriage that will pick up and drop off passengers, we know that the train should behave differently in reaction to a "discharge passengers" message depending on whether the train is stopped at a station platform or careening down a straightaway track at 90 miles per hour. Figure 22.17 illustrates a state diagram for a state variable for such a train carriage. The arrow at the top of the State = Stopped box shows that the initial state when the object is created is always Stopped. The other arrows show possible state changes, such as, from Stopped to Running or Stopped to Unloading. Notice that there is no way to change states from Running to Unloading, which logically prevents the object from discharging passengers while in motion. State diagrams are added as needed to the specification templates to document such state attributes.

Service Specification. Services are categorized as either simple or complex. Simple services involve very few conditions or operations and often apply to every Class-&-Object in the system. This category includes such services as create-object, store-object, retrieve-object, connect-object (make an instance connection), access-object (get or set values for attributes), and delete-object. Simple services are implicit, sometimes specified once in the design and never mentioned again.

Complex services involve loops, many operations, or compound conditions. These services typically apply to only one Class-&-Object. Complex services frequently entail "companion" or "private" services that are similar to subroutine modules. Private services are internal subroutines that only the object itself knows about and can trigger. Companion services are subroutines used by complex services that can also be triggered as distinct services by messages from other objects in the system. Complex services are always depicted in the service layer of the layered diagram package. The names of such services appear in the lower section of class boxes. Figure 22.18 shows three complex services: Move in the Vehicle object, Emer.Stop in the Operating Sys object, and Emer.Stop in the Database object.

We specify complex services more completely in the specification template. Almost any of the procedural specification tools discussed earlier in the chapters on software engineering (Chapter 20) and structured decision systems (Chapter 11) can be used, such as program flowcharts, Warnier-Orr diagrams, decision tables, or structured English.

Message Specification. Messages detail how one object's behavior can trigger behavior in another object; that is, messages are generated by one object with the intention of triggering a service in another object. Essentially messages document the dependence of one process on another process in a different object. Messages exist solely to communicate between services and entail both control flow and data flow.

FIGURE 22.18

Examples of complex services depicted in the Service layer of the layered diagram. The Service layer is shown in green.

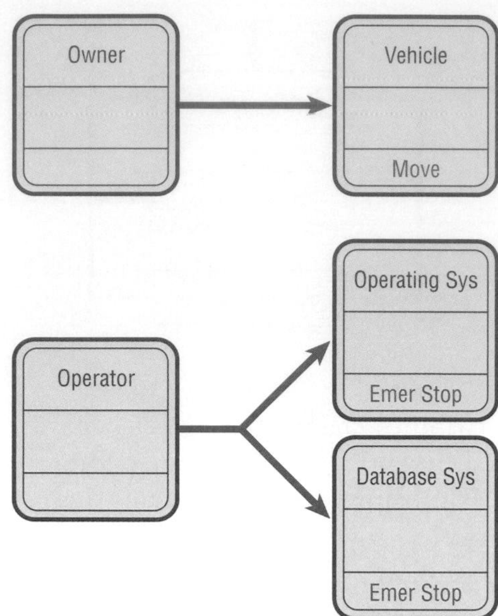

Messages regarding simple services are not documented in the layered diagram package because they are usually implicit in the simple services. Messages directed to classes, such as create-object or delete-object, are usually not diagrammed either. Consequently, most diagrammed messages end up being from object to object rather than from class to class or object to class.

Messages are shown as broad arrows in the service layer of the layered diagram package. In the lower portion of the figure, a single message is being sent from one object to two objects. Conversely, it may be the case that two services are needed independently by a single object. In this example, however, the Operator object generates a single message that triggers emergency shutdown behavior in both the Database and Operating Sys objects simultaneously.

Because messages detail complex services, they usually align with the sending service and the receiving service. Even in moderately complex systems, however, there will be very few complex services and consequently very few messages in the diagram. This fact naturally tends to highlight the really important functions of the system.

Complex services that are not triggered by messages tend to be triggered by timed events or human interaction. A class without an incoming message but with a complete service will usually need some sort of human interface component.

ASSEMBLING THE SPECIFICATION TEMPLATE

When we add service details to our O-O specification template, this portion of the analysis swells with detail. State diagrams, structured English, and flowcharts can be lengthy. The exact format of the specification can vary considerably from analysis to analysis. Figure 22.19 shows a basic specification template outline. We can add any important object details that need to be explicitly stated for the designers and programmers.

KAYJAY WORLD EXAMPLE 4

Figure 22.20 shows the analysis of the Kayjay example with the Services layer added. There are three complex services among the objects: the Yard Manager's Assemble train/Disassemble train service, the Capacity Manager's Check service (Check determines whether to add or remove trains in the system), and the Move service in the Train object. Almost everything that happens in the system results

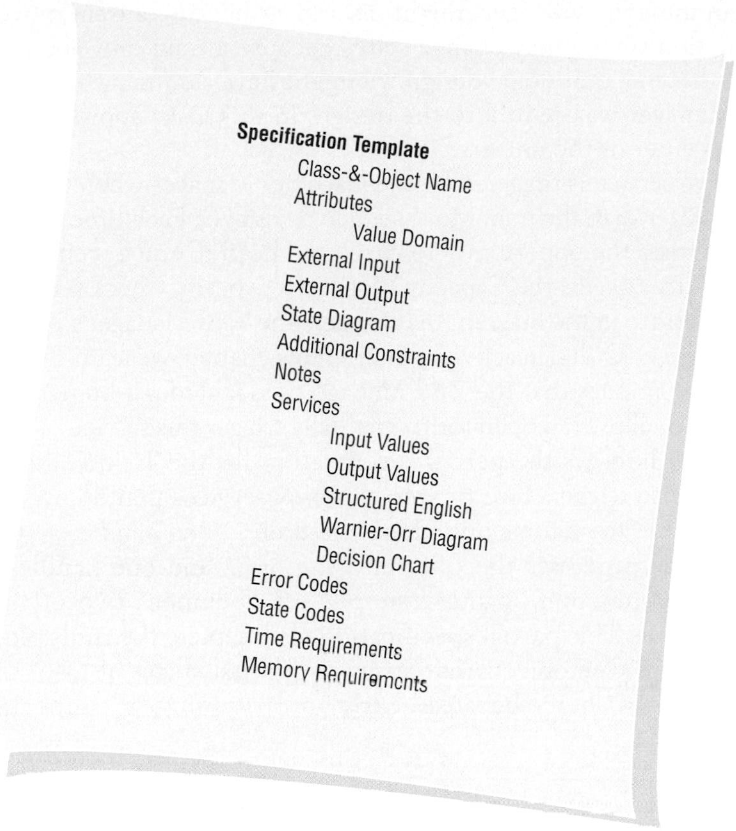

Specification Template

- Class-&-Object Name
- Attributes
  - Value Domain
- External Input
- External Output
- State Diagram
- Additional Constraints
- Notes
- Services
  - Input Values
  - Output Values
  - Structured English
  - Warnier-Orr Diagram
  - Decision Chart
- Error Codes
- State Codes
- Time Requirements
- Memory Requirements

FIGURE 22.19
The basic outline for a specification template.

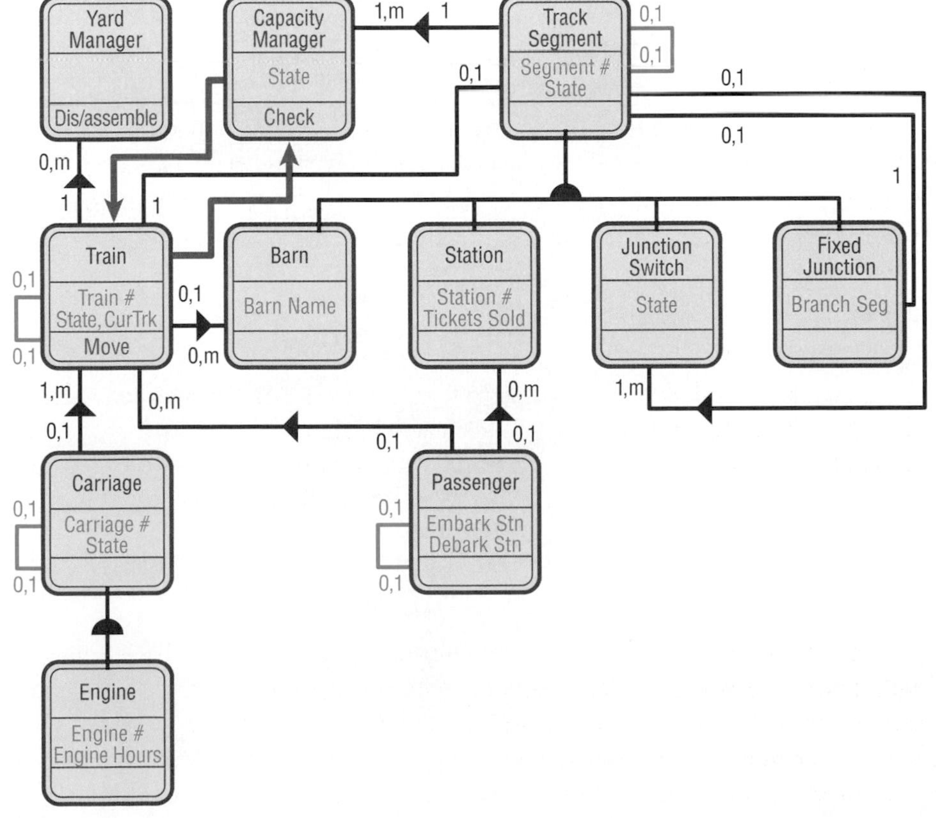

FIGURE 22.20
Layered diagram showing the Service layer of the Kayjay World monorail system. The Service layer is shown in green.

from a train moving. Passengers might disembark because a train moves into a station. A junction switch may change states because a train moves out of service. It seems odd that in a traditional design we might have assumed the Yard Manager or Capacity Manager was central to the design. In this O-O approach, however, the Train is the center of the universe.

The Move service is triggered by the Capacity Manager when the train is moved into service. After that, the train Move service takes over. Each time a train moves, the Move service uses the opportunity to check the need to add or remove trains. Thus, the Move service triggers the Capacity Manager's capacity Check service. Notice that there is no message in the diagram that triggers the Yard Manager's Dis/assemble service. During the O-O design activities later in this chapter, we learn that this "missing" trigger message implies that the Yard Manage class includes a human interface. This human interface allows an operator to manually trigger this service.

Figure 22.21 shows the partial specification for the Train Class services. The Move service also triggers four private complex services that do not appear in the layered diagram: One figures out where the train is, one handles setting the junction switch, one processes the train into the barn, and one handles the case of entering a fixed junction. In our example, we document two of these services using pseudocode. This partial specification also depicts the multiple inheritance from the List and Cell objects that arises in the first stages of O-O design activities. It is shown here because early prototyping of important objects

FIGURE 22.21
Partial specification for the Train class in the Kayjay World system.

```
Specification: Train
    Attribute: Train Number
    Attribute: Circuits completed; number of circuits
               completed since last leaving barn
    Attribute: Service State (Inop, Queued, Running,
               Barnbound, Debarking/Running, Debarking/
               Barnbound, Embarking)
    Attribute: Current Track Segment
    Attribute: Pointer to a CarList
    External_Input:
    External_Output:
    Additional_Constraints:
    Notes:          A train is a list of passengers. It is also related to the
                    CARLIST object, which is a list of carriages.
    Service: Create me
    Service: Destroy me
    Service: Access Train Number (read)
    Service: Access Circuits completed (read, incr, zero)
    Service: Access Service State (read, set states)
    State Codes: Inoperable=0, Queued =1, Running=2, Barn-Destined=3,
            Station Debarking & Barn-Destined=4, Station Debarking &
            Running=5, Station Embarking=6
    Return Codes: True=1, False=0
    Error Codes: Success=0, Failure=-1
    Train Service: enter station, debark & embark passengers
        enterStation (Station *) Service chart pseudocode
            Trigger "debark passengers" service
            IF train is "barnbound"
                THEN enterStation service is complete
        ELSE
                trigger "embark passengers" service
                enterStation service is complete

    Train Service: enter barn, join queue of ready trains (Service chart pseudocode not
        shown)
    Train Service: set switch, check if train is occupying a switch–set proper track
        branch for next move opportunity (Service chart pseudocode not shown)
    Train Service: get track, determine which track segment is under train (Service
        chart pseudocode not shown)
    Train Service: find fixed junction, finds track segment that is next for a train
        entering a fixed junction (Service chart pseudocode not shown)
    Train Service: move train, moves train to next segment when possible
        moveTrain ( ) Service chart pseudocode
            IF states are not ok
                THEN return "fails" message
        trigger get track service
        IF get track fails
            THEN return "fails" message
        trigger getNext (in list base class) service
        IF getNext is "none"
            THEN trigger find fixed junction service
                IF find junction fails
                THEN setup next track with getHead (in list base class)
                service
        IF next track segment is not free
            THEN return "error" message
        IF next track is a barn
            THEN trigger enter barn service
            AND return "ok" message
        occupy next track
        set current track value
        free previous track
        IF next track is a station
            THEN trigger enter station service
        prepare for next move by doing
            trigger setSwitch service
            trigger check system capacity service
        return "ok"
```

and services during analysis is frequently useful for establishing feasibility and proof of concept. This early prototyping required the reuse of library classes that handled list and data structures. The List and Cell classes are discussed further in the Kayjay World example of problem domain component design. It is not unusual in O-O projects for analysis and design activities to merge temporarily under opportunistic circumstances, such as prototyping of our Train Move service.

ANALYZING SUBJECTS

In the case of very large systems, we can use an additional layer in the O-O layered diagram package to organize the work of analysis, design, and implementation. This layer provides a means of dividing a complex specification into logical work units. A subject layer is only necessary in large projects involving many classes. The subjects are noted by layering a broad shaded line, one that denotes the boundaries of a particular subject, onto the O-O diagram. The name of the subject is noted in one corner of the subject box.

Usually, a subject will have an apparent "owner" class, which is a class that is centrally connected to all the classes and objects in the subject space. The subject is usually named after this class.

KAYJAY WORLD EXAMPLE 5

Because of the amount of work involved in designing the Train, perhaps we should decide to divide the project into two subjects: (1) the Train Subject, including the Yard Manager, Passengers, Trains, and their components; and (2) the Railroad Subject, including the Capacity Manager and all forms of Tracks. Figure 22.22 shows the analysis of Kayjay World with the subject layer added.

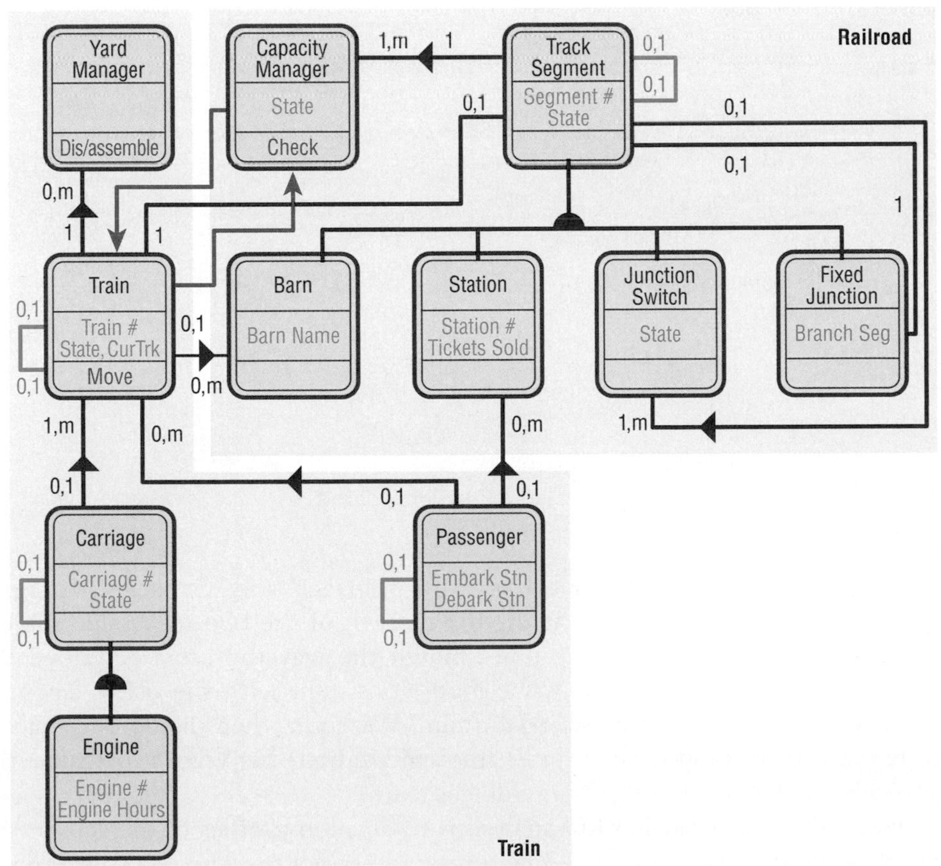

FIGURE 22.22
Subject layer for the analysis of the Kayjay World system. The Subject layer is shown in beige.

Kayjay World Monorail Control System User and Task Descriptions

1 I'm a Train Driver

1.1 Purpose: I control the deadman switch and emergency door and brake controls. I also operate the radio and make announcements on the train's public address system.

1.2 Characteristics

Age: I'm 20 years old.

Level of education: I'm a high school grad with 2 years of college.

Limitations: I don't have broad experience.

1.3 Critical Success Factors

I need to feel I am contributing.

I like to interact with people.

1.4 Skill level: Novice

1.5 Task scenario

Ride in front of train.

Monitor status indicators.

Operate deadman switch.

Make scripted announcements.

Follow emergency procedures.

2 I'm a Ticket Seller

2.1 Purpose: I explain the destinations to customers, collect cash or vouchers, and issue tickets according to customers' requests.

2.2 Characteristics

Age: I'm 17 years old.

Level of education: I'm a high school student.

Limitations: I don't have broad experience.

2.3 Critical Success Factors

I need to feel I am contributing.

I like to interact with people.

2.4 Skill level: Novice

2.5 Task scenario

Customer arrives at window from queue, and I explain the three destinations.

I ask for destination and number in party.

I ask for the proper coupons or a cash amount.

If the coupons and cash are correct, I press the destination button and give tickets to customer.

I tell the customer the next departure time and current train frequency (from the video screen).

I direct customer to the boarding queue.

FIGURE 22.23
Profiles of two employees of the Kayjay World monorail system.

We could have named the subjects "Yard Manager" and "Capacity Manager," because one of these classes was at the top of each of the two respective subject structures. These names, however, don't match the way the users described the two general subjects. We choose to be more consistent with the users' language and name the Subjects "Railroad" and "Train." We could then divide our analysis and design project (and perhaps later, implementation) between two teams: the Railroad Subject team and the Train Subject team.

One of the final tasks in O-O analysis is to develop profiles of the actors who are part of the system. Figure 22.23 features two employees who work for Kayjay:

a train driver and a ticket seller. The profile is made up of five parts: (1) purpose, (2) characteristics, (3) critical success factors, (4) skill level, and (5) task scenario.

Purpose explains why the actor is on the scene in the first place. It defines the employee's role. Characteristics describe the actor. They may include the employee's age, level of education, or even the person's limitations as an employee. Critical success factors involve an actor's performance. For example, an employee may be rewarded for good performance in these critical success factors. Skill level tells whether the employee is a novice or an expert. The task scenario is the most important part of the profile. A task scenario is a walkthrough of all the steps or procedures an actor may take. An employee may execute all or any of these tasks in the course of a day's work. Task scenarios also provide a check on the operations that are part of the O-O system.

OBJECT-ORIENTED DESIGN

Design activities in the Coad and Yourdon approach carry the analysis tools forward into the complete set of specifications for implementation. Where the analysis is reasonably technology-independent, the design activities become increasingly oriented toward a particular O-O language and development environment.

O-O design activities are grouped into the four major components of the final system: the problem component, the human interface component, the data management component, and the task management component.

All of the analysis documentation should carry directly into the design stage. Few new tools are needed at this point. The layered diagram package and the specification template remain the major components of the design. These documents are not supplemented or replaced, but instead are expanded to include the remaining implementation details during the design phase.

We frequently use prototypes (as discussed earlier in Chapter 8) during the design phase. Rough versions of the objects are created and tested in their roles within the four components. Hence, frequently the design package is sent forward to the programmers with portions of the program code already written. Designers will often use the expected implementation language (such as C++) as the mechanism to write complete specifications for the classes. For example, the designer may find it easy to copy the C++ class definition from an operational prototype into the specification. That may prove to mean less work for the designer, and it eliminates duplicate efforts on the part of the designers and implementation programmers.

DESIGNING THE PROBLEM DOMAIN COMPONENT

The problem domain component (PDC) is the basic set of functional objects that arrive from the analysis stage. These objects directly solve the problems intended to be solved by the system we are building. The other components, such as human interface and data management, are incidental functions that must be added to the PDC to "get it working."

Consequently, the design for the PDC is mostly completed in the analysis stage. Only three activities are needed to complete the design of the PDC: reuse design, implementation structures, and language accommodation.

Reuse Design. We may want to add new classes to the PDC so as to reuse objects. For example, there are commercial packages of highly generalized classes for objects. An experienced O-O programming organization usually owns a library of classes developed in-house for objects. These libraries and packages may contain classes that have attributes and services to objects similar to those required in our

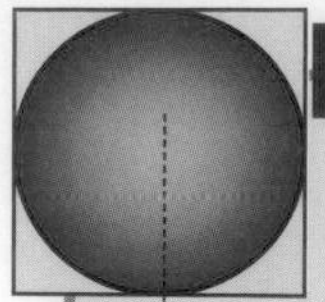

CONSULTING OPPORTUNITY 22.2

RECYCLING THE PROGRAMMING ENVIRONMENT

"I feel like I'm writing the same code over and over again," says Benito Pérez, a programmer working on a new automated warehouse design. "I have written so many programs lately that dealt with robotic-type things that control themselves: automated mailroom trolleys, building surveillance robots, automatic pool cleaners, automatic lawnmowers, monorail trains, and now warehouse trolleys. They are all variations on a theme."

Lisa Bernoulli, the project manager, had heard this sort of complaint for years. She replies, "Oh come on, Ben. These things aren't really that close. How can you compare a mailroom robot, an automated warehouse, and a monorail train? I'll bet less than 10 percent of the code is the same."

"Look," says Benito. "All three involve machines that have to find a starting point, follow a circuitous route, make stops for loading and unloading, and eventually go to a stopping point. All three have to make decisions at branches in their routes. All three have to avoid colliding with things. I'm tired of redesigning code that is largely familiar to me."

"Hmmm," Lisa muses as she looks over the basic requirements for the warehouse system and remembers the monorail system she and Benito had worked on last year. The requirements regarded a small-lot electronics manufacturing firm that was automating its warehouse and product movement system. The warehouse contains incoming parts, work in progress, and finished goods. The automated warehouse uses a flatbed robot trolley. This robot is a four-wheel electric cart, similar to a golf cart except that it has no seats. Flatbed robot trolleys have a flat, six-foot by four-foot cargo surface about three feet above ground level. These trolleys have a radio communications device that provides a real-time data link to a central warehouse computer. Flatbed trolleys have two sensors: a path sensor that detects a special type of paint and a motion sensor that detects movement. These trolleys follow painted paths around the factory floor. Special paint codes mark forks or branches in the paths, trolley start or stop points, and general location points.

The facility includes three loading dock stations and 10 workstations. Each station has a video terminal or computer connected to the central computer. When products are needed or are ready to be collected from a workstation, the central computer is informed by the worker at the station. The central computer then dispatches trolleys accordingly. Each station has a drop point and a pickup point. Flatbed trolleys move about the factory picking up work at pickup points and dropping off work at drop points. The program that will run the trolleys must interact heavily with the existing job-scheduling program that helps schedule workstation tasks.

How similar are the trolleys to the monorail trains in the Kayjay World examples? How should Lisa go about reusing Benito Pérez's work on the monorail in their current task of creating a trolley object? Explain in two paragraphs.

design. We can add these reusable classes to our design as base classes in a Gen–Spec structure. The derived classes in these Gen–Spec structures are the classes originally developed in the analysis stage.

Implementation Structures. We may want to add other structures to our design purely for implementation reasons. Also, we may want to use aggregation structures to create natural entry points for lists or queues or a Gen–Spec structure to permit several classes of objects to share a protocol or data structure. These structures use the inheritance concept to make the programming task much easier.

Language Accommodation. We may need to fix the design so that the structures can be built in the chosen programming language, because these languages may have different inheritance patterns. Some languages support multiple inheritance, others only support single inheritance, and still others support *no* inheritance. In the more restrictive cases, the inheritance patterns in the design must be modified to allow for the capabilities of the implementation language.

CRC CARDS AND OBJECT THINK

One way to begin enacting the object-oriented approach is to start thinking and talking in this new way. At first, it is natural that an analyst schooled in the structured approaches featured earlier in the text will think in those ways when attempting to perform an O-O analysis. Usually, you will then try to draw an

Class Name:			
Superclasses:			
Subclasses:			
Responsibilities	Collaborators	Object Think	Property

FIGURE 22.24
A CRC card with Object Think statements.

analogy between the traditional methods and O-O approaches. Eventually, you will use UML diagramming for O-O analysis without needing to make that conceptual translation between older methods and newer ones. This process is analogous to the process one undergoes when learning a new language, as when a native English speaker learns French. Eventually, the accomplished student will converse in French without translating statements from English first. Later in the chapter, we learn the UML approach, which helps us to diagram systems from this new perspective.

Figure 22.24 shows a template for a CRC card. CRC stands for ***C**lass*, ***R**esponsibilities*, and ***C**ollaborators* that the analyst can use when beginning to model or talk about the system from an O-O perspective. Notice that the form has blanks where the analyst can fill in the *class name, superclasses, subclasses, responsibilities*, and *collaborators*. CRC cards are used to represent the responsibilities of classes and the interaction between the classes. Analysts create the cards based on scenarios that outline system requirements. These scenarios model the behavior of the system under study. CRC cards can be written up manually on 3″ × 5″ cards for flexibility if they are to be used in a group, or a computer can be used to create them.

We have added two columns to the original CRC card template, the Object Think column and the Property column. The Object Think statements are written in plain English, and the Property or attribute name is entered in its proper place. The purposes of these columns are to clarify thinking and help move toward creating the UML diagrams.

INTERACTING DURING A CRC SESSION

CRC cards can be created interactively with a handful of analysts who can work together to identify the class in the problem domain (Beck & Cunningham, 1989; and Butler, 1996). One suggestion is to find all the nouns and verbs in a problem statement that has been created to capture the problem. Nouns usually indicate the classes in the system, and responsibilities can be found by identifying the verbs. With your analyst group, brainstorm to identity all the classes you can. Follow the standard format for brainstorming, which is not to criticize any participant's

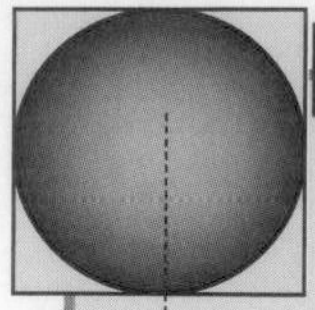

CONSULTING OPPORTUNITY 22.3

MAKING THE MAGIC REEL

Fred and Ginger, owners of the FilmMagic chain of stores (which rent videos, DVDs, and video games), have always been interested in new technology. Because they keep adding new products to rent (such as DVDs and new games for PlayStation II), their business has grown into a smash hit in several cities.

Because your home is close to their original store, you have become friends with them over the dozen years they have been in business, renting tapes as they make the move from the big screen to videos. You often swap views about which movies are "must sees" and which are "bombs."

Since you have described the new object-oriented approaches you have been learning, they would like you to analyze their business using this approach. You can find a summary of FilmMagic business activities in Figure 9.17. Notice also the series of data flow diagrams in that chapter to help you conceptualize the problem and begin making the transition to Object Think.

Because you are such good friends with Fred and Ginger and because you wouldn't mind a little practical experience using O-O thinking, you agree to apply what you know and give them a report. Once you have reread the business activities for FilmMagic, provide a timely review by completing the following tasks:

- Use the CRC cards technique to list classes, responsibilities, and collaborators.
- Use the Object Think technique to list "knows" and corresponding attributes for the objects in those classes identified in the previous stage.

Write up both steps and waltz over to FilmMagic headquarters with your report in hand. Clearly, Fred and Ginger are hoping for a rave review.

Note: Based on a problem written by Dr. Ping Zhang.

response at this point. When all classes have been identified, the analysts can then compile them, weed out the illogical ones, and write each one on its own card. Assign one class to each person in the group, who will own it for the duration of the CRC session.

Next, the group creates scenarios that are actually walkthroughs of system functions by taking required functionality for the requirements document previously created. Normal systems operations should be considered first, with exceptions such as error recovery taken up after the normal ones have been covered.

As the group decides which class is responsible for a particular function, the analyst who "owns" the class for the session picks up that card and declares, "I need to fulfill my responsibility." Notice that when a card is held in the air, it is considered an object and it can do things. The group then proceeds to refine the responsibility into smaller and smaller tasks if possible. These tasks can be fulfilled by the object if it is appropriate, or the group can decide that it can be fulfilled by interacting with other things. If there are no other appropriate classes in existence, the group may need to create one.

Examine the four CRC cards depicted in Figure 22.25 showing four classes of a car rental system. Observe that these cards represent four of the classes in a car rental system. Notice that in a class called "Rental Contract," the systems analyst is referred to three collaborators: the customer, the rental agent, and the vehicle itself. These collaborators are then described as classes of their own in the other three CRC cards.

The responsibilities listed will eventually evolve into what are called methods or operations in UML. The Object Think statements seem elementary, but they are conversational so as to encourage a group of analysts during a CRC session to describe as many of these statements as possible. As shown in the example, everything is in first person so that even the vehicle speaks, "I know my make. I know my model." These statements can then be used to describe what we call properties, or attributes, in UML. These properties can be called by their variable names, such as "make" and "model."

Class Name: Rental Contract
Superclasses:
Subclasses:

Responsibilities	Collaborators	Object Think	Property
Update customer Info	Customer	I know the customer	Customer
Select vehicle	Rental Agent	I know the rental agent	Rental Agent
Return vehicle	Rental Agent	I know the vehicle	Vehicle
Print contract	Vehicle	I know the time checkout	Out Time Stamp
		I know the time return	In Time Stamp
		I know the miles driven	Miles Driven
		I know any damage done	Damage Report
		I know the rental status	Status

Class Name: Customer
Superclasses:
Subclasses:

Responsibilities	Collaborators	Object Think	Property
Provide customer ID	Rental Agent	I know my first name	First Name
Provide driver's license	Vehicle	I know my last name	Last Name
Provide credit card		I know my address	Address
		I know my home phone	Home Phone
		I know my credit card #	CC Number
		I know my credit type	CC type
		I know my expiry data	CC Expiration
		I know my license #	License

Class Name: Rental Agent
Superclasses:
Subclasses:

Responsibilities	Collaborators	Object Think	Property
Assign vehicle	Customer	I know my first name	First Name
Check customer ID	Vehicle	I know my last name	Last Name
Check driver's license		I know my starting date	Date of Hire
Check credit card			
Process check out			
Process check in			

Class Name: Vehicle
Superclasses:
Subclasses: Car, Truck

Responsibilities	Collaborators	Object Think	Property
Provide vehicle details		I know my identity	ID
		I know my make	Make
		I know my model	Model
		I know my year	Year
		I know my class	Class
		I know my license plate	License Plate
		I know my mileage	Mileage

FIGURE 22.25
Four CRC cards for the rental car agency show how analysts fill in the details for classes, responsibilities, and collaborators, as well as for Object Think statements and property names.

UNIFIED MODELING LANGUAGE (UML)

During the 1990s, various techniques for the analysis and design of O-O systems were developed and used widely throughout the IT industry. Among the most popular is Unified Modeling Language (UML), developed by Grady Booch, Ivar Jacobson, and James Rumbaugh. Each was the primary author of independent O-O modeling methods: Booch, the "Booch" method; Jacobson, the Object-Oriented Software Engineering (OOSE) method; and Rumbaugh, the Object Modeling Technique (OMT). The three men collaborated to unify their methods, resulting in UML. The Object Management Group, an organization founded by leading corporations in the IT industry, adopted UML as a standard for modeling object-oriented systems in 1997, further solidifying the UML as the leader in O-O analysis and design methods.

Due to its wide acceptance and usage, the UML approach is well worth investigating and understanding. UML provides a standardized set of tools to document the analysis and design of a software system. The UML toolset includes diagrams that allow people to visualize the construction of an O-O system, similar to the way a set of blueprints allows people to visualize the construction of a building. Whether you are working independently or with a large development team, the documentation that you create with UML provides an effective means of communication between the development team and the businesspersons on a project.

FIGURE 22.26
An overall view of UML and its components: Things, Relationships, and Diagrams.

UML Category	UML Elements	Specific UML Details
Things	Structural Things	Classes Interfaces Collaborations Use Cases Active Classes Components Nodes
	Behavioral Things	Interactions State Machines
	Grouping Things	Packages
	Annotational Things	Notes
Relationships	Structural Relationships	Dependencies Aggregations Associations Generalizations
	Behavioral Relationships	Communicates Includes Extends Generalizes
Diagrams	Structural Diagrams	Class Diagrams Object Diagrams Component Diagrams Deployment Diagrams
	Behavioral Diagrams	Use Case Diagrams Sequence Diagrams Collaboration Diagrams Statechart Diagrams Activity Diagrams

Whereas the Coad and Yourdon OOA-OOD method focused on the five-layer approach described earlier in this chapter, UML breaks down objects and their relationships differently. An overall view of UML and its components is found in Figure 22.26.

The first components, or primary elements, of UML are called "things." You may prefer another word such as object, but in UML we refer to them as things.

Structural things are most common. Structural things are classes, interfaces, use cases, and many other elements that provide a way to create models. Structural things allow the user to describe relationships. Behavioral things describe how things work. Examples of behavioral things are interactions and state machines. Group things are used to define boundaries. An example of a group thing is a package. Finally, we have annotional things so that we can add notes to the diagrams.

Relationships are the glue that holds the things together. It is useful to think of relationships in two ways. Structural relationships are used to tie the things together in the structural diagrams. Structural relationships include dependencies, aggregations, associations, and generalizations. Structured relationships show inheritance, for example. Behavioral relationships are used in the behavioral diagrams. Behavioral relationships can show communication, includes, extends, and generalizes. We can say, for example, that an actor communicates with a use case or a use case extends another use case.

There are two main types of diagrams in UML: structure diagrams and behavioral diagrams. Structural diagrams are used, for example, to describe the relationship between classes. They include class diagrams, object diagrams, component diagrams, and deployment diagrams. Behavioral diagrams, on the other hand, can be used to describe the interaction between people (actors) and the thing we refer to as a use case as shown in Figure 22.27, a screen from Visible Analyst. Behavioral diagrams include use case diagrams, sequence diagrams, collaboration diagrams, state-chart diagrams, and activity diagrams.

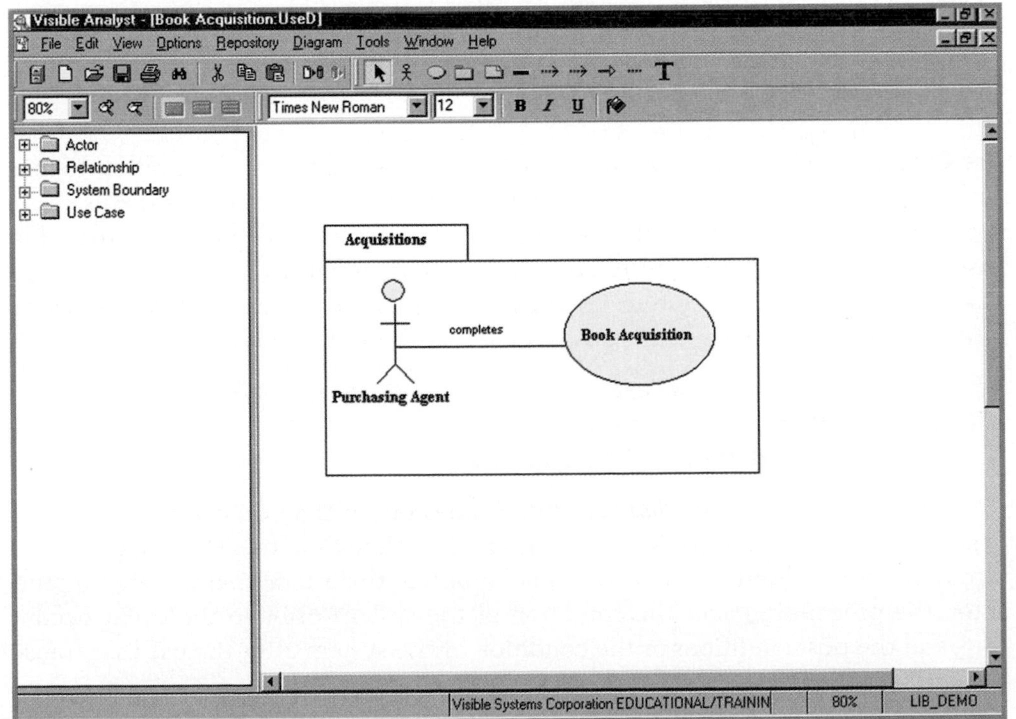

FIGURE 22.27
Visible Analyst is one of the software packages that can be used to draw UML diagrams.

In the remainder of this chapter, we first discuss use case modeling, the basis for all UML techniques. Next, we look into the fundamentals of UML: things and relationships. Then we review UML diagrams, the real toolset of UML. Finally, we discuss a proven methodology to put the UML tools to work for you.

Because entire books are dedicated to the syntax and usage of UML (the actual UML specification document is over 800 pages long), we provide only a brief summary of the most valuable aspects of UML.

USE CASE MODELING AND UML

UML is fundamentally based on an O-O analysis technique known as use case modeling. Understanding UML diagramming requires a good understanding of use case modeling. A use case model describes what a system does without describing how the system does it. The use case model reflects the view of the system from the perspective of a user outside of the system (i.e., the system requirements). UML can be used to analyze the use case model and derive system's objects and their interactions with each other and with the users of the system. Using UML techniques, you further analyze the objects and their interactions to derive object behavior, attributes, and relationships.

An analyst develops the use cases in a cooperative effort with the business experts that help define the requirements of the system. The use case model provides an effective means of communication between the business team and the development team.

A use case model partitions system functionality into behaviors (use cases) that are significant to the users of the system (the actors). Let's define some of the terminology pertaining to use case modeling. We revisit the use case model later in the chapter when we discuss UML diagramming in more depth.

Actor. The term *actor* refers to a particular role of a user of the system. The actor exists outside of the system and interacts with the system in a specific way. An actor can be a human, another system, or a device such as a keyboard or a modem. For instance, a human actor in the Kayjay World example given earlier in the chapter might play the role of a Passenger, a Capacity Manager, or a Yard Manager, whereas a Junction Switch is an example of a device actor.

Use Case. We can think of a use case, pronounced as a noun (yo͞os) rather than a verb (yo͞oz), as a sequence of transactions in a system. The purpose of the use case is to produce something of value to an actor in the system. The use case model is based on the interactions and relationships of individual use cases. Within a use case, an actor using the system initiates an event that begins a related series of interactions within the system. A use case always describes three things:

1. An actor that initiates an event.
2. The event that triggers a use case.
3. The use case that performs the actions triggered by the event.

A use case focuses on *what* the system does rather than on *how* it does it. An example of a use case from Kayjay World is "Initialize System." The use case goes into more detail about the steps taken to complete the interaction initiated by the actor, the preconditions or the condition of the system prior to the event occurring, and the postconditions or the condition of the system after the use case transactions have completed.

Use Case Name:	<the name is the behavior as a short active verb phrase>		<ID Number>
Area:	<functional area of the system that contains this use case>		
Actors:	<a role name or description for the primary actor> <other systems relied upon to accomplish this use case>		
Description:	<A brief overview of the scenario>		
Triggering Event:	<the action upon the system that starts the Use Case>		
Basic Course:	Step	Action	
	1	<List the steps for this particular scenario from the trigger of the behavior through completion of behavior, and any cleanup after>	
	2	<...>	
	3	<...>	
Preconditions:	<What must be true about the state of the system prior to this Use Case>		
Postconditions:	<State of the system as a result of the use case completing>		
Assumptions:	<Conditions that cannot be altered>		
Objective(s) Met:	<List any system objectives met by this use case scenario>		

FIGURE 22.28
A use case template that can be used to organize information for UML models.

Primary Use Case and Use Case Scenarios. The primary use case consists of a standard flow of events within the system that describes a standard system behavior. Use case scenarios are more detailed variations of that behavior; they are exceptions to the main behavior described by the primary use case. An individual use case in the use case model consists of the primary use case and all the use case scenarios, or variations, of the primary use case. The main idea in constructing the use case model is to start at a "mile-high" view of the system by developing the primary use cases and then focus in on the detailed processes of the system with the use case scenarios.

Use Case Document Template. Use cases are documented in a standardized use case document template, which makes the use cases easier to read and provides standardized information for each use case in the model. A use case template is shown in Figure 22.28.

DEVELOPING A USE CASE MODEL

Now that we have described the concepts behind the Use Case Model, we describe the steps required to create one. Start the first draft of your use case at a high level and then make several subsequent passes of analysis through the system requirements, as follows:

1. Review the business specifications and identify the actors within the problem domain.

2. Identify the high-level events and develop the primary use cases that describe those events and how the actors initiate them. Carefully examine the roles played by the actors to identify all the possible primary use cases initiated by each actor.
3. Review each primary use case to determine the possible variations of flow through the use case. From this analysis, establish the use case scenarios. Because the flow of events is usually different in either case, look for activities that could succeed or fail. Also look for any branches in the use case logic where different outcomes are possible.
4. Develop the use case documents for all primary use cases and all important use case scenarios. Be sure to review your results with the business experts to verify and refine the use cases where needed.
5. Once the verification process has been completed and all the business experts agree that the use cases are accurate, you can move to UML diagramming techniques to complete systems analysis and design.

LIMÓN CAR RENTAL AGENCY EXAMPLE

The Limón Car Rental Agency rents cars to customers at the airport. Customers either call to reserve a rental car for specific dates or come to the Limón Car Rental desk at the airport to rent a car without a reservation. The car rental agents work behind the counter, taking phone reservations and waiting on customers.

When renting a car to a customer, the rental agent fills out the customer and rental car information on a rental contract, has the customer provide a credit card number as a down payment, has the customer sign the rental contract, and provides the customer a copy of the rental contract along with the keys for the rental car.

Limón Agency also has service technicians who clean the cars when customers return them, fill cars with gas, and prepare the cars for the next rental. The service technicians also make sure that the cars receive their regular maintenance, such as oil changes, and schedule other service when the cars are in need of repair. Also, Limón has only a limited number of each type of vehicle to rent.

Figure 22.29 represents the use case "Customer Rents Car." The scenario is well described, setting the stage (front desk), the actors (customer and rental agent), the triggering event ("Customer enters Car Rental Agency and inquires about a car rental"). Each step is precisely defined from the initial action to the point where the customer locates the rental car in the parking lot. All pre- and postconditions are listed as well as a list of assumptions (for example, assuming that the customer's credit card is valid). Finally, a list of objectives that are met include basics such as printing the rental contract all the way up the organization's managerial levels to more strategic objectives such as storing information to track customers, contracts, and vehicles usage.

"THINGS": THE PRIMARY ELEMENTS OF UML

Now that we have a basic understanding of use case modeling, we can discuss the fundamental building blocks of UML. The next step is to understand the meaning and importance of "things" UML.

Houses are built with a diversity of building materials, each having its own properties and purpose in the construction and functionality of the home itself. There are innumerable types of building materials, some as simple as a nail, others as complex as a furnace. The architect and builder know the properties, functions, and uses of each building material. For more complex items, they often have manuals describing how to install, operate, and maintain them. There are similarities

Use Case Name:	Customer Rents Car	
Area:	Front Desk Activity	ACME-01.000
Actors:	– Customer – Rental Agent	
Description:	Customer interacts with Rental Agent to rent a car.	
Triggering Event:	Customer enters Car Rental Agency and inquires about a car rental	

Basic Course:

Step	Action
1	Customer initiates car rental with Rental Agent
2	Customer indicates type of car desired: Compact, Midsize, or Luxury
3	Rental Agent checks for car availability
4	Rental Agent initiates Rental Contract
5	Rental Agent asks Customer for Driver's License and verifies information
6	Rental Agent gathers Customer information and enters in Rental Contract
7	Rental Agent enters car information in Rental Contract
8	Customer provides credit card to Rental Agent for approval verification
9	Rental Agent prints copy of Rental Contract
10	Customer agrees to rental terms and signs Rental Contract
11	Rental Agent provides key and vehicle information to Customer
12	Customer finds car in Rental Agency parking lot, and drives away

Preconditions:	– Customer desires to rent car – A car is waiting on the lot for rental
Postconditions:	– Rented car is unavailable until the Customer returns car and rental is complete
Assumptions:	– Rental Agency is open – Desired car is available for rental – Customer's driver's license is valid – Customer's credit card is valid
Objective(s) Met:	– System provides fully editable online Rental Contract – Rental Contract information can be printed on multipart preprinted form – System provides means to track Customer, Rental Contract, and Vehicle information

FIGURE 22.29
Organizing material for a use case "Customer Rents Car."

here with O-O systems. The things used to build an O-O system have various levels of complexity, properties, and functionality. The more complex the object, the more documentation you should provide to describe the properties, methods, relationships, and interactions of those objects within your system.

Here you begin to see how the UML method shares similarities with Coad and Yourdon's OOA-OOD method, although the terminology is a bit different. As shown below, the "things" in UML describe the "objects" and other concepts discussed previously with OOA-OOD. The two most often used groupings of things in UML are structural things and behavioral things.

STRUCTURAL THINGS

Structural things are just that: things that define the conceptual and physical structure of an O-O system model. It may sound simplistic, but *nouns* describe structural things in UML. UML defines seven categories of structural things. The first five categories (class, interface, collaboration, use case, and control class) are conceptual or logical in nature, and the last two (component and node) are physical in nature.

Classes. The definition of a class in UML is identical to the definition of a class in OOA-OOD, that is, the class provides a common template for a group of individual objects with common attributes and common behavior. Classes are represented in UML diagrams very similar to the representation in OOA-OOD except that the class "box" has square corners but otherwise is not unlike the "bubtangle" from the earlier part of the chapter.

An example of a class is shown in Figure 22.30. Each class has a name, in this case "Rental Car." A class also has Properties or, in other words, the attributes of the rental car. These properties include the rental car's size, color, make, and model. In the lower portion of the class symbol are Methods. Methods are the operations that can be performed. In our example, a rental car can be rented as in "RentOut ( )," checked back in as in "Checkin ( )," or even serviced as in "Service ( )." Each of these operations can be performed many times.

Interfaces. An interface is the behavior of a class or component of a system that is noticeable from outside the class or component. The interface is the standard means by which external entities interact with the class or component, regardless of how the class or component executes its behavior internally. For instance, a serial port on a PC provides a standard interface for numerous devices such as a mouse or a modem. Even though the devices perform totally unrelated functions, the manufacturers know the operations provided by the standard serial port interface. The manufacturers provide software drivers that communicate with their devices via the serial port interface to provide specific functionality to the PC. In

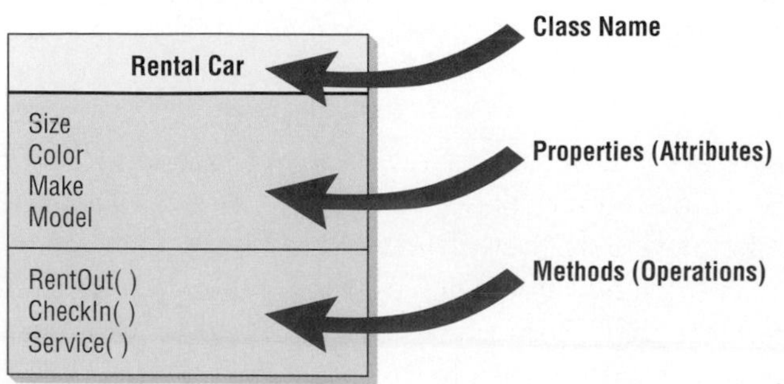

FIGURE 22.30
An example of a class in UML.

this case, the PC is the object, and the serial modem provides the interface for external devices to interact in a specific manner with the PC.

Collaborations. Collaborations describe the interactions of two or more things in the system that perform a behavior that is more than any one of the things can do alone. For instance, a car can be broken down into several thousand individual parts. The parts are put together to form the major subsystems of the vehicle: the engine, the transmission, the brake system, and so forth. The individual parts of the car can be thought of as classes, because they have distinct attributes and functions. The individual parts of the engine form a collaboration, because they "collaborate" with each other to make the engine run when the driver steps on the accelerator.

Use Cases. As explained above, a use case describes a series of actions that demonstrate a distinct behavior of the system and its interactions with the actors. In UML, each use case is treated as a "thing," an independent entity that can be shown to relate and interact with other entities within a system.

Control or Active Classes. A control class is a special type of class that can initiate and control an independent flow of activity within the system, allowing the system to perform concurrent tasks. This type of class is also known as an *active class*. In a football game, each player on the field is performing independent but concurrent activities during the game, but each team is controlled and directed by the coach. Some software systems have independent processes that run concurrently. A control class acts as the "coach" in such systems, initiating and controlling concurrent processes.

Components. A component is a physical part of a system that represents the services and interfaces implemented by the elements contained within that component, including the software code for those elements. A physical component of a software system could be, for example, a .DLL, a .COM object, a Java Bean, or a package. Components are reusable and may be used in one or more systems by other components within the system. Again, think of a car. Auto manufacturers often use the same engine component within a number of different models of vehicles. If the engine in your car breaks down, you can often have a rebuilt engine from another vehicle installed to replace your defective engine.

Nodes. A node represents a piece of hardware on which your system executes. Components are physically deployed on nodes.

BEHAVIORAL THINGS

The behavioral things are the *verbs* of a UML model and represent the behavior of the system and the states of the system before, during, and after the behaviors occur. We discussed some of these concepts in Coad and Yourdon's OOA-OOD methodology.

Interactions. Messages sent between a set of objects within the system to perform a specific task are called interactions. When you start a car, you initiate a chain of interactions within the vehicle that ultimately results in the engine running. First, you insert the key in the ignition. When you turn the ignition, it sends a message to the starter to "start the car." From that point on, the various components of your vehicle begin to interact with each other to make the car "ready to drive."

State Machines. In the first part of the chapter, we discussed the Coad and Yourdon OOA-OOD method called object state analysis. UML defines the *state machine* as the series of states that an object goes through in response to actions within the system. State machines not only document the states of an object, but also the actions and conditions that must be met to transition the object from one state to another.

GROUPING THINGS

Packages. In UML, things can be grouped together in packages. Packages can be considered as physical subsystems, .DLLs, .EXEs, ActiveX Objects, and so forth. Systems are implemented and deployed in packages.

ANNOTATIONAL THINGS

Things can also be annotational, giving developers more information about the system.

Notes. Notes are just that: notes. Notes can be attached to anything in UML: objects, behaviors, relationships, diagrams, or anything that requires detailed descriptions, assumptions, or any information relevant to the design and functionality of the system. The success of UML relies on the complete and accurate documentation of your system model to provide as much information as possible to the development team. Notes provide a source of common knowledge and understanding about your system to help put your developers "on the same page."

"RELATIONSHIPS": THE FRAMEWORK THAT HOLDS THINGS TOGETHER

Homes are not built merely by stacking bricks upon bricks or boards next to boards. The materials in a building can be grouped into subsystems, such as the framework, the plumbing, and the heating and cooling system. Each subsystem is made of individual pieces, and each piece is related to other pieces. Some pieces are connected physically, some are dependent on other pieces to function. Others are related because they share similarities and can be grouped into broader categories of materials such as nails and screws, which could both be described as fasteners.

Just as a home is made of more than just unrelated pieces of material stacked together, software systems are made of more than just unrelated and unstructured "things." Relationships provide the framework that holds the UML things. Things in a software system are related to each other in many different ways.

Because similar symbols are used in different ways in different UML diagrams, it is useful to think of UML relationships as two types, structural relationships and behavioral relationships.

STRUCTURAL RELATIONSHIPS

UML describes four primary types of structural relationships: dependencies, aggregations, associations, and generalizations. These types of relationships are used between classes, for example. Think of structural relationships as passive relationships; that is, they describe the way things are in relation to one another. Structural relationships are explained in Figure 22.31. For each relationship type, a special kind of arrow or line is used for clarity. In UML, each of the arrows and arrowheads are drawn differently.

Relationship Type	UML Notation	Example
Dependency (uses)	- - - - - - - - - - - - - - - ->	A vehicle **uses** fuel.
Aggregation (has a)	————————◇	A vehicle **has an** engine. A car **has** axles.
Association	————————	A vehicle **is licensed by** the New Jersey Department of Motor Vehicles.
Generalization (is a)	————————▷	A car **is a** vehicle.

FIGURE 22.31
Four types of UML structural relationships and the arrows and lines used to represent the relationships.

To give the reader of the UML diagram information about the number of things at either end of a relationship, we use the concept of multiplicity. For example, one car can only be assigned to one customer at a time, but a car rental agency can have many cars to rent. The kinds of relationships can be 1) one and only one; 2) zero, one, or more; 3) zero or one; or 4) one or more. The notation is shown in Figure 22.32.

Dependencies. Dependencies are behavioral connections between things. A dependency is a relationship where one thing somehow affects another thing that uses it, without necessarily being affected in return. A dependency relationship is often described as a "uses" or "used by" relationship. In a car, the electrical system provides energy to the starter to start the engine when you turn the ignition switch. The electrical system and the starter share a dependency.

Aggregations. Aggregations are often described as "has a" relationships. Aggregations provide a means of showing that the whole object is composed of the sum of its parts (other objects). In the rental car example, the rental contract "has a" customer, and the rental contract also "has a" car. The rental agency system would view the rental contract as more than a piece of paper, but rather as a binding agreement that associates a particular customer to a particular vehicle.

Associations. Associations describe structural connections between things. In a UML model, associations are usually described with a descriptive name (such as "customer has signed a rental contract") and may reflect the number of things that may be associated with the other thing (such as "one customer may have signed many rental contracts").

Generalizations. As with the nails and screws described above as fasteners, a generalization describes a relationship between a general kind of thing and a more specific kind of thing. This type of relationship is often described as a "is a" relationship. For example, a car "is a" vehicle and a truck "is a" vehicle. In this case, "vehicle" is the general thing, whereas "car" and "truck" are the more specific things. Generalization relationships are used for modeling class inheritance and specialization.

Multiplicity	UML Notation
One and only one	1
Zero, one, or more	*
Zero or one	0..1
One or more	1..*

FIGURE 22.32
Multiplicity describes the possible number of things on either end of a relationship.

Relationships	Symbol	Example
Communicates	————————	An Actor "Customer" interacts or **communicates** with a Use Case "Rent Vehicle"
Includes	<- - - < includes > - - - -	The Use Case "Verify Credit Card" and "Check Driver's License" **includes** the common use case "Rent Vehicle"
Extends	- - - < extends > - - - ->	The Use Case "Arrange for Added Insurance" **extends** the Use Case "Rent Vehicle"
Generalizes	————————▷	A "Regular Customer" **generalizes** a "Gold Card" customer.

FIGURE 22.33

Four types of UML behavioral relationships and the arrows and lines used to represent the relationships.

BEHAVIORAL RELATIONSHIPS

Relationships that are active relationships are referred to as behavioral relationships. They are used primarily in use case diagrams. There are four basic types of behavioral relationships: communicates, includes, extends, and generalizes. Notice that all these terms are action verbs. Figure 22.33 shows the symbols used to diagram each of the four types of behavioral relationships. Examples of the relationships in use are shown in Figure 22.34.

Communicates. The behavioral relationship "communicates" is used to connect an actor to a use case. Remember that the task of the use case is to give some sort of result that is beneficial to the actor in the system. Therefore, it is important to document these relationships between actors and use cases. In our example, a "Customer" communicates with "Rent Vehicle."

Includes. The "includes" relationship describes the situation where a use case contains behavior that is common to more than one use case. In other words, the common use case is included in the other use cases. A dotted arrow that points to the common use case indicates the relationship "includes." In our example, the use case "Verify Credit Card" and "Check Driver's License" includes the common use case "Rent Vehicle."

Extends. The "extends" relationship describes the situation where one use case possesses the behavior that allows the new use case to handle a variation or exception from the basic use case. For example, the extended use case "Arrange for Added Insurance" extends the use case "Rent Vehicle."

Generalizes. This relationship "generalizes" implies that one thing is more typical than the other thing. This relationship may exist between two actors or two use cases. For example, a "Regular Customer" generalizes a "Gold Card Customer." Thus, some of the customers are members of a special gold card club, but the majority of the customers are regular customers.

"DIAGRAMS": DESCRIBING THINGS AND THEIR RELATIONSHIPS

So far, we have discussed some of the concepts that form the foundation of UML. The construction of an O-O system is similar to the construction of a house in that both are constructed of interrelated things. Just as the construction of the house

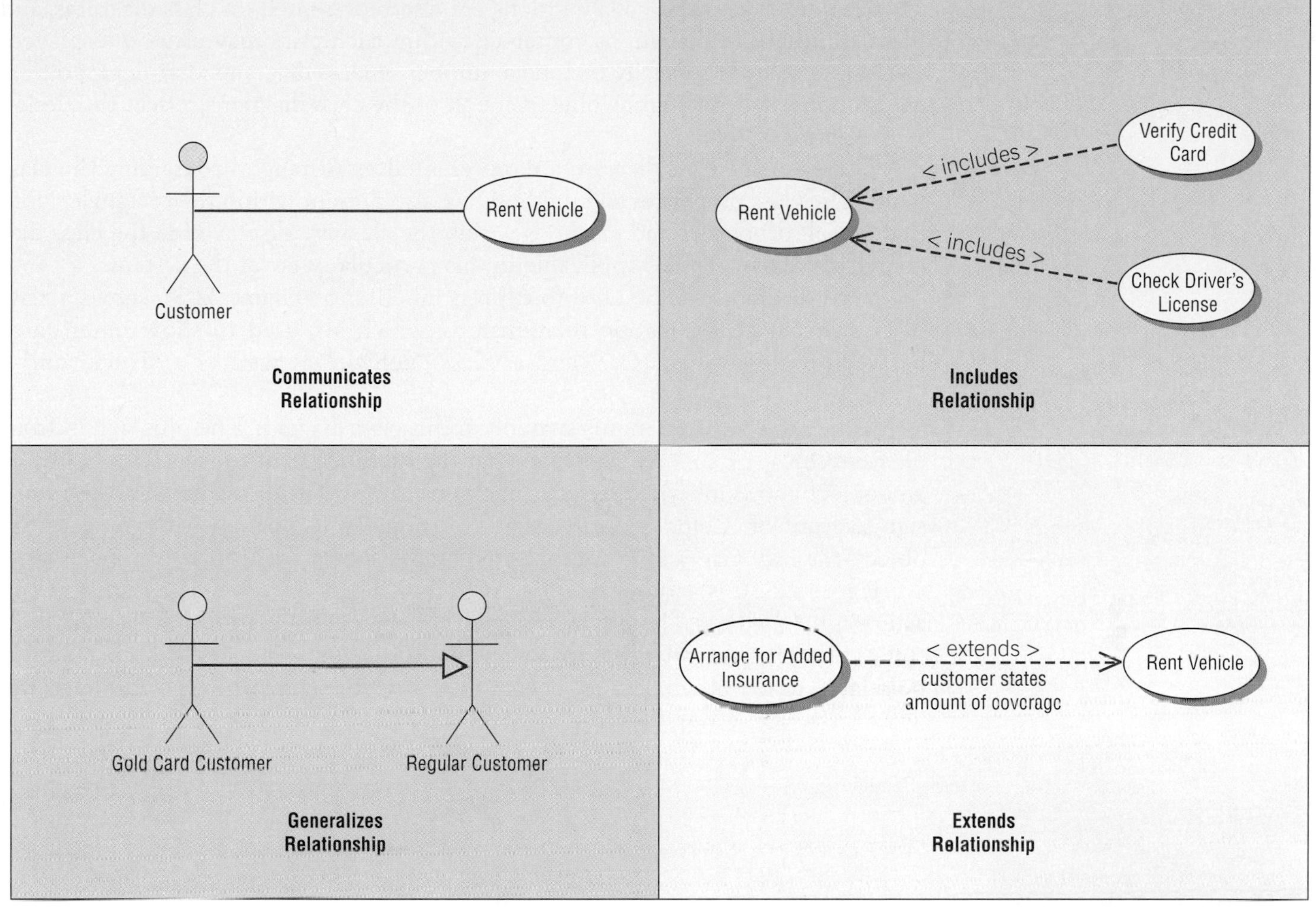

FIGURE 22.34
Examples of UML behavioral relationships in use.

can be described graphically by architectural plans prior to construction, an O-O system may be described graphically, in detail, by UML diagrams.

A good UML diagram model provides many of the same benefits to the development of an O-O system as a set of blueprints provides when building a house. Many O-O system development projects fail before a single line of source code is written. Poor systems analysis and design leads to poor system construction. Imagine the results of building a home without a set of blueprints!

UML diagrams provide the tools, when used correctly, to help ensure a robust design on paper prior to coding. In practice, redesigning a system on paper takes a fraction of the time that it would take to redesign a system that is partially developed.

Now we will describe the UML toolset, which is composed of the UML diagrams.

STRUCTURAL DIAGRAMS

UML structural diagrams include class diagrams, object diagrams, component diagrams, and deployment diagrams.

Class Diagrams. Class diagrams are used to model the static structural design view of a system. Classes along with their relationships are illustrated graphically. Class diagrams show the functional requirements of the system gathered by way of analysis as well as the physical design of the system, but they do not reflect the state of the system at a particular point in time.

Class interfaces and collaborations are also represented on class diagrams, and the attributes and operations contained within each class may also be displayed. O-O system models usually include a number of class diagrams that depict different sections of the system. Collectively, all of the class diagrams reflect the design of the entire system.

Classes in the class diagram are represented by rectangles containing the class name. The class properties and methods are also present within the rectangle. Note that not all properties and methods within the class are displayed in the class diagram, only those that are applicable to this particular view of the system.

Class diagrams can be used to express inheritance. Figure 22.35 shows a class diagram with generalization relationships, which are used to show inheritance between a class called "Car" and a Class "Vehicle" as well as a "Truck" and a "Vehicle."

Notice the plus and minus symbols in this class diagram. The plus sign in front of "RentOut ( )" in "Car" indicates that the method "Rent Out ( )" is visible to any object and is inherited by any subclasses of "Car." In the same way, the minus sign in front of "Color" indicates that the property "Color" is visible only to the objects in class "Car" and is not inherited by the subclasses of "Car."

Figure 22.36 is another example of a class diagram. Notice the numbers and asterisks on the lines. These symbols, at either end of the relationship, signify "multiplicity" within the relationships and show how many instances of each class can be related to the other class connected by the relationship arrow. For instance, the

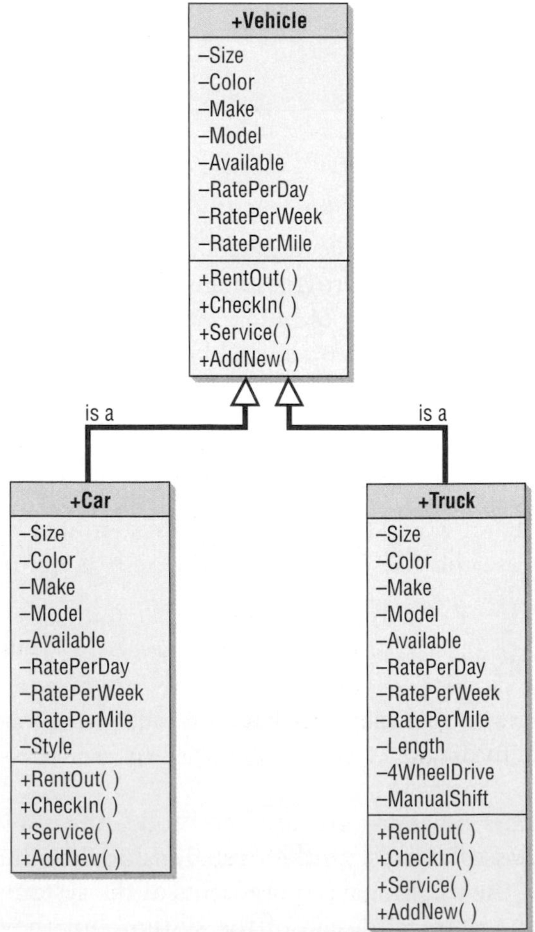

FIGURE 22.35

A class diagram showing generalization relationships from the rental car agency system. These generalization relationships are also known as inheritance.

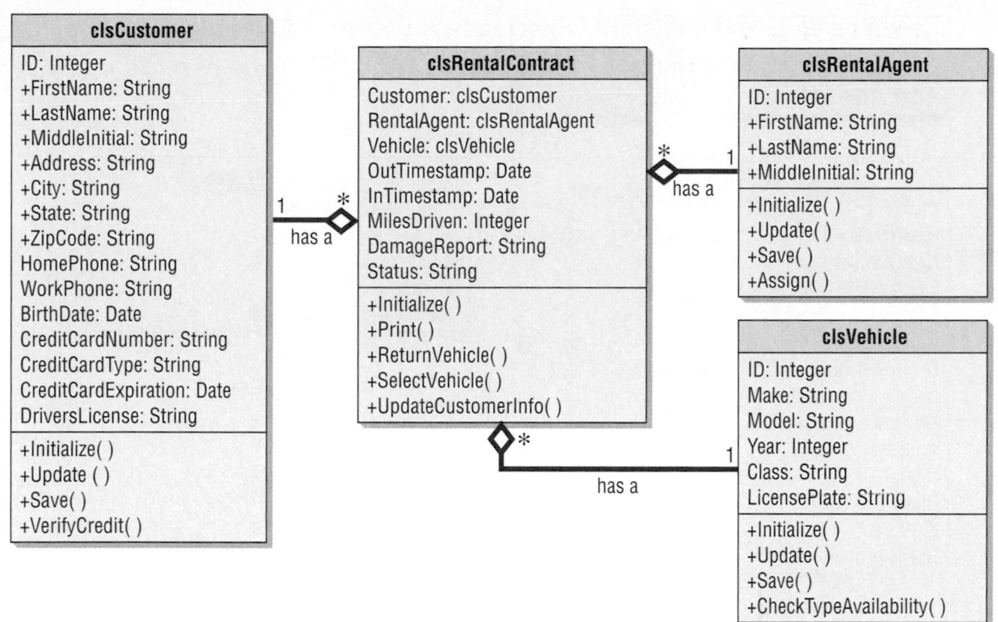

FIGURE 22.36

An example of a class diagram from the rental car agency system showing one-to-many relationships.

class clsRentalContract has a many-to-one relationship to clsRentalAgent, clsCustomer, and clsVehicle. In other words, a Rental Contract can be related to one Customer, but one Customer can have many Rental Contracts. For example, when a business rents a number of vehicles for its sales staff, we consider the business to be the customer.

Object Diagrams. Object diagrams are similar to class diagrams, but they portray the state of class instances and their relationships at a point in time. The object diagram shows objects and their relationships. The object diagram also shows optionality (Customer can have zero or more Rental Contracts) and cardinality (Rental Contract can have only one Customer).

Individual object diagrams can be related back to a specific use case scenario. The object diagram would depict the state of a set of related object instances during a specific point in time of a use case scenario, as seen in Figure 22.37.

The benefit of the object diagram is that you can visualize the state and interaction of a number of objects at a selected point in time during a complex operation. This view provides insight to processes that would normally be difficult to understand. Only use object diagrams to illustrate the most complex behaviors within your systems, where they will provide the most benefit to your project.

Component Diagrams. The component diagram is in a sense similar to a class diagram, but is more of a mile-high view of the system architecture. The component diagram shows the components of the system and how they are related to each other. The individual components in a component diagram are considered in more detail within other UML diagrams, such as class diagrams and use case diagrams.

Deployment Diagrams. The deployment diagram illustrates the physical implementation of the system, including the hardware on which the system is deployed.

FIGURE 22.37

The object diagram shows instances of classes and their relationships at a point in time.

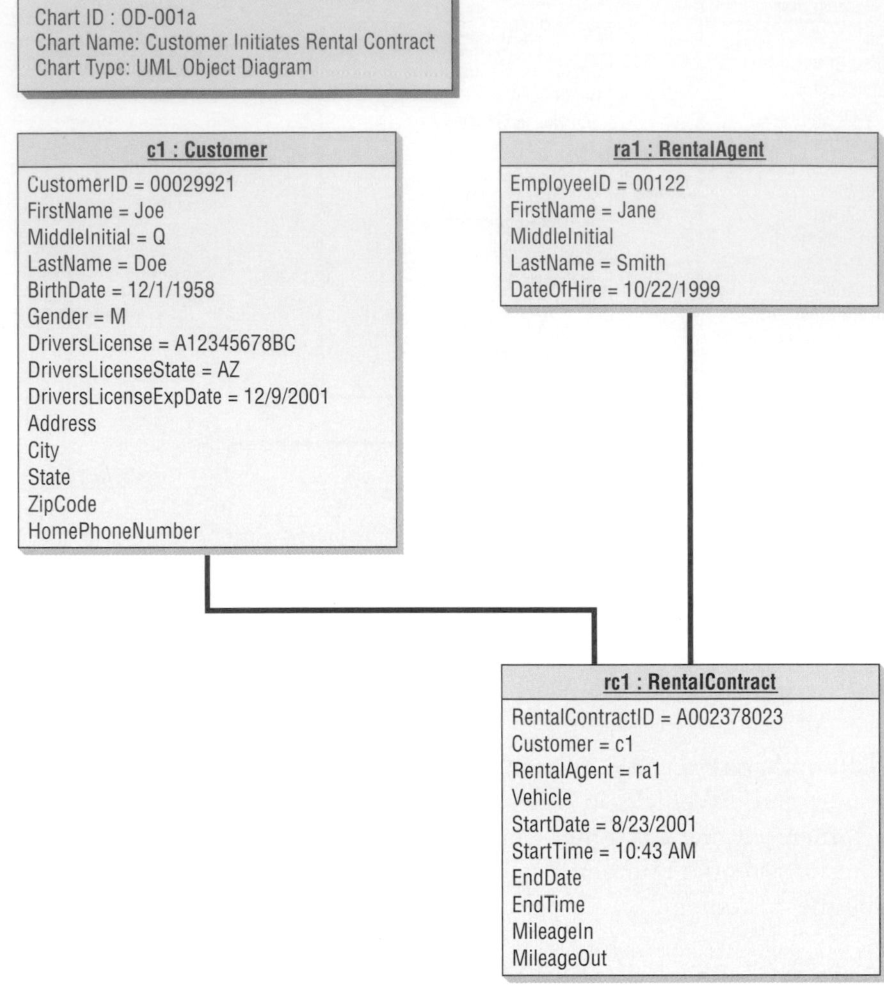

BEHAVIORAL DIAGRAMS

Behavioral diagrams include four different kinds of diagrams. They are all used to describe the interaction between people (actors) and a use case.

Use Case Diagrams. Use case diagrams show the actors and the use cases within the framework of the system. Several interrelated use cases are displayed as ovals within a box that represents the domain of the system or subsystem you are modeling. The use case diagram provides a bird's-eye view of the actors and use cases within your system and provides an overall description of the external functionality of the system. A use case diagram for the rental car example is shown in Figure 22.38.

Sequence Diagrams. Sequence diagrams can illustrate a succession of interactions between object instances over time. Sequence diagrams are often used to illustrate the processing described in use case scenarios. In practice, sequence diagrams are derived from use case analysis and are used in system design to derive the interactions, relationships, and methods of the objects in your system.

The symbols used in sequence diagrams are shown in Figure 22.39. The sequence diagram displays the actors and object instances in boxes along the top of the diagram. Lateral bars drop down from each box to the bottom of the diagram and the interactions between the objects are represented by arrows drawn from

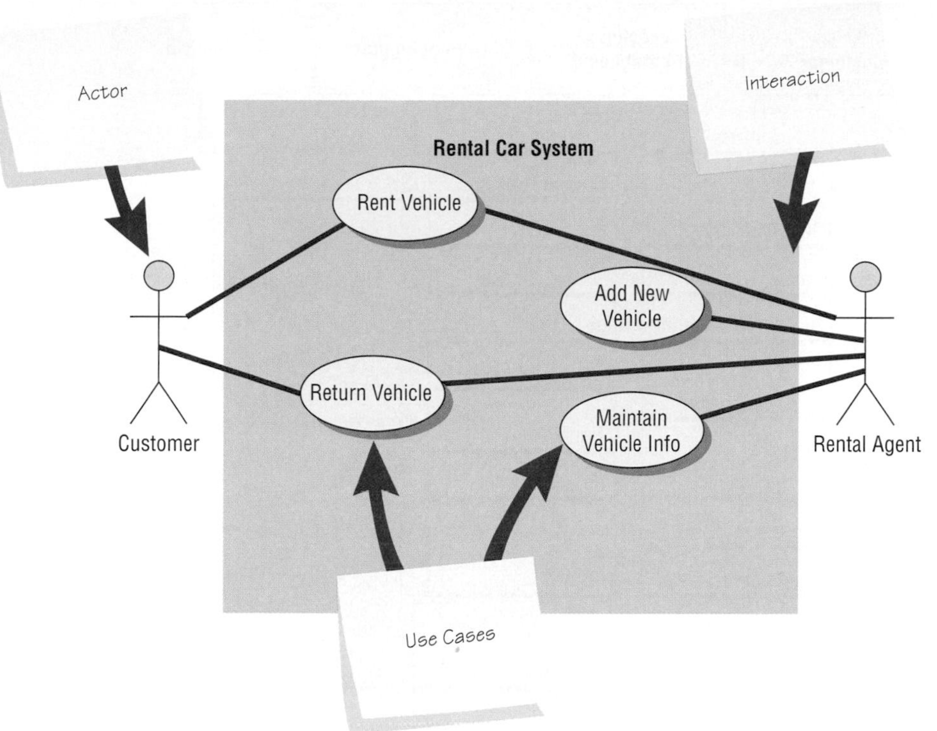

FIGURE 22.38
The use case diagram provides a basic overall view of the rental car agency system.

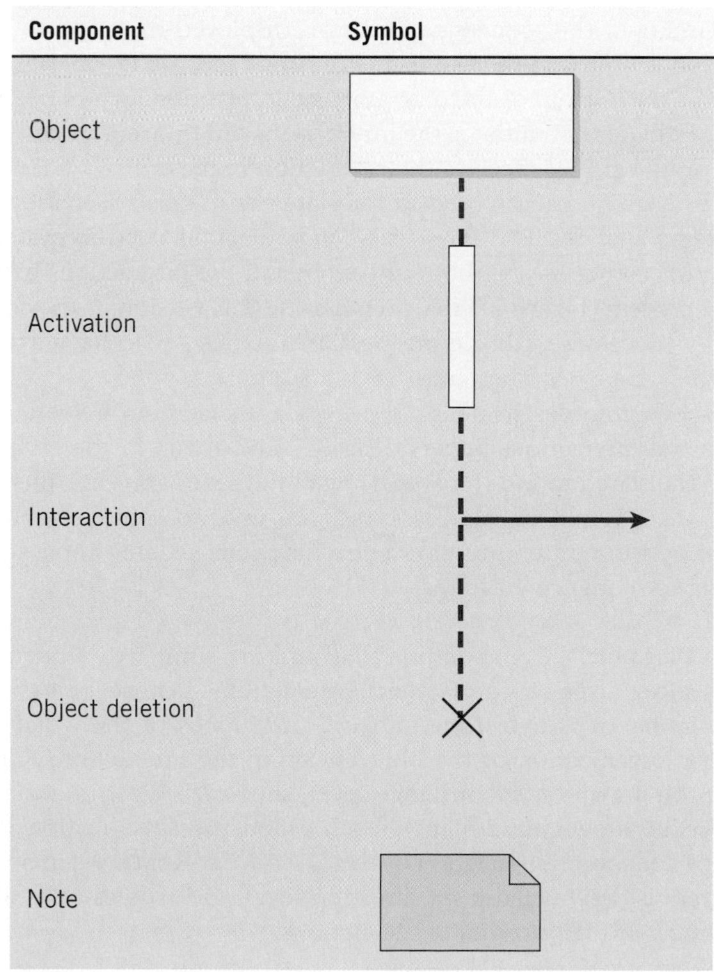

Component
Symbol
Object
Activation
Interaction
Object deletion
Note

FIGURE 22.39
The symbols used in sequence diagrams.

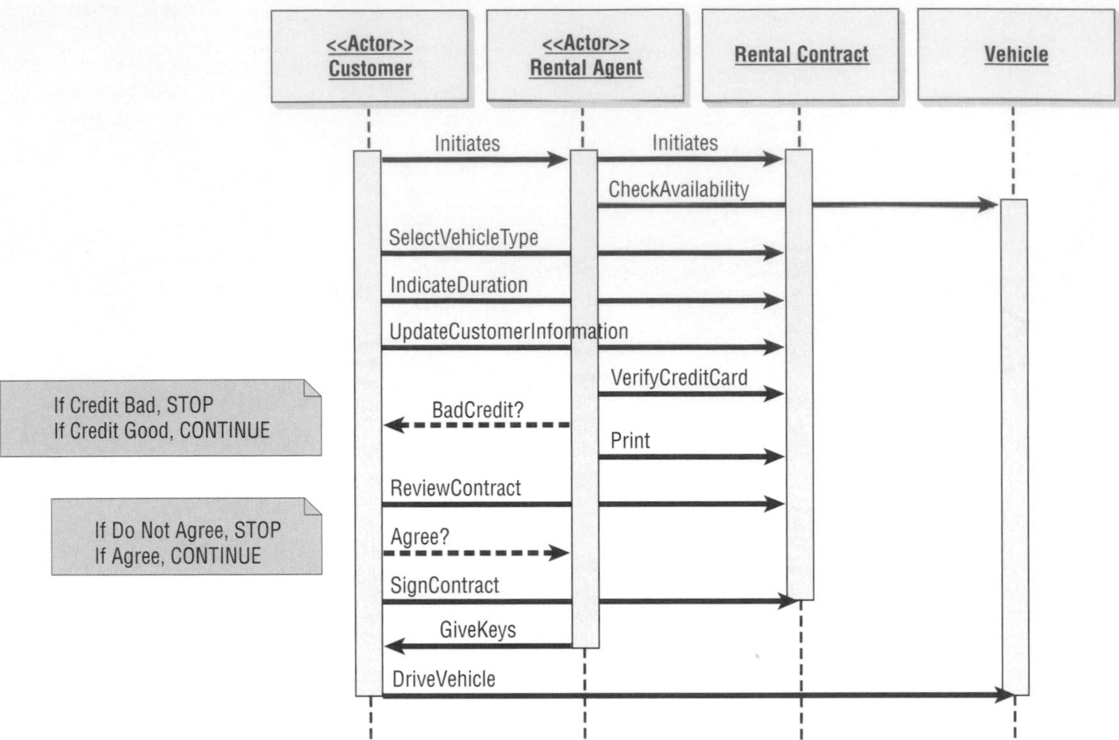

FIGURE 22.40

The initial sequence diagram used in the rental car agency system. Both actors and classes are both represented in this UML diagram.

bar to bar. Timing in the sequence diagram is displayed from top to bottom; the first interaction is drawn at the top of the diagram and the interaction that occurs last is drawn at the bottom of the diagram. The interaction arrows begin at the bar of the actor or object that initiates the interaction, and they end pointing at the bar of the actor or object that receives the interaction request.

Sequence diagrams can be used to translate the use case scenario into a visual tool for systems analysis. The initial sequence diagram used in systems analysis shows the actors and classes within the system and the interactions between them for a specific process (Figure 22.40). You can use this version of the sequence diagram to verify processes with the business area experts who have assisted you in developing the system requirements.

During the system design phase, the sequence diagrams are refined to derive the methods and interactions between classes. The actors in the earlier sequence diagrams are translated to interfaces and class interactions to class methods during system design; see Figure 22.41. Class methods used to create instances of other classes and to perform other internal system functions become apparent in the system design using sequence diagrams.

Collaboration Diagrams. Collaboration diagrams are similar to sequence diagrams in that they show a sequence of object interactions. Instead of focusing on the sequence of interactions over time, however, the focus of the collaboration diagram is on the organization of the objects during the interactions. Collaboration diagrams can be drawn on the instance level, showing objects, links, and stimuli. They can also be drawn on a higher level, called the specification level. Figure 22.42 shows a collaboration diagram of the Limón Car Rental Agency. This collaboration diagram is drawn at the specification level, so it shows roles of the actors and one-to-many or many-to-many associations.

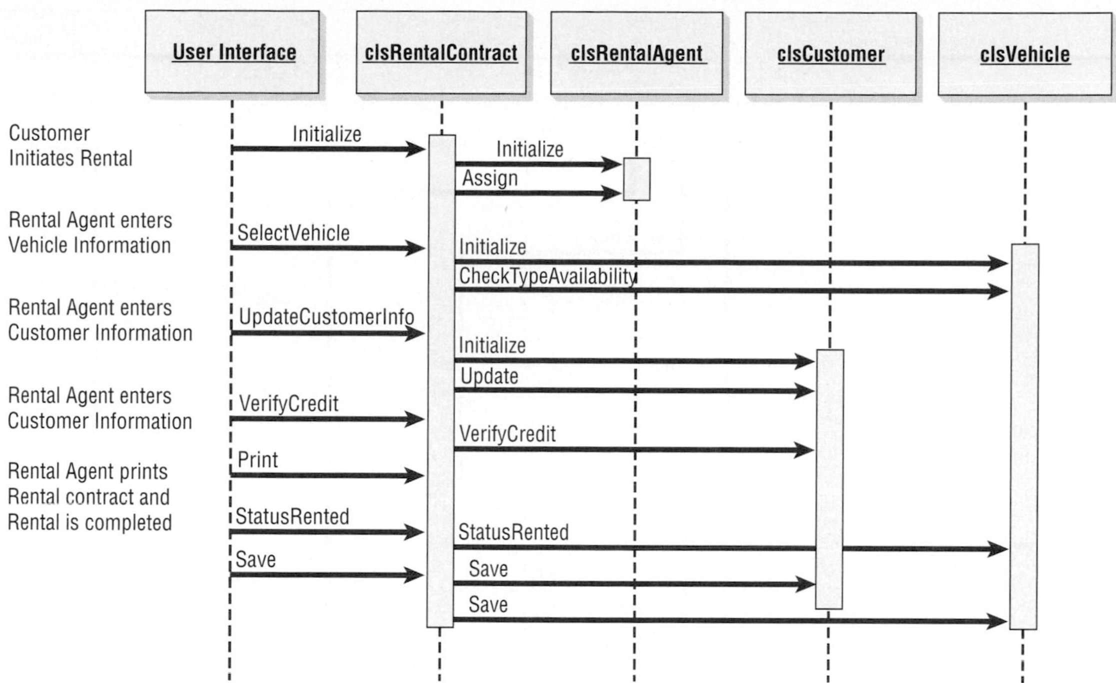

FIGURE 22.41
The refined sequence diagram used in system design shows class methods and system interfaces.

Statechart Diagrams. Statechart diagrams are similar to the state diagrams discussed in Coad and Yourdon's OOA-OOD methodology earlier in the chapter. They document a state machine, showing the various states of an object and the events and conditions that trigger a transition from one state to another. Statechart diagrams are most useful in diagramming the states of event-driven objects. An example of a statechart diagram for the car rental company is shown in Figure 22.43.

The state machine represented within a statechart diagram can be of one object, a system component, or the system itself. States are enclosed in rectangular boxes, and arrows connecting two states that are transitioning represent the transitions from state to state. Events that trigger the state transitions are listed near the transition arrows.

Activity Diagrams. An activity diagram shows the flow of activities within a process. This diagram is different than the statechart diagram, which focuses on states and state transitions instead of sequences of activities over time. Activity diagrams show transitions between state and the flow of control within a process. They also diagram concurrent activities within a process.

FIGURE 22.42
A collaboration diagram for the rental car agency system.

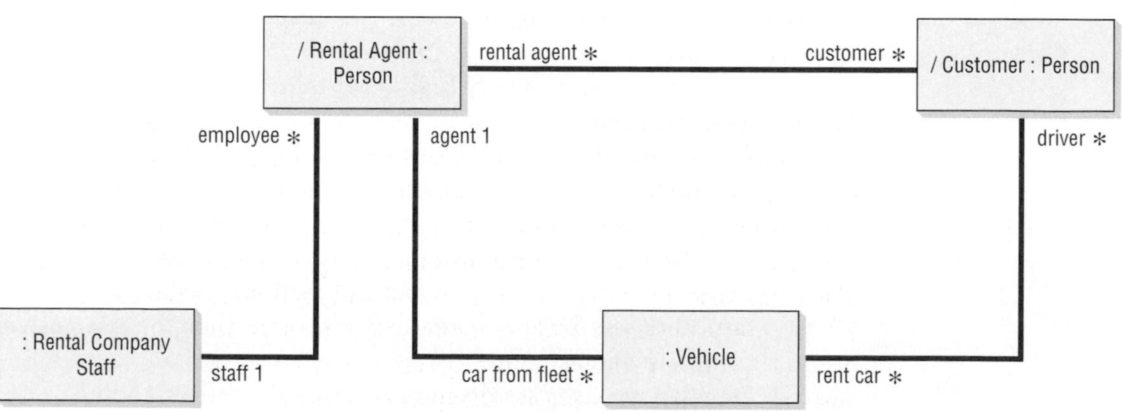

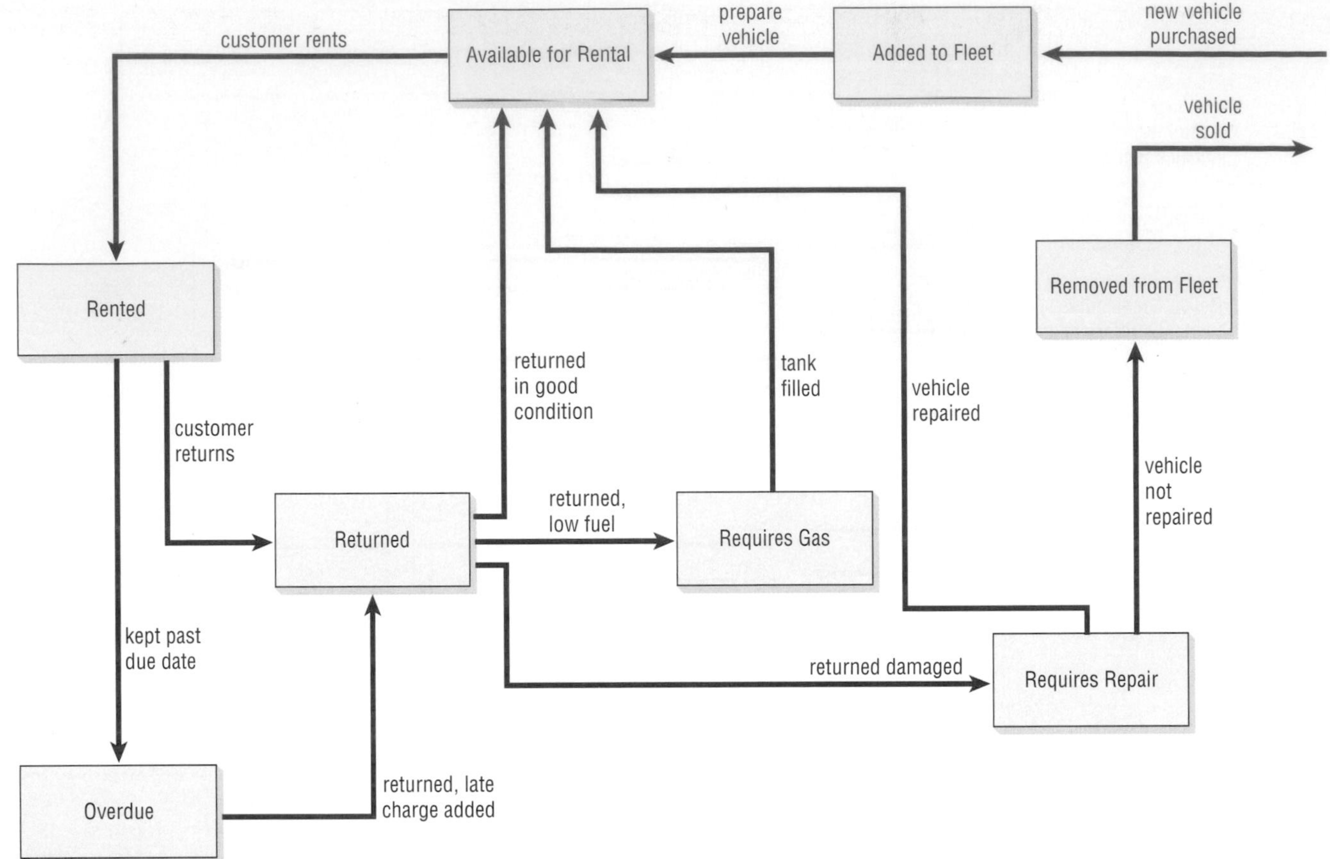

FIGURE 22.43

A statechart diagram for the rental car agency system.

A PROVEN METHODOLOGY: PUTTING UML TO WORK

UML provides a useful toolset for systems analysis and design. As with any product created with the help of tools, the value of the UML deliverables in a project depends on the expertise with which the systems analyst wields the tools. The analyst will initially use the UML toolset to break down the system requirements into a use case model and an object model. The use case model describes the use cases and actors. The object model describes the objects and object associations and the responsibilities, collaborators, and attributes of the objects.

1. Define the use case model.
   a. Find the actors within the problem domain by reviewing the system requirements and interviewing the business experts.
   b. Identify the major events initiated by the actors and develop a set of primary use cases at a very high level that describe the events from the perspective of each actor.
   c. Refine the primary use cases to develop a detailed description of system functionality for each primary use case. Provide additional details by developing the use case scenarios that document the alternate flows of the primary use cases. Review the use case scenarios with the business area experts to verify processes and interactions. Make modifications as necessary until the business area experts agree that the use case scenarios are complete and accurate.
   d. Develop the use case diagrams to provide understanding of how the actors relate to the use cases that will define the system.

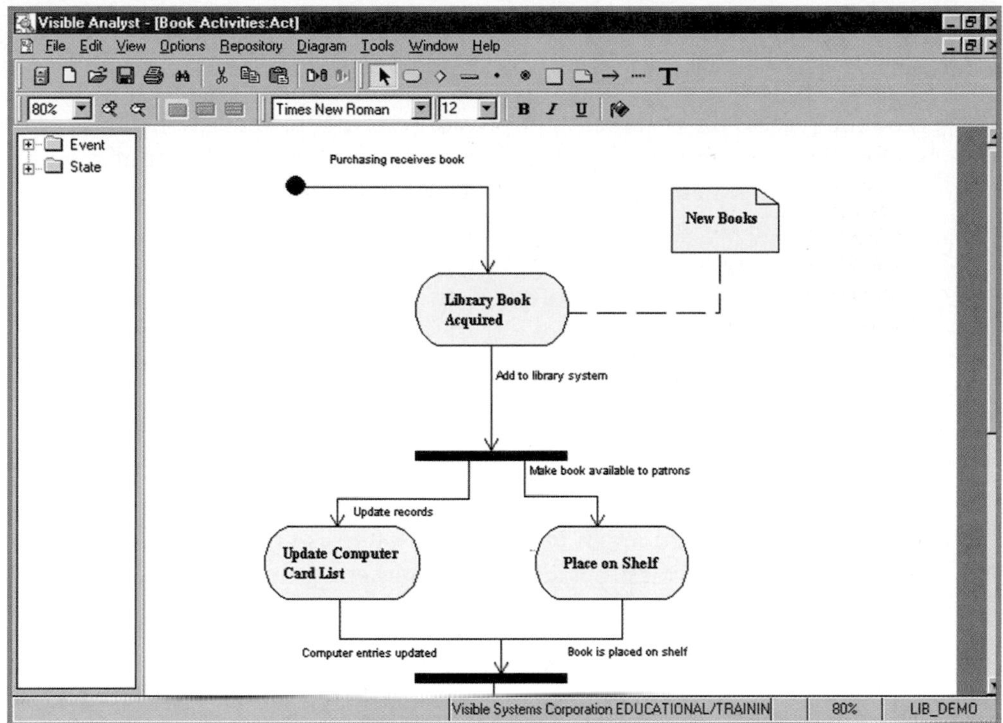

FIGURE 22.44
The example of an activity diagram drawn using Visible Analyst.

2. Define the object model.
   a. Review the use case model to discover the objects that define the problem domain. Look for nouns in use cases and list them.
   b. Once you identify the objects, look for similarities and differences in the objects due to the objects' states or behavior.
   c. Define the major relationships between the objects. Look for "has a" and "is a" relationships between objects.
   d. Beginning with the use cases that are the most important to the system design, create class diagrams that show the classes and relationships that exist within the use cases. One class diagram may represent the classes and relationships described in several related use cases.
3. Continue UML diagramming to model the system during the systems analysis phase.
   a. Derive sequence diagrams from use case scenarios and class diagrams. Review the sequence diagrams with the business area experts to verify processes and interactions. Make modifications as necessary until the business area experts agree that the sequence diagrams are complete and accurate. This additional review of the graphical sequence diagrams often provides the business area experts an opportunity to rethink and refine processes in more atomic detail than the review of the use case scenarios.
   b. Develop statechart diagrams, collaboration diagrams, and activity diagrams (an activity diagram is shown in Figure 22.44) to provide further analysis of the system at this point. Use these diagrams to aid in understanding complex processes that cannot be fully derived by the sequence diagrams.
4. Begin system design by refining UML diagrams and using them to derive classes and their properties and methods.
   a. Review all existing UML diagrams for the system. Write class specifications for each class that include the class properties, methods, and their

descriptions. Review sequence diagrams to identify class methods. Derive state (data) class properties from use cases, business area experts, and class methods. Indicate whether the methods and properties of the class are public (accessible externally) or private (internal to the class).

   b. Develop method specifications that detail the input and output requirements for the method, along with the detailed description of the internal processing of the method.
   c. Create another set of sequence diagrams (where necessary) to reflect the actual class methods and interactions with each other and the system interfaces.
   d. Analyze the class diagrams to derive the system components, that is, functionally and logically related classes that will be compiled and deployed together as a .DLL, a .COM object, a Java Bean, a package, and so forth.
   e. Develop deployment diagrams to indicate how your system components will be deployed in the production environment.

5. Document your system design in detail. This step is critical. The more complete the information you provide the development team through documentation and UML diagrams, the faster the development and the more solid the final production system.

THE IMPORTANCE OF USING UML FOR MODELING

UML is a powerful tool that can greatly improve the quality of your systems analysis and design, and it is hoped that the improved practices will eventually translate into higher-quality systems. By using UML in an iterative cycle of systems analysis, you can achieve a greater understanding between the business team and the IT team regarding the system requirements and the processes that need to occur within the system to meet those requirements. The first iteration of analysis should be at a very high level to identify the overall system objectives and validate the requirements through use case analysis. Identifying the actors and defining the initial use case model are part of this first iteration. Subsequent iterations of analysis further refine the system requirements through the development of use case scenarios, class diagrams, sequence diagrams, state chart diagrams, and so on. Each iteration takes a successively more detailed look at the design of the system until the things and relationships within the system are clearly and precisely defined within the UML documents. When your analysis and design are complete, you should have an accurate and detailed set of specifications for the classes and processes within the system.

In general, you can relate the thoroughness of the analysis and design of a system with the amount of time required to develop the system and the resultant quality of the delivered product.

Often overlooked in the development of a new system is that the further a project progresses, the costlier changes to the business requirements of a system are. Changing the design of a system on paper during the analysis and design phases of a project is easier, faster, and much less expensive than doing so during the development phase of the project.

Unfortunately, some employers are shortsighted, believing that only when a programmer or analyst is coding is that employee actually working. Some employers erroneously assume that programmer productivity can be judged solely by the amount of code produced, without recognizing that diagramming ultimately saves time and money wasted if a project is prototyped without proper planning.

The analogy with building a house is very apt in this situation. Although you hire a builder to build a house, you do not want to live in a structure built without

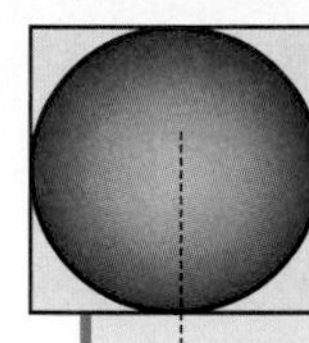

CONSULTING OPPORTUNITY 22.4

DEVELOPING A FINE SYSTEM THAT WAS LONG OVERDUE: USING OBJECT-ORIENTED ANALYSIS FOR THE RUMINSKI PUBLIC LIBRARY SYSTEM

As Dewey Dezmal enters the high-ceilinged, wood-paneled reading room of the Ruminski Public Library, a young woman, seated at a long, oak table, pokes her head out from behind a monitor, sees him, and stands, saying, "Welcome. I'm Peri Otticle, the director of the library. I understand you are here to help us develop our new information system."

Still in awe of the beauty of the old library building and the juxtaposition of so much technology amid so much history, Dewey introduces himself as a systems analyst with a small IT consulting firm, People and Objects, Inc.

"It's the first time I've been assigned to this type of project, although it's actually interesting for me, because my degree is from the Information Studies School at Upstate University. You can major in library science or IT there, so lots of my classmates went on to work in public libraries. I opted for the IT degree."

"We should work well together, then," Peri says. "Let's go to my office so we don't disturb any patrons, and I can talk you through a report I wrote."

As they pass the beautiful, winding staircase seemingly sculpted in wood, Peri notices Dewey looking at the surroundings and says, "You may wonder about the grandeur of the building, because we are a public institution. We are fortunate. Our benefactor is Valerian Ruminski. In fact, he has donated so much money to so many libraries that the staff affectionately calls him "Valerian the Librarian."

As they pass several patrons, Peri continues, "As you can see, it's a very busy place. And, regardless of our old surroundings, we don't dwell in the past."

Dewey reads the report Peri has handed him. One large section is titled "Summary of Patrons' Main Requirements," and the bulleted list states:

- A library patron who is registered in the system can borrow books and magazines from the system.
- The library system should periodically check (at least once per week) whether a copy of a book or journal borrowed by a patron has become overdue. If so, a notice will be sent to the patron.
- A patron can reserve a book or journal that has been lent out or has been in the process of purchase. The reservation should be canceled when the patron checks out the book or journal or through a formal canceling service.

As Dewey looks up from the report, he says to Peri, "I'm beginning to understand the patron (or user) requirements. I see lots of similarities between my old university library and yours. One item I didn't see covered, though, was how you decide what the library should collect and what it should get rid of."

Peri chuckles and replies, "That's an insightful question. The library staff handles the purchase of new books and journals for the library. If something is popular, more than two copies are purchased. We can create, update, and delete information about titles and copies of books and journals, patrons, loan of materials, and reservations in the system."

Dewey looks up from his note pad and says, "I'm still a little confused. What's the difference between the terms title and copy?"

Peri responds, "The library can have several copies of a title. Title normally refers to the name of a book or journal. Copies of a title are actually lent out from the library."

Based on Dewey's interview with Peri and the requirements description in her report as well as your own experience using library services, use UML to answer the following questions (*Note*: It is important to make sure your solutions are logical and workable. State your assumptions clearly whenever necessary.)

1. Draw a use case diagram to represent actors and use cases in the system.
2. For each use case, describe the steps (like we did to organize the use cases).
3. Describe scenarios for the steps. In other words, create a patron and write up an example of the patron as he or she goes through each step.
4. Develop a list of things.
5. Create sequence diagrams for use cases based on steps and scenarios.
6. Complete the class diagram by determining relationships between classes and defining the properties (attributes) and methods (operations) of each class. Use the grouping thing called package to simplify the class diagram.

Note: Based on a problem written by Dr. Wayne Huang.

planning, where rooms and features are randomly added without regard to function or cost. You want a builder to build your agreed-upon design from blueprints containing specifications that have been carefully reviewed by everyone concerned. As a member of an analyst team so accurately observed, "Putting a project on paper before coding will wind up costing less in the long run. It's much cheaper to erase a diagram than it is to change coding."

When business requirements change during the analysis phase, you may have to redraw some UML diagrams. If the business requirements change during the construction phase, however, a substantial amount of time and expense may be required to redesign, recode, and retest the system. By confirming your analysis and design on paper (especially through the use of UML diagrams) with users who are business area experts, you help to ensure that correct business requirements will be met when the system is completed.

SUMMARY

Object-oriented analysis and design techniques were developed in response to the increasing use of O-O programming languages. O-O designs can effectively specify such new O-O programming structures as inheritance and polymorphism. In this chapter, we studied two such approaches, one of which was devised by Peter Coad and Ed Yourdon. The second approach is called Unified Modeling Language, or UML.

The basic concepts underlying object-oriented analysis are objects, classes, messages, encapsulation, inheritance, and polymorphism. An object is a computer representation of some real-world thing or event. Similar objects are grouped into categories called classes. A class is the system specification for a group of similar (or related) objects. Objects communicate with each other through messages. A message is sent by one object to another object and is intended to stimulate some particular, predefined behavior by the receiver. Because all an object's internal data are buffered from other objects by internal predefined processes (the object's behavior), objects are said to encapsulate their data. The only way to change data held by an object is by sending a message to that object that will stimulate the object to make such an internal change. The inheritance concept refers to the ability to create new (derived) classes based in part on old (base) classes. Derived classes can inherit all or part of the structure and behavior of base classes and extend them with new structures and behaviors. Inheritance opens the way for polymorphic objects. Polymorphism permits a base class to take the form of any of its derived classes when the circumstances demand it. Furthermore, derived classes may substitute their own behavior in place of their base class's behavior in reaction to certain messages from other objects.

The Coad and Yourdon O-O model consists of five layers. Classes and objects are noted in the first layer. Second is a structure layer that details relationships between classes or objects. Attributes of classes are noted in the third layer. A service layer details messages and object behaviors. Finally, a subject layer may be used to divide large models into project units.

Objects are computer representations of tangible things, roles, incidents, interactions, or specifications. Objects usually embody a need to be recalled or remembered; a distinct set of several attributes, each with a meaningful value; siblings with exactly the same behavior; and independence from their implementation technology.

Structures link classes or objects. Classes may be linked together in a Gen–Spec structure for inheritance. Objects may also be linked together in a Whole–Part structure for aggregation. Aggregation involves objects that are composed or assembled from other objects.

Attributes are the data fields maintained by and related to a class. In addition, the attribute specification may include instance connections between objects. Instance connections represent some nonstructural relationship between two objects, such as the relationship between a vehicle object and an owner object.

Services are the behavior of objects and usually embody mechanisms by which the object can change its state. Simple services, such as those that merely set a data value, are generally not documented in the O-O model. Complex services that involve conditional processes or subroutines are documented in a specification template, which may also include attribute and state details. Services are typically triggered by messages created by other objects' services. Again messages that trigger simple services are generally not detailed in the O-O model.

Analysts can use CRC cards to begin the process of object modeling in an informal way. ObjectThink can be added to the CRC cards to assist the analyst in refining responsibilities into smaller and smaller tasks. CRC sessions can be held with a group of analysts to determine classes and responsibilities interactively.

Unified modeling language (UML) provides a standardized set of tools to document the analysis and design of a software system. UML is fundamentally based on an O-O technique known as use case modeling. A use case model describes what a system does without describing how the system does it. The primary components of UML are called "things." Structural things are most common; they include classes, interfaces, use cases, and many other elements that provide a way to create models. Structural things allow the user to describe relationships. Behavioral things describe how things work. Group things are used to define boundaries. Annotational things permit the analyst to add notes to the diagrams.

A use case model partitions system functionality into behaviors (called use cases) that are significant to the users of the system (called actors).

Relationships are the glue that holds the things together. Structural relationships are used to tie the things together in structural diagrams. Structural relationships include dependencies, aggregations, associations, and generalizations. Behavioral relationships are used in the behavioral diagrams to show communication, includes, extends, and generalizes. The toolset of UML is composed of the UML diagrams, of which there are two main types: structural diagrams and behavioral diagrams. Structural diagrams are used to describe the relationship between classes. They include class diagrams, object diagrams, component diagrams, and deployment diagrams. Behavioral diagrams include use case diagrams, sequence diagrams, collaboration diagrams, statechart diagrams, and activity diagrams.

UML is a powerful tool that can improve the quality of your systems analysis and design, and this improvement in practices should eventually result in a better system. Time and money can be saved by carefully planning the system on paper with the help of UML diagramming first. The resulting detailed set of specifications for the classes and the processes within the system should require only minimal recoding due to poor design.

KEYWORDS AND PHRASES

- actor
- annotational things
- assemblies
- base class
- behavioral diagrams
  - activity diagrams
  - collaboration diagrams
  - sequence diagrams
  - statechart diagrams
  - use case diagrams
- behavioral relationships
  - communicates
  - extends
  - generalizes
  - includes
- behavioral things
  - interactions
  - state machines
- cardinality
- class
- classification
- companion services
- complex services
- CRC cards
- derived class
- encapsulation

generalization–specialization
grouping things
  packages
human interface component
inheritance
instance connections
instantiate
interactions
interfaces
message
multiple inheritance
object
object-oriented
object-oriented database
object state analysis
ObjectThink
O-O analysis
primary use case
private services
polymorphism
problem component
prototype
relationships
self association
service layer
simple services
specification template
state diagrams
state machines
structural diagrams
  class diagrams
  object diagrams
  component diagrams
  deployment diagrams
structural relationships
  dependencies
  aggregations
  associations
  generalizations
structural things
  classes
  interfaces
  collaborations
  use cases
  control of active classes
  components
  nodes
subject
things
  behavioral things
  structural things
unified modeling language (UML)
use case model
use cases
use case scenario
whole–part structure

REVIEW QUESTIONS

1. What are the six basic ideas that characterize object-oriented programming?
2. Describe the difference between an object and a class.
3. How does encapsulation change the manner in which data are updated by programs?
4. What two types of classes are involved in any inheritance relationship?
5. Multiple inheritance means that there will be multiple occurrences of which type of class in the inheritance relationship?
6. Does polymorphism only occur where there is inheritance?
7. What are five general types of objects?
8. How can you tell from Coad and Yourdon's notation whether a class has been instantiated with objects?
9. What are eight criteria used to determine whether a new class is justified?
10. What are two basic types of structures that might be imposed on classes and objects?
11. What is the name of the notation used to denote a reference from one class to an unrelated class?
12. How can state changes be discovered easily in objects?
13. What are two categories of services?
14. How are messages denoted in the service layer of a Coad and Yourdon diagram?
15. What kind of project typically calls for the use of subject layers?

16. What four major components comprise the design activities?
17. What three activities take place in completing the problem domain component?
18. What does CRC stand for?
19. Describe what ObjectThink adds to the CRC card.
20. What is UML? How does it differ from the Coad and Yourdon OOA-OOD method?
21. What are the major components or primary elements of UML?
22. List what the concept of "structural things" includes.
23. List what the concept of "behavioral things" includes.
24. What are the two main types of diagrams in UML?
25. List the diagrams included in structural diagrams.
26. List the diagrams included in behavioral diagrams.
27. Describe the difference between a primary use case and use case scenarios.
28. What are the seven categories of structural things that UML defines?
29. Define what is meant by the term *class* in UML? Is it the same as the definition of class in OOA-OOD?
30. What is another name for the term *property*?
31. What is another name for the term *method*?
32. In UML, what is meant by the term *interface*?
33. Is it accurate to state that "the individual parts of the engine form collaboration"?
34. List some of the unique features of physical components.
35. Give an example of an interaction.
36. Describe what state machines do.
37. Describe the importance of UML diagramming before extensive coding.

PROBLEMS

1. The Kayjay World Station Master is usually a college student in his or her early twenties. The Station Master's duties include supervising the Ticket Seller and Platform Guards at their respective stations. The Station Master actually operates some train controls, including the "door open," "door close," and "boarding complete" switches. "I want a simple system," says Buffy Bronzebight, "one that I know will allow passengers to embark and debark safely. Confusion on the station platform is dangerous, and unexpected delays irritate the visitors. In the best of all worlds, the trains move regularly. Next best are delays you expect. You can announce them to the passengers, and then they don't get so mad. The worst delays are the unexpected ones. My job is pretty straightforward. When an arriving train comes to a complete halt, I press the 'door open' switch and allow train passengers to debark. After all passengers have debarked, I instruct the guards to allow the queued passengers to board the train. When all passengers have boarded or the train seats have filled, I instruct the guards to check the platform for safety and press the 'door close' switch. I then check the train visually according to the safety procedure and press the 'boarding complete' switch so that the train may leave."

   Based on this information, write user and task descriptions for the Station Master.
2. The Kayjay World Capacity Manager is a full-time position usually held by an experienced person with a college degree, although this individual has only rarely been someone trained in computers or engineering. Sam Spindlefold has held this position for 12 years. "I monitor the operation of the monorail system and the barn," he says, "and watch over track conditions and station

operations. It is my system: I'm in charge of starting and stopping the system and supervising all manual operations. I also have to be sure that the 'right' number of trains are operating. When the monorail was first installed, I spent three weeks with the contractor's engineers learning how it all worked. What matters most to me is that the system runs smoothly in bad conditions. Also, I should not be bothered with operational problems in typical conditions. I'm happy when the trains move smoothly around the system at peak hours and passengers are never delayed more than seven minutes when waiting for a train. I usually work the startup shift: I select 'startup system' from my screen. As soon as the Train Driver presses the deadman switch, kind of like a throttle, the first train launches from the barn. Although I rarely work a closedown shift, those people only have to select 'closedown system' from their screen. All empty trains exit the system. I dread 'alarm' conditions, such as a train failure or driver release of the deadman switch. In such cases, I may have to manually override controls by issuing radio instructions to Train Drivers or Stations Managers, by having yard workers manually switch the junction, or by sending the gasoline-powered yard engine out to tow a disabled train."

Based on this information, write user and task descriptions for the Capacity Manager.

3. The Kayjay World Yard Manager is in charge of the operating condition of the trains, monitoring their operating performance, coupling trains and placing them in service, and uncoupling trains and removing them for maintenance. Yard Managers also remove or add rolling stock when new cars are purchased or old ones are junked. Yard Managers are often trained mechanics of machinists, and most are very experienced. Billy Leroy has been working at Kayjay World since it opened 15 years ago. He says, "What I want most is a yard link that's easy to change and that works all the time so I can yank a train over to the yard fast and get another train out in its place fast. I don't want a complicated computer system that gets in my way when I need to assemble trains from yard stock. I want to turn the doggone computer control off if I need to. I spend most of my day fixing broken train parts. For example, say a train is sent to the barn with a jammed door. First I look for a quick fix. There usually isn't one, and I have to select commands from the computer menu that tells the system that the train is inoperable. Then I manually set two yard switches in the barn queue and yard and use manual controls to drive the train onto one yard siding. Next I have to uncouple the train and use a gasoline yard engine to pull the bad car onto the repair rack. I can then recouple the remaining cars and use the computer menu to enter the new train configuration. Once again, I manually set two yard switches and return the short train to the barn queue, and then I use the computer menu to tell the system that the train is again operable."

   Based on this information, write user and task descriptions for the Yard Manager.
4. Complete the Kayjay World Example 4 Specification by designing a service logic flow diagram (or other procedural specification) for the "Enter Barn" train service.
5. Complete the Kayjay World Example 4 Specification by designing a service logic flow diagram (or other procedural specification) for the "Set Switch" train service.
6. Complete the Kayjay World Example 4 Specification by designing a service logic flow diagram (or other procedural specification) for the "Get Track" train service.

7. Complete the Kayjay World Example 4 Specification by designing a service logic flow diagram (or other procedural specification) for the "find fixed junction" train service.
8. Create a series of CRC cards for World's Trend catalog division. Once an order is placed, the order fulfillment crew takes over and checks for availability, fills the order, and calculates the total amount of the order. Use five CRC cards, one for each of the following classes: order, order fulfillment, inventory, product, and customer. Complete the section on classes, responsibilities, and collaborators.
9. Finish the CRC cards in problem 8 by creating ObjectThink statements and property names for each of the five classes.
10. Draw a use case diagram for World's Trend catalog division.
11. For the FilmMagic problem in Consulting Opportunity 22.3, draw a class diagram in UML.
12. For the FilmMagic problem in Consulting Opportunity 22.3, draw statechart diagrams for (a) Customer and (b) Video.
13. Draw four pictures showing examples of the fours types of behavioral relationships for Joel Porter's BMW dealership.
14. Draw a collaboration diagram for a *student* taking a *course* from a *teacher*, who is part of the *faculty*.
15. Coleman County has a phone exchange that handles calls between callers and those receiving the call. Given these three actors, draw a simple sequence diagram for making a simple phone call.
16. You are ready to begin UML modeling for the Kirt Clinic. Draw a class diagram that includes a physician, a patient, an appointment, and a patient's bill. Do not get the insurance company involved.
17. Use UML to draw examples of the four structural relationships for the Kirt Clinic.
18. Write a sample scenario for a patient who sees a physician in the Kirt Clinic.

SELECTED BIBLIOGRAPHY

Beck, K., and W. Cunningham. "Laboratory for Teaching Object-Oriented Thinking," OOPSLA '89, as quoted in D. Butler, *CRC Card Session Tutorial*, (http://www.csc.calpoly.edu/~dbutler/tutorials/winter96/crc_b/tutorial.html) accessed Feb. 6, 2001.

Bellin, D., and S. Suchman Simone. *The CRC Card Book*. Reading, MA: Addison Wesley Longman, 1997.

Booch, G. *Object-Oriented Design with Applications*, 2d ed. Redwood City, CA: Benjamin-Cummings, 1994.

Cheesman, J., and J. Daniels. *UML Components: A Simple Process for Specifying Component-Based Software*. Reading, MA: Addison Wesley, 2001.

Coad, P., and E. Yourdon. *Object-Oriented Analysis*, 2d ed. Englewood Cliffs, NJ: Yourdon Press, 1991.

———. *Object-Oriented Design*. Englewood Cliffs, NJ: Yourdon Press, 1991.

Embley, D., B. Kurtz, and S. Woodfield, *Object-Oriented Systems Analysis: A Model-Driven Approach*. Englewood Cliffs, NJ: Yourdon Press, 1992.

Firesmith, D., B. Henderson-Sellers, I. Graham, and M. Page-Jones. *OPEN Modeling Language* (OML) *Reference Manual Version 1.0*. Cary, NC: OPEN Consortium, Knowledge Systems Corporation, December 8, 1996.

Rumbaugh, J., M. Blaha, W. Premerlani, F. Eddy, and W. Lorensen. *Object-Oriented Modeling and Design*. Englewood Cliffs, NJ: Prentice Hall, 1991.

Shlaer, S., and S. Mellor. *Object-Oriented Systems Analysis: Modeling the World in Data*. Englewood Cliffs, NJ: Yourdon Press, 1988.

———. *Object Lifecycles: Modeling the World in States*. Englewood Cliffs, NJ: Yourdon Press, 1992.

Unified Modeling Language Notation Guide ad/97–01–09. Santa Clara, CA: Rational Software Corporation, 1997.

White, I. *Using the Booch Method: A Rational Approach*. Redwood City, CA: Benjamin-Cummings, 1994.

Wilkinson, N. M. *Using CRC Cards: An Informal Approach to O-O Software Development*. New York: SIGS Books, 1995.

Wirfs-Brock, R., B. Wilkerson, and L. Wiener. *Designing Object-Oriented Software*, Englewood Cliffs, NJ: Prentice Hall, 1990.

GLOSSARY

Numbers in parentheses refer to the chapter in which that term is defined.

ACTOR In UML, a particular role of a user of the system. The actor exists outside the system and interacts with the system in a specific way. An actor can be a human, another system, or a device such as a keyboard or a modem. (22) *See also* use case.

AGGREGATION Often described as "has a" relationship when using UML for an object-oriented approach. Aggregations provide a means of showing that the whole object is composed of the sum of its parts (other objects). (22)

ALIAS Alternative name for a data element used by different users. Recorded in a data dictionary. (10)

APPLICATION SERVICE PROVIDER (ASP) A company that hosts application software that is leased by other organizations for use on the Web. Applications include traditional ones as well as collaboration and data management. (21)

ASSOCIATIVE ENTITY An entity type that associates the instances of one or more entity types and contains attributes that are peculiar to the relationship between those entity instances. (2)

ATTRIBUTE Some characteristic of an entity. There can be many attributes for each entity. (17) See also data item.

ATTRIBUTIVE ENTITY One of the types of entities used in entity-relationship diagrams. Something useful in describing attributes, especially repeating groups. (2)

BATCH PROCESSING Processing of data that are composed entirely of stored information generated and accessed by the computer, requiring no human intervention. Used for processing high volumes of data. (9)

BEHAVIOR Represents how an object acts and reacts. (22)

BIPOLAR QUESTION A subset of closed questions that can be answered in two ways only, such as yes or no, true or false, and agree or disagree. (5) *See also* closed question, open-ended question.

BROWSER Special software that runs on an Internet-connected computer enabling users to view hypertext-based Web pages on the Internet. Microsoft Internet Explorer and Netscape Communicator are examples of graphical browsers. (15)

BUBBLE DIAGRAM A simple diagram that shows data associations of data elements. Each entity is enclosed in an ellipse, and arrows are used to show the relationships. Also called a data model diagram. (17)

BUSINESS PROCESS REENGINEERING (BPR) Reorienting a business around redesigned business processes. Often involves upgrading or changing information systems to support improved processes. (1)

BUSINESS RULES Statements specific to an organization's functioning that provide a logical description of business activities. Used to help create data flow diagrams. (9)

CASE TOOLS Computer-aided software engineering tools that include computer-based automated diagramming, analyzing, and modeling capabilities. (1) *See also* lower CASE tools, upper CASE tools.

CHILD DIAGRAM The diagram that results from exploding the process on Diagram 0 (called the parent process). (9)

CLASS A common template for a group of individual objects with common attributes and common behavior in object-oriented analysis and design and UML. (22)

CLASS DIAGRAM Used to graphically model the static structural design view of a system. Class diagrams illustrate the functional requirements of the system gathered by way of analysis as well as the physical design of the system. (22)

CLIENT/SERVER ARCHITECTURE A design model that features applications running on a local-area network (LAN). Computers on the network divide processing tasks among servers and clients. Clients are networked machines that are points of entry into the client/server system. (21)

CLOSED QUESTION A type of question used in interviews or on surveys that closes the possible response set available to respondents. (5) *See also* bipolar questions, open-ended question.

CLOSED SYSTEM Part of general systems theory; a system that does not receive information, energy, people, or raw materials as input. Systems are never totally closed or totally open, but exist on a continuum from more closed to more open. (2) *See also* open system.

CODE GENERATORS Software that automatically creates all or part of a computer program. Usually a feature of a lower or integrated CASE product. (20)

COMMAND LANGUAGE INTERFACE A type of interface that allows users to control the application with a series of keystrokes, commands, phrases, or some sequence of these three methods. (18)

COMPONENT DIAGRAM A UML diagram that shows the components of the system and how they are related to each other in a mile-high view. (22)

COMPUTER-AIDED SOFTWARE ENGINEERING (CASE) Specialized software tools that include computer-based automated diagramming, analyzing, and modeling capabilities. (1) *See also* lower CASE tools, upper CASE tools.

CONTEXT DIAGRAM The most basic data flow diagram of an organization showing how processes transform incoming data into outgoing information. Also called an environmental model. (2) *See also* data flow diagram.

CONTROL FLAG Used in structure charts, they govern which portion of a module is to be executed and are associated with IF, THEN ELSE, and other similar types of statements. (20)

CONVERSION Physically converting the old information system to the new one. There are five conversion strategies: direct changeover, parallel conversion, phased conversion, modular prototype conversion, and distributed conversion. (21)

CRC CARDS The analyst creates Class, Responsibilities, and Collaborators cards to represent the responsibilities of classes and the interaction between the classes when beginning to model the system from an object-oriented perspective. Analysts create the cards based on scenarios that outline system requirements. (22)

CRITICAL PATH The longest path calculated using the PERT scheduling technique. It is the path that will cause the whole systems project to fall behind if even one day's delay is encountered on it. (3)

DATABASE A formally defined and centrally controlled store of electronic data intended for use in many different applications. (17)

DATABASE MANAGEMENT SYSTEM (DBMS) Software that organizes data in a database providing information storage, organization, and retrieval capacities. (17)

DATA COUPLE Depiction of the passing of data between two modules on a structure chart. (20)

DATA DICTIONARY A reference work of data about data (meta data) created by the systems analyst based on data flow diagrams. It collects and coordinates specific data terms, confirming what each term means to different people in the organization. (10)

DATA ELEMENT A simple piece of data. It can be base or derived. A data element should be defined in the data dictionary. (10)

DATA FLOW Data that moves in the system from one place to another; input and output are depicted using an arrow with an arrowhead in data flow diagrams. (9)

DATA FLOW DIAGRAM (DFD) Graphical depiction of data processes, data flows, and data stores in a business system. (9)

DATA ITEM The smallest unit in a file or database. Used interchangeably with the word *attribute*. (17)

DATA MINING Techniques that apply algorithms for extracting patterns from data stored in data warehouses that are typically not apparent to human decision makers. Also known as knowledge data discovery (KDD). (18)

DATA REPOSITORY A centralized database that contains all diagrams, form and report definitions, data structure, data definitions, process flows and logic, and definitions of other organizational and system components. It provides a set of mechanisms and structures to achieve seamless data-to-tool and data-to-data integration. (10)

DATA STORE Data that are at rest in the system; depicted using an open-ended rectangle in data flow diagrams. (9)

DATA STRUCTURE Structures composed of data elements, typically described using algebraic notation to produce a view of the elements. The analyst begins with the logical design and then designs the physical data structures. (10)

DATA WAREHOUSE A collection of data in support of management decision processes that is subject oriented, integrated, time-variant, and nonvolatile. (18) *See also* data mining.

DECISION SUPPORT SYSTEM (DSS) An interactive information system that supports the decision-making process through the presentation of information designed specifically for the decision maker's problem-solving approach and application needs. It does not make a decision for the user. (12)

DECISION TABLE A way to examine, describe, and document structured decisions. Four quadrants are drawn to describe the conditions, identify possible decision alternatives, indicate which actions should be performed, and describe the actions. (11)

DECISION TREE A method of decision analysis for structured decisions. It is an appropriate approach when actions must be accomplished in a certain sequence. (11)

DEFAULT VALUE A value a field will assume unless an explicit value is entered for that field. (17)

DELIVERABLES Any of the software, documentation, procedures, user manuals, or training sessions that a systems analyst delivers to a client based on specific contractual promises. (10)

DENORMALIZATION Defining physical records *not* in third or higher normal forms. It includes joining attributes from several relations together to avoid the cost of accessing several files. Partitioning is an intentional form of denormalization. (17)

DIGITAL SUBSCRIBER LINE (DSL) Protocols that allow high-speed data transmission over regular telephone wire. (21)

DISPLAY SCREEN Any one of a number of display alternatives that users employ to view computer software, including monitors, cathode-ray tubes (CRTs), liquid plasma screens, or video display terminals (VDTs). (15)

DISTRIBUTED SYSTEMS Computer systems that are distributed geographically, as well as having their processing, data, and databases distributed. One common architecture for distributed systems is a LAN-based client/server system. (21)

DOCUMENTATION Written material created by the analyst that describes how to run the software, gives an overview of the system, or details the program code that is used. Analysts can use a CASE tool to facilitate documentation. (20)

DROP-DOWN LIST BOX One of many GUI design elements that permits users to click on a box that appears to drop down on the screen and list a number of alternatives, which can be subsequently chosen. (15)

ECOMMERCE Doing business electronically including via email, Web technologies, BBS, smart cards, EFT, and EDI among suppliers, customers, governmental agencies, and other businesses to conduct and execute transactions in business, administrative, and consumer activities. (1)

ENCAPSULATION In object-oriented analysis and design, an object is encapsulated by its behavior. An object maintains data about the real-world things it represents in a true sense. An object must be asked or told to change its own data with a message. (22)

ENCRYPTION The process of converting a message into an encrypted message by using a key so that the message cannot be read by a person. The intended receiver of the message can then use a key to decode and read the encrypted message. (17)

END USERS Non-information-system professionals in an organization who specify the business requirements for and use software applications. End users often request new or modified applications, test and approve applications, and may serve on project teams as business experts. (1)

ENTITY A person, group, department, or system that either receives or originates information or data. One of the primary symbols on a data flow diagram. (2) *See also* data flow diagram, external entity.

ENTITY-RELATIONSHIP (ER) DIAGRAM A graphical representation of an E-R model. (10)

ENTITY TYPE A collection of entities that share common properties or characteristics. (10)

ENVIRONMENT Anything external to an organization. Multiple environments exist, such as the physical, economic, legal, and social environments. (2)

EXECUTIVE SUPPORT SYSTEM (ESS) A computer systems that helps executives organize their interactions with the external environment by providing graphical and communication support. (1)

EXPERT SYSTEMS (ES) A computer-based system that captures and uses the knowledge of an expert for solving a particular problem. Basic components are the knowledge base, an inference engine, and the user interface. (1)

EXTERNAL ENTITY A source or destination of data considered to be external to the system being described. Also called an entity. (9) *See also* data flow diagram.

FAVICON A small icon displayed next to any bookmarked address in a browser. Copying the bookmarked link to a desktop results in a larger version of the icon being placed there. Unique favicons can be generated with a Java icon generator or with other graphics programs. (15)

FIELD A physical part of a database that can be packed with several data items. The smallest unit of named application data recognized by system software. (17)

FIREWALL Computer security software used to erect a barrier between an organization's LAN and the Internet. Although it prevents hackers from getting into an internal network, it also stops organizational members from getting direct access to the Internet. (21)

FIRST NORMAL FORM (1NF) The first step in normalizing a relation in data used in a database so that it contains no repeating groups. (17) *See* second normal form, third normal form.

FOLKLORE A system documentation technique based on traditional methods used in gathering folklore about people and legends. (20)

FORM-FILL INTERFACE Part of GUI design elements that automatically prompt the user to fill in a standard form. Useful for ecommerce applications. (18)

GANTT CHART A graphical representation of a project that shows each task activity as a horizontal bar whose length is proportional to its time for completion. (3)

GRAPHICAL USER INTERFACE (GUI) An icon-based user interface, with several features such as pull-down menus, drop-down lists, and radio buttons. (18)

HYPERLINK Any highlighted word in a hypertext system that will display another document when clicked on by the user. (15)

ICON Small pictures that represent activities and functions that are available to the user when they activate them, often with a mouse click. Frequently used in GUI design. (18)

IMPLEMENTATION The last phase of the SDLC in which the analyst ensures that the system is in operation and then allows users to take over its operation and evaluation. (21)

INDEXED FILE ORGANIZATION A type of file organization that uses separate index files to locate records. (17)

INHERITANCE In object-oriented analysis and design, classes can have children. The parent class is known as the base class, and the child class is called a derived class. The derived class can be created to inherit all the attributes and behaviors of the base class. (22)

INPUT Any data, either text or numbers, that are entered into an information system for storage or processing via forms, screens, voice or interactive Web fill-in forms. (16)

INTANGIBLE BENEFITS Those benefits that accrue to the organization as result of a new information system such as improved decision making, enhancing accuracy, becoming and more competitive that are difficult to measure. (13) *See also* tangible benefits, intangible costs, tangible costs.

INTANGIBLE COSTS Those costs that are difficult to estimate and may not be known, including losing a competitive edge, losing a reputation for innovation, and declining company image, due to untimely or inaccessible information. (13)

INTERNET SERVICE PROVIDER (ISP) A company that provides access to the Internet and that may provide other services such as Web hosting and Web traffic analysis for a fee. (16)

INTEGRATED SERVICES DIGITAL NETWORK (ISDN) A switched network service that provides end-to-end digital connectivity for transmitting voice, data, and video simultaneously over a single line versus multiple lines. (21)

IP ADDRESS The Internet protocol address is the number used to represent a single computer on a network. The format for an IP address is 999.999.999.999. (21)

JAVA An object-oriented programming language that allows dynamic applications to be run on the Internet. (15)

JAVA APPLETS A little application program written in the Java language that can be embedded in an HTML document for use on Web pages. (15)

JOINT APPLICATION DESIGN (JAD) IBM's proprietary approach to panel interviews conducted with analysts, users, and executives to accomplish requirements analysis jointly. (5)

KEY One of the data items in a record that is used to identify a record. (17) *See also* primary key, secondary key.

LEVEL 0 DIAGRAM The explosion (or decomposition) of the context data flow diagram, showing from three to nine major processes, important data flows, and data stores of the system under study. (9)

LOCAL AREA NETWORK (LAN) The cabling, hardware, and software used to connect workstations, computers, and file servers located in a confined geographical area (typically within one building or campus). (19)

LOGICAL DATA FLOW DIAGRAM A diagram that focuses on the business and how the business operates. Describes the business events that take place and the data required and produced by each event. (9) *See also* data flow diagram, physical data flow diagram.

LOWER CASE TOOLS Those CASE tools used by analysts to generate computer source code, eliminating the need for programming the system. (1) *See also* CASE tools, upper CASE tools.

MAINTENANCE Maintaining the information system to improve it or fix problems begins in this phase of the SDLC and continues through the life of the system. Some maintenance can be done automatically through connecting to the vendor's Web site. (1)

MANAGEMENT INFORMATION SYSTEM (MIS) A computer-based system composed of people, software, hardware, and procedures that share a common database to help users interpret and apply data to the business. (1)

METHOD In UML, an operation. (22)

MNEMONIC CODE Any code (often using a combination of letters and symbols) that helps the data-entry person remember how to correctly enter data or helps the user remember how to use the information. (19)

NATURAL LANGUAGE INTERFACE An interface that permits the user to speak or write in human language to interact with the computer. (18)

NORMALIZATION The transformation of complex user views and data stores to a set of smaller, stable data structures. Normalized data structures are more easily maintained than complex structures. (17)

OBJECT In the object-oriented approach, an object is a computer representation of some real-world thing or event. Objects can have both attributes and behaviors. (22)

OBJECT CLASS A class is a category of similar objects. Objects are grouped into classes. A class defines the set of shared attributes and behaviors found in each object in the class. (22)

OBJECT DIAGRAM Diagrams similar to class diagrams, but that portray the state of class instances and their relationships at a point in time. The object diagram shows objects and their relationships. The object diagram also shows optionality (customer can have zero or more rental contracts) and cardinality (rental contract can have only one customer). (22)

OBJECT THINK Elementary statements the analyst writes on CRC cards to begin thinking in an object-oriented way. (22)

OPEN-ENDED QUESTION A type of question used in interviews or on surveys that opens up the possible response set available to respondents. (5) *See also* Bipolar question, closed question.

OPEN SOURCE SOFTWARE A development model and philosophy of distributing software free and publishing its source code, which can then be studied, shared, and modified by users and programmers. The Linux operating system is an example. (1)

OPEN SYSTEM Part of general systems theory, a system that freely receives information, energy, people, or raw materials as input. Systems are never totally closed or totally open, but exist on a continuum from more closed to more open. (2) *See also* closed system.

OUTPUT Information delivered to users through the information system by way of intranets, extranets, or the Web on printed reports, on display screens, or via audio. (15)

PACKAGE In UML, things can be grouped together in packages. Packages can be considered as physical subsystems. Systems are implemented and deployed in packages. (22)

PERT DIAGRAM A tool used to determine critical activities for a project. It can be used to improve a project schedule and evaluate progress. It stands for Program Evaluation Review Technique. (3)

PHYSICAL DATA FLOW DIAGRAM A DFD that shows how a system will be implemented, including the hardware, software, people, and files involved. (9) *See also* logical data flow diagram.

PLUG-IN A small program that augments the original application. Some plug-ins enable the use of special features on multimedia Web sites. Examples are Shockwave for Netscape Navigator and RealAudio for Internet Explorer. (15)

POLYMORPHISM Refers to alternative behaviors among derived classes in object-oriented approaches. When several classes inherit both attributes and behaviors, there can be cases where the behavior of a derived class might be different than its base class or its sibling-derived classes. (22)

PRESENT VALUE The total amount that a series of future payments is worth now. A way to assess the economic outlays and revenues of the information system over its economic life and compare costs today with future benefits. (13)

PRIMARY KEY A key that uniquely identifies a record. (17) *See also* key, secondary key.

PROBES Follow-up questions primarily used during interviews between analysts and users. (5) *See also* closed question, open-ended question.

PROCESS The activities that transform or change data in an information system. They can be either manual or automated. Signified by a rounded rectangle in a data flow diagram. (2)

PROJECT MANAGEMENT The art and science of planning a project, estimating costs and schedules, managing risk, and organizing and overseeing a team. Many software packages exist to support project management tasks. (3)

PROJECT MANAGER A person responsible for overseeing the planning, costing, scheduling, and team organization (often of a systems) project. Frequently, it is a role played by a systems analyst. (3)

PROTOTYPING A rapid, interactive process between users and analysts to create and refine portions of a new system. It can be used as part of the SDLC for requirements determination or as an alternative to the SDLC. (8) *See also* rapid application development.

PSEUDOCODE A technique that creates computer instructions that are the intermediate step between English and program code. Used to represent the logic of each module on a structure chart. (20) *See also* structure chart.

PULL-DOWN MENU One of many GUI design elements that provides an onscreen menu of command options that appear after the user selects the command name on a menu bar. (18).

QUERIES Questions users pose to the database concerning data within it. Each query involves an entity, an attribute, and a value. (18)

RADIO BUTTON One of many GUI design elements that provide a round option button in a dialog box. They are mutually exclusive, because a user can choose only one radio button option within the group of options displayed. (15)

RAPID APPLICATION DEVELOPMENT (RAD) An object-oriented approach to systems development that includes a method of development as well as software tools. (8) *See also* prototyping.

RECORD A collection of data items that have something in common with the entity described. (17)

REENGINEERING In general, redesigning the manner in which work is done and then selecting computer tools to support the redesigned process. A term used differently in engineering, programming, and business contexts. (1) *See also* business process reengineering.

RELATIONAL DATABASE MODEL Represents data in the database as two-dimensional tables called relations. As long as both tables share a common data element, the database can relate any one file or table to data in another file or table. (17)

RELATIONSHIP Associations between entities (sometimes referred to as data associations). They can take the form of one-to-one, one-to-many, many-to-one, or many-to-many. (17)

REPEATING GROUP The existence of many of the same elements within the data structure. (10) *See also* data structure.

REVERSE ENGINEERING The opposite of code generation. The computer source code is examined, analyzed, and converted into repository entities. (1)

SAMPLING The process of systematically selecting representative elements of a population. Analysts sample hard data, archival data, and people during information requirements determination. (4)

SECONDARY KEY A key that cannot uniquely identify a record. A secondary key can be used to select a group of records that belong to a set. (17)

SECOND NORMAL FORM (2NF) When normalizing data for a database, the analyst ensures that all nonkey attributes are fully dependent on the primary key. All partial dependencies are removed and placed in another relation. (17) *See also* first normal form, third normal form.

SEQUENCE DIAGRAM In UML, a sequence diagram illustrates a succession of interactions between object instances over time. Often used to illustrate the processing described in use case scenarios. (22)

SEQUENTIAL FILE ORGANIZATION A way to store data records in a file. Records must be retrieved in the same physical sequence in which they are originally stored. (17)

STATE CHART DIAGRAM In UML, a way to further refine requirements. (22)

STRUCTURE CHART A tool for designing a modular, top-down system consisting of rectangular boxes and connecting arrows. (20). *See also* control flag, data couple.

STRUCTURED ENGLISH A technique for analyzing structured decisions based on structure logic and simple English statement such as add, multiply, and move. (11)

STRUCTURED OBSERVATION OF THE ENVIRONMENT (STROBE) A systematic observational method for classifying and interpreting organizational elements that influence decision making. Based on mise-en-scène film criticism. (7)

STRUCTURED WALKTHROUGH A systematic peer review of the system's programming and overall development that points out problems and allows the programmer or analyst to make suitable changes. (20)

SUPPLY CHAIN MANAGEMENT An organization's effort to integrate their suppliers, distributors, and customer management requirements into one unified process. Ecommerce applications can improve supply chain management. (21)

SYSTEM A collection of subsystems that are interrelated and interdependent, working together to accomplish predetermined goals and objectives. All systems have input, processes, output, and feedback. One example is a computer information system; another is an organization. (2) *See also* closed system, open system.

SYSTEMS ANALYST The person who systematically assesses how businesses function by examining the inputting and processing of data and the outputting of information with the intent of improving organizational processes. (1)

SYSTEMS DEVELOPMENT LIFE CYCLE (SDLC) A seven-phase approach to systems analysis and design that holds that systems are best developed through the use of a specific cycle of analyst and user activities. (1)

SYSTEMS DEVELOPMENT METHODOLOGY Any accepted approach for analyzing, designing, implementing, testing, maintaining and evaluating an information system. (1) *See also* systems development life cycle.

SYSTEMS PROPOSAL A written proposal that summarizes the systems analyst's work in the business up to that point and includes recommendations and alternatives to solve the identified systems problems. (14)

SYSTEMS TESTING The sixth phase in the SDLC (along with maintenance). Uses both test data and eventually live data to measure error, timeliness, ease of use, proper ordering of transactions, acceptable down time, understanding procedure manuals, and so forth of the new system. (20)

T-1 LINE The T1 is a leased line Internet connection. Data can be transmitted up to 1.45 megabits per second. (21)

T-3 LINE The T3 is a leased line Internet connection. Data can be transmitted up to 45 megabits per second. (21)

TANGIBLE BENEFITS Advantages measurable in dollars that accrue to the organization through the use of the information systems. (13) *See also* intangible benefits.

TANGIBLE COSTS Those costs in dollars that can be accurately projected by the systems analyst, including the cost of computers, resources, analysts and programmer's time, and other employees' salaries to develop a new system. (13) *See also* intangible costs.

THIN CLIENT A computer or information system that accesses application and data from a server. (21)

THINGS In UML, things describe the objects of object-oriented analysis and design. The two most often used groupings of things are structural things and behavioral things. (22)

THIRD NORMAL FORM (3NF) In the third normal form, any transitive dependencies are removed. A transitive dependency is one in which nonkey attributes are dependent on other nonkey attributes. (17) *See also* first normal form, second normal form.

TRANSACTION PROCESSING SYSTEM (TPS) A computerized information system developed to process large amounts of data for routine business transactions such as payroll and inventory. (1)

UNIFIED MODELING LANGUAGE (UML) The UML provides a standardized set of tools to document the object-oriented analysis and design of a software system. (22)

UPPER CASE TOOLS CASE tools designed to support information planning and the project identification and selection, project initiation and planning, analysis, and design phases of the systems development life cycle. (1) *See also* CASE tools, lower CASE tools.

USE CASE In UML, a sequence of transactions in a system. The purpose of the use case is to produce something of value to an actor in the system. The use case model is based on the interactions and relationships of individual use cases. Within a use case, an actor using the system initiates an event that begins a related series of interactions within the system. A use case focuses on what the system does rather than on how it does it. (22)

VALIDATION SOFTWARE Software that checks whether data input to the information system is valid. Although validating input is largely done through software that is the programmer's responsibility, it is the analyst's responsibility to know what common problems might invalidate a transaction. (19)

WEBMASTER The person responsible for updating and maintenance a Web site. Often these duties initially fall to the systems analyst during development of ecommerce applications. (16)

WORKFLOW ANALYSIS An approach that analyzes and then diagrams the flow of information, paperwork, and decisions in an organization. The focus of the diagram is the work rather than the people or processes. Software tools are available to support the analyst in workflow analysis. (4)

ACRONYMS

Numbers in parentheses refer to the chapter in which that term is defined.

AI artificial intelligence (1)
ASP application service provider (21)
B2B business to business (4)
B2C business to consumer (4)
BPR business process reengineering (7)
CARE computer-assisted reengineering (1)
CASE Computer-Aided Software Engineering (4)
CD-ROM compact disk-read only memory (15)
CD-RW compact disk-read write (15)
CICS customer information control system (16)
CSCWS computer-supported collaborative work systems (1)
DBMS database management system (17)
DDE Dynamic Data Exchange (20)
DFD data flow diagram (9)
DLL Dynamic Link Library (20)
DSL digital subscriber line (21)
DSS decision support system (12)
DVD digital video disk (15)
E-R entity-relationship (2)
EIS executive information system (1)
ERD entity relationship diagram (2)
ERP enterprise resource planning (1)
ES expert system (1)
ESS executive support system (1)
FAQ frequently asked questions (15)
FIG feasibility impact grid (3)
FTP file transfer protocol (15)
GDSS group decision support system (12)
GIF graphics interchange format (15)
GSS group support system (7)
GUI graphical user interface (18)
HTML hypertext markup language (15)
ISDN Integrated Service Digital Network (15)
ISP Internet service provider (21)
JAD Joint Application Design (5)
JPEG Joint Photographic Experts Group (15)
KWS knowledge work system (1)
LAN local area network (19)
MICR magnetic ink character recognition (19)
MIS management information system (1)
OAS office automation system (1)
OCR optical character recognition (19)
OLE object linking and embedding (20)
PDA personal digital assistant (1)
PERT Project Evaluation and Review Technique (3)
PIM personal information manager (3)
PKI public key infrastructure (21)
RAD rapid application development (8)
SAN storage area network (17)
SDLC systems development life cycle (1)

SET secure electronic translation (21)

SQL structured query language (18)

SSL secure sockets layering (21)

STROBE STRuctured OBservation of the Environment (7)

TPS transaction processing system (1)

TQM total quality management (20)

UML unified modeling language (22)

URL uniform resource locator (15)

VPN virtual private network (21)

WAN wide area network (21)

WAP wireless application protocol (1)

WWW World Wide Web (15)

*Page numbers followed by an italic *f* refer to figures.

CONSULTING OPPORTUNITIES